onship

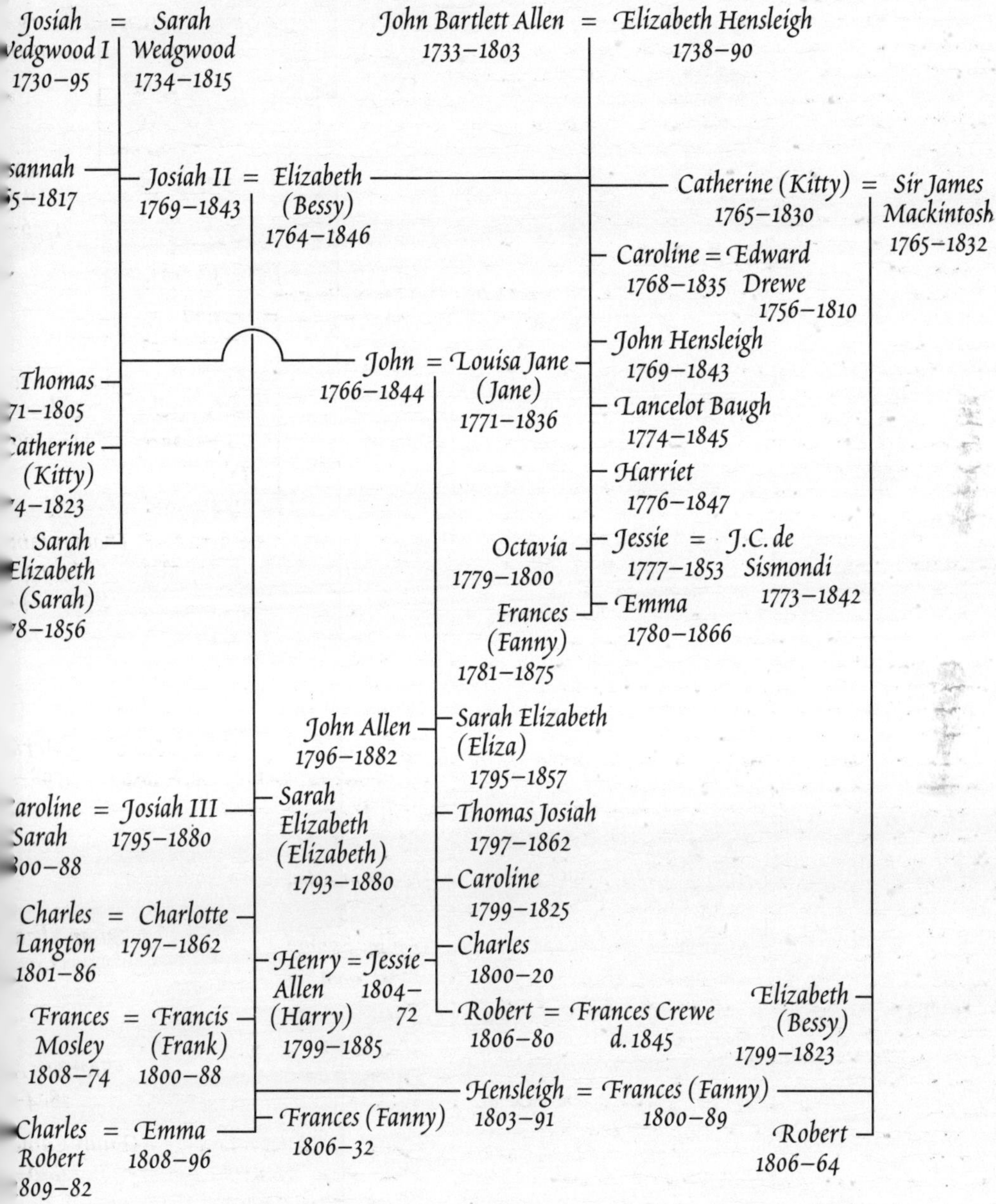

Josiah Wedgwood I 1730–95
=
Sarah Wedgwood 1734–1815
John Bartlett Allen 1733–1803
=
Elizabeth Hensleigh 1738–90
sannah
5–1817
Josiah II 1769–1843
=
Elizabeth (Bessy) 1764–1846
Catherine (Kitty) 1765–1830
=
Sir James Mackintosh 1765–1832
Caroline = Edward Drewe
1768–1835 1756–1810
John Hensleigh 1769–1843
John 1766–1844
=
Louisa Jane (Jane) 1771–1836
Thomas
71–1805
Catherine (Kitty)
74–1823
Lancelot Baugh 1774–1845
Harriet 1776–1847
Sarah Elizabeth (Sarah)
78–1856
Octavia 1779–1800
Jessie = J.C. de Sismondi
1777–1853 1773–1842
Frances (Fanny) 1781–1875
Emma 1780–1866
John Allen 1796–1882
Sarah Elizabeth (Eliza) 1795–1857
Caroline Sarah
00–88
=
Josiah III 1795–1880
Sarah Elizabeth (Elizabeth) 1793–1880
Thomas Josiah 1797–1862
Caroline 1799–1825
Charles Langton 1801–86
=
Charlotte 1797–1862
Charles 1800–20
Henry Allen (Harry) 1799–1885
=
Jessie 1804–72
Frances Mosley 1808–74
=
Francis (Frank) 1800–88
Robert 1806–80
=
Frances Crewe d. 1845
Elizabeth (Bessy) 1799–1823
Hensleigh 1803–91
=
Frances (Fanny) 1800–89
Charles Robert
809–82
=
Emma 1808–96
Frances (Fanny) 1806–32
Robert 1806–64

THE CORRESPONDENCE OF

CHARLES DARWIN

Editors

FREDERICK BURKHARDT   DUNCAN M. PORTER

JOY HARVEY   JONATHAN R. TOPHAM

Production Manager

SARAH BENTON

Assistant Editors

CHARLOTTE BOWMAN   HEIDI BRADSHAW

THOMAS JUNKER   SARAH LAVELLE

PERRY O'DONOVAN

Research Associates

ANNE SCHLABACH BURKHARDT

NORA CARROLL STEVENSON

Office Manager

HEDY FRANKS

This edition of the Correspondence of Charles Darwin is sponsored by the American Council of Learned Societies. Its preparation is made possible by the co-operation of the Cambridge University Library and the American Philosophical Society.

Support for editing has been received from the National Endowment for the Humanities, the National Science Foundation, the Andrew W. Mellon Foundation, the Alfred P. Sloan Foundation, the Royal Society of London, the British Academy, the Pilgrim Trust, the Stifterverband für die Deutsche Wissenschaft, and the Isaac Newton Trust. The National Endowment's grants (Nos. RE-23166-75-513, RE-27067-77-1359, RE-00082-80-1628, RE-20166-82, RE-20480-85, RE-20764-89, RE-20913-91, and RE-21097-93) were from its Program for Editions; the National Science Foundation's funding of the work was under grants Nos. SOC-75-15840, SES-7912492, SES-8517189, and SBR-9020874. Any opinions, findings, conclusions, or recommendations expressed in this publication are those of the editors and do not necessarily reflect the views of the grantors.

Down House. Photograph, between 1858 and 1874.
(By permission of the Syndics of the Cambridge University Library.)

THE CORRESPONDENCE OF
CHARLES DARWIN

VOLUME 10 1862

CAMBRIDGE
UNIVERSITY PRESS

PUBLISHED BY THE PRESS SYNDICATE OF THE UNIVERSITY OF CAMBRIDGE
The Pitt Building, Trumpington Street, Cambridge CB2 1RP, United Kingdom

CAMBRIDGE UNIVERSITY PRESS
The Edinburgh Building, Cambridge CB2 2RU, United Kingdom
40 West 20th Street, New York, NY10011-4211, USA
10 Stamford Road, Oakleigh, Melbourne 3166, Australia

First published 1997

Printed in the United Kingdom at the University Press, Cambridge

Typeset in TEX Baskerville 10/12 pt

A catalogue record for this book is available from the British Library

Library of Congress Cataloguing in Publication data applied for

ISBN 0 521 59032 9 hardback

CONTENTS

ILLUSTRATIONS

CALENDAR LIST OF LETTERS

The following list is in the order of the entries in the *Calendar of the correspondence of Charles Darwin*. It includes all those letters that are listed in the *Calendar* within the date range covered by this volume of the *Correspondence*. Some letters have been redated since the publication of the *Calendar*, so this list is necessary to enable users of the *Calendar* to locate such letters in the *Correspondence*. Alongside the *Calendar* numbers are the corrected dates of each letter. A date printed in italic type indicates that the item has been omitted from this volume.

Letters acquired after the publication of the *Calendar*, in 1985, appear at the end of the list. These letters are summarised in a 'Supplement' to a new edition of the *Calendar* (Cambridge University Press, 1994). Their numbers correspond to the chronological ordering of the original *Calendar* listing with the addition of an alphabetical marker. The marker 'f' denotes letters acquired after the *Calendar* went to press in 1994.

2916. [4–8 February 1862]
3144. *8 May [1866]*
3356. [1 February 1862]
3357. [28 January 1862]
3358. [before 3 July 1862]
3359. *[1848–55]*
3360. [16 July 1862]
3361. [after 29 June 1862]
3362. [November? 1862]
3363. [before 11 March 1862]
3364. [after 19 May 1862?]
3365. *[mid-February 1864]*
3366. *[after 19 May 1864]*
3367. *29 [May 1863]*
3368. [January 1862]
3369. *Part of* 4401.
3370. 2 January [1862]
3371. [3] January [1862]
3372. 2 January [1862]
3373. [1 January 1862]
3374. *[29 December 1861]*
3375. [30 December 1861 or 6 January 1862]
3376. 6 January 1862
3377. 6 January 1862
3378. 7 [February] 1862
3379. 10 January 1862
3380. [after 10 January 1862]
3381. 11 January 1862
3382. 13 January [1862]
3383. 13 January 1862
3384. 13 January 1862
3385. 13 January 1862
3386. 14 [January 1862]
3387. 14 January 1862
3388. [3–14] January [1862]
3389. 15 [and 16] January [1862]
3390. 15 January [1862]
3391. 16 January [1862]
3392. 16 January 1862
3393. 17 January 1862
3394. [25 January 1862]
3395. [19 January 1862]
3396. 20 January 1862
3397. 20 January [1862]
3398. 20 January 1862
3399. *21 [January 1863]*
3400. [21 January 1862?]
3401. 21 January 1862
3402. [before 22 January 1862]
3403. 22 January [1862]
3404. 22 January [1862]
3405. 22 January [1862]
3406. 22 January [1862]
3407. 22 January 1862
3408. [before 22 January 1862]
3409. 23 January [1862]

INTRODUCTION

As the sheer volume of his correspondence indicates, 1862 was a particularly productive year for Darwin. This was not only the case in his published output (two botanical papers and a book on the pollination mechanisms of orchids), but more particularly in the extent and breadth of the botanical experiments he carried out. While many of these remained unpublished for several years, they formed the foundation of numerous later publications. The promotion of his theory of natural selection also continued: Darwin's own works expanded on it, Thomas Henry Huxley gave lectures about it, and Henry Walter Bates invoked it to explain mimicry in butterflies. Moreover, his work was coming to be referred to routinely. In November, Joseph Dalton Hooker told him: 'you are alluded to in no less than 3 of the papers in Linn. Trans!— I do not think you are conceited, but really I do think you have a good right to be so' (letter from J. D. Hooker, [15 and] 20 November [1862]). Still taking a keen interest in the progress of his views through Europe, Darwin negotiated, in addition to a second German edition of *Origin*, a German translation of *Orchids*, and impatiently awaited the publication of *Origin* in French. His work on variation in domesticated animals and plants, the first part of the expanded version of *Origin* promised in the preface to that book, had been interrupted for months while he worked on 'Dimorphic condition in *Primula*' and *Orchids*; it suffered a further setback when illness struck the family during the summer. But towards the end of the year, Darwin was able once more to turn his attention to *Variation*.

The year began with a New Year's greeting from Huxley, triumphant over the response to his recent Edinburgh lectures (his audience applauded, sections of the Scottish press hissed). Huxley, while advocating Darwin's theory, had again espoused the view that its final proof awaited the production, by selection from a common stock, of forms that differed from one another to such an extent that they would not interbreed—the experimental production of new 'physiological' species. Darwin attempted to dissuade him from this view (letter to T. H. Huxley, 14 [January 1862]):

> no doubt you are right that here is great hiatus in argument; yet I think you overrate it— you never allude to the excellent evidence of *varieties* of Verbascum & Nicotiana being partially sterile together.

He failed. Huxley replied (letter from T. H. Huxley, 20 January 1862):

> I entertain no doubt that twenty years experiments on pigeons conducted by a skilled physiologist ... would give us physiological species sterile *inter se* from a common stock—(& in this if I mistake not I go further than you

> do yourself) and ... when these experiments have been performed I shall consider your views to have a complete physical basis[.]

The issue arose again when, through November and December, Huxley delivered a series of lectures to working men that reviewed Darwin's theory, and sent copies to Darwin. Darwin read them and pronounced them 'simply perfect', but continued (letter to T. H. Huxley, 18 December [1862]):

> you say the answer to varieties when crossed being at all sterile is "absolutely negative". Do you mean to say that Gärtner lied ... when he showed that this was the case with Verbascum & with Maize ... does Kölreuter lie when he speaks about the vars. of Tobacco.

At the end of the year, Darwin seemed resigned to their difference of opinion, but complained (letter to T. H. Huxley, 28 December [1862]):

> To get the degree of sterility you expect in recently formed varieties seems to me simply hopeless. It seems to me almost like those naturalists who declare they will never believe that one species turns into another till they see every stage in process.

This correspondence with Huxley made Darwin keener than ever to repeat the experiments by which Karl Friedrich von Gärtner had demonstrated a degree of sterility between varieties of *Verbascum*. When John Scott, foreman of the propagating department at the Edinburgh Botanic Garden, began writing long, intelligent, and informative letters, Darwin, impressed, gave him the commission (see letter to John Scott, 11 December [1862]). Darwin was altogether taken with this young protégé, telling Hooker: 'he is no common man' (letter to J. D. Hooker, 12 [December 1862]).

Darwin's views on the phenomenon of sterility were affected by, and in turn stimulated, his work on dimorphic plants, which had begun in 1861 with his study of *Primula* and escalated throughout 1862 as he searched for other cases of dimorphism. Towards the end of the year, he wrote to Hooker (letter to J. D. Hooker, 12 [December 1862]): 'my notions on hybridity are becoming considerably altered by my dimorphic work: I am now strongly inclined to believe that sterility is at first a selected quality to keep incipient species distinct.' In his private notes, he conducted a lengthy dialogue on the subject (see Appendix VI). His paper, 'Dimorphic condition in *Primula*', was read before the Linnean Society of London in November 1861, and was published in the society's journal in March 1862. The paper described the two forms of *Primula* flower, short-styled and long-styled, and demonstrated that, although each form possessed both male and female organs, the plants were only fully fertile when crossed with another plant of the reciprocal form. Darwin concluded that the two forms existed to 'favour the intercrossing of distinct individuals' ('Dimorphic condition in *Primula*', p. 92 (*Collected papers* 2: 59)). In his *Autobiography*, Darwin recalled: 'no little discovery of mine ever gave me so much pleasure as the making out the meaning of heterostyled flowers' (*Autobiography*, p. 134).

On completion of his *Primula* paper, Darwin repeated his *Primula* crosses through a second generation, both to test his previous year's results, and to observe the effects of repeated crossing with own-form pollen. He also began systematically to search for other dimorphic species, looking through botanical books for indications that a species had more than one flower form, and writing to botanists asking for information, specimens, and assistance with experiments. In January, he wrote to Asa Gray thanking him for some 'new cases of Dimorphism', and added, 'new cases are tumbling in almost daily' (letter to Asa Gray, 22 January [1862]). In a postscript, he mentioned his work on 'one of the Melastomas' and his suspicion that 'the two sets of anthers' had 'different functions'. He continued to write to Gray throughout the year about his quest for dimorphism in the Melastomataceae, Gray obligingly sending specimens and getting one of his students to make observations on American species. Hooker and George Bentham at Kew were also tapped for their knowledge. Darwin, initially hopeful, became increasingly frustrated, telling Hooker (letter to J. D. Hooker, 7 March [1862]): 'I am nearly sure that daylight is coming with respect to the melastomas'. He spent much of his valuable time on the problem: 'the labour is great', he told Gray (letter to Asa Gray, 10–20 June [1862]), 'I have lately counted one by one 6700 seeds of Monochætum!!' By October, Darwin was flagging and declared to Gray: 'I am utterly routed, beaten, "whipped" by those odious Melastomatads; yet I feel sure there is something very curious to be made out about them.' Darwin persisted with his experiments through the following year, but the Melastomataceae remained a blind alley, and no publication ever resulted from his '*enormous* labour over them' (letter to J. D. Hooker, 14 [October 1862]; see *ML* 2: 292–3).

Other species proved more profitable subjects for investigation, and Darwin was soon filling portfolios with notes on his observations and experiments. Moreover, as his work progressed, he began to appreciate the vast extent of the subject, telling Oliver: 'I can see at least 3 classes of dimorphism' (letter to Daniel Oliver, 12 [April 1862]), and experimenting to test his hypotheses about their different purposes. The number of his experiments was staggering, but by no means all were successful. He told one correspondent: 'I am contented if I get any result once out of four or five sets of experiments' (letter to M. T. Masters, 24 July [1862]). The materials that Darwin amassed on heterostyly in this year formed the foundations of several later papers, and also of his books *Forms of flowers* and *Cross and self fertilisation*.

One set of experiments led directly to publication. Many years earlier, Darwin had observed dimorphism in *Linum flavum*, but had 'at first thought it was merely a case of unmeaning variability' (*Autobiography*, p. 128). However, having made out the *Primula* case he determined to experiment on *Linum* in 1862. Soon he was enthralled, especially by the extent of the sterility of own-form crosses. He told Gray: 'Taking sexual power as the criterion of difference the two forms of this one species may be said to be generically distinct' (letter to Asa Gray, 14 July [1862]). The case was so good that he regretted having 'wasted time' in repeating his *Primula* crosses, and determined to publish on *Linum* 'at once' (letter to John Scott, 11 December [1862]), writing up his experiments in December as a paper for the Linnean Society.

One of the multi-volume treatises through which Darwin groaningly trawled seeking cases of dimorphism, produced another profitable subject for investigation—*Lythrum*, the purple loosestrife. By the summer, Darwin was experimenting. 'To day I have been looking at Lythrum & have seen the *three* forms . . . I sh^{d}. like to make out this wonderfully complex case—' (letter to Daniel Oliver, 29 [July 1862]). The three forms had different lengths of styles and stamens and differently coloured and sized pollen grains; only an elaborate system of cross-pollination between the different forms produced fertile seed. The case clearly excited Darwin, who exclaimed to Gray (letter to Asa Gray, 9 August [1862]), 'I am almost stark staring mad over Lythrum', and requested, 'For the love of Heaven', that Gray have a look at some American species, and send him 'Seed! Seed! Seed!' There was reason in his madness, he explained, as, for those who already believed in species change, the facts would 'modify . . . whole view of Hybridity'—more evidence with which to sway Huxley.

By October, Darwin had decided that the case warranted a paper for the Linnean Society (letter to J. D. Hooker, 6 October [1862]). However, it was not until 1864 that the Linnean Society heard about *Lythrum*; in 1862, it was one of the society's own specimens of the orchid *Catasetum tridentatum*, that Darwin described before the 'placid Linnæans'. This remarkable specimen produced flowers that were believed to constitute three distinct genera. Darwin explained that the three flowers represented the three sexual forms (male, female, and hermaphrodite) of a single species, differing so much from one another as to have been classified in different genera. The paper, read in April, was largely an extract from the book on orchids and their 'remarkable contrivances' that he had been preparing since the previous year.

Darwin had enjoyed observing the orchids: he described the work to Gray as a 'hobby-horse' that had given him 'great pleasure to ride' (letter to Asa Gray, 22 January [1862]). But he worried about the resulting book—whether it had been worth the effort, whether people would buy it. When he submitted the manuscript to his publisher, John Murray, he boasted: 'I can say with confidence that the M.S. contains many new & very curious facts & conclusions', but added: 'I know not **in the least**, whether the Book will sell' (letter to John Murray, 9 [February 1862]). To his son, William, his language was more blunt (letter to W. E. Darwin, 14 February [1862]): 'whether my little Book has been worth writing, I know no more than the man in the moon'. But even after the book had gone to press, Darwin could not leave the subject alone: he continued to seek out specimens of the orchids that puzzled him, and was thrilled by Gray's observations of North American species. Extending his examination of pollination mechanisms to other orders of plants, he declared: 'insects (in relation to the marriage of distinct flowers) *govern the structure of almost every* flower' (letter to Daniel Oliver, 8 June [1862]).

Darwin was careful to send out presentation copies of *Orchids* to eminent botanists in Britain and abroad, and to all those who had helped him (or might in the future) by providing specimens, information, or by doing experiments. As he had done for his *Primula* paper, he drew up a list of individuals and societies to whom the book should be sent (see Appendixes III and IV) and nervously awaited their reaction. 'I never before felt half so doubtful about anything I published', he told Hooker (letter to J. D. Hooker, 15 [May 1862]). But he did not have long to wait. 'It is a *very extraordinary* book!', wrote Daniel Oliver on 14 May, and George Bentham pronounced it 'most valuable' (letter from George Bentham, 15 May 1862). *Orchids* was published on 15 May, and by June there had been enough positive reviews for a relieved Darwin to tell Hooker that, after cursing his 'folly' in writing the book, it was, after all, 'a success' (letter to J. D. Hooker, 30 [June 1862]).

The success of *Orchids* was important: it was a follow-up to *Origin* in the sense that it was Darwin's first detailed exposition of the power of natural selection. He made the point to Hooker (letter to J. D. Hooker, 14 March [1862]): 'I have found the study of orchids eminently useful in showing me how nearly all parts of the flower are coadapted for fertilisation by insects, & therefore the result of n. selection'. The book was intended to be, as Gray put it, a '"flank-movement" on the enemy'—a way of inducing sceptics to accept the truth of natural selection through the back door (letter to Asa Gray, 23[–4] July [1862]). Moreover, it apparently worked. Gray told Darwin that George Bentham's presidential address to the Linnean Society on 24 May, in which he held up Darwin's work on orchids as an exemplar of the biological method, demonstrated that *Orchids* had 'nearly overcome his opposition to the *Origin*' (letter from Asa Gray, 2–3 July 1862).

Natural selection was also to receive support from another quarter. Henry Walter Bates, not long returned after many years in the Amazon, had invoked natural selection as the mechanism to account for the mimicry he had observed among South American butterflies. The paper in which he did so, read before the Linnean Society in November 1861, was lengthened and published in 1862. Darwin, already well-disposed towards Bates, became increasingly convinced of his worth and talent as the year progressed. Bates sent him in manuscript the first chapters of a book on his experiences in the Amazon, which Darwin declared 'excellent—style perfect—description first-rate' with 'good dashes of original reflexions' (letter to H. W. Bates, 13 January [1862]). He warmly recommended Bates and his book to Murray, who swiftly agreed to publish the work.

Hooker had met Bates and was 'much struck' with him. Early in the year, the two engaged in a lengthy correspondence about the relative effects on species of natural selection and the direct action of external conditions. Hooker sent Darwin a few of their letters; Darwin remarked (letter to J. D. Hooker, 26 [March 1862]): 'It is really curiously satisfactory to me to see so able a man as Bates (& yourself) believing more fully in nat. selection, than I think I even do myself.' The three were able to discuss the subject face to face when Bates and Hooker spent Easter at Down House.

When Bates's paper on mimetic butterflies was published towards the end of the year, Darwin praised it unreservedly: 'It is one of the most remarkable & admirable papers I ever read in my life' (letter to H. W. Bates, 20 November [1862]). He discussed it with Hooker, saying: 'To my mind the act of segregation of varieties into species was never so plainly brought forward' (letter to J. D. Hooker, 24 [November 1862]), and the two friends ended the year with an in-depth discussion of the topics that Hooker had been attempting to thrash out with Bates in the spring—what was the role of external conditions in the variation and production of species? what the role of natural selection? Hooker's peremptory criticism that in *Origin* Darwin had not conveyed with sufficient force how 'N.S. is as powerless as physical causes to make a variation' and that 'the law that "like shall *not* produce like" is at the bottom of all' (letter from J. D. Hooker, 26 November 1862) drew from Darwin a detailed explanation of his views on the nature of variation, and on how it might be affected by crossing, physical conditions, and natural selection (letter to J. D. Hooker, [after 26] November [1862]).

As well as taking pains that those already familiar with his ideas should clearly understand them, Darwin was also concerned that they should reach a wider audience, and he agreed to write an anonymous review of Bates's paper for the *Natural History Review* (see letter to John Lubbock, 16 [December 1862]). Aware of the status of his views in Europe—he told the Swiss zoologist, Edouard Claparède that they were 'more unpopular in France even than in England', though, in Germany, naturalists were beginning 'with some rapidity to adopt them' (letter to Edouard Claparède, [*c.* 16 April 1862])—he continued to interest himself in the preparation of translations of his books. When Charles Edouard Brown-Séquard informed him that he intended to write a review of *Origin* for a French periodical, Darwin replied, 'at present I am the more glad of any notice in France, as a French Translation will appear very soon' (letter to C. E. Brown-Séquard, 2 January [1862]). The translation was being prepared by Clémence Royer, a young Frenchwoman then living in Lausanne, Switzerland. She finished it in March, but, owing to printing delays, Darwin did not see a copy until June. Royer had included a long preface in which she praised Darwin's ideas as supplying an alternative to religious revelation, and had laden the text with explanatory notes. On receipt of the work, Darwin told Gray that he thought Royer must be 'one of the cleverest & oddest women in Europe'; in her additions, she had made 'some very curious & good hits'. A month later, he was expressing his regret that Royer had not 'known more of Natural History' (letter to Armand de Quatrefaces, 11 July [1862]). She had had assistance with the scientific details of *Origin* from Edouard Claparède and he too had cause for regret. In response to Darwin's thanks for the help he had given Royer, Claparède wrote an exasperated letter to Darwin detailing Royer's stubborn resistance to his suggestions and his unsuccessful attempts to remove or modify her commentary. He advised Darwin that when his larger work on species was completed, he should seek a different translator (see letter from Edouard Claparède, 6 September 1862).

In Germany, the first edition of Heinrich Georg Bronn's translation of *Origin* was nearly sold out. When Bronn wrote to Darwin in March telling him of the need for a second edition (letter from H. G. Bronn, [before 11 March 1862]), Darwin asked him to use the latest English edition, the third, for his translation, and to incorporate further additions that he would send (see Appendix VIII). Bronn complied willingly. Not only that—having read *Orchids* at Darwin's request to judge of its suitability for a German audience, he began translating it himself. He worked hard and with astonishing speed. With both tasks completed, Bronn died suddenly from a heart attack (see letter from E. Schweizerbart'sche Verlagsbuchhandlung, 11 July 1862). Yet Darwin was now finding no shortage of supporters in Germany. In September, Friedrich Rolle sent him the first part of his popular exposition of Darwin's theory (Rolle 1863; see letter to Friedrich Rolle, 17 October [1862]), and in November he received a copy of a book by the German materialist Ludwig Büchner (Büchner 1862) which included a reprint of his positive review of *Origin* (see letter to Ludwig Büchner, 17 November [1862]).

Among the foreign scientists whom Darwin hoped to interest in his work was the widely respected Swiss botanical taxonomist, Alphonse de Candolle, from whom he received a long and thoughtful reply to the gift of *Orchids* that expressed admiration but stopped short of endorsing natural selection (letter from Alphonse de Candolle, 13 June 1862). Darwin also sent presentation copies of his botanical studies to Charles Naudin, a botanist at the Muséum d'Histoire Naturelle, whose experiments on hybridisation were receiving considerable attention (see letter from C. V. Naudin, 26 June 1862). Darwin was dismissive of Naudin's methods, and his claims (see letter to J. D. Hooker, 30 [June 1862]), but still sought to enlist his help in understanding hybridity (see letter to J. D. Hooker, 24 December [1862]).

As usual, Asa Gray took care that Americans should know of Darwin's latest production. He requested the proof-sheets of *Orchids* in order to write an early review: Darwin began to send them in April. Gray was impressed, exclaiming 'What a skill & genius you have for these researches' (letter from Asa Gray, 18 May 1862). In thanking Gray for his positive comments, Darwin expressed surprise that Gray 'sh$^{d.}$ have strength of mind to care for science, amidst the awful events daily occurring' in America. Even after the news, which arrived in London early in January, that the Trent affair had been resolved (averting the danger of war between Britain and the Union states), the Civil War was still a subject of concern in Britain. Darwin followed events in *The Times*, though at one point cursing how 'detestably' the paper's special correspondent wrote on the subject, with 'not a shade of feeling against slavery' (letter to Asa Gray, 16 October [1862]). It was primarily Darwin's loathing for slavery that made him more open-minded about the Northern cause than were many other English observers, and it was his open-mindedness that permitted discussion of the subject with Gray to continue. In contrast, Hooker, who deplored Gray's national pride and anger at the British for their neutrality, avoided politics in his correspondence with Gray. Darwin, though far from relishing the conflict—a 'fearful evil to the whole world' (letter to Asa Gray, 16 October [1862])—appreciated how well Gray's letters about the war brought the distant events to life (see letter to Asa Gray, 26[–7] November [1862]).

When Darwin wrote to Gray in July that he and Emma had 'come to wish for Peace at any price' (letter to Asa Gray, 23[–4] July [1862]), the couple were in the midst of a crisis much closer to home and heart. Their son Leonard was seriously ill. He had been sent home from school in June with scarlet fever. Although at first it seemed a mild attack, there were complications, 'enlarged glands of neck, injured kidneys . . . dreadful erysipelas of the head & face' (letter to J. D. Hooker, 26 July [1862]). Darwin believed that it was only the administration of 'Port-wine every $\frac{3}{4}$ hour, night & day' that saved the boy (letter to M. T. Masters, 24 July [1862]). By August, although Lenny was convalescent, the doctors warned that all possibility of further infection must be avoided, leaving Darwin and Emma 'perplexed to death what to do' (letter to W. E. Darwin, [2–3 August 1862]). They determined on a seaside holiday in Bournemouth, setting off in mid-August. However, Leonard had a relapse and Emma caught the infection herself, forcing them to remain for the rest of August in Southampton, where they had stopped to visit Darwin's eldest son, William.

September found the invalids improved, and the whole family, divided into two households, established in Bournemouth. Here, Darwin felt the usual frustration at being separated from his work, and, moreover, the area did not at first seem to hold much of potential scientific interest. He told Hooker (letter to J. D. Hooker, 11 September [1862]): 'This is a nice, but most barren country & I can find nothing to look at'. However, a few days later, he had found something with which to amuse himself and pass the time—the insectivorous plant, *Drosera*. As he had at Eastbourne in the summer of 1860 (see *Correspondence* vol. 8), Darwin became 'wonderfully interested' in the plant's sensitive reactions. On this occasion he began investigating the effects on their reactivity of various poisons, narcotics, and anaesthetics known to affect the nervous systems of animals. The results were exciting: he told both his cousin, William Darwin Fox, and Hooker of his growing conviction that *Drosera* 'must have diffused matter . . . closely analogous to the nervous matter of animals' (letter to W. D. Fox, 20 [September 1862]; letter to J. D. Hooker, 26 September [1862]). Darwin sent an abstract of his results to Hooker, asking for his opinion as an aid in deciding 'some future year' whether and how to go on with the subject; at present, there was more pressing work to be done. And, although Darwin did some further experiments with sensitive and insectivorous plants in October, following his return to Down, the work had to be laid aside when he resumed work on his long-promised book about variation in domesticated animals and plants—the first part of his planned *magnum opus*. In May, he had lamented to Hooker (letter to J. D. Hooker, 30 May [1862]): 'what will become of my book on Variation: I am involved in a multiplicity of experiments', and, although he admitted finding experiments 'much better fun' than species (letter to J. D. Hooker, 9 February [1862]), he responded to the regular enquiries he received about the progress of his larger work by returning in October to the long-neglected collation of notes, and drafting of the manuscript.

Although Darwin's unreliable health at times interfered with his work, it also provided a protection from other potential interruptions. In June, he was diagnosed as suffering from eczema (see letter to J. D. Hooker, 30 [June 1862]), and resolved, in consequence, on growing a 'long beard' (letter from Mary Butler, [before 25 December 1862]). An 'accursed attack' of the condition prevented him from attending the Cambridge meeting of the British Association for the Advancement of Science at the beginning of October. He missed Richard Owen, one of his 'chief enemies' (letter to Asa Gray, 23[–4] July [1862]), challenging his views on transmutation in a paper on the aye-aye. However, Huxley described the event, detailing how Owen was seen in the ensuing discussion to be 'lying & shuffling', coming across as an 'innocent old sheep—being worried by . . . particularly active young wolves' (letter from T. H. Huxley, 9 October 1862).

Darwin had managed to appear in person at the Linnean Society to read his *Catasetum* paper, but regretted it, as he was confined to bed the next day; he feared he would have to 'give up trying to read any paper or speak' (letter to J. D. Hooker, 9 [April 1862]). A visit in October from three members of the old *Beagle* crew, Bartholomew James Sulivan, John Clements Wickham, and Arthur Mellersh, to prepare for which he had taken 'every possible precaution', still resulted in 'violent shaking & vomiting till the early morning'. In view of this, he prescribed strict conditions for a meeting with John Lubbock: 'if you could . . . let me go away for an hour after dinner & retire to my room at 9 o clock I do not think it would hurt me' (letter to John Lubbock, 23 October [1862]). However, his confidence was ill-founded. He told Hugh Falconer, who wondered when he might see Darwin again: 'Even talking of an evening for less than two hours has twice recently brought on such violent vomiting and trembling; that I dread coming up to London' (letter to Hugh Falconer, 14 November [1862]).

Darwin thought his children had inherited his 'poor constitution' (see letter to J. B. Innes, 24 February [1862]) and with some cause. Not only was Leonard seriously ill in 1862, but Horace too, suffering from some strange nervous malady that caused 'hysterical sobbing' and 'semi-convulsive movements'. At the height of his anxiety, Darwin made an emotional outburst to Asa Gray (letter to Asa Gray, 23[–4] July [1862]):

> Children are one's greatest happiness, but often & often a still greater misery. A man of science ought to have none, . . . then there would be nothing in this wide world worth caring for & a man might . . . work away like a Trojan.

By August he was telling Gray: 'We are a wretched family & ought to be exterminated' (letter to Gray, 21 August [1862]), and in September, with everybody's health improving, he told Fox: 'I have never passed so miserable a nine months' (letter to W. D. Fox, 12 September [1862]).

When he was not worrying about their health, Darwin enlisted his children to help with his botanical observations. George earned his father's commendation for his 'splendid work in watching Orchids' (letter to J. D. Hooker, 30 [June 1862]). William, now established in his career as a banker in Southampton, received further encouragement to use his free time to 'work a little at Botany', with Darwin assuring him that such work would make his life 'much happier' (letter to W. E. Darwin,

14 February [1862]). Darwin told William of his work on *Lythrum*, and, having ascertained that the plant grew in the vicinity of William's home, asked him to carry out some observations. William, with the help of his brothers George and Francis, who were staying with him, enthusiastically set to work (see letter to W. E. Darwin, [2–3 August 1862], and letter from W. E. Darwin, 5 August 1862).

Darwin was certainly making a reputation for himself as a botanist. Hooker, whose opinion 'on any scientific subject' Darwin valued more than that of 'any one else in the world' (letter to J. D. Hooker, 14 [October 1862]), told him (letter from J. D. Hooker, 28 June 1862): 'you are out of sight the best Physiological observer & experimenter that Botany ever saw'. Gray wrote of his gratitude to Darwin for having 'given new eyes to botanists, and inaugurated a new era in the science' (A. Gray 1862b, p. 429). Oliver joined in the chorus, telling him: 'Your late publications must surely give quite a new & most promising direction to our studies' (letter from Daniel Oliver, 14 May 1862). Darwin was thrilled by the applause, but still anxious about his audacity in daring to publish on botany. Even at the start of their correspondence he told John Scott: 'Botany is a new subject to me' (letter to John Scott, 12 November [1862]), but, impressed by his young correspondent's botanical knowledge, he later confessed: 'in fact I know only odds & ends of botany & you know far more' (letter to John Scott, 19 November [1862]).

As a geologist, Darwin's reputation was long established, and though no longer active in the field himself, his opinions on geological matters were still greatly valued by those who were. Thomas Francis Jamieson, whose work on the so-called 'parallel roads' of Glen Roy had forced Darwin to abandon his own theory on their formation, consulted him as he prepared his paper on the subject for publication. Although Darwin was glad that Glen Roy was 'settled' (letter to Charles Lyell, 22 August [1862]), he could not help once more expressing his disgust at his own mistake (letter to Charles Lyell, 14 October [1862]): 'now & for ever more I give up & abominate Glen Roy & all its belongings'. Nevertheless, Darwin agreed to referee Jamieson's paper for the Geological Society (see letter to A. C. Ramsay, 14 December [1862]).

Others were anxious for Darwin's geological approval. Andrew Crombie Ramsay, the new president of the Geological Society, sent Darwin a paper in which he argued that a number of lake basins in Europe and North America had been formed by the gouging action of glaciers; the paper had been attacked by members of the council of the Geological Society, and there was opposition to its publication. Ramsay wanted Darwin's opinion (see letter from A. C. Ramsay, 26 August 1862). Darwin gave it (letter to A. C. Ramsay, 5 September [1862]): 'As far as I can judge your theory *must* be right to a large extent, possibly wholly'. Discussing the matter with Hooker, he exclaimed (letter to J. D. Hooker, 21 [September 1862]): 'What presumption, ... in the Council of Geolog. Soc^{y}.: that it hesitated to publish the paper.' Lyell, however, rejected Ramsay's notions, prompting Darwin to admit that perhaps Ramsay pushed his theory too far (see letter to Charles Lyell, 14 October [1862]). Moreover, when the physicist John Tyndall, fresh from a summer in the

Alps with Huxley, publicised his view, inspired by Ramsay's work, that the extent of glacial action in Switzerland was so great as to have caused almost all the features of the present landscape, Darwin enjoined Hooker: 'For Heaven sake instill a word of caution into Tyndall's ears' (letter to J. D. Hooker, 10–12 November [1862]).

Another of Darwin's circle was in Switzerland in the summer: John Lubbock briefly met up with Tyndall and Huxley in August while there to examine the newly discovered prehistoric lake-dwellings (see letter from John Lubbock, 23 August 1862). Lubbock was a mine of information on the new archaeological discoveries then flooding in: earlier in the year he had been to view the prehistoric sites near Amiens (see letter from John Lubbock, 15 May 1862), and he was also anxious to have Darwin's view on his increasingly radical ideas about the antiquity of the human species (see letter from John Lubbock, 6 January 1862).

Ramsay's was not the only significant geological paper of 1862. In May, Darwin heard from Joseph Beete Jukes, director of the Irish branch of the Geological Survey, that he had a new theory to account for the topography and drainage system of southern Ireland (see letter from J. B. Jukes, 25 May 1862). In his paper, which marked the start of a revival of fluvialism in Britain, Jukes had argued that rivers not only excavate their valleys, but that they adjust their courses according to the underlying geological structures. Anxious to hear 'if any one can pick a hole in the reasoning', Jukes sought Darwin's opinion. Darwin replied with a plethora of detail (see letter from J. B. Jukes, 30 May 1862).

There was further evidence of glacial action in the antipodes, which Darwin diligently accumulated for the next edition of *Origin*. In January, William Branwhite Clarke sent new evidence from Australia, but Darwin was particularly pleased to have a new and promising correspondent in New Zealand—the provincial geologist, Julius von Haast, who sent valuable evidence of glacial action, and more besides, from the little-explored Southern Alps. This, Darwin considered, supported his view that the glacial period had affected the whole globe, and that it was the leading cause of the geographical distribution of animals and plants. Despite some worrying counter-arguments from Bates, who could not reconcile the distribution of South American insects with Darwin's hypothesis (see letter from H. W. Bates, 30 April 1862), Darwin was unshaken. He was particularly pleased with new botanical evidence from the Cameroons mountains supporting his view, telling Hooker: 'I will swear that the mundane glacial period is as true as gospel, so it must be true' (to J. D. Hooker, 9 May [1862]).

Hooker's reflections on the geographical distribution of plants led to a spirited discussion of their differing views on the subject. Darwin, reading Hooker's recently published paper on the flora of the Arctic, commented that such papers were 'the real engine to compel people to reflect on modification of species' (letter to J. D. Hooker, 25 February [1862]). Hooker had found Greenland to be 'unaccountably poor in plants', a fact that seemed to counter the view that it had been 'populated by migration since the glacial epoch' (letter from J. D. Hooker, 3 March 1862). Darwin found the case 'very curious' and, although a long-standing opponent of attempts to account for current distribution by proposing the existence of former land-bridges between one land-mass and another, he declared himself 'more willing . . . to admit a recent continental extension there than almost anywhere else' (letter to J. D. Hooker, 7 March [1862]).

Darwin was also busy with palaeontological matters. In the new second edition of Owen's *Palaeontology*, which Huxley thought 'rather more scoundrelly' than the first, Owen had committed himself more openly to evolutionary views, though not to natural selection (see letter from T. H. Huxley, 13 January 1862). Darwin told Huxley: 'he is so dishonest that I really now care little what he says' (letter to T. H. Huxley, 14 [January 1862]), but he was nonetheless irritated by Owen's lectures at the Museum of Practical Geology in May. As he told Armand de Quatrefages, the abuse commonly heaped on him did not 'in the least' hurt him, but it was different 'when the abuse comes from an old friend, like Prof. Owen, who a⟨bu⟩ses me & then advances the doctrine that all Birds are probably descended from one parent' (letter to Armand de Quatrefages, 11 July [1862]).

Much more satisfactory was the paper sent to Darwin by one of Owen's adversaries, Hugh Falconer, concerning the fossil and recent species of elephant. The paper included a section on the persistence of the specific characters of the mammoth over long periods of time. Falconer, anxious not to be thought an opponent, sought to placate Darwin, contrasting his own behaviour with Owen's—'a blackguard friend of yours and mine' (see letter from Hugh Falconer, 24–7 September [1862]). However, having read the paper, Darwin was delighted to find Falconer coming round to evolution. 'There will not be soon one good palaeontologist who believes in immutability', he told Lyell (letter to Charles Lyell, 1 October [1862]).

Huxley also wanted Darwin's palaeontological opinion: what did Darwin think of his argument, put forward in his anniversary address to the Geological Society in February, that the fossil record was 'only the skimmings of the pot of life'? (letter from T. H. Huxley, 6 May 1862). In his address, Huxley also challenged the assumption that 'geological contemporaneity' could be equated with 'chronological synchrony'. Darwin, in reply, referred him back to *Origin* (letter to T. H. Huxley, 10 May [1862]):

> I fully agree with "your skimming-of pot-theory" & very well you have put it.— With respect contemporaneity, I nearly agree with you, & if you will look to the d—d— Book . . . you will find nearly similar remarks.

The 'd—d— Book' was doing good service; Darwin was confident that the views it contained would continue to gain acceptance, that though much in it would be 'proved rubbish', the 'frame-work' would stand (letter to Hugh Falconer, 1 October [1862]). As he put it in a letter to Brown-Séquard on 16 April [1862]:

> I have not the least doubt that I have erred most seriously on many points; but now so many . . . really good judges concur *in the main* with me, that I do not fear that some such view will ultimately prevail[.]

ACKNOWLEDGMENTS

The editors are grateful to Mr George Pember Darwin for permission to publish the Darwin letters and manuscripts. They also thank the Syndics of the Cambridge University Library and other owners of the manuscript letters who have generously made them available.

The work for this edition has been supported by grants from the National Endowment for the Humanities (NEH) and the National Science Foundation (NSF). The Alfred P. Sloan Foundation and the Andrew W. Mellon Foundation provided grants to match NEH funding, and the Mellon Foundation awarded grants to Cambridge University that made it possible to put the entire Darwin correspondence into machine-readable form. Research and editorial work have also been supported by grants from the Royal Society of London, the British Academy, the Pilgrim Trust, and the Isaac Newton Trust. The Royal Society has helped meet the costs of publication. The Stifterverband für die Deutsche Wissenschaft has provided funds to edit Darwin's correspondence with German naturalists.

The Cambridge University Library and the American Philosophical Society (APS) have generously made working space and many services available to the editorial staff.

Since the project began in 1975, we have been fortunate in benefiting from the interest, experience, and practical help of many people in many places, and the editors hope that they have adequately expressed their thanks to them individually as the work proceeded. There are some, however, who have helped over so long a period that it would be ungracious not to thank them personally in the work that has profited so much from their co-operation.

Over the years, Ronald J. Overmann, former program director for History and Philosophy of Science of NSF, has given far more time and attention to the project than his formal duties required. Margaret Scrymser of NEH has also provided valuable direction. The editors would like to acknowledge with gratitude the assistance rendered to the project by Frank H. T. Rhodes, the late John E. Sawyer, and the late Shepard Stone.

Without the expert help of John L. Dawson of the Literary and Linguistic Computing Centre of Cambridge University, the computerisation of the correspondence would not have been possible, for the work on both the *Calendar* and the *Correspondence* required the solution of many novel technical problems.

The Cambridge University computer and that of the Cambridge University Library were used for initial work on this edition and for proof production. W. D. S. Motherwell, former head of automation at the Library, arranged for the installation of terminals in the project's office in the Manuscripts Department and solved the operational problems that arose. More recently, Paul Ayris and Chris Sendall have assisted us in accessing the library's laser printer. Peter Dunn and Adrian Miller of the automation department have helped with the project's numerous queries. Simon Buck has offered invaluable assistance in preparing automatic typesetting procedures and has acted as typesetting advisor to the project.

The camera-ready copy for this book was produced at the Oxford University Computing Service on a Monotype Prism Imagesetter. We are grateful to the Oxford operations staff for their advice and assistance.

Solene Morris, curator of the Darwin Museum at Down House, Downe, Kent, welcomed the editors on numerous visits and responded most generously to frequent requests for information and for material from the collections at Down. Richard Darwin Keynes kindly made available Darwin family material in his possession. Ursula Mommens has also provided letters and other materials that belonged to her grandfather, Francis Darwin. The help of Peter J. Gautrey, who, until 1989, directed the Darwin Archive at the Cambridge University Library, has been of special importance. His great knowledge of Darwiniana has always been readily shared with the editors, as with countless other Darwin scholars.

Libraries all over the world have given help that was literally indispensable by making available photocopies of Darwin correspondence and other manuscripts in their collections. The institutions and individuals that furnished copies of letters for this volume can be found in the List of provenances. The editors are extremely indebted to them all. We are also grateful to the many people who have transmitted information regarding the whereabouts of particular letters and, in many cases, have generously provided copies of such letters for the project's use.

The editors make daily use of the incomparable facilities of the Cambridge University Library and have benefited greatly from its services and from the help and expertise of its staff, particularly the staff of the Manuscripts Department. We are especially grateful to the former University Librarian, Frederick W. Ratcliffe, to his successor, Peter K. Fox, and to the Under-librarian of Manuscripts, Patrick Zutshi, for their generous support. Other members of the library's staff who have frequently responded to the editors' requests are: M. C. Allen, Gerry Bye, Kathleen Cann, Roger Fairclough, David Hall, John Hall, Brian Jenkins, Morag Law, Elisabeth Leedham-Green, David Lowe, Peter Meadows, William Noblett, Adam Perkins, Jayne Ringrose, Godfrey Waller, and Cynthia Webster. The fetchers in the Manuscripts reading room have also patiently dealt with the editors' often complex requirements.

At the American Philosophical Society Library, a splendid collection of Darwiniana and works in the history of science has been continuously available to the editorial staff since the inception of the project. Whitfield J. Bell Jr, secretary of the society until 1983, serves on the United States Advisory Committee for the project and has done his utmost to further its work. The editors have also benefited from the co-operation of Edward Carter II, Roy C. Goodman, Carl F. Miller, Elizabeth Carroll-Horrocks, and Bertram Dodelin, all of the APS Library.

The editors would like to acknowledge the assistance of the late Douglas W. Bryant, formerly of the Harvard University Library, Rodney Dennis, Jennie Rathbun, and Susan Halpert of the Houghton Library, Constance Carter of the Science Division of the Library of Congress, and Judith Warnement and Jean Cargill of the Gray Herbarium of Harvard University, who have all been exceptionally helpful in providing material from the collections in their charge.

In Britain, the editors have often received assistance from Mary Sampson, archivist of the Royal Society; Rex E. R. Banks and John Thackray of the Natural History Museum Library; Gina Douglas, librarian of the Linnean Society of London; and Sylvia FitzGerald, Cheryl Piggott, and Leonora Thompson of the Library of the Royal Botanic Gardens, Kew. We would also like to thank Richard G. Williams, Mrs Felton, and Anne Barrett (college archivist) of Imperial College of Science, Technology and Medicine; Christine Fyfe, archivist of Keele University; Salma Zabaneh of the International Bee Research Association; and Dr and Mrs Courtney of Christ's College, Cambridge. Invaluable assistance has been received from local studies librarians and county archivists, too numerous to name individually, although Elizabeth Silverthorne of Bromley Central Library and Donald Gibson of the Centre for Kentish Studies, deserve special mention for their repeated assistance. Virginia Murray of John Murray (publishers) Ltd has graciously provided the editors with copies of the Darwin correspondence in the Murray archive and with related information. Leonard G. Wilson kindly provided the editors with photocopies of manuscript material in the possession of the family of Sir Charles Lyell.

Among the others who have advised and assisted the editors in their work are Janet Browne, Nick Gill, Julius Held, Martin Rudwick, R. J. G. Savage, Jim Secord, Garry J. Tee, Hugh Torrens, and Morton N. Cohen. The editors are also pleased to acknowledge the invaluable support of the members of the British and United States Advisory Committees. Continued thanks are due to William Montgomery, former associate editor with this project, for his assistance in editing the letters in the collection of the American Philosophical Society. Gratitude is expressed to C. A. Tripp for providing the editors with an invaluable research tool by enabling them to access texts through search and retrieval software. The editors gratefully acknowledge the valuable assistance of Tim Benton, Sheila Dean, Arne Hessenbruch, Shelley Innes, Robert Lindsey, and Sarah Wilmot, and extend thanks to Jean Macqueen for providing the index to the volume.

Finally, we must acknowledge the contributions of four who left the staff during the preparation of this volume. Sarah Benton and Heidi Bradshaw ably guided the volumes through the editorial processes and production schedules; Joy Harvey edited French correspondence; and Thomas Junker edited German letters. We wish them well in all they undertake in the future.

LIST OF PROVENANCES

The following list gives the locations of the original versions of the letters printed in this volume. The editors are grateful to all the institutions and individuals listed for allowing access to the letters in their care.

American Philosophical Society, Philadelphia, PA, USA
Autographia (dealers), San Rafael, California
Bath Royal Literary and Scientific Institution, Bath, England
Bates 1892 (publication)
Bienen Zeitung (publication)
British Library, The Manuscript Collections, Great Russell Street, London, England
Cambridge University Library, Manuscripts department, Cambridge, England
Christie's (dealers)
Christ's College Library, Cambridge, England
Cleveland Health Sciences Library, Cleveland, OH, USA
CUL. *See* Cambridge University Library
Dana Family Papers, Manuscripts and Archives, Yale University, New Haven, CT, USA
DAR. *See* Cambridge University Library
Down House, Downe, Kent, England
Edinburgh University Library, Edinburgh, Scotland
Forschungsinstitut und Natur-Museum Senckenberg, Frankfurt, Germany
Milton D. Forsyth, Jr (private collection)
Gardeners' Chronicle and Agricultural Gazette (publication)
Gray Herbarium of Harvard University, Cambridge, MA, USA
Houghton Library for Rare Books and Manuscripts, Harvard University, Cambridge, MA, USA
C. H. Hughes-Johnson (private collection)
Imperial College of Science, Technology and Medicine, South Kensington, London, England
Journal of Horticulture and Cottage Gardener (publication)
La Fondation Augustin de Candolle
Leeds University Library, Leeds, England
Lehigh University Libraries, Bethlehem, PA, USA
Linnean Society of London, Burlington House, London, England
Mitchell Library, State Library of New South Wales, Sydney, Australia
John Murray (publishers), London, England
Museum of Comparative Zoology, Harvard University, Cambridge, MA, USA

National Library of Scotland, Edinburgh, Scotland
New York Botanical Garden Library, Bronx, New York, NY, USA
R. D. Pyrah (private collection)
Remember When Antiquities (dealers)
Paul C. Richards Autographs (dealers), Templeton, MA, USA
Royal Botanic Gardens, Kew, Richmond, Surrey, England
Royal College of Physicians, Regent's Park, London, England
Royal Horticultural Society, Vincent Square, London, England
Elizabeth Rütimeyer (private collection)
Smithsonian Insitution Libraries, Washington DC, USA
Sotheby (dealers)
Thomas Fisher Rare Book Library, University of Toronto, Toronto, Ontario, Canada
Mrs Ralph Vaughan Williams
Wellcome Institute for the History of Medicine Library, London, England
J. G. Zimmer (private collection)

A NOTE ON EDITORIAL POLICY

The first and chief objective of this edition is to provide complete and authoritative texts of Darwin's correspondence. In so far as it is possible, the letters have been dated, arranged in chronological order, and the recipients or senders identified. Darwin seldom wrote the full date on his letters and, unless the addressee was well known to him, usually wrote only 'Dear Sir' or 'Dear Madam'. After the adoption of adhesive postage stamps in the 1840s, the separate covers that came into use with them were usually not preserved, and thus the dates and the names of many recipients of Darwin's letters have had to be derived from other evidence. The notes made by Francis Darwin on letters sent to him for his editions of his father's correspondence have been helpful, as have matching letters in the correspondence, but many dates and recipients have had to be deduced from the subject-matter or references in the letters themselves. These tasks, together with deciphering the handwriting of Darwin and others, have presented the most troublesome editorial problems.

Whenever possible, transcriptions have been made from manuscripts. If the manuscript was inaccessible but a photocopy or other facsimile version was available, that version has been used as the source. In many cases, the editors have had recourse to Francis Darwin's large collection of copies of letters, compiled in the 1880s. Other copies, published letters, or drafts have been transcribed when they provide texts that are otherwise unavailable.

The method of transcription employed in this edition is adapted from that described by Fredson Bowers in 'Transcription of manuscripts: the record of variants', *Studies in Bibliography* 29 (1976): 212–64. This system is based on accepted principles of modern textual editing and has been widely adopted in literary editions.

The case for using the principles and techniques of this form of textual editing for historical and non-literary documents, both in manuscript and print, has been forcefully argued by G. Thomas Tanselle in 'The editing of historical documents', *Studies in Bibliography* 31 (1978): 1–56. The editors of the *Correspondence* followed Dr Tanselle in his conclusion that a 'scholarly edition of letters or journals should not contain a text which has editorially been corrected, made consistent, or otherwise smoothed out' (p. 48), but they have not wholly subscribed to the statement made earlier in the article in which he says, 'In the case of notebooks, diaries, letters and the like, whatever state they are in constitutes their finished form, and the question of whether the writer "intended" something else is irrelevant' (p. 47). The editors have preserved the spelling, punctuation, and grammar of the original, but they have found it impossible to set aside entirely the question of authorial intent. One obvious reason is that in reading Darwin's writing, there must necessarily be reliance upon both context and intent. Even when Darwin's general intent is clear, there are cases in which alternative readings are, or may be, possible, and therefore the transcription decided upon must to some extent be conjectural. When the editors were uncertain of their transcription, the doubtful text has been enclosed in italic square brackets.

A major editorial decision was to adopt the so-called 'clear-text' method of transcription, which so far as possible keeps the text free of brackets recording deletions, insertions, and other alterations in the places at which they occur. Darwin's changes are, however, recorded in the back matter of the volume, under 'Manuscript alterations and comments', in notes keyed to the printed text by paragraph and line number. All lines above the first paragraph of the letter (i.e., date, address, or salutation) are referred to as paragraph '0'. Separate paragraph numbers are used for subscriptions and postscripts. These notes enable the reader who wishes to do so to reconstruct the manuscript versions of Darwin's letters while furnishing printed versions that are uninterrupted by editorial interpolations. The Manuscript alterations and comments record all alterations made by Darwin in his letters and any editorial amendments made in transcription. For copies of Darwin letters included in the correspondence no attempt has been made to record systematically all alterations to the text, but ambiguous passages in copies are noted. The editors believe it would be impracticable to attempt to go further without reliable information about the texts of the original versions of the letters involved. Letters to Darwin have been transcribed without recording any of the writers' alterations unless they reflect significant changes in substance or impede the sense; in such cases footnotes will bring them to the reader's attention.

Misspellings have been preserved, even when it is clear that they were unintentional as, for instance, 'lawer' for 'lawyer'. Such errors often indicate excitement or haste and may exhibit, over a series of letters, a habit of carelessness in writing to a particular correspondent or about a particular subject.

Capital letters have also been transcribed as they occur except in certain cases, such as 'm', 'k', and 'c', which are frequently written somewhat larger than others as initial letters of words. In these cases an attempt has been made to follow the normal practice of the writers.

In some instances that are not misspellings in a strict sense, editorial corrections have been made. In his early manuscripts and letters Darwin consistently wrote 'bl' so that it looks like 'lb' as in 'albe' for 'able', 'talbe' for 'table'. Because the form of the letters is so consistent in different words, the editors consider that this is most unlikely to be a misspelling but must be explained simply as a peculiarity of Darwin's handwriting. Consequently, the affected words have been transcribed as normally spelled and no record of any alteration is given in the textual apparatus. Elsewhere, though, there are misformed letters that the editors have recorded because they do, or could, affect the meaning of the word in which they appear. The main example is the occasional inadvertent crossing of 'l'. When the editors are satisfied that the intended letter was 'l' and not 't', as, for example, in 'stippers' or 'istand', then 'l' has been transcribed, but the actual form of the word in the manuscript has been given in the alteration notes.

If copies of letters provide the only source, the editors have frequently retained corrections made to the text when it is clear that they were based upon comparison with the original. Francis Darwin's corrections of misreadings by copyists have usually been followed; corrections to the text that appear to be editorial alterations have not been retained.

Editorial interpolations in the text are in square brackets. Italic square brackets enclose conjectured readings and descriptions of illegible passages. To avoid confusion, in the few instances in which Darwin himself used square brackets, they have been altered by the editors to parentheses with the change recorded in the alteration notes. In letters to Darwin, square brackets have been changed to parentheses silently.

Material that is irrecoverable because the manuscript has been torn or damaged is indicated by angle brackets; any text supplied within them is obviously the responsibility of the editors. Occasionally, the editors are able to supply missing sections of text by reference to transcripts or photocopies of manuscript material made before the damage occurred.

Words and passages that have been underlined for emphasis are printed in italics in accordance with conventional practice. When the author of a letter has indicated greater emphasis by underlining a word or passage two or more times, then the text is printed in bold type.

Paragraphs are often not clearly indicated in the letters. Darwin and others sometimes marked a change of subject by leaving a somewhat larger space than usual between sentences; sometimes Darwin employed a longer dash. In these cases, and in very long stretches of text when the subject is clearly changed, a new paragraph is started by the editors without notice. The start of letters, valedictions, and postscripts are also treated as new paragraphs regardless of whether they appear as such in the manuscript. Special manuscript devices delimiting sections or paragraphs, e.g., blank spaces left between sections of text and lines drawn across the page, are treated as normal paragraph indicators and are not specially marked or recorded unless their omission leaves the text unclear.

Additions to a letter that run over into the margins, or are continued at its head or foot, are transcribed at the point in the text at which the editors believe they were intended to be read. The placement of such an addition is only recorded in a footnote if it seems to the editors to have some significance or if the position at which it should be transcribed is unclear. Enclosures are transcribed following the letter.

The hand-drawn illustrations and diagrams that occur in some letters are reproduced as faithfully as possible and are usually positioned as they were in the original text. In some cases, however, it has been necessary to reduce the size of a diagram or enhance an outline for clarity; any such alterations are recorded in footnotes. The location of diagrams within a letter is sometimes changed for typesetting reasons. Tables have been reproduced as close to the original format as possible given typesetting constraints.

Occasionally punctuation marking the end of a clause or sentence is not present in the manuscript, but the author has made his or her intention clear by allowing, for example, extra space or a line break to function as punctuation. In such cases, the editors have inserted an extra space following the sentence or clause to set it off from the following text.

Darwin consistently used the term 'fertilisation' for both fertilisation (the fusion of sperm with egg) and pollination (the transfer of pollen from anther to stigma); the first recorded usage of the term 'pollination' was in 1875 (*OED*). Darwin's use of 'fertilisation' in his letters and publications often, but not always, actually describes pollination. In the footnotes, the editors have attempted to use these terms in the context of Darwin's descriptions of experiments. When he is quoted directly, his original word usage is never altered.

Some Darwin letters and an occasional letter to Darwin are known only from entries in the catalogues of book and manuscript dealers or mentions in other published sources. Whatever information these sources provide about the content of such letters has been reproduced without change.

For every Darwin letter, the text that is available to the editors is always given in full. Some other items, however, are not printed in their entirety. Some memorandums and other documents that are not letters but are relevant to the correspondence have been summarised, as have those letters to Darwin that the editors consider can be presented adequately in shortened form. In a few cases the editors have included letters that are not to or from Darwin if they relate significantly to the published correspondence. Letters that can only be given a wide date range, in some instances spanning several decades, will be printed together in one of the concluding volumes.

The format in which the transcriptions are printed in the *Correspondence* is as follows:

1. *Order of letters*. The letters are arranged in chronological sequence. A letter that can be dated only approximately is placed at the earliest date on which the editors believe it could have been written. The basis of a date supplied by the editors is given in a footnote unless it is derived from a postmark, watermark, or endorsement as recorded in the physical description of the letter (see section 4, below). Letters with the same date, or with a range of dates commencing with that date, are printed in the alphabetical order of their senders or recipients unless their contents dictate a clear alternative order. Letters dated only to a year or a range of years precede letters that are dated to a particular month or range of months, and these, in turn, precede those that are dated to a particular day or to a range of dates commencing with a particular day.

2. *Headline*. This gives the name of the sender or recipient of the letter and its date. The date is given in a standard form, but those elements not taken directly from the letter text are supplied in square brackets.

3. *The letter text*. The transcribed text follows as closely as possible the layout of the

source, although no attempt is made to produce a type-facsimile of the manuscript: word-spacing and line-division in the running text are not adhered to. Similarly, the typography of printed sources is not replicated. Dates and addresses given by authors are transcribed as they appear, except that if both the date and address are at the head of the letter they are always printed on separate lines with the address first, regardless of the exact manuscript order. If no address is given on a letter by Darwin, the editors have supplied one, when able to do so, in square brackets at the head of the letter. Similarly, if Darwin was writing from an address different from the one given on the letter, his actual location is given in brackets. Addresses on printed stationery are transcribed in italics. Addresses, dates, and valedictions have been run into single lines to save space, but the positions of line-breaks in the original are marked by vertical bars.

4. *Physical description*. All letters are complete and in the hand of the sender unless otherwise indicated. If a letter was written by an amanuensis, or exists only as a draft or a copy, or is incomplete, or is in some other way unusual, then the editors provide the information needed to complete the description. Postmarks, endorsements, and watermarks are recorded only when they are evidence for the date or address of the letter.

5. *Source*. The final line provides the provenance of the text. Some sources are given in abbreviated form (e.g., DAR 140: 18) but are listed in full in the List of provenances unless the source is a published work. Letters in private collections are also indicated. References to published works are given in author–date or short-title form, with full titles and publication details supplied in the Bibliography at the end of the volume.

6. *Darwin's annotations*. Darwin frequently made notes in the margins of the letters he received, scored significant passages, and crossed through details that were of no further interest to him. These annotations are transcribed or described following the letter text. They are keyed to the letter text by paragraph and line numbers. Most notes are short, but occasionally they run from a paragraph to several pages, and are sometimes written on separate sheets appended to the letter. Extended notes relating to a letter are transcribed whenever practicable following the annotations as 'CD notes'.

Quotations from Darwin manuscripts in footnotes and elsewhere, and the text of his annotations and notes on letters, are transcribed in 'descriptive' style. In this method the alterations in the text are recorded in brackets at the places in which they occur. For example:

'See Daubeny ['vol. 1' *del*] for *descriptions of volcanoes in [*interl*] S.A.' *ink*

means that Darwin originally wrote in ink 'See Daubeny vol. 1 for S.A.' and then deleted 'vol. 1' and inserted 'descriptions of volcanoes in' after 'for'. The asterisk before 'descriptions' marks the beginning of the interlined phrase, which ends at the bracket. The asterisk is used when the alteration applies to more than the immediately preceding word. The final text can be read simply by skipping the material in brackets. Descriptive style is also used in the Manuscript alterations and comments.

Volumes of the *Correspondence* are published in chronological order. Each volume is self-contained, having its own index, bibliography, and biographical register. A chronology covering the period of each volume is supplied, and Appendixes give additional material as appropriate to assist the understanding of the correspondence. A cumulative index is planned once the edition is complete. References are supplied for all persons, publications, and subjects mentioned, even though some repetition of material in earlier volumes is involved.

If the name of a person mentioned in a letter is incomplete or incorrectly spelled, the full, correct form is given in a footnote. Brief biographies of persons mentioned in a volume, and dates of each correspondent's letters to and from Darwin, are given in the Biographical register and index to correspondents.

Short titles are used for references to Darwin's books and articles and to collections of his letters (e.g., *Descent*, 'Parallel roads of Glen Roy', *LL*). They are also used for some standard reference works and for works with no identifiable author (e.g., *Alum. Cantab.*, *Wellesley index*, *DNB*). For all other works, author–date references are used. The full titles of all the books referred to are given in the Bibliography.

Occasional supplements will be published containing letters that have come to light or have been redated since the relevant volumes of the *Correspondence* appeared. The first such supplement was included in Volume 7 and covered the years from 1821 to 1857.

The editors use the abbreviation 'CD' for Charles Darwin throughout the footnotes. A list of all abbreviations used by the editors in this volume is given on p. xl.

The Wedgwoo

Robert Waring Darwin 1766–1848 = Susannah Wedgwood 1765–1817

- Marianne 1798–1858 = Henry Parker 1788–1856
  - Robert b. 1825
  - Henry 1827–92
  - Francis 1829–71
  - Charles b. 1831
  - Mary Susan 1836–93
- Caroline Sarah 1800–88 = Josiah III (Jos) 1795–1880
  - Sophy Marianne 1838–9
  - Katherine Elizabeth Sophy (Sophy) 1842–1911
  - Margaret Susan 1843–1937
  - Lucy Caroline 1846–1919
- Susan Elizabeth 1803–66
- Erasmus Alvey 1804–81
- Charles Robert 1809–82 = Emma 1808–96
  - William Erasmus 1839–1914
  - Anne Elizabeth 1841–51
  - Mary Eleanor Sept.-Oct. 1842
  - Henrietta Emma (Etty) 1843–1927
  - George Howard 1845–1912
  - Elizabeth 1847–1926
  - Francis (Frank) 1848–1925
  - Leonard 1850–1943
  - Horace 1851–1928
  - Charles Waring 1856–8
- Emily Catherine (Catherine) 1810–66

win Families up to 1862

- Josiah Wedgwood II 1769–1843 = Elizabeth (Bessy) Allen 1764–1846
  - Sarah Elizabeth (Elizabeth) 1793–1880
  - Mary Anne 1796–8
  - Charlotte 1797–1862 = Charles Langton 1801–86
    - Edmund 1841–75
  - Henry Allen (Harry) 1799–1885 = Jessie Wedgwood 1804–72
    - Louisa Frances 1834–1903
    - Caroline b. 1836
    - John Darwin 1840–70
    - Anne Jane 1841–77
    - Arthur 1843–1900
    - Rowland 1847–1921
  - Frances Mosley 1808–74 = Francis (Frank) 1800–88
    - Godfrey 1833–1905
    - Amy 1835–1910
    - Cicely Mary 1837–1917
    - Clement 1840–89
    - Lawrence 1844–1913
    - Constance Rose 1846–1903
    - Mabel Frances 1852–1930
  - Hensleigh 1803–91 = Frances (Fanny) Mackintosh 1800–89
    - Frances Julia (Snow) 1833–1913
    - James Mackintosh (Bro) 1834–64
    - Ernest Hensleigh 1838–98
    - Katherine Euphemia (Effie) 1839–1931
    - Alfred Allen 1842–92
    - Hope Elizabeth 1844–1934
  - Frances (Fanny) 1806–32

ABBREVIATIONS

AL	autograph letter
ALS	autograph letter signed
LS	letter in hand of amanuensis, signed by sender
LS(A)	letter in hand of amanuensis with additions by author
Mem	memorandum
(S)	signed with sender's name by amanuensis
CD	Charles Darwin
CUL	Cambridge University Library
DAR	Darwin Archive, Cambridge University Library
del	deleted
illeg	illegible
interl	interlined
underl	underlined

TRANSCRIPTION CONVENTIONS

[some text]	'some text' is an editorial insertion
/some text/	'some text' is the conjectured reading of an ambiguous word or passage
[some text]	'some text' is a description of a word or passage that cannot be transcribed, e.g., '*3 words illeg*'
⟨ ⟩	word(s) destroyed
⟨some text⟩	'some text' is a suggested reading for a destroyed word or passage
⟨*some text*⟩	'some text' is a description of a destroyed word or passage, e.g., '*3 lines excised*'

THE CORRESPONDENCE OF CHARLES DARWIN
1862

From Ellen Frances Lubbock to Emma Darwin [January 1862][1]

Dear M^rs^. Darwin

I am writing to as⟨k if⟩ M^r^. Darwin could be pe⟨rsuaded⟩ to come here next Sunday either to early dinner at $\frac{1}{4}$ before one o'clock—or to stay all night— He would see *nobody* but John, not even me, most likely, for I am not quite strong yet—[2] & I will take care his bed is well aired, ⟨ ⟩ will tell me ⟨ ⟩ likes & has at ⟨ ⟩ I will take care ⟨ ⟩ all just the same— ⟨Jo⟩hn doesn't know I am writing this, but I have heard him lately so often wish he "could *only* have a talk with M^r^. Darwin", & I know he can't manage to get over to Downe[3] now as he used to do from High Elms—so I thought I would just *try* what I could do to persuade M^r^. Darwin to come here—[4] If you will send me a line by post I shall get it on Sunday morning.

With my best wishes for very many Happy New Years to you all I am always yours very truly | E F Lubbock

DAR 170.1: 9

[1] The date is established by the New Year's greeting, by the reference to the Lubbocks' move from High Elms, near Down (see n. 4, below), and by the relationship to the letter from John Lubbock, 6 January 1862.

[2] Ellen Lubbock had given birth to her fourth child, Norman, on 16 December 1861 (*Gentleman's Magazine* n.s. 12 (1862): 82; Hutchinson 1914).

[3] The spelling of the village name was variable. 'Down' was the commonest form up to the 1870s, and 'Downe' thereafter.

[4] John Lubbock moved from High Elms, the home of his father John William Lubbock, to Chislehurst, a village about five miles north of Down, on 19 August 1861 (John Lubbock's diary (British Library, Add. Ms. 62679: 64 r.)). See also letter from John Lubbock, 6 January 1862.

From Joseph Dalton Hooker [1 January 1862][1]

Dear Darwin

The *Heterocentron* is just out of flower, but more flowers will be open in a day or two I send *Centradenia grandiflora* & *floribunda*;[2] would you care for a tetrandrous Melast. a *Sonerila* is now in flower.[3]

I send *Eulophia viridis*.[4]

Borrer is a very nice man, but very aged indeed—[5] he will be delighted to send you seeds— he is a man of large property & a wonderfully acute British Botanist of the old school— Borrer & my Father traversed Scotland on horse-back in 1810! & explored Sutherland—[6]

Ev Yrs affec | J D H

What is the matter with your boys?[7]

DAR 101: 5

CD ANNOTATION

Top of page: '4'[8] *brown crayon, circled brown crayon*

[1] Dated by the relationship to the letter to J. D. Hooker, [30 and 31 December 1861] (*Correspondence* vol. 9), and by the reference to Hooker sending specimens of *Centradenia grandiflora* and *C. floribunda* (see n. 2, below).

[2] Hooker was assistant director of the Royal Botanic Gardens, Kew; there are observational notes, dated 'Jan 2d.', on specimens of these two species supplied from Kew, in DAR 205.8: 16. See also n. 3, below.

[3] CD had asked Hooker to arrange for specimens of *Heterocentron* and other members of the Melastomataceae to be sent to him (see *Correspondence* vol. 9, letter to J. D. Hooker, [30 and 31 December 1861]). In October 1861 CD had begun to investigate the occurrence of two different sets of stamens in the flowers of these plants, the structure and colour of the stamens facing the petals often differing from that of those facing the sepals (see *Correspondence* vol. 9, letter to J. D. Hooker, 17 November [1861] and n. 14). He suspected that the Melastomataceae might exhibit a novel form of dimorphism (see letter to George Bentham, 3 February [1862]), and continued to work on the family throughout 1862 and 1863 without ultimately being able to account for the two sets of stamens (see *Cross and self fertilisation*, p. 298 n., and *ML* 2: 292–302). CD's notes from these experiments are in DAR 205.8.

[4] Hooker offered CD a specimen of *Eulophia viridis* in his letter of [29 December 1861] (*Correspondence* vol. 9); CD accepted his offer in the letter to J. D. Hooker, [30 and 31 December 1861] (*ibid.*). The species is discussed in *Orchids*, pp. 189 and 283.

[5] CD asked about the propriety of requesting seeds or plants from William Borrer in the letter to J. D. Hooker, 18 [December 1861] (*Correspondence* vol. 9). Borrer, a wealthy botanist with an extensive plant collection, was 80 years old.

[6] William Jackson Hooker made two successive botanical excursions through Scotland for his *Flora Scotica* (W. J. Hooker 1821); he referred to his trip with Borrer in *ibid.*, p. ix. In his memoir of his father, Joseph Hooker gave the date of this excursion as 1808 (J. D. Hooker 1903, p. xiii).

[7] In the letter to J. D. Hooker, [30 and 31 December 1861] (*Correspondence* vol. 9), CD mentioned that two of his sons were 'bad'. According to Emma Darwin's diary (DAR 242), 30 December 1861, both Francis and Horace Darwin were 'feverish'.

[8] This number probably relates to one of CD's portfolios of notes.

To Charles Edouard Brown-Séquard 2 January [1862][1]

Down. | Bromley. | Kent. S.E.

Jany. 2^{d}.

My dear Sir

I cannot resist the opportunity of thanking you for your kind message through Miss Pennington.[2] I am extremely much gratified to hear that you intend noticing my Book in some French Periodical.[3]

Whether you agree with me at all I do not know, as for anything like an approach to perfect agreement, it is out of the question in so complex a subject. But I shall be truly glad to read any criticisms from one who stands so very high in one of the very highest branches of Science, as you stand. Therefore would you have the great kindness to tell me some time when & where you publish any little notice on the subject.— Just at present I am the more glad of any notice in France, as a French Translation will appear very soon.

I daresay you do not remember it, but a year or two ago I was introduced to you at the Philosophical Club, & had some little to me interesting conversation with you.—[4]

Pray believe me My dear Sir | With sincere respect | Yours very faithfully | Charles Darwin

I do not know which Edition of my "Origin" you may have seen; but the 3[d]. Edit. is in some respect considerably improved, & it is this Edition which is now nearly translated into French.—[5]

Royal College of Physicians

1 The year is established by the reference to the French translation of *Origin* and by the relationship to the letter from C. E. Brown-Séquard, 13 January 1862.
2 Miss Pennington has not been identified.
3 See letter from C. E. Brown-Séquard, 13 January 1862 and n. 2.
4 CD may be referring to a meeting of the Philosophical Club of the Royal Society that he attended on 20 May 1858, at which Brown-Séquard was a guest speaker (Philosophical Club minutes, Royal Society; Bonney 1919, p. 139).
5 Clémence Auguste Royer's French translation of *Origin* (Royer trans. 1862) was published on 31 May 1862 by Guillaumin et cie (*Journal Générale de l'Imprimerie et de la Librairie* 2d ser. 6 (pt 3): 341). CD's publisher, John Murray, at CD's request, had sent a copy of the third edition of *Origin* to Royer in Lausanne, Switzerland, in September 1861 (see *Correspondence* vol. 9, letter to John Murray, 10 September [1861]).

From John Brodie Innes[1] 2 January [1862][2]

2 Coley Hill | Reading
2nd. Jan[ry].

Dear Darwin,

I hope before this reaches you, you will have received Quiz. He left here at 10 oClock. Johnny and I saw him off, and s⟨aid⟩ ⟨go⟩od-bye very affe⟨ctionately⟩[3] ⟨ ⟩ has been used ⟨ ⟩ ⟨ ⟩m wat⟨ ⟩ ⟨ ⟩
⟨ ⟩ all parties comforted.

We are preparing for our start on Monday. I wish the journey was either over, or not coming, but it is too late to shirk now—[4]

With every good wish for a happy new year for you ⟨all⟩ (The prospects across ⟨the At⟩lantic look a little ⟨ ⟩ today than yesterd⟨ay.⟩[5]

⟨Ev⟩er Yrs | ⟨My dear⟩ Dar⟨win⟩ | ⟨ ⟩ ⟨Innes⟩

DAR 167.1: 7

[1] John Innes, the Anglican incumbent of Down, changed his name to John Brodie Innes at the end of 1861 upon inheriting an entailed estate, Milton Brodie, near Forres in Scotland, from his cousin Eliza Brodie Dunn (see *Correspondence* vol. 9, letter from John Innes, [24 December 1861], and *County families* 1864); he moved to Milton Brodie shortly afterwards (see letter from J. B. Innes, 19 February [1862]).

[2] The year is given by the relationship to the letters to J. B. Innes, 15 December [1861] and 19 December [1861], and to the letter from John Innes, [24 December 1861] (*Correspondence* vol. 9).

[3] The Darwin family had agreed to look after the pet dog 'Quiz' belonging to Innes's son John William, then 15 years old (see *Correspondence* vol. 9, letter to John Innes, 15 December [1861]).

[4] Innes and his family were preparing to move to Scotland (see n. 1, above).

[5] Innes refers to the American Civil War. Britain had recently come close to a direct involvement in the conflict as a result of the so-called *Trent* affair. In November 1861 the United States navy seized two Confederate envoys from the *Trent*, a British mail packet. News that the United States government had acquiesced to British demands for their release did not reach London until 8 January 1862. For an account of the *Trent* affair, see Ferris 1977; see also *Correspondence* vol. 9, letter to Asa Gray, 11 December [1861], and letter from Asa Gray, 31 December 1861.

From Henry Holland [3–14] January [1862][1]

Brook ⟨Street⟩
Ja⟨ ⟩ ⟨ ⟩

My dear Charles,

Though writing on ⟨ ⟩ subject, I must say a few wor⟨ds⟩ in expression of my sympathy ⟨for⟩ Mrs Darwin & yourself, in the loss y⟨ou⟩ have just sustained.[2] I was led to ex⟨pect⟩ the event as too probable, from the account that had before reached. It must be some comfort to you all, that it should have occurred without any increase of suffering. I shall be anxious now to hear of M^r^. Langton.[3]

My purpose in writing is to speak of Lord Tankerville. I believe him to be at Chillingham, whither I directed my letter, enclosing the papers[4]

⟨I⟩ have heard nothing during the ⟨ ⟩ that has since elapsed; as I ⟨ ⟩ to have done I fear it is too ⟨ ⟩able an explanation, that he ⟨ ⟩dy sent the package, ⟨ ⟩ing to the directions & address given ⟨ ⟩ may be that he is not actually at ⟨Chillin⟩gham—or that he is waiting an ⟨occa⟩sion for procuring the object required, ⟨ ⟩ of the wild cattle having been lately killed It may be again (but I unwillingly admit this idea) that he has taken offence at the *Instructions*, which were a little downright in their tone, & not perhaps suff^y^ apologetic for the trouble given:[5] A Nobleman living a good deal alone, on a property bequeathed to him in *very encumbered state*, is apt to be somewhat irritable & fastidious. Nevertheless I am not disposed to consider this the cause of delay.

If it be that the package has already been sent, I shall gladly hear of this. If not, I will let you know, when I hear of, or from, Lord Tankerville, or if I see him in Town.

There is another supposition, that my letter may have been lost *in transitu* but this is not likely.

Let me see you, if you come to London., & believe me your's ever aff^ly^ | H Holland

DAR 166.2: 238

[1] The date range is established by the reference to the death of Charlotte Langton (see n. 2, below), and by the relationship to the letter from Henry Holland, 15 January [1862] and to the letters to Ludwig Rütimeyer, 15 [and 16] January [1862] and 11 February [1862]. See also n. 5, below.
[2] Charlotte Langton, Emma Darwin's sister, died on 2 January 1862 (Emma Darwin's diary (DAR 242)).
[3] Charles Langton.
[4] The reference is to the sixth earl of Tankerville, Charles Augustus Bennet, whose Chillingham Park estate in Northumberland was home to an ancient breed of cattle (see n. 5, below).
[5] CD had asked Holland, who was an acquaintance of Lord Tankerville's, to assist him in procuring the skull of one of the Chillingham bulls to send to Ludwig Rütimeyer (see *Correspondence* vol. 9, letter to Ludwig Rütimeyer, 5 December [1861]). For a discussion of the history of this breed, see Ritvo 1992.

To J. B. Innes [3] January [1862][1]

Down Bromley Kent
Jan. 2^d.

My dear Innes

I am heartily glad to say that Quiz arrived last night safe & sound (but with a cough) & has been running about the house quite happy & very polite to every human being, *including Cats*.—[2] Hearty thanks for your present.— Also for your two notes.—[3] We most sincerely wish wish you all the happiness which can be expected in this world. Very glad we should be to see you here; but I think you will find yourself so well suited to your new career, that you will remain.[4]

We are a sick house with 3 Boys in bed with very bad feverish colds,[5] so no more at present, except kind remembrances to M^rs. Innes & Johny, the young Laird—[6]

Ever my dear Innes | Yours very sincerely | Charles Darwin

Cleveland Health Sciences Library (Robert M. Stecher collection)

[1] Dated by the relationship to the letters from John Innes, [24 December 1861] (*Correspondence* vol. 9) and 2 January [1862]. In his letter of [24 December 1861] (*ibid.*), Innes stated that Quiz would be sent on Thursday 2 January; his letter of 2 January [1862] implies that this plan was adhered to. CD refers to Quiz having arrived 'last night', and so appears to have mistaken the date.
[2] See letter from J. B. Innes, 2 January [1862] and n. 3.
[3] CD probably refers to the letters from John Innes, [24 December 1861] (*Correspondence* vol. 9) and 2 January [1862].
[4] In his letter of 19 December [1861] (*Correspondence* vol. 9), CD wished Innes happiness in his 'new career as Scotch Laird or Chieftain'. On moving to Scotland, Innes became priest in charge of Milton Brodie Mission and chaplain to the bishop of Moray; he continued to be the non-resident incumbent of Down until 1869 (*Crockford's* 1894, Freeman 1978).
[5] Emma Darwin recorded in her diary (DAR 242) on 1 January 1862 that she returned home from a trip to find the 'boys ill of influenza'. See also letter from J. D. Hooker, [1 January 1862], n. 7.
[6] CD refers to Innes's wife Eliza Mary Brodie and son, John William Brodie.

From Henry Walter Bates 6 January 1862

King St Leicester
6 Jany 1862

My Dear Sir

I have now finished the rough draught of 5 chapters of my book & send you by this post the 2nd for perusal according to your kind offer to do so. I do not know how I shall repay you for all this.[1]

Your suggestions on 1st chapter shall be all carried out.[2] Your good opinion has delighted my Father & Brother as well as myself.[3] The recommendation to alter the *arrangement* ⟨ ⟩ putting the historical sketch last is just the kind of correction that I feel most to need.[4] As to your statement that the argument on the effects of climate on dress is original & you are certainly too generous:[5] I find there is something to the same effect in your "Journal", (popular ed. p 381)[6] as my notions were suggested whilst on the Amazons the two conclusions have been derived independently & therefore my remarks will be interesting; but I must quote your passage in a note.[7]

⟨ ⟩ loose statement of the rocks being cretacean is der⟨ived⟩ from Mr Wallace's book;[8] No ⟨fossil⟩ was found & therefore it is useless to employ the word. I shall feel obliged to introduce a little Geology—so much as I observed which contributes to explain the *latest changes* & therefore connects itself with Geograph. Distribution of the Fauna. I must consult some Geologist before I do it.— One of my little theories is that part of the Delta of the Amazons was land probably an island whilst most of the Alluvial plain was under water; & that the Amazon waters have broken through it. I shall be careful however not to advance too much in the book.

As to explaining the "We" ⟨you⟩r remark shows me that there must ⟨be so⟩mething very abrupt & unsatisfactory in my commencement! I will add a straight forward preface ⟨e⟩xplaining the object of journey of Mr ⟨W⟩allace & self &c &c.—or would you incorporate the explanation in the text of 1st chapter.[9]

Please criticize this 2nd chapter without reserve. I am not satisfied with the arrangement of the matter nor with the argument on the text of Burmeister's remarks.[10] But I know you will be candid with me as before.

Yours sincerely | H W Bates

C. Darwin esq.

DAR 160.1: 64

[1] Bates was writing an account of his travels in the Amazon region, undertaken between 1848 and 1859. CD had encouraged Bates to publish an account and had offered to read the '1st Chapter or two' of the manuscript; he repeated his offer to read a second chapter after reading the first in December 1861 (see *Correspondence* vol. 9, letters to H. W. Bates, 4 April [1861], 25 September [1861], 3 December [1861], and 15 December [1861]). *The naturalist on the River Amazons* was published in 1863.

[2] Having read the first manuscript chapter of Bates 1863, CD made suggestions as to how it might be improved (see *Correspondence* vol. 9, letter to H. W. Bates, 15 December [1861]). His main point was that Bates should give the vernacular as well as the scientific names of any British animals mentioned.

[3] Bates's younger brother, Frederick, had been a keen entomologist since childhood. Their father, Henry Bates, was a hosiery manufacturer in Leicester.

[4] CD suggested that Bates place his historical description of the town of Pará, Brazil, at the end of the first chapter (see *Correspondence* vol. 9, letter to H. W. Bates, 15 December [1861]).

[5] See *Correspondence* vol. 9, letter to H. W. Bates, 15 December [1861] and n. 2. In his first chapter, Bates argued against the widespread belief that the birds and insects of the tropics were generally larger and more beautiful than those in temperate zones. From this, and from the fact that it is usually only

the males of a species that display 'brilliant dress', he concluded that, contrary to popular opinion, climate had 'little or no direct influence in the matter' (Bates 1863, 1: 18–23).

6 Bates refers to the second edition of *Journal of researches*, which was issued as part of John Murray's Colonial and Home Library series. The passage Bates cites (p. 381) discusses the generally dull coloration of the birds, plants, and insects of the Galápagos islands compared with the brilliantly coloured products of other equatorial regions. CD concluded that:

> the usual gaudy colouring of the intertropical productions, is not related either to the heat or light of those zones, but to some other cause, perhaps to the conditions of existence being generally favourable to life.

7 Bates referred to CD's conclusions (see n. 6, above) in Bates 1863, 1: 21.

8 CD had remarked in his letter to H. W. Bates, 15 December [1861] (*Correspondence* vol. 9) that without good evidence Bates should avoid identifying rocks as 'Cretaceous'. In Wallace 1853, pp. 420–1, Alfred Russel Wallace discussed his difficulties in elucidating the geology of the Amazon valley, stating: 'It is remarkable that I was never able to find any fossil remains whatever . . . We are thus unable to assign the geological age to which any of the various beds of rock belong'. Later in the text, however, Wallace stated (p. 425): 'The neighbourhood of Pará consists entirely of a coarse iron sandstone, which is probably a continuation of the rocks observed by Mr. Gardner at Maranham and in the Province of Piauhy, and which he considered to belong to the chalk formation'.

9 See *Correspondence* vol. 9, letter to H. W. Bates, 15 December [1861]. Bates explained in a preface to *The naturalist on the River Amazons* (Bates 1863, 1: iii–vi) that the expedition to the river had begun as a joint undertaking with Wallace, 'for the purpose of exploring the Natural History of its banks'. They had intended, he continued, to 'gather facts, as Mr. Wallace expressed it . . ., "towards solving the problem of the origin of species," a subject on which we had conversed and corresponded much together.'

10 Bates contrasted his positive response to the sense of the 'vastness of nature' and the 'insignificance' of man, induced by the enormous animal and plant diversity of the tropical forest, with the experience of the German zoologist Karl Hermann Konrad Burmeister, who found the 'restless selfishness' of the forest vegetation painful (Bates 1863, 1: 53). Bates gave no citation, but the reference is probably to Burmeister 1853, p. 127.

From J. D. Hooker [30 December 1861 or 6 January 1862][1]

Kew
Monday.

Dr. Darwin

I am very glad that you have given up *Acropera* ovules—[2] I could not make them out to be so; but it is so awfully difficult to pronounce (in Botany) on anothers dissection that I could not say honourably they were not. I suppose however they would not be there, were they not rude representative efforts of ovule making: & in that sense may be rudimentary ovules.

I doubt if Phanerogams will show less differences in extreme forms than Phænogams even including your d——d barnacles.[3] Compare a good dichlamydeous hermaphrodite flowered tree with the Cycas or these with a *Pine*— Why, you have nothing floral in common but *ovule*—& that physiologically & structurally different—even Pollen different— Turning again to organs of Vegetation compare *Lemna* & above— I doubt if there is more difference between a complementary male Barnacle[4] & crab than between ♀ Lemna & ♂ Oak—or ♀ & ♂ of Balanophora—

The cases are however never parallel, you forget the *constant* difference between ♀ & ♂ flowers of all dioecious plants.— Then too on other hand we have fewer organs in plants. On the w⟨hole⟩ I stick to my idea that Vegetati⟨on⟩ == at least *Annulosa* & downwards amongst animals.— My conviction is that nature has established no systematic parallelism between Animals & plants— by the year x + n plants may have branched out to greatly higher differentiation than exists between man & monad—*or* may have receded as far. There is no parallism between groups of Animals; or between groups of plants ∴ none between Animals & plants sauce for Goose =∴ sauce for Gander

If any use to you I will go over my Arctic lists— & mark colour of all the flowers, that have a really decided color— is there any use saying color of *Alchemilla myriophyllum* & such like? I should like on many accounts to do this for arctic lists, but want the stimulus of knowing that it will be used by any one—so I hope you may want it.[5] I should be curious to see how it tallies with Lecoq.[6]

Ever yrs affect | J D Hooker

I cannot understand anyone being *confident*; or *satisfied* about such awful stretches of imagination as continental extensions; Atlantises & so forth— I strongly favor them, or at least look to much greater changes than you do—but dare not even propose to bridge N Zealand & Tasmania by any *direct* continent[7]

DAR 101: 3–4

CD ANNOTATIONS
1.3 honourably] *underl pencil; cross in margin, pencil*
2.1 I doubt ... will] *cross in margin, pencil*
2.1 forms ... Phænogams 2.2] *cross in margin, pencil*

[1] The possible dates are conjectured from the relationship to the letters to J. D. Hooker, 28 [December 1861] (*Correspondence* vol. 9) and 16 January [1862] (see nn. 2 and 5, below); both dates were Mondays. This letter is published in volume 10 of the *Correspondence* rather than volume 9 because the editors' former conjectured dating, [6 January 1862], was revised subsequent to volume 9's going to press.

[2] Following his observations of the orchid *Acropera* in October and November 1861, CD had come to believe that all the specimens he had examined were male (see *Correspondence* vol. 9, letter to Daniel Oliver, 30 November [1861]). In his letter to J. D. Hooker, 1 December [1861] (*ibid.*), CD mentioned that he had requested Daniel Oliver, assistant at the Kew herbarium, to examine specimens of the 'wretched rudimentary ovules' of *Acropera*. Having received a reply from Oliver, CD told Hooker in his letter of 28 [December 1861] (*ibid.*): 'I believe the membranous fringes are placentæ with no ovules or merest rudiments'.

[3] The reference to barnacles suggests that Hooker intended to write 'Crustacea' rather than 'Phænogams'; 'phaenogam' and 'phanerogam' both mean 'flowering plant' (*OED*). In his letter to Hooker, 18 [December 1861] (*Correspondence* vol. 9), CD stated: 'The more I read about Plants, the more I get to feel that all Phanerogams seem comparable with one class, as Lepidoptera, rather than with one Kingdom as the whole Insecta'. Following a response from Hooker in a missing letter, CD explained himself further in his letter of 28 [December 1861] (*Correspondence* vol. 9):

> I wrote carelessly about the value of Phanerogams; what I was thinking of was that the sub-groups seemed to blend so much more one into another than with most classes of animals. I suspect Crustacea would show more differences in the extreme forms than Phanerogams

[4] During his study of barnacles, CD had discovered what he termed 'complemental' males – minute parasitic forms attached to hermaphrodite species of *Ibla* and *Scalpellum* (see *Living Cirripedia* (1851): 207–14, 231–43, and *Correspondence* vol. 4, Appendix II, pp. 399–400).

[5] Hooker read two papers on the Arctic flora to the Linnean Society of London on 21 June 1860. His paper on the distribution of Arctic plants (J. D. Hooker 1861a) contains extensive species lists; the other, shorter paper describes the plants collected on a specific expedition and includes lists of more localised floras (J. D. Hooker 1861b). CD had expressed an interest in the relationship between flower colour and latitude in his letter to Hooker, 28 [December 1861] (*Correspondence* vol. 9).

[6] In Lecoq 1854–8, Henri Lecoq suggested a correlation between flower colour and latitude. CD began to read Lecoq's nine-volume work in December 1861 (see *Correspondence* vol. 9, letters to J. D. Hooker, [9 December 1861] and 28 [December 1861]). There is an annotated copy of Lecoq 1854–8 in the Darwin Library–CUL (see *Marginalia* 1: 488–95).

[7] In his introductory essay to the *Flora Novæ-Zelandiæ*, Hooker had suggested that the similarity between sections of the New Zealand flora and that of parts of South America, Australia, the Antarctic, and the Pacific could be explained by postulating former land connections (J. D. Hooker 1853–5, 1: xxiii). CD had long been critical of such 'land-bridge' explanations for current geographical distribution (see *Correspondence* vol. 6, letters to Charles Lyell, 16 [June 1856] and 25 June [1856]). The existence of a sunken continent, 'Atlantis', had been invoked by Oswald Heer to explain the distribution of plants across Europe, Africa, and the Americas (Heer 1861a). In his letter to Hooker of 28 [December 1861] (*Correspondence* vol. 9), CD reported his pleasure at hearing that Daniel Oliver, who rejected Heer's theory, was to give a lecture on the distribution of northern plants at the Royal Institution of Great Britain (Oliver 1862a; see also Oliver 1862b), and he expressed the hope that 'Atlantis' would 'get a good sinking' in Oliver's lecture.

From John Lubbock 6 January 1862

11, Mansion House Street, | London, | E.C.[1]
6 Jan^y^ *1862*

My dear M^r^. Darwin

I should be very much obliged if you would look over the general remarks in the last ½ dozen pages of the paper which I enclose, as I am particularly anxious to know what you think of my notions.[2]

I often long for a good talk with you & am very sorry to live so far off.[3] If I was to come down by the 3.30 train some Saturday afternoon & drive straight over to you, so as to get to Down about 5, should that suit you? If not I must manage to drive over some afternoon earlier.

Please remember me very kindly to M^rs^. Darwin & believe me | Yours affect | John Lubbock

C Darwin Esq

DAR 170.1: 23

[1] The address is that of 'Robarts, Lubbock & Co. Bankers' in the city of London (*Post Office London directory* 1861). Lubbock travelled into London daily from his home in Chislehurst, Kent (Hutchinson 1914, 1: 52).

[2] CD was unable because of illness to read the paper Lubbock sent him (see letter to John Lubbock, 23 January [1862]). It was probably a manuscript version of Lubbock's paper entitled 'On the geologico-archæological discoveries in Denmark, Switzerland, and France' (Lubbock 1862a), which he

read before the Geological and Polytechnic Society of the West Riding of Yorkshire on 19 February 1862. The first part of this paper (pp. 238–56) was a condensed version of Lubbock's observations on ancient lake dwellings in Switzerland, which were shortly to be published in the *Natural History Review* (Lubbock 1862b). In the latter part of the paper (Lubbock 1862a, pp. 256–73), which was later expanded for publication in the *Natural History Review* (Lubbock 1862c; see also letter from John Lubbock, 15 May 1862), he described the human artefacts discovered by Jacques Boucher de Perthes in the river gravels of the Somme valley, and discussed their relevance in establishing the antiquity of the human species. The last three pages of the published paper contained Lubbock's most radical public statement on human antiquity to date, concluding that the 'cave men' of the Somme were not definitively 'the earliest human settlers in Western Europe'. He continued (p. 273):

> We have not indeed as yet the materials to decide the question, but if we were to express any opinion on the subject, it would seem more philosophical to imagine that the genus *Homo* dates back to a period as ancient as the other widely-spread genera of mammalia

[3] John and Ellen Frances Lubbock lived about five miles from Down, in Chislehurst, Kent. See also letter from E. F. Lubbock to Emma Darwin, [January 1862], n. 4.

To Edward Cresy 8 January [1862 or 8][1]

Down, | Bromley, | Kent, S.E.
Jan. 8th.

Dear Cresy,

I am much obliged to you for Theophrastas, though I did not find much which concerned me.[2] I will return it to-night by our Carrier, and I write this to notify the same to you. Many thanks, also, for the date of Boll's work[3] and for other Titles.

Yours very sincerely, | Ch. Darwin.

Copy
DAR 143: 321

[1] The years are conjectured from the form of the address printed on the stationery and by the reference to returning a book 'to-night by our Carrier'. CD used this form of printed address between 1861 and 1869 (Carroll ed. 1976, p. xiii). The Down parcel carrier, George Snow, departed for London every Wednesday night; between 1861 and 1869, the only Wednesdays that fell on 8 January were in 1862 and 1868.

[2] The reference is probably to one of the botanical works of the Greek philosopher Theophrastus, who is mentioned in *Variation* 2: 308 in a citation from A. de Candolle 1855.

[3] The reference may be to Ernst Friedrich August Boll; the work has not been identified.

From Société Impériale Zoologique d'Acclimatation[1] 10 January 1862

Paris,
le 10 janvier 1862

Monsieur et cher Confrère,

J'ai l'honneur de vous annoncer que la Société impériale Zoologique d'Acclimatation se réunira le vendredi 14 février, à 3 heures précises, rue de Lille, 19, et procédera au renouvellement annuel du Bureau et d'un tiers du Conseil.[2]

Aux termes du règlement, les Membres absents sont invités à envoyer leur bulletin de vote à M. le Secrétaire général, au siége de la Société, avant le ⟨ ⟩ ⟨fé⟩vrier (voir ci-contre).

Veuillez agréer, Monsieur et cher Confrère, l'assurance de ma considération très distinguée.

Le Secrétaire Général, | Comte D'Éprémesnil.[3]

DAR 96: 11 v.

[1] For a translation of this letter, see Appendix I.

[2] The Société d'Acclimatation had as its aim the acclimatisation of exotic species of plants and animals to Europe (Geoffroy Saint-Hilaire 1857, Osborne 1987). The society was founded by Isidore Geoffroy Saint-Hilaire in 1854. CD was admitted to membership at the meeting of the society held on 17 February 1860 (*Bulletin de la Société Impériale Zoologique d'Acclimatation* 7 (1860): 100).

[3] Jacques Louis Raoul Duval d'Eprémesnil.

To Société Impériale Zoologique d'Acclimatation [after 10 January 1862][1]

Sir

I sh[d] be much obliged if you would inform me what I owe to the Society Imperiale de Acclimatation for the annual Subscription, including the present year.[2]

Sir | Your obet sert | Ch. Darwin

Monsieur

Je vous serais bien obligé de m'informer combien je dois à la Societé Impériale d'Acclimatation pour la cotisation annuelle, en y comprenant celle de l'année actuelle

Agréez Monsieur | mes hommages respectueux[3]

Draft
DAR 96: 11 r.

[1] Dated by the relationship to the preceding letter.

[2] There is an entry in CD's Account book–cash accounts (Down House MS), dated 23 October 1862, that records a payment of £2 to 'Acclimatisation Soc[y].'

[3] The translation is in Emma Darwin's hand.

From H. W. Bates 11 January 1862

King St Leicester
11 Jan[y] 1862

My Dear Sir

It grieves me very much to hear of your illness.[1] I beg of you to throw my M.S. aside & not give a moment's thought to the subject until you are perfectly restored.[2]

I go to town on Monday to spend a few days—to study at the B.M. and also to attend the Linn. Soc. meeting, where I shall exhibit the box of mimetic butterflies.[3] I have arranged these in such a manner that any Naturalist may understand them

Yours sincerely | H W Bates

If you should wish to say anything whilst I am in town—a note will find me addressed to

John O'Groats hotel
Rupert Street
Haymarket

DAR 160.1: 65

[1] No letter containing this information has been found. Emma Darwin recorded in her diary (DAR 242) on 4 January 1862 that CD was 'ill with influ[enza]'.

[2] The reference is to the manuscript of the second chapter of Bates 1863 (see letter from H. W. Bates, 6 January 1862).

[3] Bates's collection of mimetic butterflies from the Amazon was exhibited at a meeting of the Linnean Society of London on 16 January 1862 (*Journal of the Proceedings of the Linnean Society* 6 (1862): lviii).

To H. W. Bates 13 January [1862][1]

Down Bromley Kent
Jan 13th

My dear Sir

I have been very bad for fortnight & could not read your M.S. before today & yesterday.[2] It is, in my opinion excellent—style perfect—description first-rate (I quite enjoyed rambling in forests) & good dashes of original reflexions. I must write very briefly.— Remember that large sale of a Book depends much on chance,—on whether public mind occupied—other books coming out &c &c., but I feel assured that your Book will be a permanently good one, & that your friends will always feel a satisfaction at its publication.— I will write when you like to Murray.—[3]

Could you add a notion by simile to kind of mysterious sounds heard in forest? but be vague & very brief in any simile.—for description as it now stands is grand.— Did the native look round & shrink or hide when he trembled & heard sound?—[4] This is capital.—

Matador very good—[5] Better than very good.— at p. 9. ought you not to enlarge one sentence to show why the diversity of classes of Lianas interested you? Is it because as to showing that many Families have thus become modified—something in same way as Marsupials, Edentata, Carnivora, Rodentata, & Quadrumana have here in another manner become "*climbers*".—[6]

How are moths & sphinxes in Tropics.[7] Did you sugar? Is the little Heron insectivorous?[8] State somewhere for me, whether any of the Mammals & Birds *often* & *long* kept in confinement in native home, breed?[9] I like much discussion on Burmeister.[10] I like all.— *I am very weak* & tired.— Do not think you have anything to thank me for— it has been pleasure.— Go on as you have begun & you will surely succeed.—

Yours very sincerely | C. Darwin

Hooker is much interested by what I told him about your conclusion of colours. of Butterflies & Tropics.[11]

M.S. returned by the Post.

Cleveland Health Sciences Library (Robert M. Stecher collection)

[1] The year is established by the relationship to the letters from H. W. Bates, 6 January 1862 and 11 January 1862.

[2] CD was reading the manuscript of the second chapter of Bates's book *The naturalist on the River Amazons* (Bates 1863; see letters from H. W. Bates, 6 January 1862 and 11 January 1862).

[3] CD had offered to write to his publisher, John Murray, on Bates's behalf (see *Correspondence* vol. 9, letter to H. W. Bates, 25 September [1861]).

[4] In the published version of the passage to which CD refers (Bates 1863, 1: 72–3), Bates compared a sound he sometimes heard in the Brazilian forest with 'the clang of an iron bar against a hard, hollow tree'. He described the mythical beliefs invoked by the native inhabitants of the forest to account for such sounds, and recounted the reaction to them of a native youth in his service: 'He would crouch down behind me, and beg of me to turn back'.

[5] In his second chapter, Bates described the Sipó Matador, or 'Murderer Liana', a parasitic vine that destroys its host tree as it grows, thus bringing about its own demise (Bates 1863, 1: 53–5).

[6] In the published account (Bates 1863, 1: 49), Bates included the sentence: 'The number and variety of climbing trees in the Amazon forests are interesting, taken in connection with the fact of the very general tendency of the animals, also, to become climbers.' CD repeated the analogy in *Climbing plants*, stating (p. 205): 'America, which so abounds with arboreal animals, as Mr. Bates remarks, likewise abounds . . . with climbing plants'.

[7] Bates described the moths found at Pará in his third chapter (Bates 1863, 1: 104–5).

[8] Bates 1863, 1: 82.

[9] In a later chapter (Bates 1863, 1: 191–4), Bates described the absence among Brazilian aborigines of a 'notion of domesticating animals for use', and discussed the possible influence on human culture of the 'presence or absence of domesticable animals in a country'. He noted that while there was 'no lack in the Amazonian forests of tameable animals', the animals were 'useless from not breeding in confinement'. In a section of *Variation* entitled 'Sterility from changed conditions of life', CD discussed the failure of many animals to breed in captivity, even in their native country. Citing the above, he stated: 'Mr. Bates . . . strongly insists on similar cases' (*Variation* 2: 150, 153, 156; Bates 1863, 1: 99, 2: 113).

[10] Bates had asked CD to comment on his treatment of remarks made by Karl Hermann Konrad Burmeister on the nature of tropical forests (see letter from H. W. Bates, 6 January 1862 and n. 10).

[11] In his book Bates challenged the widely held view that the brilliant colouring of many birds and insects in the tropics was due to the climate (Bates 1863, 1: 21). See also letter from H. W. Bates, 6 January 1862 and nn. 5 and 6. No correspondence between CD and Joseph Dalton Hooker on this point has been found.

From Charles Carter Blake 13 January 1862

1 Judd Street. | Brunswick Square | W.C.
13th January 1862.

Dear Sir.

I was very much gratified by the reception of your note of the 1st June last, in relation to my article on Macrauchenia in Bolivia.[1] I am thereby encouraged to address you on a different subject.

Having paid for some years careful attention to the distribution of recent and fossil Mammalia in South America, I am very anxious to obtain some information which I apprehend you alone can give me, as the materials which my friend Professor Owen[2] possesses have been already placed at my disposal.

I would ask firstly, where is any scientific anatomical description of the bones discovered at Manta, Punta St Elena, Papallacta, & Suacha to be found? Humboldt identifies some of these remains as undoubtedly Cetacean.[3] Has this been verified, and if so, where?

Secondly, whether during your many investigations, you are aware of any other species of fossil New World monkeys than

Pliopithecus antiquus.
Cebus macrognathus.
Callithrix primævus, &
Jacchus grandis.

and if any other species exist, in what work or periodical I may obtain information respecting them?[4]

I may add, that I am the author of the paper on "Distribution of Mastodon in South America" (Geologist. Novr 1861)[5] and that I have prepared a list of the recent and fossil Mammalia of Chile & Peru, which will appear in the forthcoming 2nd Edition of my friend Mr Bollaert's work on South America.[6]

I have the honour to remain, | Dear Sir, | Yours very respectfully | Charles Carter Blake

Charles Darwin. Esqre. M.A. F.R.S. | &c &c &c

DAR 160.2: 198

[1] CD's letter has not been found. In his article (Blake 1861a), Blake, who was a protégé of Richard Owen, disputed Thomas Henry Huxley's identification of a new species of fossil *Macrauchenia* (T. H. Huxley 1861a), both on a point of fact, and in respect of Huxley's 'impugning the philosophical laws of "correlation of structure" as defined by Cuvier and Owen' (Blake 1861a, p. 442). Blake claimed that the remains would probably be found to be those of *M. patachonica*, the species discovered by CD at Port St Julian and subsequently described by Owen (see *Zoology* 1: 35–56; see also Blake 1861b, p. 472, Blake 1862a, p. 325, and Rachootin 1985).

[2] Richard Owen.

[3] The reference to Alexander von Humboldt has not been traced; Blake's immediate source was probably a book by his friend William Bollaert, which contained an unreferenced account of Humboldt's opinion of these remains (Bollaert 1860, p. 80; see n. 6, below). See also Blake 1862a, pp. 323–4.

[4] Blake published an article entitled 'Fossil monkeys' in the *Geologist* for March 1862 in which he attacked the notion that there were any fossil Quadrumana that could be considered ancestors of the human species 'under any of the "derivative" theories propounded by Darwin or Lamarck' (Blake 1862b, pp. 81, 84). His list of fossil monkeys included four species of New World monkey—those listed in his letter, except that *Pliopithecus antiquus* (an Old World fossil) was replaced by *Protopithecus Brasiliensis*. These four species were discovered by the Danish naturalist Peter Wilhelm Lund (Lund

1839–40, 11: 230, 12: 208, 13: 313). CD referred to the discovery of fossil monkey remains in South America in *Origin*, p. 303.

[5] Blake 1861b.

[6] William Bollaert published his archaeological and ethnological studies of South America in 1860 (Bollaert 1860); no second edition was published (*NSTC*). Blake appended a list of the fossil Mammalia of South America to his paper, 'Past life in South America', which appeared in the *Geologist* for September 1862 (Blake 1862a, pp. 329–30).

From C. E. Brown-Séquard 13 January 1862

25, Cavendish Square. | *W.*
Jany/13/62.

My dear Sir,

I hope you will excuse the delay of my answer to your very kind letter.[1] I intended writing to you some remarks on your views and as the redaction of these remarks would have taken some little time the result has been that not having that time to dispose of I have remained without writing, up to this moment. I am now obliged to put off the writing of anything but an answer to your questions. The publication I will soon make of a review of your admirable work is to appear in my Journal (Journal de la Physiologie de l'Homme & des Animaux) It is the second edition which served me for the Review. As you are about having a translation published in France I think it will be very much better for your interests and those of the translator that I should postpone publishing the Review until the translation has appeared in France.[2] I will be obliged to you to have a copy of that translation sent to my publishers in Paris Messrs. V. Masson et fils, Place de l'Ecole de Médecine, as soon as it will have appeared.[3]

I am happy to say that few men are so near agreeing completely with you as I am, and I feel proud that my own thoughts had brought me long ago to conclusions very much similar to yours.

With much respect and sympathy, | I remain, Dear Sir, Yours faithfully, | C. E. Brown-Séquard

DAR 160.3: 327

[1] Letter to C. E. Brown-Séquard, 2 January [1862].

[2] Clémence Auguste Royer's French translation of the third edition of *Origin* appeared in the spring of 1862 (Freeman 1977, p. 102; letter to Asa Gray, 10–20 June [1862]), but the *Journal dė la Physiologie de l'Homme et des Animaux*, of which Brown-Séquard was editor, did not carry a review of the work.

[3] The publishing company V. Masson et fils was owned by Victor Masson (*Dictionnaire universel des contemporains*).

From Thomas Henry Huxley 13 January 1862

Jermyn St
Jan 13th. 1862

My dear Darwin

In the first place a new years greeting to you & yours— In the next I inclose this slip—(please return it when you have read it) to shew you what I have been doing in the North—[1]

Everybody prophesied I should be stoned & cast out of the city gate—but on the contrary I met with unmitigated applauses!!

Three cheers for the progress of liberal opinion!!

The Report is as good as any but they have not put quite rightly what I said about your views—respectg which I took my old line about the infertility difficulty—[2]

Furthermore they have not reported my statement that whether you were right or wrong—some form of the progressive development theory is certainly true—Nor have they reported here my distinct statement that I believe Man & the apes to have come from one stock.—

Having got this far I find the Lecture better reported in the "Courant" so I send you that instead[3]

I mean to publish the lectures in full by & bye (about the time the Orchids comes out)[4]

I suppose somebody her⟨e⟩ told you that Owen has gone in for progressive development in the second Edition of the 'Paleontology' which can only be described as a rather more scoundrelly book than the first—[5]

The way I am ignored & you are pooh-poohed is glorious[6]

Ever | Yours faithfully | T H Huxley

I deserved the greatest credit for ⟨not⟩ having made an onslaught on ⟨Brewster⟩ for his foolish impertinence ⟨of y⟩our views in 'Good words'—[7]but ⟨declined⟩ to stir Nationality—which you ⟨know (in him)⟩ is rather more than his Bible[8]

DAR 166.2: 290

[1] Huxley delivered two lectures under the title 'Relation of man to the lower animals' before the Philosophical Institution of Edinburgh on 4 and 7 January 1862, respectively. For an account of the lectures and their reception, see L. Huxley ed. 1900, 1: 192–4. The newspaper report of the lectures that Huxley initially intended sending has not been identified, since Huxley subsequently decided to send one from a different newspaper (see n. 3, below).

[2] In his review of *Origin*, Huxley stated that a significant if inconclusive test of whether two individuals represented distinct 'physiological' species was to attempt to hybridise them, since distinct species would often either be infertile *inter se* or produce infertile offspring, whereas varieties of the same species would give rise to fertile progeny ([T. H. Huxley] 1860, pp. 552–5). The lack of any positive evidence 'that any group of animals has, by variation and selective breeding, given rise to another group which was even in the least degree infertile with the first' was, he argued, the weakest point in CD's hypothesis of natural selection (*ibid.*, pp. 567–8). Huxley expressed the same view in the published version of his Edinburgh lectures, stating that his acceptance of CD's theory of natural selection was subject to proof that 'physiological species may be produced by selective breeding' (T. H. Huxley 1863a, p. 108).

[3] Huxley refers to the report of his second lecture that appeared in the *Edinburgh Evening Courant*, 11 January 1862 (see Appendix V); the enclosed copy has not been found.

[4] The substance of Huxley's Edinburgh lectures was published in T. H. Huxley 1863a, pp. 57–118. *Orchids* was published in May 1862 (Freeman 1977, p. 112).

[5] In the first edition of his *Palæontology*, Richard Owen's critique of all existing hypotheses to account for the origin of species was followed by a paragraph outlining principles that, he claimed, 'tended to

impress upon the minds of the closest reasoners in Biology a conviction of a continuously operative secondary creative law' (R. Owen 1860a, p. 407). In the second edition he expanded and emboldened this paragraph, heading it 'Evidence of origin of species by secondary law' (R. Owen 1861a, p. xvi) and stating (p. 444): 'The inductive demonstration of the nature and mode of operation of such secondary continuously operative species-producing force will henceforth be the great aim of the philosophical naturalist'. Owen also changed the heading of another of the concluding sections of the volume from 'Comparison of uniformitarian and progressional evidences' to 'Evidence of progress and advance in organisation'. He added a new paragraph to the section, stating (p. 449):

> the sum of what is known [from the study of extinct forms] yields the legitimate deduction, that there has been a succession of species illustrating in the main the progressive perfection of the nervous system, and the concomitant predominance of mind over matter.

There is a copy of the first edition of the work (R. Owen 1860a) in the Darwin Library–CUL, and of the second edition (R. Owen 1861a) in the Darwin Library–Down; both are lightly annotated (see *Marginalia* 1: 656).

6 In the second edition, Owen considerably reduced his already brief account of CD and Alfred Russel Wallace's theory, removing part of it to a footnote earlier in the text (R. Owen 1861a, pp. 442–3, 434–5 n.; compare R. Owen 1860a, pp. 405–6). The limited nature of Owen's response to natural selection in the first edition of his *Palæontology* is discusssed in Rupke 1994, pp. 242–3.

7 Brewster 1862.

8 The original letter is damaged. The words and letters in angle brackets have been taken from the text printed in L. Huxley ed. 1900, 1: 194.

From John Hutton Balfour 14 January 1862

27 Inverleith Row | Edinburgh
14 Jany 1862

My Dear Darwin

Many thanks for your kindness in sending me a copy of your interesting paper on the Dimorphic condition of the species of Primula.[1] I had read the notice of it in the Gardener Chronicle with great attention.[2] The facts are curious & have a most important bearing on the subject of species generally.

I hope to be able to examine some of the Primulas in the garden this year with the light which you have thrown upon him[3].

We have just had Huxley with us promulgating his views in regard to the *Zoological* relations of man & monkeys.[4] He strongly insisted on the fact that the lowest apes do not differ more in zoological structure from the highest apes than the latter do from man, & therefore they & man are in one order. If we go on in the same way taking order after order we shall find that the lowest in one order do not differ more from the highest in the same order than the latter do from the order above—& thus all animals are of one order— There will be *disorders* in place of *orders*. No doubt there is one great type throughout the orders & that is all that Huxley proves.

I still think that he must take man with all his *functions intellectual & moral* in order to determine his position. We are not entitled to leave these out of our consideration even viewing the matter zoologically. The lowest man can be *raised by education*. He is a *religious* animal & has a *conscience*. He is capable of knowing about a God & a future state. In this he differs from all animals. The tendency of man when left to himself has been to degenerate. This is shown by Humboldt & is very decidedly brought out by Whately. & Whewell.[5] Huxley was cautious & did not boldly declare that he considered man & apes to have the same origin or to be varieties of the same species.

Many of the audience however considered that as the drift of his observations.

He gave a lucid exposition of structure & he was listened to with much interest & attention

Man is the great stumbling block in regard to all *recent* theories of species. He stands by himself as a *Creation* I think, & the Records in regard to him are explicit. I cannot overlook them in considering his plan in creation.

You & others may think us in the north prejudiced in this matter.

Excuse this *yarn* which I have *spun* most unwittingly.

Browne & I often talk of you & the old Plinian Society days when as young naturalists we discussed many points of interest which have since occupied prominent places in Science.[6]

I am | Yours sincerely | J. H. Balfour

DAR 160.1: 31

[1] CD read his paper, 'Dimorphic condition in *Primula*', before the Linnean Society of London on 21 November 1861; it was published in the number of the *Journal of the Proceedings of the Linnean Society (Botany)* issued on 1 March 1862 (*General index to the Journal of the Linnean Society*, p. vi). CD had written of perhaps ordering fifty copies of the paper for distribution among individuals who had assisted his study by sending him specimens (see *Correspondence* vol. 9, letter to George Bentham, 24 November [1861]). His presentation list for these pre-prints, probably drawn up between mid-December 1861 and the end of the first week in January, is reproduced in Appendix III; Balfour's name appears on the list.

[2] *Gardeners' Chronicle and Agricultural Gazette*, 30 November 1861, pp. 1048–9.

[3] Balfour was regius keeper of the Royal Botanic Garden, Edinburgh.

[4] Thomas Henry Huxley delivered two lectures on man's relation to the lower animals at the Philosophical Institution of Edinburgh in January 1862 (see letters from T. H. Huxley, 13 January 1862 and 20 January 1862).

[5] See, for example, Humboldt and Bonpland 1819–29, 3: 208, Whately 1831, pp. 119–200, Whately 1854, and Whewell 1854, pp. 166–90 (especially p. 189). On Richard Whately's degenerationist views, see also Gillespie 1977.

[6] CD became a member of the Plinian Natural History Society in 1826 while an undergraduate at the University of Edinburgh; at the time, William Alexander Francis Browne was one of the society's five presidents. During the period that CD was a member, Browne and others read papers on the similarities between human and animal mentality; in 1827, Browne caused controversy by arguing that mind and consciousness had an entirely material basis. See Ashworth 1935, pp. 101–6 and Desmond and Moore 1991, pp. 31–3, 38.

To T. H. Huxley 14 [January 1862][1]

Down Bromley Kent

14th.

My dear Huxley

I am heartily glad of your success in the North, & thank you for your note & slip.—[2] By Jove you have attacked Bigotry in its strong-hold. I thought you would have been mobbed. I am so glad that you will publish your Lectures. You seem to have kept a due medium between extreme boldness & caution.— I am heartily glad that all went off so well.—

I hope M^rs. Huxley[3] is pretty well.— We have been a miserable family with 3 or 4 or 6 all in bed at the same time with virulent Influenza.— I have done nothing for nearly 3 weeks, & am much shaken.—[4]

I see some good hits against Owen in N. Hist. R; which, however, I have not yet read.[5]

I must say one word on Hybrid question.—[6] no doubt you are right that here is great hiatus in argument; yet I think you overrate it— you never allude to the excellent evidence of *varieties* of Verbascum & Nicotiana being partially sterile together.[7] It is curious to me to read (as I have to day) the greatest crossing *Gardener*, utterly poop-poohing the distinction which *Botanists* make on this head, & insisting how frequently crossed *varieties* produce sterile offspring.—[8] Do oblige me by reading latter half of my Primula paper in Lin. Journal for it leads me to suspect that sterility will hereafter have to be largely viewed as an acquired or *selected* character.—[9] a view which I wish I had had facts to maintain in the Origin.—

I hope to Heaven you will keep to your intention, & publish your Lectures.—[10] I do not suppose I shall see Owen's 2^d. Edit; but he is so dishonest that I really now care little what he says.—[11]

Farewell. I am poor weak wretch with trembling hands; so good night, & all good luck to you.— Ever yours | C. Darwin

Some *future* time I sh^d. like the "Three Barriers" returned; as I collect all such rubbish.[12]

I find Brown-Sequard is largely with me, & will review in France the French Translation of the Origin.—[13]

Imperial College of Science, Technology and Medicine Archives (Huxley 5: 167)

[1] The date is established by the relationship to the letter from T. H. Huxley, 13 January 1862, and by the reference to an article in the *Journal of Horticulture* of 14 January 1862 (see n. 8, below).

[2] See letter from T. H. Huxley, 13 January 1862.

[3] Henrietta Anne Huxley.

[4] Emma Darwin recorded in her diary (DAR 242) on 13 January, 'all mending'.

[5] CD refers to an article in the January issue of the *Natural History Review* (of which Huxley was the senior editor) by Jacob Ludwig Conrad Schroeder van der Kolk and Willem Vrolik (Schroeder van der Kolk and Vrolik 1862). The article discussed Richard Owen's recently published paper comparing the cerebral characters of humans and apes (R. Owen 1861b) in which he had reprinted a diagram of a chimpanzee brain taken from an earlier paper by the authors (Schroeder van der Kolk and Vrolik 1849). The authors pointed out that, while Owen used their diagram to support his view that the posterior lobe, the posterior cornu of the lateral ventricle, and the hippocampus minor were peculiar to the human brain and not found in the brains of apes, Thomas Henry Huxley and John Marshall had placed exactly the opposite interpretation on it (see T. H. Huxley 1861b, p. 75, and Marshall 1861, p. 313). They accepted, moreover, the criticism made by Louis Pierre Gratiolet (Gratiolet [1854], pp. 18–19) that their drawing reflected some distortion of the dissected brain suffered during its removal from the brain case. A recent repetition of their dissection of a chimpanzee brain had confirmed Schroeder van der Kolk and Vrolik in their view that the structures in question were indeed present in ape brains. Though the tone of the article was respectful of Owen, the authors

made the comment: 'notre illustre confrère paraît s'être laissé entraîner par son désir de combattre la théorie de M. Darwin, et, si nous ne nous trompons fortement, il s'est fourvoyé' [our illustrious colleague seems to be carried away by his desire to fight against Mr Darwin's theory, and, unless we are greatly mistaken, he has lost his way] (Schroeder van der Kolk and Vrolik 1862, p. 112). See also Rupke 1994, pp. 274–9.

6 See letter from T. H. Huxley, 13 January 1862 and n. 2.

7 In *Origin*, pp. 270–1, CD described crossing experiments carried out by Karl Friedrich von Gärtner on *Verbascum* and by Joseph Gottlieb Kölreuter on *Nicotiana*. Gärtner found that 'yellow and white varieties of the same species of Verbascum when intercrossed produce less seed, than do either coloured varieties when fertilised with pollen from their own coloured flowers.' Kölreuter observed that 'one variety of the common tobacco is more fertile, when crossed with a widely distinct species, than are the other varieties'.

8 CD refers to an article in the 14 January 1862 issue of the *Journal of Horticulture* by the gardener, Donald Beaton (*Journal of Horticulture* n.s. 2: 309–10). There is an unbound copy of this issue of the journal in the Darwin Library–CUL; CD marked several passages of this article with marginal scoring.

9 'Dimorphic condition in *Primula*'; Huxley's name does not appear on the presentation list for this paper (see Appendix III). Towards the end of the paper, having described the differential fertility of long- and short-styled plants of *Primula* species when fertilised by pollen either from the same form or from the other form, CD stated (p. 94; see also *Collected papers* 2: 61):

> Seeing that we thus have a groundwork of variability in sexual power, and seeing that sterility of a peculiar kind has been acquired by the species of *Primula* to favour intercrossing, those who believe in the slow modification of specific forms will naturally ask themselves whether sterility may not have been slowly acquired for a distinct object, namely, to prevent two forms, whilst being fitted for distinct lines of life, becoming blended by marriage, and thus less well adapted for their new habits of life.

CD's notes on this subject are reproduced in Appendix VI.

10 See letter from T. H. Huxley, 13 January 1862 and n. 4.

11 The reference is to the second edition of Richard Owen's *Palæontology* (R. Owen 1861a; see letter from T. H. Huxley, 13 January 1862 and n. 5). There is a lightly annotated copy of this edition in the Darwin Library–Down (see *Marginalia* 1: 656).

12 [Rorison] 1861. CD had drawn Huxley's attention to this book, which was critical of *Origin*, several months previously (see *Correspondence* vol. 9, letter to T. H. Huxley, 31 October [1861]).

13 See letter from C. E. Brown-Séquard, 13 January 1862.

From Henry Holland 15 January [1862][1]

Brook Street

Jany 15

My dear Charles.

I have read with very great interest your Paper on the sexual relations of the two Forms of the Primula.[2] You are right in speaking of the result of these observations as a *most surprising fact*;[3] for such in truth it is, in every sense & sort— A fact, moreover, tending to expound various anomalies, which surround this great problem of the propagation of life.

The *essential mystery* of generation,—(whether of the like or unlike)—may never be solved. I can myself see no road to it. But researches of this kind carry us closer to the point; & better define, that which our faculties may be unable to reach

I am sorry to hear from Erasmus[4] of the *subjugation* of yourself & all your family, by this Epidemic, which we call Influenza. It continues to be very rife both in London & the country. My theory of it, is the same I have more fully propounded for Asiatic Cholera.[5]

I have still no letter from Lord Tankerville.[6] But that I think he will be in Town for the meeting of Parliament, when I can see him personally, I would take the chance of sending another letter.

Ever your's very truly | H Holland

DAR 166.2: 239

[1] The year is established by the reference to CD's paper, 'Dimorphic condition in *Primula*' (see nn. 2 and 3, below).

[2] 'Dimorphic condition in *Primula*'; Holland's name appears on CD's presentation list for this paper (see Appendix III).

[3] 'Dimorphic condition in *Primula*', p. 96 (*Collected papers* 2: 63).

[4] Holland refers to CD's brother, Erasmus Alvey Darwin.

[5] Holland refers to his 'hypothesis of organic or animalcule life as the source of the Asiatic Cholera' (Holland 1855, p. 333; see also pp. 562–608). He stated that the arguments in favour of this hypothesis, 'if valid, must apply equally to the epidemic influenzas now under our notice' (p. 333). There is a copy of Holland 1855 in the Darwin Library–CUL.

[6] The reference is to Charles Augustus Bennet, sixth earl of Tankerville (see letter from Henry Holland, [3–14] January [1862]).

To Ludwig Rütimeyer 15 [and 16] January [1862][1]

Down. | Bromley. | Kent. S.E.
Jan. 15th

Dear & honoured Sir

I have been confined to my bed by illness, otherwise I should have written sooner, & thanked you for your letter of Dec. 11th.—[2]

I grieve to say that I have no good news to tell. I applied to a friend & relation, who knows Lord Tankerville[3] well, & I wrote a long letter to him with instructions, which he forwarded to L^{d}. Tankerville with a request from himself. But he has received no sort of answer!![4]

My friend says L^{d}. Tankerville leads a strange solitary life; but is much surprised at receiving no answer. I am astonished at such rudeness. My friend thinks that perhaps he may send the skulls, though he will not take the trouble to write.— My friend naturally does not choose to write again, after being treated so rudely; but if he sees him in London this Spring will ask.— I yet hope that the Skulls may be sent; but am very fearful about the result.— If the old Lord Tankerville had been alive, I feel sure that we should have got them.—[5]

I cannot tell you how *sincerely* I grieve that I have failed to aid you in your most important & interesting investigations, but I have done all that I could.—

With sincere apologies, I remain Dear Sir | Yours very faithfully | Charles Darwin

P.S. I have just had another letter from my friend, Sir H. Holland, who says he will see Lord Tankerville or write to him again;[6] & I think the noble Lord will be driven by shame to be more courteous.—

Elizabeth Rütimeyer

[1] The date is established by the relationship to the preceding letter and the letter from Henry Holland, [3–14] January [1862], and to the letter to Ludwig Rütimeyer, 11 February [1862] (see n. 6, below).
[2] The letter from Rütimeyer has not been found, but see *Correspondence* vol. 9, letter to Ludwig Rütimeyer, 5 December [1861].
[3] Charles Augustus Bennet, sixth earl of Tankerville.
[4] See letter from Henry Holland, [3–14] January]1862].
[5] The fifth earl of Tankerville, also named Charles Augustus Bennet, had a keen interest and pride in his Chillingham cattle (see Ritvo 1992).
[6] See preceding letter. The daily postal delivery to Down arrived from Bromley at 8.30 A.M., which was before the first delivery from London arrived in Bromley at 9 A.M. (*Post Office directory for the six home counties* 1862, pp. 756, 822). It is consequently unlikely that CD received Holland's letter on the same day that it was written; he would, rather, have received it on the following day. CD's postscript was, therefore, probably added on 16 January.

From William Branwhite Clarke[1] 16 January 1862

St Leonards.
16 Jany 1862.

My dear Sir,

I was extremely gr⟨at⟩ified by your letter of Octr. 25th, this day received.[2]

It will give me great pleasure, if I be enabled to gratify any of your wishes, when the season comes for experiment.[3]

I do not profess to be a *Botanist*; but it is impossible to be uninterested in the subject of Botany, especially as one is requested to dabble in the matter in connection with the f⟨ossi⟩l beds of Antiquity.[4]

The Goodeniaceæ do not blossom here now. When the season arrives we will put some under a net as you propose: I am prepared to admit that insects do to some extent assist and hinder the work ⟨of⟩ propagation. I have never seen a blossom on our Gum Trees without finding ants and beetles in abundance,—even at the very top most branch.[5]

I dare say, if one had time, the facts could be made out satisfactorily as to what plants certain insects attach themselves.

In this country the ants are the natural scavengers— they clear away whole trees—and convert the very hearts of our forest monsters in⟨to⟩ *papier maché*—

Flies too do an immense amount of labour in one way or another. When I was in camp on the flanks of the Alps in 1851–2 I was so tormented with them, that I watched their ways.[6] I found that they became active about 4 AM. and came in to my tent—so I used to get up at 3 and circumvent them—else I could not have dressed, eaten or shaved in peace.

Crowds of them got ⟨ ⟩ the top of the tent, and we beat them down at night-fall with green branches, making great slaughter. But to solve the mystery as to the morning visiting, I went out one night with a fire stick, and found by its light every blade of grass laden with them all over the space I traversed. What do they and the so-called Mosquitos live on—where no animals are found? Clearly on moisture, dew &c and doubtless on honey—and if so they probably impregnate flowers. There are certain flowers in the bush, which contain a thin sweet juice, which I often sucked in ⟨m⟩y walks in a hot day—but I have always had to bite off the end of the flower to prevent the swallowing of the *ants*.

As to Bees, I will mention only one incident.[7]

Some years ago (1844 or 5) I had charge of the country from Paramatta to and beyond the Hawkesbury.[8] I formed acquaintance with the aboriginals. In one tribe (of which none now exist) was a man called "Captain" who used to get me native honey from *[Ourinbah]* and Doural. One day I found him cutting out a large establishment from the top of a very lofty Eucalyptus: and I asked him to bring it to Paramatta to my Cottage, which he did. The hollow in the bough was filled with the harmless native bees and their honey. It lay in the yard for two or three days, till it was discovered by English bees belonging to my friend D^r. Stuart, who lived in another part some distance off.[9] In the course of a day or two these foragers killed off nearly all the little native bees and stole their honey. This I can vouch for. Yet I once saw a native hive under the same shed with English bees at the Vineyard, now "Sabiaco" Mr Hannibal McArthur's residence on the Paramatta River.[10] The English bees are very common in the bush. They make their hives in hollows in the lofty Gum trees. But the native bee is very rare now where it was once common.— I wish Mr Macleay would give the world the benefit of his researches into our insect world.—[11]

If I find any thing published on the subject mentioned in your P.S. "effect of introduced on native animal", I will let you know.—[12]

You will receive a pamphlet by this mail on the Geol^y of Australasia; and I hope you will not be offended with the mention of your notice of N.Z. & Oceania.[13]

I have always considered N. Zealand, N. Caledonia, &c as merely the summits of the ranges parallel with our Australian "Cordillera". There are soundings between this and N.Z.—[14] There is mention of the reduced state of both countries in the pamphlet that also contains a notice of supposed Glacial action in N.Z. (p 54 to 7) It brings in your doctrine of 'submergence' exemplified by the Reefs.—[15] *My paper brings up all* that is *known of Australasian Geology up to 31 D^r* 1861. and gives what is necessary to simplify the nature of my difference with M'Coy.[16]

Will you give me *your* opinion of the deductions from his paper on the *Wollumbilla fossils*. (p 48 No 5)[17] and my *Section from* Stony Ck (p 53– to 6)

The once *penal Norfolk* I^d is the island on which the greenstone boulders repose. There is basalt in Howe's Island. But I think no greenstone nearer than Tasmania or New Holland or New Zealand.[18] What should prevent ice floes or icebergs in that region if the glacial period be no fiction. Are *scratches* necessary? Are blocks *dropped* from icebergs necessarily *scratched*? Glaciers blocks would be—but ? of others. In my pamphlet I have inserted a paper on the recent journeys across this Continent. In p 60 last paragraph but two and last sentence, I define a strait. The younger Gregory boy just returned *from the point I marked* finding the same desert sands A. C. G. found on Sturt's Ck![19]

Pardon this garrulity and allow me to remain, My dear Sir, Y^rs. very truly | W B Clarke

P.S. One of the local School-masters in this one of my 5 Parishes, is a Mr Hussey who says he is known to you. He is a great fisherman and tells me wondrous things of his catchings. Can you tell me of *him*? He has been in Africa.[20]

C Darwin E^sq

DAR 161.2: 172

CD ANNOTATION

Top of last page: '16th Jan. 1862— Rev. W. B. Clarke *of St Leonards* | Sydney. Australia' *ink*

[1] This letter was apparently sent under the same cover as a postscript dated 21 January 1862. Since the postscript occupies an additional sheet and contains a separate greeting and valediction, it has been treated as a separate letter and appears in its proper chronological position (see letter from W. B. Clarke, 21 January 1862).

[2] Letter to W. B. Clarke, 25 October [1861] (*Correspondence* vol. 9).

[3] CD had asked Clarke to try an experiment that would indicate the extent to which species of Goodeniaceae, *Eucalyptus*, and *Mimosa* required the agency of insects in order to set seed (see *Correspondence* vol. 9, letter to W. B. Clarke, 25 October [1861]).

[4] Clarke carried out extensive research to determine the age of the coal formations in Australia (see n. 16, below), which necessitated the collection and examination of the fossil plants contained within them (see Clarke 1861a).

[5] CD recorded Clarke's observation in a note that is now preserved in DAR 49: 70.

[6] Clarke refers to the area known as the 'Australian Alps' which he visited during his government-sponsored survey of potential gold-bearing localities in the southern highlands of New South Wales, made between September 1851 and June 1852 (Jervis [1945], pp. 52–69).

[7] CD had asked about the effect of introduced hive-bees on native bee populations in the letter to W. B. Clarke, 25 October [1861] (*Correspondence* vol. 9).

[8] Clarke was offered charge of the new parish of North Parramatta in August 1840; he resided there for five years (Jervis [1945]).

[9] Dr Stuart has not been identified.

[10] The Vineyard, Parramatta, had been the home of the farmer and politician, Hannibal Hawkins Macarthur, who died in 1861 (*Aust. dict. biog.*). (CD had visited Macarthur while in Australia during the *Beagle* voyage (R. D. Keynes ed. 1988, p. 405).)

[11] The retired diplomat and distinguished zoologist, William Sharp Macleay, had devoted much of his time since emigrating to Sydney in 1839 to amassing a large collection of Australian insects.

[12] See *Correspondence* vol. 9, letter to W. B. Clarke, 25 October [1861].

[13] In Clarke 1861b, p. 6, Clarke quoted what he believed to be an erroneous statement by CD that 'not one oceanic island is as yet known to afford even a remnant of any palæozoic or secondary formation' (*Origin*, p. 308). Clarke pointed out that there was distinct evidence of Palaeozoic and Secondary formations in New Zealand and that CD himself had written of New Zealand as an oceanic island (*Origin*, p. 390). Clarke had previously informed CD of this discrepancy (see *Correspondence* vol. 9, letter from W. B. Clarke, [August 1861]). The fourth edition of *Origin* introduced New Zealand as an exception to the general rule (*Origin* 4th ed., pp. 372–3). There is a lightly annotated presentation copy

of Clarke 1861b in the Darwin Pamphlet Collection–CUL. The pamphlet was presumably posted so as to leave Australia on the same ship as this letter.

14 In his letter to W. B. Clarke, 25 October [1861] (*Correspondence* vol. 9), CD stated that, in his mind, he had 'excluded N. Zealand from the term oceanic Isl[d], as being connected by a Bank'.

15 Clarke refers to CD's theory that coral reefs and atolls originated as fringeing reefs adjacent to or surrounding land that subsequently subsided, the reef continuing to grow upwards, with the living coral maintaining its position near the surface of the water (*Coral reefs*).

16 Clarke and Frederick McCoy were engaged in a long-standing dispute about the age of the coal-fields of New South Wales following McCoy's description of fossil specimens sent back to England by Clarke (McCoy 1847). McCoy considered the coal deposits to belong to the Oolitic age while Clarke maintained that the deposits were Palaeozoic (Clarke 1861b, pp. 10, 17–48).

17 Clarke refers to McCoy 1861, which he reproduced in its entirety in Clarke 1861b, pp. 48–52.

18 In *Origin*, p. 373, CD, without giving details, referred to evidence of glacial action in south-eastern Australia. In response, Clarke provided CD with examples of such evidence including the existence of transported blocks of greenstone on Norfolk Island (see *Correspondence* vol. 9, letter from W. B. Clarke, [August 1861]). In his letter to Clarke of 25 October [1861] (*Correspondence* vol. 9), CD had asked for further details of the boulders in order to verify that they were erratics.

19 Clarke refers to the explorer Augustus Charles Gregory and his younger brother Francis Thomas Gregory. In his paper, Clarke stated that if there had once been a strait running from south to north across Australia, it would probably have been between the Great Australian Bight (on the south coast) and the area of the north-west coast between Point Larry and Cape Joubert now known as Eighty Mile Beach. He asserted: 'The sandy region found by Mr. A. C. Gregory, on Sturt's Creek, is on the north-east side of such a strait, the southern boundary being probably, about 500 miles to the south-west of it' (Clarke 1861b, p. 60). F. T. Gregory had just returned from an expedition to the north-west coast, where he had discovered, in the region identified by Clarke, the south-western edge of a desert plain, consisting of red drift sand, now known as the Great Sandy Desert (see F. T. Gregory 1862 and J. Bartholomew [1864?]).

20 Mr Hussey has not been identified.

To J. D. Hooker 16 January [1862][1]

Down
Jan 16th

My dear Hooker

I sh[d]. have thanked you before for valuable specimens & 3 notes,[2] but we have been in a lamentable state; with 3 or 4 or even 6 in bed at a time with virulent influenza. I have been very bad & am much shaken & have done nothing for nearly 3 weeks. The Catasetum has dropped its flowers, alas & alas; & I want to return both plants & the Bolbophyllum.—[3] Our carrier will be in London on Thursday morning;[4] *please give me address of the Kew Carrier* in a few days' time: I fear that the plants are not very healthy, though my neighbour's gardener can grow Orchids well, yet he says he cannot manage these Catasetums:[5] I hope to God they are not much injured.—

Sometime give me reference to Haast(??) Survey of Middle Is[d] *of New Zealand*??—about Glacial deposits, which interests me much.—[6]

Asa Gray is evidently sore about England: he does not say much; nor do I; but I have hitherto been able to write with some sympathy; now I must be silent; for I look at the people as a nation of unmitigated blackguards.—[7]

I have been interested by what you say about your Willy:[8] I should expect he would become all right, with all the excellent characters you specify, under advancing years. A child from such ancestry must lose the too great volatility, which seems the sole failing. I was struck years ago by what Arch^{bp} Whately[9] told me of a son of very able parents, who was thought for years a dunce & who remained so till full manhood, when he stumbled on his vocation, viz an Australian settler; & then, as W. remarked, he exhibited his inherited genius, for he became Supereminent in this line of life.—

With respect to colours of flowers; I think such an investigation as you propose would certainly be very interesting; though whether worth the labour, I cannot say.—[10] I have been excessively perplexed by opposite statements with respect to shells in deep water;[11] & now comes Bates' case to astound me.[12] He refers truly to my observations on this head at Galapagos Arch. I saw (but had forgotten) that something more than heat is required for development of colour.[13] Certainly it is an important consideration, (perhaps hardly anything more important,) to *try* to discover how much of any character stands in *direct* relation to external conditions; so that assuredly I hope you may undertake this investigation.—

Farewell— I am out of Spirits with all the illness in this house; & our youngest Boy has just failed in miserable degree, like five of our other children, with intermittent pulse.—[14]

What misery there is in this life.— God Bless you my dear old friend—

C. Darwin

Do not forget Kew Carrier.—

P.S. The letter with curious address forwarded by Mrs Hooker[15] was from a German Homœopathic Doctor—an ardent admirer of the Origin—had himself published nearly the same sort of book, but goes much deeper—explains the origin of plants & animals on the principles of Homœopathy or by the Law of Spirality— Book fell dead in Germany— Therefore would I translate it & publish it in England &c &c?![16]

DAR 115.2: 140

[1] The year is established by the reference to the incidence of influenza in the Darwin household (Emma Darwin's diary (DAR 242)) and by the relationship to the letter from J. D. Hooker, [29 December 1861] (*Correspondence* vol. 9).

[2] CD probably refers to the letters from J. D. Hooker, [29 December 1861] (*Correspondence* vol. 9), [1 January 1862], and [30 December 1861 or 6 January 1862] (this volume).

[3] Hooker had lent CD specimens of these two genera of orchids from the Royal Botanic Gardens, Kew. See *Correspondence* vol. 9, letter to J. D. Hooker, 1 December [1861] and nn. 1 and 2.

[4] George Snow, the village coal dealer, ran a carrier service to and from London every Thursday (*Post Office directory of the six home counties* 1862).

[5] John Horwood, gardener to CD's neighbour George Henry Turnbull, is thanked for his assistance in *Orchids*, p. 158.

[6] Hooker had been reading J. F. J. von Haast 1861, and told CD of Julius von Haast's allusion to glacial deposits in New Zealand in his letter of [29 December 1861] (*Correspondence* vol. 9). In the fourth edition of *Origin*, CD modified his reference to glacial action in New Zealand to read: 'we now

know, from the excellent researches of Dr. J. Haast and Dr. Hector that enormous glaciers formerly descended to a low level in New Zealand' (*Origin* 4th ed., pp. 442–3).

[7] The reference is to the aftermath of the so-called *Trent* affair (see letter from J. B. Innes, 2 January [1862] and n. 5), about which CD and Asa Gray had corresponded (see *Correspondence* vol. 9, letter to Asa Gray, 11 December [1861], and letter from Asa Gray, 31 December 1861).

[8] Hooker had mentioned his concern that his eldest child, William Henslow Hooker, was 'singularly backward & childish of his years' in his letter to CD of [29 December 1861] (*Correspondence* vol. 9).

[9] Richard Whately, archbishop of Dublin.

[10] Hooker had offered to look over his lists of Arctic plants, making a note of their colours in order to help CD determine whether there was a correlation between flower colour and latitude (see letter from J. D. Hooker, [30 December 1861 or 6 January 1862] and n. 5).

[11] CD mentioned in *Origin*, p. 133, the general observation that species of shells inhabiting deep seas tended to exhibit less colour than those inhabiting shallow seas (see also *Natural selection*, p. 282). He was citing Forbes 1850, p. 254.

[12] In Bates 1863, 1: 21, Henry Walter Bates drew on his knowledge of tropical insects to argue that, contrary to the generally held view, climate had 'little or no direct influence' on the size or coloration of animals. See also letter from H. W. Bates, 6 January 1862, and letter to H. W. Bates, 13 January [1862].

[13] See letter from H. W. Bates, 6 January 1862 and n. 6.

[14] Horace Darwin, the Darwins' youngest son, was 10 years old.

[15] Frances Harriet Hooker.

[16] The German homeopathic doctor has not been identified. He was seemingly an adherent of Johann Wolfgang von Goethe's law of spiral growth of plants. Goethe claimed in his paper, 'Über die Spiral-Tendenz der Vegetation': 'es walte in der Vegetation eine allgemeine Spiraltendenz, wodurch, in Verbindung mit dem verticalen Streben, aller Bau, jede Bildung der Pflanzen nach dem Gesetze der Metamorphose vollbracht wird.' [there is in plants a general spiral tendency, through which, in connection with a vertical tendency, every construction, every form of plant following the law of metamorphoses is achieved] (*Goethes Werke* pt 2, 7: 49). For CD's negative views on homeopathy, see *Correspondence* vol. 4, letter to W. D. Fox, 4 September [1850].

From Charles Cardale Babington 17 January 1862

Cambridge
17. Jan. 1862.

Dear Darwin

I am much obliged to you for sending the separate copy of your exceedingly interesting paper. I was looking forward to the time when it would come in the Journal, and am glad to be able to anticipate that time by reading it now.[1]

I presume that you have had no opportunity of looking at the true Oxlip, *P. elatior*, which is, I believe, confined to woods on a stiff soil in what are called the Eastern Counties. It is quite as distinct from the Cowslip and Primrose as those are from each other—at least such is my opinion—and it has the same two forms of *pin-headed* and *thumb-eyed* flowers as those possess.[2]

I noticed the existence of similar differences in the allied genus *Hottonia* as long since as the 1st edition of my "Manual" (1843).[3] But then and up to the present time had supposed that the *pin-headed* form was always barren. In fact, I have never seen seeds produced by that form of *Hottonia*. But then it is not easy to notice a quantity of such an aquatic plant. I have more than once had plants of it growing in a jar of water. It is very pretty when thus grown.

In *Stellaria graminea* two forms are found which probably correspond with those of *Primula* (See *Man*. ed. 3 or 4).[4]

Yours very truly | Charles C. Babington—

C. Darwin Esq.

May it not be species of *Hepialus* that fertilize the Primulæ. Those Moths frequent the proper places, at the proper date, and in the evening.[5]

DAR 110 (ser. 2): 58–9

CD ANNOTATIONS
1.1 I am ... now. 1.4] *crossed ink*
2.1 I presume ... existence 3.1] *crossed ink*
3.1 similar ... barren. 3.3] '(See Lecoq & Vaucher)'[6] *added in margin, red crayon*
3.1 as long ... (1843). 3.2] *scored red crayon*
4.1 In ... 4). 4.2] *scored both margins, red crayon*

1 'Dimorphic condition in *Primula*'; Babington's name appears on CD's presentation list for this paper (see Appendix III).

2 In 'Dimorphic condition in *Primula*', p. 77 (*Collected papers* 2: 45), CD referred to the terms used by florists to describe the two forms of *Primula*, namely, 'pin-headed' or 'pin-eyed' for the form displaying the stigma at the mouth of the corolla, and 'thumb-eyed' for the form displaying the stamens (according to the *OED*, the correct term is 'thrum-eyed'). CD designated the two forms 'long-styled' and 'short-styled', respectively. Although CD's paper discussed many species, it did not mention *P. elatior*. During the course of his investigations into hybridisation among *Primula* species, he had been informed of the two forms of *P. elatior* by Henry Doubleday (see *Correspondence* vol. 8, letter from Henry Doubleday, 16 May 1860). CD discussed the case in *Forms of flowers*, pp. 72–3, citing C. C. Babington 1851, p. 258.

3 C. C. Babington 1843, p. 242.

4 Babington referred to two forms of *Stellaria graminea*, one with shorter, the other with longer petals, in C. C. Babington 1851, p. 51, and in C. C. Babington 1856, p. 53. CD cited C. C. Babington 1851 on this point in *Forms of flowers*, p. 313 n. There is a lightly annotated copy of the third edition of Babington's *Manual of British botany* in the Darwin Library–CUL (see *Marginalia* 1: 27–8).

5 CD speculated about which insect species might be involved in the cross-fertilisation of *Primula* species in 'Dimorphic condition in *Primula*', p. 85 (*Collected papers* 2: 52), concluding: 'I am led to suppose that both Primroses and Cowslips are visited by moths.'

6 Lecoq 1854–8, 8: 146 and Vaucher 1841, 3: 723. There are annotated copies of both works in the Darwin Library–CUL (see *Marginalia* 1: 488–95, 812–15).

From J. D. Hooker [19 January 1862][1]

Kew
Sunday.

Dr D.

I was getting anxious to hear from you & now that I have am all the more to know again how you are all progressing my poor man, I can quite sympathize with you, a sick house is a dreadful thing—[2] We have been wonderfully well this winter considering the aggravating weather. my wife is laid up now with Neuralgia, which will I hope be better tomorrow, when the Lyells come.[3] I had Lubbock & Bates to dinner yesterday[4] the latter has done $\frac{1}{3}$ of his book: he is very earnest & I have no doubt will make a capital book as you say.—[5]

Huxley writes to me in Exuberat spirito at the reception of his lecture in "Saintly Edinburgh"—[6] he quite forgets there are sinners enough in a population of 180,000 to make him an applauding audience. I do not think H. has the smallest idea in how small a circle he makes a noise.

I dined the other day with Benthams sister—[7] a very intelligent & accomplished person who moves in the "best society" & dotes on her brother— — she did not know of what Society he was President!!! & cared not to know—[8]

Ten million thanks for Grays letter which is very interesting & very sad.[9] I was amazed with his insouciant national Egotism at the outset.— The only allusion I had then made to the war was to the effect that it would clear off the mass of scum under which I considered his nation groaned— this I intended as the only conceivable good that could come out of such a fratricidal contest.—& by George he took me for a sympathizer! not seeing that mine was the sympathy of contempt— You & I have always differed a good deal about America. I never could like but $\frac{1}{2}$ a dozen of all the Americans I had seen—but always thought that there was an element of *Grays* &c &c &c very strong in Boston &c, who, in any emergency would rise superior to the occasion, & by the might of a dignified line of conduct, gradually become the centre of a better state of things & of feelings.—

From the first of this affair I have looked for some expression of sense moderation & magnanimity from this boasted Boston party— & what do I see? **Lowell** in the chair at the Boston dinner to Wilkes[10] & Gray's rabid letters— after all why should we expect better things from a nation of upstarts— Our Aristocracy may have been (& has been) a great draw back to civilization—but on the other hand it has had its advantages—has kept in check the uneducated & unreflecting—& has forced those who have intellect enough to rise to their own level, to use it all in the struggle— There is a deal in *breeding* & I do not think that any but high bred gentlemen are safe guides in Emergencies such as these.

The moral effect of Lord Russell's despatches on the English mind has been quite astounding—[11] & I do not think you can point out a dozen men in public life—but of less breeding & culture (I do not mean by this *aristocratic* training as a specific thing) who would have been *safe* to have behaved with equal prudence dignity & consideration, & yet Gray calls *this* the pressure of a mob! If there is any thing at all in force of circumstances & Natural Selection it must arrive that the best trained, bred & ablest man will be found in the higher walks of life—true he will be rare, but then he will be obvious & easily selected by a discriminating public— When got to, he is removed above a multitude of temptations & conditions that prove the ruin of $\frac{9}{10}$ of the rising statesmen of a lower class of life— Your "Origin" has done more to enhance the value of the aristocracy in my eyes than any social political or other argument.

Now I never allude to politics in writing to Gray— it is useless I know, & furthermore whenever we did agree, it would perhaps most often be on totally different grounds—and this leads to endless misunderstandings. What puzzles me is, how are you to answer his letter? after your recent avowal that they are "unmitigated Blackguards" an opinion I heartily Echo.[12] Do tell me what you will say to him—it passes my guess-power. Have you read Spences "American Union" & what do you think of it?[13] I still like the tone of the London Review best of all the papers.[14]

I always forget to tell you that we can send you *Lythrum* plants whenever you want them.[15] Old Borrer died last week, & my Father went to the funeral.[16]

I gave Haast's book to Hector who has sailed for survey of Otago—[17] his glacial deposits was a mere mention without detail.— Hector is quite up to importance of subject.

I brought my remarkable plant before L. Soc. last Thursday with some effect—[18] it was thought quite as curious as I represented I have applied to R.S. grant for £50 to illustrate it—& of course expect to get it— I cannot afford to spend so much on my *Barnacles* as I used.[19] Genera Plantarum is passing through press i.e. 2^d^ Sheet is printing— the first part will contain DC^s^. *Thalamifloræ* & contain about 1000 genera—[20] it is an awfully laborious work.— still it is *the* great desideratum in Botany & must be done as well as we possibly can do it.— I have now Heers & Pengellys paper on Bovey Tracey to report on & Binney's on coal fossils:—so I am very busy.[21]

By the way Bates asked me last night to advise him about what was to be done about the plates for his paper for Linn: Soc:—was your offer of £10 accepted?—[22] of course the L.S. will gladly pay the engraving & coloring—but not the original drawing—but if this is done on stone at once it will save expense to all parties.

Ev yrs affec | J D Hooker

I will send Carrier on *Thursday* for Catasetum to Nagshead[23]

D^r^. Weddell. Poitiers. France[24]

DAR 101: 8–11

CD ANNOTATIONS

Top of first page: 'Glossodia' *pencil, circled pencil*; '*[*with flowers*]*' *pencil*

Margin of first page: '*[*Lebours*]*' *added pencil*

[1] Dated by the reference to Hooker's bringing his 'remarkable plant' before the Linnean Society of London (see n. 18, below), and by the relationship to the letters to J. D. Hooker, 16 January [1862] and 25 [and 26] January [1862]. In 1862, 19 January fell on a Sunday.

[2] See letter to J. D. Hooker, 16 January [1862].

[3] Hooker refers to Charles and Mary Elizabeth Lyell and Frances Harriet Hooker.

[4] John Lubbock and Henry Walter Bates.

[5] Bates 1863. CD had read the first two chapters of Bates's account of his experiences in the Amazon region (see *Correspondence* vol. 9, letter to H. W. Bates, 15 December [1861], and this volume, letter to H. W. Bates, 13 January [1862]), and told Hooker of his high expectations for the work (see *Correspondence* vol. 9, letter to J. D. Hooker, 18 [December 1861]).

[6] Thomas Henry Huxley had written to CD in the same spirit (see letter from T. H. Huxley, 13 January 1862).

[7] George Bentham had two surviving sisters, Mary Louisa de Chesnel and Sarah Jane Le Blanc (Jackson 1906).

[8] Bentham was president of the Linnean Society of London.

[9] See *Correspondence* vol. 9, letter from Asa Gray, 31 December 1861. CD sent the letter to Hooker via Francis Boott (see letter from Francis Boott, 27 January 1862).

[10] Hooker probably refers to the Boston businessman, John Amory Lowell. According to the *Boston Post*, 27 November 1861, p. 2, a dinner in honour of Charles Wilkes, captain of the *San Jacinto* and the man responsible for setting off the 'Trent affair', was held by the citizens of Boston on 26 November 1861. The report does not mention Lowell and states that John Wiley Edmands presided at the dinner.

[11] John Russell, first Earl Russell, was secretary of state at the British Foreign Office. Extracts from the diplomatic dispatches were published in the national press.

[12] See letter to J. D. Hooker, 16 January [1862].

[13] Spence 1861.

[14] The *London Review and Weekly Journal of Politics, Literature, Art and Society* began publication in July 1860, edited by Charles Mackay. Hooker had recommended this publication to CD in the previous year (see *Correspondence* vol. 9, letter to J. D. Hooker, 27 [March 1861] and n. 7).

[15] CD was interested by 'a magnificent case' of trimorphism in the genus *Lythrum* that was mentioned in Lecoq 1854–8, 6: 154–9, and had asked Hooker whether he could provide him with seeds or plants for investigation (see *Correspondence* vol. 9, letter to J. D. Hooker, 18 [December 1861] and n. 15).

[16] William Borrer died on 10 January 1862 (*DNB*). CD had asked Hooker about the possibility of approaching Borrer for plant specimens (see *Correspondence* vol. 9, letter to J. D. Hooker, 18 [December 1861], and this volume, letter from J. D. Hooker, [1 January 1862]). Borrer had been a friend of Hooker's father, William Jackson Hooker.

[17] J. F. J. von Haast 1861. James Hector was appointed geologist to the provincial government of Otago, New Zealand, in 1861. CD was interested in evidence of glaciation in Australia and New Zealand, which would support his view that there had been a worldwide glacial period (see *Correspondence* vol. 9, letter from J. D. Hooker, [29 December 1861]). Both Hector and Julius von Haast are cited in the fourth edition of *Origin* on evidence of glaciation in New Zealand (*Origin* 4th ed., p. 433). See also letter from W. B. Clarke, 16 January 1862, and letter to J. D. Hooker, 16 January [1862].

[18] Hooker had begun work on a monograph on *Welwitschia mirabilis* (J. D. Hooker 1863a); he gave an account of the plant before the Linnean Society of London on 16 January 1862 (*Journal of the Proceedings of the Linnean Society* 6: lix). He wrote to Huxley of his excitement about *Welwitschia* (L. Huxley ed. 1918, 2: 25):

> though neither Dicot, Monocot, nor Gymnosperm in flower or Exogen or Endogen in structure of axis, wood or bark . . ., it is still undoubtedly a member of the family Gnetaceae amongst Gymnosperms, as the structure of the ovule and development of the seed and embryo clearly show. It is . . . the most wonderful plant ever brought to this country—and the very ugliest. It re-opens the whole question of Gymnosperms as a class

See also *Correspondence* vol. 9, letter to J. D. Hooker, 23 October [1861] and n. 4.

[19] Hooker's investigations on *Welwitschia mirabilis*, like CD's study of barnacles, required prolonged and laborious examinations. CD had apparently remarked in a now missing letter to Hooker, 'I expect it is going to be your Barnacles' (L. Huxley ed. 1918, 2: 24).

[20] The reference is to the first part of volume 1 of Bentham and Hooker 1862–83, which covered the two botanical 'series' Thalamiflorae and Disciflorae. George Bentham had given to the printers the manuscript as far as the Resedaceae (i.e., pp. 1–112 of volume 1 of the printed work) on 2 December 1861 (see Stearn 1956, p. 129). The term 'Thalamiflorae' was first introduced by Augustin Pyramus de Candolle (Candolle and Candolle 1824–73, 1), whose name is commonly abbreviated in botanical literature to 'DC'.

[21] Heer 1862, Pengelly 1862, and Binney 1862. Hooker refers to his reviewing the papers for publication in the relevant societies' journals (see, for example, 'Papers read', Geological Society of London, archives, COM P3/1).

[22] Bates's paper on the Amazonian Heliconidae, a family of largely tropical New World butterflies (Bates 1862a), was read to the Linnean Society of London on 21 November 1861 and published in the *Transactions of the Linnean Society* 23 (1862): 495–566. In the paper, Bates invoked the theory of natural selection to account for mimetic resemblances between butterflies, arguing that the 'selecting agents' were insectivorous animals that destroyed those varieties of edible species that did not sufficiently resemble the inedible analogical form to deceive predators (pp. 511–12). CD offered financial assistance with the cost of coloured plates to accompany the paper (see *Correspondence* vol. 9, letters to J. D. Hooker, 25 November [1861] and 1 December [1861], and letter to H. W. Bates, 3 December [1861]).

[23] See letter to J. D. Hooker, 16 January [1862] and n. 4.

[24] This is the address of Hugh Algernon Weddell whose work on hybrids of *Aceras* (Weddell 1852) is cited in *Orchids*, p. 19. CD included Weddell's name on his presentation list for the volume (see Appendix III).

To C. C. Babington 20 January [1862][1]

Down. | Bromley. | Kent. S.E.
Jan 20th

Dear Babington

I thank you for your kind & very valuable letter. I shall have some future opportunity of quoting your cases, which are quite new to me.[2] I think Stellaria graminea grows here (but I know our British plant very imperfectly) & I must look sharp after it & get its seeds.— As for Hottonia I shall never see it. I know you are a great wanderer in summer: if you ever come across it, would you have the kindness to send me a few fresh spec. in a tin cannister by Post, for I should much like to see its pollen & speculate on manner of action of insects. Lecoq, I have just observed, says that Menyanthes is similarly dimorphic.[3] Perhaps where Hottonia grows Menyanthes would also grow. Koch says that Polemonium & Pyrola (according to Lecoq) are likewise dimorphic but I shall, never get to see these & still less to experiment on them, which is the really requisite thing.—[4]

As I have been begging favours, I will venture to ask you when next at your Botanic Garden to enquire whether the Curator by chance possesses seeds of any of the plants of which I will write a list, & which I much require for different experiments. I know it is a mere chance.—[5]

My health prevents me walking & that terribly interferes with my getting what I want. And now Mr. Borrer is dead, from whom I expected much.—[6] The varieties of Verbascum I want much to test Gärtners experiments.[7]

Forgive my writing at such length & believe me | Yours very sincerely | C. Darwin

P.S. | I am preparing a little Book on Orchids, which I think contains some new facts, & which I will send you when published.—[8]

I am convinced that Habenaria bifolia & chlorantha—& Ophrys apifera & arachnites are as good species as any in the world.—[9]

CUL (Add 8182)

[1] The year is given by the relationship to the letter from C. C. Babington, 17 January 1862.

[2] See letter from C. C. Babington, 17 January 1862. CD cited Babington on *Primula elatior* in *Forms of flowers*, p. 72, and on *Stellaria graminea* on p. 313 n.

[3] Lecoq 1854–8, 7: 391. There is an annotated copy of the work in the Darwin Library–CUL (see *Marginalia* 1: 488–95).

[4] CD's reference is ambiguous. He began reading Lecoq 1854–8 in December 1861 (see *Correspondence* vol. 9, letters to J. D. Hooker, [9 December 1861] and 28 [December 1861]). *Pyrola* and *Polemonium* are discussed in Lecoq 1854–8, 7: 356–62 and 413–14, respectively, but Wilhelm Daniel Joseph Koch is not cited. Lecoq identified *Pyrola* as being dimorphic in a section that CD highlighted in his copy of the work (p. 357), but did not identify *Polemonium* as being dimorphic. CD had borrowed a copy of Koch 1843–4 from Joseph Dalton Hooker in 1857 (see *Correspondence* vol. 6). Koch discussed *Pyrola* and *Polemonium* in Koch 1843–4, 2: 550–1 and 568, respectively; however, he did not identify either as being dimorphic, but remarked with respect to *Polemonium coeruleum*: 'Flores caerulei, rarius albi' [Flowers blue, more rarely white].

[5] The enclosure has not been found. As professor of botany at Cambridge, Babington directed the University Botanic Garden; the curator was James Stratton.

[6] William Borrer died on 10 January 1862 (*DNB*). CD had hoped Borrer might assist him by providing plants and seeds (see *Correspondence* vol. 9, letter to J. D. Hooker, 18 [December 1861]).

[7] In *Origin*, pp. 270–1, CD discussed experiments carried out on *Verbascum* by Karl Friedrich von Gärtner, in which crosses between differently coloured varieties of the same or of different species produced less seed than the parallel crosses between similarly coloured varieties (see Gärtner 1844, pp. 137–8, and Gärtner 1849, pp. 92, 180–1, 724–8; there are heavily annotated copies of these volumes in the Darwin Library–CUL (see *Marginalia* 1: 248–98)). These experiments were important to CD in countering an objection to natural selection based on the belief that there is 'some essential distinction between species and varieties' and that varieties always 'cross with perfect facility, and yield perfectly fertile offspring' (*Origin*, p. 268; see also letter to T. H. Huxley, 14 [January 1862] and n. 7). He had been anxious for some time to repeat Gärtner's experiments, and had sought specimens and assistance from botanical acquaintances (see *Correspondence* vol. 9). In December 1862, having failed to obtain the requisite varieties, CD persuaded John Scott to undertake the experiments on his behalf (see letters to John Scott, 19 November [1862] and 11 December [1862], and letter from John Scott, 17 December [1862]).

[8] *Orchids* was published in May 1862 (Freeman 1977, p. 112). Babington's name appears on CD's presentation list for the volume (see Appendix IV).

[9] In *Orchids*, pp. 88–9 and 72, respectively, CD questioned the classification of these plants by some botanical authorities as varieties of each other rather than as true species.

From T. H. Huxley 20 January 1862

Jermyn St
Jany 20th 1862

My dear Darwin

The inclosed article which has been followed up by another more violent more scurrilously personal & more foolish, will prove to you that my labour has not been in vain—and that your views & mine are likely to be better ventilated in Scotland than they have been[1] I was quite uneasy at getting no attack from the 'Witness' thinking I must have overestimated the impression I had made & the favourableness of the reception of what I said— But the raving of the 'Witness' is clear testimony that my notion was correct—[2]

I shall send a short reply to the 'Scotsman' for the purpose of further advertising the question—[3]

With regard to what are especially your doctrines—I spoke much more favourably than I am reputed to have done— I expressed no doubt as to their ultimate establishment—but as I particuly desire not to be misrepresented as an advocate trying to soften or explain away real dificulties—I did not in speaking enter in to the details of what is to be said in diminishing the weight of the hybrid difficulty— All this will be put fully when I print the Lecture—[4]

The arguments put in your letter are those which I have urged to other people—of the opposite side—over & over again.[5]

I have told my students that I entertain no doubt that twenty years experiments on pigeons conducted by a skilled physiologist instead of by a mere breeder—, would give us physiological species sterile *inter se* from a common stock—(& in this if I mistake not I go further than you do yourself) and I have told them that when these experiments have been performed I shall consider your views to have a complete physical basis—and to stand on as firm ground as any physiological theory whatever—[6]

This was impossible for me in the time I had to lay all this down to my Edinburgh audience—& in default of full explanation it was far better to seem to do scanty justice to you

I am constitutionally slow of adopting any theory that I must need stick by when I have once gone in for it—but for these two years I have been gravitating towards your doctrines & since the publication of your Primula paper with accelerated velocity—[7] By about this time next year I expect to have shot past you—and to find you pitching into me for being more Darwinian than yourself— However, you have set me going & must just take the consequences, for I warn you I will stop at no point so long as clear reasoning will carry me further—

My wife & I were very grieved to hear you had had such a sick house—[8] but I hope the change in the weather has done you all good— Anything is better than the damp warmth we had

I will take great care of the three 'Barriers'—[9] I wanted to cut it up in the 'Saturday'[10] but how I can fulfil my benevolent intentions—with four lectures a week—a lecture at the Royal Institution[11] & heaps of other things on my hands I don't know

Ever | Yours faithfully | T. H. Huxley

I am very glad to hear about Brown Sequard;[12] he is a thoroughly good man & told me it was worth while to come all the way to Oxford to hear the Bp. pummelled[13]

DAR 166.2: 291

[1] The enclosure has not been found. Huxley refers to the response of the Scottish press to two lectures he had recently given in Edinburgh (see letter from T. H. Huxley, 13 January 1862).

2 The *Witness*, a popular twice-weekly newspaper that served as the de facto organ of the evangelical Free Church of Scotland, carried an attack on Huxley's lectures on 14 January 1862 (see Appendix V). See also L. Huxley ed. 1900, 1: 193–4, where the report is quoted, but an incorrect date given for its publication, and Altholz 1989, pp. 90–1.

3 Huxley's letter appeared in the *Scotsman*, 24 January 1862, p. 2 (see Appendix V).

4 The substance of Huxley's lectures was eventually incorporated into T. H. Huxley 1863a, pp. 57–118.

5 See letter to T. H. Huxley, 14 [January 1862].

6 See letter from T. H. Huxley, 13 January 1862 and n. 2, and letter to T. H. Huxley, 14 [January 1862]. See also Appendix VI.

7 'Dimorphic condition in *Primula*'.

8 Many members of the Darwin household had been ill with influenza (see letter to T. H. Huxley, 14 [January 1862]).

9 [Rorison] 1861. See letter to T. H. Huxley, 14 [January 1862] and n. 12.

10 Huxley refers to the *Saturday Review of Politics, Literature, Science and Art*; no review of [Rorison] 1861 appeared in that journal.

11 Huxley gave a lecture entitled 'On fossil remains of man' at the Royal Institution on 7 February 1862 (T. H. Huxley 1862a). Huxley probably also refers to his lectures as professor of natural history at the Government School of Mines, Jermyn Street, London (*DSB*); he gave fifty lectures there in the 1861–2 session (*Medical directory* 1862, p. 243).

12 Charles Edouard Brown-Séquard. See letter to T. H. Huxley, 14 [January 1862], and letter from C. E. Brown-Séquard, 13 January 1862.

13 Huxley refers to Samuel Wilberforce, bishop of Oxford. Wilberforce's criticisms of *Origin*, made at the 1860 meeting of the British Association for the Advancement of Science at Oxford, had been vigorously answered by Huxley, Joseph Dalton Hooker, and John Lubbock (see *Correspondence* vol. 8).

From Conrad Martens 20 January 1862

St. Leonards. | —Sydney.
Jan 20th./62

To Chas. Darwin Esq., &c— .

Many thanks my old Shipmate for your kind message which I have just recd. by the padre,[1]

I thought you had quite forgotten that I was in existence, and certainly the man who voluntarily sets himself down in such a place as this has no right to grumble if he f⟨inds⟩ such to be the case.

As it appears howev⟨er⟩ you have still two of my sketches hanging up in your room.[2] I hope you will not refuse to accept another which I shall have much pleasure in preparing and will send to you by the next mail,[3]

Your "book of the season" as the reviewers have it, I must own I have not yet read, altho Mr Clarke offered to lend it me, I am afraid of your eloquence, and I don't want to think I have an origin in common with toads and tadpoles, for if there is anything in human ⟨n⟩ature that I hate it is a toady, but of course I know nothing of the subject, and they do make such microscopes now a-days— I suppose yours is one of Ross's very best,[4] by the by I got him to make two eyepieces for a reflecting telescope just before he died as I had succeeded in casting and polishing two metals of 6 and 7. feet focus, and so now I shew the good people here the mountains in the moon turned up side down, as of course they ought to be when seen from the antipodes.[5]

but I must apologize, for I suppose you don't laugh at nonsense now as you used to do in "Beagle or rather I suppose it does not come in your way.

Well, that was a jolly cruize, and I hope you have been well and happy ever since—and that you may continue so for some time to come is Believe me | the sincere wish | of your old shipmate | Conrad Martens.

PS. | I wonder whether, the Admiral, what is now,[6] I should like to send my kind regards, if you should see him, but don't if you don't like,— coffee without sugar, you remember.—[7]

DAR 171.1: 52

[1] See *Correspondence* vol. 9, letter to W. B. Clarke, 25 October [1861]. Martens, official artist on HMS *Beagle* from December 1833 until August 1834, settled in Sydney, Australia in 1835 (R. D. Keynes ed. 1979, pp. 2–3; R. D. Keynes ed. 1988, pp. 207, 263). In 1844, he moved to St Leonard's, a 'suburban township to Sydney' located in the parish of Willoughby (Whitworth ed. 1866; see L. Lindsay 1968, pp. 14–15). Following William Branwhite Clarke's arrival as incumbent of that parish in 1846, Martens became Clarke's first church-warden at the new church of St Thomas (see *Correspondence* vol. 9, letter to W. B. Clarke, [August 1861], and Jervis [1945], pp. 19–22).

[2] CD purchased two of Martens's water-colours in January 1836 when the *Beagle* visited Sydney. These water-colours are reproduced in R. D. Keynes ed. 1979 as no. 150, 'The *Beagle* in Murray Narrow Beagle Channel' (originally entitled 'View Ponsonby Sound'), and no. 193, 'River Santa Cruz' (*ibid*., pp. 116, 395, and pp. 201, 397, respectively).

[3] Martens soon afterwards sent CD a water-colour entitled 'View of Brisbane, 1862'; it is reproduced in Nicholas and Nicholas 1989, p. 128.

[4] The optician and scientific instrument maker, Andrew Ross, was one of the pre-eminent manufacturers of microscopes in Victorian London; he died in 1859 (Turner 1989, p. 154).

[5] Martens, who was a keen amateur astronomer, began constructing a six-inch reflecting telescope in 1860 (L. Lindsay 1968, pp. 13–14).

[6] Robert FitzRoy had been captain of the *Beagle* during CD's and Martens's time with the ship. He was promoted to the rank of rear-admiral in 1857 (*DNB*).

[7] According to CD's autobiography, references by the junior officers of the *Beagle* to how much 'hot coffee' had been served out in the morning indicated the state of the captain's temper (*Autobiography*, p. 73).

From W. B. Clarke 21 January 1862

St Leonard's
21 Jany 1862

My dear Sir,

I have opened this to enclose a P.S.[1]

You will find in the pamphlet which accompanies this, mention of probably *Pollicipes* among the fossils from Wollumbilla. p 52.[2]

Believing that it will be interesting to *you* to examine it, I enclose one Valve for your inspection, which I hope will reach you safely.

In your Monograph you define the oldest Cirripede to belong to the Oolite.[3] But I notice in the November number of the Quarterly Journal G.S. (XVI Part 14. p 512) Mr Moore mentions one older still.[4]

It may, therefore, serve more than one object, if you will kindly examine & compare the specimen I enclose, and favour me with a description and sketch of it. I ask this, because the Wollumbilla fossils have not yet been more than cursorily examined. In Moore's list (p 513–4) there are Molluscs, Cirripedes, Echinoderms and Annelids 13 *genera* of which I have in my first gathering. Is it possible, that the Wollumbilla collection is "*Rhætic*"?[5] You see how many Liassic and Triassic forms there are,—with Myacites, Myophoria, *Orthoceras & Belemnites* Monotis &c. M'Coy is very indistinct in his statement as to their place.[6] My views are given where I call them Triassic—i.e. upper Triassic. Can the Pollicipes help us?—

Pray let me ask you to give me such an account of it as I can use, in drawing up a further account of the fossils.

And now while writing of them—may I ask whether you know M^{r}. Moore and if he would undertake the description of the whole collection;—[7] or if you can recommend me to any one who is able and would (on terms suitable to both parties) examine and describe *all my fossils* for a work I am likely to be engaged in, on the General geology of Australasia.[8]

I hope you will pardon all this & believe me | My dear Sir, | Y^{rs}. very truly | W. B. Clarke

Mr Martens asked for your address. I think he will write to you.[9]

DAR 161.2: 173

1 This postscript was apparently sent under the same cover as the letter from W. B. Clarke, 16 January 1862. Since it occupies a separate sheet and contains an additional greeting and valediction, it has been treated as a separate letter.

2 Clarke 1861b. Clarke had received from a friend in Wollumbilla Creek, Queensland, a number of important fossils that were among the first Mesozoic specimens to be found in Australia. He forwarded the fossils to the palaeontologist Frederick McCoy for the purpose of dating, and subsequently reproduced McCoy's account and 'rough list of specimens' (McCoy 1861) in Clarke 1861b, pp. 48–52. See also Clarke 1861c. Clarke referred to sending CD his pamphlet in his letter of 16 January 1862. There is a lightly annotated presentation copy of Clarke 1861b in the Darwin Pamphlet Collection–CUL.

3 CD described *Policipes ooliticus* as 'the most ancient Cirripede as yet known' (*Fossil Cirripedia* (1851), p. v). See also pp. 8, 51–2.

4 C. Moore 1861.

5 The Rhaetic system of rocks is in the Upper Triassic, transitional to the Jurassic (*Dictionary of geological terms*).

6 McCoy dated the deposits 'to the base of the Mesozoic series, certainly not lower than the Trias, and not higher . . . than the lower part of the great Oolite' (McCoy 1861, p. 43; Clarke 1861b, p. 49).

7 CD apparently wrote for Clarke a letter of introduction to the geologist Charles Moore (see letters from W. B. Clarke, 20 June 1862 and 20 September 1862). Clarke subsequently sent his Wollumbilla fossils to Moore, who described them in C. Moore 1870 (see Jervis [1945], pp. 82–3).

8 No work on the general geology of Australasia was published under Clarke's authorship.

9 See preceding letter.

From Henry Holland [21 January 1862?][1]

Brook Street
Tuesday

My dear Charles,

At last I have received an answer from Lord Tankerville;[2] & I think a very satisfactory one.— I enclose his letter, & hope you will regard it in the same light

I trust you are all getting better.[3] Ever your's aff[ly] | H Holland

P.S | I wrote the other day to thank you for your paper on the Primula—[4]

DAR 166.2: 236

[1] The date is conjectured from the relationship to other letters of this period (see nn. 2 and 3, below). In 1862, 21 January fell on a Tuesday.

[2] For CD's interest in the Chillingham cattle, a breed kept on the Northumberland estate of Charles Augustus Bennet, sixth earl of Tankerville, see the letters from Henry Holland, [3–14] January [1862] and 15 January [1862], and the letter to Ludwig Rütimeyer, 15 [and 16] January [1862]. See also letter from C. A. Bennet, [9 February 1862].

[3] Holland had heard of the incidence of influenza in the Darwin household (see letter from Henry Holland, 15 January [1862]).

[4] Letter from Henry Holland, 15 January [1862].

From Dorothy Frances Nevill [before 22 January 1862][1]

Dangstein | *Petersfield*

My dear Sir

I have been a long time meditating whether I dared write and make the request I am about to do—but even if you cannot grant I hope you will take it as a compliment rather than in ⟨any⟩ other light— I am so gratified at having made your acquaintance[2] altho ⟨ ⟩ on paper ⟨ ⟩ ⟨w⟩ould so like if it were ⟨po⟩ssible that you would give me a *photo* or lithograph or any other portrait of yourself to hang up in my own sitting room Sir William Hooker[3] and many more have done so and it is such a pleasure to have near me the resemblance of friends to whom I am indebted for so many pleasant hours—in reading their works and listening to their conversation— Do comply with my request if you possibly can—[4] If I could I would buy one but to me that would be of no value unless it came from you and with your autograph— I shall enclose with this a list of orchids now in bloom that you may see whether you want any[5] We have 2 splendid Stangeria Paradoxa now in blossom one the fruitful one the barren flower What a curious bulbous plant it is— Do you believe it is a Cycadaceous Palm of the same species as Dion Edule—[6] I hope I have not exposed my ignorance in making these remarks—but I am so fond of all kinds of Botany and natural history that I fear I undertake too many things to do anything well— I only wish for an unframed photo or lithograph however bad I shall be most grateful for it

Y[rs] most truly | Dorothy Nevill

I am so interested in the account of your pigeons in your book[7]

Do you live ne⟨ar⟩ to Sydneys ⟨ ⟩[8]

DAR 172.1: 27

[1] Dated by the relationship to the letter to D. F. Nevill, 22 January [1862].

[2] Nevill grew orchids, pitcher plants, and other exotic species in her garden at Dangstein in Essex, and CD had asked her to assist him in procuring exotic varieties of orchids (see *Correspondence* vol. 9, letters to D. F. Nevill, 12 November [1861] and 19 November [1861]). He acknowledged her assistance in *Orchids*, p. 158.

[3] William Jackson Hooker

[4] See following letter and letter to D. F. Nevill, 22 January [1862].

[5] The enclosure has not been found.

[6] Cycadaceous plants were known to be dioecious (i.e., to bear male or female reproductive structures on different plants). See Lindley 1853, p. 223. According to Loudon 1850, p. 537, *Dion edule* had been cultivated in Britain since 1843; *Stangeria paradoxa*, which is not mentioned in Loudon 1850, was apparently a more recent introduction.

[7] An extensive account of pigeon varieties and their descent from the rock pigeon appeared in *Origin*, pp. 20–9.

[8] The reference may be to John Robert Townshend, third Viscount Sydney, whose principal residence, Frognal, was about five miles from Down at Foots Cray, Kent (*Post Office directory of the six home counties* 1862).

From D. F. Nevill [before 22 January 1862][1]

Dangstein | Petersfield

My dear Sir

I am most grateful for your most kind letter and the promise of the Photo.[2] already I am making a place for it amongst my other friends It has not arrived yet but I must be patient for I know in this dull weather they cannot make copies

I grieve to say I have at present no melastomaceous plant in flower— We had a glorious Pleroma but it is over—[3] My gardener says that he has observed that when the flower of the Cynoches is quite dry by the least touch the anther appears to go quite back with a jerk and at the same time it ejects its pollen which it throws on to the pistil[4] I do not know whether I have described it rightly or not I am most grateful for your little pamphlet[5] We have had a house full but now that I am alone I mean to give great attention to it

Thanking you again many times | believe me | most truly yours | Dorothy

DAR 172.1: 26

[1] Dated by reference to CD's interest in dimorphism (see nn. 3 and 4, below) and by the relationship to the letter to D. F. Nevill, 22 January [1862].

[2] CD's letter has not been found. It was probably a response to the preceding letter.

[3] CD was investigating dimorphism in plants and began to examine the Melastomataceae in October 1861, believing them to exhibit a novel form of this phenomenon (see letter from J. D. Hooker, [1 January 1862] and n. 3).

[4] CD had learned that, as with *Catasetum* and *Mormodes*, the pollinia of the orchid *Cycnoches* are violently ejected if the column is touched; he was anxious to study the mechanism, and had sought specimens and observations from Joseph Dalton Hooker (see *Correspondence* vol. 9, letters to J. D. Hooker,

6–7 October [1861], 11 October [1861], and 13 October [1861]). In his letters to D. F. Nevill, 12 November [1861] and 19 November [1861] (*ibid.*), CD had requested specimens of *Mormodes* and *Cycnoches*.

[5] The reference is probably to 'Dimorphic condition in *Primula*' although Nevill's name is not on CD's presentation list for the paper (see Appendix III).

To Asa Gray 22 January [1862][1]

Down Bromley Kent.—
Jan. 22[d].

My dear Gray

Your letter interested us much; for we are all curious to see how things look to you all, & a letter is something living.—[2] But first thanks for your new cases of Dimorphism: new cases are tumbling in almost daily, but I shall never have time to work a quarter of them. You will have received before this my Primula paper,[3] & will know the amount of evidence.— I have been ill with influenza (indeed we all have, for there have been 15 in bed in my household) & this has lost me 3 whole weeks, & delayed my little Orchid Book.—[4] I fear that you expect in this opusculus much more than you will find— I look at it as a hobby-horse, which has given me great pleasure to ride. I will with great pleasure send you the sheets if I can; for Clowes, my printer, often does not print off till the whole is set up.—[5] I shall be very curious to hear what you think of it; for I have no idea whether it has been worth the trouble of getting up,—though the facts, I am sure, were worth *my own* while in making out—

I am heartily glad to hear a better account of Dana, whom I much respect. What a striking looking man![6] I forwarded your letter to Boott & to Hooker, from whom I had a long & capital letter this morning.[7] He is working like a Horse. Here is a good joke, my book on Nat. Selection, he says, has made him an aristocrat in fact— he thinks breeding—the high breeding of the aristocracy—of the highest importance.—

Now for a few words on politics; but they shall be few, for we shall no longer agree, & alas & alas, I shall never receive another kind message from M[rs]. Gray.[8] I must own that the speeches & actions recently of your leading men (I regard little the newspapers), and especially the Boston Dinner have quite turned my stomach. I refer to Wilkes' being made a Hero for boarding an unarmed vessel.—to the Judges advice to him—& to your Governor triumphing at a shot being fired, right or wrong, across the bows of a British vessel.[9] It is well to make a clean breast of it at once; & I have begun to think whether it would not be well for the peace of the world, if you were split up into two or three nations. On the other hand I cannot bear the thought of the Slave-holders being triumphant; & it is really fearful to think of the difficulty of making a line of separation between the N. & the S., with armies, fortifications, & custom-houses without end with your retrograde tariff. Now I have done for myself in your eyes; & M[rs] Gray will be indignant at having sent a kind message to so false a caitiff.—

Well I can't help my change of opinion— It is all owing to that confounded Longitude.—[10] Bad man, as you will think me, I shall always think of you with affection.— Here is an insult! I shall always think of you as an Englishman.

Ever yours very sincerely | Charles Darwin

P.S | I have just performed an Herculean labour in looking through the nine big volumes of Lecoq's Bot. Geograph.—[11] it is a horrid dull Book, but I have stumbled on a few good facts; & on some cases of dimorphism,—several in Borragineæ & Labiatæ— Lythrum seems a very curious case for the two or three kinds of flowers occur on the same plant.—[12] I am now trying an experiment on one of the Melastomas; & I much suspect, that the two sets of anthers have different functions.—[13]

Hottonia is dimorphic like Primula.—[14]

Gray Herbarium of Harvard University

[1] The year is established by the relationship to the letter from Asa Gray, 31 December 1861 (*Correspondence* vol. 9).

[2] Gray discussed aspects of the American Civil War in his letter to CD of 31 December 1861 (*Correspondence* vol. 9).

[3] CD refers to 'Dimorphic condition in *Primula*'; Gray's name is on CD's presentation list for the paper (see Appendix III).

[4] *Orchids* was published on 15 May 1862 (Freeman 1977, p. 112).

[5] In his letter to CD of 31 December 1861 (*Correspondence* vol. 9), Gray had asked CD to send him the sheets of *Orchids* 'one by one—as soon as they come out', in order that he might write an early review of it. William Clowes was head of the London printing firm William Clowes & Sons, printers to John Murray.

[6] James Dwight Dana suffered a nervous breakdown in 1859 (see *Correspondence* vol. 7, letter to Charles Lyell, 29 [December 1859] and n. 10). In his letter of 31 December 1861 (*Correspondence* vol. 9), Gray enclosed a photograph of Dana and informed CD of his improved condition.

[7] The references are to Francis Boott and Joseph Dalton Hooker, and to the letter from J. D. Hooker, [19 January 1862]. See also letter from Francis Boott, 27 January 1862.

[8] In his letter of 31 December 1861 (*Correspondence* vol. 9), Gray included a message from his wife, Jane Loring Gray: 'my wife . . . bids me send her *love* to Mr. Darwin, & say that his is the only Englishman whose letters do not give her a shock to read'.

[9] In November 1861, Charles Wilkes, captain of the Union vessel, *San Jacinto*, had ordered the seizure of two Confederate envoys from the British mail packet, the *Trent*. A dinner in honour of Wilkes was held in Boston on 26 November, the details of which were reported in *The Times*, 10 December 1861, p. 9. Several speeches were made, praising Wilkes's action. The judge, George Tyler Bigelow, called upon to give a legal opinion on the case, advised: 'England, unless she falsifies her own conduct, unless she falsifies the statements of her own statesmen, unless she sets aside the judgments of her own judiciary, cannot undertake to make an issue with us upon this great act.' He went on to state that it was acceptable to regard the envoys as 'contraband of war'. John Albion Andrew, governor of Massachusetts, also gave a eulogy, in which he said: 'And that there might be nothing left to crown the exultation of the American heart, Commodore Wilkes fired his shot across the bows of the ship that bore the British lion at its head'. See also letter from J. D. Hooker, [19 January 1862] and n. 10.

[10] This is a reference to the suggestion made by the historian Henry Thomas Buckle that there was a statistical relationship between the nature of a country's climate and the progress of its civilisation (Buckle 1857–61, 1: 38 *et seq.*). In his letter to Asa Gray, 11 December [1861] (*Correspondence* vol. 9), CD had remarked: 'Buckle might write a chapter on opinion being entirely dependent on Longitude!'

[11] CD had begun reading Lecoq 1854–8 in December 1861 (see *Correspondence* vol. 9, letters to J. D. Hooker, [9 December 1861] and 28 [December 1861]). There is an annotated copy of the work in the Darwin Library–CUL (see *Marginalia* 1: 488–95).

[12] Lecoq 1854–8, 6: 154–9. CD mentioned his interest in the 'magnificent case' of trimorphism in *Lythrum* in the letter to J. D. Hooker, 18 [December 1861], and asked if Hooker could provide him with seeds or plants for experimentation. He carried out numerous crossing experiments in 1862 and 1863 with help from his son, William Erasmus Darwin (see the experimental notes by CD in DAR 27.2 and by W. E. Darwin in DAR 117). His paper on the sexual relations of the three forms of *Lythrum salicaria* ('Three forms of *Lythrum salicaria*') was read before the Linnean Society of London in June 1864.

[13] CD had recently begun to investigate what he believed might be a novel form of dimorphism in the Melastomataceae (see letter from J. D. Hooker, [1 January 1862] and n. 3). The reference in this letter is probably to a pollination experiment on *Heterocentron roseum* that CD began in October 1861; the resulting seed pods were gathered at the end of January 1862 (see *Correspondence* vol. 9, letters to J. D. Hooker, 17 November [1861] and [30 and 31 December 1861]; see also the dated experimental notes in DAR 205.8: 45–6). CD had also begun to examine *Monochaetum ensiferum*, but did not begin experiments on this species until early in February (see the dated notes in DAR 205.8: 22–4, 26).

[14] Charles Cardale Babington pointed out the occurrence of dimorphism in *Hottonia* in his letter to CD of 17 January 1862. CD discussed the case in *Forms of flowers*, pp. 50–4, 252, and 254.

To T. H. Huxley 22 January [1862][1]

Down Bromley Kent
Jan. 22^{d}.

My dear Huxley

I have been much amused at the Witness.— Such abuse is as good as praise. What fools these Bigots are.—[2]

I have been much pleased at what you say about Sterility & Hybridity. I did not suppose that you would enter on the general question of modification in your published Lectures, but am pleased to hear that you will touch on it.[3] It will be a good joke if ever I come to cry "hold hard"— I well know that you will go as far, but no further, than your reason tells you.— It is really odd how most of the objectors, (as Sir D. Brewster)[4] never allude to the arguments, which alone have much weight in favour of such views, as affinities rudimentary organs &c &c—

Your whole letter has interested me, & many thanks for it.

We are all better; but we have had 16 ill in the House!

I cannot think how on earth you will find time to write out your Lectures with all your work. I wish to God you had more spare time.

Ever yours very sincerely | C. Darwin

I have read about $\frac{1}{3}$ of the N. Hist. R. it is a capital number;[5] how well Lubbock writes, not that I have finished his article.—[6]

Imperial College of Science, Technology and Medicine Archives (Huxley 5: 252)

[1] The year is established by the relationship to the letters from T. H. Huxley, 13 January 1862 and 20 January 1862.

[2] Huxley had recently delivered two lectures at the Philosophical Institution of Edinburgh on man's relation to the lower animals. The lectures were fiercely criticised in an article that appeared in the Presbyterian newspaper, the *Witness*, on 14 January 1862 (see Appendix V). Huxley sent CD a copy of the article with his letter of 20 January 1862.
[3] See letters from T. H. Huxley, 13 January 1862 and 20 January 1862, and letter to T. H. Huxley, 14 [January 1862]. See also Appendix VI.
[4] Brewster 1862. See letter from T. H. Huxley, 13 January 1862.
[5] Huxley was the senior editor of the new series of the *Natural History Review*. There is a lightly annotated copy of the January 1862 number in the Darwin Library–CUL.
[6] Lubbock 1862b.

From Leonard Jenyns 22 January [1862][1]

Darlington Place | Bath
Jan. 22. 1861.

My dear Darwin,

With this I send the proof-sheet containing the notice you were kind enough to contribute to my memoir of Henslow.[2] You expressed a wish to see it.— I have not sent your MS, as I myself have compared it with that, & found it correct;—but if you wish to make any alteration or addition you can now do so.—[3] Perhaps you will be good enough to return the sheet the same day you receive it, if possible,—either to myself or the printers;—to myself if much altered,—but otherwise to the printers direct, whose address I have given on the other side.—

Very sincerely your's | L. Jenyns.

P.S. Your contribution begins at p. 51.—

"Mess[rs] Woodfall & Kinder
Angel Court
Skinner Street
London". E.C.

DAR 168: 55

[1] The year is established by the reference to Jenyns 1862 (see n. 2, below). Jenyns wrote '1861' in error.
[2] John Stevens Henslow died on 16 May 1861. Jenyns's memoir of his brother-in-law was published in May 1862 (*Publishers' Circular* 25: 223). CD's contribution appears in Jenyns 1862, pp. 51–5 (see also *Collected papers* 2: 72–4 and *Correspondence* vol. 9, Appendix X).
[3] CD had added a note to his manuscript asking to see a proof of the piece (see *Correspondence* vol. 9, letter from Leonard Jenyns, 31 December 1861).

To D. F. Nevill 22 January [1862][1]

Down. | Bromley. | Kent. S.E.
Jan. 22[d]

Madam

I received the Orchids safely & the other flowers, & I ought to have thanked your Ladyship earlier for your kindness, but I happened to be much engaged & deferred writing.

with my best thanks | I beg leave to remain | Your Ladyships | Truly obliged | Charles Darwin

P.S. Since the above was written I have received your Ladyships most kind note.—[2] I am infinitely obliged for all your great kindness.—

No doubt the Photograph will be sent in a few days; but assuredly it will not ornament your room.—[3]

Wellcome Institute for the History of Medicine Library, London

[1] The year is conjectured from CD's having first written to Nevill the previous November, requesting her help in procuring orchid specimens (see *Correspondence* vol. 9, letters to D. F. Nevill, 12 November [1861] and 19 November [1861]).

[2] CD probably refers to the second of the two letters from Nevill dated [before 22 January 1862].

[3] In the first letter dated [before 22 January 1862], Nevill requested a photograph or lithograph of CD to hang in her sitting room. The photograph to which CD refers may have been one taken the previous year by his son, William Erasmus Darwin (see *Correspondence* vol. 9, letter to Asa Gray, 11 April [1861] and n. 19; the photograph is reproduced as the frontispiece to that volume). Since, however, it appears that CD did not send the photograph to Nevill personally, it may be that he had ordered her a copy of the photograph taken by Maull and Polyblank in about 1857 (*Correspondence* vol. 8, frontispiece). See also letter from E. A. Darwin, [April – May? 1862].

From John Rogers 22 January 1862

16. Eversfield Place | St. Leonards. | Hastings
Jany. 22 1862

Dear Sir

I have never succeeded in growing Cycnoches.— & I do not recollect ever to have had Mormodes,[1] but as this last is one of the aberrant forms of Catasetum, I suspect it follows its habits— You are doubtless aware that Myanthus & Catasetum are *identical* though totally unlike.—[2] Twenty years & more ago, I flowered a most beautiful specimen of Myanthus barbatus.—two full spikes of flower, exactly like the plant represented in an early volume of Paxtons Mag^{e}.[3]

It was an *imported* plant from Demarara: & I was greatly delighted with it & prized it highly.— Judge of my dismay when the plant flowered the next year, & was a simple & pure Catasetum.—[4] There could be no mistake for I had no other plant like it. & besides I watched my own plants *then* almost day by day.— The transmogrification of that genus is enough to shake my belief in all floral identity.— I regret that I made no drawings of either form—& I cannot recollect any thing definite about the structure of the anther.—

I am spending the winter here & have no access either to my books or my plants—

Soon after I last wrote to you,[5] a Catasetum—which had escaped notice up in the roof of my Orchid house flowered, but with only 2 flowers & these had shed their pollen masses before I saw them—so that it was useless to send it to you.—

I wish we were nearer neighbours[6] as I should have much pleasure in showing you any thing which I have, though my plants through neglect now of some years standing are not what they once were—

Believe me, Dear Sir | Faithfully yours | John Rogers

PS. | Tropical orchids are very capricious in their growth & often require some slightly peculiar treatment which it is very difficult to find out— One or two species will sometimes flourish, where all others languish, & a slight alteration in the construction of a house will sometimes cause failure where success had previously been uniform—

DAR 176.1: 194

[1] CD was particularly interested in the Catasetidae, which he called 'the most remarkable of all Orchids' (*Orchids*, p. 211), because their pollinia are violently ejected, thus becoming attached to the insect disturbing the flower. While he had observed this phenomenon in *Catasetum*, CD had been unable to obtain mature and undamaged flowers of *Cycnoches* and *Mormodes*, despite many requests to acquaintances (see *Correspondence* vol. 9 and this volume, second letter from D. F. Nevill, [before 22 January 1862] and n. 4).

[2] CD had already encountered the remarkable case of trimorphism in *Catasetum tridentatum* (see *Correspondence* vol. 9, letter to Daniel Oliver, 7 December [1861]), and he subsequently discussed it in his paper 'Three sexual forms of *Catasetum tridentatum*'. There he explained that the female and hermaphrodite forms of this species were so different from the male form and from each other as to have been classified in different genera (as *Monachanthus viridis* and *Myanthus barbatus*, respectively).

[3] *Paxton's Magazine of Botany* 2: 124.

[4] CD cited Rogers's anecdote about *Myanthus* in 'Three sexual forms of *Catasetum tridentatum*', p. 151 n. (*Collected papers* 2: 69 n. 3), and in *Orchids*, p. 236 n. (In *Orchids*, CD misspelt his correspondent's name as 'Rodgers'.)

[5] No further correspondence between CD and Rogers has been found.

[6] Rogers lived at Sevenoaks Weald in Kent, about eight miles from Down.

To John Lubbock 23 January [1862][1]

Down
Jan. 23.

My dear Lubbock

At last I am pretty well after having lost three weeks & having had 16 in my household ill! It is an age since we have met, & I sh^d. much enjoy seeing you.[2] I would come over any day to your luncheon & not stay so long as last time & then I am sure it would not tire me; though my movements must always be doubtful, but I would come *punctually* to your luncheon (my dinner) or not at all.— Or how would it suit you to come & dine & sleep here on Saturday or any day & take your chance of my being pretty well. Meet we must & it really makes to me no difference: settle whichever plan suits you, who have so little time to spare, best.— By the way I cannot come till after next Tuesday; any day after that would suit me to come to you or most gladly to see you here—

I was very sorry I could not read your paper which you sent,[3] but I was downright ill with fever.—

I have been just reading your paper in N. H. Review, & have been much interested in it.—[4] How well you write & how you find time to do all that you do, is simply marvellous.—

M^rs. Darwin, though very little fit for exertion, has gone to London for a couple of days.—[5]

I hope Mrs Lubbock is strong again.[6]

Yours most truly | C. Darwin

DAR 263

[1] The year is established by the relationship to the letter from John Lubbock, 6 January 1862, and by the reference to Emma Darwin's trip to London (see n. 5, below).

[2] In 1861, John and Ellen Frances Lubbock moved from High Elms, near Down, to Chislehurst, Kent. Lubbock had been trying to arrange a meeting with CD for several weeks (see letter from E. F. Lubbock to Emma Darwin, [January 1862], and letters from John Lubbock, 6 January 1862 and 7 [February] 1862).

[3] Lubbock had asked CD for comments on the manuscript of one of his papers, probably Lubbock 1862a (see letter from John Lubbock, 6 January 1862 and n. 2).

[4] Lubbock 1862b.

[5] According to her diary (DAR 242), Emma Darwin stayed in London from 23 to 25 January 1862 in order to take Leonard Darwin to the dentist. She stayed at the home of Erasmus Alvey Darwin, CD's brother.

[6] Ellen Frances Lubbock gave birth to the Lubbocks' fourth child, Norman, on 16 December 1861 (Hutchinson 1914; *Gentleman's Magazine*, n.s. 12 (1862): 82). See also letter from E. F. Lubbock to Emma Darwin, [January 1862].

To Leonard Jenyns 24 January [1862][1]

Down Bromley Kent
Jan. 24th

My dear Jenyns

I have to thank you for two kind notes.[2] I received the proof-sheets this morning—altered only a few words & forwarded by todays post to the Printers. I am heartily glad that you have undertaken this memoir of poor dear Henslow, & that your labours are over & that it will soon appear.—[3] I shall read it with deep interest. His death has indeed been a cruel loss to many.—

I hope you keep well & tolerably happy. We lately have had a wretched household of invalids.—[4] I often look back to the happy & unanxious days of Cambridge.—[5]

My dear Jenyns | Yours very sincerely | Ch. Darwin

Do you ever see F. Hope at Bath;[6] if you do please give him my very kind remembrances.—

Bath Royal Literary and Scientific Institution

[1] The year is established by the reference to Jenyns's memoir of John Stevens Henslow (Jenyns 1862; see n. 3, below).

2 Letters from Leonard Jenyns, 31 December 1861 (*Correspondence* vol. 9) and 22 January [1862].

3 Jenyns's memoir of his brother-in-law, John Stevens Henslow, (Jenyns 1862) was published in May 1862 (*Publishers' Circular* 25: 223). CD's contribution to this memoir is reprinted in *Correspondence* vol. 9, Appendix X (see also *Collected papers* 2: 72–4).

4 Many members of CD's household had been ill with influenza (Emma Darwin's diary (DAR 242)).

5 CD was an undergraduate at Cambridge when he first met Jenyns, at that time vicar of Swaffham Bulbeck, Cambridgeshire. They had a shared interest in entomology, occasionally going on entomological excursions together (see *Correspondence* vol. 1).

6 Frederick William Hope, the son of one of the Darwins' Shropshire neighbours, had assisted CD with his early entomological collecting (see *Correspondence* vol. 1). He lived in London, but travelled for the sake of his health (*Journal of the Proceedings of the Linnean Society of London* 6 (1862): xci). Jenyns settled in Bath in 1860.

From H. W. Bates 25 January 1862

King St Leicester
25 Jany 1862

My Dear Sir

I found the returned M.S & your kind letter when I came home after 8 days stay in town chiefly occupied by studies relative to the travels.[1] I hope your good opinion of the 2nd chapter will be equally merited by the rest. I will now trouble you to write to Mr. Murray, as there is quite enough ready for him to form an opinion of the work.[2]

I will not write more at the present time— The complaint which I heard in London has afflicted your family has extended to here— I am now suffering from a kind of influenza with some symtoms of an old ague but it has not upset me completely at present[3]

Yours sincerely | H W Bates

DAR 160.1: 65a

1 Bates refers to the manuscript of the second chapter of his book describing his experiences as a naturalist in the Amazon region of South America, which he had sent to CD for comment (see letter to H. W. Bates, 13 January [1862]).

2 CD had offered to write to his publisher, John Murray, recommending Bates's manuscript (see *Correspondence* vol. 9, letter to H. W. Bates, 25 September [1861]). Murray published the work the following year (Bates 1863).

3 According to Henry Holland, influenza was 'very rife both in London & the country' (see letter from Henry Holland, 15 January [1862]).

To J. D. Hooker 25 [and 26] January [1862][1]

Down Bromley Kent
Jan. 25th

My dear Hooker.

Many thanks for your last Sunday's letter, which was one of the pleasantest I ever received in my life.[2] We are all pretty well redivivus, & I am at work again.

I thought it best to make a clean breast to Asa Gray & told him that the Boston dinner &c &c had quite turned my stomach,—that I almost thought it would be good for peace of world if U. States were split up: on other hand I said that I groaned to think of the Slaveholders being triumphant, & that the difficulties of making a line of separation were fearful. I wonder what he will say: I shall never have love sent me again from M^rs. Gray.—[3]

Your notion of the aristocrats being ken-speckle,[4] & the best men of a good lot being thus easily selected is new to me & striking.[5] The Origin having made you, in fact, a jolly old Tory, made us all laugh heartily. I have sometimes speculated on this subject: primogeniture is dreadfully opposed to selection,—suppose the first-born Bull was necessarily made by each farmer the begetter of his stock! On other hand, as you say, ablest men are continually raised to peerage & get crossed with the older Lord-breeds—& the Lords continually select the most beautiful & charming women out of the lower ranks; so that a good deal of indirect selection improves the Lords. Certainly I agree with you, the present American row has a very toryfying influence on us all.—

I am very glad to hear you are beginning to print the "Genera":[6] it is a wonderful satisfaction to be thus brought to bed,—indeed it is one's chief satisfaction, I think, though one knows that another bantling will soon be developing. I feel infinite satisfaction in knowing that in 2 or 3 weeks I shall be correcting press of my little orchid Book: whether this book is worth printing I have no more idea than the man in the moon—[7]

It is very true what you say about Huxley having no idea how little science is generally regarded;[8] & he deceives himself in wonderful manner about writing down that beggar Owen.[9] By the way Huxley tells me that Owen goes in for progressive development in 2^d Edit. of his Palæontology, pooh-poohing natural selection.—[10] Those Dutchmen in N. Hist. Review give Owen a good setting down.—[11] I am quite ashamed how demoniacal my feelings are towards Owen.— I had the other day the feeblest letter from Balfour against Huxley's conclusion on man & monkeys;[12] & he says that H. was very cautious in affirming that they were codescendants.—

I have read your interesting paper on Cedars.—[13] what a good case of doubtful species & of broken distribution.

With respect to Bates I repeated by letter the offer of assistance for any scientific requirement;[14] & it would look presumptuous in me to say anything more; so he must apply if he wants aid.— I wish I could have seen the specimens which he exhibited at Linn. Soc^y.—[15]

I am so glad to hear that you can give me Lythrum:[16] a little later in Spring, I will send list of my desiderata.[17] A bad job poor old M^r. Borrer's death.—[18]

Farewell | Ever yours affect^ly | C. Darwin

P.S.[19] I shall be very glad indeed to see an Arethusa[20] & then I must stop. I have just received such a Box full from Mr Bateman with the astounding Angræcum sesquipedalia with a nectary a foot long—[21] Good Heavens what insect can suck it— I will write about Dimorphism.—[22]

DAR 115.2: 141

[1] The year is established by the reference to Schroeder van der Kolk and Vrolik 1862 (see n. 11, below). The postscript was probably written on the day following the rest of the letter (see n. 19, below).
[2] Letter from J. D. Hooker, [19 January 1862].
[3] See letter to Asa Gray, 22 January [1862] and nn. 8 and 9. The reference is to Jane Loring Gray.
[4] Kenspeckle: 'easily recognisable; conspicuous' (*OED*).
[5] See letter from J. D. Hooker, [19 January 1862].
[6] See letter from J. D. Hooker, [19 January 1862]. The first part of volume 1 of *Genera plantarum* (Bentham and Hooker 1862–83) was published on 7 August 1862 (Stearn 1956, p. 130).
[7] *Orchids* was published in May 1862 (Freeman 1977, p. 112).
[8] In his letter of [19 January 1862], Hooker, commenting on Thomas Henry Huxley's delight at the reception accorded to lectures he had recently given in Edinburgh, stated: 'I do not think H. has the smallest idea in how small a circle he makes a noise'.
[9] Richard Owen.
[10] See letter from T. H. Huxley, 13 January 1862 and n. 5.
[11] CD refers to Schroeder van der Kolk and Vrolik 1862. See also letter to T. H. Huxley, 14 [January 1862] and n. 5.
[12] Letter from J. H. Balfour, 14 January 1862.
[13] J. D. Hooker 1862a appeared in the January issue of the *Natural History Review*. CD annotated the article in his copy of this issue (Darwin Library–CUL).
[14] Hooker had inquired whether or not Henry Walter Bates had accepted CD's offer to pay towards the cost of plates for Bates 1862a (see letter from J. D. Hooker, [19 January 1862] and n. 22). The paper was published with two coloured plates.
[15] Bates exhibited specimens of 'Lepidopterous Insects' from South America at a meeting of the Linnean Society of London on 16 January 1862 (*Journal of the Proceedings of the Linnean Society* 6: lviii).
[16] See letter from J. D. Hooker, [19 January 1862] and n. 15.
[17] See letter to J. D. Hooker, 25 February [1862] and n. 10.
[18] William Borrer died on 10 January 1862 (see letter from J. D. Hooker, [19 January 1862]). CD had hoped to obtain plants and seeds from him.
[19] This postscript, which is on a separate sheet, was written in response to the following letter, and was thus probably added on 26 January.
[20] See following letter.
[21] CD thanked James Bateman for his assistance in supplying plants in *Orchids*, pp. 114 n. and 158 n.; he discussed *Angraecum sesquipedale* on pp. 197–203. See also letter from Robert Bateman, [28 January 1862] and Kritsky 1991.
[22] See the following letter, in which Hooker provided CD with a list of dimorphic plant families and offered to procure specimens for him. CD responded to Hooker in his letter of 30 January [1862].

From J. D. Hooker [25 January 1862][1]

Kew
Saturday.

Dear Darwin

I write chiefly to say that we have an *Arethusoid* coming into flower, & that I hope next week to send 1 flower—& if all goes well with the plant some more afterwards.[2]

I have always forgotten to mention if amongst your dimorphous orders.—You have *Malpighiaceæ*. *Leguminosæ*, *Oxalideæ*, *Caryophyllaceæ*, *Balsaminæ* (of course) *Campanulaceæ*—& do you want a bit of all I can get?[3]

I fear you have no better report to give of your own & families health—

Falconer I hear snubs & pooh poohs & contradicts all the Sumatra + Ceylon Elephant story. & he is so accurate that I suppose he must be believed[4]

Ev Yrs affec | J D Hooker

Plants have come all right.[5]

Lyells dined here last Monday, & are as rabid as ever for America—that I do not care about—but I cannot stand her abuse of *our* conduct.[6]

Busk has broken his arm above wrist both bones.— —fell down on Monday night coming from Burlington House.[7]

DAR 101: 6–7

[1] Dated by the reference to the Lyells' having dined with the Hookers (see n. 6, below) and by the reference to George Busk's accident (see n. 7, below, and letter from J. E. Gray, 29 January 1862). In 1862, 25 January fell on a Saturday.

[2] CD had long been anxious to examine a specimen of John Lindley's 'great Division of Arethuseae', as described in Lindley 1853 (see *Correspondence* vol. 9, letter to J. D. Hooker, [28 July – 10 August 1861]), and had repeatedly requested specimens from Hooker and from other botanical acquaintances (see *Correspondence* vol. 9). However, in *Orchids*, p. 269, CD wrote that he had 'not seen any living flowers' of the Arethuseae.

[3] Following the publication of 'Dimorphic condition in *Primula*', CD was investigating the occurrence of dimorphism in other plant families (see, for example, *Correspondence* vol. 9, letter from Asa Gray, 31 December 1861, and this volume, letter to Asa Gray, 22 January [1862]).

[4] An article on the elephants of Sumatra and Ceylon, by the German naturalist Hermann Schlegel, appeared in the January issue of the *Natural History Review* (Schlegel 1862). Schlegel claimed that these elephants constituted a species distinct from that found in mainland India. Hugh Falconer, an authority on fossil and living elephants, subsequently published his objections to Schlegel's claim (Falconer 1863, pp. 81–96).

[5] CD had returned the *Catasetum* plants he borrowed from the Royal Botanic Gardens, Kew (see letter to J. D. Hooker, 16 January [1862]).

[6] Charles and Mary Elizabeth Lyell were supporters of the Union cause in the American Civil War and intimates of Charles Francis Adams, the United States' minister in London (Adams 1918). The Lyells dined with the Hookers on 20 January 1862 (see letter from J. D. Hooker, [19 January 1862]).

[7] George Busk was the zoological secretary of the Linnean Society of London. The society maintained offices and held its meetings at Burlington House, Piccadilly.

To William Walmisley Baxter 26 January [1862][1]

Down. | Bromley. | Kent. S.E.
Jan 26th

Dear Sir

I am much obliged for your note.[2] In early part of year some Cod Liver oil was used by myself, viz part of 2 or 3 Bottles; but to be sure that I have not deducted unfairly, I have deducted for 9 out of the 13 Bottles sent.—[3] I am sorry I troubled you about the Friar's Balsam, which I did not know was the same with T. of Benzoin—

I send cheque[4] & remain, Dear Sir | Yours faithfully | Ch. Darwin

Thomas Fisher Rare Book Library, University of Toronto

[1] The year is established by an entry in CD's Account book (Down House MS) (see n. 4, below).

[2] Baxter's note has not been found.

[3] Baxter ran an apothecary's shop in Bromley, Kent. The previous year, Joseph Dalton Hooker had recommended the external application of cod-liver oil as something that might assist Henrietta Emma Darwin in her recuperation from illness (see *Correspondence* vol. 9, letter to J. D. Hooker, 8 [February 1861]).

[4] An entry in CD's Account book–cash accounts (Down House MS) for 27 January 1862 records a payment to Baxter of £10. 2*s.* 6*d*.

From Francis Boott 27 January 1862

London
Jany 27. 1862

My Dear Darwin

I have sent what I have published of my Work to Your Brother for Your acceptance & only wish it were more worthy.[1] If I live, & the American War does not ruin the Country, & what I have in it, I hope to give about 200 additional figures, & a general view of the genus.[2] This last portion of the work is above my powers, & I often wish I had never attempted that which every one has failed in. Willdenow, Wahlenberg, Kunth, Sprengel Torrey, Tuckerman have left abortive attempts, & mine will be of no other value than shewing their short coming—[3]

The old observation that "there is nothing stabile but change",[4] is especially true in Carex—& it is the absence of any fixed character that renders the definition of groups so perplexing. Carey & Drejer insist on the orifice of the perigynium, as entire, bidentate—or bicuspidate,[5] but I find in plants otherwise affiliated the orifice exhibiting each form, & no selection of words is adequate to express clearly specific (so called) differences. If you take groups or parts of a large group as variable species, the definition is equally embarassing, & I think no general clavis is possible—that is, affording a ready means for ascertaining a species. I believe the more satisfactory way will be to give first a view of the species of different countries, & then to write the whole as intelligibly as may be, giving indications of such variations as may be found in species widely diffused over the world.

With a view to geographical distribution we want an arrangement of Countries. I have thrown the 600 odd species into such an arrangement—first into the *5 quarters* of the World Europe, Asia, Africa America Australia—& the different groups into more definite portions of these large divisions of the world— Asia for instance alone means nothing definite— Then Islands sometimes puzzle me.

I sent Grays note to Hooker, & trust he has returned it to you.[6] You seem to have conciliated Dear Gray. I send The Times & Saturday Review to my sister who is a neighbour of Grays & I come into the disgrace of those Journals.[7] I see no issue to the war & can only lament the animosity against England—to me unaccountable—

Yrs sincerely | F. Boott

C. Darwin Esq—

DAR 160.2: 252

[1] The reference is to Boott's study of the sedge genus *Carex* (Boott 1858–67). There are unannotated copies of the first two parts of this work, published in 1858 and 1860, in the Darwin Library–Down; the third part was published in 1862. CD's brother, Erasmus Alvey Darwin, lived at 6 Queen Anne Street, London.

[2] Boott, who was the son of a wealthy New England merchant, had inherited money and was financing the publication of Boott 1858–67 himself, including the drawing and engraving of the many plates, which cost him 'a very large sum of money' (*Journal of the Proceedings of the Linnean Society of London* 8 (1865): xxv). His health had long been poor and he died in December 1863, before the completion of his work; the fourth and final part, containing 189 plates, was prepared for publication by Joseph Dalton Hooker. The volume contained no general discussion of the genus.

[3] Karl Ludwig Willdenow, Göran Wahlenberg, Karl Sigismund Kunth, Kurt Polycarp Joachim Sprengel, John Torrey, and Edward Tuckerman all wrote extensively on the genus *Carex*. See Willdenow 1805 and Willdenow ed. 1797–1824, Wahlenberg 1802–3, Kunth 1815 and Kunth 1833–50, K. P. J. Sprengel ed. 1825–8, Torrey 1836 and Torrey ed. 1824, and Tuckerman 1843.

[4] The phrase is an American proverb (see Mieder *et al.* eds. 1992, p. 90).

[5] John Carey had written the section on *Carex* for Asa Gray's *Manual of the botany of the northern United States* (A. Gray 1848). Salomon Thomas Nicolai Drejer wrote two important papers on the genus (Drejer 1840–1 and Drejer 1844).

[6] See letter from Asa Gray, 31 December 1861 (*Correspondence* vol. 9), and letter from J. D. Hooker, [19 January 1862].

[7] Both *The Times* and the *Saturday Review* attacked the Union cause in the American Civil War. Boott was born in Boston, Massachusetts, but left the United States for England after completing his education at Harvard University (*DAB*).

From John Lubbock 27 January 1862

11, Mansion House Street, London, E.C.
27 Jan *1862*

My dear M^r. Darwin

Many thanks for your kind letter, I am delighted to hear that you are better.[1]

When children are offered a choice of two good things they always ask for both & I feel inclined to do the same, so I shall let you know which day I shall be at home next week & shall then perhaps hope to see you. We are expecting the Carpenters on Saturday, but have not yet their answer,[2] if they do not Come I will write & offer myself for that day as you suggest, knowing that you will not hesitate to tell me if in the mean time it should have become inconvenient.

Willy called in here the other day & seems to like his business which you can easily believe pleased me much.[3]

I am also very glad that you liked my "Lake habitations" & approve of my Style of writing.[4]

I often long to be nearer to you as I used so much to enjoy my little visits to Down.[5]

Hoping you will all keep well I remain | Your aff | John Lubbock

DAR 170.1: 26

[1] See letter to John Lubbock, 23 January [1862].

[2] The reference is to Louisa and William Benjamin Carpenter. Carpenter, a physiologist and microscopist was, like Lubbock, one of the contributing editors of the *Natural History Review*.

[3] William Erasmus Darwin, CD's eldest son, was a partner in the Southampton and Hampshire Bank, Southampton. John Lubbock had suggested William for the position and had helped CD to complete the partnership agreement (see *Correspondence* vol. 9).

[4] In his letter to John Lubbock, 23 January [1862], CD praised Lubbock's article reviewing recent archaeological finds made at the sites of ancient lake-dwellings in Switzerland (Lubbock 1862b).

[5] In August 1861, John and Ellen Frances Lubbock moved from the home of John William Lubbock, High Elms, near Down, to their new house, 'Lamas', in Chislehurst, Kent (John Lubbock's diary (British Library, Add. Ms. 62679: 64 r.)).

From Robert Bateman [28 January 1862][1]

Biddulph Grange, | Congleton.
Tuesday.

Dear Sir.

My father being obliged to leave home early this morning was unable to answer your note which he hopes to do tomorrow.[2] Meanwhile he bids me to send you three other species which he hopes will be of service, namely a Phaius, Leptotes & Goodyera.[3] The names of the orchids of which you send notes are the following.

1. "Labellum pale spotted with purple, sepals & petals green with purple spots" Zygopetalum crinitum.
2. Odontoglossum Bictoniense.
3. Odontoglossum pulchellum
4 Calanthe vestita.
5. Purple Cattleya-like flower Laelia anceps.
6 The Angræcum is a A. Sesquipedale & is from Madagascar

Believe me | Dear Sir | Yrs very truly | Robert Bateman.

to Dr Darwin.

DAR 160.1: 60

CD ANNOTATION

Verso of first page: 'Phaius like Cattleya | Leptotes like Cattleya | I shall have to add several genera to Foreign Orchids' *pencil*[4]

[1] Dated by reference to CD's commenting in the letter to J. D. Hooker, 25 [and 26] January [1862] that he had just received a box of orchids from James Bateman, Robert Bateman's father, that included a specimen of *Angraecum sesquipedale*. The following Tuesday was 28 January.

[2] See n. 1, above. CD's letter to James Bateman has not been found. For Bateman's reply, see the letter from James Bateman, [1 February 1862].

[3] James Bateman's assistance in providing orchid specimens is acknowledged in *Orchids*, pp. 114 n., 158 n., and 197).

[4] CD mentioned the similarity between the general manner of fertilisation of *Leptotes*, *Phaius*, and *Cattleya* in *Orchids*, p. 164.

From John Edward Gray 28 January 1862

B M.
28 Jan 62.

My Dear Darwin

I am going to read a curious paper at the Society this eveng.[1]

The discovery of a New *Species of Domestic Animal*

This species is curious as being only known in the domestic state! no wild prototype of it having as yet been sent to our Collection

The Animal is the Japan Pig It is very different in its osteology from any of the Wild or Domestic Swine of Europe or Asia

The skull being much more like a Potamochœrus of Africa than a sus in general outline with the Characters of a *Sus*. yet there a[re] characters enough to make it form a peculiar section of the related genus sus characterized by the symetrical Plicature of the face the flatness & margins of the nose of the skull and the width of the Palate

Ever Yours Sincerely | J. E Gray

You will find a figure of the head in the July issue Proceedings for 1861.[2] & of the Entire Animal in the Illustrated News of this Month.[3] I think the 11th I call it *Sus pliciceps*

DAR 165: 204

[1] Gray read his paper on the skull of the Japanese pig, *Sus pliciceps* (J. E. Gray 1862a), before the Zoological Society of London on 28 January 1862. CD referred to Gray's paper in his discussion of variation in domestic pigs in *Variation* 1: 69–70.

[2] The *Proceedings of the Scientific Meetings of the Zoological Society of London* for each year were published in three parts, in the months of May, August, and February (the part published in February completing the volume for the preceding year). The woodcut to which Gray refers appeared in the issue for August 1861, p. 263. A copy appears in *Variation* 1: 69.

[3] An illustration of 'Japanese pigs' appeared in the *Illustrated London News*, 11 January 1862, p. 49.

To John Murray 28 January [1862][1]

Down. | *Bromley.* | *Kent. S.E.*
Jan. 28th

My dear Sir

M^{r}. W. H. Bates of King St, Leicester has travelled in wild parts of region of Amazon &c during eleven years, as a collecting Naturalist.[2] He has great knowledge of Nat. History, & what is far rarer is a capital reasoner & generalizer. He has seen much of the natives; & has attended to the habits of Monkeys & higher animals; but Entomology & Botany are his forte.— I am sure that he is no common man.— He wrote me several long letters, so well expressed & showing such powers of mind, that I urged several months ago to write a Book of Natural History Travels, such as would suit the general reader;[3] but not to fear to go occasionally into pretty deep questions. He has sent me his two first chapters.[4] His style & his powers of description seem to me first rate. I do not pretend to be a Critic; but my *deliberate* opinion is that of the class of the Books, such as my Journal, his will be the best ever published.— That is of course judging from the two Chapters which I have read.—

He has five or six Chapters well copied out & ready for inspection.— He applied to me to whom he had better ask to publish.— I, of course, named you.[5] He wishes to publish soon.— He is a poor man & tells me he must look to money for his work. Now, if what I have said with entire truth, according to my judgment, should make you willing to enter into negotiation with him; will you write me a note, which I can forward to him.—[6] I should have said that I have taken the liberty to write all this, as he asked me to give him a common note of introduction to you.[7] And I thought it better to give you my opinion of the man.— Dr. Hooker knows him, & if you see Dr. H. you can ask his opinion of Mr. Bates' talents.[8]

My dear Sir | Yours very sincerely | C. Darwin

John Murray Archive

[1] The year is established by the relationship to the letter from H. W. Bates, 25 January 1862.

[2] Henry Walter Bates and Alfred Russel Wallace travelled to South America in 1848. Wallace returned to England in 1852; Bates continued to explore the Amazon area until 1859 (Bates 1863, 1: iii).

[3] See *Correspondence* vol. 9, letter to H. W. Bates, 4 April [1861].

[4] Bates sent CD the manuscript of the first chapter of *The naturalist on the River Amazons* (Bates 1863) in December 1861 (see *Correspondence* vol. 9, letter to H. W. Bates, 15 December [1861]). The second chapter followed at the beginning of January 1862 (see letter from H. W. Bates, 6 January 1862, and letter to H. W. Bates, 13 January [1862]).

[5] See *Correspondence* vol. 9, letter from H. W. Bates, [before 25 September 1861], and letter to H. W. Bates, 25 September [1861].

[6] See letter from John Murray, 30 January [1862].

[7] See *Correspondence* vol. 9, letter from H. W. Bates, 30 September 1861, and this volume, letter from H. W. Bates, 25 January 1862.

[8] Joseph Dalton Hooker had spent a day with Bates in November 1861 and had formed a favourable opinion of him (see *Correspondence* vol. 9, letters to J. D. Hooker, 1 December [1861] and [9 December 1861], and letter to H. W. Bates, 3 December [1861]).

From J. E. Gray 29 January 1862

My Dear Darwin

Are you in existence? if you are I am glad, but I think Owen thought he had snuffed you & your theory out, last night in some observation tagged to a Paper on the Aye Aye[1]

This paper is a long one ocupped more than an hour at a preceding meeting & 2 hours last night. and ends in shewing what every one who has seen the Animal knew that it is a Lemur with anomalous teeth as it has all the other characters of the Primates.[2]

The Paper no doubt is good contribution to Science but why it was all to be read is another question The Secretary went fast asleep[3] and I fear the Chairman your humble servant felt very much inclined to do the same but duty prevented him. However the Methaphysic and the religious? observations suited the minds of some of the Audience & they applauded. There was no discussion as Huxley & Busk (who has broken his arm)[4] could not be there & it was so late (11. oclock) that I was not inclined to sit & listen to the observations of Capt Altherly the only person present who seemed inclined to say anything[5]

Ever Yours Sincerely | J. E Gray

29 Jan 1862

DAR 165: 205

[1] R. Owen 1862a. Richard Owen read his paper on the aye-aye at meetings of the Zoological Society of London held on 14 and 28 January 1862. In his conclusion, Owen endorsed the idea 'of the origin of species by a continuously operative secondary cause or law' (*ibid.*, p. 91) but used the case of the aye-aye, as he put it, to 'test' some of the 'different hypotheses, guesses, and beliefs . . . propounded in explanation of the way of the origin of species' (*ibid.*, p. 92). After denying the validity of George Louis Leclerc, comte de Buffon's and Jean Baptiste Pierre Antoine de Monet de Lamarck's theories, Owen attacked natural selection in harsh terms, arguing that it was inadequate to explain the origin of the aye-aye, and concluding: 'Darwin seems to be as far from giving a satisfactory explanation . . . as Lamarck' (*ibid.*, p. 96). Owen later repeated the substance of these remarks in a paper with the title 'On the characters of the aye-aye as a test of the Lamarckian and Darwinian hypotheses of the transmutation and origin of species' (R. Owen 1862b), which was read before the meeting of the British Association for the Advancement of Science, held at Cambridge in October 1862.

[2] The aye-aye had formerly been classified by some naturalists as a rodent (R. Owen 1862a, pp. 33–7).

[3] Philip Lutley Sclater was secretary of the society (Scherren 1905, pp. 104, 143).

[4] Thomas Henry Huxley and George Busk. See also letter from J. D. Hooker, [25 January 1862] and n. 7.

[5] The reference is probably to Francis Henry Atherley.

From John Lubbock 29 January 1862

My dear Mr. Darwin

The Carpenters[1] are not coming this Saturday, so if still quite convenient to you I should be very glad to spend the evening with you.[2]

I shall come down by the 3.30 train & drive straight over to down.

If however you do not feel so well, or for any reason would rather I came some other day you will I am sure not hesitate to say so.

Believe me | Yours affec | John Lubbock

Lamas
Wednesday 29 Jan./62

DAR 170.1: 27

[1] The reference is to William Benjamin Carpenter and his wife Louisa (see letter from John Lubbock, 27 January 1862).

[2] In the letter from Emma Darwin to William Erasmus Darwin, [3 February 1862] (DAR 219.1: 48), Emma reported that Lubbock visited Down on 1 February.

From C. C. Babington 30 January 1862

Cambridge
30. Ja⟨n. 186⟩2

Dear Darwin,

I send a few seeds; all that we have.[1]

I am told that Le Coq⟨,⟩ who has published a voluminous work on the Geograph. Distrib. of Europæan plants mentions not a few instances of apparent dimorphism.[2]

Yours truly | Charles C. Babington—

DAR 160.1: 2

[1] See letter to C. C. Babington, 20 January [1862] and n. 5.

[2] Lecoq 1854–8. CD had just finished 'looking through' this nine-volume work, and had derived from it a number of cases of apparent dimorphism (see letter to Asa Gray, 22 January [1862]); he had cited some of them in his letter to Babington of 20 January [1862]. There is an annotated copy of the work in the Darwin Library–CUL (see *Marginalia* 1: 488–95).

From Henry Holland 30 January [1862][1]

25 Brook Street
Jan$^{ry.}$ 30

My dear Sir,

I again write a few lines to you,—but on another topic.[2]

I have been asked at different times heretofore, & again now, to reprint in a volume some of the Articles (chiefly on subjects of science) which I have contributed to the Ed^{n} & Q^{y} Reviews.[3] I shall probably write no others, as I now have one or more companions in my distant journies to America, Asia, &c, & have therefore no need of that occupation in writing these Articles, which was a great comfort to me in the frequent solitariness of unfrequented places of travel.

I am therefore about to collect some nine or ten of the Articles into a volume;[4] & among the first of these will be one on *Life & Organization*, published almost a year before your work on the *Origin of Species*.[5] In this article some 8 or 10 pages are occupied on this *crucial question*, & mainly (though I hope without any irrational partiality) in vindication of the old doctrine[6]

But I wish to add a postscript in relation to your work; & I have just put down a few lines on the enclosed paper, which I send for your perusal, remarks, & emendation. Suggest any change that you desire, or think good;—attending especially to the close of the first paragraph, where I have written what *perhaps* is not wholly consistent with fact, or with your wish that it should be so understood—But, I wrote it, with full design to submit it to you, before sending it to the printer.

Perhaps you will send me back the paper, with any comments or changes you may suggest?

In haste, but | Ever your's affect^{ly}, | H Holland

[Enclosure][7]

The foregoing article was published a year before the appearance of Mr. Darwin's remarkable work on the 'Origin of Species'. Whatever be thought of some of the conclusions at which Mr. Darwin has arrived, the value of his work as a contribution to the Natural History sciences, and as a guide to their future successful pursuit, cannot be too highly estimated. Nor is it possible to speak too strongly of the candid spirit manifest throughout this volume; leading its author to anticipate objections and acknowledge all difficulties; and to state, even perhaps beyond his own belief in them, the conclusions to which his doctrine might be liable, if carried to its extreme inferences.

Thinking it well that the argument on the other side should be fully and fairly weighed, I have made only a few very slight alterations in the article now reprinted; one of these referring to the changes in the animal world due to the principle of natural selection, operating amidst the general struggle for existence. This principle, which Mr. Darwin has so largely and effectively used in his researches, may be considered in great degree a new path of inquiry; and one which, steadily pursued by exact and patient observation, cannot fail of conducting to important results. It is certain that at some future time a revision and reform will be required of all existing catalogues of genera and species. How far the needful curtailment may carry us is yet doubtful; but there is reason to presume that it will stop very far short of anything like unity of origin, even with unlimited concession of time for the process and progress of change. An original act of creation, *in time and under design*, being assumed in every hypothesis, the conception of any primordial unity, capable of evolving and multiplying itself into all the actual forms of life, is infinitely more difficult than that of many distinct primitive forms, brought into simultaneous or successive existence by one designing and creating Power. Numbers, in truth (and this must ever be kept in mind), admit of no line or limitation, when applied to problems which pass so far beyond all human comprehension. Like in this to Time, similarly applied, their only boundary becomes that of Infinity.

DAR 166.2: 240

1 The year is established by reference to the publication of Holland 1862 (see n. 4, below).

2 See the letters from Henry Holland, [3–14] January [1862], 15 January [1862], and [21 January 1862?].

3 The *Wellesley index* attributes seven essays in the *Edinburgh Review* and nine essays in the *Quarterly Review* to Holland in the period up to January 1862.

4 Holland 1862. The publication eventually included twelve essays. For CD's negative judgment of Holland's scientific writings, see the letter to J. D. Hooker, 23 [June 1863] (*Correspondence* vol. 11).

5 Holland's paper, 'Life and organisation' ([Holland] 1859), was published in January 1859. *Origin* was published in November 1859 (Freeman 1977, p. 75).

6 Holland discussed the 'nature and permanence of Species' in [Holland] 1859, pp. 244–52.

[7] CD approved Holland's original version of the 'postscript', though expressing reservations about one section (see letter from Henry Holland, [1 or 8 February 1862]). The original manuscript that Holland sent to CD has not been found; the text published here is that given in Holland 1862, pp. 98–9. There is an unannotated copy of the book in the Darwin Library–Down.

To J. D. Hooker 30 January [1862][1]

Down
Jan 30th

My dear Hooker

I received the Bletia this morning, rather dry from paper-Box.— I suppose that you are sure it is a Bletia for its pollen is very different from that of any of the Epidendreæ which I have seen, & agrees with description of that of the Arethuseæ. If it be a Bletia, by Jove, Lindley's grand divisions are not a little fanciful.[2] I shall be astounded if the distinction from state of pollen alone can make good main divisions.

I have not written since your note of Saturday, in which you offer to collect cases of Dimorphism:[3] I shd. not wish you to take trouble on purpose; but if you would make a note of any cases on which you stumble, I shd. be very glad. The cases of Balsamineæ, Violaceæ &c, I believe to be widely different from that of Primula.—

You allude to Caryophyllaceæ; I shd. in this one instance like soon to hear to what you allude.—[4]

I have been very sorry to hear about Busk.—[5]

Bateman has just sent me a lot of orchids with the Angræcum sesquipedale: do you know its marvellous nectary 11½ inches long, with nectar only at the extremity. What a proboscis the moth that sucks it, must have! It is a very pretty case.[6]

Farewell | C. Darwin

DAR 115.2: 142

[1] The year is established by the reference to George Busk's accident (see n. 5, below).

[2] John Lindley divided the Orchidaceae into seven tribes distinguished principally by the nature of the anthers and pollen-masses (Lindley 1853, p. 179). CD had expressed dissatisfaction with Lindley's classification system in his letter to J. D. Hooker, 18 [December 1861] (*Correspondence* vol. 9); he made similar comments in *Orchids*, p. 159. CD referred to the pollen-masses of *Bletia* in *Orchids*, p. 162 n.

[3] See letter to J. D. Hooker, 25 [and 26] January [1862], and letter from J. D. Hooker, [25 January 1862].

[4] For Hooker's reply, see the letter from J. D. Hooker, [31 January – 8 February 1862].

[5] George Busk broke his arm on the evening of 20 January 1862 (see letter from J. D. Hooker, [25 January 1862], and letter from J. E. Gray, 29 January 1862).

[6] CD had already told Hooker about the box of orchids sent by James Bateman (see letter to J. D. Hooker, 25 [and 26] January [1862]). See also letter from Robert Bateman, [28 January 1862].

From John Murray 30 January [1862][1]

Albemarle St
Jany. 30

My Dear Sir

I thank you for your very obliging note—introducing to me Mr W H. Bates. What you tell me of him interests me much[2] I shall be very happy to make his personal acquaintance if he be in London—& am ready to receive as much of his MS. as he has prepared—as soon as convenient. I cannot of course proceed farther until I have seen a considerable portion of it—to form an opinion of its character & attractiveness. The description you give leads me to form high expectations & if & an actual inspection will enable me to make proposals for the publication wch, I hope, may meet the Authors views

Will you add to your kindness by informing Mr Bates that I am at home daily with the rarest exceptions from 10 to 4 or 5. (Sat^{y} only until 2) but that if it be at all inconvenient to call he has only to send his MS. wch I will take the utmost care of & treat as a confidential communication

I thought it wd never do to keep the orchids for your weekly carrier & I hope they arrived in good order[3] I take it you & Lyell will be out about the same time He has only 2 Chapters to finish[4]

I am My Dear Sir | Your obliged & faithful | John Murray

Charles Darwin Esq

American Philosophical Society (B/D25.268a)

[1] The year is established by the relationship to the letter from H. W. Bates, 25 January 1862 and to the letter to John Murray, 28 January [1862].

[2] See letter to John Murray, 28 January [1862]. The reference is to Henry Walter Bates.

[3] Murray may refer to proofs of the woodcuts for *Orchids*, which CD had requested in his letter to Murray of 21 October [1861] (*Correspondence* vol. 9). George Snow, the Down coal merchant, operated a carrier service to and from London every Thursday (*Post Office directory of the six home counties* 1862).

[4] Charles Lyell was preparing *Antiquity of man* (C. Lyell 1863a), which was published by John Murray on 6 February 1863 (C. Lyell 1863b, p. vii). *Orchids* was published on 15 May 1862 (see 'Journal' (Appendix II)).

To H. W. Bates 31 January [1862][1]

Down Bromley Kent
Jan. 31.

My dear Sir

I hope the enclosed note will be satisfactory to you, as far as it goes.[2] You see that he wishes to see as much M.S. as you can let him have.— When it comes to negotiation, remember my advice about disposing only of 1st. Edit.[3] He will probably offer to take risk & give you a certain share of profit of sales. I believe it is universally thought that Mr Murray can be quite trusted. You must remember that as yet your name is not known to the general public, which necessarily always makes a Publisher cautious.—

I heartily wish you all the success, which I am sure you deserve. If I can do anything whatever for you, it will give me real pleasure.—

In Haste | Yours very sincerely | C. Darwin

Cleveland Health Sciences Library (Robert M. Stecher collection)

1 The year is established by the relationship to the preceding letter and to the letter from H. W. Bates, 25 January 1862.
2 CD evidently enclosed the preceding letter (see letter to John Murray, 28 January [1862]).
3 See *Correspondence* vol. 9, letter to H. W. Bates, 25 September [1861].

From J. D. Hooker [31 January – 8 February 1862][1]

Dear Darwin

I wrote you a frightful screed the other day about the development of an Aristocracy being the necessary consequence of Natural Selection—& then burnt it—so you must take the will for the deed & be thankful![2] If ever we meet again we will talk it over—

I have a capital letter from Bates who is the only man I know that is "thinking out" your doctrines to any purpose— he tells me he is drawing the butterflies himself, I am glad of it— I did not want you to write to offer him again, but only to *tell me* whether he had arranged any thing with you as we shall be glad to get the drawings in hand. & engraved early.[3] We have got the L.S. up to a very decent pitch of prosperity & hope to keep it so.— I think I have driven Bates back to Nat. Selection as the only way of solving his difficulties.—[4] I do not know when I have met a more interesting thing than his mimetic butterflies— I wish I had time to do the same thing with plants, which is quite feasable to a very considerable degree.

What the deuce can keep you so irritable about Owen:[5] how I wish I could soothe you, I suppose it is the effect of your isolated life, & yet I dare say I am as insane upon some far less worthy score. My only care is to avoid owen— I can see that he *hates me* now with an intense *hate*— he fell foul of me at the Linnæan the other night in a most contemptible manner,[6] & in so foolish a one that in half a dozen words of answer I set the whole society laughing at him. My God what an eye he fixed on me— Won't I catch it— of course I shall, but no worse than if I had not— what do I care

On the back of this you will find the case of dimorphous *Stellaria* flowers[7]

Huxley has got into a most contemptible squabble with the Edinburgh newspapers, I really am astonished that he should notice such rubbish as they fulminate—[8] the beauty of it is that no one in Edinburgh who reads either side sees the other & no one out of Edinburgh reads either! It is not like a Times controversy which every one reads

Ever Yrs | J D Hooker

Stellaria bulbifera—[9] a Siberian form of this has the *apparently* fertile flowers at upper part of plant (as in other Stellarias) these are said by Maximowitch never to ripen seeds.—[10] At base of plant are flowers apetalous, or with very short petals, and barren stamens or 0, a succulent ovary with 1 instead of 3 styles, & very numerous ripe seeds.

DAR 101: 14, DAR 111: 93

CD ANNOTATIONS
0.1 Dear . . . Hooker 6.1] *crossed ink*
7.1 *Stellaria* . . . seeds. 7.5] 'very doubtfully [Cleistogamic]'[11] *red crayon*
Top of letter: 'Heterocentron mexicanum | *Bletia* or [Arethusa]'[12] *pencil*

[1] Dated by the relationship to the letters to J. D. Hooker, 30 January [1862] (see n. 9, below) and 9 February [1862].
[2] Hooker and CD had light-heartedly discussed the relationship between natural selection and the system of hereditary aristocracy (see letter from J. D. Hooker, [19 January 1862], and letter to J. D. Hooker, 25 [and 26] January [1862]).
[3] CD had offered Henry Walter Bates financial assistance with the cost of coloured plates to accompany his account of South American mimetic butterflies (Bates 1862a; see *Correspondence* vol. 9, letters to H. W. Bates, 25 November [1861] and 3 December [1861]). The Linnean Society of London was willing to pay for the engraving and colouring but not for the original drawing (see letter from J. D. Hooker, [19 January 1862]). The plates published with Bates 1862a were drawn and lithographed by Edward W. Robinson (*Transactions of the Linnean Society of London* 23 (1862): tabs. 55 and 56).
[4] Hooker and Bates initiated a flourishing correspondence in 1861, which is partly reproduced in Edward Clodd's memoir of Bates (Bates 1892, pp. xvii–lxxxix). Hooker wrote to Bates at length about natural selection on 2 February 1862 (see *ibid.*, pp. l–liii, and the enclosure to the letter from J. D. Hooker, [23 March 1862]). Many of the extant letters between Bates and Hooker are now in the Archives of the Royal Botanic Gardens, Kew.
[5] In his letter to J. D. Hooker, 25 [and 26] January [1862], CD confessed shame at the 'demoniacal' nature of his feelings towards Richard Owen.
[6] There were meetings of the Linnean Society of London on 16 January and 6 February 1862 (*Journal of the Proceedings of the Linnean Society* 6 (1862): lvii, lix).
[7] See n. 9, below.
[8] See letter from T. H. Huxley, 20 January 1862, and letter to T. H. Huxley, 2 February [1862].
[9] This note is on the verso of the last page of the letter. In his letter to J. D. Hooker, 30 January [1862], CD requested information concerning a case of dimorphism in the Caryophyllaceae that Hooker had mentioned; the genus *Stellaria* belongs to the Caryophyllaceae. CD thanked Hooker for his information on *Stellaria* in the letter to J. D. Hooker, 9 February [1862].
[10] Maximowicz 1859, p. 58; see also Bentham 1862a, p. 67.
[11] This annotation was apparently added at a later date, since the word 'cleistogamic' was not in use before 1867 (see *Forms of flowers*, p. 310, and *OED*).
[12] See letter to J. D. Hooker, 9 February [1862].

From Charles Kingsley 31 January 1862

Eversley Rectory, | Winchfield.

Private

My dear M^r. Darwin

I have just returned from Lord Ashburton's, where the Duke of Argyle, the Bp of Oxford, & I, have naturally talked much about you & your book.[1] As for the Bp. you know what he thinks—& more important, *you know what he* **knows**.[2] The Duke is a very difft. mood; calm, liberal, & ready to hear all reason; though puzzled as every one must be, by a hundred new questions w^h. you have opened.[3]

What struck us on you & your theory, was, the shooting in the park of a pair of "blue Rocks",[4] w^{h}. I was called to decide on. There were several Men there who knew blue Rocks. The Duke said that the specimen was $diff^{t}$ from the Blue Rock of the Hebrides— Young Baring[5] that it was difft from the B. R. of Gibraltar, & of his Norfolk Rabbit warrens (w^{h}. I don't believe from the specimens I have seen, to be a B. R. at all, but a stunted Stock dove, w^{h}. breeds in rabbit holes.), & I could hardly swear that this was a B. R. (as the keeper held) till I saw, but very weakly developed, the black bars on the wing coverts.

Do you care engh about the matter to have a specimen of the bird? He comes in 2 & 3^{s}. (from the Isle of Wight, I suppose) to the heart of S. Hants, & feeds on dry berries—

My own view is—& I coolly stated it, fearless of consequences—that the specimen before me was only to be explained on your theory, & that Cushat, Stock doves & Blue Rock, had been once all one species—& I found—to shew how your views are steadily spreading—that of 5 or 6 men, only one regarded such a notion as absurd. If you want a specimen, I can get you one at once.

I want now to bore you on another matter. This great gulf between the quadrumana & man; & the absence of any record of species intermediate between man & the ape. It has come home to me with much force, that while *we* deny the existence of any such, the legends of most nations are full of them. Fauns, Satyrs, Inui, Elves, Dwarfs—we call them one minute mythological personages, the next conquered inferior races—& ignore the broad fact, that they are always represented as more bestial than man, & of violent sexual passion.

The mythology of every white race, as far as I know, contains these creatures, & I (who believe that every myth has an original nucleus of truth) think the fact very important.

The Inuus of the old Latins is obscure: but his name is from *inire*—sexual violence

The Faun of the Latins (or Romans, I dont know w^{h}.) has a monkey face, & hairy hind legs & body— the hind feet are traditionally those of a goat, the goat being the type of lust.

The Satyr of the Greeks is completely human, save an ape-face & a short tail—

The Elves Fairies & Dwarfs puzzle me, the 2 first being represented, originally, as of great beauty, the Elves *dark*, & the Fairies *fair*; & the Dwarfs as cunning magicians, & workers in metal— They may be really conquered aborigines.

The Hounuman, monkey God of India, & his monkey armies, who take part with the Brahminæ invaders, are now supposed to be a slave negro race, who joined the new Conquerors against their old masters. To me they point to some similar semi-human race. That such creatures sh^{d}. have become divine, when they became rare, & a fetish worship paid to them—as happened in *all* the cases I have mentioned, is consonant with history—& is perhaps the only explanation of fetish-worship. The fear of a terrible, brutal, & mysterious creature, still lingering in the forests.

That they should have died out, by simple natural selection, before the superior white race, you & I can easily understand.

That no sculls, &c. of them have been found, is a question wh. may bother us when the recent deposits of Italy & Greece have been as well searched as those of England. Till then, it concerns no man.

I hope that you will not think me dreaming— To me, it seems strange that we are to deny that any Creatures intermediate between man & the ape ever existed, while our forefathers of every race, assure us that they did— As for having no historic evidence of them—How can you have historic evidence in pre-historic times? Our race was strong enough to kill them out while it was yet savage— We are not niggers, who can coexist till the 19th. century with gorillas a few miles off. I do not say that this notion is true, as a fact: but I do say that it has to be looked to, & weighed patiently quantum valeat.[6]

At least, believe me | Ever, differing now, & now agreeing— | Yours most faithfully | C Kingsley

Eversley
January 31/62

DAR 169.1: 29

[1] Kingsley refers to William Bingham Baring, Lord Ashburton, George Douglas Campbell, duke of Argyll, and Samuel Wilberforce, bishop of Oxford. Baring's principal seat was The Grange, Alresford, Hampshire (*Burke's peerage* 1862).

[2] Wilberforce reviewed *Origin* anonymously in the *Quarterly Review* ([Wilberforce] 1860). CD thought the review 'uncommonly clever' but 'not worth anything scientifically' (see *Correspondence* vol. 8, letter to Asa Gray, 22 July [1860]). More famously, Wilberforce had been fiercely critical of the *Origin* during a debate at the British Association meeting at Oxford in 1860 (see *Correspondence* vol. 8).

[3] The duke of Argyll had discussed *Origin* in his presidential address, delivered before the Royal Society of Edinburgh on 3 December 1860 (G. D. Campbell 1860, pp. 371–6). He praised CD's 'most curious and original' observations, and his capacity for 'arranging and co-ordinating physical phenomena' (*ibid.*, p. 376), but raised a number of objections to what he described as 'essentially another form of the old theory of development' (*ibid.*, p. 375). CD had been impressed with the liberal tone of the duke's discussion, but did not value his arguments (see *Correspondence* vol. 9, letter to T. H. Huxley, 1 April [1861]).

[4] The duke of Argyll had argued in his presidential address that CD's work on the artificial selection of pigeons only served to illustrate the action of 'a restraining law of reversion to type' (G. D. Campbell 1860, p. 373). He had earlier told Charles Lyell: 'As regards the effects of breeding, I think the facts he gives in respect to pigeons tell more against than for his theory' (I. E. Campbell ed. 1906, 2: 482).

[5] W. B. Baring had no surviving sons, but the reference may be to his nephew, Alexander Hugh Baring, who was a member of parliament for Thetford, Norfolk. Alexander Baring's father, Francis, lived at Buckenham Hall, near Brandon, Norfolk, which stood in extensive parkland (*Post Office directory of Cambridge, Norfolk, and Suffolk* 1858).

[6] Kingsley wrote a letter to CD on 23 March 1862 expressing more of his notions about the human species; he described what he thought to be the physical imperfections of the human body. However, Kingsley did not send the letter until 1867 (see the enclosure to the letter from Charles Kingsley, 1 November 1867 (*Calendar* nos. 3482 and 5664)).

To C. C. Babington 1 February [1862][1]

Down. | Bromley. | Kent. S.E.
Feb 1.

Dear Babington

I received the seeds this morning, which are real treasures to me, & I am sincerely obliged.—[2] I have already looked through Lecoq's gigantic & tedious Book.—[3]

With many thanks, | Yours very sincerely | C. Darwin

CUL (Add 8182)

[1] The year is established by the relationship to the letter from C. C. Babington, 30 January 1862 (see nn. 2 and 3, below).
[2] See letter to C. C. Babington, 20 January [1862], and letter from C. C. Babington, 30 January 1862.
[3] CD's copy of Lecoq 1854–8 (Darwin Library–CUL) bears extensive annotations (see *Marginalia* 1: 488–95). See also letter from C. C. Babington, 30 January 1862.

From James Bateman [1 February 1862][1]

Biddulph Grange, | Congleton.
Saturday

My Dear Sir,

I was very glad indeed to hear that the Orchid flowers were so acceptable.[2]

Pray forgive my gardener's carelessness in omitting the names;—[3] when I charged him with the misdemeanour he defended himself on the ground that he could never have supposed *you* could have been ignorant of them!!

I am sorry to say that my knowledge of Orchids almost ends where yours—according to your own modest representation—would seem to begin i.e. I know them much better *systematically* than *structurally*. Indeed, even as a young man, my eyes were too weak to bear the strain of the microscope without which nothing of course could be done.

I believe however I can answer in the affirmative your question as to whether the pollinia of *Mormodes* are ever projected;—of *Cycnoches* I cannot speak positively but I have scarcely a doubt that its pollinia have learned the same trick.[4]

You should study these organs in *Chysis*, the flowering season of which will soon arrive (N.B. if you want specimens I shall be most happy to send them) and I cannot help thinking that it might be useful to investigate the hybrid between Limatodes & Calanthe that is almost constantly in flower at Veitchs nursery.[5]

I have never actually *seen* insects gnawing away at the labella of Orchids but several, minute ickneumon-like things often lurk in the recesses of the more complicated forms.[6]

I much wish you would take up the subject of the marvellous changes—I might almost call them metempsychosis—to which Orchids are prone. Though by no means a convert to your theory as to the 'Origin of Species' I wish the matter to be thoroughly ventilated and cannot but think that facts of great significance may be gathered in the direction I have indicated. If you put me in the witness-box I shall be happy to tell all I know.

I have to thank you very much for the interesting little pamphlet on Primula that you were good enough to send me, & I need not say how anxious I am to see yr. more important work on Orchids.[7] I was interrupted when about to write you before, for which, or rather the consequent delay—I beg to apologise & am very truly | Jas. Bateman

DAR 160.1: 59

[1] Dated by the reference to CD's *Primula* paper (see n. 7, below), and by the relationship to the letter from Robert Bateman, [28 January 1862], in which Bateman stated that his father hoped to write to CD the following day. The first Saturday after 28 January 1862 fell on 1 February.

[2] CD's letter to Bateman has not been found. CD received a box of orchids from Bateman on 25 January 1862 (see letter to J. D. Hooker, 25 [and 26] January [1862]).

[3] James Bateman's son Robert identified at least some of the orchids his father had sent to CD in his letter of [28 January 1862].

[4] CD had learned that, as with *Catasetum*, the pollinia of *Mormodes* and *Cycnoches* are violently ejected when the column is touched in a certain way. He was anxious to observe the mechanism, and had requested specimens from a number of correspondents (see *Correspondence* vol. 9). The genera are discussed in *Orchids*, pp. 249–69; the ejection of pollen is described on pp. 256–62.

[5] The Veitch family owned nurseries in London and Exeter (R. Desmond 1994). The assistance of James Veitch Jr in procuring orchid specimens is acknowledged in *Orchids*, pp. 158 and 214; see also *Correspondence* vol. 9. *Limatodes* is a synonym for *Calanthe* (Willis 1973); Bateman presumably refers to *Calanthe dominii*, a hybrid between *C. furcata* and *C. masuca*, first raised in the Veitches' Exeter nursery in 1856 (see *Gardeners' Chronicle* 3d ser. 46 (1909): 328). CD discussed *C. dominii* and *C. masuca* in *Orchids*, pp. 195–7.

[6] CD was interested in the possibility that species of orchid that lacked nectaries might attract insects by possessing either a thick and fleshy labellum or excrescences on the labellum that might serve the insect as food (see *Orchids*, pp. 283–5).

[7] The references are to CD's paper, 'Dimorphic condition in *Primula*' and to *Orchids*. Bateman's name appears on CD's presentation lists for both of these publications (see Appendixes III and IV).

From J. E. Gray 1 February 1862

B M

1 Feb 1862

My Dear Darwin

Thanks for your note[1] we agree in the estimate of the Man.[2] I wish you would use your influence with your pupil Lubbock to get him to use his endeavours to prevent the Council (or certain members of it) of the Zoological Society inflicting a great injury on Science and Scientific men by proposing much less procuring a man whose conduct has been so eratic as the head of the Society & making the world beleive that we scientific men consider such conduct as of no importance[3]

I know that the old non Scientific Fellows who have been on & off the Council ever since the formation of the society have lost all confidence in him & dislike the idea of his being ever proposed as they say did he not misuse his influence when he came down & proposed a gratuity of £500 to Mitchell who shortly after proved a defaulter[4] and enquire did you ever see him attend regularly & pay any permanent attention to the wellbeing of the Society?

Ever Yours Sincerely | J. E Gray

DAR 165: 206

CD ANNOTATIONS
1.4 Scientific] *del pencil*; 'Scientific' *interl above, pencil*
2.6 attention] 't's *crossed pencil*; 'i' *dotted pencil*

[1] The note from CD has not been found; it was probably a response to the letters from J. E. Gray, 28 January 1862 and 29 January 1862.
[2] The reference is to George Clerk, who was to be invited to succeed Prince Albert as president of the Zoological Society of London. See also letter from John Lubbock, 7 [February] 1862.
[3] John Lubbock was on the council of the society. Clerk was elected president at a council meeting on 5 February 1862, and remained in office until his death in December 1867 (Scherren 1905, p. 126).
[4] Upon his resignation as secretary to the Zoological Society in 1859, David William Mitchell was presented with a gratuity of five hundred pounds (Scherren 1905, p. 104).

From Henry Holland [1 or 8 February 1862][1]

2 Brook Street
Saturday

My dear Sir,

Thank you cordially for your letter.[2] While thus kindly acquiescing in what I have written, I fancy I see that you are not quite satisfied with my mode of using the word *path*—[3] This then I will change, & in some such fashion as follows

"This must be considered a new path of research in the great field of Life; & one which, steadily pursued by exact & patient observation, cannot fail of conducting to many important results. It is certain that some future revision & reform will be required, of all existing catalogues of species & genera.— How far this may carry us is yet doubtful," &c

I write, as you will see, in great haste to secure the post, & put down the above lines as suggested to me while writing them. But something like these words I shall take instead of the other sentences of which I before sent you.

Ever yours affect^ly | H Holland

DAR 166.2: 235

[1] The possible dates are conjectured by the relationship to the letter from Henry Holland, 30 January [1862] (see also n. 2, below).
[2] CD's letter in response to the letter from Henry Holland, 30 January [1862], has not been found. Holland had asked CD to comment on a postscript he had written to one of the essays that he was preparing to republish as a collection (Holland 1862), since it contained a discussion of CD's work.
[3] Holland refers to the enclosure sent with the letter from Henry Holland, 30 January [1862]; the enclosure has not been found, but the published note has been reproduced with that letter.

To T. H. Huxley 2 February [1862][1]

Down
Feb. 2

My dear Huxley

I return the enclosed sent me by Lyell.[2] It is a *capital* letter; you certainly are a dead hand at writing; but whether it is worth wasting your precious time, I cannot help feeling rather doubtful. I must however confess that a man must be utterly blind with prejudice, who would not take up your Lectures, when published, with a more impartial feeling after reading this letter.[3] Can you spare me the article to read which you praise as written in a candid and good spirit by a man of science??[4]

With respect to your reference to me;—it is not a case for thanks; it is quite beyond my deserts or claims.[5]

Most truly yours | Charles Darwin

DAR 145: 223

[1] The year is established by the reference to a letter written by Huxley to the *Scotsman* (see n. 3, below).

[2] Charles Lyell apparently sent CD a copy of the *Scotsman*, 24 January 1862, in which a letter from Huxley was published (see n. 3, below).

[3] In his letter to the *Scotsman*, Huxley pointed out that his recent Edinburgh lectures had achieved their object, namely to state his conclusions, to avoid giving offence, and to provoke unfair attacks in the Scottish press to ensure wide dissemination of his views (see Appendix V and L. Huxley ed. 1900, 1: 196–7). See also letter from T. H. Huxley, 20 January 1862.

[4] Huxley had cited an article published in the *Week*, 17 January 1862 (see Appendix V), as an example of reasoned criticism given in the proper spirit, contrasting it with the unreasoned attacks that had appeared in the *Witness*.

[5] Huxley wrote of 'my revered friend Mr Darwin' and described himself as a 'mere disciple of that great naturalist'.

To George Bentham 3 February [1862][1]

Down Bromley Kent
Feb. 3^d.

Dear Bentham

As you so kindly helped me before on Dimorphism, will you forgive me begging for a little further information, if in your power to give it.[2]

The case is that of the Melastomads with 8 stamens, on which I have been experimenting.[3] I am perplexed by opposed statements: Lindley says the stamens which face the *petals* are sterile;[4] Wallich says in Oxyspora paniculata, that the stamens which face the *sepals* are destitute of pollen:[5] I find plenty of apparently good pollen in **both** sets of stamens in Heterocentron, Monochætum & Centradenia.[6] Can you throw any light on this?

But there is another point on which I am more anxious for information: please look at enclosed miserable diagram:[7] I find that the pollen of the yellow petal-facing stamens produce more than twice as much seed as the pollen of the purple sepal-facing stamens.[8] This is exactly opposed to Lindley's statement, viz that the petal-facing stamens are sterile. But I cannot at present believe that the case has any relation to abortion; it is hardly possible to believe that the longer & very curious stamens, which face the sepals in this Heterocentron are tending to be rudimentary, though their pollen applied to their own flowers produce so much less seed.— It is conformable with what we see in Primula that the sepal-facing anthers which in the plant seen by me, stood quite close on each side of the stigma should have been rendered less fitted to fertilise the stigma, than the stamens on the opposite side of the flower. Hence the suspicion has crossed me, that if many plants of the Heterocentron roseum were examined, half would be found with the pistil nearly upright, instead of being rectangularly bent down, as shown in the Diagram. Or, if position of pistil is fixed, that in half the plants the petal-facing stamens would bend down & in the other half of the plants the sepal-facing stamens would bend down as in the diagram.— I suspect the former case, as in Centradenia I find the pistil nearly straight.—[9] Can you tell me? Can the name **Heterocentron** have any reference to such diversity? Would it be asking too great a favour to ask you to look at dried specimens of Heterocentron roseum (which would be best) or of Monchætum or any 8-stamened Melastomatid, of which you have specimens from several localities (as this would ensure specimens having been taken from distinct plants) & observe whether the pistil bends differently, or stamens differently in different plants.— You will at once see that if such were the fact, it would be a new form of dimorphism, & would open up a large field of enquiry with respect to the potency of the pollen in all plants which have two sets of stamens, viz longer & shorter.

Can you forgive me for troubling you at such unreasonable length? But it is such waste of time to experiment without some guiding light. I do not know whether you have attended particularly to Melastomas— if you have not, perhaps Hooker or Oliver may have done so.[10]

I should be very grateful for any information, as it will guide future experiments.

Yours very sincerely | C. Darwin

Do you happen to know when there are only 4 stamens, whether it is the petal- or sepal facers which are preserved? & whether in the 4-stamened forms the pistil is rectangularly bent or is straight?

Royal Botanic Gardens, Kew (Bentham letters: 694–6)

[1] The year is established by the relationship to the letter to J. D. Hooker, 9 February [1862], and to an observational note on *Heterocentron roseum* in DAR 205.8: 46, dated 3 February 1862. See also nn. 6 and 8, below.

[2] Following the presentation of CD's account of the two forms in *Primula* at a meeting of the Linnean Society of London on 21 November 1861, Bentham had provided CD with additional instances of dimorphism (see *Correspondence* vol. 9, letters from George Bentham, 26 November 1861 and 29 November 1861).

[3] CD had begun crossing experiments on the Melastomataceae in October 1861, believing that plants of this family might exhibit a novel form of dimorphism. Although he continued to work on the family throughout 1862 and 1863, making extensive notes (see DAR 205.8), he reached no definite

conclusion and did not publish on the subject (see *Cross and self fertilisation*, p. 298 n., and *ML* 2: 292–3).

4 Lindley 1836, p. 41.

5 N. Wallich 1830–1, 1: 78. There is a note by CD recording this reference in DAR 205.8: 61.

6 There are notes by CD recording his observations of the pollen of *Heterocentron roseum* (dated 28 October 1861), *Monochaetum ensiferum* (dated 15 January 1862), and *Centradenia grandiflora* and *C. floribunda* (dated 2 January [1862]) in DAR 205.8: 44, 22, and 16 v. He summarised his findings in the three genera mentioned here, together with Nathaniel Wallich's findings in *Oxyspora paniculata*, on the reverse of an experimental note dated 3 February 1862 (DAR 205.8: 46; see n. 8, below).

7 The diagram has not been located, but a diagram of *Heterocentron roseum* included in a note by CD dated 29 October 1861 labels petal- and sepal-facing anthers (DAR 205.8: 45). See also the diagram in the letter to Asa Gray, 16 February [1862].

8 CD refers to the results of his first pollination experiment with a melastomaceous plant, carried out on *Heterocentron roseum* between October 1861 and January 1862. The results of this experiment are recorded in a note dated 3 February 1862 (DAR 205.8: 46). See also letter to Asa Gray, 22 January [1862] and n. 13.

9 There is a note recording this observation, dated 2 January [1862], in DAR 205.8: 16 r.

10 CD refers to Joseph Dalton Hooker and Daniel Oliver. See letter to J. D. Hooker, 9 February [1862], and letter to Asa Gray, 16 February [1862].

From Daniel Oliver [4–8 February 1862][1]

M^r^. Darwin to consult Naudin. Ann. Sc. Nat. Ser. 3 Vol. xii to–xviii He says *Dissochæta* sometimes has but 4 stamens, then 'cum petalis alterna'.— Stylus '*magis minusve sigmoideus*'. (xv. 69).— In *Aplectrum* with 8 stamens, the 4 opposite the *petals are* '*saepius steriles*'. (p. 303) *Sonerila* has "*Stamina* 3 cum pet. alternantia, rarissime 6 et tunc alternatim inæqualia.— He figures *Omphalopus* XV. t. 4. with 4 stamens & nearly straight style—[2]

DAR 205.8 (Letters)

CD ANNOTATIONS

1.1 M^r^. ... xviii] '(both inclusive)' *added ink*

1.2 4 stamens,] '& these are sepal-facers' *interl ink*

1.2 'cum petalis alterna'] 'ie petal-facers aborted' *added ink*

1.4 *Sonerila* ... inæqualia.— 1.5] 'ie. one sepal facer, & all petal facers aborted' *added ink*

1 Dated by the relationship to the letter to George Bentham, 3 February [1862] (see n. 2, below), and to the letter to J. D. Hooker, 9 February [1862], in which CD asked Hooker to thank Oliver for his 'short extract & references'. See also letter to Asa Gray, 16 February [1862].

2 In his letter to George Bentham, 3 February [1862], CD had requested information concerning the relative fertility of the two sets of stamens found in octandrous species of the Melastomataceae and concerning the floral anatomy of tetrandrous species. Bentham declined to help (see letter to J. D. Hooker, 9 February [1862], and letter to Asa Gray, 16 February [1862]), but apparently communicated the enquiries to Oliver and Joseph Dalton Hooker, his associates at the Royal Botanic Gardens, Kew (see letter from J. D. Hooker, [8 February 1862] and n. 2). Oliver cites Naudin 1849–52. CD's notes and comments on this work are in DAR 205.8: 62–3; he also appended notes based on Charles Naudin's work to an earlier summary of his experimental findings dated 3 February 1862 (DAR 205.8: 46 v.).

To T. H. Huxley 6 February [1862][1]

Down
Feb. 6th

My dear Huxley

I return by this Post "The Week" which I have been glad to see.[2]

I entirely agree with all that you say in your published Letter on its excellent spirit. The writer must be a good man, whether as good a Logician as he thinks himself *we* at least shall doubt.

Ever yours | C. Darwin

American Philosophical Society (B/D25.293)

[1] The year is established by the reference to the *Week*, and by the relationship to the letter to T. H. Huxley, 2 February [1862].

[2] See letter to T. H. Huxley, 2 February [1862] and n. 4.

To Charles Kingsley 6 February [1862][1]

Down. | Bromley. | Kent. S.E.
Feb. 6th

My dear Mr Kingsley

I thank you sincerely for your letter.—[2] I have been glad to hear about the Duke of Argyle, for ever since the Glasgow Brit. Assoc. when he was President, I have been his ardent admirer.[3] What a fine thing it is to be a Duke: nobody but a Duke, the first time he geologised would have found a new formation; & the first time he botanised a new lichen to Britain.—[4]

With respect to the pigeons, your remarks show me clearly (without seeing specimens, though I thank you for the kind offer) that the birds shot were the Stock Dove or C. Oenas, long confounded with the Cushat & Rock-pigeon.[5] It is in some respects intermediate in appearance & habits; as it breeds in *holes* in trees & in rabbit-warrens. It is so far intermediate that it quite justifies what you say on all the forms being descendants of one.—

That is a grand & almost awful question on the genealogy of man to which you allude.[6] It is not so awful & difficult to me, as it seems to be most, partly from familiarity & partly, I think, from having seen a good many Barbarians. I declare the thought, when I first saw in T. del Fuego a naked painted, shivering hideous savage, that my ancestors must have been somewhat similar beings, was at that time as revolting to me, nay more revolting than my present belief that an incomparably more remote ancestor was a hairy beast.[7] Monkeys have downright good hearts, at least sometimes, as I could show, if I had space. I have long attended to this subject, & have materials for a curious essay on Human expression, & a little on the relation in mind of man to the lower animals.[8] How I shd. be abused if I were to publish such an essay! I hope & rather expect that Sir C. Lyell will enter in his new Book on the relations of men & other animals; but I do not know what his recent intentions are.[9]

It is a very curious subject, that of the old myths; but you naturally with your classical & old-world knowledge lay more stress on such beliefs, than I do with all my profound ignorance.[10] Very odd those accounts in India of the little hairy men! It is very true what you say about the higher races of men, when high enough, replacing & clearing off the lower races. In 500 years how the Anglo-saxon race will have spread & exterminated whole nations; & in consequence how much the Human race, viewed as a unit, will have risen in rank. Man is clearly an old-world, not an American, species; & if ever intermediate forms between him & unknown Quadrumana are found, I should expect they would be found in Tropical countries, probably islands. But what a chance if ever they are discovered: look at the French beds with the celts, & no fragment of a human bone.—[11] It is indeed, as you say absurd to expect a history of the early stages of man in prehistoric times.—

I hope that I have not wearied you with my scribbling & with many thanks for your letter, I remain with much respect— | Yours sincerely | Charles Darwin

As you seem to care for all departments of n. History, I send a pamphlet with a rather curious physiological case.—[12]

Cleveland Health Sciences Library (Robert M. Stecher collection), Ralph Colp

[1] The year is established by the relationship to the letter from Charles Kingsley, 31 January 1862.

[2] Letter from Charles Kingsley, 31 January 1862.

[3] CD was in Glasgow in September 1855 and heard George Douglas Campbell, eighth duke of Argyll deliver the presidential address at the meeting of the British Association for the Advancement of Science; CD reported that he spoke '*excellently*' (*Correspondence* vol. 5, letter to W. D. Fox, 14 October [1855]). Argyll subsequently discussed *Origin* in an address delivered to the Royal Society of Edinburgh (G. D. Campbell 1860). CD told Thomas Henry Huxley that, though the address had been 'highly complimentary', he did not think much of Argyll's argument (see *Correspondence* vol. 9, letter to T. H. Huxley, 1 April [1861]).

[4] CD refers to the discovery by one of the duke of Argyll's tenants on the Isle of Mull of Tertiary fossil leaf-beds intercalated among basalt lavas. Argyll communicated the discovery to the British Association meeting in 1850 (G. D. Campbell 1850), and he described the deposits in his first and, according to contemporaries, his most important scientific paper (G. D. Campbell 1851; see also I. E. Campbell ed. 1906, 1: 349–54). The reference to a 'new lichen' has not been traced.

[5] See letter from Charles Kingsley, 31 January 1862. CD had finished his chapters on doves and pigeons for *Variation* in June 1860 (see *Correspondence* vol. 8, Appendix II); he described *Columba oenas* in *Variation* 1: 183.

[6] See letter from Charles Kingsley, 31 January 1862.

[7] For CD's earliest descriptions of the Fuegians encountered on the *Beagle* voyage, see *'Beagle' diary*, pp. 121–43, and *Correspondence* vol. 1, letter to Caroline Darwin, 30 March – 12 April 1833, and letter to J. S. Henslow, 11 April 1833. See also *Journal of researches*, pp. 227–30.

[8] See also *Autobiography*, p. 130. CD recorded observations on the expression of the emotions in humans and on the relationship of the human mind to that of animals in his early notebooks (see *Notebooks*), and had also recorded the emotional reactions of his own children (see *Correspondence* vol. 4, Appendix III). He published his detailed observations on human and animal mental processes and emotional expression in *Descent* and *Expression*. CD's notes for these works are in DAR 80–6 and DAR 191, and in DAR 53, DAR 189, and DAR 195, respectively.

[9] Charles Lyell was preparing a book on the antiquity of the human species (C. Lyell 1863a).

[10] See letter from Charles Kingsley, 31 January 1862.

[11] The reference is to the gravel beds in the vicinity of Abbeville, France, in which stone implements or celts had been found in association with the fossil bones of extinct animals. Their discoverer, Jacques Boucher de Perthes, had claimed that both the artefacts and the beds in which they were found were of great antiquity, and many eminent geologists, including Hugh Falconer, Joseph Prestwich, and Charles Lyell, had visited the site in recent years to investigate his claims (see *Correspondence* vol. 8 and Grayson 1985, pp. 185–90). CD cited these discoveries in *Origin* 3d ed., p. 18, as evidence for the great age of the human species.

[12] CD probably refers to his paper 'Dimorphic condition in *Primula*', although Kingsley's name does not appear on the presentation list (see Appendix III).

From John Lubbock 7 [February] 1862[1]

11, Mansion House Street, | *London,* | *E.C.*
7 Jan *1862*

My dear M^r. Darwin

Sir G. Clerk is to be the new Zool. Pres. My feeling in the matter was quite the same as yours.[2]

The Busks[3] are coming to us on Friday 14th. & if you can come over on the Saturday I will stay at home to meet you. Indeed I will do so unless you write to the Contrary.

Hoping you will be able to come I remain | Very sincerely yours | John Lubbock

C. Darwin. Esq | F.R.S.

DAR 170.1: 25

[1] It appears from Lubbock's mention of 'Friday 14th' that he mistakenly dated his letter January instead of February: 14 January 1862 did not fall on a Friday, whereas 14 February did. See also letter from John Lubbock, 13 February 1862.

[2] In February 1862, George Clerk was elected to succeed Prince Albert as president of the Zoological Society of London (Scherren 1905, p. 126). There was some opposition to the selection of Clerk who was 75 years old and did not have a scientific background (see letter from J. E. Gray, 1 February 1862).

[3] George and Ellen Busk were close friends of the Lubbocks (Hutchinson 1914, 1: 39).

From J. D. Hooker [8 February 1862][1]

Kew
Saturday.

Dr Darwin

We have sent you a load of dried specimens of tetramerous Melastomaceæ—[2] they go today to Bromley station "to be left till called for. When you cut up a flower please put the 'disjecta membra', if worth keeping, into one of the little paper capsules of which a bundle is in the box, & leave it on the sheet. Do not do this except when the flowers are few, or the dissection otherwise worth keeping—

Ever yours | J D Hooker

DAR 101: 12

[1] Dated by the relationship to the letter to George Bentham, 3 February [1862], and to the letter to J. D. Hooker, 9 February [1862] (see n. 2, below); the intervening Saturday was 8 February.

[2] CD had requested specimens of octandrous Melastomataceae in the letter to J. D. Hooker, [30 and 31 December 1861] (*Correspondence* vol. 9). In his reply, Hooker offered to send also a fresh tetrandrous specimen (see letter from J. D. Hooker, [1 January 1862]), but it appears that CD did not receive this specimen, since he asked George Bentham to describe the floral anatomy of tetrandrous Melastomataceae in his letter of 3 February [1862]. Hooker probably sent the 'dried specimens' in response to CD's request to Bentham for information (see letter to J. D. Hooker, 9 February [1862]). There is a description of CD's observations on a dried specimen from Kew of the melastomad *Heterocentron subtriplinervium* in DAR 205.8: 48, dated 13 February 1862.

From Charles Augustus Bennet, Lord Tankerville [9 February 1862][1]

We had more fighting this last year than usual,[2] seven Bulls contending for the mastery;[3] they endeavoured to drive the Old Bull out of the herd, and Hardy was witness of the fight.[4] Two of the younger ones attacked him at the same moment and overthrew him & so disabled him that he walked away discomfited to small wood where he stayed alone for some days & was supposed to be mortally wounded, but this could not be ascertained as under such circumstances they are very dangerous to approach. A few days afterwards the herd was grazing near the wood, and the leader of the opposition singled himself out, and approached the wood where the old "Monarch of the Chase" had been lashing himself up for vengeance (as Virgil describes so truthfully).[5] This was exactly what he wanted, out he came, and in two minutes they were in mortal combat; when he got him singly he soon killed his antagonist & then quietly walked back to the herd where he has resumed undisputed sway ever since.

I am very much disposed to adopt Sir Walter Scott's theory with regard to them. In days comparatively not so very remote, the old primæval Forests extended, as he maintains, from the borders to nearly as far as Glasgow, and the aboriginal breed of Cattle (w^{h}. he calls[6]

Incomplete
DAR 83: 157–8

CD ANNOTATIONS
2.1 I am ... he calls 2.4] *crossed ink*
Top of letter: 'Lord Tankerville | Feb. 9th. /62/' *ink*

[1] Dated by CD's annotation.

[2] The first seven words of the letter are in CD's hand, evidently copied from the preceding page of the original. CD retained this portion of the letter because of its bearing on sexual selection; he recounted Lord Tankerville's description of the Chillingham bulls fighting in *Descent* 2: 240–1. The folded sheet on which this part of the letter was written was marked '2' by Tankerville; details from some of the missing sections are given in the letter to Ludwig Rütimeyer, 11 February [1862].

[3] Lord Tankerville was the owner of Chillingham Park, Northumberland, home to the Chillingham cattle. CD had been attempting to procure the skull of a bull from the herd for Ludwig Rütimeyer,

through his friend, the physician Henry Holland, who was an acquaintance of Tankerville's (see *Correspondence* vol. 9, letter to Ludwig Rütimeyer, 5 December [1861], and this volume, letter from Henry Holland, 15 January [1862], and letter to Ludwig Rütimeyer, 11 February [1862]).

4 William Hardy was Lord Tankerville's bailiff (*Post Office directory of Northumberland* 1858).

5 The Roman poet Virgil described the vengeful preparations of a defeated bull to challenge his rival in *Georgics* 3: 209–42.

6 In his poem 'Cadyow Castle', Walter Scott depicted the 'wild' white cattle of Chillingham Park as descendants of the original wild ox that had continued to survive in what remained of the ancient Scottish forests (see Ritvo 1992).

To J. D. Hooker 9 February [1862][1]

Down Bromley Kent
Feb. 9th Sunday

My dear Hooker

I hear that the Box with Melastomas is at Bromley & I will send for it tomorrow.[2] As I have never looked at dryed flowers, Heaven knows what I shall make of them: it is wonderfully kind in you to send them.— I particularly beg you to tell Bentham (as he may think me the most unreasonable of mortals) that I never for a moment thought of his investigating points; I asked only in case he had the subject at his fingers' ends.[3] I thought he could have told about the pistil of Heterocentron, by just looking at a sheet of dried specimens.— When Heterocentron Mexicanum flowers I shd. rather like to see a flower.—[4] Also please thank Oliver for his *most valuable* short extract & references:[5] pistil "plus minusve sigmoides"[6] tells a whole story; I do not doubt they will turn out a new form of dimorphism, & I can see my way to meaning of odd positions & forms of the two sets of stamens.

I am now crossing Monochætum.[7] Have you this genus in flower? The case seems worth working out; as I shd. not be at all surprised if most flowers with 5 or 4 long & short stamens turned out dimorphic.— I wish I had more time. I am sometimes half tempted to give up Species & stick to experiments; they are much better fun.— Many thanks for the curious Stellaria case.—[8] If you had time you ought to look at the pollen of Bletia hyacinthina: it is quite unlike other Bletias & exactly the same as in Epipactis.[9] From the few *drawings* which I have seen of Arethuseæ, the group looks to me like a refuge for the destitute.[10]

You allude to a "long frightful scrawl" on Aristocracy which you tore up, & we all **most heartily** abused you for it; for we should all have liked beyond anything to have seen it.[11] How I shd. have chuckled with my demoniacal feelings to have heard you raise a laugh against Owen:[12] I can well imagine how savage & revengeful his eye must have glared at you.

Bates & Murray are now negotiating about his Book:[13] I am very glad you have been so much struck with the man: he seems to me quite out of the common way. I cannot help rather groaning over your Genera Plantarum, though I can quite believe it the most useful & difficult job a man could undertake—[14]

Farewell | my dear Hooker | Ever yours | C. Darwin

DAR 115.2: 143

[1] The year is established by the references to Bates 1863 and Bentham and Hooker 1862–83 (see nn. 13 and 14, below).
[2] See letter from J. D. Hooker, [8 February 1862].
[3] CD had sent a diagram of *Heterocentron* to George Bentham and asked for his explanation of some details about its structure, but Bentham had declined to assist (see letter to George Bentham, 3 February [1862], and letter to Asa Gray, 16 February [1862]).
[4] CD had asked for flowers of this plant in his letter to J. D. Hooker, [30 and 31 December 1861] (*Correspondence* vol. 9), but Hooker had replied that it was 'just out of flower' (see letter from J. D. Hooker, [1 January 1862]).
[5] Daniel Oliver had supplied CD with a reference to a monograph by Charles Naudin on the Melastomataceae (see letter from Daniel Oliver, [4–8 February 1862]).
[6] Plus minusve sigmoides: 'more or less sigmoid'.
[7] CD began a crossing experiment with *Monochaetum ensiferum* on 7 February 1862, using a plant belonging to his neighbour George Henry Turnbull (DAR 205.8: 24). He continued to work on this species until May 1863. See the dated experimental and observational notes in DAR 205.8: 22–41.
[8] See letter from J. D. Hooker, [31 January – 8 February 1862].
[9] In his letter to J. D. Hooker, 30 January [1862], CD had discussed the apparently atypical pollen of a specimen of *Bletia* sent him by Hooker. He noted the resemblance of the pollen-masses of *Bletia* to those of *Epipactis* in *Orchids*, p. 162 n.
[10] See letter to J. D. Hooker, 30 January [1862] and n. 2. In *Orchids*, p. 269, CD reported that he had not seen any living flowers of this orchid tribe.
[11] See letter from J. D. Hooker, [31 January – 8 February 1862].
[12] In the letter from J. D. Hooker, [31 January – 8 February 1862], Hooker mentioned an incident involving himself and Richard Owen that occurred during a meeting of the Linnean Society of London.
[13] Bates 1863. See letter from H. W. Bates, 25 January 1862, letter to H. W. Bates, 31 January [1862], letter to John Murray, 28 January [1862], and letter from John Murray, 30 January [1862].
[14] Bentham and Hooker 1862–83. CD regretted the length of time the project would take to complete, saying he would not be able to 'enter on & understand' Hooker's work (see *Correspondence* vol. 9, letter to J. D. Hooker, 18 [December 1861]). The first part of the first volume of *Genera plantarum* went to press in December 1861 (see Stearn 1956, p. 129; see also letter from J. D. Hooker, [19 January 1862]).

To John Murray 9 [February 1862][1]

Down. | Bromley. | Kent. S.E.
Sunday 9th.

My dear Sir

I will send up to you by a servant tomorrow (Monday) the M.S of my Orchid Book, excepting the last Ch. which can be fully completed before I get first proofs.—[2]

For Heavens sake be careful of the M.S. for I have no copy of three of the Chapters.—

Inside parcel you will see letter of Instructions to Messrs. Clowes;[3] please read & modify as you think fit.—

Urge Messrs. Clowes to print *quickly*, as I am incapable of changing my work & want to get on with my other Books.—[4]

Now I have finished the Orchids, I can say with confidence that the M.S. contains many new & very curious facts & conclusions.— I have done my best to make the facts striking & clear. I think they will interest enthusiasts in Nat. History; but I fear will be too difficult for general public. In short, I know not **in the least**, whether the Book will sell.[5] If it prove a dead failure, I shall hold myself to a large extent responsible for having tempted you to publish with your eyes shut.— Perhaps there may be enough enthusiasts to prevent a dead failure⟨.⟩

My dear Sir | Yours very sincerely | Ch. Darwin

Endorsement: '*1862*'
John Murray Archive (Darwin 114–15)

[1] The month is established by the relationship to the letter to W. E. Darwin, 14 February [1862]. See also letter to H. W. Bates, 27 February [1862].
[2] *Orchids* was published on 15 May 1862 (see 'Journal' (Appendix II)).
[3] William Clowes & Sons was the London printing firm used by John Murray.
[4] CD refers to his planned three-part work on natural selection, of which *Origin* was an 'abstract'. In the summer of 1861, he had interrupted work on *Variation*, the first part of the projected work, to write *Orchids* (see *Correspondence* vol. 9, Appendix II).
[5] Though well-reviewed, *Orchids* sold slowly (Freeman 1977, p. 112). In first suggesting the book to Murray, CD had used the most tentative of terms, warning that interest in the book was likely to be limited (see *Correspondence* vol. 9, letters to John Murray, 21 September [1861] and 24 September [1861]).

From Maurice Alberts[1] 10 February 1862

Prussian Legation | London
Feb. 10. 1862

Sir

The University of Breslau has transmitted to this Legation a letter and a Doctor Diploma for you,[2] and I request you to let one know whether you expect to come to London soon so that you might call for the same or whether I shall forward it to you by rail.—[3]

I am | Sir | Your ob^d^ servt | M. Alberts. | Secretary

DAR 229, Down House (MS 3: 13–14)

[1] The name of Alberts, the secretary of the Prussian embassy in London, is wrongly transcribed in *Correspondence* vol. 9 and in the *Calendar* (Supplement) as 'H. Albertof'.
[2] At the celebration of the 50th anniversary of the foundation of the University of Breslau (2–4 August 1861), CD was awarded an honorary doctorate in medicine and surgery by the Faculty of Medicine (see *Correspondence* vol. 9, letter from the University of Breslau, 4 August 1861). The diploma is in DAR 229 (see Appendix IX).
[3] CD's reply has not been found, but see the letter from Maurice Alberts, 13 February 1862.

To Ludwig Rütimeyer 11 February [1862][1]

Down. | Bromley. | Kent. S.E.
Feb 11th

Dear Sir

I received some time ago your last kind note.[2] I forwarded your thanks & wish for the Leg-Bones to the Earl at his old castle of Chillingham.—[3] This morning I received a very long, kind & interesting letter from him, saying that the leg-bones shall be sent with the other. But he adds that perhaps they will not be sent off very soon, as they are forced to be very careful about slaughtering them. The Earl remarks with truth on the strong probability of these cattle being remnants of the wild cattle, preserved from time immemorial in the Parks of unknown antiquity at Chillingham & the Duke of Hamilton's.—[4]

Have you heard of a most extraordinary domestic Pig, now living in the Zoological Gardens of London from Japan. It is most extremely unlike any other Pig in appearance; & Dr Gray of the British Museum has just read a paper (as he tells me) before the Zoolog. Soc. on its skull, which differs much from all common pigs.—[5] It certainly seems to be quite new & a distinct species.—

I hope & expect to hear when the skulls are sent off from Chillingham, & when I hear I will let you know.—

With much respect I remain Dear Sir | Yours very faithfully | Ch. Darwin

Elizabeth Rütimeyer

[1] The year is established by the reference to John Edward Gray's publication about a Japanese variety of domestic pig (J. E. Gray 1862a) (see n. 5, below).

[2] Rütimeyer's letter has not been found; it may have been a response to the letter to Ludwig Rütimeyer, 15 [and 16] January [1862].

[3] Chillingham Castle, Northumberland, was the estate of the sixth earl of Tankerville, Charles Augustus Bennet. CD had initially offered to assist Rütimeyer in procuring the skull and some upper neck vertebrae of a bull from the Chillingham herd of cattle (see *Correspondence* vol. 9, letter to Ludwig Rütimeyer, 5 December [1861]).

[4] Only a part of the letter referred to is extant (letter from C. A. Bennet, [9 February 1862]). Hamilton Park, seat of William Alexander Anthony Archibald Douglas, the eleventh duke of Hamilton, included within its boundaries part of the Cadzow forest, thought to be a remnant of the ancient Caledonian forest, and boasted a herd of white cattle similar to the Chillingham herd (see Auld 1888, pp. 507–9).

[5] J. E. Gray 1862a. Gray had told CD of this paper in his letter of 28 January 1862.

From Maurice Alberts 13 February 1862

Prussian Legation
February 13. 1862.

Sir

In reply to your letter of the 11th inst I beg to inform you that I have forwarded this day the Diploma in question to 6 Queen Ann St.[1] and request you to let me have a receipt for the same which I may transmit to the University of Breslau.[2]

I am, | Sir, | your obd^t^. serv^t^. | M. Alberts.

DAR 96: 2 v.

[1] CD's letter has not been found. He had been awarded an honorary degree by the University of Breslau (see letter from Maurice Alberts, 10 February 1862). CD's brother, Erasmus Alvey Darwin, lived at 6 Queen Anne Street, London.

[2] See letter to Maurice Alberts, [after 13 February 1862].

From John Lubbock 13 February 1862

11, Mansion House Street, London, | E.C.
13 Febr *1862*

My dear M^r. Darwin

I write one line to remind you to come if possible to Lunch with us on Saturday at half past one.[1]

The Busks & Hooker will be with us.[2] Hoping you have all kept well, I remain always | Your affec | John Lubbock

DAR 170.1: 28

[1] See letter from John Lubbock, 7 [February] 1862.
[2] George and Ellen Busk and Joseph Dalton Hooker.

To Maurice Alberts [after 13 February 1862][1]

Sir

I beg leave to acknowledge your letter & the the receipt of the Diploma of Doctor's Degree from the Un. of B. & take this opportunity of expressing my most sincere thanks for the great Honor, which this distinguished Univ has been pleased to confer on me[2]

I beg leave to remain | Sir | Your most obed sv | C. Darwin

Draft
DAR 96: 2 r.

[1] Dated by the relationship to the letter from Maurice Alberts, 13 February 1862.
[2] CD had been made an honorary doctor of medicine and surgery by the Faculty of Medicine at the University of Breslau (see *Correspondence* vol. 9, letter from the University of Breslau, 4 August 1861). The diploma is in DAR 229 (see Appendix IX).

To William Erasmus Darwin 14 February [1862][1]

Down
Feb. 14th

My dear William

Lubbock tells me that the new microscope by Smith & Beck is to be Stereoscopic, with 3 object-glasses turning round, fixed to a circular plate;[2] so that it sound stunning & all for 10 Guineas (if I am not mistaken) so it is rather aggravating that you have got yours.— Nevertheless I am pleased to hear that you have got one;, for the new one may not be on sale for many months to come; & then you can exchange,, if you find that you want it, for something still better. I am delighted to hear that you are beginning to work a little at Botany.[3] You seem to me to have a taste for original research, & I believe such work will make your life much happier. I can well believe it must be a great relief having the mathematicks off your mind, & you must be glad that you did not shirk your degree.[4] No doubt the Banking business will get to feel much easier, & after a year or two you will relish some extra work. Lord how proud I shall be if ever you write a Botanical paper![5]

By the way, here is a trifling subject I wish in the summer you would look to: I asked Prof. Oliver whether there were any rudiments of ovules in such *male* plants as in the male of Lychnis dioica, & he looked for me at, I suppose, *dried* specimens,; & could see no difference in ovules of male & female!![6] This seems to me very odd. Begonia & some others might be looked at.— You seem to be leading a jolly dissipated life, & that was a jolly letter, which George forwarded to us.[7] You seem to be well established in the best Society of the Place.— To return for moment to microscope; I know nothing about the polarising instrument; perhaps it would tell something of difference in contents or structure of cells.— Remember to buy some Phlox seed, mentioned by Lindley, & look at spiral ejected threads.[8] When you have looked at some British forms, I could, if you cared for subject, get seeds of out-of-the way forms from Hooker & lots of seeds of the order could be bought at London seedsmen. I dare say Carter & Co could sell you 50 kinds of Compositæ all named.—[9]

We have been very miserable, & I keep in a state of almost constant fear, about poor dear little Skimp, who has oddest attacks, many times a day, of shuddering & gasping & hysterical sobbing, semi-convulsive movements, with much distress of feeling.[10]. These semi-convulsive movements have been less during these few last days, & are never accompanied by any loss of consciousness. Do you remember his being pitched out of the Truck: Mr Headland thinks his Brain probably suffered a little concussion;[11] but I cannot help thinking that it is all due to some extreme irritation of stomach.— Miss Ludwig is unspeakeably kind to him, & he will remain with her all day & night.[12] We shall have no peace in life till the poor dear sweet little man gets better.—

Tomorrow we are going to lunch with the John Lubbocks & Hooker will be there which will be a real pleasure.[13] On Wednesday I have another lark to London to a Dentist, & be hanged to it, though it is only for stopping. Elizabeth is here & Mary Parker, who though dull is a nice girl.[14]

I have sent my Orchid M.S. to the Printers, & shall soon be hard at work correcting:[15] whether my little Book has been worth writing, I know no more than the man in the moon.— I am now working & shall continue all this summer, a little at Dimorphism.—

Good night, my dear old fellow: I often rejoice that you are, I hope, fairly well & comfortably settled; but it is a dreadful loss that we cannot see you oftener, & for longer times.—

Goodnight. Your affect. Father | C. Darwin

DAR 210.6: 95

[1] The year is established by CD's reference to the publication of *Orchids* and by the relationship to the letter from John Lubbock, 13 February 1862 (see n. 13, below).

[2] John Lubbock had apparently recommended a microscope for botanical research made by the London instrument makers, Smith, Beck & Beck, of Coleman Street and Holloway Road, London (*Post Office London directory* 1861).

3 William's letter to CD has not been found.

4 In 1861, William left the University of Cambridge without a degree to take up the offer of a partnership with the Southampton and Hampshire Bank, Southampton. However, as he had kept the required number of terms, he was able to sit the mathematical tripos in January 1862 and receive his degree (*Cambridge University Calendar* 1861, p. xxx; F. Darwin 1914). See also *Correspondence* vol. 9.

5 Although William Darwin began extensive studies in botany and conducted observations for CD over a number of years (see his botanical notebooks (DAR 117 and 234), and his botanical sketchbook (DAR 186)), he never published independently.

6 There are observational notes on male *Lychnis dioica*, dated 11 May 1862, in William's botanical sketchbook (DAR 186: 43). CD had, for a number of years, been interested in the sexual relations of *Lychnis dioica*, and believed that he had 'observed female Lychnis dioica seeded without pollen' (letter to John Scott, 19 November [1862]; see also *Correspondence* vol. 9, letter to Daniel Oliver, 7 December [1861] and n. 5). In November 1861 he sought information from Daniel Oliver and Joseph Dalton Hooker concerning the rudimentary sexual organs found in flowers of this plant (see *ibid.*, letter to J. D. Hooker, 25 November [1861]). No letter from Oliver containing the information referred to has been found, but CD acknowledged Oliver's reply in the letter to Daniel Oliver, 30 November [1861] (*ibid.*). Following his own investigation of the male plants of *Lychnis*, CD concluded that Oliver was mistaken (see letter to J. D. Hooker, 30 May [1862]). In *Cross and self fertilisation*, pp. 410–11, CD used the rudimentary stamens and pistils found, respectively, in the female and male plants of this species to argue that some diclinous plants evolved from hermaphroditic species.

7 George Howard Darwin, aged 16, attended Clapham Grammar School, near London (Darwin 1916). The letter has not been found.

8 Lindley 1853, p. 635. In an attempt to further his son's interest in botany, CD had advised William to obtain a copy of John Lindley's *Vegetable kingdom* (see *Correspondence* vol. 9, letter to J. D. Hooker, 19 June [1861]).

9 The reference is to James Carter & Co. of 237 and 238 High Holborn, London (*Post Office London directory* 1861). For other letters concerning CD's attempts to stimulate William's interest in botany, see *Correspondence* vol. 7, letter to W. E. Darwin, [3 May 1858], and *Correspondence* vol. 9.

10 CD refers to 11-year-old Horace Darwin whose periods of illness during early 1862 are recorded in Emma Darwin's diary (DAR 242).

11 CD had begun to consult the London physician, Edward Headland in 1860 (see *Correspondence* vol. 8, letter to J. D. Hooker, [22 January 1860]). Emma Darwin took Horace to Headland on 11 February 1862, and recorded in her diary the commencement of an acid treatment on 14 February (DAR 242).

12 Camilla Ludwig was governess to the Darwin children.

13 CD refers to the lunch party at John Lubbock's on 15 February 1862 to which, in addition to Joseph Dalton Hooker, George and Ellen Busk had been invited (see letter from John Lubbock, 13 February 1862).

14 The references are to Sarah Elizabeth Wedgwood, Emma Darwin's eldest sister, and CD's niece, Mary Susan Parker.

15 See letter to John Murray, 9 [February 1862]. *Orchids* was published on 15 May 1862 (see 'Journal' (Appendix II)).

To J. D. Hooker [before 15 February 1862][1]

My dear Hooker

Will you give me Mr W. C. *Crockers* (**is this spelt right?**) address in enclosed envelope:[2] I am half mad about Mormodes perhaps he may know. [3]

C.D.

DAR 96: 7 r.

[1] Dated by the relationship to the following letter.
[2] Charles William Crocker had retired from his post as foreman of the propagating department at the Royal Botanic Gardens, Kew, and lived in Chichester (see following letter, and letter from C. W. Crocker, 17 February 1862).
[3] CD had for some time been anxious to obtain reliable information concerning the mechanism by which the pollinia of this orchid are ejected, and had unsuccessfully sought observations and specimens from various of his botanical correspondents. See *Correspondence* vol. 9, and this volume, second letter from D. F. Nevill, [before 22 January 1862], letter from John Rogers, 22 January 1862, and letter from James Bateman [1 February 1862]. CD discussed *Mormodes* in *Orchids*, pp. 249–69.

From J. D. Hooker [before 15 February 1862][1]

I should think—

Mr W. C. Crocker[2]
Verger
Chichester Cathedral

would do

I doubt if he can help you with Mormodes What is it you want to know—

I am to see you at Lubbocks on Saturday.[3]

DAR 96: 7 v.

[1] Dated by the relationship to the letter from C. W. Crocker, 17 February 1862, and by Hooker's reference to seeing CD 'at Lubbocks' on the coming Saturday (see n. 3, below).
[2] Charles William Crocker. See preceding letter and n. 2.
[3] The reference is to the lunch at John and Ellen Frances Lubbock's on 15 February 1862 (see letter from John Lubbock, 13 February 1862).

To Asa Gray 16 February [1862][1]

Down Bromley Kent
Feb. 16th

My dear Gray

I have been trying a few experiments on Melastomatads;[2] & they seem to indicate that the pollen of the two curious sets of anthers (ie the petal-facers & the sepal-facers) have very different powers; & it does not seem that the difference is connected with any tendency to abortion in the one set.— Now I think I can understand the structure of flower & means of fertilisation, if there be two forms,—one with the pistil bent rectangularly out of the flower, & the other with it nearly straight. Study the enclosed magnificent diagram![3] Our hot-house & green-house plants have probably all descended by cuttings from a single plant of each species; so I can make out nothing from them. I applied in vain to Bentham & Hooker;[4] but Oliver picked out some sentences from Naudin, which seem to indicate differences in the position of the pistil.[5]

I see that Rhexia grows in Massachusetts; & I suppose has two different sets of stamens. Now, if in your power, would you observe the position of the pistil in different plants, in lately-opened flowers of the same age (I specify this because in Monochætum I find great changes of position in the pistils & stamens, as flower gets old).[6] Supposing that my prophecy should turn out right, please observe whether in both forms the passage into the flower is not in the *upper* side of the pistil, owing to the basal part of the pistil lying close to the ring of filaments on the under side of the flower.— Also I should like to know the colour of the two sets of anthers.—[7]

This would take you only a few Minutes & is the only way I see that I can find out whether these plants are dimorphic in this peculiar way, i.e. only in the position of the pistil & in its relation to the two kinds of pollen.— I am anxious about this, because if it should prove so, it will show that all plants with longer & shorter or otherwise different anthers, will have to be examined for dimorphism.—

Will you keep this note as a memorandum? You see that I do not scruple to trouble you.— I am tired tonight, so will write on no other subject.—

Ever yours most truly | C. Darwin

[Enclosure]

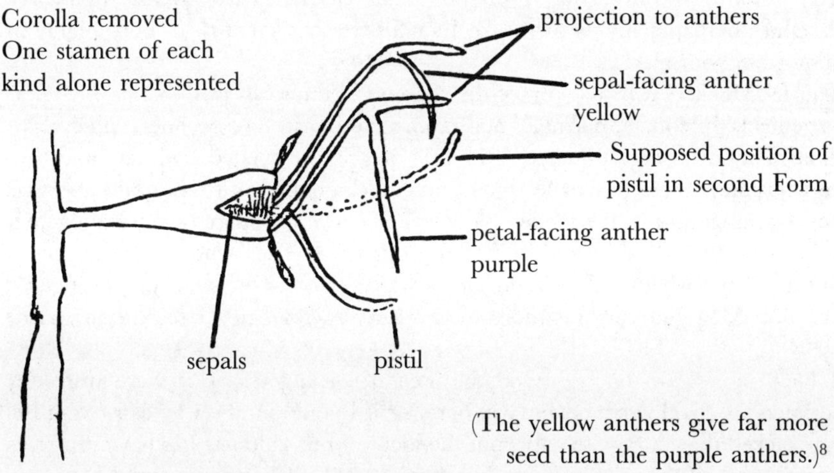

(The yellow anthers give far more seed than the purple anthers.)[8]

Gray Herbarium of Harvard University (63)

[1] The year is established by the relationship to the letter from Asa Gray, 6 March [1862].

[2] CD began to experiment with the melastomaceous plant *Heterocentron roseum* in October 1861 (see *Correspondence* vol. 9, letter to J. D. Hooker, 17 November [1861], and this volume, letter to George Bentham, 3 February [1862] and n. 8). He had also recently started crossing experiments with *Monochaetum ensiferum* (see letter to J. D. Hooker, 9 February [1862] and n. 7). There are dated notes from these experiments in DAR 205.8.

[3] See enclosure.

[4] See letter to George Bentham, 3 February [1862], letter from J. D. Hooker, [8 February 1862], and letter to J. D. Hooker, 9 February [1862].

[5] See letter from Daniel Oliver, [4–8 February 1862].
[6] CD recorded this observation in a note dated 12 February 1862 (DAR 205.8: 23).
[7] For Gray's agreement to carry out observations on *Rhexia*, see the letter from Asa Gray, 6 March [1862].
[8] CD refers to the results of a crossing experiment with *Heterocentron roseum*, begun in October 1861 and harvested in February 1862. There are dated notes from this experiment in DAR 205.8: 44–6. See also letter to George Bentham, 3 February [1862] and n. 8. This drawing is apparently a composite based on the drawing of *Heterocentron roseum*, dated 29 October 1861, in DAR 205.8: 45, and the drawings of *Monochaetum ensiferum*, dated 15 January and 12 February 1862, in DAR 205.8: 22–3. See also letter to Asa Gray, 15 March [1862] and n. 10. The diagram has been reduced to 75% of its original size.

From Charles William Crocker 17 February 1862

28 South St. Chichester
Feby 17th. /62.

Dear Sir

I cannot sufficiently thank you for your kindness in sending me a copy of your paper on the Dimorphic states of Primula.[1] I had eagerly listened to all I could hear on this subject last year, and had devoured the long report of the paper in the Gard: Chron:[2] transferring to my note book all the leading points. Highly as I shall value the paper for its own sake I shall prize it a great deal more highly as coming from yourself.

Did Dr. Hooker[3] ever tell you of the division of the sexes in Bilbergia bivittata? This genus is, of course, normally hermaphrodite but in the specimen which flowered at Kew the sexes alternated—i.e. only one flower opened each day and they were alternately male & female. In the male flower the pistil was quite abortive, almost wanting, and in the female blossoms I believe the stamens were altogether wanting and the style exserted. I pointed it out to Mr. Fitch who made sketches of both states,[4] but whether the circumstance was noticed when the plant was figured in the Bot: Mag: (tab 5270) is more than I can say, for I never see the magazine now.[5]

I have more than once formed the idea of writing to ask if there were any experiments I could carry on for you here, or if I could in any way assist you; but always lacked the courage to carry out the idea— your kindness has however given me confidence. If there be any way in which I can be of service to you I hope and beg you will do me the favour of telling me. I have now a little garden of my own, but no glass-houses at present, and I am in hopes that a little out-of-door exercise will aid me in recovering my strength. While at Kew there were many little things which, under the direction of Dr. Hooker or Prof. Oliver,[6] I was enabled to observe. As far as circumstances admit I am now at your service and if there be anything which I can observe for you, or any experiments which I am competent to perform I shall be most delighted to do it. It will indeed be a labour of love with me, as you will easily understand if, as you doubtless have, you have heard my character from my friends at Kew.—

There are many British Orchids in our neighbourhood but I suppose you are now satisfied with your observations upon them as I see by one of the Reviews that your work upon them is now ready for the press.[7]

Incomplete
DAR 161.2: 254

[1] CD read his paper, 'Dimorphic condition in *Primula*', to the Linnean Society of London on 21 November 1861. Crocker's name appears on CD's list of people to whom copies of the paper were to be sent (see Appendix III). See the letter to J. D. Hooker, [before 15 February 1862], for CD's intention to write to Crocker about *Mormodes*; however, no such letter has been found.

[2] An abstract of the papers read at the meeting of the Linnean Society on 21 November 1861 appeared in the *Gardeners' Chronicle and Agricultural Gazette*, 30 November 1861, pp. 1048–9.

[3] Before his retirement Crocker had worked under Joseph Dalton Hooker as foreman of the propagating department at the Royal Botanic Gardens, Kew.

[4] Walter Hood Fitch was a botanical artist at the Royal Botanic Gardens, Kew (R. Desmond 1994).

[5] *Billbergia bivittata* is figured in *Curtis's Botanical Magazine* 2d ser. 17 (1861): tab. 5270. The journal comprised plates of the plants in the Royal Botanic Gardens, Kew and in other British botanical establishments, together with descriptions provided by William Jackson Hooker, J. D. Hooker's father. The description of *B. bivittata* refers to its having six stamens, 'three attached to the petals, short, with sterile(?) anthers in the pistillate flowers', and to the pistil being 'absent in some flowers'.

[6] Daniel Oliver, who was the librarian and assistant in the herbarium at the Royal Botanic Gardens, Kew, became professor of botany at University College London in 1861.

[7] CD sent most of the manuscript of *Orchids* to his publisher, John Murray, on 10 February 1862 (see letter to John Murray, 9 [February 1862]).

From Andrew Crombie Ramsay 17 February 1862

Jermyn S^t^
17 Feby 1862

My dear Sir

I have a paper coming on at the Geol: Soc: on Wednesday the 26th. which I will tell you the subject of, as it will interest you, though I do not expect you will be able to attend.[1] I am going to prove (I believe) that all the lake basins of the Alps have been scooped out by glacier ice. After two re-examinations of the country, with Lyell I have come to the conclusion that the old glaciers did extend as far as the Jura on the North, & quite into the Plains of Lombardy & Piedmont on the South.[2] The vast moraines of Ivrea &c completely prove the latter.

Further both in America & Europe the number of lakes (true rock basins) increase in number just in proportion as the country has been glaciated, & therefore I apply my theory to glaciated regions in the widest sense, showing, I conceive, that nothing but a solid such as ice could scoope out deep true rock-basins.

Of course I also allow special areas of subsidence, but these are not what I deal with. North of the St Lawrence you would require them every few miles & of all sizes from a few yards in diameter, which is I think impossible.

Ever truly | And^w^ C Ramsay

DAR 176: 8

[1] Ramsay's paper, entitled 'On the glacial origin of certain lakes in Switzerland, the Black Forest, Great Britain, Sweden, North America, and elsewhere', was read to the Geological Society of London on 5 March 1862 (Ramsay 1862).

[2] Both Charles Lyell and CD had formerly propounded the belief that the erratic boulders of the Jura were transported from the Alps in icebergs that floated across a vast submerged tract of Switzerland (see C. Lyell 1855, p. 151). After travelling in Switzerland to study glacial phenomena in the summer of 1857, Lyell decided that the erratics had been carried from the Alps to the Jura by glaciers extending across the intervening region (K. M. Lyell ed. 1881, 2: 243–62). CD informed Ramsay of Lyell's new opinion in the letter to A. C. Ramsay, 24 June [1859] (*Correspondence* vol. 7). See also Davies 1969, pp. 291–2, 310–11.

From Asa Gray 18 February 1862

Cambridge, Mass.
18: Feb. 1862

Dear Darwin

Accept a hasty line at this present, when I am busy above measure.

Thanks for the *Primula* paper, which I have barely looked over.[1] I do hope that you and the other 14 of your household are out of bed and done with influenza.[2] As I have not given you up ⟨ ⟩ notwithstanding your very shocking principles and prejudices against *design in nature*,[3] so we shall try to abide your *longitudinarian* defection.[4] I suppose it *is* longitude, and I am sorry to see that their is a wide and general desire in that meridian that we (U.S.) should fall to pieces. But the more you want we should, the more we w'ont, & the more important it appears to us that we should be a strong and unbroken power. God help us, if we shall not keep strong enough, at whatever cost now it may be, to resist the influence of a country which looks upon the continuation of our steady policy to protect and diversify our domestic industry, as a *wrong and sin* against them. No, no, we must have our own way. But the triumph of the Republicans was the political destruction of the very people who were always making trouble with England and, if you would only let us, and have some faith in the North, we should have been permanently on the best of terms.

What you complain of in ⟨th⟩e Boston dinner, was indeed ⟨l⟩amentable.[5] Such men should not have talked *bosh*, even at a little private ovation. And we have reason to know some of them were heartily ashamed of it as soon as they saw it in print.. It was immediately spoken of here, by influential people (some of whom *refused* to attend the dinner), and in at least one paper, in a tone like your own. It was really as bad as the speeches of some members of Parliament, and worse because it was foolish.

The fact is, a set of cunning fellows on both sides of the water,—(but here utterly characterless) have contrived to make both English & Yankees believe that ⟨e⟩ach was bent up⟨on⟩ quarrelling with the othe⟨r.⟩

Your thinking of me "as ⟨an⟩ Englishman",[6] would once ha⟨ve⟩ been a compliment, and is what from my well known feelings & expressions I have passed for among my friends here. Had the North gone on giving in to the South as for years past, *I should have been one*, at least in residence just as soon as I could have got out of the country. I thank God, it has been otherwise, and that I have a country to be proud of, and which I will gladly suffer for, if need be. With all its weakness & follies—and I know them well, I go ⟨ ⟩ my country, and friendly to those ⟨ ⟩ we *ought* to be on good terms wit⟨h.⟩ I am cured of some illusions ⟨ ⟩ We shall do very well, and ⟨the⟩ two countries will be on the best of terms *when we are strong*,— till then we must not expect it.

If it is the old question of struggle for life,—good feeling has not much to do with it:—the weak must go to the wall, because it cant help it. "Blessed are ⟨the⟩ ⟨*s*⟩*trong*, for they shall ⟨inher⟩it the earth".

⟨My⟩ wife, who is loath to strike ⟨you⟩ from her books,[7] begs you ⟨to m⟩ake allowances for the people here, who were so very *cocky* at having caught two such ineffable scamps as Mason & Slidell—[8]whom we have reason to hate with perfect hatred, that they thought of nothing else, and did not mean to be saucy to England. But you have made us sore, there is no denying it. We did not allow enough for longitude.

Her former message did not refer to Boott,—tho' he is unfortunately influenced by longitude; but is a Yankee born,[9]—nor to Hooker, who, Gallio-fashion cares for none of these things,—thinks us ⟨un⟩wise for fighting, I presume.—[10] ⟨ ⟩ we perfectly agree to say nothing ⟨abou⟩t such matters. It is odd ⟨how⟩ you all fail to appreciate that ⟨it⟩ is simply a struggle for existence ⟨on o⟩ur part, and that men will persist in thinking their existence of some consequence to themselves—tho' you prove the contrary ever so plain,—and will strike or grasp or kick, right & left, in an undignified way some times,—which the safe & sound bystander, cooly looking on, may not appreciate, not sharing his feelings. Telling him the world will get on quite as well without him; yet he some how does not quite like it.

Ever, Yours, | A. Gray.

DAR 165: 106

CD ANNOTATION

Top of first page: 'Pamphlet | Rhexia'[11] *ink*

[1] Gray refers to 'Dimorphic condition in *Primula*'. CD wrote of having fifty copies of his paper printed for private distribution in his letter to George Bentham, 24 November [1861] (*Correspondence* vol. 9); Gray's name appears on the presentation list CD drew up for these pre-prints (see Appendix III).

[2] CD told Gray about the influenza that had swept through his household in the letter to Asa Gray, 22 January [1862].

[3] For the disagreements between Gray and CD on design in nature, see their extensive correspondence on this topic in *Correspondence* vols. 8 and 9.

[4] CD joked in his letter to Gray of 22 January [1862] that his difference of opinion with Gray about the *Trent* affair in the American Civil War was 'all owing to that confounded Longitude', a reference to Henry Thomas Buckle's suggestion of a statistical relationship between a country's climate and the progress of its civilisation (Buckle 1857–61, 1: 38 *et seq.*). Gray is also making a pun on the word 'latitudinarian', meaning broad-minded or liberal in religious belief. For CD and Gray's earlier correspondence about the *Trent* affair, see *Correspondence* vol. 9, letter to Asa Gray, 11 December [1861], and letter from Asa Gray, 31 December 1861.

[5] The reference is to the banquet in Boston on 26 November 1861 given in honour of Charles Wilkes for his action in the *Trent* affair (see letter to Asa Gray, 22 January [1862] and n. 9).
[6] See letter to Asa Gray, 22 January [1862].
[7] Jane Loring Gray. In his letter to Asa Gray, 22 January [1862], CD wrote that his political opinions might mean that he would 'never receive another kind message from M[rs]. Gray.'
[8] James Murray Mason and John Slidell were the two Confederate envoys removed from the British mail-packet, the *Trent*.
[9] Gray probably refers to Jane Loring Gray's message, relayed in the letter from Asa Gray, 31 December 1861 (*Correspondence* vol. 9), that CD was 'the only Englishman' whose letters did not 'give her a shock to read'. Francis Boott, though a resident in England since early adulthood, had been born and brought up in Boston, Massachusetts. He told CD of his feelings about the American Civil War, and of his regret at the animosity between the United States and Great Britain, in his letter to CD of 27 January 1862.
[10] The reference appears to be to the Roman administrator, Junius Annaeus Gallio, who refused to pass judgment on the apostle Paul (Acts 18: 12–17). For an indication of Joseph Dalton Hooker's views on the American Civil War, see the letter from J. D. Hooker, [19 January 1862].
[11] See letter to Asa Gray, 15 March [1862].

To A. C. Ramsay 18 February [1862][1]

Down Bromley Kent
Feb. 18th

Dear Ramsay

Many thanks for your note. I should much like to hear your paper & new view on the mountain Lakes,[2] but I can seldom stand the hot & late Scientific meetings. I have been for some time wishing to pay you a visit at the Museum & have a little talk, & when next in London I hope to get time to put my wish into execution.—[3]

Yours very sincerely | C. Darwin

DAR 261 (DH/MS 9: 3)

[1] The year is established by the relationship to the letter from A. C. Ramsay, 17 February 1862.
[2] Ramsay 1862. See letter from A. C. Ramsay, 17 February 1862.
[3] Ramsay was on the staff of the Geological Survey of Great Britain, which had its headquarters at the Museum of Practical Geology, Jermyn Street, London.

From Searles Valentine Wood 18 February 1862

Brentwood | Essex

My dear Sir

I am much obliged by your very friendly note & for your commendatory remark upon my monog[h].[1] I have been waiting to see if I co[d] procure any information you require but I fear after all my delay I shall not be able to give a very satisfactory answer to your question.[2]

As a general rule among the Mollusca I may say that the Genus most abundant in species & the species most abundant in individuals will present the greatest variation as we might naturally expect altho this rule like most others is not without exceptions

I have found some species of fossils which are but rare or sparingly exhibited present very great diversity of character this possibly may have arisen from a paucity of individuals in one locality—while another locality which has not been examined or perhaps removed might have furnished them more abundantly & the apparently rare species might not in reality have been so altho no doubt some animals have a greater tendency to change than others or possess a less adaptability to altered conditions by which of course greater permanence of character will be maintained. In regard to the the special genus Lucina there is I believe every possible variation between what are called sectional divisions & the line of demarkation is extremely shadowy fading away in a most evanescent manner. The same may I believe be said of any other extensive Genus such as Helix, Unio, Cerithium, Conus Cypræa &c &c which have been divided into sections but which merge imperceptibly the one into the other

I presume that you are still employed collecting & arranging data for your work of which the essay published is you tell us but an outline I hope that I may live to see it & that you may have health & strength to complete the work you have begun[3] I may add in reference to your remark about the obloquy thrown upon it that it has been a great surprise to me that so much toleration has been accorded to you as seems to me to have been the case I admired your courage in so boldly avowing your opinions

The mistake into which Authors like that of the vestiges[4] fell by treating the organic world as a chain of developement in a continuous line instead of as an ever diverging ramification of being had furnished the opponents of the Theory of the origin of beings by the natural process of production out of a preexisting form with the means of an easy victory & it was in the midst of the general gratulation at this "scotching of the snake" that your sounder views came upon the world & backed by the reputation that you had previously so justly acquired they co^{d} not be so easily poohpoohed as might have been the case had they emanated from a less known man. I think that your thus boldly coming forwar⟨d⟩ is a merit of itself (apart from that attaching to the lucid developement of your argument⟨)⟩ & I rather suspect that there are other naturalists who have by their studies been forced to regard this natural origin of being as the true one but have not the confidence to avow their conviction in the face of the opprobrium which has been attached to such opinions

Believe me Dear Sir | Yours very truly | Searles Wood

Feb. 18. 62
Chas Darwin Esq

DAR 181: 144

CD ANNOTATIONS
2.1 Genus . . . without 2.3] *scored brown crayon*
3.1 I presume . . . production 4.4] *crossed brown crayon*
Top of first page: 'S. Wood | Feb 18 | 1862' *ink*

[1] CD's letter has not been found. The reference is to the first part of Wood's monograph on Eocene bivalves, which was published in December 1861 (S. V. Wood 1861–77).

[2] Evidently CD's question concerned whether species belonging to large genera of molluscs were more variable and wider ranging than those from small genera; this had a bearing on his principle of divergence, first outlined in *Origin*, pp. 111–30 (see also Kohn 1985). In order to demonstrate the connection between large genera and variable species, CD had already given the results of his extensive tabulation of the genera listed in the floras of various countries in *Origin*, pp. 54–8. For a discussion of CD's objective in relation to his botanical work, see Browne 1980 and 1983.

[3] In the introduction to *Origin* (pp. 1–6), CD explained that he regarded the work as an abstract of a larger work yet to be completed. The only part of this larger work that was published during CD's lifetime was *Variation* (1868). The remaining draft chapters of what CD called his 'big book' were published posthumously as *Natural selection*.

[4] The first edition of *Vestiges of the natural history of creation* was published anonymously in 1844 ([Chambers] 1844). It argued for a designed progressive evolution of life, and aroused a storm of protest and criticism. Robert Chambers was thought by many to be the author of the work, but this was not confirmed publicly until 1884, well after his death. See Secord 1989 and Secord ed. 1995.

From J. B. Innes 19 February [1862][1]

Milton Brodie | Forres. | NB
Feb^{ry}. 19th.

Dear Darwin,

You must not suppose we only think of you and yours when some fact of natural history turns up, for indeed we often think and speak of our kind friends in the South, and some times Stephens gives us a bulletin[2] We were sorry the last reported some of your party indisposed[3] I hope you have forgotten all about this long ago.

My gardener has got a bird the offspring of a male mule between a canary and green finch, and a hen canary. He says he is quite sure that papa was a mule, though he is not quite sure whether it was half greenfinch or chaffinch. It was reared by a labourer who was then in this garden, and he persisted in putting it with the canary in spite of all assurances that they would not breed, and this bird is the result. Probably you know plenty such cases, but it is new to me— If you want any thing looked after up here in Earth air or water tell me and we will do our little utmost.

We have had very mild weather no frost to Johnny's sorrow as he wants to skate and has only had them on once for a short morning when rain came—[4] today it has been quite warm.

We saw the announcement of Mrs Langton's death.[5] I know you were prepared for and expecting it and believe she had been in much suffering. We have been all as well as usual. Johnny has not tired of his home pursuits yet, and looks forward to some swimming in the sea when hot weather comes He likes his tutor and works pretty willingly. Eliza is much as usual and has been once out to dinner, a mighty feat for her, but I fear she will not repeat it very often.[6]

You will be all gay with the Exhibition. We hear so much of it, that I suppose some of us at least must struggle up to see it before it closes.[7]

With all our best regards to your circle | Believe me Dear Darwin | Yours faithfully | J. B. Innes

DAR 167.1: 8

CD ANNOTATION
End of letter: '11 10 $\frac{1}{4}$ | 2' *ink*

[1] The year is established by the reference to the death of Charlotte Langton, Emma Darwin's sister (see n. 5, below).
[2] Thomas Sellwood Stephens was the curate at Down (*Post Office directory of the six home counties* 1862). Innes, who was the incumbent of the parish, moved to his family's ancestral home, Milton Brodie, near Forres, Scotland, in January 1862 (see letter from J. B. Innes, 2 January [1862], and letter to J. B. Innes, [3] January [1862]).
[3] CD reported that three of his sons were ill in bed in the letter to J. B. Innes, [3] January [1862]. Subsequently, many more members of the household were afflicted with influenza (see letter to John Lubbock, 23 January [1862], and letter to J. B. Innes, 24 February [1862]).
[4] Innes refers to his son, John William Brodie Innes.
[5] According to Emma Darwin's diary (DAR 242), Charlotte Langton died on 2 January 1862. Her death was announced in *The Times*, 6 January 1862, p. 1.
[6] Innes refers to his wife, Eliza Mary Brodie Innes.
[7] The International Exhibition opened at South Kensington on 1 May 1862 (*The Times*, 2 May 1862, pp. 11–12). Regular descriptions of the plans for the exhibition appeared in the London papers throughout the first half of 1862.

From J. E. Gray 21 February 1862

21 Feb 1862

My Dear Darwin

I do not know if the enclosed has any interest to you when read please return it.[1]

A curious incident has occurred here *Gerrard* has compiled a catalogue of the Bones of mammalia[2] he also aded where these skeletons had been figured & gave a list of those that had been figured we had not

I printed the Catalogue & put Gerrard name in the title

My report had to go through Owen as usual[3] *He* objected to the name being there because the list was on a system now *made by Gerrard himself* there[fore] it could not be catalogue[d] & his name should only be mentioned in the Preface I replied that the system adopted was that on which the Bones & stuffed skins were arranged & any other system would not be so usefull to the Museum He persisted in his objection & the Trustees have ordered the name to be errased![4]

Adieu | Ever Yours sincerely | J. E Gray

DAR 165: 207

[1] The enclosure has not been identified.
[2] J. E. Gray ed. 1862b. Edward Gerrard, curator of the galleries and storerooms of the British Museum natural history collections, was Gray's 'right hand man' (Gunther 1975, pp. 268–9). As there was not

room in the zoological department for Gerrard to have a desk, he wrote the catalogue on a board across his knees (Gunther 1975, p. 275 n. 11).

3 Richard Owen was superintendent of the natural history collections at the British Museum.

4 Only Gray's name appeared on the title-page.

To J. B. Innes 24 February [1862]

Down Bromley Kent
Feb. 24th

Dear Innes.

Many thanks for your friendly note.[1] You seem all very prosperous, & we are very glad to hear of it.— I have heard of the mule from the canary & other finches occasionally breeding; but it is very rare (except with the siskin where the case is not so rare) & there is hardly one quite well authenticated case of two such mules breeding together.[2] I will not forget your offer if I should wish for any observations or enquiries made in the north.

Life rolls on, as you know, very uniformly in Down, & we have no news. Yes, we have, the Butcher has jilted his old love, & is going to be married to a new one![3]

We went a few days ago to lunch with the John Lubbocks & they evidently seem thoroughily to enjoy their new home & freedom.[4] They gave us a good account of poor Montague.[5]

We have had the Influenza here very badly— 16 were sick in this house, & at one time six in bed. Etty keeps capital;[6] but now we have Horace failing badly with intermittent weak pulse, like four of our other children previously.[7] It is a curious form of inheritance from my poor constitution, though I never failed in exactly that way.— I am glad to hear that Mrs. Innes (to whom pray give our kind remembrances) has been out to dinner;[8] she beats me, for I have not ventured on such a bold step for an age.

Believe me Dear Innes | Yours very sincerely | C. Darwin

Postmark: FE 24 62
Cleveland Health Sciences Library (Robert M. Stecher collection)

1 Letter from J. B. Innes, 19 February [1862].

2 In *Variation* 2: 154, CD wrote of finches kept in confinement that 'more than a dozen species could be named which have yielded hybrids with the canary; but hardly any of these, with the exception of the siskin . . . have reproduced their own kind.'

3 The name of the butcher's shop in Down was Osborne & Whitehead (*Post Office directory of the six home counties* 1862); the reference is to Alfred James Osborne (Census returns 1861 (Public Record Office, RG9/462: 71)).

4 CD refers to a lunch party held by the Lubbocks on 15 February 1862 (see letter from John Lubbock, 13 February 1862). In 1861, John and Ellen Frances Lubbock moved from the Lubbock family home, High Elms, near Down, to Chislehurst, several miles north of Down (Hutchinson 1914, 1: 52).

5 John Lubbock's younger brother, Montagu, had been seriously injured in a carriage accident in 1861 (Hutchinson 1914, 1: 178). See also *Correspondence* vol. 9.

6 Henrietta Emma Darwin had been ill throughout 1861 (see *Correspondence* vol. 9).

7 For an account of Horace Darwin's symptoms, see the letter to W. E. Darwin, 14 February [1862].

8 Eliza Mary Brodie Innes (see letter from J. B. Innes, 19 February [1862]).

To J. D. Hooker 25 February [1862][1]

Down Bromley Kent
Feb. 25th

My dear Hooker

I have almost finished your Arctic paper, & I must tell you how I admire it.[2] The subject treated, as you have treated it, is really magnificent. Good Heavens what labour it must have cost you! And what a grand prospect there is for the future.— I need not say how much pleased I am at your notice of my work;[3] for you know that I regard your opinion more than that of all others. Such papers are the real engine to compel people to reflect on modification of species: anyone with an enquiring mind could hardly fail to wish to consider the whole subject after reading your paper.— By Jove you will be driven,, nolens volens, to a cooled globe—[4] think of your own case of Abyssinia & Fernando Po, & S. Africa[5] & of your Lebanon case;[6] grant that there are high lands to favour migration, but surely the lowlands must have been somewhat cooled.— I hope I blunder but I fear there is serious erratum at p. 258, compare whole table of "233 Arctic Asiatic species" with p. 264:[7] the **"O"** to "Tropical Mountains of Asia" gave me a shudder: I hope I am wrong; but if it be an erratum, ought it not to be corrected in next vol. of Transactions.?

What a splendid new & original evidence & case is that of Greenland: I cannot see how, even by granting bridges of continuous land one can understand the existing Flora. I should think from state of Scotland & America & from isothermals, that during the coldest part of Glacial period, Greenland must have been quite depopulated. Like a dog to his vomit, I cannot help going back & leaning to accidental means of transport by ice & currents.[8] How curious also is the case of Iceland. What a splendid paper you have made of the subject. When we meet I must ask how much you attribute richness of Flora of Lapland to mere climate: it seems to me very marvellous that this point should have been a sort of focus of radiation: if, however, it is unnaturally rich, ie contains more species that it ought to do for its latitude, in comparison with the other arctic regions, would it not thus falsely seem to a focus of radiation? But I shall hereafter have to go over & over again your paper; at present I am quite muddy on subject. How very odd on any view, the relation of Greenland to the mountains of E. N. America; this looks as if there had been wholesale extinction in E. N. America.— But I must not run on.— By the way I find Link in 1820 speculated on relation of Alpine & Arctic plants being due to former colder climate, which he attributed to higher mountains cutting off the warm southern winds.[9]

I enclose list of specimens much wanted for experiment. Aid me, if you can.—[10] Do not send off in a frost; otherwise soon, or transplanting may check seeding.—

I am much troubled in mind about Masdevallia fenestrata;[11] I shd. like to make it out: is it a large plant or very precious? if not, could you lend me the specimen whenever this next summer it is near flowering?—

When at Lubbock's[12] you said you shd. ask Bentham about my reading some extracts to Linn. Soc: about the odd sexual orchids.— I shd. doubt the propriety,

& am quite indifferent on subject.— I mention it only because if wished I must *at once* get the M.S. from printers & have a few pages copied & get a few diagrams made. If I do not hear I will understand it is not desired.[13]

Ever my dear Hooker | Yours most truly | C. Darwin

DAR 115.2: 144

[1] The year is established by the relationship to the letter from J. D. Hooker, 27 February 1862.

[2] J. D. Hooker 1861a. Having divided the polar zone into five areas 'characterized by differences in their vegetation', Hooker traced the distribution of Arctic plants into temperate and alpine regions in an attempt to show how far their present distribution could be accounted for by 'slow changes of climate during and since the glacial period' (*ibid.*, p. 251). The issue of the *Transactions of the Linnean Society of London* in which Hooker's paper appeared was published on 14 December 1861 (Raphael 1970); CD's annotated copy of this issue is in the Darwin Library–CUL.

[3] J. D. Hooker 1861a, pp. 253–4. The reference is to CD's discussion of the southern, transtropical migration of northern types during the glacial period (see *Origin*, pp. 365–82).

[4] In 1856, CD sent Hooker part of the manuscript of his 'big book' on species that discussed geographical distribution (a section of chapter 11 of *Natural selection*). Hooker was initially unconvinced by CD's proposal of a cooling of the whole globe, including the tropics, during the glacial period (see *Correspondence* vol. 6, letter from J. D. Hooker, 9 November 1856, and *Correspondence* vol. 7, letter to Charles Lyell, 26 April [1858]).

[5] J. D. Hooker 1862b.

[6] J. D. Hooker 1862a.

[7] As explained in the letter from J. D. Hooker, 27 February 1862, the erratum is on page 264 of J. D. Hooker 1861a. In the discussion of the distribution of the Arctic Asiatic species, the figure given for how many species occurred on the mountains of the two Indian peninsulas was '4' instead of '0' and the number of species on the mountains of Australia and New Zealand was recorded as '8' instead of '5' (see letter from J. D. Hooker, 27 February 1862).

[8] CD had consistently opposed the ad hoc invocation of land-bridges to explain the geographical distribution of plants and animals.

[9] Link 1821, p. 102. There is an annotated copy of this work in the Darwin Library–CUL (see *Marginalia* 1: 501).

[10] The enclosure has not been found, but see the letter from J. D. Hooker, 3 March 1862 and n. 5, and the letter to J. D. Hooker, 14 March [1862] and n. 13.

[11] The structure of the flower of *Masdevallia fenestrata*, with its small window opening, appeared to prevent both the withdrawal and insertion of pollinia by insects (see *Orchids*, pp. 168–9).

[12] John Lubbock held a luncheon party at his house on 15 February 1862, which both CD and Hooker attended (see letter to W. E. Darwin, 14 February [1862]).

[13] See letter from J. D. Hooker, 27 February 1862 and nn. 6 and 7. George Bentham was president of the Linnean Society of London.

From J. D. Hooker [26 February 1862?][1]

Kew
Wednesday

Dr. Darwin

The box of Melastomads has arrived all safe.[2] I hope you paid nothing on this—for carriage but I see 2/6 in *red ink* on one direction, & there is 1/10 charged here. We are so awfully cheated here by carriers &c &c that I think it best to tell you of this. There is no railway ticket on the box, though you have directed it to go "*by rail*". Can you find out what you paid for the box when you received it.—

I had a talk with the D of Argyll about your book on species, he seems to be between two stools—Owen & Lyell[3]

Ever yours | J D Hooker

DAR 101: 13

[1] The date is conjectured from the relationship to the letter from J. D. Hooker, 27 February 1862; the preceding Wednesday was 26 February.

[2] See letter from J. D. Hooker, [8 February 1862], and letter to J. D. Hooker, 9 February [1862].

[3] The references are to George Douglas Campbell (the eighth duke of Argyll), Richard Owen, and Charles Lyell. Charles Kingsley also provided an account of Campbell's reaction to *Origin* at this time (see letter from Charles Kingsley, 31 January 1862).

To Maxwell Tylden Masters 26 February [1862][1]

Down Bromley
Feb. 26th.

My dear Sir

I am much obliged to you for sending me your article, which I have just read with much interest.[2] The History and a good deal besides was quite new to me. It seems to me capitally done, and so clearly written. You really ought to write your larger work. You speak too generously of my Book; but I must confess that you have pleased me not a little; for no one, as far as I know, has ever remarked on what I say on classification,—a part, which when I wrote it, pleased me.[3]

With many thanks to you for sending me your article, pray believe me | My dear Sir | Yours sincerely | C. Darwin

DAR 146: 339

[1] The year is established by the reference to Masters 1862 (see n. 2, below).

[2] Masters 1862.

[3] Masters considered *Origin* a 'wonderful book' and argued that CD had 'done a good service', regardless of whether his conclusions were true. Masters particularly recommended chapter 13, entitled 'Mutual affinities of organic beings: morphology: embryology: rudimentary organs', as 'a clear exposition . . . of the rules and methods employed by systematists' (Masters 1862, p. 216).

To H. W. Bates 27 [February 1862][1]

Down
27th

My dear Sir

I will write again; but I do not want to lose a post to say that the terms are **very** favourable: I never heard of such terms being offered for first work.— You may depend he thinks very highly of your Book.—[2] I have always agreed for each edition separately, (except for my Journal which I disposed of all at once for much worse terms);[3] but I think the terms are so good, I would let Murray do as he likes.—

You might say you should prefer agreeing for only one Edition.—
I am **heartily** glad.—
Yours | C Darwin

Cleveland Health Sciences Library (Robert M. Stecher collection)

[1] Dated by the relationship to the letter from John Murray, 30 January [1862], the letter to H. W. Bates of 31 January [1862], and to the following letter, all of which relate to negotiations for the publication of Bates 1863.
[2] CD refers to the publisher John Murray, to whom he had recommended Bates's travel writings (Bates 1863) for publication. See letter from H. W. Bates, 25 January 1862, letter to John Murray, 28 January [1862], letter from John Murray, 30 January [1862], and letter to H. W. Bates, 31 January [1862]. No letter from Bates describing the terms offered by Murray has been found.
[3] CD had earlier recommended that Bates should initially sell only the first edition of his book (see *Correspondence* vol. 9, letter to H. W. Bates, 25 September [1861]). Owing to his dissatisfaction with his dealings with Henry Colburn over the publication of the first edition of *Journal of researches*, CD sold the copyright for the work to John Murray in 1845 for £150 (see *Correspondence* vol. 3, letters to John Murray, 7 March [1845], 12 April [1845], and 2 September [1845]).

To H. W. Bates 27 February [1862][1]

Down Bromley Kent
Feb. 27th

My dear Sir

I have nothing to add to my morning's scrawl,[2] except to thank you heartily for giving me what information you could on domestic animals of Indians.— I am very glad to have the list & your remarks on the breeding of the animals tamed by the Indians.—[3] Case of the Fowl will be useful to me.—[4]

I am very sincerely glad that Murray took same view of your M.S. that I did. I told him that I believed it would be the best Book of the class ever published.—[5] Whether public approve highly or not, I am sure I shall not change my opinion.— If I can be of any use about Geology or about anything, pray ask me.—

My dear Sir | Yours very sincerely | C. Darwin

I am hard-at-work at proof-sheets of my little Orchid Book.—[6]

Leeds University Library (Brotherton collection)

[1] The year is established by the reference to the negotiations for the publication of Bates 1863 (see n. 5, below and preceding letter).
[2] See preceding letter.
[3] The letter from Bates has not been found. In his letter to Bates of 13 January [1862], CD had asked whether any South American animals would breed when '*long* kept in confinement'. Bates commented on this subject in Bates 1863, 1: 99, 191–4, and 2: 113. Bates's observations on the sterility of South American animals kept in captivity by Indian tribes were cited by CD in *Variation* 2: 150–6.
[4] In Bates 1863, 1: 193 and 2: 112, Bates stated that curassow birds (large, turkey-like birds from South America) apparently did not breed when kept in captivity by South American Indians. CD cited this observation in *Variation* 2: 156.
[5] See letter to John Murray, 28 January [1862].

[6] CD sent most of the manuscript of *Orchids* to John Murray on 10 February 1862 (see letter to John Murray, 9 [February 1862]). See also letter to W. E. Darwin, 14 February [1862] and n. 15. *Orchids* was published on 15 May 1862 (Freeman 1977, p. 112).

From J. D. Hooker 27 February 1862

Kew
Feb 27/62

Dear Darwin

I am greatly pleased & indeed relieved by your letter.[1] for no one but Oliver (who can judge) has pronounced any opinion on my Greenland paper,[2] & I find that one is so easily deceived as to the value of such researches, that I was any thing but sanguine of your approval. You have caught me up in a great blunder at p. 264. where the 4 should be 0.—[3] on turning to the proof sheets (which I have kept) I find both this & the fig 8. of same line (which should have been 5) **marked** by myself **for correction**: but the correction not inserted by some accident of my own. I must correct it in the forthcoming number. it is very awkward & stupid.

I do not attach much importance to there being no *Arctic* **Asiatic** species in mts. of trop. Asia, both considering how low latter are & because Arctic Asia is itself so wretchedly poor a flora.— & it is quite met by the fact of such an Arctic *European* species being in the Mts of the Peninsula (which is not in Himalaya!—) as *Alchemilla vulgaris*, which seems to indicate some current of migration from Europe or W. asia down S.E. to Peninsula & Ceylon. & is an awful staggerer to bridge migrations.

I will send you *Lythrum* &c. as soon as weather improves I fear *Masdevallia* is all out of flower but will see & let you have the plant if worth sending on[4]

I have a hazy recollection of Link's speculation. where is it?[5]

I think with you that it is perhaps better not to bring a resume of sexual orchids before Linnæan; but wish indeed you will give us a note on our *trimorphous* Gongora of Schomburgk,[6] which Bentham & I certainly should like to see noticed, however shortly, in our Journal or Transactions.[7] It has long been considered as one of the most wonderful things in the Society's possession: & the explanation of it should certainly appear in our publications.

Do not bother any more about the box, the irregularity was of our Rail-way here in not sending a Rl. ticket—[8]which reminds me that you should not *prepay* every thing as you do: there is not the least occasion for it.

I suppose you will hardly be able to get up to Oliver's Lecture on Tertiary Floras, at R.I. next Friday. Evg.[9]

Ever yours affec | J D Hooker

DAR 101: 15–16

CD ANNOTATIONS
0.3 Dear . . . stupid. 1.8] *crossed brown crayon*

1.5 where ... o.—] *two crosses in margin, pencil*
1.6 fig ... 5)] *two crosses in margin, pencil*
2.4 as ... *vulgaris*, 2.5] *two crosses in margin, pencil*
3.1 I will ... *Lythrum*] *underl brown crayon*
3.1 I will ... Hooker 8.1] *crossed brown crayon*
4.1 I have ... speculation.] *two crosses in margin, pencil*
Top of letter: 'Primula | Glycerine | Crocker' *pencil*

1 See letter to J. D. Hooker, 25 February [1862].
2 J. D. Hooker 1861a. Hooker refers to Daniel Oliver, his colleague at the Royal Botanic Gardens, Kew.
3 See letter to J. D. Hooker, 25 February [1862] and n. 7.
4 See letter to J. D. Hooker, 25 February [1862], and letter from J. D. Hooker, 3 March 1862.
5 Link 1821. CD mentioned Heinrich Friedrich Link's speculation that a 'former colder climate' could account for the similarities between Alpine and Arctic plants in the letter to J. D. Hooker, 25 February [1862].
6 Hooker refers to an apparently monstrous form of orchid found by Robert Hermann Schomburgk in British Guiana, that bore three distinct flowers, at the time considered to represent three separate genera, namely *Catasetum tridentatum*, *Monachanthus viridis*, and *Myanthus barbatus* (Schomburgk 1837).
7 CD read a paper on the trimorphic *Catasetum* before the Linnean Society of London on 3 April 1862 ('Three sexual forms of *Catasetum tridentatum*'); it was published in the *Journal of the Proceedings of the Linnean Society (Botany)* 6 (1862): 151–7. George Bentham was president of the Linnean Society and Hooker vice-president.
8 See letter from J. D. Hooker, [26 February 1862?].
9 Daniel Oliver's lecture, 'On the distribution of northern plants', was delivered at the Royal Institution of Great Britain on 7 March 1862 (Oliver 1862a). CD did not attend.

To J. D. Hooker[1] [1 March – 15 May 1862][2]

Evelyna caravata
Steudel has *capitata*[3]
Pleurothallis *ligulata*[4]
Stelis racemiflora
Steudel has *racemosa*[5]
Bolbophyllum rhizophoræ[6]

Memorandum
DAR 101: 87

1 This may have been an enclosure with one of the letters written to Hooker at this time. Hooker's annotations are recorded in nn. 3–6, below.
2 The date range is established by the reference to Lindley 1862a (see n. 6, below). The number of the botanical section of the *Journal of the Proceedings of the Linnean Society of London* in which Lindley 1862a appeared was published on 15 May 1862; the preceding number was published on 1 March 1862 (*General index to the Journal of the Linnean Society*, p. vi).
3 Hooker wrote '=Caravata, Lindl.' next to these two lines. CD refers to Steudel 1841, a copy of which is in the Darwin Library–CUL. CD mentioned *Evelyna caravata* in *Orchids*, p. 164.
4 Hooker added ', Lindley. Peru. I think you had better put a ? to this, as there is a little difference between the original wild specimens & our garden ones.' CD referred to 'Pleurothallis prolifera and ligulata (?)' in *Orchids*, p. 166.

[5] Hooker added 'Should I think be *Pleurothallis racemiflora*, Lindl.' next to these two lines. CD cited *Stelis racemiflora* in *Orchids*, p. 167.

[6] Hooker added 'Lindl. A new W. African species, will be described by Lindley in Linn. Journal. next number'. An article by John Lindley on West African tropical orchids was published in the 15 May 1862 number of the *Journal of the Proceedings of the Linnean Society* (Lindley 1862a). Lindley described *Bolbophyllum rhizophorae* in *ibid.*, p. 125. CD discussed the species in *Orchids*, p. 170.

From George E. Harris 3 March 1862

20, Glo'ster Street, | Queen Square, | Bloomsbury, W.C.
Mar: 3/62.

Sir,

Being but a poor working man with a wife and four children to support, and further, being one of those "*thirsty souls*" who would drink deep at the fountain of science—when accessible, I have taken the liberty to obtrude upon you and to ask of you the favor of a *presentation* copy of your great work on the "*Origin of Species*" the price (12/-) being far above my purchasing means. (power)[1]

As I am but a poor tailor, I am desirous not to ask the labor of others for nothing, and shall, in return, be but too glad to do work for the am[t]. "My soul thirsteth after Knowledge" is the apology I make for my obtrusion, and remain, Sir, | yours faithfully, | Geo: E. Harris.

C. Darwin, Esq[r].

DAR 166.1: 107

[1] There is no known answer to this letter, nor any record that CD sent Harris a copy of *Origin*.

From J. D. Hooker 3 March 1862

Kew
March 3[d]/62

My dear Darwin

I think my last letter must have crossed yours, in wh I made it clear that all B. cares about is a notice however short of the trimorphous plant in the Societys possession appearing in Linn. Journal or Transactions—[1] a verbatim copy of your obs[ns]. in your book would do quite well.[2] You will never be able to give up orchids once you have begun them & I have no doubt we shall have plenty papers on them from you— I am mighty curious to see your little book & prepared to devour it with eagerness.

I am really sorry about the blunder in my Arctic paper (& in anticipation for the others you will find) but it is of mighty little consequence, you being the only one who has found it out.—[3] it is well this should be so— I should never have written such papers but for you: & the evulgation of your views is the purest pleasure I derive from them—

I am staggered equally with you by the idea that Greenland ought to have been depopulated during the Glacial period: but if so how is it that its *temperate* flora is no richer than its arctic— if it had been populated by migration since the glacial epoch, surely some species suited to the south end would have got over there— there are plenty such in Iceland.— then again the absence of *Caltha*, **anywhere** in Greenland & other plants that *swarm* elsewhere all round the circle, is as fatal as any indirect evidence can be to the population of the whole by chance migration— If you intend to ask me when we meet how I account for richness of Lapland Flora, I will take care to flee your presence.—[4] I am utterly at sea when I attempt to jog out of the quiet locus standens—of Lapland being the focus for the lattermost migration. I grant that the idea may be utterly false, of its being the centre I have some vague notion that the preglacial focus of Scandinavian plants was a terra polaris, that united Greenland, Iceland & Scandinavia (not perhaps in latitude but somehow). What it may have embraced to the N. of America & Asia I neither know nor care: for it is quite clear that there have been very great modern changes of level amongst the Polar American Islands, which I suppose are rising. I only call this vegetation Scandinavian because it is now represented best in Scandinavia & this partly because of present climate of Scandinavia & partly because of its mts. having afforded a favoring climate to said plants during post glacial warm period.

I cannot too strongly impress the fact that Greenland is unaccountably poor in plants.— its comparatively equable (for an Arctic) climate is singularly favorable for a northern Flora. In summer the line of perp. snow in Disco is above *4000 feet* I am told. Just look again at the list of Arctic species at p. 272, found in Europe & America but *not in Greenland*.

I have not a shadow of doubt about wholesale extinction in E.N. America.

Masdevallia has not a single flower. I shall try to recollect it next year.— the Lythrum &c shall go whenever the weather admits.[5]

If you take up G. Lewis Astronomy—I advise you to read only the Chapter VI.[6]

Has Etty read Miss Rogers "Domestic Life in Palestine[7] it is charming to my mind.

Ever yours "by George" (only an expletive) | Jos D Hooker

DAR 101: 17–19

CD ANNOTATIONS

1.1 I think . . . yours,] *cross in margin, pencil*
1.1 I think . . . eagerness. 1.7] *crossed brown crayon*; '(1' *added in corner of page, brown crayon*
1.6 I am . . . eagerness. 1.7] *cross in margin, pencil*
2.3 I should . . . you: 2.4] *cross in margin, pencil*
3.3 by migration . . . centre 3.11] '(2' *added in corner of page, brown crayon*
3.5 then . . . migration— 3.7] *cross in margin, pencil*
3.11 I have . . . somehow). 3.14] *double scored pencil*
3.16 I only . . . Scandinavia 3.17] *cross in margin, pencil*
4.1 I cannot . . . whenever 6.2] '(3' *added in corner of page, brown crayon*
6.1 Masdevallia . . . whenever 6.2] *crossed pencil*

6.2 the weather ... Hooker 9.1] *crossed brown crayon*
End of letter: 'Greenland good evidence that Arctic forms *can* live further South.' *ink*

[1] See letter from J. D. Hooker, 27 February 1862; CD's letter has not been found. Hooker refers to George Bentham, president of the Linnean Society of London.
[2] CD's paper, 'Three sexual forms of *Catasetum tridentatum*', was largely extracted from *Orchids*, pp. 236–48.
[3] J. D. Hooker 1861a. See letter to J. D. Hooker, 25 February [1862], and letter from J. D. Hooker, 27 February 1862.
[4] See letter to J. D. Hooker, 25 February [1862].
[5] CD requested a specimen of the orchid *Masdevallia fenestrata* and enclosed a list of other species he needed for experiment in the letter to J. D. Hooker, 25 February [1862]. See also letter from J. D. Hooker, 27 February 1862. He had come across the 'magnificent case' of trimorphism in the genus *Lythrum* in December 1861, and had arranged for Hooker to send him specimens for investigation (see letter from J. D. Hooker, [19 January 1862], and letter to J. D. Hooker, 25 [and 26] January [1862]).
[6] Lewis 1862. Chapter 6 is entitled 'Early history and chronology of the Egyptians'.
[7] Rogers [1861]. The reference is to Henrietta Emma Darwin, CD's eldest daughter. Emma Darwin listed this book at the end of her diary for 1863 (DAR 242).

From Asa Gray 6 March [1862]

Cambridge. [Massachusetts]
March 6, 18⟨62⟩

My Dear Darwin

I have your note of Feb. 16, about *Melastomaceæ*[1] The test of a good theory is said to be its power of predicting. If your speculations lead you to predict style curved to one side in *Melastomaceæ*, and the prediction is verified,—that will be a great matter in your favor. Why you are coming out so strong in *final causes* that they should make a D.D. of you at Cambridge.[2]

I shall be pleased if I can help you about *Rhexia*. *R. Virginica* grows not far from here, and I will set to watching it next summer.[3] But I fear ⟨I⟩ may not help you, as it is stated (in our Flora of N. America) to have "anthers uniform". I see, however, the phrase: "style somewhat declined" in the character,—which must be looked to.[4] The character was drawn wholly from dried specimens. I have good details from the fresh drawn by Mr. Sprague,[5] but cannot just now lay hands on them

Freely point out any thing else you want looked at. I have now a very zealous pupil, who will be glad to be entrusted with looking up plants & observing,[6]

Ever Yours | Cordially | Asa Gray

There is some jolly science in the Saturday Review, now and then—as in Dec. 28, p. 665, where we are informed that icebergs "are formed by the splashing of the waves on the coast of Labrador".[7]

Mill, being "the greatest logician in England", I send you an American reprint of a specimen of his logic, which I know you will like.[8]

We are very sad here at the death of the President of our University, who had also many warm friends in England.[9]

Postmark: MR 21 62
DAR 165: 107

CD ANNOTATION
End of first paragraph: 'Holly-trees'[10] *added pencil*

[1] Letter to Asa Gray, 16 February [1862].
[2] For the extensive correspondence between CD and Gray on questions of teleology, see *Correspondence* vols. 8 and 9.
[3] Asa Gray provided CD with further information on *Rhexia* later in the year (see letter from Asa Gray, 4 August 1862).
[4] Torrey and Gray 1838–43, 1: 476–7.
[5] Isaac Sprague was a botanical illustrator who had worked for Gray since 1845 (Dupree 1959, p. 166).
[6] Gray probably refers to his assistant and student, Joseph Trimble Rothrock, whose observations on *Houstonia* were sent to CD later in the year (see letters from Asa Gray, 15 July [1862] and 4 August 1862, *DAB*, and Dupree 1959, p. 326).
[7] An article published in the *Saturday Review of Politics, Literature, Science and Art*, 28 December 1861, pp. 665–7, examined the conditions that British forces could expect to encounter during a Canadian winter; additional troops were dispatched to Canada in the tense period following the *Trent* affair in the American Civil War.
[8] The enclosure has not been found, but the reference is probably to John Stuart Mill's article entitled 'The contest in America', which discussed the recent tensions between Britain and the Federal states in America following the *Trent* affair (Mill 1862); the article, first published in *Fraser's Magazine*, was subsequently reprinted in Boston (MacMinn *et al.* 1990, p. 94). Mill argued that English sympathies in the American Civil War should lie with the northern states since, whatever their initial motives had been, their purpose in the conflict was increasingly the abolition of slavery.
[9] Cornelius Conway Felton, president of Harvard University, died on 26 February 1862 (*DAB*).
[10] See letter to Asa Gray, 21 April [1862] and n. 11.

To J. D. Hooker 7 March [1862][1]

Down Bromley Kent
March 7th

My dear Hooker.

I have thought you would like to see enclosed from Gray: without having seen mine, you will not perceive what nice little sneers it contains; & there is a slap at you.[2] I heartily wish I cd. sympathise more with so excellent a man. Some time, return the note to me.—

I will soon prepare & modify some extracts for Linn. Soc about the three Orchid forms; why I wrote again was because I did not think that you understood they would be chiefly mere extracts from my book.[3]

You will be disappointed in my little book: I have got to hate it, though the subject has fairly delighted me: I am an ass & always fancy at the time that others will care for what I care about: I am convinced its publication will be bad job for Murray. Well it won't be a big concern.—[4]

Your last note will be very useful when I come to reconsider your Arctic paper[5] (by the way I will never believe that naturalists are so dull, as not sooner or later to appreciate this paper); your notion of a preglacial centre of dispersal far north, seems good. I have often speculated that during eocene period, there could hardly have been any strictly Arctic Flora & Fauna; & consequently their curious poverty, from want of time for great modification in strictly Arctic genera.— Greenland is indeed very curious; I do not feel quite so sure as you (considering direction of currents of sea, & greater proximity of land far north) that chance migration would have brought to there *temperate* forms. I am more willing, considering Geolog. nature of Spitzbergen & Bear Is^d. to admit a recent continental extension there than almost anywhere else.—[6]

"Link Die Urwelt & das Alterthum &c 1821. p. 102",—on Alpine plants & change of climate.—[7]

I have had a most obliging letter from M^r. Crocker;[8] who offers & wishes to experiment, so I have given him some things to do;[9] it will be grand if he will work.— I am at work on Dimorphism; in Primula & am finding out some very odd & perplexing facts; including a third form in the Chinese Primrose;[10] & I am nearly sure that daylight is coming with respect to the melastomas.— Can you tell me whether anything is better than Spirits & Water to preserve flowers in, as I have to preserve all, as I cannot draw.—

Have you read Buckle's 2^d. Vol: it has interested me greatly; I do not care whether his views are right or wrong; but I sh^d. think they contained much truth.[11] There is a noble love of advancement & truth throughout; & to my taste he is the very best writer of the English Language that ever lived, let the other be who he may.—

Yours affect | C. Darwin

DAR 115.3: 185

[1] The year is provided by the reference to CD's paper on the Linnean Society's trimorphic orchid specimen (see n. 3, below).

[2] See the letter from Asa Gray, 18 February 1862, which is a reply to the letter to Asa Gray, 22 January [1862].

[3] CD's letter has not been found, but see the letter from J. D. Hooker, 3 March 1862. CD's paper, 'Three sexual forms of *Catasetum tridentatum*', was read before the Linnean Society of London on 3 April 1862.

[4] CD's publisher, John Murray, issued 1500 copies of the first edition of *Orchids* on 15 May 1862. Although favourably reviewed, the book sold slowly. See letter to John Murray, 9 April [1862], and Freeman 1977, p. 112.

[5] J. D. Hooker 1861a. See letter from J. D. Hooker, 3 March 1862.

[6] See letter to J. D. Hooker, 25 February [1862], and letters from J. D. Hooker, 27 February 1862 and 3 March 1862.

[7] Link 1821. See letter from J. D. Hooker, 27 February 1862 and n. 5.

[8] See letter from C. W. Crocker, 17 February 1862.

[9] CD's letter has not been found. The letters from C. W. Crocker, [before 13 March 1862] and 13 March 1862 give an indication of the experiments Crocker agreed to carry out for CD.

[10] CD had carried out crossing experiments with the Chinese primrose, *Primula sinensis*, in 1861 ('Dimorphic condition in *Primula*', pp. 87–8; see also *Collected papers* 2: 54–6), and had subsequently raised plants from the resulting seeds. With this second generation of plants CD carried out a second series of crosses in late January and February 1862, some of the results of which were later given in 'Illegitimate offspring of dimorphic and trimorphic plants', pp. 410–18 (see also the experimental

notes in DAR 108: 15–18, 26–8, 34–9). The experiments were designed to investigate further the connection between dimorphism and sterility (see 'Illegitimate offspring of dimorphic and trimorphic plants', pp. 431–7, letter to T. H. Huxley, 14 [January 1862] and n. 9, and Appendix VI). See also letter to Alphonse de Candolle, 17 June [1862] and n. 2, letter to J. D. Hooker, 23 June [1862] and n. 4, and letter to John Scott, 11 December [1862]. In 'Illegitimate offspring of dimorphic and trimorphic plants', p. 414, CD reported that his attention was first drawn to the equal-styled variety of *Primula sinensis* in 1862 by observing some anomalous flowers in a long-styled plant derived from a self-fertilised long-styled parent. As a result, he reported (p. 415), he had examined the plants 'in several small collections', and had discovered that 'the equal-styled variety is not rare'. CD's notes from these observations and experiments, dated 27 February – 30 March, 24 April 1862, and 30 May 1862, are in DAR 108: 38 v., 56–66.

11 CD had read and greatly admired the first volume of Henry Thomas Buckle's *History of civilization in England* (Buckle 1857–61) in 1858 (see *Correspondence* vol. 7, letter to J. D. Hooker, 23 February [1858]). In the second volume, published in 1861, Buckle used the examples of Spain and Scotland to defend his four leading propositions concerning the history of civilisation: first, that human progress depends on the prosecution and diffusion of science; second, that scientific investigation requires a prior 'spirit of scepticism', itself subsequently fostered by science; third, that the discoveries thus made increase the influence of intellectual as compared with moral truths; and, finally, that the 'great enemy' of such progress, and thus of civilisation, is 'the protective spirit' found in the constant interference of church and state in human affairs (Buckle 1857–61, 2: 1).

From J. D. Hooker [10 March 1862][1]

Kew
Monday

Dr Darwin

I return Grays letter with many thanks. I need not say that he is quite wrong in thinking I "care for none of these things," but I told you before why I ceased to write to him about them, & am very glad I did so—[2] he writes to you & to Boott & did to me as an angry man—visibly—& there is no use arguing with such.[3] I confess moreover to be miserably disappointed in him as with the North generally. I did think that there were some at any rate who would have preserved a calm & dignified bearing, let their feelings be ever so much outraged; but the Boston dinner & Gray's letters have put all that out of the question;[4] his whole letters are steeped in the most inordinate self esteem as a Yankee. he can allow of no difference of opinion, is blind to everything, & what is worse brags like the greatest bullies amongst them. It is the total want of self-respect that I so deeply deplore in Gray.— I mean of course in his capacity as Citizen—for I have the same high opinion of him as a man as ever—

What folly he talks of 2 such nations as England & America ever being on the best terms— What is there in the whole history of the human race to quote for such a state of things as "*best of terms*" between two nations of the same blood & bone, & with the same aims & prospects. Nothing but the power of despising us, or we them, ever can or ever will bring one of us to look amicably on the other. It is not in the bounds of possibility that two nations so powerful, so ambitious, so *like* should love one another, & it will be bad day for one or both when they do. A Gray knows no more of the philosophy of the "struggle for life" than the Bp of Oxford does.[5] You might as well talk of High X loving low X,[6] God knows they are each powerful enough, &c *like* enough to form one body religious with a common aim & object—, *if* they would sink differences & agree *each to be nothing*, or one to be everything & the other nothing.

It always amuses me to hear political noodles regret that we as a nation are *hated* abroad, God forbid we should be any thing else (till the Millenium comes)— so sure as we are loved we shall be done for— Fancy your credit as a Naturalist if Houghton loved you as one or Owen! or Sedgwick! you must be reduced to a nonentity first.[7] Rival nations may fear & hate—never love & Gray must be utterly demented not to know that.

Poor Grays account of apology for the Boston dinner is a very lame one & it is lamentable to see him try to cloak it—by sneering allusions to our "members of Parliament."— How lame too the attempt to put all on the shoulders of a few cunning fellows on both sides the water—. as if the general feeling here in favor of giving Jonathan[8] a licking were not spontaneous & heartfelt—(right or wrong I do not say)— I can quite see that the poor fellow is most hurt at the loss of your cordial sympathy & there is an uneasy state of mind running through his letter. Again many thanks for it, I was most anxious to see it.

I find nothing so good as Alcohol & water to preserve plants in for after dissection—[9] it hardens enough & not too much, leaves tissues tolerably transparent, & quite clean. Vinegar will preserve pollen masses if I remember right but renders other tissues far too soft. Other fluids are apt to clog or discolor, as far as my experience goes.

Olivers lecture went off very well he will have a capital paper in next Review—[10] would you care to see it before publication?

I send you a capital letter of Bates, which please return as I have not answered it.[11]

Lubbock spent a day here last week & I am as much charmed as you are with him. We hope he will bring Mrs L. to Kew, I like her so much[12]

I hope much to get down to you at Easter week, if you can take me in: but my Father's health is so uncertain that I cannot be sure of my movements.[13] This an examination week at Chelsea for Asst Surgeons Army.[14]

Ever yours affec | J D Hooker

P.S. Reading over this I fear you will think my politics very Mephistophelean—

I am curious to hear of your trimorphous Primula— I suspect it will prove a very complicated case.[15]

[Enclosure]

King St Leicester
March 5

My Dear Sir

I have not been quick to answer your last not only because most of my writing time is just now taken up by my book of Travels[16] but because your letter was so full of new observations on the most difficult of all Nat. Hist. subjects that it required long consideration.[17] I hope you will attend to mine only when you have nothing else to do. The view you propound of the origin of species—by the nat. Selection of varieties which always occur as a universal condition of reproduction, in organic species, independent of the direct action of external conditions; is simple & grand. As you say it recommends itself by its simplicity & clearness & is amply sufficient to explain the origin of species. But what if the real state of things is not so simple as this? Are there not some phenomena in species which seem to show that local conditions & use & disuse have some direct effect on individuals the effect being propagated to the offspring & so complicating the question of the origin of species. Whilst I was travelling I used to attribute like many others the production of distinct local varieties or races to the direct action of the local conditions & when I found such coming in contact with their supposed parents without intermarrying or reverting, I supposed the cause was the gradual & slow change in their constitutions which had brought them to a point when they were incapable of going back. But since I read Mr Darwin's book I believe that Nat. Sel. has effected this change operating both on structure & functions. But still the belief is strong that Climatal & other causes have some slight effect; so slight however, that of themselves they are unable to produce a race. Your letter shows me plainly that the effect must be yet slighter. I think the strongest case of the inability of local conditions acting directly to produce a race is that of the two races of man inhabiting the Tropics of America & Africa. All facts go to show that the American race must have lived in Brazil many thousands of years yet there is no approach to the negro. Yet the effect of exposure to the sun is to blacken the skin. For the nomade tribes of Amazon Indians are darker coloured than the agricultural & the families of chiefs in Africa are lighter col^r^. than the rest of the tribe &c &c. My studies of those curious mimetic butterflies gave me convincing proof that local physical conditions do not directly produce a race for there I see a species exhibiting nearly a dozen varieties in one limited spot & 4 or 5 of them as it were in the act of being selected & segregated; the motive for such a process being also quite plain to detect.[18]

A little more in corroboration of your views.— On testing them by my case of the mimetic butterflies I find they hold good. There is really no proof that the variable species originally sported in a different manner in one locality from what it did in another. Inappreciable variations or very small variations similar in all the localities & Nat. Sel. *drawing them out* as it were would be sufficient to explain the whole case. The elimination of intermediate vars. formerly, & of the occasional offspring of cross marriages between the divergent selected forms in one limited district must have been & must be now very stringent & perpetually acting; probably however there is a *moral* barrier preventing frequent intermarriages although there is not a *geographical* one; in other words the selected & now somewhat widely divergent vars. refuse to intermarry.

Now for a few cases which seem to show that the modifications induced by direct action, are propagated. A few sheep have been imported into the Amazons region from Portugal. The wool nearly all falls off & the young at about 2 or 3 months old (when alone I have noticed them) have a thin coat of straight wool more like hair than wool. I know this is an imperfect observation. I did not notice that *hair* or a tendency to hair was produced in the lifetime of a sheep & I did not see lambs when first born, to prove that they have this modified coating at their birth. The sheep are only kept by twos & threes as pets & have not been in the country for many generations so that I do not think selection artificial or Natural has operated.

Dogs become lopeared in consequence of the relaxation of muscles from disuse puppies are born lop-eared.

Mr Darwin's case of tame Ducks having less developed wing bones from disuse seems to come under this head[19] but unconscious or Nat. Sel. may have come into play here. For when ducks were first tamed strong flyers would less readily submit than weak ones & so forth.

The general blindness of animals in deep caverns seems to be partly owing to direct effect of external cond. on individuals leading to disuse of the organs & collapse of muscles & nerves reproduced in offspring. But this belongs to the same class of facts as those cases which show a general similarity in some features amongst widely different species of animals & plants living under the same conditions; such as fleshiness of plants on sea coast brassy tints of beetles ditto & some similar cases which I observed in S. America such as transparency of wings in Heliconia butterflies living near either tropic; very suffused colouring on E. of Andes under the equator &c.

Some part of this general similarity is owing to Nat. Sel. adapting one sp. to another in external dress; but I think there are some cases where this explanation is inapplicable. It is curious in the case of negro that the colour produced by direct action should be also the one selected as most suitable by nature. Of course the whole argument depends on this question which you put & which I thank you for having placed so vividly before me. Are the peculiarities induced in the lifetime of an organism transmitted to its offspring? We may allow several generations for the operation. I cannot think of any facts giving affirmation to this question except the poor ones given in this letter. Even if they prove it they would not show that a race & new species could be produced by direct action; Nat. Sel. is always acting & would act of course on the offspring of these species

I am much obliged to you for recalling me to order in the matter of forgetting the thousands of generations & millions of individuals extinguished without offspring in the case of insect species. I really was not giving the full force to that.

I have not fully answered your letter but will finish another time. Would you oblige me by getting from Mr Bentham the names of the following trees— I think Mr Spruce may have sent them under these same native names[20]

Tapiribá		a stone-fruit tree (very acid— com^n^)
Acutitiribá	—	round yellow mealy fruit
Jabuti-puhé	—	anonaceæ? delicious wild fruit

Massarandúba	—	cow tree
Umarí or Marí and Uixí	}	two similar fruits Upper Amazon
Pamá		wild stone fruit flavour of cherry
Sucu-úba		of Santarem[21]

Please give my kind regards to your family

Yours sincerely | H W Bates

I shall send tomorrow a small box per post with a specimen of leaves that have a fine woolly substance on their under surface to get information from you on its nature.— a certain ant (Polyrhachis bispinosa) collects this material & makes large nests of it after being elaborated by them it is collected & used as tinder & sold in all shops on the Amazons. Latreille gave an interesting account of the ant & its work but no one up to present time has explained the matter.[22] | Yrs | H W B

Will you please give me the reference to your remarks on colours of plants & fur of artic animals not being dependent on climate. I must quote you in my book where I have given similar generalization[23]

DAR 101: 20–2; Royal Botanic Gardens, Kew

CD ANNOTATIONS

6.1 Olivers . . . publication? 6.2] *cross in margin, pencil*
7.1 I send . . . it. 7.2] *cross in margin, pencil*
8.1 Lubbock . . . much 8.2] *cross in margin, pencil*
9.1 I hope . . . in:] *cross in margin, pencil*
11.1 P.S. . . . Mephistophelean—] *cross in margin, pencil*
Top of first page: 'Photograph'[24] *pencil, circled pencil*
Above last paragraph: 'Remember Lythrum'[25] *ink*

1 Dated by the relationship to the letters to J. D. Hooker, 7 March [1862] and 14 March [1862]. The only Monday between these dates fell on 10 March 1862.

2 See letter from Asa Gray, 18 February 1862, and letter from J. D. Hooker, [19 January 1862].

3 See letter from Francis Boott, 27 January 1862, and *Correspondence* vol. 9, letter from Asa Gray, 31 December 1861.

4 Hooker refers to the dinner held in honour of Charles Wilkes in Boston in November 1861 (see letter to Asa Gray, 22 January [1862], n. 9).

5 The bishop of Oxford, Samuel Wilberforce, was well-known for his outspoken opposition to natural selection (see *Correspondence* vol. 8, and this volume, letter from Charles Kingsley, 31 January 1862).

6 The reference is to High and Low Church Anglicans.

7 Samuel Haughton, Richard Owen, and Adam Sedgwick had all written hostile reviews of *Origin* (see *Correspondence* vol. 8, Appendix VII).

8 Jonathan: 'a generic name for the people of the United States, and also for a representative United States citizen' (*OED*).

9 CD had asked Hooker about methods for preserving plant specimens in the letter to J. D. Hooker, 7 March [1862].

10 Daniel Oliver had delivered a lecture entitled 'On the distribution of northern plants' at the Royal Institution of Great Britain on 7 March 1862 (Oliver 1862a). In an article published in the April issue of the *Natural History Review*, Oliver challenged the proposal made by Oswald Heer (Heer 1861a, pp. 213–20) and others that the geographical distribution of plants could be explained by postulating

a continental link between Europe and America during the Miocene period, the so-called 'Atlantis' hypothesis (Oliver 1862b).

11 Henry Walter Bates. See enclosure.

12 Hooker refers to John and Ellen Frances Lubbock.

13 William Jackson Hooker had been seriously ill since the summer of 1861 (Allan 1967, pp. 207, 208).

14 Hooker refers to examinations for admission to the Army Medical Service, held at Chelsea Hospital in March 1862. Hooker served for many years as a scientific examiner for medical officers in the armed services (L. Huxley ed. 1918, 1: 387); he was examiner for the paper on natural history, held on 11 March 1862 (*Statistical, Sanitary, and Medical Reports* 1860, p. 488).

15 See preceding letter and n. 10.

16 Bates 1863.

17 For Hooker's letter to H. W. Bates of 2 February 1862, see the enclosure to the letter from J. D. Hooker, [23 March 1862].

18 Bates displayed his collection of Amazonian mimetic butterflies to the Linnean Society on 16 January 1862 (*Journal of the Proceedings of the Linnean Society of London (Botany)* 6 (1862): lviii). His paper on the butterflies (Bates 1862a), read before the society on 21 November 1861, was published in the society's *Transactions* in November 1862 (see Raphael 1970, p. 76).

19 *Origin*, p. 11.

20 The botanist Richard Spruce had been collecting plants in South America since 1849; George Bentham received, named, and distributed the plants Spruce sent back to England (*DNB*). Hooker, in his reply to Bates of 18 March 1862, informed him that Bentham had not kept a record of Spruce's native names, but had reclassified the specimens according to their natural orders (Bates 1892, pp. lv–lvi).

21 Bates described several of these fruits in Bates 1863, 2: 216–18.

22 Latreille 1802, pp. 133–7. Pierre André Latreille referred to this ant as *Formica bispinosa*.

23 Hooker had reported in his letter to Bates of 2 February 1862 that he had long denied 'that tropical heat or light *produces* the bright colouring of plants, or that arctic climates *produce* woolly covering' (see the enclosure to the letter from J. D. Hooker, [23 March 1862]). In his reply to Bates of 18 March 1862, Hooker stated: 'I do not think I have anywhere published any notices of colours of plants in relation to climate, but I will see and let you know' (Bates 1892, p. lvii). In his discussion of the independence of colour and climate (Bates 1863, 1: 19–22), Bates did not cite any work of Hooker's, but did refer to CD's similar conclusions (see letter from H. W. Bates, 6 January 1862 and nn. 5–7).

24 Hooker apparently enclosed a photograph of himself in this letter (see letter to J. D. Hooker, 14 March [1862] and n. 4).

25 Hooker had repeated his promise to send CD specimens of *Lythrum* in the letter from J. D. Hooker, 3 March 1862. See also letter to J. D. Hooker, 14 March [1862] and n. 13.

From Heinrich Georg Bronn[1] [before 11 March 1862][2]

⟨ ⟩
⟨ ⟩

Hochverehrter Herr!

Nachdem ich im Laufe des letzten Jahres ang⟨estrengt⟩ mit zoologischen Arbeiten beschäftigt gewesen ⟨bin, brachte⟩ mich ein von Herrn Schweizerbart erhaltener ⟨Brief⟩ zur Natural Selection zurück, indem mich dersel⟨be be⟩nachrichtigt, dass die Deutsche Auflage von "Origin of Species" (welche übrigens nicht sehr stark gewesen ist) wohl in Jahres-Frist vergriffen seyn werde;[3] so daß dann eine neue Auflage veranstaltet werden würde. Ich erlaube mir deshalb die ergebene Anfrage an Sie zu richten, ob nicht auch eine Englische *neue* Auflage bevorstehe,—ob dieselbe Abänderungen und Zusätze erhalten werde—oder ob (auch wenn Solches nicht der Fall) Sie wünschen würden, Abänderungen oder Zusätze zur Deutschen Ausgabe zu machen?[4] Vielleicht steht auch das Erscheinen des größeren Werkes in mehren Bänden bevor, welches Sie dem Kleineren folgen zu lassen die Absicht gehabt haben?[5]

⟨Ich ha⟩be wohl ein Dutzend Kritiken Ihres ⟨Wer⟩kes in Deutschen, Holländischen, Englischen und ⟨A⟩merikanischen Zeitschriften gelesen, günstige und ⟨v⟩iele ungünstige. Indeßen haben sie an *meinem* Urtheile nichts geändert.[6]

1) Ich sehe in Ihrer Theorie den einzigen natürlichen Weg zur endlichen Auflösung des Räthsels der Schöpfung.

2) Ihre Theorie steht aber in Widerspruch mit dem *jetzigen* Stande unsrer Kenntniße von Bildung *organischer Materie* aus unorganischer, von *Belebung* organischer Materie und deren Gestaltung zu organischer *Form* **ohne** *älterlichen Einfluss*

3). Unsre Kenntniße in dieser Beziehung sind aber negative;—und es lässt sich *nicht behaupten*, daß nicht spätre Entdeckungen uns in Besitz solcher positiver Kenntnisse bringen werden welche Ihre Theorie anzunehmen gestatten. Wenn man Ihre Theorie auch jetzt nicht zulassen kann, so darf man sie doch nicht *für alle Zeit* verwerfen!

Ich habe unter den vielen Urtheilen über Ihre **Theorie** keines gefunden, was mit dem meinigen übereinstimmte; aber ich halte das meinige für richtig. Aber alle Urtheile über Ihr **Buch** waren günstig

Sie haben mir unlängst einen Aufsatz über Primula gesendet, wofür ich herzlich danke.[7] Ich habe ihn mit vielen Interesse gelesen, wie ich überhaupt alle auf Ihre Theorie bezüglichen Beobachtungen und Erscheinungen sorgfältig zu beachten gewohnt bin; viele erregen meine Theilnahme die ich sonst übersehen haben würde.

Indem ich Sie bitte, die Versicherung meiner ausgezeichnetesten Hochschätzung zu genehmigen, habe ich die Ehre zu verbleiben | Ihr | ergebenster Diener | H. G. Bronn

DAR 160.3: 319

CD ANNOTATION

Top of letter: '*New Edit*— | *Orchid Book*' *ink*

1 For a translation of this letter, see Appendix I.

2 Dated by the relationship to the following letter.

3 Bronn had translated *Origin* into German soon after its publication in November 1859. The German edition (Bronn trans. 1860) was published in Stuttgart by the firm of E. Schweizerbart'sche Verlagsbuchhandlung, under the direction of Christian Friedrich Schweizerbart. See *Correspondence* vol. 8, letter to T. H. Huxley, 2 [February 1860], and letters to H. G. Bronn, 4 February [1860] and 5 October [1860], and letter from H. G. Bronn, 13 [or 15] October 1860.

4 The first German edition of *Origin* (Bronn trans. 1860) had been translated from the second English edition (see *Correspondence* vol. 8, letter to H. G. Bronn, 4 February [1860], and this volume, letter to H. G. Bronn, 25 April [1862]). A third English edition was issued in April 1861, and it was from this that the second German edition (Bronn trans. 1863) was prepared (see following letter).

[5] CD had originally planned to publish his account of the origin of species by means of natural selection as a multi-volume treatise (see *Correspondence* vols. 6 and 7), but was persuaded to postpone this project in favour of producing an 'abstract' of his material, *Origin* (see, for example, *Correspondence* vol. 7, letter to Charles Lyell, 18 July [1858]). The only part of the projected larger work that was published during CD's lifetime was *Variation* (1868).

[6] For a list of the reviews of *Origin* that appeared in 1860, see *Correspondence* vol. 8, Appendix VII. Bronn included a final chapter in his translation of *Origin* that presented his own views of CD's theory (Bronn trans. 1860, pp. 495–520). See also Bronn 1860, and *Correspondence* vol. 8, letter to H. G. Bronn, 14 July [1860], and letter from H. G. Bronn, 13 [or 15] October 1860. For a discussion of Bronn's attitude towards CD's theory, see Junker 1991.

[7] The reference is to CD's paper, 'Dimorphic condition in *Primula*'. Bronn's name is included on CD's list of people to whom presentation copies of the paper were to be sent (see Appendix III).

To H. G. Bronn 11 March [1862][1]

Down | Bromley. | Kent S.E.
March 11th.

Dear & much honoured Sir.

I thank you for your very kind letter received this morning.[2] I am surprised & pleased to hear that a new Edition of the Origin will be wanted— The last Edition in England contains a considerable number of small corrections & a few of importance; & I should like to make a few more corrections on clean sheets of the last English Edition which I will send you— I hope the Publisher will employ some one to compare the German Edition with this last English Editn. & make the additions in the new German Edition—[3] I am, however at present extremely busy, & it would be a great convenience if I could wait 5 or 6 weeks before sending the English Edition, with the new corrections—[4] If I do *not* hear I will assume that this will be time enough. I have not made much progress in my larger work for I have been tempted away by other subjects— I have however made some progress.—[5]

In about a months time I shall publish a little book on the Fertilisation of Orchids & on their Homologies,—of which I will send you a copy, as a mark of my sincere gratitude, for I do not suppose that the subject will interest you—[6] I may add that if M Schweizerbart should like to publish (but this is very improbable) a translation I would try & procure stereotype plates of the several woodcuts at no expense beyond the casting[7] But I doubt whether the Book would be worth translating though it contains I believe some new & curious facts—

I have lately been reading the French Copy (for I find the German very difficult) of your great work crowned by the French Academy—[8] I have not finished it, but admire, & am profoundly interested as far as I have gone— I regret deeply that I did not know this book before I wrote the Origin—[9]

With sincere respect & gratitude I remain—Dear Sir. | Yours truly obliged. | Ch. Darwin.

Copy
DAR 143: 153

[1] The year is established by the reference to the publication of *Orchids* (see n. 6, below).
[2] See preceding letter.
[3] See preceding letter and n. 4. CD refers to the third English edition of *Origin*, published in April 1861.
[4] CD sent Bronn the corrections and additions for the new German edition of *Origin* at the end of April (see letter to H. G. Bronn, 25 April [1862]).
[5] The reference is to CD's 'big book' on species, from which *Origin* had been abstracted (see *Correspondence* vol. 7). CD made some progress on the work during the first half of 1861, but had since been working on *Orchids* and 'Dimorphic condition in *Primula*' (see *Correspondence* vol. 9, 'Journal' (Appendix II), and this volume, 'Journal' (Appendix II)).
[6] *Orchids* was published on 15 May 1862 (Freeman 1977, p. 112). CD included Bronn's name on his presentation list for the volume (see Appendix IV).
[7] The Stuttgart publishing house E. Schweizerbart'sche Verlagsbuchhandlung, headed by Christian Friedrich Schweizerbart, published a translation of *Orchids* by Bronn later in the year (Bronn trans. 1862; see letter from H. G. Bronn, 27 March 1862, and letter from E. Schweizerbart'sche Verlagsbuchhandlung, 7 June 1862).
[8] In 1857, Bronn had won a prize offered by the Académie des Sciences of Paris to the author of the best essay discussing the question of how the distribution of fossils in geological deposits helped to provide an explanation for the appearance and extinction of species through time. A German translation of the essay was published the following year (Bronn 1858), but the original French version was not published until 1861 (Bronn 1861). There are lightly annotated copies of both the German and French versions of the essay in the Darwin Library–CUL (see *Marginalia* 1: 91).
[9] In his essay (Bronn 1858), Bronn discussed the origin of species and argued that species were always created anew by some unknown creative force rather than originating from other species (see, for example, *ibid.*, pp. 232–5).

From C. W. Crocker [before 13 March 1862][1]

28 South St. Chich⟨ester⟩
Ma⟨r.⟩ ⟨ ⟩

Dear Sir

I cannot tell you how much pleasure your long letter gave me—[2] I shall have much real enjoyment in trying the experiment upon Hollyhocks.[3] I shall also try the Larkspurs if possible, I have seeds of several kinds— I had also promised myself that I would try Beaton's long and short stamened experiment on Pelargoniums—[4] I always like to prove these kinds of things for myself.

Mr. Black from the Herbarium at Kew was stopping at Shanklin for his health last week, so taking advantage of a leisure day I ran down to see him.[5] Walking along the cliff near the landslip we saw what we took to be a cluster of wild carrots. I took up a few roots, would you like to have one or two? The celery and beet, too, grow wild here— would they be of any service to co⟨mpare⟩ with the cultivated vars.

⟨ ⟩ ⟨flo⟩wer find the two forms of Linum[6] ⟨ ⟩ but then we never raised it from seed, ⟨ ⟩ ⟨possi⟩bly all I observed might have originated ⟨from⟩ ⟨c⟩uttings from one individual. I must try if I can get seed of it.

We have still a few bits of marshy ground left undrained near here—some inland towards the downs and some near the shore with brackish water. Pray let me know what plants growing in such places are likely to be dimorphous.

I do not know for certain whether Mormodes &c eject their pollen-masses to a distance;[7] it is better to keep silence than to state anything upon which there is a doubt in one's mind. To be candid I must admit that I have less practical acquaintance with Orchids than, perhaps, with any family of plants.

Do not fear that I shall take less interest in any experiment you suggest than in any thing I might myself originate. It has ⟨alw⟩ays been my opinion that persons situated ⟨as I⟩ am should be satisfied with observing ⟨and rec⟩ording little facts—leaving the Master-minds to show what is their practical use and bearing; and at the same time to suggest what sort of facts are most worthy of being sought for, and where most likely to be found. There is yet so much to be learned in the way of experiment that I have often thought it would be only right for the Government, or better still one of the great scientific societies, to furnish the means for carrying them out in a systematic manner. A little bit of ground, a director, and one or two earnest enquirers after truth, would be all that would be required. There is too much to be done at Kew, the men are too busy to be able to follow out an experiment with the necessary patience and constant attention. At least that was my case while there.

I am very sorry to hear that you are unwell— as for myself I am by degrees regaining strength and hope as the sprin⟨g⟩ advances I shall be myself again. C⟨lose⟩ confinement in the hot-houses at Ke⟨w tried⟩ me severely; had I taken more out-of-doors exercise perhaps I should not have felt it so much, but I was too anxious about my plants and even on a Sunday I was afraid to lose sight of them. In the hands of a careless man so much damage may be done in so short a time.

Believe me, dear Sir, to remain | ever yours very respectfully | C. W. Crocker

C. Darwin Esq[e]

DAR 161.2: 255

CD ANNOTATIONS
2.3 I took . . . vars. 2.5] *scored brown crayon*
4.1 We . . . dimorphous. 4.3] *scored brown crayon*
4.2 Pray . . . dimorphous. 4.3] '—*Epipactis palustris*—'[8] *pencil*
5.1 I do . . . plants. 5.4] *crossed pencil*
6.6 There is . . . would 6.7] *double scored brown crayon*
6.7 be only . . . manner. 6.9] *scored brown crayon*
Top of first page: 'Cut off petals of Hollyhock' *pencil*

[1] Dated by the relationship to the following letter.

[2] CD's letter has not been found, but see the letter to J. D. Hooker, 7 March [1862].

[3] CD told Crocker of some points about hollyhocks on which he wished for more information in the letters to C. W. Crocker, 18 May [1861] and 1 June [1861] (*Correspondence* vol. 9).

[4] Donald Beaton contributed a regular column to the *Journal of Horticulture and Cottage Gardener.* Beaton referred to the occurrence of both long and short stamens in geranium flowers (the term 'geranium' is often applied to cultivated varieties of the genus *Pelargonium*) in the *Journal of Horticulture* n.s. 1

(1861): 312. In *ibid.*, p. 355, he stated:

in Pelargoniums . . . the pollen of the shortest stamens has been proved to produce seedlings of more dwarf habit than the parents; and by applying from the shorter stamens to the stigmas of the dwarf race, again and again, the seedlings at last will be so faint . . . that no art can grow them.

[5] Allan A. Black was the curator of the herbarium at the Royal Botanic Gardens, Kew, where Crocker had been foreman of the propagating department before retiring to Chichester, West Sussex. Shanklin is on the Isle of Wight.

[6] Having studied dimorphism in *Primula*, CD was interested in investigating similar cases in other plant genera (see letters to Asa Gray, 22 January [1862] and 16 February [1862]). He referred to the incidence of dimorphism in two species of the genus *Linum* in his paper 'Dimorphic condition in *Primula*', p. 96 (*Collected papers*, 2: 62–3), stating his intention to carry out further crossing experiments on the two forms during the following summer; he had begun experimenting on the genus in September 1861 (see *Correspondence* vol. 9, letter to Asa Gray, 16 September [1861]). CD's experiments with *Linum* are described in his paper 'Two forms in species of *Linum*', written in December 1862 (see 'Journal' (Appendix II)).

[7] CD had contacted Crocker to inquire about the orchid genus *Mormodes* earlier in the year (see letter to J. D. Hooker, [before 15 February 1862], and letter from J. D. Hooker, [before 15 February 1862]). *Mormodes* is discussed in *Orchids*, pp. 249–69.

[8] CD inquired about this orchid in a subsequent letter to Crocker, now missing (see following letter).

From C. W. Crocker 13 March 1862

South St. Chichester
Mar. 13th 62

Dear Sir

I thank you much for your good opinion of me, I shall try to deserve it.[1]

Epipactis palustris has been found near us, but is rare—[2] If you want a supply of it write to Kew it grows plentifully in the Willow beds between Kew and Mortlake by the river-side. It is *very* common there but from the colour is often overlooked. Unfortunately Menyanthes does not grow here-abouts. But I shall keep my eyes open.[3]

Don't forget that there is a Cypripedium at Kew which belongs to you— I should think it would flower this spring. It was under my charge and before leaving I pointed it out to Gower—[4]but in case it is forgotten—it is (or was) in the second light of the pit nearest the Propagating house.

Beaton talks too confidently for me— I never could put much faith in him.[5] The more I see the more careful I become in my statements. and I think this is the natural effect upon any properly

Incomplete
DAR 161.2: 256

CD ANNOTATION
2.2 it grows . . . Mortlake] *scored brown crayon*

[1] CD's letter has not been found.

[2] *Epipactis palustris*, or the marsh epipactis, is discussed in *Orchids*, pp. 95–102; CD had asked about this orchid in a missing letter to Crocker (see preceding letter, CD annotations). He thanked Alexander Goodman More for providing him with fresh specimens of the plant in *Orchids*, p. 95 n., but referred to remaining difficulties in understanding the 'peculiar structure of the labellum' in *Orchids*, pp. 101–2 n.

[3] CD had recently learned that *Menyanthes* was dimorphic, and was keen to see a specimen of the genus (see letter to C. C. Babington, 20 January [1862]).

[4] Crocker had recently retired from his post as foreman of the propagating department at the Royal Botanic Gardens, Kew. William Hugh Gower was also a foreman at the gardens.

[5] The reference is to Donald Beaton (see preceding letter and n. 4). CD held a similar opinion of Beaton (see *Correspondence* vol. 9, letter to J. D. Hooker, 14 May [1861], and letter to C. W. Crocker, 18 May [1861]).

To J. D. Hooker 14 March [1862][1]

Down Bromley Kent
March 14th

My dear Hooker

Thanks for your letter:[2] I agree with much of what you say about the amiable reciprocal feelings of nations; but Emma agrees with your last sentence that you wrote in a Mephistophelerian spirit. I think you are a bit too hard on Asa Gray; but he evidently tried to be as severe as he civilly could. I knew he was quite wrong about your indifference.—[3]

Thanks, also, for Photograph, which about a fortnight ago we were wishing for; but it does not give your expression & so by no means does you justice.—[4]

What a capital letter of Bates':[5] he is evidently a true thinker; it has made me very curious to see your letter; if it contains nothing personal relating to Bates or yourself, might I see it? If so, & you are writing, would you ask him to send it; or I would write; but I thought he might feel scruples without your permission in sending it.[6]

The point which you have been discussing is most difficult: I always come, after doubt, to your side. There is one pretty clear line of distinction;— when many parts of structure as in woodpecker show distinct adaptation to external bodies, it is preposterous to attribute them to effect of climate &c—but when a single point, alone, as a hooked seed, it is conceivable that it may thus have arisen. I have found the study of orchids eminently useful in showing me how nearly all parts of the flower are coadapted for fertilisation by insects, & therefore the result of n. selection,—even most trifling details of structure. I have just, by the way, been studying Mormodes ignea—; it is a prodigy of adaptation; but I had to examine 12 flowers in all sorts of ways, before I made out its mechanism.[7]

I should like to read Oliver's paper, but I am so hard-worked with proofs &c, that I must give it up, till it appears in print.—[8]

It is real good news that you will try & come here in Easter;[9] Emma desires to join me in hoping that Mrs. Hooker[10] will come also; I fear we cannot take in your children, as all our Boys, & perhaps others, will be at home.

I am pleased to hear that you like Lubbock & Mrs. L.; he is a real good fellow & she is a charmer.—[11]

Farewell, my dear old fellow | Yours affect^{ly}.— | C. Darwin

Wallace will be home in a month or two.—[12]

Do not forget Lythrum, Saxifrages &c. Avoid Saxifrages with flexuous or woolly hair; but choose a plant with longest straight hairs.[13]

DAR 115.2: 150

[1] The year is established by the reference to Alfred Russel Wallace's return to England (see n. 12, below).

[2] Letter from J. D. Hooker, [10 March 1862].

[3] See letter from Asa Gray, 18 February 1862.

[4] The photograph was apparently enclosed with the letter from J. D. Hooker, [10 March 1862]: CD wrote 'Photograph' at the top of that letter (see CD annotations). For Hooker's negative opinion of his photographic likeness, see the letter from J. D. Hooker, 17 March 1862.

[5] See the enclosure to the letter from J. D. Hooker, [10 March 1862].

[6] See the enclosure to the letter from J. D. Hooker, [23 March 1862].

[7] CD had for some time been anxious to discover the mechanism by which the pollinia of this orchid are ejected, and had unsuccessfully sought specimens from various of his botanical correspondents (see letter to J. D. Hooker, [before 15 February 1862] and n. 3). In *Orchids*, p. 249 and n., CD thanked Sigismund Rucker for having lent him a plant of *M. ignea*, making reference to the fact that he had examined twelve flowers before making out 'the meaning and action of the several parts.' See also following letter and n. 3.

[8] Oliver 1862b. See letter from J. D. Hooker, [10 March 1862] and n. 10.

[9] See letter from J. D. Hooker, [10 March 1862]. According to Emma Darwin's diary (DAR 242) and the letter to H. W. Bates, 16 April [1862], Hooker stayed with the Darwins from 17 to 21 April 1862.

[10] Frances Harriet Hooker.

[11] John and Ellen Frances Lubbock. See letter from J. D. Hooker, [10 March 1862]. In her Autobiography (DAR 246), Henrietta Emma Darwin recalled that CD was 'fascinated' by E. F. Lubbock, who was reported to be a 'beautiful and fascinating creature'.

[12] Alfred Russel Wallace arrived back in England in the spring of 1862, having spent eight years in the Malay Archipelago (Wallace 1905, 1: 385).

[13] CD had enclosed with the letter to J. D. Hooker, 25 February [1862], a list of specimens he needed for experiment; he wished to see specimens of *Lythrum* in order to investigate what he believed to be 'a magnificent case' of trimorphism in the genus, and of *Saxifraga* for his research into insectivorous plants. Having observed that minute flies were caught in the hairs of *S. umbrosa*, CD had concluded that, while this species apparently did not derive nutriment from the dead flies, some of the more hairy saxifrages might do so (see *Correspondence* vol. 9, letter to J. D. Hooker, 30 May [1861]).

From D. F. Nevill [*c.* 14 March 1862][1]

Dangstein | Petersfield

My dear Sir

I ought long ere this to have acknowledged the photo which has been for some time hung up in my own sitting room opposite my esteemed friend Sir W Hooker.[2] When in London—for a few days I went down to Mr Ruckers to see his gardens which are wonderful and he told me he had sent you a plant of a Mormodes—[3] I am watching our Vanilla lutescens— it is a magnificent plant full 20 feet long as yet no flowers—but I never forget your wants when I enter the house—[4] I hope you will never scruple to ask me for anything I can give for it will be a real pleasure to me to do so—

believe me | most truly yours | Dorothy Nevill

DAR 172.1: 28

[1] The date is conjectured from the reference to CD's having sent Nevill a photograph of himself (see n. 2, below), and from the reference to Sigismund Rucker's having sent CD a specimen of *Mormodes*; in the letter to J. D. Hooker, 14 March [1862], CD described his examination of a specimen of *M. ignea* sent to him by Rucker (see n. 3, below).

[2] William Jackson Hooker. In the first of two letters dated [before 22 January 1862], Nevill asked CD for a photograph or portrait of himself. In his letter to D. F. Nevill, 22 January [1862], CD stated that the photograph would be sent 'in a few days'.

[3] See preceding letter and n. 7. Nevill was among those from whom CD had sought specimens of *Mormodes* (see *Correspondence* vol. 9, letters to D. F. Nevill, 12 November [1861] and 19 November [1861]). In mid-March 1862, he was lent a plant of *M. ignea* by Rucker, an East and West India broker residing in West Hill, Wandsworth, whose assistance in this regard is recorded in *Orchids*, p. 249. Rucker also subsequently, on 8 June 1862, provided CD with a specimen of *M. luxata*, CD's description of which is in DAR 70: 99–102.

[4] CD had wished to examine specimens of the orchid tribe Arethuseae, to which the genus *Vanilla* belongs, before completing *Orchids*, and had asked Nevill if she could provide him with specimens (see *Correspondence* vol. 9, letter to D. F. Nevill, 12 November [1861]). He eventually acquired some *Vanilla* flowers from Joseph Dalton Hooker (see letter from J. D. Hooker, [29 May 1862]).

To Asa Gray 15 March [1862][1]

Down. | Bromley. | Kent. S.E.
March 15th

My dear Gray

Thanks for the newspapers (though they did contain digs at England) & for your note of Feb. 18th—[2] It is really almost a pleasure to receive stabs from so smooth, polished & sharp a dagger as your pen.— I heartily wish I could sympathise more fully with you, instead of merely hating the South. We cannot enter into your feelings; if Scotland were to rebel, I presume we should be very wrath, but I do not think we should care a penny what other nations thought. The Millenium must come before nations love each other; but try & do not hate me. Think of me, if you will, as a poor blinded fool. I fear the dreadful state of affairs must dull your interest in Science.—

Two days ago, I heard from Trübner, who says of the 250 copies of your Pamphet, he has only 38 in hand; so that he will, I suppose, soon transmit to you a few pounds, enough to cover all your expences.[3] I believe that your pamphlet has done my book *great* good; & I thank you from my heart for myself; & believing that the views are in large part true, I must think that you have done natural science a good turn. Natural Selection seems to be making a little progress in England & on the

Continent; a new German Edition is called for & a French one has just appeared.[4] There has even been a Dutch Edition![5] One of the best men, though at present unknown, who has taken up these views, is Mr Bates;[6] pray read his Travels in Amazonia, when they appear;[7] they will be very good, judging from M.S. of two first chapters.[8]

I wrote some little time ago about Rhexia:[9] since then I have been carefully watching & experimenting on another genus, Monochætum; & I find, that the pistil is first bent rectangularly, (as in the sketch sent)[10] & then in a few days becomes straight; the stamens also move. If there be not two forms of Rhexia, will you compare the position of the part in young & old flowers. I have suspicion (perhaps will be proved wrong when seed-capsules are ripe) that one set of anthers are adapted to pistil in early state, & the other set for it in its later state.— If Bees visit the Rhexia, for Heavens sake watch exactly how the anthers & stigma strike them, both in old & young flowers . . . & give me a sketch.—

I have got lots of seeds planted for experiment this summer, including Amsinckia spectabilis![11]

Again I say, do not hate me.

Ever yours most truly | C. Darwin

Gray Herbarium of Harvard University (64)

[1] The year is established by the relationship to the letter from Asa Gray, 18 February 1862.

[2] Letter from Asa Gray, 18 February 1862.

[3] CD refers to Gray's pamphlet on natural selection and natural theology (A. Gray 1861); CD and Gray had shared the cost of having the pamphlet printed (see *Correspondence* vol. 8, letter to Asa Gray, 11 December [1860], and *Correspondence* vol. 9, letter to Asa Gray, 17 February [1861]). Nicholas Trübner was head of the publishing firm, Trübner and Co., which acted as the London agent for distribution of the pamphlet in Britain. See also *Correspondence* vol. 9, Appendix III.

[4] CD refers to Bronn trans. 1863 (see letter from H. G. Bronn, [before 11 March 1862]). Clémence Auguste Royer's French translation of *Origin* was not in fact published until 31 May 1862 (*Journal Générale de l'Imprimerie et de la Librairie* 2d ser. 6 (pt 3): 341. See also letter to J. D. Hooker, 26 [March 1862], and letter to Asa Gray, 10–20 June [1862].

[5] Winkler trans. 1860. Tiberius Cornelius Winkler sent CD a copy of his Dutch translation of *Origin* in 1861 (see *Correspondence* vol. 9, letter from T. C. Winkler, 7 July 1861).

[6] Henry Walter Bates argued that mimicry in Amazonian butterflies offered 'a most beautiful proof of the truth of the theory' of natural selection (Bates 1862a, p. 513).

[7] Bates 1863.

[8] See letter to H. W. Bates, 13 January [1862].

[9] See letter to Asa Gray, 16 February [1862].

[10] CD probably refers to the diagram enclosed with the letter to Asa Gray, 16 February [1862]. He had begun crossing experiments with *Monochaetum ensiferum* on 7 February 1862, and continued to work on the species until May 1863. See the dated experimental and observational notes in DAR 205.8: 22–43.

[11] In 'Dimorphic condition in *Primula*', p. 95, CD mentioned that his study of dried specimens sent him by Gray indicated that *Amsinckia spectabilis* was probably heterostyled (see also *Collected papers* 2: 62). However, in *Forms of flowers*, p. 110, he reported: 'on raising many plants from seed, I soon became convinced that the whole case was one of mere variability.'

From J. D. Hooker 17 March 1862

Kew
March 17/62

Dear Darwin

The Saxifragas &c shall go on Wednesday for your carrier to take on on Thursday.— just look at the Saxifrages & let me know at once if they will suit,— as if not I will send others[1]

As regards my photograph, I believe I have very little expression[2] I have often remarked that I am not recognized except by those who know me tolerably well; that I have often to introduce myself—added to which all my photographs & portraits make me look either silly or stupid or affected— artists find nothing salient, nothing to idealize upon— Poor Richmond who generally knocks off his chalk heads in 2 sittings gave me 8 I think & grumbled all the time, & has turned me out a very lackadaisaical young gentleman.[3]

I am so glad you have asked for my letter to Bates,[4] to tell the truth he has roused my curiosity to see it myself, I have not the slightest recollection of what I can have said that has so much impressed him— When I write I shall ask him to send it. I have done nothing more that I know of than ram you down his throat.— probably like some others he did not realize your views until I enforced them by pointing out their applicability to his cases.—

I am greatly puzzled just now in my mind by a very prevalent difference between animals & vegetables—in as much as the *individual* animal is certainly changed materially by external conditions, the latter (I think) never except in such a coarse way as stunting or enlarging—& this is because in animals there is a direct relation between stimulated function & consequent change in organs concerned in that function—Eg. no increase of cold on the spot, or change of *individual* plant from hot to cold, will induce said individual plant to get more woolly covering.— but I suppose that a series of cold seasons would bring about such a change in an *individual* quadruped: just as rowing will harden hands, &c. The cases are not parallel because the parts of plants that could be so changed are annually lost—& the only conceivable parallel is afforded by Bark;—would a cycle of cold seasons cause the bark of a tree to thicken more than it otherwise would?[5]

I cannot suppose that the buds of the individual would get thicker, or more scales, or more resinous scales; or that its successive leaves can become *annually* more hairy: except indeed we assume the annual death of a large proportion of the buds, & that those alone are preserved that have most woolly leaves—when no doubt the woolly tendency would be inherited by the successive phytons of that bud as by successive generations from seeds.

Be all this as it may, in neither plant or animal would the induced character be of necessity inherited by the offspring by seed of the individual to any greater extent than if it had not been changed— At least so far as the animal is concerned; though with regard to the plant it might be, the seed being that of the phyton, not of the whole tree, or average tree— Thus a wild complication is introduced into the whole subject that perplexes me greatly.

Berkeleys article on acclimatization is very unclear I think (see last Saturdays Gard. Chronicle).—[6]

I cannot conceive what you say, that climate could have effected even such a single character as a hooked seed. You know I have a morbid horror of 2 laws in nature for obtaining the same end. hence I incline to attribute the smallest variation to the inherent tendency to vary; a principle *wholly* independent of physical conditions—but where effects on the race are absolutely dependent on physical conditions—for their conservation—

Huxley is rather disposed to think you have overlooked "*saltus*"—[7]but I am not sure that he is right. Saltus quod *individuals*, is not saltus quod *species*—as I pointed out in the *Begonia* case.[8] though perhaps that was rather special-pleading in the present state of science.

I hear Falconer is indignant at the base idea of abolishing a tertiary Atlantis—[9] what a queer mixture he is of dogma & fancy.— What do you think of **3** of Heers Madeira fossils referred to *Ulmus*, *Corylus* & *Leguminosæ*, all turning out to be the *leaflets* of *one* Bramble,[10] & this the commonest plant in Europe, & found in Madeira too—! I must confess it rather turns my stomach, for I was beginning to waive many of my objections against Bot. Palæontology in favor of O. Heer.[11] I wrote to Lyell about it 10 days ago, but he does not answer my letter!—[12] I suppose he is what is called *flabbergasted*—

Do you know anything of Earl Powis— he is a Shropshire man I know—[13] I have to *make out* a visit to him this week, having been asked so often that I cannot refuse without appearing churlish I do hate such visits, feeling generally thoroughly out of my element. Walcot Hall Shropshire is the place— I go from Thursday till Saturday or Monday, so please direct there if you have anything to say—or whether or no— His brother lives next house to us here[14] & I like what I have seen of the Earl; but really know nothing about him.

Many thanks to Mrs Darwin for her kind thoughts of my wifes going to Down with me:[15] but I am sure she will not get away, & as to the children it is *out of the question*, even if you could have taken them in. 2 of my children are equal to a round dozen of another mans. The baby is just the same "nunquam otiosus"—[16] Except Charlie[17] there is not one that can sit still *one moment* & he is never idle. Yet I should like my children to know your's one day.

Ever Yours affec | J D Hooker

DAR 101: 23–6

CD ANNOTATIONS

4.2 in as ... conditions, 4.3] *scored brown crayon*
8.1 I cannot ... end. 8.3] *scored brown crayon*
8.3 smallest ... conditions— 8.6] *scored brown crayon*
9.1 but I ... case. 9.3] *scored brown crayon*
10.2 of **3** ... Bramble, 10.4] *scored brown crayon*
11.1 Do ... churlish 11.3] *scored brown crayon*
12.2 but I ... in. 12.3] *double scored brown crayon*

[1] See letter to J. D. Hooker, 14 March [1862] and n. 13. The carrier service between Down and London was provided by George Snow.

[2] See letter to J. D. Hooker, 14 March [1862].

[3] George Richmond made a portrait in chalk of Hooker in 1855 (see Lister 1981, p. 162). The portrait is reproduced as the frontispiece to the first volume of L. Huxley ed. 1918.

[4] See letter to J. D. Hooker, 14 March [1862] and nn. 5 and 6. The reference is to Henry Walter Bates.

[5] The possibility of species being modified by the direct action of external conditions, as well as by natural selection, formed part of the discussion contained in the recent correspondence between Hooker and Bates (see enclosure to the letter from J. D. Hooker, [23 March 1862], and enclosure to the letter from J. D. Hooker, [10 March 1862]).

[6] Miles Joseph Berkeley wrote a short article on acclimatisation of plants that appeared in the *Gardeners' Chronicle and Agricultural Gazette*, 15 March 1862, pp. 235–6. CD annotated this article in his copy of the magazine, which is at the Cory Library, Cambridge Botanic Garden. The number of the magazine in which the article appears is also listed on the 'List of the numbers of special interest to Darwin and kept by him in separate parcels' (DAR 222).

[7] One of Thomas Henry Huxley's earliest criticisms of *Origin* was that by advocating gradual change through the selection of small variations and by adhering 'so unreservedly' to the principle that 'Natura non facit saltum', CD had burdened himself with an unnecessary difficulty (see *Correspondence* vol. 7, letter from T. H. Huxley, 23 November 1859). Huxley repeated this criticism in his review of *Origin* ([T. H. Huxley] 1860, p. 569).

[8] The reference is to Hooker's article in the *Gardeners' Chronicle and Agricultural Gazette*, 25 February 1860, pp. 170–1, entitled 'The monstrous Begonia frigida at Kew, in relation to Mr. Darwin's theory of natural selection'. The article was a response to a piece by William Henry Harvey (*ibid.*, pp. 145–6), which argued that the plant was an illustration of the abrupt formation of a new species. Hooker's response stated that, on the contrary, the plant demonstrated how 'slow and partial' was change effected by natural selection: 'We cannot indeed conceive the new form replacing the old till after the lapse of many generations, and a long course of . . . operation of natural selection'. See *Correspondence* vol. 8, letter to Charles Lyell, 18 [and 19 February 1860], and letters to J. D. Hooker, [20 February 1860] and 26 [February 1860].

[9] The reference is to Hugh Falconer. Daniel Oliver had delivered a lecture at the Royal Institution of Great Britain on 7 March 1862 questioning the assertion made by Oswald Heer and Franz Unger, that the geographical distribution of plants provided evidence for the existence of a land link between Europe and America during the Tertiary period (Oliver 1862a). Oliver was about to have a paper on the subject published in the *Natural History Review* (Oliver 1862b). See also letter from J. D. Hooker, [10 March 1862] and n. 10.

[10] In his *Manual flora of Madeira* (Lowe 1857–68, pp. 249, 251), Richard Thomas Lowe had challenged Heer's identification (in Heer 1857) of several fossil plants found in Madeira. Oliver noted Lowe's challenge in Oliver 1862b, p. 170.

[11] Hooker was able to view Heer's botanical collections himself during a visit to Zürich later in the year (see letter from J. D. Hooker, [24 July 1862]).

[12] Charles Lyell invited Heer, whom he considered 'the best botanist in Europe for fossil tertiary plants', to England in 1861 to study the fossil plants found at Bovey Tracey in Devon (K. M. Lyell ed. 1881, 2: 346). Lyell subsequently communicated Heer's paper on the fossils (Heer 1862) to the Royal Society of London (see *Proceedings of the Royal Society of London* (1861): 449–55). CD mentioned Lowe's doubts about Heer's work on the fossil plants of Madeira in the letter to Charles Lyell, [15 September 1861] (*Correspondence* vol. 9).

[13] Hooker refers to Edward James Herbert, third earl of Powis. The family was known to the Darwins (see *Correspondence* vol. 2, letter to Susan Darwin, [26 April 1838]). CD's father, Robert Waring Darwin, lent substantial sums of money to Edward Clive, first earl of Powis and his son Edward Herbert, Viscount Clive (see R. W. Darwin's Account book (Down House MS) and Brent 1981, p. 17).

14 Hooker refers to Percy Egerton Herbert, deputy quartermaster-general at the Horse Guards (*DNB*, *Post Office directory of the six home counties* 1862).

15 See letter to J. D. Hooker, 14 March [1862]. The references are to Emma Darwin and Frances Harriet Hooker.

16 The Latin designation indicates that the baby, Brian Harvey Hodgson Hooker, then nearly 2 years old, was 'never inactive'.

17 Charles Paget Hooker.

From M. T. Masters 17 March 1862

Rye Lane | Peckham
March 17. 1862.

My dear Sir

I can only speak from memory as to the Primrose with ten stamens as I have not the specimen[1] My impression is that the upper stamens alternated with the lobes of the corolla. the lower on the contrary were opposite to them but you must take this impression for what it is worth—not much— Should I ever come across another such I will assuredly bear you in mind— I think too I told you that I had a sketch of the late Prof Henslow's[2] showing a double row of stamens this I find on examination is an error— My sketch is merely a diagram showing the different position of the stamens thus—

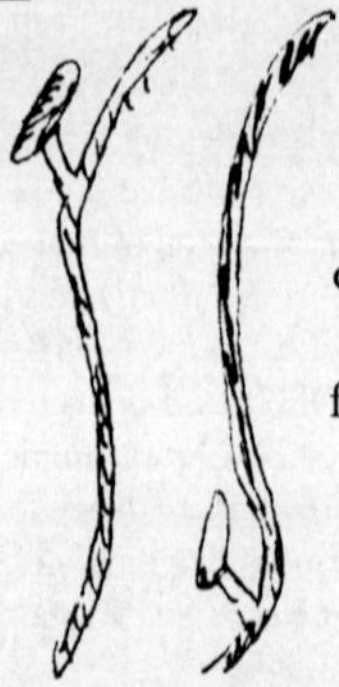

evidently referring to two different flowers not to the same.

Believe me | dear Sir | Yours faithfully | Maxwell. T. Masters.

DAR 171.1: 67

1 Neither Masters's letter mentioning the specimen nor CD's letter seeking further information about it has been found.

2 John Stevens Henslow, professor of botany at the University of Cambridge, died on 16 May 1861.

To J. D. Hooker 18 March [1862][1]

Down Bromley Kent.
March 18th

My dear Hooker

I am astonished at your larkiness in going to Walcot; I wish it had been Powis Castle, which, as far as I remember, is very grand. I saw your host, as a child; & have heard since that he is reckoned stiff & proud, but very respectable.—[2]

Your letter discusses lots of interesting subjects,[3] & I am very glad you have sent for your letter to Bates.[4] What do you mean by "individual plant"? I fancied a bud lived only a year & you could hardly expect any change, in that time; but if you call a tree or plant an individual, you have sporting buds. Perhaps you mean that the whole tree does not change. Tulips in "breaking" change. Fruit—seems certainly affected by the stock. I think I have got cases of slight changes in alpine plants transplanted. All these subjects have rather gone out of my head owing to Orchids; but I shall soon have to enter on them in earnest, when I come again to my volume on Variation under Domestication.[5] In the life time of an animal you would, I think, find it very difficult to show effects of external conditions on animals, more than shade & light, good & bad soil, produce on a plants.[6]

You speak of "an inherent tendency to vary wholly independent of physical conditions". This is a very simple way of putting the case (as D^r. Prosper Lucas also puts it);[7] but two great classes of facts make me think that all variability is due to changes in the condition of life. (1) that there is more variability & more monstrosities (& these graduate into each other) under unnatural domestic conditions, than under nature. And secondly that changed conditions affect in an especial manner the reproductive organs,—those organs which are to produce a new being.— But why one seedling out of thousands presents some new characters transcends the wildest powers of conjecture. It was in this sense that I spoke of "*climate, &c*" possibly producing *without selection* a hooked seed; or any not great variation.[8] I have for years & years been fighting with myself not to attribute too much to Nat. selection,—to attribute something to direct action of conditions; & perhaps I have too much conquered my tendency to lay hardly any stress on conditions of life. I am not shaken about "saltus";[9] I did not write without going pretty carefully into all the cases of normal structure in animals resembling monstrosities which appear per saltus.—

I saw about Heer, & thought of you; but may not Lowe be wrong?[10] I shall be so glad to see you here at Easter, if you are able to come; if Willy is at home, why not bring him?[11] this is Emma's message as well as mine; but perhaps he will be at school.— We have been very anxious for 6 weeks about our boy Horace, who three or four times a day has spasmodic attacks, something like Chorea, yet different.[12] Our country Doctor thinks it certainly caused only by irritation in alimentary canal;[13] but I can see that Sir H. Holland thinks it serious.[14] All that one can do, is to hope

Farewell my dear old friend | C. Darwin

DAR 115.2: 145

[1] The year is established by the relationship to the letter from J. D. Hooker, 17 March 1862.

[2] Edward James Herbert, third earl of Powis, had invited Hooker to spend the weekend at Walcot Hall, Shropshire (see letter from J. D. Hooker, 17 March 1862 and n. 13).

[3] Letter from J. D. Hooker, 17 March 1862.

[4] With his letter of [10 March 1862], Hooker enclosed a letter from Henry Walter Bates. CD had asked to see the letter from Hooker to which Bates's was a reply (see letter to J. D. Hooker, 14 March [1862]).

[5] See letters to J. D. Hooker, 24 [November 1862] and [after 26] November [1862], and letter from J. D. Hooker, 26 November 1862.

[6] CD devoted a whole chapter in *Variation* to a discussion of the possibility that external conditions such as climate or food could alone cause species modification (*Variation* 2: 271–92).

[7] Lucas 1847–50. There is a heavily annotated copy of this work in the Darwin Library–CUL (see *Marginalia* 1: 513–23).

[8] See letter to J. D. Hooker, 14 March [1862].

[9] See letter from J. D. Hooker, 17 March 1862 and n. 7.

[10] See letter from J. D. Hooker, 17 March 1862 and n. 10.

[11] See letter to J. D. Hooker, 14 March [1862] and n. 9. CD refers to Hooker's eldest child, William Henslow Hooker.

[12] According to Emma Darwin's diary (DAR 242), Horace Darwin had been ill since January. Emma took him to London in February for a consultation with Edward Headland (see letter to W. E. Darwin, 14 February [1862]).

[13] Stephen Paul Engleheart was the Down surgeon (*Post Office directory of the six home counties* 1862).

[14] Henry Holland, a London physician, was frequently consulted by the Darwin family. See also letter from Henry Holland, 26 March [1862].

To Richard Kippist[1] 18 March [1862][2]

March 18th

My dear Sir

I send by this Post, a paper for Linn. Socy. as has been requested of me.—[3] You will already have, or will soon receive, some large Diagrams to hang up.— Please endeavour to hang the *Section* in some place, where I could point to it, & indicate the parts.—

Will you kindly inform me, *three or four days* before whatever time the paper is to be read, & I will endeavour to come up & read it myself.—[4]

Will you be so good as to attend to separate note about Refereeship.—[5]

My dear Sir | Yours sincerely | C. Darwin short

Down Bromley Kent
March.—18th

Linnean Society of London

[1] The recipient is identified by the relationship to the following letter. Richard Kippist was librarian and clerk to the Linnean Society of London (Gage 1938).

[2] The year is established by reference to CD's paper, 'Three sexual forms of *Catasetum tridentatum*', read before the Linnean Society on 3 April 1862.

[3] Joseph Dalton Hooker and George Bentham, respectively vice-president and president of the Linnean Society, had encouraged CD to contribute a paper on an orchid in the possession of the society that produced three types of flower (see letter from J. D. Hooker, 27 February 1862).

[4] CD read his paper on *Catasetum tridentatum* before the Linnean Society on 3 April 1862 (see letter to Charles Lyell, 1 April [1862]); according to Emma Darwin's diary (DAR 242), he travelled to London on 2 April and returned on 4 April.

[5] See following letter.

To Richard Kippist 18 March [1862][1]

Down
March 18th.—

My dear Sir

Will you be so good as to inform the **Referee**, to whom my paper may be referred, that, if it be ordered to be printed, I will endeavour to borrow from Mr Murray the two or three woodcuts, which would be almost necessary.—[2] If, however, the Referee thinks fit, I will withdraw the paper, as nearly the whole will appear in my small work on the Fertilisation of Orchids.[3] Or I could strike out as much as the Referee should think fit.— I am, however, not willing to spare the time to condense the whole.—

My dear Sir | Your's sincerely | C. Darwin

R. Kippist Esq—

Linnean Society of London

[1] The year is established by the reference to CD's paper on *Catasetum tridentatum*, read before the Linnean Society of London on 3 April 1862.

[2] The woodcuts used in CD's paper, 'Three sexual forms of *Catasetum tridentatum*', were those of *C. tridentatum*, *Myanthus barbatus*, and *Monachanthus viridis* prepared for publication in *Orchids* (see *Orchids*, pp. 232 and 239). John Murray, CD's publisher, had possession of the woodcuts, since *Orchids* was then in press.

[3] *Orchids* was published on 15 May 1862 (Freeman 1977, p. 112).

To William Bernhard Tegetmeier 18 March [1862][1]

Down. | *Bromley.* | *Kent. S.E.*
March 18th

My dear Sir

Many thanks for your note.[2] Your name has been on my list to write to, for, I am ashamed to say, months. I have been extra busy in preparing & getting through the press a small Botanical work on orchids; & this has taken up my time & driven all other subjects out of my head. I hope now in three or four weeks to resume Domestic Animals.—[3] Thank you much for keeping my M.S. I shall not want it for some time, for I have much to do, before going a second time over my M.S.—[4]

The case about the Hen-feathered Cock, is very interesting, & I am very glad to hear of it.—[5] It is very kind in you to offer me a bird, but I have so many irons in the fire that I could not properly attend to it, so I must decline with hearty thanks. I hope we shall meet some day at one of the Shows or elsewhere.—

I long to be at work again at Fowls, Rabbits & such small cattle.—

My dear Sir | Yours very sincerely | C. Darwin

New York Botanical Garden Library (Charles Finney Cox collection)

[1] The year is established by the reference to the publication of *Orchids*, which occurred in May 1862.

[2] Tegetmeier's letter has not been found.

[3] CD had suspended work on *Variation* at the end of May 1861 in order to write up his researches on the fertilisation of orchids (see *Correspondence* vol. 9, 'Journal' (Appendix II)).

[4] CD had sent Tegetmeier his manuscript chapter on domestic fowls, intended for inclusion in *Variation* (see *Correspondence* vol. 9, letter to W. B. Tegetmeier, 14 June [1861]).

[5] At a meeting of the Zoological Society of London on 26 March 1861, Tegetmeier exhibited a live male specimen of *Gallus domesticus* that had assumed female plumage; he had been attempting to ascertain how far the disposition to undergo this change was hereditary (see *Proceedings of the Zoological Society of London* (1861), p. 102, and *Correspondence* vol. 9, letters to W. B. Tegetmeier, 3 February [1861] and 2 April [1861]). CD discussed the phenomenon, citing Tegetmeier's case, in *Variation* 1: 252–3.

To J. D. Hooker 22 [March 1862][1]

Down
22[d]

My dear Hooker

This is the last time, I fully believe, that I shall trouble you to correct names of two Calanthes.—[2] I wrote to you at L[d]. Powis'.[3] I suppose you forwarded to me this morning a note from Asa Gray:[4] it contained nothing, but ends "Yours cordially" so I hope to Heaven I am forgiven.—

My Catasetum paper is to be read by me on April 3[d]. I have had some diagrams made.—[5]

I am so consumedly sick of correcting my orchid Book.— | Farewell. C. Darwin

The plants & seeds arrived all safe & will be of greatest use; especially Lythrum, which seems from Vaucher a grand case of Trimorphism.—[6] What sort of man is Vaucher? I have just got 4 awful vols.— phys. Hist. of Plants of Europe.—[7]

DAR 115.2: 146

[1] The date is established by the reference to Hooker's visit to the home of Lord Powis (see n. 3, below) and by CD's references to his publications (see nn. 2 and 5, below).

[2] For Hooker's reply, see the letter from J. D. Hooker, [23–5 March 1862]. CD was checking the species names of orchids in the process of correcting proofs of *Orchids*. See also letter to J. D. Hooker, [1 March – 15 May 1862].

[3] Letter to J. D. Hooker, 18 March [1862]. Hooker was staying at Walcot Hall in Shropshire as the guest of Edward James Herbert, third earl of Powis (see letter from J. D. Hooker, 17 March 1862).

[4] See letter from Asa Gray, 6 March [1862].

[5] See letters to Richard Kippist, 18 March [1862].

[6] Vaucher 1841, 2: 370–1.

[7] There is an annotated copy of Vaucher 1841 in the Darwin Library–CUL (see *Marginalia* 1: 812–15).

From J. D. Hooker [23 March 1862][1]

Kew
Sunday

My dear Darwin

I returned last night & found Bates' letters which I send herewith,[2] I have no time to compare them— I hope I have not abused you unmercifully in my letter to Bates— you must take your chance!

I had a very profitable stay at Walcot considering all things & came away with food for much reflection.[3] I could not make up my mind to stay over Sunday, though kindly pressed with real English hospitality— Some of the family are very nice—all the Ladies particularly so, the Servants perfection (such Nat: selection of flunkies)—the food good & plenty. The country beautiful—the weather detestable—& the habits & hours of the house quite intolerable— it would take a letter from you every morning to have supported me under such a system of killing time & outraging the stomach— However it does one good to go to such places rarely, gives one much food for reflection & will add a chapter to my posthumous work "On the principles which regulate the development of an Aristocracy".[4] The principle part of this work will consist of 4 chapters, each headed with a B. viz. Blood, Blunt,[5] Brains, Beauty.— These are all good things, of use to the organism possessing them, & hence sought after by all human organisms, & their accumulation, by Natural Selection, must culminate in an Aristocracy—or there is no truth in darwinism. The better these are blended, the better will be your Aristocracy—the more they are seperated the worse. & it is hard to say which is worst per se, or which is best when all are mixed.

You have the Aristocracy purely of B1 in Germany; of B.2 in America, of B.3 in France, of B.4 everywhere. but of 4B in England only: where indeed we have $4B^4$ in the highest nobility. I met nothing beyond B1 & B2 at Walcot, however, perhaps with ever so small an element of the two others I might have been induced to stay Sunday, for I do maintain that the union of all must be irresistable, in every degree & condition of life, from Fuegia to London.

I have no time to answer your kind long letter received at Walcot.—[6] There must be, as you say something effective in the alteration of the reproductive system under variation—but I do not see why, that is not a form of variation, not necessarily *induced* by domestication but accompanying some variety artificially selected. I cannot however forget that it is through *marriage* alone that the 4B^{s}. are usually recruited in after life & so there may be something in what you say!!! That's my philosophy— make the best of it, till we meet.

We are concerned to hear of Horace,[7] but after Ettys recovery "nil desperandum"—[8] does he get regular exercise 2–3 times a day & **very** regular nutritious food— I do not give much for Hollands opinion.[9]

My wife likes the idea of my taking Willie down with me.[10] pray thank Mrs Darwin very much— I have only one condition—that Mrs Darwin would let him sleep with me— I have seen *nothing* of him by myself since he was 6 years old & I should like to have him beside me when I am dressing. I find Charlie's intellect singularly active in the morning.[11]

Ever yours affec | J D Hooker

[Enclosure]

Kew
Feby 2^{d}/62

My dear Mr Bates

I have been thinking much of your extremely interesting letter before answering it,[12] & shall be only too glad if I can say any thing that can tend to remove any difficulties. I need hardly commence by telling you that my opinions are nothing, that the whole question wants working out by observation as you are doing, & by experiments which no one has attempted or even suggested that I know of— What I have said & shall now say you must take *not* as opinions of mine, but as my tendencies of thought.— Certainly I *incline* to believe that variation is sufficient to ensure any amount of divergence & that it (the principle or fact called variation) is independent **wholly** as to amount & kind of local circumstances.

My reasons for thus thinking are 1.) That it has never been shown experimentally that induced habits of the individual are propagated 2) That no such effects of local circumstances on the *individual* is necessary.— there being variation enough without it. 3) It seems more philosophical to suppose that the principle of variation is one thing, immutable, & that local circumstances are the secondary causes acting through Nat. selection on the varieties killing some sparing some directing intermarriages &c &c. 4) that it would simplify matters very much if we could thus disentangle the two phenomena of variation & Nat selection. I grant that not one of these reasons has a leg to stand upon in a strictly scientific point of view—they are a string of hypotheses, & further there are some seemingly opposed facts. 1. hereditary diseases might have been originally induced by local causes; though I think they may be explained without. It is said that a peculiar form of teeth becomes hereditary in syphilitic families.—but I suspect that this could not be perpetuated—it could only be propagated by intermarriage of children of syphilitic parents, which would surely die out.— Then too I have heard that D^{r}. Brown-Sequard has induced epilepsy in rabbits, & this has happened in the brood.—[13] if really so this is a strong point; but here again any attempt to perpetuate the brood of epileptics must end in extinction. All these are cases of induced *diseases* of the individual being propagated;—could induced innocuous peculiarities be so propagated?— Englishmen show no tendency to beardless faces after 4 generations of shaving—& a thousand similar instances may be quoted—the oldest custom of all—circumcision has had no effect on the organ of the race, after hundreds of generations: which I have always regarded as the most wonderful fact.— Again though the habit of inducing varieties in plants & animals has gone on from time immemorial no one has ever supposed that one place is better than another causing the *first brood* to vary— Whoever plants most gets most varieties i.e secures most new sorts by after selection &c &c &c— To put the question in its simplest form—no one ever supposed that of 12 peas in one pod, 6 grown in England would be more different *inter se* or from their parents than the other 6 grown in Australia. Again as I have remarked in the Introd: Essay to New Zealand, the *true*ness with which seeds from all parts of the world come up in our gardens is astounding—[14] Whoever heard of a new species thus raised *by the first sowing*— Were it otherwise—did individual seeds *come up* differently in different localities—you would surely in extreme cases have different species & different genera coming up at once in our gardens— On the other hand—Rivett wheat seed produces Rivetts in Australia the first year; but whereas even Rivett wheat has its varieties & and some of them are more suited to Australia than England, *after several generations* the Rivetts of Australia will differ from those of England, but through Natural Selection.—

In all these considerations we must carefully exclude mere stunting or the effect of overgrowth & undergrowth, which are not variations in the sense I allude to, though I confess to my inability of pointing out any scientific distinction that is irrefragable.

Now all your apparent exceptions are I grant very strong cases at first sight; but may all be explained by assuming more time than you do, & correcting more than you do for the millions of lost individuals. That "species will be constant under one set of conditions & variable under another" is quite true—but this is no proof that even extreme external physical conditions have acted *on the pregnant female* so as to have produced greater differences amongst her ova.—or on the sperm cells of the male previously.—to an extent *perceptible in the first brood*.

I have long on other grounds denied that tropical heat or light *produces* the bright coloring of plants, or that arctic climates *produce* woolly covering.—[15] these phenomena are far too partial to be attributed to such cosmical phenomena. You say "you are convinced that inorganic conditions have some effect"— I am quite ready to believe it when I see any one experimental case, or such an accumulation of arguments *pro*, as I think can be adduced *con*. It is an open question, I grant, quite so.—but you must bring your case to the point. You say "you cannot see that the species has produced the same variations in all the stations" &c—how should you ever see whether or no, except you compare progress of individuals of same brood in each?. Again you ask "if variety A succeed in locality I. why should it not succeed at locality II. if it had ever been existent there"—my answer is that no locality II. is identical with a loc. I. & that Nat. Selection will act on an imperceptible difference—it **searches** where no faculty of man can follow. Consider again & ponder well the number of individuals of a brood that die for every one that lives, multiply these by the countless generations that it must have taken to have established that amount of change wh[ich] we call specific & then reflect that if besides all this we are to have **direct** effects of local conditions, which (*remember*) vary in kind & amount from year to year—& you will have such an accumulation of change effecting forces that there could be no such things as recognisable species & genera. Not only every place would have different varieties but *every year* in *each place*, & in the case of temperate plants grown in a hothouse, wet country plants in a dry house, &c &c &c, we should have startling changes, ever occurring.

Another objection to my line of argument is the changes wrought in bees by either feeding or heat (as the case may be).— but this again is change of individual, & is not propagated, for the Queen after all lays again males & drones, not Queens. Darwin I believe holds with you as to the influence of external conditions on the variation of the brood.[16] I have however failed to be convinced by him of it, & I do not think he recognises the *facts* of variation to the extent I do. Indeed I think his book would have been more convincing had he treated variation somehow so as to have impressed the unaccustomed reader & thinker to regard it as the origin of species, & Nat Selection as the *fixer* of these. As it is, in most minds the two are confused, or Nat. Sel. is supposed to *make* the varieties as well as to fix them.[17] At other times no one more stanchly denounces the effects of external circumstances in producing variation than Darwin does.— Darwin also believes in some reversion to type which is opposed to my view of variation.[18] You may have a single character persist or reappear, but the sum of differences goes on increasing as you depart from the parent.— Variation I hold to be centrifugal;[19] if it were not so how could it go on making species, which are only the preserved forms of each brood which circumstances favoured Remove the circumstances which *kill* the others as man does when he cultivates, (who kills those that nature would have spared), & you have what you call a variety, & fancy *you* made it, whereas you only prevented nature killing it.

After all Darwins axiom that man has never failed in getting varieties of *any species he has* fairly tried,[20] is in favor of my view that the abstract principle called variation is enough *with time* to beget any amount of change; & by means of Nat: Selection to retain only such as may present any amount of difference. Finally I have come to look upon the law of variation as I do on gravitation: local circumstances may mask its effects, but upon *itself* they cannot act. The power that drives the stone to the ground is *the same* whether the stone permeates water or air or vacuum.—whether you let it fall straight or throw it forwards, or even upwards— So, A certain amount of centrifugal force of variation is distributed in certain proportions amongst the 12 peas in the pod; & except to arrest or retard the progress & amount of development of the individuals or their organs (I incline to think) local circumstances are powerless. Give more Nitrogen to Pea No 1. & you will have more & greener leaves; but its seeds again will not be as green as they too are not supplied with nitrogen.

I think my long letter will disgust you with asking any more questions; but I should be greatly obliged if you have time & would write me again on the subject— I know no one but yourself who is really thinking out Darwins views to any purpose in zoology I am sure that with you, as with me, the more you think, the less occasion you will see for anything but time & N.S. to effect change:— & that this view is the simplest & clearest in the present state of science, is one advantage at any rate. Indeed I think that it is in the present state of the enquiry, the legitimate position to take up; it is time enough to bother our heads with a 2d. secondary cause when there is some evidence of it, or some demand for it—at present I do not see one or other, & so I feel inclined to renounce any other *for the present*. It is not so very long since I thought differently & that variation was the first effect of circumstances on the individual.

We are sorry to hear of your Influenza—I hope it is gone now. We are all quite well & the children returned to school—

I hope when you next come to town you will let me know, & that we may have some more boxes of butterflies at the Linnean[21] & get the many curious facts you name well impressed in the languid circulation of the Entomologist.

Ever most truly yrs | Jos D Hooker

DAR 101: 27–9; American Philosophical Society Library (Hooker papers, B/H76.2)

[1] The date is established by the reference to Hooker's weekend stay at Walcot Hall, Shropshire (see n. 3, below).

[2] The reference is to Hooker's letter to Henry Walter Bates, 2 February 1862 (see enclosure), which CD had asked to see (see letter to J. D. Hooker, 14 March [1862]), and probably to Bates's letter to Hooker, 21 March 1862 (printed in Bates 1892, pp. lvii–lx).
[3] Hooker stayed at Walcot Hall, Shropshire, as the guest of Edward James Herbert, third earl of Powis, from 20 to 22 March 1862 (see letter from J. D. Hooker, 17 March 1862).
[4] See letters from J. D. Hooker, [19 January 1862] and [31 January – 8 February 1862], and letter to J. D. Hooker, 25 [and 26] January [1862].
[5] In this context, 'blunt' is used to mean 'ready money' (*OED*).
[6] Letter to J. D. Hooker, 18 March [1862].
[7] Horace Darwin. See letter to J. D. Hooker, 18 March [1862] and n. 12.
[8] Nil desperandum: 'an exhortation to have hope in difficult circumstances and not to despair', deriving from Horace, *Odes*, 1.7.27 (*OED*). Henrietta Emma Darwin had been seriously ill for much of 1860 and 1861 (see *Correspondence* vols. 8 and 9).
[9] Henry Holland. See letter to J. D. Hooker, 18 March [1862] and n. 14.
[10] Hooker planned to spend Easter at Down House and CD had suggested that he bring his eldest child, William Henslow Hooker, with him (see letters to J. D. Hooker, 14 March [1862] and 18 March [1862]).
[11] Hooker refers to his son Charles Paget Hooker.
[12] Bates's letter has not been located; it is not printed in Bates 1892.
[13] Charles Edouard Brown-Séquard reported his successful attempts to transmit experimentally induced epilepsy through several generations of guinea-pigs in Brown-Séquard 1860.
[14] J. D. Hooker 1853, p. x.
[15] Hooker had expressed the view that climate had little direct influence on the form of plants during correspondence with CD, in April and May 1857, about the hairiness of alpine plants (see *Correspondence* vol. 6, letter to J. D. Hooker, [29 April 1857]). However, he subsequently stated that climate was a cause of variability, describing it as 'an active handmaid influencing its mistress most materially' (see *Correspondence* vol. 8, letter to J. D. Hooker, 5 June [1860]). See also this volume, letter from J. D. Hooker, 17 March 1862.
[16] CD mentioned the idea that external conditions might have a greater effect on larval than on adult insect forms in the letter to J. D. Hooker, 1 July [1857] (*Correspondence* vol. 6).
[17] CD and Hooker had corresponded extensively on this point in 1860 (see *Correspondence* vol. 7, letter from J. D. Hooker, [20 December 1859], and *Correspondence* vol. 8, letter from J. D. Hooker, 8 June 1860, and letters to J. D. Hooker, 29 [May 1860], 5 June [1860], and 12 [June 1860]).
[18] CD discussed his ideas on reversion with Hooker during the preparation of his 'big book' on species in 1857 (see, for example, *Correspondence* vol. 6, letter to J. D. Hooker, 5 June [1857]).
[19] Hooker described his concept of centrifugal variation in J. D. Hooker 1844–7, p. 315. See also *Correspondence* vol. 3, letter from J. D. Hooker, [16 April 1846].
[20] CD discussed the effect of artificial selection on domesticated animals and plants in *Origin*, pp. 29–43.
[21] Bates had exhibited specimens of Amazonian butterflies at the Linnean Society of London on 16 January 1862.

From J. D. Hooker [23–5 March 1862][1]

Dr Darwin

Calanthe Masuca. C. *Dominii* is said to be a hybrid between C. *furcata* & C. *Masuca*, & certainly is quite intermediate—[2] see Gard Chron: 1858 p. 4 & Bot. Mag. 5042.[3]

I have no idea that A Gray would quarrell with either of us, under any provocation, & that a good deal of snubbing from us would have done him more good than our sympathy.[4]

Which Vaucher do you ask about, the old Vaucher who published in Geneva & Paris[5] if so he was considered a very accurate acute & able observer— his first work on Conferva appeared in 1800,[6] his later on Devellopment down to 1841.[7] I have never seen the latter's books, & speak from hearsay.

Ever yours affec | J D Hooker.

DAR 101: 30

[1] Dated by the relationship to the letters to J. D. Hooker, 22 [March 1862] and 26 [March 1862].

[2] CD was correcting the proofs of *Orchids* and had asked Hooker to check the names of two species of *Calanthe* (see letter to J. D. Hooker, 22 [March 1862]).

[3] In the *Gardeners' Chronicle and Agricultural Gazette*, 2 January 1858, p. 4, John Lindley reported the raising of a hybrid between *Calanthe furcata* and *C. masuca* in the Exeter nursery of James Veitch Jr by Veitch's foreman, John Dominy. Lindley proposed the name *C. dominii* for the plant. The hybrid was described in *Curtis's Botanical Magazine* 3d ser. 14 (1858): 5042. *C. dominii* was the first known man-made orchid hybrid (R. Desmond 1994). CD's copy of the *Gardeners' Chronicle* is at the Cory Library, Cambridge Botanic Garden.

[4] See letter to J. D. Hooker, 22 [March 1862]. The friendliness of the correspondence between Hooker and Asa Gray had been adversely affected by the *Trent* affair; Hooker had expressed his disappointment with Gray in the letter from J. D. Hooker, [10 March 1862].

[5] See letter to J. D. Hooker, 22 [March 1862]. Hooker refers to the Swiss botanist, Jean Pierre Etienne Vaucher.

[6] Vaucher 1803.

[7] Vaucher 1841.

From Thomas Francis Jamieson 24 March 1862

Ellon Aberdeen[sh].

24 March 1862

My Dear Sir,

I have just lighted upon what I consider an important fact with reference to the parallel roads of Glen Roy & w[h]. may probably interest you.[1]

In the letter I wrote you after my return from Lochaber I think I mentioned that owing to the impossibility of tracing the lowermost line along the shores of Loch Laggan with any degree of certainty I felt somewhat in doubt as to the fact of the outlet at Makoul coinciding with that line—[2] I afterwards wrote to both M[r]. Chambers & M[r]. Milne Home on the subject but failed to elicit anything certain on the subject.[3]

I had however little doubt in my own mind that Loch Laggan had formerly discharged its waters E,ward by that channel from the circumstance of the great old delta of the Gulban which fills the S.W. corner of the lake & rises high above its present level.

—But by the greatest good fortune I find that the ordnance survey had run a line of spirit levelling up Glen Spean on to Dalwhinnie. and from the record of the engineers it appears that the water shed at Makoul just coincides with the lowermost of the Glen Roy lines— Thus, the height of said line found by the spirit levelling of an engineer employed for Mr. R. Chambers is 847 feet above the sea it is not stated whether this refers to high water or mean sea level, but most of these surveyors usually take high water mark. and the summit level of the road at the water-shed of Makoul according to the Ordnance surveyors is 850 feet above the *mean sea level*.

—Now if the sea level taken by Mr. Chambers' surveyor was high water mark it wd. make the Glen Roy line up somewhat over 850 feet— something also would depend upon what part of the line the surveyor took as the line of terrace is generally several feet in vertical height.

Anyhow the coincidence is too close to be accidental—

I am inclined to think however that the level of the outflowing water at Makoul had been at one time fully 860 feet. judging from some measurements wh. I made of the height of the old delta at the S. W. corner of L. Laggan. & also from my remembrance of the beds of waterworn pebbles &c at the Makoul outlet wh. rise some feet I think above the summit level of the road. —Whether this is owing to some inequality in the level of the land that has occurred since the period of the Glen Roy roads—or whether it is owing to some discrepancy in the levellings I cannot say—

But I think it is evident that over a stretch of 20 miles from E. to W. this lowermost of the Glen Roy lines is wonderfully near if not altogether parallel to the present sea level, and I think there can be now very little doubt that it coincides with this outlet & has been determined by it.

If you think Sir C. Lyell wd. be interested in this communication I should feel much obliged if you wd. show it him as it wd. save me writing it over again.[4]

I am | My dear Sir | Your very obed servt | Thos. F. Jamieson

C. Darwin Esq | F. R. S | &c &c &c

Edinburgh University Library (Gen. 112/2834–5)

[1] The so-called 'parallel roads' of Glen Roy are three terraces that run parallel to one another along the sides of Glen Roy in Lochaber, Scotland. In the nineteenth century several attempts were made to account for their formation; in a paper published in 1839 ('Parallel roads of Glen Roy'), CD argued that the roads were the remains of beaches formed by the sea as the landmass of Scotland rose in graduated steps. By contrast, Jamieson believed that during a great 'Ice Age' ice-flows trapped a series of lakes in the glen and that the roads represented the shorelines of three of these former lakes. M. J. S. Rudwick's account of what CD regarded as his 'gigantic blunder' in relation to Glen Roy details the major explanations proposed during the first half of the nineteenth century to provide this celebrated phenomenon with a natural history (Rudwick 1974).

[2] See *Correspondence* vol. 9, letter from T. F. Jamieson, 3 September 1861. Loch Laggan is in Glen Spean, a glen adjacent to Glen Roy. The lowermost of the Glen Roy roads continues out of Glen Roy and around Glen Spean.

[3] Robert Chambers and David Milne-Home.

[4] Jamieson conducted an extensive correspondence with Charles Lyell on the subject of Glen Roy in 1861 (see *Correspondence* vol. 9, Appendix IX).

From Henry Holland 26 March [1862][1]

25 Brook Street
Wednesday | March 26

My dear Charles,

I have some difficulty in answering your letter of yesterday.[2] The circumstances of the case are so far peculiar (especially those you state in this letter) that on present knowledge I can hardly hazard an opinion as to their cause & exact import.

That there is a *cause of physical irritation* present, affecting at times the brain & nervous system, must be deemed certain. The intermittent nature of the attacks, & the freedom from obvious disorder between, may be fairly admitted in proof that the cause is transient in kind; & therefore, as I trust, not depending on any organic disease.[3]

Though I know (& I believe told you) that worms are often a cause of very irregular symptoms in young Children, yet I cannot readily suppose, that at the age of eleven, *Ascarides* should be capable of producing the peculiar effects you have described to me. Still it is important to bring the treatment to act upon this ascertained fact; & to remove what would add to irritation of nerves, if not producing it.

If the Salt injection should not be of avail, one including Castor Oil might be used, to clear the Rectum of these worms—Or other injections still might be suggested, if these fail of effect.

Does he pick or rub his nose or lips frequently? Irritation from Ascarides often shews itself in this way. Even without reference to Ascarides, I should consider that the use of Enemas, small in quantity but somewhat stimulating to the mucous membranes of the Rectum, would tend to relieve the irritation elsewhere.

Without going further at present, I would ask that if you come to Town again, you would let me have the satisfaction of seeing you—to talk further as to the details of the case, upon having of which I should desire to put questions more minutely than can be done or answered by letter.

Ever yours affect[ly] | H Holland

It is important, (as far as may be possible) to ascertain the state of the *Heart action*, during the attacks

DAR 166.2: 241

[1] The year is established by the discussion of Horace Darwin's illness (see n. 2, below).

[2] CD apparently outlined Horace Darwin's symptoms in a letter to Holland that has not been found. Emma Darwin kept a detailed record of Horace's symptoms and treatment in her diary (DAR 242). See also letter to W. E. Darwin, 14 February [1862].

[3] Emma Darwin noticed a connection between Horace's symptoms and his 'great fondness' for the children's governess, Camilla Ludwig. The local doctor, Stephen Paul Engleheart, advised that Horace be separated from her (see the letters from Emma Darwin to William Erasmus Darwin, [27 May 1862] and [19 November 1862] in DAR 219: 57 and 68).

To J. D. Hooker 26 [March 1862][1]

Down Bromley Kent
26th

My dear Hooker.

Thanks about Calanthe: it must be the old Vaucher, so I am answered.—[2] Thanks, also, for the aristocratic note about the four B^s.[3] Also for your own & Bates' letter, now returned.[4] They are both *excellent*; you have, I think said all that can be said against direct effect of conditions & capitally put. But I still stick to my own & Bates' side. Nevertheless I am pleased to attribute little to conditions; & I wish I had done, what you suggest, started on the fundamental principle of variation being an innate principle; & afterwards made a few remarks, showing that hereafter perhaps this principle would be explicable.— Whenever my Book on poultry, pigeons, Ducks & Rabbits is published, with all the measurements & weighings of bones, I think you will see that "use & disuse" at least have some effect.—[5] I do not believe in perfect reversion.—[6]

I rather demur to your doctrine of "centrifugal variation";[7] I suppose you do not agree with, or do not remember my doctrine of the good of diversification; this seems to me amply to account for variation being centrifugal— if you forget it, look at this discussion (p. 117 of 3d edit); it was the last point, which, according to my notions, I made out, & it has always pleased me.—[8] It is really curiously satisfactory to me to see so able a man as Bates (& yourself) believing more fully in nat. selection, than I think I even do myself. By the way I always boast to you, & so I think owen will be wrong that my Book will be forgotten in 10 years,[9] for a French Edit is now going through the press[10] & a 2nd German Edit. wanted.—[11] Your long letter to Bates has set my head working & makes me repent of the nine months spent on Orchids;[12] though I know not why I should not have amused myself on them, as well as slaving on bones of Ducks & pigeons &c. The Orchids have been splendid sport, though at present I am fearfully sick of them.

I enclose a waste copy of woodcut of Mormodes ignea; I wish you had a plant at Kew; for I am sure its wonderful mechanism & structure would amuse you. Is it not curious the way the labellum sits on the top of the column; here insects alight & are beautifully shot, when they touch a certain sensitive point, by the pollinia.—[13] How kindly you have helped me in my work. Farewell my dear old fellow.—

We have been miserably anxious about Horace, but he has been a little better for two days.[14] One day I expected every minute he would go into convulsions or become insane.

Farewell | C. Darwin

DAR 115.2: 147

[1] The date is established by the relationship to the letter to J. D. Hooker, 22 [March 1862] (see n. 2, below).

[2] Jean Pierre Etienne Vaucher. See letter to J. D. Hooker, 22 [March 1862], and letter from J. D. Hooker, [23–5 March 1862].

[3] Letter from J. D. Hooker, [23 March 1862].

[4] CD refers to Hooker's letter to Henry Walter Bates, 2 February 1862 (enclosed with the letter from J. D. Hooker, [23 March 1862]), and probably to the letter from Bates to Hooker, 5 March [1862] (enclosed with the letter from J. D. Hooker, [10 March 1862]).

[5] *Variation* was published in 1868. One of the chapters on the laws of variation included a section entitled 'On the effects of the increased use and disuse of organs' (*Variation* 2: 295–303).

[6] CD discussed the difficulty in establishing the validity of the commonly held view that domestic varieties 'when run wild' would gradually revert to an ancestral form in *Origin*, pp. 14–15, 160–2.

[7] See enclosure to the letter from J. D. Hooker, [23 March 1862] and n. 19.

[8] CD discussed his 'principle of divergence' in *Origin* 3d ed., pp. 117–32. See also *Correspondence* vol. 6, letter to J. D. Hooker, 22 August [1857], and letter to Asa Gray, 5 September [1857]. For discussions of CD's formulation of the principle of divergence, see Browne 1980 and 1983 and Ospovat 1981.

[9] The occasion on which Richard Owen made this remark has not been identified. CD first mentioned Owen's comment in March 1860 (see *Correspondence* vol. 8, letter to J. D. Hooker, 3 March [1860]).

[10] The first edition of Clémence Auguste Royer's French translation of *Origin* (Royer trans. 1862) was published on 31 May 1862 (*Journal Générale de l'Imprimerie et de la Librairie* 2d ser. 6 (pt 3): 341).

[11] See letter from H. G. Bronn, [before 11 March 1862].

[12] CD worked on *Orchids* from July 1861 until the end of April 1862 (see *Correspondence* vol. 9, 'Journal' (Appendix II), and this volume, 'Journal' (Appendix II)).

[13] *Mormodes ignea* is described in *Orchids*, pp. 249–65; a woodcut of the flower is printed on p. 250.

[14] Horace Darwin had been ill since January, but, according to Emma Darwin's diary (DAR 242), showed a slight improvement towards the end of March.

From Charles Lyell [26–31 March 1862][1]

My dear Darwin

I have had what is called the "Shingles" a troublesome though not serious complaint & which seems to come & go no one knows how or wherefore.

I was very glad to hear of you & envy you being so near out. I have printed more than half.[2] Just as I got your letter I was upon Glen Roy & glad to get Jamieson's last but I have got myself into a puzzle which I should be much indebted to you to help me out of as I suspect you will be able to do & make me wonder why I did not see the explanation before—[3]

Assuming ice-blockages, they may alter in height from year to year without that variation affecting the permanency of the level of the shelves because the level of each shelf is determined not by the height of the ice-dam, but by the "col" or parting ridge—

Now I perfectly understand that there may first have been a blockage which caused the lowest of all the shelves common to Glen Roy & Spean, then a separate ice-dam for Glen Roy which made the waters stand higher & escape over a higher "col".

But how in one & the same Glen like Glen Roy could there be two shelves one above the other & coextensive caused by the blocking up by ice of the waters of the Roy? How can it happen that the lower of the two cols does not always prevent the water from rising to the level of the upper one?

Is there not this advantage in the marine theory over that of ice-blockages, that the sea sinking leaves the uppermost & thus the 2^{d}. & then the 3^{d} shelves & does not meddle with them any more, whereas in the glacier lake hypothesis the lowest shelf must be made first? & thus the water must rise without injuring the beach first made & afterwards sink again from the uppermost level & not damage the inferior ones as it goes down?

But this I could grant if my other difficulty were removed— If it be said that the two upper shelves of Glen Roy were caused the highest by a "col" & the 2^{d}. by the ice-dam which was lowered I should have a difficulty in supposing a long permanent level due only to ice—[4]

Please return this note as if I have to write to Jamieson after getting your answer it may save me trouble as I shall be then on some quite different subject.

I have heard from your brother of your child's illness & was very sorry for it—[5]

believe me | ever affect^{y} yrs | Cha Lyell

American Philosophical Society (274)

[1] Dated by the relationship to the letter from T. F. Jamieson, 24 March 1862 (see n. 3, below), and to the letter to Charles Lyell, 1 April [1862].

[2] CD completed the manuscript of *Orchids* in February 1862, and thereafter spent some weeks correcting proofs (see letter to John Murray, 9 [February 1862], and letter to H. W. Bates, 27 February [1862]). *Orchids* was published on 15 May 1862 (Freeman 1977, p. 112). Lyell's *Antiquity of man* (C. Lyell 1863a) was published on 6 February 1863 (C. Lyell 1863b, p. vii).

[3] CD's letter to Lyell has not been found; with it was apparently forwarded the letter from T. F. Jamieson, 24 March 1862. Lyell refers to the so-called 'parallel roads' of Glen Roy, a series of parallel terraces running along the sides of a glen in Lochaber, Scotland. Thomas Francis Jamieson had visited the area in August 1861, since which time, CD, Jamieson, and Lyell had pursued a three-way correspondence discussing the various theories about the formation of the roads (see *Correspondence* vol. 9). Lyell discussed Glen Roy in C. Lyell 1863a, pp. 252–64.

[4] There were a variety of competing explanations of the origin of the parallel roads of Glen Roy. The first published geological accounts postulated that the roads were the shorelines of a former lake impounded by debris. In a paper on the phenomena, published in 1839 ('Parallel roads of Glen Roy'), CD argued that the roads were marine beaches formed during intervals in the rise of the landmass. Jamieson rejected CD's marine theory and adopted the lake hypothesis, though invoking glaciers as the containing barriers rather than debris. For a detailed discussion of the various interpretations of Glen Roy, see Rudwick 1974.

[5] Lyell refers to CD's brother, Erasmus Alvey Darwin, and to CD's youngest child, Horace Darwin, who had been ill since the beginning of the year (see letter to W. E. Darwin, 14 February [1862]).

From J. D. Hooker [after 26 March 1862?][1]

the old type as away from it— this is conceivable of course, but I maintain it is not so in nature— it is opposed to my idea of centrifugal variation.—[2] Now variation is not centrifugal because of any repulsive or repellent power, but simply because in fact it is a million chances to one that identity of form once lost will be returned to. Each character we estimate is a compound of lesser characters each too minute to estimate in the gross, but missed if with-drawn. A petal if further varying after it has once varied has a thousand characters of form color consistence nervation &c &c &c to chuse amongst (so to speak) besides those its grandparents bore. After all, experience is our best guide, & we do not find in the human race any reversion so strong as would lead us to confound a man with his ancestor, a Yankee with an Englishman— Whoever saw a Grandfathers portrait that would really pass for his Grandsons—& if we do not find reversion amongst individuals so close in kin & in time, how can we expect them in organisms that have reached the specific term of divergent development—

True you may have a 10th removed lineal descendent (who supposing the name of Wedgwood to be lost to genealogists—except by a stray portrait) may so resemble the Wedgwood portrait that an Ethnologist would call him an anomalous Darwin representing the lost Wedgwood type.—but the chances are a hundred to one in favor of both Wedgwood & Darwin physiognomy being wiped out long before the 10th. generation—[3] How often do we consider 2 people strikingly alike till we see them together & then consider them wholly unlike; this is because **one** minute character common to both was **alone** carried away, in the mind, but which, though no doubt present, is not perceived when *again* sought for amongst the thousand other characters of the 2 faces.

Incomplete
DAR 47: 214

CD ANNOTATIONS
1.1 the old ... with-drawn. 1.6] *crossed brown crayon*
1.9 & we ... Grandsons— 2.12] *scored brown crayon*
1.12 & if ... faces. 2.10] *crossed brown crayon*

1 The date is conjectured from the discussion of 'centrifugal variation', which relates to the letter to J. D. Hooker, 26 [March 1862].

2 See letter from J. D. Hooker, [23 March 1862] and n. 19, and letter to J. D. Hooker, 26 [March 1862].

3 Hooker may have been aware of the Wedgwood family joke that the Darwins were 'more Wedgwood than the Wedgwoods', since CD was the son of Susannah Wedgwood, and had married his cousin, Emma Wedgwood (Wedgwood and Wedgwood 1980, p. 269).

From H. G. Bronn[1] 27 March 1862

Heidelberg
den 27 März 1862.

Dear Sir!

Ich danke Ihnen verbindlich für die gütige Aufnahme und wohlwollende Beantwortung meines letzten Briefes,[2] worin Sie mir schreiben, dass Sie auch die neue Übersetzung Ihres "Origin of Species" mit Beiträgen unterstützen wollen, wenn Ihnen dafür einige Zeit gewährt würde. Diess kann ganz wohl geschehen; denn auch ich selbst kann vor dem 1. May nicht an die Arbeit kommen,—welche übrigens nicht schwer seÿn würde und vielleicht nicht viele Zeit in Anspruch nehmen dürfte, wenn ich erst die Stellen kennen werde, die in der Original-Schrift seit der ersten Übersetzung geändert worden sind.

Ich schreibe Ihnen jetzt hauptsächlich in der Absicht, um Ihnen zu melden, dass Herr Schweizerbart auch bereit seÿn würde, Ihre Arbeit über die Befruchtung der Orchideen zu übersetzen[3] und daher mit Dank es annehmen würde wenn Sie Ihm die Abdrücke der dazu gehörigen Abbildungen, oder die Holz-Blöcke mit den Holzschnitten, oder die *Clichés* davon zu verschaffen die Güte haben würden. Mit Vergnügen habe ich auch aus Ihrem Briefe ersehen, dass Sie an Ihrem grösseren Werke noch immer fortarbeiten.[4]

Indem ich bitte die Versicherung meiner aufrichtigsten Hochschätzung zu genehmigen habe ich die Ehre zu verbleiben | Ihr | ergebenster Diener | H. G. Bronn

DAR 160.3: 320

[1] For a translation of this letter, see Appendix I.
[2] See letter from H. G. Bronn, [before 11 March 1862], and letter to H. G. Bronn, 11 March [1862].
[3] In the letter to H. G. Bronn, 11 March [1862], CD had asked whether Christian Friedrich Schweizerbart, head of the publishing firm E. Schweizerbart'sche Verlagsbuchhandlung, might be interested in publishing a translation of *Orchids*.
[4] See letter to H. G. Bronn, 11 March [1862] and n. 5.

To George Bentham 30 March [1862][1]

Down. | Bromley. | Kent. S.E.
March 30th

Dear Bentham

Many thanks for your note.[2]

I will certainly come to London & be at Linn. Soc. if possible, but I have been just lately rather extra headachy; but I hope I shall not fail.[3]

I fear that my paper will by no means be worth Lindley's[4] attendance.

Yours very sincerely | C. Darwin

Royal Botanic Gardens, Kew (Bentham letters: 699)

[1] The year is established by CD's reference to reading a paper before the Linnean Society of London (see n. 3, below). CD used the form of address printed on this stationery from 1861 to 1869 (see Carroll ed. 1976, pp. xxii–xxiii); during this time, the only paper by CD read before the Linnean Society in the months March, April, or May, was read in 1862.
[2] The letter from Bentham has not been found.
[3] CD read his paper, 'Three sexual forms of *Catasetum tridentatum*', before the Linnean Society on 3 April 1862. According to Emma Darwin's diary (DAR 242), he travelled to London on 2 April and returned home on 4 April.
[4] John Lindley.

From Asa Gray 31 March [1862]

Cambridge. [Massachusetts]
⟨M⟩arch 31. | ⟨1862⟩

My D⟨ear Darwin⟩

⟨Yours of⟩ the 15 came this evening.[1] ⟨To⟩-morrow I am busy all day in College (where I began my course this year with lectures on Fertilization, developing your views on Orchid-Insect. fertilization, dimorphism, &c,—&c—to an interested class!)—[2]so I must drop a line for you into a letter for Boott, for Wednesday's post.[3]

A friend has just handed me Morell's new book, which—looking at psychology from the physiological side I see brings up several notions which have been turning ov⟨er⟩ in my mind for some years⟨.⟩[4] He is coming out a good D⟨ar⟩winian, I see, and is qu⟨ite of⟩ my way of thinking ab⟨out⟩ design. You see ⟨I am⟩ determined to *baptize* ⟨ ⟩ng nolen⟨s vo⟩lens, which ⟨will⟩ be its ⟨salvation⟩ But if you won't ⟨have it⟩ done it will be *damn⟨ed⟩*, I fear.

A few days ago I had a letter from Trübner, informing me that he had ordered his agent in Boston to pay me £8. from the sale of 200 copies of my pamphlet![5]

Now this should have been paid to you, and it shall be presently, when I shall have funds on your side of the water. For, you are to remember that I took about £50— I cannot find the mem. ⟨ ⟩ the exact amount now—which ⟨Ap⟩pleton & Co sent me for ⟨your⟩ book, to pay the printers with, as far as it would ⟨ ⟩[6]

How much Ticknor ⟨ha⟩s sold here, I know not.[7] ⟨ ⟩ not much, and that ⟨ ⟩ no pains. I have asked for an account,—thus far in vain. When I get it I will communicate with you. Doubtless I could send you lots of copies to give away, if you like, and any body now cares for such matters.

Things move on here—on the whole very well.

Yes, I will promise not to *hate* you:—quite the contrary![8] Our sensitiveness as to England was the natural result of the strong *filial* feeling on our part. It was very undignified I dare say, But I think we are getting bravely over it, and getting really not to care what the Old country may think or say,—so it lets us alone.

As to rebeldom, there is now hardly any State that we have not got some foothold in.

I do not do so much scientific work as before the war,—but still I keep pottering away. From now till July, I can expect to do little besides my College duties.—

Ever, dear Darwin | Your cordial friend | and true Yankee | A. Gray

Postmark: AP. 16 | 62
DAR 165: 108

CD ANNOTATION
2.1 Morell's] *underl brown crayon*

[1] Letter to Asa Gray, 15 March [1862].

[2] Asa Gray was Fisher Professor of natural history at Harvard University and lectured at the Lawrence Scientific School (Dupree 1959).

[3] Francis Boott.

[4] Morell 1862. See also letter from Asa Gray, [late June 1862].
[5] Nicholas Trübner. See letter to Asa Gray, 15 March [1862] and n. 3.
[6] Gray had arranged for the New York publishing firm D. Appleton & Co. to publish a revised American edition of *Origin*, for which CD was to receive £50 royalties (see *Correspondence* vol. 8, letter from Asa Gray, 20 February 1860, and letter to Asa Gray, 8 March [1860]). Gray had used a portion of this money to pay for the printing costs of A. Gray 1861 (see n. 5, above, and *Correspondence* vol. 9, letter to Asa Gray, 11 April [1861]).
[7] Gray refers to the Boston firm Ticknor and Fields, the publishers of A. Gray 1861.
[8] See letter to Asa Gray, 15 March [1862].

From Erasmus Alvey Darwin [April – May? 1862][1]

Dear Charles

Will you send me an order to buy your photograph at Polyblank. for Mr Tait.[2]

Polyblank says that for some he has a general order to sell & for others he requires special permission so I should think you might as well give a general order as it is a good photograph.

I hav'nt seen any of your world but I dare say I shall at the great Ex.[3]

ever yours | E. D

DAR 105 (ser. 2): 9–10

[1] The date range is conjectured from the reference to the International Exhibition (see n. 3, below).
[2] The reference may be to the painter Robert Tait, who was Erasmus's near neighbour in Queen Anne Street, Cavendish Square, London (*Post Office London directory* 1861). Tait was associated with Erasmus's circle of friends, and had a keen interest in photography (see Bliss ed. 1953, pp. 322–3). George Henry Polyblank, in partnership with Henry Maull, was the proprietor of one of the leading photographic studios in London, Maull & Polyblank, located at 187a Piccadilly and 55 Gracechurch Street (Pritchard 1986). The photograph referred to may be that taken by Maull & Polyblank *circa* 1857 (see frontispiece to *Correspondence* vol. 8; the copy reproduced there is taken from a print prepared by Maull & Fox from the original negative).
[3] The 1862 International Exhibition opened in South Kensington, London, on 1 May (*The Times*, 2 May 1862, p. 11).

From Henry Holland [*c.* April 1862][1]

Brook Street
Monday

My dear Charles

I write a few lines to thank you for your letter of this mg. & to express my satisfaction in the account you give me of your little boy. Pray let me hear of his progress from time to time[2]

Pasteur's Memoir, to which you allude, is a very able & convincing one—[3] He completely pushes Pouchet from the field.[4] I am rather surprized that he does not refer (or, I think, does not) to Ehrenberg's papers in the Transs. of the Berlin Academy 10 or 12 years ago; cataloguing some 250 to 300 forms of organic life, which he found *at different altitudes* in the atmosphere above Berlin.[5] If I recollect rightly, he collected these matters by processes analogous in kind to those employed by Pasteur. He sent me his Memoirs at the time, but I scarcely know what has become of them; as I keep very few papers

Ever yours very sincerely | H Holland

DAR 166.2: 237

[1] The date is conjectured on the basis of the references to Pasteur 1861 and to Horace Darwin's health (see nn. 2–4, below).

[2] CD's letter has not been found. Horace Darwin had been ill since the beginning of 1862, but showed signs of improvement in April (see Emma Darwin's diary (DAR 242)). CD wrote to Holland, with a description of Horace's symptoms, on 25 March 1862 (see letter from Henry Holland, 26 March [1862]).

[3] Pasteur 1861.

[4] In 1858, Félix Archimède Pouchet revived the French debate regarding spontaneous generation with a paper in which he described the appearance of micro-organisms in boiled hay infusions, kept under mercury, after the introduction of artificially produced oxygen (Pouchet 1858). Pouchet expanded his claims in a major work on the subject (Pouchet 1859), but his conclusions were opposed by Louis Pasteur in a series of five notes presented to the Académie des Sciences Naturelles, and subsequently brought together in his prize-winning essay, 'Mémoir sur les corpuscules organisés qui existent dans l'atmosphère' (Pasteur 1861). Pasteur argued that contaminated mercury was the source of error in Pouchet's work (Pasteur 1861, p. 79). For an account of the French debate about spontaneous generation, see Farley 1977.

[5] Holland probably refers to Ehrenberg 1848 and 1849.

From George Busk 1 April 1862

15 Harley St
April 1 1862—

My dear Darwin

There seems to be more difficulty in the matter of the Army returns than we anticipated, from the enclosed letter from Dr Parkes you will see that the usual official nonsense or obstructiveness has got into the Director General's head.[1] Parkes however is so zealous & efficient a man, that I have little doubt, the matter with his assistance may yet be achieved. I have written to him to say that you will communicate with him on the subject—[2]

Ever Yours very truly | Geo Busk

[Enclosure]

Frindsbury | Rochester
30 March 1862.

My dear Busk

I brought M^r^. Darwins communication under the notice of the Director General of the Army Medical Department, yesterday.

The Director General wished me to say that while he is always anxious to do what he can to promote any scientific object, he has yet to consider that the Army

Surgeons have their own duties to perform, and that additional labour should not be thrown on them without consideration. After looking over M^r. Darwins memorandum he was of opinion that a very careful & probably elaborate investigation would alone give results of any value. Before therefore taking the subject into further consideration, or pledging himself to any course, he would wish to know the exact form, in which the information will have to be put, and if M^r. Darwin would not mind the trouble of preparing some definite scheme, the Director General will then consider whether or not he can carry it out.

It has occurred to me that if M^r. Darwin's wishes could be made known in a non-official way, there would be some Army Surgeons who would be glad to assist in the collection of these facts, and that this voluntary aid would be more likely to be well don⟨e⟩ than if compulsory returns were called for, which might be hastily and incorrectly filled up.

However on this point I can talk to the Director General when we see the exact form the investigation should assume.

Believe me | Very sincerely yours | E. A. Parkes.

George Busk Eqre

DAR 160.3: 377, DAR 174.1: 22

[1] See enclosure. Busk refers to Edmund Alexander Parkes, professor of hygiene at the Army Medical School, Chatham, and James Brown Gibson, director-general of the Army Medical Department. CD was seeking information from colonial army medical staff to help him establish whether there was a relationship between hair colour and disease susceptibility among British soldiers serving in tropical climates. In *Descent* 1: 244 n. 48, CD wrote:

> In the spring of 1862 I obtained permission from the Director-General of the Medical department of the Army, to transmit to the surgeons of the various regiments on foreign service a blank table, with the following appended remarks, but I have received no returns.

No copy of the printed document survives (Freeman 1977, pp. 111–12), but see the letter from E. A. Parkes, 8 April 1862, n. 1, for a transcription of the 'appended remarks' that accompanied the table. The correlation between complexion and resistance to disease was of long-standing interest to CD, who had, over a number of years, written 'several letters to various parts of the world' seeking information on the subject (see *Correspondence* vol. 7, letter to Asa Gray, 18 November [1858]). See also *Correspondence* vol. 6, letter from W. F. Daniell, 8 October – 7 November 1856, and *Correspondence* vol. 8, letter to Jeffries Wyman, 3 October [1860]. The material CD was able to collect on this point was used in *Descent* in a chapter on human races (*Descent* 1: 214–50).

[2] See letter from E. A. Parkes, 8 April 1862.

To Charles Lyell 1 April [1862][1]

Down

April 1st

My dear Lyell.

I am not quite sure that I understand your difficulty, so I must give what seems to me the explanation of the glacial-lake-theory at some little length.[2] You know that there is a rocky outlet at the level of *all* the shelves.— Please look at my map:[3]

I suppose whole valley of Glen Spean filled with ice; then water would escape from outlet at Loch Spey & the highest shelf would be **first** formed. Secondly ice began to retreat, & water would flow for short time over its surface; but as soon as it retreated from behind hill marked Craig Dhu, where the outlet on level of 2^d^ shelf was discovered by Milne, the water would flow from it, & the *second* shelf would be formed.[4] This supposes that a vast barrier of ice still remains, under Ben Nevis, along all the lower part of the Spean. Lastly I suppose the ice disappeared, everywhere along L. Laggan, L Treig & Glen Spean, except close under Ben Nevis, where it still formed a Barrier, the water flowing out at level of *lowest* Shelf, by the pass of Muckul at head of L. Laggan.— This seems to me to account for everything. It presupposes that the shelves were formed towards close of Glacial period.—

I come up to London to read on Thursday short paper at Linn. Soc.^y^[5] Shall I call on Friday morning at $9\frac{1}{2}$ & sit half an hour with you? **Pray** have no scruple to send a line to Q. Anne St to say "no"—if it will take anything out of you.—[6] If I do *not* hear, I will come.—

Yours affect | C. Darwin

American Philosophical Society (275)

[1] The year is established by the reference to CD's reading a paper at the Linnean Society (see n. 5, below).
[2] See letter from Charles Lyell, [26–31 March 1862].
[3] CD refers to the map accompanying his paper, 'Parallel roads of Glen Roy'.
[4] Milne 1849, p. 398.
[5] CD read his paper, 'Three sexual forms of *Catasetum tridentatum*', before the Linnean Society of London on 3 April 1862 (*Journal of the Proceedings of the Linnean Society of London (Botany)* 6: lxiv). According to Emma Darwin's diary (DAR 242), CD left for London on Wednesday 2 April and returned home on Friday 4 April 1862.
[6] CD's brother Erasmus Alvey Darwin lived at 6 Queen Anne Street, London.

From Edward Newman 6 April 1862

York Grove | Peckham
6 April 1862

My dear Sir

You expressed an interest in the existence of dimorphism among insects;[1] I have several illustrations of this, and should you intend being in London on the evenings of either the Linnean or Entomological meetings I shall have great pleasure in taking them for exhibition and will read a few very brief notes thereon with the view of eliciting observations of more value from yourself.[2]

Should this plan be inconvenient to you, I will leave the specimens with Mr Kippist or elsewhere for your examination.[3]

I may just observe that in the families of ants which belong to the old genera Formica and Atta, the two forms are equally abundant a fact that has obtained the notice of all entomologists. What may be called a "guess solution" of this phenomenon has been attempted by an untenable hypothesis namely that the larger workers are undeveloped females, the smaller undeveloped males but this is opposed to fact, both sets or forms of workers are aculeates and aculeates are invariably female: in Atta Mr Smith tells me the sting is always present,[4] in Formica the poison is present but not the sting; the skin or flesh of the object attacked is perforated by the mandibles and the poison injected into the wound no male insect possesses poison: the poison is called formic acid

Believe me dear Sir | *[illeg]* faithfully yours | Edward Newman

Charles Darwin Esq

Position of the common wood ant
when it attacks on object

DAR 172.2: 38

[1] CD may have spoken with Newman at the meeting of the Linnean Society of London on 3 April 1862.

[2] According to Emma Darwin's diary (DAR 242), CD had recently returned home from a trip to London, and did not visit again until 6 May, so did not attend the next meetings of either the Linnean Society or the Entomological Society.

[3] Richard Kippist was librarian and secretary to the Linnean Society.

[4] Frederick Smith, an entomologist with the British Museum, specialised in the Hymenoptera.

From J. D. Hooker [7 April 1862][1]

Royal Gardens Kew

Dr. Darwin

I shall I hope be able to send you Vanilla Flowers in a day or two.—[2] How are you after your tremendous effect on the placid Linnæans?[3]

Ever yrs affec | J D Hooker

Kew Monday.

DAR 101: 32

[1] Dated by the reference to CD's having recently read a paper at a meeting of the Linnean Society (see n. 3, below), and by the relationship to the letter to J. D. Hooker, 9 [April 1862]. The first Monday following the meeting in question was 7 April 1862.

[2] CD had been anxious to examine specimens of the orchid tribe Arethuseae, to which the genus *Vanilla* belongs, before completing *Orchids*, but had been unable to procure any living flowers (see *Correspondence* vol. 9, letters to J. D. Hooker, 27 July [1861] and 13 [August 1861], and this volume, letter to J. D. Hooker, 9 [April 1862]). See also *Orchids*, p. 269.

[3] CD read his paper, 'Three sexual forms of *Catasetum tridentatum*', before the Linnean Society of London on 3 April 1862 (see letter to Charles Lyell, 1 April [1862] and n. 5).

From Alfred Russel Wallace 7 April 1862

5 Westbourne Grove Terrace. W
April 7th. 1862

My dear Mr. Darwin

I was much pleased to receive your note this morning.[1] I have not yet begun work but hope to be soon busy.[2] As I am being doctored a little I do not think I shall be able to accept your kind invitation at present but trust to be able to do so during the summer.

I beg you to accept a wild honeycomb from the island of Timor, not quite perfect but the best I could get. It is of a small size but of characteristic form & I think will be interesting to you. I was quite unable to get the honey out of it, so fear you will find it somewhat in a mess but no doubt you will know how to clean it.

I have told Stevens to send it to you.[3]

Hoping your health is now quite restored & with best wishes | I remain | My dear Mr Darwin | Yours very sincerely | Alfred R. Wallace

C. Darwin Esq.

DAR 106/7 (ser. 2): 1

[1] CD's letter has not been found.

[2] Wallace had recently arrived in England, having spent eight years in the Malay Archipelago (Wallace 1905, 1: 385).

[3] Samuel Stevens was Wallace's London agent, responsible for handling the natural history specimens that Wallace collected. In a drawer of the rent table in CD's study at Down House is a pill box marked: 'Bees: Timor Wallace, of which I have comb.—'

From Edmund Alexander Parkes 8 April 1862

Frindsbury | Rochester
8 April 1862.

My dear Sir

I think if the Memorandum, which you sent to M^r. Busk & which I returned to him were attached to the form or table to be filled up it would be an advantage, as explaining the object of the enquiry, & leading men to take an interest in it.[1]

Do you not think it might be as well in the first instance to restrict the enquiry to Malarious & Yellow Fevers and to Dysentery. These are definite & easily recognized Complaints; if "liver diseases" are added there would be more difficulty about diagnosis.

A tabular form would perhaps be the Easiest to fill up. I enclose one for your consideration.[2] The object of having columns for 1st. & 2nd or subsequent attacks is to prevent the fallacy of one man going into Hosp^l. many times with the same disease & so giving an apparent preponderance of attacks to a particular colour.

Did you ever happen to read a little paper by D^r. John Beddoe on the colour of the hair & eyes of the Scotch & of some of the Continental nations?[3] He seems to think the colour of the Iris less trustworthy than that of the hair.

But it would be easy to introduce columns for the colour of the Irides if you thought it desirable.

If you thought some table of the kind I enclose would do, I would show it to the Director General[4] & if he approves he might send it out to Calcutta, Bombay, Madras & the West Indies with a request that the attention of the Army Surgeons be drawn to the subject, but without making the return compulsory. There are sure to be some who will take with interest to the work if they know what it means.

If this course be adopted it might be well to lithograph or print a few of your memoranda and tables to be sent out.

Incomplete
DAR 174.1: 23

[1] See letter from George Busk, 1 April 1862 and n. 1. The memorandum has not been found; however, the text of the memorandum 'explaining the object of the enquiry' is given in *Descent* 1: 244–5 n. 48:

> As several well-marked cases have been recorded with our domestic animals of a relation between the colour of the dermal appendages and the constitution; and it being notorious that there is some limited degree of relation between the colour of the races of man and the climate inhabited by them; the following investigation seems worth consideration. Namely, whether there is any relation in Europeans between the colour of their hair, and their liability to the diseases of tropical countries. If the surgeons of the several regiments, when stationed in unhealthy tropical districts, would be so good as first to count, as a standard of comparison, how many men, in the force whence the sick are drawn, have dark and light-coloured hair, and hair of intermediate or doubtful tints; and if a similar account were kept by the same medical gentlemen, of all the men who suffered from malarious and yellow fevers, or from dysentery, it would soon be apparent, after some thousand cases had been tabulated, whether there exists any relation between the colour of the hair and constitutional liability to tropical diseases. Perhaps no such relation would be discovered, but the investigation is well worth making. In case any positive result were obtained, it might be of some practical use in selecting men for any particular service. Theoretically the result would be of high interest, as indicating one means by which a race of men inhabiting from a remote period an unhealthy tropical climate, might have become dark-coloured by the better preservation of dark-haired or dark-complexioned individuals during a long succession of generations.

[2] The enclosure has not been found.

[3] Beddoe 1862. There is a heavily annotated copy of this article in the Darwin Pamphlet Collection–CUL.

[4] The reference is to James Brown Gibson, director-general of the Army Medical Department.

To J. D. Hooker 9 [April 1862][1]

Down
9th

My dear Hooker

One line to say that I shall *very* much like to examine Vanilla (give specific name if you can), but it will be too late for my Book.[2] I am told that Vanilla flowers drop in 3 or 4 hours, which is a queer thing for an orchid.— The menyanthes which you sent me is going to flower in my greenhouse soon![3] If you can remember, when your plants bloom, just look whether you have both the long- & short-styled form.— Also look some day whether you have any Saxifrages with long hairs **glandular** at the tip.—[4]

I by no means thought that I produced a "tremendous effect" on Linn. Soc;[5] but by Jove the Linn. Soc. produced a tremendous effect on me for I vomited all night & could not get out of bed till late next evening, so that I just crawled home.— I fear I must give up trying to read any paper or speak. It is a horrid bore I can do nothing like other people.—

My dear old friend | Ever yours | C. D.

DAR 115.2: 148

[1] The date is established by the references to the publication of *Orchids* and to CD's paper on *Catasetum* (see nn. 2 and 5, below).

[2] See letter from J. D. Hooker, [7 April 1862]. *Orchids* was about to be printed (see following letter). CD recorded finishing work on the volume on 28 April (see 'Journal' (Appendix II)).

[3] CD had recently learned that *Menyanthes* was dimorphic, and had sought specimens from a number of correspondents (see letter to C. C. Babington, 20 January [1862], and letter from C. W. Crocker, 13 March 1862); he probably included the genus on the list of specimens required for experiment enclosed with the letter to J. D. Hooker, 25 February [1862]. In DAR 110 (ser. 2): 52 there is a note dated '1862' that states: 'Menyanthes from Kew.— Short-styled 〈 〉 set its own pods in plenty.— but none germinated'. This observation is also recorded in *Forms of flowers*, p. 115.

[4] CD had observed that minute flies often became caught in the hairs of *Saxifraga umbrosa*, and while he believed that this species derived no nutriment from the flies, he wished to know 'whether some of the more hairy Saxifrages may not profit by caught flies' (see *Correspondence* vol. 9, letter to J. D. Hooker, 30 May [1861]). Hooker sent CD specimens of the genus in March 1862, with the promise that he would send others if they were not suitable (see letter from J. D. Hooker, 17 March 1862).

[5] See letter from J. D. Hooker, [7 April 1862]. CD read his paper, 'Three sexual forms of *Catasetum tridentatum*', before the Linnean Society of London on 3 April 1862.

To John Murray 9 April [1862]

Down. | *Bromley.* | *Kent. S.E.*
April 9th

My dear Sir

The enclosed slip has apparently been sent to me by you by mistake.[1]

Thanks for your note.[2] You are bold man to print 1500 copies, & I hope to Heaven you will not repent it.—[3] I shall want about 50 copies for myself.—[4]

Yours very sincerely | C. Darwin

Endorsement: '*1862—Ap. 9*'
John Murray Archive (Darwin 116–17)

[1] The enclosure has not been found.

[2] Murray's letter has not been found.

[3] *Orchids* was offered for sale on 15 May 1862 (Freeman 1977, p. 112). In the letter to John Murray, 21 September [1861] (*Correspondence* vol. 9), CD warned Murray that *Orchids* might have limited appeal.
[4] For CD's presentation list for *Orchids*, see Appendix IV.

To Andrew Murray 10 April [1862][1]

Down. | Bromley. | Kent. S.E.
April. 10th

My dear Sir

Pray excuse me for writing to you on a wild-goose chase. About a year ago I received from author a pamphlet "Scudder on blind cave-insects of N. America".[2] I much want it, & resember lending it to some one. I have *no reason* to suppose that I lent it to you, excepting that you are the most likely man, for caring about the subject.—[3] Will reflect for a moment & if by any extraordinary piece of good fortune, you have it return it to me.—

If I do **not** hear, I will understand that I did not lend it to you.—

Pray excuse me & believe me yours very sincerely | C. Darwin

R. D. Pyrah

[1] The year is established by the relationship to the letter from Andrew Murray, 12 April 1862.
[2] Scudder 1861. There is an annotated copy of this article in the Darwin Pamphlet Collection–CUL.
[3] CD and Murray had corresponded at length about whether CD's theory of natural selection could adequately account for the existence of blind cave insects (see *Correspondence* vol. 8).

From Daniel Oliver 10 April 1862

Richmond, SW.
10. IV. 1862

My dear Sir.

It is very atrocious of me to trouble you espec^{ly}. that I gather from D^{r}. Hooker that he thinks you may be unwell.[1] But you must not on any acct. reply to me until it is *perfectly convenient*. Surely you were observing *Fumariaceæ* some time ago.[2] Now I wonder however if (the flower of Fumar^{æ}) can be fertilized by foreign pollen. By other than the pollen of its own stamens. Many other plants might be named in same cate⟨gory⟩

⟨ ⟩ common one & I saw a few days ago that the insects had bored a hole at x thus getting at the nectar without disturbing the anthers which instead fit very closely under the stigma x x x

They wished me to notice your case of *Primula* dimorphism in Nat. Hist. Review & I thought of giving a *short* acct. of some cases of the phenomenon so far as we know anything abt. it,—as in Viola, Oxalis, &c.—[3] And the case of such plants as the *Corydalis*, &c. occurred to me apropos of the thing.— I was showing Dr. Hooker & Mr. Lubbock the 2 forms of Primrose the other day which confirmed the admirable accuracy of yr. obs. upon them[4] Besides the differences in pollen, stigma, &c.—you might notice (perhaps indeed you say so, but I do not recollect it) that the ovules of the *long-styled* are considerably larger (& probly. less numerous) *in the bud* than in the *short-styled* plant even when expanded.—[5]

I have made within the last few days a rather interesting obs. on dimorphous *Campanula*. I wished to allude to the case of the order (mentioned by Hooker & Thomson. Journ. Linn Soc. ii. p. 7[6] & previously by Linnæus in Praelectiones (of C. *perfoliata*))[7]—& examined some of the smaller flowers. I find them thus,—

No Corolla, no
apparent stamens
No—style or stigma,—
but with ovary &
numerous ovules.

x is a hairy mammilla sticking slightly up in middle of disk Dissecting further I found the disc membrane *double* in the middle the interspace enclosing a stigma-like body:—thus—

In the minute flowers,—(size of coriandras to pea), it was at first difficult to understand, but I now find that the stigma-like body enclosed by *imperforate membrane*—is both stigma & anthers. Looking down upon it after dissecting away upper membrane I find it, in the large flowers 5-lobed—with slender rays from the tips of the lobes to base of each calyx-lobe thus—

the ends of the rays are attached.— Further examining & comparison shows satisfly. that there are stamens closely connate over a stigmatic body. In flowers of larger size,—approaching normal conditions—I find the stigma raised in short style (it is rather *long* normally) & the membrane (*which represents the corolla*) ruptured—or rather not closed over it.— Now I have not means at hand to ascertain certainly just now but *do not doubt* but that the *small* flowers do *perfect* seed, tho' the 'flowering' takes place (one w$^{d.}$ think) where 'crossing' was out of the question. I thought of giving a little drawing or two of it to Linn: Soc. as I can make time before my lectures,—calling attention to it.—[8]

By the way I think I c$^{d.}$ get you some *Primula farinosa* this season if it w$^{d.}$ be useful to you.— If you w$^{d.}$ care *in the least* for it,—if you'd kindly spare me a duplicate of your *Primula* paper I w$^{d.}$ lend it to a person (whom I do *not* know personally) living very near *heaps* of the plant, & dont doubt he w$^{d.}$ try to send a box.—[9]

Yours very Sincerely | D$^{l.}$ Oliver

DAR 173.1: 13

CD ANNOTATIONS
1.4 By ... cate⟨gory⟩ 1.6] *cross in margin, brown crayon*
1.7 ⟨ ⟩ ... stigma 1.9] *cross in margin, brown crayon*
2.3 as in ... thing.— 2.4] *cross in margin, brown crayon*
2.8 the ovules ... expanded.— 2.9] *cross in margin, brown crayon*
2.8 *long-styled*] 'short-styled'[10] *added brown crayon*
4.11 but *do* ... question. 4.12] *cross in margin, brown crayon*
5.1 By ... you.— 5.2] *cross in margin, brown crayon*
Top of letter: 'Your Paper | F. Water Production'[11] *pencil*; 'Diagrams' *ink*

[1] CD had been unwell following his recent trip to London to read a paper before the Linnean Society (see letter to J. D. Hooker, 9 [April 1862] and n. 5).

[2] CD and Oliver had corresponded the previous year about the pollination mechanisms of Fumariaceae (see *Correspondence* vol. 9, letter to Daniel Oliver, 9 April [1861]).

[3] [Oliver] 1862c. CD made several annotations on the review in his copy of this issue of the *Natural History Review*, which is preserved in the Darwin Library–CUL.

[4] Oliver refers to Joseph Dalton Hooker and John Lubbock. CD read his paper, 'Dimorphic condition in *Primula*' before the Linnean Society of London on 21 November 1861 (see *Correspondence* vol. 9).

[5] CD's paper does not describe differences in the size of the ovules of the long- and short-styled forms of *Primula*, but CD included Oliver's observations on this point when he revised the paper for inclusion in *Forms of flowers* (see *Forms of flowers*, p. 17). See also letter to Daniel Oliver, 12 [April 1862].

[6] Hooker and Thomson 1858.

[7] Linnaeus 1792.

[8] In addition to his duties at the Royal Botanic Gardens, Kew, Oliver was professor of botany at University College London.

[9] The person to whom Oliver refers has not been identified; *Primula farinosa*, the bird's-eye primrose, is common in damp woods in northern England and southern Scotland. CD had requested specimens of this species from Oliver the previous season, but Oliver had been unable to send any (see *Correspondence* vol. 9, letters to Daniel Oliver, 1 May [1861], 27 May [1861], and 11 September [1861]).

[10] See letter to Daniel Oliver, 12 [April 1862] and n. 4.

[11] See letter to Daniel Oliver, 12 [April 1862].

From Andrew Murray 12 April 1862

Royal Horticultural Society, | South Kensington, W.
12 April *1862*

My dear Sir

No! I am not the guilty person.—[1]

Since I came here, there has been such a multitude of things passing incessantly thro' my mind, that I should speak less positively were it not for one Circumstance & that is that I never heard of Mr Scudders pamphlet before.— I *could not* have forgot the Existence of a work upon a subject on which I was interested, particularly if I had borrowed it from you—

No! you must throw your net again into the world & see if you cannot fish out the right man—

I have been promised & I believe it is on the way, a specimen of the representative of our Aepus which lives under shale between low & high water in Australia— I think that should interest us both[2]

Yours Ever Sincerely | And^W^. Murray

DAR 171.2: 325

[1] CD was trying to locate a pamphlet (Scudder 1861) that he thought Murray might have borrowed from him (see letter to Andrew Murray, 10 April [1862]).

[2] *Aepus* is a beetle found on the seashore towards mid-tide level; when covered by the sea at high tide, it probably survives in air pockets within cracks in rocks (G. Evans 1975, p. 106).

To Daniel Oliver 12 [April 1862][1]

Down Bromley Kent
12th

My dear Sir

I seldom see anyone, so it is a great pleasure to me to receive a scientific letter & I thank you for your very interesting one.[2]

First for Primula; I should like to see (pollen of) P. farinosa; but it really is of little importance, so tell your correspondent not to give himself much trouble; I send copy of my paper.—[3] I had noticed difference in size of ovules, but slurred the case over, & concluded that I had taken buds of different ages: I have just looked at a couple of large flower-buds of Primrose; in these the ovules were largest in the *short*-styled, just the reverse of what you say! Have you written "long"-styled for "short"-styled? or is the size a variable point?[4] The short-styled certainly produce a greater number of seed; but I did not attend to number of ovules: I will look at a few more of buds of equal age.—

Your discovery about Campanula seems to me *extremely* interesting; & I am specially surprised at the gradation of the two states on the same(?) plant; I am very glad you will read a paper on this.[5] Twice I have had Campanulas & Violets for experiments, & twice I have been prevented observing.[6] I can see at least 3 classes of Dimorphism,—Primula & co:—Thymus & co:—Campanula. Violets & co. For the latter cases I have theorised from what I have seen & read on Violets that the final object of the imperfect flowers is to produce seed safely, *without any crossing*. (so your remarks on the structure please me); the perfect flowers being adapted for getting an occasional cross; but I have not space here to give my reasons for coming to this provisional conclusion.— You will see that according to my notion, the final object of the dimorphism in Violets &c is exactly the reverse of what it is in Primula.—

With respect to Fumaraceæ, I had attended a little to them, when 3 or 4 years ago Asa Gray (& so does Vaucher) advanced this order as one with perpetual self-fertilisation;[7] this made me attend more carefully to it, & I am convinced the whole structure of the flower is beautifully adapted to favour an *occasional* cross.— I have many recorded observations, but God knows whether I shall ever have time to make use of them.[8] I write this letter only from memory.

Insects are not indispensable, for all that I have tried set seed without insect aid. It is really pretty to watch a Humble-bee visit Dielytra (& allied forms) Adlumia &c &c & see how neatly its hind-legs rest on the projecting plates on each side of the hood over the stamens & pistil, & push it to one or the other side, & get its body covered with pollen, against which the stigma is rubbed.— This tube has nectary on both sides & the "hood" can be pushed either way & the pistil is straight— Next look at Corydalis *tuberosa*? & allies (purple flower with white tube & now in bloom) here secreting nectary on only **one** side, & the hood can be pushed off only on one side, & the pistil is curved (it is straight in Dielytra) towards the gangway into the nectary.—[9]

Pray look at Corydalis lutea, with nearly similar structure; but with additional curious contrivance; when Bee (as I have often seen) visits flower, it pushes back the hood, & then the pistil moves with a jerk or spring forward.— This movement determines at once the act of fertilisation, though it will take place if the flower is never visited & the pistil never snaps off. (but flowers in this case are less fertile) But I must not run on; & I have written so briefly that you will not understand me.—

With respect to Bees biting holes; this occurs in almost all flowers *occasionally*, which are difficult to enter; I have never seen it occur in any Fumariacieæ, at least I think not.— But I know of no single instance of this always occuring in any plant— When it does occur, there can be no crossing as you say, & I have some reason to believe that the fertility of flower is diminished; certainly in some cases it would be quite stopped. Oddly this biting of holes occurs *chiefly* when large masses of the same species are grown together; I sh^d. like to hear what species of Fumaria you have seen with bitten corolla.—

I am pleased to hear that you are going to discuss Dimorphism in Nat. Hist R. for I have no doubt you will make a very valuable paper.—[10] By the way I have been much interested by your Atlantis paper, & rejoice at your conclusion;[11] I think, however, you might have made it more popularised, & I will add another criticism, viz that you ought (as I think) to have laid much more stress on migration having taken place during former warmer period.— I wish that your Lectures did not take up so much time, so that you might write on various subjects.[12] I do not know whether the diagrams of Catasetum would be worth your taking away from Linn. Soc; but if so they are at your service; & there was a diagram of Primula there.—[13] This reminds me that Mr Fitch drew for me cuts of Primula, would you be so very kind as to ask him what I am indebted to him for his kind assistance; though perhaps the Linn. Soc paid for the drawings on the wood.—[14]

I am nearly certain that Asa Gray (perhaps also in his Manual) told me that the imperfect flowers of some N. American Campanula produced *more* seed than the perfect.—[15] But I keep my notes in such a stupid manner that I cannot refer to single points, only to whole large subjects.

I have written you a frightfully long letter not worth the trouble of decyphering. Another thing just occurs to me which I meant to say whenever we met: viz that if ever you had leisure, you might make a grand essay on the charater of Fr. Water plants, taking, I presume, those whose roots were always covered with water. I presume no one could treat of more than Europe & N. America, though some Tropical country ought to be included.— It would be interesting to see how many had sent one or two colonists into the water: but what would be most curious would be to consider the nature of affinities, & degree of organisation of those groups, in which all the species were aquatic; I strongly suspect they would come out a queer set.— Do think of this.—

Forgive me scribbling at such length, & believe me Yours very sincerely
C. Darwin

DAR 261 (DH/MS 10: 1)

[1] The date is established by the relationship to the letter from Daniel Oliver, 10 April 1862, and to the following letter.

[2] Letter from Daniel Oliver, 10 April 1862.

[3] See letter from Daniel Oliver, 10 April 1862. CD refers to 'Dimorphic condition in *Primula*'.

[4] On the letter from Daniel Oliver, 10 April 1862, CD wrote 'short-styled' above Oliver's '*long-styled*' (see CD annotations). Oliver repeated his conclusions without modification in [Oliver] 1862c, p. 237. See also following letter, and letter to Daniel Oliver, 15 April [1862].

[5] Oliver had planned to report to the Linnean Society of London on an observation he had made on *Campanula* (see letter from Daniel Oliver, 10 April 1862), but subsequently decided against it (see following letter).

[6] CD had intended to carry out experiments on the cleistogamic flowers of *Viola* and *Campanula* in the summer of 1860, but his attempts were frustrated (see *Correspondence* vol. 8, letters to J. D. Hooker, 7 June [1860] and 12 [June 1860], and letter to Daniel Oliver, 24 [September 1860]).

[7] CD refers to his correspondence with Asa Gray during 1857 and 1858 (see *Correspondence* vol. 6, letter from Asa Gray, 7 July 1857, and letter to Asa Gray, 29 November [1857], and *Correspondence* vol. 7, letter from Asa Gray, 21 June 1858). CD also refers to Vaucher 1841, 1: 143; there is an annotated copy of this work in the Darwin Library–CUL (see *Marginalia* 1: 812–15).

[8] CD's initial notes on various members of the Fumariaceae, dated May and June 1858, are in DAR 76: 13–18; further notes, dated 1861 and 1863, are in DAR 76: 17–21. Some of CD's observations on species of *Fumaria* were eventually published in 1876, in *Cross and self fertilisation*, as part of a discussion of plants that are fertile without insect aid (p. 366). CD also published a notice on pollination in Fumariaceae in *Nature* 9 (1874): 460 (*Collected papers* 2: 182–3). See also *Correspondence* vol. 9, letter to Daniel Oliver, 9 April [1861].

[9] CD described this phenomenon in *Variation* 2: 58–9.

[10] [Oliver] 1862c. See letter from Daniel Oliver, 10 April 1862.

[11] Oliver 1862b. Oliver argued against Oswald Heer and Franz Unger's suggestion that the relationship between the Tertiary flora of Europe and the present flora of eastern North America indicated the existence of an Atlantic land connection between Europe and America during the Miocene period (Heer 1857 and 1861a, and Unger 1860). Oliver concluded that the botanical evidence did not favour

'the hypothesis of an Atlantis'; he went on to say (Oliver 1862b, pp. 164–5):

> it strongly favours the view that at some period of the Tertiary epoch, North-eastern Asia was united to North-western America, perhaps by the line where the Aleutian chain of islands now extends, since there is sufficient ground to believe that the temperature in that latitude was high enough to allow the migration of types, which at the present period, are characteristic of lower latitudes

CD annotated this essay in his copy of this issue of the *Natural History Review*. CD's copies of the journal are in the Darwin Library–CUL.

12 In addition to his duties at the Royal Botanic Gardens, Kew, Oliver was professor of botany at University College London.

13 CD refers to the illustrations for his papers 'Three sexual forms of *Catasetum tridentatum*' and 'Dimorphic condition in *Primula*' (see letters to Richard Kippist, 18 March [1862]).

14 CD marked this request with two vertical lines in the left margin. Walter Hood Fitch was a botanical artist at the Royal Botanic Gardens, Kew. The Linnean Society paid Fitch's 10*s*. bill (see following letter).

15 North American species of *Campanula* are described in A. Gray 1856, pp. 243–4, but the point to which CD refers is not mentioned.

From Daniel Oliver 14 April 1862

Kew G.
Monday. 14. IV. 1862

My dear Sir

How very kind you are to write me at such length!—[1] The *Linn: Soc*y. *will pay* Mr. Fitch. His bill amounts only to 10/.—[2] It is curious about the Primrose ovules. I have repeated the obs. with same result & got the boy downstairs who knows nothing which I thought largest to compare bud of one with open flower of other & he wholly confirms me.[3] However it must be variable & I ought to delete its mention or modify it in the M.S.S. which I send you by this post?[4] that the No. of seeds matured may vary relatively in individual capsules tho' as you shew more in the short-styled on the whole.— So with the ovules?

The whole paper however you may think so stupid & obscure & devoid of interest that I may have to burn it.— It is what I have hastily sketched out for N.H.R. You understand these things so *infinitely* better than I that you perhaps think me very foolish. You see it is anonymous!— Pray tell me if you think it worth putting in;—that is if you can find time to look at it. If you do not burn it wd. you kindly post it again to me.— The Atlantis Paper did not satisfy me.[5] But with regard to greater prominence to early migration during warm period I merely must have taken it for granted that no body could have thought of any thing else.—unless it be plurality-of-centres-of-creation people.[6]

To write popularly I must learn how.— I am so stupid with a pen, that I believe one ought to sit down & attempt not to "write a paper"—but to "*talk* one"

The *Corydalis* with hole bit in at top we thought might be C. *Marschalliana*, but it was a doubtful plant rather. I shall write about the *Primula farinosa*.[7]

The *Campanula* I mention in the M.S.S.— as Brongiart had previously described similar structure I shall not write about it to Linn. Soc.[8] The *closed* membrane was new to me over the sexual organs where I found it & *extremely interesting*. You see I have stopped at 2 kinds of dimorphism doubtless this may be my ignorance. The thing is difficult to put in words—*defining* the kinds.—[9] Now I must beg you excuse all this trouble.

I am not clear that I wholly apprehend your observat^{n}. about water plants just yet.—[10] I must read again.

Very sincerely Yours | Da^{l}. Oliver

DAR 101: 54–5

CD ANNOTATIONS

2.6 But ... people. 2.9] *cross in margin, brown crayon*
4.1 The ... rather. 4.2] *cross in margin, brown crayon*
5.2 The *closed ... interesting*. 5.3] *cross in margin, brown crayon*

[1] See preceding letter.

[2] Walter Hood Fitch, a botanical artist at the Royal Botanic Gardens, Kew, had drawn the illustrations for CD's paper, 'Dimorphic condition in *Primula*' (see preceding letter).

[3] Oliver had observed a difference in the size of the ovules of long- and short-styled forms of *Primula*; on repeating the observation, CD had found the size difference to be the reverse of that described by Oliver (see letter from Daniel Oliver, 10 April 1862, and preceding letter).

[4] Oliver refers to the manuscript of his review of CD's paper on *Primula* ([Oliver] 1862c; see letter to Daniel Oliver, 15 April [1862]). Oliver repeated his observations, apparently without modification, in *ibid*., p. 237.

[5] Oliver 1862b.

[6] See preceding letter. Oliver refers to those who, like Louis Agassiz, advocated multiple centres of creation to account for sporadic or disjunct species (see Rehbock 1983, p. 152, and Browne 1983, pp. 138–44).

[7] Oliver had offered to acquire for CD specimens of *Primula farinosa* (see letter from Daniel Oliver, 10 April 1862 and n. 9).

[8] See letter from Daniel Oliver, 10 April 1862. Oliver refers to Brongniart 1839.

[9] In [Oliver] 1862c, p. 236, Oliver stated:

> We may ... rudely group the kinds of dimorphism exhibited in the flower under two heads. First, a dimorphism, apparently favourable to variation, marked primarily by a partial or complete separation of the sexes, which may be accompanied or not by alteration in the form or arrangement of the parts of the perianth surrounding them; and, second, a dimorphism, conservative, and unfavourable to variation, marked primarily by an alteration in the form or arrangement ... of the outer whorls of the flower, which more or less completely enclose and seal up the sexual organs, which are never wholly separated.

[10] See preceding letter.

From J. D. Hooker [15 April 1862][1]

D^{r}. Darwin

Will it be convenient to Mrs Darwin to receive Willy & me from Thursday till Monday? if so we will find our way down that afternoon.[2] I shall not want dinner, so do not arrange or disarrange for us.

Next week would suit us equally well.

Ever yours affec | Jos D Hooker

Kew Tuesday.

DAR 101: 31

[1] The date is established by the relationship to the letter to H. W. Bates, 16 April [1862].

[2] Hooker and his eldest son, William Henslow Hooker, had been invited to spend Easter at Down House (see letter to J. D. Hooker, 18 March [1862]); according to Emma Darwin's diary (DAR 242), they stayed at Down from Thursday 17 April to Monday 21 April 1862.

To Daniel Oliver 15 April [1862][1]

Down Bromley Kent
Ap. 15th

Dear Oliver

How can you doubt about publishing your paper?[2] I may be, but I do not think I am, blinded by the too much praise you give me.— Why the *lowest* merit of the paper, the number of references, would alone make it worth publishing. Why do you not attach your name; pray remember it more than doubles the value of every remark?— It is just the paper, with your original remarks & suggestions, to set men observing. Your observation of possible tendency to hermaphroditism is quite new to me & on *ancient* prevalence of diœcious forms.—[3]

It is, as I *now* see, certainly quite possible; but I cannot call to mind, at present, any case supporting this view. Yet we entirely agree, I see, in *probable* function of the imperfect flowers of Viola &c, ie to make sure of a stock of seed.— I am almost certain that the Frenchmen are wrong about the perfect flowers not producing seed;[4] but I have not time to hunt up my notes. I shd not be surprised that they observed plants in pots in greenhouse or in chamber! As number of seeds is different in 2 forms of Primula, you have a right to put this case in your first class;[5] but I doubt (I say only in my paper *possibly* tending to diœcious condition) whether the tendency is real—whether the difference in number of seed is not merely some odd case of correlation like size of pollen-grains.— Would it not be well to keep your hypothetical or rather possible class of dimorphism altogether in the note: the sentence at present is not clear.—[6]

I have been looking at ovules again; one case seemed to agree with you, but others did not.[7] Had you not better have one more look; the case, as you put it seems probable, but certainly in some buds of *apparently* exact same age, did not hold; I got one of my Boys to look, as you did, & he gave verdict, without knowing which with me.—[8]

I am forced to write in great hurry.— For Heaven sake publish & append your name: if you do not publish, pray let me make M.S. copy of your paper.— The time will come I suspect, when *all* animals & plants will have to be viewed as primordially hermaphrodite; though confervæ are opposed to this.—

I am disgusted that your pretty discovery about Campanula has been forestalled;[9] you are quite right about insects & Campanula— C. K. Sprengel long ago observed & proved same fact, ie on necessity of insects in this genus.[10] Do not forget to look at Corydalis lutea; it will interest you: Vaucher observed the fact, but blundered greatly about all the details.—[11]

It makes me laugh at myself hearing you say that migration into N. America so manifestly depends on old warmer period;[12] I believe I meditated for four years on the whole case before this occurred to me!!! I then wrote to Asa Gray who used it in his paper.—[13]

Farewell— I hope to Heaven my vile hand-writing does not break your heart—

Yours very sincerely | C. Darwin

I return your paper by this same Post.—

DAR 261 (DH/MS 10: 45)

[1] The year is established by the relationship to the letter from Daniel Oliver, 14 April 1862.

[2] [Oliver] 1862c. See letter from Daniel Oliver, 14 April 1862.

[3] In considering the apparent tendency in *Primula* towards a separation of the sexes, Oliver raised the question, 'why did they ever become hermaphrodite?' He went on to say ([Oliver] 1862c, p. 238):

> While we may . . . suggest that certain species are tending to a separation of the sexes, we must not forget that arguments may be advanced to shew that it is not impossible but that they may be striving towards more perfect hermaphroditism, especially if we bring to mind the evidence . . . furnished by the 'Geological Record.' This evidence does certainly appear in favour of a greater predominance of unisexual forms at an early period than obtains at the present day.

[4] In his discussion of the fertility of Violaceae, Oliver cited Gingins-La-Sarra 1823, Monnier 1833, Müller 1857, and Michalet 1860 ([Oliver] 1862c, pp. 238–9).

[5] See letter from Daniel Oliver, 14 April 1862 and n. 9.

[6] Oliver apparently followed CD's suggestion (see letter from Daniel Oliver, 14 April 1862, n. 9, and letter from Daniel Oliver, 23 April 1862). The note in the published paper reads ([Oliver] 1862c, p. 236 n.): 'This second group we have not framed to include a dimorphic condition of the male flower, or of the female flower, of a unisexual plant. We are not aware, however, that such exist. If there be none, the circumstance is worth noticing.'

[7] See letters from Daniel Oliver, 10 April 1862 and 14 April 1862, and letter to Daniel Oliver, 12 [April 1862].

[8] See letter from Daniel Oliver, 14 April 1862.

[9] See letter from Daniel Oliver, 14 April 1862 and n. 8.

[10] C. K. Sprengel 1793, pp. 110–11. There is an annotated copy of this work in the Darwin Library–CUL (see *Marginalia* 1: 774–85).

[11] Vaucher 1841, 1: 151. There is an annotated copy of this work in the Darwin Library–CUL (see *Marginalia* 1: 812–15).

[12] See letter from Daniel Oliver, 14 April 1862.

[13] A. Gray 1858–9. See *Correspondence* vol. 7, letter to Asa Gray, 11 August [1858].

To H. W. Bates 16 April [1862][1]

⟨ ⟩
April 16th

My dear Mr Bates

Dr. Hooker is coming here tomorrow night to stay till Monday,[2] is there any chance of your being able to spare time to come here during this time:[3] it would give me great pleasure if you could. If you can spare the time or want a little rest, come without writing. You must come to Bromley Kent, by Railway; but I ⟨am⟩ sorry to say Bromley is six miles from this House.— On Monday I could send you back to Bromley with Hooker.— I hope to get M^r. Lubbock over on Sunday or Saturday evening—[4]

I only heard this morning when Hooker was coming or I would have given you longer notice.—[5] A bit of rest would do you good.—

Yours very sincerely | C. Darwin

Bromley is 10 miles from London & you can start from the Victoria Station Pimlico, or from London Bridge.— Do **not** get out at "Shortlands for Bromley", but at the proper Bromley Station.—

Cleveland Health Sciences Library (Robert M. Stecher collection)

1 The year is established by the references to Bates and Joseph Dalton Hooker visiting Down House (see nn. 2 and 3, below).
2 See letter from J. D. Hooker, [15 April 1862]. According to Emma Darwin's diary (DAR 242), Hooker was a guest at Down House from 17 to 21 April 1862.
3 Emma Darwin recorded Bates's arrival at Down House in her diary for 18 April 1862 (DAR 242).
4 See letter from John Lubbock, 17 April 1862.
5 See letter from J. D. Hooker, [15 April 1862].

To C. E. Brown-Séquard 16 April [1862][1]

Down. | Bromley. | Kent. S.E.
April 16th

My dear Sir

I have taken a very long time in thanking you for your extremely kind note of Jany. 13th,[2] but I have been waiting in hopes of being able to tell you that the French Translation of the Origin was published.[3] There has been great delay in the printing, but at last more than half is actually printed so that I suppose that in a month or two, it will appear. The Translator will have a copy sent to your Publishers, & is delighted to hear that you intend noticing it. If in your power to give it, I sh^d. very much like to have a copy of your Review, for I am sure that it will be very valuable to me.—[4]

You say in your note, "that few men are so near agreeing with me as I am".— I cannot tell you how this has pleased me; for I look with profound interest for the judgment of such men as yourself. I have not the least doubt that I have erred most seriously on many points; but now so many (yet few) really good judges concur *in the main* with me, that I do not fear that some such view will ultimately prevail, notwithstanding all the abuse & ridicule so freely poured on me.— On the Continent the Reviewers have treated me more fairly than in this country; & an excellent, but too favourable, Review has lately been published by Claparéde in the Revue Germanique.—[5]

Pray excuse me scribbling so long a note & believe me, my dear Sir | With sincere respect | Yours very faithfully | Ch. Darwin

Royal College of Physicians of London

[1] The year is provided by the reference to the imminent publication of the French translation of *Origin* (see n. 3, below).

[2] Letter from C. E. Brown-Séquard, 13 January 1862.

[3] Royer trans. 1862.

[4] See letter to C. E. Brown-Séquard, 2 January [1862], and letter from C. E. Brown-Séquard, 13 January 1862. Brown-Séquard intended to publish a review of *Origin* in the *Journal de la Physiologie de l'Homme et des Animaux*; however, no such review appeared.

[5] Claparède 1861. Edouard Claparède had sent CD a copy of the review (see following letter); it had in fact been published in the numbers of the *Revue Germanique* for August and September 1861.

To Edouard Claparède[1] [*c.* 16 April 1862][2]

Dear Sir

I thank you sincerely for sending me your excellent Review;[3] & for the generous & most kind manner in which you have spoken of my views.— I have read a great many critiques & abstracts of my Book; but I cannot remember one written with so much vigour & clearness & with so full an appreciation of the bearing of all the leading points. It is admirably done; & could have been so done only by one who entered on the subject with spirit, as well as with knowledge.—[4] I have been particularly struck by the eloquence of the latter portion of your article.—[5] You will have helped much in spreading in France sound views (as I hope & believe they are) on the descent and modification of species; & I have been told that my views were more unpopular in France even than in England; in Germany naturalists are beginning with some rapidity to adopt them.—[6]

Pray accept my sincere thanks & respects & believe me Dear Sir | Yours very faithfully

Draft
DAR 96: 17

[1] The recipient is identified by CD's description of the review, which corresponds to Claparède 1861 (see nn. 3–5, below).

[2] The date is conjectured from the relationship to the preceding letter (see n. 3, below), and to the letter from Edouard Claparède, 6 September 1862.

[3] Claparède 1861. See preceding letter and n. 5. There is an annotated copy of the review in the Darwin Pamphlet Collection–CUL.

[4] In the letter to George Bentham, 15 April [1863] (*Correspondence* vol. 11), CD singled out Claparède's review of *Origin* as one of the best, reporting that it was 'important as coming from so good an observer, as Claparède.'

[5] In the first part of his review, Claparède described the main tenets of CD's theory of natural selection, showing how the theory solved a number of problems in natural history; the latter part discussed CD's responses to possible objections to the theory.

[6] On the reception of natural selection in Germany, see, for example, *Correspondence* vol. 9, letter from Hugh Falconer, 23 June 1861, and letter to John Murray, 10 September [1861].

From H. W. Bates [17 April 1862][1]

Gregory's Hotel | Rupert Street [London]
Thursday evening

My Dear Sir

I could not resist your invitation[2] & the beautiful weather, although I had made other arrangements for Easter, intending to have a trip South about the middle of next week. Nothing could be more delightful to me than to visit you. I have arrived so far, en route, & intend to take an early train for Bromley tomorrow[3] I see a train from Pimlico at 9.55— I will try that & arriving at Bromley I shall get a boy to carry my bag & walk to Down having a great desire to loiter about the Kentish lanes— the county is quite new to me

I have been very poorly lately & shall be poor company

Yours sincerely | H W Bates

DAR 160.1: 66

[1] The date is established by reference to Bates's visit to Down House from 18 to 21 April (see n. 3, below); the previous Thursday was 17 April 1862.

[2] See letter to H. W. Bates, 16 April [1862].

[3] According to Emma Darwin's diary (DAR 242), Bates arrived at Down House on 18 April 1862.

From John Lubbock 17 April 1862

Lamas, Chiselhurst. S.E.
17 April 62

My dear M^r. Darwin

I wish the world had been so arranged that one could be in two places at once. We might I am sure enjoy ourselves more; now I am off today with Prestwich & Evans on a little geological excursion, & cannot therefore come to you though I should have enjoyed it so much.[1] Nelly is very glad that you approve of her review.[2]

Hoping that Horace is better & that you yourself are well,[3] I remain | Yours affec^{ly} | John Lubbock

DAR 170.1: 29

[1] In a letter that has not been found, CD had invited Lubbock to visit him at Down House on 19 or 20 April (see letter to H. W. Bates, 16 April [1862]). Between 17 and 28 April 1862, John Lubbock travelled with Joseph Prestwich and John Evans to Amiens and Abbeville in France to examine the Somme Valley sites at which flint implements considered to be of great antiquity had been discovered (Lubbock 1862c, p. 248, and Prestwich 1899, p. 170).

[2] Ellen Frances Lubbock; the reference has not been identified.

[3] Horace Darwin, CD's youngest child, had been ill since the beginning of the year.

To Daniel Oliver 20 [April 1862][1]

Down
20th

Dear Oliver

Hooker says you pass daily some Oxalis acetosella.[2] Will you oblige me by gathering a dozen or score of flowers **from different plants**, & if possible plants *growing a little apart.*— I find some evidence of dimorphism in the plants here, as in Primula, & much want to see plants from some other station.[3] The plant does not grow here within my walking distance.— Would you send me the flowers by Post in a little tin box, with a bit of damp blotting paper.—

Yours very sincerely | C. Darwin

DAR 261 (DH/MS 10: 56)

[1] The date is established by the relationship to the letter from Daniel Oliver, 23 April 1862.

[2] Emma Darwin's diary (DAR 242) records that Joseph Dalton Hooker spent the Easter weekend, 17 to 21 April 1862, at Down House. In his review of CD's paper, 'Dimorphic condition in *Primula*', Oliver cited Louis Eugène Michalet's observation that flowers of *Oxalis acetosella* produced early in the season differed from those produced in the summer ([Oliver] 1862c, p. 239). Oliver had sent CD the manuscript of his review (see letter from Daniel Oliver, 14 April 1862).

[3] There are notes on *Oxalis acetosella*, dated 17–27 April 1862, in DAR 109 (ser. 2): 4–5. See also *Forms of flowers*, pp. 181–3.

To Asa Gray 21 April [1862][1]

Down Bromley Kent
Ap. 21st

My dear Gray.—

I am in your debt for two pleasant notes.[2] First for business: I should *greatly* prefer your not returning any of Trübners remittance;[3] but you really must not return more than half, as otherwise I shall have gained an immense advantage in having given away many copies of your Pamphlet, gratuitously.[4] So add to all your kindness by letting matters remain as they are. I have settled with Trübner.— Trübner has not sold quite all; if copies are quite superfluity rich I shd. certainly like $\frac{1}{2}$ a dozen or dozen to give away. I was asked for one but yesterday. I have never met one person who was not delighted with your writing.—

Secondly, in a week or 10 days I shall send $\frac{1}{2}$ of my vol. on Orchids, as you desired;[5] & other half will soon follow for they set up the whole, before they printed off a sheet.— I fear it can never be popular; but do not judge too severely by first half; for, if I do not deceive myself the two last chapters are better.—[6] I believe I have been *very* foolish in publishing in popular form.—[7] When I told Murray that I wanted clean sheets to send you; he thought of some arrangement for American republication, as he said Lyell's new Book is to appear in America; but with my incomparably less important book it seems to me, as things now are, quite out of the question; but I have thought it as well just to mention what Murray said.—[8]

The North seems going on grandly victorious;[9] & thank God there is distinct ground broken on the Slavery question;[10] but we *stupid* English cannot yet believe that you will ever be a single Union again.—

I hope that you will ask your pupil to look carefully to gradation in sexes in your Hollies.—[11] As far as I can yet judge, I am not only wrong, but diametrically wrong about Melastomas, or at least about some of them;[12] if a Rhexia grew in a Garden, it would be good to cover up a plant under net & see if it seeded as well as uncovered plants.—[13]

Thanks for Mill's pamphet, which is very good & I had not seen it;[14] indeed I see hardly any Reviews or Periodicals.—

Hooker has been here for 3 days & we had lots of pleasant talk:[15] I am always full of admiration & love for him: I wish he had not so tremendous & dry a job in hand, as the Genera Plantarum.—[16]

Yours affectionately | C. Darwin

Gray Herbarium of Harvard University (65)

[1] The year is established by the reference to the publication of *Orchids* (see n. 5, below).

[2] See letters from Asa Gray, 6 March [1862] and 31 March [1862].

[3] The reference is to the London publisher Nicholas Trübner (see letter from Asa Gray, 31 March [1862]). See also n. 4, below.

[4] CD and Gray shared the cost of reprinting and publishing as a pamphlet Gray's reviews of *Origin* from the *Atlantic Monthly* (A. Gray 1861). CD privately distributed over 100 copies of the pamphlet. Trübner acted as London agent for the sale of the remainder of the pamphlets sent to Britain. See *Correspondence* vol. 9, especially Appendix III.

[5] Gray had asked CD to send him the sheets of *Orchids* as soon as they were printed because he wished to write an early review (see *Correspondence* vol. 9, letter from Asa Gray, 31 December 1861). *Orchids* was published on 15 May 1862 (Freeman 1977, p. 112).

[6] The penultimate chapter of *Orchids*, chapter 6, was devoted largely to the Catasetidae, described by CD as 'the most remarkable of all Orchids' (*Orchids*, p. 211). The final chapter discussed homologies of orchid flowers and the diversity of flower structure, and presented CD's general conclusions about the functions of the various contrivances for pollination.

[7] CD originally intended to publish his account of orchid adaptations as a paper in the *Proceedings* or *Transactions* of the Linnean Society of London (see *Correspondence* vol. 9, letter to John Murray, 21 September [1861]).

[8] CD refers to his publisher, John Murray, who also published Charles Lyell's *Antiquity of man* (C. Lyell 1863a). Lyell's book was first published in America in 1863 (*NUC*); an American edition of *Orchids* did not appear until 1877 (Freeman 1977, p. 113).

[9] The Union army secured a number of victories in the American Civil War during the spring of 1862—at Fort Henry and Fort Donelson, Tennessee, in February, Pea Ridge, Arkansas, in March, and Shiloh, Tennessee, in April (McPherson 1988, pp. 392–414).

[10] On 6 March 1862, President Abraham Lincoln asked the United States Congress to support a resolution offering Federal compensation for voluntary emancipation of slaves. Congress adopted Lincoln's resolution on 10 April 1862 (Curry 1968).

[11] In the letter to Asa Gray, 17 September [1861] (*Correspondence* vol. 9), CD asked whether, unlike the 'English Holly (& all the cultivated vars.)', any of the American species of holly showed signs of gradation indicating 'the steps by which it became dioicous'. In reply, Gray sent some observations, but reported that further observations would be necessary (see *ibid.*, letter from Asa Gray, 11 October 1861, and letter to Asa Gray, [after 11 October 1861]). CD refers to Gray's

'zealous pupil', probably Joseph Trimble Rothrock (see letter from Asa Gray, 6 March [1862] and n. 6).

12 See letter to Asa Gray, 16 February [1862].

13 CD had asked Gray to carry out some observations on *Rhexia* in the letters to Asa Gray, 16 February [1862] and 15 March [1862].

14 Mill 1862. See letter from Asa Gray, 6 March [1862] and n. 8.

15 Joseph Dalton Hooker was a guest at Down House from 17 to 21 April 1862 (Emma Darwin's diary (DAR 242)).

16 Bentham and Hooker 1862–83.

From C. W. Crocker 22 April 1862

South St. Chichester
April 22nd. 62

Dear Sir

I fear you will have thought me neglectful ere this—but I have been in sad grief and trouble lately. There has been nothing but illness and death in our family for the last six months.[1] I hardly expect my poor Mother to get through the night— It is impossible she can last long.[2] With so many troubles and anxieties one's thoughts are seldom free.—

After receiving your first note with regard to Prim: sin: I went to all the gardens in our neighbourhood. But it was then too late.[3] Where ten or twelve dozen plants are grown annually I could only find four or five exhausted plants—almost bloomed to death and not fair specimens to judge by. That there are *three* forms I am quite certain. I took notes of the numbers but as no garden furnished more than 2 or 3 *good* i.e. healthy plants I dont think it is any use.[4]

My own little garden has been terribly neglected of late, for as every hour during the last fortnight, might have been my Mother's last I hardly like to leave the house. And my garden is some little distance away. However brighter times must surely come at last, and I try to look forward hopefully to the time when this tide of trouble which has has been running in so long will once more ebb

Yours very respectfully | C. W. Crocker

C. Darwin Esq.

DAR 161.2: 257

1 Crocker's father, the poet and former shoemaker, Charles Crocker, died on 6 October 1861. See *DNB* and *Journal of Horticulture* n.s. 14 (1868): 206–7.

2 Crocker's mother, Mrs Charles Crocker, died on 27 April 1862 (see letter from C. W. Crocker, 17 May 1862).

3 CD's letter about *Primula sinensis* has not been found, but, regarding Crocker's willingness to carry out experiments or observations for CD, see the letters from C. W. Crocker, 17 February 1862 and [before 13 March 1862]. *P. sinensis*, the Chinese primrose, flowers in England in the winter and early spring (*EB*).

4 CD had been carrying out crossing experiments on *P. sinensis*, which he had hitherto considered dimorphic, since the end of January. In the course of his experiments he had come across what he

believed to be an 'equal-styled' form of the flower, in addition to the familiar long- and short-styled forms, and he had examined plants in a number of collections in order to establish the prevalence or otherwise of this third form (see letter to J. D. Hooker, 7 March [1862] and n. 10). There is a note recording Crocker's observation in DAR 108: 64. See also letter from C. W. Crocker, 17 May 1862 and n. 6.

From Daniel Oliver 23 April 1862

Kew.
Wednesday | 23. Apr. 1862

Dear Sir

Here are the flowers of *Oxalis* as requested. I do not perceive anything distinctly dimorphic.—[1]

My examining of the plant had reference chiefly to the aestival small flowers: they are very remarkable.— I altered a little the "definition" of the two groups of dimorphism in the paper which you so kindly looked over (& tho't worth printing!).—[2] Making one group with the Dimorphism manifest in, primarily, a separation more or less of the sexual organs, accompanied or not by alteration in the outer whorls.— (Thus including all wholly or partially diclinous plants,—*Catasetum*, *Primula*, &c) & the other group marked primarily by alteration primarily in envelopes of the flower without separation of the sexes.

Of course this is only the *morpholog*l. definition

After discussing their function &c. we may class them in corresponding group by other characters.

Very sincerely yours | Dan!. Oliver

DAR 173.1: 14

[1] See letter to Daniel Oliver, 20 [April 1862] and n. 2. There are observational notes relating to these specimens, dated 24 April 1862, in DAR 109 (ser. 2): 5. CD subsequently concluded that *Oxalis acetosella* was not dimorphic (see *Forms of flowers*, pp. 181–3).

[2] [Oliver] 1862c. See letter to Daniel Oliver, 15 April [1862].

From Thomas Gold Appleton 24 April [1862][1]

Dear Mr Darwin

I venture to send this to you by my brother Mr William S. Appleton, who leaves us for a summer trip to Old England.[2] Though very young he has mature tastes & ways. He is passionately fond of antiquities and if he has the luck to see your Roman camp will be in delight.[3] But he chiefly will devote himself to our national & family records with you.[4] He is sure to have ⟨a⟩ charming summer, and I ⟨do⟩ not know a better way to see England than just the one he proposes, to hunt up odd bits & ends in out of the way places.—

We are here all impatiently waiting for the continuation of your wonderful book and I think you will find it difficult to excite a greater interest now than did the first. It was something wonderful and Endures yet—[5]

I am hoping to find some of the "sweet sap" to send the no longer very young people I fear. It will remind them of long ago, and of our maple fruits.[6] I ⟨do⟩ not venture to write you about the war. It is a most painful calamity, but we all now firmly believe we could not have avoided it. The South in infatuation and evil-mindedness was determined to ruin us if we did not resist. We shall whip them and probably break up slavery but it requires very philosophic feelings to witness such "a struggle for life" without dismay.[7]

We feel we have gained at the North whatever betide a higher character and a loftier principle by our sacrifices and endurance.

My best remembrances to Mrs Darwin and the children & believe me | faithfully, | T. G. Appleton—

24th April
Boston.

DAR 159: 111

[1] The year is established by the reference to William Sumner Appleton's visit to Britain (see n. 2, below).

[2] Appleton refers to his half-brother William Sumner Appleton who left the United States for Europe in April 1862; he spent the summer and autumn in Britain before returning home in November 1862 (C. C. Smith 1903).

[3] The reference is probably to the earthwork known as Cæsar's Camp, which is situated about two miles from Down at Holwood Hill, near Keston (see Page ed. 1908–32, 1: 398–9 and 3: 119–21). The earthwork was supposed by some to be the Roman station Noviomagus, and a club had been formed, called the Noviomagian Society of Antiquaries, for its exploration (*Post Office directory of the six home counties* 1862). The site was located in Holwood Park, the estate of Robert Monsey Rolfe, Baron Cranworth, with whom the Darwins were on visiting terms (see letter from R. M. Rolfe, 28 November 1862).

[4] Appleton's sister, Mary, was married to Robert Mackintosh, Emma Darwin's cousin (F. W. Gregory 1975).

[5] In the introduction to *Origin*, CD had explained that the book was an abstract of a larger work that he expected to complete in 'two or three more years' (*Origin*, p. 1).

[6] Appleton subsequently sent the Darwins a present of some American maple sugar (see letter from Emma Darwin to T. G. Appleton, 28 June [1862]). He had formerly visited Down House in October 1849 (Emma Darwin's diary (DAR 242)), and had sent a box of maple sugar for the Darwin children in 1852 (see the letter from Emma Darwin to William Erasmus Darwin, [23 April 1852], in DAR 219.1: 4).

[7] The reference is to the American Civil War; Appleton alludes to *Origin* (see *Concordance*, pp. 725–6).

To Daniel Oliver 24 April [1862][1]

Down. | Bromley. | Kent. | S.E.
April. 24th

Dear Oliver.

Many thanks for Oxalis; of 38 flowers sent 24 are long-styled & 14 short-styled, just as in the flowers here; but whether this depends on mere useless variability or on useful dimorphism, I suspect experiment will alone show:[2] I have got plants in pots for next spring.

I am quite aware the other æstival flower is a very much more curious case. I return 6^d stamp which I suppose is a Kew affair.—

Yours very sincerely | C. Darwin

DAR 261 (DH/MS 10: 47)

[1] The year is established by the relationship to the letter from Daniel Oliver, 23 April 1862.
[2] See letter to Daniel Oliver, 20 [April 1862], and letter from Daniel Oliver, 23 April 1862.

From Berthold Carl Seemann 24 April 1862

22. Canonbury Square | London N.
April 24, 1862

Dear Sir.

That passage to which you allude in your letter I have copied out for your information. It is found in Botany of the Voyage of H.M.S. Herald. p. 67. (4to. London. 1852–57.).[1]

I also enclose a prospectus of a new work I am going to bring out, if a sufficient number of subscribers be forthcoming.[2] I understood Dr. Hooker to say that you were deeply interested in an exploration of the Sandwich or Hawaiian Islands; and had offered to give £50 to anyone attempting it in a thoroughly scientific manner.[3] It has always been my ardent wish to investigate the Flora of the South sea, and publish a good synopsis of it, and I have made up my mind to do, as soon as I have placed all the information accumulated during my last trip upon record, which I find some difficulty in doing as I have no assistance from Government, and was so ill paid during the time I was engaged in the Fiji's that I lost a good deal of money by my trip.[4]

As a preliminary to a Flora of Polynesia I am preparing a list of all the plants known from there.[5]

Yours very truly | Berthold Seemann.

More than five years ago I adopted as my motto a passage from your book. 'A traveller should be a botanist, for in all views *plants* form the chief embellishment'.[6]

I adopted it my Botany of the Herald,[7] and many people have agreed with you in thinking that without botanical knowledge it is impossible to describe scenery with any approach to correctness. | B S

DAR 177: 130

[1] Neither CD's letter nor the copied passage has been found. The passage in which CD had expressed an interest was probably the following, taken from a description of the flora of the isthmus of Panama (Seemann 1852–7, p. 67):

> Mountains, exceeding 2000 feet in elevation, situated principally in Western Veraguas, possess a vegetation which resembles in many respects that of the Mexican highlands; one in which the forms of the torrid region are harmoniously blended with those of the temperate.

In the chapter on geographical distribution in the larger work on species that CD was preparing, he quoted a very similar phrase from Seemann's narrative of his voyage (Seemann 1853, 1: 253) in support of the argument that temperate northern forms could have migrated across the isthmus to South America (*Natural selection*, p. 550 and n. 1).

[2] The enclosure has not been found, but probably related to Seemann's *Flora Vitiensis*, which was published by subscription (see Seemann 1865–73, p. ii, and n. 4, below).

[3] See *Correspondence* vol. 7, letter to J. D. Hooker, 26 [December 1859].

[4] In 1860, Seemann had travelled to Fiji as naturalist with a mission of inquiry to advise the British government about an offer of cession by the islands' king. His official report, 'On the resources and vegetable products of Fiji', was presented to parliament and later published (Seemann 1862). In addition, Seemann had been led to expect government assistance with the costs of publishing a flora of the islands, but none was forthcoming, and he was obliged to finance his *Flora Vitiensis* from his own resources (Seemann 1865–73, pp. i–ii).

[5] In an 'Advertisement' on the paper wrapper of the first part of Seemann 1865–73, Seemann stated:

> In working up this 'Flora Vitiensis' I have gone through the whole Polynesian plants found between lat. 30° N and 30° S., and it is my intention to publish the results of this investigation in a separate form.

Seemann died before the last part of the *Flora Vitiensis* was published, and his projected 'Flora of Polynesia' never appeared.

[6] *Journal of researches*, p. 604.

[7] The quote to which Seemann refers is printed on the title-page of Seemann 1852–7.

To H. G. Bronn 25 April [1862][1]

Down. | *Bromley.* | *Kent. S.E.*
April 25th

Dear & Honoured Sir

At last I have found time to correct the Origin.[2] I have compared the sheets of the Third English Edition with the Second which was translated into German, & have marked with a pencil line all the additions & corrections. Some of these additions were sent to you in M.S formerly,[3] but as I do not know which, I have marked all. As I am so poor a German scholar, it would take me a long time to compare all, but it will lose you very little time. Where merely a few words have been altered I have underlined them with pencil: where a sentence has to be omitted I have marked "dele". I also send a few new M.S. additions & corrections.—[4] You will, perhaps, be surprised at some of the additions, & will think them trifling.; I could of course have amplified many parts, but I have been guided almost solely in enlarging the parts which have been most criticised. I much regret to think that the additions will cause you some trouble: but I very much hope you will add to the load of kindness already conferred on me by looking through the English Sheets & correcting the new German Edition by them. As corrected sheets, cannot go by Post, I send them addressed to you to the care of Messrs. Schweizerbart, through Messrs Williams & Norgate.[5] I must add that some of the M.S. additions to *1st* German Edition, have since been a little modified.

In your very kind letter of March 27th, you say that M. Scheweizerbart is willing to publish a translation of my Book on Orchids;[6] but I shd. not be easy at his undertaking this until you or some one had read part & had expressed to him whether you thought it worth translation. I am quite doubtful. Will you have the great kindness to read Ch. I & tell M. Sweizerbart what you think. If I do not deceive myself the two last chapters VI & VII are a good deal the best in the Book.[7] In the hope that you will do this I will send by Post to you in about a week, the first half of volume in sheets. There are 35 wood cuts. With respect to electrotype copies, I have had some little difficulty. M^r. Murray, who publishes at his own risk, demurred rather, & said he thought M. Schweizerbart ought to pay a little more than the mere cost of the copies. M^r. Murray, however, at last agreed: he said the copies on average would cost 5^d each; therefore altogether £8:^s15:0.[8] It would be pleasanter to me, if M. Scheweizerbart would pay a little more, say £10, to M^r. Murray. But the cost perhaps will stop his wishing to produce a Translation. When you, or anyone M. Schweizerbart may employ, has read part, *or the whole*, & has decided, if he would write to me I would get the copies of the woodcuts made.—

The English Edition is printed in much larger type than is necessary. I will soon send the second half of volume, as it will soon be printed. Can you forgive me causing you so much trouble.

With sincere respect | Yours truly obliged | Ch. Darwin

Special Collections, Lehigh University Libraries

[1] The year is established by the reference to the second German edition of *Origin* (Bronn trans. 1863) (see n. 2, below).

[2] Christian Friedrich Schweizerbart, head of the firm E. Schweizerbart'sche Verlagsbuchhandlung, publishers of the first German translation of *Origin* (Bronn trans. 1860), wanted to bring out a second edition (see letter from H. G. Bronn, [before 11 March 1862], and letter to H. G. Bronn, 11 March [1862]).

[3] CD had sent Schweizerbart some 'corrections & additions' and a historical preface for inclusion in the first German edition of *Origin* (Bronn trans. 1860) (see *Correspondence* vol. 8, letter to H. G. Bronn, 14 February [1860]). The historical preface was subsequently expanded and included in the third English edition.

[4] The manuscript version of these additions and corrections has not been found. However, the changes incorporated in the second German edition of *Origin* that do not occur in the third English edition, and that presumably, therefore, correspond to the additional corrections CD sent with this letter, are given in Appendix VIII.

[5] Edmund Sydney Williams and Frederick Norgate were partners in the London booksellers and publishers, Williams and Norgate, which specialised in foreign scientific literature. British postal regulations stipulated that printed papers sent to Germany should 'not contain any writing or other manuscript marks besides the name and address of the person to whom they are sent' (*British postal guide*, 1 January 1862, p. 57). The regulations of the German Postal Union prohibited the circulation by post of letters above 50 g in weight, classifying heavier items as freight (*ibid.*, pp. 62–3). Williams and Norgate could presumably send CD's proofs to the Stuttgart-based publisher as part of a larger consignment of book-related materials.

[6] See letter from H. G. Bronn, 27 March 1862.

[7] See also letter to Asa Gray, 21 April [1862] and n. 6.

[8] The correspondence between CD and his publisher John Murray on this point has not been found, but see letter to H. G. Bronn, 11 March [1862] and n. 7.

To W. E. Darwin 26 April [1862][1]

Down
Ap. 26th

My dear William

What an incomparably good fellow you are to send me the eye-glass: I have now got it with a string round my neck, & practise every now & then, making horrible contortions, to keep it to my eye. I believe with practise I shall at last succeed, & that it will be *very* useful; & again I say that you are an incomparable good fellow.—

The Boys have all had a prosperous time of it, & return on Monday Evening to school:[2] they all went up with Mamma (who has got her usual very bad headach) & went to play to see Ld. Dundreary, & are now constantly repeating the good jokes, with old Jingo throwing his head back with laughter.—[3]

Poor dear little Skimp has been rather better for these few days, but has had some attacks of the involuntary movements:[4] he & Miss Ludwig go next week to Aunt Elizabeth's at Hartfield to see what a change of air may do for him.[5] He is a real little darling, so patient with all his discomfort.—

To day, thank Heavens, I finished last revise of my accursed little orchid-Book, of which a copy shall be sent you when it is out; but it will be stiff reading.—[6] Tim (i.e. Alfred) arrived here last night; not an atom altered in any way, except in having an untidy stumpy beard.—[7] We had another arrival, the night before last, of a school-fellow & friend of Franks, who told us an unintelligible story of losing his way & purse:[8] he has just started, which is a good job. I wonder when you will come here again; I saw nothing of you last time; but I am very glad that you saw Hooker work on plant.—[9]

Farewell, dear old fellow | Yours affecty. | C. Darwin

DAR 210.6: 96

[1] The year is established by CD's reference to having finished work on the proofs of *Orchids* (see n. 6, below).

[2] Emma Darwin's diary (DAR 242) records that the 'Boys went to school' on Monday 28 April 1862. George Howard Darwin and Francis Darwin both attended Clapham Grammar School in south-west London (see *DNB* s.v. Darwin, G. H., and F. Darwin 1920, p. 63). Leonard Darwin had been tutored privately by George Varenne Reed since summer 1859 (see *Correspondence* vol. 7, letter to G. V. Reed, 1 July [1859], and CD's Classed account book (Down House MS)), but apparently started to attend Clapham Grammar School in January 1862. According to CD's Classed account book, CD made no payment to Reed for the period January to June 1862, and made a proportionate increase in his half-yearly payment to Clapham Grammar School. See also letter from Emma Darwin to W. E. Darwin, [3 February 1862] (DAR 219.1: 48). Leonard was sent home from Clapham with scarlet fever in June 1862 and was again tutored by Reed during his convalescence (see CD's Classed account book (Down House MS), letter from Emma Darwin to W. E. Darwin, [13 December 1862] (DAR 219.1: 69), and *Correspondence* vol. 11, letter to G. V. Reed, 12 January 1863).

[3] Emma Darwin wrote in her diary (DAR 242) that she 'Went to London with 3 boys & Lizzy to the play' on Wednesday 23 April. Lord Dundreary was a comic character in Tom Taylor's popular farce, *Our American cousin* (*OED*). 'Jingo' or 'Gingo' was a family nickname for George (see, for example, the letter from Emma Darwin to W. E. Darwin, [26 March 1858] (DAR 219.1: 33), which begins 'My dear Georgy', but later continues 'Since beginning this letter I remember it is you I want to write to & not Gingo').

[4] CD refers to Horace Darwin who had been ill since the beginning of the year.
[5] According to Emma Darwin's diary (DAR 242), Camilla Ludwig, the governess at Down House, accompanied Horace Darwin to the home of his aunt Sarah Elizabeth Wedgwood on 1 May 1862.
[6] CD recorded in his journal (see Appendix II) that he 'finished Orchis Book' on 28 April 1862. William's name appears on CD's list of presentation copies for this work (see Appendix IV).
[7] The reference is to Alfred Allen Wedgwood, the son of Hensleigh and Frances Mackintosh Wedgwood.
[8] There is an entry in Emma Darwin's diary (DAR 242) on 25 April 1862 that reads: 'young Sayer here'.
[9] William had been home for Easter when Joseph Dalton Hooker was also visiting Down House (Emma Darwin's diary (DAR 242)).

From E. A. Parkes 28 April 1862

Frindsbury | Rochester
28 April 1862

My dear Sir

I saw the Director General on Saturday.[1] He thought the best way of carrying out your wishes would certainly be to send out copies of your Memorandum & printed forms to India & the West Indies & to see what the voluntary efforts of the Army Surgeons would do in the matter.[2] I enclose a copy (wh. please return) of a circular letter from the Director General which would perhaps meet the case.[3] But please to alter or modify it in any way. The Director General has not yet seen it.

I think the Memorandum & tables should both be printed, & that these should be sent to each Presidency & to the W. I.[4]

10 copies of the Memor.
20 " " table.

So that a total number of 50 copies of the Memorandum & 100 of the table would do.

Or perhaps it would be still better to let the Memorandum head the table & be on the same sheet.

With many apologies for so untidy a form I forward one for your consideration, but probably you will be able to arrange one much better—

I enclose a paper by D^r^. Beddoe of Bristol which may perhaps interest you— it has just reached me— would you kindly return it to me.[5]

Pray excuse this very hurried note | & Believe me | very sincerely yours | E A. Parkes.

We must put your name in. Without it nothing will be done— When the Med. Officers know you wish the investigations, they are much more likely to do it.

DAR 174.1: 24

CD ANNOTATION

3.1 Or … sheet. 3.2] '*Letter* on one side | Dedication— | Is the D. Gen letter to be printed' *added pencil*

[1] James Brown Gibson was the director-general of the Army Medical Department.
[2] See letter from George Busk, 1 April 1862, and letter from E. A. Parkes, 8 April 1862.
[3] The enclosure has not been found.

[4] West Indies. The term 'presidency' was applied to each of the three divisions of the East India Company's territory, Bengal, Madras, and Bombay; the administration of these territories was transferred to the Crown in 1858 (*OED*).

[5] Beddoe 1862. See letter from E. A. Parkes, 8 April 1862. CD's heavily annotated copy of Beddoe 1862 is in the Darwin Pamphlet Collection–CUL.

From H. W. Bates 30 April 1862

King St. Leicester
30 April 1862

My Dear Mr Darwin

I arrived here late on Friday night much fatigued. I wish to tell you what I learnt at the British Museum relative to the insects of South Temperate S. America & New Zealand. The *Carabi* of Chili & Tierra del Fuego are a remarkable case. There are 11 species known & I examined 8 of them, I have sent for a German Monograph which will tell me all about them.[1] They form a subgenus Ceroglossus of Solier[2] & as a group they are quite distinct from all other Carabi their nearest relationship being with S. European species. The genus Carabus is absent from Tropical America, N. Zealand, Australia & the Malay archipelago. I will not be quite sure before seeing the Monograph that none have any near resemblance with species of N. Temperate zone; I believe however there is no near affinity & therefore that no Carabus crossed the tropics during the recent Glacial epoch. It is inconceivable that these Carabi ⟨sh⟩ould have crossed & that their near allies in the North (2 or 300 species) should have undergone an entire modification amounting to subgeneric value, since the crossing, as Dr Hooker suggested.[3] Could the genus have originated in Chili independently? The great genus Calosoma nearly allied to Carabus but sharply distinct is almost cosmopolitan & appears of higher antiquity than Carabus for O. Heer (über die fossile Calosomen, just out)[4] finds many in tertiary strata of Europe & N. America & no⟨t⟩ a single Carabus. Could Carabi have segregated from Calosoma in the North & in the South independently? I think it highly improbable & have no doubt you think so.

What do other groups say? I was surprised at the poverty of the British Museum in Chilian Butt⟨erfl⟩ies I thought it was rich. In your journal you mention flocks of many species off Patagonia[5] I only find 3 from E. side of S. America from Buenos Ayres to Falkland islands. 46 species in all have been described. All accounts agree that Chili is poor in butterflies but still there must be more than 46 for Britain has 66. Of these 46 I found only 12 in B. M. but I referred to descriptions & was able to get a good knowledge of 13 more. These teach us much the same lesson as the Carabi with a few other things in addition. There are species *very closely* allied to European & Californian ones for instance a "meadow brown" Epinephile Janiroides very near our Janira. The genus or subg. Epinephile is quite unknown in tropical America but is present in California, Canada (& U.S.?) & N. Temperate zone old world, but the Chilian sp. comes nearest European sp. The genera are generally the same as N. Temperate, but the species in 6 cases form groups peculiar to Chili. In one genus the species are *very closely allied* to species of *mountainous Tropical* America.

One solitary sp. is common to Chili, S. Brazil (30^{d} S. lat.) & Venezuela & is totally absent, genus & species, from Amazon region. 2 or 3 quite tropical species exist in middle chili as local varieties.

The Cicindelidæ of which 8 are recorded confirm the above with the exception that there appears to be no species more nearly related to N. Temperate than to Tropical American sp.

The New Zealand insect fauna is wonderfully scanty I find only 7 Butterflies & 3 Cicindelidæ recorded all of which I examined. The Cicindelæ belong to a group which occurs in New Guinea but I do not know where else. All the Butterflies but one (a com^{n}. Australian sp.) are peculiar to the country; one forms a peculiar N. Zealand genus The rest have a generic resemblance to those of N. Temperate zone; two of them come nearer to European species than they do to any Australian or equatorial Asian but they are of such a nature that it is inconceivable how they could be modifications of Northern species which crossed the equator so recently as during the Glacial epoch.

It would be very desirable to publish in some journal a complete analysis of the insect fauna of these S. Temperate countries; but the great deficiency of our English collections makes me afraid of undertaking it. I never felt more painfully the confused state of the B.M. collections & the loose manner in which additions have been made. Nothing w^{d}. be easier than to obtain a very large set of Chilian Insects for there are resident collectors & Continental Museums seem to get supplied. Do you think it would be worth while to analyse the Chilian Carabi & adduce confirmations from other groups?

I send you the references which I promised.[6] At the B.M. one day Mr Pascoe shewed me the case of Dimorphism he announced.[7] It is in the males only— two forms of male very different in *structure*. of course this has nothing to do with dimorphism in flowers but does it not throw light on the first origin of neuter ⟨ ⟩ts,—2 forms of female?—

I set an artist to work on the plates of mimetic butterflies which ⟨it⟩ appears will be very expensive.[8]

Please give my kind regards to Mrs Darwin & family

Yours sincerely | H W Bates.

I find this precious bit in a Vienna periodical just to hand. "Dedicamus hocce genus eximium cel. Domino Henrico Bates, Darwinianæ doctrinæ propugnatori acerrimo &c"[9] I don't know how the author got to know I was a Darwinian.

I am told that the vacant place at B.M. will be filled by a young man who has Prof. Owen's protection & that it will be little or no use my becoming a candidate[10]

Wingless Coleoptera on the Caucasus

D^{r}. Fr. Kolenati, Meletemata Entomologica, Fascic. 1.2. Petrop. 1845—[11] (The work is advertised by Williams & Norgate— it is not in Ent. Soc. library London— I have seen a very good abstract & it appears Kol. goes very deeply into the geograph. distrib. of insects in the Caucasian lands. It was sent out by the Imp. Botanic

Gardens, S[t] Petersburg.) "The snow zone begins at 1400 Klafter (8,400 ft) & ends at 2000 Kl. (12,000 ft.?) The insects are few, nearly all beetles; *all peculiar* to the snow zone & *all wingless*. The genera are as follows:

Platychrus Feronia (omaseus Pseudomaseus & Platyoma) Nebria Calathus Amara Trechus Tachys	} Carabici—feeding on the Onisci, Myriapoda & Molluscs of the snow region	Philonthus Lathrobium Xantholinus Hister Helops Elmis Hydræna ochthebius Meriones otiorhynchus"

It w[d]. be desirable to know whether arctic beetles of these genera are wingless, but I cannot at present think of any treatise on the subject.

Coleoptera of the Azores

H. Drouet has given an account of these in Revue & Magazin de Zoologie 1859.[12] I could not find this work in London. I have seen only a short abstract— 57 species of beetles are given all except one, have a striking European character & many are common to the Azores & Europe; one (Calathus fulvipes) I have hunted out & find to be a *N.* European insect. A thorough analysis is desirable & could easily be made by a good European Coleopterist. The solitary tropical beetle is a common American one found throughout Brazil & W. Indies but *not* in N. America; it presents local varieties in America it appears well established on Azores, found on fig trees in 3 islands.

⟨ ⟩ Lucanus cervus L. cervus varies in the number of leaf⟨l⟩ike expanded joints of the antennæ, from 4 to 6 This feature was always considered of specific importance & many species were f⟨ounde⟩d on it *other differences accompanying*—viz. L. pentaphyllus—L. Fabiani—L. Pont-Crianti & L. turcicus— They are now believed to be all one sp. by Kraatz Berliner Entom. Zeitschrift 1860 p 68–9.—[13] I notice these varieties are local, being found in S. & S.E. Europe

DAR 47: 175, DAR 160.1: 67–8

CD ANNOTATIONS

6.4 in flowers . . . candidate 11.2] *crossed ink*

8.1 Please . . . Bates. 9.1] 'in Portfolio Ch. 7. extract on character of Wingless Beetles. Northern forms'[14] *added in margin, ink; circled ink*

12.1 *Wingless* . . . islands. 13.10] 'Ch 7' *brown crayon*; 'H. W. Bates. May 1862' *ink*

13.9 on fig . . . islands. 13.10] *in CD's hand, ink*

14.1 Lucanus . . . Europe 14.6] 'H. W. Bates | May 1862' *ink*

Top of first page: 'Gl⟨ac⟩ial' *ink, circled brown crayon*; 'Insects.' *ink*

CD note:[15]

Do not believe Carabi could have segregated— very strange not crossed during glacial period— Birds just same puzzle & sea['s' *over* 'S']-shells. Only caution about land-shells altering so much, i.e. rapidly in small details, yet in grand details so constant.— Do not consider Glacial as too short or recent: Forbes believed that it probably corresponded to *a sub [*interl*] stage in Tertiary Geology[16] I think it would be worth while— what are differences of Carabi in other parts of world; is the Chilian ⟨ ⟩ more distant say than Siberian or N. American, or Himalayan if there found.—

*A [*two words illeg*] B. Mus. [*illeg*] very odd males different— Hearty thanks for 3 most curious references. [*crossed ink*]

1 Gerstaecker 1858. See letter from H. W. Bates, 19 May 1862.

2 Solier 1848, p. 58

3 Bates and Joseph Dalton Hooker had recently spent several days at Down House (see letter to H. W. Bates, 16 April [1862], and letter from H. W. Bates, [17 April 1862]). During 1861, Bates and CD had corresponded about Bates's work on the geographical distribution of South American insects, and its implications for CD's views on glaciation (see *Correspondence* vol. 9). CD held the view that the glacial period had simultaneously affected the whole world, or a large part of it, probably causing much extinction among tropical forms and allowing northern temperate forms to migrate southwards towards the equator and even across it (*Origin*, pp. 376–9). Bates's study of butterflies in the Amazon led him to question this view, the region being still 'rich in endemic species' (see Bates 1861, pp. 351–3). In addition, Bates had told CD: 'there are genera peculiar to the high temperate zones of both hemispheres which present in South America compact groups of species so very dissimilar to those of North America that it is out of the question that they could have been derived thence so lately as the glacial period.' In evidence he cited the 'strong and very intelligible case' of the beetle genus *Carabus*, his observations on which CD considered 'very important' (see *Correspondence* vol. 9, letter from H. W. Bates, 18 March 1861, and letter to H. W. Bates, 4 April [1861]).

4 Heer 1861b.

5 *Journal of researches* 2d ed., pp. 158–9.

6 See nn. 11–13, below.

7 At the meeting of the Entomological Society of London on 7 April 1862, Francis Polkinghorne Pascoe, an entomologist at the British Museum, exhibited specimens of *Xenocerus semiluctuosus* and read a short paper on the dimorphic condition of the males (Pascoe 1862).

8 Bates refers to the plates to be published with his paper on the butterflies of the Amazon valley (Bates 1862a); the artist was Edward W. Robinson. CD had offered Bates £10 towards the cost of including coloured plates; in the event, the Linnean Society of London paid for the production of the plates, but not for the drawings (see *Correspondence* vol. 9, letter to J. D. Hooker, 25 November [1861], and letter to H. W. Bates, 3 December [1861]).

9 'We dedicate this rare genus to Master Henry Bates, keen and excellent defender of the doctrine of Darwin' (Felder and Felder 1862, p. 113).

10 Richard Owen was superintendent of the natural history collections at the British Museum. The vacancy for which Bates considered applying was created by the retirement of Adam White, one of the assistants in the zoological department (Bates 1892, p. lx). With Owen's support, Albert Charles Lewis Günther was appointed to the position (Gunther 1975, pp. 270–3, 286, and 297).

11 Kolenati 1845–6.

12 Drouet 1859.

13 Kraatz 1860.

14 The reference is to chapter 7 of the manuscript of CD's 'big book' on species, entitled 'Laws of variation: varieties and species compared', in which CD discussed the frequently wingless condition of the beetles on the island of Madeira (*Natural selection*, pp. 291–3).

15 See letter to H. W. Bates, 4 May [1862].

16 Forbes 1846, pp. 402–3. There is an annotated copy of the volume of the *Memoirs of the Geological Survey* in which this paper appears in the Darwin Library–CUL.

To T. H. Huxley 30 April [1862][1]

Down
Apr. 30th

My dear Huxley

It is late at night & I have this minute finished your Address to Geolog. Soc^{y}. & I must just thank you for sending it me.—[2] I have read it with uncommon interest. It is a wonderful condensed & original summary on Palæontology, & I should think & hope will do much good.— I hope you are not killing yourself with too much work—

Good night— | yours very sincerely | C. Darwin

Paul C. Richards Autographs (catalogue 183)

[1] The year is established by the reference to Huxley's anniversary address to the Geological Society of London (see n. 2, below).

[2] In the absence of the president, Huxley, as one of the secretaries of the Geological Society, delivered the anniversary address on 21 February 1862 (T. H. Huxley 1862d). The president, Leonard Horner, was in Italy with his convalescent wife.

To J. D. Hooker 1 May [1862][1]

Down.
May 1.

My dear Hooker,

If you can screw out time, do look at stigma of blue Leschenaultia biloba; I have just examined large bud with indusium not yet closed, & it seems to me *certain* that there is no stigma within.[2] The case would be very important for me, & I do not like to trust solely to myself. I have been impregnating flowers, but it is rather difficult. Singular case of highly peculiar structure, now **re**modified into the functional condition of a Campanula.—[3]

Ever yours | C. Darwin

How odd it is that pollen sh^{d} be enclosed within the indusium & then have to be removed out of it & put on the stigma by insects.

DAR 115.2: 153

[1] The year is provided by the relationship to the letter from J. D. Hooker, [5 May 1862].

[2] CD began to experiment with *Leschenaultia* in April 1860 in regard to his work on pollination (see *Correspondence* vol. 8, letter to J. D. Hooker, 12 March [1860] and n. 1). *L. formosa* seemed to have an effective mechanism for preventing cross-fertilisation, the pollen being 'shed in the early bud' and 'there shut up round the stigma within a cup or indusium' (see *Correspondence* vol. 9, letter to *Journal of Horticulture*, [17 May 1861]). CD initially suspected that the two viscid surfaces on the outside of the indusium might in fact be stigmatic surfaces, thus requiring insect agency to effect pollination (*Correspondence* vol. 8, letter to J. D. Hooker, 26 April [1860]), but Hooker had convinced him that the stigma was *inside* the indusium (*Correspondence* vol. 8, letter from J. D. Hooker, [28 April 1860]). However, CD carried out further investigations that encouraged him to maintain that pollination involved removal of the pollen from the indusium by insects, thus allowing *cross*-pollination (see *Correspondence* vol. 8, and *Correspondence* vol. 9, letter to *Journal of Horticulture*, [17 May 1861]). In CD's Experiment book (DAR 157a), there is a diagram of the indusium in a flower of the blue *L. biloba* var. *splendens*, dated 18 April 1862, showing the stigma located externally. CD recorded the observations

given here in a note dated 1 May 1862 (DAR 265). CD also attempted to effect pollination in several flowers, but did not succeed until later in the year (see the letter to J. D. Hooker, 11 June [1862], and the notes dated 2 June and 29 October 1862 in DAR 265)). CD subsequently published an account of his observations on *L. formosa* and *L. biloba*, made during 1860 and 1862, in a letter to the *Gardeners' Chronicle and Agricultural Gazette*, 9 September 1871 (*Collected papers* 2: 162–5).

[3] In species of *Campanula* the pollen is shed at an early stage in the maturation of the flower, adhering to collecting hairs that surround the pistil beneath the stigma. As a result the flowers cannot be pollinated without mechanical aid, as CD believed to be the case in *Leschenaultia*. See *Cross and self fertilisation*, p. 174.

To J. B. Innes 1 May [1862][1]

Down. | *Bromley.* | *Kent. S.E.*
May 1st

Dear Innes

I have bad news about Quiz: perhaps you had better not tell your son for a time.[2] He has been killed; it was done instantaneously by a gun. We were forced to do this, for he would fly at poor people, & one day bit a child & two days after a beggar woman & we had an awful row about it. There was another reason we could not stop him having fearful battle with Tartar; I had such a job one day in separating them both streaming with blood; & this was *incessantly* happening. Poor little Quiz had, also, got so asthmatic that he could not run, so that altogether we had no choice left us, though we were very sorry about it.

I hope the world goes well with you all; it has not of late with us, for we have had our youngest boy strangely ill, with singular involuntary movements, for two months;[3] but at last he is decidedly better. We feared much that there was mischief in the Brain, but it now seems clear that it was all sympathetic with irritation of stomach.

I know of no news of any sort here; we all go on exactly as usual as quiet as mice.

Dear Innes | Yours very sincerely | Charles Darwin

Cleveland Health Sciences Library (Robert M. Stecher collection)

[1] The year is established by the reference to Horace Darwin's illness (see n. 3, below).

[2] 'Quiz' was the pet dog of Innes's son John William Brodie Innes. When the Innes family moved to Scotland at the turn of the year they left Quiz with the Darwins (see letter from J. B. Innes, 2 January [1862] and n. 3).

[3] Emma Darwin's diary (DAR 242) records that Horace Darwin was seriously ill throughout the early part of 1862.

To John Murray 2 May [1862]

Down. | *Bromley.* | *Kent. S.E.*
May 2d

My dear Sir

I returned this morning to Printers last page of Index.[1] This last page is 365 so that the Book will not be quite so big as we feared. I hope you will reconsider the price: 10^{s} seems to me high.[2] About lettering the back of volume; I can think only of "Fertilisation of Orchids: Darwin,"; but this, I fear is too long. Please let me hear on this head. I have got rather to think that Red Cloth would look too gaudy for a grave volume.—[3]

You will, of course, settle what Reviews to send to; but I may mention that Editor of London Review would be inclined to be favourable.—[4] Also the Editors of old "Cottage Gardener", now called "Journal of Horticulture" are very civil to me,[5] & I have contributed little articles for them; so that if the keep any Reviewers, they would wish to review me favourably, & this Journal has very large circulation amongst men who cultivate Orchids.—

Lastly, will you be so very kind as to distribute by Post & otherwise the long enclosed list (which please keep safely):[6] I fear the copies for Foreigners will cause you some trouble; but I beg you kindly to do the best for me.— I hope you will give me a few copies, as heretofore.— God knows whether my little book will succeed at all, but I am frightened when I think what a large Edition is published.—[7]

My dear Sir | Yours very sincerely | Ch. Darwin

Endorsement: '1862—May 2'
John Murray Archive (Darwin 118–19)

[1] CD refers to the page-proofs of *Orchids*, and to the London printing firm of William Clowes & Sons.
[2] *Orchids* was published in May 1862 at the price of 9*s*. (*Publishers' Circular* 25 (1862): 247).
[3] The lettering on the spine of *Orchids* reads: 'Fertilization | OF | Orchids | Darwin.' The volume was published with a plum-coloured cloth binding (see Freeman 1977, p. 112).
[4] A favourable review of *Origin* had been published in the *London Review and Weekly Journal of Politics, Literature, Art and Society* 1 (1860): 11–12, 32–3, 58–9 (see *Correspondence* vol. 8, letter to J. D. Hooker, 29 July [1860]). The journal began publication in July 1860, and was edited for the first six months by Charles Mackay (*Men of the time*); his successor has not been identified. CD subscribed to the journal from May 1861 (see *Correspondence* vol. 9, letters to J. D. Hooker, 18 [May 1861] and 24–5 May [1861]). See also letter from J. D. Hooker, [19 January 1862].
[5] The *Journal of Horticulture, Cottage Gardener and Country Gentleman*, which changed its name from *The Cottage Gardener, Country Gentleman's Companion, and Poultry Chronicle* in 1861, was edited by George William Johnson and Robert Hogg.
[6] CD refers to his presentation list for *Orchids* (see Appendix IV).
[7] The first edition of *Orchids* had a print-run of 1500 copies (see letter to John Murray, 9 April [1862]).

To H. W. Bates 4 May [1862][1]

Down
May 4th

My dear Mr Bates.

Hearty thanks for your most interesting letter & three very valuable extracts.[2] I am very glad that you have been looking at the S. temperate insects. I wish that the materials in B. Museum had been richer, but I should think the case of the S. American Carabi, supported by some other case, would be worth a paper. To us who theorise I am sure the case is very important. Do the S. American Carabi differ more from the other species, than do, for instance, the Siberian & European & N. American & Himalayan (if the genus exists there); if they do, I entirely agree with you that the difference would be too great to account for by the recent Glacial period. I agree, also, with you in utterly rejecting an independent origin for these Carabi.—

There is a difficulty, as far as I know, in our ignorance whether insects change quickly in time; you could judge of this by knowing how far closely allied Coleoptera generally have much restricted ranges, for this almost implies rapid change. What a curious case is offered by Land-Shells, which become modified in every sub-district, & have yet retained the same general structure from very remote geological periods. When working at glacial period, I remember feeling much suprise how few Birds, no mammals & very few sea-mollusca seemed to have crossed, or deeply entered, the intertropical regions during the cold period. Insects, from what you say, seem to come under the same category. Plants seem to migrate more readily than animals.[3] Do not underrate the length of Glacial period; Forbes used to argue that it was equivalent to the whole of the pleistocene period in the warmer latitudes.—[4] I believe with you that we shall be driven to an older Glacial period.—[5]

I am very sorry to hear about B. Museum; it would be hopeless to contend against any one supported by Owen.[6] Perhaps another chance might occur before very long. How would it be to speak to Owen, as soon as your own mind is made up? From what I have heard, since talking to you, I fear strongest personal interest with Minister is requisite for Pension.—

Farewell & may success attend the "acerrimo propugnatori"[7] Yours very sincerely | C. Darwin

I deeply wish you could find some situation in which you could give your time to Science; it would be a great thing for science & for yourself.—

Cleveland Health Sciences Library (Robert M. Stecher collection).

[1] The year is established by the relationship to the letter from H. W. Bates, 30 April 1862.

[2] See letter from H. W. Bates, 30 April 1862.

[3] See letter from H. W. Bates, 30 April 1862 and n. 3. In *Origin*, p. 378, CD stated:

> as I believe, a considerable number of plants, a few terrestrial animals, and some marine productions, migrated during the Glacial period from the northern and southern temperate zones into the intertropical regions, and even crossed the equator.

In his 'big book' on species, published posthumously as *Natural selection*, CD made more detailed remarks about the migration during the glacial period of beings other than plants (*Natural selection*, pp. 554–5):

> In mammals & reptiles, I know of no cases of the same or representative species being found in the opposite hemispheres & not in the intermedial Tropics. In Australia, ... there are some striking cases of Birds, ... which represent northern forms, & are not known to occur within the Tropics ... In regard to insects, I carefully collected the beetles of Tierra del Fuego, ... but none are identical with, or closely representative of, northern forms; ... In southern Australia & New Zealand, there are only a few very doubtful cases of representatives of northern forms. ... In regard to sea-shells, ... I am surprised that there is not more identity or very close representation between north & south.

[4] See Forbes 1846, p. 403. There is an annotated copy of the volume of the *Memoirs of the Geological Survey* in which this paper appeared in the Darwin Library–CUL.

[5] Bates believed that the extent of the differences between species of the beetle genus *Carabus* in the temperate regions of North and South America was too great to have arisen since the *recent* glacial

period (see letter from H. W. Bates, 30 April 1862). However, when Bates first described this case to CD, in 1861, CD had reminded him that some geologists had 'speculated on a Permian & even on a Chalk Glacial period' (see *Correspondence* vol. 9, letter from H. W. Bates, 28 March 1861 and CD annotations, and letter to H. W. Bates, 4 April [1861]). CD included a brief discussion of this case in his much-revised discussion of the mundane glacial period in *Origin* 4th ed., p. 454.

6 The reference is to the competition for the position of zoological assistant at the British Museum (see letter from H. W. Bates, 30 April 1862 and n. 10). Richard Owen was superintendent of the natural history collections at the British Museum and supported the application of Albert Charles Lewis Günther for the post.

7 CD refers to a dedication to Bates, describing him as a 'keen and excellent defender' of CD's doctrines, given in Felder and Felder 1862, p. 113 (see letter from H. W. Bates, 30 April 1862 and n. 9).

From J. B. Innes to Thomas Sellwood Stephens[1] [before 5 May 1862][2]

J. B Innes.

If you see Darwin will you ask him if he can send me particulars of Tegetmeier's bee hives.[3] I want some improvement on the Scotish common ones. I find Milton who made your straw wooden top one has given up.[4] It was the best I had. Send me any particulars you can about hives ⟨ ⟩ want are supplied.

I see Tromer lodge is up for the 27th.[5] I fear the title is only such as to entitle the fortunate purchaser to a chancery suit with his Bleak Hall[6]

Incomplete
DAR 167: 2

1 Stephens was Innes's curate at Down from 1859 (*Crockford's* 1860), and was left in charge of the parish when Innes moved to Scotland early in 1862.

2 Dated by the relationship to the letter from J. B. Innes, 5 May [1862].

3 Innes probably refers to the 'observation hives' described by William Bernhard Tegetmeier in a paper he read at the 1858 meeting of the British Association for the Advancement of Science (see Tegetmeier 1858, and Richardson 1916, p. 47). The hives were constructed with 'each side formed of four parallel plates of glass, with thin strata of air between', thus preventing heat loss and obviating the need to cover the hives with opaque material; such hives were always open to observation without the risk of disturbing the bees by a sudden admission of light (Tegetmeier 1858, p. 133).

4 Innes refers to the London tea-dealer and grocer, John Milton (see letter from J. B. Innes, 5 May [1862]).

5 Tromer Lodge is a property located near Down House. Innes considered purchasing the estate in 1860 (see *Correspondence* vol. 8, letters to John Innes, 18 July [1860], 6 September [1860], and 28 December [1860]).

6 A reference to Charles Dickens's novel *Bleak House*, published in 1853.

From J. D. Hooker [5 May 1862][1]

Kew
Monday

Dear Darwin

I have not a *Leschenaultia* to examine, so pray send me some flowers. I should like to dissect them—[2]

We have had a regular kick up, & been in a troublous state for some time. We have had **all** our spoons forks &c to the tune of 80 pieces of silver walked off with; by a nice young man who introduced himself to our maids, & was made so much of that he could not make enough of us without & I have had tears, groans, hysterics, Police inspectors & all the other evidences of civilization in the house. It is all our own faults, wholly & entirely, for not looking better after our servts, doors & establishment. I don't care a brass farthing for the silver, which I have replaced already (with electro-plate!) but I must confess to a feeling of shame at finding out that my establishment has for some months had the reputation of being not a little **disreputable**—

Oddly enough, your lovely Teapot, 2 solid silver Candlesticks (which I have long wanted my wife to sell) & various other very expensive articles were left standing **by** the plate-basket which was emptied. I am disgusted at their not taking the candlesticks, which are of no use to me a bit, & at their assuming your tea-pot to be plated!—or they surely would have taken it.— So "there is no pleasing some people" you will say

The Cameroons' plants are most interesting, lots of temperate forms descending to 4000 feet, for your private satisfaction— I am drawing up an account of them for Lin Journal.—[3] More temperate Abyssinian species than ever.[4]

The upper region of Cameroons 7–13000 ft, consists of Volcanic cones, exactly like Auvergne according to the sketches our collector has sent home:[5] he has sent a fair Journal too— he spent some weeks at 7000 feet.

I had hoped to have had your Willy up ere this, but have been put out by this ugly affair: please tell him I will write very soon.[6]

Ever yours affec | J D Hooker

DAR 101: 33, 134a

CD ANNOTATIONS

0.3 Dear . . . say 3.6] *crossed ink*
6.1 I had . . . soon. 6.2] *crossed ink*
7.1 Ever . . . Hooker] 'May 6th 1862' *added in margin, ink*
Top of first page: 'Corrections | Orchid Book' *ink*; '*Miss Pugh*'[7] *pencil*

[1] Dated by reference to CD's endorsement (see annotations), presumably recording the date of the letter's arrival; 5 May was a Monday.

[2] See letter to J. D. Hooker, 1 May [1862].

[3] Hooker read a paper entitled 'On the vegetation of the Cameroons' at a meeting of the Linnean Society of London on 5 June 1862 (*Journal of the Proceedings of the Linnean Society of London (Botany)* 6 (1862): cvi). This paper was not published by the society; a further paper by Hooker on the topic was read the following year and published in 1864 (J. D. Hooker 1863c).

[4] In a paper read to the Linnean Society in 1861 (J. D. Hooker 1862b), Hooker had highlighted the affinity of the vegetation of Clarence Peak, Fernando Po, off the coast of West Africa, with that of Abyssinia 1800 miles distant on the other side of Africa, and of islands east of Africa. See also letter to J. D. Hooker, 9 May [1862] and n. 6.

[5] Hooker's work on the West African flora was based on botanical collections made by Gustav Mann (see J. D. Hooker 1862b and 1863b).

[6] Hooker had been helping CD to foster the botanical interest of his eldest son William Erasmus Darwin (see *Correspondence* vol. 9, letters to J. D. Hooker, 22 June [1861] and 17 [July 1861]). William had moved to Southampton in 1861 to begin a banking career (see *Correspondence* vol. 9).
[7] See letter to J. D. Hooker, 9 May [1862] and n. 4.

From J. B. Innes 5 May [1862][1]

Milton Brodie—
May 5th.

Dear Darwin,

I was sorry to find by your letter that you have had so much trouble with Quiz who I hoped would have been an acquisition.[2] Thanks however for the trouble and patience you have had with him and for so entirely doing as I asked you. I did not anticipate his biting humans improperly, & believed he would have at once succumbed to Tartar's prowess, very likely as he had pluck enough to fight his temper got soured. Any how I am glad to know the end of the little animal and that he had so good a chance of a happy life had he been virtuous.

We are very sorry to hear of your boy's indisposition. Now he is on the mend it is to be hoped he will be all right again very shortly.[3] I told Stephens the other day to ask you if you have tried any of Tegetmier's bee-hives and if you find them answer.[4] The general trouble with boxes is to get them not to split and warp. There was a good straw hive made by Milton G^t Marylebon St.[5] but I find by a letter from him that he has given up apiarian dealings, and he does not tell me who makes on his pattern. My views are not scientific but regard honey for which I want an improvement on the common cottage which is the only article in use here.

We are all pretty well & have real spring. The trees are in full leaf and bloom and all looking well at present. I am in much want of a good pony, which seems a hard thing to find here abouts.

M^{rs}. Innes[6] unites with me in kindest regards to you and Mrs Darwin—

Believe me, | Dear Darwin | Faithfully Yours | J. B Innes

DAR 167: 9

CD ANNOTATION
Top of first page: 'Forres | *N.B*' *ink*

[1] The year is established by the relationship to the letter to J. B. Innes, 1 May [1862].
[2] See letter to J. B. Innes, 1 May [1862] and n. 2.
[3] Innes refers to Horace Darwin.
[4] See letter from J. B. Innes to T. S. Stephens, [before 5 May 1862] and n. 3. The reference is to William Bernhard Tegetmeier.
[5] The reference is to John Milton, a tea-dealer and grocer of 10 Great Marylebone Street (*Post Office London directory* 1861).
[6] Eliza Mary Brodie Innes.

From T. H. Huxley 6 May 1862

Jermyn St
May 6th. 1862

My dear Darwin

I was very glad to get your note about my address—[1] I profess to be a great stoic you know, but there are some folks from whom I am glad to get a pat on the back— Still I am not quite content with that and I want to know what you think of the argument—whether you agree with what I say about contemporaneity or not and whether you are prepared to admit as I think your views compel you to do—that the whole Geological Record is only the skimmings of the pot of life—[2]

Furthermore I want you to chuckle with me over the notion I find a great many people entertain—that the address is dead against your views— The fact being, as they will by & bye wake up see that yours is the only hypothesis which is not negatived by the facts.— One of the great merits being that it allows not only of indefinite standing still but of indefinite retrogression[3]

I am going to try to work the whole argument into an intelligible form for the general public as a chapter of my forthcoming 'Evidence' (one half of which I am happy to say is now written) so I shall be very glad of any criticisms or hints[4]

Since I saw you indeed from the following Tuesday onwards—I have amused myself by spending ten days or so in bed.—[5] I had an unaccountable prostration of strength which they called influenza—but which I believe was nothing but more obstruction in the liver—

Of course I can't persuade people of this—and they will have it that it is over-work— I have come to the conviction however that steady work hurts nobody— the real destroyer of hard working men being not their work—but dinners late hours—and the universal humbug & excitement of Society

I mean to get out of all that & keep out of it—

Ever | Yours faithfully | T. H. Huxley

DAR 166.2: 293

CD ANNOTATIONS

1.6 the whole ... life— 1.7] *cross in margin, pencil*
4.2 I had ... influenza 4.3] *cross in margin, pencil*
5.3 but dinners ... Society 5.4] *two crosses in margin, pencil*

[1] Letter to T. H. Huxley, 30 April [1862]. The reference is to Huxley's anniversary address to the Geological Society of London on 21 February 1862 (T. H. Huxley 1862d); the address was later republished in the *Quarterly Journal of the Geological Society of London* (T. H. Huxley 1862b).

[2] In his address (see n. 1, above), Huxley challenged the assumption that 'geological contemporaneity is the same thing as chronological synchrony', arguing that 'neither physical geology nor palaeontology' possessed a method by which the 'absolute synchronism of two strata' could be demonstrated; all that

geology could prove was the 'local order of succession'. He continued (T. H. Huxley 1862b, p. xlvi):

> the moment the geologist has to deal with large areas or with completely separated deposits, then the mischief of confounding that "homotaxis" or "similarity of arrangement," which *can* be demonstrated, with "synchrony" or "identity of date," for which there is not a shadow of proof, under the one common term of "contemporaneity" becomes incalculable, and proves the constant source of gratuitous speculations.

'Skimmings of the pot of life' was Huxley's metaphor for the imperfection of the geological record; in *Origin*, CD had adopted Charles Lyell's metaphor of the remains of a damaged volume from 'a history of the world imperfectly kept, and written in a changing dialect' (*Origin* 3d ed., pp. 336–7).

3 Huxley emphasised the persistence of forms in the geological record and criticised the view that the fossil record indicated the progressive modification of forms towards less embryonic or less generalized types (T. H. Huxley 1862b, pp. xlvii–liv). He concluded that 'any admissible hypothesis of progressive modification must be compatible with persistence without progression through indefinite periods'. For a discussion of Huxley's views on this point, see A. Desmond 1982, pp. 85–112.

4 T. H. Huxley 1863a.

5 According to Emma Darwin's diary (DAR 242), Huxley visited CD at Down House on 6 April 1862.

From Hugh Falconer 7 May [1862][1]

21 Park Crescent N.W. | Portland Place
7th. May

My Dear Darwin,

There is no measure to my regrets— I literally lament having missed seeing you— I am just come home to find, that you had called and I out. This is the second time—that this evil luck has befallen me. Why did you not write and tell you were to be in town?[2] I would have had a constable put over me—to stop my going out, for a chat with you is a green spot in the vista of my most agreeable remembrances.

and this time, I wanted so much to have talked to you about the new results, from the *Pliocene* Fauna of N. America. If Leidy is to be trusted—there were no fewer than 6 genera! & 8 species of *Equid*⟨*æ*⟩ besides a Rhinoceros like the Existing Indian species.[3] Among the Horses he affirms that there was one which had the *milk* dentition of *Anchitherium*—and the permanent teeth of the Horse!! I would willingly believe it if I could—as a fact of great interest to your views.[4] *Anchi* was founded on the *Palæotherium aurelianense* of Cuvier—from the middle or lower miocenes. You will see it figured in the "ossements fossiles".[5]

I address this note at once to your brother—on the chance that you are staying in town today.[6] I would go over to see

Incomplete
DAR 205.9 (Letters)

CD ANNOTATIONS
1.1 There ... results, 2.1] *crossed ink*
3.1 I ... see 3.2] *crossed ink*
Top of first page: 'Falconer' *ink*; '21'[7] *brown crayon, circled brown crayon*

[1] The year is established by the reference to CD's trip to London (see nn. 2 and 6, below).
[2] Emma Darwin's diary (DAR 242) records that CD travelled to London on Tuesday 6 May 1862 and returned to Down House on Friday 9 May.
[3] Leidy 1858, 26–8. See also *Proceedings of the Academy of Natural Sciences of Philadelphia* (1858): 11.
[4] Falconer refers to *Merychippus*, described in Leidy 1858, p. 27. See also the letter to Joseph Leidy, 4 March [1861] (*Correspondence* vol. 9), in which CD expressed delight at Leidy's statement that he had some interesting facts 'in support of the doctrine of selection' that he would report at a suitable opportunity.
[5] Cuvier 1812.
[6] On his visits to London, CD stayed with his brother, Erasmus Alvey Darwin, at 6 Queen Anne Street. CD remained in London until Friday 9 May, when he met Falconer, before returning to Down House (see letter to Hugh Falconer, [8 May 1862], and letter to J. D. Hooker, 9 May [1862]).
[7] This is the number of CD's portfolio of notes on palaeontology and extinction.

To W. E. Darwin [8 May 1862][1]

6. Queen Anne St
Thursday

My dear William

a note from Hooker has been forwarded to me here from Down, in which he says he will soon write to you about your coming there.[2] I had heard nothing of this; but the Hookers have just had all their Plate stolen & in consequence much bother with Police &c &c & this has delayed Hooker writing. I have been doing odds & ends in London & am now just starting for the Exhibition!![3]

Yours | C. Darwin

a good account of Horace last night.—[4]

DAR 210.6: 97

[1] The date is established by reference to CD's trip to London in May 1862 (see nn. 2 and 3, below); 8 May 1862 was a Thursday.
[2] Letter from J. D. Hooker, [5 May 1862]. According to Emma Darwin's diary (DAR 242), CD travelled to London on 6 May 1862 and returned to Down on 9 May.
[3] The International Exhibition opened in South Kensington, London, on 1 May 1862 (*The Times,* 2 May 1862, p. 11).
[4] Horace Darwin was ill throughout the early part of 1862 (see Emma Darwin's diary (DAR 242)).

To Hugh Falconer [8 May 1862][1]

6 Queen Anne St.

My dear Falconer

Sincere thanks for your very kind note.[2] Whenever I come to London, which is seldom enough, I *always* look forward to a talk with you as a treat. I had intended trying my luck with another call. I doubt whether I shall be able this afternoon, but to-morrow (i.e Friday) I will call, perhaps about eleven but anyhow before one o'clock. If I do not appear you may understand that I have one of my attacks of sickness.

What you tell me about the horses, and about your readiness to believe (though this is a dreadful fib) makes me very eager to come,[3] not that this was necessary to draw me to Park Crescent.

My dear Falconer | Ever yours | C. Darwin

Copy
DAR 144: 24

[1] Dated by the relationship to the letter from Hugh Falconer, 7 May [1862]. The letter was written on a Thursday; 8 May 1862 was a Thursday.
[2] Letter from Hugh Falconer, 7 May [1862].
[3] See letter from Hugh Falconer, 7 May [1862].

To H. W. Bates 9 May [1862][1]

Down. | *Bromley.* | *Kent. S.E.*
May 9th

Dear Bates

I have been talking in London with Sir C. Lyell & he told me facts showing that there almost certainly must have been a Glacial **Miocene** period![2] In N. Italy he has seen huge boulders embedded with Miocene shells!—[3]

I saw at B. Museum the admirable mimetic case of Philippine Longicorn & Curculia— Really wonderful.— The gentlemen at the Museum sucked in your doctrine in explanation as easily as mother's milk.[4] I saw, also, there Mr Baily(?) & he showed me the generative organs of closely allied Chrysomelæ— on my word it is a very curious & unpleasant case, for he vows that these parts never vary.—[5] Dr. Günther, I find, is the candidate for place in B. Museum, & I apprehend no one would have a chance against him.—[6] I hear also with much surprise that a candidate would have to pass a civil-service examination.—[7]

Good Night | I am tired— | Yours sincerely | C. Darwin

Cleveland Health Sciences Library (Robert M. Stecher collection)

[1] The year is established by the relationship to the letter from H. W. Bates, 30 April 1862.
[2] See letter to H. W. Bates, 4 May [1862] and n. 5. CD was in London from 6 to 9 May 1862 (Emma Darwin's diary (DAR 242)), during which time he met with Charles Lyell (see letter to J. D. Hooker, 9 May [1862]).
[3] Lyell recorded observations of this phenomenon while travelling in northern Italy during September 1857 (see K. M. Lyell ed. 1881, 2: 265–8).
[4] CD refers to the application to mimetic beetles of Bates's explanation of mimetic resemblances in Amazonian butterflies as given in Bates 1862a. Bates argued that such resemblances could not be explained as purely the result of adaptations to similar conditions. He proposed that the imitated form was probably 'unpalatable to insect enemies', and, invoking natural selection, suggested that the 'selecting agents' were insectivores that destroyed any varieties not sufficiently resembling the unpalatable analogical form (Bates 1862a, pp. 510–12). In his review of Bates 1862a, CD stated: 'Mr. Waterhouse has noted an excellent instance (and we have seen the specimens) of a rare beetle inhabiting the Philippine Archipelago, which most closely imitates a very common kind belonging

to a quite distinct group' ('Review of H. W. Bates on mimetic butterflies', p. 221; see also *Collected papers* 2: 89); the reference is to the entomologist George Robert Waterhouse, who was keeper of the geology department at the British Museum.

5 Bates had told CD in 1861 of Joseph Sugar Baly's claim that the male organs of generation of Chrysomelidae could be used to distinguish one species from another (see *Correspondence* vol. 9, letter from H. W. Bates, 18 March 1861).

6 The reference is to the vacancy for a zoological assistantship at the British Museum for which Bates hoped to be considered (see letter from H. W. Bates, 30 April 1862 and n. 10, and letter to H. W. Bates, 4 May [1862]). Albert Charles Lewis Günther was appointed to the post in July 1862 (Gunther 1975, p. 308 n. 1).

7 Examinations for appointment to the civil service had been introduced in 1855 (*EB*). However, Günther did not have to sit the examination: at the instigation of the principal librarian, Anthony Panizzi, the Superannuan Act was invoked to provide Günther with a level of seniority that rendered him unaffected by established requirements (Gunther 1975, p. 297).

To J. D. Hooker 9 May [1862][1]

Down Bromley Kent
May 9th

My dear Hooker

I am so sorry for all the annoyment & loss which you have had.[2] But I firmly believe that care on your part of your household would have made no difference; I do not find that those who look most after their servants succeed by any means best. I know that we let matters take their course & upon the whole get on very well. It must have been a horrid bother & I am sure would have greatly annoyed me. I have given your message to our William: I had not heard that you had asked him; it is very kind in you & I *am sure* he will value it; but you must not give yourself much trouble with him.—[3]

How goodnatured, also, you & Mrs. Hooker have been to poor Miss Pugh:[4] she writes to Emma with great pleasure about it.—

I will most gratefully try & get & send a couple of flowers of Leschenaultia.—[5]

What a grand case that of the Cameroons; the 4000 ft has been much to my "private satisfaction".— I will swear that the mundane glacial period is as true as gospel, so it must be true.—[6]

In a few days you will receive my orchid-book.—[7] whenever you read it, will you kindly mark with pencil any errors,—for a German publisher wants to bring out a translation at once, but I have refused till he has got some one to read, that he may not be entrapped;[8] so I could correct any glaring error, which is likely enough to have crept in or rather to have walked in.

I have just returned from London & saw old Falconer, very jolly & *not* at all bitter against modification of species!!![9] He is a good old fellow.

Ever my dear old friend | Yours | C. Darwin

I saw, also, Lyell[10] very flourishing & very pleasant.—

DAR 115.2: 149

[1] The year is established by CD's reference to the publication of *Orchids* (see n. 7, below).
[2] Hooker's home had recently been burgled (see letter from J. D. Hooker, [5 May 1862]).
[3] In his letter of [5 May 1862], Hooker mentioned that he hoped to invite William Erasmus Darwin to the Royal Botanic Gardens, Kew, in the near future. See also letter to W. E. Darwin, [8 May 1862].
[4] From January 1857 until January 1859, Miss Pugh had been the governess of the Darwin children (Emma Darwin's diary (DAR 242)). According to Henrietta Emma Litchfield's autobiography (DAR 246: 36–7), she was 'neurotic and semi-mad', and was finally admitted into an asylum. See also Freeman 1978. At this time Miss Pugh resided in Kew, Surrey (see the letter from Emma Darwin to W. E. Darwin, [14 May 1862] in DAR 219.1: 231), and was thus a neighbour of Hooker's; CD also refers to Hooker's wife, Frances Harriet Hooker.
[5] See letter from J. D. Hooker, [5 May 1862].
[6] See letter from J. D. Hooker, [5 May 1862]. The discovery of temperate plants at heights as low as 4000 ft above sea level in the Cameroon mountains was further evidence in favour of CD's views on migration during a mundane glacial period. He proposed that the onset of a cold climate enabled temperate forms to migrate into tropical regions and across the equator, their progress facilitated by the weakened condition of the tropical forms, the greater adaptability of temperate forms to a changed climate, and the presence of high land (*Origin*, pp. 377–8). CD added Hooker's evidence from the Cameroon mountains to the discussion on this topic in the fourth edition of *Origin* (*Origin* 4th ed., pp. 445–50).
[7] *Orchids* was offered for sale on 15 May 1862 (Freeman 1977, p. 112). For CD's presentation list for *Orchids*, see Appendix IV.
[8] See letter from H. G. Bronn, 27 March 1862, and letter to H. G. Bronn, 25 April [1862].
[9] See letter from Hugh Falconer, 7 May [1862], and letter to Hugh Falconer, [8 May 1862]. Emma Darwin's diary (DAR 242) records that CD was in London from 6 to 9 May 1862.
[10] Charles Lyell.

To T. H. Huxley 10 May [1862][1]

Down Bromley Kent
May 10th

My dear Huxley

I have been in London, which has prevented my writing sooner.[2] I am very sorry to hear that you have been ill;[3] if influenza I can believe in any degree of prostration of strength; if from over-work for God's sake do not be rash & foolish.

You ask for criticisms, I have none to give only impressions.—[4] I fully agree with "your skimming-of pot-theory" & very well you have put it.—[5] With respect contemporaneity, I nearly agree with you, & if you will look to the d—d— Book 3d Edit p. 349, you will find nearly similar remarks.[6] But at p. 22 of your Address in my opinion you push your ideas too far: I cannot think that future geologists would rank the Suffolk & St. Georges strata as contemporaneous, but as successive sub-stages;[7] they rank N. American & British Stages as contemporaneous, *notwithstanding* a percentage of different species (which they, I presume, would account for by geographical difference) owing to the parallel succession of the forms in both countries.

For terrestrial productions I grant that great errors may creep in; but I shd. require strong evidence before believing that in countries at all well known so-called

Silurian, Devonian & Carboniferous strata could be contemporaneous.[8] You seem to me on the third point, viz on non-advancement of organisation, to have made a very strong case.[9] I have not knowledge or presumption enough to criticise what you say. I have said what I could at p. 363 of origin.[10] It seems to me that the whole case may be looked at from several points of view. I can add only one miserable little special case of advancement in cirripedes. The suspicion crosses me that if you endeavoured your best, you would say more on the other side. Do you know well Bronn in his last Entwickelung (or some such word) on this subject;[11] it seemed to me very well done: I hope before you publish again you will read him again & consider the case as if you were a Judge in a court of Appeal: it is a very important subject: I can say nothing against your side, but I have an "inner consciousness" (a highly philosophical style of arguing!) that something could be said against you; for I cannot help hoping that you are not quite as right as you seem to be.—

Finally I cannot tell why, but when I finished your Address, I felt convinced that many would infer that you were dead against change of species; but I clearly saw that you were not.—[12]

I am not very well—so good night & excuse this horrid letter. Ever yours | C. Darwin

Imperial College of Science, Technology and Medicine Archives (Huxley 5: 171)

[1] The year is established by the relationship to the letter from T. H. Huxley, 6 May 1862.

[2] According to Emma Darwin's diary (DAR 242), CD was in London from 6 to 9 May 1862.

[3] See letter from T. H. Huxley, 6 May 1862.

[4] CD refers to Huxley's anniversary address to the Geological Society of London (T. H. Huxley 1862d), later republished as T. H. Huxley 1862b. See letter to T. H. Huxley, 30 April [1862], and letter from T. H. Huxley, 6 May 1862.

[5] See letter from T. H. Huxley, 6 May 1862 and n. 2. Huxley's metaphor for the imperfection of the geological record did not feature in his address.

[6] CD refers to a section with the sub-heading 'On the forms of life changing almost simultaneously throughout the world' (*Origin* 3d ed., pp. 349–56), which begins, in contrast to Huxley's views: 'Scarcely any palæontological discovery is more striking than the fact, that the forms of life change almost simultaneously throughout the world.' CD subsequently restricted this observation to marine forms, arguing that insufficient data existed to judge whether land and freshwater forms 'change at distant points in the same manner', and stating: 'We may doubt whether they have thus changed' (*ibid.*, p. 350). He then qualified his position further, stating:

> When the marine forms of life are spoken of as having changed simultaneously throughout the world, it must not be supposed that this expression relates to the same thousandth or hundred-thousandth year, or even that it has a very strict geological sense; . . . Nevertheless, looking to a remotely future epoch, there can, I think, be little doubt that all the more modern *marine* formations, namely, the upper pliocene, the pleistocene and strictly modern beds, of Europe, North and South America, and Australia, from containing fossil remains in some degree allied, and from not including those forms which are only found in the older underlying deposits, would be correctly ranked as simultaneous in a geological sense.

See also letter from T. H. Huxley, 6 May 1862 and n. 2.

[7] Huxley, in a *reductio ad absurdam* of the commonly held view that similarity of organic contents afforded proof of the synchrony of the deposits that contain them, had stated (T. H. Huxley 1862b, p. xlv):

> suppose that, a million or two of years hence, when Britain has made another dip beneath the sea and has come up again, some geologist applies this doctrine, in comparing the strata laid bare by the upheaval of the bottom, say, of St. George's Channel with what may then remain of the Suffolk Crag. Reasoning in the same way, he will at once decide the Suffolk Crag and the St. George's Channel beds to be contemporaneous; although we happen to know that a vast period (even in geological sense) of time, and physical changes of almost unprecedented extent, separate the two.

[8] In his address to the Geological Society, Huxley stated (T. H. Huxley 1862b, p. xlvi):

> For anything that geology or palæontology are able to show to the contrary, a Devonian fauna and flora in the British Islands may have been contemporaneous with Silurian life in North America, and with a Carboniferous fauna and flora in Africa. Geographical provinces and zones may have been as distinctly marked in the Palæozoic epoch as at present, and those seemingly sudden appearances of new genera and species, which we ascribe to new creation, may be simple results of migration.

[9] In T. H. Huxley 1862b, pp. xlviii–liv, Huxley emphasised the persistence of forms from all sections of the animal and plant kingdoms in the geological record, and presented what he described as an 'impartial survey of the positively ascertained truths of palæontology', which, he claimed, refuted 'the common doctrines of progressive modificaton, which suppose that modification to have taken place by a necessary progress from more to less embryonic forms, or from more to less generalized types, within the limits of the period represented by the fossiliferous rocks' (*ibid.*, p. liii). For an account of Huxley's opposition to the idea of progressive divergence and specialisation, see A. Desmond 1982, pp. 85–112. See also M. Bartholomew 1975 and 1976.

[10] In a section under the sub-heading 'On the state of development of ancient compared with living forms' (*Origin* 3d ed., pp. 363–7), CD stated:

> natural selection will constantly tend . . . to render the organisation of each being more specialised and perfect, and in this sense higher; not but that it may and will leave many creatures with simple and unimproved structures fitted for simple conditions of life, and in some cases will even degrade or simplify the organisation, yet leaving such degraded beings better fitted for their new walks of life.

He continued:

> on the theory of natural selection the more recent forms will tend to be higher than their progenitors; for each new species is formed by having had some advantage in the struggle for life over other and preceding forms. . . . modern forms ought on the theory of natural selection to stand higher than ancient forms.

As to whether this was in fact the case, CD wrote that, although a 'large majority of palæontologists' believed so, he himself 'could concur only to a limited extent'. CD's copy of the third edition of *Origin* is at CUL; the pages carrying this discussion are heavily marked and annotated for revision.

[11] Bronn 1858.

[12] See letter from T. H. Huxley, 6 May 1862.

To William Darwin Fox 12 May [1862]

Down Bromley Kent
May 12th

My dear Fox.

I am going to bother you. Looking over some of your old notes, I see that you have kept the wild breed of Turkeys from Ld. Leicester & Powis.[1] You know that they now say that the common Turkeys have descended from a southern so-called species.[2] Have you ever crossed intentionally or accidentally your wild & common; & did you ever cross the hybrids inter se or with either pure parent & were they quite fertile? Have you ever given half-bred birds to other people, & did they with them become mingled with common Turkeys? Can you recognise the half-breds by their appearance? I sh^{d} be grateful for any information, which I might quote on your authority.—[3]

When you write tell me how *you* & all are. We were very glad to see your son at Torquay.[4]

I am much as usual, always grumbling & complaining. We have of late had much anxiety about our youngest Boy, who has failed in same way, but worse than, four other of our children.[5]

This is very shabby note, but I am tired with having written a heap of letters.—

My dear old friend.— | Yours affect^{ly}. | C. Darwin

Do you know anything of so-called Japanned Peacocks suddenly appearing from the common Peacock?[6]

Endorsement: '/62'
Christ's College Library, Cambridge (Fox 132)

[1] No letter from Fox containing this information has been found. The then earl of Leicester was Thomas William Coke and the earl of Powis was Edward James Herbert.

[2] CD refers to a paper by John Gould describing a new species of turkey, *Meleagris mexicana*, a native of Mexico, which Gould believed to be distinct from the wild turkey of North America (J. Gould 1856). He believed the domesticated turkey to be descended from the former and not, as commonly supposed, from the North American form. CD cited Gould on the origin of the turkey in *Variation* 1: 292.

[3] CD was working on chapter 8 of *Variation* (see 'Journal' (Appendix II)), which includes a discussion of the turkey (*Variation* 1: 292–4). In *Variation* 1: 292, he mentioned, on Fox's authority, that the wild turkey of the United States would cross freely with the common domestic kind, the turkeys in the area showing traces of their crossed parentage for many years afterwards. CD offered this as 'an instance of a domestic race being modified by a cross with a distinct species or wild race'.

[4] The Darwins had spent July and August 1861 in Torquay (see *Correspondence* vol. 9). CD may refer to Fox's eldest son, Samuel William Darwin Fox.

[5] Horace Darwin had been seriously ill since the beginning of the year (see Emma Darwin's diary (DAR 242)).

[6] In chapter 8 of *Variation*, on which CD was then working (see 'Journal' (Appendix II)), he described several cases of the 'japanned' or 'black-shouldered' peacock 'suddenly appearing in flocks of the common kind kept in England', and expounded his belief that the phenomenon was 'evidence of the first appearance of a new variety' (*Variation* 1: 290–2). See also following letter and n. 1.

To Philip Lutley Sclater 12 May [1862]

Down. | Bromley. | Kent. S.E.
May 12th

Dear Sclater

Mr Bartlett told me at the Gardens that the Japanned Peacock (I forget at moment your name) appeared amongst Mr Gurney's birds.[1] As he is an ornithologist perhaps you know him & if so would you have the great kindness to write & ask particulars; or if not can you tell me his address & I would write as a stranger. The chief point would be to know whether his birds appeared pure & whether any Japanned Peacocks lived anywhere near, so that there could have been a recent cross. I sh^d^. be very much obliged if you would kindly aid me & the point no doubt will interest you.—

Yours very sincerely | C. Darwin

Endorsement: 'May 1862'
American Philosophical Society (276)

[1] Philip Lutley Sclater was secretary to the Zoological Society of London. Abraham Dee Bartlett was superintendent of the society's gardens in Regents Park, London. CD was working on material for chapter 8 of *Variation* (see preceding letter, n. 6). In *Variation* 1: 291, CD stated that, through Sclater, he had received information from Hudson Gurney about his having reared 'a pair of black-shouldered peacocks from the common kind'. Sclater believed, in contrast to CD's view (see preceding letter, n. 6), that the 'japanned' peacock should be regarded as a distinct species from the common form, and named it *Pavo nigripennis* (Sclater 1860).

From Daniel Oliver 14 May 1862

Richmond, S.W.
14. May 1862

My dear Sir

I must not defer thanking you for *your very kind present*,—sent me by post from M^r^. Murray's—[1] I received it this morning that I have only had time to look over it. It is *a very extraordinary* book!— Your late publications must surely give quite a new & most promising direction to our studies. I am just now, you know, busy with my Class that I have very little time to spare, but shall not fail to inform them of your general results.[2]

With many thanks | Yours very truly | Danl: Oliver

DAR 173.1: 15

[1] CD's publisher John Murray offered *Orchids* for sale on 15 May 1862 (Freeman 1977, p. 112). Oliver's name appears on CD's list of presentation copies (see Appendix IV).
[2] Daniel Oliver was professor of botany at University College London.

To P. L. Sclater 14 May [1862]

Down. | Bromley. | Kent. S.E.
May 14th

Dear Sclater

Very sincere thanks for your two notes.[1] It will not be fair to plague you or Mr Gurney with another note, but when you see him, will you ask him whether he can remember at the time when the P. nigripennis appeared he had any white or pied Birds.[2] In two of the three cases mentioned by Sir R. Heron there were white & pied birds in the lot.—[3] With four cases now recorded I would wager the P. nigripennis will prove a variety,—[4]hardly more surprising in its origin, than the so-called Himalayan rabbit.[5] It is a very curious case. Have you a white Peacock in the Gardens; if so do match a white & common for the chance of P. nigripennis appearing.— The effects of crossing are sometimes marvellous in bringing out old & lost characters or in producing new characters—

Yours very sincerely | C. Darwin

Endorsement: 'May 1862'
American Philosophical Society (277)

[1] The letters from Sclater have not been found; evidently they were in response to CD's queries on peacocks (see letter to P. L. Sclater, 12 May [1862]).
[2] CD refers to Hudson Gurney (see letter to P. L. Sclater, 12 May [1862] and n. 1).
[3] Heron 1835. CD cited Robert Heron's observations in *Variation* 1: 290.
[4] In his account of variation in the peacock (*Variation* 1: 290–2), CD presented five cases of japanned birds suddenly appearing in common peacock flocks. CD regarded the japanned form as a variety of the common form (see letter to W. D. Fox, 12 May [1862], n. 6); Sclater regarded it as a distinct species (see letter to P. L. Sclater, 12 May [1862] and n. 1). CD concluded his discussion by stating (*Variation* 1: 291–2):

> On the view that the black-shouldered peacock is a variety, the case is the most remarkable ever recorded of the abrupt appearance of a new form, which so closely resembles a true species that it has deceived one of the most experienced of living ornithologists.

[5] CD described the origin of the Himalayan breed of rabbit in *Variation* 1: 108–9. He stated that these rabbits, white except for the ears, nose, feet, and upper side of the tail, which are brownish-black, were initially thought to be specifically distinct from other domestic breeds and given the name *Lepus nigripes*. CD then recounted how it was subsequently discovered that the Himalayans resulted from a complex cross: the offspring of 'chinchillas' crossed with the common black breed were crossed with the offspring of chinchillas that had been crossed with silver-greys. The Himalayan progeny, 'notwithstanding their sudden origin, if kept separate, bred perfectly true' (*ibid.*, p. 109).

From John Obadiah Westwood 14 May 1862

University Museum | Oxford
14 May 1862

My dear Sir

Thankyou very much for a Copy of your highly interesting book which arrived this morning[1]

I have only had time just to dip into it but I do not see any other bees or humble bees recorded as having been observed with pollen except those I sent you—[2] It will therefore perhaps interest you to learn that yesterday I captured on the alighting board of one of my hives, a bee with the front of its head invested with a large patch of pollen masses, 8 or 10,— Another bee had seized hold of this one & was trying to drag & fly away with it from the hive as an obnoxious individual. I intended to have sent it off to you alive but it died in the Course of the day So I put it into Spirits If you like to look at it I shall be most happy to send it to you for your Examination

But if I have not had time to devour your book my wife[3] has & she says she now quite understands the structure of the pollen &c. of the Orchids which she has in vain tried to do in Lindleys different books—[4]

Believe me | My dear Sir | Yours very truly | Jno O Westwood

This letter was misdirected to Down Hatherley[5] & has been rambling about England for the last fortnight | J.O.W.

DAR 181: 90

[1] Westwood's name appears on CD's presentation list for *Orchids* (see Appendix IV). Among others, CD sent presentation copies to those who, like Westwood, had assisted him with his study (see n. 2, below).

[2] Westwood had sent CD bees with orchid pollen-masses attached to them in July 1860 (see *Correspondence* vol. 8, letters to J. O. Westwood, 25 June [1860] and 9 July [1860]). CD wrote to Westwood again in 1861 requesting specimens of moths with orchid pollinia attached; Westwood sent him some wasps, but the pollinia were found not to be from orchids (see *Correspondence* vol. 9, letters to J. O. Westwood, 15 August [1861] and 4 September [1861]). Westwood's assistance is acknowledged in *Orchids*, p. 35.

[3] Eliza Westwood.

[4] John Lindley was the leading authority on the classification of orchids, and published three major books on the family, which constituted his chief botanical speciality. Lindley was, however, a prolific author, and he also wrote a number of introductory and reference works on botany, most of which passed through many editions (see *DSB* and *Taxonomic literature*).

[5] Down Hatherley is a village near Gloucester.

From George Bentham 15 May 1862

25. Wilton Place, S.W.
May 15 1862

My dear Sir

Many thanks for the copy of your most valuable book which I received last night[1] I have only had time as yet to go through the first two chapters but they are quite enough to show what a new field for observing the wonderful provisions of Nature you have opened—as the notice you gave of a portion of it at a recent Linnean meeting put us on a new and unexpected track to guide us in the explanation of phenomena which had before that appeared so irreconcileable with the ordinary prevision and method shown in the organised world.[2] I can only hope you will continue to give the world the benefit of your invaluable observations.

I sincerely trust your health did not suffer after your evening at the Linnean Society and that we may again occasionally have the great advantage of seeing you there

Ever yours most sincerely | George Bentham

C. Darwin Esq

DAR 160.1: 152

[1] Bentham's name is on CD's presentation list for *Orchids* (see Appendix IV). George Bentham helped CD to acquire specimens of *Orchis pyramidalis* (see *Correspondence* vol. 9, letters to George Bentham, 17 June [1861] and 22 June [1861]).

[2] CD read his paper, 'Three sexual forms of *Catasetum tridentatum*', before the Linnean Society of London on 3 April 1862.

To J. D. Hooker 15 [May 1862][1]

Leith Hill Place, Dorking (where we stay till this day week)
Thursday 15

My dear Hooker

I would not have sent Leschenaultia, had I known that Sir William was away, & you so busy;[2] for Emma has only just shown me M^rs. Hooker's note.[3] I am going to beg M^rs. Hooker to have the great kindness to send me here two answers from you: (1) address where I can get best paper for drying plants for Henrietta[4] (2) what must I call the *simple* microscope made by *Ross*?? (where does he live?) which you recommend for young Surgeons, & about what does it cost?—[5]

The remainder of this letter read at any time. You stated at Linn. Soc. that different sets of seedling Cinchona grew at very different rate, & from my Primula case you attributed it probably to two sorts of pollen;[6] I confess I thought you rash, but I now believe you were quite right. I find the yellow & crimson anthers of **same** flower in the Melastomatous Heterocentron roseum, have different powers; the yellow producing on the same plant thrice as many seeds as the crimson anthers;[7] I got my neighbours most skilful gardener[8] to sow both kinds of seeds & yesterday he came to me & said it is a most extraordinary thing that though both lots have been treated exactly alike one lot all remain dwarfs, & the other lot are all rising high up. The dwarfs were produced by the pollen of the crimson anthers.[9] In Monochætum ensiferum the facts are more complex & still more strange; as the age & position of the pistil comes into play in relation to the two kinds of pollen.[10] These facts seem to me so curious, that I do not scruple to ask you, (*mind, when you have a little leisure*) to see whether you can **lend** me any Melastomatad, just before flowering, with a not very small flower, & which will endure for a short time a greenhouse or sitting room; when fertilised & watched, I could send it to Mr Turnbulls[11] to a cool Stove to mature seed. I fully believe the case is worth investigation.

Farewell, my dear old fellow. Yours affect. | C. Darwin

You will not have time at present to read my orchid book:[12] I never before felt half so doubtful about anything which I published: when you read it, do not fear "punishing" me, if I deserve it. Adios.— I am come here to rest, which I much want.—[13]

Whenever you have ocasion to write pray tell me whether you have Rhododendrum Boothii from Bhootan with a smallish yellow flower & *pistil bent the wrong way*; if so I would ask Oliver to look for nectary, for it is an abominable error of nature, that must be corrected.— I could hardly believe my eyes, when I saw the pistil.—[14]

DAR 115.2: 151

[1] The date is established by reference to the Darwins' stay at Leith Hill Place, home of Josiah Wedgwood III (see n. 13, below).

[2] CD had asked Hooker to examine the stigma of *Leschenaultia biloba* (see letter to J. D. Hooker, 1 May [1862]), and had apparently sent him specimens at Hooker's request (see letter from J. D. Hooker, [5 May 1862], and letter to J. D. Hooker, 9 May [1862]). Hooker's father, William Jackson Hooker, was director of the Royal Botanic Gardens, Kew.

[3] Frances Harriet Hooker's letter to Emma Darwin has not been found.

[4] Henrietta Emma Darwin.

[5] The reference is to the optician and scientific instrument maker, Thomas Ross, whose father, Andrew Ross, had been one of the pre-eminent manufacturers of microscopes in Victorian London (Turner 1989, p. 154). Ross's factory was at 2 and 3 Featherstone Buildings, High Holborn, London (*Post Office London directory* 1861). See also letter from J. D. Hooker, [17 May 1862].

[6] Hooker's comments may have been made either following CD's presentation of his paper, 'Dimorphic condition in *Primula*', before the Linnean Society of London on 21 November 1861, or during the Linnean Society meeting of 3 April 1862.

[7] The reference is apparently to the results of pollination experiments that CD carried out on *Heterocentron roseum* between October 1861 and January 1862. The results of these experiments, showing that crosses with pollen from the yellow anthers produced *twice* as much seed as those with pollen from the crimson anthers, are recorded in a note dated 3 February 1862 (DAR 205.8: 46; see also letter to George Bentham, 3 February [1862]).

[8] The reference is probably to John Horwood, gardener to CD's neighbour, George Henry Turnbull; Horwood had supplied CD with specimens of *Heterocentron roseum* and *Monochaetum ensiferum* (see DAR 205.8: 24 and 47). He had previously assisted CD with his study of orchids (see *Correspondence* vol. 9), and is acknowledged in *Orchids*, p. 158.

[9] There is a note recording this observation, dated 14 May 1862, in DAR 205.8: 49. CD recorded similar observations on 30 May and 9 August 1862 (see DAR 205.8: 49–50).

[10] CD had observed that the position of the pistil in flowers of *Monochaetum ensiferum* changed over time (see letters to Asa Gray, 16 February [1862] and 15 March [1862]). For CD's attempts to understand the significance of these changes, see the dated experimental notes relating to this species in DAR 205.8: 22–43.

[11] George Henry Turnbull (see n. 8, above).

[12] CD had sent Hooker a presentation copy of *Orchids*, and had asked Hooker to inform him of any errors that he found (see Appendix IV, and letter to J. D. Hooker, 9 May [1862]).

[13] The Darwins stayed at the home of Emma's brother, Josiah Wedgwood III, from 15 to 22 May 1862 (see Emma Darwin's diary (DAR 242) and 'Journal' (Appendix II)).

[14] CD refers to Daniel Oliver. CD had previously observed that in rhododendrons, the pistils were bent so as to be in the path of insects seeking to reach the nectar (see *Correspondence* vol. 8, letters to J. D. Hooker, 26 April [1860] and 27 April [1860]). There is a note, dated May 1862, describing CD's observation on *R. boothii* in DAR 49: 72, which is marked 'Dichogamy'. See also the more detailed note on this subject, dated 16 May 1862, in DAR 49: 77–8. Hooker had collected many species of *Rhododendron* during his expedition to the Himalayas, 1848–50, and his *The rhododendrons of Sikkim-Himalaya* (J. D. Hooker 1849) had widened the interest of horticulturalists in using rhododendrons as garden plants. For Hooker's reply to CD's request, see the letter from J. D. Hooker, 23 May 1862.

From John Lubbock 15 May 1862

Lamas, | *Chiselhurst.* | *S.E.*
15 May 62 | Evening

My dear M^r. Darwin

Many thanks for your Orchis book which I have received this afternoon.[1] I have already read 50 pages; they are almost as good as animals, & it will be a great treat to me.

Prestwich & Evans & I had a delightful trip in the French valleys, & I am writing an account of the Somme implements for our next number.[2] You are I hope pretty well & all your party. I have heard nothing of any of you for ever so long, & am anxious to know how Horace has been getting on.[3] The big book will no doubt go on again now the Orchise's are over.[4] I must write to Oliver to get a review of them for our next number; he is doing one on Dimorphism, & it is sure to be good, but perhaps a little heavy.[5] Sir Charles Lyell is they say not coming out till Autumn, but of course you know all about that.[6]

Hoping to see you soon & with kindest regards to M^rs. Darwin believe me | Yours very affec^tely | John Lubbock

C Darwin Esq

DAR 170.1: 30

[1] Lubbock's name is on the presentation list CD drew up for *Orchids* (see Appendix IV).

[2] Lubbock refers to Joseph Prestwich and John Evans with whom he had travelled to France in April 1862 to investigate the sites in the Somme valley, near Amiens, at which alleged prehistoric implements had been found (Hutchinson 1914, 1: 56; Prestwich 1899, p. 170). Lubbock's essay 'On the evidence of the antiquity of man, afforded by the physical structure of the Somme valley' (Lubbock 1862c), was published in the July 1862 number of the *Natural History Review*; Lubbock was one of the journal's editors.

[3] Horace Darwin had been ill since the beginning of the year (see Emma Darwin's diary (DAR 242)).

[4] The reference is to CD's plan to complete an extended account of his views on species; the only part of the projected three-volume work to be published in CD's lifetime was *Variation*, which appeared in 1868.

[5] Daniel Oliver's review of 'Dimorphic condition in *Primula*' was published in the July 1862 number of the *Natural History Review* ([Oliver] 1862c); an anonymous review of *Orchids* appeared in the October number, but it was written by Joseph Dalton Hooker ([J. D. Hooker 1862d]; see letter from J. D. Hooker, 7 November 1862 and nn. 6 and 7). See also letter to Daniel Oliver, 13 October [1862] and n. 8.

[6] Charles Lyell was preparing *Antiquity of man* (C. Lyell 1863a). John Murray, the publisher of both *Orchids* and *Antiquity of man*, had expected the volumes to be published at about the same time (see letter from John Murray, 30 January [1862]). In the event, *Antiquity of man* was not published until 6 February 1863 (C. Lyell 1863b, p. vii).

From M. T. Masters [*c.* 15 May 1862][1]

Rye Lane | Peckham

My dear Sir

I have to thank you greatly for a copy of your new work on orchid fertiliz^n.[2]

I was so struck with your paper at the Linnæan that I promise myself very great pleasure and interest from the perusal of your book[3]

I see another book is promised on the Variation of plants & animals etc[4] I need hardly say that if it is in my power to give you any information as to the natural & abnormal variations in plants (to w^h. I have paid some little attention)—I shall be delighted to do so[5]

Believe me dear Sir | With many thanks | Your's faithfully | Maxwell T Masters

Ch.[as] Darwin Esq

DAR 171.1: 66

[1] The date is conjectured from the reference to the publication of *Orchids* (see n. 2, below).
[2] *Orchids* was published on 15 May 1862 (Freeman 1977, p. 112). Masters's name appears on CD's presentation list for the volume (see Appendix IV).
[3] CD read his paper, 'Three sexual forms of *Catasetum tridentatum*', before the Linnean Society of London on 3 April 1862.
[4] A notice announcing that CD was preparing *Variation* for publication appeared in *Orchids*, p. ii.
[5] Masters, a lecturer in botany at St George's Hospital, London, and his father, the nurseryman William Masters, were frequently cited in *Variation*.

From George Chichester Oxenden 15 May 1862

Dear Sir

Mr. *Malden*[1] & *myself* were on the Hill-Side, for some hours—last Tuesday, *with splendid sport*— —Even of *O. Aranifera* we did not find far short of a **thousand**, in **full** *flower*—

I am sure you would like Mr. Malden greatly— A *thousand thanks* for the beautiful, useful, charming Book—[2] I am **delighted** with it—

—Can I do aught for *you*, in revenge—?

—This means to be a *famous* year for Wild Orchids—

—Never before have I seen—"*fusca*" so beautiful or so *very* abundant—

There will also be good Harvest of "**Epipactis palustris**"—[3]

—Even in March last, when Snipe-shooting, I found fully *200*, well up above the ground; about 4 inches high—

—Of "Ep: **purpurata**" I observe you take no notice—

—Is it because it is *over-abundant* with you?

—The Earliest of my "*Lizards*" will bloom in about 3 weeks—

—Shall I cut some blooms off & send them to you—? or have you them, at command, in your own Neighbourhood?[4]

—With many kind thanks & regards | G: Chichester Oxenden

Broome
May 15 | 1862—

DAR 173.2: 48

[1] Bingham Sibthorpe Malden.
[2] Oxenden and Malden were both included on CD's presentation list for *Orchids* (see Appendix IV). Both men had supplied CD with orchid specimens; their assistance is acknowledged in *Orchids*, pp. 31 n., 43, and 78. In DAR 70: 10–11 there are notes dated 24 May 1862, recording observations on specimens of *Orchis ustulata* sent to CD by Malden.

[3] In the summer of 1861, CD had sent Oxenden a memorandum of inquiry concerning *Epipactis palustris*, requesting him to observe the behaviour of the insects visiting the flowers, and the resulting movements of the parts of the flowers (see *Correspondence* vol. 9, letter to A. G. More, 4 June 1861). In his discussion of the species in *Orchids*, pp. 94–102, CD mentioned outstanding problems in understanding the 'peculiar nature of the labellum' because of a shortage of specimens for observation at the end of the previous flowering season. CD was keen to continue these observations, and, earlier in the year, had made inquiries of Charles William Crocker regarding the location of specimens of this plant (see letter from C. W. Crocker, 13 March 1862 and n. 2). See also letter to A. G. More, 18 May 1862, and letter from G. C. Oxenden, 4 June [1862].

[4] CD had hoped to examine a specimen of the rare lizard orchid, *Orchis hircina*, before finishing *Orchids*. In the letter to B. S. Malden, 15–16 June [1861] (*Correspondence* vol. 9), CD wrote: 'Now I have seen & carefully every British orchid, . . . except the Lizard, which I still hope either from your or Mr Oxenden's kindness to see.' However, the plant is not mentioned in *Orchids*. Oxenden did subsequently send CD a specimen, a description of which CD included in the second edition (*Orchids* 2d ed., pp. 25–6 and n.).

From George Henry Kendrick Thwaites 15 May 1862

Peradenia, Ceylon
15 May 1862

Dear Mr. Darwin,

In a conversation I had a few evenings ago with our excellent Governor Sir Charles MacCarthy,[1] who takes great interest in all that you have published, he mentioned having met with a passage in Plato's Timæus which reminded him of your theory, and, curiously enough, I had read some short time previously, but I cannot discover where, a remark that Plato had anticipated you in the views you had enunciated in your work on the Origin of Species.

Sir Charles has kindly transcribed for me the passage he referred to, and I have great pleasure in sending it to you, with the note he has written to me with it—[2] the "private" in the corner of which only indicating its being sent me by the Governor in his private, not public, capacity; & so it is no breach of confidence my sending it for your perusal: indeed I shall tell Sir Charles I have done so.

Though the ζῶον ἓν of Plato will not mean the original primordial germ of all creation, still it is a fine expression for the aggregate of all created things, is it not?

I have just been enjoying the perusal, in the new number of the Linnean Journal, of your very interesting paper on *Dimorphism*[3] You will see in my little Enum. Pl. Zeyl. pp. 54 & 205 the same structure noticed in the genera *Sethia* and *Limnanthemum*.[4]

Trusting that your health is improved | I am always | Dear Mr Darwin | Very sincerely your's | G. H. K. Thwaites.

[Enclosure]

Private

Pavilion.
13th May.

Dear Mr Thwaites,

Many thanks for the books, which I shall be very glad to look into. —

I have transcribed the passage from Plato which I spoke to you about, but on looking at it again, I find that tho' the concluding words (under-lined) are curiously co-incident, the whole substance could only by a most uncritical process of *torture*, be construed so as to consider it as foreshadowing the Darwinian theory. The "one animal having (or containing) in itself all animals, mortal & immortal" is evidently the preceding "πᾶν, τόδε", "the universe", not a single primordial Being containing the germs of all other beings, to be hereafter developed out of it, but the one universal world containing & embracing all kinds of beings, already created & developed.

His whole Cosmogony is fanciful enough, & more of the realm of poetry than philosophy.

I don't know whether you are a Greek Scholar, but the words I have translated are the essential ones in the passage.

Yrs. ever faithfully | C. J. Mac Carthy

τότε γὰρ οὔτε τούτων ὅσον μὴ τύχῃ τι μετεῖχεν,
οὔτε τὸ παράπαν ὀνομάσαι τῶν νῦν ὀνομαζομένων
ἀξιόλογον ἦν οὐδέν, οἷον πῦρ καὶ ὕδωρ καὶ εἴ τι
τῶν ἄλλων. ἀλλὰ πάντα ταῦτα πρῶτον διεκόσμησεν,
ἔπειτα ἐκ τούτων πᾶν τόδε ξυνεστήσατο, <u>ζῶον ἓν</u>
<u>ζῶα ἔχον πάντα ἐν αὑτῷ θνητὰ ἀθανατά τε.</u>

Timaeus. $44.[5]

DAR 110 (ser. 2): 79–80, DAR 171.1: 3

CD ANNOTATIONS

0.3 Dear ... not? 3.2] *crossed blue crayon*

4.2 You will ... *Limnanthemum*. 4.4] 'Villaria dioicus. same sub. [*interl*] fam. as Menianthes, so ascertain fertility.—'[6] *added in margin, ink*

4.3 *Sethia*] 'Erythroxylacea—little order allied to Malpighiaceæ' *interl ink*

4.3 *Limnanthemum*. 4.4] 'allied to Menyanythes.' *added ink*

1 Charles Justin MacCarthy was appointed governor of Ceylon in 1860 (*Modern English biography*).

2 See enclosure and n. 5, below.

3 'Dimorphic condition in *Primula*'.

4 Thwaites 1858–64. CD referred to Thwaites's observations in *Forms of flowers*, pp. 116 and 122.

5 Plato, *Timaeus*, 69B–C. The passage translates: 'For at that time nothing partook thereof, save by accident, nor was it possible to name anything worth mentioning which bore the names we now give them, such as fire and water, or any of the other elements; but He, in the first place, set all these in order, and then out of these He constructed this present Universe, one single Living Creature containing within itself all living creatures both mortal and immortal. (Plato, *Timaeus*, Loeb Classical Library (London: Heinemann. Cambridge, Mass.: Harvard University Press, 1966), p. 179).

6 See letter to G. H. K. Thwaites, 20 June [1862].

From J. D. Hooker [16 May 1862][1]

Royal Gardens Kew | Kew
Friday.

My dear Darwin

I have dissected the *Leschenaultia* Flowers very carefully—[2] I can find no trace whatever of stigma within the indusium, there is certainly no naked surface there such as I saw in the former plant I dissected—[3] I find the cuticle continuous every where down to the chink at the base.

The tissue under the cuticle of the indusium inside is not stigmatic, though it is of very large delicate utricles such as lead the way to stigmatic tissue, & this tissue on the side next the glandular surface is continuous to the latter & thus deliquesces into a viscid surface quite stigmatic in many characters, though I do not see the long utricles so characteristic of most stigmatic tissues.

There is no doubt then that the external viscid surface is the apparent stigma. I have failed to prove it to be *the real* one—for—I find no pollen tubes in it, nor any protruded from the pollen grains, nor do I find any pollen grain either inside or outside indusium that has protruded a tube.

Then again I cannot find any stigmatic tissue in the style— There is a large hollow canal all the way down, communicating with the chink at the bottom of the indusium—but I have failed to trace any communication between this & the tissues of the *apparent* stigma.— To our shame, we have hardly a *Goodeniaceous* plant in the Garden, & none in good state. Poor Smith, our always wholly inefficient curator, is now half blind.—[4]

Ever yrs aff | J D Hooker

DAR 261 (DH/MS 11: 27)

CD ANNOTATIONS
1.1 *Leschenaultia*] 'biloba' *interl pencil*
3.1 the external ... stigma.] *scored pencil*
4.4 To our ... blind.— 4.6] *crossed pencil*

[1] Dated by the relationship to the letters to J. D. Hooker, 1 May [1862], 9 May [1862], and [18 May 1862]. The Friday following 9 May 1862 fell on 16 May 1862.

[2] CD had asked Hooker to examine the stigma of *Leschenaultia biloba* in the letter to J. D. Hooker, 1 May [1862], and, at Hooker's request, had sent him flowers for the purpose (see letter from J. D. Hooker, [5 May 1862], and letters to J. D. Hooker, 9 May [1862] and 15 [May 1862]).

[3] See letter to J. D. Hooker, 1 May [1862] and n. 2. Hooker had examined *Leschenaultia formosa* at CD's request in April 1860, and had reported: 'There is a perfectly good & normal stigmatic surface ... where it ought to be, at the base of the cup *inside*, on each lip ... These surfaces are bona fide structurally, i.e. histologically "stigmatic"' (*Correspondence* vol. 8, letter from J. D. Hooker, [28 April 1860], and Hooker's diagram, dated 29 April 1860, in CD's Experiment book (DAR 157a)). Hooker had also stated his conviction 'that the external viscid surfaces ... have nothing whatever to do with the stigmatic surfaces'.

[4] John Smith had been the curator of the Royal Botanic Gardens, Kew, for twenty years (R. Desmond 1994).

From D. F. Nevill [16? May 1862][1]

29 Upper Grosvenor St.
Friday

Oh my dear Sir

How can I thank you enough for your great kindness in sending me your book which I only received this day from Dangstein—[2] What I value even more is the writing at the beginning which renders it doubly valuable in my eyes Pray always remember what a real pleasure it is to me to hear from you and I shall be grateful—if I can ever at any time do anything for you

ever Y^s^ most truly | Dorothy

DAR 172.1: 25

[1] The date is conjectured from the distribution date of the presentation copies of *Orchids* (see n. 2, below); the following Friday fell on 16 May 1862.

[2] *Orchids* was published on 15 May 1862 (Freeman 1977, p. 112). Nevill, who had supplied CD with orchid specimens from her estate, Dangstein, in Sussex, is included on CD's presentation list for *Orchids* (see Appendix IV). The publisher, John Murray, appears to have sent the presentation copies by post on 13 May 1862 (see, for example, letter from Daniel Oliver, 14 May 1862).

From C. W. Crocker 17 May 1862

South St. Chichester
May 17th. 62

Dear Sir

Could you see the pleasure your book has already given me, for I have just dipped into it, and the continued pleasure I shall receive in perusing it and trying all the points within my reach, it would repay your kindness better than any words of mine could do.[1] I shall give up the attempt to try and thank you, for it would be impossible to do it properly; all I can say is that you have furnished me with a great amount of pleasure of the highest and purest kind. That I appreciate your present I need hardly say.

I had already begun to examine the structure of Orchis maculata before the book reached, before it had been in my possession ten minutes I understood its peculiarities. clearly.— Have you ever examined Sobralia I do not find the name in the index and have not yet read the whole book.[2] A friend of mine sent me a flower of S. macrantha which I kept in water for some days for I did not like to sacrifice the pleasure of seeing it even to the curiosity of examining its structure. It yesterday began to shrivel and I then tried taking away the sepals and petals I saw there was a passage leading to the nectary. An insect going in would disturb nothing but in *coming out* would lift up a cup, hinged at the top, and covering the pollinia. These so quickly set that after about half a minute I had some little difficulty in removing them from the object to which they were attached. If the insect failed in removing them the lid fell back into its former position to keep all moist.— I had only one flower to examine and know the danger of judging from such meagre materials but if you examine this plant yourself I think you will find this the plan of operation.

I am terribly afraid I shall be obliged to lose a twelvemonth in the case of the Hollyhock experiment. I cannot get the necessary plants.[3] Everybody discards the plants which have a tendency to be single and the very double ones would be useless for the purpose. It would be of no use to write to the nurserymen for they would grow none but what they term first class plants. I have tried every garden near us with the same result. I shall however do the best I can with the plants such as they are and should the result be unsatisfactory will make sure of single plants from someb⟨ody's⟩ seed-bed for the following year.

I have been trying to make myself acquainted with the history of Ranunculus Ficaria, and you will see the result in the little paper I post with this.[4] I mentioned some points connected with it in the Chronicle hoping to get light especially upon the point as to what enemies it has to fight against in the struggle for life. Two correspondents said they had found the young tubers to be eaten by the common wood pigeon.[5] Th⟨is is un⟩satisfactory to me; the tubers ⟨ ⟩ ⟨ ⟩ those formed the preceeding ⟨ ⟩ upon the surface of the ground. This was not the point I wished for information upon— I wanted to know what killed them in the *second* year— there are hundreds produced every spring from last summer's tubers for every single plant which grows on the second year. What kills them after they have once made a good start? They dont rot away because they keep very well even under water. It is not due to the pigeons because I dont believe they would dig them out (in the second year the tubers are buried an inch or so below the surface) and because the wholesale destruction is going on where pigeons are rarely if ever seen. When a root does manage to live on it grows much stronger and flowers more early and more plentifully than the plants from last year's tubers.

I enclose a copy of my notes upon Primula sinensis but look upon them as very unsatisfactory because the plants were quite exhausted with flowering.[6] In other places ⟨ ⟩ I visited I found only two or three ⟨ ⟩ blooms.

Primula sinensis

	Garden N°. 1	—Garden N° 2.
Long styled	— 8	— 4
Short styled	— 4	— 5
Medial	— 2	— 1?

Since I wrote to you last I have lost my poor Mother after a most trying illness. She only survived my father for a little more than six months.[7]

With many thanks for your kind present I remain, dear Sir, | yours very respectfully | Chas. W. Crocker

C. Darwin Esqre.

Incomplete
DAR 108: 133, DAR 161.2: 258

CD ANNOTATIONS
5.4 Primula ... respectfully 7.1] *crossed pencil and ink*
End of letter: 'These are nearly last remnants of flowers' *ink*

[1] Crocker's name is included on CD's presentation list for *Orchids* (see Appendix IV).

[2] In *Orchids*, p. 269, CD reported that he had been unable to obtain any living flowers of the Arethuseae, the orchid tribe to which *Sobralia* belonged. He included an account of *S. macrantha* in *Orchids* 2d ed., pp. 91–2.

[3] See letter from C. W. Crocker, [before 13 March 1862].

[4] Crocker 1862. There is a copy of the number of the *Gardener's Weekly Magazine* in which Crocker's article appeared in the Darwin Pamphlet Collection–CUL. CD cited Crocker 1862 on the non-production of seeds in England by *Ranunculus ficaria* in *Variation* 2: 170 n., where the paper's publication date is mistakenly given as 1852.

[5] Crocker's query concerning *Ranunculus ficaria* was printed in the *Gardeners' Chronicle and Agricultural Gazette*, 7 December 1861, p. 1070. The two responses were printed in the issues for 21 December (p. 1117) and 28 December (p. 1136).

[6] Crocker had been making observations for CD on the prevalence of an equal-styled form in this species, which CD had previously considered dimorphic (see letter from C. W. Crocker, 22 April 1862 and n. 4). There are notes recording Crocker's and other observations on the proportions of the three forms in DAR 108: 66 and 29; these findings were later incorporated into 'Illegitimate offspring of dimorphic and trimorphic plants', p. 415 (a paper not included in *Collected papers*).

[7] Crocker's mother, Mrs Charles Crocker, died at his home on 27 April 1862 (*Sussex Advertiser*, 6 May 1862); his father died on 6 October 1861 (*DNB*).

To W. D. Fox [17 May 1862]

Leith Hill Place | Dorking
Saturday

My dear Fox

Your letter has been forwarded here, where we remain till Thursday evening.[1] I am so very sorry to miss your visit. Our house is utter confusion with painters & we shall, when we return have to live in a bed-room; but I think by Monday or Tuesday week one room will be ready; is there any chance of your being in London so long as that?—

Very many thanks for your letter about Turkeys, which interests me much.[2] If without very great trouble you could get me any more information about the fertility &c of the hybrids or mongrels, I sh[d]. be very much obliged.—

In Haste | My dear Fox | Ever yours | C. Darwin

Postmark: MY 17 62
Christ's College Library, Cambridge (Fox 133)

[1] In his 'Journal' (see Appendix II), CD recorded spending 15 to 22 May 1862 at Leith Hill Place, the Surrey home of his sister Caroline Sarah Wedgwood and her husband Josiah Wedgwood III, Emma Darwin's brother.

[2] Fox's letter has not been found. CD had asked Fox for information about crosses between wild and domestic turkeys in the letter to W. D. Fox, 12 May [1862].

From J. D. Hooker [17 May 1862][1]

Royal Gardens Kew | Kew
Saturday

My dear Darwin

I have been at the red *Leschenaultia* this morning.[2] Not the same species as last year. I find *no stigmatic surface whatever* in the indusium—but as before a hard dense very unstigmatic tissue under the glandular surface. No pollen tubes any-where. A very stigmatic tissue underlies a pulvinus within the indusium at *b* but it is covered perfectly with cuticle.[3]

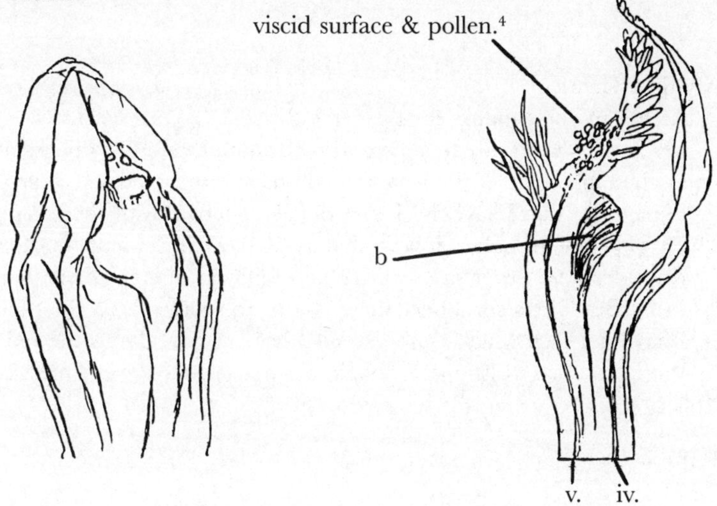

Best paper for drying plants can be got at Pamplin 45 Frith Street Soho London.[5]

Ask Ross for the simple microscope price £4" 4 which Dr Hooker orders. his address is Featherstone buildings Holborn[6]

Bentham & Oliver are quite struck up in a heap with your book & delighted beyond expression[7]

Incomplete
DAR 261 (DH/MS 11: 28)

CD ANNOTATIONS

4.1 Bentham ... expression 4.2] *crossed pencil*

Top of first page: 'Red Leschenaultia not formosa'[8] *pencil*; 'West Strand Eye glass' *pencil, del pencil*; 'Address for Paper' *pencil, del pencil*

Diagram: 'A Red Leschenaultia not formosa.' *pencil*

[1] Dated by the relationship to other correspondence with Hooker concerning *Leschenaultia* (see n. 2, below) and by reference to the publication of *Orchids* (see n. 7, below); 17 May 1862 fell on a Saturday.

[2] See letters to J. D. Hooker, 1 May [1862], 9 May [1862], and 15 [May 1862], and letters from J. D. Hooker, [5 May 1862] and [16 May 1862].

[3] See letter from J. D. Hooker, [16 May 1862] and n. 3.

[4] The diagram has been reduced to 65% of its original size.

5 See letter to J. D. Hooker, 15 [May 1862]. Hooker refers to the botanical bookseller William Pamplin (*Post Office London directory* 1861).

6 The reference is to the London optician and instrument-maker Thomas Ross (see letter to J. D. Hooker, 15 [May 1862] and n. 5).

7 CD sent presentation copies of *Orchids* to George Bentham and Daniel Oliver (see letter from Daniel Oliver, 14 May 1862, and letter from George Bentham, 15 May 1862, and Appendix IV).

8 See letter to J. D. Hooker, [18 May 1862].

From Georgina Tollet 17 May [1862][1]

30 Q. A. S^{t}
May 17

My dear Charles

Accept my best thanks for the Orchid book.[2]

My head is in too weak a state to read more than 2 or 3 pages at a time, but I enjoy these glimpses of the wonders you have discovered, exceedingly.

I much fear that I have no chance of talking it over with you this summer. Even if I were to get better before June 5, that awful day will certainly make me bad again.[3]

I wonder if you ever saw a bird's nest built on the top of another.

In Ld Bradford's shrubbery there is a hedgesparrow's on the roof of a wren's.[4]

I have not seen Mackintosh yet; all I hear of him alarms me much.[5]

With best love to Emma & thanks for her letter | Believe me | Y^{rs}. affectly | G Tollet

DAR 178: 128

1 The year is established by reference to the publication of *Orchids* (see n. 2, below).

2 Tollet's name appears on CD's presentation list for *Orchids* (see Appendix IV). *Orchids* was published on 15 May 1862 (Freeman 1977, p. 112).

3 Tollet refers to the marriage, on 5 June 1862, of her niece, Marianne Caroline Clive, to John Robert Orlando Bridgeman, son of the second earl of Bradford (see n. 4, below).

4 The reference is to George Augustus Frederick Henry Bridgeman, second earl of Bradford.

5 The reference is probably to Emma Darwin's nephew, James Mackintosh Wedgwood, who was suffering from an incurable cancer (Wedgwood and Wedgwood 1980, pp. 274 and 279).

From Asa Gray 18 May 1862

Cambridge. U.S.
May, 18. 62

My Dear Darwin,

Yesterday came by post the sheets—B–I of your Orchid book.[1]

This evening (Sunday) I have opened the parcel and read *Introduction* and Chapt. I. What a charming book it is.! You are right in issuing it in this form. It would be a sin not to do so.[2]

I fear, tho', that no publisher would reprint it here.; though I may, on reading farther conclude to offer it to the Appleton's—who should have the refusal.[3] But it will surely be popular in England, where *Orchids* are popular and the species known to most intelligent and educated people.

—I hope soon to get the other sheets. I am perfectly delighted with O. pyramidalis, and must extract the whole account of its fertilization for Sill. Jour.[4]

But now for a request, Sill. Jour. does not really pay its cost, and so cant well pay for making wood cuts. Would it be too much bother to order for me (at my expense) electrotype or stereotype casts of the illustrations of *O. pyramidalis*—and perhaps of *O. mascula* also.[5]

Indeed, if I get your little book reprinted here—and I am not sure that I ought not—it would save much of the cost and risk to have casts from all your cuts. Those of *O. pyramidalis*, &c could be sent thro. Trübner,[6] ⟨no⟩t mounted or backed.

Indeed, if the *electrotyped copper* casts alone were sent, in a proper box by post, they would hardly exceed a single 1/ postage. And I could have them backed & mounted here.

Our only *Orchis*—i.e. *O. spectabilis* I brought last summer from W. New York, & planted. I shall in a week have 3 or 4 spikes coming into flower, and I will cover one and leave the others exposed. They are in a wooded part of the garden, like their natural habitat. The rest of our Ophrydeæ are *Habenarias*, (*Platanthera*)

⟨I⟩ must recur to your letter about *Cypripedium* and see what you wanted of ⟨it⟩ i.e. what observation.[7]

If there be any adaptation—be it ever so pretty—I shall never see it without your direction. What a skill & genius you have for these researches. Even for the structure of the flower of Ophyrideæ I have to-night learned more than I ever knew before.

I think I have somewhere an older letter of yours unanswered, but at this moment I find only yours of April 21.[8] I have sent to Trubner for you a package of my pamphlet from Atlantic, for you to give away— Can send as many more as you like, for I find—owing to my own folly in giving it to Ticknor & Field, & not looking after ⟨it⟩—that it has not sold at all here.[9] ⟨In⟩deed, I had no care that it should. ⟨Th⟩e £8 is strictly yours: do not think of returning a penny of it.[10]

Hollies are 20 miles off, but I can send for them. *Rhexias* I hope to get young live plants of, soon.[11]

I shall be continuously overworked now till 10th July,

Ever yours | A. Gray.

Incomplete[12]
DAR 165: 109

1 Asa Gray had asked CD to send him the printed sheets of *Orchids* as they appeared so he could write an early review (see *Correspondence* vol. 9, letter from Asa Gray, 31 December 1861). See also letter to Asa Gray, 21 April [1862]. Gray's name also appears on CD's presentation list for *Orchids* (see Appendix IV).

2 CD initially considered publishing his study of orchid pollination as a paper in one of the journals of the Linnean Society of London (see *Correspondence* vol. 9, letter to John Murray, 21 September [1861]).

3 The New York publishing house D. Appleton & Co., established by Daniel Appleton and continued by his son William Henry Appleton, had published the first American edition of *Origin* (see *Correspondence*

vol. 7, letter to Asa Gray, 21 December [1859], and *Correspondence* vol. 8, letter from Asa Gray, [10 January 1860]).

4 The *American Journal of Science and Arts* was commonly known as 'Silliman's journal' after its founder Benjamin Silliman. Gray's review of *Orchids* appeared in the July issue of the journal; the review included extracts from *Orchids* on *Orchis pyramidalis*, a species not found in the United States (A. Gray 1862a, pp. 140–2).

5 Gray refers to *Orchids*, figures 3 and 1, facing pages 22 and 18, respectively. He used the illustrations in a follow-up article to his review published in the November issue of Silliman's journal (A. Gray 1862b, pp. 421, 423). See also letter to John Murray, 13 June [1862] and n. 6.

6 The London publisher Nicholas Trübner frequently acted as Gray's London agent.

7 See *Correspondence* vol. 9, letters to Asa Gray, 12 March [1861], 5 June [1861], and 21 July [1861]; see also letter to J. D. Hooker, [28 July – 10 August 1861].

8 Gray probably refers to CD's letter of 15 March [1862]; although Gray notes its receipt in his letter to CD of 31 March [1862], CD had asked Gray to carry out some observations on *Rhexia* and let him know the results, and Gray had yet to do so.

9 CD and Gray shared the cost of reprinting and publishing as a pamphlet Gray's reviews of *Origin* that had appeared in the *Atlantic Monthly* (A. Gray 1861). CD had already given away more than 100 copies of the pamphlet (see *Correspondence* vol. 9); in his letter to Gray of 21 April [1862], he asked for further copies to give away. Gray refers to the Boston publishing company conducted by William Davis Ticknor and James Thomas Fields.

10 See letter from Asa Gray, 31 March [1862], and letter to Asa Gray, 21 April [1862].

11 See letter to Asa Gray, 21 April [1862] and nn. 11 and 13.

12 See letter to Asa Gray, 10–20 June [1862] and n. 2.

To J. D. Hooker [18 May 1862][1]

Leith Hill Place
Sunday

My dear old friend.

I cordially thank you for your two letters; & for your answers to all my questions. Leschenaultia seems very odd:[2] I will try with pollen left on for 48 hours; for I I am sure I could trace the tubes for some way. It seems to me very curious that anthers should shed the pollen, & then the indusium collect it, ultimately to be removed again & placed on stigma. It is a good illustration of a point which I discuss & endeavour to explain in the orchid-book, namely the diversity of structure, acquired by successive modifications, for same purpose.[3] You have pleased me much by what you say in regard to Bentham & Oliver approving of my book;[4] for I had got a sort of nervousness & doubted whether I had not made an egregious fool of myself, & concocted pleasant little stinging remarks for Reviews,—such as "Mr Darwin's head seems to have been turned by a certain degree of success, & he thinks that the most trifling observations are worth publication" &c &c &c &c.—

You say that you have no Goodeniaceæ;[5] I suppose this refers to having none to dissect; or did you think that I asked for them? It is any *Melastomatads*, going into flower & which would stand my greenhouse for a short period, which I am so anxious to experiment on, & which I feel nearly sure will give important results.—[6] I fear it will be no satisfaction to you, but upon my soul, I do feel deeply guilty at the degree to which I trouble you; & you never in all your life have given me a rebuf. Now mind do not answer on purpose, but at some time tell me what the enclosed Sikkim R[h]od. is;— it has a reflexed pistil like R. Boothii & secretes nectar *within* the stamens all round the germen![7] R. Dalhousianum secretes nectar on lower side of flower; so that I do not fear the usual law will hold with respect to bending of pistil & position of nectar.—

We heard from William this morning; he asked me, supposing that you asked him to Kew within a fortnight, whether he might put it off for a week or so, till another Sunday was convenient to you, & I told him that I was sure you would let him do so.—[8]

Emma sends her love to M^rs. Hooker & desires me to say that she quite forgot to thank M^rs. H. for a very nice photograph.—

Good Bye my dear old fellow | C. Darwin

P.S. Are you rich in Leschenaultia formosa I sh^d. like to try whether the visid surface outside the indusium can be fertilised; that inside can be in this species be fertilised.—[9]

Many thanks for Saxifrages.—[10]

DAR 115.2: 154

[1] Dated by the relationship to the letters from J. D. Hooker, [16 May 1862] and [17 May 1862] (see nn. 2, 4, and 5, below), and by reference to the Darwins' stay at Leith Hill Place, home of Caroline Sarah Wedgwood and Josiah Wedgwood III. The Darwins stayed at Leith Hill Place from 15 to 22 May 1862 (see 'Journal' (Appendix II)). In 1862, 18 May fell on a Sunday.

[2] See letters from J. D. Hooker, [16 May 1862] and [17 May 1862].

[3] This point is discussed in *Orchids*, pp. 346–51, in a section entitled 'Cause of the vast diversity of structure for the same general purpose'.

[4] See letter from J. D. Hooker, [17 May 1862]. See also letter from Daniel Oliver, 14 May 1862, and letter from George Bentham, 15 May 1862.

[5] See letter from J. D. Hooker, [16 May 1862].

[6] See letter to J. D. Hooker, 15 [May 1862].

[7] CD mentioned his interest in *Rhododendron boothii* in the letter to J. D. Hooker, 15 [May 1862]. For Hooker's identification of CD's specimen, see the letter from J. D. Hooker, 19 [June 1862]. There is a note describing the bent pistils of rhododendrons in DAR 49: 77–8, dated 'May 16— 62.'

[8] The letter from William Erasmus Darwin has not been found. Hooker had invited William to the Royal Botanic Gardens, Kew (see letter to J. D. Hooker, 9 May [1862]). See also letter from J. D. Hooker, [5 May 1862], and letter to W. E. Darwin, [8 May 1862].

[9] See letter from J. D. Hooker, [17 May 1862] and CD annotations.

[10] See letter to J. D. Hooker, 9 [April 1862].

To Alexander Goodman More 18 May 1862

Down, Bromley, Kent. [Leith Hill Place]
May 18, 1862

My dear Sir

If you should have leisure and your health should be good, and if again you should feel any little interest on subject, perhaps you would try the experiment on Ep. palustris. Flower should be chosen in middle of spike, and marked by a thread, and when the capsules are nearly ripe, each should be *separately* folded up in paper; and another capsule from an unmutilated flower close by from same spikes should be gathered and folded up in paper as standard of comparison. I would then endeavour to estimate whether there was any difference in fertility in the flowers possessed of the distal portion of the labellum, and those deprived of it. I should much like to see this point ascertained;[1] but of course you must not think of troubling yourself unless you should feel some little curiosity in the subject. Many thanks for your note.[2] Pray do not think of acknowledging this

My dear Sir | Yours sincerely | Ch. Darwin

Of course the distal portion of labellum would have to be removed from flowers not fully expanded.

Copy
DAR 146: 404

[1] More lived on the Isle of Wight where *Epipactis palustris*, the marsh epipactis, was common. CD and More tried this experiment (designed to ascertain the importance of the structure of the labellum) the previous summer, but, as CD reported in *Orchids*, p. 101 n., the attempt was made too late in the season and the seed capsules More did manage to send shed most of their seed before reaching Down House. See *Correspondence* vol. 9, letters to A. G. More, 4 June 1861, 7 July 1861, and 19 July 1861. See also this volume, letter from C. W. Crocker, 13 March 1862 and n. 2.

[2] The letter from More has not been found; he may have written to thank CD for his presentation copy of *Orchids* (see Appendix IV).

From H. W. Bates 19 May 1862

King St Leicester
19 May /62

My Dear Mr Darwin

The discovery of signs of a glacial period in Miocene times is very remarkable as I dare say you have thought, in as much as the whole tertiary ⟨ep⟩och has been always considered a time of greater warmth than the present in the Northern Hemisphere. But if it turns out a cold epoch intervened in middle of the tertiary it would only compel us to enlarge the lapse of time allowed to the whole.[1]

I have just received the German monograph on Chilian Carabi.[2] It is very complete. The author is of high reputation. He concludes that the 11 species form only a section of the genus although very closely allied amongst themselves & distinct as a body from all Carabi of the Northern Hemisphere. He anticipates no future discovery to modify ⟨current⟩ conclusions on the Geographical distribution of the genus. No Carabus has been found within the tropics & none beyond the Southern tropic in Eastern Hemisphere He goes into many inquiries arising out of the subject but strange to say does not start the question "Whence came these isolated Chilian Carabi?"

I never thought of modifications in horny genital apparatus of closely-allied Chrysomelæ being a difficulty ⟨for the⟩ theory.[3] Quite the contrary it see⟨ms that⟩

they were a strong support of it. M^r^ ⟨Baly⟩ has found that an English Chr⟨ysomela⟩ believed on other grounds to be a ⟨ ⟩ in these organs from its supp⟨ ⟩ ⟨ ⟩[4] He therefore again separates it & ⟨there⟩ is an end of the matter. Like many other ⟨natur⟩alists who aim only at separating species *neatly* in their collections & monographs, Mr Baly never thinks of the possibility of gradual modification & consequently never looks in Nature for it. I believe, as regards the total specific form, $\frac{3}{4}^{\text{ths}}$ of the species in all large groups will be found, when specimens are collected over a wide area, to be incapable of sharp definition from their nearest allies.

⟨ ⟩ modifications have scarcely ever been ⟨called for⟩, even in the whole specific char⟨acter or⟩ special organs. Mr Janson,[5] an ⟨entom⟩ologist tells me there is a difference ⟨in genital⟩ apparatus between two English ⟨Carabi⟩ C. hybrida & C. maritima. Now ⟨modern⟩ ⟨en⟩tomologists, on the most conclusive ⟨evidenc⟩e have re-united these two. It appears ⟨h⟩owever the two are more distinct in England than on the Continent! This is what I should expect from observations on S. American insects. There is a gradual divergence amongst varieties of a species over a wide area. Who would think of examining the genital apparatus through all the graduated series of vars. of these Cicindelæ?

Thanks for the copy of "Orchids". I have read it through with great pleasure.[6] It is very clear.— In Ann. Nat. Hist. for June you will see a note of mine on a new way of regarding local varieties[7] I have sent in an application for British Museum. situation but it will be of little use.[8] Something else may turn up soon

Yours sincerely H W Bates

I go tomorrow to London, for three days chiefly to see the artist about finishing the plates for Linnean Transactions.[9] I shall call on Murray.[10] Can I do anything for you. The address 43 Harwood St Hampstead road will find me.

DAR 160.1: 69

CD ANNOTATIONS
1.1 The ... whole. 1.5] *crossed pencil*
2.2 He concludes ... Carabi?" 2.9] *crossed pencil*
5.1 Thanks ... Bates 6.1] *crossed pencil*
7.1 I go ... me. 7.3] *crossed pencil*
Top of first page: 'Ch. IV'[11] *added brown crayon*

1 See letters to H. W. Bates, 4 May [1862] and 9 May [1862].
2 Gerstaecker 1858.
3 See letter to H. W. Bates, 9 May [1862] and n. 5.
4 The entomologist Joseph Sugar Baly was a leading authority on phytophagous Coleoptera (*Modern English biography*).
5 Edward Westley Janson was the curator of the Entomological Society of London.
6 Bates's name appears on CD's list of presentation copies of this work (see Appendix IV).
7 Bates refers to the final section of Bates 1862b. In *ibid.*, p. 451, having described some local varieties of *Anisocerus onca*, Bates stated:

> It is the custom of naturalists, when they subordinate varieties to a species, to fix upon one of the forms as the original, to which the rest are referred: this original is generally the one first

described or best known . . . but, strictly speaking, no form can be said to be a variety of *another existing* form unless it can be proved or shown to be highly probable that the one *descended* from the other, this other itself remaining meanwhile unchanged.

[8] The reference is to an assistantship in the department of zoology at the British Museum (see letter from H. W. Bates, 30 April 1862, and letters to H. W. Bates, 4 May [1862] and 9 May [1862]).

[9] Edward W. Robinson drew the coloured illustrations for Bates 1862a.

[10] John Murray was to publish Bates's account of his travels in Brazil (Bates 1863). See the letters to H. W. Bates, 31 January [1862] and 27 [February 1862] concerning Bates's arrangement with Murray.

[11] CD probably refers to chapter 4 of his 'big book' on species, on 'Variation under nature'.

From H. G. Bronn[1] 19 May 1862

Heidelberg
le 19 Mai 1862

Monsieur!

J'ai eu l'honneur de recevoir les paquets, que Vous avez bien voulu m'addresser soit directement soit par l'intermédiaire de Mr Schweizerbart;[2]—et je vous remercie sincèrement non seulement pour les additions à la dernière édition anglaise de vôtre Origin of Species, mais particulièrement encore pour la manière, dont vous m'avez su faciliter de retrouver les pages et les lignes, où un changement ou une addition doit avoir lieu dans la nouvelle édition allemande, dont l'impression va commencer dans ces jours mêmes.[3] Il faut cependant me permettre une observation sur un passage dans la dernière édition Anglaise, qui ne s'est pas trouvé dans la précédente.

Pag. 142: Geology shows us, at least within the whole imm⟨ense⟩ tertiary period, that the number of species of shells and probably, *of mammals*, has not greatly or at all increased.

N'ÿ a-t-il pas là quelque faute d'impression? L⟨es⟩ ⟨ ⟩ n'existent au commencement de la période tertiaire ⟨ ⟩ nombre et de peu d'ordres seulement; ils deviennent toujo⟨urs⟩ plus nombreux, de sorte que dans l'étage Miocène ils paraissent être même plus variés et plus abondants, qu'aujourdhui. Au moins le nombre d'espèces trouvées ensemble dans une même couche d'une même localité (à Mayance, Sansan etc)[4] est quelquefois plus grand, que celui qu'on m'en trouveroit à présent dans toute l'Europe. Il n'ÿ a pas de doute, que le nombre des Mammifères, dont on ne connoît qu'une douzaine d'espèces à peu près dans les terrains pré-tertiaires, n'ait augmenté plus considérablement peutêtre que celui d'aucune autre classe justement dans la période tertiaire. Je vous prie donc, Monsieur, de bien vouloir me faire savoir comment il faut entendre ce passage?[5]

Les feuilles de Vôtre livre sur les Orchidées m'ont fait reconnoître, qu'il sera d'une grande importance pour la science, qu'il va enrichir de nouvaux faits relativement à une famille de plantes qui sont des plus interèssantes de toutes, et dont il corrigera la théorie.[6] Si peût-être pas tous les lecteurs de "l'Origin" seroient assez botanistes, pour s'ÿ intéresser il ÿ aura sans doute un grand nombre de botanistes qui, ne possedent pas l'Origin, mais achèteront les Orchidées. C'est ce que j'ai écrit à Mr. Schweizerbart en lui communiquant les épreuves; cependant je l'ai prié également d'en consulter encore un botaniste même, comme p.e. Mr. *v. Mohl*.[7] Ensuite il aura à faire son compte rélativement à ses frais; et je crois qu'il sera à même à Vous communiquer sa decision en 18–24 jours auplus!—car on vient de me prévenir qu'il est parti pour la foire de Leipsic, voÿage que' chaque libraire éditeur doit faire tous les ans et qui conge 2–3 semaines.[8]

Je Vous ai dit, que je n'adhère pas à Vôtre théorie, parce-que, malgré tous les avantage qu'elle auroit pour la science, elle est encore en opposition avec des faits fondamentaux de la science (le premier developpement d'un animal sortant de matière inorganique), mais que néanmoins je suis convaincu, qu'*elle* nous conduira enfin sur la route de la verité.[9] Peût-être, si je trouve assez de temps, j'ajouterai à la traduction quelques observations intéressantes sur la variabilité des espèces relativement à des charactères, qu'on a cru être les plus constants.[10]

Veuillez aggréer, Monsieur, l'expression de la considération toute particulière, avec laquelle j'ai, l'honneur d'être, Monsieur | Vôtre | très devoué | H. G. Bronn

DAR 160.3: 321

CD ANNOTATION
5.5 Peût-être, . . . constants. 5.7] *two crosses in margin, pencil*

1 For a translation of this letter, see Appendix I. Although Bronn had previously corresponded with CD in German, he began writing in French after learning that CD found it more difficult to read German than French (see letter to H. G. Bronn, 11 March [1862]).

2 See letter to H. G. Bronn, 25 April [1862]. Christian Friedrich Schweizerbart was the head of E. Schweizerbart'sche Verlagsbuchhandlung, the publishing firm that published the German translation of *Origin*, and was preparing to publish a translation of *Orchids*.

3 Bronn was preparing a second German edition of *Origin* (Bronn trans. 1863) from the third English edition. The manuscript alterations and corrections sent by CD with his letter of 25 April [1862] have not been found. However, those changes incorporated in the second German edition of *Origin* that do not occur in the third English edition are given in Appendix VIII. Bronn trans. 1863 was initially published in three parts, whose publication was announced on 6 October, 17 November, and 19 December 1862, respectively (*Börsenblatt für den Deutschen Buchhandel* 29 (1862): 2083, 2447, and 2735).

4 Mayence is the French name for the German city of Mainz. Bronn refers to the rich Miocene fauna found at both Eppelsheim, a village south of Mainz, and at Sansan, a village near Auch in the south of France (see Bronn 1858, p. 473).

5 No reply from CD to this query has been found, and in Bronn trans. 1863, p. 152, the sentence is translated without modification. However, Bronn added a footnote that reads: 'Bekanntlich hat sich die Säugthier-Welt fast ganz erst im Laufe der Tertiär-Zeit entwickelt.' [As is commonly known, mammals developed almost exclusively during the Tertiary period.] CD annotated the sentence in his copy of the third edition of *Origin*, and in the fourth edition published in 1866, he amended it to read: 'But geology shows us, that from an early part of the long tertiary period the number of species of shells, and that from the middle part of this same period the number of mammals, has not greatly or at all increased' (*Origin* 4th ed., p. 151). CD's copies of his own books are in the Rare Books Room–CUL.

6 CD sent Bronn the first half of the proof-sheets of *Orchids* at the end of April (see letter to H. G. Bronn, 25 April [1862]); Bronn's name also appears on CD's list of presentation copies for this work (see Appendix IV).

7 See letter to H. G. Bronn, 25 April [1862]. Hugo von Mohl, professor of botany at Tübingen, was a specialist on plant anatomy.

[8] See letter from E. Schweizerbart'sche Verlagsbuchhandlung, 7 June 1862. Leipzig was a major centre for the book trade in the nineteenth century; the book fair formed part of the annual Easter fair (*Börsenblatt für den Deutschen Buchhandel* 29 (1862): 1172–3, and *EB*).

[9] See also letter from H. G. Bronn, [before 11 March 1862]. For a discussion of Bronn's attitude towards CD's theory, see Junker 1991.

[10] Bronn had appended a chapter commenting on *Origin* to his translation of the first edition (see *Correspondence* vol. 8). He added no further remarks to the second German edition, stating that, as the latest English edition (*Origin* 3d ed.) contained several responses to his original comments, he felt obliged to leave his chapter unchanged in the second German edition (Bronn trans. 1863, p. 525 n.).

From Edward Cresy 19 May 1862

Metropolitan Board of Works | Spring Gardens
19 May '62

My dear Sir,

Pray accept my very best thanks for your wonderful book on the fertilization of Orchids—[1] I dont pretend to have mastered it yet for it wants very careful reading but what I have read has given me the greatest pleasure— I knew how singular & striking was the mechanism of many foreign orchids, but had no idea of the extent and variation of contrivance in the British— I am truly glad you have yourself explained so strong a class of cases for the advocates of separate creation— No one can accuse you of suppressing anything in their favor—& I dare say they will profit by some of your labors— I confess I am altogether puzzled by the Bee Ophrys— Why so much pains should be taken with all the others to ensure intercrossing and this one go on for ever self fertilising is a regular teazer—[2] Is there any solution to be found in multiplication by the roots— I believe some orchids do and some do not— I think your anticipation by analogy of a Madagascar moth with a probiscis ten inches long equals Adam's & Leverrier— What a triumph it will be to find him—[3]

I am very grateful for your few remarks on the secretion of nectar I own to having felt it to be a difficulty in the path and it never occurred to me 'as matter excreted to free the system from superfluous or injurious substances' seized upon by natural selection as a means for working out an end—[4] I had never noticed the glands secreting nectar in the laurel leaves— I confess myself greatly astonished at the prolificness of the orchids you cite— I thought they must be shy bearers from their comparative scarceness— What can keep them down?— You do not mention the source of the very peculiar smell of most of them. I suppose it to be important in attracting insects—or has it not been worked up by natural selection— the common orchis I have noticed to be almost scentless by day & remarkably fetid at night. it makes a room stink as if no end of cats had passed the night in it—

The chapter on homologies is to me and will be to many others who have not access to the great monographs peculiarly instructive—[5] From having your attention so constantly directed to the subject & therefore at your finger's ends you can hardly conceive the extreme value of such a diagram as fig 32—[6] Half the story in fact is incomprehensible to our weak minds without it— with it we feel quite cleared up— I never thanked you for your interesting brochure on dimorphism in Primula—[7] The beauty of natural selection is the immense variety of new thoughts it suggests—

Yours very truly & very gratefully | E Cresy

I hope your daughter continues to gain health & strength[8] & that Mrs Darwin is well Pray give my very kind remembrances—

C Darwin Esq.

DAR 161.2: 239

[1] Cresy's name appears on CD's presentation list for *Orchids* (see Appendix IV). Cresy had assisted CD over the previous two years with his studies of insectivorous plants (see *Correspondence* vols. 8 and 9).

[2] CD was also puzzled by the fact that *Ophrys apifera*, the bee-orchis, appeared to be adapted to favour perpetual self-fertilisation (see *Correspondence* vol. 9, letters to A. G. More, 17 June 1861 and 17 July 1861). In *Orchids*, p. 63, he described the plant as having 'widely different means of fertilisation as compared with the other species of the genus, and, indeed, . . . , with all other Orchids'.

[3] *Angraecum sesquipedale*, an orchid with 'large six-rayed flowers' and a 'whip-like green nectary' nearly a foot long, is discussed in *Orchids*, pp. 197–203. The nectar being held at the lower end of the nectary, CD surmised that in the orchid's natural habitat in Madagascar 'there must be moths with probosces capable of extension to a length of between ten and eleven inches!' (*Orchids*, p. 198). For a discussion of CD's view, see Kritsky 1991. In 1845, John Couch Adams and Urbain Jean Joseph Leverrier independently postulated the existence and location of the planet Neptune from the perturbations in the orbit of Uranus (*DSB*).

[4] *Orchids*, p. 278.

[5] The homologies of orchids are discussed in *Orchids*, pp. 286–307.

[6] Cresy refers to the illustration, showing an orchid flower in section, in *Orchids*, p. 292.

[7] Cresy refers to CD's paper, 'Dimorphic condition in *Primula*'. For CD's presentation list for this paper, see Appendix III.

[8] CD's eldest daughter, Henrietta Emma Darwin, had been ill throughout much of 1861 (see *Correspondence* vols. 8 and 9).

From Henry Holland 19 May [1862][1]

Brook Street
May 19

My dear Charles

Though dating from Brook S^t^., I am beginning this letter from the house of a patient— Its purport is to thank you for the very interesting volume on the orchids, which you have been good enough to send me;[2] the major half of which I have already perused; & the remainder of which I look to with the same expectation of pleasure

There are two or three comments which occur to me on the subject, as it bears upon the principle of "natural selection". But these matters lie beyond the limits of a note, & I will not enter upon them now. I earnestly wish your time, & your health, gave us more frequent opportunities of meeting & talking over these subjects.

I should gladly hear of any improvement in the state of your little Boy, as respects this singular form of malady.[3] Let me know if there is anything which can be considered a change in the character of the symptoms

Ever, my dear Sir | Yours affec^{ly} | H Holland

DAR 166.2: 242

[1] The year is provided by reference to the publication of *Orchids*.
[2] Holland's name appears on the presentation list CD drew up for *Orchids* (see Appendix IV).
[3] Holland refers to Horace Darwin (see letter from Henry Holland, 26 March [1862]).

From Thomas Campbell Eyton [after 19 May 1862?][1]

Dear Darwin

I was persuaded to sit the other day for a photograph a copy of which I enclose in hopes that you will do the same thing and send me a copy[2] I have got lots of chicken by the jungle fowl from game hens[3]

Yours truly | Th C Eyton

Eyton
Saturday

DAR 163: 40

CD ANNOTATION
Top of first page: '1862' *pencil*

[1] The date is conjectured from CD's annotation and from Eyton's use of stationery with a mourning border: his father-in-law, Robert Aglionby Slaney, died on 19 May 1862 (*DNB*).
[2] Eyton's photograph has not been found in the Darwin Archive–CUL.
[3] In 1861, CD had sought assistance from Eyton while preparing chapter 7 of *Variation* on domesticated fowl (see *Correspondence* vol. 9, letters to T. C. Eyton, 3 May [1861], 6 May [1861], and 14 May [1861]).

From C. C. Babington 22 May 1862

Cambridge
22. May. 1862

Dear Darwin

I ought to have written sooner to thank you for a copy of your exceedingly interesting and valuable book upon Orchids.[1]

It is highly satisfactory in all respects. The results are most curious and the skill shown in discovering them equally so.

Yours very truly | Charles C. Babington—

DAR 160.1: 3

[1] Babington's name is on the presentation list CD drew up for *Orchids* (see Appendix IV).

Dorothy Frances Nevill.
Photograph, 1865.
(From Nevill [1910]. By permission of the Syndics of the Cambridge University Library.)

Charles Cardale Babington.
Photograph by Ernest Edwards, *c.* 1863.
(From L. Reeve ed. 1863–6. By permission of the Syndics of the Cambridge University Library.)

Henry Holland. Portrait by Thomas Brigstocke, *c.* 1860.
(Courtesy of the National Portrait Gallery, London.)

From J. D. Hooker 23 May 1862

Dr Darwin

I do not know *Rhod. Boothii*: this is *R. Keysii*, a remarkable form.[1] I will send a Melastomad or two anon—[2]

In haste Ever Yrs | J D Hooker

Rhod. Boothii is a Bhotan species of which I have seen dried leaves only

Kew May 23/62.

DAR 101: 34–6

[1] See letters to J. D. Hooker, 15 [May 1862] and [18 May 1862].
[2] See letter to J. D. Hooker, [18 May 1862].

From A. R. Wallace 23 May 1862

5, Westbourne Grove Terrace, W.
Friday May 23rd. 1862

My dear Mr. Darwin

Many thanks for your most interesting book on the *Orchids*.[1] I have read it through most attentively & have really have been quite as much *staggered* by the wonderful adaptations you shew to exist in them as by the *Eye* in animals or any other complicated organs..[2] I long to get into the country & have a look at some Orchids guided by your *new lights*—, but I have been now for 10 days confined to my room with what is disagreeable though far from dangerous;—*boils*.

I have been reading several of the Reviews on the "Origin", & it seems to me that you have assisted those who want to criticise you by your *overstating* the difficulties & objections— Several of them quote your *own words* as the strongest arguments against you.

I think you told me Owen wrote the article in the "Quarterly". This seems to me hardly credible as he speaks so much *of Owen* quotes him as such a *great authority* & I believe even calls him a *profound philosopher*. &c. &c.. Would Owen thus speak of himself?[3]

Trusting your health is good | I remain My dear M^r. Darwin | Yours very faithfully | Alfred R. Wallace

C. Darwin Esq.

DAR 106/7 (ser. 2): 2–3

[1] Wallace's name appears on CD's list of presentation copies for *Orchids* (see Appendix IV).
[2] In chapter 6 of *Origin*, entitled, 'Difficulties on theory', CD included a discussion of 'Organs of extreme perfection and complication', chief among which was the eye. He stated (*Origin*, pp. 186–7):

> To suppose that the eye, with all its inimitable contrivances for adjusting the focus to different distances, for admitting different amounts of light, and for the correction of spherical and chromatic aberration, could have been formed by natural selection, seems, I freely confess, absurd

in the highest possible degree. Yet reason tells me, that if numerous gradations from a perfect and complex eye to one very imperfect and simple, each grade being useful to its possessor, can be shown to exist; if further, the eye does vary ever so slightly, and the variations be inherited, which is certainly the case; and if any variation or modification in the organ be ever useful to an animal under changing conditions of life, then the difficulty of believing that a perfect and complex eye could be formed by natural selection, though insuperable by our imagination, can hardly be considered real.

3 The anonymous, critical review of *Origin* published in the *Quarterly Review* was in fact by Samuel Wilberforce, bishop of Oxford ([Wilberforce] 1860), though CD believed that Owen had assisted Wilberforce with its composition (see *Correspondence* vol. 8, letter to J. D. Hooker, [20? July 1860], letter to Asa Gray, 22 July [1860], and letter to W. E. Darwin, [30 July 1860]). There is no mention of Wilberforce's review in the previous extant correspondence between CD and Wallace, but CD did tell Wallace of Owen's anonymous, negative review of *Origin* for the *Edinburgh Review*, in which Owen cited his own works favourably ([R. Owen] 1860b; see *Correspondence* vol. 8, letter to A. R. Wallace, 18 May 1860). See also letter to A. R. Wallace, 24 [May 1862].

To Leonard Jenyns 24 May [1862][1]

Down. | Bromley. | Kent. S.E.
May 24th

Dear Jenyns

I thank you most sincerely for your kind present of your memoir of Henslow.[2] I have read about half & it has interested me much. I did not think that I could have venerated him more than I did; but your Book has even exalted his character in my eyes. From turning over the pages of the latter half I shd. think your account would be invaluable to any Clergyman who wished to follow poor dear Henslow's noble example. What an admirable man he was.

I hope that you are yourself pretty well. I cannot say much for my own health.

With sincere thanks, believe me dear Jenyns | Yours very truly | Ch. Darwin

Bath Royal Literary and Scientific Institution

1 The year is established by the reference to Jenyns's memoir of his brother-in-law, John Stevens Henslow (see n. 2, below).

2 Henslow died on 16 May 1861. CD had contributed his own recollections of Henslow to Jenyns's memoir (Jenyns 1862, pp. 51–5; see also *Correspondence* vol. 9 and Appendix X).

To A. R. Wallace 24 [May 1862][1]

Down. | Bromley. | Kent. S.E.
24th

My dear Mr Wallace

I write one line to thank you for your note & to say that the B. of Oxford wrote the Quarterly R (paid £60), aided by Owen. In the Edinburgh Owen no doubt praised himself.[2] Mr. Maw's Review in Zoologist is one of the best; & staggered me in part, for I did not see the sophistry of parts—[3] I could lend you any which you might wish to see; but you would soon be tired. Hopkins in Fraser & Pictet are two of the best.—[4]

I am glad you approve of my little Orchid Book; but it has not been worth, I fear, the 10 months it has cost me: it was a hobby horse & so beguiled me.—[5]

I am sorry to hear that you are suffering from Boils; I have often had fearful crops: I hope that the Doctors are right in saying that they are serviceable.—[6]

How puzzled you must be to know what to begin at.[7] You will do grand work, I do not doubt.— My health is, & always will be, very poor: I am that miserable animal a regular valetudinarian.—

Yours very sincerely | C. Darwin

British Library (Add 46434: 25)

[1] Dated by the relationship to the letter from A. R. Wallace, 23 May 1862.

[2] [Wilberforce] 1860 and [R. Owen] 1860b. See letter from A. R. Wallace, 23 May 1862 and n. 3.

[3] Maw 1861. Soon after the review's publication, CD had corresponded with George Maw about some of the points it raised (see *Correspondence* vol. 9, letter to George Maw, 19 July [1861], and letter from George Maw, 27 August [1861]).

[4] Hopkins 1860 and Pictet de la Rive 1860. For CD's correspondence about these reviews, see *Correspondence* vol. 8.

[5] CD had sent Wallace a presentation copy of *Orchids* (see letter from A. R. Wallace, 23 May 1862, and Appendix IV). According to his journal, CD began work on *Orchids* in July 1861 and completed it at the end of April 1862 (see *Correspondence* vol. 9, Appendix II, and this volume, Appendix II).

[6] Henry Holland, one of the physicians consulted by the Darwin family, believed that abscesses or other 'affections of the skin' heralded the end of a disease, giving 'relief to the system' (see Holland 1855, p. 143).

[7] Wallace had recently returned to England having spent eight years in south-east Asia (Wallace 1905, 1: 385).

From Joseph Beete Jukes 25 May 1862

Geological Survey of Ireland, | *Office, 51, Stephen's Green, Dublin,* | Nenagh
May 25 *1862*

My dear Darwin

Many thanks for your orchid book which I received some time ago & have been carrying about with me in hopes of being able to read it.[1] Hitherto I have only dipped into it and it seems excessively interesting.— Field work and reading are I find almost incompatible, after a blow on the hills, or as yesterday a climb about a mine, one has such a fatal facility for falling asleep after dinner if one attempts to read, that one is off into the land of nod without being aware of it & though I am an early bird yet I find letters, geological notes &c &c take all ones time before breakfast.—

While you are working away at the Organics I am chiefly laying the physical substratum for future observers to work on. I have however hit upon one conclusion lately which I think will interest you.—

Many of the rivers of Ireland after running over a low limestone plain with water sheds not exceeding 300 feet, escape to the sea by deep gorges thro' hills of Slate & Old Red &c more than double that height.—

Some in the S. after running down long limestone valleys between ORS[2] ridges, those valleys & ridges running straight out to the sea with same features, suddenly turn from the valleys & run through deep transverse ravines across the ridges. These latter cases I have been enabled to link on to some lateral brooks coming down from the loftier ridges on the N opposite the points where the transverse ravines commence & to show that the ravines were formed by these brooks, being commenced on a higher surface & always cutting down by running water faster than the surface of the valleys sank. Therefore it was always dry land & the valleys have been worn by *atmospheric* degradation alone.—[3] The Limestone has in fact been dissolved; & the limestone plains & valleys have lowered & sunk down past the other rocks like glaciers sinking in their beds under a hot sun.

The other rocks have also suffered by atmospheric degradation valleys being worn in the softer parts of them.

Supposing my explanation be right in Ireland it must be applicable elsewhere. The Weald for instance, which after all is a mere flea bite compared to the denudation of the Palæozoic rocks.— I suspect that the Chalk was bared of the Tertiary rocks by marine denudation as the rock rose above the Sea, that brooks commenced to run down the chalk slopes along the courses of those which now cut ravines through the Chalk escarpments, & that those ravines have been worn by those brooks continually cutting deeper than the ground inside, that the Chalk which has been removed has been merely *dissolved* off the crown of the arch by atmospheric action & the hills & valleys inside worn by the *rain* only & the *weather*. Your 300,000,000 of years is not nearly enough for the denudation of the Weald by this process.[4]

I have sent in a paper to the Geol: Soc: Lond: on the Irish valleys w.^{h} is to be read on June 18th.—[5]

I intend to be there with large maps and sections and I shall be anxious to hear if any one can pick a hole in the reasoning or give another explanation for the phenomenon

I have long been considering the "Form of Ground" as a geological problem as yet unsolved and think my idea will give us a help.—[6]

Agrarian outrages beginning to sprout again all over this Tipperary country.—[7] Much wider emigration wanted. Only fancy! I was told by Cap^{tn}. King at Silvermines yesterday, that one man had given another £80 for the possession of 6 acres of land on the hill side there: both being mere tenants at will of Lord Dunally's.[8] When the peasants *buy* the land in this way from one another no wonder they fancy it is their own.[9]

Believe me | yours very truly | J. Beete Jukes.

DAR 168: 90

[1] Jukes's name appears on the presentation list CD drew up for *Orchids* (see Appendix IV).
[2] Old Red Sandstone.
[3] Jukes was studying the mode of formation of regional drainage patterns. In his paper on the topic (Jukes 1862), Jukes introduced the concept of drainage superimposition, and propounded the idea that rivers not only excavate their valleys but adjust their courses according to underlying structures (Davies 1969, p. 330). In his history of geomorphology, G. L. Davies identified this 'pioneer study' as a seminal event in the 'fluvial revival' witnessed in Britain in the 1860s (Davies 1969, p. 330).
[4] In *Origin*, pp. 285–7, CD calculated that the erosion of the Weald, at what he estimated to be a reasonable rate, would have taken 306,662,400 years. In response to criticisms in the *Saturday Review*, 24 December 1859, however, CD halved his estimate in the second edition of *Origin*, and removed the discussion altogether from the third edition. When, in the spring of 1860, CD informed Jukes of his intention to remove the passage, Jukes had sought to persuade him not to do so, expressing his conviction that the estimate 'was not at all exagerrated' (see *Correspondence* vol. 8, letter to Asa Gray, 3 April [1860], and letter to Charles Lyell, 10 April [1860]).
[5] Jukes 1862.
[6] In 1855, the director-general of the Geological Survey, Roderick Impey Murchison, ordered that maps be accompanied by explanatory booklets, including an account of the 'form of ground' or topography of the area; as director of the Irish branch of the survey, Jukes's duties included editing these explanatory notes (Jukes 1862, p. 379). See also Davies 1969, p. 323.
[7] Violent action by combinations formed by tenant farmers wanting reforms in the system of land tenure was on the increase in rural Ireland in the 1860s (Clark 1979, p. 211). 'Agrarian outrage' was a term used by the Irish constabulary to refer to any offences concerning the occupation of land (Warwick-Haller 1990, p. 22).
[8] Henry Sadleir Prittie, third Baron Dunalley, was proprietor of 21,000 acres of land in Tipperary, Ireland (Dun 1881). Captain King has not been identified.
[9] Jukes refers to the custom of selling one's 'interest' in a holding (the 'Ulster custom' or 'tenant right'). This 'right' had no legal status, but was a non-contractual privilege for which the tenant was usually required to obtain permission from the landlord (see Clark 1979, pp. 166–7, 180–1).

From G. C. Oxenden 26 May 1862

Broome | Canterbury
May 26. 1862.

Dear Sir

Each fresh page of the *beautiful Book* you sent to me does but **humble** me more & more—supplying, as it does, the true measure of my own ignorance[1]

—And *you yourself*, far-seeing as you are, do you not constantly feel humbled & abashed, before the awful Analogies of Creation?

—Mr Knight (so long *President* of the Hort: Soc:) *told me* that he was once speaking to *Sir Humphry Davy* of our own *very* narrow insight into *the mysteries of Creation*—[2] "Aye (said Davy) it reminds one of the *Nursery song*

I, said the Fly,
With **my little** Eye!—

I used to spend much of my life with Andrew Knight, at Downton Castle—[3] he, then an old Man, & I about 18—[&c&c] & wonderfully he used to amuse & astonish me— In regard to "*Sexualism*", I well remember hearing him say once that he had known an instance of a Male Yew Tree going 2 Miles to find a female—

—He once said to me "Mr Oxenden, you are a young Man, & fond of Horticulture— take the *White Currant* as your *Starting Point*— keep *on breeding* from **the very largest** *White Currant* you can grow—and in *twenty years* you will have them as large as *Grapes* & *as sweet*"—

I *began* this experiment—but the accidents of existence cast it aside—

—Do *forgive me*, in *the past*, for having often thrust upon you my own *most crude views*— I see & feel now that the *seeing*, *finding*, *comparing*, such things as meet my *bodily Eyes*, constitute *my* "Speciality"—& that I have *no business* with *aught beyond*—& for this mere "*Eye-knowledge*", strong walking powers & a boundless range of wild country especially adapt me— I will **not** fail to send you *good blossoms* of "*Lizard Orchis*"—nor aught else that you will trust me to get for you—[4] Can you send me either a *coloured Drawing*, or a *dried specimen*, of "*Malaxis paludosa*"—?[5] I have discovered the *peculiarity* which *governs* the "*habitat*" of the "*Lizard*"—

Believe me heartily & truly yours | G. Chichester Oxenden

DAR 173.2: 49

[1] Oxenden's name appears on CD's presentation list for *Orchids* (see Appendix IV). See also letter from G. C. Oxenden, 15 May 1862.

[2] Oxenden refers to Thomas Andrew Knight, president of the Horticultural Society between 1811 and 1838, and Humphry Davy, president of the Royal Society between 1820 and 1827.

[3] In 1809, Knight took over the management of his brother's estate at Downton Castle, Herefordshire, where he raised new varieties of cultivated plants (*DNB*).

[4] See letter from G. C. Oxenden, 15 May 1862 and n. 4.

[5] *Malaxis paludosa* is discussed in *Orchids*, pp. 130–9, and illustrated in *ibid.*, p. 132.

From Leonard Jenyns 28 May 1862

1. Darlington Place | Bath
May 28th. 1862

My dear Darwin,

It was a great pleasure to get your letter yesterday, & to have your testimony in favour of my memoir of our departed friend.—[1] You are thoroughly qualified to judge of its faithfulness as a correct portrait, as well as of any value it may possess in a scientific & literary point of view,—from your intimate acquaintance with Henslow, & with the subjects to which he devoted so much of his attention.—[2] I had previously received letters from Berkeley & others equally satisfactory in respect of their judgment,—[3]so I am inclined to hope the book will not be altogether a failure, as I was afraid it might turn out.— Some might wish for a longer biography,—but there scarce materials for it,—were it desirable.

I may take this opportunity of stating that I made many inquiries after Hope in Bath some time back, from what you stated in a former letter,—but I could hear nothing of him.[4] None of the letter carriers at the Post Office knew any one of that name in the town, at least a clergyman.— I think it must have been a mistake of yours—supposing he was here. I know him very well, & should have been glad to see him: I think also he would have found me out, had he been in the place.—

I am sorry to hear you are still in indifferent health.— I thank you for your inquiries after myself,[5] & am thankful to report myself quite well at this present time,—though not quite so strong as formerly, nor equal to taking the same long Nat. Hist. rambles, in which I delighted.—

With best wishes, believe me, | My dear Darwin, | Very Sincerely Yours | L. Jenyns.

DAR 168: 57

[1] See letter to Leonard Jenyns, 24 May [1862]. Jenyns refers to his memoir of John Stevens Henslow (Jenyns 1862).

[2] Henslow was professor of botany at Cambridge University from 1827 to 1861; CD was a member of Christ's College, Cambridge, from 1827 to 1831 and studied under Henslow. The two formed a close association that was maintained until Henslow's death in May 1861 (see *Correspondence* vol. 9 and Appendix X; see also Barlow ed. 1967).

[3] The botanist Miles Joseph Berkeley, who had known Henslow in his early days at Cambridge and before, also contributed recollections to Jenyns's memoir of Henslow (Jenyns 1862, pp. 55–7).

[4] CD mistakenly believed that the entomologist Frederick William Hope had moved to Bath (see letter to Leonard Jenyns, 24 January [1862]). Hope died in London on 15 April 1862 (*DNB*).

[5] See letter to Leonard Jenyns, 24 January [1862].

From J. D. Hooker [29 May 1862][1]

Kew
Thursday

Dear Darwin

I send 2 flowers of Vanilla from Sion House.[2] I can get more if wanted. I ordered 3 biggish plants of *Melastomaceæ* to go to Down Carrier, Nag's head—today—[3] We have no smaller ones in the flowering way at present.

I have done nothing of interest since writing last, but prepare the Cameroons list,—[4] do "Genera Plantarum",[5] Jury work[6] & see an everlasting round of visitors, who I (for the most part) wish at Jericho. I broke 3 solemn engagements yesterday

Ever yours affec | J D Hooker

DAR 101: 37

CD ANNOTATION
Top of first page: 'Henslow | Violets'[7] *pencil*

[1] Dated by the relationship to the letter from J. D. Hooker, 23 May 1862, and to the letter to J. D. Hooker, 30 May [1862]; the intervening Thursday fell on 29 May.

[2] Although *Orchids* had been published, CD continued to study orchids. He had been anxious for some time to see a living specimen of a member of the tribe Arethuseae to which the genus *Vanilla* belongs (see letter from J. D. Hooker, [7 April 1862] and n. 2); CD's notes on the specimens sent by Hooker are in DAR 70: 94–5. Syon House, near Isleworth, Middlesex, was the property of Algernon Percy, fourth duke of Northumberland, who further developed its notable gardens (*Gardeners' Chronicle and Agricultural Gazette*, 18 February 1865, p. 148); its grounds are located in view of the Royal Botanic Gardens, Kew, on the opposite side of the River Thames (Turrill 1959, p. 223).

[3] The carrier service between Down and London, operated every Thursday by George Snow, returned from the 'Nag's Head' public house in Borough High Street, south-east London (Freeman 1978, p. 261, and *Post Office directory of the six home counties* 1862). Earlier in May, CD had asked Hooker for specimens of Melastomataceae that were about to flower (see letters to J. D. Hooker, 15 [May 1862] and [18 May 1862]). There are observational notes on specimens from Kew of the melastomaceous species *Rhexia glandulosa* (dated 1 June 1862) and *Centradenia floribunda* (dated 31 May 1862) in DAR 205.8: 14 r. and 19 r.

[4] Hooker was compiling a list of the plants recently collected by Gustav Mann in the Cameroon mountains and on islands off the West African coast; the list was published in J. D. Hooker 1863c and summarised in J. D. Hooker 1863b. See also letter from J. D. Hooker, [5 May 1862] and nn. 3 and 5.

[5] Bentham and Hooker 1862–83.

[6] The International Exhibition opened in South Kensington, London, on 1 May 1862. Hooker was a member of the jury for class 4, section C, 'Vegetable substances used in manufactures, &c.', and an associate juror for class 3, section B, 'Drysaltery, grocery, and preparations of food as sold for consumption' (*Reports by the juries*).

[7] See letter to J. D. Hooker, 30 May [1862] and nn. 7 and 8.

From G. C. Oxenden [before 30 May 1862][1]

Dear Sir

I hunted some splendid Ground for "**Arachnites**", this day—

And, in the whole long day, only found *five plants*, of which only **three** were in flower—[2]

It is too Early for them, by two or three weeks—or a Month—

—On the very same ground, two years ago, I found fully 50—of which some were 9 and 10 inches high—

—When the wretched *ruminating* Animals bite them back, they come up again, it is true, but shorn of half their size & beauty—

—Of the three which I *did* find, I have enclosed *two* to you, by Post, this day—[3] The *third* flower I gave to Lady Dundonald,[4] whom I met on the Hills—a lover of Wild flowers—but with even *less knowledge* of Plants *than myself*—

Sincerely | G: C: Oxenden

DAR 173.1: 47

CD ANNOTATION
Top of first page: 'May 62' *pencil*

[1] Dated by CD's annotation and by the reference to Oxenden's enclosing specimens of *Ophrys arachnites* (see n. 3, below).

[2] In his letter of 15 May 1862, Oxenden asked whether there was any way in which he could assist CD; CD's reply has not been found, but he apparently asked Oxenden for specimens of *Ophrys arachnites*. In *Orchids*, pp. 72–3, CD reported that, while some botanists regarded *O. arachnites* as 'a mere variety of the varying Bee Ophrys' (*O. apifera*), his examination of specimens of *O. arachnites* sent him by Oxenden had led him to conclude that it was functionally very different from *O. apifera*; he stated: 'until these forms can be shown to be connected by intermediate varieties, we must rank O. arachnites as a good species'. Following the publication of *Orchids*, CD continued to investigate the relationship between the two forms (see n. 3, below, and letter to A. G. More, 7 June 1862, letters from G. C. Oxenden, 21 June 1862 and 8 July 1862, and letter to Daniel Oliver, 24 July [1862]).

[3] In DAR 70: 25 there is an observational note, dated 30 May 1862, that states:

2 spikes of Arachnites sent me by Mr Oxenden, kept in water (& young flowers opened) for more than week & intentionally jarred strongly & often, the [*interl*] pollinia did not fall out— The Shiny lateral knobs at base of Lab[ellum]. are [*interl*] larger than in Bee; if these be perforated *by insects, then [*interl*] ['then' *del*] this wd be important difference.— A more important one, is that the stigmatic surface is more perpendicular & therefore more open with respect to base of Labellum in the Bee than in Arachnites; evidently so as [to] allow surface to be struck by ['own' *del*] *its own [*interl*] pollinia in the Bee.

CD reported some of these findings in 'Fertilization of orchids', p. 145, a paper comprising revised and additional notes on the subject keyed by page number to the first edition of *Orchids* (see also *Collected papers* 2: 142).

[4] Louisa Harriet Cochrane was married to Thomas Barnes Cochrane, eleventh earl of Dundonald (*Burke's peerage* 1980).

From George Dickie 30 May 1862

Aberdeen
May 30, 1862

Dear Sir,

In accordance with your wish I procured a supply of L. cordata & made some observations on it—[1] About the time of full expansion & when the pollen is matured the small hairs on the lower surface of rostellum are directed perpendicularly downwards & their points touch the central line of the labellum. I carefully—under a lense—pushed a bristle along central line of labellum towards the rostellum; it or any other small body cannot penetrate the space between rostellum & labellum without coming in contact with hairs & front margin of the former; when such contact occurs there is a sudden explosion & the *whole* pollen mass instantly adheres to the touching body, the rostellum at the same time spreading out & covering the stigmatic surface. After a time the rostellum becomes retracted & the stigma exposed; so that the pollen cannot possibly touch the stigma of its own flower, but when applied to another—whose pollinia are removed—fragments adhere. *Small* Diptera and Hymenoptera have been seen on the flowers but as yet no pollinia on them. I send you bristles with adhering pollinia, I fear however they may become detached during their long journey by post.

I have read your work with much satisfaction,[2] & while subscribing to your remarks—page 28—respecting adaptations,[3] I frankly confess that I cannot comprehend how they can be explained by "natural selection" or what relation they have to that view.

Truly yours | G Dickie

DAR 162.2: 178

CD ANNOTATION

Top of letter: '*Listera*' *red crayon*; 'I noticed that R. was very much turned down in Mr Jamiesons specimen.—'[4] *ink*

[1] George Dickie was professor of botany at the University of Aberdeen. He had provided CD with specimens of the orchid *Listera cordata* in 1861 (see *Correspondence* vol. 9, letter to George Dickie, [5 July

1861], and letter to George Gordon, 6 July [1861]). Dickie's assistance is acknowledged in *Orchids*, p. 152, where CD reported that he had requested the specimens 'rather too late in the season'. CD incorporated the observations given here in the German edition of *Orchids* (Bronn trans. 1862, pp. 94–5n; see the second enclosure to the letter to H. G. Bronn, 30 June [1862]).

2 Dickie's name appears on the presentation list CD drew up for *Orchids* (see Appendix IV).

3 In *Orchids*, p. 28, CD concluded a minute account of the perfect adaptation of form to function in *Orchis pyramidalis*, saying:

> in no other plant, or indeed in hardly any animal, can adaptations of one part to another, and of the whole to other organised beings widely remote in the scale of nature, be named more perfect than those presented by this Orchis

4 In 1861, CD had asked the agriculturalist and geologist Thomas Francis Jamieson, who, like Dickie, resided in Aberdeenshire, to procure for him specimens of the orchid *Listera cordata* (see *Correspondence* vol. 9, letter from T. F. Jamieson, 13 June 1861). There is a note, dated 31 May 1862, in DAR 70: 77, recording observations on a specimen of this orchid supplied by Jamieson.

To J. D. Hooker 30 May [1862][1]

Down
May 30th

My dear Hooker

Infinite thanks for 3 grand plants of Melastomas;[2] if anything can be safely made out, I have now good opportunity. I saw again today the 2 sets of plants of Heterocentron raised from the two pollens from same flower, & you never could imagine what a marvellous difference in stature.[3] Also many thanks for Vanilla:[4] it is very different indeed from any orchid which I have seen, & has pollen like Cypripedium(!) & very curious stigma. But oh Lord what will become of my book on Variation:[5] I am involved in a multiplicity of experiments.— I have been amusing myself by looking at the small flowers of Viola. If Oliver has had time to study them,[6] he will have seen the curious case (as it seems to me) which I have just made clearly out, viz that in these flowers the *few* pollen-grains are never shed or never leave the anther-cells, but emit long pollen-tubes, which penetrate the stigma. Today I got the anther with the included pollen-grains (now empty) at one end, & a bundle of tubes penetrating the stigmatic tissue at the other end; I got the whole under the microscope without breaking the tubes: I wonder whether the stigma pours some fluid into the anther so as to excite the included grains. It is a rather odd case of correlation that in the double sweet Violet, the little flowers are double; i.e. have a multitude of minute scales representing the petals. What queer little flowers they are.—[7]

Have you had time to read poor dear Henslow's life:[8] it has interested me for the man's sake, & what I did not think possible, has even exalted his character in my estimation. But I much fear that the Public will think it dull. It will, however, be useful to any clergyman trying to follow, longo intervallo, Henslow's footsteps.—

How bothered with visitors & hard worked you are.[9]

Good Night | Charles Darwin

P.S. Tell Oliver, I think he must by mistake, when I asked him, have opened the ovarium of female Lychnis dioica: anyhow I can find no vestige of ovules—[10]

Tell him that my fancied dimorphisms, like that of Primula, of Oxalis acetosella is all a confounded mistake; only great variability in length of pistils.—[11]

I have just looked again at V. canina—the case odder;[12] only 2 stamens which embrace the stigma have pollen; the 3 other stamens have no anther-cells & no pollen.— These 2 fertile anthers are of different shape from the 3 sterile others; & the scale representing lower lip is larger & differently shaped from the 4 other scales, representing 4 other petals.—

In V. odorata, (single flower) all 5 stamens produce pollen. But I daresay all this is known.—

DAR 115.2: 152

[1] The year is established by the reference to Jenyns 1862 (see n. 8, below).
[2] See letter from J. D. Hooker, [29 May 1862] and n. 3.
[3] See letter to J. D. Hooker, 15 [May 1862] and n. 7.
[4] See letters from J. D. Hooker, [7 April 1862] and [29 May 1862], and letter to J. D. Hooker, 9 [April 1862].
[5] CD refers to the account of variation in domesticated plants and animals (*Variation*) that he had begun to prepare in January 1860 after the publication of *Origin*. He suspended work on *Variation* in July 1861 to produce *Orchids*, and, because of ill health, had only recently turned his attention once more to *Variation* (see *Correspondence* vols. 8 and 9, and 'Journal' (Appendix II)).
[6] Daniel Oliver had talked of including an account of dimorphism in *Viola* in his review of 'Dimorphic condition in *Primula*' ([Oliver] 1862c) (see letter from Daniel Oliver, 10 April 1862). See also letters to Daniel Oliver, 12 [April 1862] and 15 April [1862].
[7] There are notes by CD on his experiments with *Viola* in DAR 111: 3–5, dated 28 May and 30 May 1862.
[8] Jenyns 1862. John Stevens Henslow, Hooker's father-in-law, died in May 1861. See also letter to Leonard Jenyns, 24 May [1862] and n. 2.
[9] See letter from J. D. Hooker, [29 May 1862].
[10] In the letter to J. D. Hooker, 25 November [1861] (*Correspondence* vol. 9), CD had asked Hooker whether he or Oliver knew whether 'the **male** of Lychnis dioica' had 'rudimentary ovules in the ovarium'. Oliver either had observed or subsequently did observe such ovules (see *ibid.*, letters to Daniel Oliver, 30 November [1861] and 7 December [1861]).
[11] See letters to Daniel Oliver, 20 [April 1862] and 24 April [1862], and letter from Daniel Oliver, 23 April 1862.
[12] See n. 7, above.

From J. B. Jukes 30 May 1862

Geological Survey of Ireland, | Office, 51, Stephen's Green, Dublin,
May 30th *1862*

My dear Darwin

Many thanks for your letter.[1] It is of course impossible for any one to see the proof of my case without maps & sections & even then difficult without knowing the ground.—[2] In my paper I confine myself to Ireland & merely ask if the principle be not applicable to the Weald & all other cases.[3] Your account of the form of residuum of the insoluble matter of the Chalk however is very interesting.

Concerning marine and atmospheric denudation I am inclined to lay down the following canons.—

Marine denudation acts horizontally, like a wide plane, & cuts back into the land so as to make long lines of vertical cliffs if the land be stationary, or broad slopes if it be rising or any combination of the two according to rates of movement &c.

It does not make narrow winding valleys, except as passes on the crests of rising mountain chains, just while these passes are straits between islands & the arm of the sea made to run back & forward with a river-like action.—

(I do not see how the Sea can make mere inlets or Fiords, or even wide valleys if they be closed at the end, though it might cut a little at the mouths of such valleys).—

Atmospheric denudation acts vertically by weathering over the whole surface of land degrading its whole surface at different rates according to the nature of the rock. The matter removed is carried off ultimately by the rills, brooks & rivers either in solution or mechanical suspension.

The rivers (of all sizes) also erode the rocks by their own action, continually cutting their own channels deeper, and also widening them where those channels wind much from side to side. Now I think the question whether the river make a wide channel or a ravine depends on the ratio between the rate of the motion of its water and the nature of the rock.— If the river does not wear the rock very fast its channel may shift very much & eventually make a wide open valley. If however the river once cut a deep channel it can only make a ravine & that can only be widened by the slipping of the sides of the cut into the bottom & therefore its form depends on the nature of the rock whether it will stand perpendicular or at what slope.

I take these as tolerably obvious & certain rules.—

All that I propose to do is to push the amount of atmospheric & river action to a greater extent than most people have propos⟨ed⟩ to do, because I can, I think, prove it to have gone to that extent in the S. of Ireland.

As to the valleys you mention in Australia I looked at some of them & have often considered your notions respecting them & on the whole agree with Dana.—[4] Look at Sydney Harbour with its numerous arms all successively joining and going out by one narrow entrance.— I do not see how the Sea can possibly have made those ravines. Lay them down on a map & they have exactly the form of a number of brooks uniting to form a river—

Look at the slope of the ascent of the Blue Mountains above the Hawkesbury it is all gullied by similar ravines successively meeting, often emptying their brooks into one another through the narrowest & most precipitous gorges. I cannot but think they have all been cut by brooks when the land was higher & Sydney Harbour above sea level, perhaps a long way, & much more rain pouring down perhaps than there is now.

But if so then those huge circular valleys you describe so well must nevertheless have been formed also on dry land however inconceivable it may be.— If the others are furrows worn in a sheet of sandstone those are only bigger holes worn perhaps by the union of some of the furrows. It is however quite possible that the erosion may have been helped a bit by the sea after the land had been down again & during its last rise.—

So in S. Ireland the Sea during the Glacial period doubtless has left its mark, but the main modelling of the "form of ground" below the original surface of marine denudation has been done on the *dry land*.

Who is that goose that reviews your book in the Athenæum.[5] I had a good read at it yesterday coming up in the train.

Yours very truly | J. Beete Jukes

DAR 168: 91

[1] CD's letter has not been found, but it was presumably a reply to the letter from J. B. Jukes, 25 May 1862.

[2] Jukes had formerly believed that the topography of modern Ireland resulted from marine denudation of Upper Carboniferous strata. However, in a paper first delivered before the Geological Society of Dublin in May 1862 (Jukes 1862), Jukes announced his rejection of the marine erosion theory of landscapes, adopting instead the view that Ireland's drainage system had been formed by the action of fluvial processes. See letter from J. B. Jukes, 25 May 1862, and Davies 1969, p. 326.

[3] In a postscript to his paper (see n. 2, above), Jukes suggested that the Weald district in Kent might lend itself to a form of explanation similar to that he had provided for the drainage pattern of the south of Ireland (Jukes 1862, p. 400). See also letter from J. B. Jukes, 25 May 1862.

[4] CD, in attempting to explain the formation of the 'grand valleys' penetrating the mountains of New South Wales, dismissed as 'preposterous' the idea that they could have been formed by fluvial action (*Volcanic islands*, p. 136), arguing instead that they resulted from the action of the sea during a period of slow elevation of the land. James Dwight Dana took issue with CD's conclusions and argued in favour of fluvial action as the main agent involved (Dana 1850).

[5] According to the publisher's marked copy of the *Athenæum* (City University Library, London), John R. Leifchild was the author of the anonymous review of *Orchids* that appeared in the issue of 24 May 1862, pp. 683–5 ([Leifchild] 1862). CD's copy of this review is in DAR 226.1: 11 (Scrapbook of reviews).

From G. C. Oxenden 30 May [1862][1]

Broome
May 30

Dear Sir

I will send early *Lizard Blooms* **any** day you will write & appoint—[2] —for instance, on *Monday next*, to a **certainty**—but I **forewarn** you that these early flowers will *not* be so handsome or so *vivid* as the *later* Lizards, from the double effects of the *last severe frost*, and from the destruction of the lower leaves by *Snails*— —However, I will with pleasure send you some July Blooms *also, which are finer*—

—Unless you send other & better instructions **by return**, I purpose placing the blooms in clean *damp Moss*, & then in a *small deal Box*—

(By *Kentish Lines of Rail*, all goods suffer injury which are not "*Merrimacs*" or "*Monitors*"—)[3]

"*Malaxis paludosa*"

—I am pretty sure there is a coloured plate of this Bog-Orchis in "*Sowerby*"—[4]

—I have *not* got him—& he is now out of Print— Doubtless *you* know many Persons who possess this Work— If so, please send me the Volume which contains "*Malaxis*"

—I merely want to know *him by sight*, and also to know the *Month* in which *he is in flower*, and I am almost sure to come upon him—

—My present **impression** is that he is *low in Stature* (like **Ophrys Monophrys**) & the *Flower* of the *same tint*, namely a **Yellowish Green**—

You talk of *thousands* of "*Arachnites*" being seen in a good year—[5]

I know only 4 spots in Kent where it is found— —And in the three out of the four—he is only found **sparsely**—

With best regards | G. Chichester Oxenden

C. Darwin Esq—

DAR 173.2: 50

CD ANNOTATION
6.1 —I . . . upon him— 6.2] *marked with cross, brown crayon*

[1] The year is established by the relationship to the letter from G. C. Oxenden, 26 May 1862 (see n. 2, below).

[2] Oxenden had offered to send CD specimens of the lizard orchis, *Orchis hircina*, in his letters to CD of 15 May 1862 and 26 May 1862. See also letter from G. C. Oxenden, 4 June [1862] and n. 2.

[3] The reference is to the American warships *Merrimack* and *Monitor*. The *Merrimack*, a frigate salvaged by the Confederate navy, was coated with iron plating and renamed the *Virginia*. In response, at the close of 1861, the Union navy commissioned the construction of the iron-clad warship, the *Monitor*. In March 1862, the two ships engaged in a famous encounter at the mouth of the James River, Virginia, an event that ended in deadlock (McPherson 1988, pp. 373–7).

[4] Smith and Sowerby 1790–1814, vol. 1, tab. 72. In his letter of 26 May 1862, Oxenden had asked CD for a coloured drawing or dried specimen of this orchid.

[5] CD's letter has not been found, but it was presumably a response to the letter from G. C. Oxenden, [before 30 May 1862], in which Oxenden described his difficulty in finding specimens of *Ophrys arachnites*.

To W. E. Darwin [31 May 1862][1]

Down
Saturday night

My dear William

Have you dried Malaxis; if & you can spare a good spec. please put it between 2 good cards & post it to,

G. Chichester Oxenden Esq[e]
Broome Park
Canterbury.

He much wishes for it, & I am *bound* to do anything I can for him: if you cannot send it, please tell me, as I must get an artist to make coloured drawing for him.[2]

I shamefully never thanked you for my Photographs: please make mamma pay you for them.[3]

I expect it will do Horace much good coming to you: he looks forward with much pleasure to it—[4]

I have been dissecting the minute green little flowers now on Viola canina & hirta & odorata; the pollen-tubes are emitted from the pollen-grains still included in the anther, & penetrate the stigma.—[5]

The 3 Boys ⟨are⟩ here & George is hard at bugging—[6] Good night my dear old man | C. D.

DAR 210.6: 98

[1] Dated by the relationship to the preceding letter (see n. 2, below); in 1862, 31 May fell on a Saturday.

[2] George Chichester Oxenden had asked CD for a dried specimen or a coloured drawing of the orchid *Malaxis paludosa* in the letter to CD of 26 May 1862. See also preceding letter.

[3] William had for some time been a keen amateur photographer (see *Correspondence* vol. 6, letter to W. E. Darwin, [before 11 September 1857], and *Correspondence* vol. 7, letter to W. E. Darwin, [26 April 1858]). The reference may be to copies of the photograph of CD that William had taken in April 1861 (see *Correspondence* vol. 9, frontispiece, and letter to Asa Gray, 11 April [1861]).

[4] According to Emma Darwin's diary (DAR 242), she and Horace visited William in Southampton from 3 to 12 June 1862. Emma added a note to William at the end of CD's letter, giving her expected time of arrival in Southampton.

[5] See also letter to J. D. Hooker, 30 May [1862] and n. 7.

[6] The entry in Emma Darwin's diary (DAR 242) for 31 May 1862 reads: 'Boys came from school'. She refers to George Howard Darwin, Francis, and Leonard Darwin, who all attended Clapham Grammar School (see letter to W. E. Darwin, 26 April [1862], n. 2). 'Bugging' refers to hunting or collecting insects (*OED*).

From G. C. Oxenden 31 May [1862][1]

Broome
May 31

Dear Sir

I have just sent you two or three specimens of—Aceras. A. &—Hypopithys—which I found this Morning,[2] whilst seeking for "*Lizard Orchis*"—of which last however I only found 3 Plants—[3]

—In regard to the Mysterious disappearance of rare plants from localities, wherein they may previously have been abundant, I humbly submit that it admits of satisfactory explanation—[4]

Your's most truly | G. Chichester Oxenden

DAR 173.2: 51

CD ANNOTATION

1.1 I have ... Hypopithys—] '*Monotropa?*'[5] *added in margin, pencil*

[1] Dated by the relationship to the letters from G. C. Oxenden, 26 May 1862 and 30 May [1862].

[2] Oxenden refers to *Aceras anthropophorum*, the man orchid, and *Monotropa hypopitys*, the yellow bird's nest, a saprophytic dicotyledon that resembles an orchid. See also letter from G. C. Oxenden, 4 June [1862] and n. 8.

[3] See letter from G. C. Oxenden, 30 May [1862].

[4] Oxenden probably refers to his correspondence with CD about *Ophrys arachnites*, CD's side of which has not been found, but see the letters from G. C. Oxenden, [before 30 May 1862] and 30 May [1862].

[5] See letter from G. C. Oxenden, 4 June [1862] and n. 8.

From W. E. Darwin June 1862

Southampton
June 1862

Found Orchis Latifolia with 27 flowers. in 16 flowers a small fly (of one kind) was killed entering sideways, apparently just at the stigma[1]

in 2 more flowers parts of flies
in 9 of 16 pollen not gone
in 7— — gone
in other flowers of same plant pollen nearly all gone
W. E D.

DAR 162.1: 89

[1] Following the publication of *Orchids* in May 1862, CD continued, with assistance from his sons, William, Francis, and George Howard Darwin, to observe, experiment, and keep notes on orchid adaptation and pollination. In a letter to A. G. More, 17 July 1861 (*Correspondence* vol. 9), CD had expressed the opinion that *Orchis latifolia* and the closely allied *O. maculata*, which some botanists believed to be one species, might be distinguishable by the manner of their insect pollination. He was unable to find conclusive evidence of this before the publication of *Orchids* (see *Orchids*, p. 42). However, there is a note in DAR 70: 13–14, dated 20 June 1862, which describes observations by CD's sons George and William on the pollination of *O. maculata* by the flies *Empis livida* and *E. pennipes*; the note also reads: 'A D^r. from Southampton sent me 3 flowers of O. latifolia with the smaller black fly [*E. pennipes*] within the nectary, when he killed it.—' CD incorporated these observations in the German translation of *Orchids* (Bronn trans. 1862, p. 22 n.; see the enclosure to the letter to H. G. Bronn, 30 June [1862]). He later published them in 'Fertilization of orchids', p. 142, a paper comprising revised and additional notes on orchids, keyed by page number to the first edition of *Orchids* (see also *Collected papers* 2: 139).

From Asa Gray [2 June 1862][1]

Long-styled *Houstonia* has shorter, stouter, & far more hispid stigmas—*smaller anthers*, & *smaller pollen* than short styled,—evidently a step towards the separation of the sexes.[2]

I am kept very hard at work all this month in College, &c— But after 6th. July, I can do something.[3]

Viola (V. cucullata, & V. lanceolata) is very pretty for cross-fertilizing.

Ever Yours | A. Gray

Do I bore you, with these scattering notes?

I jot things down here—keeping no other records, & send them off to you.

I enclose some flower-buds of Arethusa—.[4] You can soak them.

Incomplete[5]
DAR 110 (ser. 2): 66

CD ANNOTATIONS
2.1 I am . . . to you. 6.1] *crossed red crayon*
4.1 Ever . . . Gray] 'June 2^d^ 62' *added in margin, ink; circled ink*
5.1 Do I . . . them. 7.1] *crossed ink*

1 Dated by CD's annotation.

2 While preparing his paper, 'Dimorphic condition in *Primula*', CD had asked Gray for information on analogous cases (see *Correspondence* vol. 9, letter to Asa Gray, 16 September [1861]). In response, Gray mentioned, among other things, dimorphism in *Houstonia*, stating his intention to observe the differences in the pollen of the two forms the following spring, and his unsuccessful attempts to procure seedlings of *H. carerulea* to send to CD (see *Correspondence* vol. 9, letters from Asa Gray, 11 October 1861 and 9 November 1861). In addition, Gray had encouraged his pupil, Joseph Trimble Rothrock, to carry out further observations and crossing experiments on the two forms (see letters from Asa Gray, 2–3 July 1862 and 4 August 1862). CD mentioned the possibility that dimorphism might in some cases represent a stage in the development of dioeciousness in 'Dimorphic condition in *Primula*', p. 95 (*Collected papers* 2: 62).

3 Gray was Fisher Professor of natural history at Harvard University and lectured at the Lawrence Scientific School (Dupree 1959).

4 In July 1861, Gray had recommended this orchid genus (or any other member of the Arethuseae) to CD's attention (see *Correspondence* vol. 9, letter to J. D. Hooker, 17 [July 1861]). However, although CD made repeated requests for specimens from his botanical correspondents (see *Correspondence* vol. 9, and this volume), he was unable to see any 'living flowers' of the Arethuseae before the publication of his study (see *Orchids*, pp. 269–70).

5 Some indication of the subjects discussed in the missing portion of the letter is given by CD's response in the letter to Asa Gray, 10–20 June [1862].

From G. C. Oxenden 4 June [1862][1]

Broome
June 4—

Dear Sir

This Evening, by Rail, I send you two or three very bad Lizard Blooms—[2] A few days ago the leaves of the forward Lizards got **horribly** bitten by snails & slugs, almost killed by them— The *scent* of Lizard seems to attract them— *Did* I, or did I **not**, ever send to you the *dried stem* of my immense Lizard of 1860—**?** It is a real wonder—[3]

These wretched blooms, sent this day, are from the very same plant—

—In a few more days I will send you half a dozen more—& (I hope) some "*Arachnites*" also[4]

I enclose, with the Lizard, a very pretty, rather sparse specimen, of of "*Conopsea albida*—"[5]

Later in the Summer, I will send good **spikes** of

Epipactis latifolia
—— purpurata
—— palustris[6]

Do you care to have "*Epipactis grandiflora*"—which is so **very common**? Do *not* trouble yourself about the "*Bog-Orchis*"— My recollection of him has *returned completely*—& I feel pretty sure of finding him—[7]

—Here, in our *benighted* state, we do **not** call that brown Orchid, which I sent you the other day, the "*Birds nest Orchis*"— That which we **do** call "Nidus Avis" is nearly ⟨of⟩ the *same colour*, but a **much larger**, **stronger**, ⟨*mo*⟩*re warlike* Plant, *parasitic* upon **decayed** ⟨Ha⟩zel Roots—[8]

—At this very time of writing, I could shew to any one *200* O. **apifera**, in full bloom—

With most kind regards | G. C. Oxenden

DAR 173.2: 53

[1] The year is established by the relationship to the letters from G. C. Oxenden, [before 30 May 1862] and 30 May [1862].

[2] In his letters to CD of 26 May 1862 and 30 May [1862], Oxenden had offered to send CD specimens of the lizard orchis, *Orchis hircina*; see also letter from G. C. Oxenden, 31 May [1862]. CD's notes on these specimens, dated 6 and 10 June 1862, are in DAR 70: 27–9. CD acknowledged Oxenden's assistance in providing specimens of this plant in *Orchids* 2d ed., p. 25 n.

[3] CD did not discuss *O. hircina* in the first edition of *Orchids*. No correspondence between Oxenden and CD on this specimen has been found.

[4] See letters from G. C. Oxenden, [before 30 May 1862] and 30 May [1862].

[5] Oxenden refers to the species that CD called *Gymnadenia albida* in *Orchids*, pp. 83–4.

[6] See letter from G. C. Oxenden, 15 May 1862 and n. 3.

[7] Oxenden had asked CD to provide him with a coloured drawing or a specimen of *Malaxis paludosa*, the bog orchis (see letters from G. C. Oxenden, 26 May 1862 and 30 May [1862]; see also letter to W. E. Darwin, [31 May 1862]).

[8] Oxenden had sent CD a specimen of *Monotropa hypopitys*, the yellow bird's nest, a saprophytic dicotyledon that resembles an orchid (see letter from G. C. Oxenden, 31 May [1862]). The orchid *Neottia nidus-avis* is commonly called the bird's nest orchid; there are observational notes on this species, dated 17 and 21 May 1862, in DAR 70: 73–4. CD later reported these observations in 'Fertilization of orchids', p. 152 (*Collected papers* 2: 148–9).

To A. G. More 7 June 1862

Down, Bromley, Kent. S.E.
June 7, 1862.

My dear Sir

If you are well and have leisure will you kindly give me one bit of information Does Ophrys arachnites occur in the Isle of Wight? or do the intermediate forms, which are said to connect abroad this species and the Bee-O., ever there occur?[1]

Some facts have led me to suspect that it might just be possible, though improbable in the *highest degree*, that the Bee might be the self-fertilising *form* of *O. arachnites* which requires insect's aid, something (in the same way) as we have self-fertilising flowers of the Violet and others requiring insects.[2] I know the case is widely different as the Bee is borne on a separate plant and is incomparably commoner. This would remove the great anomaly of the Bee being a perpetual self-fertiliser.[3] Certain Malpighiaceæ for years produce only one of the two forms. What has set my head going on this is receiving to-day a Bee having one alone of the best marked characters of O. arachnites.[4]

Pray forgive me troubling you and believe me | Yours sincerely | C. Darwin

Copy
DAR 146: 405

[1] More had assisted CD with his study of orchid pollination since 1860 by carrying out observations and sending him specimens (see *Correspondence* vols. 8 and 9). See also letter to A. G. More, 18 May 1862.

[2] On CD's interest in the relationship between *Ophrys arachnites* and *O. apifera*, see the letter from G. C. Oxenden, [before 30 May 1862], nn. 2 and 3. See also letter to C. C. Babington, 20 January [1862] and n. 9. CD had been interested over a number of years in the occurrence in *Viola* of small unopening flowers in which self-pollination occurred, a phenomenon later called cleistogamy (see letter to Daniel Oliver, 12 [April 1862] and n. 6); he had recently made observations and notes on the genus (see letter to J. D. Hooker, 30 May [1862] and n. 7).

[3] One of CD's two objectives in *Orchids* was to show that the contrivances by which orchids are pollinated had 'for their main object the fertilisation of each flower by the pollen of another flower' (*Orchids*, p. 1). In the conclusion, CD stated: 'In the Bee Ophrys alone, as far as I have seen, there are special and perfectly efficient contrivances for self-fertilisation' (*Orchids*, p. 359).

[4] The specimen of the bee-orchis, *Ophrys apifera*, may have been sent to CD by George Chichester Oxenden (see preceding letter).

From E. Schweizerbart'sche Verlagsbuchhandlung[1] 7 June 1862

Stuttgart
den 7 Juni 1862

Verehrtester Herr!

Sie werden von Herrn Hofrath Bronn vernommen haben, dass das Exemplar Ihrer neuesten Ausgabe des "Origin", welches Sie die Güte hatten an mich absenden zu lassen, mir richtig zugekommen ist und von mir Herrn Bronn gesendet wurde.[2] Ich habe bei diesem Anlasse von Neuem Ihre grosse Gefälligkeit zu rühmen, mit welcher Sie uns bei der Veranstaltung eines neuen Abdruckes von 1000 Ex der deutschen Ausgabe Ihres berühmten Buches entgegen zu kommen die Freundlichkeit haben, wozu Sie auch mich zu grossem Danke verpflichten.[3]

Herr Bronn hat mir auch angezeigt dass Sie die Gewogenheit hatten ihm 1 Ex des Orchideen Buches zu senden so wie dass so gütig seyn wollen, es zu vermitteln durch Ihren Verleger Herrn Murray dass er mir für den Preis von £10— die Clichés der in den Text eingedruckten Abbildungen liefere;[4] ich nehme dieses Offert gerne an und ersuche Sie H^rn^ Murray zu beauftragen mir in guten Abgüssen recht bald die Clichés per Post zu senden und mir anzugeben an wen ich den Betrag sodann bezahlen soll, oder ob er eine Rimesse auf London wünscht.

Der 2ten. deutschen Ausgabe werde Ihr Portrait beifügen, in Photographie, die ich mir zu diesem Zwecke von London kommen liess, ich glaube damit allen Freunden des Buches eine erwünschte Gabe zu bringen.[5]

Genehmigen Sie, verehrtester Herr, die Versicherung vollster Hochachtung, | Ihres ergebensten | E Schweizerbart[6]

DAR 177: 68

[1] For a translation of this letter, see Appendix I. It was written by Christian Friedrich Schweizerbart, who was head of the Stuttgart publishing company E. Schweizerbart'sche Verlagsbuchhandlung (see n. 6, below).

[2] See letter to H. G. Bronn, 25 April [1862], and letter from H. G. Bronn, 19 May 1862.

[3] Schweizerbart refers to CD's offer to assist in incorporating new material from the third English edition of *Origin* into the new (second) German edition of the book (Bronn trans. 1863). See letters to H. G. Bronn, 11 March [1862] and 25 April [1862].

[4] Schweizerbart had agreed, through Bronn, to CD's suggestion that he might publish a German translation of *Orchids* (see letter to H. G. Bronn, 11 March [1862], and letter from H. G. Bronn, 27 March 1862); he refers to John Murray, CD's British publisher.

[5] The photograph included as the frontispiece to the second German edition of *Origin* (Bronn trans. 1863) is that taken by Maull & Polyblank *circa* 1857 (see the frontispiece to *Correspondence* vol. 8; the photograph reproduced there is of a print prepared by Maull & Fox from the original negative).

[6] The publishing firm E. Schweizerbart'sche Verlagsbuchhandlung was founded by Wilhelm Emanuel Schweizerbart; in 1841 he sold the firm to his nephew, Christian Friedrich Schweizerbart, who continued using the signature 'E. Schweizerbart' in business communications (*Jubiläums-Katalog*, pp. x–xi).

To Daniel Oliver 8 June [1862][1]

Down. | *Bromley.* | *Kent. S.E.*
June 8th

Dear Oliver

Very many thanks for the orchid, which was new to me & interested me, but by Jove I must stop & go on with confounded dull old subjects. The orchid must be a Catasetum (allied to C. tridentatum) & has no doubt its own Monchanthus.[2] The stigmatic surface was more viscid than in the other species examined by me but not viscid enough to break the caudicles. The utriculi & ovules after spirits showed also very little contained pulpy matter: An examination of the tissue or utriculi of stigmas of utterly sterile Hybrids after being kept for 24 or 48 hours in spirits, in comparison with the utriculi of the pure & fertile parent species, would be a point worth attention. But time time time, as you no doubt exclaim with your lectures,[3] & as I often exclaim, with my wretched stomach, though having no lectures or other disturbance.

That is a curious monster which you sent with its 2 anthers & 2 rostellums.—

I am glad that you have read my orchis book & seem to approve of it;[4] for I never published anything which I so much doubted whether it was worth publishing & indeed I still doubt.[5]

The subject interested me beyond what, I suppose, it is worth.— Almost every day I get more convinced that insects (in relation to the marriage of distinct flowers) *govern the structure of almost every* flowers: I have been led, from crossing, to look to Pelargonium, & see how well the 7 anthers stand & face, so that an insect visiting the nectary may take them all; & see the open stigmas in an older flower.—[6]

Yours very truly | C. Darwin

DAR 261 (DH/MS 10: 32)

[1] The year is established by reference to the publication of *Orchids*, of which Oliver received a presentation copy (see letter from Daniel Oliver, 14 May 1862, and Appendix IV).

[2] Oliver's letter has not been found; however, a note in DAR 70: 96–7, dated 'June 7th 62.', describes 'A Catasetum sent . . . from Kew, . . . Species closely allied to tridentatum'. CD described this specimen in *Orchids* 2d ed., p. 193, under the heading '*Catasetum planiceps (?)*'. The reference to *Monachanthus* concerns the explanation provided in *Orchids* of an apparent anomaly: flowers of *Catasetum tridentatum*, *Monachanthus viridis*, and *Myanthus barbatus* had been observed to grow on the same plant. As CD explained it, *Myanthus barbatus* should be considered the hermaphrodite form of the same species of which the *Catasetum* was the male and the *Monachanthus* the female form (see *Orchids*, pp. 231–48).

[3] Oliver was professor of botany at University College London. On the demands of his lectures, see the letters from Daniel Oliver, 10 April 1862 and 14 May 1862, and the letter to Daniel Oliver, 12 [April 1862].

[4] Oliver's letter has not been found, but see the letter from Daniel Oliver, 14 May 1862.

[5] When CD suggested to John Murray that he might publish *Orchids*, he expressed some of his own doubts about the advisability of doing so (see *Correspondence* vol. 9, letters to John Murray, 21 September [1861] and 24 September [1861]); he wrote to Joseph Dalton Hooker (*Correspondence* vol. 9, letter to J. D. Hooker, 24 September [1861]): 'Heaven knows whether it will not be a dead failure'.

[6] CD had begun crossing experiments with different varieties of pelargoniums on 11 May 1862, in an attempt to produce seed from the normally sterile, central peloric flowers of these plants (see the notes from these experiments, dated 11 May – August 1862, in DAR 51: 4–9, 12–13, and *Variation* 2: 167). In DAR 51 (ser. 2): 10–11, there is a note dated 6 June 1862 that discusses the number and position of the anthers in pelargoniums, and their relationship to the activities of insects.

From J. D. Hooker 9 June 1862

Royal Gardens Kew | Kew
June 9/62

My dear Darwin

Oliver has written what appears to me to be an able review of "Dimorphism" apropos of your *Primula paper* for the forthcoming No. of the N.H Review.[1] Your account of *Viola* impregnating direct from the anther is to me novel & *most interesting* it is half-way to the structure of Asclepias &c.[2] I have finished the Cameroons Mt plants, are you ready for the gross results? as to the proportions &c of temperate forms.[3]

I am daily at Exhibition Jury work except Monday— it is just as well to be slaving there as at Kew, & as amusing.[4]

We are still in domestic perplexity— My wife[5] is very thin & watery, lacks energy, blood & muscle. & though she does her best honestly & heartily with the children, she lacks energy & method & does not get on.— We have agreed to a plan of housekeeping which will I hope answer better—to get a middle aged cook who will be sort of housekeeper in as far as keeping an absolute control over who comes in & out of the kitchen—[6] Mrs Darwin is not likely to know of such a person, but perhaps you will kindly mention it to her—quite a plain cook is all we want, who can roast, boil & bake, but she must be beyond the age of flirtation— I can promise her a quiet place, a most indulgent mistress & good wages.— Then we shall return to the Nursery Governess plan—but endeavour to get an older & more practised one than we had before—

I am getting over my dispiriting feelings of annoyance & anxiety. I wish I could add that my wife was better— she complains of palpitation of heart & shortness of breath & she has hardly a perceptible pulse—that she looks very ill every one says. I want above all to take her away, but neither of us can leave home till our household is arranged.

Ever yrs affec | J D Hooker

DAR 101: 40–1

CD ANNOTATION

Top of first page: 'This letter **Private**; to be returned to me You will see why I send it.'[7] *pencil*

[1] [Oliver] 1862c. See letter from Daniel Oliver, 10 April 1862.

[2] See letter to J. D. Hooker, 30 May [1862].

[3] Hooker had been preparing a list of the plants collected in the Cameroon mountains by Gustav Mann (see letters from J. D. Hooker, [5 May 1862] and nn. 3 and 5, and [29 May 1862] and n. 4); he read a paper entitled 'On the vegetation of the Cameroons' before the Linnean Society of London on 5 June 1862 (*Journal of the Proceedings of the Linnean Society of London (Botany)* 6 (1862): cvi). The paper was not published in the society's journal, but in 1863 Hooker read before the society a further paper on the subject that was published together with his list of the plants (J. D. Hooker 1863c; see also J. D. Hooker 1863b). CD was interested in the evidence these collections provided concerning the historical causes of the prevailing geographical distribution of plant species in tropical areas (see letter to J. D. Hooker, 9 May [1862] and n. 6).

[4] See letter from J. D. Hooker, [29 May 1862] and n. 6.

[5] Frances Harriet Hooker.

[6] The Hookers' home had recently been broken into, an incident for which Hooker felt they themselves were to blame for not taking better care of their 'servts, doors & establishment' (see letter from J. D. Hooker, [5 May 1862]).

[7] CD appears to have sent the letter to Emma Darwin, who was in Southampton (see letter to J. D. Hooker, 11 June [1862] and n. 5).

To *Journal of Horticulture* [before 10 June 1862][1]

I should feel much obliged if the "Devonshire Beekeeper" or any of your experienced correspondents would have the kindness to state whether there is any sensible difference between the bees kept in different parts of Great Britain.[2] Several years ago an observant naturalist and clergyman, as well as a gardener, who kept bees, asserted positively that there were certain breeds of bees which were smaller than others, and differed in their tempers.[3] The clergyman also said that the wild bees of certain forests in Nottinghamshire were smaller than the common tame bees. M. Godson, a learned French naturalist, also says that in the south of France the bees are larger than elsewhere, and that in comparing different stocks slight differences in the colour of their hairs may be detected.[4] I have also seen it stated that the bees in Normandy are smaller than in other parts of France. I hope that some experienced observers who have seen the bees of different parts of Britain will state how far there is any truth in the foregoing remarks.[5] In the Number of your Journal published May 15, 1860, Mr. Lowe gives a curious account of a new grey or light-coloured bee which he procured from a cottager.[6] If this note should meet his eye I hope he will be so good as to report whether his new variety is still propagated by him.— Charles Darwin.

Journal of Horticulture, Cottage Gardener, and Country Gentleman n.s. 3 (1862): 207

1 The letter was printed in the issue of 10 June 1862. As well as publishing it in their own journal, the editors of the *Journal of Horticulture* sent CD's letter on to the *Bienen Zeitung* (see letter to *Bienen Zeitung*, 18 June 1862).

2 Thomas White Woodbury, who was one of the contributors to the beekeeping section of the *Journal of Horticulture*, always signed his pieces for the journal 'A Devonshire bee-keeper' (*Journal of Horticulture* 28 (1862): i). According to CD's 'Journal' (Appendix II), he was writing up the chapter on 'Silk-worms Geese &c' for *Variation*. This chapter (chapter 8) also included a section on hive-bees (*Variation* 1: 297–9).

3 This individual has not been identified.

4 Godron 1859, 1: 459; Dominique Alexandre Godron's name was misspelled in the published letter. There is an annotated copy of Godron 1859 in the Darwin Library–CUL (see *Marginalia* 1: 331–5).

5 Five responses to CD's letter were published in the *Journal of Horticulture*: Woodbury and Henry Wenman Newman published replies in the issue for 17 June 1862 (p. 225); replies from a 'Surrey Highlander' and from J. Lowe of Edinburgh appeared in the 24 June 1862 issue (pp. 242–3); and Sylvanus Bevan Fox of Exeter published a response in the issue for 8 July 1862 (p. 284). In *Variation* 1: 297–8, CD stated:

> It is frequently asserted that the bees in different parts of Great Britain differ in size, colour, and temper; . . . The best authorities concur that, with the exception of the Ligurian race or species, . . . , distinct breeds do not exist in Britain or on the Continent. There is, however, even in the same stock, some variability in colour.

In evidence of this CD cited the above replies to his letter (*ibid.*, p. 298 n. 57).

6 The account by J. Lowe of Edinburgh was published in *Cottage Gardener, Country Gentleman's Companion, and Poultry Chronicle* 24 (1860): 110; the journal was renamed the *Journal of Horticulture and Cottage Gardener* in 1861. CD's unbound copies of this journal are in the Darwin Library–CUL; CD's copy of Lowe's article is annotated.

To Asa Gray 10–20 June [1862][1]

Down Bromley Kent
June 10th.

My dear Gray,

Your generous sympathy makes you overestimate what you have read of my orchid Book. But your letter of May 18th + 26th has given me an almost foolish amount of satisfaction.[2] The subject interested me, I know, beyond its real value; but I had lately got to think that I had made myself a complete fool by publishing in a semi-popular form. Now I shall confidently defy the world. I have heard that Bentham & Oliver approve of it;[3] but I have heard the opinion of no one else, whose opinion is worth a farthing. What strange creatures these orchids are, for instance Mormodes, of which I have this morning examined another species, & which supports all that I have said, but which has completely puzzled me.[4]

I thank you most heartily for your notes on several American species. I am not surprised as no true Orchis grows near you, that the pollinia of O. spectabilis were not removed; I sh^{d}. expect that it would take probably a long time before new insects would learn the dodge.—[5] You probably pushed too hard against the viscid disc & crumpled the contracting atom of membrane, which, I know, interferes with the proper movement.

I will write to Murray about casts of 3 first woodcuts; but I doubt whether he will send the casts, for I believe that there is to be set to be sent to Germany for German Edition.—[6] I will do my best, but by Jove you shall not pay for them. If there be (which is very improbable) an American Edit, Murray will expect a little more than simple cost. But I will keep back this letter till I hear from him.[7]

Enough & too much about my orchids, which are now again become beloved in my eyes, & which were quite lately accursed. Many thanks about copies of your Pamphlet.[8] Do not trouble about Hollies; I thought they grew near; the case is not important.[9] Nothing will be made out, I fear, about Rhexias, unless indeed a plant or plants could be protected from insects. I have now a Rhexia glandulosa under trial, but there is little difference in stamens & little to be made out.[10] I am working at several Melastomas; but am at fault; I am, however, *certain* there is something very remarkable; the pollen of one set of anthers produce less seed & to my amazement their seedlings are dwarfs compared to the other set, all produced from the same plant.—[11] The labour is great: I have lately counted one by one 6700 seeds of Monochætum![12] Mr Meeham has sent me his paper on parallel differences in trees of N. America & Europe; *pray be so kind as to remember* to tell me whether this can be approximately trusted; for the case interests me much, as best case I have seen of apparently direct action of conditions of life.—[13] Forgive me for one bit more trouble: I have a Boy with the collecting mania & it has taken the poor form of collecting Postage stamps: he is terribly eager for "Well, Fargo & Co Pony Express 2^{d} & 4^{d} stamp", & in a lesser degree "Blood's 1. Penny Envelope, 1, 3, & 10 cents". If you will make him this present you will give my dear little man as much pleasure, as a new & curious genus gives us old souls.[14]

Since this was written the above little man has been struck down with scarlet-fever; but thank God this morning the case has taken a mild form.—[15]

I have just received your long notes on Cypripedium; you may believe how profoundly interesting they are to me. Will you not publish them, either in noticing my Book in Silliman, or otherwise?[16] But your notes are more interesting than you will suppose, for since publishing I saw at Flower show, C. hirsutissimum, but could not touch it, but it seemed to me that the sterile anther entirely covered the passages by the anthers. I was amazed & saw clearly that there must be some quite distinct manner of fertilisation. But I did not think of insects crawling into flower; still less of different kind of pollen & in somewhat concave & viscid stigma.[17] By Jove it is wonderful. You have hit on the same very idea which latterly has overpowered me, viz the exuberance of contrivances for same object: you will find this point discussed & attempted to be partly explained in the last Chapter.[18] No doubt my volume contains much error: how curiously difficult it is, to be accurate, though I try my utmost. Your notes have been interested me beyond measure. I can now afford to d—d. my critics with ineffable complacency of mind. Cordial thanks for this benefit.—

It is surprising to me that you sh^d. have strength of mind to care for science, amidst the awful events daily occurring in your country. I daily look at the Times with almost as much interest as an American could do. When will peace come: it is dreadful to think of the desolation of large parts of your magnificent country; & all the speechless misery suffered by many. I hope & think it not unlikely that we English are wrong in concluding that it will take a long time for prosperity to return to you. It is an awful subject to reflect on.— Good Bye my dear friend.— I will keep this open till I hear from Murray, which I sh^d. think must be tomorrow.[19]

I am keeping back this letter till I hear from Murray, who, I fear is absent. I have now received your interesting notes of June 2^d.[20] How can you ask whether your letters bore me? I never in my life received a letter from you that was dull. Your letters are a very great pleasure & profit. I seldom see or hear from a soul on Science. Most of my scientific friends (See p. 8 at back of p. 5.) (This page has got in wrong place). are so busy that I scruple to write to them.

Arethusa is very pretty: I sh^d. *conjecture* its fertilisation was effected nearly as described under Cattleya; for so it seems to be with Vanilla, which I have lately seen.—[21] How well you are attending to Cypripedium. I can at any time return you (making copy for self) your notes on this genus or other notes..[22] How *very very* kind it is in you, overworked as you are, to send me so many notes.— Hearty thanks about Houstonia: that subject, I am working at hard & interests me much.—[23] By the way did you ever look at the little (so-called imperfect) flower of Viola & Oxalis; they are very curious, the pollen-grains emit their tubes whilst within the anthers; & it is curious to see these tubes travelling up in straight lines from the *lower* anthers in Oxalis, right to stigmas; it is like spermatozoa finding their way to ovules.[24]

I received 2 or 3 days ago a French Translation of the Origin by a Mad^elle. Royer, who must be one of the cleverest & oddest women in Europe:[25] is ardent Deist & hates Christianity, & declares that natural selection & the struggle for life will explain all morality, nature of man, politicks &c &c!!!. She makes some very curious & good hits, & says she shall publish a book on these subjects, & a strange production it will be.

Good Bye—till I hear from that wretch Murray.

(I have had another look at your Arethusa; structure seems very like Vanilla & unlike that of other orchids. In Vanilla, the Labellum is furnished with a compound curious comb, which would compel an insect in retreating to rub its back against rostellum; but the papillæ in Arethusa seem very different. How beautifully clear the spiral ducts are visible in wings of Clinandrium & colum.)

If you come across Specularia do look & tell me whether pollen-grains emit tubes direct from anthers or are grains collected on collecting hairs.—[26]

I have just had letter from Alp. De Candolle about Primula & he gives me facts & his queries show he appreciates the case, & about nat. selection.[27] He says he goes as far as you about change of species, & he laughs at Linnæus' old definition "Species tot numerasmus quot . . . . . . sunt creatæ".—[28] But I think from his letter you go further; he says he wants direct proof of nat. selection & he will have to wait a long time for that. Opticians do not wait for direct proof of undulation of ether. But Good Heavens what a higglety-pigglety letter I am scribbling to you, who have hardly a minute to spare.— It is a horrid shame, so I will stop.—

20th. At last I have heard from Murray that he will instantly send the 3 casts & will let, if wanted, a publisher have whole set "on easy terms"[29]

Yours cordially C. Darwin

Gray Herbarium of Harvard University (66)

[1] The year is established by reference to the publication of *Orchids*.

[2] In the letter from Asa Gray, 18 May 1862, Gray gave his first reactions to the proof-sheets of part of *Orchids* sent to him by CD. The portion of this letter dated 26 May 1862 has not been found.

[3] See letter from J. D. Hooker, [17 May 1862], and letter to J. D. Hooker, [18 May 1862]. See also letter from Daniel Oliver, 14 May 1862, and letter from George Bentham, 15 May 1862.

[4] CD refers to *Mormodes luxata*; there are notes on a specimen sent to him by Sigismund Rucker on 8 June 1862 in DAR 70: 99–102. The only species of *Mormodes* described in *Orchids* is *M. ignea* (*Orchids*, pp. 249–65). CD concluded his notes on *M. luxata* indicating his uncertainty about the mechanism of ejection of the pollinia: 'I do not clearly understand whole mechanism.— . . . If an insect entered & [*del illeg*] climbed up face of column till it touched the filament then disc would stick to abdomen & I could understand all, but I doubt.—' CD added a passage about *M. luxata* to the second edition of *Orchids* (*Orchids* 2d ed., pp. 219–20), and stated: 'If an insect were to gnaw the terminal cup [of the labellum], it could hardly fail to touch the apex of the column, and then the pollinium would swing upwards and adhere to some part of the insect's body.'

[5] The notes referred to have not been found, but see the letter from Asa Gray, 18 May 1862, in which Gray mentioned having transplanted some *Orchis spectabilis* from the state of New York.

[6] The reference is to CD's publisher, John Murray. Gray planned to review *Orchids* in the *American Journal of Science and Arts* and wanted to use some of the original illustrations from the book (see letter from Asa Gray, 18 May 1862). Christian Friedrich Schweizerbart was planning to publish a German edition of *Orchids* and also wanted a set of electrotype plates of the illustrations (see letter from E. Schweizerbart'sche Verlagsbuchhandlung, 7 June 1862).

[7] See letters to John Murray, 13 June [1862] and 20 [June 1862].

[8] In his letter to CD of 18 May 1862, Gray mentioned that he had sent some copies of his pamphlet on *Origin* (A. Gray 1861) for CD 'to give away'.

[9] CD had sought information regarding 'gradation in sexes' in American species of holly (see letter to Asa Gray, 21 April [1862]). Gray replied that the nearest hollies grew twenty miles away, but that he could 'send' for some (see letter from Asa Gray, 18 May 1862).

[10] Suspecting that *Rhexia*, a member of the Melastomataceae, might be dimorphic, CD had asked Gray to observe the position of the pistil and colour of the anthers in different *Rhexia* plants (see letter to Asa

Gray, 16 February [1862]); if Gray found that there were not two forms, CD asked him to compare the position of the pistil in young and old flowers (see letter to Asa Gray, 15 March [1862]). Gray offered to 'set to watching' *R. virginica* in the summer, but doubted whether the plant was dimorphic (see letter from Asa Gray, 6 March [1862]). See also letter to Asa Gray, 21 April [1862] and n. 13. CD had recently received a specimen of *Rhexia glandulosa* from Joseph Dalton Hooker, on which he had begun to experiment (see letter from J. D. Hooker, [29 May 1862], and the experimental notes, dated 1 June – 2 July 1862, in DAR 205.8: 14–15).

[11] CD had been experimenting since the end of 1861 on several members of the Melastomataceae that he believed might be dimorphic, namely *Heterocentron*, *Monochaetum*, and *Centradenia*. CD refers in particular to the results of his crossing experiment with *H. roseum* (see letters to J. D. Hooker, 15 [May 1862] and 30 May [1862]).

[12] CD's notes on the crossing experiments with *Monochaetum ensiferum* that he began in February and harvested in April and May 1862 are in DAR 205.8: 22–33, 37, and 39. They include tabulations of the numbers of seeds produced following crosses using pollen from the two different kinds of anther (DAR 205.8: 30, 31).

[13] Meehan 1862; there is an annotated copy of the paper in the Darwin Pamphlet Collection–CUL. Thomas Meehan, having noticed that (Meehan 1862, pp. 10–11):

> European willows, oaks and other trees retained their green leaves in the autumn much longer than closely allied American species growing near them, and that this could not be owing to immediate climatic influences, ... was led to believe it was rather the result of inherent specific peculiarities, which further investigation tended to confirm.

CD discussed Meehan's results in *Variation* 2: 281–2, in a chapter entitled: 'Direct and definite action of the external conditions of life'. See also Baker 1965.

[14] CD refers to his twelve-year-old son Leonard Darwin. Wells, Fargo & Co. was an American stage-coach company set up by Henry Wells and William George Fargo. In 1861, the company became the San Francisco agents for the Pony Express mail service and issued stamps inscribed 'Pony Express'. Daniel O. Blood & Co. was a mail company based in Philadelphia. (Sutton 1966, pp. 41, 237, 332.)

[15] According to Emma Darwin's diary (DAR 242), Leonard Darwin 'came home with sc. fever' on 12 June 1862.

[16] The notes referred to have not been found; CD asked Gray to observe and experiment on *Cypripedium* the previous year (see *Correspondence* vol. 9, letters to Asa Gray, 12 March [1861] and 5 June [1861]). In *Orchids*, pp. 274–5, CD had suggested that *Cypripedium* must be pollinated by an insect inserting its proboscis into one of the two lateral entrances at the base of the labellum, directly over one of the two lateral anthers, and thus either placing the pollen onto the flower's own stigma, or carrying it away to another flower. In 'Fertilization of orchids', pp. 155–6 (*Collected papers* 2: 152), CD stated:

> Prof. Asa Gray, after examining several American species of *Cypripedium*, wrote to me ... that he was convinced that I was in error, and that the flowers are fertilized by small insects entering the labellum through the large opening on the upper surface, and crawling out by one of the two small orifices close to either anther and the stigma.

Gray included his observations on American species of *Cypripedium* in a follow-up article to his review of *Orchids* for the *American Journal of Science and Arts* (also known as 'Silliman's journal'). See A. Gray 1862b, pp. 427–8.

[17] There is an undated note in DAR 70: 119–20, which states:

> I must correct my saying that insects could reach end of Labellum with greatest ease by open end of the toe of slipper— Cypripedium hirsutissimum seen very *[aberrant]* with rudimentary anther joined to soldered edge of labellum.—

CD refers to his statement in *Orchids*, p. 274 that 'An insect could reach the extremity of the labellum, or the toe of the slipper, through the longitudinal dorsal slit'. CD extensively revised his discussion of *Cypripedium* in the second edition of *Orchids*. See also letter to J. D. Hooker, 4 November [1862] and n. 10.

[18] *Orchids*, pp. 346–51. In the table of contents CD headed this section: 'Cause of the vast diversity of structure for the same general purpose'.

[19] See nn. 6 and 7, above.

[20] See letter from Asa Gray, [2 June 1862].

[21] Gray enclosed some flower-buds of the orchid *Arethusa* with his letter of [2 June 1862]. CD described the manner of pollination of *Cattleya* in *Orchids*, pp. 160–4. Hooker had sent CD some specimens of *Vanilla* at the end of May (see letter from J. D. Hooker, [29 May 1862] and n. 2).

[22] See n. 16, above.

[23] See letter from Asa Gray, [2 June 1862] and n. 2.

[24] There is a note dated 16 June 1862, recording CD's observations on the 'little imperfect flower' of *Oxalis acetosella* in DAR 111: 44. For CD's observations on *Viola*, see the letter to J. D. Hooker, 30 May [1862] and n. 7.

[25] Royer trans. 1862. The first edition of Clémence Auguste Royer's French translation of *Origin* was published on 31 May 1862 (*Journal Générale de l'Imprimerie et de la Librairie* 2d ser. 6 (pt 3): 341). CD's copy of the work has not been found; however, there is a lightly annotated copy of Royer's preface to her translation in the Darwin Pamphlet Collection–CUL.

[26] Gray had told CD that the occurrence in *Specularia perfoliata* of small unopening flowers in which self-fertilisation occurred (a phenomenon later called cleistogamy) had 'long been known' (see *Correspondence* vol. 9, letter from Asa Gray, 11 October 1861). CD had recently been investigating this phenomenon in *Viola* and *Oxalis* (see n. 24, above).

[27] See letter from Alphonse de Candolle, 13 June 1862.

[28] The reference is to Linnaeus 1751, aphorism 157: 'We count as many species as there were forms created in the beginning' (Stafleu 1971, p. 63).

[29] Murray's letter has not been found, but see the letter to John Murray, 20 [June 1862].

To Daniel Oliver [before 11 June 1862][1]

Dear Oliver.

I have been much pleased by the Zeitung for Müller describes exactly what I have seen, except that he overlooked the funny minute rudiments of petals.[2] V. hirta belongs to type of V. odorata. Why I write is that you have marked "your Regards" on copy;[3] but you do not mean me to keep copy? Does it not break set? I will keep it safe, & return it whenever you may have occasion to write

with many thanks.— yours very truly | C. Darwin

It is a pity that Müller did not know that the perfect flowers are fertile only when visited by Bees

P.S. By the way I must tell you that I had long letter from Asa Gray this morning, approving of my orchis book, & what is better comparing what I say with structure of living American allied orchids, & finds, as yet, all true.[4] This pleases me much.— He likes the book, incomparably more than I ever ventured to hope.—

P.S. What a wonderful scheme of reference you must have, always to know, what & where has been written on any subject.—

DAR 261 (DH/MS 10: 33)

[1] Dated by the reference to CD's having received a letter from Asa Gray, approving of *Orchids*, on the morning on which this letter was written (see n. 4, below); CD received Gray's letter of 18 May 1862, on or before 10 June (see letter to Asa Gray, 10–20 June [1862]). See also n. 2, below.

[2] Müller 1857; there is an annotated copy of this paper in the Darwin Pamphlet Collection–CUL. Oliver cited Daniel Ernst Müller's article on *Viola hirta* and *V. odorata* in his review of 'Dimorphic condition in *Primula*' ([Oliver] 1862c, p. 239). See also letter from Daniel Oliver, 10 April 1862, and letter to J. D. Hooker, 30 May [1862].

[3] CD's copy of the issue of the *Botanische Zeitung* in which Müller 1857 appears is marked: 'With DO's kind regards'.

[4] For Gray's reaction to the first half of *Orchids* (of which he read the proof-sheets), see the letter from Asa Gray, 18 May 1862; the portion of this letter written on 26 May (see letter to Asa Gray, 10–20 June [1862]) is missing. After sending him the proof-sheets, CD sent Gray a presentation copy of *Orchids* (see Appendix IV).

To H. W. Bates 11 June [1862][1]

Down Bromley Kent
June 11th

Dear Bates.

The writer of the enclosed is my Brother-in-law who studies & writes on Language.[2] Can you answer his question & tell me? I hope that your Book progresses & all goes well.—[3] I have been not so well lately—

Yours very sincerely | C. Darwin

American Philosophical Society (284)

[1] The year is established by the relationship to the letter from H. W. Bates, 14 June 1862.

[2] The enclosure has not been found. Emma Darwin's brother, Hensleigh Wedgwood, was probably collecting material for his book on the origin of language (Wedgwood 1866). Wedgwood cited Bates 1863 regarding the language of Amazonian Indians in Wedgwood 1866, pp. 29–30 and 132.

[3] Bates 1863. Following Bates's return to Britain in 1859 after 11 years spent exploring the Amazon region, CD had encouraged him to write an account of his travels in South America (see *Correspondence* vol. 9, and this volume, letter from H. W. Bates, 6 January 1862, and letter to H. W. Bates, 13 January [1862]).

To J. D. Hooker 11 June [1862]

Down
June 11th

My dear Hooker

I was glad to see your handwriting this morning.[1] I am very sorry to hear an indifferent account of Mrs. Hooker:[2] I heard from Miss Pugh (who is here) that she was looking not well.[3] How kind Mrs. H. was to Miss Pugh, who by the way is charmed with your Charlie & has repeated to us some of his capital little speeches.[4] Emma is at Southampton with Horace (who has got *much* good from the change) but I have sent your message about trustworthy oldish cook;[5] I fear she cannot help. We might have recommended the very woman; but she is going to marry: I wish to Heaven Nat. Selection had produced "neuters", who would not flirt or marry; I am sure that they would be as useful as neuter Bees. How the Exhibition works you![6] I have had only a glance, & I was not well, & saw nothing, & was dispirited.[7] Any time that you are *actually* inclined to write, I sh^{d}. of course be eager to hear about Cameroons.—[8]

I wish you had time to discuss a little the mundane Glacial period: I still believe it will be the turning point of all recent Geographical distribution. But the case, as I have put it is infinitely weaker than if I had published all the direct evidence of glacial action.[9]

Did I tell you that pollen placed for 65 hours on the apparent (& I still think real) stigma of Leschenaultia had not protruded a vestige of a tube!! I won't be beaten & will get a plant.—[10]

Oliver, the omniscient, has sent me a paper in Bot. Zeitung with most accurate description of all that I saw in Viola.—[11]

I have had long letter from Asa Gray, (*with not one allusion to politicks*.(!)) about my orchid book:[12] he gives red-hot praise; & did I not know he would not write falsely, I sh^{d}. think he was humbugging me; but I am sure he is honest, & his kind sympathy has run away with him.—

Farewell my dear old friend. | C. D.

Endorsement: '/62'
DAR 115.2: 155

[1] Letter from J. D. Hooker, 9 June 1862.

[2] In his letter of 9 June 1862, Hooker had written of Frances Harriet Hooker: 'My wife is very thin & watery, lacks energy, blood & muscle'.

[3] Miss Pugh had been governess at Down House between 1857 and 1859 (Emma Darwin's diary (DAR 242)); she was currently resident in Kew (see the letter from Emma Darwin to William Erasmus Darwin, [14 May 1862], in DAR 219.1: 56) and was thus a neighbour of the Hookers. According to Emma Darwin's diary (DAR 242), 'Miss Pugh came to Down' on 5 June 1862.

[4] CD also mentioned the Hookers' kindness to 'poor Miss Pugh' in the letter to J. D. Hooker, 9 May [1862]. Charles Paget Hooker, the Hookers' third child, was 7 years old.

[5] See letter from J. D. Hooker, 9 June 1862. Emma and Horace Darwin were in Southampton from 3 to 12 June 1862 (Emma Darwin's diary (DAR 242); see also letter to W. E. Darwin, [31 May 1862] and n. 4). Horace had been unwell earlier in the year (see letters to W. E. Darwin, 14 February [1862] and 26 April [1862]).

[6] See letter from J. D. Hooker, 9 June 1862. CD refers to the International Exhibition, which opened in South Kensington, London, on 1 May 1862. Hooker was a member of the jury for class 4, section C, 'Vegetable substances used in manufactures, &c.', and an associate juror for class 3, section B, 'Drysaltery, grocery, and preparations of food as sold for consumption' (*Reports by the juries*).

[7] CD was in London from 6 to 9 May 1862 (Emma Darwin's diary (DAR 242)), and visited the International Exhibition on 8 May (see letter to W. E. Darwin, [8 May 1862]).

[8] See letter from J. D. Hooker, 9 June 1862 and n. 3.

[9] CD put forward his ideas about migration during a worldwide glacial period in explanation for current patterns of geographical distribution in *Origin*, pp. 365–82. Henry Walter Bates's assertion that there was no evidence for the extinction of indigenous butterfly species from the Amazon region during the glacial period, and that the differences between North and South American temperate species of some insect genera were too great to have occurred since the glacial period, had led CD to reconsider his arguments in relation to whether the glacial period had affected all parts of the globe simultaneously and whether equatorial regions had been affected at all. CD had discussed this question with Bates and Hooker when they visited Down House in April 1862 (see letter from H. W.

Bates, 30 April 1862 and n. 3). However, Hooker's work on the flora of Cameroon and the islands off the coast of West Africa provided support for CD's view that temperate plants had migrated into tropical regions during the glacial period (see letter to J. D. Hooker, 9 May [1862] and n. 6).

10 CD refers to his investigations into fertilisation in *Leschenaultia* (see letter to J. D. Hooker, [18 May 1862] and n. 2). CD's notes on the observation reported here, which are dated 2 June 1862, are preserved in DAR 265. See also *ML* 2: 257–63.

11 Müller 1857. See letter to Daniel Oliver, [before 11 June 1862].

12 See letter from Asa Gray, 18 May 1862. CD's allusion to politics refers to the strained correspondence between Gray and some of his English correspondents in the wake of the so-called '*Trent* affair' (see letter from J. D. Hooker, [19 January 1862], letters to J. D. Hooker, 16 January [1862] and 25 [and 26] January [1862], and letter to Asa Gray, 22 January [1862]).

From Frederick Currey 12 June 1862

3, New Square, Lincolns Inn. W.C.
June 12. 1862

My dear Sir,

I have a spike of the rare Irish Orchid Spiranthes gemmipara which has been preserved in spirits for about 3 years & seems in good condition— I dont know whether it would be a matter of any interest to you to examine it, but if you would care to see it I shall be very happy to send it to you— It is *possible* I may be able later in the season to get a fresh spike of Orchis hircina— If so perhaps might like to see it— Your interesting book has induced me to write to you on the subject

Your's faithfully | Fred^k Currey

C. Darwin Esqr.

DAR 161.2: 305

CD ANNOTATION
1.2 for about . . . condition—] *double scored ink*

From George Howard Darwin [12 June 1862][1]

My dear Papa

L. is unwell & the doctor says it is Scarlet fever so he asked me if it would be best to send home or to lodging so I said home.[2]

I remain | G. H D

As I thought you would like M^r. Williams to see him I have telegraphed to M^r. W. to go to Down so don't send for him[3]

DAR 251

1 Dated by the reference to Leonard Darwin's illness (see n. 2, below).

2 George and Leonard Darwin attended Clapham Grammar School in south-west London (see letter from Charles Pritchard, 17 June [1862] and n. 2). Emma Darwin's diary (DAR 242) records on 12 June 1862: 'Lenny came home with sc. fever'; George's letter was probably sent with Leonard.

[3] Edward Augustus Williams was a surgeon in practice with Charles Morgan in Bromley, Kent (*Post Office directory of the six home counties* 1862); he attended various members of the Darwin household between 1845 and 1863 (see *Correspondence* vols. 7, 8, and 9, and CD's Classed account book (Down House MS)). In 1829, Williams had contributed articles on the treatment of scarlet fever to the *London Medical Gazette* (*London Medical Gazette* 4 (1829): 297, 432).

From Alphonse de Candolle[1] 13 June 1862

Genève
13 Juin 1862.

Mon cher Monsieur

je vous remercie extremement de m'avoir adressé votre article *On the dimorphic condition of Primula*.[2] Il est très curieux et bien propre à faire réflechir, comme tout ce que vous imprimez. J'ai engagé mon fils à en faire un extrait pour la bibliothèque universelle (archives des sciences) et il l'a rédigé effectivement pour un des prochains numéros.[3]

Le fait principal, celui de la fécondité supérieure, par le croisement des deux formes les moins semblables, ne m'a paru se rattacher en physiologie qu'à un seul fait connu, celui de la fecondité supérieure et des produits plus vigoureux d'individus non parents les uns des autres, tandis que le breeding in and in est défavorable. Cela est bien mystérieux—en théorie on aurait plutot supposé le contraire—mais c'est un fait.

Je n'ai pas vu si vous avez semé les graines obtenues par les divers croisements de vos Primula. On aimerait savoir si les deux formes reparaissent indifferemment des graines de divers croisements, ou si telle catégorie de graines produit une des formes. Vous dites bien que le même pied conserve sa forme de fleurs d'année en année, mais qu'arrive-t-il de génération en génération?

La double forme des Borraginées m'avait frappé en les rédigeant pour le *Prodromus*.[4] Je l'ai constaté dans cet ouvrage pour quelques espèces par ex. *l'Alkanna hispidissima* DC. Prodr. X p. 94, mais je n'avais point fait les recherches intéressantes que vous avez faites. On a aussi remarqué dans ⟨ ⟩ de la famille des Campanulacées des fleurs tardiv⟨es⟩ ⟨ ⟩ presque nulle, qui sont bien fertiles, et dans ⟨ ⟩ ⟨ ⟩munes autant quil m'en souvient deux espèces de fleurs, où les moins po⟨ ⟩ ⟨ ⟩ sont les plus fertiles, mais ce ne sont pas des ⟨ ⟩ analogues et l'on n'a pas expérimenté sur les ⟨ ⟩

Aurons-nous bientot le grand ouvrage que vous annonciez comme donnant les preuves detaillées des faits dont vous parlez dans le volume sur l'origine des espèces?[5] Je l'attends avec une vive impatience. En somme après vous avoir lu trois ou quatre fois, tantot de suite, tantot partiellement, je suis assez comme Asa Gray.[6] J'aime votre théorie. Elle plait à mon esprit. C'est la seule qui rende compte de questions très obscures, inabordables par d'autres voies—mais il lui faudrait des preuves, surtout quant à le *natural selection*. L'hypothèse générale d'une transmission indéfinie des formes au travers des siècles, avec des modifications plus ou moins graves, semble préférable à toute autre, mais que le natural selection soit le mode,

Joseph Beete Jukes.
(From Woodward 1907. By permission of the Syndics of the Cambridge University Library.)

Andrew Crombie Ramsay.
Photograph by G. C. Wallich.
(From G. C. Wallich 1870. By permission of the Syndics of the Cambridge University Library.)

Asa Gray. Portrait, 1862.
(Courtesy of the Archives, Gray Herbarium, Harvard University.)

voila ce qui est vague dans mon esprit. Il y a tant de causes qui maintiennent longtemps les mêmes formes de génération en génération ou qui les ramènent! Il est si rare qu'une forme nouvelle paraissant, se conserve sans la protection de l'homme! De ce dernier cas je ne connais aucun exemple constaté. Il y en a probablement, mais pas un seul n'a été prouvé, à ma connaisance. Ces graves questions, sur lesquels vous avez jeté un si grand jour, viennent de m'occuper à l'occasion d'une revue du genre *Quercus* pour le Prodrome[7] et d'une analyse des travaux de Heer sur la Flore tertiaire, que j'ai faite pour la Bibliothèque ⟨universelle⟩ (Mai 1862).[8] J'ai voulu traiter les Quercus ⟨ ⟩ sur l'espèce, au moyen de materiaux ⟨ ⟩t aux formes. Les variations sur la ⟨ ⟩ sont étonnantes. Il en découle une ⟨ ⟩ très grande de prétendues espèces des auteurs. Cela m'a conduit de nouveau à la question des origines des formes. J'ai ri de bon coeur en relisant la définition de Linné (Philos. bot. p. 157): *Species tot numeramus, quod diversæ formæ in principio sunt creatæ.*[9] Combien le bon homme croyait savoir de choses que nous ne savons pas cent ans après! Je n'ai découvert nulle part la date et le lieu précis d'origine d'un millier peut-etre de formes des Quercus que je groupe en 200 espèces environ et que je réduirais encore—si j'avais pour les formes étrangères autant d'échantillons que pour celles d'Europe et des Etats-Unis. Les formes principales du Q. Robur L. (que je considère comme *une* espèce) sont antérieures à la séparation de l'Irlande d'avec la Grande Bretagne et par consequent de celle-ci d'avec le Continent. Voila bien du temps que les formes pedunculata et sessiliflora luttent dans toutes les forêts de toute l'Europe, sans que l'une chasse l'autre. Les américains sont moins acharnés. Si je me décide à publier ce que j'ai commencé de rediger sur les Quercus j'aurai l'honneur de vous le faire passer.[10]

L'achevement des Dicotyledones pour le Prodromus m'a occupé depuis quelque temps. J'espère ou plutot je desire avoir assez de force pour reprendre la géographie botanique lorsque j'aurai publié les volumes XV et XVI avec l'aide de divers collaborateurs. Malheuresement pour moi les années avancent et les forces diminuent.

Agréez, mon cher Monsieur, l'assurance de ma haute considération et de tout mon dévouement | Alph. de Candolle

DAR 161.1: 10

[1] For a translation of this letter, see Appendix I.

[2] Candolle's name is on CD's presentation list for his paper 'Dimorphic condition in *Primula*' (see Appendix III).

[3] C. de Candolle 1862. Candolle refers to his son, Casimir de Candolle.

[4] Candolle and Candolle 1824–73. There is a note, dated 13 June 1862, recording this observation and reference, in DAR 110 (ser. 2): 25a.

[5] Candolle refers to CD's explanation in *Origin*, p. 2, that the work was an abstract, without references or authorities. CD stated: 'No one can feel more sensible than I do of the necessity of hereafter publishing in detail all the facts, with references, on which my conclusions have been grounded; and I hope in a future work to do this.' The only part of the planned three-part work that was published during CD's lifetime was *Variation* (1868).

[6] CD had sent Candolle a copy of Asa Gray's pamphlet on *Origin* in which Gray argued that natural selection and natural theology were not inconsistent with one another if one took the view that natural

selection operated in accordance with divine purpose (A. Gray 1861; for CD's presentation list for this pamphlet, see *Correspondence* vol. 9, Appendix III). In the discussion of CD's theory given in A. de Candolle 1862b, p. 60 (see n. 10, below), Candolle noted that many apparent anomalies that were repugnant and embarrassing ('répugnent et embarrassent') for the theory of special creation (like the rudimentary nipples of male mammals), were, by contrast, brought under a general law by the theory of evolution from a common ancestor. He continued (p. 61):

> je trouve naturel que des hommes fort éloignés des idées matérialistes, ayant même une tendance prononcée vers d'autres opinions, comme le docteur Hooker, M. Asa Gray et le professeur Heer, préfèrent la théorie de l'évolution et s'attachent plus ou moins aux doctrines ou aux études par lesquelles on s'efforce de la démontrer. [I find it natural that men who distance themselves from materialist ideas, and who may even have a definite tendency towards other opinions, like Dr Hooker, Mr Asa Gray and Professor Heer, prefer the theory of evolution and attach themselves more or less to those doctrines and studies by which attempts are made to demonstrate it.]

The references are to Joseph Dalton Hooker and Oswald Heer. On Gray's response to CD's theory, see also Dupree 1959.

[7] Candolle and Candolle 1824–73, 16, pt 2: 1–109.

[8] Candolle's paper (A. de Candolle 1862a) discussed Heer 1855–9 and Heer 1861a.

[9] Linnaeus 1751, aphorism 157: 'We count as many species as there were forms created in the beginning' (Stafleu 1971, p. 63).

[10] Candolle published two papers relating to the oak genus (*Quercus*) in 1862 (A. de Candolle 1862b and 1862c); there are annotated copies of these publications in the Darwin Pamphlet Collection–CUL. The first paper (A. de Candolle 1862c) described a newly identified defining characteristic in acorns, and proposed a new division of the genus. Candolle probably refers to the second paper (A. de Candolle 1862b), entitled 'Étude sur l'espèce à l'occasion d'une révision de la famille des Cupulifères' [Study on species occasioned by a revision of the family of Cupuliferae], in which he discussed the grouping and historical origins of the members of the oak family, conluding with an extensive commentary on CD's theory (A. de Candolle 1862b, pp. 57–68). Although Candolle considered that evolution from a common ancestor was 'l'hypothèse la plus naturelle' [the most natural hypothesis] (p. 66), which explained otherwise inexplicable facts, he also observed that there was no direct proof of the theory, and recommended a cautious approach, especially in view of the vast periods of time that would be required to accomplish species change by such a process.

To W. E. Darwin 13 [June 1862][1]

Down

13th

My dear William

Mamma returned with Horace very brisk yesterday.[2] An hour before she came, Lenny arrived ill in carriage & this morning has decided Scarlet Fever.—[3] So the House is in miserable state; but he is not bad.—

Please return my watch in some little Box, registered, & please return "Journal of Horticulture".— I am so very sorry that you yourself are not very well.—

My dear old fellow | C. Darwin

DAR 210.6: 99

[1] The date is established by the reference to Leonard Darwin becoming ill with scarlet fever (see n. 3, below).

[2] According to Emma Darwin's diary (DAR 242), she and Horace Darwin visited Southampton, where William lived, from 3 to 12 June 1862. Horace had been ill earlier in the year.
[3] Twelve-year-old Leonard Darwin attended Clapham Grammar School in south-west London (see letter from Charles Pritchard, 17 June [1862]). Emma Darwin's diary (DAR 242) records, on 12 June 1862: 'Lenny came home with sc. fever'.

To Leonard Horner 13 June [1862][1]

Down. | Bromley. | Kent. S.E.
June 13th

My dear Mr. Horner

I hope that you will forgive me troubling you with this note, but I wish to say how sincerely sorry we are at your great loss & at all the grief which you must have suffered.[2] This event has recalled vividly to my mind the great kindness which I received, & the many pleasant hours which I have spent, in old days in Bedford Place.[3] Emma desires me to thank you for having remembered us and sent her the Memorial paper; she feels that she has lost a most kind friend; and she most sincerely joins me in sympathising with you & all your family.

Pray believe me my dear M[r] Horner | Yours very sincerely | Ch. Darwin

National Library of Scotland (MS 2216: 167)

[1] The year is established by reference to the death of Anne Susan Horner (see n. 2, below).
[2] Horner's wife, Anne Susan Horner, died on 22 May 1862 (K. M. Lyell ed. 1890, 2: 352).
[3] During the 1830s and 1840s, the Horners lived at 2 Bedford Place, London (see *Correspondence* vol. 2, letter from Leonard Horner, 7 April [1838], and *Post Office London directory* 1846, 1848). CD had visited the Horners as a bachelor; he and Emma lived in London from their marriage in 1839 until their removal to Down House in 1842 (see *Correspondence* vol. 2).

To Patrick Matthew 13 June [1862][1]

Down. | Bromley. | Kent. S.E.
June 13th

Dear Sir

I presume that I have the pleasure of addressing the Author of the work on Naval Architecture & the first enunciator of the theory of Natural Selection.[2] Few things would give me greater pleasure than to see you; but my health is feeble & I have at present a son ill & can receive no one here, nor leave home at present.—[3]

I wish to come up to London as soon as I can; if, therefore you are going to stay for more than a week, would you be so kind as to let me hear, & if able to come up to London, I would endeavour to arrange an interview with you, which afford me high satisfaction;

with much respect, I remain Dear Sir | Yours very faithfully | Ch. Darwin

National Library of Scotland (Acc. 10963)

[1] The year is established by the relationship to the letter from Patrick Matthew, 3 December 1862, and by reference to Leonard Darwin's illness (see n. 3, below).

[2] In a letter to the *Gardeners' Chronicle and Agricultural Gazette*, 7 April 1860, Matthew claimed to have formulated a principle of natural selection in his book *On naval timber and arboriculture* (Matthew 1831). CD conceded the claim in a letter published in the *Gardeners' Chronicle and Agricultural Gazette*, 21 April 1860, pp. 362–3 (see *Correspondence* vol. 8, letter to *Gardeners' Chronicle*, [13 April 1860]). Matthew was mentioned in the 'Historical sketch' published in the third edition of *Origin* as one of those who had, prior to the publication of *Origin*, enunciated ideas on species change (*Origin* 3d ed., pp. xiv–xv).

[3] Leonard Darwin had been sent home from school on 12 June 1862 suffering from scarlet fever (see letter to W. E. Darwin, 13 [June 1862]). Matthew's letter suggesting a meeting with CD has not been found; however, see the letter from Patrick Matthew, 3 December 1862.

To John Murray 13 June [1862]

Down. | Bromley. | Kent. S.E.
June 13th

My dear Sir

You were so kind as to say that, I might have electrotypes of woodcuts for a German edition.[1] I wrote to M. Schweizerbart of Stuttgart to say that if he continued to wish for German Edition, he must pay to you £10 for them, which will leave a few pounds profit, I believe, over the cost.[2] There are 34 (several very small); for the embellishment I shd. think need not be sent.[3] He wishes to have them *immediately*, but does not say how they are to be sent; you will probably know; & asks how he is to pay you the £10; whether by a note from London? You will have to send to Taylor & Francis for the 2 they have.—[4]

I am heartily sorry to give so much trouble, & here follows more.—

Prof. Asa Gray doubts whether at present an American Edit. can be published; but he wishes to know what you would charge for electrotypes of all.[5] Fix what you think right, but remember if you fix high there will be no chance of Edition.— Please answer all my queries soon.— Again Asa Gray wishes to review the Book in Sillimans Journal (an admirable Journal) & wishes for cuts of Nos I., II., & III.:[6] if you are inclined to grant me this favour (which would please me) I will pay for these 3 cuts. Asa Gray says he wishes for them immediately. I do not understand their manufacture; but he says "*not to be mounted*" & sent through Trübner;[7] or what what would be better for the "electrotype *copper casts* to be put in light Box & sent by Post, & postage would hardly be more than 1s.".—" But I fancy the "casts" would be required for the German cuts; but I do not in the least understand the subject.— Please act as you think fit. Again I say I am sorry to give so much trouble.—

The Botanists praise my orchid-book to the skies.—[8] Some one sent me (perhaps you) the Parthenon with good Review.—[9] The Athenæum treats me with very kind pity & contempt; but the Reviewer knew nothing of his subject.[10]

Yours very sincerely | C. Darwin

I wish I had asked Schweizerbart a higher sum; for he evidently is glad to pay the 10£; but it is too late now.—

Please write soon.—

Endorsement: '*1862*'
John Murray Archive (Darwin 120–2)

[1] Murray's letter has not been found, but see the letter to H. G. Bronn, 25 April [1862].
[2] See letter to H. G. Bronn, 25 April [1862], and letter from E. Schweizerbart'sche Verlagsbuchhandlung, 7 June 1862.
[3] *Orchids* was offered for sale in a plum-coloured cloth binding with a gilt orchid embossed on the front cover.
[4] Taylor & Francis of Red Lion Square, Fleet Street, London, printed the *Journal of the Proceedings of the Linnean Society of London*, in which CD's paper, 'Three sexual forms of *Catasetum tridentatum*' was published. The paper included two illustrations from *Orchids*, *Catasetum tridentatum* (*Orchids*, fig. 27, p. 232) and *Myanthus barbatus* and *Monachanthus viridis* (*Orchids*, fig. 28, p. 239). See also letters to Richard Kippist, 18 March [1862].
[5] See letter from Asa Gray, 18 May 1862, and letter to Asa Gray, 10–20 June [1862].
[6] The illustrations referred to are those of *Orchis mascula* (*Orchids*, facing p. 18), the pollinia of *O. mascula* (*Orchids*, p. 15), and *O. pyramidalis* (*Orchids*, p. 22). Asa Gray followed up his review of *Orchids* (A. Gray 1862a), which appeared in the July number of the *American Journal of Science and Arts* (also known as 'Silliman's journal' after its founder, Benjamin Silliman), with a further article on the pollination of orchids in the November number (A. Gray 1862b); the three illustrations from *Orchids* were reproduced in the latter paper (pp. 421–3).
[7] The publisher Nicholas Trübner frequently acted as Gray's London agent (see *Correspondence* vols. 8 and 9).
[8] CD was particularly pleased at the response to *Orchids* of Gray, George Bentham, and Daniel Oliver (see letter to Asa Gray, 10–20 June [1862]).
[9] *Parthenon*, 7 June 1862, pp. 177–9. The author of the review has not been identified. There is a copy of it in the Scrapbook of reviews (DAR 226.1: 129).
[10] According to the publisher's marked copy of the *Athenæum* (City University Library, London), the author of the review of *Orchids* that appeared in the *Athenæum*, 24 May 1862, pp. 683–4, was John R. Leifchild. There is a copy of this review in the Scrapbook of reviews (DAR 226.1: 11).

From H. W. Bates 14 June 1862

King St Leicester
14 June 1862

My Dear Mr Darwin

Adjoined is a sort of answer to Mr Wedgwood's query & I shall be glad if it prove of any use.[1] I am truly sorry to hear of your being again more unwell than usual: but this under the circumstances of the interruption to your work which it causes, is a national concern.

My book "progresses" it is true, but slowly.[2] How thoroughly ashamed I am to have bragged to you how quickly I could write it.[3] I am at it every day as many hours as I can stand; but what takes me one day to write, takes 5 to alter.

Please do not make use of the facts about generative organs in beetles which I gave you.[4] Conversing in London again with those gentlemen I find such a chaos of statements that the facts are not to be depended upon—

Yours sincerely | H W Bates

DAR 160.1: 70

[1] See letter to H. W. Bates, 11 June [1862] and n. 2. The enclosure has not been found.
[2] Bates 1863.
[3] No letter from Bates containing such a claim has been found. However, when CD wrote to John Murray asking whether he would be interested in publishing Bates's travel narrative, he emphasised that Bates wished 'to publish soon' (see letter to John Murray, 28 January [1862]).
[4] See letter from H. W. Bates, 19 May 1862.

To G. H. K. Thwaites 15 June [1862][1]

Down. | Bromley. | Kent. S.E.
June 15th

My dear Mr Thwaites

If you see Linn. Journal, I hope that you will read my paper on the fertilisation of Primula.[2] Why I hope that you will do this, is because I have been told that Dr. Weddell states that Cinchona presents the same case of some individual trees with long pistils & some with short.[3] Now if this be the case, it is almost certain that in order to get good seed & good seedlings there must be reciprocal fertilisation between the two forms. It is quite likely, but by no means certain, that Ceylon insects may do this work effectually; but the subject, I am sure, is worth your attention.

If there be any difficulty in getting seed in Ceylon, *or in raising strong plants from Ceylon seed*, do I beg you, try artificial fertilisation on a few dozen flowers, making the cross between the two forms.— There is no need of castration The growth of Cinchona is so important for mankind, that I am sure you will excuse my making this suggestion.[4] It is a subject I am still working at.[5] This form of dimorphism seems common with the Rubiaceæ;[6] & if you have observed any analogous facts I shd. be grateful for information.—

I sincerely hope that you are well & in all ways prosperous.—

Pray believe me | Yours very sincerely | Ch. Darwin

American Philosophical Society (278)

[1] The year is established by the relationship to the letter to G. H. K. Thwaites, 20 June [1862].
[2] CD read his paper, 'Dimorphic condition in *Primula*', before the Linnean Society of London on 21 November 1861 (see *Correspondence* vol. 9); the paper was published in the number of the society's *Journal* issued on 1 March 1862 (*General index to the Journal of the Linnean Society*).
[3] A 'rather tall man with upturned eyebrows' gave CD this information, after hearing CD's paper on *Primula*, at the Linnean Society meeting on 21 November 1861 (see *Correspondence* vol. 9, letter to J. D. Hooker, 25 November [1861]). The reference is to Hugh Algernon Weddell, a specialist on the flora of South America who had made a detailed study of *Cinchona* (see, for example, Weddell 1849).
[4] *Cinchona* trees were valued on account of their bark, the source of quinine, an effective febrifuge or antipyretic medicine. The enormous demand for the bark and the wasteful manner in which it was procured from South America led to fears that supplies would soon be exhausted. From the middle of the nineteenth century, attempts were made to introduce the most valuable *Cinchona* species into Asian localities likely to support their cultivation; in the 1850s, the Dutch introduced the trees to Java, and, in the 1860s the British established plantations in India and Ceylon (*EB*).

[5] CD was investigating the possible occurrence of dimorphism in a number of species, particularly in several members of the Melastomataceae (see, for example, letter to Asa Gray, 10–20 June [1862]).

[6] *Cinchona* is a member of the Rubiaceae. Asa Gray told CD of the dimorphic nature of many Rubiaceae in his letter of [27 and 29 August] and 2 September 1861 (*Correspondence* vol. 9), and mentioned *Houstonia* as a good example on which to experiment (see *ibid.*, letter from Asa Gray, 11 October 1861). Gray had recently examined plants of this genus for CD (see letter from Asa Gray, [2 June 1862]). See also *Correspondence* vol. 9, letter from George Bentham, 26 November 1861.

To Alphonse de Candolle 17 June [1862]

Down. | Bromley. | Kent. S.E.
June 17th

My dear Sir

I am extremely much obliged for your kind & very interesting letter.[1] I am pleased that you are interested by the Primula case. Your questions & remarks show that you have gone to the root of the matter. I am now trying various analogous experiments on several plants & on the seedlings raised from the so-called heteromorphic & homomorphic unions; & the results (as far as I have yet seen; for the capsules are gathered, but not yet examined) are interesting; Whenever I publish I will do myself the pleasure of sending you a copy.[2] I am particularly obliged for your information on Alkanna. I have examined the small imperfect flowers of Viola & Oxalis: the case is very different both functionally & structurally from that of Primula.—[3]

You kindly enquire about my larger work; it does make progress, but very slowly owing to my own weak health & ill-health in my family.[4] I have, also, been seduced to publish a small work on the Fertilisation of Orchids, which has taken up nearly ten months. As Mr Bentham & Asa Gray think well of this Book, I have sent by this post a copy for you.[5] One main object has been to show how wonderfully perfect the structure of plants is; another regards close breeding in & in, to which I see you have attended.— I am not at all surprised that you are not willing to admit natural selection: the subject hardly admits of direct proof or evidence.[6] It will be believed in only by those who think that it connects & partly explains several large classes of facts: in the same way opticians admit the undulatory theory of light, though no one can prove the existence of ether or its undulations.—

I hope you will publish on Quercus, & I shall be most grateful for a copy; the genus has long appeared to me preeminently interesting under the point of view to which you refer.[7] I am, also, rejoiced to hear that you have the intention of again returning to Geographical Distribn. I believe few, or no one, can have read your truly great work with more care than I have;[8] & no one can feel more respect & admiration for it & its author.—

Pray believe me, my dear Sir | Yours sincerely & respectfully | Ch. Darwin

Endorsement: '17 Juin 1862'
La Fondation Augustin de Candolle

[1] Letter from Alphonse de Candolle, 13 June 1862.
[2] In January and February 1862, CD carried out crossing experiments with *Primula sinensis* using plants raised the previous year from the seed produced by homomorphic and heteromorphic crosses (see letters to J. D. Hooker, 7 March [1862] and n. 10, and 23 June [1862] and n. 4). (By 'homomorphic unions', CD meant crosses in which fertilisation was effected by own-form pollen; by 'heteromorphic unions', those in which it was effected by different-form pollen.) In April and May, he performed experiments on the common cowslip, *P. veris*, not only repeating the homomorphic and heteromorphic crosses that he had carried out in 1860 and 1861 (see letter to J. D. Hooker, 23 June [1862] and n. 4), but also investigating the possible prepotency of different-form pollen from a polyanthus (which he considered a variety of the same species) over own-form pollen from a cowslip (see CD's notes from these experiments in DAR 157a: 76–7 and DAR 108: 70). Some of the results of CD's experiments with *P. sinensis* are recorded in 'Illegitimate offspring of dimorphic and trimorphic plants' (a paper not included in *Collected papers*), which was published in 1869; the results of his experiments with *P. veris* are recorded in *Cross and self fertilisation*, p. 28, and 'Three forms of *Lythrum salicaria*', p. 189 (see also *Collected papers* 2: 121–2). In the letter, CD interlined the phrase 'on several plants &' (see Manuscript alterations and comments); he was engaged in crossing experiments with several members of the Melastomataceae, which he believed might exhibit dimorphism (see, for example, letter to George Bentham, 3 February [1862], letters to J. D. Hooker, 9 February [1862], 15 [May 1862], and 30 May [1862], and letters to Asa Gray, 15 March [1862] and 10–20 June [1862]).
[3] See also letters to Daniel Oliver, 12 [April 1862] and 15 April [1862], letter to J. D. Hooker, 30 May [1862], and letter to Asa Gray, 10–20 June [1862].
[4] See letter from Alphonse de Candolle, 13 June 1862 and n. 5. CD and many of his household had had influenza early in the year (see letter to J. D. Hooker, 16 January [1862], and letter to Asa Gray, 22 January [1862]). Horace Darwin became ill during January and had suffered bouts of ill health ever since (see letter to W. E. Darwin, 14 February [1862]). On 12 June, Leonard Darwin was sent home from school suffering from scarlet fever (see letter to W. E. Darwin, 13 [June 1862]). In his 'Journal' (see Appendix II), CD recorded: 'Much time wasted June & July from Leonard's illness'.
[5] See letter from George Bentham, 15 May 1862, and letter from Asa Gray, 18 May 1862. See also letter from J. D. Hooker, [17 May 1862], and letter to Asa Gray, 10–20 June [1862]. Candolle's name is included on CD's presentation list for *Orchids* (see Appendix IV).
[6] See letter from Alphonse de Candolle, 13 June 1862.
[7] See letter from Alphonse de Candolle, 13 June 1862 and n. 10.
[8] A. de Candolle 1855. There is an annotated copy of this work in the Darwin Library–CUL (see *Marginalia* 1: 106).

From Charles Pritchard 17 June [1862][1]

Clapham
June 17.

My dear Sir,

You see I have taken the precaution of breaking up a very few days earlier than we intended— I feared possibilities.

I rejoice to hear that your boy is nearly convalescent.[2]

I shall send some work for the holidays, part of which is to be remitted to me in about a month. The more formal Examination I must defer until August 12 when the school re-assembles.

I am | dear Sir | faithfully yours | C. Pritchard

I desire my respects to Mrs Darwin.

I shall hunt for Pollen tubes in the I. Wight: but not under -/10—but try what I can do with an erecting eye-piece and inch object glass giving me a power *[lumina]* of about 90.[3]

DAR 174.2: 77

[1] The year is established by the reference to Leonard Darwin's illness (see n. 2, below).

[2] Pritchard was headmaster of Clapham Grammar School, the school then attended by George Howard Darwin, Francis, and Leonard Darwin (see letter to W. E. Darwin, 26 April [1862], n. 2). Emma Darwin's diary (DAR 242) records that Leonard had been sent home from school on 12 June with scarlet fever, that George returned home on 16 June, and that Francis, having come home for the weekend on 31 May, did not return to school as he became 'feverish' on 2 June.

[3] Pritchard, whose main interests were mathematics and astronomy, was also interested in botany, having at Clapham, 'one of the finest fernaries in England' (*DNB*). Although CD and Pritchard evidently corresponded on botanical topics, these letters have not been found. In his memoirs (A. Pritchard ed. 1897, p. 60), Pritchard recalled that George Darwin had, while a pupil at Clapham, asked him 'to read with him certain portions of his illustrious father's book on *The Fertilisation of Orchids*.' He continued: 'More than gladly I did as he desired, and in due time we succeeded in fertilising and ripening the seeds of Oncidium Papilio; but we did not succeed in our attempts to induce the seeds to germinate, as his father challenged us to do: the necessary environments were wanting.' Pritchard left his post at Clapham Grammar School in 1862, and moved to Freshwater on the Isle of Wight (*DNB*).

To *Bienen Zeitung*[1] 18 June 1862

Es sollte mir sehr angenehm sein, wenn Herr Pfarrer Dzierzon,[2] oder ein anderer erfahrener Correspondent der Bienenzeitung, die Güte haben würde, zu erklären, ob bei den ordinären Bienen (apis mellifica), welche in den verschiedenen Gegenden Deutschlands gehalten werden, ein merklicher Unterschied stattfindet oder nicht. Ein aufmerksamer Naturforscher und Geistlicher sowohl als Gärtner behauptete vor einigen Jahren, daß gewisse Brut der Bienen kleiner sei als andere und daß in der Gemüthsart der Bienen ein Unterschied stattfinde.[3] Dieser Geistliche erklärte ferner, daß die wilden Bienen in gewissen Wäldern Englands kleiner als die gewöhnlichen zahmen Bienen seien. Mons. Godson, ein gelehrter französischer Naturforscher, sagt ebenfalls, daß im Süden Frankreichs die Bienen grösser als anderswo seien und daß beim Vergleich gewisser Stöcke ein geringer Unterschied in der Farbe des Haars entdeckt werden kann.[4]

Ich hoffe, daß einige erfahrene Beobachter, welche die Bienen in den verschiedenen Orten Deutschlands gesehen haben, darthun werden, inwieweit die vorstehenden Bemerkungen begründet sind.[5]

Bromley, Kent, England 18/6 62.
Charles Darwin.

Bienen Zeitung 18 (1862): 145

[1] For a translation of this letter, see Appendix I. CD sent a slightly different version of this inquiry to the *Journal of Horticulture* (letter to *Journal of Horticulture*, [before 10 June 1862]). It was Thomas White

Woodbury, one of the editors of the bee section of the *Journal of Horticulture*, who forwarded CD's letter to the editors of the *Bienen Zeitung* (*Journal of Horticulture and Cottage Gardener* n.s. 3 (1862): 463).

2 The reference is to the renowned German beekeeper, Johannes Dzierzon.

3 The reference has not been traced.

4 'Godson' is a misspelling; the reference is to Godron 1859, 1: 459. There is an annotated copy of this work in the Darwin Library–CUL (see *Marginalia* 1: 331).

5 A reply from Dzierzon was published in the *Bienen Zeitung* 18 (1862): 145–6. CD's query also elicited a response from Georg Kleine (*Bienen Zeitung* 18 (1862): 206–7). These replies were translated into English and published in the *Journal of Horticulture and Cottage Gardener* n.s. 3 (1862): 463–4 and 642–3. In *Variation* 1: 298, referring to Dzierzon's reply to this letter, CD stated: 'The great apiarian Dzierzon . . . says that in Germany bees of some stocks are decidedly dark, whilst others are remarkable for their yellow colour. Bees also seem to differ in habits in different districts'. CD also summarised Kleine's reply (*ibid.*, p. 298 n. 59): 'though there is some variability in colour, no constant or perceptible differences can be detected in the bees of Germany'.

To John Murray 18 [June 1862][1]

Down
18th

My dear Sir

I wish much for an answer about the Woodcuts. I am very sorry to be so troublesome.— Ask one of your clerks to answer me.—[2]

Yours very sincerely | in great Haste | C. Darwin

There is a superb, but I fear exaggerated, Review in "London Review".[3] But I have not been a fool, as I thought I was, to publish; for Asa Gray, about the most competent judge in world, thinks almost as highly of book as does London Review.—[4]

The Athenæum will hinder sale greatly.—[5]

John Murray Archive (Darwin 123)

1 Dated by the relationship to the letter to John Murray, 13 June [1862], and to the letter to Asa Gray, 10–20 June [1862] (see n. 2, below).

2 CD had asked Murray whether he could provide three plates from *Orchids* for a review proposed by Asa Gray (see letter to John Murray, 13 June [1862], and letter to Asa Gray, 10–20 June [1862]).

3 Miles Joseph Berkeley wrote an anonymous review of *Orchids* for the *London Review and Weekly Journal of Politics, Arts and Science* ([Berkeley] 1862). CD wrote 'Rev J. M. Berkeley' on a copy of the review now in the Scrapbook of reviews (DAR 226.1: 87).

4 See letter from Asa Gray, 18 May 1862, and letter to Asa Gray, 10–20 June [1862].

5 CD refers to the critical review of *Orchids* by John R. Leifchild in the *Athenæum*, 24 May 1862, p. 691.

From J. D. Hooker 19 [June 1862][1]

Royal Gardens Kew | Kew
19th.

Dr Drn.

Your Rhododendron is *R. glaucum*— it has a very peculiar scent.—[2] this was sent me by you a month ago under some other name.

I saw Lubbock yesterday who tells me that you were in bed last Sunday & that one of your boys has come home ill from school with Scarlet fever[3] I am truly sorry to hear of all this.

We are in statu quo—no Cook yet, though we hope we have heard of one who will suit— We have also I hope found the right sort of person for Nursery Governess—[4] Meanwhile we are carrying on the war as best we can & have a party of 12 persons chiefly Italian Botanists coming to a Tea dinner tomorrow. How glad we shall be to have Willy tomorrow to meet them—[5] one is Mr. Watson Taylor of Monte Cristo notoriety & his wife.[6] Leonard Henslow my eldest brother in law is engaged to a very nice girl who has been here today, a Miss Wall, niece of Dean Pellew.[7] apparently a very nice girl indeed with a little money— she is only 19. *very* large, quite plain, but most amiable & pleasing-looking. we are rejoicing at it. Leonard Jenyns is to be married next week[8]

My Jury work is nearly over & a pretty time I have had of it![9]

My wife is a trifle better I think, but very pale and bloodless & cannot sleep at night—[10] I wish I could take her to Switzerland

On Monday I go down to Ross' place to look over some of his things previous to the sale—[11] Sabine goes after some Meteorological &c books of observation—[12] I for a whole lot of things chiefly glacial & Kerguelens Land that I never could find in the British Museum.

This is a weary life, I am daily expecting to hear of the death of a dear little niece, 8 years old, at school at Bury, of inflammation of lungs,—a daughter of my eldest sister Mrs M^c^Gilvray.[13]

Ever yours affec | J D Hooker

DAR 101: 38–9

CD ANNOTATION
Top of first page: 'London Review | Orchids'[14] *pencil*

[1] Dated by the reference to Leonard Jenyns's impending marriage to Sarah Hawthorn (see n. 8, below).
[2] See letter to J. D. Hooker, [18 May 1862].
[3] Hooker refers to John Lubbock. Leonard Darwin was sent home from school with scarlet fever on 12 June 1862 (see letter to W. E. Darwin, 13 [June 1862]).
[4] The governess has not been identified. For Hooker's concern about the state of his household, see the letters from J. D. Hooker, [5 May 1862] and 9 June 1862.
[5] Hooker had been helping CD to foster William Erasmus Darwin's interest in botany (see *Correspondence* vol. 9, letters to J. D. Hooker, 22 June [1861] and 17 [July 1861]). Hooker had been planning for some time to have William, who lived in Southampton, visit Kew (see letter from J. D. Hooker, [5 May 1862], letter to W. E. Darwin, [8 May 1862], and letter to J. D. Hooker, 9 May [1862]).
[6] George Graeme Watson-Taylor was a British subject who owned the Mediterranean island of Monte Cristo. The island came under Sardinian control after the Italian War of 1859, and in 1860 Watson-Taylor and his wife, Victorine, were charged with sedition following altercations with the troops stationed there. Escaping to Britain, they were found guilty *in absentia*, but were granted remission of their penalties by the Sardinian king, Victor Emmanuel II. In their absence from Monte Cristo the island was sacked by members of Giuseppe Garibaldi's volunteer army, the 'Red shirts', who had commandeered a British steamer, the *Orwell*, in pursuance of their military campaign for a united Italy

under the rule of the Sardinian king. In 1861, Watson-Taylor unsuccessfully sought compensation for the damage to his property from the new Italian government of Victor Emmanuel. The case attracted much publicity in Britain, and on 13 June 1862 it was the subject of a lengthy debate in the House of Commons (*EB*, *Hansard's Parliamentary Debates* 3d ser. 167: 545–608).

7 Frances Harriet Hooker's eldest brother, Leonard Ramsay Henslow, married Susan Wall on 9 September 1862 (*Gentleman's Magazine* n.s. 13 (1862): 489). George Pellew was dean of Norwich.

8 Leonard Jenyns married Sarah Hawthorn of Stapleford, Cambridgeshire, on 24 June 1862 (*Gentleman's Magazine* n.s. 13 (1862): 222).

9 See letter to J. D. Hooker, 11 June [1862], n. 6.

10 See also letter from J. D. Hooker, 9 June 1862.

11 The polar explorer, James Clark Ross, died in Aylesbury, Buckinghamshire on 3 April 1862; Hooker had been assistant surgeon on Ross's Antarctic expedition of 1839–43 (*DNB*).

12 Edward Sabine, president of the Royal Society and a major-general in the Royal Artillery, had been for many years a friend and colleague of Ross's. The two had sailed together on the 1818 Arctic expedition in search of a North-west Passage, commanded by Ross's uncle, John Ross. During the 1830s, J. C. Ross had assisted Sabine in the first systematic magnetic survey of the British Isles, and Ross's Antarctic expedition of 1839–43 was organised largely to make magnetic observations for Sabine's work on the subject (Cawood 1979, *DNB*).

13 Anne McGilvray, daughter of Maria McGilvray, died on 15 August 1862 (see 'Hooker pedigree' in Allan 1967).

14 For CD's comments on Miles Joseph Berkeley's review of *Orchids* for the *London Review and Weekly Journal of Politics, Arts and Science* ([Berkeley] 1862), see the letter to J. D. Hooker, 23 June [1862].

From W. B. Clarke 20 June 1862

St Leonard's | N.S.W
20th June 1862

My dear Sir,

I am glad you got my tin box with the crammed contents, which I did not know how to send otherwise.[1]

I thank you very much for your kind introduction to Mr. Moore.[2]

I write to him by this Mail, and will send to him the Fossils much increased by new findings, in July; among others huge Ammonitidæ.

I have also written to Mr. Rupert Jones about getting my whole collections of Tertiary, Secy and Palæozoic fossils described and figured.[3] I want to find some one or more who would undertake the service. The Parliament here have in addition to the 2000£ I had before from N.S.W and Victoria, put me into a pension of £*[200]* per an. and also propose this year to give me 5000£ in addition for my past and *future* services, if I will write a book for them containing all I know about the Geology of these parts.[4] at the same time the Radical Members are hoping here to cut down all *State Aid* to the Clergy. I may, perhaps, therefore be not able to do what I would in the money way—but I doubt not I should be able to make some remuneration for the Palæontological work I require. Can you assist Mr Jones in telling me what it is best to do. I have *thousands* of fossils—of which *hundreds* are distinct species. I can work up M'Coy's description of those I sent to Cambridge by re-printing his paper.[5] And as the book is for the Colony, I intend to have *figured* every Colonial fossil I can get—if I find the means.

I am sorry you broke the supposed Cirripede.[6] But I have myself serious doubts now, if it was such. I never recognised such a structure in the wings of Pecten or Monotis:—but I begin to think it belonged to one or other. Mr. Moore will have another and then you can determine it.

It was the . . . . . . line of dots which partly induced me to think the thing was a valve of a Cirripede.

I will try and get you some native Comb.—[7]

I have applied to some persons on whom I can rely, with respect to the Gardenias. They grow here—but *not* in gardens: so there is a difficulty.— Last year the Eucalypti did not seed much— this year the seeds are forming abundantly. We have had a mild dry autumn and thus far winter. Perhaps moisture has something to do with it. For the ants are as usual every where at all times up and down.

Every now & then I am hearing of people who find *trees partly* living or recent partly converted into stone. I⟨s⟩ that a fact known out of ⟨Aus⟩tralia? It might account for stone trees lying on the surface of our plains—without calling in the aid of a post mortem silicification.

I hope you will soon recover your health and strength. '*Serus in cælum redeas*".[8] I have also been ill of late, but I effected a partial cure by taking a hammer and walking up and down the railway lines, examining sections, in *[illeg]* and Illawarra. I am going next week to Mount *[illeg]*, the top of wh. you saw when you went to Bathurst.[9]'

Believe me, My dear Sir, Yours very truly | W. B. Clarke

C. Darwin Esq^{e}. F.R.S.

DAR 161.2: 174

[1] In January 1862, Clarke had sent CD a copy of Clarke 1861b and a specimen of a fossil cirripede (see letter from W. B. Clarke, 16 January 1862, and n. 6, below); the box may have contained these items.

[2] Clarke, seeking someone to describe his important collection of fossils from Wollumbilla Creek in Queensland, had asked CD whether he knew the geologist Charles Moore, or whether there was anyone else he could recommend to undertake the description of his entire collection of Australasian fossils (see letter from W. B. Clarke, 21 January 1862).

[3] Thomas Rupert Jones was an assistant secretary, librarian, and curator at the Geological Society of London. He also edited the *Quarterly Journal of the Geological Society* (*DNB*).

[4] The legislative councils of Victoria and New South Wales had each voted Clarke £1000 for his researches into the gold resources of the southern highlands of New South Wales, made between September 1861 and June 1862 (Jervis [1945], p. 66). Following a campaign by Clarke's friends in 1860–1, the legislative council of New South Wales appointed a committee to examine his entitlement to further recognition for his services in developing the mineral resources of the colony; the committee's report was positive, and he was granted an allowance of £200 per annum in 1861. In May 1862, a deputation to the Colonial Secretary, Charles Cowper, urged that Clarke be presented with a sum in acknowledgment of his services, that would allow him to publish the results of his work in England. However, while £5000 was placed on the estimates, that amount was reduced to £3000 when the matter was brought before the legislative council (*ibid.*, pp. 67–8).

[5] McCoy 1847. Frederick McCoy, professor of zoology and natural history at the University of Melbourne, had worked at the Woodwardian Museum in the University of Cambridge between 1846 and 1850, where he described the Australasian fossils sent to Britain by Clarke.

[6] CD's letter has not been found. The reference is to a fossil cirripede specimen that Clarke sent to CD in January (see letter from W. B. Clarke, 21 January 1862).

[7] The reference is to native Australian bee-combs (see letter from W. B. Clarke, 16 January 1862). In the letter to W. B. Clarke, 25 October [1861] (*Correspondence* vol. 9), CD had asked Clarke for information on the effects of introduced species on native species. CD was writing a draft of the chapter on 'Silk-worms Geese &c' for *Variation* (see 'Journal' (Appendix II)); this chapter also included a section on hive-bees (*Variation* 1: 297–9).

[8] Horace, *Odes*, 1.2.45: 'Late be thy journey home'.

[9] CD visited Bathurst, New South Wales, in 1836 during the *Beagle* voyage (*Journal of researches*, p. 442).

From E. A. Darwin 20 June [1862][1]

Dear Charles.

I dont know if it will be in your power to give Carlyle a helping hand about a book. On some former occasion he borrowed a book from the Jermyn St Museum he does not remember thro' whom or the exact title. It was an 8v^{o} with a col^{d} geological map on the 'Saxon Switzerland'.[2]

Do you know the Librarian well enough to give him the trouble of ascertaining the tittle of the Book & if such things are done to lend the book for a few days to Carlyle.[3]

E D—

June 20

DAR 105 (ser. 2): 4–5

[1] Dated by the reference to material cited in the fourth volume of Carlyle 1858–65, published in February 1864 (see n. 2, below).

[2] CD's brother, Erasmus, was a close friend of Thomas Carlyle (see *Collected letters of the Carlyles* and Fielding 1978). In June 1862, Carlyle was attempting to complete volume 4 of his life of Frederick II of Prussia (Carlyle 1858–65; see Dyer 1928, pp. 12–13, and Wilson 1929, pp. 464, 470). In his account of Frederick's military campaign in Saxony at the commencement of the Seven Years' War, Carlyle quoted a description of the picturesque highlands of Saxony (known as 'Saxon Switzerland') from Wilhelm Lebrecht Götzinger's *Schandau und seine Umgebungen, oder Beschreibung der Sächsischen Schweiz* (Götzinger 1812; Carlyle 1858–65, 4: 595–6). This book meets the physical description given by Carlyle, being in octavo format and containing a coloured geological map with the title 'Topo- und petrographische Reisekarte durch die Sæchsische Schweiz und umliegende Gegend'. However, it was not in the library of the Museum of Practical Geology, Jermyn Street, London, when the first printed catalogue was made in 1878 (White and Newton 1878). Carlyle later reported to Erasmus that the book he was seeking was not in the library of that museum, and that, therefore, he must on a previous occasion have borrowed it from the Geological Society of London (see letter from E. A. Darwin, 1 July [1862]); the Geological Society possessed Götzinger 1812 (Dallas 1881).

[3] Trenham Reeks was the curator and librarian at the Museum of Practical Geology.

To John Murray 20 [June 1862][1]

My dear Sir

One word to make clear (as you are so busy) that it is not at all certain that cuts are wanted by an American Publisher;[2] & secondly that I fixed £10 for price for M. Scheweizerbart.—[3]

Hearty thanks for the 3 cuts for A. Gray[4]

C. Darwin

Down— 20th—

Please remember that M. Scheweizerbart is anxiously wishing for the Electros of all the cuts.—

John Murray Archive (Darwin 124)

[1] Dated by the relationship to other letters in May and June 1862 between CD and John Murray and CD and Asa Gray (see nn. 2–4, below).

[2] CD refers to the possibility of an American edition of *Orchids* (see letter from Asa Gray, 18 May 1862, and letter to John Murray, 13 June [1862]).

[3] Christian Friedrich Schweizerbart wished to publish a German edition of *Orchids*; CD had told him that he should pay Murray £10 for electrotype plates of the woodcuts from the English edition, but he subsequently told Murray that he regretted not having asked for more (see letter from E. Schweizerbart'sche Verlagsbuchhandlung, 7 June 1862, and letter to John Murray, 13 June [1862]).

[4] Asa Gray had asked CD whether he could use a number of the illustrations from *Orchids* for a review of the work (see letter from Asa Gray, 18 May 1862).

To W. B. Tegetmeier 20 June 1862

Down. | Bromley. | Kent. S.E.
June 20th. 1862

My dear Sir

I have real pleasure in sending you my opinion, as a testimonial, of your varied acquirements, & acute powers of observation.[1] I have repeatedly been struck at the extent of your knowledge in many departments of Science,—at the originality of your ideas,—at your powers of clear expression in writing, & especially at your pure & disinterested love of Science.— Your paper on Bees' cells, read before the British Association was highly useful & suggestive to me.—[2] You know that I have often applied to you for information, & I beg permission to state that I have rarely in my whole life met anyone more thoroughily obliging & willing to give aid of every kind.— I hope that you may be successful in your application, for I am convinced that the Hartley Institution will obtain in you a really good & zealous Curator.—[3]

My dear Sir | Yours sincerely | Ch. Darwin

To | W. B. Tegetmeier Esqre—[4]

American Philosophical Society (279)

[1] Tegetmeier and CD had corresponded extensively since 1855. In addition to facilitating CD's study of the habits of bees and bee-cell construction, Tegetmeier introduced CD to pigeon fanciers at meetings of London clubs and fairs, and assisted CD in acquiring live birds, skins, and skeletons (see *Correspondence* vols. 5–9 and Secord 1981).

[2] Tegetmeier 1858.
[3] The Hartley Institution, Southampton (later Hartley University College), opened in October 1862 (*History of Hampshire* 2: 385–6). Tegetmeier apparently did not succeed in his application to become curator (see letter to W. B. Tegetmeier, 12 August [1862], and Richardson 1916).
[4] There is a draft of this letter in DAR 96: 8.

To G. H. K. Thwaites 20 June [1862][1]

Down. | *Bromley.* | *Kent. S.E.*
June 20th

My dear Mr Thwaites

By an odd chance, two days before receiving your letter of May 15th I wrote to you on Primula.—[2]

I am particularly glad to hear of Sethia. Menyanthes is said to be dimorphic like Primula; so I am not surprised at Limnanthemum;[3] it will be a curious point to compare Villarsia (I have been blundering, I fancied Villarsia was diœcious.) with Menyanthes, if I can make out any difference in fertility in the two of Menyanthes.[4] Have you any Malpighiaceæ? if so, I very much wish you would mark the imperfect flowers & see if they set seed.— Also whether they are closed, & whether the pollen-tubes are emitted from the pollen-grains within the anthers & then penetrate the stigma.— This is the case in the imperfect flowers of Viola & Oxalis.—[5]

Many thanks for your Governor's letter: you do not say whether I am to return it, so I will keep it till I hear.—[6]

In Haste, pray believe me | yours very sincerely | Ch. Darwin—

I suppose it would be too troublesome for you to mark ½ a dozen plants of the two forms Limnanthemum & count the capsules, & compare the produce of seed by weighing or counting.— I suspect the dimorphism of Primula is often, (though not at all necessarily) the high-road to diœciousness.[7]

American Philosophical Society (280)

[1] The year is established by the relationship to the letter from G. H. K. Thwaites, 15 May 1862.
[2] Letter from G. H. K. Thwaites, 15 May 1862, and letter to G. H. K. Thwaites, 15 June [1862].
[3] In his letter of 15 May 1862, Thwaites reported that he had just read CD's paper, 'Dimorphic condition in *Primula*', and that he had noticed the same phenomenon in the genera *Sethia* and *Limnanthemum.*
[4] In *Forms of flowers*, p. 116, CD noted that the genera *Menyanthes*, *Limnanthemum*, and *Villarsia* constituted 'a well-marked sub-tribe of the Gentianeæ' and that all the species, as far as was then known, were 'heterostyled'. CD had been anxious to see specimens of *Menyanthes* since he had learned earlier in the year that it was dimorphic (see letter to C. C. Babington, 20 January [1862], and letter from C. W. Crocker, 13 March 1862); he had recently acquired a short-styled specimen from the Royal Botanic Gardens, Kew (see letter to J. D. Hooker, 9 [April 1862] and n. 3). See also letter from G. H. K. Thwaites, 15 May 1862, CD annotations.
[5] For CD's interest in *Viola* and *Oxalis*, see also the letters to Daniel Oliver, 12 [April 1862] and 15 April [1862], the letter to J. D. Hooker, 30 May [1862], and the letter to Alphonse de Candolle, 17 June [1862].
[6] See the enclosure to the letter from G. H. K. Thwaites, 15 May 1862. Charles Justin MacCarthy was the governor of Ceylon.
[7] See 'Dimorphic condition in *Primula*', p. 95 (*Collected papers* 2: 61–2).

From G. C. Oxenden [before 21 June 1862][1]

Dear Sir

That you are *quitting Orchideous* life, & launching into realms of Scientific Space, whither I have no hope to follow you, is to me infinite grief—[2]

I feel like a Servant just discharged from a Good Place, & thrown upon his own resources—

& half the pleasure I have had, in hunting these Chalky Wildernesses, is now lost to me for Ever—

—As a last proof of fidelity, I have *just* discovered quite a new & *rich* mine of O. *Arachnites* & also of "*Epipactis palustris*"—

—So tell me, if you still *care* to have either—[3]

—The latter will not be in full flower for fully three more weeks—[4]

—The former are now in their **very prime**—[5]

With most kind regards | G. Chichester Oxenden

Chas. Darwin Esqr

DAR 173.2: 52

[1] Dated by the relationship to the letters from G. C. Oxenden, 4 June [1862] and 21 June 1862, and by the references to the flowering season of *Ophrys arachnites* and *Epipactis palustris* (see nn. 4 and 5, below).

[2] CD's letter has not been found, but Oxenden probably refers to CD's intention to return to his work on *Variation*, which had been interrupted for ten months while he prepared *Orchids* (see letter to J. D. Hooker, 30 May [1862] and n. 5). Oxenden had assisted CD's study of orchids by providing him with a number of specimens (see *Correspondence* vol. 9 and *Orchids*, p. 31 n.).

[3] In his letter of 4 June [1862], Oxenden promised to send CD 'good **spikes**' of *Epipactis palustris* later in the summer. He had already provided CD with specimens of *Ophrys arachnites* (see letter from G. C. Oxenden, 31 May [1862]), but in his letter of 4 June [1862], he expressed the hope of soon being able to send more.

[4] *Epipactis palustris* flowers from early July to mid-September (Lang 1989, p. 50). See also letter from G. C. Oxenden, 8 July 1862.

[5] *Ophrys arachnites* flowers from late June to mid-July (Lang 1989, p. 97). In his letter of [before 30 May 1862], Oxenden reported that it was too early for flowers of this species 'by two or three weeks—or a Month'.

From H. G. Bronn[1] 21 June 1862

Heidelberg
le 21 Juin 1862

Monsieur!

Je viens de recevoir une lettre de Mr. Schweizerbart, où il m'avertit, que Monsieur Treviranus est disposé à traduire vôtre excellent volume sur les Orchidées.[2] Vous n'auriez pû trouver un meilleur traducteur pour introduire Vôtre livre en Allemagne,[3] et je Vous en aurois felicité de tout mon coeur. Mais je suis désesperé de Vous dire, que la traduction *est faite* jusqu'à *la page* **306**.[4]

Monsieur Schweizerbart m'avoit prié de la faire. Je lui ai repondu, que j'ÿ serois disposé, s'il ne trouveroit pas un traducteur, dont le nom en qualité de botaniste pourroit être avantageux pour le succès du livre, et sous la condition, qu'il pourroit me confier le travail si-tôt, que je pourrois encore en finir avant la fin du mois de Juin, parceque j'avois déjà disposé des mois de Juillet et d'Août. Il me repondit aussitôt, qu'il désireroit de recevoir la traduction sitôt que possible, et 'ainsi j'ÿ ai travaillé jour et nuit, pour repondre à *Vôtre* voeu de voir une traduction Allemande de vos Orchidées et à *ses* intentions de publier bientôt cette traduction![5]

J'avois consenti à faire la traduction, parceque je considérois ce travail comme un complément de "l'Origin";[6] j'esperois de connoître en cette manière plus complêtement vos vues, que par la lecture passagère du livre, et je pensois ainsi instruire moi même d'une manière plus fondamentale. Je fis m'apporter du hot-house de nôtre jardin botanique quelques espèces d'Orchidées, qui ètoient en fleur pour les étudier et comparer avec Vos descriptions. Après avoir commencé la traduction, je sentis bientôt tendu mon ésprit au plus haut degré par l'interêt que m'apportoit continuellement la lecture de vos observations et des conclusions que vous en avez tirées. Et dejà j'allois vous écrire pour quelques petites difficultées, qui s'étoient offertes à ma traduction et dont Vous trouverez la liste si-jointe.[7] Mais avec tout l'interêt, que m'ont inspiré Vos recherches sur les rélations sexuelles des végétaux hermaphrodites et qui sont également importantes pour mes études zoologiques, j'avoue que je sens parfaitement bien comme je ne saurois pas remplacer un traducteur-botaniste du mérite de M. Treviranus, sous les mains duquel l'édition Allemande auroit dû gagner considérablement vis-à-vis de la mienne.

Monsieur Schweizerbart auroit dejà commencé l'impression, s'il n'auroit été en attente des copies des xylographies.[8] "L'origin" est sous presse.[9]

Veuillez bien excuser, Monsieur, si cette fois mes intentions n'ont pas été si avantageuses non seulement pour Vos publications, mais encore, il faut l'avouer, pour la science même, comme je l'avois désiré et espéré en me chargeant d'un travail, qui au moins pour moi même est des plus attirants et des plus instructifs.

Veuillez aggréer, Monsieur, l'expression de la considération distinguée, avec laquelle j'ai l'honeur d'être | Vôtre | très devoué | H. G. Br⟨onn⟩

[Enclosure]

Darwin Orchids

pag. 22, fg. III: (**l'**) manque dans la figure[10]

pag. 171: **india-rubber-band**: qu'est ce que c'est, je ne trouve nulle part ce terme

pag. 254: **cocket-hat**: non plus![11]

pag. 272, fg. XXXI: (**a'**) manque dans la figure.[12]

pag. 285 **Homopterous Insects**. nous comprenons (*en Allemagne*) tout généralement sous ce nom une partie des Hemiptera ou Rhynchota, et quoique il ÿ ait là quelques formes assez particulières (Fulgora), je ne sais pas, si les formes souvent phantastiques des *Orthoptères* tropiques ne serviroient pas beaucoup mieux à Votre comparaison?[13]

Anther-case: y-a-t-il un terme botanique pour cette expression? Je ne sais pas toujours, si c'est le *propre* paroi des anthères, ou la cavité, qui les entoure?

pag. 104, ligne 4: **renewed** doit ètre **removed**??[14]

(pag 292, fg. XXXII. et suiv.) **group of spiral-vessels**: cette expression, peût-elle être rendue par *Spiralgefäss-Bündel*, c.à.d. *bundle of spiral vessels*,—ou est ce qu'*un group* peût contenir plusieurs *bundles*? (nous sommes très accoutumés à l'usage du mot Gefäss-*Bündel*)

—Dans la description page 292 et suivantes, y est il toujours soutenu la distinction entre

group of vessels et *vessels* ?	ce qui seroit très important à savoir pour le "general reader"

—Enfin il me paroit, qu'il seroit de beaucoup d'utilité, également pour le general reader, d'avoir un tout petit tableau (remplissant une page seulement) de la composition de la famille entière des Orchids, de ses tribus, sub-tribus et des genres les plus importants (suivant Lindley?[15]) pour le faire une juste idée des affinités des genres, dont il rencontre les noms isolés. Le botaniste n'en a pas besoin.

DAR 70: 2, DAR 160.3: 318

CD ANNOTATIONS

Enclosure

1.1 *pag. 22*, ... figure] *cross in margin, brown crayon*
4.1 *pag. 272*, ... figure.] *cross in margin, brown crayon*
5.1 **Homopterous** ... comparaison? 5.5] *crossed brown crayon*
7.1 *pag. 104*, ... **removed**??] *cross in margin, brown crayon*
10.1 —Enfin ... besoin. 10.6] *scored brown crayon*

[1] For a translation of this letter, see Appendix I.

[2] Bronn refers to Christian Friedrich Schweizerbart, head of E. Schweizerbart'sche Verlagsbuchhandlung, the Stuttgart publishing firm that was preparing to publish a German edition of *Orchids* (see letter from H. G. Bronn, 27 March 1862, and letter from E. Schweizerbart'sche Verlagsbuchhandlung, 7 June 1862). CD had apparently written on 18 June 1862 to ask Ludolph Christian Treviranus, professor of botany at the University of Bonn, whether he would be prepared to translate *Orchids* for this edition. However, having learned that Bronn had nearly completed a translation of the book, CD wrote to inform Treviranus of this on 24 June. See letter from F. H. G. Hildebrand, 14 July 1862, and *Correspondence* vol. 11, letter from L. C. Treviranus, 12 February 1863. CD's letters to Treviranus of June 1862 have not been found.

[3] Treviranus was well known in Germany for his botanical publications. He was familiar with CD's views as they pertained to botanical problems. In his review of Joseph Dalton Hooker's *Flora Tasmaniæ* (J. D. Hooker 1855–60), where Hooker discussed CD's views on species as they related to the question of the geographical distribution of plants, Treviranus was critical of CD's ideas about the advantages to plants of cross-pollination (see Treviranus 1861; see also Junker 1989, pp. 143–4). CD was apparently unaware of Treviranus's review: there is no copy in the Darwin Pamphlet Collection–CUL, and CD made no mention of it in his extant correspondence. For Treviranus's views on *Orchids*, see Treviranus 1863a, and *Correspondence* vol. 11, letter from L. C. Treviranus, 12 February 1863.

4 The English edition of *Orchids* has 365 pages. Since CD did not send Bronn the sheets of the first half of the book until the beginning of May, Bronn apparently translated more than 300 pages in only six weeks (see letter to H. G. Bronn, 25 April [1862]).

5 Bronn's German translation of *Orchids* (Bronn trans. 1862) was published on 20 October 1862 (*Börsenblatt für den Deutschen Buchhandel* 29: 2195). See also letter from E. Schweizerbart'sche Verlagsbuchhandlung, 16 October 1862.

6 Bronn translated both the first and second German editions of *Origin* (Bronn trans. 1860 and 1863).

7 See enclosure.

8 CD had arranged for Schweizerbart to purchase electrotype plates of the illustrations for *Orchids*, made from John Murray's original woodcuts (see letters to John Murray, 13 June [1862] and 20 [June 1862]).

9 Bronn refers to the second German edition of *Origin* (Bronn trans. 1863; see letters from H. G. Bronn, [before 11 March 1862], 27 March 1862, and 19 May 1862). It was published in three parts, appearing on 6 October, 17 November, and 19 December 1862 (*Börsenblatt für den Deutschen Buchhandel* 29: 2083, 2447, and 2735); however, the title-page bore the date '1863'. There is a copy of the work in the Darwin Library–CUL.

10 In the key to figure 3, given in *Orchids*, p. 23, the label 'l' is said to stand for 'labellum', and 'l′' for 'guiding plate on the labellum'; there is, however, no l′ shown on the figure. The error was corrected for the second edition (*Orchids* 2d ed., p. 18).

11 Bronn mistranscribed the term: it read 'cocked-hat'.

12 In the key to figure 31 (*Orchids*, p. 272), the label 'a' is said to stand for 'anther', and 'a′' for 'rudimentary shield-like anther'; there is, however, no a′ shown on the figure. The error was corrected for the second edition (*Orchids* 2d ed., p. 227).

13 In the last paragraph of his chapter on the Catasetidae (*Orchids*, p. 285), CD stated:

> The flowers of Orchids in their strange and endless diversity of shape, may be compared with the great vertebrate class of Fish, or still more appropriately with tropical Homopterous insects, which seem to us in our ignorance as if modelled by the wildest caprice.

In his own copy of *Orchids* (Darwin Library–CUL), CD marked this final paragraph for deletion. However, the paragraph including the comparison between orchids and homopterous insects is maintained in the second edition (*Orchids* 2d ed., pp. 224–5).

14 The sentence Bronn refers to reads:

> In England the flowers are much visited by insects: during the wet and cold season of 1860 a friend in Sussex examined five spikes bearing eighty-five expanded flowers: of these, fifty-three had the pollinia removed, and thirty-two had them in place; but as many of the latter were immediately beneath the buds, ultimately a larger proportion would almost certainly have been renewed.

In his own copy of *Orchids* (Darwin Library–CUL), CD deleted the word 'renewed' in pencil and wrote 'removed' in the margin. In the second edition, the word was again misprinted, this time as 'remowed' (*Orchids* 2d ed., p. 102).

15 John Lindley was the leading authority on orchid taxonomy. Bronn refers to the table listing the various tribes of orchids and their genera in Lindley 1853, pp. 181–3, which he included in his translation (Bronn trans. 1862, p. vi).

To E. A. Darwin 21 June [1862][1]

Down. | Bromley. | Kent. S.E.
June 21st

Dear Erasmus

My friend Mr. Trenham Reeks, the Secretary at the Museum Jermyn St, would, I am sure, gladly give any information or aid to so distinguished a person as Mr.

Carlyle.—[2] I do not believe that Books are ever lent out, & Sir R. Murchison could alone give permission.—[3] Carlyle could, if he thinks fit, take this note as an introduction to Mr Reeks.—[4]

Yours affecty | Ch. Darwin

National Library of Scotland (MS 553: 440 (241))

[1] Dated by the relationship to the letter from E. A. Darwin, 20 June [1862].

[2] Thomas Carlyle wanted to borrow a book from the library of the Museum of Practical Geology, Jermyn Street, London, where Trenham Reeks was registrar, librarian, and curator (*Post Office London directory* 1861, p. 59). Erasmus had sought CD's help on Carlyle's behalf (see letter from E. A. Darwin, 20 June [1862]).

[3] Roderick Impey Murchison was director of the Government School of Mines and Museum of Practical Geology, and director-general of the Geological Survey of the United Kingdom, all of which were located in the same building in Jermyn Street, London (*Post Office London directory* 1861, p. 59). The rules of the library stated that 'under no circumstances' was any book 'allowed to be taken out of the building' (White and Newton 1878, p. [iii]).

[4] Erasmus enclosed CD's note in a letter to Carlyle dated 23 June 1862 (National Library of Scotland, MS. 1768: 34–5), for use as an introduction to Reeks. Carlyle in turn gave the letter to Joseph Neuberg, who was assisting him with the research for his life of Frederick II of Prussia (Carlyle 1858–65). At the top of the letter Carlyle wrote: '*Erasmus Darwin Esqr.*— in helping me (T. Carlyle)'; at the end of the letter he added: 'If Mr Reeks would extend his charity to Mr Neuberg on my behalf?— | T. C.'

From G. C. Oxenden 21 June 1862

Broome
June 21. 1862

Dear Sir

I hope your Son is *Convalescent*— —If you were to place him in a **warm** bath, containing *seven parts water*, and *one* part of "*Condy's Ozonised Fluid*", & if he remained therein for $\frac{3}{4}$ of an hour, it is likely that infinite benefit would accrue, by **absorption**—[1]

Notwithstanding your prohibition, I this day send you some fine specimens of "*Arachnites*" caught yesterday—[2]

—Also (enclosed in *an Envelope*) a curious Pink Plant, with small blue flowers, concerning which I am anxious for your *verdict*, as to *what it is*—

I myself entertain *no doubt*—& that his *Initials* are

"P. C."

I yesterday hunted a splendid range, for **many miles**—which I had *never tried before*—

—We found, at the lowest Computation, 500 Bees in Flower—& enough of "**Arachnites**" to drive any Plant-hunter stark staring *distracted mad*— —Certainly not *less* than *160*, in full flower—

—In nineteen cases out of Twenty, "Arachnites" & "Apifera" stand aloof from each other—

—The former Especially is apt to Congregate in little *clusters* of from 3 to 16 Plants—*wide* away from any Bees—which they *clearly do not seem* to like— both anatomically, & as to *dress* & *markings*, the two Plants stand widely apart—[3]

—I took up 2 plants only, out of the *multitude*—& the *single spiders* which I have sent you were taken, here & there, wherever a stem was *richly covered* with blooms— Altogether it was fine a days sport as I ever saw—

Sincerely | G. C. Oxenden

DAR 173.2: 54

[1] Leonard Darwin was ill with scarlet fever (see letter to W. E. Darwin, 13 [June 1862]). 'Condy's fluid' was a solution of alkaline permanganates, developed by Henry Bollmann Condy, and used 'for the purification of air and water, and sanitary objects in general' (Condy 1862, p. 69).

[2] Oxenden had sent CD specimens of *Ophrys arachnites* at the end of May, and at the start of June expressed a hope that he would be able to send more in 'a few more days' (see letters from G. C. Oxenden, [before 30 May 1862] and 4 June [1862]). CD's replies to these letters have not been found, but in his letter of [before 21 June 1862], Oxenden expressed his regret at CD's claim to be '*quitting Orchideous* life', and offered to send further specimens of *O. arachnites* 'as a last proof of fidelity'.

[3] CD doubted the claim of some botanists that *Ophrys arachnites* was a variety of the bee-orchis, *O. apifera*, but was exploring the possibility that 'the Bee might be the self-fertilising *form* of *O. arachnites* which requires insect's aid' (see letter to A. G. More, 7 June 1862 and nn. 2 and 3). CD had asked Alexander Goodman More whether he had seen any of the 'intermediate forms' that were said to connect the two, and had evidently made a similar enquiry of Oxenden. See also letter from G. C. Oxenden, 8 July 1862.

To J. D. Hooker 23 June [1862][1]

Down Bromley Kent
June 23^d

My dear Hooker

Thank you for telling me about Mrs Hooker & yourself, which was just what I wanted to hear.[2] How hard-worked you are; do not stretch the string too much; I most cordially hope you may soon get a holiday right away from Kew.— I am sure I have lately agreed with you that it is a weary world: we have had lots of sickness (thank God the Scarlet F. did not turn out bad)[3] & I have been extra bad with violent skin inflammation— I have done hardly anything lately, except just to attend to my experiments.— I have tried over again Primula crosses & former results most fully verifyed & some very curious facts on the sterility of the homomorphic seedlings.[4] Did you see the review of my Orchis Book in London R. By Jove it was too strong & made me feel modest & that was a wonderful feeling. I wonder who wrote it.[5] I have had several letters from Asa Gray, who seems about as much infatuated over orchids, as I have been.—[6]

I asked Emma about a cook; but she knew no one whom she could recommend. I do hope your household will soon be comfortable; but I long to hear of your starting somewhere.—[7]

This is a very dull letter; but my hands are burning as if dipped in hell-fire.—
Good Night my dear old fellow.— | C. Darwin
William enjoyed his visit to you.—[8]

DAR 115.2: 156

[1] The year is established by reference to the publication and reception of *Orchids* (see n. 5, below).
[2] See letter from J. D. Hooker, 19 [June 1862].
[3] The apparently terminal illness of his eight-year-old niece, Anne McGilvray, had caused Hooker to declare, in his letter of 19 [June 1862]: 'this is a weary life'. Leonard Darwin had been sent home from school on 12 June 1862 suffering from scarlet fever (see letter to W. E. Darwin, 13 [June 1862]).
[4] CD apparently refers to repeating the crossing experiments with *Primula sinensis* and *P. veris* that he had carried out in 1860 and 1861 while preparing 'Dimorphic condition in *Primula*'. In January and February 1862, he carried out similar crossing experiments on plants of *P. sinensis* raised from the seed of the earlier experiments. He crossed the offspring of the homomorphic crosses (i.e., those in which long- or short-styled parents were impregnated with own-form pollen) in order to investigate the extent of their sterility (see Appendix VI), and he crossed the offspring of the heteromorphic crosses (i.e., those in which the parents were impregnated with different-form pollen) 'for comparison . . . as these would of course be fully fertile'. He noted also that the latter experiments would serve to test the previous year's experiments (DAR 108: 35; see also letter to J. D. Hooker, 7 March [1862] and n. 10). CD's results from the crosses with homomorphic plants are given in 'Illegitimate offspring of dimorphic and trimorphic plants', pp. 410–18 (CD came to use the word 'illegitimate' in place of 'homomorphic'). CD repeated his crosses with *P. veris* on 1 May 1862; his notes on these experiments are in DAR 157a: 77 and DAR 108: 70, and the results were later published in *Cross and self fertilisation*, p. 28. See also letter to Alphonse de Candolle, 17 June [1862] and n. 2, and letter to John Scott, 11 December [1862].
[5] The review of *Orchids* published in the *London Review and Weekly Journal of Politics, Arts, and Science* 4 (1862): 553–4 was written by Miles Joseph Berkeley (see letter from J. D. Hooker, 28 June 1862).
[6] See letters from Asa Gray, 31 March [1862] and 18 May 1862.
[7] See letters from J. D. Hooker, 9 June 1862 and 19 [June 1862], and letter to J. D. Hooker, 11 June [1862].
[8] William Erasmus Darwin visited Hooker at the Royal Botanic Gardens, Kew, on 20 June 1862 (see letter from J. D. Hooker, 19 [June 1862]).

To Nicholas Trübner 23 June [1862][1]

Down. | Bromley. | Kent. S.E.
June 23^{d}

Dear Sir

I have received a letter from D^{r}. Asa Gray this morning, who says "pray ask M^{r}. Trübner to send at once $\frac{1}{2}$ a dozen copies of the (Darwin's) Orchis Book." (Murray).—[2]

Will you be so good as to attend to this—

Yours faithfully | C. Darwin

Remember When Antiquities (catalogue 26)

[1] The year is established by reference to the publication of *Orchids*. See also letter to Asa Gray, 1 July [1862].

[2] The letter from Gray has not been found. Trübner, a publisher and bookseller, frequently acted as Gray's London agent (see *Correspondence* vols. 8 and 9). In addition to reviewing *Orchids* for the *American Journal of Science and Arts* (A. Gray 1862a), Gray was attempting to find an American publisher for the book, which had been published in Britain by John Murray (see letter from Asa Gray, 18 May 1862 and n. 3).

From Charles Victor Naudin[1] 26 June 1862

Muséum | d'Histoire naturelle, | Culture, | Paris,
le 26 Juin 1862.

A Monsieur Ch. Darwin, Membre de la Société royale des Sciences &c.
Monsieur,

J'ai reçu, par l'entremise de M. J. Murray, éditeur, l'exemplaire de votre savant ouvrage, sur la fertilisation des Orchidées par les insectes, que vous avez bien voulu m'adresser;[2] Je l'ai lu avec le plus grand intérêt, et je viens vous en faire mes vifs et très sincères remerciments.

Vous savez que, depuis long-temps, je m'occupe d'un sujet analogue, l'observation des hybrides dans le règne végétal.[3] Je crois être arrivé, si non à la solution définitive de cette question, du moins à quelque chose qui en approche de bien près. Je serai en mesure, vers la fin de l'année, de publier de nombreuses observations qui, je l'espère, porteront la conviction dans l'esprit des lecteurs, relativement à la terminaison *ordinaire* des générations hybrides. Je me ferai un devoir et un plaisir de vous offrir, en retour, une exemplaire de mon travail.[4]

En attendant, Veuillez agréer, Monsieur, avec mes remerciments, l'expression de ma considération et de mes sentiments les plus distingués, | Votre bien dévoué serviteur, | Ch. Naudin | aide-naturaliste au Muséum

DAR 172.1: 6

CD ANNOTATION
Top of first page: 'Naudin' *ink*

[1] For a translation of this letter, see Appendix I.

[2] *Orchids* was published by John Murray on 15 May 1862 (Freeman 1977, p. 112). Naudin's name appears on CD's presentation list for the volume (see Appendix IV).

[3] In the 1840s, Naudin had become interested in hybridisation as a means of establishing taxonomic relationships between different forms. In the 1850s he became increasingly interested in its significance in regard to species formation, and from 1854 he carried out extensive crossing experiments in the botanic garden of the Muséum d'Histoire Naturelle in Paris (Olby 1985, pp. 47–8, and Rheinberger 1983, p. 201). CD owned a copy of one of Naudin's papers on hybridisation (Naudin 1858), which is now in the Darwin Pamphlet Collection–CUL, and had read two further papers by Naudin that dealt with the subject (Naudin 1852 and 1856; see *Correspondence* vol. 4, Appendix IV, *128: 155, 157, 167). In Naudin 1852, which CD had apparently read by November 1855 (see *Correspondence* vol. 5, letter to J. D. Hooker, 6 November [1855] and n. 2), the author maintained that the present range of species had been derived from a smaller number of ancestral forms by a process analogous to that of artificial selection, by which domestic varieties had been derived. After the publication of *Origin*, Naudin's colleague at the Muséum, Joseph Decaisne, claimed that Naudin had anticipated CD's

views in this paper (see *Correspondence* vol. 7, letter to J. D. Hooker, 23 [December 1859]); CD subsequently corresponded on the subject with another of Naudin's colleagues, Armand de Quatrefages (see *Correspondence* vol. 9, letter to J. L. A. de Quatrefages de Bréau, 25 April [1861] and n. 7).

4 In December 1861, Naudin submitted an account of his researches into plant hybridisation to the Académie des Sciences for consideration for the *prix des sciences physiques*. The essay was awarded the prize in December 1862, and the second part of it, containing his conclusions, was published the following year (Naudin 1863). The material on Cucurbitaceae used in this paper was covered in greater detail in a separate paper (Naudin 1862). There are offprints of both papers in the Darwin Pamphlet Collection–CUL. See also *Correspondence* vol. 11, letter to C. V. Naudin, 3 February 1863. Naudin's prize-winning essay was not published in full until 1865 (Naudin 1865); there is an annotated reprint of this paper in the Darwin Library–CUL (see *Marginalia* 1: 638–9). In the essay, Naudin described the results of hybridisation experiments with sixty different species of plant, the offspring from which he had propagated for as many as five generations. In his conclusions, Naudin discussed the significance of his discovery that whereas the first generation from each of his crosses consisted of individuals that all resembled each other, the subsequent generations often comprised individuals that were markedly different from each other, and that, over a number of generations, the hybrid forms reverted to the parental types. Naudin explained this in terms of his theory of specific essences and non-blending heredity, arguing: 'Tous ces faits vont s'expliquer naturellement par la *disjonction des deux essences spécifiques* dans le *pollen* et dans les *ovules* de l'hybride' [All these facts are naturally explained by the disjunction of the two specific essences in the pollen and ovules of the hybrid] (Naudin 1865, p. 150). He also stated that he had not been studying hybrids for long enough to have formed a settled opinion on the question of whether there were any exceptions to his law of reversion, by which certain hybrids might become new species, but that he doubted this was so (Naudin 1865, pp. 156–7). See also Guédès 1975.

From August Wilhelm von Hofmann 27 June 1862

R. College of Chemistry
June. 27. 1862.

My dear Sir

I have to apologize for transmitting to you only now the specimens of Carbonate of Ammonia and Gelatine, which I had promised to you.[1] I have examined both; they are free from chlorine and may be used for your experiment.[2] It was not quite easy to get a sample of gelatine without chlorine; a very considerable number of specimens, which I tested contained abundance of this element. In fact hydrochloric acid is the agent by which gelatine is generally separated from the earthy phosphates. At last I procured a sample from the french exhibition which I found to be pure and which I forward to you.[3]

Hoping that the chemicals may not arrive too late for your experiment, I remain | My dear Sir | Yours very sincerely | A W Hofmann

The Specimens are sent in a box forwarded by Rail

DAR 166.2: 232

1 No letter from Hofmann containing this promise has been found. Hofmann was the director of the Royal College of Chemistry; in 1860, when CD was investigating the sensitivity of the hairs of *Drosera*, an insectivorous plant, to certain chemical substances, Hofmann provided CD with information about the sensitivity of chemical tests (see *Correspondence* vol. 8, letters from A. W. von Hofmann to Edward

Cresy, 13 October 1860 and 27 October 1860, and letters to Edward Cresy, 14 October [1860] and 2 November [1860]).

[2] CD had become interested in the sensitivity of insectivorous plants to different chemical substances in the summer of 1860, and had continued to experiment on the subject (see *Correspondence* vols. 8 and 9); the results of this research were ultimately published in *Insectivorous plants*. He had recently carried out experiments to determine the effects of carbonate of ammonia on *Saxifraga platypetala* (see the experimental note in DAR 54: 27, dated 22 April 1862) and on *Drosera* (see the note in DAR 54: 75, dated 21 May 1862). The effects of carbonate of ammonia on the hairs of *Drosera rotundifolia* are detailed in *Insectivorous plants*, pp. 141–8. In *Insectivorous plants*, pp. 110–12, CD described experiments on the digestive powers of *Drosera*, in which he had used 'pure gelatine in thin sheets' given him by Hofmann.

[3] The International Exhibition ran from 1 May to 1 November 1862 at South Kensington, London. Of the foreign nations represented, France had the largest number of exhibitors, several of whom were awarded medals for their exhibits of glues and gelatines (*Reports by the juries*, class 4, section B, p. 4).

From Frederick Smith 27 June [1862][1]

British Museum
27th. June

My dear Sir

I did not answer directly as I wished to show your insects to Mr. Walker—thinking as they belonged to a family he has monographed—I should get you the name generic as well as specific—[2] Mr. Walker however does not come as usual—so I write to say that the insect belongs in my opinion—to the genus Tetrastichus— Fam. Chalcididæ ⟨ ⟩ I do not know— when I see Mr. Walker I will try and get the name—or names for I think the specimen with mottled or banded wings is distinct— you shall have all I can ascertain in a day or two

Yours very truly | Fredk Smith

Chas. Darwin Esqre.

DAR 177: 194

[1] The year is established by the relationship to the letter from Frederick Smith, 28 June 1862.

[2] Smith refers to his colleague at the British Museum, the entomologist Francis Walker, and to Walker's *Monographia Chalciditum* (Walker 1839). The letter in which CD asked Smith to identify these insects has not been found; however, see the letter from Frederick Smith, 28 June 1862 and n. 3.

From Emma Darwin to T. G. Appleton 28 June [1862][1]

Down. | Bromley. | Kent. S.E.
June 28.

My dear Mr Appleton

It was very pleasant to find ourselves so kindly remembered by you at such a distance of time & Mr Darwin begs me to thank you most cordially for the beautiful book—[2] (He has not at all lost his taste for maple sugar any more than the children have).[3] I am writing for him as he is very unwell at present but he begs me to say that the book is one of the most beautiful specimens of works of Nat. History he has ever seen.

I wish we could have had the pleasure of seeing your brother here but Mr Darwin has been too unwell to see any visitors.[4]

We shall rejoice at the termination of the war & if we cannot hope to see Slavery abolished I think it must at all events be prevented from Spreading. If you sh^{d} be coming to England I hope you will let us know as we should be so glad to see you again. Mr Darwin begs to be remembered to you. He has the pleasantest recollection of the days you spent here[5]

very truly yours | E. Darwin

J. G. Zimmer (private collection)

[1] The year is established by the reference to William Sumner Appleton's visit to Britain (see n. 4, below).

[2] The book referred to has not been identified. Appleton visited Down House in October 1849 in company with his sister, Mary, the wife of Emma Darwin's cousin, Robert Mackintosh (Emma Darwin's diary (DAR 242)).

[3] Appleton had sent a gift of maple sugar to the Darwin family (see letter from T. G. Appleton, 24 April [1862]).

[4] Appleton's half-brother, William Sumner Appleton, spent the summer of 1862 in Britain pursuing antiquarian and historical interests (see letter from T. G. Appleton, 24 April [1862]).

[5] See n. 2, above.

From J. D. Hooker 28 June 1862

Royal Gardens Kew
June 28/62

My dear Darwin

I am distressed indeed to hear of your suffering. & can only hope it may be transient, & long of returning.[1]

Berkeley was author of the London Review article, I thought it very well done indeed.[2] I have read a good deal of the Orchid book & echoe all he says: you are out of sight the best Physiological observer & experimenter that Botany ever saw. & I do sometimes feel most confoundedly humiliated when I think how much I missed when doing the *Listera*— But for your loving praise of that paper I should wish it withdrawn—[3] you do warm the cockles of one's heart.

I am still in perplexity— We have found no cook yet at all to suit us & I have determined to send the children to Worthing with the Governess, who seems a capital person,[4] & if possible take my wife to Switzerland for a fortnight— She seems to go down hill steadily, & complains of shortness of breathing & palpitation of heart— Whether it is all weakness (as the Doctors say) or symptoms of the affliction her father died of, God only knows, & I dare not ask myself—[5] I *must* get her away but my Father is again laid up with Eczema on the legs,[6] & I do not see how I am to go.

Switzerland is the only place she has the smallest fancy for, I *talk* of going next Friday or Saturday & joining Lubbock,[7] but I never felt less disposed or able to go anywhere than now, when Gen. Plant. Part. I—is all but, but not quite, through the press & I must leave it at the last sheets—[8] I wish I knew of any one with whom I could send my wife.

More plants from Fernando Po & more Europæans![9] Sibthorpia *Europæa*, *Bromus giganteus*, & *Myosotis arvensis*, or one of its vars (*stricta* of authors)

My Jury work is over, & I am comparatively free.[10]

Ever yours affec | Jos D Hooker

DAR 101: 42–3

CD ANNOTATIONS

3.1 I am … heart— 3.5] *crossed pencil*

Top of letter: 'Masdevallia | Bonatea' *pencil*; '*Naudin*[11] | complexion Queries'[12] *ink*

[1] See letter to J. D. Hooker, 23 June [1862].

[2] [Berkeley] 1862. See also letter to John Murray, 18 [June 1862], and letter to J. D. Hooker, 23 June [1862].

[3] J. D. Hooker 1854a. In *Orchids*, p. 139, at the beginning of his discussion of *Listera ovata*, CD mentioned Hooker's 'highly remarkable' study of the structure of the rostellum, but noted that Hooker had not attended to the role that insects play in the fertilisation of this flower. Hooker's name is on CD's presentation list for *Orchids* (see Appendix IV).

[4] The governess has not been identified; she had only recently been appointed (see letter from J. D. Hooker, 19 [June 1862]).

[5] Frances Harriet Hooker's father, John Stevens Henslow, died in May 1861 as the result of long-standing heart disease, aggravated by bronchitis and 'congestion of the lungs' (L. Huxley ed. 1918, 2: 60).

[6] William Jackson Hooker.

[7] John Lubbock visited Switzerland in the summer of 1862 in order to examine the sites of prehistoric lake-dwellings, the recent discovery and examination of which he had described in a paper for the January number of the *Natural History Review* (Lubbock 1862b; see also Lubbock 1862a). See Hutchinson 1914, 1: 56, John Lubbock's diary (British Museum, Add. Ms. 62679: 64 r.), and letter from John Lubbock, 23 August 1862).

[8] The first part of *Genera plantarum* (Bentham and Hooker 1862–83) was published on 7 August 1862 (Stearn 1956, p. 130).

[9] In a paper read before the Linnean Society of London in 1861 (J. D. Hooker 1862b), Hooker had described the plants sent by Gustav Mann from the mountainous interior of the West African island of Fernando Po, noting that of the forty-eight temperate genera represented, only twelve were *not* European, and that nine of the species were European. This provided CD with further evidence for his claim, made in *Origin*, pp. 365–82, that during the glacial period the whole world, or large parts of it, had been simultaneously much colder than at present, allowing northern temperate species to migrate into tropical regions. CD added Hooker's evidence from Fernando Po to his discussion of this topic in the fourth edition of *Origin* (p. 445). See also letter to J. D. Hooker, 9 May [1862] and n. 6.

[10] At the International Exhibition of 1862, Hooker was a member of the jury for class 4, section C, 'Vegetable substances used in manufactures, &c.', and an associate juror for class 3, section B, 'Drysaltery, grocery, and preparations of food as sold for consumption' (*Reports by the juries*).

[11] Charles Victor Naudin.

[12] See letter to J. D. Hooker, 30 [June 1862].

From Frederick Smith 28 June 1862

Brit Museum
June 28th. 1862.

My dear Sir

I have seen M[r]. Walker this morn[g]. and he gives me the names of the two species—⟨*Tetr*⟩*astichus Diaphantus* and the little fellow with dark fasciæ on ⟨the⟩ wings is *Coleothrips fasciata*—[1] Many thanks for the Volume on the Fertilization of Orchis which I have just received—[2] what ⟨a⟩ most interesting subject. I ⟨ ⟩ just had from the ⟨Journal⟩

believe me | Yours very Sincerely | Fred[k] Smith

DAR 177: 195

CD ANNOTATIONS
1.2 *Diaphantus*] 'this is the common one' *interl ink*
Top of page: 'Hymenoptera caught on Musk Orchis' *ink*[3]

CD note:[4]
*M[r]. Waterhouse [*circled ink*][5] This is minute Beetle with pollinia of Musk Orchis—

[1] The reference is to the entomologist Francis Walker (see letter from Frederick Smith, 27 June [1862]). See also n. 3, below.
[2] The names of Frederick Smith and Francis Walker are both on CD's presentation list for *Orchids* (see Appendix IV).
[3] George Howard Darwin had recently begun observing, on his father's behalf, the insects that entered the flower of the green musk orchis, *Herminium monorchis* (see CD's notes of these observations, dated 22–7 June 1862, in DAR 70: 32–6; see also DAR 70: 31, 37). CD reported George's findings in the German translation of *Orchids* (Bronn trans. 1862, pp. 47–8 n.; see the second enclosure to the letter to H. G. Bronn, 30 June [1862]). He later published them in 'Fertilization of orchids', p. 145 (*Collected papers* 2: 142), stating that *Tetrastichus diaphantus* was the most common of the minute Hymenoptera that visited *H. monorchis*, and that the Hymenoptera identified in the paper had all been named for him 'by our highest authority, Mr. Frederick Smith' ('Fertilization of orchids', p. 142; *Collected papers* 2: 139).
[4] CD's note is on a scrap of paper that was pinned to the letter.
[5] CD probably refers to the entomologist, George Robert Waterhouse, who was keeper of the geology department at the British Museum.

From E. A. Parkes 29 June 1862

Frindsbury | Rochester
29 June 1862

My dear Sir

There has been a little delay in the transmission of your papers, as I was unable to go to town for some days after receiving them, & then the Director General had to obtain the sanction of the War Office to send them to India &c, as the rules are very strict against sending any but official letters in the Gov[t]. bag.[1]

However you will see by the enclosed that the authorities have sanctioned the papers being sent out.[2]

On talking the matter over with the Director General (Dr. Gibson C.B.) he thought it better that the papers should come back to him instead of to you direct. The great reason for this is that then the Surgeons abroad can send them home officially & without expense to themselves.

I hope some good returns will come of it.

Dr. Gibson has been so kind in this matter & so desirous of meeting your wishes, that I venture to suggest you should write a note to him to thank him for his compliance with your request.[3]

Excuse my suggestion | & Believe me | My dear Sir | Very sincerely yours | E A. Parkes.

Charles Darwin Esqre

DAR 174.1: 25

[1] James Brown Gibson was director-general of the Army Medical Department. CD had prepared a questionnaire for army surgeons serving abroad as part of an inquiry about the association between hair colour and susceptibility to disease among the European colonial military (see letter from George Busk, 1 April 1862, and letters from E. A. Parkes, 8 April 1862 and 28 April 1862).

[2] The enclosure has not been found; it may have been one of the printed questionnaires. No copy of this questionnaire has been found; however, the text of the memorandum 'explaining the object of the enquiry' is given in *Descent* 1: 244–5 n. 48 (see letter from E. A. Parkes, 8 April 1862, n. 1). See also letter from George Busk, 1 April 1862 and n. 1.

[3] See following letter.

To James Brown Gibson [after 29 June 1862][1]

Sir

I hope that you will permit me to return to you my sincere & cordial thanks for the kind manner with which you have acceded to my wishes in the endeavour the endeavour to find out whether there exists any re— *[between]* complexion & liability & tropical diseases.—[2] Without your aid nothing whatever could have been done.—

The result may not *[illeg]* turn out a complete failure; but I have not ventured to make my request without *[mature]* consideration.—

With my sincere thanks for all your kindness I *[hope]* to remain | Your oblig Servt, | C. Darwin

Draft
DAR 96: 4

[1] Dated by the relationship to the preceding letter.

[2] See preceding letter. See also letter from George Busk, 1 April 1862, and letters from E. A. Parkes, 8 April 1862 and 28 April 1862.

To H. G. Bronn 30 June [1862][1]

Down. | Bromley Kent.
June 30

Dear & Honoured Sir.

I have been very unwell otherwise I should have written sooner.[2] I enclose answers. & some few corrections & additions—[3] I fear that the latter on account of my bad handwriting will give you trouble; but I hope that you will insert them, as they are of some importance It would have been better to have worked them into the text, but I really have not time.

When you write to Mr. Schweizerbart will you say that I was in such a hurry when I wrote to him that I believe I forgot to thank him for his kind letter of June 7th.[4] Please to say that I should very much like to have a Copy of the 2nd. German Edit of the Origin & of the Orchis Book, & I hope he will send me them through Messrs. Williams & Norgate.[5]

Believe me | Dear Sir | Yours truly obliged | Ch. Darwin.

I assure you that I feel most fully sensible of the great honour & kindness which you have done me in translating my Orchis-book.

What powers of work you must have!—[6]

[Enclosure 1]

Answers

p. 22. fig iii. p. 272 fig XXXI. I know that the letters were omitted. The reader must guess.—

p. 171 indian-rubber = "Gummi" = caoutchouc— The strap or hinge of Labellum is elastic when extended, as well as when bent.—

p. 254. Cocked-hat—an old fashioned hat worn formerly by gentlemen & Officers.

p. 285. I was struck in Brazil & in British Museum with the astonishing diversity of prominences, &c &c on small *Homopterous* insects: I marvel whenever I think of them.—

Anther-case, I only mean, the walls of the anther, which form the case or receptacle for the pollen.—

p. 104. line 4. for renewed, read removed.

p. 292 "group of spiral vessels" is well translated by your "Spiral gefass-Bündel". There is no distinction between "groups of vessels" & "vessels". Sometimes there is only a single vessel in a petal. As the vessels collect from the several organs, the groups get bigger & bigger low down towards, or in, the ovarium, & form the six large ovarian groups or bundles. Generally even a single organ as one petal or one anther has a small bundle or group of vessels.—

[Enclosure 2]

[p. 34 (3 lines from bottom) Orchids*]

[*] With respect to Orchis maculata, my son George Darwin, who is an entomologist and careful observer, has clearly made out the manner of its fertilisation.[7] He saw many specimens of a fly (Empis livida) inserting their proboscises into the nectary. He brought home six specimens, with pollinia attached to their spherical eyes on a level with the bases of the antennae. The pollinia had undergone the movement of depression, and stood a little above and parallel to the proboscis: hence they were in a position excellently adapted to strike the stigma. Six pollinia were thus attached to one specimen, and three to another. My son also saw another and smaller species (Empis tenuipes) inserting its proboscis into the nectary; but this species did not act so well or so regularly as the other in fertilising the flowers. One specimen of this latter Empis had five pollinia, and a second had three pollinia, attached to the dorsal surface of the convex thorax. It is very probable that Orchis masculata, O. latifolia, and O. morio are fertilised by Diptera. (Postscript, June 1862.)[8]

[p. 75 (bottom line) removed*]

[*] My son has observed that Herminium is visited by various minute Hymenoptera, which are so small that I had overlooked them.[9] They belong to at least two groups, the largest being only 1/20 of an inch in length. Each time he brought home 1-2-3, together at last twenty-four specimens all with pollinia attached to them. It is an extraordinary fact that in all the specimens, with only one exception, the viscid discs were attached to the same peculiar spot, namely, to the outer side of the front leg, to the projection formed by the articulation of the femur with the coxa. In one instance alone a pollinium was attached to the inside of the femur. In addition I received a very small beetle from the family of Serricornia, which also carried a pollinium at the same spot of the front legs. These facts demonstrate how magnificently precise the form and depth of these flowers must be calculated to force various different kinds of insects to assume the same position so that they touch the helmet-like viscid disk with the projecting point of the femur. Answering my remark that these insects have to crawl in with their backs turned directly or obliquely towards the labellum he confirmed that this usually is the case. He, however, had seen several that had begun to crawl into the flower in a different position; but they came out, changed their position and crawled in again. One individual had been caught in the flower with the femur glued to the disc which had not yet been pulled off. This instance made me realise that the insects enter on both sides between the labellum and the upper petals and the discs (with rare exceptions) are glued to the outer surface of the femur. It is probable that when the insect retreats, the femur and hip joint brush against the lower side of the disc and pull out the pollinia. I know of hardly any other case where the pollinia stick to the legs of insects, although this occurs in Asclepiadeae as I have seen myself. (Postscript, June 1862.)[10]

[p. 82 (13 lines from bottom) nectary*]

[*] My son went at night to a bank where G. conopsea grows and soon caught

several moths, like Plusia Chrysites with six pollinia, Plusia gramma with three, Anaitis plagiata with five, and Triphena pronuba with five pollinia attached to their proboscises.[11] The two viscid discs in the flower form an arched roof over the nectary and are large, compared with their diameter. That is why they become attached on the sides of the proboscis and, after a vertical movement of depression, occupy a proper position for striking the lateral stigma on the same side. Now if the moths are sucking the nectar they are resting on the labellum and there are no guiding-ridges. That is why it can frequently occur that the proboscises are inserted obliquely. In this case one pollinium alone is usually removed, as I convinced myself by repeated experiments with a bristle. This explains how it happens that so many pollinia are attached to the proboscises of these moths. (Postscript, June 1862.)[12]

[p. 88 (top line) other*]

[*] Professor Asa Gray has written to me that he had compared Platanthera hookeri of North America with my description of Chlorantha and found that they correspond in most respects; he, however, also found some curious differences.[13] The two viscid discs stand widely separated from each other; consequently a butterfly, unless of gigantic size, would be able to suck the copious nectar without touching either disc; but this risk is avoided in a most remarkable manner. The central line of the stigma is prominent, and the labellum, instead of hanging down, is curved upwards, so that the front of the flower is divided into two halves. Thus the moth is compelled to go either to one or the other side to suck nectar, and its head will be brought into contact with one of the two discs. The drum of the pollinium, when removed, contracts in the same manner as in Pl. chlorantha. (Postscript, June 1862.)[14]

[p. 152 (1 line from bottom) insects*]

[*] Prof. Dickie has been so good as to observe the flowers on living plants.[15] He informs me that, when the pollen is mature, the crest of the rostellum is directed towards the labellum, and that, as soon as touched, the viscid matter explodes, the pollinia becoming attached to the touching object; after the explosion, the rostellum bends downwards and spreads out, thus protecting the virgin stigmatic surface; subsequently the rostellum rises and exposes the stigma; so that everything here goes on as I have described under L. ovata. These flowers are frequented by minute Diptera and Hymenoptera. (Postscript, June 1862.)[16]

[p. 156 (bottom line) Listera*]

[*] I have recently had better opportunity to observe this orchid,[17] and found that the rostellum lost its power of explosion in about four days, the viscid matter then turning brown within the loculi of the rostellum. This fact does not correspond with some earlier observations, but the weather in that year was unusually hot. After the four days had elapsed, the pollen had become very incoherent and some had fallen on the two corners of the stigma, which was penetrated by the pollen-tubes. Hence, if insects should fail to remove the pollinia by causing the explosion of the rostellum, and carry them to their destination, this orchid certainly seems capable

of occasional self-fertilisation. But the scattering of the incoherent pollen was largely aided by the presence of small insects of the genus Thrips, insects that cannot be excluded by any net, and that are numerous in these flowers and scatter the pollen in all parts of the flowers.[18]

[p. 276 (3 lines from top) convex*]

[*] Professor Asa Gray carefully examined several American species of Cypripedium after reading these sheets and has found various major differences of structure and several beautiful adaptations.[19] He infers from these observations that in several cases small insects are entering the labellum and he completely agrees with me that these insects are necessary for the fertilisation of all species. With regards to the state of the stigma he confirms that in C. acaule, where the pollen is much more granular than in other species, that, also, the stigma is slightly concave and viscid. (Addition made in June 1862.)[20]

Copy[21]

Bronn trans. 1862; DAR 143: 155; Houghton Library for Rare Books and Manuscripts, Harvard University

[1] The year is established by the relationship to the letter from H. G. Bronn, 21 June 1862.

[2] CD had been suffering from eczema for three weeks (see following letter).

[3] Bronn was translating *Orchids* into German (Bronn trans. 1862). For Bronn's questions concerning the translation, see the letter from H. G. Bronn, 21 June 1862; for CD's answers, see enclosure 1. Although CD's 'corrections and additions' have not been located, Bronn trans. 1862 includes some additions by CD in the form of footnotes, most of which are dated June 1862. As these printed additions presumably incorporate at least some of the additions to which CD refers in his letter, they are transcribed here as enclosure 2, and keyed to the relevant passages in Bronn trans. 1862. See also letter to H. G. Bronn, 11 July 1862.

[4] Letter from E. Schweizerbart'sche Verlagsbuchhandlung, 7 June 1862. CD's letter to Christian Friedrich Schweizerbart has not been found.

[5] The booksellers Williams and Norgate specialised in foreign scientific literature. There are two copies of the second edition of the German translation of *Origin* (Bronn trans. 1863) in the Darwin Library–CUL, and two copies of the German translation of *Orchids* (Bronn trans. 1862) in the Darwin Library–Down.

[6] CD refers to the speed with which Bronn had prepared a new German edition of *Origin* and translated *Orchids*; Bronn had worked 'night and day' for two months (see letters from H. G. Bronn, 21 June [1862], and from E. Schweizerbart'sche Verlagsbuchhandlung, 11 July 1862).

[7] George Howard Darwin's observations on the insects visiting *Orchis maculata* are detailed in a note dated 20 June 1862, which is in DAR 70: 13–14. See also the letter from W. E. Darwin, June 1862.

[8] Translated from Bronn trans. 1862, p. 22 n. A similar passage is included in 'Fertilization of orchids', p. 142 (*Collected papers* 2: 139–40).

[9] George's observations on the insects visiting *Herminium monorchis* are detailed in a series of notes dated 22–7 June 1862, which are in DAR 70: 32–6. See also the letters from Frederick Smith, 27 June [1862] and 28 June 1862.

[10] Translated from Bronn trans. 1862, pp. 47–8 n. A similar passage is included in 'Fertilization of orchids', p. 145–6 (*Collected papers* 2: 142–3).

[11] George's observations on the insects visiting *Gymnadenia conopsea* are detailed in a series of notes dated 21–4 June 1862, which are in DAR 70: 30.

[12] Translated from Bronn trans. 1862, p. 52 n. A similar passage is included in 'Fertilization of orchids', pp. 146–7 (*Collected papers* 2: 143–4).

[13] The letter from Asa Gray has not been found; however, see the letter to Asa Gray, 1 July [1862] and nn. 3 and 4.
[14] Translated from Bronn trans. 1862, pp. 55–6 n. A similar passage is included in 'Fertilization of orchids', pp. 147–8 (*Collected papers* 2: 144–5).
[15] George Dickie was professor of botany at Aberdeen University. He communicated his observations on *Listera cordata* in his letter to CD of 30 May 1862.
[16] Translated from Bronn trans. 1862, pp. 94–5 n. This passage occurs in 'Fertilization of orchids', pp. 151–2 (*Collected papers* 2: 148).
[17] The reference is to *Listera ovata* (see *Orchids*, p. 149).
[18] Translated from Bronn trans. 1862, p. 97 n. A similar passage is included in 'Fertilization of orchids', p. 151 (*Collected papers* 2: 148).
[19] Gray's observations on American species of *Cypripedium* have not been found; however, see the letters to Asa Gray, 10–20 June [1862] and n. 16, and 1 July [1862] and nn. 3 and 4.
[20] Translated from Bronn trans. 1862, p. 169 n. *Cypripedium* is discussed in 'Fertilization of orchids', pp. 155–6 (*Collected papers* 2: 152–3).
[21] The text of the letter to Bronn is taken from the copy in DAR 143.

To J. D. Hooker 30 [June 1862][1]

Down
30th.

My dear old friend.

You speak of my "warming the cockles of your heart", but you will never know how often you have warmed mine.[2] It is not your approbation of my scientific work (though I care for that more than for any one's); it is something deeper. To this day I remember keenly a letter you wrote to me from Oxford, when I was at the water-cure, & how it cheered me, when I was utterly weary of life.—[3] Well my orchis-book is a success (but I do not know whether it sells) after cursing my folly in writing it. I saw that the London R. was not written by a common Reviewer; but by Jove I must now read it again.—[4] I am better today than I have been for 3 weeks: the doctors told me it was eczema that I have had; so Sir William & I are fellow sufferers; & suffering it is.—[5]

I am grieved at the poor account of Mrs. Hooker. I shd. think it could be a great rest to get the children away; but there certainly is a strange influence in change of place for a patient. It is folly in me to have an opinion; but is not Switzerland too great an exertion? Does not Mrs. Hooker rely too much on it having done her good formerly? it seems to me a frightful thing to go so far as Switzerland.[6]

How marvellous it is about the European forms in Fernando Po!—[7] Do try & have some rest after you have printed 1st. vol. of Genera.—[8] No one can stand such wear & tear as you.— I hear Huxley is failing.[9] Why on earth should Brain-work take so much out of every man? I had a note from Naudin yesterday; he is going to publish a Book this autumn on Hybridity.[10] From some of his paper I have much fear that he has underrated the distribution of pollen by insects.[11]

Your Melastomatous plants are setting pods splendidly, & if I can make out the meaning of the two sets of anthers, I have every chance.[12] You may remember about complexion & Tropical diseases, & your & Busks aid; well, they are printed & gone to all quarters of world, through Dr. Parkes kindness—[13]

I long to hear of your schemes being settled & of M^{rs}. Hooker being stronger.— Yours affet. | C. Darwin

P.S. | Could you not give H. Gower a memorandum about Masdevallia, to let me know when near flowering & I could write to him how to send it.—[14] It would be off your mind & mine.— I am more curious about it than any other Orchis. Have you Bonatea speciosa— I cannot buy it at Veitchs.[15] When I have seen these two, I will not be seduced to look into more.— By the way my son George has been doing splendid work in watching Orchids. The case of Herminium beats in close adaptation almost every other orchid: in 24 insects the pollinia were attached to one point on one limb!—[16]

DAR 115.2: 157

1 Dated by the relationship to the letter from J. D. Hooker, 28 June 1862.

2 See letter from J. D. Hooker, 28 June 1862.

3 CD refers to the letter in which Hooker reported on the proceedings at the 1860 meeting of the British Association for the Advancement of Science held in Oxford from 26 June to 3 July 1860 (see *Correspondence* vol. 8, letter from J. D. Hooker, 2 July 1860). CD had been undergoing treatment at Edward Wickstead Lane's hydropathic establishment at Sudbrook Park in Richmond, Surrey.

4 The reference is to Miles Joseph Berkeley's review of *Orchids* in the *London Review and Weekly Journal of Politics, Arts, and Science* ([Berkeley] 1862). Hooker identified Berkeley as the author in the letter from J. D. Hooker, 28 June 1862.

5 CD refers to Hooker's father, William Jackson Hooker (see letter from J. D. Hooker, 28 June 1862).

6 Hooker planned to take Frances Harriet Hooker to Switzerland for a period of rest and recuperation (see letter from J. D. Hooker, 28 June 1862; see also Allen 1967, p. 208). Switzerland had been the Hookers' favoured holiday destination over a number of years (L. Huxley ed. 1918, 2: 431).

7 See letter from J. D. Hooker, 28 June 1862 and n. 9.

8 The first part of *Genera plantarum* (Bentham and Hooker 1862–83) was published on 7 August 1862 (Stearn 1956, p. 130).

9 Thomas Henry Huxley was 'crippled by neuralgic rheumatism' in his arm and shoulder. Accompanied by John Tyndall, he left for a holiday in Switzerland on 1 July 1862 (L. Huxley ed. 1900, 1: 234). See also letter from T. H. Huxley, 6 May 1862.

10 Naudin 1863. See letter from C. V. Naudin, 26 June 1862.

11 In the account of some of Naudin's hybridisation experiments given in Naudin 1858 (of which there is an annotated copy in the Darwin Pamphlet Collection–CUL), the author sought to forestall objections to one of his experiments on the grounds that he had not guarded sufficiently against the experimental plant, *Datura stramonium*, being pollinated by the same species, instead of by *D. ceratocaula*. He reported that he had provided for this eventuality by removing the stamens from the experimental flowers, and that, on another occasion, having left stamenless flowers open to the air and observing that they set no seed he had assured himself that they could not be accidentally pollinated by the agency of wind or insects, or by any other conceivable cause. In his copy of the paper, CD wrote in the margin at this point (*ibid.*, p. 6) 'But this was done at different date [i.e., from the hybridising experiment], perhaps no insect abroad—'. On the previous page, where Naudin described the very close resemblance of his supposed hybrids between *D. stramonium* and *D. ceratocaula* to the former species, CD wrote in the margin 'I do not believe were hybrids.— plants not protected.—' (*ibid.*, p. 5). See also CD's comments on this point in his copy of Quatrefages de Bréau 1861, p. 161 (the annotated copy is in the Darwin Library–CUL; see *Marginalia* 1: 693–4).

12 At the end of May, Hooker had complied with CD's request to send him specimens of Melastomataceae that were about to flower (see letter from J. D. Hooker, [29 May 1862], and letter to J. D. Hooker, 30 May [1862]). CD's notes from crossing experiments with specimens of the melastomaceous

plants *Rhexia glandulosa* (dated 1 June – 2 July) and *Centradenia floribunda* (dated 31 May – July), sent from the Royal Botanic Gardens, Kew, are in DAR 205.8: 14 v.–15, and 19–20.

13 CD refers to George Busk and Edmund Alexander Parkes (see letter from E. A. Parkes, 29 June 1862).

14 William Hugh Gower was a foreman at the Royal Botanic Gardens, Kew. CD refers to the orchid *Masdevallia fenestrata*; in his study of orchid pollination CD conceded that he had 'failed to understand' this species. *M. fenestrata* is discussed in *Orchids*, pp. 168–9.

15 CD refers to the Chelsea nursery run by James Veitch and his son James Veitch Jr. In his study of orchid pollination, CD examined only one specimen of *Bonatea speciosa*. In this species, the course of the so-called spiral vessels represented an anomaly for CD's account of the homologies of orchids (see *Orchids*, pp. 302–5; see also *Correspondence* vol. 9, letters to J. D. Hooker, 10 November [1861] and 14 November [1861]).

16 George Howard Darwin's observations on the insects visiting the orchid species *Orchis maculata*, *Gymnadenia conopsea*, and *Herminium monorchis* are detailed in a series of notes made between 20 and 27 June 1862; these notes are in DAR 70: 13–14, 30, and 32–6. CD incorporated the observations in the German translation of *Orchids* (Bronn trans. 1862, pp. 22 n., 47–8 n., and 52 n.; see the second enclosure to the preceding letter). He later published them in 'Fertilization of orchids', pp. 142, 145–7 (*Collected papers* 2: 139, 142–4).

From George Maw 30 June 1862

Benthall Hall | *Broseley*
30th. June 62

Dear Sir

Some time ago you asked me to send you specimens of the Pelargonium leaves I referred to in a former note as presenting a singular relation in their form to the characters of the respective flowers of the two varieties—[1] of the enclosed, No 1 is a leaf of the original "Tom Thumb" No 2 of "Brilliant" which is a monstrous variety or rather condition of "Tom Thumb" The symmetrical leaf of Tom Thumb is always as you are doubtless aware accompanied by a perfect and symmetrical truss of flowers but "Brilliant" with the deformed leaves very rarely perfects a good truss— The individual flowers are thin with a tendency to wither & petals reflexed. in fact the distortion of leaf is accompanied by a corresponding deformity in the flower.— As far as I have observed, this relation between leaf and flower holds good in all the cultivated varieties of the Scarlet Pelargonium those with flat symmetrical leaves always have a good truss of flowers & those in which the leaf is much crumpled (as in "mountain-of-light" "Mrs Lennox" &c) the truss is poor & imperfect—[2] I believe you consider these relationships between parts as evidence of their being modified forms of a common organ but do we not see similar phenomena in the relations of organs as to character, time of development, &c that cannot possibly be homologous take as a familiar example the development of *hair* on the face at the time the generative organs are perfected[3]

If the relationship in character of leaf and flower in Pelargonium is from their being modified forms of a common organ—why not on the same principle assume that the generative organs & the hair on face are homologous? & on the supposition that the occasional leafy development of a petal is an instance of reversion to an original common organ. I think we might with equal justice assume that the Castration of male producing the contours & general aspect of female (in those races where the aspect of the sexes is notably distinct as in the ox tribe) as evidence of the sexes having been developed from an original a-sexual form, resembling the female— It is however very remarkable that the destruction of the generative organs in either Sex does not produce reversion to a *common neuter* form, but in many instances it causes the assumption of the special attributes of the opposite sex for instance the castration of the bull produces in the ox the physical contours of the cow & many instances are recorded of the destruction of the ovary in hen birds producing cock plumage—the voice of the male & general physical aspect of male bird— The appearance of hair on the chin of old women is perhaps a parallel case I cannot help thinking that these cases show how very much may depend on the mere relationship of parts, & that two organs *may* be subject to similar influences without their necessarily being homologous— with respect to the cotemporary phenomena at birth I noticed as presenting a difficult case for Natural Selection. I can understand it as perfectly logical that the several phenomena as the production of milk by the mother—the organization of heart—the altered Circulation—respiration of fetus on the cotemporary operation of which depends the life of the offspring, may be *kept together* at birth on the principle of Natural Selection but my difficulty is as to how these functions which would be useless either separately or in an imperfect state could be accumulated *gradually* because the offspring unless absolutely perfect in every one of these functions (with a perfected supply of milk from its mother) could not survive & accumulate & advance on their progenitors—

The perfect readyness with which you have discussed objections to your theory induces me to take the liberty of mentioning these difficulties that have occurred to me but pray do not feel under any obligation to reply as I know you must be overwhelmed with correspondence.

Believe me Dr Sir | yours truly | George Maw.

Charles Darwin Esqr

DAR 99: 5–9

CD ANNOTATIONS

1.1 Some . . . imperfect— 1.14] *crossed ink*

1.12 leaves . . . perfected 1.18] 'These remarks show he does not understand— case of Hair is correlation in period alone' ['I never suppose' *del*] 'It is rash to speak of correlation except when organs are homological.' *added in margin, ink*

2.2 common . . . sexes 2.8] '*Is this* so. Ox? I think it is in some birds' *added in margin, ink*

2.11 instances . . . women 2.15] *scored brown crayon*

Top of letter: 'See his Review *of me*'[4] *blue crayon*

Cover: 'G. Maw on correlation cases & difficulties' *ink, del brown crayon*; 'On castrated males & females not returning to neuter state, but to opposite state; as if must take on one or other state neuter state last' *ink*; Bears *[*on*]* Selection' *ink, underl brown crayon*; 'Ch VII'[5] *brown crayon, del brown crayon*

CD note:[6]

Why simultaneous? Young fish with yolk milk need not be selected till long after.— Mucus in *marsupial* Batrachians— diffused mammary glands in marsupial Ornithorhynchus.

[1] See *Correspondence* vol. 9, letter to George Maw, 31 August [1861], and letter from George Maw, 27 August [1861].

[2] CD cited Maw's observations on this subject in *Variation* 2: 330–1.

[3] CD had elaborated his views on the 'correlation of parts' in *Origin*, pp. 143–50, stating that 'the whole organisation is so tied together during its growth and development, that when slight variations in any one part occur, and are accumulated through natural selection, other parts become modified' (p. 143). In particular, he argued that '[t]he several parts of the body which are homologous, and which, at an early embryonic period, are alike, seem liable to vary in an allied manner' (p. 143), asserting that these laws of correlation were thus responsible for 'modifying important structures, independently of utility and, therefore, of natural selection' (p. 144), and citing in evidence 'a striking case of correlation' in pelargoniums (p. 145). In his review of the third edition of *Origin*, Maw criticised CD's views on this subject, arguing that such morphological relationships were perfectly consistent with special creation, and that there were cases of 'correlation of organization' in which genealogical affinity was 'out of the question', but in which it was 'not unreasonable to suppose that the uniformity of model, upon which the more important parts are formed' resulted from 'special creation working out a consistent proportion of the integral parts' (Maw 1861, p. 7587). CD and Maw had subsequently corresponded on this question (see *Correspondence* vol. 9, letters to George Maw, 19 July [1861] and 31 August [1861], and letter from George Maw, 27 August [1861]).

[4] Maw 1861. See n. 3, above.

[5] CD refers to chapter 7 of his 'big book' on species, entitled 'Laws of variation: varieties and species compared' (see *Natural selection*, pp. 275–338), which contained his original discussion of 'Correlation of growth'. CD later addressed the subject of this chapter at greater length in *Variation*, chapters 24–6, including a revised discussion of 'Correlated variability' (*Variation* 1: 319–38; see also letter to George Maw, 3 July [1862] and n. 7).

[6] The note is in DAR 99: 10. See also letter to George Maw, 3 July [1862].

From Asa Gray [late June 1862][1]

[Cambridge, Massachusetts]

I have now read (by snatches) the greater part of Morell's Outlines of Mental Philosophy on the Inductive method.—[2]a book which has interested me very much from its full acquaintance with, and good use of scientific facts (so rare with metaphysicians),—and I think it is a book to please you.

A fortnight more will essentially relieve me from this wearying work in College, which not only consumes all my time, but unfits me for anything else.[3]

I ought to go and look up *Rhexia*; but I cannot find time yet.[4]

We are getting on quietly with our war; and now that we are used to it, we can keep it up two years longer as well as not, if our rebels choose not to yield. The longer they defer, the worse for slavery.

So some of our wisest and coolest people say that they have no great desire to have it end too soon.

Ever, dear Darwin | Yours cordially | Asa Gray

Incomplete
DAR 165: 110

[1] The date is established by Gray's references to Morell 1862 (see n. 2, below), to his duties at Harvard University, where he was Fisher Professor of natural history (see n. 3, below), and to the genus *Rhexia* (see n. 4, below).

[2] Morell 1862; Gray had told CD that he had been given a copy of this book in his letter of 31 March [1862].

[3] In May, Gray had told CD that he would be fully occupied with college duties until 10 July (see letter from Asa Gray, 18 May 1862). See also letter from Asa Gray, 2–3 July 1862.

[4] CD was investigating what he thought might be a novel form of dimorphism in the Melastomataceae, and had asked Gray to examine plants of the genus *Rhexia* for signs of this phenomenon (see letter to Asa Gray, 16 February [1862]; see also letters to Asa Gray, 15 March [1862] and 21 April [1862]). Gray had agreed to carry out this task in the summer (see letter from Asa Gray, 6 March [1862]); in his letter of 2–3 July 1862, he told CD that he would shortly 'try to look up & bring home living *Rhexia Virginica*'.

From E. A. Darwin 1 July [1862][1]

July 1

Dear Charles

I am sorry to bother you, but Carlyle took your letter to the Museum & found that the book was not there & therefore supposes that he had it from the Geolog. Soc:[2] Will you give me a letter of introduction just stating that it is to enquire the title of a book for Carlyle & if possible to borrow it—

yours | E D

If you are unwell there is no hurry

DAR 105 (ser. 2): 6

[1] The year is established by the relationship to the letter from E. A. Darwin, 20 June [1862].

[2] Erasmus was assisting his friend Thomas Carlyle in obtaining a publication containing a topographical account of the highlands of Saxony (see letter from E. A. Darwin, 20 June [1862]). The references are to the library of the Museum of Practical Geology, Jermyn Street and to that of the Geological Society at Somerset House, London.

To Asa Gray 1 July [1862][1]

Down Bromley Kent

July 1st.

My dear Gray

I have been baddish for 2 or 3 weeks, but am better tonight[2] & mean to amuse myself (& have no other object) by scribbling a few lines on orchids. Since writing last, I have received your notes on Platanthera Hookeri & on diversity of forms of Cypripedium. On the latter you ought, if you can spare time, to write a paper.[3] Platanthera Hookeri is really beautiful & quite a new case. It is almost laughable the viscid discs getting so far apart that the front of the flower has to be divided into two bridal chambers! I have added a note to the German Edition about this & a few words on Cypripedium on your authority.[4] On getting your letter I wrote to Trübner to send the $\frac{1}{2}$ dozen copies:[5] I wish you would let me pay for them; but you are so punctilious that you would fling without permission first granted the money across the Atlantic in the same shameful manner in which you did the £8.—[6]

My son George, who is an entomologist, has been watching orchids with enthusiasm & indomitable patience.[7] He has made out clearly that it is a Fly, (Empis) which fertilises O. maculata, & probably other Diptera fertilise the closely allied species; it was pretty to see the pollinia affixed to their spherical eyes, & after the act of depression parallel to & rather above the probosces.[8] But the most remarkable case is that of Herminium monorchis; he has brought me 24 specimens of very minute Hymenoptera with pollinia attached to all, & always to the same *exact* spot—viz to the exterior base of femur of front legs. Nothing has given me such an idea of close adaptation of form of whole flower: the labellum hangs *obliquely* downwards & the minute insects enter between its Edge & the huge viscid disc on one side; & in retreating they hit their prominent femora against the under side of the disc. So closely fitted is the flower to the insect, that my son saw several times insects after entering in a wrong position come out, change their position & reenter.[9] So much for orchids: I am fairly astonished at the success of my book (not that I know whether it sells) with Botanists: Berkeley has reviewed it in London R. rather egregiously, & Hooker writes strongly.[10] So that now I can dismiss the subject & stick to other work.

Lately I have done very little, except some crossing of plants. I have made a great series of crosses on the peloric flowers of Pelargonium: but I doubt whether I shall get such good results, as I at first hoped with respect to sterility of hybrids.[11] By the way Naudin writes that he is going to publish on this subject this autumn: his papers give me the idea that he does not know what has been done in Germany.—[12] Rhexia glandulosa, I does require insect agency to set seed; but I see as yet no probability of dimorphism.[13] You mentioned some genus (name forgotten without searching your letters) in which you found two forms like Primula, & *a third* **form** *with both pistil* & *stamen* short.[14] Can you tell me whether *all flowers* on the specimen were thus characterised? I much want to know because Lythrum is trimorphic.—[15]

Yours ever most sincerely | C. Darwin

Gray Herbarium of Harvard University (69)

[1] The year is established by the reference to the publication of the German translation of *Orchids* (see n. 4, below).

[2] CD had been suffering from eczema (see letters to J. D. Hooker, 23 June [1862] and 30 [June 1862]).

[3] The notes referred to have not been found; they may have been included with the letter from Asa Gray, [late June 1862]. Gray included information on *Platanthera Hookeri* in his review of *Orchids* (A. Gray 1862a, p. 143) and published observations on various species of *Cypripedium* in a follow-up article (A. Gray 1862b, pp. 427–8). See also letter to Asa Gray, 10–20 June [1862] and n. 16.

[4] CD had sent Heinrich Georg Bronn, the German translator of *Orchids*, a list of corrections and additions to be included in the German edition (Bronn trans. 1862; see the second enclosure to the letter to H. G. Bronn, 30 June [1862]).

[5] Gray had asked CD to arrange for six copies of *Orchids* to be sent to him by the publisher and bookseller, Nicholas Trübner, who frequently acted as Gray's London agent (see letter to Nicholas Trübner, 23 June [1862]). Gray's letter requesting the copies has not been found.

[6] Gray had returned a payment resulting from the sale of his pamphlet on *Origin* (A. Gray 1861), the publication costs of which he and CD had shared (see letter from Asa Gray, 31 March [1862], and letter to Asa Gray, 21 April [1862]).

[7] George Howard Darwin, then 16 years old, was on holiday from Clapham Grammar School. CD's notes, dated 20–7 June 1862, recording George's observations of the insects visiting several orchid species, are in DAR 70: 13–14, 30, 32–6 (see also nn. 8 and 9, below).

[8] CD's note recording George's observations on flies visiting *O. maculata*, dated 20 June 1862, is in DAR 70: 13–14. CD reported these findings in the German edition of *Orchids* (Bronn trans. 1862, p. 22 n.; see the second enclosure to the letter to H. G. Bronn, 30 June [1862]). He later published them in 'Fertilization of orchids', p. 142 (*Collected papers* 2: 139–40).

[9] The records kept by CD of George's observations on insects visiting *H. monorchis*, dated 22–7 June 1862, are in DAR 70: 32–6 (see also letter from Frederick Smith, 28 June 1862); these observations were incorporated in the German edition of *Orchids* (Bronn trans. 1862, pp. 47–8 n.), and were later described in 'Fertilization of orchids', pp. 145–6 (*Collected papers* 2: 142–3). CD's notes on his own observations of the species, dated 2 July 1862, are in DAR 70: 37.

[10] See letter from J. D. Hooker, 28 June 1862. CD refers to Miles Joseph Berkeley's review of *Orchids*, which appeared in the *London Review and Weekly Journal of Politics, Arts and Science* 4 (1862): 553–4.

[11] CD had begun a series of crossing experiments with the normally sterile central peloric flowers of different varieties of pelargonium on 11 May 1862 (see the experimental notes in DAR 51 (ser. 2): 4–9, 12–13; see also letter to Daniel Oliver, 8 June [1862]); in 1861 he had attempted to encourage readers of the *Gardeners' Chronicle* to carry out such experiments (see *Correspondence* vol. 9, letter to *Journal of Horticulture*, [before 18 June 1861]). The results of CD's experiments are given in *Variation* 2: 167, as part of a discussion on the relationship between abnormal plant structure and the incidence of sterility. For CD's interest in the causes of sterility, see Appendix VI.

[12] See letter from C. V. Naudin, 26 June 1862 and n. 4. CD probably refers to the experiments of the German hybridisers Karl Friedrich von Gärtner and Joseph Gottlieb Kölreuter.

[13] Suspecting that plants of the melastomaceous genus *Rhexia* might exhibit a novel form of dimorphism, CD asked Gray to make observations for him; Gray promised to do so in the summer (see letters to Asa Gray, 16 February [1862] and 15 March [1862], and letter from Asa Gray, 6 March [1862]). Later, CD asked Gray if he would cover his plant under a net '& see if it seeded as well as uncovered plants' (letter to Asa Gray, 21 April [1862]). However, having obtained a specimen of *R. glandulosa*, CD began a series of crossing experiments on 1 June 1862 (see the experimental notes, dated 1 June – 2 July 1862, in DAR 205.8: 14–15, and letter to Asa Gray, 10–20 June [1862]).

[14] See *Correspondence* vol. 9, letter from Asa Gray, 9 November 1861. See also this volume, letter from Asa Gray, 15 July [1862].

[15] CD had discovered that *Lythrum salicaria* and *L. thymifolia* were trimorphic in December 1861, on reading Lecoq 1854–8 (see *Correspondence* vol. 9, letter to J. D. Hooker, 18 [December 1861]), and had obtained specimens for experiment from the Royal Botanic Gardens, Kew (see letter to J. D. Hooker, 22 [March 1862]). He had previously informed Gray of the case in his letter of 22 January [1862].

From E. A. Darwin 2 July [1862][1]

July 2

Dear Charles

If you have not already written me the letter, never mind, as the obliging sub Lib of the Museum went to the Geolog: & did the necessary.[2]

It is really wondeful the trouble Carlyle takes for such a minute point— I have had 4 or 5 letters & he has probably written half a dozen besides

Yours | E D

DAR 105 (ser. 2): 7–8

[1] The year is established by the relationship to the letters from E. A. Darwin, 20 June [1862] and 1 July [1862].

[2] See letters from E. A. Darwin, 20 June [1862] and 1 July [1862]. The assistant librarian at the Museum of Practical Geology was Thomas W. Newton (Flett 1937, p. 255).

From Asa Gray 2–3 July 1862

[Cambridge, Mass.]
July 2, 1862.

My dear Darwin.

Many thanks for your long and interesting letter of June 10th, & later.[1] Pray write upon thinner paper, and then one of the long letters which I find so enjoyable and so stimulating, will cost you only 1/ in postage.

The word *postage* reminds me of your young son's request (I do hope he has quite recovered his health), which shall surely be attended to.[2] Some young people here, of Mrs. Gray's family take to stamp-collecting, and will help. They s⟨ay⟩ Wells, Fargo & Co. Express ⟨are⟩ most rare. I never ⟨saw⟩ them. But we will ⟨ ⟩ *Blood*,—a Philadelph⟨ia⟩ ⟨ ⟩ penny-post carrier, is more common. I used to see his stamp upon my Philadelphia letters, and I think I may find or procure them. And for the rest, is it our U.S. stamps on letter envelopes your boy wants? I enclose a 3 cent, and will lay hold of the first one & two cent ones that I see. I am glad if my off-hand orchid notes interest you, or prove of the least use..[3] I am daily expecting a copy to send you of my notice of the early chapters of your book. I will continue in the ensuing number.[4] And whatever of the notes I send you seem ⟨to⟩ you worth touching upon, ⟨you⟩ have only to indicate ⟨ ⟩ and send back my ⟨memor⟩anda, and I will take ⟨ ⟩. But as to *Cypripedium*, I should like to have an opportunity of examining them (except C. *acaule*) more at large, and growing.

A week from to-morrow, I expect to be able to leave Cambridge.—to go down, with my examination-papers to read, to my *beau-pere's* place on the shore for a few days.[5] There I will try to look up & bring home living *Rhexia Virginica*,[6] and also I expect to have a look at *Calopogon pulchellus* with its strong bearded labellum. And I hope it will not be too late to get plenty of *Mitchella repens* which my pupils do not bring in as they ought.[7] I want to see if long-styled stigma & short, differ, and also the pollen of the two, as they do in *Houstonia*,—of which I hope I sent you Rothrock's observations. At least I will send when he has completed them.[8]

Meehan—a good gardener—send me his ms. before printing.[9] I tried to find exceptions to his rule, and thought I had; but he beat me down.

If any body comes out with a new empirical law, I always disbelieve him *prima facie*. But Meehan is an honest and I suppose very good observer, and you may "*approximately*" trust him, I should think.[10] He may have got hold of something.

Precocious fertilization in the bud was much noticed here very long ago, by Torrey, in *Viola*, *Specularia*, &c, &c—[11]also in Impatiens—about which see my Genera Fl. vol. 2.[12] I once mentioned it to you as good evidence of *close-fertilisation*.[13] As to the pollen-tubes of such, I have no observations of my own, but a memory, or fancy, that they were shown to me by Torrey. I will ask him, and have him look at *Specularia*.[14]

As to the French Lady's translation and commentary on the Origin, I am not so much surprised.[15] As I view it there are only two sides to the main question. Very likely she takes one side in a thorough-going and consistent manner; and either she is right, or I am right. I.e. there is design in nature or there is not. The *no-design* view, if one can bring himself to entertain it may well enough lead to all she says, and we may very much admire how *collission*, and destruction of least favored brings about apparently orderly results,—*apparent* contrivances or adaptations of means to ends. On the other hand, the implication of a designing mind must with it a strong implication of design in matters where we could not directly prove it. If you grant an intelligent designer anywhere in Nature, you may be confident that he has had something to do with the "contrivances" in your Orchids.

I have just received and glanced at Bentham's address, and am amused to see how your beautiful *flank*-movement with the Orchid-book has nearly overcome his opposition to the Origin.[16]

The military simile above leads me to speak of your wonder that I can think of science at all in the midst of war.[17] Well, 1st we get used to it. 2d, We need something to turn to, and happy are they who, forbidden to engage personally in the ⟨war⟩ (as I am ever itching to do,⟨)⟩ have something to turn to.[18] 3d. I do not do much.—do nothing in fact except my college duties now for months.—and that is the reason I have time to write to you, and be interested in all your doings.[19]

If you suppose everything is paralysed and desolate here, and country greatly put back, read a very sensible letter of an Englishman in the *Spectator* of June 7.[20] It is very just & true. We shall recuperate fast enough, and be better off than ever, as much prosperity as is good for us, and more sol⟨id,⟩ more independent, more sel⟨f-⟩contained,—which is our g⟨reat⟩ desideratum. *Free-tr⟨ade⟩ be blowed*; we must nee⟨ds have⟩ high duties on imports; an⟨d⟩ ⟨ ⟩ that we should. By ⟨th⟩ese and by direct taxes—the ⟨t⟩ax-bill just passed—we shall have to *pay over* largely.[21] Very well.

Just at present our prospects (viz. evening of July 3) are looking badly enough. Mc'Clellan has clearly been over matched and driven to the wall, after very obstinate fighting with very heavy loss on both sides.[22] Whether it is retrievable with reinforcements, or whether the whole campaign has to be begun again against Richmond is not yet clear. Anyway we have got to put shoulder to the wheel anew, and it may be done, we suppose, more easily, and far more ⟨p⟩romptly than last year. ⟨All⟩ we ask is that Europe shall ⟨leave us⟩ alone.

⟨Enoug⟩h for today. Ever Yours | cordially | A. Gray

Note | *Utricularia vulgaris* is about as neatly contrived for cross-fertilization by insects as almost any *orchid*.[23]

Postmarks: Jul 5; JY 18 | 62
DAR 165: 110a, 112–12a

[1] Letter to Asa Gray, 10–20 June [1862].

[2] Gray refers to Leonard Darwin. In the letter to Asa Gray, 10–20 June [1862], CD requested Gray's assistance in procuring a number of North American postage stamps for Leonard's collection, and told him that Leonard had become ill with scarlet fever.

[3] CD had assured Gray, who was concerned that his 'scattering notes' were tiresome, that he took 'very great pleasure' in Gray's letters (see letter to Asa Gray, 10–20 June [1862], and letter from Asa Gray, [2 June 1862]). It is evident from CD's reply to this letter (letter to Asa Gray, 23[–4] July [1862]) that Gray had sent with it 'a bundle of notes' on orchids; these notes have not been found.

[4] Gray's review of the first part of *Orchids* (A. Gray 1862a) was published in the July number of the *American Journal of Science and Arts*; according to the paper wrapper in which the journal was issued, numbers were published on the first day of the month. Gray wrote a follow-up article to his review for the November number of the journal (A. Gray 1862b). In the List of reviews (DAR 262 (DH/MS* 8: 6–18)) that served as CD's index to his collection of reviews of his own books, there is an entry that reads '*Asa Gray* On Orchids, Review of'. However, the offprint corresponding to the index number is absent from the Darwin Pamphlet Collection–CUL. See also letter from Asa Gray, 15 July [1862] and n. 16, and letter to Asa Gray, 28 July [1862].

[5] Gray's father-in-law, Charles Greely Loring, had an estate in Beverly, Massachusetts, situated on the Atlantic coast (Dupree 1959, pp. 179–80).

[6] In the letter from Asa Gray, 6 March [1862], Gray had agreed to assist CD with his investigation of the possible occurrence of dimorphism in the Melastomataceae by observing specimens of *Rhexia virginica* in the summer. See also letter from Asa Gray, [late June 1862] and n. 4.

[7] Gray had told CD that *Mitchella repens* was dimorphic in his letter of [27 and 29 August] and 2 September [1861] (*Correspondence* vol. 9), and had offered to make observations for CD the following spring on this or any other dimorphic species.

[8] Gray had informed CD in October 1861 that *Houstonia* was dimorphic, and had promised to look for any differences in the pollen of the two forms the following spring (see *Correspondence* vol. 9, letter from Asa Gray, 11 October 1861); he sent CD observations of the pollen and stigmas of *Houstonia* in the letter from Asa Gray, [2 June 1862]. Joseph Trimble Rothrock was one of Gray's students (Dupree 1959, p. 326), whose services as an observer Gray apparently recommended to CD in his letter of 6 March [1862]; Rothrock's observations on *Houstonia caerulea* are included in the letter from Asa Gray, 4 August 1862.

[9] Meehan 1862. Thomas Meehan had sent CD a copy of his paper, which described characteristic differences between various American trees and allied European species grown in proximity to each other and under similar conditions (see letter to Asa Gray, 10–20 June [1862]).

[10] See letter to Asa Gray, 10–20 June [1862]. A summary of Meehan's findings is given in *Variation* 2: 281–2; CD concluded that the differences could not have been the result of natural selection, but were caused by the 'long-continued action of the different climate of the two continents on the trees'.

[11] The reference is to the botanist John Torrey (see, for example, Torrey 1843, 1: 428). In the letter to Asa Gray, 10–20 June [1862], CD discussed the occurrence in several species of small flowers that did not open and in which self-pollination occurred (a phenomenon later known as cleistogamy); CD had recently been experimenting on this phenomenon in *Viola* and *Oxalis*.

[12] A. Gray 1848–9, 2: 134–5. There is a note recording this reference in DAR 111: 31.

[13] See *Correspondence* vol. 9, letter from Asa Gray, 11 October 1861.

[14] In the letter to Asa Gray, 10–20 June [1862], CD asked Gray to observe the behaviour of the pollen-tubes in the cleistogamic flowers of *Specularia*. For Torrey's observations on *Specularia*, see the letter from Asa Gray, 18–19 August 1862.

[15] Royer trans. 1862. See letter to Asa Gray, 10–20 June [1862].

[16] As president of the Linnean Society of London, George Bentham delivered the society's anniversary address on 24 May 1862 (Bentham 1862b). Bentham concluded that while 'Biology', defined strictly as 'the history of animal and vegetable life' (*ibid.*, p. lxviii), was 'a field of inquiry comparatively untrodden', the 'remarkable success' attending CD's work 'should stimulate others to follow in the

same track', (*ibid.*, p. lxxxi). He continued:

> I do not refer to those speculations on the origin of species, which have excited so much controversy; for the discussion of that question, when considered only with reference to the comparative plausibility of opposite hypotheses, is beyond the province of our Society. . . . But we must all admire that patient study of the habits of life, with that great power of combining facts, which has revealed to us so much of surprising novelty in the economy of nature. The wonderful contrivances for the cross-fertilization of Orchids, so graphically detailed in Mr. Darwin's new work, and which rival all that had been previously observed in the singular economy of insect life, had been hitherto unsuspected even by those botanists who had specially devoted themselves to that family.

Bentham had at first been 'greatly agitated' by *Origin* (see *Correspondence* vol. 7, letter from J. D. Hooker, [20 December 1859]), and CD had been concerned at Bentham's lack of support for his views (see *Correspondence* vol. 8).

17 See letter to Asa Gray, 10–20 June [1862].

18 As Gray was over 50, he was too old for military service (Dupree 1959, p. 306).

19 Gray was Fisher Professor of natural history at Harvard University.

20 Gray refers to a letter from the new town of Lanark, Illinois, signed 'An English Traveller', claiming that 'the great fact of American history' was the progress westwards across the prairie represented by towns like Lanark, and that such progress was continuing, despite the American Civil War. The writer contrasted this with the 'popular English view of American affairs', that the 'whole country' was 'in revolution', that trade was bankrupt, and 'the entire progress of the country stopped for years to come' (*Spectator*, 7 June 1862).

21 The Internal Revenue Act was passed on 1 July 1862. The legislation introduced taxes on alcohol and tobacco, luxury items, professions, banks and corporations, and manufactured goods, as well as an inheritance tax and new income taxes (McPherson 1988, pp. 447–8).

22 Gray refers to the Seven Day's battles, waged in the Virginia peninsula campaign between 25 June and 1 July 1862, which claimed 30,000 casualties. The Union forces, commanded by George Brinton McClellan, had unsuccessfully attempted to lay siege to the Confederate capital, Richmond, Virginia (McPherson 1988, pp. 461–71).

23 This note is written on the inside of the envelope (DAR 165: 110a).

From J. D. Hooker 2 July 1862

Kew
July 2/62

My dear Darwin

One line to say, that I have given instructions to Gower about *Masdevallia*,[1] & will write to the Cape for *Bonatea*-roots.[2]

We have got a Cook, who I hope will suit, she was Sir F. Palgrave's, during all his widowerhood & is a most trustworthy person, too old, but that is a fault on the right side in our case:[3] & at any rate she will enable me to get things belowstairs put on a proper footing— you cannot conceive the relief it is to us to have found a suitable person! Children & Governess go to Worthing tomorrow—we on Friday to Dover, we shall travel slowly. I also fear length &c of journey to Switzerland, but her heart is there, & just see what weather we have here![4] I saw Dr Walshe yesterday who examined her & assures me she has no organic complaint & that it is all functional[5]

I saw Lady Bell[6] on Monday who gave a very poor account of Lyell I am grieved to say. I called in Harley street yesterday (of course did not ask to see anyone). the servant told me he was *not so well* that day, but that he & all would go to Eastbourne on Saturday.[7]

I heard of Lyell having had to be put in a warm bath after arrival in England.[8]

I do hope you will unravel the *Melastoma* mystery.[9]

Your Eczema must be very different from my father's, which gives little or no pain.[10] Paget told me that Eczema was a sort of nom de guerre for any skin complaint that had no other recognized name,—a sort of "suppressed gout:" I suppose, w^{h}. means any thing but Gout!.[11]

Ever yrs affec | J D Hooker

P.S. We have just heard from Lady Lyell— Sir Charles is not well enough to go to Eastbourne on Saturday— she says she sees no one

DAR 101: 44–5

1 See letter to J. D. Hooker, 30 [June 1862] and n. 14; William Hugh Gower was a foreman at the Royal Botanic Gardens, Kew.

2 See letter to J. D. Hooker, 30 [June 1862] and n. 15; CD was hoping to obtain a specimen of *Bonatea speciosa*, an orchid found in the Cape of Good Hope (*Orchids*, p. 87).

3 Hooker refers to his uncle, the historian, Francis Palgrave. For Hooker's concern about the state of his household, see the letters from J. D. Hooker, [5 May 1862], 9 June 1862, 19 [June 1862], and 28 June 1862.

4 The Hookers were planning a trip to Switzerland in the hope that Frances Harriet Hooker might recover her health (see letter from J. D. Hooker, 28 June 1862, and letter to J. D. Hooker, 30 [June 1862]).

5 Walter Hayle Walshe was professor of the principles and practice of medicine at University College London (*DNB*). For Hooker's account of his wife's ill health, see the letters from J. D. Hooker, 9 June 1862 and 19 [June 1862].

6 Marion Bell.

7 Charles and Mary Elizabeth Lyell resided at 53 Harley Street, London.

8 The Lyells had travelled to Florence in May 1862, following the death there of Mary Lyell's mother, Anne Susan Horner (K. M. Lyell ed. 1881, 2: 343).

9 See letter to J. D. Hooker, 30 [June 1862] and n. 12.

10 See letter to J. D. Hooker, 30 [June 1862]. Hooker refers to his father, William Jackson Hooker.

11 James Paget was a leading London surgeon. In 1849, Henry Holland had diagnosed CD as suffering from 'suppressed gout' (see *Correspondence* vol. 4, letter to W. D. Fox, 6 February [1849] and n. 2).

To John St Barbe[1] [before 3 July 1862][2]

Dear Sir—

I wish to invest through the Union Bank about £17,000 in Railway Shares.[3] if I can find shares that suit me. Under these circumstances, I trust you will not think it unreasonable in me to request, through you, from the gentleman, who acts as your Broker, some careful information.— I wish for Guaranteed or Preference shares which divide with ordinary shares, when the latter rise above the Guarantee—Lancaster & Carlisle, some of Newcastle & Carlisle. & of Grt. N. of Scotland &c are of this nature.[4] Will the Brokers reflect what shares there are of this kind which can be purchased. Or whether any Railways acts likely soon to pass include such shares?[5] I sh^{d} perhaps purchase some Lancaster & Carlisle (these I now hold a good number) & sh^{d} be glad to know at what price I could purchase.—[6] Secondly will the Brokers inform me what Preference or Guaranteed shares there are, with *perpetual* power of changing into ordinary stock; (though I sh^{d} prefer the former kind of stock) N.E. Berwick 4 per cent Prefer is said to be of this nature; & I sh^{d} be glad to know price. But I wish to know what other shares there are of this nature.—[7] I hope you will be so kind as to request the best information, which your Broker can give, & you can forward this letter to them—

Dear Sir | Yours &c

P.S. I have been told that the N. Staffordshire guarantees the G.T. Canal, with a promise of sharing profit; but I know not if this be accurate or scale of guarantee &c &c.[8]

P.S.

Draft
DAR 96: 5

[1] John St Barbe was manager of the Charing Cross branch of the Union Bank of London.

[2] Dated by reference to entries in CD's Investment and Account books (Down House MSS) (see nn. 3, 6, and 7, below).

[3] According to CD's Account book–banking account (Down House MS), CD made purchases of railway stock to a total of £17,140 5*s.* in July and August 1862. See nn. 5 and 6, below.

[4] According to CD's Investment book (Down House MS), pp. 84, 57, 49, and 92, CD owned preference shares of this nature in the Great North of Scotland Railway Company, and had purchased such shares in trust for Emma Darwin in the Newcastle and Carlisle Railway Company; he also had a number of investments in the Lancaster and Carlisle Railway Company.

[5] CD refers to the passage through Parliament of the private bill legislation by which railway companies were granted powers to build new lines. On the relationship between the railway companies and Parliament in this period, see Parris 1965, pp. 56–62.

[6] An entry in CD's Investment book (Down House MS), p. 92, dated 3 and 9 July 1862, states that he purchased '5000 Stock' of the Lancaster and Carlisle Railway at a cost, including brokerage and stamps, of £10,150 4*s.* 9*d.*

[7] CD had purchased 9000 4% preference stock of the North Eastern Berwick railway in July 1859, at a cost of £8707 10*s.*; an entry in CD's Investment book (Down House MS), p. 74, dated 9 July 1862, records his purchase of a further 1000 stock at a cost of £980. An entry for 23 August (p. 53) shows that he also bought £6000 of London and Brighton Railway 4% debenture stock.

[8] CD refers to the North Staffordshire Railway Company and to the Grand Trunk Canal (another name for the Trent and Mersey Canal) in the founding of which both of CD's grandfathers, Erasmus Darwin and Josiah Wedgwood I, had been instrumental (J. Lindsay 1979). Both CD and Emma Darwin had received shares in the company from their respective fathers, Robert Waring Darwin and Josiah Wedgwood II (see CD's Investment book (Down House MS), pp. 3 and 27, and *Correspondence* vol. 3, letter to Solicitor?, 1 October 1844). In 1845, the canal company was taken over by the New North Staffordshire Railway Company. The terms of the take-over were that, when the whole line had been opened, the dividend on the Trent and Mersey shares would be 22.5%, and that on the 15 January 1847, the canal shares would be exchanged for 5% railway preference shares; thereafter the profits 'were to be divided in proportion to the capital, until the canal shareholders had received a total of

£30 per share' (J. Lindsay 1979, p. 116). In his Investment book (Down House MS), CD noted: Jos[iah]. W[edgwood II]. tells me each Canal share was converted into 22½ preference shares to pay beleive 1£ per share when whole R. Way open & afterwards, if dividend be sufficient to pay 5 per cent to other shareholders, then preference share holders to share same as at present, but never more.—

From Frederick Currey 3 July 1862

3, New Square | Lincolns Inn
July 3. 1862

My dear Sir,

My friend M^r. Wollaston of Chiselhurst has a plant of Spiranthes gemmipara which has thrown up a fine spike—[1] The flowers will be in perfection in a few days— I dont know whether it would be worth your while to go as far as Chiselhurst, but if so M^r. W. would be glad to submit it to your examination—

He has paid great attention to the cultivation of British Orchids, although not specially to their mode of fertilization—

I see you consider Ophrys arachnites a species—[2] M^r. Wollaston entertains a strong opinion that it is a hybrid between ophrys apifera & O. aranifera— If you do not happen to have seen his remarks upon the subject I think they would interest you— You will find them in the Volume of the Phytologist for 1855—[3]

Believe me | Yours sincerely | Fred^k Currey

Chas Darwin Esqr.

P.S

G. B. Wollaston Esqr.
Chiselhurst | S.E

is my friend's address—

DAR 161.2: 306

CD ANNOTATION

3.2 If you ... them in 3.4] *double scored brown crayon*

[1] George Buchanan Wollaston. See letter from Frederick Currey, 12 June 1862.

[2] In *Orchids*, pp. 72–3, CD discussed the claim that *Ophrys arachnites* was 'a mere variety of the varying Bee Ophrys' (*O. apifera*); he concluded: 'it seems to me that, until these forms can be shown to be connected by intermediate varieties, we must rank O. arachnites as a good species, more closely allied to O. aranifera in its manner of fertilisation than to O. apifera.'

[3] G. B. Wollaston 1855.

To George Maw 3 July [1862][1]

Down Bromley Kent
July 3^d

Dear Sir

Your criticisms are profoundly interesting to me.[2] A downright good objection or difficulty is very nearly as interesting as a favourable fact.— Your Review always struck me as very able (though I am not now quite so much shaken by some remarks as I was at first)[3] & I have given some few of your objections in a new German edition.[4] Very sincere thanks for the Pelargonium leaves.[5] I sh^{d}. certainly rank this as a case of true correlation; because the parts are homologous & it seems to me inherently probable that parts, which are quite alike in an early stage of development on the same individual would be apt to be similarly affected by the unknown causes of variation. If you will look at my discussion in Ch. V. Origin (p. 162–*164* 3^{d}. Edit) you will see that I expressly guard against the assumption that parts going together throughout whole classes are necessarily correlated.[6] But I now see that I ought more plainly to have said that "correlation of growth" contains a most heterogenious set of facts.[7] If you had asked me yesterday about hair & sexes, I sh^{d}. have said that it was only a case correlation so far as that the Hairy character had become attached to the male sex, in the same *accidental* (to speak loosely) manner in which large wattles have become correlated with male Carrier pigeons.— But your remarks show me that there is something more. I sh^{d}. hardly have said that the ox was quite like a cow; or an old woman like a man; but I think I have read of female birds quite like males in their plumage, & the case seems very strange & inexplicable. I must remember & attend to this; but the subject is far too large for me.—[8] I cannot help looking confidently to all vertebrata as having descended from an Hermaphrodite form; (though only one hermaphrodite genus still exists) & that males & in lesser degree females have acquired new characters in "correlation" (to use my favourite term) or in "accidental correlation" with sex.[9] But a man must be a fool to hope even to conjecture (supposing for the moment that my views are in the main correct) how all the wondrous changes have been effected. I had not forgotten your good case of coadapted structure at or before birth.[10] I do not quite see necessity of such close simultaneous development as you do. Some Young fish are born with bag of yolk, & this might last long till milk was secreted. Those young Batrachians which are hatched in a quasi marsupial pouch on mothers back are believed to feed on mucus secreted by whole surface; concentration of glands would make a mamma, & in the Ornithorhynchus (*marsupial* order) the mammary glands are still diffused. But this is all conjectural rubbish. Pray however (I must add) reflect on the quasi-placenta in cartilaginous fishes.

I have lately published a small Book on Orchids (small portions of which alone bear on the general question of Species) parts of which I think would interest you. I cannot conceive how I forgot your name when I made out a list of copies to send away.—[11] As I hope you will permit me to send you a copy, I will today write to my publisher.—[12] I sh^{d}. be much pleased, if any criticisms occur to you, to hear them, or read them if ever you review semi-botanical works.—

I assure you that I feel most sincerely grateful for your remarks & for the very kind manner in which you make them.—

Pray believe me | My dear Sir | Yours sincerely | C. Darwin

Royal Horticultural Society

[1] The year is established by the relationship to the letter from George Maw, 30 June 1862.
[2] See letter from George Maw, 30 June 1862.
[3] Maw 1861. For CD's reaction to Maw's review of *Origin*, see *Correspondence* vol. 9, letters to George Maw, 13 July [1861] and 19 July [1861].
[4] In Bronn trans. 1863, p. 445, CD added a reference to Maw's objection to his statement that 'the grand fact in natural history of the subordination of group under group' was explained by the theory of descent from a common ancestor and of divergence. See letter to H. G. Bronn, 25 April [1862] and Appendix VIII.
[5] See letter from George Maw, 30 June 1862.
[6] The section to which CD refers, headed 'Correlation of growth' (*Origin* 3d ed., pp. 161–4), concludes:

> We may often falsely attribute to correlation of growth structures which are common to whole groups of species, and which in truth are simply due to inheritance; for an ancient progenitor may have acquired through natural selection some one modification in structure, and, after thousands of generations, some other and independent modification; and these two modifications, having been transmitted to a whole group of descendents with diverse habits, would naturally be thought to be correlated in some necessary manner.

[7] At the start of his discussion in the third edition of *Origin* (p. 161), CD stated: 'This is a very important subject, most imperfectly understood'; in the fourth edition he added: 'and no doubt totally different classes of facts may be here easily confounded together: we shall presently see that simple inheritance often gives the false appearance of correlation' (*Origin* 4th ed., p. 170). CD gave a more detailed discussion of this subject in *Variation* 2: 319–38, under the heading 'correlated variability'; he noted that he had formerly used 'the somewhat vague expression of correlation of growth, which may be applied to many classes of facts', and sought to distinguish 'correlated variability' from what in this letter he calls 'accidental correlation'. See also letter from George Maw, 30 June 1862 and n. 5.
[8] In the chapter on reversion in *Variation*, CD included a discussion of what he called 'latent characters', the 'most obvious illustration' of which, he claimed, was 'afforded by secondary sexual characters' (*Variation* 2: 51). He then stated:

> It is well known that a large number of female birds . . . when old or diseased, or when operated on, partly assume the secondary male characters of their species . . . On the other hand, with male animals, it is notorious that the secondary sexual characters are more or less completely lost when they are subjected to castration . . . But characters properly confined to the female are likewise acquired . . . We thus see that in many, probably in all cases, the secondary characters of each sex lie dormant or latent in the opposite sex, ready to be evolved under peculiar circumstances.

[9] CD discussed the origin of the sexes in vertebrates in *Descent* 1: 207–11, arguing that 'some extremely remote progenitor of the whole vertebrate kingdom appears to have been hermaphrodite or androgynous'. He also thought it 'quite possible' that as each sex 'gradually acquired the necessary organs proper to it, some of the successive steps or modifications were transmitted to the opposite sex', and he presented 'innumerable instances of this form of transmission' (p. 208).
[10] See letter from George Maw, 30 June 1862, and *Correspondence* vol. 9, letter from George Maw, 27 August [1861].
[11] Maw's name is included on the presentation list for *Orchids* (see Appendix IV); it was probably added as an afterthought.
[12] The letter to John Murray has not been found.

To W. E. Darwin 4 [July 1862][1]

Down
4th

My dear William.

I have been looking at the fertilisation of Wheat, & I think, *possibly*, you might find something curious.— I observed in almost every one of the pollen-grains, which had become empty & adhered to (I suppose the viscid) branching hairs of stigma, that the pollen-tube was *always*(?) emitted on opposite side of grain to that in contact with the branch of stigma.— This seems very odd.

The branches of stigma are very thin, formed apparently of 3 rows of cells, of hardly greater diameter than pollen-tube:: I am astonished that the tubes sh^d. be able to penetrate the walls. The specimen examined (not carefully by me) had pollen only during few hours on stigma; & the mere **suspicion** has crossed me that the pollen-tubes crawl down these branches to the base & there penetrate the stigmatic tissue. The paleæ open for a short period for stigma to be dusted & then close again, & such travelling down of tubes would take place under protection. High powers & good adjustment are necessary—[2] Ears expel anther, when kept in water in room; but the paleæ *apparently* do not open & expose stigma;— but the stigma could easily be artificially impregnated. If I were you I w^d. keep memoranda of points worth attending to. Years ago I asked Hooker how stigma of grasses were impregnated & he seemed to know nothing.—[3]

That d——d. Verbascum will do nothing—[4]

The House is not very cheerful— Lenny is not well, & has glands swollen on neck, which is sometimes serious after Scarlet F.—[5] Horace is not very well today.—[6] Miss Pugh went today,[7] & good old Brodie came.[8] Miss Rendle came yesterday for dancing—[9]

We have got a new grey mare from M^r. Edwards on trial (he is the man for horses) & if she answers we shall sell that beast that fell down.—[10] She is a beautiful canterer. We got all our Hay in splendidly & we have a gigantic stack, worth £140. I have bought £10,000 Lancaster & Carlisle shares.—[11] Here is a dry catalogue of facts. I wish you might be able to come before very long.—

Good night | My dear old fellow | Your affect. Father | C. D.

Mamma says I am to wear a beard.— I am better.

DAR 210.6: 100

[1] Dated by the reference to CD's investment in the Lancaster and Carlisle Railway Company (see n. 11, below).

[2] William used a microscope in his botanical studies; see also letter to W. E. Darwin, 14 February [1862].

[3] See letter to J. D. Hooker, 13 July [1856] (*Correspondence* vol. 6). William's botanical sketch-book (DAR 186: 43) contains a note dated 10 July 1862, describing pollen-masses and pollen tubes of 'Grasses'.

[4] CD had been impressed by Karl Friedrich von Gärtner's experiments on *Verbascum*, in which crosses between differently coloured varieties of the same or of different species were found to produce

less seed than the parallel crosses between similarly coloured varieties, and he discussed the case in *Origin*, pp. 270–1. In September 1861, he told Joseph Dalton Hooker that he had decided to test Gärtner's experiments, stating: 'I do not think any experiment can be more important on Origin of species; for if he is correct, we certainly have what Huxley calls new physiological species arising' (*Correspondence* vol. 9, letter to J. D. Hooker, 28 September [1861]; see also Appendix VI). CD unsuccessfully sought the requisite specimens from a number of botanical acquaintances (see *Correspondence* vol. 9, letters to J. D. Hooker, 18 October [1861], 23 October [1861], 1 November [1861], and this volume, letter to C. C. Babington, 20 January [1862]), and on 28 June 1862 began crossing experiments with a specimen of *Verbascum* transplanted from a nearby field (see the experimental notes dated 28 June and 2 July 1862 in DAR 108: 2). He was, however, unable to obtain fertile seed-pods from these crosses, later concluding that the experimental plant was a sterile hybrid (see letters to J. D. Hooker, 6 October [1862] and 14 [October 1862], and letter to Asa Gray, 16 October [1862]). CD later discussed this experiment in 'Specific difference in *Primula*', pp. 451–4.

5 Leonard Darwin became ill with scarlet fever on 12 June 1862 (Emma Darwin's diary (DAR 242)).

6 Horace Darwin.

7 Miss Pugh had been governess to the Darwin children from January 1857 to January 1859 (see Emma Darwin's diary (DAR 242) and CD's Classed account book (Down House MS)). The Darwins' current governess, Camilla Ludwig, made an extended visit to her family in Hamburg between early June and early November 1862. She was apparently sent away on full pay in order to separate her from Horace Darwin. Stephen Paul Engleheart, the Down surgeon, was concerned that Horace's attachment to her might be exacerbating the illness from which Horace had been suffering since early in the year (see letters from Emma Darwin to W. E. Darwin, [2 March 1862], [27 May 1862], and [6 November 1862] (DAR 219.1: 49, 57, 64), letter to Camilla Ludwig, 26 August [1862], Emma Darwin's diary (DAR 242), and CD's Classed account book (Down House MS)). Miss Pugh was employed as a replacement governess for Elizabeth Darwin from 5 June to 4 July 1862; on 5 July she was paid £5 for 'education' (see CD's Classed account book (Down House MS), Emma Darwin's diary (DAR 242), and letter from Emma Darwin to W. E. Darwin, [27 May 1862] (DAR 219.1: 57)).

8 Brodie had been the nurse at Down House between 1842 and 1851 (Freeman 1978). She apparently remained with the family until the return of Camilla Ludwig in November (see letters from Emma Darwin to W. E. Darwin, [9 August 1862] and [6 November 1862] (DAR 219.1: 61, 64), and *Emma Darwin* 2: 178).

9 Miss Rendle has not been identified. There are entries in CD's Account book–cash account (Down House MS), dated 31 March and 6 December 1861, that state: 'Miss Rinaldi Dancing'.

10 George Edwards was a farmer living in Bromley, Kent (*Post Office directory of the six home counties* 1862); he is listed in CD's Address book (Down House MS) as a 'Horse-Breeder'. In CD's Account book–cash account (Down House MS), CD recorded payments of £42 to 'Edwards' on 31 May 1862 and £37 16*s.* on 14 July 1862.

11 The reference is to the Lancaster and Carlisle Railway Company; according to CD's Investment book (Down House MS), the stock was purchased on 3 July and 9 July 1862. CD's Account book–banking account (Down House MS) records the settlement of the account on 16 July 1862. See also letter to John St Barbe, [before 3 July 1862].

From Charles Giles Bridle Daubeny 5 July 1862

Botanic Garden. | *Oxford.*
July 5th. 1862

My dear Sir,

I took occasion last week in a lecture on Orchids to notice your interesting researches on the agency of Insects in distributing the Pollen of that class of plants, and as it was the first time that your views had been publickly noticed at Oxford, since the famous discussion in which the Bishop of Oxford & Huxley played so prominent a part,[1] took advantage of the opportunity to set the Academical public ⟨righ⟩t as to the general tendency of your former work, and its bearings on Theology.

As I have no intention of publishing my Lecture which indeed contained little or nothing which has not been much better explained by yourself and other original investigators of the subject of orchids, I have got the few concluding remarks transcribed, under the impression, that it might be satisfactory to you to see the light in which your enquiries were represented in one of the great centres of clerical influence.—[2]

The account you give of your own researches is so clear and compleat, that the only point upon which I am somewhat puzzled, is as to Insects visiting the same species of orchis from which they had abstracted the Pollen. Are they attracted to one Species more than to others? for if not, the fertilisation of the flower must be a matter of accident; and although the same remark may apply to other plants, which have their pollen distributed by the same agency, yet the rarity of certain orchises renders the difficulty greater in their case than in most others.—

Owing to my having been absent in Italy all the winter and part of the spring, it is long since I have met you at the Philosophical Club,[3] or elsewhere in London, but I hope I may augur favorably of your health from the activity with which you have pushed your researches

Believe me | ever | dear Sir | y^{rs} faithfully | C Daubeny

DAR 162.1: 115

CD ANNOTATIONS

2.4 it might . . . influence.— 2.6] *double scored brown crayon*
3.2 Insects . . . Pollen. 3.3] *double scored brown crayon*
3.6 yet . . . others.— 3.8] *double scored brown crayon*
4.2 it is . . . researches 4.4] *double scored brown crayon*

[1] Daubeny held the chairs of botany and of rural economy at the University of Oxford. He refers to the verbal exchange between the bishop of Oxford (Samuel Wilberforce), Thomas Henry Huxley, and others, on 30 June 1860 at the Oxford meeting of the British Association for the Advancement of Science. For accounts of this incident, see *Correspondence* vol. 8, letter from J. D. Hooker, 2 July 1860, and *ibid.*, Appendix VI. At the same meeting Daubeny read a paper on plant sexuality in regard to CD's views, in which he expressed 'dissent from the Darwinian hypothesis in the full extent to which the Author seems disposed to carry it' but denied 'that any higher considerations than those of pure science need enter into the field of discussion' (Daubeny 1860, p. 26); CD considered the paper '**very** liberal & candid' (see *Correspondence* vol. 8, letter to Charles Lyell, 11 August [1860]).

[2] The manuscript has not been located, but see Daubeny's comments on this subject in Daubeny 1867, 2: 192–8.

[3] Daubeny refers to the Philosophical Club of the Royal Society.

From William Alexander Wooler 5 July 1862

Halliwell | Heighington | Darlington
5th. July 62

Dear Sir

In reading your very interesting work on "Fertilization of orchids you will I am sure excuse me in offering an opinion contra to that which you express in reference to the oppening of flowers—[1] From some experiment I made some years ago my impression is that when the anthers arise from the petals that the amputation of part of the petals effects the development of the pollen and as we well know that fertilization is followed by decay of the flower I fancy there is considerable connection between the flower & the production of the pollen & seed I state this with defference as it needs much & careful attention &c before an opinion is stated; for we know how much we have all had to unlearn from mear fances having been stated so positively that they for a while have stoped further observations—being taken as authoritive—

If I am right then there would be an analogy between the vegetable & animal kingdom—

Nature taking so much & beautifll provision for the fertilization of the orchids yet seems to work vastly to no purpose when she produces so many 1000s of seeds but yet these form so extremely few plants—[2] This may indicate that like many other objects of nature they are in the slow progress of passing away— My notion of inter breeding is as Mr Jonas Webb with his sheep[3] & Booths &c &c, with the short horns[4] that within certain limits when there are certain strengths of constitution &c &c without any defect of healthy action that inter breeding combines the strength & other desired properties in the offspring but I quite think that it is a very small limit that this inter breeding can be carried without increasing some tendancy to deteriotion—[5]

My orange hose in hose polyanthus plant*s* planted together have produced flowering seedlings this year as good hose in hose as the parent plant but of a lighter yellow colour— also the pollens fertilizing common polyanthus have produced some few good hose-in-hose the seedling varying in colour from yellow to that of the seed bearing plant— also the orange hose in hose fertilized by a large course & dark polyanthus (for the purpose of marking its effect upon the progeny) has resulted in some seedlings without any appearance of hose in hose but some hose in hose one of which the calyx is as perfectly changed as the mother plant but the colour & form of the flower almost precisely like the plant used as the father.[6]

I have also some seedlings from a cowslip which I fertilized with pollen from the polyanthus & altho some of the seedlings come cowslip some have come coloured but only a little enlarged in size of flower—[7] I had not however planted them out of the seed place & therefore they not only flowered poorly but many did not do so at all

I met a man who has asses & mules to carry coals from Potto in Yorkshire up the Cleveland Hills & think he might have some knowledge of these dorsal marks with the bars across the legs of horses[8] I learnt from him that he knew a bay pony which often breeding a mule bred ponies & these latter had the line along the back and the bars across the legs— I have sent a few observations to the field on this subject[9] & the fact appears to me to present that the variation of species may be produced by this influence—left of the male of a different species upon the female— I have many young Ducks pure Aylesbury ducks & muscovy drakes but next year I intend breeding first from the aylesbury ducks & muscovy drake & then breeding from the same ducks with an aylsbury drake, to see if the other progeny of these ducks the products of eggs obtained after the intercourse with the muscovy vary from the pure aylesbury—and to what extent this influence may have upon the descent of these ducks

The recent work upon the horse by the Editor of the field[10] points out that the Americal trotting breed is a pure bred English racer only the taste & ⟨the⟩ adaptation of the Country made trotting races preferable to the galloping ones of this country & therefore selection to breed for trotting was followed & has produced so much a superior trotting breed[11] Norfolk was well known for a certain breed of trotters & this arose from that County being a great breeder of poultry for the London market which ⟨it⟩ was desirable to deliver fresh & thus the practice of sending it up in vans ⟨&c⟩ drawn by trotting horses begetting a want for trotters and they were thus produced

—The Editor of the field also in his work upon the dog gives product of the dogs bred between the Grey Hound & bull dog with the descendants[12]

I shall I hope give more minute particulars of the results of my polyanthus in the Journal of Horticulture shortly.[13] I am strongly inclined to concur with you in your paper upon the fertilization of the poly⟨anthus⟩[14] and did intend to have ⟨carried⟩ out some experiments this ye⟨ar⟩ for altho I am in wonderfully productive cowslip locality yet it is a little too high for my other objects & I am looking out for an estate to build upon for myself—& I am afraid I could not carefully pursue such experiments[15] In my cowslip seedlings some have come quite as cowslips altho the plant was grown in the garden & may have been impregnated from pollen from the fields— This plant is strongly endowed with an idiosyncrasy which may ⟨make⟩ it not such a ready seeder if shaded by net— I would prefer that a plant should be obtained with a large ball just coming into flower from the Lowlands taken to a warm muck in the hills & then made the object of the experiments as the cowslip grows better upon our hills but are much later in flowering & so would be free from any interference from other plants growing perhaps within a radius of miles & in early spring bees &c are not likely to come from the warm shelter of the low lying lands to the cold & windy hills—

I fear I tire you so beg to remain Yours truly | W Woo⟨ler⟩

C Darwin Esq | am FRS

DAR 181: 157

CD ANNOTATION

Verso of last page: 'Fert & seeding of Orchids | Cross fert | Petals | Pony | Aylesbury Ducks | Polyanthus [*above del* 'Polyanthus'] experiments | I have reported many I think I sent you a copy— Beaton did not [*illeg*] my paper'[16] *pencil*

[1] CD discussed the 'uses of the petals and sepals' in *Orchids*, pp. 339–43; the exact reference has not been identified.

[2] CD described the vast superabundance of seed in orchid species in *Orchids*, pp. 344–6.

[3] Jonas Webb of Babraham, Cambridgeshire, was a renowned breeder of pedigree livestock. The celebrated Babraham flock of short-woolled sheep had been maintained by selective interbreeding between five families within the flock (*Gardeners' Chronicle and Agricultural Gazette*, 6 July 1861, pp. 631–2).

[4] Thomas Booth and his sons John and Richard Booth of Killerby and later Warlaby, Yorkshire were the originators of the 'Booth' strain of shorthorn cattle, first bred in 1790 (*DNB*, *EB*).

[5] One of CD's objectives in *Orchids* had been to demonstrate that cross-fertilisation was the 'main object' of the contrivances by which orchids were pollinated (p. 1). He concluded the book by stating:

> Nature thus tells us, in the most emphatic manner, that she abhors perpetual self-fertilisation. . . . For may we not infer as probable, in accordance with the belief of the vast majority of the breeders of our domestic productions, that marriage between near relations is likewise in some way injurious,—that some unknown great good is derived from the union of individuals which have been kept distinct for many generations?

[6] CD's earlier correspondence with Wooler about polyanthuses has not been found; however, see 'Dimorphic condition in *Primula*', p. 79 (*Collected papers* 2: 47). CD cited Wooler's observations in *Variation* 1: 365:

> In the hose and hose primulæ, the calyx becomes brightly coloured and enlarged so as to resemble a corolla; and Mr. W. Wooler informs me that this peculiarity is transmitted; for he crossed a common polyanthus with one having a coloured calyx, and some of the seedlings inherited the coloured calyx during at least six generations.

[7] Having carried out a number of crosses between polyanthuses and common cowslips (*Primula veris*), CD had concluded that the former were a cultivated variety of the latter (see *Correspondence* vol. 9, letter to Daniel Oliver, 23 March [1861], this volume, letter to Alphonse de Candolle, 17 June [1862], n. 2, and 'Three forms of *Lythrum salicaria*', p. 187 (*Collected papers* 2: 121).

[8] CD was seeking information on colour variation and 'striping' in horses for inclusion in *Variation*. See also *Correspondence* vol. 9, letter to the *Field*, [before 27 April 1861], and letter to W. E. Darwin, 9 May [1861]. The subject is discussed in *Variation* 1: 55–63.

[9] The journal, the full name of which was *The Field, the Farm, the Garden. The Country Gentleman's Newspaper*, does not appear to have published Wooler's observations on striping in ponies.

[10] John Henry Walsh.

[11] Lupton and Walsh 1861, p. 31.

[12] Walsh 1859, pp. 177–86.

[13] Wooler had published a notice on polyanthus in the *Cottage Gardener and Country Gentleman* in 1860 (Wooler 1860); the journal was renamed the *Journal of Horticulture and Cottage Gardener* in 1861, but Wooler appears to have published nothing further on the subject.

[14] Wooler refers to CD's paper, 'Dimorphic condition in *Primula*'. Wooler's name is included on CD's presentation list for the work (see Appendix III).

[15] Wooler lived at Halliwell House, Heighington, County Durham; by May 1863 he had moved to Sadberge Hall, near Darlington (see *Correspondence* vol. 11, letter from W. A. Wooler, 5 May 1863; see also *North Eastern Daily Gazette*, 5 May 1891).

[16] Donald Beaton's name appears on CD's presentation list for 'Dimorphic condition in *Primula*'; no letter from Beaton on this subject has been found.

From George Maw 7 July 1862

Benthall Hall. | *Broseley.*
7th. July 62.

My dear Sir.

I am sincerely obliged for your kind letter of the 4th. inst & for the very liberal spirit in which you receive my criticisms[1]

It is exceedingly kind of you to think of sending me your new work on orchids[2] it reached me safely yesterday but I have not yet had time to do more than just look into it

I was about ordering it of my bookseller—but to receive it as a gift from the author is certainly a much more agreeable way of becoming possessed of it—

An Uncle of mine Mr. Jno Frdk Johnson[3] who resides at Araquipa was mentioning to me a few days ago the case of a female mule *breeding* by a horse.— he has also *heard* of the male mule gendering, but the case of the female he can vouch for as it came under his own observation— it occurred in the valley of Vitor about 12 leagues from Araquipa. I mention it because I think it is a fact bearing on the subject of hybridism which may interest you

My Uncle who will return to S' America in a few months has promised to make careful enquiries respecting the alleged cases of the fertile male mules.[4]

Believe me Dear Sir | Sincerely & obliged | George Maw.

Charles Darwin Esqr

DAR 171.1: 96

[1] Letter to George Maw, 3 July [1862].

[2] Maw's name is included on the presentation list for *Orchids* (see Appendix IV). See also letter to George Maw, 3 July [1862].

[3] John Frederick Johnson has not been further identified.

[4] No further correspondence either from Johnson or Maw regarding fertile male mules has been found. However, in 1865, CD transmitted to the *Natural History Review* a report from Egypt of a mule producing a foal ('Notice of a mule breeding', *Natural History Review* n.s. 5: 147–8).

To M. T. Masters 8 July [1862][1]

Down. | *Bromley.* | *Kent. S.E.*
July 8th

My dear Sir

You wrote me some little time since an extremely kind note,[2] which makes me believe you will excuse me now troubling you.—

I have been experimenting largely on fertility of central peloric flower of Pelargonium, but do not yet know result;[3] but I much wish to try some experiments on other peloric flowers. Have you a garden & have you by *chance* any peloric plants in it? If so I would ask you to try a few simple experiments for me.— Or can you tell me what seed I could sow another spring with a fair chance of getting peloric flowers? The wild yellow Linaria is, I believe, often peloric; but I do not know how I could get seed. Does any cultivated Linaria produce often peloric flowers.— Does the wild colombine produce a whole circle of nectaries? I suppose so.

Any suggestions would be valuable. My object is to see when a flower undergoes any great change of structure, whether its fertility with others of the same species & unaltered, remains the same.—

You will see it is with the forlorn hope of illustrating sterility of Hybrids.—[4]

My dear Sir | Yours very sincerely | C. Darwin

American Philosophical Society (Getz 3645)

[1] The year is established by the relationship to the letter from M. T. Masters, 12 July 1862.

[2] Letter from M. T. Masters, [*c.* 15 May 1862].

[3] On 11 May 1862, CD had begun a series of crossing experiments with different varieties of pelargoniums, in an attempt to fertilise the normally sterile peloric flowers (see the experimental notes in DAR 51 (ser. 2): 4–9, 12–13; see also letter to Daniel Oliver, 8 June [1862], and letter to Asa Gray, 1 July [1862]). The results of these experiments are given in *Variation* 2: 167. Masters was making a special study of plant morphology and teratology; in April 1860, he and CD had discussed CD's claim in *Origin*, p. 145, that 'in irregular flowers, those nearest to the axis are oftenest subject to peloria, and become regular' (see *Correspondence* vol. 8, letter to M. T. Masters, 13 April [1860], and *Correspondence* vol. 9, letter to *Journal of Horticulture*, [before 18 June 1861]).

[4] In *Variation* 2: 166–7, CD discussed the sterility of peloric flowers in pelargoniums and other plants in a section entitled 'Monstrosities as a cause of sterility'. He noted that, while '[g]reat deviations of structure . . . sometimes cause plants to become sterile . . . in other cases plants may become monstrous to an extreme degree and yet retain their full fertility', and concluded: 'no general rule can be laid down; but any great deviation from the normal structure, even when the reproductive organs themselves are not seriously affected, certainly often leads to sexual impotence'. In the following chapter, CD sought to relate his findings regarding the factors affecting fertility and sterility to the case of hybrids (*ibid.*, pp. 178–91). For CD's interest in hybrid sterility, see Appendix VI.

From G. C. Oxenden 8 July [1862][1]

Dear Sir

This post brought me many letters—& when an hour afterwards I sought for *your's*, that I might reply to it, it was missing, & I have not yet found it—[2]

Thus, I forget the *address* of Mr Wollaston, & *cannot* write to you *there*—[3]

for this reason also, I have this day sent you a few spikes of "Epipactus palustris" addressed to *Bromley as usual*— I hope they will *reach* you 'ere entirely faded—

Lastly, I sent, in a separate parcel, a little Book which I have just published—which I trust may make you *laugh*—[4]

Sincerely | G. C. Oxenden

Broome July 8—

DAR 173: 55

[1] The year is established by the reference to CD's visiting George Buchanan Wollaston (see n. 3, below), and by the reference to Oxenden 1862 (see n. 4, below).
[2] CD's letter has not been found, but see the following letter.
[3] George Buchanan Wollaston, a fern collector of Chislehurst, Kent, had invited CD to visit him in order to examine a specimen of the orchid *Spiranthes gemmipara* (see letter from Frederick Currey, 3 July 1862). No such visit is recorded in Emma Darwin's diary (DAR 242), and CD did not refer to this species in the second edition of *Orchids*.
[4] Oxenden 1862. *Railway Horace* was a parody of Horace, the Roman satirist; the copy sent to CD has not been located in the Darwin Library.

From G. C. Oxenden 8 July 1862

Broome
July 8. 1862

Dear Sir

I do assure you, you expressed a desire to know if Insects visited "*Epipactis palustris*" *at Night*—[1] We have not seen *even One instance* of any such Visits—but I feel *sure* that they **are** thus visited—& that their *absence* on the nights in question was due to *the very* **disturbed** state of the *atmosphere*—

—As regards the access of Insects in the **day time**— I have spent *two whole days* in a Marsh Containing these plants in *flower*, without detecting so much as *one insect upon them*—But then, the days were each *damp* & *lowering*—

—In this splendid Marsh, I yesterday found *four Mowers* hard at work— —I tried to save the *flower crop* by a *very large Money Offer*—but the Farmer assured me he really **needed** the rough rush & reed for *thatching* purposes—

If I live 'till next year, I will spare no pains to secure this Marsh—

As it is, all that I could do was to cause them to leave (*un*touched) certain *patches*, here & there, where these lovely Plants most abounded—

—You say you suppose that I do not ⟨ ⟩ "*Fens*"— I do believe I have *h*⟨ ⟩ *explored* **more** of the *principal* ⟨ ⟩ *Europe*, from the *Arctic Circle* ⟨ ⟩ dangerous Morasses of *Southern* ⟨ ⟩ than *any Man living*—

—Botanists seem determi⟨ned⟩ that "*Arachnites*" shall not exist ⟨ ⟩ independent *Country Gentleman*

—**You** accuse him of **dup⟨licity⟩** and of being one year a true Be⟨e⟩ in the next, a *Spider*—[2]

—And Mr Woollaston asserts the ⟨ ⟩ Existence of Intrigue between "Ar⟨achnites⟩ and Aranifera"—[3]

⟨*four lines excised*⟩

—But, even if there were sufficient accordance in the flowering-times, to render the thing *possible*, is it *probable* that the Union of two *very delicate Orchids* (of which One is of *very low* small habit, & the other small as to flower, & moderate as to stem) sh^{d}., by their Union, produce a flower of the *size and strength* and *ferocity* of "*Arachnites*"—& which stands aloof from its Congeners in the proudest Isolation?

—In like manner, if you meet a "*Birdologist*" tomorrow, he will tell you, that "*Scolopax Major*" (the Solitary Snipe) is *not* a species but a *Cross*—

—In all these matters, the *dissecting knife* has *large powers of deciding the dispute*—but a much larger power resides in the Man who is incessantly face to face with these objects in their living & natural habitats—

—The fight about the Origin of *Tetrao medius*" is interminable—[4]

Sincerely | G. C. Oxenden

Incomplete
DAR 173: 56

[1] CD's letter has not been found. In 1861, CD had sent Oxenden a 'memorandum' containing queries relating to *Epipactis palustris*, but although this included a question about the kind of insects that visited the flowers, it did not make reference to nocturnal visits by insects (see *Correspondence* vol. 9, letter to A. G. More, 4 June 1861).

[2] Having previously disputed the claim of some botanists that the late spider orchid, *Ophrys arachnites*, was a variety of the bee orchid, *O. apifera*, CD had recently speculated that these two forms might be crossing and self-fertile forms of the same species (see letters from G. C. Oxenden, [before 30 May 1862] and 21 June 1862, and letter to A. G. More, 7 June 1862). CD's letters to Oxenden on this subject have not been found, but in the letter to J. T. Moggridge, 13 October [1865] (*Calendar* no. 4914), CD reported that some years previously he had written 'to an acquaintance asking him to mark some Spider orchises and observe whether they retained the same character'. CD recalled: 'he evidently thought the request as foolish as if I had asked him to mark one of his cows with a ribbon to see if it would turn next Spring into a horse.'

[3] Wollaston 1855. 'Woollaston' is a misspelling; the reference is to George Buchanan Wollaston.

[4] *Tetrao medius* is the name that was given in 1811 by the German naturalist, Bernhard Meyer, to a form of grouse subsequently widely regarded as a naturally occurring hybrid between *Tetrao urogallus* and *Lyrurus tetrix* (see *Index animalium*, and Elliot 1865).

To W. E. Darwin 9 July [1862][1]

Down
July 9th.

My dear William

Lenny is very ill, but Mr Engleheart does not think there is danger:[2] his kidneys hardly act & his urine is tinged with blood— His liver is much disordered & he vomits. Poor dear little man, he is so patient.—

I have not much heart for Botany. I forget whether I said that pollen-tubes of wheat are same diameter as the cells of stigma—[3] it is very improbable that they run down it, but if they do, they could be distinguished, I think, from the cells of stigma by not having transverse septa— It is much the most probable that they penetrate to hairs near where they adhere.— But it is very odd that tubes shd. come out on opposite side to point of adhesion— A rather oldish flower of wheat with anthers gone would be best; put in water & under thin glass.—

The Valerian (what species?) does seem case for me; it is, as in Primula, that some plants should have, all flowers with longer pistil than others.[4] In Primula there is no gradation. Are stamens shorter in those flowers with long pistils? It would, I think, be better to look at a good many flowers on same two plants, if *all* on one have long pistil & all on other short pistil, I sh^{d}. very much like to see specimens. This would perhaps be better than marking plant & comparing at different ages—certainly in several plants pistil does elongate with age.—

Vaucher asserts that in all parts of Europe, plants of Lythrum salicaria occur of 3 forms;[5] some with long, some with mid- & some with short pistil, & that all flowers on same plant are alike.— I have so many young plants growing up, that I hope I shall get all 3 forms next year.—[6]

Mamma, thank God, keeps pretty well with all our anxiety. All the rest are pretty well

Good Bye | My dear old fellow | C. Darwin

DAR 185: 11

[1] The year is established by reference to Leonard Darwin's illness (see n. 2, below).

[2] Stephen Paul Engleheart was a surgeon in Down (*Post Office directory of the six home counties* 1862); Leonard Darwin had been suffering from scarlet fever since 12 June 1862 (Emma Darwin's diary (DAR 242)).

[3] In the letter to W. E. Darwin, 4 [July 1862], CD had encouraged William to make observations on the pollination of wheat.

[4] William's letter has not been found. CD had learned from Asa Gray that some species of *Valeriana* were dimorphic (see *Correspondence* vol. 9, letter from Asa Gray, 31 December 1861). For William's observations on *Valeriana*, see the letter from W. E. Darwin, 14 July 1862.

[5] Vaucher 1841, 2: 371. There is a heavily annotated copy of this work in the Darwin Library–CUL (see *Marginalia* 1: 812–15).

[6] CD had noted the occurrence of trimorphism in *Lythrum salicaria* and *L. thymifolia* on reading Lecoq 1854–8 in December 1861, and had asked Joseph Dalton Hooker to provide him with seeds or plants of the genus for experiment (see *Correspondence* vol. 9, letter to J. D. Hooker, 18 [December 1861]). Hooker did so in March 1862 (see letter from J. D. Hooker, [19 January 1862], and letter to J. D. Hooker, 22 [March 1862]).

From J. D. Hooker 10 July 1862

Samaden Enghedien Valley
July 10th. 62.

Dr Darwin

We arrived here last night after a fairly prosperous Journey & having got into a quiet clean little German Hotel we propose remaining en pension for a week & exploring the neighbourhood.[1] My wife has been pretty well on the whole. At Dover she did not sleep & she had a bad palpitation &c on ascending the Louvre Stairs.— At Paris we staid a night (good) & took the following *night*-express to Basle, where we walked about & proceeded the same afternoon to Zurich by rail without fatigue— There we spent $1\frac{1}{2}$ days quietly enough with O. Heer,[2] an old friend of mine, whose name you know & he recommended us to come here for the rest of our Swiss stay, because of the purity of the air & beauty & interest of the scenery— We accordingly took Rail to Chur (Coir) a most magnificent route, by the Wallen See, & then up the Rhine— At Chur she had no sleep & unfortunately we had to leave by Diligence at 5 next morning for this place, a 15 hours drive.— The first part of the day was intensely hot, the latter, crossing a pass, at 7000 f^{t},—was as cold & a violent fit of Neuralgia was the consequence— We arrived at 8 last night & this morning she is all right again—& will I hope continue so.

The views yesterday were as grand as any thing I have any where seen & except for the little piece called *via Mala* on the Splugen, this route beats that hollow,—it is only a few miles East of the Splugen route, of which I caught a glimpse— this, the valley of the Inn, appears to me to combine the beauty of the Tyrol with the savage grandeur of Switzerland in a remarkable degree— In Science I have seen little but Heer's fossils, he showed me a leaf apparently *Dicotyledonous* from the **Lower** Lias in Jura—which please tell Lyell of.[3] He has a wonderful collection of fossil insects & crustacea from the same, beside which the fossil plants are as nothing, in point of absolute value of characters for systematic determination— I am as always impressed with the identity of physical features & wonderful analogy of *biological*, between Alps & Himalaya, the former we can suppose we understand, because physical causes are the same every where & the sequence of these is probably the same in Alps & India— The representation of allied species too we can now (thanks to you) account for largely, but the repetition of forms in plants & animals in no way allied is always a puzzle—especially when accompanied by startling contrasts between allied forms. These latter can best no doubt be accounted for by the *indirect action* of physical causes, i.e. Nat. selection & I think there are already many reliable facts to be quoted in illustration of this & that after the *course of alternatives* you have administered, I could write a suggestive chapter, comparing the vegetation of Alps Andes & Himalaya, my (never to be begun) book on *Plants*.[4]

I cannot yet give up my dream of meeting you in Switzerland one day;—if you ever did come here, & I could see you for 5 minutes a day, I should be the happiest man alive. These rocks plants & insects teem with thoughts of you & reminiscences of your writing.

Your Orchid book which I have not read through has suggested to me that Insects &c may have had a wonderful deal more to do with *checking* migration than climate or geographics, & that the absence of whole genera may thus one day be accounted for by absence of genera of Insects: in short that the Cat; & Clover story is capable of immediate expansion by any one having sufficient knowledge of Plants Insects & Geography—[5] I shall be home on 23^{d}. so do not write to me. I have seen & heard 0 of Lubbock, Huxley & Tyndall—[6]

With united regards to you & yours | Ever yr affec | J. D. Hooker.

DAR 101: 46–7

1 The Hookers had travelled to Switzerland in the hope that Frances Harriet Hooker might recover her health (see letters from J. D. Hooker, 19 [June 1862], 28 June 1862, and 2 July 1862).

2 Hooker refers to the Swiss botanist, Oswald Heer.

3 No angiosperms had ever been found in rocks older than the Cretaceous system (see *Correspondence* vol. 6, letter from Charles Lyell, [16 January 1857]); the Lower Lias is the oldest formation of the

Jurassic system. As a long-standing opponent of what he called the 'Theory of progressive development of organic life at successive geological periods' (C. Lyell 1853, p. 130), Lyell had rejected the suggestion that the dicotyledonous angiosperms, widely considered to be the most advanced plants, were necessarily absent from the older rocks. While noting the 'remarkable' fact that no such fossils had yet been found in the Carboniferous system, he emphasised the great paucity of plant fossils generally from that era. Likewise with the older Secondary rocks he argued that the data were 'too scanty as yet to affirm whether the vegetation of this second epoch was or was not on the whole of a simpler organization than that of our own times', and he pointed to 'the important fact of the co-existence of a large number of angiosperms with cycadeæ' in the Lower Cretaceous formation (pp. 132–3). However, in noting the absence 'as yet' of all signs of dicotyledonous angiosperms from the Jurassic rocks in the fifth edition of *A manual of elementary geology* (Lyell 1855, p. 329), Lyell conceded that:

> The leaves of such plants are frequent in tertiary strata, and occur in Cretaceous, though less plentifully ... The angiosperms seem, therefore, to have been at the least comparatively rare in these older secondary periods, when more space was occupied by the Cycads and Conifers.

On Lyell's non-progressivist views, see Bartholomew 1973.

4 In 1860, CD and others, including the publisher John Murray, encouraged Hooker to write what he called a 'Darwinian book on Botany' (see *Correspondence* vol. 8, letters to J. D. Hooker, 8 February [1860] and 18 [March 1860], and L. Huxley ed. 1918, 1: 535).

5 In *Origin*, CD discussed the 'web of complex relations' that binds plants and animals together; he continued (pp. 73–4): 'the presence of a feline animal in large numbers in a district might determine, through the intervention first of mice and then of bees, the frequency of certain flowers in that district!'

6 Thomas Henry Huxley and John Tyndall had been on holiday in Switzerland since early July, exploring the glacier at Grindelwald (L. Huxley ed. 1900, 1: 234, and Eve and Creasy 1945, p. 92). John Lubbock met up with them on 13 July, on their way to the Rhône glacier, and stayed with them for a week before travelling to examine the remains of prehistoric lake-dwellings, the recent discovery and examination of which he had described in a paper for the January number of the *Natural History Review* (Lubbock 1862b; see Hutchinson 1914, 1: 56, John Lubbock's diary (British Museum, Add. Ms. 62679: 64 r.), and letter from John Lubbock, 23 August 1862).

To H. G. Bronn[1] 11 July 1862

Down Bromley Kent
July 11th.

Dear & Honoured Sir

I am very sorry to be troublesome, but I hope you will insert following note to "Mormodes"* p. 265. (7 lines from bottom) instead of the short one before sent.[2]

*I have now examined perfect flowers of this orchid. It is the Cynoches ventricosum.[3] I have erred to a certain extent in my conjectures on the action of the parts. The sensitive point lies in some part of the filament of the anther, between two little leaf-like appendages on the summit of the column. The movement of the pollinium is nearly the same as in Mormodes ignea, & the anther is torn off. But the power of ejection is more feeble; & the viscid surface of the disc after the movement projects at right angles to the anther. There can be no doubt that insects either alight on, or touch, the anther or end of the column, which hangs downwards, & then the disc is flirted out, & sticks probably to their heads; but the whole pollinium is not shot to a distance as in Catasetum. In about quarter of an hour the pedicel of the pollinium slowly straightens itself, as in the case of Mormodes ignea.

July 1862

p. 324 (3 lines from bottom) Orchids*

*I now find that in several, perhaps in most of the Arethuseæ,—a tribe which as stated I had until lately no opportunity of examining,—the pollen-grains are simple, that is are not compounded of three or four granules.[4]

July 1862

Dear sir | Yours truly obliged | Ch. Darwin

Houghton Library for Rare Books and Manuscripts, Harvard University (Lowell Autograph)

[1] CD had not yet learned that on 5 July 1862 Bronn had died suddenly of a heart attack (see letter from E. Schweizerbart'sche Verlagsbuchhandlung, 11 July 1862).

[2] Immediately prior to his death, Bronn had been working on the German translation of *Orchids* (Bronn trans. 1862), and had sent CD a list of queries (see letter from H. G. Bronn, 21 June 1862). In his reply (letter to H. G. Bronn, 30 June [1862]), CD sent a number of corrections and additions to be incorporated into the translation. Despite Bronn's death, the two notes given here were included in Bronn trans. 1862, pp. 163 n. and 198 n.

[3] CD misspelled 'Cycnoches'. In *Orchids*, pp. 265–9, CD had described an unnamed species of *Mormodes*, sent to him by James Veitch Jr. Veitch subsequently sent CD further flowers from the same plant, which CD described in notes dated 9 July 1862 (DAR 70: 103–4); these notes were headed 'Cycnoches Ventricosum | named by Lindley'. See also 'Fertilization of orchids', p. 155 (*Collected papers* 2: 151–2).

[4] Before publishing *Orchids*, CD had been anxious to examine a specimen of the 'great Division of Arethuseæ' as described in Lindley 1853 (see *Correspondence* vol. 9, letter to J. D. Hooker, [28 July – 10 August 1861]), but he reported in *Orchids*, p. 324, that he had not been able to do so. Since publishing *Orchids*, he had received specimens and descriptions of several of the genera ascribed to this tribe in Lindley 1853. CD's notes, dated 30 May 1862, on a specimen of *Vanilla* sent to him by Joseph Dalton Hooker, describe the pollen as consisting of '*single* grains' (DAR 70: 94–5). He also received a dried specimen of *Arethusa* from Asa Gray, the structure of which he considered 'very like Vanilla' (letter to Asa Gray, 10–20 June [1862]), and a description from Gray of *Pogonia ophioglossoides*, CD's notes on which, dated 23 June [1862], describe the pollen as consisting of a '**single** grain' (DAR 70: 80).

To Armand de Quatrefages 11 July [1862][1]

Down. | Bromley. | Kent. S.E.
July 11th

Dear Sir

I thank you cordially for so kindly & promptly answering my questions.[2] I will quote some of your remarks.— The case seems to me of some importance, with reference to my heretical notions, for it shows how larvæ might be modified.[3] I shall not publish, I daresay for a year, for much time is expended on experiments;[4] if within this time you should acquire any fresh information on the similarity of the moths of distinct races, & would allow me to quote any facts on your authority I should feel very grateful.—[5]

I thank you for your great kindness with respect to the Translation of the Origin;[6] it is very liberal in you, as we differ to a considerable degree.— I have been atrociously abused by my religious countrymen; but as I live an independent life in the country, it does not in the least hurt me in any way,—except indeed when the abuse comes from an old friend, like Prof. Owen, who a/bu/ses me & then advances the doctrine that all Birds are probably descended from one parent.—[7]

I wish the Translator had known more of Natural History; she must be a clever, but singular Lady; but I never heard of her, till she proposed to translate my Book.—[8]

Dear Sir | Yours sincerely obliged | Ch. Darwin

Wellcome Institute for the History of Medicine Library

[1] The year is established by the reference to the French translation of *Origin* (see n. 6, below). See also letter from Armand de Quatrefages, [after 11 July 1862].

[2] CD's letter to Quatrefages and the reply referred to have not been found.

[3] The reference may be to Quatrefages's views on larval development, later published in Quatrefages 1862, p. 129. In *Variation* 2: 269, CD wrote:

> Insects sometimes have their antennæ or legs in a monstrous condition, and yet the larvæ from which they are metamorphosed do not possess either antennæ or legs; and in these cases, as Quatrefages believes, we are enabled to see the precise period at which the normal progress of development has been troubled.

There is an annotated presentation copy of Quatrefages 1862 in the Darwin Library–CUL (see also letter from Armand de Quatrefages, [after 11 July 1862]).

[4] *Variation* was not published until 1868.

[5] Quatrefages was a leading authority on silk moths, and CD frequently cited him in the account of the moths in *Variation* 1: 300–4; according to his 'Journal' (Appendix II), CD finished writing that section of *Variation* in the summer of 1862. In *Variation* 1: 303, CD referred to a statement he had received from Quatrefages confirming that there were no constant differences in moths 'of the most distinct races'.

[6] CD refers to the French translation of *Origin* (Royer trans. 1862), which was published in June 1862 (see letter to Asa Gray, 10–20 June [1862] and n. 25). (CD had unsuccessfully sought Quatrefages's help in finding a French translator for *Origin* in December 1859 (see *Correspondence* vol. 7, letter to Armand de Quatrefages, 5 December [1859] and n. 5).)

[7] The reference is probably to Richard Owen's lectures at the Museum of Practical Geology on the 'Characters, Organisation, Geographical Distribution, and Geological Relations of Birds'. The series of six lectures ran from 14 to 30 May 1862 (*Athenæum*, 10 May 1862, p. 613). The accounts of these lectures given in the *Parthenon* 1 (1862): 82, 115, 148, 179, 211–12, and the *Medical Times and Gazette* (1862), pt 1: 513, 537–8, 563–4, 590–1, 616–7, 645–6, do not report the claim that CD mentions; nor does the claim appear in the manuscript text which survives for some of the lectures (Natural History Museum, London, OC38.3/286–331). Owen also gave a series of four lectures on birds between 7 and 28 April 1862 at the London Institution, but no reports of these lectures have been found (see the syllabus in the Natural History Museum, London, OC38.3/285). CD apparently learned of the content of the lectures from one of his acquaintances (see letter to Charles Lyell, 22 August [1862]), possibly Miles Joseph Berkeley, who, in his review of *Orchids* ([Berkeley] 1862, p. 553) stated of Owen:

> one of the most ardent opponents of Mr. Darwin's views at the great Oxford meeting two years ago, was so far fallen in with them as apparently to allow the possibility of the whole ornithological world being derived from a single type

See also letter to Asa Gray, 23[–4] July [1862].

[8] Clémence Auguste Royer's letter to CD proposing to translate *Origin* into French has not been found, but see *Correspondence* vol. 9, letter to John Murray, 10 September [1861]. In her preface to

the translation, Royer expressed regret at her lack of detailed scientific knowledge but added that she had sought to convey the intent of the author (Royer trans. 1862, p. xxxv). See also letter from Edouard Claparède, 6 September 1862.

From E. Schweizerbart'sche Verlagsbuchhandlung[1] 11 July 1862

Stuttgart
den 11 Juli 1862.

Verehrtester Herr!

Ich weiss nicht ob Sie schon die Kunde erhielten, dass die deutsche Wissenschaft einen grossen Verlust zu beklagen hat, der auch Sie und mich besonders in hohem Grade trifft,—unser verehrter Freund Hofrath Bronn ist vor einigen Tagen plötzlich am Herzschlag gestorben, während er wie gewöhnlich im Hotel Schrieder seinen Nachmittags Kaffé trank!—[2]

Diese betrübte Nachricht wurde mir von Hause abwesend gemeldet; ich eilte sogleich nach Heidelberg um der Familie persönlich meine Theilnahme zu bezeigen.[3] Herr Bronn hinterlässt eine Wittwe, eine Tochter und 2 Söhne, welche eben von Marseille ankamen wo sie in Handlungshäusern angestellt sind.

Bronn hatte eine immense Arbeits-Kraft, daher war nicht nur die neue Auflage der Uebersetzung vom "Origin" für den Druck vorbereitet,[4] sondern auch das M.S. der Uebersetzung des Buches über die Orchideen war in der Kürzesten Zeit fertig, so dass wir nur auf die Clichées warteten um Satz und Druck zu beginnen. Ich wäre Ihnen recht verbunden wenn Sie Herrn Murray zur schleunigsten Absendung der Clichées veranlassen wollten.[5]

Was die Beobachtungen des Herrn Treviranus betrifft, so liesse sich vielleicht die Bekanntmachung derselben gleichzeitig damit bewerkstelligen, wenn er überhaupt deren Publikation beabsichtigt und sein M.S. fertig ist.[6] Wissen Sie etwas von ihm?

Herr Bronn sandte neulich noch das M.S. der Zusätze welche Sie ihm lieferten und welches Sie für die amerik. Ausgabe wieder zu erhalten wünschten;[7] ich lege solches diesem Briefe bei.

Ich werde nicht verfehlen Ihnen sobald die Uebersetzungen gedruckt sind, Ex. davon zu schicken;[8] leider kann nun die Correctur nicht mehr von Br. gelesen werden, doch werde ich einen tüchtigen Corrector dafür anstellen.

Wenn Sie das unter der Presse sich befindende neue Werk von Ihnen mir später auch zusenden wollten, wäre ich Ihnen sehr verpflichtet.[9]

Mit grösster Hochachtung und Verehrung | Ihr ganz ergebenster
E. Schweizerbart[10]

DAR 177: 69

[1] For a translation of this letter, see Appendix I. The letter was written by Christian Friedrich Schweizerbart, who was head of the Stuttgart publishing firm E. Schweizerbart'sche Verlagsbuchhandlung (see n. 10, below).

[2] Heinrich Georg Bronn died on 5 July 1862, aged 62 (*Quarterly Journal of the Geological Society of London* 19 (1863): xxxiii).

[3] Bronn was professor of zoology at the University of Heidelberg.

[4] The second German edition of *Origin* (Bronn trans. 1863) was initially issued in three parts, whose publication was announced on 6 October, 17 November, and 19 December 1862, respectively (*Börsenblatt für den Deutschen Buchhandel* 29 (1862): 2083, 2447, and 2735).

[5] Bronn trans. 1862. CD arranged for Schweizerbart to buy electrotype plates of the illustrations for *Orchids* from John Murray, CD's British publisher (see letter from E. Schweizerbart'sche Verlagsbuchhandlung, 7 June 1862, and letter to John Murray, 13 June [1862]). The German translation of *Orchids* was published on 20 October 1862 (*Börsenblatt für den Deutschen Buchhandel* 29 (1862): 2195).

[6] CD had apparently written to ask the German botanist Ludolph Christian Treviranus whether he would be prepared to translate *Orchids* into German, only to discover that Bronn had already almost completed a translation (see letter from H. G. Bronn, 21 June 1862 and n. 2). The observations referred to by Schweizerbart have not been identified; Treviranus included a detailed critique of *Orchids* in his review of CD's views on dichogamy, published in the *Botanische Zeitung* in January 1863 (Treviranus 1863a), and in his 'Supplementary observations on the fertilisation of some orchids' published in August 1863 (Treviranus 1863b). Treviranus sent CD a copy of Treviranus 1863a with his letter of 12 February 1863 (*Correspondence* vol. 11); there are lightly annotated copies of Treviranus 1863a and 1863b in the Darwin Pamphlet Collection–CUL.

[7] See letter to H. G. Bronn, 30 June [1862].

[8] See letter to H. G. Bronn, 30 June [1862].

[9] The reference is to *Variation*, on which CD had been working intermittently since January 1860 (see *Correspondence* vol. 8, 'Journal' (Appendix II)); the work was not published until 1868.

[10] Although C. F. Schweizerbart purchased the publishing firm of E. Schweizerbart from his uncle Wilhelm Emanuel Schweizerbart in 1841, he continued to use the signature 'E. Schweizerbart' in business communications (*Jubiläums-Katalog*, pp. x–xi).

From Armand de Quatrefages[1] [after 11 July 1862][2]

Mon cher confrè⟨re⟩

Il est trop tard cett⟨e⟩ ⟨année⟩ pour revoir les nouvelle⟨s⟩ informations relativement ⟨aux⟩ diverses races de vers à ⟨soie⟩ Mais l'an prochain v⟨ers⟩ le mois d'Avril les ecl⟨ores⟩ recommenceront et alor⟨s⟩ je tacherai de répondre ⟨a⟩ vos desirs—[3] Seulemen⟨t⟩ ⟨ ⟩ *avoir la bonté de ⟨me⟩* **rappeler ma prome⟨sse.⟩** ⟨Je⟩ me méfie de ma ⟨mémoire⟩ et serais désolé que ⟨vous⟩ prissiez pour négligenc⟨e⟩ ⟨ou⟩ mauvais vouloir ce ⟨qu'est⟩ le résultat d'un défaut d'organisation. Voilà même pourquoi je vous écris sur le champ. Mes distractions pouvant me faire oublier de vous faire une recommendation que je sais malheureusement etre nécessaire.*[4]

Vous dites vrai; nous diffèrons beaucoup. Mais peut etre moins que vous en pensez. Je crois que depuis la periode géologique actuelle les espèces sont restées a qu'elles sont. Je trouve que nous manquons de données suffisantes pour juger de ce qui existait avant. Je m'abstiens donc plus que je ne combats. Mais avant tout—je puis le dire—je cherche la verité et la science. Tout ce qui tend a etendre ce double domaine a mes sympathies profondes. Dut-il y avoir un peu de poesie dans les efforts tentés je ne m'en allarme pas. En somme tout ⟨en⟩ préferant Cuvier je mets Geoffroy tres haut et Buffon est à mes yeux un fort grand homme.[5] J'applique à mes contemporains la même nature de jugement— Voila tout.

Je vous adresserai prochainement un petit volume sur les Metamorphoses.[6] La conclusion exprime à peu près les idées que j'indiquais tout à l heure. Elle sera donc en désaccord avec les votres. Mais soyez bien certain qu'un désaccord de doctrines ne suivra jamais chez moi a l estime méritée par tout effort fait pour agrandir le domaine de l'esprit humain. Quant à mon indépendance personelle je crois que l'introduction de mon Unité de l'espèce humaine en témoigne. On m'a même dit que c'etait la ce qui avait empêché la traduction en Anglais.[7]

Adieu mon cher confrere. Si je puis vous etre utile disposez de moi

De Quatrefages

Incomplete[8]
DAR 175: 8

[1] For a translation of this letter, see Appendix I.

[2] Dated by the relationship to the letter to Armand de Quatrefages, 11 July [1862], and by the references to Quatrefages 1861 and 1862 (see nn. 6 and 7, below). See also *Correspondence* vol. 11, letter from Armand de Quatrefages, 29 March 1863.

[3] See letter to Armand de Quatrefages, 11 July [1862].

[4] The original letter included a symbol like an asterisk at this point. It was apparently intended to refer CD to a marginal reference; the broad left margin of this page of the letter has been excised.

[5] Quatrefages refers to the prolonged controversy earlier in the century between Etienne Geoffroy Saint-Hilaire and Georges Cuvier, both professors at the Muséum d'Histoire Naturelle in Paris, that culminated in 1830 with a famous debate at the Académie des Sciences. Part of the substance of the controversy related to the question of species change, Geoffroy Saint-Hilaire arguing for environmentally driven transmutation, while Cuvier argued for the permanence of species. Geoffroy Saint-Hilaire drew heavily on the philosophical approach of Georges Louis Leclerc, comte de Buffon, who, in his later writings, classified mammals and birds into natural families, each of which he believed had descended by 'degeneration' from a single parent stock, modifications being the result of adaptations to changes in the environment (Appel 1987).

[6] Quatrefages 1862. There is an annotated copy of this publication in the Darwin Library–CUL (see *Marginalia* 1: 691).

[7] Quatrefages 1861, pp. 1–10. The introduction to the volume details the religious and political basis of arguments for and against the doctrine that all human races belong to one species. Quatrefages proposed that the question was settled by the scientific evidence of complete interfertility between the different races. There is an annotated copy of this work in the Darwin Library–CUL (see *Marginalia* 1: 693–4).

[8] See n. 4, above.

From M. T. Masters 12 July 1862

Rye Lane | Peckham
July 12. 1862

My dear Sir

I shall be happy to give you a few memoranda I have made with reference to peloria in a day or two—[1]and thinking my Father might be able to communicate some information respecting comparative fertility of Peloria Gloxinias & other cultivated pelorias—I have written to him and will let you know the result—[2] at present I have no peloria in my little garden— I write this now lest you should think I had overlooked your letter but for the reason I have mentioned and also because I hope in a day or two to have more leisure than I have at this moment I know you will excuse the delay—

Believe me Dear Sir | Yours faithfully | Maxwell T. Masters

DAR 171.1: 68

CD ANNOTATION
Verso of last page: '36
1—16
———
37c 16' *ink*[3]

[1] CD was carrying out crossing experiments on the normally sterile flowers of pelargoniums, in order to explore the relationship between changes in plant structure and the incidence of sterility; he had written to Masters to ask him for assistance with experiments and for information on other peloric flowers (see letter to M. T. Masters, 8 July [1862]). No memoranda from Masters on this subject have been found, but see the letter to M. T. Masters, 24 July [1862].

[2] At the end of the letter to M. T. Masters, 8 July [1862], Masters had written 'fertility of Gloxinias *peloriated* with others'. His father, William Masters, was an expert hybridiser and owned a nursery in Canterbury, Kent (R. Desmond 1994). See also letter to M. T. Masters, 24 July [1862] and n. 6.

[3] The annotation apparently records the payment, on 14 July 1862, of £37 16*s*. to a local farmer, George Edwards, for the purchase of a horse (CD's Account book–cash account (Down House MS)); see also letter to W. E. Darwin, 4 [July 1862] and n. 10.

From Down Friendly Society to Bromley Savings Bank [before 14 July 1862][1]

We being the Trustees of the Down F. Soc beg leave to give notice that we wish to withdraw the sum of . . from the Brmly Savings Bank on . . July 14th. & the above amt to be paid to Mr C R Darwin the Treasurer of the Society.[2]

——— }
——— } ———

witness the signature of C. Darwin
——— of G. F. Duck[3]

156: 13: 10[4]

Draft
DAR 96: 3 v.

[1] Dated by the request to withdraw money from the Bromley Savings Bank on 14 July, and by an entry in CD's Account book–banking account (Down House MS), dated 16 July 1862, recording receipt of funds from the Bromley Savings Bank for the Down Friendly Club (see n. 4, below).

[2] CD had been treasurer of the Down Friendly Club since 1852 (Freeman 1978).

[3] George Duck was a carpenter and a grocer, and landlord of the 'George' public house in Down (Freeman 1978, *Post Office directory of the six home counties* 1862).

[4] The entry in CD's Account book–banking account (Down House MS) gives the sum as £153, stating: 'Cheque from Bromley Saving Bank, for Down Friendly Soc. ['for' *del*] together with cash for the

160 on opposite page'. On the opposite (debit) page CD had entered £160 on 16 July 1862, stating: 'Bank of England; invested for Down Friendly Soc'.

From W. E. Darwin 14 July 1862

Southampton & Hampshire Bank, | Southampton
July 14 *1862*

My Dear Father,

I believe there is something in Valerian after all. A day or two ago I gathered parts of flowers pairing them as to their apparent age.[1] And this morning I examined one pair taking 6 flowers of each, as they are withered rather & I had not time to send them off this morning. I send you an analysis I made of them; I send a fair copy as I doubt if you could make out my hieroglyphics.[2]

This was the first pair I examined so that I was lucky.

The species is Officinalis, I call them A and B, A has a decidedly pinker tint than B.

You will see by *Elbow* what I mean in the rough sketch I send.[3]

I judge length of pistil, by turning back a petal and then using the tube of Corolla down to the Elbow as standard of comparison

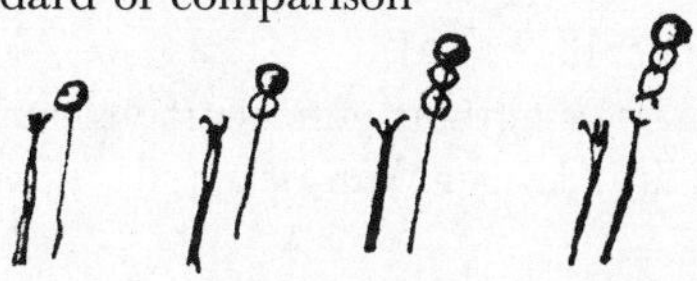

(1) (2) (3) (4)

1.2.3.4. is the pistil and a stamen in (1) I call the stamen longer by an anther, in (2) by two anthers and so on.

I am not sure whether you will read my fair copy— at all events I will send you some actual flowers.

Did you know or am I wrong that Centaury is dimorphic? on back of my rough copy, I send a rough sketch I made this morning with naked eye,[4] these two forms run thro' the whole of their particular plants as I opened buds— The sketch is not exaggerated the pistil has the odd side twist but I dare say I shall find it a mistake on looking at other plants—

I have found plenty of Lithrum not yet in flower—[5]and a large bed of Lysimachia Thyrsiflora where I mean to look for dimorphism[6]

What a wonderful thing the Scarlet fever coming back.[7]

I am your affect | son | W E Darwin

[Enclosure 1]

A.—*pinker variety*

(1) pistil $\frac{3}{4}$ down to Elbow.—pollen & anther gone, and stamens curled back

(2) p.$\frac{3}{4}$—Elbow—do—&—do—

(3). p. almost $\frac{3}{4}$—. E—pollen & Anther not gone— stamens upright longer than pistil by one *anther*— pistil a little beyond petals.

(4.) p—$\frac{3}{4}$—E— rest same as (3)
(5.) p— $\frac{3}{4}$—E—pollen & anther gone stamens curled back, stamens apparently same length as others.
(6.) p— $\frac{3}{4}$ to E.— pollen & anthers not gone stamens upright— longer by nearly 2 anthers than pistil stigmas rather above petals in these 6 of A some were old enough to have lost anther—but the pistils in the older flowers did not seem longer than the other.

[Enclosure 2]
B. (1). pistil not $\frac{1}{2}$ down to Elbow. pollen & anther gone stamens upright from 3 to 4 anthers longer than pistil—pistil *hardly* longer than petals.
(2). p. not $\frac{1}{2}$— to E. pollen & anthers not gone stamens upright nearly 4 anthers longer than pistil—pistil not so *long* as petals.—
(3). p about $\frac{1}{2}$ to E.— pollen &c not gone stamens upright— nearly $2\frac{1}{4}$ anthers longer than pistil pistil a little longer than petals.
(4). p. hardly $\frac{1}{2}$ to E.— pollen & anthers gone stamens curled, about $2\frac{1}{2}$(;) Anthers above pistil pistil hardly any longer than petals.
(5). about $\frac{3}{4}$ to E. pollen not gone stamens upright about $2\frac{1}{4}$ anthers longer than pistil, pistil little longer than petal
(6.) p about $\frac{1}{2}$ to E. pollen & anthers gone stamens curled about 2 Anthers longer than pistil— pistil a little longer than petals

[Enclosure 3]

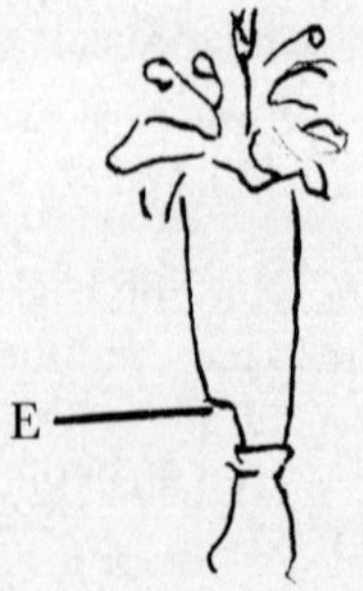

Valeriana officinalis
E. Elbow.

[Enclosure 4]

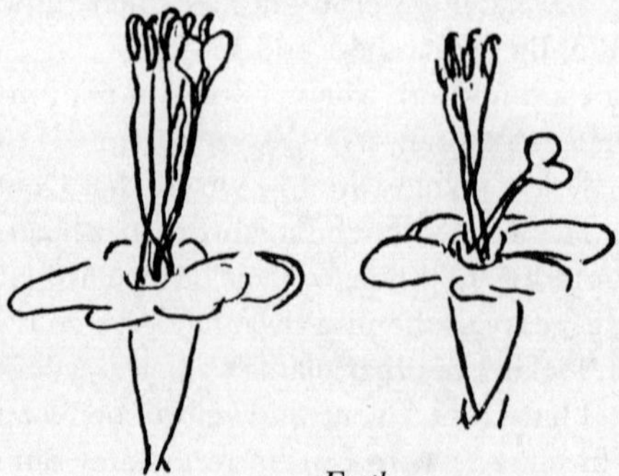

Erythræa

DAR 110 (ser. 2): 23, 41–2, 81–2

CD ANNOTATION
8.1 Did . . . back. 10.1] *crossed pencil*

[1] William had apparently offered to examine plants of the genus *Valeriana* for signs of dimorphism (see letter to W. E. Darwin, 9 July [1862]); his botanical notebook includes a note dated 13 July 1862 stating that he had 'got some Valerian to look at—' (DAR 117: 1).

[2] For the fair copy of William's notes, see enclosures 1 and 2; William also sent a rough copy of his notes, crossed out, on the reverse of the diagram in enclosure 4.

[3] See enclosure 3.

[4] See enclosure 4; the diagram is on the reverse of the piece of paper on which William wrote a rough copy of his notes on *Valeriana* (see n. 2, above). There is a series of observations and drawings of the pistils and stamens of *Erythraea centaurium*, dated 'July 1862', in William's botanical notebook (DAR 117: 16–31).

[5] On 13 July 1862, William recorded in his botanical notebook: 'Looked at *Lithrum*, for two or 3. length of pistil' (DAR 117: 1).

[6] William detailed his observations on '*Lysimachia vulgaris*', dated 20 July 1862, in his botanical notebook (DAR 117: 2–11).

[7] Leonard Darwin had been sent home from school with scarlet fever on 12 June 1862 (Emma Darwin's diary (DAR 242)); he had suffered a relapse at the beginning of July (see letter to W. E. Darwin, 9 July [1862]).

To Asa Gray 14 July [1862][1]

Down Bromley Kent
July 14th

My dear Gray

I shd. have returned the enclosed sooner;[2] but we have had had a miserable ten days. Our Boy (the Postage Stamp collector) has had a return of Scarlet fever, with all sorts of mischief, kidneys, glands of neck &c & still has much fever.[3] But he is now made up for night & I will try & forget the misery of this weary world, & write a bit about science.—

With respect to Pogonia, it would be a very great anomaly, if insects open the anther for nectar: you say nothing about the rostellum;[4] from Vanilla I shd. expect that viscid matter would be forced under lip of anther. Insects ought to be watched at work. In Australia Bees open the indusium of Goodeniaceæ for *pollen*.—[5] Platanthera fimbriata is a pretty case.[6] There is no end to the adaptations. Ought not these cases to make one very cautious when one doubts about the use of all parts?— I fully believe that the structure of all irregular flowers is governed in relation to insects. Insects are the Lords of the floral (to quote the witty Athenæum) world.—[7] How well you have worked the N. American Orchids! I am heartily glad by this time that your harassing Lectures are over.[8]

Hooker is very anxious about Mrs Hooker & has started on health-tour to Switzerland.[9] Today I heard of the sudden death of poor old Prof. Bronn, just as he finished translating my orchid-book.[10] There is misery & anxiety everywhere.— Poor dear Hooker seemed very anxious.—

A few days ago I made an observation which has surprised me more that it ought to do— it will have to be repeated several times, but I have scarcely a doubt of its accuracy.— I stated in my Primula paper that the long-styled form of Linum grandiflorum was utterly sterile with own pollen;[11] & I have lately been putting the pollen of the two forms on the division of the stigma of the *same* flower; & it strikes me as truly wonderful, that the stigma distinguishes the pollen;

& is penetrated by the tubes of the one & not by those of the other; nor are the tubes exserted. Or, (which is the same thing) the stigma of the one form acts on & is acted by (for the papillæ become colourless) pollen, which produces not the least effect on the stigma of the other form. Taking sexual power as the criterion of difference the two forms of this one species may be said to be generically distinct.—[12]

Farewell my dear friend | C. Darwin

Gray Herbarium of Harvard University (70)

[1] The year is established by CD's reference to the death of Heinrich Georg Bronn (see n. 10, below).

[2] The enclosure has not been found, but CD was probably returning some of Gray's notes on American species of orchids (see letter to Asa Gray, 1 July [1862]).

[3] Entries in Emma Darwin's diary (DAR 242) for 5 and 6 July 1862 state that Leonard's 'neck swelled' and that he was suffering from 'bad symptom of kydney'. See also letters to W. E. Darwin, 4 [July 1862] and 9 July [1862]. In the letter to Asa Gray, 10–20 June [1862], CD had asked Gray to provide Leonard with several kinds of American postage stamp for his collection.

[4] The observations by Gray on *Pogonia* to which CD refers have not been found; they were probably among the notes on American species mentioned in the letter to Asa Gray, 10–20 June [1862]. In DAR 70: 80, there is a note in CD's hand that reads:

> June 23^{d}. Asa Gray— Pogonia ophioglossoides— Pollen powdery *loose* grains, no threads— **single** grains Pollen within a lid, he thinks insects lift this to get at nectar within! Does not mention Rostellum— At least it is clear that aid of some kind required.—

See also letter to H. G. Bronn, 11 July 1862 and n. 4.

[5] In 1860, CD had asked James Drummond, a botanical collector and one-time superintendent of the government gardens, Western Australia, to observe whether bees visited the flowers of *Leschenaultia formosa*, and, if so, whether they opened the lips of the indusium, or cup, at the top of the style; he also asked for similar information regarding any other members of the Goodeniaceae that possessed an indusium (see *Correspondence* vol. 8, letter to James Drummond, 16 May 1860; on *Leschenaultia*, see also this volume, letter to J. D. Hooker, 1 May [1862]). Drummond supplied CD with one instance of this sort (see *Correspondence* vol. 8, letter from James Drummond, 8 October 1860, and *Correspondence* vol. 9, letter to *Journal of Horticulture*, [17 May 1861]), by which CD confessed himself '*extremely* much interested' (*Correspondence* vol. 8, letter to James Drummond, 20 December [1860]).

[6] Gray included an account of the adaptation for insect pollination of the floral anatomy of this species in the follow-up article to his review of *Orchids* (A. Gray 1862b, p. 424).

[7] The reference may be to the review of *Orchids* that appeared in the *Athenæum* on 24 May 1862, pp. 683–5 ([Leifchild] 1862). The reviewer contrasted CD's pursuits with the fear of war between Britain and the United States, suggesting that 'in the gardens of green and gladsome Kent', with a 'philosopher's peaceful diligence', CD was principally occupied with his search for floral nectar while 'half the world' was 'experiencing or dreading the bitterness of war'. The phrase 'Lords of the floral world' does not appear in the review; this may be CD's paraphrase of the reviewer's comments on the fertilisation of orchids (*Athenæum*, 24 May 1862, p. 685):

> the greatest care is taken throughout this vast order, . . ., that the pollen shall not be wasted; and yet, if we admit all Mr. Darwin's views, the act of fertilization is, with few known exceptions, left to insects.

[8] Gray was Fisher Professor of natural history at Harvard University and lectured at the Lawrence Scientific School (Dupree 1959); for Gray's comments on his college duties, see the letters from Asa Gray, [2 June 1862], [late June 1862], and 2–3 July 1862.

[9] Joseph Dalton Hooker had taken his wife, Frances Harriet Hooker, to Switzerland in the hope that she might recover her health (see letters from J. D. Hooker, 9 June 1862, 19 [June 1862], 28 June 1862, 2 July 1862, and 10 July 1862).

[10] See letter from E. Schweizerbart'sche Verlagsbuchhandlung, 11 July 1862. In May and June 1862, Heinrich Georg Bronn had translated *Orchids* in preparation for a German edition (Bronn trans. 1862; see letter from H. G. Bronn, 21 June 1862); he died on 5 July 1862 (*Quarterly Journal of the Geological Society of London* 19 (1863): xxxii).

[11] 'Dimorphic condition in *Primula*', p. 96 (*Collected papers* 2: 63).

[12] CD described these observations in detail in 'Two forms in species of *Linum*', pp. 73–5 (*Collected papers* 2: 96–8), concluding:

> Taking fertility as the criterion of distinctness, it is no exaggeration to say that the pollen of the long-styled *Linum grandiflorum* (and conversely of the other form) has been differentiated, with respect to the stigmas of all the flowers of the same form, to a degree corresponding with that of distinct species of the same genus, or even of species of distinct genera.

For CD's interest in the sterility of own-form crosses in dimorphic plants, see Appendix VI.

From Friedrich Hermann Gustav Hildebrand 14 July 1862

Bonn
July 14th | 1862.

Sir,

Professor Treviranus informed me, that after having asked him to translate your work on the fertilisation of Orchids you had made an agreement with the late Professor Bronn on the same matter.[1] Very likely Professor Bronn has not finished the translation or even begun it before his death, and I should feel very honoured if I could be of any use to you in the same way. Professor Treviranus thinks me fit to make the translation, but I must say that I should not be able to begin it before the end of September.— I have presumed to send you a copy of a little work of mine on the Distribution of Coniferous Plants, hoping that it might interest you in some way.[2]

I have the honour to be | Sir | yours | respectfully | Dr Hildebrand | Privat Docent of Botany | at the University.[3] | Bonn.

DAR 166: 199

[1] CD had apparently written to Ludolph Christian Treviranus, professor of botany at the University of Bonn, asking him whether he would be willing to translate *Orchids* into German, only to discover that Heinrich Georg Bronn, through whom he had arranged for the edition to be published by the Stuttgart publishing company, E. Schweizerbart'sche Verlagsbuchhandlung, had already almost completed a translation (see letter from H. G. Bronn, 21 June 1862 and n. 2). Bronn died on 5 July 1862, shortly after completing his translation (Bronn trans. 1863; see letter from E. Schweizerbart'sche Verlagsbuchhandlung, 11 July 1862 and n. 2).

[2] Hildebrand 1861. In this paper, Hildebrand stated that CD's doctrine of common descent helped to explain the geographical distribution of the Coniferae (Hildebrand 1861, p. 381); it was one of the first botanical papers published in Germany to support CD's theory. There is a copy of Hildebrand 1861 in the Darwin Library–Down (see *Marginalia* 1: 380).

[3] Hildebrand had been a student of Treviranus at the University of Bonn. In 1860, he became a *Privatdozent* (lecturer) in botany at the University (see Correns 1916).

To W. E. Darwin [after 14 July 1862][1]

My dear William

I write in hurry to say Lenny so ill (but not I hope dangerously) that I cannot attend to anything:[2] so do not send off Valerian, till you hear.—[3]

How capitally you have been working— The Erythræa seems grand new case, if it does not fail.—[4] Look to roughness of stigmas & size of pollen-grains in the 2 forms.—

I sh^d. enjoy quoting your observations.— I have hardly considered Valerian yet—

It is a fearful illness of Lenny | yours | C. Darwin

DAR 185: 12

[1] Dated by the relationship to the letter from W. E. Darwin, 14 July 1862.

[2] Leonard Darwin was suffering from scarlet fever (Emma Darwin's diary (DAR 242); see also letters to W. E. Darwin, 4 [July 1862] and 9 July [1862], and letter to Asa Gray, 14 July [1862] and n. 3).

[3] William had proposed to send CD specimens of *Valeriana* (see letter from W. E. Darwin, 14 July 1862). For CD's request that William send specimens, see the letter to W. E. Darwin, [24 July 1862].

[4] In the letter from W. E. Darwin, 14 July 1862, William had suggested that *Erythraea centaurium* might offer an additional instance of dimorphism.

To *Journal of Horticulture* [before 15 July 1862][1]

I am very much obliged to your several correspondents for their information in regard to the supposed differences in the bees of Britain.[2] Possibly some few of your readers may be interested in the following case:— The hive bee was introduced many years ago into Jamaica. Having seen it stated that the cells were larger, I procured (through the kindness of Mr. R. Hill, of Spanish Town), some bees and comb.[3] The bees have been carefully examined by Mr. F. Smith,[4] of the British Museum, and pronounced to be the common species. I also secured the hind and front legs, the antennæ and jaws of worker bees from Jamaica and my own stock, and could detect no trace of difference in size or other character. But here comes the remarkable point—the diameter of the cells is conspicuously greater in about the proportion of 60 to 51 or 52 than in our English combs. The wax seems tougher, and the walls, I think, are thicker. The cells in parts of the comb were much elongated, and the whole hive contained a great quantity of honey. It certainly appears as if the instinct of the bee had become modified in relation to its new, hot, and rich home. But it seems to me an astonishing fact that the cells should have been made larger without a corresponding increase in the size of the body of the architect.—

I am well aware that when any person dips into a subject with which he is not familiar he is apt to make great blunders. And this conviction leads me to ask whether the excellent observer *[Mr]* Woodbury[5] would like to look at the pieces of combs, the queen, drone & *[worker]*-bees. If he would, I would most gladly & gratefully send the specimens, (carriage paid) to his residence; for I should either get a curious fact established or my error would be corrected. If he will consent to examine them, I hope he will publish a brief note of the result in your Journal[6]

Charles Darwin, *Down, Bromley, Kent.*

Christie's East (catalogue 26 April 1995: the Philip M. Neufeld collection, pt 2); *Journal of Horticulture, Cottage Gardener, and Country Gentleman* n.s. 3 (1862): 305

[1] CD's letter was published in the issue of 15 July 1862.
[2] See letter to *Journal of Horticulture*, [before 10 June 1862].
[3] See *Correspondence* vol. 7, letters from Richard Hill, 10 January 1859 and 26 November 1859, and letter to Richard Hill, 8 August [1859]. On CD's interest in the comb-building instinct of bees, see Prete 1990.
[4] See *Correspondence* vol. 8, letter from Frederick Smith, 3 April 1860.
[5] Thomas White Woodbury was one of the regular contributors to the beekeeping section of the *Journal of Horticulture* (see letter to *Journal of Horticulture*, [before 22 July 1862] and n. 3).
[6] The text of this paragraph, which was not published in the *Journal of Horticulture*, is reportedly 'neatly crossed through by Darwin'. However, it is probable that the text was in fact crossed out by the editors of the *Journal of Horticulture* to indicate that it was not to be printed. See letter to *Journal of Horticulture*, [before 22 July 1862].

From Bernard Philip Brent 15 July 1862

Dallington | nr. Robertsbridge | Sussex
July 15th. 1862

My dear Sir,

Ever since I last wrote I have been endeavouring to repeat the experiment with the Guinea Pigs, only one has bred, the same Sow that, was kept out of doors last spring went to the Boar this year on the afternoon of Friday May the second— I placed her in a hutch and kept her by the kitchen fire until the 10th. of July when I removed her not wishing her to have young before the children, Saturday morning July the 12th. I found she had kindled being exactly the same time namely a few hours over the ten weeks which she went when exposed to the cold and I hope you will consider this quite conclusive[1] her young two in number were dead one apparently had been long dead the other died at the birth.

I think we may set down the period of gestation as ten weeks and that warmth has no influence I wish to part with the Guinea Pigs shall I send them to you? there is the Boar and one Sow that Messrs. Baker sent,[2] an old Sow I got here, and a young one bred last year four in all, and though I have watched them closely I have not seen any desire in the other Sows to breed

I have a Rat's tooth that forms a complete circle the rat was blind and very thin both the upper teeth, had grown in a circle the points growing into the roof of the mouth and almost touching the roots. one was broken by the dog killing the rat the other is perfect and at your service if you think it worth acceptance,

I have also the skull of my old dog a bull terrier 12 years old, would you like it,? I have not yet seen any notice of your work on Variation under domestication so I conclude it is not yet published[3] I trust that your health and also your daughters has improved since I last heard,[4] I do not call to mind that I have anything else to communicate,

I am sorry to say I am not yet free from my law suit I hope however for the final settlement in November unless any unforseen hinderance prevents,[5]

With best wishes believe me | My dear Sir | Your's truly | B P Brent

To C Darwin Esqre

DAR 160: 301

CD ANNOTATIONS
2.3 there is . . . sent,] *scored brown crayon*
3.1 I have . . . published 4.3] *scored brown crayon*
5.1 I am . . . suit] *scored brown crayon*

1 In 1861, CD had apparently asked Brent to examine the effect of temperature on the gestation period in guinea-pigs (see *Correspondence* vol. 9, letter from B. P. Brent, 15 June 1861).
2 Brent may refer to Samuel C. and Charles N. Baker, dealers in birds and live animals, with premises in London and Paris (*Post Office London directory* 1861).
3 *Variation* was published in 1868.
4 Henrietta Emma Darwin was seriously ill throughout 1861 (see *Correspondence* vol. 9).
5 The particulars of Brent's law suit are not known; it came to an end in 1864, with expenses of over £900. Brent subsequently told CD: 'had I not defended myself I have no doubt I should have lost everything' (letter from B. P. Brent, 2 September 1864 (*Calendar* no. 4607)).

From Asa Gray 15 July [1862][1]

[Cambridge, Mass.]
July 15. Evening.

Dear Darwin.

On my way into Boston to mail the present *envoi*,[2] I received yours of July 1..[3] I open my envelope to acknowledge it.—

First, I am very sorry to hear that your health has suffered.[4]

2. Platanthera orbiculata has discs still wider apart than P. Hookeri, but no division into "2 bridal chambers"[5]

I am curious to know what you will say of my notes on P. hyperborea.[6] I hope to be able to repeat the observations—in the field. Your son's obs. on *minute* insects fertilising is to be noted.[7] My pupil, Rothrock, catches *Thrips*, and only Thrips—in *Houstonia*.[8]

About "a third form in some genus—both stamens & pistil short"—[9] I think you have a *Mertensia* in mind which you have referred to in your *dimorphic* paper;[10] but it was noticed only on one or 2 specimens,—of a rare plant—and I do not think much of it. You want facts which can be verified and re-examined. I doubt you have made quite enough of it already,—unless it jibes in with some other better-observed facts.

A. G.

See end of my over sheet.[11]

[Enclosure][12]

Mitchella repens, L

Two sorts of flowers,

I have had no time to look at them, fresh or dry.[13]

If young enough the differences would doubtless be as evident (between the pollen & stigmas of the two) as in Houstonia, of which I have sent, or will send, details.[14]

I have failed to get *Rhexia*,[15]

Incomplete
DAR 110 (ser. 2): 116, DAR 165: 113

CD ANNOTATIONS
1.1 On my . . . suffered. 2.1] *crossed ink*
4.1 I am . . . A. G. 6.1] *crossed ink*
Verso: 'Stamps | Review—Separate notice[16] | Mitchella | Pogonia | Houstonia— your pupil experiment | Lythrum | Trübner'[17] *ink*

CD note:[18]
The ['pistil' *del*] stigma will project in one form & the anther hidden— Anther projects in other & stigma hidden— Pollen in both, nearly same size; perhaps little larger in short-styled. Pollen $\frac{11}{7000}$ + $\frac{12}{7000}$ in short-style long diameter— *ink*

1 The year is established by the relationship to the letter to Asa Gray, 1 July [1862].

2 The letter from Gray to which this is a postscript (see letter from Asa Gray, 21 July 1862) has not been found; it was apparently returned with the letter to Asa Gray, 28 July [1862]. Some indication of its content is conveyed by CD's annotations, and by his reply (letter to Asa Gray, 28 July [1862]).

3 Letter to Asa Gray, 1 July [1862].

4 In the letter to Asa Gray, 1 July [1862], CD complained that he had felt 'baddish for 2 or 3 weeks'.

5 In the letter to Asa Gray, 1 July [1862], CD acknowledged receipt of Gray's notes on *P. Hookeri*, stating that the species was 'really beautiful & quite a new case', and that it was 'almost laughable the viscid discs getting so far apart that the front of the flower has to be divided into two bridal chambers!'

6 Gray's notes on *P. hyperborea*, which were sent with the letter from Asa Gray, 2–3 July 1862, have not been found; however, see CD's response in the letter to Asa Gray, 23[–4] July [1862] and n. 23.

7 Gray refers to George Howard Darwin's observations of insects pollinating *Orchis maculata* and *Herminium monorchis* (see letter to Asa Gray, 1 July [1862] and nn. 8–9).

8 In the letter from Asa Gray, 2–3 July 1862, Gray had promised to send Joseph Trimble Rothrock's observations on the dimorphic plant, *Houstonia*, when completed. See also letter from Asa Gray, 4 August 1862.

9 See letter to Asa Gray, 1 July [1862] and n. 14.

10 'Dimorphic condition in *Primula*', pp. 95–6 (*Collected papers* 2: 62).

11 This portion of Gray's package has not been found; it was possibly a sheet protecting Gray's several enclosures. See also n. 12, below.

12 This enclosure has been identified by reference to CD's annotation on the letter respecting *Mitchella* (see n. 13, below), and to his comments on the subject in the letter to Asa Gray, 28 July [1862]. The paper on which the enclosure is written is different from that of the rest of the letter, and may be part of the 'over sheet' referred to by Gray.

13 Gray had suggested that the dimorphic plant *Mitchella repens* would be a good case for CD to experiment upon (see *Correspondence* vol. 9, letter from Asa Gray, 11 October 1861). In the letter from Asa Gray, 2–3 July 1862, Gray expressed a hope that he would be able to collect specimens of the

plant during his stay in Beverly, Massachusetts, beginning on 10 July 1862; he apparently enclosed such specimens with this letter (see CD's annotations, and the letters to Asa Gray, 28 July [1862] and 9 August [1862]).

14 Gray sent CD observations on the pollen and stigmas of *Houstonia* in the letter from Asa Gray, [2 June 1862]. See also letters from Asa Gray, 2–3 July 1862 and 4 August 1862.

15 See letter from Asa Gray, 2–3 July 1862 and n. 6.

16 Gray had evidently enclosed a copy of his review of *Orchids* (A. Gray 1862a; see letter from Asa Gray, 2–3 July 1862, and letter to Asa Gray, 28 July [1862] and n. 3). Although the review is listed in the List of reviews (DAR 261 (DH/MS* 8: 6–18)) that served as CD's index to his collection of reviews of his own books, it is absent from the Darwin Pamphlet Collection–CUL.

17 The annotation refers to Nicholas Trübner, Gray's London agent (see also letter to Asa Gray, 28 July [1862] and n. 18).

18 The note is on the verso of the enclosure.

To John St Barbe[1] [16 July 1862][2]

(Cheque Book)

Dear Sir

I enclose cheque Mr Wight[3] formerly was so good as to invest for me, as Treasurer of the Down Friendly Soc.[4] sum of money with the Comm.. of National Debt— I hope, therefore that you will be so kind as to take the trouble to pay into B. of E . . . . . ., (as soon as enclosed cheque for 122 has been paid into my account) sum of . . . . . .—(figures & writing, with the Commission, in accordance with enclosed paper which must be transmitted to the Commission; & transmit me the Receipt of the Commission[5]

Hoping that you will grant this favour, I remain | Dear Sir, Yours faithfully, C. D.

Draft
DAR 96: 3 r.

1 John St Barbe was manager of the Charing Cross branch of the Union Bank of London. The recipient is identified by the reference to the former manager of the bank (see n. 3, below).

2 Dated by an entry in CD's Account book–banking account (Down House MS), on 16 July 1862, recording a payment of £160 to the Bank of England 'invested for Down Friendly Soc'.

3 Alexander Wight had been the manager of the Charing Cross branch of the Union Bank of London until 1860 (*Banking almanac*). CD had banked with the branch since 1853 (see *Correspondence* vol. 5, letter to John Higgins, 11 April [1853]).

4 CD had served as treasurer of the Down Friendly Society since its establishment in 1850 (see *Correspondence* vol. 4, letter to J. S. Henslow, 17 January [1850] and n. 6).

5 Neither the enclosure nor the receipt has been found.

From Asa Gray 21 July 1862

Cambridge [Massachusetts]
21. July 1862.

My Dear Darwin

The enclosed I trust will please your boy,—who I hope is now well, and his Father also.[1]

Since my last, in the P.S. to which I hastily acknowledged yours of July 1,[2] I have received the casts of *cuts* of *Orchis*.[3] Very many thanks, I shall use them—to illustrate my 1st part of notice of your book.[4] I have received invoice from Trübner & Co for 6 copies—at £3·3. What is the retail price of your book? Is it not half a guinea?[5] It would be wrong you should pay for these copies,—since I suppose Murray would supply you at half the price.

No time to write now.— No more Orchid-matter has come to hand.

Ever Yours Cordially | Asa Gray.

DAR 165: 114

CD ANNOTATIONS

3.1 No time ... hand.] '*[*Malaxis*]*' *brown crayon*

Verso of letter: 'Mitchella. (rec'd)[6] | Echinum | Lythrum— | Trübner 6/10=£2:s1:d0' *ink*

1 In the letter to Asa Gray, 10–20 June [1862], CD had told Gray that Leonard Darwin was suffering from scarlet fever, and had asked Gray to send Leonard several kinds of American postage stamp for his collection. In his letter to Gray of 1 July [1862], CD complained that he had felt 'baddish for 2 or 3 weeks'.

2 See letter to Asa Gray, 1 July [1862], and letter from Asa Gray, 15 July [1862].

3 At Gray's request, CD had arranged for John Murray to send electrotype plates of three of the illustrations from *Orchids*, figuring *Orchis mascula* and *O. pyramidalis*, for reproduction in Gray's review of the book in the *American Journal of Science and Arts* (A. Gray 1862a; see letter from Asa Gray, 18 May 1862, letter to Asa Gray, 10–20 June [1862], and letters to John Murray, 13 June [1862] and 20 [June 1862]).

4 Gray apparently wrote '1st part' in error; he had sent CD a copy of his review of *Orchids*, which was published in the July number of the *American Journal of Science and Arts* (A. Gray 1862a), with the letter from Asa Gray, 15 July [1862]. The illustrations were used in Gray's follow-up article to his review (A. Gray 1862b, pp. 421–3).

5 Gray had asked CD to arrange for six copies of *Orchids* to be sent to him by the London publisher and bookseller, Nicholas Trübner, who frequently acted as Gray's London agent (see letter to Nicholas Trübner, 23 June [1862]). *Orchids* was offered for sale at a retail price of 9*s*. (*Publishers' Circular* 35 (1862): 247).

6 See letter from Asa Gray, 15 July [1862] and n. 13.

To *Journal of Horticulture* [before 22 July 1862][1]

I am sorry that you did not append the closing sentence to my communication about the Jamaica bees, as it would have shown your readers that I was doubtful on the subject.[2] I have now to confess that I have made a gross blunder. The cells which I measured were drone-cells, as I am informed by the "Devonshire Bee-Keeper."[3] I could offer some explanation and apology to your readers for making so great a mistake; but it is a personal matter and would not interest them. How the statement in French works arose that the cells in West Indian combs are larger than those in European combs, I cannot conceive.—[4]

C. Darwin

Journal of Horticulture, Cottage Gardener, and Country Gentleman n.s. 3 (1862): 323

[1] CD's letter was published in the issue of 22 July 1862.
[2] See letter to *Journal of Horticulture*, [before 15 July 1862].
[3] Thomas White Woodbury contributed to the bee-keeping section of the *Journal of Horticulture*, writing under the pen-name 'A Devonshire Bee-keeper' (*Journal of Horticulture* 28 (1862): i). No letter from Woodbury containing this information has been found; however, see the letter from T. W. Woodbury, 9 August 1862. According to his 'Journal' (Appendix II), CD was writing up the chapter of *Variation* that dealt with 'Silk-worms Geese &c' (chapter 8); this chapter also included a section on hive-bees (*Variation* 1: 297–9). In *ibid.*, p. 298, CD wrote:

> I procured a hive full of dead bees from Jamaica, where they have long been naturalised, and, on carefully comparing them under the microscope with my own bees, I could detect not a trace of difference.

[4] The reference has not been traced.

To Asa Gray 23[–4] July [1862][1]

Down Bromley Kent
July 23^d

My dear Gray

I received several days ago two large packets, but have as yet read only your letter;[2] for we have been in fearful distress & I could attend to nothing. Our poor Boy had the rare case of second rash & sore throat, besides mischief in kidneys; & as if this was not enough a most serious attack of erysipelas with typhoid symptoms.[3] I despaired of his life; but this evening he has eaten one mouthful & I think has passed the crisis. He has lived on Port-wine every $\frac{3}{4}$ of an hour day & night. This evening to our astonishment he asked whether his stamps were safe & I told him of the one sent by you,[4] & that he sh^d. see it tomorrow. He answered "I should awfully like to see it now"; so with difficulty he opened his eyelids & glanced at it & with a sigh of satisfaction said "all right".— Children are one's gretest happiness, but often & often a still greter misery. A man of science ought to have none,—perhaps not a wife; for then there would be nothing in this wide world worth caring for & a man might (whether he would is another question) work away like a Trojan.— I hope in a few days to get my Brains in order & then I will pick out all your orchid letters (& read by & bye your last)[5] & return them in hopes of your making use of them—[6] Planthanthera would be eminently well worth giving & as much as feel safe about Cypripedium;[7] in part I am not sure that I understand the passages by which insects crawl in & out. Could you give a diagram?[8] I have such an arrear of letters & such a number of experiments,[9] all going to the dogs, that I have not time to make abstract of your letters. Will you return me such, as you do not use: but I hope you will be led to use all some time or another.—[10] I shall be very glad to hear of *Rosmacks**? observations on Houstonia; you only just alluded to them.—[11] You did formerly tell me about Specularia:[12] in viola & oxalis the case seems to me to be much too remarkable to be called "precocious flowering".[13]

*I hope he will publish note; I hear the French say that my paper on Primula is all pure imagination; but I cannot hear that this is grounded on any observations—[14]

You will never read my horrid writing, if I write on both pages, of thin paper which I have taken in obedience to orders.—[15] Of all the carpenters for knocking the right nail on the head, you are the very best: no one else has perceived that my chief interest in my orchid book, has been that it was a "flank movement" on the enemy.[16] I live in such solitude that I hear nothing, & have no idea to what you allude about Bentham & the orchids & Species.[17] But I must enquire.—

By the way one of my chief enemies (the sole one who has annoyed me) namely Owen, I hear has been lecturing on Birds, & admits that all have descended from *one*, & advances as his own idea that the oceanic wingless Birds have lost their wings by gradual disuse.[18] He never alludes to me or only with bitter sneers & coupled with Buffon, & the Vestiges.—[19]

Well it has been an amusement to me this first evening scribbling as egotistically as usual about myself & my doings; so you must forgive me, as I know well your kind heart will do.— I have managed to skim the news-paper, but had not heart to read all the bloody details. Good God what will the end be; perhaps we are too despondent here; but I must think you are too hopeful on your side of the water. I never believed the "canard" of the army of the Potomac having capitulated.[20] My good dear wife & self are come to wish for Peace at any price.

Good Night my good friend. I will scribble no no more— C. D.

One more word. I sh[d] like to hear what you think about what I say in last Ch. of Orchid Book on the meaning & cause of the endless diversity of means for same general purpose.— It bears on design—that endless question—[21]

Good Night Good Night.

P.S. Last night after writing the above, I read the great bundle of notes.[22] Little did I think what I had to read. What admirable observations! You have distanced me on my own hobby-horse! I have not had for weeks such a glow of pleasure as your observations gave me.— Plat. hyperborea is indeed a most curious case & especially interesting to me. How like the Bee ophrys.[23] Does it live in arctic regions where insects may be scarce? It would be *very good* to ascertain whether there actually is any occasional crossing, or removal of pollinia in this species.[24] How curious about the nectary. See my *note* p. 324 about Aceras.[25] Aceras, I now find, leads, also, most closely into the rare O. hircina.[26] How organic beings are connected! How excellently you have worked Cyp. spectabilis. I daresay I may be altogether wrong, & fertilisation may always be by small insects bodily crawling in: I wish you could get some 2 youths to watch on warm day for 2 or 3 hours a fine plant of some Cypripedium.—[27] What diversity in Platanthera— Your observations seem to me *much* too good to be sunk in any review of my Book; they won't be noticed.—[28] But I am so very sorry I did not return your M.S. earlier: I shall be so grieved if I thus cause you inconvenience; but in truth it was physically impossible for me before last night to read or attend to anything.

Farewell my good Friend | C. Darwin

Gray Herbarium of Harvard University (76)

[1] The year is established by the reference to Leonard Darwin's illness (see n. 3, below).

[2] Gray had sent a 'great bundle of notes' with the letter from Asa Gray, 2–3 July 1862 (see n. 22, below); the envelope to that letter bears a London postmark: 'JY 18 | 62'.

[3] Leonard Darwin had been sent home from school on 12 June 1862 suffering from scarlet fever; his recovery was interrupted by a recurrence of symptoms in July (see Emma Darwin's diary (DAR 242), and the letters to W. E. Darwin, 4 [July 1862], 9 July [1862], and [after 14 July 1862]).

[4] Gray had sent a three-cent postage stamp in the letter from Asa Gray, 2–3 July 1862.

[5] Letter from Asa Gray, 2–3 July 1862 (see n. 2, above).

[6] Since starting to read *Orchids* in May, Gray had sent CD a number of notes on American species of orchids (see letters from Asa Gray, 18 May 1862, [2 June 1862], [late June 1862], and 2–3 July 1862). CD had urged Gray to publish his observations, either in reviewing *Orchids*, or separately, and had offered to return the notes to Gray to enable him to do so (see letters to Asa Gray, 10–20 June [1862], 1 July [1862], and 14 July [1862]). In the letter from Asa Gray, 2–3 July 1862, Gray asked CD to indicate those observations that seemed to him 'worth touching on', and to send back the appropriate notes.

[7] Gray had expressed a wish to examine most of the species of *Cypripedium* further before publishing on the subject (letter from Asa Gray, 2–3 July 1862). He incorporated an account of American species of *Cypripedium* and *Platanthera* in the follow-up article to his review of *Orchids* (A. Gray 1862b), stating in regard to *Cypripedium* that it was a subject on which he would 'hazard a few remarks' (p. 427).

[8] In *Orchids*, pp. 274–5, CD had suggested that *Cypripedium* must be pollinated by an insect inserting its proboscis into one of the two lateral entrances at the base of the labellum, directly over one of the two lateral anthers, and thus either placing the pollen onto the flower's own stigma, or carrying it away to another flower. In 'Fertilization of orchids', pp. 155–6 (*Collected papers* 2: 152), CD stated:

> Prof. Asa Gray, after examining several American species of *Cypripedium*, wrote to me . . . that he was convinced that I was in error, and that the flowers are fertilized by small insects entering the labellum through the large opening on the upper surface, and crawling out by one of the two small orifices close to either anther and the stigma.

Gray detailed his observations in A. Gray 1862b, but did not provide any illustrations, concluding: 'The beauty of these adaptations can be appreciated only by actual inspection of the parts or of a series of figures.'

[9] On CD's experiments, see, for instance, the letters to J. D. Hooker, 30 May [1862] and n. 7, and 23 June [1862] and n. 4, the letter to Alphonse de Candolle, 17 June [1862] and n. 2, and the letter to M. T. Masters, 8 July [1862] and n. 3.

[10] Gray's notes to CD on American species of orchids have not been found in the Darwin Archive–CUL, or in the Gray Herbarium Archives.

[11] 'Rosmack' is a misspelling; Joseph Trimble Rothrock was a student of Gray's (see letter from Asa Gray, 2–3 July 1862). Gray forwarded Rothrock's observations on *Houstonia* in the letter from Asa Gray, 4 August 1862.

[12] See letter from Asa Gray, 2–3 July 1862, and *Correspondence* vol. 9, letter from Asa Gray, 11 October 1861.

[13] See letter from Asa Gray, 2–3 July 1862.

[14] The reference has not been traced.

[15] Gray advised CD to use thinner writing paper, thereby reducing postage charges, in the letter from Asa Gray, 2–3 July 1862.

[16] See letter from Asa Gray, 2–3 July 1862.

[17] See letter from Asa Gray, 2–3 July 1862 and n. 16. The reference is to George Bentham's presidential address to the anniversary meeting of the Linnean Society of London on 24 May 1862 (Bentham 1862).

18 CD probably refers to Richard Owen's lectures at the Museum of Practical Geology on the 'Characters, Organisation, Geographical Distribution, and Geological Relations of Birds'. The series of six lectures ran from 14 to 30 May 1862 (*Athenæum*, 10 May 1862, p. 613). Neither the printed accounts of these lectures, nor the manuscript text which survives for some of them, report the details cited by CD (see letter to Armand de Quatrefages, 11 July [1862], n. 7). However, CD apparently learned of the content of the lectures from one of his acquaintances (see letter to Charles Lyell, 22 August [1862] and n. 7).

19 CD refers to the evolutionary views of George Louis Leclerc, comte de Buffon (see letter from Armand de Quatrefages, [after 11 July 1862], n. 5) and to the anonymous evolutionary work, *Vestiges of the natural history of creation* ([Chambers] 1844). Owen discussed CD's theory in common with the views of Buffon and Robert Chambers in R. Owen 1861a, pp. 442–3. See also R. Owen 1862a. In his fifth lecture, on the geographical distribution of birds (Natural History Museum, London, OC38.3/318; see also *Medical Times and Gazette* (1862), pt 1: 617), Owen stated:

> we must remember that by the word 'creation' we mean "a process we know not what." We have not yet ascertained the secondary causes which operated when "the earth brought forth grass & herb yielding seed after its kind" and when "the waters brought forth abundantly the moving creature that hath life." And if the 'spontaneous generation' of a fruit-bearing tree, or of a fish, were conceivable, & the whole process demonstrable, we should still retain as strongly the idea which is the chief of the 'mode' or group of ideas called 'creation'; viz., that the process was ordained by and had originated from an . . . all-wise First Cause of all things.

On Owen's views concerning evolution, see Rupke 1994, pp. 220–58.

20 The reference is to the failed attempt made by the Union army of the Potomac in June 1862 to seize the Confederate capital, Richmond, Virginia. Between 25 June and 1 July 1862 the Confederates drove the Union forces away from Richmond; the ensuing battles, resulting in 30,000 casualties, were the bloodiest yet seen in the conflict (McPherson 1988, pp. 461–71).

21 In chapter 7 of *Orchids*, pp. 348–9, CD stated:

> In my examination of Orchids, hardly any fact has so much struck me as the endless diversity of structure,—the prodigality of resources,—for gaining the very same end, namely, the fertilisation of one flower by the pollen of another. The fact to a certain extent is intelligible on the principle of natural selection.

Gray and CD had corresponded at length on the question of design in nature (see *Correspondence* vols. 8 and 9). For Gray's response to CD's question, see the letter from Asa Gray, 22 September 1862 and n. 11.

22 CD refers to the notes on orchids sent with the letter from Asa Gray, 2–3 July 1862 (see n. 2, above, and the letter from Asa Gray, 15 July [1862] and n. 6); the notes have not been found.

23 Gray's notes on this subject have not been found; however, in the portion of A. Gray 1862c that was published in the September number of the *American Journal of Science and Arts*, Gray disputed Joseph Dalton Hooker's claim that *Platanthera hyperborea* and *P. dilatata* constituted a single species, reporting that he had recently observed that 'while *P. dilatata* . . . can rarely if ever self-fertilise, *P. hyperborea* readily does so, much in the manner of *Ophrys apifera* as recently illustrated by Darwin' (p. 259). (Gray gave further details of pollination in *P. hyperborea* in the follow-up article to his review of *Orchids* (A. Gray 1862b, p. 426).) In *Orchids*, pp. 72–3, CD similarly distinguished *Ophrys arachnites* as a species separate from *O. apifera* (the bee ophrys) on the basis that the former was adapted for cross-fertilisation, while the latter was adapted for self-fertilisation. One of CD's objectives in *Orchids* was to demonstrate that cross-fertilisation was the 'main object' of the contrivances by which orchids were pollinated (p. 1), and in the conclusion, he noted that the bee ophrys was anomalous in being adapted for self-fertilisation (p. 359). He had subsequently been speculating about possible explanations for this apparent anomaly (see letters from G. C. Oxenden, [before 30 May 1862], 21 June 1862, and 8 July 1862, and letter to A. G. More, 7 June [1862]).

[24] In A. Gray 1862b, p. 426, Gray described how it would be possible for an insect to cross-pollinate *P. hyperborea*, continuing:

If the rule holds here as elsewhere, that a stigma is more sensitive to the pollen of another flower than to that of its own, there will be no lack of sufficient crossing in this species, wherever proper insects abound; where they do not, it will be prolific without them.

[25] Gray did not mention the nectary of *P. hyperborea* in either of his published accounts (A. Gray 1862b and 1862c). CD's account of the monstrous flowers of *Aceras* in *Orchids*, p. 324 n., referred to the pollinia not having viscid discs, and to the two anther cells being widely separated; this resembles Gray's statement that *P. hyperborea* had smaller viscid discs than *P. dilatata*, with the anther-cells more divergent, and the stalks of the pollinia being 'very attenuated and weak' (A. Gray 1862c, p. 260).

[26] CD had recently received specimens of *Orchis hircina* from George Chichester Oxenden (see letter from G. C. Oxenden, 4 June [1862] and n. 2). See also *Orchids* 2d ed., pp. 25–6.

[27] See n. 8, above. In his account of insect pollination in *Cypripedium* (A. Gray 1862b, p. 428), Gray stated that CD's theory might account for fertilisation in the genus, 'but hardly in *C. spectabile*'. He went on to note that the 'rigid, sharp-pointed papillæ, all directed forwards', that were particularly striking on the stigma of *C. spectabile*, offered 'no slight confirmation' of his own hypothesis, since they would act like a 'wool-card' in removing pollen from any insect 'working its way upwards to the base of the labellum'. However, Gray noted with respect to his hypothesis that he had 'not been able to detect insects actually at work'.

[28] Gray's observations on American species of *Platanthera* and *Cypripedium* were incorporated into the follow-up article to his review of *Orchids* (A. Gray 1862b).

To W. E. Darwin [24 July 1862][1]

Down
Thursday night

My dear William

Lenny is going on capitally.[2] He took a little tea & powdered meat this evening! His poor dear little face is something like itself again.

Any time that you can send me a *few* valerian flowers & of Erythræa, I sh^{d}. like to look at them.—[3] But if the cases hold good you shall draw up a little account, which I will look over, & send it to the Linnean Soc^{y}.

I don't quite understand your scale of measurement of pistil in Valerian: you speak of it as $\frac{3}{4}$ &c & yet it projects beyond the corolla, if I understand rightly.—[4]

The Erythræa almost seems a more curious case: is there any difference in nectary or in secretion of nectar or position of flowers in the two forms?[5]

Good Night— I am tired.— | C. Darwin

DAR 210.6: 101

[1] Dated by the relationship to the letter from W. E. Darwin, 14 July 1862, and to the letter to W. E. Darwin, [after 14 July 1862], and by reference to a postscript, written in pencil, in Emma Darwin's hand: 'Enquire a little about quiet sea places near Southampton I expect G & F will come to you at the end of next week'. George Howard Darwin and Francis stayed with William in Southampton from 2 August (see letter from G. H. Darwin, [after 5 August 1862] and n. 3). The letter was written on a Thursday; 24 July was the Thursday prior to the week ending 2 August 1862.

[2] Leonard Darwin was recovering from scarlet fever (Emma Darwin's diary (DAR 242); see also preceding letter, letters to W. E. Darwin, 4 [July 1862], 9 July [1862], and [after 14 July 1862], and letter to Asa Gray, 14 July [1862]).

[3] See letter from W. E. Darwin, 14 July 1862, and letter to W. E. Darwin, [after 14 July 1862].
[4] See letter from W. E. Darwin, 14 July 1862.
[5] See letter from W. E. Darwin, 14 July 1862. There is a series of observations and drawings of the parts of *Erythraea centaurium*, made in July, in William's botanical notebook (DAR 117: 16–31).

From J. D. Hooker [24 July 1862][1]

University of London, | Burlington House, W.
Thursday

Dr Darwin

We arrived last night at 7pm. Having left Zurich on the previous morning only at 10 am. stopping 3 hours at Basle & as many at Paris. The journey has done my wife a wonderful deal of good. The day before leaving Zurich she mounted 6 times in the day to her room in the Hotel **83** steps up each time without the slightest palpitation or loss of breath. she enjoyed our 8 days in the Enghadien valley immensely & walked about a great deal there.[2] I suppose there must be some Hygienic effect in a diminished atmospheric pressure— there ought to be at any rate. Before leaving I found another hybrid orchid— between *Gymnadenia odoratissima* & *Nigritella*— the former one was between *G. Conopsea* & the same *Nigritella*:— I have brought some plants of the latter hybrid & have left them at Kew this morning for Fitch to draw—[3] (I am here at L.U. Examinations).[4]

I was delighted with Heer & went over all his collections which are grand & good, they serve to convince me that the Miocene vegetation was Himalayan not American as H. supposed.[5] Heers error was very natural, for no one knows from any published works what the real nature of the Himalayan vegetation is.[6] Heers works on Insects seem excellent too[7] & I should hope from what I heard & saw that Zurich will become a good Nat Hist. School: besides a good staff of Professors they have good paid Curators of Botanical, Zoological, Entomological & Geological Cabinets, which will all be united in the new Polytechnicon, Except the Herbaria which will all go to the Bot. Garden. They have also good departmental libraries.

There is a capital review of the "Nat. Hist. Review" in the Parthenon, I wonder who wrote it, it expresses my opinion exactly.[8] The last number of the Review is a sad falling off & the last page is disgraceful for errors & misprints: poor Oliver is quite down-hearted about it—[9] they all seem afraid of Huxley who has undertaken sole responsibility of Editorship, which he is not up to & has not time for.[10]

I saw Oliver for a moment this morning when rushing up here & he told me you are again in trouble about your boy,[11] my dear old friend I do grieve to hear it. I have nothing more to say till I am settled again | Ever Yrs affec | J D Hooker

Endorsement: 'July | 62'
DAR 70: 171, DAR 101: 48–9

[1] Dated by the reference to the Hookers' holiday in Switzerland; the Hookers were abroad from 4 to 23 July 1862 (see letters from J. D. Hooker, 2 July 1862 and 10 July 1862). The letter was written on a Thursday; 24 July 1862 fell on a Thursday.

[2] Hooker had taken his wife, Frances Harriet Hooker, to Switzerland in the hope that she might recover her health (see letters from J. D. Hooker, 28 June 1862, 2 July 1862, and 10 July 1862).

[3] Walter Hood Fitch was a botanical artist at the Royal Botanic Gardens, Kew.

[4] Hooker was an examiner in botany at the University of London until 1864 (L. Huxley ed. 1918, 1: 537; *Medical directory* (1862): 200). University College London was one of several organisations to have its offices at Burlington House, Piccadilly (*Post Office London directory* 1861).

[5] The Hookers spent some of their holiday in Switzerland as guests of Oswald Heer (see letter from J. D. Hooker, 10 July 1862, and L. Huxley ed. 1918, 1: 401–2). In attempting to explain the resemblance between the Tertiary flora of Europe and Madeira, and the present flora of Atlantic North America, Heer had argued that, during the Miocene era, there must have existed an Atlantic land-bridge between Europe and North America, which was subsequently submerged, with the exception of the various Atlantic islands (Heer 1857 and 1860). CD had long been critical of attempts to explain geographical distribution by postulating former land-bridges (see, especially, *Correspondence* vol. 6, letters to Charles Lyell, 16 [June 1856] and 25 June [1856]).

[6] Hooker had gained an extensive first-hand knowledge of Himalayan botany during his travels in the region in the period 1847–50.

[7] In addition to his work in palaeobotany, Heer had published extensively on both recent and fossil insects; he studied the latter in Miocene deposits in Germany and Croatia, Oligocene deposits in France, and Liassic deposits in Switzerland (*DSB*).

[8] Hooker refers to the review of the first volume of the new series of the *Natural History Review* in the *Parthenon* 1 (1862): 373–5. The reviewer welcomed the journal as partially meeting a long-felt need for 'a serial devoted to the advancement of natural science', but was critical of the editorial arrangements, arguing that, while the name of any of the eleven editors might seem 'sufficient warranty' for the quality of the contents, the review suffered from the 'absence of a directing mind'. He continued:

> It is as though every author and reviewer had edited his own production, and that the printer had been left to do the remainder. It follows that there are numerous faults of style and endless obscurities, which would have been avoided by a more perfect unity of supervision, and an independent critical reading of manuscript matter.

Thus, while many of the articles were 'of supreme interest to men of science', the journal was of no use to 'the great world, which requires earnestly to be instructed'.

[9] The *Natural History Review* was published quarterly; Hooker refers to the number for July 1862. Daniel Oliver was one of the botanical editors of the review.

[10] In July 1860, Edward Perceval Wright, one of the then editors of the *Natural History Review* had offered Thomas Henry Huxley 'effectual control' of the new series of the journal, due to start publication in January 1861, if he would become one of the editors. Huxley had organised a 'commissariat' comprising eleven co-editors, but by June 1861 he was increasingly taking the responsibility of editing on himself, writing: 'It is no use letting other people look after the journal. I find unless I revise every page of it, it goes wrong' (see L. Huxley ed. 1900, 1: 209–10).

[11] Leonard Darwin's recovery from scarlet fever had been interrupted by a recurrence of symptoms in July (see Emma Darwin's diary (DAR 242); see also preceding letter, letters to W. E. Darwin, 4 [July 1862], 9 July [1862], and [after 14 July 1862], and letters to Asa Gray, 14 July [1862] and 23[–4] July [1862]).

To M. T. Masters 24 July [1862][1]

Down. | Bromley. | Kent. S.E.
July 24th

My dear Sir

My poor Boy rallied last night & is now out of danger. He had recurrent scarlet-fever with every sort of mischief in the glands; & this followed by dreadful erysipelas of head with typhoid symptoms.[2] The Doctors never saw such a complication of illness. But thank God Port-wine every $\frac{3}{4}$ of hour, night & day, seems to have saved him.

I thank you cordially for taking the trouble of writing at such length: your letter is in many ways of great value to me.[3] The distinction of the two sorts of Peloria, though so excessively obvious when pointed out, never occurred to me.—[4] I shall now know what flowers to look to. It is quite likely I may make nothing of these peloric gentlemen; but I am contented if I get any result once out of four or five sets of experiments.[5]

Pray give my compliments & best thanks to your Father for his kind information.[6] The seeds are not ripe, but *apparently* I have got some few from a few of the peloric Pelargoniums; but perhaps the seed will prove bad.—[7] Many thanks for references to Bull. Bot. Soc;[8] as until within a few weeks I did not see that Peloric flowers would have any bearing on my subjects, I never attended to them. That is a curious case of hereditariness, which you mention: I think Prosper Lucas gives an analogous case.[9]

I am glad to hear that you are continuing your work on malconformations in Plants.[10]

With sincere thanks for your valuable aid. Believe me, my dear Sir | Yours sincerely | C. Darwin

American Philosophical Society (Getz 3663)

[1] The year is established by the reference to Leonard Darwin's illness (see n. 2, below).

[2] See also Emma Darwin's diary (DAR 242), letter to Asa Gray, 23[–4] July [1862], and letter to W. E. Darwin, [24 July 1862].

[3] CD had written to Masters requesting information on peloric flowers (letter to M. T. Masters, 8 July [1862]); in the letter from M. T. Masters, 12 July 1862, Masters promised to send 'a few memoranda' on the subject when he had more time, but no such correspondence has been found.

[4] Masters published an account of his observations on two forms of peloric flowers the following year (Masters 1863); he classified the apparent malformations as follows (Masters 1863, p. 260):

> As the word Peloria itself merely signifies something strange and out of the common way, there can be no objection, I think, to the introduction of the terms Regular and Irregular Peloria. "Regular or Congenital Peloria" would include those flowers which, contrary to their usual habit, retain throughout the whole of their growth their primordial regularity of form and equality of proportion. "Irregular or Acquired Peloria", on the other hand, would include those flowers in which the irregularity of growth that ordinarily characterizes some portions of the corolla is manifested in all of them.

CD cited Masters 1863 in his discussion of this point in *Variation* 2: 58; his annotated copy of the number of the *Natural History Review* in which the article appeared is in the Darwin Library–CUL.

[5] In the letter to M. T. Masters, 8 July [1862], CD had asked for suggestions as to what plants he might grow for experimentation the following season, with an expectation of their producing peloric flowers. CD described crossing experiments on the peloric form of *Antirrhinum majus* in *Variation* 2: 70; his notes from these experiments, dated 1863–5, are in DAR 51 (ser. 2): 18–23.

[6] In the letter from M. T. Masters, 12 July 1862, Masters told CD that he had written to his father, the nurseryman William Masters, to ask for information on the fertility of the peloric flowers in *Gloxinia* and other cultivated plants. In *Variation* 2: 167, CD noted William Masters's observations on

the sterility of peloric flowers in pelargoniums; CD had told M. T. Masters of his interest in this question in the letter to M. T. Masters, 8 July [1862].

7 CD refers to the crossing experiments with the normally sterile peloric flowers of pelargoniums that he had begun in May 1862 (see letter to Daniel Oliver, 8 June [1862], letter to Asa Gray, 1 July [1862], and letter to M. T. Masters, 8 July [1862]). CD's notes from these experiments are in DAR 51 (ser. 2): 4–9, 12–13. In *Variation* 2: 167, CD reported that he had made 'many vain attempts' to set seed from these peloric flowers, but that he had 'sometimes succeeded in fertilising them with pollen from a normal flower of another variety' and, conversely, had 'several times fertilised ordinary flowers with peloric pollen.' Only once, he reported, had he 'succeeded in raising a plant from a peloric flower fertilised by pollen from a peloric flower borne by another variety'.

8 The references have not been identified, but may have included an article on peloric flowers in the genus *Zingiber* by Arthur Gris, in the *Bulletin de la Société Botanique de France* (Gris 1859), cited by Masters in his article on peloric flowers (Masters 1863, p. 262).

9 The reference has not been traced. The work of the French physician Prosper Lucas on inheritance (Lucas 1847–50) is extensively cited in *Variation*; there is a heavily annotated copy of this work in the Darwin Library–CUL (see *Marginalia* 1: 513–23).

10 Masters was making a special study of plant morphology and teratology, and, in 1869, published *Vegetable teratology* (Masters 1869).

To Daniel Oliver 24 July [1862][1]

Down. | Bromley. | Kent. S.E.
July 24th

Dear Oliver

My poor Boy rallied last night & the Doctors think he has passed the crisis & is out of danger.—[2] I have had a miserable month, & many of my experiments on Dimorphism are gone to the dogs.[3]

I am very glad that your Lectures are nearly over, & that you will be a somewhat free man.—[4] It must be very hard work. Asa Gray seems to be able to do nothing else when he is lecturing.[5] He is now free & has been making some capital observations on orchids. He has got a self-fertilising Platanthera just like the Bee, with thin caudicle, but still more plainly than the Bee with adaptations for an occasional cross.[6] I think, in opposition to you, that some day the Bee-ophrys will be explained; I have been speculating, *against* evidence, that arachnites may be the crossing form & the Bee the self-fertilising form of same species; but it won't do.—[7]

Many thanks for your reference to Duchartre.[8] The Vanda must be something curious.[9] By the way I find that the orchid mentioned by me as *Mormodes unnamed sp.* is Cycnoches ventricosum:[10] it is hermaphrodite & I wish much to see a dark Cynoches, of which I once received a flower from Kew, for I believe it to be a male.[11] Will you ask Mr H. Gower[12] whether he could spare me one, when it flowers again.

Many thanks for Bot.-Zeitung: it will be very useful.—[13] Dont take trouble about the Primula; if you stumble on it, I sh^d. like to see it.[14] Cytisus adami is a strange puzzle; I have failed in fertilising C. purpureus by pollen of common Laburnum.—[15]

If you can spare the time to notice my orchid Book, I have not the least doubt you will do it right well. There was capital matter in your Review of my Primula paper.—[16] I am always astonished at your knowledge.

Farewell— I am tired so no more. I hope you will enjoy your holidays & be idle | Yours very sincerely | C. Darwin

If you can remember, please remember that I want any peloric plants for experiments; ie any in pots, which could be lent me from Kew.— I have been working at peloric pelargoniums; but whether I shall get any good result, I know not.[17]

DAR 261 (DH/MS 10: 34)

[1] The year is established by the references to Leonard Darwin's illness (see n. 2, below) and to [Oliver] 1862c.

[2] Leonard Darwin was recovering from scarlet fever (see Emma Darwin's diary (DAR 242); see also preceding letter, letter to W. E. Darwin, [24 July 1862], and letter to Asa Gray, 23[–4] July [1862]).

[3] In his 'Journal' (see Appendix II), CD wrote, 'Much time wasted June & July from Leonards illness'. On CD's experiments, see, for instance, the letters to J. D. Hooker, 30 May [1862] and n. 7, and 23 June [1862] and n. 4, the letter to Alphonse de Candolle, 17 June [1862] and n. 2, and the letter to M. T. Masters, 8 July [1862] and n. 3.

[4] Oliver was professor of botany at University College London.

[5] See letters from Asa Gray, [2 June 1862], [late June 1862], and 2–3 July 1862. Gray was Fisher Professor of natural history at Harvard University and lectured at the Lawrence Scientific School (Dupree 1959).

[6] Gray sent CD notes on this subject with the letter from Asa Gray, 2–3 July 1862; the notes have not been found, but see the letter to Asa Gray, 23[–4] July [1862] and nn. 23–4.

[7] See letters from G. C. Oxenden, [before 30 May 1862], 21 June 1862, and 8 July 1862, and letter to A. G. More, 7 June 1862.

[8] Duchartre 1862. Oliver's letter has not been found; however, there is a note in CD's hand, recording this reference, in DAR 70: 161. The note is pasted on a second piece of paper, which bears the annotation: 'M. D— has given a full historical account of the form of Catasetums | (I doubt whether worth geting)'. CD cited this paper in *Orchids* 2d ed., p. 196 n.

[9] Pierre Etienne Simon Duchartre's paper, which discussed the different kinds of polymorphism in orchids, included an account of *Vanda Lowei*, a Sumatran orchid in which the flowers at the base of each spike are consistently different in colour and form from the rest of the flowers on the spike (Duchartre 1862, pp. 118–22). Duchartre suggested that this phenomenon might be explained by supposing *V. Lowei* to be a wild hybrid, with the two sorts of flowers representing another example of the dissociation of parental types well documented in *Cytisus Adami* (see n. 15, below). On *Vanda Lowei*, see also the letter from J. D. Hooker, 20 August 1862 and n. 12.

[10] CD had described an unnamed species of *Mormodes*, sent to him by James Veitch Jr, in *Orchids*, pp. 265–9. Veitch subsequently sent CD further flowers from the same plant, which CD described in notes dated 9 July 1862 (DAR 70: 103–4); these notes were headed 'Cycnoches Ventricosum | named by Lindley'. See also letter to H. G. Bronn, 11 July 1862, and 'Fertilization of orchids', p. 155 (*Collected papers* 2: 151–2).

[11] CD acknowledged receipt of a specimen of *Cycnoches* from the Royal Botanic Gardens, Kew, in the letter to J. D. Hooker, 6–7 October [1861] (*Correspondence* vol. 9). In his notes on *C. ventricosum*, dated 9 July 1862 (DAR 70: 103–4), CD described the observations that led him to conclude that the specimen he had received from Veitch (see n. 10, above) was hermaphroditic. He then noted: 'See reference to Mr Bateman. C. Egertonianum must be male or female—'. The reference is to CD's notes (DAR 70: 152) on Bateman [1843], recording that:

> Cycnoches Egertonianum . . . in Guatemala produced & once in England a scape dark purple & totally different species of Cycnoches, but generally in England scapes of the common Cycnoches ventricosum yellow & much larger— . . . It w^{d} appear as if in this country it *reverted to C. ventricosum.*

CD had originally noted: 'C. ventricosum seeds itself, so not sexual distinction', but in *Orchids* 2d ed., p. 224, he reported that he was inclined to believe, by analogy with *Catasetum tridentatum*, that the various forms were to be explained by the existence of 'male, female, and hermaphrodite forms of the same species of Cycnoches.'

12 William Hugh Gower was a foreman at the Royal Botanic Gardens, Kew.

13 Oliver had sent CD a copy of Müller 1857 (see letter to Daniel Oliver, [before 11 June 1862] and n. 2, and letter to J. D. Hooker, 11 June [1862]).

14 CD probably refers to the bird's-eye primrose, *Primula farinosa*; having failed to supply CD with specimens of this species during the previous season (see *Correspondence* vol. 9), Oliver had expressed a hope in April 1862 that he would be able to do so in the coming season, and had promised to write to let CD know (see letters from Daniel Oliver, 10 April 1862 and 14 April 1862; see also letter to Daniel Oliver, 12 [April 1862]). *P. farinosa* flowers in Britain in June and July (Lindley 1859, p. 184).

15 CD had for many years been interested in the origin of the hybrid laburnum, *Cytisus Adami*, branches of which had reportedly been found to bear the leaves and flowers of both the parent species (*C. purpureus* and *C. laburnum*, the common laburnum) in addition to those of the hybrid form (see *Correspondence* vols. 4–6). In *Variation* 1: 387–97, CD discussed at length the evidence for and against the different theories that had been advanced to account for the hybrid form, dwelling particularly on the possibility that it was either an ordinary hybrid, formed by seed, or that it was what he called a 'graft-hybrid'. In regard to the former he noted that he had 'tried in vain to cross *C. laburnum* and *purpureus*' (p. 389), continuing: 'when I fertilised the former with pollen of the latter, I had the nearest approach to success, for pods were formed, but in sixteen days after the withering of the flowers they fell off.'

16 [Oliver] 1862c. At the end of his review of 'Dimorphic condition in *Primula*' in the *Natural History Review*, Oliver expressed his hope to be able to discuss dimorphism in the Orchidaceae 'in connection with Mr. Darwin's new work on the 'Fertilisation of Orchidaceae' in a future number.' In the event, Joseph Dalton Hooker wrote the review of *Orchids* for the *Natural History Review* (see letter from J. D. Hooker, 7 November 1862).

17 See letter to M. T. Masters, 8 July [1862] and n. 3.

To J. D. Hooker 26 July [1862]

Down
July 26th

My dear Hooker

I sincerely rejoice that your tour answered so well for Mrs. Hooker, & that you have returned safe. I hope it did you good & rested you.—[1]

We have been utterly miserable; but it over now: for now patience alone is wanted; & when he is strong enough we shall take him to the sea.—[2] Did you ever hear of such a catalogue of evil. Scarlet fever, enlarged glands of neck, injured kidneys—recurrent scarlet fever with fresh & bad sore-throat & eruption—dredful erysipelas of the head & face—fever with typhalid petechiæ.[3] Port-wine alone saved him.

I have not done a stroke of work for weeks & it has played old Harry with my experiments.—[4]

Your Hybrid orchids are interesting to me, as I never heard of but one case before.[5]

I was struck also with review of Nat. Hist R. in the Parthenon (which I take in):[6] now you point it out that last page is astounding.[7] I remember being surprised at "tubular" stems & wondering what "squarrose cymes" were. What an odd case that of the Calluna.—[8] I wrote to you Poste Restante in the Swiss Valley;[9] but there was nothing in my note worth sending.

Goodnight my dear old friend. | C. Darwin

It is surprising that many hybrids orchids are not produced, when clearly allied species grow & flower together. George caught a moth sucking G. conopsea, with the pollen-mass of a Habenaria bifolia attached to its face.—[10]

Endorsement: '/62'
DAR 115.2: 159

[1] See letter from J. D. Hooker, [24 July 1862] and n. 1.

[2] Leonard Darwin was recovering from scarlet fever (see Emma Darwin's diary (DAR 242); see also preceding letter, letter to M. T. Masters, 24 July [1862], letter to W. E. Darwin, [24 July 1862], and letter to Asa Gray, 23[–4] July [1862]).

[3] CD may have intended to write 'typhoidal' (see letter to Asa Gray, 23[–4] July [1862]); petechiae are red or purple spots on the skin that may accompany fever (*OED*).

[4] On CD's experiments, see, for instance, the letters to J. D. Hooker, 30 May [1862] and n. 7, and 23 June [1862] and n. 4, the letter to Alphonse de Candolle, 17 June [1862] and n. 2, and the letter to M. T. Masters, 8 July [1862] and n. 3.

[5] See letter from J. D. Hooker, [24 July 1862].

[6] CD refers to the review of the first volume of the new series of the *Natural History Review* published in the *Parthenon* 1: 373–5 (see also letter from J. D. Hooker, [24 July 1862] and n. 8).

[7] See letter from J. D. Hooker, [24 July 1862]. The reference is to the last page of the April issue of the *Natural History Review*, which contained numerous typographical errors.

[8] The last page of the April 1862 number of the *Natural History Review* concluded with an unsigned article recording the occurrence of *Calluna vulgaris* (misspelled '*Callema*' throughout) in North America; the author reported Asa Gray's opinion that this might be one of those 'species of the old world so sparingly represented in the new, that they are known only at single stations' (*Natural History Review* n.s. 2 (1862): 346).

[9] CD's letter has not been found. The Hookers had been on holiday in Switzerland from 4 to 23 July 1862 (see letters from J. D. Hooker, 2 July 1862, 10 July 1862, and [24 July 1862]).

[10] George Howard Darwin had made observations in June on the insects visiting several orchid species, including *Gymnadenia conopsea* (see letter to J. D. Hooker, 30 [June 1862], and letter to Asa Gray, 1 July [1862]). In DAR 70: 30 there is a note, dated 21 June 1862, that states that George had seen the moth *Plusia chrysites* 'with one pollinia of Butterfly! [the butterfly orchid, *Habenaria bifolia*] it was settled in conopsea & sucking; but had none of conopsea attached!'

To Asa Gray 28 July [1862][1]

Down Bromley Kent
July 28th

My dear Gray.—

I hardly know what to thank for first. Your stamps gave infinite satisfaction. I took him first one lot & then an hour afterwards another lot.[2] He actually raised himself on one elbow to look at them. It was the first animation he showed. He said only "You must thank Prof. Gray awfully".— In evening after long silence, there came out the oracluar sentence "He is awfully kind". And indeed you are; overworked as you are, to take so much trouble for our poor dear little man.—

And now I must begin the "awfullys" on my own account: what a capital notice you have published on the orchids![3] it could not have been better; but I fear that you overrate it. I am very sure that I had not the least idea that you or anyone would approve of it so much.— I return your last note[4] for chance of your publishing any notice on subject; but after all perhaps you may not think it worth while; yet in my judgment *several* of your facts, especially P. hyperborea are *much* too good to be merged in a Review.[5] But I have always noticed that you are prodigal in originality in your Reviews.— I wish with all my heart I could understand use of the common vibratile labellum in foreign orchids;[6] insects would have to watched at work.—

You speak of two Pogonias always growing together;[7] see my account of Aropera, a *male* orchid.[8] Cannot you get a pupil to watch Calopogon? I sh^d. be very much surprised if insect suck stigma: the speculation crossed my mind & was rejected from want of any analogous case.[9] I received two copies of your Review; so I sent one to Hooker,[10] as if you have not sent it, he would surely like to see your excellent remarks.— You exactly express my conclusion about Greenland. Hooker, I think, underrates occasional means of transport.[11] Till I proved the contrary, he used to maintain vehemently without a fact that the sea would kill all seeds.[12] I have not had time yet to look at Mitchellia.[13] Cannot you persuade your pupil to protect under fine net & experiment on some plants.—[14] Perhaps Houstonia is visited by moths:[15] I have reason to *suspect* that many Galiaceæ are so visited.

I was looking at Lythrum salicaria this morning; it is beautifully dimorphic like Primula, but with addition in both forms (I believe there is a third form) of six short stamens: it will be curious to make out use of them.[16]

But I must hold hard, otherwise I shall spend my life over dimorphism: what I said about pollens & stigmas of Linum grandiflorum, knowing each other is confirmed by further experiment.[17]

I have written to Trübner about the copies of Orchid Book.[18]

Farewell. This is a dull letter & is sent merely to give you my most cordial thanks.— Adios | C. Darwin

Gray Herbarium of Harvard University (75)

[1] The year is established by the references to Leonard Darwin's illness and to Gray's review of *Orchids* (A. Gray 1862a).

[2] Gray had evidently sent a selection of American postage stamps for Leonard Darwin, who was recovering from scarlet fever, with the letter from Asa Gray, 15 July [1862] (see also CD's annotations to that letter).

[3] Gray's review of *Orchids* (A. Gray 1862a) appeared in the July number of the *American Journal of Science and Arts*; Gray had sent CD a copy of the review (see letter from Asa Gray, 15 July [1862] and n. 16).

[4] CD probably refers to the main body of the letter from Asa Gray, 15 July [1862], which has not been found.

[5] In the letter to Asa Gray, 23[–4] July [1862], CD tried to persuade Gray to publish a separate account of his observations on American species of orchids, rather than merely adding them to his review of *Orchids*, and sent back all Gray's notes to enable him to do so. Gray detailed some of his observations, including those on *Platanthera hyperborea* in the follow-up article to his review of *Orchids*

(A. Gray 1862b); he also gave an account of *P. hyperborea* in A. Gray 1862c, pp. 259–60. See also letter to Asa Gray, 23[–4] July [1862] and nn. 23–5.

6 In *Orchids*, p. 171, CD conjectured that the extreme flexibility and liability to movement of the labellum in some orchids might be designed to attract insects 'in the same manner as the bright colours and strong odours of many other Orchids apparently serve to attract insects' (*Orchids*, p. 171).

7 CD discussed Gray's observations on *Pogonia* in the letter to Asa Gray, 14 July [1862]. From CD's annotation to the letter from Asa Gray, 15 July [1862], it appears that Gray sent CD further notes on the subject with that letter; these notes have not been found.

8 CD misspelled *Acropera*; in *Orchids*, pp. 206–10, CD noted that, 'although no instance of the separation of the sexes was known in Orchids', he had become convinced that the specimens of *Acropera luteola* he had examined were from a male plant.

9 Gray's notes on this species have not been found, but see the letter from Asa Gray, 2–3 July [1862], in which Gray reported his intention of looking at *Calopogon pulchellus*.

10 Joseph Dalton Hooker.

11 The reference is to Gray's review of Hooker's paper on the distribution of Arctic plants (J. D. Hooker 1861a), which followed Gray's review of *Orchids* (A. Gray 1862a) in the July issue of the *American Journal of Science and Arts*. In his paper, Hooker sought to explain the general paucity of the Greenland flora on the hypothesis that 'the polar region was once occupied by the Scandinavian flora', and that the cold of the glacial epoch drove the vegetation southwards, causing the temperate flora of southern Greenland to be 'driven into the sea'. Following 'the return of the heat', Hooker argued, Greenland would have been repopulated by the migration of Arctic species surviving in the south of the peninsula (J. D. Hooker 1862a, p. 254). On reading Hooker's paper, CD concluded that 'during the coldest part of Glacial period, Greenland must have been quite depopulated', suggesting that it must then have been repopulated by 'accidental means of transport by ice & currents' (see letter to J. D. Hooker, 25 February [1862]; see also letter from J. D. Hooker, 3 March 1862, and letter to J. D. Hooker, 7 March [1862]). In his review (A. Gray 1862d, p. 145), Gray reported Hooker's hypothesis, but stated:

> Considering the present frigid climate of Greenland, . . . its moderate summer and low autumnal temperature, we should rather have supposed the complete extermination of the Greenland ante-glacial flora; and have referred the Scandinavian character of the existing flora . . . directly to subsequent immigration from the eastern continent.

See also letter to J. D. Hooker, 4 November [1862].

12 Starting in March 1855, partly prompted by Hooker's scepticism on the subject, CD had carried out a series of experiments to ascertain the fertility of seeds after immersion in salt water (see *Correspondence* vol. 5, letter to J. D. Hooker, 7 April [1855]).

13 Gray sent CD a specimen of *Mitchella repens* with the letter from Asa Gray, 15 July [1862].

14 CD refers to one of Gray's students, Joseph Trimble Rothrock (see letter from Asa Gray, 15 July [1862]).

15 See letter from Asa Gray, 15 July [1862] and n. 8.

16 CD recorded the first of a series of observations on *Lythrum salicaria* on 29 July 1862 (DAR 27.2 (ser. 2): 1). See also letter to Asa Gray, 1 July [1862], and letter to W. E. Darwin, 9 July [1862].

17 See letter to Asa Gray, 14 July [1862] and n. 12.

18 Gray had asked CD to arrange for six copies of *Orchids* to be sent to him by the London publisher and bookseller Nicholas Trübner (see letter to Nicholas Trübner, 23 June [1862], and letter to Asa Gray, 1 July [1862]).

From Daniel Oliver 28 July 1862

Monday morn[g]. 28.vii.62

My dear Sir,

I have seen Gower & he gives me the accompanying which he thinks *may* be the "dark *Cycnoches*" of your letter.—[1] He also gives me a few flowers of the two spp. of *Acropera*, *A. Loddigesii* & "*A. luteola*".[2]

I must *try* & notice your extraordinary work, tho' I am very doubtful as to the mess I shall make.—[3]

Very sincerely yours | D^l. Oliver.

DAR 173: 16

CD ANNOTATION

End of letter: '8^d'[4] *ink*

[1] See letter to Daniel Oliver, 24 July [1862]; William Hugh Gower was a foreman at the Royal Botanic Gardens, Kew.

[2] Oliver had assisted CD with his work on *Acropera* the previous autumn (see *Correspondence* vol. 9, letters to Daniel Oliver, 30 November [1861] and 7 December [1861], and letter to J. D. Hooker, 28 [December 1861]). In *Orchids*, pp. 203–10, CD explained that he had to some extent removed the 'opprobrium' of being unable to elucidate the mechanism of fertilisation in these two species by his discovery that *A. luteola* was apparently dioecious, and that the plant at the Royal Botanic Gardens, Kew, from which he had seen specimens was male. He confessed, however: 'the use of some important parts remains quite unintelligible'. In *Orchids* 2d ed., p. 166, CD noted that he had made observations on these two species 'during several seasons'.

[3] In the letter to Daniel Oliver, 24 July [1862], CD had encouraged Oliver to write a review of *Orchids*.

[4] The annotation refers to the cost of postage of the plant specimens (see letter to Daniel Oliver, 29 [July 1862] and n. 7).

From Philip Henry Stanhope 28 July 1862

Chevening
July 28. 1862

My dear M^r. Darwin

Would you & M^rs. Darwin do us the pleasure of coming over here early on the afternoon of Thursday the 7th. & staying to dinner.[1]

If instead of returning at night you would sleep at Chevening we should be all the better pleased[2]

y^rs. very sincerely | Stanhope

DAR 177: 244

[1] No reply to this letter has been found; however, Emma Darwin's diary (DAR 242) records for 7 August 1862: 'went to London to see children and returned'. (Horace, Elizabeth, and Henrietta Emma Darwin were at that time staying in London because of Leonard Darwin's illness (*Emma Darwin* 2: 178); Francis and George Howard Darwin were staying with William Erasmus Darwin in Southampton (see letter from G. H. Darwin, [after 5 August 1862]).)

[2] Chevening, near Sevenoaks, Kent, was the seat of the Stanhope family.

From Asa Gray 29 July 1862

Providence R.I.
July 29, 1862

My Dear Darwin.

No more news in the Orchis line. I am making 2 or 3 days of holiday, and yesterday I found a few specimens of *Gymnadenia tridentata*. But the flowers are too small to examine well with a hand lens. If they keep, I will take them back to Cambridge in a day or two, and see what is to make of them.[1] I write a line to say that I have just received the 6 copies of Orchis-book from Trübner. And I wish to ask you *not* to pay Trübner for them.[2] Leave it for me to do at my leisure. I find—supposing the book is a *half-guinea* book, that he has charged me £3·3. for the six,—i.e. full retail price, instead of treating my order as he would have treated one from a bookseller.—which is what I was fairly entitled to.

It would be gross, therefore that *you* should pay £3·3. for what your own publisher would have supplied you for a little more than two.

On my return home I must sit down and write a further notice of your book. But I hope that, meanwhile, I shall learn from you how you like my first notice.[3] You ought to be satisfied with it, as it is mainly a string of extracts from the book itself.

As to the country, you will see by this time that we have not the least idea of abandoning the struggle. We have learned only, that there is no use trying any longer to pick up our eggs gently, very careful not to break any. The South force us at length to do what it would have been more humane to have done from the first,—i.e. to act with vigor,—not to say *rigor*.[4]

We shall be complained of for our savageness, no doubt,,—whereas we feel that our error has been all the other way. But the independence, the total indifference to English feeling which you recommended last year, has come at length. Now we care nothing what Mrs. Grundy *says*.[5]

Ever, dear Darwin. | Yours faithfully | Asa Gray

Postmark: AU 11 62
DAR 165: 115

[1] For Gray's observations on this species, see the letters from Asa Gray, 4 August 1862 and 18–19 August 1862.

[2] Gray had asked CD to arrange for six copies of *Orchids* to be sent to him by the London publisher and bookseller, Nicholas Trübner, who frequently acted as Gray's London agent (see letter to Nicholas Trübner, 23 June [1862]). CD had offered to pay for these copies (see letter to Asa Gray, 1 July [1862]), but Gray declined CD's offer in the letter from Asa Gray, 21 July 1862.

[3] A. Gray 1862a; the review appeared in the July number of the *American Journal of Science and Arts*, of which Gray had sent CD a copy (see letter from Asa Gray, 15 July [1862] and n. 16). Gray followed the review with an article in the November issue of the journal, consisting of observations on a number of North American orchids (A. Gray 1862b).

[4] Following the failure of the Union army to take the Confederate capital, Richmond, Virginia, at the start of July 1862, Union policy 'took a decisive turn toward total war' (McPherson 1988, p. 490; see also letter from Asa Gray, 2–3 July 1862 and n. 22).

[5] 'Mrs Grundy' is a character in Thomas Morton's play, *Speed the plough* (1798); fear of her sneering causes her neighbour, Dame Ashton, repeatedly to ask 'What will Mrs. Grundy say?' (*OED*).

To Daniel Oliver 29 [July 1862][1]

Down Bromley Kent
29th

Dear Oliver

Thanks for Orchids.—[2] I liked to see the strange labellum of the Catasetum new to me.—[3] Cycnoches is considerably different; there is a truly wretched sketch of the genus at the back of my orchid Book.—[4] Do not take any more trouble about it: I only thought that as I had carefully examined C. ventricosum, I sh^d. like to see its male.—[5]

Yours very truly | C. Darwin

I care more for Dimorphism now than orchids, & I think it is a more important subject. To day I have been looking at Lythrum & have seen the *three* forms;—ie long-styled—mid-styled & short-styled:[6] as each form has two sets of anthers, 18 *different* crosses are practicable within the limits of this one species!! As I ought to cross 10 of each, this would make 180 fertilisations & markings & counting of seeds. A nice job, Heaven knows whether my patience will last; but I sh^d. like to make out this wonderfully complex case—

I enclose stamps not to cheat Kew.—[7]

DAR 261 (DH/MS 10: 55)

[1] Dated by the relationship to the letter from Daniel Oliver, 28 July 1862.

[2] See letter from Daniel Oliver, 28 July 1862.

[3] There is a note in DAR 70: 98, dated 29 July 1862, that reads:

> a curious Catasetum sent me from Kew. with Lab[ellum] upwards, & thus is formed into a bucket with small oral orifice, by the base— lateral lobes meeting each other in front of column. The two antennae project in middle of the dark cavity & their tips can be seen through the oral orifice. Like a jar with narrowed in mouth.

[4] CD had discovered that the specimen of which he provided a drawing in *Orchids*, p. 267, under the title 'Section of the flower-bud of a Mormodes', was actually *Cycnoches ventricosum* (see letter to Daniel Oliver, 24 July [1862] and n. 10; see also *Orchids* 2d ed., p. 223).

[5] See letter to Daniel Oliver, 24 July [1862] and n. 11.

[6] Having learned in December 1861 that *Lythrum salicaria* was trimorphic, CD had obtained plants for experiment from the Royal Botanic Gardens, Kew, and elsewhere (see letter to J. D. Hooker, 22 [March 1862], and DAR 27.2 (ser. 2): 1–7). There are a number of notes in DAR 27.2 recording the details of crossing experiments carried out by CD on this species in the summer of 1862; the earliest is dated 31 July 1862 (DAR 27.2 (ser. 2): 7). CD's paper on the three forms in *Lythrum* ('Three forms of *Lythrum salicaria*') was read before the Linnean Society of London on 16 June 1864.

[7] CD enclosed stamps worth 8*d.* to cover the cost of postage of the plant specimens sent with the letter from Daniel Oliver, 28 July 1862 (see CD's annotation to that letter).

From W. E. Darwin 1 August 1862

Southampton & Hampshire Bank, Southampton
Aug 1 *1862*

My Dear Father,

I have got your Lythrum letter, and will send you the 3 kinds tonight. it certainly will be most awfully complicated work with 18 possible crosses.[1] there is as much Lythrum as you like to be got here. when I first found a bed, or rather *beds* along a stream, I gathered haphazards parts of 27 different plants, and examined them when I got home.[2] it is very odd the symmetry the the division had out of the 27 plants

11	of them were	long pistilled		=	Lp
9	—	short	—	=	Sp
7	—	middle	—	=	Mp.

You see the short are exactly a $\frac{1}{3}$ the Lp—2 above the $\frac{1}{3}$, the Mp—2 below the $\frac{1}{3}$.

If you liked I could gather 90 or 120 or 150 or 300 plants and class them. there is one odd thing if *true*, that would I should make it less complicated; all the Lp that I have looked at yet are less ripe or advanced than the Sp or the Mp.

This was plain in the 27 plants, (unless of course I have made some hideous mistake)

when the pollen is ripe and the anthers are opening, the filament is crimson and the pollen green— in these 27 plants the filaments were crimson and the pollen green in all the long stamens (only of course of the quite open flowers) both of the Sp and of the Mp.— while in all the 11 heads of the Lp there was not a single red filament or green pollen to be seen—and I looked tolerably carefully through them all.

I have looked at the *two* pollen of the Lp, and drawn and measured them by Camera Lucida and there is a decided difference.[3]

there is also difference I think between the pollens of the different kinds, but you will see all that.

I had gathered a lot more yesterday to have another look, but with the family luck my mare tumbled crossing her legs and cut her knee very badly, and in the scrummage all the plants tumbled out of my case

Incomplete
DAR 162: 90

CD ANNOTATIONS
1.1 I have . . . home. 1.5] *crossed pencil*
2.2 '$\overline{27}$' *added under column of numbers, pencil*
3.1 If you . . . my case 8.3] *crossed pencil*

[1] CD's letter has not been found; however, in the preceding letter, CD described the crosses he planned to make with the three forms of *Lythrum salicaria*. See also letter to W. E. Darwin, 9 July [1862].

[2] William's observations on the twenty-seven *Lythrum* plants, dated 26 July 1862, are detailed in his botanical notebook (DAR 117: 36–7). See also William's earlier observations on *Lythrum*, dated 13 July and 20 July, in DAR 117: 1, 12–13.

[3] In the letter from W. E. Darwin, 5 August 1862, William enclosed camera lucida drawings, dated 1 August 1862, of the two sets of pollen grains from the long and short stamens of the long-styled form of *Lythrum salicaria*.

From J. O. Westwood 1 August 1862

Oxford
1 Augt 1862

My dear Sir

I lent my bee to Dr Daubeny for his lecture on your Book,[1] & unfortunately the pollinia got shaken off in transitu.[2] If you like to see the insect & the detached appendage, I will send you the bottle. I suppose it was Orchis Maculata as it was in flower & I had one or two plants in my garden

Yours very sincerely | Jno O Westwood

DAR 181

[1] Westwood, who was professor of zoology at Oxford University, refers to the lecture on *Orchids* given at the end of June by the professor of botany, Charles Giles Bridle Daubeny (see letter from C. G. B. Daubeny, 5 July 1862).

[2] In the letter from J. O. Westwood, 14 May 1862, Westwood offered to send CD a preserved bee specimen with pollen masses attached to its head.

From W. E. Darwin 2 August [1862][1]

Southampton
Aug 2.

My Dear Father,

I send off the 3 kinds of Lythrum, but I foolishly forgot moss, till I had not time to get any, so if they come withered I must send some more.[2] I got these at a different place to the ones I had examined & I find the proportion of *M*p's much greater than before, at least 3 or 4 to 1 of either of the other, I only judge this from hunting about for Lps and Sps. the Mps seemed to grow in large clumps together.

Incomplete
DAR 162: 91

CD ANNOTATION
1.1 I send . . . examined 1.3] *crossed pencil*

[1] The year is established by the relationship to the letter from W. E. Darwin, 1 August 1862.

[2] See letter from W. E. Darwin, 1 August 1862. There is a note, dated 'Aug 3^{d}', recording CD's observations on specimens of the three forms of *Lythrum salicaria* sent from Southampton, in DAR 27.2 (ser. 2): 8.

To W. E. Darwin [2–3 August 1862][1]

Down
Saturday

My dear William.—

I am very sorry to hear about your mare. I never knew what an unfortunate family we are in horse-flesh. It really is most provoking.—[2] Mamma orders me to compliment you on what a capital hand you have taken to write: it is very good. Aunt Susan was much pleased with your very nice note to her, which she sent on to us.—[3]

Now for business: Lythrum is a really wonderful case & will be worth an enormous amount of labour; & labour, by Jove, there will be, for I find I must castrate every flower. It will be a grand ending to my work on dimorphism.— The enclosed splendid diagram shows what I have hardly a doubt is the case: you must study it; the red lines show which pollen will fertilise which stigmas;[4] all the other crosses I expect to be sterile or little fertile. In the long-styled (your long pistilled)[5] the longer anthers are *always* yellow: has not this deceived you about maturity of flowers? *Please let me hear*.— I had overlooked difference in colour of filaments;—thanks.—

It w^d. be enormous advantage if you could have patience to gather at least 100 twigs & class them, & then I sh^d. give fact on your authority:[6] employ the Boys;[7] but caution them to gather only one twig from each plant.— I noticed slight difference in pollen of long & short stamens in the long-styled; & *considerable* difference between green & yellow pollen in the mid- & the short-styled plants. I wanted 3 good specimens. (not crushed with *little* damp, not wet, moss) to compare pollens: I suspect those that go in pairs in diagram are alike or nearly alike.—

I dare not trust specimens in my garden for some are monstrous. I also wanted specimens from same locality to get relative heights of stamens & pistils, & to put in spirits for Artist. You will not have time to do it; but if you could make me *with camera in same position* outlines of all *six* pollens, I would have them engraved from your outlines: these outlines would show whether there are three or more than three sizes & shapes. Pollen had better be **dry**; it sticks to glass I find; but whether when perpendicular I know not. But I do not think you will have time for the outlines.—[8]

If my diagram is the truth, is it not very curious & complex case? nothing like it, is known in whole organic world. See how well corresponding heights will carry pollen on proboscis from flower to flower. Do Humble or Hive-bees visit the flowers? ask George to see to this.—

Did you say that the other sp. of Lythrum grows with you?—Vaucher says it has only 2 forms;[9] I am *intensely* curious to see them; & if you can I would get you in autumn to send me 3 or 4 plants of each kind.

GoodBye my dearest old man.— Lenny is going on very well.[10] We are perplexed to death what to do. The Doctors urge in strongest terms to avoid all chance of infection; so did my Father in bad cases, & this has been very bad. All the children are now perfectly safe, if we can avoid future infection.—[11]

C. Darwin

I think I have plants enough of Lythrum salicaria, but could you mark two of *mid-styled*.

Sunday Evening

P.S. I kept back my letter, because I received this morning the specimens.[12] They have been **very** useful: I return stamps. I am perplexed about pollen, & I believe I must measure all again distended by water: the breadth of dry grains seem to vary much from some cause.— Will you mark with string (3 strings for long-styled, 2 for mid & 1 for short-styled) 3 plants of Lythrum, as I much want a few pods, (wrapped up separately when ripe) to count seeds in each, to know natural product, & to compare shape of pods.—

George is a rare hand at watching insects[13] & beg him to observe carefully what part of body of bee the knobbed stigma of *long* & *mid*-styled rubs against, when Bee sucking: no doubt the short anthers & stigma of short-styled rubs against the proboscis. After a Bee has visited several flowers of *mid*- & *short*-styled & long-styled flowers, it would be grand to catch him & kill him & see with lens whether one part of body did not abound with the green pollen & another part with yellow. Ask George to study my diagram & he will see what I mean; it is like the two pollens of Cowslip on proboscis of moth & Bees.

George ought to look out for Epipactis latifolia now in flower, & see what insects visit it.[14] Good Night— Lenny sat in arm-chair for $\frac{1}{2}$ an hour.— Goodnight my three dear Boys.— | C. D.

I shall weary you about Lythrum.

[Enclosure][15]

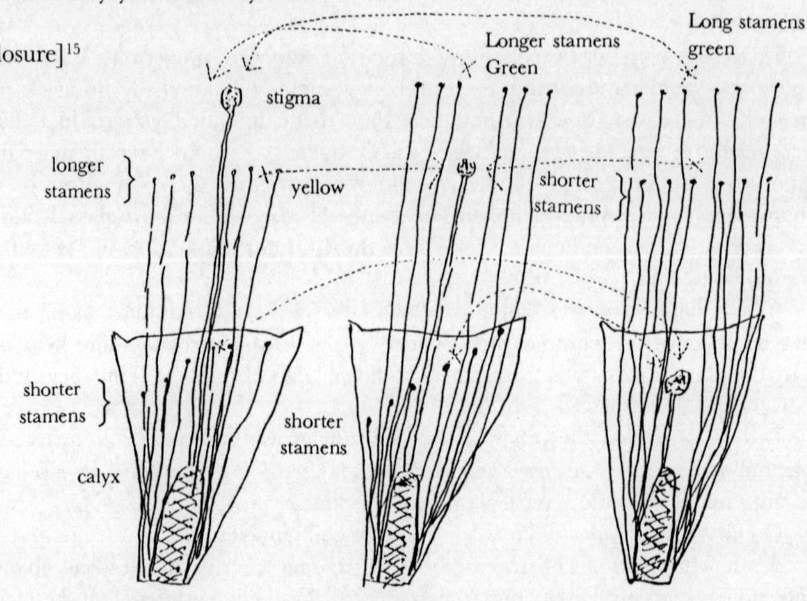

Long-styled Mid-styled Short-styled

DAR 185: 70, DAR 210.6: 102

1 Dated by the relationship to the letters from W. E. Darwin, 1 August 1862, 2 August [1862], and 5 August 1862. The letter was written on Saturday and Sunday; 2 August 1862 fell on Saturday.

2 See letter from W. E. Darwin, 1 August 1862.

3 The reference is to CD's sister, Susan Elizabeth Darwin. In a letter to William of [late July 1862] (DAR 219.1: 59), Emma Darwin reported that 'Aunt Susan' had 'a great fancy for larking down to Southampton with the boys', and suggested that William should write to invite her.

4 See enclosure.

5 See letters from W. E. Darwin, 1 August 1862 and 2 August [1862].

[6] William examined and classified specimens from 102 plants of *Lythrum salicaria* on 6 August 1862 (see letter from W. E. Darwin, 5 August 1862); CD cited his results, along with those given in the letter from W. E. Darwin, 1 August 1862, in 'Three forms of *Lythrum salicaria*', p. 174 (*Collected papers* 2: 110).

[7] Francis and George Howard Darwin arrived in Southampton on a visit to William on 2 August 1862 (see letter from G. H. Darwin, [after 5 August 1862] and n. 3).

[8] There are camera lucida drawings of the two sets of pollen from each of the three forms of *Lythrum salicaria*, all with August 1862 dates, in DAR 27.2 (ser. 2): 27–32 bis. William enclosed the first of these with the letter from W. E. Darwin, 5 August 1862. Although CD referred to William's drawings in 'Three forms of *Lythrum salicaria*', p. 172 (*Collected papers* 2: 108–9), he did not reproduce them as engravings.

[9] Vaucher 1841, 2: 371. CD refers to *Lythrum hyssopifolia* (see letter to H. C. Watson, 8 [August 1862]). See also 'Three forms of *Lythrum salicaria*', pp. 190–1 (*Collected papers* 2: 124).

[10] Leonard Darwin was recovering from scarlet fever (see Emma Darwin's diary (DAR 242), and letter to W. E. Darwin [24 July 1862]).

[11] During the latter part of Leonard's illness, the other Darwin children had been sent away with their former nurse, Brodie, who was at Down at the time (*Emma Darwin* 2: 178; see also letter to W. E. Darwin, 4 [July 1862] and n. 8).

[12] See preceding letter.

[13] In June, George Darwin had made a series of observations on the insects visiting orchids (see letter to J. D. Hooker, 30 [June 1862] and n. 16, and letter to Asa Gray, 1 July [1862] and nn. 7–9).

[14] The following year, CD made his own observation of insects visiting *Epipactis latifolia* on plants that grew near his home, and concluded that they were pollinated by a species of wasp (see 'Fertilization of orchids', p. 150 (*Collected papers* 2: 146–7)). He remarked in a note dated 13–14 August 1863: 'We had all often watched and saw no insect visiting though Bees abounded!' (DAR 70: 56).

[15] In the original, the red lines were drawn in crayon; they are represented in the diagram by dashed lines. The diagram is reduced to 50% of its original size.

From Margaret Susan Wedgwood [before 4 August 1862][1]

The Manor | Llandudno

Dear Uncle Charles

Of 256 specimens of Lythrum gathered this morning from different plants, we find

94 with long pistil
95 — middle length pistil
69 — shortest pistil.

These plants were all in one large field or near it but tomorrow we will go if we can manage it in a different direction for more—[2] We find it rather difficult in gathering to know what are distinct plants and what only offsets. At Criccieth & Aber we thought the different sorts usually grew in plots together mixed with a few of the others but here they are all together. We have found the Hottonia, and find the pistils of different lengths as I think you told us.[3]

Your affec niece | M. S. Wedgwood

DAR 181

CD ANNOTATIONS
2.2 We find . . . niece 3.1] *crossed pencil*
Top of letter: 'Aug 4th 1862'

[1] Dated by CD's annotation, and by reference to CD's reply to this letter (letter to K. E. S., L. C., and M. S. Wedgwood, 4 [August 1862]).

[2] CD wanted to know the proportions of the three forms of *Lythrum salicaria* in different localities (see, for example, the letter to W. E. Darwin, [2–3 August 1862]), and had evidently asked his nieces, Katherine Elizabeth Sophy, Margaret Susan, and Lucy Caroline Wedgwood, to examine specimens during their visit to North Wales. Margaret provided a further tally of the three forms in the letter from M. S. Wedgwood, [6 August 1862]; CD quoted the totals in 'Three forms of *Lythrum salicaria*', p. 174 (*Collected papers* 2: 110).

[3] CD had learned from Charles Cardale Babington that *Hottonia* was dimorphic, and was anxious to see specimens (see letter from C. C. Babington, 17 January 1862, and letter to C. C. Babington, 20 January [1862]). See also letter to K. E. S., L. C., and M. S. Wedgwood, 4 [August 1862], and letter from M. S. Wedgwood, [6 August 1862].

From Asa Gray 4 August 1862

Cambridge. [Massachusetts]
4th. August, 1862

My Dear Darwin

My pupil, Rothrock, now away in vacation, has sent me a brief abstract of his Observations on *Houstonia cærula*.[1] The following are culled from them.

Long-styled: stigmatic hairs are in length			—	·04 mm.
Short-styled —	"	"		·023 "
	long		wide	
Long-styled pollen	·020.	×	·017	
Short-styled "	·036	×	·02 :	

in the fresh plants, but dry. Distended with water became round, by increase of the shorter axis alone, the long diameter unaltered.

Long-styled	becoming	·026
short-styled	"	·036.

(In *Mitchella*, however, the long diameter was shortened as the other enlarged.)

Pollen of *Houstonia* not wet, seen endwise is quite strongly 3-lobed. In water the reentering angles come out, and so increase the width.

Short-styled had the smaller stigmas, and the largest and best filled anthers.

The above results I overlooked enough to verify substantially. The following I did not. They may be taken as approximately good

A patch of *long-styled*—a good deal choked by other vegetation, yielded 34 capsules. The 6 largest of them gave 86 seeds,—average of $14\frac{1}{3}$. Another patch of same better placed 6 best capsules gave 123 seeds

average $21\frac{1}{2}$.

A patch of short styled: out of 44 capsules, the six best gave gave an average of $15\frac{1}{2}$ each. Out of the 44, 4 were wholly abortive; 8 averaged 2 seeds each. Most of the rest gave about 6 seeds each.

The above was all from wild plants.

Transplanted specimens in the garden.

Long-styled, average no of seeds		
of best capsules	—	13
Short-styled	—	??

Of 13—short-styled pods
11 were wholly sterile
1 had 4 seeds
1 " 8 "

This was from specimens placed far off by themselves, so that they were not likely to be visited by insects which had left long-styled flowers.

Long-styled flowers fertilised artificially by short-styled pollen.: a patch in what proved to be a most unfavorable situation, much dwarfed. But 6 best capsules gave 77 seeds. In flowers of plants covered with fine netting (coarse gauze) a species of *Thrips* abounded, also a larva of some small beetle—completely dusted with pollen.[2] In many flowers a species of *Podura*.

I have been looking at the flowers of *Rhexia Virginica* (I suppose you have examined no true *Rhexia*)[3]

Style declined to lower side of the flower. Stamens with their anthers also declined. The small subapical pore of, latter facing inwards. A little pressure on the base of the anther causes puffs of pollen to be blown out through the pore.

In one flower only have I observed the style change its position, in that it bent over towards the upper side of the flower;—accidental?[4] A clump in water now 4 days: no stigma has been detected with pollen on it.[5]

As far as I can see, the likeliest way is that an insect approaching the flower from below, and searching the tube of calyx prolonged above the ovary into a cup (where if he finds any nectar he has sharper eyes than I have),—as he knocks his head against the enlarged bases of the anthers, will get puffs of pollen against his body, I should think on the sides of a humble bee &c. or abdomen underneath.— If he approaches the next flower from the front, below, he will brush against the subcapitate stigma.[6]

No Orchid examined since my last, except *Gymnadenia tridentata*,—on which I have a few obs.— It is a congener of Platanthera dilatata,—with the discs formed in two large shallow *saucers* occupying the whole breadth of the stigma; and the pollen pockets most readily, detached from the caudicle, some pulling off at a touch. It must be fertilised by a very fine proboscis.[7]

Several more Orchids will soon be sent me by sharp-sighted youngster up in Maine,—to whom I have just sent a copy of your book, to stimulate him.

My latest from you is July 14.[8] It leaves me in a state of much anxiety for your boy.[9] I will hope for a better account in your next.

I looked to-day for seeds of the little *Houstonia* for you—in vain. I told Rothrock to gather or save some & hope he has, but know not.— I am worth little now for any commission,—and—now that I am to set down to systematic work, shall be worth still less to you.

Ever Yours most cordially | Asa Gray

DAR 110 (ser. 2): 67–9

CD ANNOTATIONS

1.1 My pupil, ... from them. 1.2] *crossed blue crayon*
1.8 Distended ... ·036. 1.11] *closing square bracket, red crayon*
1.12 (In ... enlarged.)] *closing square bracket, red crayon*
5.1 A patch ... each. 6.3] *crossed red crayon*
5.1 patch] *underl red crayon*
5.2 $14\frac{1}{3}$] *underl red crayon*
5.4 $21\frac{1}{2}$] *underl red crayon*
6.2 $15\frac{1}{2}$] *underl red crayon*
7.6 Of 13–short-styled ... flowers. 9.11] *enclosed in square brackets, red crayon*
8.1 Long-styled ... 77 seeds. 8.3] '6/77
6
—
17 12·8 sq 13 [*illeg calculation*]' *ink*
8.1 fertilised artificially] *underl red crayon*
8.1 short-styled] *underl red crayon*
8.1 pollen.:] 'Long' *interl red crayon*
8.2 6 best capsules gave 77 seeds. 8.3] *underl red crayon*
9.1 I have ... on it. 11.3] *crossed red crayon*
12.1 the likeliest ... stigma. 12.7] *crossed red crayon*
13.1 No Orchid ... to you. 16.4] *crossed ink*
Top of first page: 'Rhexia' *ink*; 'Houstonia' *blue crayon*; 'Rubiaceae' *pencil*; 'Rothrock' *pencil*

[1] The reference is to Joseph Trimble Rothrock. Gray had informed CD that *Houstonia* was dimorphic in October 1861, and had promised to look for any differences in the pollen of the two forms the following spring (see *Correspondence* vol. 9, letter from Asa Gray, 11 October 1861); he sent CD his own observations on differences between the pollen and stigmas of the two forms in his letter of [2 June 1862], and subsequently promised to send further observations by Rothrock (see letter from Asa Gray, 2–3 July 1862). CD included Rothrock's observations and experimental results in *Forms of flowers*, pp. 132, 254.

[2] In his letter to CD of 15 July [1862], Gray mentioned that Rothrock found only Thrips in *Houstonia.*

[3] In the letter to Asa Gray, 16 February [1862], CD asked Gray to help him in his investigation of the possible occurrence of dimorphism in the Melastomataceae, by making observations of the floral anatomy of *Rhexia*. Gray promised to do so in the summer (see letter to Asa Gray, 16 February [1862], and letter from Asa Gray, 6 March [1862]), but experienced difficulties in obtaining specimens (see letter from Asa Gray, 2–3 July 1862). Meanwhile, CD informed Gray that he was making a number of crosses with a plant of *Rhexia glandulosa* from the Royal Botanic Gardens, Kew (see letter to Asa Gray, 10–20 June [1862] and n. 10, and letter to Asa Gray, 1 July [1862] and n. 13).

[4] In the letter to Asa Gray, 15 March [1862], CD asked Gray to compare the position of the pistil in young and old *Rhexia* flowers, noting that in a related genus he had observed that the pistil and stamens changed position over time, and reporting his suspicion that one set of anthers was 'adapted to pistil in early state, & the other set for it in its later state'.

[5] CD had asked Gray to observe whether *Rhexia* could be fertilised if insects were excluded (see letter to Asa Gray, 21 April [1862] and n. 13). See also letters to Asa Gray, 10–20 June [1862] and 1 July [1862].

[6] In the letter to Asa Gray, 15 March [1862], CD asked Gray to watch how the anthers and stigma touched bees that visited *Rhexia* flowers.
[7] Gray had gathered flowers of *Gymnadenia tridentata* while on holiday in July (see letter from Asa Gray, 29 July 1862); for Gray's further observations on the species, see the letter from Asa Gray, 18–19 August 1862.
[8] Letter to Asa Gray, 14 July [1862].
[9] Leonard Darwin was suffering from scarlet fever (see letter to Asa Gray, 14 July [1862]).

From G. C. Oxenden 4 August [1862?][1]

Crowhurst Park | Battle— Sussex
Augt. 4.—

Dear Sir

At this place,[2] I find many of this charming Epipactis—very erect & showy—& the leaves *beautiful*—[3]

Sincerely | G. C. Oxenden—

DAR 173: 57

[1] The year is conjectured from Oxenden's reference to *Epipactis* (see n. 3, below).
[2] Crowhurst Park, Sussex, was the estate of Thomas Papillon, the husband of Oxenden's sister, Frances Margaret Papillon (*Burke's landed gentry* 1862).
[3] In letters written earlier in the summer of 1862, Oxenden discussed various species of *Epipactis*, and promised to send CD specimens later in the season (see letters from G. C. Oxenden, 15 May 1862, 4 June [1862], and [before 21 June 1862]); Oxenden sent specimens of *E. palustris* with his letter of 8 July 1862. See also letter from G. C. Oxenden, 17 September [1862].

To K. E. S., L. C., and M. S. Wedgwood 4 [August 1862][1]

Down
4th

My dear Angels!

I can call you nothing else.—

I never dreamed of your taking so much trouble; the enumeration will be invaluable.[2] I will write this evening if possible & explain what I have very little doubt is the case with Lythrum, & which I am daily working to prove by most laborious crosses.—[3]

But I write now to ask whether you will be more angelic than angels & send me in tin, not tightly packed, with *little* damp (*not wet*) moss (perhaps tied round stems??) 2 or 3 flowers of *both* forms of Hottonia: I much wish to measure pollen & compare stigmas.—[4]

GoodBye | My dear Angels | C. Darwin

Mrs Ralph Vaughan Williams

[1] The date is established by the relationship to the letters from M. S. Wedgwood, [before 4 August 1862] and [6 August 1862].
[2] See letter from M. S. Wedgwood, [before 4 August 1862] and n. 2.

[3] CD apparently wrote a further letter (see letter from M. S. Wedgwood, [6 August 1862] and n. 3), but it has not been found. On CD's crossing experiments with *Lythrum salicaria*, see, for example, the letter to Daniel Oliver, 29 [July 1862] and n. 6.
[4] See letter from M. S. Wedgwood, [6 August 1862] and n. 2.

From W. E. Darwin 5 August 1862

Southampton & Hampshire Bank, Southampton
Aug 5 *1862*

My Dear Father,

I have got your fine long Lythrum letter, and have packed off the boys at once to watch and gather 99 pieces cautioning particularly about taking different plants.[1] I quite understand the diagram, it is quite beautiful how they all fit each other, and if that is not the way they cross, more fools they is all I can say. It is so elegantly symmetrical, as well as its being so probable that insects should impregnate stigmas of the same height as the stamens from which they got the pollen; I mean to look if the shorter stamens on all have yellow pollen, because then you see each pistil will be impregnated by one sort of pollen either green or yellow, and it will be so odd that the pollen of long stamens of the long styled plants should be yellow on purpose to suit the middle length styled plants.—

I will look again about the maturity of the L. styled. I think it is quite certain that the yellow pollen deceived me; but I remember thinking that the anthers seemed less open, and the pollen softer; but when I class the plants I will look to that; I am not quite certain that the filaments of L. styled don't become red when ripe.[2]

I can easily mark some plants. I am sorry to say the rare Lythrum does not grow about here,[3] it seems to be a rare plant.

I send you drawings by Camera Lucida and measured by micrometer of the pollen of L. styled[4] you will see there is a decided difference in size. I could easily draw all the others; tell me if that is too large a size; the difficulty is to know when to draw them[5] I drew the enclosed just as the anthers open which seems a good *middle* time— tell me what time? as they vary so in shape at different ages; in the bud they are spherical, and get elliptical as they get older. I shall not finish this till I have classed the kinds this afternoon. I got Mama's letter, & Orley[6] please send July Orley and the shirt-front, Franky is not sure whether he has got one or two shirts at all events he can't spare one to be mended.

I had a very nice visit from Aunt Susan[7] she went off yesterday afternoon, we luckily had a beautiful Sunday.

We are off to Christ Church tomorrow to look at Muddyford for houses,[8] which is said to be *very* nice, but does not sound it.

Tuesday

I examined 102 plants this morning[9]

42. Long styled }
31 M. — } 102.
29 Short — }

I think I found a Long styled with green pollen this morning—but will look at it again, anyhow it was only one. I did not quite know whether you wanted more specimens at once, send me a line if you do.

I have just written an account of Bournemouth to Mama in London from where (Bournemouth) I and the two boys have just returned.[10]

Your affect son | W. E Darwin.

[Enclosure][11]

drawn Cam. Lucida. Aug 1. 62.

the measurings by the micrometer seem unequal—and make it more difficult to judge of the sizes

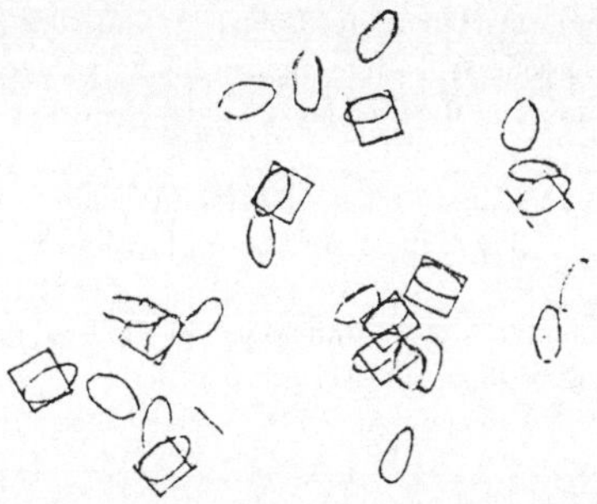

Short stamen's pollen from same flower under same power—

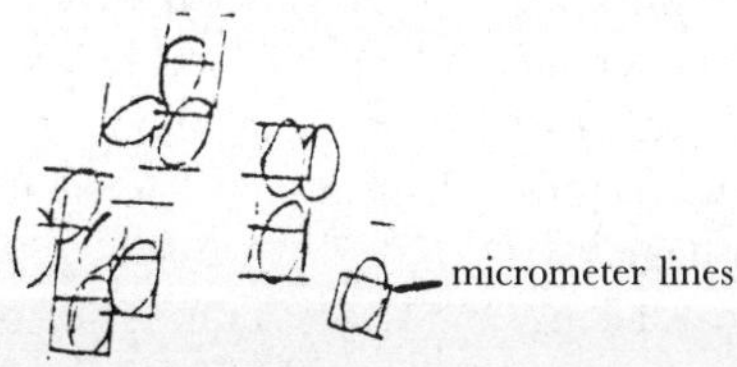

Long stamens pollen from long styled Lythrum.—
open flower— anther valves just beginning to open
under strongest object glass and eye piece

DAR 27.2 (ser. 2): 32 bis, DAR 162: 92

CD ANNOTATIONS
1.1 I have ... Sunday. 5.2] *crossed pencil*
7.1 *Tuesday* ... Short — 7.5] *double scored pencil*
8.1 I think ... returned. 10.2] *crossed pencil*

[1] See letter to W. E. Darwin, [2–3 August 1862] and nn. 6 and 7. CD refers to Francis and George Howard Darwin, who arrived on a visit to their brother on 2 August (see letter from G. H. Darwin, [after 5 August 1862] and n. 3).

2 CD had questioned William's assertion that the flowers of the long-styled form of *Lythrum salicaria* matured later than the other two forms (see letter from W. E. Darwin, 1 August 1862, and letter to W. E. Darwin, [2–3 August 1862]).

3 The reference is to *Lythrum hyssopifolia* (see letter to W. E. Darwin, [2–3 August 1862] and n. 9).

4 See enclosure.

5 See letter to W. E. Darwin, [2–3 August 1862] and n. 8.

6 *Orley Farm*, by Anthony Trollope, was published in monthly shilling parts between March 1861 and October 1862 (J. N. Hall 1991, p. 560); it was published in book form in 1862 (Trollope 1862). The letter from Emma Darwin has not been found.

7 Susan Elizabeth Darwin visited William in Southampton on 2 and 3 August, bringing Francis and George Howard Darwin with her (see following letter). See also letter to W. E. Darwin, [2–3 August 1862] and n. 3.

8 Muddiford is near Bournemouth in Hampshire. Emma Darwin added a postscript to the letter to W. E. Darwin, [24 July 1862], asking William to inquire about 'quiet sea places near Southampton'; the family planned to rent a house in the area for a holiday (see letter to H. C. Watson, 8 [August 1862]).

9 In the letter to W. E. Darwin, [2–3 August 1862], CD asked William to examine at least 100 plants of *Lythrum salicaria* and to produce a tally of the occurrence of the three sexual forms. See also following letter and n. 7.

10 William's letter has not been found; however, Emma Darwin discussed the 'account of Bournemouth', in a letter to William of [9 August 1862] (DAR 219.1: 61).

11 The diagrams in the enclosure have been reduced to 32.5% of their original size.

From G. H. Darwin [after 5 August 1862][1]

too much of a job & so F.[2] & I got the tickets for Thursday night gratis. It was most awful fun & they were very good places. On Saturday we came down here with Aunt Sue.[3] On Monday Birch[4] & a lot of men (Ronnald)[5] from his yatch came & had dinner here. Ernest couldn't stop but had to go off to London early in the morning.[6] They went back to their yatch at about 12 & sailed very early in the morning to the channel islands. On Tuesday I went to watch the Lythrums—& had a very wet scrummage.[7]

I caught 3 hive bees sucking & saw any number of others & also caught 5 flies sucking & saw lots of others; saw a butterfly (Pieris Rapæ) suck several flowers but could not catch it.

Incomplete
DAR 162: 90.1

CD ANNOTATION
1.5 They went ... islands. 1.6] *crossed pencil*

1 The date is established by the reference to George observing *Lythrum salicaria* (see n. 7, below) and by the reference to Susan Elizabeth Darwin (see n. 3, below).

2 Francis Darwin.

3 Susan Elizabeth Darwin evidently joined George and Francis in London, and accompanied them on their visit to William Erasmus Darwin in Southampton. See also letter to W. E. Darwin, [2–3 August 1862], and preceding letter.

[4] This individual has not been identified.

[5] The reference is probably to Ronald Mackintosh, cousin of Ernest Hensleigh Wedgwood (see n. 6, below).

[6] George's cousin, Ernest Hensleigh Wedgwood, was a clerk in the Colonial Office, London (*Post Office London directory* 1861).

[7] In the letter to W. E. Darwin, [2–3 August 1862], CD asked George to observe the insects visiting *Lythrum salicaria*. CD discussed the necessity of insect pollination in this species, and described the insects involved, in 'Three forms of *Lythrum salicaria*', pp. 175–6 (*Collected papers* 2: 111–12). See also preceding letter and n. 9.

From M. S. Wedgwood [6 August 1862][1]

Shrewsbury
Wednesday

Dear Uncle Charles

We made a mad rush this morning after the Hottonia before we started, but I am sorry we could only find 1 specimen of long pistil and we had no time to hunt for more and they were rather withered as we could not pack them at once—[2]

We are very much obliged to you for sending us such a full account of the Lythrums but I am afraid it must have been a great deal of trouble—[3] We went out Lythrum-hunting again but could only find 8 plants of which 1 was long style 2 mid-style & 5 short style[4]

We go home tomorrow after sleeping this one night here, Aunt Susan comes home today or tomorrow,[5] it is very provoking missing her—

I am very glad Lenny will soon be strong enough to move to the sea[6]

Yr affec | Margaret S. Wedgwood

DAR 181

CD ANNOTATIONS

1.3 they were . . . trouble— 2.2] *crossed pencil*

3.1 We go . . . sea 4.1] *crossed pencil*

[1] Dated by the relationship to the letter to K. E. S., L. C., and M. S. Wedgwood, 4 [August 1862], and to CD's notes on *Hottonia* of 8 August 1862 (see n. 2, below); the intervening Wednesday fell on 6 August 1862.

[2] Having learned from Charles Cardale Babington that *Hottonia* was dimorphic (see letter from C. C. Babington, 17 January 1862), CD had been anxious to see flowers of this genus; he requested specimens in his letter to K. E. S., L. C. and M. S. Wedgwood, 4 [August 1862]. CD's notes on the flowers sent with this letter, dated 8 August 1862, are in DAR 110: 16 (see also letter to Asa Gray, 9 August [1862]). CD later gave a copy of his observations to John Scott, for publication in Scott's paper on the structure and functions of the reproductive organs in the Primulaceae (Scott 1864c, pp. 78–9; see *Correspondence* vol. 11, letter to John Scott, 20 [April 1863]).

[3] In the letter to K. E. S., L. C., and M. S. Wedgwood, 4 [August 1862], CD stated his intention to write a further letter, that evening, explaining what he believed to be the relationship between the three sexual forms of *Lythrum salicaria*. This letter has not been found.

[4] CD reported these findings, along with the information given in the letter from M. S. Wedgwood, [before 4 August 1862], in 'Three forms of *Lythrum salicaria*', p. 174 (*Collected papers* 2: 110).

[5] Susan Elizabeth Darwin had been visiting Southampton (see letter from W. E. Darwin, 5 August 1862). Following their holiday in North Wales (see letter from M. S. Wedgwood, [before 4 August 1862]), Margaret and her sisters (Katherine Elizabeth Sophy and Lucy Caroline Wedgwood) had apparently broken their journey home to Surrey at the Mount, Shrewsbury, the home of their aunt.
[6] Leonard Darwin was recovering from scarlet fever (see letter to W. E. Darwin, [2–3 August 1862] and n. 10). The Darwins intended taking a holiday on the south coast, near Southampton (see letter to W. E. Darwin, [24 July 1862], and letter to H. C. Watson, 8 [August 1862]).

From A. R. Wallace 8 August 1862

5, Westbourne Grove Terrace, W.
August 8th. 1862

My dear Mr. Darwin

I sincerely trust that your little boy is by this time convalescent,[1]—& that you are therefore enabled to follow your favourite investigations with a more tranquil mind.

I heard a remark the other day which may not perhaps be new to you, but seemed to me a "*fact*" if true, in your favour. Mr. Ward[2] (I think it was,) a member of the MicroscoScopical Society mentioned as a fact noticed by himself with much surprise, that "the muscular fibres of the whale were no larger than those of the bee"! an excellent indication of community of origin.

While looking at the ostriches the other day at the Gardens[3] it occurred to me that they were a case of special difficulty, as, inhabiting an ancient continent, surrounded by numerous enemies how did their wings ever become abortive, & if they did so before the birds had attained their present gigantic size strength & speed, how could they in the transition have maintained their existence? I see Westwood in the "Annals" brings forward the same case,[4] arguing that the ostriches shd. have acquired better wings within the historic period;—but as they are now the swiftest of animals they evidently do not want their wings, which in their present state may serve some other trifling purpose in their economy such as fans, or balancers which may have prevented their being reduced to such rudiments as in the Cassowaries'— The difficulty to me seems to be, how if they once had flight could they have lost it, surrounded by swift & powerful carnivora against whom it must have been the only defence.

This probably is all clear to you but I think it is a point you might touch upon as I think the objection will seem a strong one to most people.[5]

In a day or two I go to Devonshire for a few weeks & hope to lay in a stock of health to enable me to stick to work at my collections during the winter— I begin to find that large collections involve a heavy amount of manual labour which is not very agreeable.[6]

Present my compliments to Mrs. & Miss Darwin[7] & believe me Yours very faithfully | Alfred R. Wallace

C. Darwin Esq.

DAR 106/7 (ser. 2): 4–5

[1] Leonard Darwin had been suffering from scarlet fever since mid-June (see letter to W. E. Darwin, 13 [June 1862]). Wallace had visited Down House earlier in the summer, apparently in late June or early July (see letter to A. R. Wallace, 20 August [1862] and n. 1).

[2] Wallace refers to the physician and botanist Nathaniel Bagshaw Ward, who was a founding member of the Microscopical Society (R. Desmond 1994, *List of the Microscopical Society*).

[3] The reference is to the gardens of the Zoological Society in Regent's Park, London.

[4] Westwood 1860. John Obadiah Westwood's article, 'Mr. Darwin's theory of development', appeared in the April 1860 number of the *Annals and Magazine of Natural History*, having been reprinted from the *Gardeners' Chronicle*, 11 February 1860, p. 120. Westwood was dismissive of CD's theory on the grounds that species had 'remained permanent during the whole historic period'. He noted that the modern form of the ostrich was 'faithfully represented' in ancient Egyptian records, and continued:

> Now, there can be no doubt that it would have been beneficial to this bird, both specifically and individually, if its coveted plumes could have been shortened and its wings lengthened, so as the better to escape from its pursuers. Moreover ... when driven to their fullest speed they stretch out their short stumps of wings in order to assist in their attempts to escape. But all their efforts to acquire by such means the additional power of flight have been unavailing, and the type of the species remains as it was in this respect 3000 years ago.

[5] For CD's response to Wallace's queries concerning the ostrich, see the letter to A. R. Wallace, 20 August [1862].

[6] In the spring of 1862, Wallace had returned home, in 'a very weak state of health', from an eight-year collecting expedition in the Malay Archipelago. His private collections comprised 'about three thousand bird skins of about a thousand species, and, perhaps, twenty thousand beetles and butterflies of about seven thousand species'; he spent the next five years writing papers based on these specimens. See Wallace 1905, 2: 385–6, 395.

[7] Emma and Henrietta Emma Darwin.

To Hewett Cottrell Watson 8 [August 1862][1]

Down. | Bromley. | Kent. S.E.
July 8th

My dear Mr Watson

You are accustomed to me applying to you when in want. And now I am in great want. I am trying a most laborious series of experiments on Lythrum salicaria, & I think if the result can be proved, which I fully expect to be the case, you will think that your aid will be worth bestowing.[2]

what I want intensely is a few fresh flowers of the rare Lythrum hyssopifolia. Vaucher says it presents two forms like Primula,[3] & these would be invaluable to me; why they would be so is too long a story for a note. Can you give me address (*& allow me to use your name*) to some one or two Botanists, who may live anywhere near this plant, & who would aid me.—[4] If too late for fresh plants, perhaps I could then get seed.— If you can, will you aid me?—

I have had a sick House for last 8 weeks, with one of my poor Boys terribly ill, whom we must take to sea-side next week.[5] All this has cut up my work in a cruel degree.— How is your poor patient, whom you mentioned a year or so ago:[6] you spoke then as if there was little hope. For five years we have never been a whole month without some anxiety. It is a weary world.—

Yours very sincerely | Ch. Darwin

Endorsement: '8 Aug[t] 1862'[7]
Museum of Comparative Zoology, Harvard University

[1] The date given by CD has been changed on the basis of Watson's endorsement and CD's reference to Leonard Darwin's having been ill for '8 weeks' (see nn. 5 and 7, below).
[2] For CD's experiments on *Lythrum salicaria*, see, for example, the letter to Daniel Oliver, 29 [July 1862] and n. 6.
[3] Vaucher 1841, 2: 371.
[4] No letter from Watson on this subject has been found; however, in 'Three forms of *Lythrum salicaria*', p. 190, CD thanked Watson for having provided him with dried specimens of *L. hyssopifolia* (*Collected papers* 2: 124).
[5] CD refers to Leonard Darwin, who had been ill with scarlet fever since mid-June (see letter to W. E. Darwin, 13 [June 1862]).
[6] This individual has not been identified.
[7] Watson wrote 'Aug[t]' over 'July'; he also wrote 'Aug[t].' above CD's date.

To Asa Gray 9 August [1862][1]

Down Bromley Kent
Aug 9th

My dear Gray.—

It is late at night & I am going to write briefly & of course to beg a favour. But first let me thank you most cordially for the stamps in letter of July 21, of *superlative* value to my Boy, who has gloated over them once.—[2] We hope in bed-carriage to get him to sea on Wednesday next.—[3]

I have settled with Trübner; I presume he spoke to Murray for he has charged me reduced price 6[s]·8[d], instead of 9[s].).—[4] If you have spare copy pray present one from me to your pupil (*Rosback?*)[5]

The Mitchella very good, but pollen apparently equal-sized.[6] I have just examined Hottonia grand difference in pollen.—[7] Echium vulgare a humbug, merely case like Thymus.[8] But I am almost stark staring mad over Lythrum; if I can prove what I fully believe, it is grand case of **Tri**morphism with 3 different pollens & 3 stigmas; I have castrated & fertilised above 90 flowers, trying all 18 distinct crosses which are possible within limits of this one species! I cannot explain, but I feel sure you would think it grand case.—[9]

I have been writing to Botanists to see if I possibly can get L. hyssopifolia,[10] & it has just flashed on me that you might have Lythrum in N. America.—& I have looked to your manual[11] For the love of Heaven have a look at some of your species, & if you can get me seed, do; I want much to try species with few stamens, if they are dimorphic: Nesæa verticillata I sh[d] expect to be trimorphic.[12] Seed! Seed! Seed!

I sh[d]. rather like seed of Mitchella— But oh Lythrum!—

Your utterly mad friend | C. Darwin

There is reason in my madness, for I can see that to those who *already believe in change of species*, these facts will modify to certain extent whole view of Hybridity—

*Homomorphic grand*children (from two homorphic unions) of Primula are become **more** sterile.—[13]

Gray Herbarium of Harvard University (71)

[1] The year is established by the relationship to the letter from Asa Gray, 21 July 1862.

[2] Gray had enclosed some American postage stamps for Leonard Darwin with the letter from Asa Gray, 21 July 1862.

[3] According to Emma Darwin's diary (DAR 242), the Darwins travelled to Southampton on 13 August 1862. Before they could continue their journey to Bournemouth, however, Leonard suffered a 'slight relapse' in his recovery from scarlet fever, and Emma contracted the disease (see letter to A. R. Wallace, 20 August [1862]); consequently they spent the remainder of August in Southampton, moving on to Bournemouth in September (see 'Journal' (Appendix II)). Emma discussed details of the holiday plans in a letter to William Erasmus Darwin, [9 August 1862] (DAR 219.1: 61).

[4] Gray had asked CD to arrange for six copies of *Orchids* to be sent to him by the London bookseller and publisher Nicholas Trübner (see letter to Nicholas Trübner, 23 June [1862]). In the letter from Asa Gray, 21 July 1862, Gray mentioned having received an invoice for the copies, and stated that it would be wrong for CD to pay the retail price for them, since he supposed CD's publisher, John Murray, would supply him with copies at half price. CD's Account book–cash accounts (Down House MS) records a payment of £2 1*s.* to Trübner on 3 August 1862.

[5] Joseph Trimble Rothrock had offered to provide CD with information on *Houstonia* (see letters from Asa Gray, 2–3 July and 15 July [1862]). CD had not yet received Rothrock's observations, detailed in the letter from Asa Gray, 4 August 1862.

[6] Gray had sent with his letter to CD of 15 July [1862], specimens of *Mitchella repens* to examine for heterostyly.

[7] See letter from M. S. Wedgwood, [6 August 1862] and n. 2.

[8] No previous correspondence with Gray on *Echium vulgare* has been found. However, in *Forms of flowers*, p. 111, CD reported that 'from statements made by Vaucher [Vaucher 1841, 3: 461]', he had thought at first that this species was heterostyled, but had soon seen his error. Later in the volume (pp. 305–6), CD described the existence of two forms of flower in this species, namely, an ordinary hermaphrodite form and a female form. CD had already observed this phenomenon in *Thymus* (see *Correspondence* vol. 9, letter to George Bentham, 30 November [1861], and letter to Asa Gray, 11 December [1861]; see also this volume, letter to Daniel Oliver, 12 [April 1862]).

[9] CD had told Gray of his interest in the occurrence of trimorphism in *Lythrum salicaria* in the letters to Asa Gray, 1 July [1862] and 28 July [1862]. There are a number of notes in DAR 27.2 (ser. 2), recording the details of crossing experiments carried out by CD on this species in the summer of 1862; the earliest is dated 31 July 1862 (DAR 27.2 (ser. 2): 7). CD's paper on the subject ('Three forms of *Lythrum salicaria*') was read before the Linnean Society of London on 16 June 1864.

[10] See letter to W. E. Darwin, [2–3 August 1862], and letter to H. C. Watson, 8 [August 1862].

[11] A. Gray 1856. There is an annotated copy of this work in the Darwin Library–CUL (see *Marginalia* 1: 348–51).

[12] *Nesaea*, like *Lythrum*, is a member of the Lythraceae.

[13] In January and February 1862, CD carried out crossing experiments on plants of *Primula sinensis* raised from the seed produced by similar crosses made in 1861 (see letter to J. D. Hooker, 23 June [1862] and n. 4). He crossed the offspring of the homomorphic crosses (i.e., those in which long- or short-styled parents were fertilised from own-form pollen) both homomorphically and heteromorphically, confirming his expectation that they would be sterile, and crossed the offspring of the heteromorphic crosses 'for comparison ... as these would of course be fully fertile' (DAR 108: 35). CD made notes on the rates of growth and health of the offspring from these experiments (i.e., the 'grandchildren') on 28 July and 9 August (DAR 108: 127, 129; see also DAR 108: 125). His overall results are given in 'Illegitimate offspring of dimorphic and trimorphic plants', pp. 410–18. In his conclusion (p. 436), CD asserted: 'the

illegitimate [i.e., homomorphic] offspring from an illegitimate union are hybrids formed within the limits of one and the same species'; consequently 'the lessened fertility of the first union of the offspring of two forms is no sure criterion of specific distinctness'. CD also expounded these views in *Origin* 4th ed., p. 323, and in *Variation* 2: 184. For CD's interest in cross and hybrid sterility, see Appendix VI.

From Thomas White Woodbury 9 August 1862

[17 Lower Mount Radford Terrace, Exeter]
9th Augt 1862

Dear Sir

Accept my warmest thanks for your kind present of combs and workers of apis testacea.[1] I have examined them carefully and with the greatest interest The cells appear precisely the same as those of A. mellifica with this important difference—the breeding cells are *one-fourth deeper* than those of A mellifica, proving that the insect is *one-fourth longer* in the body when it arrives at maturity. It is very remarkable that an insect so much larger than the common bee should build cells of no greater diameter, and satisfies me that I am right in believing A. Ligustica to be a larger bee than A mellifica although it makes cells of the same diameter

What an immense comb yours must have been! Could you ascertain by any means if A testacea is a hive bee? If so I believe it would be invaluable in this country. Its very large wings augur enormous powers of flight and its capacity for honey-gathering is I should think, immense. I really cannot get it out of my head and shall not rest until I have attempted by some means to obtain a living colony.

With regard to living bees from central Africa, I see no very great difficulty in the matter; if the gentleman you mention[2] would attempt it I should be happy to give instructions pay expenses of freight &c and if successful and a new species would gladly present the sender with £10 or £20 according to their value, although I can scarcely think they would turn out the extraordinary honey-gatherers which I fancy A testacea to be

One thing appears remarkable, and that is that the melting-point of the wax of A testacea seems *lower* than that of A. mellifica

Again thanking you most sincerely I am | My dear Sir | Yours very truly | T W Woodbury

C. Darwin Esq

P.S.— The bees of Jamaica are I believe A mellifa and I doubt not their combs are the same as ours[3]

DAR 181

CD ANNOTATIONS

3.1 With regard ... think 3.5] *crossed pencil*

Top of first page: 'A. R. Wallace 5. Westbourne Grove Terrace.—'[4] *ink*; '(Keep Bees)' *pencil*, *square brackets in MS*

[1] CD had sent Woodbury a portion of the honeycomb from the island of Timor, sent to him by Alfred Russel Wallace (see letter from A. R. Wallace, 7 April 1862, and letter to A. R. Wallace, 20 August

[1862]). Woodbury was one of the contributors to the bee-keeping section of the *Journal of Horticulture and Cottage Gardener*, to which CD had recently directed several queries concerning variability in hive-bees (see letters to the *Journal of Horticulture*, [before 10 June 1862] and [before 15 July 1862]). Woodbury had written privately to CD in response to the latter of these (see letter to *Journal of Horticulture*, [before 22 July 1862]); however, Woodbury's letter has not been found. According to his 'Journal' (Appendix II), CD was writing up the chapter of *Variation* on 'Silk-worms Geese &c' in the summer of 1862; this chapter included a section on hive-bees (*Variation* 1: 297–9).

2 The reference is to Gustav Mann, who was collecting in West Africa for the Royal Botanic Gardens, Kew (see letter to J. D. Hooker, 22 [August 1862] and nn. 5 and 6). No letter from CD to Woodbury on this subject has been found; however, CD had evidently been attempting to obtain specimens of bees and honeycomb from West Africa (see letter from J. D. Hooker, 20 August 1862 and nn. 8 and 9).

3 See letter to *Journal of Horticulture*, [before 22 July 1862] and n. 3.

4 See n. 1, above.

To W. B. Tegetmeier 12 August [1862]

Down. | Bromley. | Kent. S.E.
Aug. 12th.

My dear Sir

I think yours is a capital but too kind & flattering an article on my orchid Book.—[1] You are an excellent hand at writing & I know that you once attended much to Botany

Many thanks for Green feathers, which I am very glad to see.—

I much fear, from your not mentioning subject, that you have not got the Curatorship at Southampton.[2] I write in hurry for I start tomorrow for a month to "Sea Villa, Bournemouth", on account of one of my Boys, who has had frightfull illness of 8 weeks.[3] All my work has gone of late to the dogs.—

My dear Sir | Yours very sincerely | C. Darwin

Endorsement: '1862—'
New York Botanical Garden Library (Charles Finney Cox collection)

1 Tegetmeier's review of *Orchids* appeared in the August 1862 number of *Weldon's register of facts and occurrences relating to literature, the sciences, and the arts* (Tegetmeier 1862). There is a copy of this review in the Scrapbook of reviews (DAR 226.1: 140).

2 Tegetmeier had applied for a position as curator of the museum at the new Hartley Institution, Southampton; CD had written a testimonial in support of his application (see letter to W. B. Tegetmeier, 20 June 1862).

3 Although the Darwins travelled as far as Southampton on 13 August (Emma Darwin's diary (DAR 242)), their onward journey to Bournemouth was delayed because Leonard Darwin suffered a relapse in his recovery from scarlet fever, and Emma Darwin contracted the disease (see letter to A. R. Wallace, 20 August [1862]).

From Asa Gray 18–19 August 1862

Cambridge. [Massachusetts]
Aug. 18. 1862

My Dear Darwin

The dawn of hope in yours of July 23d, so brightened in your last, of the 28th gave us heart-felt joy.[1] Thank God your dear boy is convalescent, by this time, we trust so decidedly so as to give you full relief from all anxiety. Really, if one can give so much satisfaction at so cheap a rate, one would become a stamp collector for the purpose of supplying the good fellow.[2]

When he gets quite well he is to write me a note and let me know what American stamps he lacks, and I will see that some of our young people help him. Tell him that so far, he is not so much indebted to the Harvard professor as to a venerable French gentleman of Philadelphia, M^r. Durand.—aged something like 75.[3]

I have lost the time in which I was to write you this week.—and can send only a hasty line. My old friend Dr. Torrey[4] is expected to-morrow to visit us for several days, and I must clear up matters before his coming. But I shall scold him for not having supplied me with *Specularia perfoliata*, with the "*precociously fertilized flowers*". (It seems a good term,—it express the fact.)[5]

Since my last a few Orchids have come in, e.g. our White Fringed Orchis *P. blephariglottis*.—but rather too old,—and while I was out, after sunset, a friend has brought me *P. ciliaris*—the Yellow Fringed O.— which I shall look at to-morrow.

I will record notes, and try to put on record the most important,—so it will not be worth while to send you so many details in the raw state.[6]

If you so wish it, perhaps I may make a note of some obs. elsewhere— But I like to stick such things into my Notices & Reviews.[7] No matter if they are overlooked by the outsiders. If *you* know of them, or any one else who really wants to use such observations, that is all I care. I rather like to stick these farthing candles under the bushel,—and the same of larger lights, to a degree.

The Sept. no. of Sill. Jour is so delayed that your enclosures were back in time. I was sitting down to use them, when Silliman wrote me that I had already sent him more matter than he could print.[8]

And so, I was not sorry to let the Orchids go by till the Nov. no.—where I hope to use the cuts you sent me,[9] and give some gossiping observations on several American orchideæ. But these must be regarded as only rapid *reconnaisance*.— All should be held subject to confirmation or the reverse by more careful & reiterated observations—

But I have *two points*, and both, I know, will please you.

1. *Goodyera repens*. Before reading your remark on p. 114 line 11, —i.e. having forgotten it.—I have confirmed it.[10] It is very distinct, the difference between the *early position* when the proboscis must hit gland & remove *pollinia*, but cannot thrust pollinia down to stigma.—and the later position, as seen in flowers lower down the spike. Where there is room, the stigma is in sight on looking down the flower, and the pollinia on a pin go down to stigma as sure as can be.— *But the* difference is only slightly, if at all owing to any downward movement of the labellum.; it comes from a backward movement of the column, which becomes *more erect*.

2. I have another, and a different case of *close-self-fertilization*,—in *Gymnadenia tridentata*. The arrangement different from P. hyperborea,—[11] and indeed, I do not see just how the pollen so surely gets on to the stigmas—on to an arm of stigma carried up each side of anther, & one between the cells. But they get pollen packets

in the *bud*, and pollen-tubes are emitted abundantly.— It is most interesting case yet.—such determination to self-fertilize and yet I suspect pollinia are often removed by insects, & cross-fertilize occasionally. I will describe this.[12] But it must be looked at more particularly next year.

19th | The charm of *Platanthera ciliaris* & *blephariglottis* is—the very long & narrow arms to the stigma, which, with the anterior portion of the anther containing the long caudicle, are like a stalk bearing the (small) disk on the extremity.— these *thrust forward* to meet the head of insect, who cannot get in the whole length of proboscis in to the long ($1\frac{1}{4}$ inch) straight spur, before his face will come in to contact with the glands.

Unfortunately for my experiments, the little glands, so far projecting, dry up sooner than usual, & loose their viscidity, & the stalk its power of depression.

The slight differences between the two species—the white and the yellow—are interesting; but I will not trouble you now, as I keep notes on them.[13]

Torrey is here. He tells me he could not find any pollen-tubes emitted from *Specularia* perfoliata— the early fertilised flowers—while the pollen was still in the anthers.[14]

Good bye, in haste, Ever Yours | A. Gray

DAR 165: 111, 116

CD ANNOTATIONS

1.1 The dawn ... *reconnaisance.*— 8.3] *crossed pencil*
10.1 1. *Goodyera*] *opening square bracket, pencil*
15.1 Torrey ... anthers. 15.3] *double scored pencil*
Top of first page: 'Keep' *circled pencil*; 'Orchids' *pencil*; 'Goodyera' *red crayon, enclosed in parentheses, double underl*

1 Gray refers to Leonard Darwin's gradual recovery from scarlet fever (see letters to Asa Gray, 23[–4] July [1862] and 28 July [1862]).

2 At CD's request, Gray had sent a number of stamps from the United States for Leonard's collection (see letter to Asa Gray, 10–20 June [1862], and letters from Asa Gray, 2–3 July 1862, 15 July [1862], and 21 July 1862). See also letters to Asa Gray, 23[–4] July [1862] and 28 July [1862].

3 Gray refers to the botanist and pharmacist, Elias Durand. Gray was Fisher Professor of natural history at Harvard University.

4 Gray refers to John Torrey, his former mentor and botanical collaborator.

5 In his letter to CD of 2–3 July 1862, Gray promised to obtain specimens of this species from Torrey, and to observe the behaviour of the pollen-tubes in those flowers that underwent 'precocious fertilization' (later known as cleistogamy). In reply, CD commented that the phenomenon seemed 'too remarkable to be called "precocious flowering"' (letter to Asa Gray, 23[–4] July [1862]).

6 *Platanthera blephariglottis* and *P. ciliaris* are described in A. Gray 1862b, p. 424.

7 In the letters to Asa Gray, 23[–4] July [1862] and 28 July [1862], CD argued that Gray's observations on American species of orchids were too good to be reported only in a review of *Orchids*, and that Gray should publish at least some of them separately.

8 Gray refers to his notes on American species of orchids, returned by CD with the letter to Asa Gray, 23[–4] July [1862]. Gray intended to include some of his observations in a follow-up article to his review of *Orchids* (A. Gray 1862b), to be published in the monthly *American Journal of Science and Arts*, edited by Benjamin Silliman Jr, and commonly referred to as 'Silliman's journal'.

[9] CD had arranged, at Gray's request, for John Murray to send Gray electrotype plates of three of the illustrations from *Orchids*, figuring *Orchis mascula* and *O. pyramidalis*, for reproduction in Gray's review of the book (A. Gray 1862a; see letter from Asa Gray, 18 May 1862, and letter to Asa Gray, 10–20 June [1862]). The plates arrived too late for use in Gray's review (see letter from Asa Gray, 21 July 1862 and nn. 3 and 4), but appeared in the follow-up article, published in the November number of the *American Journal of Science and Arts* (A. Gray 1862b).

[10] In *Orchids*, p. 114, CD stated that in *Goodyera repens* the passage into the flower between the rostellum and labellum was contracted. He reported that, by analogy with *Spiranthes autumnalis*, he suspected the labellum moved 'further from the column in mature flowers, in order to allow insects, with the pollinia adhering to their heads or probosces, to enter the flower more freely.' Gray gave his observations on this point in A. Gray 1862b, p. 427; CD cited Gray's confirmation of his view in 'Fertilization of orchids', p. 151 (*Collected papers* 2: 148).

[11] Gray sent CD notes on *Platanthera hyperborea* with his letter of 2–3 July 1862; although the notes have not been found, it is clear from CD's response in the letter to Asa Gray, 23[–4] July [1862], that Gray had told him that the species was often self-pollinated. In *Orchids*, one of CD's purposes was to demonstrate that the 'main object' of the various 'contrivances by which Orchids are fertilised' was *cross-fertilisation* (p. 1), and he noted only one exception (p. 359). CD included *P. hyperborea* and *Gymnadenia tridentata* on an undated list of 'self-fertilisers' that is now in DAR 70: 167; he also included a modified discussion of the occurrence of 'self-fertilisation' in orchids in *Orchids* 2d ed., pp. 288–93.

[12] Gray discussed *Gymnadenia tridentata* in A. Gray 1862c, p. 260 n. and A. Gray 1862b, p. 426. CD made undated notes referring himself to the latter account (DAR 70: 8, 17); he cited Gray's observations in 'Fertilization of orchids', p. 147 (*Collected papers* 2: 144).

[13] See n. 6, above.

[14] See nn. 4 and 5, above.

From J. D. Hooker 20 August 1862

Kew
Aug. 20/62

My dear Darwin

I can well appreciate your frightful suspense— There is nothing my wife dreads so much, or I for her, as an attack of scarlet fever.[1] You know perhaps that Belladonna is a prophylactic in opinion of good medical men—a drop or 2 of tincture in water, 3 times a day. Poor little Lenny, I did not know *he* had been ill— I thought it was Horace alone—who had made you uneasy lately.[2]

We are all pretty well, my wife remarkably so, though she has a little palpitation & short breath occasionally—[3] her aunt (Miss Henslow) is here now suffering much from it, so it is completely hereditary,[4] & I recognize a tendency to it in one of my boys. I have had a great deal to do since my return with Examinations—all over now,[5] & had some very hard work with this *Welwitschia* & its pollen tubes, corpuscula, Embryo-sacs, & all that horrid complexity of Gymnospermous Embryology.[6] I have sat 5 hours together at microscope at least 6 times lately, besides all the odd days & hours I have spent over it, & am very far from finished yet. Every part is so curious, how the pollen gets to nucleus of ovule is absolutely unintelligiable—dozens of pollen **grains**, get down a microscopic tube nearly $\frac{1}{2}$ inch long & settle on top of nucleus— they must get in before tube elongates, but if so the development of ovule is very difft from other plants. By good luck just as I am at work on it, I receive

5 splendid specimens from a Mr Monteiro of Luando,[7] to whom I wrote 5 months ago asking him to send down the Coast to Cape Negro for it, & like a trump he has done so! I have just written to thank him, sent him a few books, & asked him for Honey comb & bees.[8] I will also ask Mann for these—to whom I am writing—[9]

We (wife Willy[10] & self) go to Scotland on Saturday, & shall be back 15th. Sept.

Lindley[11] sent me yesterday two totally difft. flowers from one spike of *Vanda Loweii*—[12] I have sent them to be drawn & bottled.

What do you think of Ramsay's glacial Lake origin?[13] I like it in print, but did not on hearsay—but am not mechanic enough to *rely* on my own opinion— We are quite overrun with visitors, & never a day alone.

Ever Yrs affec | J D Hooker

Huxley is about to publish a *[curious]* amusing as well as clever book on Monkey Man[14] it will be a great success. I am hugely pleased with A Grays Review of my Arctic Essay[15]

DAR 101: 52–3

CD ANNOTATION

Top of first page: 'Coloured pollen | (Your microscope) | Lythrum | (Lindley) | Lythrum' *ink*

1 According to her diary (DAR 242), Emma Darwin became ill with scarlet fever on 13 August 1862. Hooker refers to Frances Harriet Hooker, who had been ill in June and July (see letters from J. D. Hooker, 9 June 1862, 19 [June 1862], 28 June 1862, and 2 July 1862).

2 Horace Darwin had been ill during the spring of 1862 (see 'Journal' (Appendix II)). Leonard Darwin had been suffering from scarlet fever since June (see letters to W. E. Darwin, 13 [June 1862] and 9 July [1862], and letter to A. R. Wallace, 20 August [1862]).

3 The Hookers had travelled to Switzerland in July so that Frances might recover her health (see letters from J. D. Hooker, 28 June 1862, 2 July 1862, 10 July 1862, and [24 July 1862]).

4 Frances Hooker's late father, John Stevens Henslow, was survived by four sisters (Jenyns 1862, p. 5); the reference is probably to the eldest, Anne Frances Henslow, who died unmarried on 30 August 1863 (*Gentleman's Magazine* n.s. 15 (1863): 520); see also letter from J. D. Hooker, 25 October 1862.

5 Hooker had recently been involved in examinations at the University of London, where he was examiner in botany (see letter from J. D. Hooker, [24 July 1862]). He was also examiner for the Society of Apothecaries' prize in botany, and a botanical examiner for medical officers in the armed services and the East India Company (L. Huxley ed. 1918, 1: 385–7 and 537).

6 The reference is to the plant *Welwitschia mirabilis*, from Angola, which Hooker described as the 'most wonderful' and the 'very ugliest' plant ever brought to Britain (L. Huxley ed. 1918, 2: 24–5). Hooker read his account of *Welwitschia* before the Linnean Society of London on 18 December 1862 and 16 January 1863 (J. D. Hooker 1863a).

7 The reference is to the mining engineer and zoologist, Joachim John Monteiro, who resided in Loanda, Angola (see J. D. Hooker 1863a, p. 4, and Monteiro 1875, 2: 228–31).

8 CD was writing up the chapter of *Variation* on 'Silk-worms Geese &c' (see 'Journal' (Appendix II)), which included a section on hive-bees (*Variation* 1: 297–9). He intended to compare specimens of *Apis mellifica* from different localities, in order to investigate whether differences in climate produced any effect on their form (see letters to the *Journal of Horticulture*, [before 10 June 1862] and [before 15 July 1862]). See also letter from T. W. Woodbury, 9 August 1862.

9 Gustav Mann was botanical collector for the Royal Botanic Gardens, Kew, on the Niger expedition led by William Balfour Baikie (see R. Desmond 1995, p. 433). See also letter from T. W. Woodbury, 9 August 1862.

10 William Henslow Hooker.

11 John Lindley.

12 In *Vanda Lowii*, the flowers at the base of each spike are different in both colour and form from the rest of the flowers. A note in DAR 70: 156 states: 'Vanda Loweii— lowest flowers different from others Gardeners Chronicle 1862. p. 791. Specimen in Spirits at Kew'. The article, published in the 23 August 1862 issue, described a specimen belonging to Sigismund Rucker. CD's annotated copy of this number of the *Gardeners' Chronicle*, now at the Cory Library, Cambridge Botanic Garden, was kept separately by him (see the 'List of the numbers of special interest to Darwin and kept by him in separate parcels' in DAR 222). See also letter to Daniel Oliver, 24 July [1862] and n. 9.

13 Ramsay 1862. Andrew Crombie Ramsay read his controversial paper on the glacial origin of various lake basins before the Geological Society of London on 5 March 1862. For an account of the controversy initiated by Ramsay's paper, see Davies 1969, pp. 303–9.

14 T. H. Huxley 1863a.

15 CD had sent Hooker a copy of A. Gray 1862d, which was a review of J. D. Hooker 1861a (see letter to Asa Gray, 28 July [1862] and n. 11).

From Charles Lyell 20 August 1862

Freshwater Gate, Isle of Wight:
August 20, 1862.

My dear Darwin,—

Mr. Jamieson of Ellon has been again to Lochaber, and confirms his former theory of the glacier lakes.[1] The chief new point is a supposed rise at the rate of a foot per mile of the shelves as we proceed from the sea inland. It seems to me to require many more measurements, before we can rely on it. He found some splendid moraines opposite the mouth of Glen Trieg. He found some shells of Arctic character in the forty feet high raised beach of the Argyllshire coast, and has asked me to learn about one of them, of which he sends a drawing.

I fell in yesterday in my walk with Mr. A. G. More, whom you cite in your orchid book.[2] He considers you the most profound of reasoners, to which I made no objection, only being amused at remembering that, such being the case, you had performed a singular feat, as the Bishop of Oxford assured me, of producing 'the most illogical book ever written.'[3]

We shall be here for a week longer. I have been with my nephew Leonard[4] to Alum and Compton Bays.

Ever most truly yours, | Charles Lyell.

Incomplete[5]
K. M. Lyell ed. 1881, 2: 358

1 Thomas Francis Jamieson visited the Scottish district of Lochaber in August 1861, to examine the so-called 'parallel roads' of Glen Roy (see *Correspondence* vol. 9). In 1839, CD had published a paper in which he argued that the 'roads', a series of terraces running parallel to each other along the sides of the glen, were the remains of beaches, formed by the sea as the landmass of Scotland gradually rose ('Parallel roads of Glen Roy'). Following his own observations, however, Jamieson concluded that during a great 'Ice-Age', ice-flows had trapped a series of lakes in the glen, and that the 'roads' represented the shorelines of those former lakes. While CD at first appeared to have conceded defeat

on the question, stating that his paper had been 'one long gigantic blunder' (see *Correspondence* vol. 9, letter to Charles Lyell, 6 September [1861]), subsequent letters indicate that he was reluctant to abandon his own explanation (see *Correspondence* vol. 9, Appendix IX). Jamieson made a second visit to the site in July 1862 (see Jamieson 1863, p. 240); Lyell subsequently sent CD Jamieson's letter describing his visit (see the enclosure to the letter to Charles Lyell, 14 October [1862]).

2 In addition to assisting CD with a number of experiments in 1861, Alexander Goodman More had supplied CD with orchid specimens from the Isle of Wight (see *Correspondence* vol. 9). His assistance is acknowledged several times in *Orchids* (see *Orchids*, pp. 67, 95 n., 99, and 101 n.).

3 In his letter to CD of [13–14 February 1860] (*Correspondence* vol. 8), Lyell reported that Samuel Wilberforce, the bishop of Oxford, had maintained that *Origin* 'was the most unphilosophical [book] he had ever read.'

4 Leonard Lyell.

5 From CD's reply, it appears that this letter included a postscript that has not been found (letter to Charles Lyell, 22 August [1862] and n. 15).

To A. R. Wallace 20 August [1862]

1. Carlton Terrace | Southampton
Aug. 20th

My dear Mr Wallace

You will not be surprised that I have been slow in answering, when I tell you that my poor [boy] became frightfully worse after you were at Down;[1] & that during our journey to Bournemouth he had a slight relapse here & my wife took the Scarlet Fever rather severely.[2] She is over the crisis. I have had a horrid time of it & God only knows when we shall be all safe at home again. Half my Family are at Bournemouth.—[3]

I have given a piece of the comb from Timor to a Mr Woodbury,[4] (who is working at subject) & he *extremely* interested by it (I was sure the specimen would be valuable) & has requested me to ascertain whether the Bee (A. testacea) is domesticated & when it makes it combs? Will you kindly inform me?

Your remarks on ostriches have interested me, & I have alluded to case in 3d. Edition.—[5] The difficulty does not seem to me so great as to you.— Think of Bustards which inhabit wide open plains, & which so seldom take flight: a very little increase in size of body would make them incapable of flight.— The idea of ostriches acquiring flight is worthy of Westwood;[6] think of the food required in these inhabitants of the Desert to work the Pectoral muscles! In the Rhea the wings seem of considerable service in the first start & in turning. The distribution & whole case of these birds is, however, very interesting: considering their apparently real affinities to mammals, I have sometimes speculated whether we do not here get an obscure glimpse of

Endorsement: '1862'
Incomplete
British Library (Add. 46434: 28)

1 See letter from A. R. Wallace, 8 August 1862. Wallace's visit to Down House is not recorded in either Emma Darwin's diary (DAR 242) or CD's 'Journal' (Appendix II). However, Wallace later recalled:

'Soon after I returned home, in the summer of 1862, Mr. Darwin invited me to come to Down for a night, where I had the great pleasure of seeing him in his quiet home, and in the midst of his family' (Wallace 1905, 2: 1). This visit probably took place in late June or early July 1862: Leonard Darwin became ill with scarlet fever in June (see letter to W. E. Darwin, 13 [June 1862]), and suffered a relapse early in July (see Emma Darwin's diary (DAR 242), and letter to W. E. Darwin, 9 July [1862]).

2 According to Emma Darwin's diary (DAR 242), the Darwins travelled to Southampton on 13 August 1862; she recorded that on the same date, she became ill with scarlet fever.

3 During the latter part of Leonard's illness, the other Darwin children had been sent away with their former nurse, Brodie, who was at Down at the time (*Emma Darwin* 2: 178; see also letter to W. E. Darwin, 4 [July 1862] and n. 8). The family had planned to be reunited in Southampton, before moving on to Bournemouth for a holiday (see the letter from Emma Darwin to William Erasmus Darwin, [9 August 1862], in DAR 219.1: 61). As a result of Emma Darwin's illness, however, the children were apparently sent on to Bournemouth, and were joined by the rest of the family on 1 September (see 'Journal' (Appendix II), and *Emma Darwin* 2: 178).

4 See letter from T. W. Woodbury, 9 August 1862.

5 See letter from A. R. Wallace, 8 August 1862. CD's remarks appear in *Origin* 3d ed., pp. 151–2.

6 CD refers to John Obadiah Westwood (see letter from A. R. Wallace, 8 August 1862 and n. 4).

From A. R. Wallace [after 20 August 1862][1]

⟨*half a page excised*⟩

I have only the 1st. Edition of the "*Origin*"— I suppose the one you kindly sent to Singapore must have reached after I left & they did not think it worth while to return it.[2] I should be much pleased to have the 3rd. Ed^{n}.[3]

I have never considered the Ostriches to have any *real* connection with Mammals,[4] & should imagine ⟨*half a page excised*⟩ difficult ⟨*3 words excised*⟩ of anomalous semiaquatic animals which should easily admit of modification into these three forms.—

I find myself very delicate & dread the winter, as I have been suffering here from inflamation of the *pleura*, & bad cough through a little exposure to night air.

I remain | My dear Mr Darwin | Yours very faithfully | Alfred R. Wallace

C. Darwin Esq.

Incomplete
DAR 181

CD ANNOTATION

2.2 difficult ... air. 3.2] 'Malacca | ova cells | domesticated Bees' *added in margin, ink*

1 Dated by the relationship to the preceding letter. See also n. 3, below.

2 Wallace had returned to Britain in the spring of 1862, having been on a collecting expedition in the Malay Archipelago since 1854 (Wallace 1905). His name appears on CD's list of presentation copies for *Origin* (see *Correspondence* vol. 8, Appendix III).

3 Wallace acknowledged receipt of a copy of the third edition of *Origin* in his letter to CD of 30 September 1862.

4 See preceding letter.

To Asa Gray 21 August [1862][1]

Southampton
Aug 21.

My dear Gray.

We are a wretched family & ought to be exterminated. We slept here to rest our poor Boy on his journey to Bournemouth, & my poor dear wife sickened with Scarlet-fever & has had it pretty sharply, but is recovering well.[2] Our Boy suffered sadly from the Journey, though we took it on the advice of two Doctors. I fear he will be an invalid for months, if not years.— There is no end of trouble in this weary world.— I shall not feel safe till we are all at home together, & when that will be I know not. But it is foolish complaining.

I received a few days ago your letter of Aug. 4th, with all the interesting details on Houstonia.[3] It seems a grand case. I hope that Mr Rothwick will surely publish them.[4] The simple fact of two pollens in the same species, & the reciprocal action of two hermaphrodites seems to me well worth establishing; & till any first account is confirmed, nothing can be considered as established. I feel no sort of doubt after repeating my experiments on Primula; but I shall probably not publish till winter, (even if then) & so Mr. R. could first establish the case.—[5]

I must just recur to stamps; my little man has calculated that he will now have six stamps which no other Boy in the school has.[6] Here is a triumph. Your last letter was plaistered with many coloured stamps & he long surveyed the envelope in bed with much quiet satisfaction.

I wrote you a mad letter the other day about Lythrum;[7] but the case is worth some madness. Thanks for remarks about Rhexia; what you say about pollen flirting out agrees with what I have seen.— My Rhexia glandulosa seems very different, & I believe offers nothing odd.[8] Heterocentron will, I suspect, turn out, as I prophecyed something marvellous like Lythrum.—[9] I know *almost* as well as you, that systematic work is the foundation of *everything*; yet in your case & Hooker's case, I perpetually feel inclined to d——n systematic work.—[10] I had a note from Hooker this morning giving a pretty fair account of Mrs Hooker; but it almost seems that her heart is slightly affected.[11] He tells me that he has got two wonderfully different flowers on same spike of a Vanda.

Huxley is going to bring out a very curious Book on man & monkey.—[12]

I can see no Honey in Melastoma;[13] but secretion of Honey depends on most delicate combination of circumstances. The common Polygala will go on for many days & secret none, & then will suddenly all commence— I am scribbling away at a great rate.

Affairs seem to be getting with you more & more terrible.[14] What will the end be. It seems to us here far more fearful, than it apparently does to you.

Farewell my dear Friend | C. Darwin

I shd. very much like, if time permits to hear what you think of my last chapter in Orchid-book.—[15]

Gray Herbarium of Harvard University (67)

[1] The year is established by the reference to Emma and Leonard Darwin's illnesses (see n. 2, below).

[2] The Darwins had planned to take a holiday in Bournemouth to assist Leonard Darwin's convalescence from scarlet fever (see letter from W. E. Darwin, 5 August 1862 and n. 8). According to Emma Darwin's diary (DAR 242), CD, Emma, and Leonard travelled to Southampton on 13 August 1862; on the same date, Emma recorded that she became ill with scarlet fever.

[3] Letter from Asa Gray, 4 August 1862.

[4] CD refers to Joseph Trimble Rothrock, one of Gray's students who had carried out observations and experiments for CD on the dimorphic plant *Houstonia caerulea* (see letter from Asa Gray, 4 August 1862). Rothrock does not appear to have published his results; however, CD included Rothrock's findings in *Forms of flowers*, pp. 132, 254.

[5] Rothrock had confirmed and expanded upon Gray's observations of differences in the pollen from each of the two flower forms of *Houstonia caerulea* (see letter from Asa Gray, 4 August 1862 and n. 1). He had also demonstrated experimentally that each kind of pollen was more fertile when applied to the stigma of a flower of the other form than when applied to the stigma of a flower of the same form. This paralleled CD's findings in *Primula*, published in March 1862 ('Dimorphic condition in *Primula*'; see *General index to the Journal of the Linnean Society*, p. vi). CD repeated his experiments on *P. sinensis* in January and February 1862, with similar results (see letter to Asa Gray, 9 August [1862] and n. 13); however, he did not publish his findings until 1869 ('Illegitimate offspring of dimorphic and trimorphic plants').

[6] Since June, Gray had sent North American postage stamps for Leonard Darwin (see letters to Asa Gray, 10–20 June [1862], 23[–4] July [1862], 28 July [1862], and 9 August [1862], and letters from Asa Gray, 2–3 July 1862, 15 July [1862], and 21 July 1862).

[7] Letter to Asa Gray, 9 August [1862].

[8] CD suspected that plants of the melastomaceous genus *Rhexia* might be dimorphic, and had asked Gray to make observations for him (see letter from Asa Gray, 4 August 1862 and nn. 3–6). Having obtained a specimen of *Rhexia glandulosa*, CD carried out crossing experiments, the results of which he reported in his letter to Gray of 10–20 June [1862]. Detailing his observations on *Rhexia Virginica* in his letter to CD of 4 August 1862, Gray wrote: 'I suppose you have examined no true *Rhexia*.'

[9] In October 1861, CD had begun to investigate the occurrence of two different sets of stamens in each flower of *Heterocentron roseum*, those facing the petals differing in structure and colour from those facing the sepals (see *Correspondence* vol. 9, letter to J. D. Hooker, 17 November [1861]). Suspecting that this might be a novel form of dimorphism (see letter to George Bentham, 3 February [1862]), he carried out pollination experiments between October 1861 and January 1862, in which the crosses made with pollen from the yellow anthers produced more seed than those made with pollen from the crimson anthers. In addition, the seedlings produced from the former crosses grew much larger than those resulting from the latter (see letter to J. D. Hooker, 15 [May 1862] and nn. 7–9). On 9 August, CD again compared the two sets of seedlings, noting: 'the bigger ones were still twice as tall; the smaller being more compact habit & more purplish leaves' (DAR 205.8: 50). However, despite continuing to work on the family throughout 1862 and 1863, CD was not able to account for the two sets of stamens (see *Cross and self fertilisation*, p. 298 n., and *ML* 2: 292–302).

[10] In his letter to CD of 4 August 1862, Gray commented that, since he was preparing to do systematic work, he would be unable to carry out observations and experiments for CD. For CD's concern at the amount of time Joseph Dalton Hooker spent preparing his *Genera plantarum* (Bentham and Hooker 1862–83), see the letter to J. D. Hooker, 9 February [1862], and the letter to Asa Gray, 21 April [1862].

[11] Letter from J. D. Hooker, 20 August 1862.

[12] T. H. Huxley 1863a; see letter from J. D. Hooker, 20 August 1862.

[13] In his letter to CD of 4 August 1862, Gray mentioned that he could see no nectar in the melastomaceous plant *Rhexia Virginica*.

[14] CD apparently refers to Gray's comments in his letter to CD of 29 July 1862, on the progress of the American Civil War.

[15] CD refers to chapter 7 of *Orchids*, which examined the homologies of the parts of orchid flowers, and ended with concluding remarks on natural selection and on the importance of cross-fertilisation (*Orchids*, pp. 286–360). See also letter to Asa Gray, 23[–4] July [1862] and n. 21.

To John Lubbock 21 August [1862][1]

1. Carlton Terrace | Southampton
Aug. 21.

My dear Lubbock

How long it is since we have had any communication. I want very much to hear some news of you & of M[rs] Lubbock.—[2] I heard indirectly that your tour was very successful.—[3] Do write me a note about yourself.—

We have been a very unhappy family, since you went, with Leonard fearfully ill from the effect of Scarlet-fever; & now, here, (on our Journey to Bournemouth where our other children are) M[rs]. Darwin has been struck with Scarlet-Fever.[4] Her attack has been pretty severe, but she is now recovering. When we shall move I hardly know. We are lucky to be in William's House. William seems a comfortable old Man of Business & has sole charge for a fortnight of the Bank!!! M[r] Atherley being in London.—[5] "Atherley & Darwin" in big letters on the Bank looks very grand.—

I have done hardly anything in science for an age; except beginning to make out a marvellous case of trimorphism like Primula, but far more complex, with two kinds in each flower of three kinds.—[6]

Farewell, my dear Lubbock, I suppose our troubles will some day end.

Yours affect[y] | C. Darwin

I enclose a prospectus.[7] I formerly knew the man & a very clever fellow he was. It is a real case of distress & I have helped him occasionally during several years—[8]

Down House (MS 7)

[1] The year is established by the relationship to the letter from John Lubbock, 23 August 1862.
[2] Ellen Frances Lubbock.
[3] Lubbock visited Switzerland in July and early August 1862, spending ten days mountaineering with Thomas Henry Huxley and John Tyndall, before travelling to examine the remains of prehistoric lake-dwellings, the recent discovery and examination of which he had described in a paper for the January number of the *Natural History Review* (Lubbock 1862b). See Hutchinson 1914, 1: 55–6, John Lubbock's diary (British Museum, Add. Ms. 62679: 64 r.), and the letter from John Lubbock, 23 August 1862.
[4] See letter to A. R. Wallace, 20 August [1862] and nn. 2 and 3.
[5] In 1861, John Lubbock had helped to arrange William Erasmus Darwin's partnership with George Atherley in the Southampton and Hampshire Bank, Southampton (see *Correspondence* vol. 9).
[6] CD refers to his work on *Lythrum salicaria* (see letter to Asa Gray, 9 August [1862] and n. 9).
[7] The enclosure has not been found.
[8] The reference is apparently to William Charles Linnaeus Martin (see letter from John Lubbock, 23 August 1862 and n. 10). As superintendent of the museum of the Zoological Society of London between 1830 and 1838, Martin had described some of CD's mammal specimens from the *Beagle* voyage, presented by CD to the society (Martin 1837). In later years, he was reported to have 'suffered

many severe afflictions' (*Gentleman's Magazine* n.s. 16 (1864): 536). According to CD's Classed account book (Down House MS), CD made a charitable donation of £15 to 'Mr Martin' on 20 August 1862. CD had previously made two donations to Martin, each of £5 5*s*., on 9 June and 19 December 1857.

To J. D. Hooker 22 [August 1862]

1. Carlton Terrace | Southampton
22d

My dear Hooker.

Hearty thanks for your note just received.[1] I am very glad Mrs. Hooker feels so well & that you are off to Scotland so soon.[2] I can give a good account of my patients.[3] Poor Leonards kidneys are certainly in some degree organically injured; & it will be months before he will be strong. I hope we shall get to Bournemouth, where we must take separate House in 10 days or fortnight.— We are staying here at William's house.—[4]

My chief object in writing is to ask for Mr Mann's address.—[5] My Bee friend, Mr Woodbury—a good man in his way, wants to offer him £8 or £10 to bring home live Bees, sending him instructions.[6] It is quite hopeless. But I suppose there wd. be no impropriety in making Mr Mann the offer.? Please answer this, as soon as you are established in Scotland.— Is Oliver at Kew?[7] when I am established at Bournemouth; I am completely mad to examine any fresh flowers of any Lythraceous plant & I would write & ask him if any are in bloom. Hardly any case has interested me so much as Lythrum salicaria.—[8] You must let me some time examine this wonderful Vanda.[9] Good Heavens what work you have had over Wellwitschia![10] How mortal man can work 5 hours with high power passes my understanding. By the way, perhaps you did not know the fact, but a young man at Ross' told me that no one can dissect with a $\frac{1}{10}$th inch focal glass!![11] I saw your microscope, & was a little disappointed at it for zoological purposes.[12] It seems to me an easily remedied, but great fault that the wheel for bringing lens nearer & further is on the *cheek*-side.—

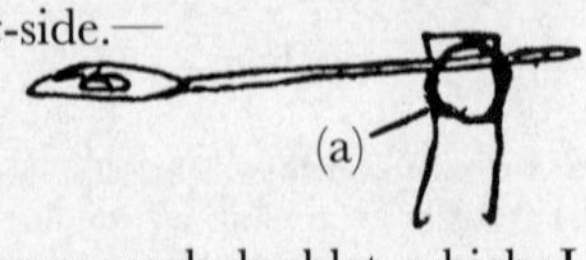

Ross does not give a huge weak doublet, which, I find, almost the most useful glass. With all necessary apparatus Smith & Beck charge 11£ for my microscope!![13]

But they are going to improve & I daresay spoil it. I find that slips of glass held by spring on stage of simple microscope, invaluable for quick transference of dissected object to compound. You see, God help you, by my scribbling that I am idle & am amusing myself; as my patients want nothing.—

One other question.— Can you think of plants, which have differently coloured anthers or pollen in same flowers, as in Melastomas or on same & in different plants as in Lythrum. It would be safe guide to dimorphism.—[14] Do just think of this.—

Did I tell you of one curious observation which I have made on action of pollen in Linum grandiflorum: the long-styled form is sterile with its own pollen & by Jove the pollen does not even emit tubes:[15] it is very curious to put pollen of long-styled & of short-styled on separate divisions of same stigma of long-styled, & after about 10 hours, to mark the wonderful difference both in state of pollen & stigma. In function, but not in appearance, the pollen of these two forms, as tested by their action may be said to be generically distinct.[16]

Now I have driven all care for half-an-hour out of my head; so farewell my dear old friend.— Yours affect | C. Darwin

I heartily hope that Huxley's book will be very successful; he will be well abused.—[17]

Endorsement: 'August/62.'
DAR 115.2: 162

1 Letter from J. D. Hooker, 20 August 1862.

2 Having recently returned from a holiday in Switzerland, intended to improve Frances Harriet Hooker's health, the Hookers were planning to leave for a three-week trip to Scotland on 23 August 1862 (see letter from J. D. Hooker, 20 August 1862).

3 Emma and Leonard Darwin were both recovering from scarlet fever (see preceding letter).

4 While travelling to Bournemouth for a holiday, CD, Emma, and Leonard had been obliged to remain at William Erasmus Darwin's house in Southampton, following the onset of Emma's scarlet fever (see preceding letter). They did not continue their journey until 1 September (see 'Journal' (Appendix II)).

5 Gustav Mann was botanical collector for the Royal Botanic Gardens, Kew, on the Niger expedition led by William Balfour Baikie (see R. Desmond 1995, p. 433).

6 See letter from T. W. Woodbury, 9 August 1862.

7 Daniel Oliver.

8 For CD's interest in this species, see, for example, the letter to Daniel Oliver, 29 [July 1862], the letter to W. E. Darwin, [2–3 August 1862], and the letter to Asa Gray, 9 August [1862].

9 See letter from J. D. Hooker, 20 August 1862 and n. 12.

10 See letter from J. D. Hooker, 20 August 1862 and n. 6.

11 CD refers to the establishment of the optical instrument maker, Thomas Ross, at 2 and 3 Featherstone Buildings, High Holborn, London (*Post Office London directory* 1861).

12 In the letter to J. D. Hooker, 15 [May 1862], CD asked for details of 'the *simple* microscope made by *Ross*' that Hooker recommended for young surgeons. See also letter from J. D. Hooker, [17 May 1862].

13 CD refers to Smith, Beck & Beck, instrument makers of 6 Coleman Street and Pear Tree Cottage, Holloway Road, London. In 1847, the company (then trading as Smith & Beck) had built for CD a simple microscope to his own design. They subsequently sold copies of the original model under the name 'Darwin's Single Microscope' (see *Correspondence* vol. 4, letter to Richard Owen, [26 March 1848] and n. 2). A full description of the microscope is given in Beck 1865, pp. 102–4.

14 In October 1861, CD had begun to investigate the occurrence of what he considered might be a novel form of dimorphism in the Melastomataceae, the structure and colour of the stamens facing the petals differing from that of the stamens facing the sepals in the same flower (see *Correspondence* vol. 9, letter to J. D. Hooker, 17 November [1861], and this volume, letter to George Bentham, 3 February [1862]). In *Lythrum salicaria*, the filaments and anthers of the long-styled form are different in colour from those in the other two forms (see letter to W. E. Darwin, [2–3 August 1862]). In *Forms of flowers*, p. 244, CD noted that he had been 'often deceived' by judging the occurrence of heterostyly on the basis of the length of stamens and pistils alone, and had decided that 'the more prudent course' was not to rank any species as heterostyled unless there was 'evidence of more important differences between the forms'. He concluded that 'absolutely conclusive evidence' of heterostyly could only be derived from experiments demonstrating differences in fertility (p. 245).

[15] CD had told Hooker that the long-styled pollen of *Linum grandiflorum* was sterile with its own stigma, in the letter to J. D. Hooker, 28 September [1861] (*Correspondence* vol. 9). See also 'Dimorphic condition in *Primula*', p. 96 (*Collected papers* 2: 63). CD also mentioned his more recent observations on the failure of the pollen to emit pollen-tubes, in the letters to Asa Gray, 14 July [1862] and 28 July [1862]. He described these observations in detail in 'Two forms in species of *Linum*', pp. 73–5 (*Collected papers* 2: 96–8).

[16] In 'Two forms in species of *Linum*', p. 75 (*Collected papers* 2: 98), CD stated:

> Taking fertility as the criterion of distinctness, it is no exaggeration to say that the pollen of the long-styled *Linum grandiflorum* (and conversely of the other form) has been differentiated, with respect to the stigmas of all the flowers of the same form, to a degree correspending with that of distinct species of the same genus, or even of species of distinct genera.

For CD's interest in this question, see also the letter to Asa Gray, 9 August [1862] and n. 13, the letter to Asa Gray, 21 August [1862] and n. 5, and Appendix VI.

[17] See letter from J. D. Hooker, 20 August 1862. The reference is to T. H. Huxley 1863a.

To Charles Lyell 22 August [1862]

1. Carlton Terrace | Southampton
Aug 22d.

My dear Lyell

I thank Lady Lyell (whose kind note to William I answer) & yourself for all your sympathy.[1] Emma is going on very well, but she has had a sharpish attack.[2] Lenny has been fearfully ill, but is perhaps as strong as one could expect; he now gets up for some hours every day.[3] We long to get to Bournemouth, where though in a separate house, we shall be close to our other children;[4] & Horace is far from strong.[5] The last two months misery has been enough to try anyone; but I suppose better times will come.— The Lubbocks have had for nearly a year (*& have enjoyed*) a separate house, "Lamas, Chiselhurst."[6]

I am glad Glen Roy is settled;[7] the moraines opposite L. Treig are obviously very important: if the slope inland can be proved, it will indeed be an important fact.— I heartily hope that you will be out in October;[8] I fancy Huxley will be out sooner;[9] Hooker speaks as if the book would be very interesting.[10] You say that the Bishop & Owen will be down on you;[11] the latter hardly can, for I was assured that Owen in his Lectures this Spring, advanced as a new idea that wingless Birds had lost their wings by disuse.[12] Also that magpies stole spoons &c from a *remnant* of some instinct like that of the Bower-bird, which ornaments its playing passages with pretty feathers.[13] Indeed I am told that he hinted plainly that all Birds are descended from one. What an unblushing man he must be to lecture thus after abusing me so & never to have openly retracted or alluded to my Book.[14]

Your P.S. touches on, as it seems to me, very difficult points.[15] I am glad to see in Origin, I only say that naturalists generally consider that low organisms vary more than high;[16] & this I think certainly is the general opinion. I put the statement this way to show that I considered it only an opinion probably true. I must own that I do not at all trust even Hooker's contrary opinion, as I feel pretty sure that he has not tabulated any result.[17] I have some materials at home, & think I attempted to make this point out, but cannot remember result.[18]

Mere variability, though the necessary foundation of all modifications, I believe to be almost always present enough to allow of any amount of selected change; so that it does not seem to me at all incompatible, that a group which at any one period (or during all successive periods) varies less, should in the long course of time have undergone more modification than a group which is generally more variable. Placental mammals e.g. might be at each period less variable than Marsupials, & nevertheless have undergone more differentiation & development than marsupials, owing to some advantage, probably Brain development.—

I am surprised, but do not pretend to form an opinion, at Hooker's statement that higher species, genera &c are best limited.—[19] It seems to me a bold statement.

Looking to the Origin I see that I state that the productions of the land seem to change quicker than those of the sea (Ch X. p. 339 3^d. Edit) & I add there is some reason to believe that organisms considered high in the scale change quicker than those that are low. I remember writing these sentences after much deliberation; but cannot now remember why I did not more fully adopt & quote your axiom of 1832.—[20] I remember well feeling much hesitation about putting in even the guarded sentences which I did. My doubts, I remember related to the rate of change of the Radiata in the Secondary formation & of the Foraminifera in the oldest Tertiary beds.

I daresay, however, your axiom may be quite true: I only remember considerable perplexity on subject; I sh^d. think mammals & molluscs rather too remote from each other for fair comparison.

I am tired with writing this long letter (though it has amused me writing it) & I fear that you will be tired with reading it, & that it will be much too vague to be of any service—

I was very glad to get your note.— I hope to goodness your Book won't be delayed—

With kindest remembrances to Lady Lyell— Good Night— | C. Darwin

Postmark: AU 23 62
American Philosophical Society (281)

[1] The note from Mary Elizabeth Lyell to William Erasmus Darwin, which may have been enclosed with the letter from Charles Lyell, 20 August 1862, has not been found.

[2] According to her diary (DAR 242), Emma Darwin became ill with scarlet fever on 13 August 1862.

[3] Leonard Darwin had been suffering from scarlet fever (see letters to W. E. Darwin, 13 [June 1862] and 9 July [1862], and letter to A. R. Wallace, 20 August [1862]).

[4] In consequence of their illnesses, Emma and Leonard, together with CD, were staying at William Erasmus Darwin's house (see letter to A. R. Wallace, 20 August [1862] and n. 3); CD's 'Journal' (Appendix II) records that on 1 September, they joined the rest of the family on holiday in Bournemouth.

[5] Horace Darwin had been ill earlier in the year (see letter to W. E. Darwin, 14 February [1862]).

[6] John and Ellen Frances Lubbock had lived with John Lubbock's father (John William Lubbock) at High Elms, near Down, until August 1861, when they moved to a separate house, which they called 'Lamas', at Chislehurst, a village about five miles north of Down (John Lubbock's diary (British Library, Add. Ms. 62679: 64 r.)).

[7] See letter from Charles Lyell, 20 August 1862 and n. 1.

[8] CD refers to Lyell's *Antiquity of man* (C. Lyell 1863a), which was not published until 6 February 1863 (C. Lyell 1863b, p. [vii]).

[9] Huxley's *Evidence as to man's place in nature* (T. H. Huxley 1863a) was also published in February 1863 (*Publishers' Circular* 26 (1863): 112).

[10] See letter from J. D. Hooker, 20 August 1862.

[11] Lyell's comments on this subject were probably made in the missing postscript to the letter from Charles Lyell, 20 August 1862 (see n. 15, below). The references are to the bishop of Oxford, Samuel Wilberforce, and to Richard Owen, both of whom had opposed CD's views (see especially *Correspondence* vol. 8).

[12] CD probably refers to Richard Owen's lectures at the Museum of Practical Geology, London, on the 'Characters, Organisation, Geographical Distribution, and Geological Relations of Birds'. The series of six lectures ran from 14 to 30 May 1862 (*Athenæum*, 10 May 1862, p. 613). See also letter to Armand de Quatrefages, 11 July [1862], and letter to Asa Gray, 23[–4] July [1862].

[13] Owen's observations on this point, made in his third lecture, were reported in the *Medical Times and Gazette* (1862), pt 1: 563–4.

[14] Owen had written a critical review of *Origin*, which was published anonymously in the *Edinburgh Review* in April 1860 ([R. Owen] 1860b). For CD's reaction to Owen's review, see *Correspondence* vol. 8.

[15] The postscript to the letter from Charles Lyell, 20 August 1862, was not reproduced in K. M. Lyell ed. 1881; the original letter has not been found. In the postscript, Lyell had evidently discussed the differences in the longevity of species exhibited by different groups of organism (see C. Lyell 1863a, pp. 441–3).

[16] *Origin* 3d ed., p. 167. See also *Origin*, p. 149, and *Origin* 2d ed., p. 149.

[17] In J. D. Hooker 1859, p. vii n., Joseph Dalton Hooker stated that the higher plants manifested their physical superiority in a 'greater extent of variation'. Lyell cited Hooker's observation in C. Lyell 1863a, p. 442. CD's annotated copies of both works are in the Darwin Library–CUL (see *Marginalia* 1: 394–8, 525–7).

[18] CD's notes on this subject have not been found. However, a single sheet of his discussion of this question in the 'Big book' on species, survives in DAR 47: 95; it is transcribed in *Natural selection*, pp. 567–8.

[19] CD refers to Hooker's statement in J. D. Hooker 1859, pp. v–vi, that in the vegetable kingdom, as a general rule,

> the varying species are relatively most numerous in those classes, orders, and genera which are the simplest in structure. Complexity of structure is generally accompanied with a greater tendency to permanence in form: thus Acotyledons, Monocotyledons, and Dicotyledons are an ascending series in complexity and in constancy of form.

[20] CD refers to a passage in the third volume of Lyell's *Principles of geology* (C. Lyell 1830–3, 3: 140), which states that:

> the longevity of *species* in the mammalia is, upon the whole, inferior to that of the testacea ... Their more limited duration depends, in all probability, on physiological laws which render warm-blooded quadrupeds less capable, in general, of accommodating themselves to a great variety of circumstances, and consequently, of surviving the vicissitudes to which the earth's surface is exposed in a great lapse of ages.

See also C. Lyell 1830–3, 3: 48. CD had presented evidence in support of this statement, which he described in *Journal of researches*, p. 97, as 'the remarkable law so often insisted on by Mr. Lyell'; see also *Correspondence* vol. 2, letter to Charles Lyell, 30 July 1837 and nn. 1 and 6. The third volume of C. Lyell 1830–3 was published in 1833; CD wrote '1832' in error. Lyell referred to his earlier statement in C. Lyell 1863a, p. 441:

> It has since been found that this generalisation can be carried much farther, and that, in fact, the law which governs the changes in organic beings is such, that the lower their place in a graduated scale, or the simpler their structure, the more persistent are they in form and organisation.

From John Lubbock 23 August 1862

15, Lombard Street. E.C.
23 August 1862

My dear M^r^. Darwin

I am very sorry to hear so sad an account of your family, but hope that you are now getting out of your troubles.[1] Atherley called here the other day & gave a flourishing account of William, but leaving him for a fortnight was even more complimentary.[2]

My Swiss tour was most successful;[3] besides spending a week in the mountains with Tyndall & Huxley,[4] which was capital fun, I visited nearly all the collections of Lake antiquities & saw five of the Pfahlbauten themselves.[5] Three of them, those at Nernier, Thonon, & Morges in the Lake of Geneva I saw from a boat. The water was from 8 to 12 f^t^. deep, but so clear that I could see quite well, the piles & other things at the bottom.

We thought we saw a hatchet and I undressed & dived for it. After two or three ineffectual attempts we poked at it with a pole, & it turned out to be only a bit of wood. At Wauwyl the old floor is covered by three or four feet of peat, and the lake has been drained. We spent several hours in digging & got three hatchets, three or four implements in bone & a great many bits of pottery & more or less broken bones. There also the beams forming the floor are preserved in the peat, & one could stand, as it were, on the old floor.

I have got a hymenopterous insect which uses its wings to swim with *in the water*! Probably it is looking for a victim on which to lay eggs; but it strikes me as a most curious adaptation & one which would interest you much.[6]

I have also a case of dimorphism in Psocus, but have not yet been able to make out much about it. It had been already suspected by Westwood.[7]

Your case of trimorphism makes ones mental mouth water for more information.[8]

Let me know when you return as I am very anxious for a talk with you.[9]

Yours affec | John Lubbock

P.S. I send a subscription to Martin[10]

DAR 170: 31

[1] See letter to John Lubbock, 21 August [1862].

[2] See letter to John Lubbock, 21 August [1862]. Lubbock refers to George Atherley, William Erasmus Darwin's partner in the Southampton and Hampshire Bank; in 1861, Lubbock had helped to arrange William's partnership with Atherley (see *Correspondence* vol. 9).

[3] Lubbock had visited Switzerland in July and early August (Hutchinson 1914, 1: 55–6; John Lubbock's diary (British Library, Add. Ms. 62679: 64 r.)).

[4] According to John Lubbock's diary (British Library, Add. Ms. 62679: 64 r.), Lubbock spent ten days with Thomas Henry Huxley and John Tyndall, who were in Switzerland for a holiday (Eve and Creasy 1945, p. 92; L. Huxley ed. 1900, 1: 234).

[5] According to his diary (British Library, Add. Ms. 62679: 64 r.), Lubbock spent the latter part of his visit to Switzerland with the Swiss archaeologist, Charles Adolphe Morlot, studying the recently

discovered remains of prehistoric lake-dwellings or 'Pfahlbauten'. Lubbock had described the remains from written reports, in a paper for the January 1862 number of the *Natural History Review* (Lubbock 1862b). He gave details of his own observations in his book *Pre-historic times* (Lubbock 1865, pp. 119–70).

[6] In October 1862, Lubbock read an account of this insect, which he proposed to name *Polynema natans*, and of a second, unrelated, aquatic species of Hymenoptera, before the meeting of the British Association for the Advancement of Science in Cambridge (Lubbock 1862d). He reported that he had found the insects 'on one of the early days in August' in water from a pond near his house (Lubbock 1862d, p. 110). Lubbock described the species in greater detail in Lubbock 1863e.

[7] Lubbock refers to the entomologist John Obadiah Westwood; the reference has not been further identified.

[8] See letter to John Lubbock, 21 August [1862].

[9] CD, Emma, and Leonard Darwin were staying with William in Southampton (see letter to John Lubbock, 21 August [1862]). According to CD's 'Journal' (Appendix II), they did not return to Down until 30 September.

[10] William Charles Linnaeus Martin; see letter to John Lubbock, 21 August [1862] and n. 8.

To John Murray 24 August [1862]

☞ 1. Carlton Terrace | Southampton[1]
Aug. 24th

My dear Sir

I enclose a cheque for your account, which please return to above address.[2]

I do not know what you think, but I am well contented with the sale of 768 copies; I shd. hope & expect that the remainder will ultimately be sold.—[3]

Pray believe me | My dear Sir | Yours sincerely | C. Darwin

You will have to charge me for the 2 or 3 electrotypes sent to America.[4] Has M. Scheweizerbart of Stuttgard paid you the 10£ for the cuts for German Edition?[5] this sum, I presume, will be in excess of actual cost.—

Endorsement: '*1862— Aug 24*'
John Murray Archive (Darwin 125–6)

[1] CD, Emma, and Leonard Darwin were staying with William Erasmus Darwin in Southampton (see 'Journal' (Appendix II)).

[2] In his Account book–cash accounts (Down House MS), CD recorded a payment of £17 16*s*. to Murray on 24 August 1862, marking the entry 'J Murray Orchid Book'. The sum probably represents the cost of some or all of the presentation copies of *Orchids*, charged at a reduced price (see letter to Asa Gray, 9 August [1862] and n. 4). For CD's presentation list for *Orchids*, see Appendix IV.

[3] The first edition of 1500 copies of *Orchids* was published on 15 May 1862 (see Freeman 1977, p. 112, and letter to John Murray, 9 April [1862]); CD had been concerned that the book might not sell (see letter to John Murray, 9 [February 1862]).

[4] CD had arranged for Murray to send electrotype plates of three of the illustrations from *Orchids* for reproduction in Asa Gray's review of the book in the *American Journal of Science and Arts* (A. Gray 1862a; see letters to John Murray, 13 June [1862] and 20 [June 1862]).

[5] CD had arranged for Christian Friedrich Schweizerbart, head of the publishing company E. Schweizerbart'sche Verlagsbuchhandlung, to purchase from Murray electrotype plates of the illustrations for the German translation of *Orchids* (Bronn trans. 1862); see letter to John Murray, 13 June [1862].

From J. D. Hooker [26–31 August 1862][1]

Address Care of Mrs Hooker[2]
Westbank Terrace | Hillhead | Glasgow

Dear Darwin

Yours of 22^d. arrived just as I was leaving Kew.[3] We took train to Glasgow, staid there Sunday & came down Clyde on Monday; we are now with our friends the Smiths Jordan hill[4] at his son in laws Mr Buchanan of Shandon a most lovely spot on the Gare Loch,[5] & remain till Monday. I am amusing myself wandering over the hills & yachting. I hear from Huxley on Loch Fyne[6] & perhaps shall join him after the marriage, (on the 4th)[7] but we are uncertain as to our movements after that. Are you sure it is Lennys kidneys that are hurt?[8] heaps of kidney symptoms are nothing but rheumatism + vitrated secretions

Mann's address is[9]

Mr Gustav Mann
Govt. Botanist
Care of HRM Consul
Fernando Po

he will gladly do any thing he can & is a capital collector & packer but nothing else. He returns to England next spring. Oliver returned to Kew after I left & is there now.[10] I bottled the two Vanda flowers after Fitch had sketched them—[11] By the way I have a bottle of the Gymnadenia hybrids for you, tell me when you want them.[12]

I do not wonder you are disappointed with the microscope in *the abstract* & for Zoological purposes.—[13] the **great** fault you mention of the wheel on wrong side I remedy myself at once with screw driver. (it is a horrible arrangement) The doublet is a necessity easily added.— The fact was that Ross[14] alone would supply a good microscope at £4·40. suitable for **ordinary** botanical purposes.—firm, steady, portable, good glasses, cheap—& above all capable of any & every further appliance that a simple microscope is capable of having.— so a young student of moderate means, may buy his essential at once, & add to it, as he can afford.— Smith & Beck would do nothing of the kind.[15] I know from long experience how essential it is that something efficient should be cheap.—especially for young men going abroad with all their outfit to procure & an expensive set of medical & surgical books & instruments to purchase besides.

I cannot remember any plants but melastomads with differently colored *polliniferous* anthers in same flower,[16] but I shall

Incomplete
DAR 101: 50–1

[1] The date range is established by the reference to the letter to J. D. Hooker, 22 [August 1862], and on the basis of Hooker's departure from London for Scotland on 23 August 1862 (see letter from J. D. Hooker, 20 August 1862). The letter was written after the Monday following 22 August (25 August) and before the subsequent Monday (1 September).

[2] The reference is to Isabella Whitehead Hooker, widow of Hooker's brother, William Dawson Hooker; she lived at 11 West Bank Terrace, Hillhead, Glasgow, with her daughter Willielma (Census returns 1861 (Mitchell Library, Glasgow, 646^2/11a: 6)). See also n. 7, below.

[3] Letter to J. D. Hooker, 22 [August 1862].

[4] Hooker refers to the antiquarian and geologist, James Smith, known as 'Smith of Jordanhill', and probably to his two surviving unmarried daughters, Jane Charlotte and Sabina Douglas Clavering Smith.

[5] The reference is to Walter Buchanan, widower of James Smith's daughter, Christina Laura.

[6] Huxley was at Loch Fyne in his capacity as a member of the Royal Commission on the Operation of the Acts relating to Trawling for Herring on the Coasts of Scotland (L. Huxley ed. 1900, 1: 200–1; *Report of the Royal Commission on Herring*, pp. 141, 143).

[7] Hooker refers to the wedding of his niece Willielma Hooker to James Campbell at St Mary's Episcopal Church, Glasgow, on 4 September 1862 (*Gentleman's Magazine* n.s. 13 (1862): 488).

[8] The reference is to Leonard Darwin (see letter to J. D. Hooker, 22 [August 1862]).

[9] Gustav Mann was botanical collector for the Royal Botanic Gardens, Kew, on the Niger expedition led by William Balfour Baikie (see R. Desmond 1995, p. 433). CD wanted Mann's address for Thomas White Woodbury (see letter from T. W. Woodbury, 9 August 1862, and letter to J. D. Hooker, 22 [August 1862]).

[10] The reference is to Daniel Oliver, librarian and assistant in the herbarium at the Royal Botanic Gardens, Kew (R. Desmond 1994; *List of the Linnean Society of London* 1862). See letter to J. D. Hooker, 22 [August 1862].

[11] Hooker refers to a spike of *Vanda Lowii* bearing two different types of flower, which he had mentioned in his letter to CD of 20 August 1862. Walter Hood Fitch worked as a botanical artist at the Royal Botanic Gardens, Kew.

[12] See letter from J. D. Hooker, [24 July 1862].

[13] See letter to J. D. Hooker, 22 [August 1862].

[14] The reference is either to the optician and scientific instrument maker, Thomas Ross, or to his father, Andrew Ross, who had been one of the leading microscope manufacturers in London.

[15] Hooker refers to Smith, Beck & Beck, instrument makers at 6 Coleman Street, and Pear Tree Cottage, Holloway Road, London (*Post Office London directory* 1861).

[16] CD had suggested that the occurrence of differently coloured anthers or pollen in the same flower, as in the Melastomataceae, would prove to be a 'safe guide' to dimorphism and asked Hooker if he knew of any instances other than those he mentioned (see letter to J. D. Hooker, 22 [August 1862]).

To Camilla Ludwig 26 August [1862][1]

1. Carlton Terrace | Southampton
Aug. 26th.

My dear Miss Ludwig

We thank you sincerely for your heart-felt sympathy.[2] We have had a bad time of it; but Emma is nicely recovering, & came down to the drawing room for a little bit to day.[3] She had considerable fever. Leonard travelled too soon, & was injured by the journey.[4] Poor little fellow, he has been so patient during all his terrible illness; he will be months before he will be strong. We hope to move to Bournemouth in about a week, if both patients go on well.[5] Elizabeth Wedgwood is here, & has been, as usual, the most unselfish & devoted of nurses.—[6]

Horace is going on well & only occasionally has a baddish day.[7] Etty[8] is accustoming him to have no one to sit with him at night; & she has so much judgment

& kindness, that she will do it well. It is a horrid bore, but we have been forced to engage a second house at Bournemouth, & so shall not be all together.— Emma sends her love & will write when she is strong: she hopes that Horace will soon be able to do some lessons, & then it will be capital for him to have you to return. But I hope we shall then have got him out of his invalid habits. Poor little man he has often cried, when he has tried & failed to write to you. And no wonder for nothing could have possibly exceeded your kindness to him.— What a wretched summer & spring we have had! We had a very nice letter lately from Louisa, written with so much feeling.[9] I fear some things are very uncomfortable at the school; but she seems determined to bear them with excellent spirit. I sincerely hope that you have been happy & enjoyed yourself at home with as few drawbacks as this weary world permits.

My dear Miss Ludwig— Yours very sincerely | C. Darwin

American Philosophical Society (B/D25.272)

[1] The year is provided by reference to CD's stay in Southampton and Bournemouth (see n. 5, below).

[2] Ludwig's letter has not been found. Ludwig, the Darwins' governess, was on an extended visit to her family in Hamburg, having apparently been sent away in early June, on full pay, in order to separate her from Horace Darwin. The Down surgeon, Stephen Paul Engleheart, was concerned that Horace's attachment to her might be exacerbating the illness from which he had been suffering. See the letters from Emma Darwin to William Erasmus Darwin, [2 March 1862], [27 May 1862], and [6 November 1862], in DAR 219.1: 49, 57, and 64; see also Emma Darwin's diary (DAR 242), and CD's Classed account book (Down House MS).

[3] According to her diary (DAR 242), Emma Darwin had become ill with scarlet fever on 13 August 1862.

[4] The Darwins had planned to take a holiday in Bournemouth to assist Leonard Darwin's convalescence (see letter from W. E. Darwin, 5 August 1862 and n. 8). CD, Emma, and Leonard began their journey on 13 August 1862, but were obliged to remain in Southampton as a result of Emma and Leonard's illness (see letter to A. R. Wallace, 20 August [1862] and n. 2).

[5] CD, Emma, and Leonard stayed with William in Southampton from 13 August until 1 September 1862, when they joined the other Darwin children in Bournemouth (see 'Journal' (Appendix II), and letter to A. R. Wallace, 20 August [1862] and n. 3).

[6] CD refers to Emma's sister, Sarah Elizabeth Wedgwood.

[7] Horace Darwin had been ill earlier in the year (see n. 2, above).

[8] Henrietta Emma Darwin.

[9] The reference is to Louisa Ludwig, apparently Camilla's sister. During Camilla's absence, she had assisted the family in looking after Horace (see the letter from Emma Darwin to W. E. Darwin, [27 May 1862], in DAR 219.1: 57). CD's Classed account book (Down House MS) records, under the heading 'Governess', payments to Louisa totalling £8, made on 1 and 8 August 1862.

From A. C. Ramsay 26 August 1862

London
26 Augt 1862

My dear Sir

By this post I send a paper[1] which I hope you may find time to read, for you are one of the few whose opinion I specially care about on such a subject.

When I read it Falconer[2] made a 40 minutes onslaught on it, & I accidentaly heard (what did not surprise) me that the Council had some difficulty about passing it at all.[3]

Falconer was of opinion that had I known the Himalayah I would not have propounded such a theory. Hooker, however, who was not present, writes me that "the great standing puzzle of the Himalayah, *its wanting lakes*, is explicable on your hypothesis & on no other that I ever heard propounded".—[4] He then shows cause on my hypothesis for their absence on the South & the immense number of them in Cashmire & Thibet where the valleys are wide & of gentle slope.

Ever sincerely | And^w C Ramsay

DAR 176: 9

[1] Ramsay 1862.
[2] Hugh Falconer.
[3] Ramsay, who became president of the Geological Society of London in 1862, read his controversial paper on the glacial origin of rock basins before the society on 5 March 1862. For an account of the controversy initiated by Ramsay's paper, see Davies 1969, pp. 303–9.
[4] For Joseph Dalton Hooker's reaction to Ramsay's paper, see the letter from J. D. Hooker, 20 August 1862.

To C. C. Babington 2 September [1862][1]

Cliff Cottage | Bournemouth | Hants
Sept 2^d.

Dear Babington

As you seemed a little interested in the Primula case,[2] I venture to trouble you, to ask whether you think you could get me by any of your correspondents some seed of the rare Lythrum hyssopifolia.— It would be of *great* importance for my work; as I have lately been working hard at Lythrum salicaria.—[3]

I have seen Hottonia, of which you told me:[4] it is splendidly dimorphic with widely different pollen in the two forms.

We shall be here about a month.—

I believe that your kindness will forgive me troubling you.—

In Haste | Yours very sincerely | C. Darwin

CUL (Add 8182)

[1] The year is established by reference to CD's interest in *Lythrum*, and by reference to the Darwin family's holiday in Bournemouth (see 'Journal' (Appendix II)).
[2] See letter from C. C. Babington, 17 January 1862.
[3] At the end of July 1862, CD had begun crossing experiments on the trimorphic plant, *Lythrum salicaria* (see letter to Daniel Oliver, 29 [July 1862] and n. 6). Having learned from his reading of Vaucher 1841 that *L. hyssopifolia* was dimorphic (see Vaucher 1841, 2: 371), CD was anxious to confirm the observation, and had sought specimens from several correspondents (see following letter, letter to H. C. Watson, 8 [August 1862], and letter to Asa Gray, 9 August [1862]).

[4] In his letter to CD of 17 January 1862, Babington had noted that *Hottonia* was dimorphic; CD obtained specimens of the plant early in August (see letters from M. S. Wedgwood, [before 4 August 1862] and [6 August 1862], letter to K. E. S., L. C., and M. S. Wedgwood, 4 [August 1862], and letter to Asa Gray, 9 August [1862]).

To W. E. Darwin [2–3 September 1862][1]

[Bournemouth]

My dearest William

Thanks for your pleasant letter.[2] Your account made us all laugh.—

Dare you in morning steal 2 flowers & one leaf & one pod, & post same day in enclosed.—[3]

The patients are very well & mamma better.[4] I have not yet had strength to dissect the Bee's mouth.—[5]

Good Bye my dear old fellow | C. D

DAR 210.6: 103

[1] The date range is established by the relationship of this letter to CD's correspondence with John Lubbock concerning bees (see n. 5, below), and by reference to the letter to William from Emma Darwin, written on the verso of this letter. CD, Emma, and Leonard Darwin stayed with William in Southampton until 1 September 1862, when they joined the rest of the family in Bournemouth (see 'Journal' (Appendix II)). In her letter to William, Emma described the house where they were staying, and suggested that William should visit them in Bournemouth on his first free Saturday; according to Emma Darwin's diary (DAR 242), William first visited them on 6 September.

[2] William's letter has not been found.

[3] The reference has not been traced.

[4] Leonard Darwin had been ill with scarlet fever since 12 June 1862 (see letter to W. E. Darwin, 13 [June 1862]); Emma contracted the fever on 13 August 1862 (Emma Darwin's diary (DAR 242)).

[5] At the end of August 1862, during his stay with William in Southampton, CD noted that whereas some of the bees in a field of red clover sucked nectar at the mouths of the flowers, others did so at the bases of the flowers, through holes bitten in the corollas. CD speculated that there might be two types of bees, with those that bit holes in the corollas possibly having shorter probosces than the others, or possibly working with the others according to a 'division of labour principle' (DAR 49: 149); on 31 August, he gathered two specimens of each type, but apparently did not examine them until the evening of 3 September (see following letter, letter to Asa Gray, [3–]4 September [1862], and letter to John Lubbock, [3 September 1862]).

To John Lubbock 2 September [1862][1]

Cliff Cottage | Bournemouth | Hants
Sept 2^{d}

Dear Lubbock.—

Hearty thanks for your note.[2] I am so glad that your Tour answered so splendidly.—[3] My poor patients got here yesterday & are doing well;—& we have a second House for the well ones.—[4]

I write now in great Haste to beg you to look (though I know how busy you are, but I cannot think of any other naturalist who w^{d} be careful) at any field of common red clover (if such a field is near you) & watch the Hive Bees: probably (if not too late) you will see some sucking at the mouth of the little flowers & some few sucking at the base of the flowers, at holes bitten through the corollas—[5] All that you will see is that the Bees put their Heads deep into the head & rout about.— Now if you see this, do for Heaven sake catch me some of each & put in spirits & *keep them separate*.— I am almost certain that they belong to two castes, with long & short probosces. This is so curious a point that it seems worth making out.— I cannot hear of a clover field near here.—

Pray forgive my asking this favour, which I do not for one moment expect you to grant, unless you have clover field near you & can spare $\frac{1}{2}$ hour

In Haste— Yours most sincerely | C. Darwin

DAR 263

[1] The year is established by the relationship to the letter from John Lubbock, 23 August 1862 (see nn. 2 and 3, below).

[2] Letter from John Lubbock, 23 August 1862.

[3] CD refers to Lubbock's recent tour of Switzerland (see letter from John Lubbock, 23 August 1862).

[4] Emma and Leonard Darwin were recovering from scarlet fever; together with CD they travelled from Southampton to Bournemouth on 1 September 1862. The other Darwin children had been sent away with their former nurse, Brodie, during the latter part of Leonard's illness (see *Emma Darwin* 2: 178, and letter to W. E. Darwin, 4 [July 1862] and n. 8); they had been in Bournemouth since mid-August (see letter to A. R. Wallace, 20 August [1862] and n. 3, and letter to John Lubbock, 21 August [1862]).

[5] CD had observed this phenomenon at Southampton on 29, 30, and 31 August (see preceding letter and n. 5).

To Daniel Oliver 2 September [1862][1]

Cliff Cottage | Bournemouth | Hants.

Sept. 2^d.

Dear Oliver.

I want very much to beg a favour. I have been working hard at Lythrum salicaria, which offers a most curious case (beating Primula all to bits) of **tri**morphism, & I hope my numerous crosses at home will explain the functional meaning of all the differences.[2] This work has given me an intense wish to see fresh flowers of any member of the Lythraceæ. Have you anything in bloom at Kew?[3] If so would you be so very kind as to send me anything in little tin by Post—tying something damp round cut off stems.— When you hear the case of Lythrum, I really think you will not think the trouble wasted.—[4]

We are here on account of a long miserable illness of one of my Boys from Scarlet Fever; & now my poor wife has caught it, but has almost recovered.—[5] I have in consequence done hardly anything this summer.

Can you tell me of any plants, which bear differently coloured anthers? I wrote to Hooker & he could not remember any besides the Melastomas & Lythrums,[6] & I have just seen in a garden a *small* bush, which seems to me either a Clarkia or Epilobium with crimson tall anthers & short white ones.—[7] This difference I suspect would be good guide to functional dimorphism.—[8]

I hope you have enjoyed your holidays.— Hooker tells me you are at Kew, where I wrote to him about Lythaceæ.—[9]

Dear Oliver | Yours very sincerely | in Haste | C. Darwin

DAR 261 (DH/MS 10: 35)

[1] The year is established by reference to the incidence of scarlet fever among members of the Darwin family (see n. 5, below).

[2] In his letter to Oliver of 29 [July 1862], CD had described the crosses he planned to make with *Lythrum salicaria.* The earliest of his notes from these experiments is dated 31 July 1862 (DAR 27.2 (ser. 2): 7).

[3] Oliver was librarian and assistant in the herbarium at the Royal Botanic Gardens, Kew.

[4] In a subsequent letter, which has not been found, CD described what he believed to be the relationship between the three sexual forms in this species; for Oliver's response, see the letter from Daniel Oliver, 13 September 1862.

[5] The Darwin family were on holiday in Bournemouth to assist with the convalescence of Leonard Darwin, who had been ill with scarlet fever since mid-June (see letter to W. E. Darwin, 13 [June 1862], and letter to H. C. Watson, 8 [August 1862]). Emma Darwin became ill with the disease on 13 August 1862, on the first stage of their journey to Bournemouth (Emma Darwin's diary (DAR 242)).

[6] See letter to J. D. Hooker, 22 [August 1862], and letter from J. D. Hooker, [26–31 August 1862].

[7] In a note dated 25 September 1862 (DAR 48: 50), CD recorded the existence of two differently coloured sets of stamens in *Clarkia pulchella*, describing the shorter ones as 'aborted with little shrivelled anther with no pollen'. He noted that the species presented a 'good [case of] gradation', marking his notes '*Transition*': he apparently considered that the species was transitional between an ordinary hermaphrodite plant and a dimorphic plant. For CD's later views on the origin of heterostyly, see *Forms of flowers*, pp. 260–8. See also letter from Daniel Oliver, 4 September 1862.

[8] See letter to J. D. Hooker, 22 [August 1862] and n. 14.

[9] See letter to J. D. Hooker, 22 [August 1862], and letter from J. D. Hooker, [26–31 August 1862].

To Asa Gray [3–]4 September [1862][1]

Cliff Cottage. Bournemouth. Hants

Sept. 4th.

My dear Gray.

I have just received your most kind letter of Aug 18 & 19th[2] & will amuse myself, now my patients are in bed,[3] by beginning a letter to you. My poor Boy (whose waxen face blushed up to the eyes at the thought of writing to a live Professor)[4] has this day made a marked step & has taken several walks of a few hundred yards; & my wife is recovering well & her skin well peeling.— We have taken two houses here, so I hope & trust this dredful fever will not spread.—[5] I am very glad that at present you intend to publish some separate notes on orchids, which you have so capitally worked out.[6] No doubt I am not a fair judge; but I must think that it is worth your while.— I am pleased to hear about Goodyera & about Gymnadia tridentata. Your account makes me think the latter like the Bonatea speciosa & often & often have I speculated what on earth could be meaning of its wonderful horn-like stigmas & projecting anthers.[7] I suspect its structure may have been arrived at by a process somewhat analogous to that which apparently has produced the wondrous nectary of Angræcum sesquipedale.[8] It would appear that self-fertilisation is commoner than I thought: since publishing I have found that Neottia nidus-avis fertilises itself, *if insects fail to do the job.*—[9] Many thanks for Houstonia seed.—[10]

I am glad to hear, but disappointed, about the Specularia pollen-tubes.[11] I cannot resist sending you a diagram about Lythrum (why should I resist? does not prescription give a right? & have I not for long years written & bothered you about all sorts of things?) Lythrum has 3 kinds of stigmas & 3 kinds of pollen; the stamens of same height on two of the three forms producing the same sort of pollen, & I cannot doubt are fitted to fertilise the stigma of that height. I conclude so from watching the Bees; but hope to prove it by my crosses. So that we have three hermaphrodite forms each depending on half the stamens of either one of the two other forms. This is shown by the lines with the arrows. This strikes me as a very curious case. The three forms coexist in about equal numbers. By the way it seems to me that Lindley in Veg. Kingdom describes the homological structure of flower wrong:[12] the so called calyx, with its 12 bundles of spiral vessels, appears to consist of 6 narrow sepals & 6 modified petals, all cohering; & that the coloured petals belong to an inner whorl & are modified stamens. By the way can you tell me of any flowers, with fertile anthers of different colours; I believe that this would be pretty sure guide to dimorphism or trimorphism.—[13]

All my semibotanical work, as you know, has been connected with insects, & now I am almost sure (but I find it a disgusting truth that with me first observations are generally all a blunder) that flowers have led me to a curious little discovery with respect to the best-known insect in the world, the Hive Bee. I saw the other day to my dismay (see Origin) Hive Bees sucking the common red clover, but it was a second crop, which I am *told* produces shorter flowers;[14] but many of the Bees never attempted this, but always inserted their heads between the flowers & sucked at holes bitten through the corolla.— The same bee always followed the same practice. And apparently those which suck at the mouth of the flower have a longer proboscis than the other bees, which suck through the holes.

Farewell my dear Gray, forgive me scribbling about my own work; for I seldom see a soul to talk with on natural history.

Good night— Ever yours most truly | C. Darwin

Since writing the above by Jove I have found I have as usual at first blundered about the probosces; but if you had seen the Bees, the blunder was almost excusable—[15] What an ass I was to scribble all the above.—

Progress of Education.— one of my little Boys Horace[16] said to me, "there are a terrible number of adders here; but if everyone killed as many as they could, they would sting less".— I answered "of course they would be fewer" Horace "Of course, but I did not mean that; what I meant was, that the more timid adders, which run away & do not sting would be saved, & after a time none of the adders would sting".—

Natural selection!!

Lythrum salicaria[17]

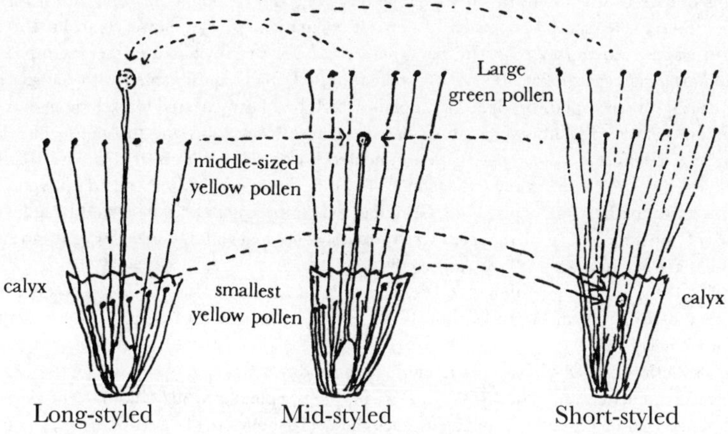

Gray Herbarium of Harvard University (68)

[1] The year is established by the relationship to the letter from Asa Gray, 18–19 August 1862. CD's date was written in the top left-hand margin of the first page of the letter, and apparently represents the date on which the letter was completed. The relationship of this letter to the following letter indicates that it must have been started before the letter to Lubbock, on the evening of 3 September 1862, and completed after that letter (see nn. 14 and 15, below).

[2] Letter from Asa Gray, 18–19 August 1862.

[3] Emma and Leonard Darwin were recovering from scarlet fever (see letter to John Lubbock, 2 September [1862] and n. 4).

[4] Gray had sent a number of North American postage stamps for Leonard Darwin's stamp collection (see letter from Asa Gray, 21 July 1862), and in his letter to CD of 18–19 August 1862, Gray suggested that Leonard should write to let him know what stamps he still required. A letter from Leonard was apparently enclosed with this letter (see letter from Asa Gray, 22 September 1862 and n. 3); however, the enclosure has not been found.

[5] See letter to John Lubbock, 2 September [1862] and n. 4.

[6] See letter from Asa Gray, 18–19 August 1862 and n. 8.

[7] In the letter from Asa Gray, 18–19 August 1862, Gray described the occurrence in *Gymnadenia tridentata* of 'an arm of stigma carried up each side of anther, & one between the [anther] cells'.

[8] The reference is to the long nectary of *Angraecum sesquipedale*; in the specimens examined by CD, the nectary had been 'eleven and a half inches long' (*Orchids*, p. 197). In *Orchids*, pp. 202–3, CD provided the following explanation of the origin of the nectary:

> As certain moths of Madagascar became larger through natural selection in relation to their general conditions of life, either in the larval or mature state, or as the proboscis alone was lengthened to obtain honey from the Angræcum and other deep tubular flowers, those individual plants of the Angræcum which had the longest nectaries . . . , and which, consequently, compelled the moths to insert their probosces up to the very base, would be best fertilized. These plants would yield most seed, and the seedlings would generally inherit longer nectaries; and so it would be in successive generations of the plant and moth.

In *Orchids*, p. 330, CD had suggested that the length of the caudicle carrying the pollinia in *Bonatea speciosa* had probably been 'largely increased in length by natural selection', rather than merely by the abortion of the lower pollen-grains.

[9] In *Orchids*, CD had sought to demonstrate that the primary object of the various pollination mechanisms of orchids was to ensure cross-pollination (p. 1), and he noted only one case (*Ophrys apifera*) in which there were 'special contrivances' for self-pollination (p. 359); however, in his letter to CD of 18–19 August 1862, Gray reported that *Gymnadenia tridentata* provided a further example. Suspecting that *Neottia nidus-avis* might also be self-pollinating, CD had experimented with the species in May, comparing the seed-pods produced by a plant which had been covered to exclude insects, with those from an uncovered plant (see the notes, dated 17 and 21 May 1862, and 10 August 1862, in DAR 70: 73–4, 81). CD included this species on an undated list of 'self-fertilisers' that is now in DAR 70: 167, and also discussed it in his revised account of 'self-fertilisation' in *Orchids* 2d ed., p. 290.

[10] In his letter to CD of 4 August 1862, Gray had expressed a hope that he would be able to send CD seeds of 'the little *Houstonia*'; the seeds were probably enclosed with Gray's letter of 18–19 August 1862.

[11] See letter from Asa Gray, 18–19 August 1862 and n. 14.

[12] Lindley 1853, pp. 574–5. Although CD's copy of John Lindley's *Vegetable kingdom* has not been found, this is the edition of the work listed in CD's Library catalogue (DAR 240), and frequently referred to in *Orchids*.

[13] See also letter to J. D. Hooker, 22 [August 1862] and n. 14.

[14] See following letter, and letter to W. E. Darwin, [2–3 September 1862] and n. 5. In *Origin*, pp. 94–5, CD stated that hive-bees were unable to suck the nectar out of the flowers of the common red clover, which he claimed were 'visited by humble-bees alone' (see also *Origin*, p. 73). However, on the basis of information provided by Charles Hardy (see *Correspondence* vol. 8, letter from Charles Hardy, 23 July 1860), CD added the following statement to the third edition (p. 100): 'I have been informed, that when the red clover has been mown, the flowers of the second crop are somewhat smaller, and that these are abundantly visited by hive-bees.' In the fourth edition (p. 107), CD expressed some doubt about the accuracy of this statement, noting:

> That this nectar is much liked by the hive-bee is certain; for I have repeatedly seen, but only in the autumn, many hive-bees sucking the flowers through holes in the base of the tube which had been bitten by humble-bees.

[15] See following letter.

[16] Horace Darwin.

[17] The diagram has been reduced to 50% of its original size.

To John Lubbock [3 September 1862][1]

Cliff Cottage | Bournemouth
Wednesday

My dear Lubbock

I beg a million pardons. Abuse me to any degree but forgive me— it is all an illusion (but almost excusable) about the Bees.[2] I do so hope that you have not wasted any time for my stupid blunder.— I hate myself I hate clover & I hate Bees—

In Haste to catch evening post | Yours sincerely | C. Darwin

DAR 263

[1] Dated by the relationship to the letters to John Lubbock, 2 September [1862] and 5 September [1862]; the intervening Wednesday was 3 September.

[2] See letter to John Lubbock, 2 September [1862]. See also preceding letter, and letter to W. E. Darwin, [2–3 September 1862] and n. 5.

From Daniel Oliver 4 September 1862

Royal Gardens Kew
4. IX. 1862

Dear Sir

I have been looking amongst the herbaceous beds & enclose a few Lythraceae & Onagraceae which may interest you a little.[1]

Our *Lythrums*—forms of *L. salicaria* I daresay most of them—are going back. *Lopezia* (Onagr.) is a curious thing, but I never studied its economy. I do not recollect any additional plants with 2-colored anthers tho' I think they might be looked for amongst tetramerous genera with 8 stamens & pentam^s. with 10—as in the latter—*Saxifrages*—in former Melast^æ. or Onagrac^æ.—

I *fancy* the colour of anthers may be used by some botanists in discriminating critical species—as in *Drosera rotundifolia* ("white") & *D. intermedia* ("yellow").

Clarkia elegans—of which a scrap is put in may be the same with your plant.—[2]

I am busy examin^g. wood structure of D^r. Hooker's *Welwitschia*.[3] It is very curious & puzzling. I did hope to have visited Mull with D^r. H. to see the place where the Duke Argyle found the Tertiary leaf-prints, but have had to give it up,—having left too early[4]

Very sincerely yours | Danl. Oliver

Chas. Darwin, Esq.

DAR 173: 17

CD ANNOTATION
Top of letter: 'seeds of L. Hyssopifolium[5] | pay 10^d.—'[6] *ink*

[1] CD was working on trimorphism in *Lythrum salicaria*. In the letter to Daniel Oliver, 2 September [1862], he asked Oliver to send him fresh flowers of any of the Lythraceae, and also asked if Oliver knew the names of any plants possessing differently coloured anthers (see letter to Daniel Oliver, 2 September [1862]). In DAR 27.2 (ser. 2): 17, there are notes, dated 5 September 1862, describing a fresh flower of *Lythrum hyssopifolia* from the Royal Botanic Gardens, Kew. The same sheet records notes on specimens of *L. pubescens* and *L. acuminatum* with the comment: 'These are perhaps *Vars.* of L. salicaria.—' On 5 September, CD also recorded observations on other specimens from Kew: *Cuphea lanceolata* and *C. ignea*, which are members of the Lythraceae (DAR 109 (ser. 2): 2), and *Lopezia*, which belongs to the Onagraceae (DAR 205.8: 3).

[2] See letter to Daniel Oliver, 2 September [1862] and n. 7. CD initially identified the specimen as *Clarkia elegans*, but later corrected it to *C. pulchella* (see DAR 48: 50).

[3] Joseph Dalton Hooker was preparing a monograph on the Angolan plant, *Welwitschia mirabilis* (J. D. Hooker 1863a; see letter from J. D. Hooker, 20 August 1862). Oliver was an assistant in the herbarium at the Royal Botanic Gardens, Kew.

[4] Oliver refers to the discovery of Tertiary fossil leaf-beds on the Isle of Mull by one of the tenants of George Douglas Campbell, eighth duke of Argyll (G. D. Campbell 1851). Hooker had left London for Scotland on 23 August 1862 (see letter from J. D. Hooker, 20 August 1862).

[5] See letter from Daniel Oliver, 13 September 1862 and n. 1.

[6] CD apparently intended to pay Oliver the cost of postage for the specimens that he had sent (see n. 1, above).

From Asa Gray 5 September 1862

Cambridge. [Massachusetts]
Sept. 5. 1862

My Dear Darwin

Thanks for yours of Aug. 9.[1] *Lythrum Hyssopifolia* is not in my reach.[2] Get it from France, where it is common. I may, *perhaps*, get you L. *lineare* & *alatum* thro. correspondents. *Nesæa verticillata* I will send you seeds of. Probably a very hopeful plant for your purpose[3]

Mitchella—you will hardly be able to raise from seed.[4] But I mean to send you this autumn patches of it—hoping to get better sorts,—and live roots of Cypripedium.[5] Let them rest in the cold till Xmas, & then they will flower in your room very well.

So you have paid for the Orchid-Books sent by Trübner![6] Well, I had given away a part, and shall so arrange for the rest. I hope to get observations & fresh orchids in return.

Rothrock suddenly enlisted from Pennsylvania, & is before the enemy![7]

I am *so* glad to hear of your boy's recovery.[8]

Of observations, I have little to say this time. Impatiens sends off pollen tubes out of the anthers of the precociously fertilized blossoms. I could not get *specularia*. But Torrey says he could not find the pollen-tubes coming out.[9]

No *Lythrum* within my reach here.

In Ammannia, I have distinguished 2 species by length of style, *dimorphic*—no doubt, though the 2 species are good.

Goodyera repens & I believe *pubescens* also. The *column* of the flower turns back a little & exposes the stigma in the older flowers.— *not the labellum drum.*[10]

Spiranthes cernua. The difference between the older flowers & those first opened is striking. The latter presents the disc—the former the stigma. But here too, I am very confident the change not in the position of the *labellum*—but in that of the *column*![11]

You have opened a *rich mine* indeed among dimorphic & *trimorphic*? flowers.

Ever— in haste— Yours | most cordially | A. Gray

DAR 111: 81, DAR 165: 117

CD ANNOTATIONS
0.3 My Dear ... enemy! 4.1] *crossed ink*
6.1 Impatiens ... *specularia*. 6.2] *double scored ink*
8.1 In ... good. 8.2] *double scored ink*
9.1 *Goodyera* ... change 10.3] *scored ink*; 'Orchids' *opposite page ink*
Top of first page: 'Dimorphism' *ink*; 'Orchids' *del ink*

[1] Letter to Asa Gray, 9 August [1862].
[2] CD was working on trimorphism in *Lythrum salicaria*, and had asked Gray if he could obtain specimens of the rare *L. hyssopifolia* (see letter to Asa Gray, 9 August [1862]).
[3] In his letter to Gray of 9 August [1862], CD commented that he expected *Nesaea verticillata*, a member of the Lythraceae, to be trimorphic.
[4] CD asked for seed of *Mitchella* in his letter to Gray of 9 August [1862].

[5] CD had been impressed with Asa Gray's observations on American species of *Cypripedium* (see letter to Asa Gray, 1 July [1862]).

[6] See letter to Asa Gray, 9 August [1862] and n. 4. Gray had asked CD to arrange for six copies of *Orchids* to be sent to him by the London bookseller and publisher Nicholas Trübner (see letter to Nicholas Trübner, 23 June [1862]).

[7] Joseph Trimble Rothrock was Gray's assistant and student at the Lawrence Scientific School, Harvard University; Gray had sent Rothrock's observations on *Houstonia caerula* in his letter to CD of 4 August 1862. Rothrock enlisted in the 131st Pennsylvania Infantry on 1 July 1862 (*DAB*).

[8] Gray refers to Leonard Darwin, who was recovering from scarlet fever.

[9] Gray refers to his former mentor and botanical collaborator, John Torrey (see letter from Asa Gray, 18–19 August 1862 and n. 5).

[10] See letter from Asa Gray, 18–19 August 1862 and n. 10.

[11] Gray included his observations on this point in A. Gray 1862b, p. 427, noting: 'we suspect that it is the column, rather than the labellum, which changes its position, but we have not been able to demonstrate it'. CD cited Gray's findings in 'Fertilization of orchids', p. 151 (*Collected papers* 2: 148).

To John Lubbock 5 September [1862][1]

Cliff Cottage | Bournemouth
Sept 5th

My dear Lubbock

Many thanks for your pleasant note in return for all my stupid trouble.—[2] I did not fully appreciate your insect-diving-case before your last note;[3] nor had I any idea that the fact was new, though new to me. It is really very interesting. Of course you will publish an account of it.[4] You will then say whether the insect can fly well through the air. My wife asked how did he find out that it stayed 4 hours under water without breathing; I answered at once "Mrs. Lubbock[5] sat four hours watching". I wonder whether I am right—

I long to be at home & at steady work, & I hope we may be in another month.[6] I fear it is hopeless my coming to you, for I am squashier than ever, but hope two shower baths a day will give me a little strength:[7] so that you will, I hope, come to us. It is an age since I have seen you, or any scientific friend.—

I heard from Lyell the other day in Isle of Wight & from Hooker in Scotland.[8] About Huxley I know nothing; but I hope his Book progresses, for I shall be very curious to see it—[9]

I do nothing here except occasionally look at a few flowers; & there are very few here, for the country is wonderfully barren.

Ever dear Lubbock | Yours very sincerely | C. Darwin

See what it is to be well trained. Horace[10] said to me yesterday, "if everyone would kill adders they would come to sting less". I answered "of course they would, for there would be fewer". He replied indignantly "I did not mean that; but the timid adders which run away would be saved, & in time they would never sting at all" Natural selection of cowards!

DAR 263

[1] The year is established by the relationship to the letter from John Lubbock, 23 August 1862, and to the letters to John Lubbock, 2 September [1862] and [3 September 1862] (see nn. 2 and 3, below).
[2] Lubbock's letter has not been found. In the letter to John Lubbock, 2 September [1862], CD asked Lubbock to observe how hive-bees sucked nectar from the flowers of common red clover, but he wrote again in his letter of [3 September 1862], explaining that he had made the request in error.
[3] Lubbock told CD in his letter of 23 August 1862 of his discovery of a swimming hymenopterous insect; no subsequent letter has been found.
[4] Lubbock read a paper on the subject in October 1862 at the British Association for the Advancement of Science meeting in Cambridge (Lubbock 1862d); he described the species in Lubbock 1863a.
[5] Ellen Frances Lubbock.
[6] The Darwins were on holiday in Bournemouth; they returned to Down House on 30 September 1862 (see 'Journal' (Appendix II)).
[7] CD had attended hydropathic establishments periodically since 1849 (see *Correspondence* vols. 4, 6, 7, and 8). He also had a 'douche' erected in the garden at Down House in order to continue the cold-water treatment at home, but by 1853 he apparently no longer used it (see *Correspondence* vol. 5, letter to Edward Cresy, 29 April [1853]). There were public baths and a sanatorium in Bournemouth (*Post Office directory of Hampshire, Dorsetshire, and Wiltshire* 1859, 1867).
[8] See letter from Charles Lyell, 20 August 1862, and letter from J. D. Hooker, [26–31 August 1862].
[9] T. H. Huxley 1863a. In his letter of [26–31 August 1862], Joseph Dalton Hooker told CD that Thomas Henry Huxley was at that time also in Scotland, at Loch Fyne.
[10] Horace Darwin.

To A. C. Ramsay 5 September [1862][1]

Cliff Cottage, Bournemouth
Sept. 5

Dear Ramsay.—

Absence from home & severe illness in my family have prevented me from sooner answering your note & thanking you for your paper.—[2] It has interested me much, & I am surprised, if you are not, that the Council sh^d. hesitate about publishing it—[3] I dispute that the Council have any right to set up their opinion against yours.— I have been the more interested in your paper from years ago marvelling what could be the meaning of so many lakes in Finland, Scandinavia & N. America.— I could form no conjecture.— As far as I can judge your theory *must* be right to a large extent, possibly wholly. Would it not be worth while for you to look at maps of rocky mountainous countries within the Tropics. I cannot remember lakes in Brazil. How is it in Ceylon, Sumatra, the Neilgherries? Years ago I worked through all the reported cases of erratic boulders within the Tropics, & all seemed to be mere weathering of granitic rocks in situ, as shown in appendix to 1st Edit of my Journal.—[4] Would not something of same kind be worth your consideration? As no doubt you will attend for future to all lakes; it may possibly be worth your notice, that I always heard in T. del Fuego that sealers &c always searched for anchorage at the mouths of the deep fiords which penetrate so deeply the land; & that if they passed the bank at the mouth, no soundings could be obtained. This, I believe, is simply due to detritus there alone being formed & accumulated, from the wear of the *exposed* coasts on each side of entrance:[5] during upheaval, such vast mounds of detritus might possibly bank up the water & form a lake.—

No doubt the great depth of the Italian Lakes is rather a staggerer; but I think your theory must be true to very great extent & seems to me very ingenious & satisfactory. The only doubt which occurs to me is the a priori probability in a much troubled country, that some areas would be lifted up less than others & depressions thus left. I sh$^{d.}$ have doubted whether such irregular depressions could be detected by the lines of stratification, & perhaps in part be due to faults.— I remember years ago being struck with frequency of large lakes at base of volcanos, which fact, conjoined *with frequency in all parts of world* of interstratification of *volcanic* & *lacustrine* deposits, led me (together with a few other facts) to believe that *very generally* large areas subside at the base or near active volcanos.[6] Might there not be same tendency near points or ranges, into which much fluid rock has been *in*jected, instead of *e*jected? Could such depressed areas be detected by stratification? An examination of several mountainous countries within the Tropics would throw much light on this doubt.—

I was pleased to see your concluding sentence on cause of Glacial Period:[7] it is an old opinion of mine, over which I have fought battles with Hooker, but never dared with Lyell.[8] In M.S. I have even gone into details, in attempting to show that there has been no such vast recent geographical change as could account for such vast climatal change.—[9] As for Hopkin's Gulf-stream change; it is, in my opinion, an empty hypothesis—[10]

Excuse this scribbling paper worthy of the scribble written on it.—

Again I thank you for your valuable paper & remain | Dear Ramsay | Yours very sincerely | C. Darwin

Jamieson has smacked my marine view of Glen Roy in splendid & most satisfactory style: he seems a real good observer—[11] The shelves are a magnificent record of the Glacial period—

DAR 261 (DH/MS 9: 7)

[1] The year is established by the relationship to the letter from A. C. Ramsay, 26 August 1862 (see n. 2, below).

[2] CD refers to the letter from A. C. Ramsay, 26 August 1862, and Ramsay 1862. CD left Down on 13 August 1862, travelling via Southampton to Bournemouth, before returning home on 30 September 1862. Leonard Darwin had been suffering from scarlet fever since June 1862; Emma Darwin developed the disease on 13 August (see 'Journal' (Appendix II), and Emma Darwin's diary (DAR 242)).

[3] See letter from A. C. Ramsay, 26 August 1862. In his paper (Ramsay 1862), Ramsay argued for the glacial origin of a number of lake-filled European and American rock-basins, and rejected former explanations of them as structural features associated with synclines, as the results of local subsidence, or as fissures along fault lines. The paper was strongly criticised by Hugh Falconer and others when it was read on 5 March 1862 before the Geological Society of London, of which Ramsay was president, and the council considered not passing the paper for publication in the society's journal. Two members of the council, Roderick Impey Murchison and Charles Lyell, subsequently attacked Ramsay's theory in print. For an account of the controversy, see Davies 1969, pp. 303–9.

[4] CD refers to the Addenda to his *Journal of researches*, pp. 609–29; the discussion of erratic blocks is on pages 612–25.

[5] In *South America*, p. 24 n., CD described these 'fiords' as being much shallower at their mouths than further inland, attributing the phenomenon to the accumulation of sediment formed by the wearing down of the rocks exposed to the open sea.

[6] See *Volcanic islands*, p. 96 n.

[7] Ramsay concluded his paper by rejecting the theory advocated by Charles Lyell that the elevation and subsidence of landmasses were responsible for large-scale climatic change (Ramsay 1862, p. 204). He stated:

> I find it difficult to believe that the change of climate that put an end to this [the glacial period] could be brought about by mere changes of physical geography. The change is too large and too universal, having extended alike over the Lowlands of the Northern and the Southern Hemispheres.

[8] CD had told Joseph Dalton Hooker of his disbelief in 'the Lyellian view that [the] Global Epoch is connected with [the] position of continents', in his letter of 15 March [1859] (*Correspondence* vol. 7). CD had opposed Hooker's hypothesis of former land-bridges as an explanation for the geographical distribution of animals and plants (see, for example, *Correspondence* vol. 6, letters to Charles Lyell, 16 [June 1856] and 25 June [1856]). Hooker had linked the elevation of such land-bridges to Lyell's explanation of the causes of climatic change (see J. D. Hooker 1853, pp. xxiii–xxv, and Browne 1983, pp. 133–4).

[9] CD may refer to the manuscript of his projected 'big book' on species, written between 1856 and 1858, and published posthumously as *Natural selection*. CD discussed the Pleistocene glacial period in his draft chapter on the geographical distribution of animals and plants, written in 1856 (*Natural selection*, pp. 534–66). It was in discussing the abstract of this chapter, drawn up for publication in *Origin*, that CD broached with Hooker his disagreement with Lyell's view of the causes of the climatic changes of the glacial period (see n. 8, above). In his letter to Hooker of 30 March [1859] (*Correspondence* vol. 7), CD explained:

> I think if I had space & time, I could make pretty good case against any great *continental* changes since Glacial epoch, & this has mainly led me to give up the Lyellian doctrine as insufficient to explain all mutations of climate.—

[10] The Cambridge mathematician, William Hopkins, had argued that the European glacial period may have been caused by the Gulf Stream being diverted from its course (Hopkins 1852a and 1852b).

[11] In 1839, CD had published 'Parallel roads of Glen Roy', in which he argued that the series of parallel terraces running along the sides of Glen Roy in Lochaber, Scotland, had been formed by the sea as the landmass of Scotland gradually rose. Rejecting CD's marine theory, Thomas Francis Jamieson believed that the so-called 'roads' represented the shorelines of a series of former lakes, trapped in the glen by ice-flows during a great 'Ice-Age' (Jamieson 1863). Jamieson corresponded with CD on the subject in 1861 (see *Correspondence* vol. 9), and CD had recently been informed by Charles Lyell that Jamieson had confirmed his theory following a second visit to the site (see letter from Charles Lyell, 20 August 1862 and n. 1).

From Edouard Claparède[1] 6 September 1862

Cologny près Genève
6 Sept. 62.

Monsieur!

J'ai vraiment honte de prendre si tardivement la plume pour vous remercier de l'aimable lettre dont vous avez bien voulu m'honorer.[2] J'ai été malade pendant sept à huit mois à la suite d'une fièvre typhoïde et j'ai dû laisser en conséquence une foule de choses en arrière. Une de mes premières lectures a cependant été votre ouvrage sur la fécondation des Orchidées dont les admirables *contrivances* ont excité à un haut degré mon intérêt.—[3] Je suis heureux que l'analyse de votre théorie de l'élection naturelle, insérée par moi dans la Revue Germanique, ait reçu votre approbation.[4] Je l'ai rédigée avec tout *l'amour* que peut faire naître une conception, qui est à mes yeux le plus grand pas en avant dont les Sciences Naturelles puissent se vanter à notre époque et pourtant je pense avoir examiné vos idées sans passion, laissant à la porte tout enthousiasme.

Vous me remerciez aussi, Monsieur, du concours que j'ai prêté à Mlle. Royer.[5] J'aurais préféré que ce détail vous fût resté inconnu, car, je dois le dire, j'ai regretté de voir votre ouvrage traduit par cette personne pour laquelle je professe d'ailleurs beaucoup d'estime. Sa traduction est lourde, indigeste, parfois incorrecte et les notes qui l'accompagnent ne seront certainement point de votre goût. J'ai usé de toute mon influence auprès de Mlle. Royer pour la décider à se borner au simple rôle de traducteur, mais mes efforts n'ont pas été couronnés de succès. Je dois dire cependant à l'éloge de Mlle. Royer qu'elle a supprimé sans exception toutes les notes que j'ai qualifiées d'absurdes et de contre sens scientifiques. En revanche elle en a imprimé un très grand nombre (la majeure partie de celles qui illustrent son traduction) qui ne m'avaient point été soumises.

Mlle. Royer est une personne singulière, dont les allures ne sont point celles de son sexe. Toutefois l'éducation semi-masculine qu'elle s'est donnée à force de travail a été puisée avant tout à une école philosophique exclusivement déductive et sa manière de penser s'en ressent.[6] Elle avait imaginé, en traduisant votre ouvrage, d'y introduire des corrections de son propre chef, corrections qui vous auraient étrangement et désagréablement surpris. J'ai cependant réussi à la détourner de cette manière de faire en lui montrant que ⟨ ⟩ manquer de délicatesse à votre égard.— ⟨La⟩ nature de ces corrections était vraiment interessante en montrant combien les méthodes d'un esprit comme celui de Mlle. Royer sont opposées à la marche des Sciences naturelles. Je vous en citerai deux exemples.

Dans le chapitre sur l'instinct des abeilles, Mlle. Royer avait remplacé partout dans sa traduction les termes de pyramide trièdre (pour la base des alvéoles) par celui de pyramide hexaèdre, parceque affirmait-elle les abeilles ne pouvaient pas terminer un prisme hexagone autrement que par un point hexagonal.[7] L'idée ne lui était point venue, avant d'introduire une modification aussi capitale, de jeter elle-même un coup d'oeil sur un rayon de miel.

Le second exemple est de même force. Mlle. Royer n'avait imaginé rien de mieux que de faire descendre dans la traduction tous les poissons électriques d'un ancêtre commun ayant un organe électrique. Comme elle n'a pas de notions de Zoologie non plus que d'anatomie comparée, j'ai eu beaucoup de peine à lui faire comprendre que vous aviez eu vos raisons pour ne pas émettre une idée aussi simple. J'ai cependant réussi à la convaincre tant bien que mal par une description des organes électriques de la torpille, du gymnote, du malaptérure, du mormyre et des nerfs que s'y rendent que ces organes bien qu'identiques au point de vue du tissu ne sont cependant point morphologiquement homologues.[8]

⟨Q⟩uelqu'imparfaite que soit donc la traduction d⟨e⟩ Mlle. Royer, quelque déplacées que soient certaines parties de sa préface et de ses notes, je m'applaudis cependant d'avoir empêché qu'elle défigurât plus complètement votre œuvre. Mais si le grand ouvrage sur les espèces dont vous nous annoncez la publication pour un avenir un peu éloigné vient, comme je l'espère, à être publié, je lui souhaite un traducteur plus versé dans les sciences naturelles et moins désireux de faire remarquer sa propre personalité![9]

J'ai pris la liberté de vous adresser il y a quelques temps un Mémoire sur l'évolution des Araignées et un peu plus tard un autre sur les vers Oligochètes.[10] Tous deux, le premier surtout traitent des questions d'homologies morphologiques qui je l'espère auront quelque intérêt pour vous

Votre bien sincèrement dévoué | Ed. Claparède

DAR 161.1: 149

[1] For a translation of this letter, see Appendix I.

[2] See letter to Edouard Claparède, [*c.* 16 April 1862]. However, either the extant draft of that letter is incomplete, or CD must have sent a subsequent letter to Claparède (see n. 4, below).

[3] Claparède's name appears on CD's presentation list for *Orchids* (see Appendix IV).

[4] CD praised Claparède 1861 in his letter to Edouard Claparède, [*c.* 16 April 1862]. There is an annotated copy of this review in the Darwin Pamphlet Collection–CUL.

[5] Clémence Auguste Royer translated and edited the first French edition of *Origin*, which was published on 31 May 1862 (Royer trans. 1862 and *Bibliothèque Générale de l'Imprimerie et de la Librairie* 2d ser. 6 (pt 3): 341).

[6] As a young woman, Royer had obtained certificates to teach music and languages, and, after reading works by radical Enlightenment authors, devoted herself to acquiring a knowledge of science and philosophy; having educated herself through a rigorous programme of reading, she instituted in 1858 a course in philosophy for women, extending it in 1859 to include science (Harvey 1987, pp. 150–2).

[7] CD discussed the cell-making instinct of hive-bees in chapter seven of the third edition of *Origin*, entitled 'Instinct' (pp. 245–56); he referred to 'the three-sided pyramidal bases of the cell of the hive-bee' on page 246. In Royer trans. 1862, the passage is translated accurately (p. 323).

[8] CD discussed the origin of electrical organs in fish in a chapter of *Origin* entitled 'Difficulties on theory', pointing out that they occurred 'in only about a dozen fishes', several of which were 'widely remote in their affinities' (*Origin* 3d ed., p. 212). Consequently, CD argued, the occurrence of electrical organs in these fish could not be explained as a product of inheritance from a common ancestor, but might be explained as the result of independent adaptations (*Origin* 3d ed., p. 213). Royer added a lengthy footnote to CD's discussion of this question in Royer trans. 1862, pp. 277 n.–279 n. She began by suggesting that all the ancient ancestral forms from which the existing species of electric fish were descended, themselves possessed electrical organs, but that the organs had only been preserved in certain lines. However, Royer also noted that a number of serious objections could be raised against this hypothesis, including the considerable differences between the various electrical organs and the very distant affinities of the existing species of electric fish. In consequence, she proposed an explanation for the independent origin of electrical organs, based on the localised concentration of the capability of all muscle tissue to produce weak electric currents.

[9] In the introduction to *Origin*, CD described the book as an abstract of a larger work on natural selection, which he anticipated would take him 'two or three more years to complete' (p. 1). He intended to publish the work in three parts (see *Correspondence* vol. 7, letter to John Murray, 22 December [1859]); however, the only part ever completed was *Variation*, published in 1868. The remainder of the first draft of CD's 'big book' on species was published posthumously as *Natural selection*.

[10] Claparède 1862a and 1862b. There are lightly annotated copies of these works in the Darwin Pamphlet Collection–CUL.

To W. E. Darwin [10? September 1862][1]

Cliff Cottage [Bournemouth]
Wednesday

My dear William

As I am not a metereologist, my advice is not good for much. Observations made in so inland a county would, I sh^d. think be valuable.— on the other hand, observations have been accumulated in very many cases to a useless extent, chiefly because they have been made without reference to any theory.— I sh^d. fear if the instruments were not compared with some *standard* instruments, they would be of no scientific value.

The observations would be of much more value, if the necessary corrections have been calculated, e.g. the height of Barometer corrected for temperature.— The address of the London Metereological Soc^y. might be found in the Directory, & a letter might be written to Secretary,[2] asking him whether the Soc^ty. would like to possess the observations, describing their nature.

Possibly Mr Glaisher of the Royal. Observ. Greenwich would answer a letter on this subject.—[3] This is all the advice which I can give

The patients are going on capitally.—[4] I hope you will come soon here again[5] | Good Bye | C. Darwin

DAR 210.6: 104

[1] The date is conjectured from the references to the health of Emma and Leonard Darwin (see n. 4, below), and to William having visited Cliff Cottage, Bournemouth, where CD, Emma, and Leonard stayed in September 1862. According to Emma Darwin's diary (DAR 242), William first visited them there on 6 September, and again on 12 September; the intervening Wednesday was 10 September. However, William visited them again on 20 and 27 September.

[2] The British Meteorological Society was located at 25 Great George Street, London; the society's secretaries were James Glaisher and Charles Vincent Walker (*Post Office London directory* 1861).

[3] James Glaisher was head of the magnetic and meteorological department at the Royal Observatory, Greenwich (*DNB*).

[4] Leonard and Emma Darwin were recovering from scarlet fever (see letter to John Lubbock, 2 September [1862] and n. 4).

[5] See n. 1, above.

To J. D. Hooker 11 September [1862]

Cliff Cottage Bournemouth
Sept 11.

My dear Hooker

You once told me that cruciferous flowers were anomalous in alternation of parts & had given rise to some theory of dedoublement.[1] Having nothing on earth to do here I have dissected all spiral vessels in a flower, & instead of burning my diagrams, I send them to you, you miserable man. But mind I do not want you to send me a discussion, but just some time to say whether my notions are rubbish & then burn the diagrams. It seems to me that all parts alternate beautifully by fours (!) on the hypothesis that two short stamens of outer whorl are aborted; & this view is perhaps supported by there being so few, only two, sub-bundles in the two lateral main bundles, where I imagine two short stamens have aborted; but I suppose there is some valid objection against this notion. The course of the *side* vessels, (*not* of the midrib) in the sepals is curious just like my difficulty in Habenaria.[2] I am surprised at the 4 vessels in the ovarium. Can this indicate 4 confluent pistils? anyhow they are in right alternating position. The nectary within the base of the shorter stamens seems to cause the end sepals apparently, but not really, to arise beneath the lateral sepals.—

I think you will understand my diagrams in five minutes, so forgive me for bothering you. My writing this to you, reminds me of a letter which I received yesterday from Claparède, who helped the French Translatress of the Origin, & he tells me he had difficulty in preventing her (who never looked at a bee-cell) from altering my whole description, because she affirmed that an hexagonal prism must have an hexagonal base![3] Almost everywhere in Origin, when I express great doubt, she appends a note explaining the difficulty or saying that there is none whatever!! It is really curious to know what conceited people there are in the world, (people for instance after looking at *one* cruciferous flower, explain their homologies!!!).—

This is a nice, but most barren country & I can find nothing to look at.[4] Even the brooks & ponds produce nothing— The country is like Patagonia.— My wife is almost well, thank God, & Leonard is wonderfully improved:[5] his kidneys excrete less blood & the albumen decreases. Good God what an illness scarlet-fever is. The Doctor feared rheumatic fever for my wife; but she does not know her risk. It is now all over. I do not know whether you have returned; so I have marked this "not to be forwarded".—[6] I saw in paper the marriage of your niece.[7] I heartily hope that you have enjoyed yourself, & that it has done M^rs. Hooker good,[8] & that you will return to your work refreshed. Thanks for your note from Scotland, with address of Mann &c &c.—[9]

I was delighted to see by Lindley's review that your 1st. vol. of Genera is out.[10] I have ordered a copy, for I daresay it will be useful even to me.[11] The more I potter over flowers, the more I get delighted with them. I expect we shall return home in 3 weeks.[12] | Farewell my dear old friend— | Yours affec^y | C. Darwin

Ideal section, horizontal, of flower.

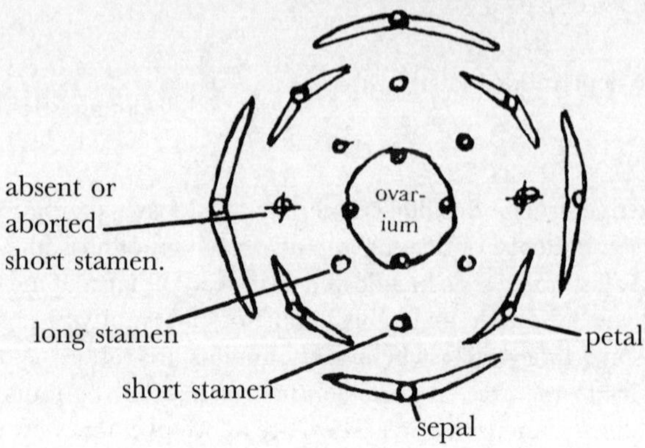

The little circles represent the bundles of spiral vessels, actually seen (except the two with X) in their proper position, except that they are all crowded to centre.

Longitudinal section of flower of Crucifer, laid flat open, showing by dotted lines the course of spiral vessels in all the organs.[13] Sepals & petals shown on one side alone, with the stamens on one side above, with course of vessels indicated, but *not* prolonged. Near side of pistil with one spiral vessel cut away.

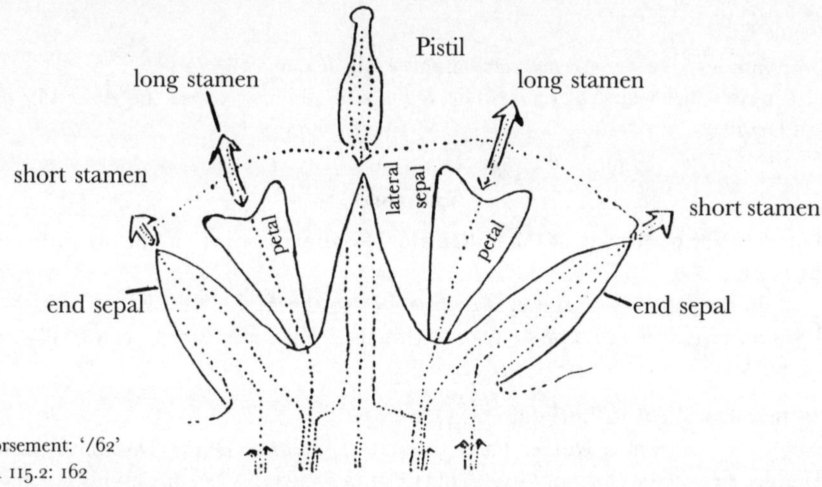

Endorsement: '/62'
DAR 115.2: 162

[1] The reference has not been traced.

[2] CD discussed the 'anomalous' course of the spiral vessels in *Habenaria* and in the allied genus *Bonatea* in *Orchids*, pp. 297 and 302–5. See also *Correspondence* vol. 9, letters to J. D. Hooker, 10 November [1861] and 14 November [1861].

[3] See letter from Edouard Claparède, 6 September 1862 and n. 7. The reference is to Clémence Auguste Royer, who translated and edited the first French edition of *Origin* (Royer trans. 1862).

[4] The Darwins spent September 1862 on holiday in Bournemouth (see 'Journal' (Appendix II)).

[5] Emma and Leonard Darwin were recovering from scarlet fever (see letter to John Lubbock, 2 September [1862] and n. 4).

[6] The Hookers left London for Scotland on 23 August 1862, and returned on 16 September 1862 (see letters from J. D. Hooker, [26–31 August 1862] and 16 September 1862).

[7] Hooker had attended the wedding of his niece, Willielma Hooker, to James Campbell, which took place in Glasgow on 4 September 1862 (see letter from J. D. Hooker, [26–31 August 1862] and n. 7). CD probably refers to *The Times*, to which he subscribed at this time (see letter to Asa Gray, 10–20 June [1862], and CD's Classed account book (Down House MS)), and which carried an announcement of the wedding in the issue of 9 September 1862.

[8] Frances Harriet Hooker had been ill for several months (see letters from J. D. Hooker, 9 June 1862 and 20 August 1862 and n. 3).

[9] Letter from J. D. Hooker, [26–31 August 1862]. The reference is to the botanical collector, Gustav Mann.

[10] [Lindley] 1862b. The first part of the first volume of *Genera Plantarum*, which Hooker co-authored with George Bentham (Bentham and Hooker 1862–83), was published on 7 August 1862 (Stearn 1956, p. 130).

[11] There is a lightly annotated copy of Bentham and Hooker 1862–83 in the Darwin Library–CUL (see *Marginalia* 1: 52–3).

[12] CD returned to Down on 30 September 1862 (see 'Journal' (Appendix II)).
[13] The diagram has been reduced to 50% of its original size.

From G. C. Oxenden 11 September 1862

Broome | nr Canterbury—
Sept. 11. 1862

Dear Sir

I wonder whether the *following Anecdote* will *interest* you—

I passed the year **1821** in Arctic regions, & chiefly, *between* the *Arctic Circle, and Spitzbergen*—

At the *End* of *June* in that year, I was at a place called

"**Bossecop**"[1]

Lat: 70 North—which is the **extreme Northern point** of the **Mainland of Europe**—

—In the Pine Forest there, (which is the Northernmost Pine Forest *in the World*) I found, in full bloom, one or two specimens of that very rare Cryptogs. Plant

"*Splachnum luteum*"—

& these I brought to England—

—*Last Spring*, at a Dinner Party in Kent, I heard some one say that a Kentish Gentleman, *Major Dyne*, of Gore Court, Sittingbourne,[2] was just about starting for the Mouth of the "*Alten River*", for *Salmon Fishing*—which River Alten *meets the North Sea* at about 2 miles from the *aforesaid* place "*Bossecop*"

—& it struck me that Major Dyne might possibly bring me a specimen of this **most** beautiful "*Splachnum luteum*"—

—I accordingly sent to him as *close and minute* a description of the *precise spot* where I found it, as it was possible to give of a plant not 3 inches high, extremely rare, inhabiting an Arctic Forest 1600 miles from England—& finally, after the lapse of *40 years* from the time when I myself found it—

—Nevertheless—this morning, by the Open Post, I received a mere letter from Major Dyne, enclosing two charming specimens of the much coveted Splachnum—

—I enclose to you his letter, in which he gives me the credit of having furnished him with extremely accurate data to work upon—[3] But *I* think he has shewn *rare intelligence*—

—Have you ever seen this "Splachnum? & have you any *desire* to see it— If you have this desire, I will try & manage it for you—if you will tell me that you are likely to be at home—

I remain, with my kind regards, Your's | G. Chichester Oxenden

Please to **return** to me Major Dyne's letter—

DAR 173: 58

[1] Bossekop is a fishing village on the Alta Fjord in Norway (Seltzer ed. 1952).

[2] Musgrave James Bradley Dyne.
[3] The enclosure has not been found.

To W. D. Fox 12 September [1862]

Cliff Cottage | Bournemouth
Sept. 12th

My dear Fox

Very sincere thanks for all the trouble which you have so kindly taken for me about the Turkeys: your information will be of use to me, whenever circumstances will permit me to finish my half completed volume.—[1] Your ancient case of the woman with a hairy face is curious;[2] for, as perhaps you know, there has of late occurred an instance in the Malay archipelago; & the peculiarity was hereditary & accompanied by peculiarities in the teeth.[3]

Thanks for all your sympathy about us: we have been most unfortunate; Horace[4] seriously ill, in a strange manner, all the Spring; & then Leonard came home from school with Scarlet Fever, had recurrent fever, with serious mischief in his kidneys & then bad erysipelas.[5] At last we started for this place; but he suffered much from the journey & at Southampton Emma had Scarlet-fever pretty sharp: we have been here for about 10 days & both my patients are going on admirably, & we have two Houses so that I trust the other children will escape. I have never passed so miserable a nine months.—[6] I hope we shall all get safe home in about 3 weeks. I had thought of going to Cambridge;[7] but now I feel very doubtful & shall not make up my mind till the time arrives & I see how I am. If I do go it will be only for 2 or 3 days. It would indeed be most pleasant to meet you there; but I never know what I can do. All this misery has shaken me a good deal; but I am righting now.— Emma sends her kind remembrances to you.

My dear old friend | Yours affectionly | C. Darwin

If you go to Cambridge please tell me.

Endorsement: 'Sep 15/62'
Postmark: SP 13 62
Christ's College Library, Cambridge (Fox 134)

[1] CD refers to *Variation*, which was eventually published in 1868. In the letter to W. D. Fox, 12 May [1862], CD asked Fox whether he could provide him with information about crosses between the wild and the common turkey. Fox's reply has not been found; however, in *Variation* 1: 392, CD reported Fox's observation that wild turkeys crossed freely with the common domestic kind, and that 'during many years afterwards ... the turkeys in his neighbourhood clearly showed traces of their crossed parentage'.

[2] CD refers to the case of Barbara Van Beck; he acknowledges Fox's assistance in providing information about this case in *Variation* 2: 4.

[3] CD may be referring to the case, reported in 1855, and cited in *Variation* 2: 327, of a Burmese family in which superfluous hair growth and abnormal dental structure persisted through three generations.

[4] Horace Darwin.

[5] Leonard Darwin was sent home from Clapham Grammar School on 12 June 1862, suffering from scarlet fever (see letter to W. E. Darwin, 13 [June 1862] and n. 3); his condition deteriorated in early

July (see, for example, the letter to Asa Gray, 14 July [1862] and n. 3, and Emma Darwin's diary (DAR 242)).

[6] CD, Emma, and Leonard Darwin had started their journey to Bournemouth on 13 August 1862, but were delayed at Southampton until 1 September by Leonard's relapse and by Emma's becoming ill with scarlet fever (see 'Journal' (Appendix II)).

[7] In 1862, the British Association for the Advancement of Science held its annual meeting in Cambridge during the first week of October.

From Edward Cresy 13 September 1862

Riverhead— Kent
13 Sept '62—

My dear Sir.

As we are making our annual sojourn with my mother[1] I should certainly have availed myself of the opportunity of paying my respects to you but learn with regret that Mrs Darwin has been Seriously ill with fever—& have therefore not ventured to walk over until I should have heard from you—[2] Not that I have the slightest fear of contagion or infection— my experience has shewn me that a well fed healthy person runs very little danger when coming immediately from & returning shortly to the fresh air—but I know that when you have illness in the house even a casual call sometimes deranges the invalid—but I sincerely hope to hear a favorable account as well as of your 3^d^ son who if I remember rightly you told me was at Southampton for his health—[3]

I am greatly luxuriating in the simple luxury of having nothing to do— Last session was a very laborious one for me[4] and I felt regularly done up without having anything the matter with me—but having been to Frant near Tunbridge W⟨ells⟩ ⟨ ⟩ Hastings I feel like a giant refreshed— How glad I should be to hear you give the same account of yourself— I fear the amount of sickness you & your family have suffered the last few months can have left you but little time for work—but trust you are now able to resume your pen.

I have long been looking for your memoir on Drosera—[5] Walter White has recently introduced me to your Linnæan Secretary—[6] how is it he has advanced so little towards the Origin of Species? I should have thought you & Dr Hooker[7] irresistable in that quarter—

Pray tell Mrs Darwin how much we regretted to learn of her attack & assure her of our sincere wishes for her recovery— We hope your daughter maintains the wonderful improvement we saw in her—[8]

Yours very truly | E Cresy—

Charles Darwin Esq—

DAR 161.2: 240

[1] Eliza Cresy lived at Riverhead, near Sevenoaks, Kent, about seven miles south-east of Down (*Post Office directory of the six home counties* 1862). Edward Cresy had stayed with CD in September 1860, on his way home to Ham Moor, Surrey, from his mother's house (see *Correspondence* vol. 8, letter to Edward

Cresy, 25 August [1860], and 'Journal' (Appendix II)), and had visited CD again on 30 August 1861 (see *Correspondence* vol. 9, 'Journal' (Appendix II)).

2 Emma Darwin became ill with scarlet fever on 13 August 1862 (Emma Darwin's diary (DAR 242)).

3 The reference is probably to CD's *fourth* son, Leonard Darwin, who had been ill with scarlet fever since June 1862. See letter to H. C. Watson, 8 [August 1862], and letter to A. R. Wallace, 20 August [1862].

4 Cresy was principal assistant clerk at the Metropolitan Board of Works.

5 At the time of Cresy's visit to Down House in September 1860, CD had been observing the response of the insectivorous plant *Drosera rotundifolia* to various substances (see *Correspondence* vol. 8, letters to Daniel Oliver, 11 September [1860] and 15 [September 1860]). Cresy assisted CD's research by putting him in contact with the organic chemist, August Wilhelm von Hofmann (see *ibid.*, letters from A. W. von Hofmann to Edward Cresy, 13 October 1860 and 27 October 1860, and letter from Edward Cresy, 30 October 1860). Although CD reported his preliminary results at a meeting of the Philosophical Club of the Royal Society of London on 21 February 1861, he postponed undertaking experiments to confirm his results (see *Correspondence* vol. 9, letter to Daniel Oliver, 11 September [1861], and Appendix IV). CD carried out further experiments in September 1862 (see letter to J. D. Hooker, 26 September [1862]); however, his work on this subject was not published until 1875 (*Insectivorous plants*).

6 Walter White was librarian and an assistant secretary at the Royal Society of London (*DNB*). The botanical and zoological secretaries at the Linnean Society of London were, respectively, Frederick Currey and George Busk; however, Cresy apparently refers to the society's librarian, Richard Kippist, who performed some administrative duties (see, for example, *List of the Linnean Society of London* 1862). See also letter to Edward Cresy, 15 September [1862].

7 Joseph Dalton Hooker.

8 Cresy refers to Henrietta Emma Darwin, who was seriously ill when Cresy visited Down House in September 1860 (see *Correspondence* vol. 8, 'Journal' (Appendix II)); when Cresy visited in August 1861, Henrietta's condition was improved (see *Correspondence* vol. 9, 'Journal' (Appendix II), and letter to J. D. Hooker, 30 August [1861] and n. 6).

From Daniel Oliver 13 September 1862

Royal Gardens Kew
13. IX. 1862

My dear Sir

I have given directions about saving seeds of *Lythrum hyssopifolium*.[1]

Your Diagram of *L. Salicaria* is very remarkable[2] I think it will be a very serious labour & difficult too satisfactorily to settle such a complication.

I wonder did you ever think about Compositae[3]

&c.

Rolle is new to me,—the name I mean.—[4] *Bolle* is a botanist who has worked on Atlantic Island botany.[5]

Pray excuse the dirty blot.— | Sincerely yours | Dan! Oliver.

Charles Darwin Esq

DAR 111 (ser. 2): 60

CD ANNOTATION
Top of first page: 'Keep. | *Dimorphism*' *ink*

[1] At CD's request, Oliver had sent him a number of fresh specimens of *Lythrum* with his letter of 4 September 1862, including the rare *L. hyssopifolia*, plants and seeds of which CD had attempted to obtain from several correspondents (see letter to C. C. Babington, 2 September [1862] and n. 3). CD had written again, in a letter that has not been found, apparently to ask Oliver to obtain seed of this species for him (see CD's annotations to the letter from Daniel Oliver, 4 September 1862).

[2] CD's letter has not been found; however, see CD's diagrams of *Lythrum salicaria* in the letter to W. E. Darwin, [2–3 August 1862], and in the letter to Asa Gray, [3–]4 September [1862].

[3] Oliver refers to CD's work on dimorphic and trimorphic plants. His diagrams represent cross-sections of the inflorescences of the Compositae, indicating the differences between the ray flowers and the disc flowers. From left to right they indicate: ray flowers bisexual, disc flowers bisexual; ray flowers female, disc flowers bisexual; ray flowers sterile, disc flowers bisexual; and ray flowers female, disc flowers male (all these combinations occur in Compositae).

[4] The German geologist and palaeontologist, Friedrich Rolle, had apparently sent CD a copy of the first part of his popular exposition of CD's theory (Rolle 1863; see letter to Daniel Oliver, [17 September 1862] and n. 10).

[5] The German dendrologist and ornithologist, Carl August Bolle, had collected and identified plants in the Canary and Cape Verde islands (*Taxonomic literature*).

To John Lindley 14 September [1862][1]

Cliff Cottage | Bournemouth
Sept. 14th

My dear Lindley

I suppose that it is a sin against propriety, but I really cannot resist expressing my cordial thanks for your most kind review of my book.[2] One quarter of the praise which you have bestowed on it, coming from you to whom I have long looked up, would I assure you have much more than satisfied me. Considering that you are the great authority on orchids the cordial tone of your article strikes me as something much more than merely kind. I was led to publish the book almost by accident; for whilst I made all the earlier observations I had not the least intention of publishing even a separate paper on the subject, but was led to it by finding that some persons were interested by what I showed them of our British species. And you in one of your letters when I had looked at only two or 3 foreigners, encouraged me, & to a great extent led me to go on.—[3]

Accept my cordial thanks & forgive me for troubling you.

Yours very sincerely | Charles Darwin

Royal Botanic Gardens, Kew (Lindley letters, A–K: 192)

[1] The year is established by the reference to [J. D. Hooker] 1862c (see n. 2, below).

[2] CD refers to the first part of a review of *Orchids* that appeared in the *Gardeners' Chronicle and Agricultural Gazette* of 23 August 1862 ([J. D. Hooker] 1862c). Lindley was the editor of the *Gardeners' Chronicle*; however, the review was actually written by Joseph Dalton Hooker (see letter to J. D. Hooker, 3 November [1862], and letter from J. D. Hooker, 7 November 1862). CD's annotated copy of the

Francis Darwin.
(By permission of the Syndics of the
Cambridge University Library.)

Henrietta Emma Darwin.
Photograph by S. J. Wiseman, Southampton, *c.* 1862.
(By permission of the Syndics of the
Cambridge University Library.)

George Howard Darwin.
Photograph by S. J. Wiseman, Southampton, *c.* 1862.
(By permission of the Syndics of the Cambridge University Library.)

Horace Darwin.
(By permission of the Syndics of the Cambridge University Library.)

Gardeners' Chronicle is in the Cory Library, Cambridge Botanic Garden; CD kept in a separate parcel his copies of the numbers in which [J. D. Hooker] 1862c appeared (see DAR 222 and DAR 75: 1–12).

3 Lindley's letter has not been found; however, the two corresponded at length on the subject of orchids in 1861 (see *Correspondence* vol. 9).

To Daniel Oliver 14 September [1862][1]

Cliff Cottage | Bournemouth
Sept. 14th.

Dear Oliver

My sister-in-law sent me several specimens dried of Lythrum hyssopifolium to compare with the fresh specimen, which you kindly sent me,[2] & amongst them was the enclosed: it is clearly not L. hyssopifolium or L. salicaria: it has 12 stamens, large petals, smooth calyx, & flowers not in whorls. Could you find out its name?[3] I fancy the genus is not large. It is a European specimen. The specimen sent answers to the "short-styled" in L. salicaria, but differs in many important respects. The stigma of "mid-styled" would not project beyond the calyx, & this perhaps led old Vaucher (who always blunders when that is possible) to assert that some species are dimorphic like Primula.[4] It would be a very interesting aid to me if you could name this species for me, & at same time, when you find the specimens in the Herbarium, (if the species be not rare) pluck off a single **young** *unopened* flower from a few specimens, as I shd. very much wish to compare the pollen of the two sets of anthers in the "long-styled" or "mid-styled" form of this new species.[5]

I hope you will not think me very unreasonable to ask all this; for I hope & believe that the species of Lythrum are not numerous; & I have been much perplexed, how any of the species could be **di**morphic, as old Vaucher says.—

Yours very sincerely | C. Darwin

DAR 261 (DH/MS 10: 37)

1 The year is established by CD's reference to having received dried specimens of *Lythrum hyssopifolia* from Sarah Elizabeth Wedgwood (see n. 2, below).

2 The reference is to Sarah Elizabeth Wedgwood; in DAR 27.2 (ser. 2): 14, there is a note by CD, dated 8 September 1862, which records:

> Elizabeth sent me 3 specs of *L. hyssopifolia* from different countries these all agreed with Kew specimens; so that I have seen 4 specimens. *very improbably [*del*] *This species is probably not [*interl*] dimorphic.—

These specimens are preserved in DAR 142 in a wrapper marked 'Lythrum hyssopifolium from Elizabeth from 3 countries.—' Oliver had sent CD specimens of *L. hyssopifolia* from the Royal Botanic Gardens, Kew, with his letter of 4 September 1862; CD's notes on those specimens are in DAR 27.2 (ser. 2): 17.

3 Oliver's reply to this letter has not been found. However, in DAR 21.2 (ser. 2): 14, CD noted:

> E. sent me a 4th spec. *of L Graefferi (named at Kew) [*interl*] with larger leaves & larger petals & longer flowers than L. hyssopifolia; & differing quite in narrow *smooth* *& longer [*interl*] calyx from L. salicaria.—

The specimen is preserved in DAR 142 in a wrapper marked 'Elizabeth W'; this was preserved, together with several other specimens of *Lythrum Graefferi*, in an envelope addressed to CD by Oliver, postmarked 16 September 1862, and annotated by CD 'Lythrum Graefferi' (*Calendar* no. 13891f). See also letter to Daniel Oliver, [17 September 1862] and n. 2.

4 Vaucher 1841, 2: 371.

5 See letter to Daniel Oliver, [17 September 1862] and n. 3.

To Edward Cresy 15 September [1862][1]

Cliff Cottage, Bournemouth
Sept. 15th.

Dear Cresy

I have just received your kind note.—[2] You will see where we are. My third son (not the former invalid, who is better) had the Scarlet Fever dreadfully badly[3] and on our road here, at Southampton, Mrs. Darwin sickened;[4] but both our patients are at last going on capitally. We shall remain here some weeks longer.— This incessant illness has utterly stopped all work, except a few miscellaneous observations. I intend to give up my beloved Drosera till I have got out a separate volume on Variation; and Heaven knows when that will be.—[5] I am very glad to hear that you have had a good holiday and feel yourself rested; for I am sure your incessant work is enough to tire out anyone.[6] One week of your work would send me to bed for half a year.— Kippist to whom you allude is a good-hearted little man; but not one whose opinion one would value.[7]

Mrs. Darwin sends her kind remembrances to Mrs. Cresy[8] and yourself.— I am sorry that I shall miss seeing you.

Pray believe me | Yours very sincerely | C. Darwin

Copy
DAR 143: 322

1 The year is established by the relationship to the letter from Edward Cresy, 13 September 1862.

2 Letter from Edward Cresy, 13 September 1862.

3 CD refers to his *fourth* son, Leonard Darwin, who had been ill with scarlet fever since 12 June 1862 (Emma Darwin's diary (DAR 242)). The 'former invalid' referred to is Horace Darwin, who was seriously ill during the early months of 1862 (see, for example, letter to W. E. Darwin, 14 February [1862]).

4 Emma Darwin became ill with scarlet fever on 13 August 1862 (Emma Darwin's diary (DAR 242)).

5 See letter from Edward Cresy, 13 September 1862 and n. 5. *Variation* was published in two volumes in 1868; CD's work on *Drosera rotundiflora* was not published until 1875 (*Insectivorous plants*).

6 See letter from Edward Cresy, 13 September 1862. Cresy was principal assistant clerk at the Metropolitan Board of Works.

7 See letter from Edward Cresy, 13 September 1862 and n. 6. Richard Kippist was librarian of the Linnean Society of London.

8 Mary Cresy.

From C. C. Babington 16 September 1862

Cambridge
16 Sept. 1862

Dear Darwin,

I hope to be able to get seeds of Lythrum hyssopifolium from our Botan. Garden in a few days.[1] The Curator[2] has undertaken to look for them and I will send them when received.

You know that the British Association is to meet here on *Oct 1.* next and it would give great pleasure to many of us to see you here at that time.[3] Could we have calculated upon your convenience and health allowing you to take the office we should have advised your being placed in the chair of the Sec. of Zool. & Botany.[4] But, altho not in the position in which we would have placed you, nevertheless you might perhaps come for a few days as a private member of the Association. We want all eminent Cambridge-men to attend, and amongst these we place C. Darwin in the first class.[5]

Yours very truly | Charles C. Babington—

DAR 160: 4

[1] Babington was professor of botany at the University of Cambridge. In his letter to Babington of 2 September [1862], CD asked if Babington could obtain seeds of *Lythrum hyssopifolia* for him.
[2] The curator of the Cambridge University Botanic Garden was James Stratton (Walters 1981).
[3] In 1862, the British Association for the Advancement of Science held its annual meeting in Cambridge during the first week of October.
[4] In the event, the president of Section D of the meeting, encompassing 'zoology and botany, including physiology', was Thomas Henry Huxley (*Report of the 32d meeting of the British Association for the Advancement of Science*, p. xxix).
[5] CD matriculated at Christ's College, Cambridge, in 1828; he did not compete for honours, but was tenth in the list of candidates for the ordinary BA in January 1831. He was admitted BA in April 1831, and MA in October 1836 (*Alum. Cantab.*, *Cambridge University Calendar* 1832 and 1837, Pass lists–mathematical examinations 1822–84 (Cambridge University Archives, Exam. L. 6: 178)).

From J. D. Hooker 16 September 1862

Kew
Sept. 16/62.

My dear Darwin

We returned this morning to Kew all well,[1] I do not think that the damp relaxing air of W. Scotland did my wife either harm or good, she is remarkably well for her, though she still has occasional attacks of palpitation.[2] I was delighted with Staffa.[3] I saw a good deal of the D. of Argyll who has been pottering over the Orchid book, to small purpose, though he evidently tries hard to understand it.[4] He & the D^s.[5] are great Northerns, in Yankee matters[6] I was wonderfully struck with the humidity of Inverary, the cryptogamic vegetation was like Fuegia for rank luxuriance.[7]

I am rejoiced to hear such good news of your party— what a trial you have had of it.![8]

Thank you 1000 times for the Cruciferous diagram,[9] which I shall ponder over, my impression always was that the arrangement of the parts was quaternary arguing however only from appearances & from analogy with Papaveraceæ & Capparideæ—for I have never gone into the thing. Asa Gray is my opponent in this & in *dieclytra* which I wish very much you could find time to examine.[10] The 4 bundles to the pistil is anomalous I should have expected that each bundle sent to replum was a double one.— I have no objection to offer to the quaternary composition of the pistil in Cruciferæ except the simply bilobed of the stigma, & that in Tetrapoma in others that have 4 valves the repla are identical with those that have only two.

I am staggered with the intricacy of Welwitschia, Oliver (who is a real blessing) has been examining the tissues, where I left off on going to Scotland & I have a pretty job before me. he has made me some charming drawings that will save me a world of trouble.[11]

The marriage went off well & has pleased us all.[12] We had rather bad weather all along, but I enjoyed visiting old faces & places.[13]

I hope to go to Cambridge for a day to show Welwitschia—[14] my wife will stay with Mr & Mrs Liveing Prof. of Chemistry.[15]

A torrent of visitors hangs over us.

I saw Huxley on Loch Fyne with the Herring Fishery Commissioners, entre nous I doubt their inspiring much confidence, & I find every man's verdict is against their bias, which is in *favor* of trawling.[16]

Write to me when you have nothing better to do.

Ever Yours affec | J D Hooker

How are the *Lyells*?[17]

DAR 101: 56–7

[1] The Hookers had left Kew for Scotland on 23 August 1862 (see letter from J. D. Hooker, [26–31 August 1862]).

[2] See letter to J. D. Hooker, 11 September [1862] and n. 8. Frances Harriet Hooker had been ill for several months (see letter from J. D. Hooker, 9 June 1862).

[3] Staffa is an uninhabited island, in the Inner Hebrides, renowned for its basaltic caverns, the principal of which is Fingal's Cave (J. Bartholomew n.d.).

[4] The eighth duke of Argyll, George Douglas Campbell, reviewed *Orchids*, together with five other books, in an article entitled 'The supernatural' that appeared in the *Edinburgh Review* for October 1862 ([G. D. Campbell] 1862). There is a copy of Argyll's review in the Darwin Pamphlet Collection–CUL. Argyll's name appears on CD's presentation list for *Orchids* (see Appendix IV).

[5] The duchess of Argyll, Elizabeth Georgiana Campbell.

[6] Hooker refers to the Union states in the American Civil War.

[7] The duke of Argyll's family seat (Inveraray Castle) was located at Inveraray, the county town of Argyllshire, located on Loch Fyne (J. Bartholomew n.d.). Both CD and Hooker had visited Tierra del Fuego, CD during the *Beagle* voyage, and Hooker while on board HMS *Erebus*.

[8] See letter to J. D. Hooker, 11 September [1862].

[9] See letter to J. D. Hooker, 11 September [1862].

[10] The reference has not been traced.

[11] Hooker was preparing a monograph on the Angolan plant *Welwitschia mirabilis* (J. D. Hooker 1863a; see letter from J. D. Hooker, 20 August 1862). Daniel Oliver was an assistant in the herbarium at the Royal Botanic Gardens, Kew. See also letter from Daniel Oliver, 4 September 1862.

[12] Hooker refers to the wedding of his niece, Willielma Hooker, to James Campbell, in Glasgow on 4 September 1862 (*Gentleman's Magazine* n.s. 13 (1862): 488).

[13] Hooker spent most of his childhood in Glasgow, where from 1820, his father was professor of botany at the university (Allan 1967). See also letter from J. D. Hooker, [26–31 August 1862].

[14] In 1862, the British Association for the Advancement of Science held its annual meeting in Cambridge during the first week of October; Hooker did not attend the meeting (see letter from J. D. Hooker, [12 October 1862]).
[15] Catherine and George Downing Liveing.
[16] Thomas Henry Huxley was in Scotland between 8 August and 16 September 1862, in his capacity as a member of the Royal Commission on the Operation of the Acts relating to Trawling for Herring on the Coasts of Scotland (L. Huxley ed. 1900, 1: 234). Since the 1840s a number of fishermen on the west coast of Scotland had been using trawl nets for herring fishing, instead of the traditional drift nets, as the former generally required lower capital investment and smaller boats. Trawl nets were rendered illegal by acts of Parliament of 1851, 1860, and 1861, the last two acts suppressing trawling almost entirely (*Report of the Royal Commission on Herring*, p. 145). In their report of 1862, the commissioners rejected the claims advanced by the drift-net fishermen, and stated that fish stocks were not damaged by the trawlers' practices, that trawling acted as 'an important means of cheapening fish to the consumer', and that the legislation already introduced was 'repressive' and 'altogether unnecessary', constituting 'Acts for protecting class interest', and serving only to interfere with 'the invention and application of new and more productive forms of industry' (*Report of the Royal Commission on Herring*, pp. 167–70).
[17] Charles and Mary Elizabeth Lyell.

From Philip Gidley King 16 September 1862

Goonoo Goonoo. New South ⟨Wales⟩
16 Sepr. 1862

My dear M^{r}. Darwin

My brother John[1] is now on a visit to England & I have asked him if he has time to call on you. I am sure he would not regret making your acquaintance & I should rejoice on his return to the Colony to know that he had seen you. His address is

care of Messrs. Stillwell & Co.[2]
22 Arundel Street
Strand.

I should so much like him to see you that he might bring me a description of, as Jonathan says, your "location"[3] and of yourself and doings. Your work the Origin of Species has a prominent place in my Library & was read with much interest. I think you are thought by many to be right who will hardly allow it. I feel in the small scope of my Experience that there is much truth in y^{r}. deductions, but the question is where do they lead us to—or what is their limit. I have recently joined in a Newspaper correspondence relative to breeding Sheep in the Colony, upholding the idea that we can breed as good sheep as can be imported

I remember a very old letter you once wrote me asking why it was necessary to go on importing.[4] I was too green then to know much about y^{r}. queries, but I now ask the same question—. One of our most eminent breeders a M^{r}. Bayly[5] had written that Australia is the finest country for producing Sheep (ie wool and Mutton especially wool.) yet he continued unless occasional importations are made, a degeneracy ensues.— with these assertions he coupled another that he had sheep without a fault. I thought his logic so bad that I pressed him for a cause for the degeneracy—but he could not explain at all what it proceeded from. I believe it is caused (for that there is degeneracy no one can doubt) by carelessness feeding sheep in large flocks, hunting them with dogs, obliging them to travel every day 4 or 5 miles for food & keeping them 14 hours out of the 24 without it— If we could afford with cheaper labor to keep Sheep as they are kept in Germany we might soon export Rams there

If you ever publish your promised large work on y^r. favorite Subject I shall look out for it eagarly—.[6] We have now grown up families— I think we should never recognize each other I wish you would send me your carde de Visite if you ever indulge in such matters

With kindest remembrances Ever My dear Mr Darwin | Y^r very sincere old friend | Philip Gidley King

If you ever see Admiral FitzRoy will you remember me to him—[7]

Charles Darwin Esq.

DAR 169: 26

[1] John King was Philip Gidley King's younger brother.

[2] John G. & Thomas Stilwell were naval agents (*Post Office London directory* 1861).

[3] Jonathan: 'a generic name for the people of the United States, and also for a representative United States citizen' (*OED*). In the sense of 'place of settlement or residence', 'location' is chiefly United States usage (*OED*).

[4] CD's letter has not been found. After leaving HMS *Beagle*, King had become a farmer in New South Wales (*Aust. dict. biog.*).

[5] Nicholas Paget Bayly.

[6] CD stated in the introduction to *Origin* that the book was an abstract of a larger work that was 'nearly finished', but would take 'two or three more years to complete' (p. 1). He intended to publish the larger work in three parts (see *Correspondence* vol. 7, letter to John Murray, 22 December [1859]); however, the only part he completed was *Variation*, published in 1868. The remainder of CD's first draft of his 'big book' on species was published posthumously as *Natural selection*.

[7] Robert FitzRoy was captain of HMS *Beagle* during the voyage of 1831–6, when CD and King were on board.

To Daniel Oliver [17 September 1862][1]

Cliff Cottage | Bournemouth
Wednesday Evening

Dear Oliver

I am sorry you had trouble of writing two notes,[2] but I thank you *sincerely* for the flowers of Lythrum, which I shall carefully examine with great interest.[3] Unless L. hyssopifolium presents 2 or 3 forms (which I do not believe) I am astounded at D. C. speaking of L. Graefferi as possibly the same as L. hyssop.[4] They are totally unlike!— I am truly obliged for the specimens, for I feel a strange interest in this case. As you say it will be laborious to prove the case;[5] but before I left home, I castrated marked, & fertilised in 18 methods above a hundred flowers, & protected them from insects.—[6]

I am much obliged for your Photograph.;[7] I have none to *match* of myself, but I have one made by my son, which I will send when at home if you care to have it.[8] Thanks, also, for diagram of Compositæ, in which the "&c." means I fear that all cases are not represented.[9] By chance the day before I was looking at the marygold & was puzzled by the apparently different stigma of the ray florets which, do not seem to produce seed.— But a man must be a Botanist to think about so gigantic an order: moreover it would be hardly possible to experiment by crossing.

Thanks again about Bolle or Rolle; which I will ascertain, when I get my copy from Lyell:[10] I fancied from some remarks that he might be a Botanist; but my wife[11] read it Rolle as well as I did.—

Having nothing on earth to do here, I have been working a little bit at the never-ending Drosera: as the glands absorb so readily & the Hairs move so rapidly under certain stimulants, it seemed a good opportunity to test how far this plant was sensitive to various vegetable substances, which are known to act energetically on the nervous system of animals.[12] As yet I can make out no sort of rule; but the difference in action is very great. Thus, strychnine produces no effect; belladonna causes movement, as does veratrine; Henbane does *not* cause movement, but does not in the least check subsequent & immediate action of meat.— Opium on the other hand, does not cause movement, but afterwards meat instead of causing movement in less than one minute, does not act for 2, or 3 or 4 hours—it puts the plant to sleep! I wonder whether analogous experiments have been tried on other plants; but anyhow I shall go on, as it amuses me & passes the time.

But why do I waste your time?— I suppose for same reason it amuses me & passes the time, so forgive me & believe me, Yours very sincerely | C. Darwin

DAR 261 (DH/MS 10: 36)

[1] The date is established by the relationship to the letter from Daniel Oliver, 13 September 1862; the Wednesday following 13 September 1862 was 17 September.

[2] Only one of the two letters referred to has been found (letter from Daniel Oliver, 13 September 1862). However, there is in DAR 142 an envelope addressed to CD in Oliver's hand, which is postmarked 'SP 16 62' (London) and 'SP 17 62' (Poole and Bournemouth) (*Calendar* no. 13891f). This probably contained a letter and specimens sent in response to CD's letter of 14 September [1862] (see n. 4, below); CD's letter and Oliver's letter of 13 September 1862 had apparently crossed in the post.

[3] With his letter to Oliver of 14 September [1862], CD sent a dried specimen of the genus *Lythrum* for identification, and requested further dried specimens of the flower buds of that species. In a note in DAR 27.2 (ser. 2): 14, CD recorded that the specimen had been identified at the Royal Botanic Gardens, Kew, as *L. Graefferi.* The envelope of Oliver's missing letter of 16 September 1862 (*Calendar* no. 13891f; see n. 2, above) is annotated in CD's hand 'Lythrum Graefferi', and was preserved as a receptacle for several dried specimens of *L. Graefferi*, including that originally sent to Oliver by CD; the descriptions on the wrappers of most of the remaining thirteen specimens are in Oliver's hand. However, since Oliver sent further specimens of the flowers of *L. Graefferi* in October (see letter to J. D. Hooker, 6 October [1862], and letter to Daniel Oliver, 13 October [1862]), it is impossible to identify which of the specimens were sent on this occasion. CD's notes on the first set of specimens sent by Oliver are in DAR 27.2 (ser. 2): 15.

[4] Candolle and Candolle 1824–73, 3: 82. Having learned that the Swiss botanist, Jean Pierre Etienne Vaucher considered *Lythrum hyssopifolia* to be dimorphic, CD had sought specimens of the plant from a number of correspondents (see letter to C. C. Babington, 2 September [1862]), obtaining fresh specimens from Oliver and dried specimens from Sarah Elizabeth Wedgwood early in September 1862 (see letter to Daniel Oliver, 14 September [1862] and nn. 2 and 3).

[5] See letter from Daniel Oliver, 13 September 1862.

[6] See letter to Daniel Oliver, 29 [July 1862] and n. 6.

[7] The photograph of Oliver has not been found. The photograph may have been the same as the carte-de-visite by Maull & Polyblank that is reproduced in this volume; that photograph has been dated to the years 1856–65 by reference to the period during which Maull & Polyblank were in partnership.

[8] William Erasmus Darwin's portrait of CD is reproduced as the frontispiece to *Correspondence* vol. 9.

[9] See letter from Daniel Oliver, 13 September 1862.

[10] CD had apparently asked Oliver whether he had heard of the German geologist and palaeontologist, Friedrich Rolle; Oliver replied in his letter of 13 September 1862 that he had not, but wondered whether CD was referring to the German dendrologist and ornithologist, Carl August Bolle. CD's question was probably prompted by his having received a copy of the first part of Rolle 1863, which he had subsequently lent to Charles Lyell (see letter to Charles Lyell, 1 October [1862] and n. 2).

[11] Emma Darwin.

[12] CD began to observe the effects of various substances on the insectivorous plant, *Drosera rotundifolia*, in 1860 (see *Correspondence* vol. 8). He had hoped to continue and complete the experiments in the summer of 1861, but subsequently decided to postpone them (see *Correspondence* vol. 9, letter to J. D. Hooker, 4 February [1861], and letter to Daniel Oliver, 11 September [1861]); he carried out a small number of experiments in May 1862 (see the observational notes dated 18 and 21 May 1862 in DAR 54: 74–5). There are notes detailing his extensive experiments with the species between 14 and 26 September 1862 in DAR 54: 29–49; on 17 September, CD carried out a set of experiments with the substances described in this letter. See also letter to J. D. Hooker, 26 September [1862]. CD's work on this subject was not published until 1875 (*Insectivorous plants*).

From G. C. Oxenden 17 September [1862][1]

Broome
Sept. 17

Dear Sir

When I am after "*Epipactis palustris*", in July, I see **Meneanthes** in profusion—& in bloom—[2]

—Whether I shall as easily recognize it *now*, I cannot tell—but I fully believe that I should—& I will see to it the first spare day—

—Typhus, & Scarlet Fever, are each as bad as true Oriental "*Plague*"—only we handle them somewhat better than does the stupid Turk—[3]

—But even in England, whenever any treatment *out* of the common course—*any* departure from old worn rut & groove, is suggested to a Medical Man, he becomes as Mad as a Hatter; & had much rather see all his patients, & all his own family, *die*, than depart in any way from the old stereotyped formula—

—It is my firm belief that there *does* exist a Medecine which is not only a "**prophylactic**" against the *access* of either of those two fevers; but also possessed of the power of *mitigating, assuaging, & subduing* them, when they **have** got hold of the human frame—

—In the last year or two "*Ozone*" has not merely *knocked* off their *pedestals* those nauseous & *dangerous* disinfectants

{ Chloride of Soda
Chloride of Lime
Chloride of *Zinc*

but has itself come to *the very front* as a disinfectant, in a *pure*, *sweet*, & even *enjoyable shape*—

A patient suffering from *Scarlet Fever*, ought to drink (**ad libitum**, & *whenever the Tongue is parched*) of

Condy's pure *Ozonised Fluid*"

largely diluted with Water—

—It is **impossible** for a *Fever patient* to drink **too much** of it—& the Mouth sh^{d}. be *rinsed* with it *incessantly*—

—It is so entirely **harmless** also, that any person in sound health, may drink it to *any amount*, as a pleasant daily diet drink, at Meals instead of Beer, Wine, &c &c—but it *should be largely* diluted— & **all** its **effects** are **beneficial**—without **any** strain upon the *constitution*—[4]

2dly— in fever, the patient should be placed in a bath of tepid Water, into which should be poured several bottles of the "*Ozonised fluid*"—or, (*better still*) *one large bottle* of "Condy's *Ozonised Disinfectant*"—(which is the **same** Medecine, in a *stronger* shape)—

Every Chemist sells each of these Medecines

By this Mode of treatment—the patient rapidly Convalesces—& the Medical Man becomes *infuriate*—

Your's most sincerely | G. Chichester Oxenden

DAR 173: 59

[1] The year is established by the references to *Menyanthes*, scarlet fever, and Condy's 'Ozonised Fluid' (see nn. 2–4, below).

[2] CD had apparently asked Oxenden, probably in a reply to Oxenden's letter of 11 September 1862, if he could send him some specimens of the marsh plant *Menyanthes trifoliata*. Having learned that *Menyanthes* was dimorphic (see letter to C. C. Babington, 20 January [1862] and n. 3), CD had obtained a specimen of the genus from Joseph Dalton Hooker (see letter to J. D. Hooker, 9 [April 1862]). In the summer of 1861, CD had sent Oxenden a memorandum of inquiry concerning the orchid *Epipactis palustris*, which is also a marsh plant (see *Correspondence* vol. 9, letter to A. G. More, 4 June 1861); for details of Oxenden's attempts to obtain specimens of *E. palustris*, see the letters from G. C. Oxenden, 15 May 1862, 4 June [1862], [before 21 June 1862], and 8 July [1862].

[3] Oxenden apparently refers to the illnesses of Emma and Leonard Darwin, both of whom became ill with scarlet fever in 1862 (see 'Journal' (Appendix II)). See also letter from G. C. Oxenden, 21 June 1862.

[4] Condy's 'Ozonised Fluid' was a solution of alkaline permanganates, developed by Henry Bollman Condy, and used 'for the purification of air and water, and sanitary objects in general' (Condy 1862, p. 69). CD took the fluid himself later in the year (see letter to J. D. Hooker, 10–12 November [1862]).

From Alphonse de Candolle[1] 18 September 1862

Mon cher Monsieur

j'ai recu et lu avec un grand interet le volume sur les Orchidées que vous avez bien voulu m'envoyer.[2] Il est tellement rempli de faits qu'on voudrait pouvoir le lire à côté de chaque plante mentionnée pour comprendre mieux les détails, mais il y a souvent des idées générales qui se rattachent à l'ensemble de vos théories et c'est ce qui m'a le plus intéressé. Je viens de remettre au Directeur de la Bibliothèque universelle pour la partie scientifique deux pages environ pour annoncer l'ouvrage.[3] Je n'ai pas osé entrer dans plus de détails, mais chaque lecteur doit être averti voir le livre si cela l'intéresse.

Les masses polliniques implantées sur les insectes ont donné lieu jadis à de singulières erreurs. Je ne sais plus quels botanistes les avaient prises pour des cryptogames! Wydler, autant qu'il me souvient, en a fait l'objet d'un article pour relever l'erreur, dans quelque journal allemand—il y a 25 ou 30 ans.[4]

Les derniers cahiers du Bulletin de la Société botanique de France renferment des articles sur l'hybridation par Lecoq[5] et sur le dimorphisme dans les Orchidées par Duchartre[6] qui touchent un peu aux sujets dont vous vous occupez.

Je fais bien des vœux pour votre santé car avec un si grand tresor accumulé de bonnes observations il serait facheux de ne pas continuer à publier.[7]

Mon travail sur les Quercus est fini.[8] La partie description dans le Prodrome sera la principale,[9] mais je compte publier auparavant deux petits opuscules que j'aurai l'avantage de vous adresser.[10]

Agréez, je vous prie, en attendant l'assurance de mes sentiments très dévoués | Alph. de Candolle

Genève 18 Sept. 1862.

DAR 161.1: 11

[1] For a translation of this letter, see Appendix I.
[2] Candolle's name appears on the presentation list for *Orchids* (see Appendix IV).
[3] A. de Candolle 1862d.
[4] The article by Heinrich Wydler has not been identified. See also Baehni 1955, p. 119 n. 13.
[5] Lecoq 1862.
[6] Duchartre 1862.
[7] See letter to Alphonse de Candolle, 17 June [1862] and n. 4.
[8] See letter from Alphonse de Candolle, 13 June 1862.
[9] Candolle and Candolle 1824–73, 16, pt 2: 1–109.
[10] A. de Candolle 1862b and 1862c; there are annotated copies of these works in the Darwin Pamphlet Collection–CUL. See also letter from Alphonse de Candolle, 13 June 1862 and n. 10.

To J. D. Hooker [18 September 1862][1]

Cliff Cottage | Bournemouth
Thursday Evening

My dear Hooker

Thanks for your pleasant note, which told me much news, & upon the whole good of yourselves.[2] You will be awfully busy for a time; but I write now to say that if you think it really worth while to send me a *few* Dielytra or other Fumariaceous plant (which I have already tried in vain to find here) in a little tin Box, I will try & trace vessels;[3] but please observe, I do not know that I shall have time, for I have just become wonderfully interested in experimenting on Drosera with poisons &c.—[4] **If** you send any Fumariaceous plant, send, if you can, also 2 or 3 single Balsams.— After writing to you[5] I looked at vessels of ovary of a Sweet Pea, & from this & other cases, I believe that in ovary the mid-rib vessel alone gives homologies & that the vessels on the edge of the carpel-leaf often run into wrong bundle, just like those on *sides* of sepals. Hence I look in Crucifers that the ovarium consists of two pistils;

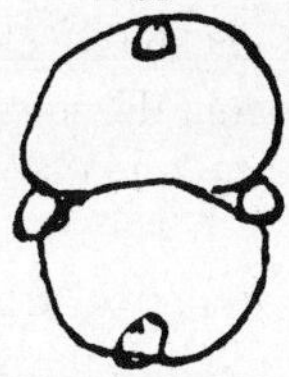

A A being the midrib vessels; & B B being those formed of the vesses on edges of the two carpels run together & going to wrong bundles.— I came to this conclusion before receiving your letter. I wonder why Asa Gray will not believe in the quaternary arrangement;[6] I had fancied that you saw some great difficulty in the case, & that made me think that my notion must be wrong.

I have been thinking of Cambridge, for a few days, & your going is an immense temptation, but I very much fear I shall not be strong enough;[7] I have had headach half every day with my stomach intolerably bad.—

Farewell; hearty thanks for your note | C. Darwin

Endorsement: 'Sept 20/62'
DAR 115.2: 160

[1] The date is established by Hooker's endorsement; the Thursday preceding 20 September 1862 was 18 September.
[2] Letter from J. D. Hooker, 16 September 1862.
[3] See letter from J. D. Hooker, 16 September 1862 and n. 10.
[4] See letter to Daniel Oliver, [17 September 1862] and n. 12.
[5] Letter to J. D. Hooker, 11 September [1862].
[6] See letter from J. D. Hooker, 16 September 1862.
[7] CD refers to the meeting in Cambridge in October 1862 of the British Association for the Advancement of Science.

From W. B. Clarke 20 September 1862

St. Leonard's | N.S.W.
20 Sepr. 1862.

My dear Sir,

I beg you to accept my best thanks for your book on the Orchids[1]—which as yet I have been too busy to read—but from which I anticipate a great deal of instruction.

I have been so much occupied of late with unavoidable business of all kinds, that I have had less time for scientific pursuits than usual. I have, however, managed to send off a box of specimens to M^{r}. T Rupert Jones—which if you are in town in the end of Novr. or beginning of December he will be able, perhaps, to show you.[2]

I took the liberty of writing to M^{r}. Moore on the strength of your kind letter.[3] I intended that the Mesozoic fossils should have come home before this: but they will so next month.

Can you advise me what to do about the description &c of all my fossils. There is a sum of 5000£ down on the Estimates for me this year—but the Parliament will expect me to do something in the way of a general Geol^{l}. work on Australia as an acknowledgment.[4] How can I get the fossils described and figured, best?—

My experiments here on Goodenia do not satisfy me.[5] The season has not been favourable—and blossoms not abundant. What we had covered *under glass—did not seed*.

M^{r}. C. Moore Director of the Botanical Gardens will seek out for me as many specimens as possible and subject them to the most rigid trial.[6] I will report thereon to you.—

There is a prospect of a good Native Bee Comb for you—[7] I have a person on the lookout for me—(they are scarcer than they used to be)—and when I get it, I will forward it you, with a letter.—

I hope you are now better than when you wrote last.

Since Burke and Wills—Landsborough and McKinley as well as Walker have been across the Continent to the Gulf! Stuart's is the only party now out.—[8]

Believe me | My dear Sir, | Y^{rs}. very truly | W. B. Clarke.

C. Darwin Esqre. &c—

DAR 161: 175

[1] *Orchids* was published in May 1862; Clarke's name appears on CD's presentation list for the work (see Appendix IV).

[2] Clarke had asked Thomas Rupert Jones to assist him in finding someone to describe and draw his entire collection of Australasian fossils (see letter from W. B. Clarke, 20 June 1862 and nn. 2 and 3).

[3] CD's letter has not been found. He had apparently written for Clarke a letter of introduction to the geologist Charles Moore, who Clarke hoped would describe his collection of fossils from Wollumbilla Creek, Queensland (see letters from W. B. Clarke, 21 January 1862 and 20 June 1862).

[4] See letter from W. B. Clarke, 20 June 1862 and n. 4.

[5] In 1861, CD had asked Clarke to cover with a net any species of Goodeniaceae just before the flowers opened, so as to exclude insects, and observe whether or not the plant set seed (see *Correspondence* vol. 9, letter to W. B. Clarke, 25 October [1861]). See also letter from W. B. Clarke, 16 January 1862.

[6] Charles Moore was director of the Botanic Gardens, Sydney (R. Desmond 1994).

[7] In his letter to CD of 20 June 1862, Clarke promised to attempt to procure a native Australian bee comb for CD. See also letter from W. B. Clarke, 16 January 1862.

[8] A reward of £10,000 had been offered by the legislature of South Australia for the first successful crossing of the Australian continent from south to north, and subsequently, a number of parties had attempted the crossing. Robert O'Hara Burke and William John Wills succeeded early in 1861, but perished on the return journey. William Landsborough, John McKinlay, and Frederick Walker each

commanded search parties for Burke and Wills. Beginning in 1858, John McDouall Stuart made two unsuccessful attempts at the crossing, before eventually succeeding in 1862 (*Aust. dict. biog.* and *EB*, s.v. Australia).

To W. D. Fox 20 [September 1862]

Cliff Cottage | Bournemouth
Saturday night 20th.

My dear Fox

Your affectionate & pleasant letter has interested, pleased & grieved me.—[1] I know that your lungs have often troubled you; but I greatly fear that this last attack has been more serious & has left mischief behind. If the sea-air suits you, it seems a thousand pities that you should not make everything bend to it & stay some time there. Your life is a precious one. Four days is nothing for the sea-air. When I read of you & your dog I exclaimed to Emma, "how like Fox", but I did not know of Erasmus' saying.[2] I suppose you have given up all idea of Cambridge;[3] I have begun to turn tail & have resolved to go home on Oct 1st.; but if I can screw my courage up, I may go there on Saturday or Monday; but it will depend on how my stomach is. I shd like to see the old place once again, where I have spent so many happy hours; but I am not sure whether it will not be more melancholy than pleasant; for I know I shall feel knocked up & unable to ramble about & see the old haunts. What pleasant hours we have spent together at our alternate breakfast & teas.[4] There were no fears & anxious looking forward in those days. And poor dear Henslow is gone.[5] About two years ago I stumbled at Down on a Panagæus crux major: how it brought back to my mind Cambridge days!

You did me a great service in making me an entomologist: I really hardly know anything in this life that I have more enjoyed that our beetle-hunting expeditions;[6] Prince Albert told Lyell,[7] that he looked back with more pleasure to collecting insects, than he had ever found in stag-shooting. I am *much* pleased & somewhat surprised at your liking my orchid-book:[8] the Botanists praise it beyond its deserts,[9] but hardly anyone, not a Botanist, except yourself, as far as I know, has cared for it. The subject interested me much, & was written almost by accident; for it was half written as a mere paper & then I found it too long, & thought I would risk publishing it separately.[10] What you say about it, is very pleasant; for at one time I agreed with Lyell that I was an ass to publish it.[11]

I have lately been making some curious observations on the "dimorphic" fertilisation of other plants[12] & likewise on their sensibility; & upon my life I am coming to the conclusion, that they must have something closely analogous to diffused nerve-matter.[13] But as you most truly say what a mystery life is; & a mystery one feels the more, the more one knows.— As soon as I get home, if we all can but keep well, I must return to variation under domestication.[14]

I do not wonder that you found the jelly-fishes puzzles to dissect; it would take weeks to get even a glimmering of their structure,—mere organised water.—

I did not know that you were doubly a grandfather.[15] I will send your messages to my sister.[16] Perhaps Erasmus, who has never stirred out of London all this summer, will come here for our last week.[17] He is very far from strong. All Darwins ought to be exterminated.

Farewell my dear old friend; I do most truly hope that your health may improve & your lungs recover. Farewell | Yours affectionately | Charles Darwin

Endorsement: 'Sep 24/62'
Postmark: SP 22 62
Christ's College Library, Cambridge (Fox 135)

[1] The letter from Fox has not been found.

[2] The reference has not been traced; CD refers to his brother, Erasmus Alvey Darwin.

[3] In 1862, the British Association for the Advancement of Science held its annual meeting in Cambridge during the first week of October. See also letter to W. D. Fox, 12 September [1862].

[4] CD and Fox had both been undergraduates at Christ's College, Cambridge, where their periods of residence overlapped in 1828; they were second cousins and had shared an interest in entomology (see *Correspondence* vol. 1, and Desmond and Moore 1991, pp. 58–73).

[5] John Stevens Henslow was professor of botany at the University of Cambridge for thirty-six years, and had been a friend to both CD and Fox; he died in May 1861.

[6] See n. 4, above.

[7] CD refers to Charles Lyell and Albert Francis Charles Augustus Emmanuel, prince-consort of England.

[8] *Orchids* was published in May 1862.

[9] CD had been particularly pleased by the responses to *Orchids* of George Bentham, Daniel Oliver, and Asa Gray (see letter to Asa Gray, 10–20 June [1862]).

[10] Originally conceived as a paper for one of the journals of the Linnean Society of London, CD began preparing his manuscript on the pollination mechanisms of various species of orchids in July and August 1861. He first thought of publishing it as a book in September 1861 (see *Correspondence* vol. 9, letter to John Murray, 21 September [1861], and letter from John Murray, 23 September 1861).

[11] The occasion on which Charles Lyell expressed this view has not been identified; however, see letter to J. D. Hooker, 7 March [1862] and n. 4.

[12] After reading his paper, 'Dimorphic condition in *Primula*', before the Linnean Society in November 1861, CD had continued in 1862 to work on the phenomenon of heterostyly.

[13] For details of CD's experiments with insectivorous plants, see the letter to Daniel Oliver, [17 September 1862], and the letters to J. D. Hooker, [18 September 1862] and 26 September [1862].

[14] In the introduction to *Origin*, CD described the book as an abstract of a larger work he was preparing on natural selection (p. 1). He intended to publish the larger work in three volumes, the first of which was to be entitled 'Variation under domestication' (see *Correspondence* vol. 7, letter to T. H. Huxley, 16 December [1859], and letter to John Murray, 22 December [1859]). CD had been working on the manuscript intermittently since January 1860, and had reached chapter eight on 'Silk-worms Geese &c' by the summer of 1862 (see *Correspondence* vols. 8–10, 'Journal' (Appendix II)).

[15] Fox had two married daughters: Eliza Anne Sanders gave birth to her first child, Charles Henry Martyn Sanders, on 21 March 1862 (*Gentleman's Magazine* n.s. 12 (1862): 638); Harriet Emma Overton also gave birth in 1862 to her first child, Frederick Arnold Overton (*Alum. Oxon.*, s.v. Overton, Frederick Arnold; *Darwin pedigree*, pp. 15–16).

[16] CD's surviving sisters were Caroline Sarah, Emily Catherine, and Susan Elizabeth Darwin. CD may be referring to Susan Elizabeth Darwin, as she was a close friend of one of Fox's own sisters (see *Correspondence* vol. 5, letter to W. D. Fox, [27 March 1851], n. 5). See also *Correspondence* vol. 8, letter to W. D. Fox, 17 December [1860] and n. 5.

[17] Erasmus Alvey Darwin lived at 6 Queen Anne Street, Cavendish Square, London (*Post Office London directory* 1861). According to Emma Darwin's diary (DAR 242), the Darwins stayed with him on 29 September 1862, on their way home from Bournemouth.

From Emile M. J. M. P. Goubert 20 September 1862[1]

Londres,
20 7bre. 1862

Monsieur,

J'ai pris la liberté de me présenter chez vous, sans avoir l'honneur et l'avantage de vous rencontrer. J'aurais été fort aise en effet de visiter votre collection paléontologique, si vous aviez bien voulu le permettre.

Quoiqu'il en soit, je suis aise de profiter de l'occasion pour vous proposer d'entrer avec vous en rapporte d'échange. Contre des fossiles anglais du terrain qu'il vous plaisait de m'envoyer, pourvu qu'ils fussent déterminés et avec localité indiquée, je pourrais vous envoyer de très belles suites d'espèces françaises ou allemandes de l'étage qu'il vous plaira de m'indiquer. Je vous signalerai notamment mes nombreux doubles tertiaires de Paris et Mayence, des moindres niveaux déterminés presque tous par M. Deshayes.[2] Je puis encore vous offrir spécialement du crétacé de Picardie, du jurassique de Normandie, du permien et des schistes houillers d'allemagne.

J'ai publié quelques très jolies espèces coralliennes et autres que je me ferais un plaisir de vous addresser aussi, avec les brochures les concern⟨ant.⟩[3]

Veuillez, quoiqu'il en soit, mon ⟨cher⟩ et très honoré confrère, agréer l'hommage de ma considération la plus distinguée | Emile Goubert

à Paris, rue du cherche-midi 23 | professeur au collège Louis le grand | membre de la Société géologique et de la Soc. botanique de France, de l'académie de Moscou et de celle de Hanau, etc

DAR 165: 80

[1] For a translation of this letter, see Appendix I.
[2] Gérard Paul Deshayes.
[3] The paper on corals to which Goubert refers is probably Goubert and Zittel 1861.

From J. D. Hooker 20 September 1862

Kew
Sept 20/62

Dr. Darwin

What do you think of Ramsays Lake glacial theory?—[1] I think the enclosed may interest you as it does me.—[2] if not pray take the will & forget the deed.

I am frightfully busy & inundated with d——d visitors.— there goes the bell—just as I wrote!

Ever yours affec | J D Hooker

[Enclosure][3]

Geological Survey Office Christchurch NZ.
9 June 1862.

My dear Sir!

I can not tell you, how thankful I am for your kind letter of the 22 Decb,[4] which I received a fortnight ago with all my mails of three months in a lonely shepherd hut at the Alpine lake Pukaki.[5] Your judgement on my Nelson Rapport is very kind and the same complaint, that there is no map with it, has been made to me from different sides, but the Nelson Government was afraid of the costs.[6] Although I had no instructions from you, I have collected quite after your wishes, and you will find many interesting undescribed plants in the collection, which I shall forward as soon as the box arrives in which it is contained; probably in three weeks.[7] Of course I have not forgotten the grasses, amongst which are many remarkable Alpine forms.[8] Our mountains are unfortunately so very inaccessible and my occupations so varied, that it is only with great difficulty, that I can find the time for collecting.

I shall accompany the plants with a general description of the country, which, as I hope, will give you a clear insight into the general distribution of the vegetation. There are some very interesting observations which I was able to make.[9]

Dr Hector is safely arrived. He sent me your letter and wrote me.[10] Of course I offered him the hand of friendship and can assure you that I am very glad, that another member of the hammer and such a distinguished one is in New Zealand, so as to be able to consult with him about any difficult geological question. If I find the time I shall go over to Dunedin and see him. I have to thank you very much indeed for your great kindness in giving me such manifold information on scientific subjects of the day and if your time allows you I should be delighted, if in future letters you would do the same

A few days ago I came back from my wanderings; the winter set in very soon, so that having the snow feet deep on the mountains, I was obliged to retreat. My journey has been a very interesting one and as I am so very much pressed with official business, I send you this time only the small accounts, which I published in one of our newpapers.[11] You will find an account of my Exploration of last year, when Poor Dr Sinclair lost his life in the Transactions of the Royal Society of Victoria.[12] Being now naturalized,[13] I think it my duty to offer my work to the scientific societies in England. And as I know, that you are also a fellow of the Geological Society, I shall take the liberty to send you in a few months as soon as my map is finished an account of the geological features of the Alps, on deposits of the glacial periods, accompanied by Maps, sections and sketches, which perhaps you will have the kindness to lay before the Geological Society.[14] In October last I sent a Copy of my Nelson Map to Arrowsmith, requesting him to lay it before the R. G. Society, but I got no answer.[15] If you see him by chance in one of the meetings, would you perhaps be so kind as to ask him, if he has received my letter with this map?

Our Alps are really superb; the glaciers all of such large dimensions and splendour that it is quite impossible to give an adequate description of it. Why do you not come and make a journey through our Alps?![16] It would amply repay costs and trouble and I should be highly delighted, were I allowed to act as your Cicerone.

Hoping that you have safely received the box of plants with the Chrysolishe and to hear from you, I remain | my dear Sir | Yours very sincerely | Julius Haast.

Dr J. D Hooker | FRS. | etc etc.
Royal Botanical Gardens | Kew.—

DAR 101: 58, Royal Botanic Gardens, Kew (Director's Correspondence 174 (New Zealand letters, 1854–1900): 273)

CD ANNOTATIONS
Top of first page: '*Home* | Brit. assoc. | Mrs French' *ink*
End of letter: '*Rats.*—' *ink*

1 Hooker had previously asked CD what he thought of Ramsay 1862 in his letter of 20 August 1862. See also letter from A. C. Ramsay, 26 August 1862, letter to A. C. Ramsay, 5 September [1862], and following letter.

2 Hooker sent CD a letter from the German-born New Zealand geologist, Julius von Haast, together with some newspaper reports that had been enclosed with his letter (see letter from Julius von Haast, 9 December 1862). The letter is reproduced here (see enclosure), but the newspaper reports have not been identified; however, it is evident from the following letter that they related, among other subjects, to glacial remains observed during Haast's 1862 expedition in the Southern Alps of New Zealand. Hooker had told CD about Haast's earlier observations on glacial phenomena in New Zealand in his letter of [29 December 1861] (*Correspondence* vol. 9); for CD's response, see the letter to J. D. Hooker, 16 January [1862] and n. 6.

3 The enclosure has been identified by Hooker's statement, in a letter to Haast of 18 September 1862 (Alexander Turnbull Library, Wellington, New Zealand, MS. papers 37.96), that he had just received Haast's letter of 9 June 1862.

4 Hooker began to correspond with Haast in 1861; his letters to Haast are preserved in the Alexander Turnbull Library, Wellington, New Zealand, MS. papers 37.96–105. His letter of 22 December 1861 is partly reproduced in H. F. von Haast 1948, pp. 248, 250–1.

5 Following the discovery of gold in the province of Otago in 1861, Haast, who was made provincial geologist for Canterbury in that year, set out at the end of January 1862 to search for gold-bearing deposits in the Southern Alps of New Zealand, and to continue there the regular work of the geological survey of the province. He returned to Christchurch at the end of May 1862 (H. F. von Haast 1948, pp. 199–219).

6 J. F. J. von Haast 1861. In November 1859, Haast had been appointed by the provincial government of Nelson province, New Zealand, to conduct a topographical and geological survey of its western district between January and August 1860. The provincial government had originally intended to illustrate the report with maps and plants, but in the event, declined to do so on the grounds of expense, 'much to the disgust of Haast and others' (H. F. von Haast 1948, pp. 47, 105). See H. F. von Haast 1948, pp. 42–9, 69–112.

7 In his letter to Haast of 22 December 1861 (H. F. von Haast 1948, p. 248), Hooker had expressed a hope that Haast would be able to collect botanical specimens on his expedition in the little-explored Southern Alps. Having collected many new plants, Haast sent them to Hooker in August 1862, with a friend who was travelling to Europe 'by the overland route'. See the letter from Haast to Hooker, dated 9 August 1862, which is in the Royal Botanic Gardens, Kew (Director's Correspondence 174 (New Zealand letters, 1854–1900: 273)). Hooker, who had published a flora of New Zealand following his own visit there (J. D. Hooker 1853–5), used them in preparing his government-sponsored *Handbook of the New Zealand flora* (see J. D. Hooker 1864–7, p. 12*). See also H. F. von Haast 1948, pp. 247–51, 294–8.

[8] In his letter to Haast of 22 December 1861 (H. F. von Haast 1948, p. 248), Hooker had especially requested Haast to collect some of the 'alpine grasses, *junci*, *cyperaceae* and small plants, especially on the west coasts, and near the snow', stating that it was amongst these that the representatives of Australian and Fuegian plants and 'most of the interesting novelty' were to be found.

[9] See the letter from Haast to Hooker, dated 9 August 1862, which is in the Royal Botanic Gardens, Kew (Director's Correspondence 174 (New Zealand letters, 1854–1900: 273)).

[10] The Scottish geologist James Hector was appointed provincial geologist of the province of Otago, New Zealand, in 1861. In his letter to Haast of 22 December 1861 (H. F. von Haast 1848, p. 250), Hooker had mentioned that he intended to give Hector a letter of introduction to Haast, continuing: 'I am sure you will find him a most agreeable and worthy coadjutor'. Hector wrote to Haast on 19 April 1862, enclosing Hooker's letter of introduction.

[11] The newspaper articles have not been identified, but see n. 2, above.

[12] Haast refers to the former colonial secretary, Andrew Sinclair, who was drowned while accompanying Haast on his geological expedition to the headwaters of the Rangitata River, Canterbury province, in 1861 (H. F. von Haast 1948, pp. 179–81). Haast apparently refers to his paper, 'On the physical geography and geology of New Zealand, principally in reference to the Southern Alps', which was read before the Royal Society of Victoria on 26 May 1862 (see *Transactions and Proceedings of the Royal Society of Victoria* 6 (1865), Proceedings, p. xv); however, the paper was subsequently lost, and was consequently not published (*Transactions and Proceedings of the Royal Society of Victoria* 6 (1865): vi). Hooker and Sinclair had collected plants together in New Zealand in 1841, and Hooker dedicated his *Flora Novæ-Zelandiæ* (J. D. Hooker 1853–5) to Sinclair. Having retired as colonial secretary, Sinclair returned to New Zealand in 1858 in order to collect plants for Hooker's *Handbook of the New Zealand flora* (J. D. Hooker 1864–7). On Sinclair's death, Hooker wrote: 'His loss has been a very great one, whether as a botanist or as an enthusiastic and liberal patron of science' (*DNZB*).

[13] Haast, who was born in Germany, was naturalised as a British subject in February 1861 (H. F. von Haast 1948, p. 123).

[14] Haast's map of the province of Canterbury, showing 'the glaciation during the Pleistocene and recent periods as far as explored', was communicated to the Geological Society of London by Roderick Impey Murchison in 1864 (J. F. J. von Haast 1864a). See also J. F. J. von Haast 1867.

[15] Haast refers to the map-maker John Arrowsmith, who was a member of the council of the Royal Geographical Society. Although a paper by Haast on the Southern Alps of New Zealand appeared in the *Journal of the Royal Geographical Society* in 1864 (J. F. J. von Haast 1864b), it was accompanied by a composite map of the provinces of Otago and Canterbury, illustrating two other additional papers.

[16] In his letter to Haast of 22 December 1861 (H. F. von Haast 1948, p. 248), Hooker stated: 'To me the Middle Island of New Zealand is still the most interesting spot botanically in the globe, and I would rather botanize those mountains you traverse than any others.'

To J. D. Hooker 21 [September 1862][1]

Cliff Cottage | Bournemouth
Sunday 21st.

My dear Hooker

I am specially obliged to you for sending me Haasts communications.[2] They are very interesting, & grand about Glacial & drift or marine glacial.— I see he alludes to whole southern hemisphere.[3] I wonder whether he has read the Origin.—[4] Considering your facts on the Alpine plants of N. Zealand & remarks, I am particularly glad to hear of the geological evidence of glacial action.[5] I presume he is sure to collect & send over the mountain "Rat" of which he speaks; I long to know what it is.[6] A Frog & Rat together would to my mind prove former connection of N. Zealand to some continent;[7] for I can hardly suppose that the Polynesians introduced the Rat as game, though so esteemed in the Friendly Islands.[8]

Ramsay sent me his paper & asked my opinion on it.[9] I agree with you & think highly of it.[10] I cannot doubt that it is to large extent true: my only doubt is that in a much disturbed country, I sh$^{d.}$ have thought that some depressions, & consequently lakes, would have almost certainly have been left— I suggested a careful consideration of mountainous Tropical countries, such as Brazil, peninsula of India &c; if lakes are there very rare, I sh$^{d.}$ fully subscribe to Ramsay's views.[11] What presumption, as it seems to me, in the Council of Geolog. Soc$^{y.}$: that it hesitated to publish the paper.—[12]

We return home on the 30$^{th.}$— I have made up my mind, *if I can keep up my courage*, to start on the Saturday for Cambridge & stay the last few days of the Assoc.^{n} there.[13] I do so hope that you may be there then.—

Farewell | C. Darwin

DAR 115.2: 161

1 The date is established by the relationship to the letters from J. D. Hooker, 20 August 1862 and 20 September 1862.

2 Hooker had sent CD a letter from Julius von Haast, together with some newspaper reports of his recent explorations in the Southern Alps of New Zealand, that had been enclosed with his letter (see enclosure to preceding letter and n. 2).

3 Haast's newspaper reports have not been identified; they apparently related, among other subjects, to glacial remains observed during his expedition in the Southern Alps of New Zealand (see J. F. J. von Haast 1862a, and H. F. von Haast 1948, pp. 199–219). In *Origin*, p. 373, CD referred to the 'evidence of former glacial action in New Zealand' as a point in favour of his argument that the Pleistocene glacial epoch had affected large portions of the world. Hooker told CD about Haast's earlier observations of glacial remains in New Zealand in his letter of [29 December 1861] (*Correspondence* vol. 9). See also letter to J. D. Hooker, 16 January [1862] and n. 6. In *Origin* 4th ed., pp. 442–3, CD stated: 'we now know, from the excellent researches of Dr. J. Haast and Dr. Hector, that enormous glaciers formerly descended to a low level in New Zealand'.

4 Haast had read *Origin*, and held it in high esteem (see letter from Julius von Haast, 9 December 1862 and n. 8).

5 CD refers to Hooker's flora of New Zealand (J. D. Hooker 1853–5), and to the remarks in the introductory essay concerning the possibility of a former cool period, occasioned by the existence at that time of a greater amount of landmass in the southern seas (J. D. Hooker 1853–5, 1: xxiii–xxv). There is a copy of Hooker's introductory essay in the Darwin Library–CUL (*Marginalia* 1: 394).

6 Haast probably referred to the rat in the newspaper reports that have not been identified. The reference may have been to the 'large Norwegian grey species' of rat that had been introduced to New Zealand, and that Haast, on his 1862 expedition, found to have reached 'the very heart of our Alps' (H. F. von Haast 1948, p. 215). However, CD evidently believed Haast was referring to the supposed native species of rat, *Mus rattus* (now called *Rattus exulans*). See n. 7, below; see also letter from Julius von Haast, 9 December 1862. On *Mus rattus*, see Buller 1870.

7 CD had long regarded the presence of frogs and rats in New Zealand as evidence supporting the possibility of a continental connection for these islands, especially since he found frogs' spawn to be very easily killed by salt water (see *Correspondence* vol. 5, letter to J. D. Hooker, 10 June [1855], and L. G. Wilson ed. 1970, p. 53). However, in *Origin*, p. 393, after stating that frogs are never found on oceanic islands, CD suggested that their presence in New Zealand might be explained by glacial

transport. CD had previously attempted to obtain specimens of the New Zealand frog and rat (see *Correspondence* vol. 6, letter to W. B. D. Mantell, 5 June [1856–9]). See also *Correspondence* vol. 9, letter to W. B. Clarke, 25 October [1861].

8 CD refers to present-day Tonga.

9 Ramsay 1862; see letter from A. C. Ramsay, 26 August 1862.

10 See letter from J. D. Hooker, 20 August 1862 and n. 13.

11 See letter to A. C. Ramsay, 5 September [1862].

12 See letter from A. C. Ramsay, 26 August 1862, and letter to A. C. Ramsay, 5 September [1862].

13 In 1862, the British Association for the Advancement of Science held its annual meeting in Cambridge during the first week of October. See also letter to W. D. Fox, 20 [September 1862].

From Asa Gray 22 September 1862

Cambridge. [Massachusetts]
Sept. 22, 1862.

My Dear Darwin.

Your pleasant epistles of Aug. 21 & Sept. 4. are to be acknowledged, with thanks.[1] But I have nothing in particular to communicate, except our hearty congratulations that your boy and Mrs. Darwin are recovering so well.[2]

Tell Leonard that I was pleased both with his attention in writing, and with the occular proof of his convalescence in his being able so soon to use a pen.[3]

His requests shall be kept in view,—the 5^{cts} stamp I send now,—dare say I shall some time pick up the 30 and 90, though I never saw the latter, nor the 12, 20 & 24 on envelopes. (the 24^{ct} per se he must have already, as it is often used on my envelopes to you.

Bravo for Horace, whose illustration of Natural Selection as to the *adders* is capital.[4] A chip of the old block, he evidently is.

I told you that *Rothrock* had gone to the war, and perhaps has already been under—probably not.[5] I had intended that next spring he should do up *Houstonia* more perfectly, and work up this and some related matters for his thesis when he comes up for examination. But all this is broken up by his enlistment.[6]

I may take occasion just to allude briefly to the case in next Silliman apropos to something.[7] But if you think it of the least use, *dress up a note of the facts and send it to Gard. Chronicle*. I should be pleased if you would.[8]

I have been lazy about all my writing,—working all day at *dry* & *dull* systematic Botany, which you anathematise.[9] But *if I get time to turn it over* I will say a few words on the last chapter of your Orchid book.[10] But it opens up a knotty sort of question about *accident or design*, which one does not care to meddle with much until one can feel his way further than I can.[11]

You make too much far of my casual obs. on Orchids last summer.[12] Most of them should be worked over anew before I should put much confidence in them. But I am convinced there is something *mighty queer* about *Gymnadenia tridentata*.[13]

Just you come over here next summer—bringing all your sickly ones to this drier climate—and work up these Orchids, &c.—

Here are good quarters for you and I will have the orchids poured in upon you from all the country round.

The change of position in *Spiranthes cernua* is striking enough. Once seen it can never be overlooked. I cannot however, tell whether it is the labellum or the column that moves, but I suppose the latter, resting much on *Goodyera*, where I had no doubt.[14]

Well, it is very pretty, the triple forms of Lythrum Salicaria.[15] Being in the Garden I will look to it next year.

I doubt if I get seeds of any N. Amer. sp. But *Nesæa* verticillata I expect to supply seeds—not yet ripe. Ten to one you will find triple state in it.—[16]

I have not seen it growing since your letters came.

So, to spite the book, the honey-bees will suck clover![17]

The extract from M^c^Millan's Mag, which I send you in an English newspaper, herewith, is a just and fair account—will let you know how we feel & think as to the civil war.[18]

We find it a far tougher job than we supposed. But we have no more notion of giving it up than—the English nation would have, under a similar case. For my part, I would fight till $\frac{4}{5}$ of our property and half our men were destroyed before I would give up!—and I think that is the general opinion. Cambridge volunteered to fill up her quota under both calls long ago, and a number over.[19]

Ever, dear Darwin— | Yours | A. Gray

DAR 165: 118, 119

CD ANNOTATIONS
9.1 Just ... round 10.2] *crossed ink*
12.1 Well, ... year. 12.2] *crossed ink*
13.1 I doubt ... over. 17.5] *crossed ink*
Top of first page: 'Seeds— | Pollen' *pencil*

1 Letters to Asa Gray, 21 August [1862] and [3–]4 September [1862].

2 In his letter to Gray of [3–]4 September [1862], CD reported that Emma and Leonard Darwin were recovering from attacks of scarlet fever.

3 In the summer of 1862, Gray sent a number of North American postage stamps for Leonard's collection; Gray suggested that Leonard write to let him know what stamps he still required. Leonard's letter has not been found, but was apparently enclosed with CD's letter to Gray of [3–]4 September [1862] (see letter to Asa Gray, [3–]4 September [1862] and n. 4).

4 Horace Darwin. See letter to Asa Gray, [3–]4 September [1862].

5 Joseph Trimble Rothrock was Gray's assistant and student at the Lawrence Scientific School, Harvard University; he enlisted in the 131st Pennsylvania Infantry on 1 July 1862 (*DAB*). Gray told CD this news in his letter of 5 September 1862.

6 In the summer of 1862, Rothrock carried out experiments and observations on CD's behalf with the dimorphic plant *Houstonia caerulea* (see letter from Asa Gray, 4 August 1862). Rothrock was eventually wounded at Fredericksburg, Virginia, and honourably discharged from the Union army in June 1864; he took his degree from Harvard University in July 1864 (*DAB*).

7 Gray mentioned that *Houstonia* was dimorphic in a brief notice on the subject of dimorphism that appeared in the November 1862 number of the *American Journal of Science and Arts*, commonly referred to as 'Silliman's journal', after its founder Benjamin Silliman (A. Gray 1862e, p. 419).

[8] CD regularly published brief notices in the *Gardeners' Chronicle and Agricultural Gazette*; however, he did not publish an account of Rothrock's experiments with *Houstonia caerulea* until 1877 (*Forms of flowers*, pp. 132 and 254).

[9] See, for example, the letter to Asa Gray, 21 August [1862] and n. 10.

[10] Gray's review of *Orchids* (A. Gray 1862a) was published in the *American Journal of Science and Arts* in July; his follow-up article (A. Gray 1862b) appeared in the November issue.

[11] CD had asked Gray for his opinion of the last chapter of *Orchids*, noting: 'It bears on design' (see letters to Asa Gray, 23[–4] July [1862] and n. 21, and 21 August [1862] and n. 15). In the follow-up article to his review of *Orchids* (A. Gray 1862b, pp. 428–9), Gray praised CD for having 'brought back teleological considerations into botany', and concluded:

> we *faithfully* believe that both natural science and natural theology will richly gain, and equally gain, whether we view each varied form as original, or whether we come to conclude, with Mr. Darwin, that they are derived:— the grand and most important inference of *design in nature* being drawn from the same data, subject to similar difficulties, and enforced by nearly the same considerations, in the one case as in the other.

For CD's extended correspondence with Gray on the question of design in nature, see *Correspondence* vols. 8 and 9.

[12] Gray had made a number of observations on American species of orchids in the summer of 1862, sending several sets of notes to CD, who encouraged Gray to publish them. In his letter to Gray of [3–]4 September [1862], CD expressed pleasure at Gray's intention to include some of his observations in the follow-up article to his review of *Orchids* (A. Gray 1862b).

[13] See letter to Asa Gray, [3–]4 September [1862] and n. 7. In A. Gray 1862b, p. 426, Gray described in this species an instance of self-pollination that was 'apparently so remarkable that we hesitate to bring forward our too scanty observations until another summer affords an opportunity to test them'. Gray noted that the pollen-masses could not independently 'fall upon or reach the stigma beneath', and argued that they must therefore be conveyed on an insect's proboscis. However, he also noted that the pollen-masses were frequently found lying on the three narrow processes of the rostellum, which, since the pollen freely sent down pollen-tubes into them, appeared to 'act as stigmas'. Gray confirmed this view after re-examining the species the following summer (A. Gray 1863a, p. 293). CD cited the case in 'Fertilization of orchids', p. 147 (*Collected papers* 2: 144).

[14] See letter from Asa Gray, 5 September 1862 and n. 11.

[15] See letters to Asa Gray, 9 August [1862] and [3–]4 September [1862].

[16] See letter to Asa Gray, 9 August [1862], and letter from Asa Gray, 5 September 1862.

[17] See letter to Asa Gray, [3–]4 September [1862] and nn. 14 and 15.

[18] D[icey] 1862; the reference is to the American Civil War.

[19] Two proclamations calling for volunteers had been made by the Union, the first in April 1861 and the second in July 1862 (McPherson 1988, pp. 274 and 491).

From Hugh Falconer 24–7 September [1862][1]

21 Park Crescent N.W. | Portland Place.
24th.–27th Sept.

My Dear Darwin.

Do not be frightened at the enclosure.[2] I wish to set myself right by you—before I go to press. I am bringing out a heavy memoir on Elephants—an Omnium gatherum affair—with observations on the fossil & recent specs. One section is devoted—to the *persistence* in time of the specific characters of the mammoth.[3] I trace him—from before the Glacial period—through it—and after it—unchangeable & unchanged, so far as the organs of digestion (teeth) & locomotion are concerned.

Now the glacial period was no joke— it would have made ducks and drakes of your dear pigeons & doves.[4]

With all my Shortcomings, I have such a sincere & affectionate regard for you—& such admiration of your work, that I should be pained to find that I had expressed my honest convictions, in a way that would be open to any objection by you. The reasoning may be very stupid—but I believe that the observation is sound. Will you therefore, look over the few pages which I have sent, and tell me whether you find any flaw—or whether you think I should change the form of expression.[5] You have been so unhandsomely and uncandidly dealt with by a blackguard friend of yours and mine,[6] that I should be sorry to find myself in the position of an opponent to you—and more particularly with the chance of making a fool of myself.

I met your brother yesterday who tells me you are coming to town.[7] I hope you will give me a hail. I long for a jaw, with you, and have much to speak to you about.

You will have seen the eclaircissement—about the Eocene monkeys of England. By a touch of the conjuring wand they have been metamorphosed—a la Darwin—into *Hyracotherian* Pigs.[8] —Would you believe it— this even is a gross blunder. They are not pigs

Kindly return the Enclosed & believe me, my Dear Darwin | Yours ever Sinc^{ly} | H. Falconer

I hope this will find you at Bournemouth. I have not got your exact address.[9] | H F

27th. Your note just come to hand—and I dispatch the Cover.[10]

DAR 164: 6

1 The year is established by the reference to the manuscript of Falconer 1863 (see n. 2, below).

2 Falconer enclosed with his letter a portion of the manuscript of a paper on fossil and living species of elephant, that was later published in the January 1863 number of the *Natural History Review* (Falconer 1863; see letter to Charles Lyell, 1 October [1862]).

3 Falconer 1863, pp. 77–81.

4 Between 1855 and 1858, CD had made a detailed study of artificial selection in domestic pigeons, concluding that all the breeds kept by pigeon-fanciers were descended from a single ancestral species (see *Origin*, pp. 20–9 and Secord 1981). Falconer had assisted CD with the osteological comparison of the various breeds (see *Correspondence* vol. 7, letter to T. C. Eyton, 4 October [1858] and n. 4).

5 Falconer sought to show that European fossil elephants from successive geological strata displayed a marked persistence of specific characters. Conceding that his conclusions did not oppose the 'leading propositions of Darwin's theory' (Falconer 1863, p. 80), Falconer, nevertheless, argued:

> The whole range of the Mammalia, fossil and recent, cannot furnish a species, which has had a wider geographical distribution, and at the same time passed through a longer term of time, and through more extreme changes of climatal conditions, than the Mammoth. If species are so unstable, and so susceptible of mutation through such influences, why does that extinct form stand out so signally a monument of stability?

For CD's response, see the letter to Hugh Falconer, 1 October [1862].

6 Richard Owen.

[7] CD stayed in London with his brother, Erasmus Alvey Darwin, on the night of 29 September 1862, on the way home from his holiday in Bournemouth (see 'Journal' (Appendix II)).

[8] The reference is to Richard Owen's retraction, in the September number of the *Annals and Magazine of Natural History* (R. Owen 1862b), of his earlier identification of certain fossil remains from the Eocene sand at Kyson, Suffolk, as being those of a species of the monkey genus *Maeacus* (R. Owen 1844); Owen concluded that the remains were in fact probably those of *Hyracotherium cuniculus*.

[9] CD spent September 1862 at Cliff Cottage, Bournemouth (see 'Journal' (Appendix II)).

[10] CD's letter has not been found.

To J. D. Hooker 26 September [1862]

Cliff Cottage | Bournemouth
Sept. 26th

My dear Hooker

Do not read this till you have leisure, if that blessed moment ever comes. I sh^{d}. be very glad to have your opinion on the subject of this letter. I am led to the opinion that Drosera must have diffused matter, in organic connection, closely analogous to the nervous matter of animals.[1] When the gland of one of the papillæ or tentacles, in its natural position (A.), is supplied with nitrogenised fluid & certain other stimulants, or when loaded with an extremely slight weight, or when struck

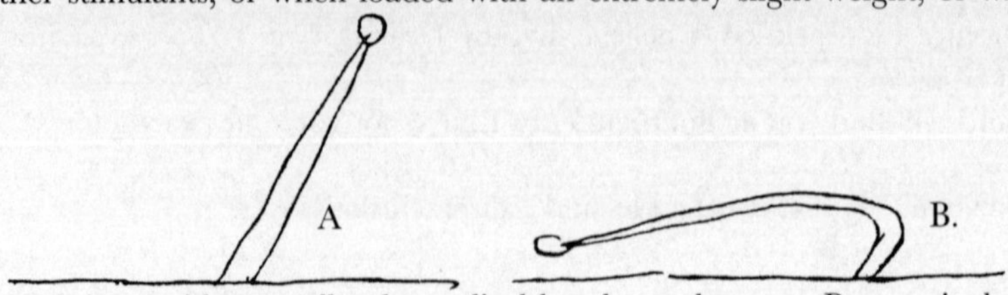

several times with a needle, the pedicel bends, as shown at B, near its base, in under one minute. These varied stimulants are conveyed down the pedicel by some means: it cannot be vibration for drop of fluid put on quite quietly causes the movement: it cannot be absorption of fluid from cell to cell, for I can see rate of absorption, which though quick, is far slower, & in Dionæa the transfusion is instantanous. Analogy from animals would point to transmission through nervous matter. Reflecting on rapid power of absorption in the gland; the extreme sensibility of the whole organ, & the conspicuous movement caused by varied stimulants, I have tried a number of substances, which are not caustic or corrosive, & which I believe do not act on such organic compounds, as sugar gluten cellulose &c, but most of which are known to have a remarkable action on the nervous matter of animals. You will see the result in enclosed paper.[2] As the nervous matter of different animals are differently acted on by the same poisons, one would not expect the same action on plants & animals; only if plants have diffused nervous matter, some degree of analogous action. And this is partially the case. Considering these experiments, together with the previously made remarks on the function of the parts, I cannot avoid the conclusion, that Drosera possesses matter at least in some degree analogous in constitution & function to nervous matter.

Now do tell me what you think, as far as you can judge from my abstract. Of course many more experiments would have to be tried; but in former years I tried on whole leaf, instead of on separate glands, a number of innocuous substances such as sugar, gum, starch &c & they produced no effect.— Your opinion will aid me in deciding some future year in going on with this subject.—[3]

I shd. not have thought it worth attempting, but I had nothing on Earth to do.—[4]

My dear Hooker | Yours very sincerely | Ch Darwin

We return home on Monday 28th thank Heaven—[5]

[Enclosure]

☞ Please return this, as it will aid me hereafter as short abstract[6]

Vapours

Small dose of *Chloroform,* causes rapid, even spasmodic movement; but if very small dose be given & the tentacles do not reach centre, they are paralysed & do not move with meat on gland under from 2 to 4 hours.

Nitric Ether, sometimes causes a few tentacles to move; & thus all are paralysed for some hours.

Sulphuric Ether, caused no movement, & tentacles were paralysed for one or two hours

Alcohol caused no movement; & tentacles were paralysed for about one hour.

Carbonic Acid: plant exposed for one hour; caused no movement, but I *think* slightly dulled power of movement.

Turpentine, vapour of for 15′ caused no movement, & entirely stopped subsequent action of meat.—

All substances used in fluid state

Stramonium Tobacco Henbane Hemlock Strong infusion of Tea	produced no effect; caused no movement, & subsequently meat caused usual rapid movement; I think tea rather dulled power of movement.—
Prussic Acid diluted Belladonna Digitalis	caused the tentacles to move rapidly.
Strychnine acetate of	caused no movement, & entirely stopped the subsequent action of meat & killed the glands. When a tentacle with an atom of meat had commenced moving the strychnine arrested the movement.
Veratria acetate of	caused no movement; & stopped for many hours any movement from meat; but some tentacles after several hours regained power of movement

Morphia acetate of, & common Opium	caused no movement, & stopped for one or two hours subsequent action of meat, & then the movement was excessively slow
Poison from adders' fang	caused rapid movement; & produced no subsequent injury; perhaps acted merely like saliva, which from having some nitrogen causes movement
Poison from an Ant	—caused no movement, but stopped subsequent action of meat: concentrated acetic acid acts in same manner.

Endorsement: '/62'
DAR 60.2: 88, DAR 115.2: 163

[1] CD had begun his experiments with the insectivorous plant *Drosera rotundifolia* in 1860 (see *Correspondence* vol. 8). He had hoped to continue and complete the experiments in the summer of 1861, but subsequently decided to postpone them (see *Correspondence* vol. 9, letter to J. D. Hooker, 4 February [1861], and letter to Daniel Oliver, 11 September [1861]). There are notes detailing CD's experiments with this species between 14 and 26 September 1862, in DAR 54: 29–49. See also letter to Daniel Oliver, [17 September 1862] and n. 12.

[2] See enclosure.

[3] CD did not again work extensively on the subject of insectivorous plants until 1872 (*LL* 3: 322); his findings were published in 1875 as *Insectivorous plants.*

[4] CD was on holiday in Bournemouth (see 'Journal' (Appendix II)); he had begun to study *Drosera* while 'idling and resting' at the Sussex home of his sister-in-law, Sarah Elizabeth Wedgwood, in July 1860 (*Autobiography*, p. 132; see also *Correspondence* vol. 8).

[5] This appears to be an error: 28 September 1862 fell on a Sunday. However, CD did not return to Down House until Tuesday 30 September (see 'Journal' (Appendix II)).

[6] These notes represent an abstract of CD's experimental notes, dated 14–26 September 1862, which are in DAR 54: 29–49.

To A. G. More 26 September 1862

Cliff Cottage | Bournemouth
Sept. 26, 1862

My dear Sir

I thank you cordially for the specimens received this morning; and for all the trouble which you have so kindly taken.[1] Putting on one side the 3 imperfect capsules, 6 are left, of which the labellum had been removed.[2] In comparison with the 5 naturally fertilised capsules, two of the mutilated produced as much seed and seemed to have been fully fertilised; the four others produced very much less seed. So that the result in a *limited* degree supports my belief in use of labellum; but I can *now* see that a dozen other capsules from mutilated flowers ought to be compared with a dozen naturally fertilised. But this would not be worth your time and trouble.

I should, however, be very much obliged if you would answer me two questions (writing to Down where I return on Monday),[3] viz., whether you chose average capsules of the naturally fertilised; for if you picked out the finest, in this case I should doubt the result; and secondly whether you removed only the further or distal or articulated part of the Labellum, as you speak in your letter of removing the labellum.

Pray believe me | My dear Sir | Yours very truly obliged | Ch. Darwin

I sincerely hope that your health is reestablished.

Copy
DAR 146: 406

[1] More's letter has not been found. The reference is to specimens of the orchid *Epipactis palustris* (see n. 2, below).

[2] In the summer of 1861, More had carried out an experiment for CD on the orchid *Epipactis palustris*, in order to test CD's ideas about the importance of the labellum in the process of insect pollination in that species. More removed the distal half of the labellum from two flowers before they opened, and compared the seed-capsules produced with those from undamaged flowers (see *Orchids*, pp. 101 n.–102 n., and *Correspondence* vol. 9, letters to A. G. More, 7 July 1861, 19 July 1861, 23 September 1861, and 1 October 1861); the capsules from the damaged flowers were smaller, and, in 1862, CD asked More to confirm these results under more rigorous conditions (see letter to A. G. More, 18 May 1862). CD published More's later results in 'Fertilization of orchids', pp. 149–50 (*Collected papers* 2: 146).

[3] Although CD left Bournemouth on Monday 29 September 1862, he stayed in London that evening, returning to Down House on Tuesday 30 September (see 'Journal' (Appendix II)).

From G. C. Oxenden 27 September [1862][1]

Broome
Sept 27—

Dear Sir

I rather think that the "*Ozonised Fluid*" is a very clean, pure, neat solution of the "*Per-manganate of Soda*—"[2]

So small a dose as *10 drops*, could not, by possibility, produce **any** appreciable effect whatever, save & except a sensation of *purity* in the *mouth & Tongue*—

—You should, at the Outset, take a **Teaspoonful** of the Fluid in a full **Wine** glass of Water, *four* or *five times a day*, or even *more* often—[3]

—Then, in a fortnight, increase the dose of Fluid to a **Dessert Spoonful**, in a **large** *Wine*-glass of Water—& stick at this for a Month—by which time you will surely know & feel whether you are the better for it—

—In **no** case, can it by possibility do **any** harm— —& ultimately, it might be used as daily diet Drinks, (in lieu of Beer or claret,) well *diluted with Water*—

—It does not in any way impede digestion—but, as a *rule*, it is best to take it upon an Empty, or half-empty Stomach— —(which rule applies to **every** description of Medecine)—

—I myself take it with avidity, as a matter of **luxury**, just as I w[d]. use Eau de Cologne—or Soda Water—

—I conceive this Ozonised Fluid to be quite capable of arresting, & *preventing*, *Diarrhœa*, but I have never tried it—nor had need—

—Lastly, I believe the *cheaper*, *stronger*, *preparation* of the *same* (sold by every Chemist) will be found eminently sanative & prophylactic, (by **absorption**) if largely added to the Water of a tepid Bath, & especially in, & after, Fever—.

—If anything farther strikes me, I will write— | Your's sincerely | G. Chichester Oxenden

DAR 173.2: 60

[1] The year is established by the references to Condy's 'Ozonized Fluid' (see n. 3, below).

[2] Condy's 'Ozonized Fluid' was a solution of alkaline permanganates developed by Henry Bollmann Condy, and used 'for the purification of air and water, and sanitary objects in general' (Condy 1862, p. 69).

[3] Oxenden first mentioned Condy's 'Ozonized Fluid' to CD in his letter of 21 June 1862, recommending it for the external treatment of Leonard Darwin, then suffering from scarlet fever. However, in his letter of 17 September [1862], Oxenden also recommended its use as a 'pleasant daily diet drink'. In his letter to J. D. Hooker, 10–12 November [1862], CD mentioned that he had been taking 'a tea-spoon in water 3 or 4 times. a day' with, he thought, '*extraordinary* advantage'.

From Bartholomew James Sulivan 27 September [1862][1]

Board of Trade. S.W.
Sept[r]. 27—

My dear Darwin

It is long since I have heard any thing of you or yours. I hope "no news" is "good news"—

Our old shipmate Wickham will be with me for some days after the sixth: when I last saw him he expressed a wish to see you again;[2] would there be any chance of your coming to Town about that time? if so we could easily arrange a meeting—and if you could manage to come to us for one night it would give us great pleasure—& I could get Mellersh—who is in Town—to come down also.[3] I know you do not like being away from home, but if you have to come to Town at all about then and could stay for a night it might not interfere much with your arrangements. I see a mid named 'Darwin' just appointed to a ship[4] have you a boy in the Navy? I think not.

I hope M[rs]. Darwin is well and that your daughter's health has gone on improving.[5] | with very kind remembrances | Believe me dear Darwin— | Yours very sincerely | B. J. Sulivan

DAR 177: 275

[1] The year is established by reference to the visit to Down House of Sulivan, John Clements Wickham, and Arthur Mellersh, on 21 October 1862 (see nn. 2 and 3, below).

[2] John Clements Wickham was second in command on board HMS *Beagle* during the surveying voyage of 1831–6; he emigrated to Australia in 1842, but moved to southern France in 1862 (*Aust. dict. biog.*). Sulivan was a lieutenant on board the *Beagle* during the same voyage; in 1856, he was appointed as one of the naval officers in the marine department at the Board of Trade (*DNB*, *Post Office London directory* 1861).

[3] Arthur Mellersh was midshipman and then mate on board HMS *Beagle* between 1825 and 1836; in 1862, after serving on various vessels, he returned to Britain to take command of HMS *Forte*, for a posting to South America (*Modern English biography*; *The Times*, 28 September 1894, p. 4).

[4] Sacheverel Charles Darwin, who was CD's first cousin, once removed, was appointed midshipman to HMS *Phœbe* in the autumn of 1862 (*Colburn's United Service Magazine* (1862), pt 3: 447).

[5] Sulivan refers to Henrietta Emma Darwin, who had been seriously ill in 1860 and 1861 (see *Correspondence* vols. 8 and 9).

From Henry Johnson 30 September 1862

Shrewsbury
Sept 30. | 1862

Dear Darwin,

Perhaps you have seen by the Transactions that the R. Society has published a Paper of mine about the crushed skulls found at Wroxeter.[1]

In consequence of this one of the Fellows has asked me if I would be proposed as an F.RS.[2]

Much as I should like to be so honoured—I have always heard that those three letters *are the dearest in the alphabet*.

Will you kindly tell me what is the cost—exactly for if it is what I fancy I must decline.[3] I hope you are stronger & better in health than when I last heard of you.[4]

If I could know how to procure a photograph or carte de visite of you for love or money I should be extremely glad,

Yours very truly | Henry Johnson

C. Darwin Esq

DAR 168: 66

[1] An abstract of Johnson's paper, entitled 'On the distorted skulls found at Wroxeter (Salop), with a mechanico-chemical explanation of the distortion', was published in the *Proceedings of the Royal Society of London* (Johnson 1862). Excavation of the site of the Roman city of Uriconium, at Wroxeter in Shropshire, began in 1859 under the auspices of the Shropshire Natural History and Antiquarian Society, of which Johnson was honorary secretary; Johnson had been appointed to the committee responsible for the excavation (Wright 1872, p. iii).

[2] Johnson may refer to Erasmus Wilson, who had communicated Johnson's paper to the Royal Society (Johnson 1862, p. 149).

[3] Johnson did not become a fellow of the Royal Society of London (*Record of the Royal Society of London*, pp. 517-66). The admission fee for the Royal Society of London was £10 and, thereafter, £1 per quarter; alternatively, one could pay a 'composition' fee of £40 (*Record of the Royal Society of London*, p. 100).

[4] CD and Johnson had been contemporaries at Shrewsbury School and, later, at the University of Edinburgh (*Correspondence* vol. 1, *Medical directory*); no earlier correspondence between them has been found.

From A. R. Wallace 30 September 1862

5, *Westbourne Grove Terrace, W.*
Sept.r 30th. 1862

My dear Mr. Darwin

Many thanks for the third Ed. of the "Origin" which I found here on my return from Devonshire on Saturday.[1]

I have not had time yet to read more than the "Historical Sketch" which is very interesting & shews that the time had quite come for your book.[2]

I am now reading Herbert Spencer's *First Principles*, which seems to me a truly great work, which goes to the root of everything.[3]

I hope you will be well enough to come to Cambridge.[4]

I remain | My dear Mr. Darwin | Yours very faithfully | Alfred R. Wallace

DAR 106/7 (ser. 2): 6

[1] See letter from A. R. Wallace, [after 20 August 1862] and n. 3.

[2] The third English edition of *Origin*, published in 1861, was the first to include the historical preface that CD had appended to the first American and German editions; the preface was entitled 'An historical sketch of the recent progress of opinion on the origin of species' (*Origin* 3d ed., pp. xiii–xix).

[3] Spencer 1860–2. There is a partly uncut copy of this work in the Darwin Library–CUL (*Marginalia* 1: 769).

[4] In 1862, the British Association for the Advancement of Science held its annual meeting in Cambridge during the first week of October. See also letter to W. D. Fox, 20 [September 1862], and letter to J. D. Hooker, 21 [September 1862].

To Charles Lyell 1 October [1862]

Down Bromley Kent
Oct 1.

My dear Lyell

You will find the long discussion on the "Règne Humain" in Ch. VII. of *Tom. 2* p. 167 of Isid. G. Hist. Nat. Générale.[1] I quite forgot to ask whether you have done with Rolle, as I could have taken it away with me.—[2]

I found here a short & very kind note from Falconer with some pages of his Elephant memoir, which will be published, in which he treats admirably on long persistence of type.[3] I thought he was going to make a good & crushing attack on me; but to my great satisfaction he ends by pointing out a loophole & adds "with him I have no faith that the mammoth & other extinct elephants made their appearance suddenly . . . . the most rational view seems to be that they are the modified descendants of earlier progenitors &c"[4]

This is capital. There will not be soon one good paleontologist who believes in immutability. Falconer does not allow for the Proboscidean Group being a failing one & therefore not likely to be giving off new races.—

He adds that he does not think natural selection suffices; I do not quite see the form of his argument, & he apparently overlooks that I say over & over again that

N. Selection can do nothing without variability, & that variability is subject to the most complex fixed laws.

I much enjoyed my little chat with you.—[5]

Very sincerely yours | C. Darwin

Endorsement: 'Sept. 1. 1862'
American Philosophical Society (282)

[1] CD refers to Geoffroy Saint-Hilaire 1854–62, 2: 167–261, in which the author concluded that, while in regard to its physical organisation, the human species constituted a family in the order of primates, in regard to its moral and intellectual faculties, it constituted a separate kingdom, of equal significance with the animal and plant kingdoms. CD's annotated copy of this work is in the Darwin Library–CUL (see *Marginalia* 1: 316–20). Lyell was preparing a major work on the antiquity of the human species (C. Lyell 1863a), in which he referred to Geoffroy Saint-Hilaire 1854–62 as providing 'an elaborate and faithful summary' of 'those systems of classification which have been proposed by the most eminent teachers of natural history', especially in regard to the human species (p. 473).

[2] CD apparently refers to the first part of Rolle 1863, the publication of which was announced on 8 September 1862 (*Börsenblatt für die Deutschen Buchhandel* 29 (1862): 1862). In his letter to Friedrich Rolle of 17 October [1862], CD reported that he had received the first part of Rolle's work on returning home 'after a long absence'; however, the work had apparently been forwarded to CD in Bournemouth, and subsequently to Lyell. See also letter to Daniel Oliver, [17 September 1862]. CD's annotated copies of the four parts of Rolle 1863 are in the Darwin Pamphlet Collection–CUL. See also n. 5, below).

[3] Falconer 1863. See letter from Hugh Falconer, 24–7 September [1862], and following letter.

[4] Falconer 1863, p. 80. See also following letter.

[5] CD met Lyell in London on 30 September 1862 (see following letter). According to Emma Darwin's diary (DAR 242), CD had spent the night of 29 September at the London home of his brother, Erasmus Alvey Darwin.

To Hugh Falconer 1 October [1862][1]

Down
Oct. 1,

My dear Falconer

On my return home yesterday I found your letter and MS. which I have read with extreme interest.[2] Your note and every word in your paper are expressed with the same kind feeling which I have experienced from you, ever since I have had the happiness of knowing you.[3] I value scientific praise, but I value incomparably higher such kind feeling as yours. There is not a single word in your paper to which I could possibly object: I should be mad to do so; its only fault is perhaps its too great kindness. Your case seems the most striking one which I have met with of the persistence of specific characters. It is very much the more striking as it relates to the molar teeth, which differ so much in the species of the genus, and in which consequently I should have expected variation. As I read on, I felt not a little dumb-founded, and thought to myself that whenever I came to this subject, I should have to be savage against myself; and I wondered how savage you would be. I trembled a little. My only hope was that something could be made out of the

B-o-g N. American forms, which you rank as a geographical race;[4] and possibly hereafter out of the Sicilian species.[5] Guess then my satisfaction when I found that you yourself made a loophole which I never of course could have guessed at; and imagine my still greater satisfaction at your expressing yourself as an unbeliever in the eternal immutability of species.[6] Your final remarks on my work are too generous; but have given me not a little pleasure.[7] As for criticisms, I have only small ones: when you speak of "moderate range of variation",[8] I cannot but think that you ought to remind your readers (though I daresay previously given) what the amount is, *including case of the* American Bog Mammoth. You speak of these animals as having been exposed to vast range of climatal changes from before to after the Glacial period;[9] I should have thought from analogy of sea-shells, that by migration (or *local* extinction when migration not possible) these animals might and would have kept under nearly the same climate.

A rather more important consideration, as it seems to me, is that the whole Proboscidean Group may, I presume, be looked at as verging towards extinction: anyhow the extinction has been complete as far as Europe and America are concerned. Numerous considerations and facts have led me in the Origin to conclude that it is the flourishing or dominant members of each order which generally give rise to new races, sub-species and species;[10] and under this point of view I am not at all surprised at the constancy of your species. This leads me to remark that the sentence at the bottom of p. 60 is not applicable to my views, though quite applicable to those who attribute modification to the direct action of the conditions of life.[11] An elephant might be more individually variable than any known quadruped (from the effects of the conditions of life or other innate unknown causes), but if these variations did not aid the animal in better resisting all hostile influences, and therefore making it increase in numbers, there would be no tendency to the preservation and accumulation of such variations, i.e., to the formation of a new race. As the proboscidean group seems to be from utterly unknown causes a failing group in many parts of the world, I should not have anticipated the formation of new races.

You make important remarks versus natural selection,[12] and you will perhaps be surprised that I do to a large extent agree with you. I could show you many passages, written as strongly as I could in the Origin, declaring that Natural Selection can do nothing without previous variability; and I have tried to put equally strongly that variability is governed by many laws, mostly quite unknown. My title deceives people, and I wish I had made it rather different.[13] Your phyllotaxis will serve as example; for I quite agree that the spiral arrangement of a certain number of whorls of leaves (however that may have primordially arisen; (and whether quite as invariable as you state?)) governs the limits of variability, and therefore governs what natural selection can do.[14] Let me explain how it arose that I laid so much stress on Natural Selection, and I still think justly: I came to think from Geographical Distribution &c. &c. that species probably change; but for years I was stopped dead by my utter incapability of seeing how every part of each creature (a wood-pecker

Daniel Oliver. Photograph by Maull & Polyblank, *c.* 1862.
(Courtesy of the Linnean Society of London.)

Clémence Royer.
Photograph, 1865.
(Courtesy of the Bibliothèque Nationale, Paris.)

Edouard Claparède.
Photograph by Runcaldier & Artus, 1863.
(Courtesy of the Bibliothèque Publique et Universitaire, Geneva.)

or swallow for instance) had become adapted to its conditions of life. This seemed to me and does still seem the problem to solve, and I think natural selection solves it, as artificial selection solves the adaptation of domestic races for man's use. But I suspect that you mean something further,—that there is some unknown law of evolution by which species *necessarily* change; and if this be so, I cannot agree.[15] This, however, is too large a question even for so unreasonably long a letter as this.

Nevertheless just to explain by mere valueless conjectures how I imagine the teeth of your elephants change; I should look at the change, as *indirectly* resulting from changes in the form of the jaws, or from development of tusks, or in case of the "primigenius" even from correlation with the woolly covering; in all cases natural selection checking the variation. If indeed an elephant could succeed better by feeding on some new kinds of food, then any variation of any kind in the teeth, which favoured their grinding power would be preserved. Now I can fancy you holding up your hands and crying out what bosh! To return to your concluding sentence; far from being surprised, I look at it as absolutely certain that very much in the Origin will be proved rubbish; but I expect and hope that the frame-work will stand.[16]

I had hoped to have called on you on Monday evening, but was quite knocked up.[17] I saw Lyell yesterday morning; he was very curious about your views, and as I had to write to him this morning I could not help telling him a few words on your views.[18] I suppose you are tired of the Origin and will never read it again; otherwise I should like you to have the 3rd edit., and would gladly send it rather than you should look at the 1st or 2nd editions.[19]

With cordial thanks for your generous kindness, believe me | My dear Falconer | Very truly yours | Ch. Darwin

Copy
DAR 144: 25

[1] The year is established by the reference to the manuscript of Falconer 1863 (see n. 2, below).

[2] CD refers to the portion of a manuscript paper on fossil and living species of elephant that Falconer had enclosed with his letter to CD of 24–7 September [1862]. The paper was published in the January 1863 number of the *Natural History Review* (Falconer 1863); in his letter to CD of 3 January [1863] (*Correspondence* vol. 11), Falconer told CD that although he had made some additions to the manuscript in response to CD's comments, he had not altered the existing text.

[3] CD had known Falconer since at least November 1845 (see *Correspondence* vol. 3, letters to J. D. Hooker, [21 November 1845] and [25 November 1845]). See also *Correspondence* vol. 2, letter to Richard Owen, [March 1843 – 15 May 1846].

[4] Falconer 1863, p. 79.

[5] In his account of the 'persistence in time of the distinctive characters of the European fossil elephants', Falconer excluded from consideration the fossil elephants of Sicily or Italy (*Elephas Armeniacus*), 'in order to relieve the argument of any elements, which may not be considered as being at present established on sufficient evidence' (Falconer 1863, p. 78 n.).

[6] In the conclusion to this section of his paper (Falconer 1863, p. 80), Falconer stated:

> The inferences which I draw from these facts, are not opposed to one of the leading propositions of Darwin's theory. With him I have no faith in the opinion, that the Mammoth and other

extinct Elephants made their appearance suddenly, after the type in which their fossil remains are presented to us. The most rational views seem to be, that they are in some shape the modified descendants of earlier progenitors. But if the asserted facts be correct, they seem clearly to indicate that the older Elephants of Europe . . . were not the stocks from which the later species . . . sprung, and that we must look elsewhere for their origin.

7 Falconer concluded the section by stating (Falconer 1863, pp. 80–1):

By his admirable researches and earnest writings, Darwin has, beyond all his cotemporaries, given an impulse to the philosophical investigation of the most backward and obscure branch of the Biological Sciences of his day; he has laid the foundations of a great edifice; but he need not be surprised, if, in the progress of erection, the superstructure is altered by his successors . . . to a different style of architecture.

8 Falconer 1863, p. 79.

9 Falconer 1863, p. 79.

10 See especially *Origin*, pp. 52–5, 116–7, and 325–9.

11 The reference is to the manuscript pagination; CD probably refers to the following statement, which appeared on p. 80 of the published paper (Falconer 1863):

The whole range of the Mammalia, fossil and recent, cannot furnish a species, which has had a wider geographical distribution, and at the same time passed through a longer term of time, and through more extreme changes of climatal conditions, than the Mammoth. If species are so unstable, and so susceptible of mutation through such influences, why does that extinct form stand out so signally a monument of stability?

12 Having stated that he accepted CD's proposition of the evolution of species by modified descent, Falconer continued: 'Another reflexion is equally strong in my mind, that the means which have been adduced to explain the origin of species by 'Natural Selection,' or a process of variation from external influences, is inadequate to account for the phenomena' (Falconer 1863, p. 80). Noting the regular laws limiting or determining the variation in form of plants and animals, he concluded: 'it is difficult to believe, that there is not in nature, a deeper seated and innate principle, to the operation of which 'Natural Selection' is merely an adjunct'.

13 On the title-page of *Origin*, CD defined 'Natural Selection' as the 'preservation of favoured races in the struggle for existence'. In the text of the work he defined it as the 'preservation of favourable variations and the rejection of injurious variations' (*Origin*, p. 81). The laws of variation are the subject of chapter 5 of *Origin*.

14 One of Falconer's examples of the laws of variation (see n. 12, above) was 'the law of Phyllotaxis, which governs the evolution of leaves around the axis of a plant', which he considered 'nearly as constant in its manifestation, as any of the physical laws connected with the material world' (Falconer 1863, p. 80).

15 See n. 12, above.

16 See n. 7, above.

17 See letter from Hugh Falconer, 24–7 September [1862]. Emma Darwin's diary (DAR 242) records that CD spent the night of Monday 29 September 1862 at the London home of his brother, Erasmus Alvey Darwin.

18 See preceding letter.

19 The third edition of *Origin* was published in March 1861.

To A. G. More 1 October 1862

Down, Bromley, Kent. S.E.
Oct. 1, 1862

My dear Sir

I am very much obliged for your satisfactory answers to my questions.[1]

I am ashamed to say so, but if you are in the Isle of Wight I should certainly be very glad to see the result of some more trials.[2] The result would probably throw light on the use of the labellum in many exotic orchids.

My dear sir | Yours sincerely and obliged | Ch. Darwin.

Copy
DAR 146: 407

[1] More's letter has not been found. In his letter to More of 26 September 1862, CD asked for further details of the experiments carried out by More on the labellum of the orchid *Epipactis palustris*.

[2] More lived at Bembridge on the Isle of Wight. According to his biographer, 'unforeseen circumstances' prevented the fulfilment of CD's wish (Moffat ed. 1898, p. 158).

From B. J. Sulivan 2 October [1862][1]

Board of Trade. S.W.
Oct 2.

My dear Darwin

I am indeed sorry to hear such an account of M^rs. Darwin & your family.[2] I can answer for myself and (I think for Wickham & Mellersh) that I should be very glad indeed to run down to you for a night, probably the week *after next* would suit Wickham best as he will then be in Lodgings in Town.[3] I will ask them both and let you know—but it will be after *Sixth* of this month as Wickham does not come to us till that day and I do not know where a letter would catch him.

I had intended volunteering to run down to you this autumn for a day or two after the other people returned here to work, & I should be free for a day when I wished to take one. But when I knew Wickham was coming I thought it a nice chance for us to meet.

Believe me my dear Darwin | Yours very sincerely | B. J. Sulivan

DAR 177: 276

[1] The year is established by reference to the visit to Down House on 21 October 1862, of Sulivan, John Clements Wickham, and Arthur Mellersh (see n. 3, below).

[2] CD's letter has not been found. Horace Darwin had been ill early in 1862, and Emma and Leonard Darwin had been ill with scarlet fever during the summer (see Emma Darwin's diary (DAR 242)).

[3] Sulivan, Mellersh, and Wickham had served with CD on board HMS *Beagle* during the surveying voyage of 1831–6. In his letter to CD of 27 September [1862], Sulivan suggested that CD and Mellersh might visit him in London early in October while Wickham, who had emigrated to Australia in 1842, was staying with him. According to Emma Darwin's diary (DAR 242), Sulivan, Mellersh, and Wickham visited Down House on 21 October 1862.

From Hugh Falconer 3 October 1862

British Association For The Advancement of Science. | Cambridge Meeting, 1862.
3^d. Oc^r.

My Dear Darwin.

On Saturday last—the 27th—on the receipt of your note,[1] I posted to you—

addressed to you—to *Down* near Bromley—a heavy letter containing some folio m.ss. in a large *paid* cover—4^{d} postage.[2]

Not having heard from you, I am afraid it may have miscarried.

Did any such letter reach you. | Yours very Sin^{ly} | H Falconer

DAR 164: 7

[1] CD's letter has not been found.

[2] Falconer refers to his letter to CD of 24–7 September [1862], with which he sent a portion of the manuscript of a paper on fossil and living species of elephant (Falconer 1863). For CD's favourable reaction to Falconer's work, see the letter to Hugh Falconer, 1 October [1862].

From Hugh Falconer 4 October 1862

British Association For The Advancement of Science. | Cambridge Meeting, 1862.
4 Oc^{tr}.

My Dear Darwin.

I seize a stray moment—to thank you for your kind reply to my note.[1] Under the alarm, that it might have miscarried I wrote my note of yesterday—that I might trace it up if lost.[2]

How differently you and I think—from the honest results of hard earned Conviction—upon certain points—and yet what an ineffable charm—in the feeling, that we can exchange our differences of thought—without the risk of offending! The virtue of the position is all on your side—because your views are such an integral part of your present existence—and of your future fame—that you must be more than angelic—to tolerate dissent with equanimity.

What I sent you was only—a section of results from the evidence. You must come and have a jaw and we will talk over it. I was so disappointed, at not seeing you. I yearned for it—a couple of days—and sent to Queen Anne S^{t}. on Tuesday to enquire—when the messenger said you had been there and were gone.[3]

Excuse a hurried line & believe me | my Dear Darwin | Yours Ever | H Falconer

DAR 164: 8

[1] Letter to Hugh Falconer, 1 October [1862].

[2] See preceding letter.

[3] CD stayed at 6 Queen Anne Street, Cavendish Square (the London home of his brother, Erasmus Alvey Darwin) on Monday 29 September 1862, on the way home from his holiday in Bournemouth (Emma Darwin's diary (DAR 242)). See also letter from Hugh Falconer, 24–7 September [1862], and letter to Hugh Falconer, 1 October [1862].

From Asa Gray 4 and 13 October 1862

Cambridge. [Massachusetts]
Oct. 4, 1862

My Dear Darwin

I have just been reading Max Müllers Lectures on the Science of Language, with much interest.[1] But perhaps what has interested me most is, after all, his perfect appreciation and happy use of Natural Selection, and the very complete *analogy* between diversification of species and diversification of language.[2]

I can hardly think of any publication which in England could be more useful to your cause than this volume is, or should be. I see also with what *great effect* you may use it in our occasional discussion about *design*,—indeed I hardly see how to avoid conclusion adverse to *special design*.—tho' I think I see indications of a way out.[3]

Depend on it, Max Müller will be of real service to you.

Oct. 13.

I have been so much occupied that I deferred to the last moment to write out my 2d notice of your Orchid Book for Silliman's Journal. I wrote out Saturday evening what I could, and to-day have finished and sent of my MSS. to New Haven.[4] The greater part consists of a record of some of my observations last summer.—very hurriedly penned, & sent off. I trust you will be pleased, and will think that my little contributions cannot be better hatched than under your wings.[5]

Hoping that my young correspondent is fast recovering strength, tell him that I have no more stamps for him yet, but shall pick up his desiderata one of these days.[6]

I have some nice live roots of Cypripedium, 2 or 3 sp. to send you,—and mean to send *Mitchella*.[7]

How Hooker does praise up your book,—in Gard. Chron.[8]

Ever Yours | A. Gray

DAR 165: 120

[1] Max Müller 1861. In his lectures, originally delivered at the Royal Institution in the summer of 1861, Friedrich Max Müller sought to establish the study of language as a science operating according to the same epistemology as the natural sciences (Stocking 1987, pp. 56–62).

[2] In his final lecture, on the 'theoretical stage' of the science of language, and on the 'origin of language' (Max Müller 1861, pp. 329–78), Max Müller argued that human language originated in the instinctive faculty of giving 'articulate expression to the rational conceptions' of the mind (p. 370). Although he considered that the nature of these first '*phonetic types*' was entirely determined by natural instincts, the 'almost infinite' number of them was, he claimed, subsequently reduced by a 'process of elimination, or natural selection', so that 'clusters of roots, more or less synonymous, were gradually reduced to one definite type.' The combining of the fundamental roots into different languages was, he claimed, 'the work of man, not in his individual and free, but in his collective and moderating capacity' (pp. 374–5). He concluded (p. 377):

> The first natural and instinctive utterances, if sifted differently by different clans, would fully account both for the first origin and for the first divergence of human speech. We can understand not only the origin of language, but likewise the necessary breaking up of one language into many; and we perceive that no amount of variety in the material or the formal elements of speech is incompatible with the admission of one common source.

On Max Müller's views on evolution, see, for instance, Schrempp 1983 and Knoll 1986.

[3] Gray probably refers to Max Müller's use of natural selection as an analogy for the development of language. Denying that 'the various forms of development in language must be explained . . . as *necessary* evolutions, founded in the essence of human speech', Max Müller argued that, after its first

origin, the evolution of language was entirely a matter of historical contingency (Max Müller 1861, pp. 373–4). This suggestion could be seen as undermining Gray's view, first given in [A. Gray] 1860, that the variations on which natural selection operated were specially designed. See also *Correspondence* vol. 11, letter from Asa Gray, 20 April 1863, letter to J. D. Hooker, [9 May 1863], and letter to Asa Gray, 11 May [1863]. Gray and CD had discussed intermittently since 1860 the question of design in nature (see *Correspondence* vols. 8 and 9).

[4] Having published a review of *Orchids* in the June number of the *American Journal of Science and Arts* (A. Gray 1862a), Gray had planned to write a follow-up article (A. Gray 1862b). The *American Journal of Science and Arts*, founded by Benjamin Silliman, was commonly known as 'Silliman's journal'; it was published in New Haven, Connecticut.

[5] In the summer of 1862, Gray had made numerous observations of the mechanisms for effecting pollination in several American species of orchid. CD had initially suggested that Gray might publish some of his observations with his review of *Orchids*, but subsequently sought to persuade him to publish them separately (see letters to Asa Gray, 10–20 June [1862], 1 July [1862], 23[–4] July [1862], 28 July [1862], and [3–]4 September [1862]).

[6] Gray refers to Leonard Darwin, who had been ill with scarlet fever during the summer. A letter from Leonard, requesting American postage stamps for his collection, was apparently enclosed with the letter to Gray of [3–]4 September [1862] (see letter from Asa Gray, 22 September 1862 and n. 3).

[7] See letter from Asa Gray, 5 September 1862 and nn. 4 and 5.

[8] [J. D. Hooker] 1862c. CD was unaware that Joseph Dalton Hooker was the author of this anonymous review (see letter to J. D. Hooker, 3 November [1862], and letter from J. D. Hooker, 7 November 1862).

To J. D. Hooker 6 October [1862]

Down Bromley Kent
Oct 6th.

My dear Hooker.

Sincere thanks for opinion on Drosera, which will be of real use as guide & your note shall be put in the Drosera Portfolios.[1] I do not know whether others know the feeling, but if I work for a time hard on any subject, I become *absolutely* incapable of judging of its value.

I enclose 2 Queries, which you can answer by a word.

But here is a rather more bothersome affair, on which I am really ashamed to trouble Oliver without your aid.[2] If you will ask Oliver, I think he will tell you I have got a real odd case in Lythrum.[3] It interests me extremely; & seems to me the strongest case of propagation recorded amongst plants or animals, viz a necessary triple alliance between 3 hermaphrodites.— I feel sure I can now prove the truth of case, from a multitude of crosses made this summer.—[4]

Now Oliver at my request sent me a set of *buds* of L. Graefferi; but I most stupidly forgot that I shd. require open flowers (i.e. with petals expanded) to establish comparison with L. salicaria: the buds answered *capitally* & I got pollen from all the anthers; but I cannot complete case without open flowers.—[5] I do not care for localities, if I have single flower from 6 or 7 *distinct* plants. This species is trimorphic like L. salicaria. Some flowers of L. thymifolia would be extraordinarily interesting to me, as Vaucher says it is **di**morphic;[6] & I am most curious to see how a **tri** morphic form passes or graduates into **di**-morphic. Is it very much trouble to turn to these plants in the Herbarium?— No case has so much interested me. & I shall write paper for Linnean.—[7]

I hope that you may have gone to Cambridge & read your Wellwitschia paper;[8] it does seem a most grand case to connect two such groups; & I presume you will leave the Gymnosperms, which I am rather glad of.— Oh for your chart of vegetable orders to hang up & study!— I failed in going to Cambridge from another accursed attack of Eczema. I sh^{d}. so much like to pay you a visit of an hour or two at Kew, that I must try; but I get to dread more & more fatigue. I grieve to hear about Miss Henslow.[9]

Pray thank Oliver for his clear & favourable notice of my Orchid Book.[10] I have been going through the Bibliography & picking out references: by Heavens what labour; I sh^{d}. not have thought any mortal man could have done it.[11]

Farewell. This is a horridly dull & troublesome letter. Farewell | Yours affect^{ly} | C. Darwin

Queries, which you can answer by a word on this paper.—

(1) In two forms of Linum perenne, I find in one, the stigmatic surface faces the axis of flower; in the other form twist of style the five stigmas face the circumference of flower. Is this a character of any importance? For instance, if you found this difference in a form, which you doubted whether to rank as a species or variety, would this difference decide you?[12]

(2) Shall I return the 3 Melastomatads by Railway; must I mark by any Railway?

(N.B They came here much broken, & I fear will suffer in their return.)[13]

P.S. Here is a fact which may possibly interest you. In a field here I find many Verbascum thapsus & lychnitis; & lots of varieties making an almost perfect series between these two distinct forms. I am sure many species have been run together on less perfect evidence. But lo & behold every one of these intermediate forms are absolutely sterile! & no doubt are natural hybrids. I found 33 of these hybrids in one field!![14]

Endorsement: '/62'
DAR 115.2: 164

1 Hooker's response to CD's letter of 26 September [1862] has not been found. CD's work on the insectivorous plant *Drosera rotundifolia* was eventually published in 1875 in his book *Insectivorous plants*; the extant portfolios of notes on the subject are in DAR 54–61.

2 Daniel Oliver was librarian and assistant in the herbarium at the Royal Botanic Gardens, Kew.

3 See letters to Daniel Oliver, 29 [July 1862], 2 September [1862], and [17 September 1862].

4 There are dated notes from CD's experiments with *Lythrum salicaria* during the summer of 1862 in DAR 27.2.

5 See letters to Daniel Oliver, 14 September [1862] and n. 3, and [17 September 1862] and n. 3.

6 Vaucher 1841, 2: 371.

7 'Three forms of *Lythrum salicaria*' was read before the Linnean Society of London on 16 June 1864.

8 Hooker was preparing a monograph on the Angolan plant *Welwitschia mirabilis* (J. D. Hooker 1863a), and had intended to display the plant at the 1862 meeting of the British Association for the Advancement of Science, held in Cambridge from 1 to 8 October (see letter from J. D. Hooker, 16 September 1862).

[9] The reference is to one of Frances Harriet Hooker's aunts, probably Anne Frances Henslow (see letter from J. D. Hooker, 20 August 1862 and n. 4).

[10] CD had suggested to Oliver that he should write a review of *Orchids* for the *Natural History Review* (see letter to Daniel Oliver, 24 July [1862], and letter from Daniel Oliver, 28 July 1862). In the event, Hooker wrote the review ([J. D. Hooker] 1862d; see letter from J. D. Hooker, 7 November 1862).

[11] CD refers to the botanical bibliography in the latest number of the *Natural History Review*, apparently compiled by Oliver, who was an editor of the journal ([Oliver] 1862d). CD's annotated copies of the *Natural History Review* are in the Darwin Library–CUL.

[12] Having discovered in 1861 that certain species of *Linum* were dimorphic, CD carried out crossing experiments in the summer of 1862 on *L. perenne* and *L. grandiflorum*; the results from these experiments are given in 'Two forms in species of *Linum*', which was read before the Linnean Society on 5 February 1863. In the paper (pp. 75, 78–80), CD noted the torsion of the style in the long-styled form of *L. perenne*, and discussed its importance in effecting cross-pollination (*Collected papers* 2: 98, 100–3). CD's query apparently reflects his interest in the parallels between the sterility of a dimorphic plant with its own-form pollen and interspecific sterility (see, for instance, letter to J. D. Hooker, 22 [August 1862] and n. 16). See also Appendix VI.

[13] The reference is probably to the three melastomaceous plants that Hooker sent from the Royal Botanic Gardens, Kew, in May 1862, for use in CD's experiments on the possible occurrence of a novel form of dimorphism in the Melastomataceae (see letter from J. D. Hooker, [29 May 1862] and n. 3). There are notes dated 2 October 1862 in DAR 205.8: 54, which record the results of crossing experiments carried out in June and July 1862 on a specimen from Kew of the melastomaceous plant *Heterocentron Mexicanum*.

[14] Having transplanted a wild specimen of *Verbascum* into his garden in order to repeat the hybridisation experiments of Karl Friedrich von Gärtner, CD had been unable to obtain fertile seed-pods from the plant (see letter to W. E. Darwin, 4 [July 1862] and n. 4). In consequence, he searched the field in Cudham Valley, near Down House, from which the plant had been removed, and obtained the results reported here. CD's notes on this subject, dated 4, 6, and 16 October 1862, are in DAR 108: 3–4; he reported his observations in 'Specific difference in *Primula*', pp. 451–4.

To John Murray 7 October [1862?][1]

Down, Bromley
7 October

My eye has just caught sight of a curious misprint in your announcement of my Book, viz., the good affects of *intercussing* for intercrossing.[2] It might do a man good to swear a little occasionally, but a formal treatise on the good effects of intercussing is a novelty

Incomplete
Christie's 27 March 1985

[1] The year is conjectured from the date of publication of the first edition of *Orchids* (see n. 2, below).

[2] The announcement has not been traced. The first edition of *Orchids* was published in May 1862 with the subtitle 'on the good effects of intercrossing'. The subtitle was not used in the second edition.

From Henry Johnson 8 October 1862

Shrewsbury
Oct 8 | 1862

My dear Darwin,

I am very much obliged to you for your very kind letter,[1] and also for the

Photograph, which is a capital likeness of you—& which I shall always greatly value.[2] I should know it anywhere.

I did not know that the difficulty of getting into the R. Society was so great. There is no other scientific distinction except the F.R.S. that I should care to add to my name—but, I confess to have had always rather a longing for that but the sum paid by my brother always frightened me.[3]

From what you tell me about the number of Candidates on the list, I think I should hardly *live* to be elected for I am already 58.!![4]

I dont think it is impossible that I might be able to get leave to send you a skull of the Wroxeter ox.[5]

Whenever we have a meeting of the Council I will remember to moot the question.[6]

Yours very sincerely | Henry Johnson

DAR 168: 67

[1] CD's letter has not been found. Johnson had written to CD inquiring about the costs entailed in becoming a fellow of the Royal Society of London (see letter from Henry Johnson, 30 September 1862).

[2] See letter from Henry Johnson, 30 September 1862. CD probably sent Johnson a copy of the photograph taken by William Erasmus Darwin in 1861, which is reproduced as the frontispiece to *Correspondence* vol. 9 (see also letter to Daniel Oliver, [17 September 1862] and n. 8).

[3] Johnson's brother was George Henry Sacheverell Johnson, dean of Wells and vicar of St Cuthbert's, Wells (*DNB*, *Salopian Shreds and Patches* 5 (1882): 2). The fee for admission as a fellow of the Royal Society of London was £10; thereafter, the annual fee was £4. Alternatively, one could pay a 'composition' fee of £40 (*Record of the Royal Society of London*, p. 100).

[4] Johnson did not become a fellow of the Royal Society of London (*Record of the Royal Society of London*, pp. 517–66).

[5] The reference has not been traced; however, among the remains of extinct animal species found at Wroxeter were 'crania of the *Bos longifrons*, one of which, now in the Museum [of the Shropshire Natural History and Antiquarian Society], bears . . . the mark of the blow of the butcher's axe by which it was slaughtered', and also bones of another species of ox which was thought to be extinct (Wright 1872, p. 321). The reference is to the museum of the Shropshire Natural History Society, under whose auspices the remains of the Roman city of Uriconium, at Wroxeter in Shropshire, were being excavated. CD did not refer to the Wroxeter finds in his account of the history of the different species of cattle in *Variation* 1: 79–93.

[6] Johnson was honorary secretary of the Shropshire Natural History Society, he had also been appointed to the committee responsible for the excavation of the remains at Wroxeter (Wright 1872, p. iii).

From T. H. Huxley 9 October 1862

26 Abbey Place
Oct. 9th. 1862

My dear Darwin

It is a source of much pleasure to me to learn that anything I can say or do is a pleasure to you and I was therefore very glad to get your letter at that whirligig of an Association meeting the other day—[1] We all missed you but I think it was as well you did not come—for though I am pretty tough, as you know, I found the pace rather killing— Nothing could exceed the hospitality & kindness of the University people—and that, together with a great deal of speaking, on the top of a very bad cold which I continued to catch, just before going down—has somewhat used me up

Owen[2] came down with the obvious intention of attacking me on all points— Each of his papers was an attack & he went so far as to offer stupid & unnecessary opposition to proposals of mine in my own Committee—[3] However—he got himself sold at all points— Not a soul seconded him in the Committee & how the mendacious audacity of the man was shewn up at the discussion—is printed in the 'Times'—[4]The Polypterous paper & Aye-Aye paper fell flat—[5] The latter was meant to raise a discussion on your views—but it was all a stale hash and I only made some half sarcastic remarks which stopped any further attempt at discussion—[6]

All the people present who could judge saw that Owen was lying & shuffling—: the other half saw he was getting the worst of it but regarded him I think, rather as an innocent old sheep, being worried by three particularly active young wolves—[7] He rolled his eyes about & smiled so sweetly every time the teeth set sharp into his weasand![8]

My wife & belongings are at Felixtow— I was so alarmed about my wife that I came back rather hastily from Scotland—[9] I am glad to say however, that she is very much better— I think she had weakened herself by over-working & the grief which September always opens afresh[10]

I trust you are getting better & that Mrs Darwin & the children are all well again[11]

I took my book to Scotland but did nothing I shall ask leave to send you a bit or two as I get on[12]

Ever | Yours | T H Huxley

A "Society for the propagation of common honesty in all parts of the world" was established at Cambridge— I want you to belong to it, but I will say more about it by & bye[13]

DAR 166.2: 294

[1] CD's letter has not been found. In 1862, the British Association for the Advancement of Science held its annual meeting in Cambridge from 1 to 8 October. CD may have written to thank Huxley for the praise he had accorded CD in his inaugural address as president of the sectional committee for 'zoology and botany, including physiology' (Section D), as reported in *The Times*, 3 October 1862, p. 5. Speaking on the condition and prospects of biological science, Huxley had 'passed a graceful encomium on the labours of Mr. Darwin, whose name was received with a burst of applause', and had 'emphatically affirmed that Mr. Darwin's work was as perfect in its logical method as it was accurate in its scientific facts.'

[2] Richard Owen.

[3] See n. 1, above.

[4] Huxley refers to the discussion of Owen's paper, 'On the zoological significance of the cerebral and pedial characters of man' (R. Owen 1862c), which was read in Section D of the British Association meeting in Cambridge on 3 October 1862. The substance of the paper and the ensuing discussion were

reported in *The Times*, 4 October 1862, p. 7. Owen's paper was a defence of his view, first enunciated in 1858, that certain anatomical characters of the human brain required that the species be classified in a sub-class separate from the anthropoid apes. Since its publication, Owen's classification, and the cerebral anatomy on which it was based, had been repeatedly attacked, particularly by Huxley, who saw it as an opportunity to impugn both Owen's scientific reputation and his honesty (see Gross 1993 and Rupke 1994). In the discussion following the reading of Owen's paper in Cambridge, Huxley had been Owen's most vocal opponent, disputing both Owen's 'facts' and his 'reasoning'.

5 R. Owen 1862d and 1862e.

6 Owen read his paper, 'On the characters of the Aye-aye, as a test of the Lamarckian and Darwinian hypotheses of the transmutation and origin of species' (R. Owen 1862e), before Section D of the British Association for the Advancement of Science on 3 October 1862 (*The Times*, 4 October 1862, p. 7). His argument was that the remarkable adaptation of parts of the Aye-aye, and their correlation, could not be explained by natural selection or the Lamarckian theory of evolution, but depended on a guiding intelligence. Owen had previously affirmed this view in a paper read before the Zoological Society of London in January 1862 (R. Owen 1862a; see letter from J. E. Gray, 29 January 1862).

7 According to the report in *The Times*, 4 October 1862, p. 7, Owen's paper on the human brain was criticised chiefly by Huxley, George Rolleston, and William Henry Flower, all of whom had previously attacked Owen's views on this subject (see Gross 1993). *The Times* reported that Rolleston had concluded by saying that:

> if he had expressed himself with any unnecessary vehemence he was sorry for it, but he felt there were things less excusable than vehemence, and that the laws of ethics and love of truth were things higher and better than were the rules of etiquette or decorous reticence.

8 Weasand: 'the throat generally' (*OED*).

9 Huxley refers to Henrietta Anne Huxley. Huxley was in Scotland from 8 August to 16 September 1862 in his capacity as a member of the Royal Commission on the Operation of the Acts relating to Trawling for Herring on the Coasts of Scotland (L. Huxley ed. 1900, 1: 234).

10 The Huxleys' first child, Noel Huxley, died from scarlet fever on 15 September 1860 (L. Huxley ed. 1900, 1: 151–2 and 216).

11 Horace Darwin had been seriously ill at the beginning of 1862. Emma and Leonard Darwin had both become ill with scarlet fever during the summer (see 'Journal' (Appendix II)).

12 T. H. Huxley 1863.

13 An account of the formation of this association was given by the zoologist Alfred Newton, in a letter to his brother Edward Newton, dated 8 October 1862 (A. F. R. Wollaston 1921, pp. 123–4):

> I had meant to have an "Ibis" dinner, but the last was the only evening we could have it, and then a lot of others wanted to dine together, so it ended in establishing a new "Club for Promoting Common Honesty" and we had a feed at the "Lion" under the presidency of Huxley, with Kingsley as *vice*. . . . This club, I believe, was founded with one rule only, and that was that any one drinking Sclater's health was to be expelled (this was Sclater's stipulation in his nervous juxta-matrimonial state, and the only condition under which he would allow the dinner to take place), so that as soon as Sclater left, which he did early, I proposed his health and every one drank it; where by it is difficult to say whether the association did not thereupon dissolve itself!

The references are to Charles Kingsley and Philip Lutley Sclater, editor of *Ibis*, the British Ornithologists' Union journal. Sclater married Jane Anne Eliza Hunter-Blair on 16 October 1862 (*Gentleman's Magazine* n.s. 13 (1862): 630).

To *Natural History Review* [before 10 October 1862][1]

In my work on Cirripedes I have described an orifice, previously unobserved, beneath the first pair of cirri, on each side of the body, including a very singular elastic sack, which I considered to be an acoustic organ. Furthermore I traced the oviduct from the peduncle to a mass of glands at the back of the mouth, and these glands I called ovarian.[2] Dr. Krohn has recently stated that these glands are salivary, and that the oviduct runs down to the orifice, which I had thought to be the auditory meatus.[3] It is not easy to imagine a greater mistake with respect to function than that made by me; but I expressly stated that I could never succeed in tracing the oviducts into actual union with these glands; nor the supposed nerve from the so-called acoustic sack to any ganglion.[4] As Dr. Krohn is no doubt a much better dissector than I am, I fully admitted my error and still suppose that he is right. Nevertheless, several facts can hardly be reconciled with his view of the function of the several parts. To give one instance: if any one will look at the figure of the Anelasma (Lepadidae, Pl. iv.),[5] he will see how extremely difficult it is to understand by what means the ova coming out of the orifices (e) above referred to, could be arranged in the symmetrical lamellae which extend up to the summit of the capitulum: it must be observed that the ova are united together by a delicate membrane enclosing each ovum; moreover the cirri in this animal are in atrophied condition, without regular articulations, so that it is inconceivable how the ova can be transported and arranged by their agency.

I have lately received from an eminent naturalist, Prof. F. de Filippi, a paper (Estratto dall' Arch. per la Zoolog. 31st Dec. 1861), chiefly devoted to the development of the ova of Cirripedes, in which the following passage occurs:—[6]

"The small size of *Dichelaspis Darwinii* has not enabled me to verify the relationship discovered by Krohn between this problematical organ and the termination of the oviduct; but on the other hand the transparency of the tissues has enabled me to perceive a peculiarity of structure which may help to elucidate the question. Fig. 13 represents what I persist in calling a hearing organ. Within a cavity, the walls of which are united to the surrounding tissues, there is a pear-formed sack or ampulla. On the neck of this ampulla, at *a*, are numerous minute lines parallel to each other and to the axis of the ampulla. I doubted at first whether the appearance of these lines arose from folds in the membrane, and therefore I separated some of the sacks, and I could then better convince myself that these lines correspond with true nervous fibres, thin and simple, embedded in the rather thick, resisting, and transparent substance which forms the walls of the ampulla. This circumstance seems to me to show clearly the sensitive nature of the organ, and hence to favour Darwin's opinion, who considers them to be organs of hearing."

My object in asking you to publish this note, is to induce some one to attend to this curious organ; to endeavour to discover ova within the so-called auditory sack; for as each cirripede produces so many eggs, assuredly this might be effected without great difficulty. It is, however, possible (as I believe was suggested by Mr. R. Garner at the British Association, but whose paper I have mislaid,) that cirripedes, like certain Entomostraca, may lay two kinds of eggs;[7] one set passing out through the problematical orifices; and another set coming out of the body in sheets, in the manner suggested by me;—namely, the ova collecting under the lining membrane of the sack before the act of exuviation, with a new membrane formed beneath them; so that the layer of eggs becomes external after the act of exuviation.[8] If this view, to which I was led by many appearances, be correct, improbable as it may seem, it ought not to be difficult to find a specimen with the old membrane of the sack loose and ready to be moulted, with the new underlying membrane almost perfect, and with the layer of ova between them. Or a specimen might be found which had lately moulted, with its skin still soft, (and this I believe that I saw) with a layer of eggs still loosely attached to the new lining membrane of the sack.

Natural History Review n.s. 3 (1863): 115–16

[1] The date is established by the relationship to the letter from T. H. Huxley, 10 October [1862]. CD's letter was published in the January 1863 number of the *Natural History Review* ('"Auditory-sac" of Cirripedes' (*Collected papers* 2: 85–7)).
[2] *Living Cirripedia* (1851), pp. 53–5, 57–8.
[3] Krohn 1859.
[4] In *Living Cirripedia* (1851), p. 57, CD stated: 'I was not able to ascertain whether the two main ducts, coming from the peduncle, expanded to envelope them, or what the precise connection was'. He also reported, with regard to the supposed auditory nerve, that he had 'not yet succeeded in tracing this nerve' (*ibid*., p. 54); CD noted that the salivary function of these glands had been asserted previously by Georges Cuvier (*ibid*., p. 57).
[5] *Living Cirripedia* (1851).
[6] The reference is to the Italian zoologist Filippo De Filippi; the passage is translated from De Filippi 1861, p. 205. There is a presentation copy of this paper in the Darwin Pamphlet Collection–CUL, together with a manuscript translation, annotated by CD, of the section quoted here.
[7] Garner 1861. Robert Garner read a paper entitled 'On the structure of the *Lepadidae*' at the 1860 meeting of the British Association for the Advancement of Science, held in Oxford (Garner 1861). However, in the report of the paper, no reference was made to the production of two kinds of eggs.
[8] CD's lapse of memory (see n. 7, above) may be explained by the fact that Garner referred not only to ova passing out through the disputed acoustic apertures, but also to 'broad sheets' of ova, as described by CD in this letter. Unlike CD, Garner attributed the existence of such sheets to the ovipository action of a 'so-called proboscis', which arranged the ova into sheets after their emergence from the supposed acoustic apertures.

From T. H. Huxley 10 October [1862][1]

My dear Darwin

'The smallest contribution thankfully received' from you—

You shall receive a proof in due time[2]

Ever | Yours truly | T H Huxley

Jermyn St
Octr. 10th

I can't find anything to talk to the working men about this year but your book—I mean to give them a commentary à la Coke upon Lyttleton.—[3]

DAR 166.2: 295

[1] The year is established by the reference to T. H. Huxley 1863b (see n. 3, below).
[2] The reference is apparently to CD's notice, '"Auditory-sac" of Cirripedes', which was published in the January 1863 number of the *Natural History Review* (see preceding letter). Huxley was editor-in-chief of the *Natural History Review.*
[3] The Government School of Mines in Jermyn Street, London, at which Huxley was professor of natural history, ran a popular series of lectures for artisans every winter. In 1862, Huxley's lecture series was entitled 'On our knowledge of the causes of the phenomena of organic nature' (T. H. Huxley 1863b; see Bibby 1959, pp. 97–100 and Montagu ed. 1968, pp. vi–vii). The reference is to the commentary by Edward Coke (Coke 1628) on Thomas Littleton's fifteenth-century treatise on English law (Littleton *c.* 1481), commonly referred to as 'Coke upon Littleton' (B. E. Smith 1894, D. M. Walker 1980).

From J. D. Hooker [12 October 1862][1]

Kew
Sunday

Dr. Darwin

There is an article in Bull. Bot. Soc. France VIII. 519 by A. de Lassus on irritability of leaves of Aldrovanda a kinsman of Drosera.[2]

Thanks for your long letter on Linum &c &c.[3] I sent you last Thursday a box with two good species of Impatiens flowers, & can send a third if you care for it. I shall be most curious to know what you make of the Floral whorls & their vasc. bundles[4]

Cassia is another genus that has different anthers in same flower. I have also thought of some Comelyneæ but am not sure.[5] Oliver[6] has brought in a Tray of Drosera rotundifolia, but I doubt either of us having time to look at it.

My wife went to Cambridge & enjoyed it— I stayed at home! (& enjoyed it) working away at Welwitschia every day & almost every night—[7] I entirely agree with you by the way, that after long working at a subject, & after making something out of it, one invariably finds that it all seems dull flat stale and unprofitable—[8] this feeling however you will observe only comes (most mercifully) after you *really* have made out something worth knowing— I feel as if every body must know more of *Welwitchia* than I do, & yet I cannot but believe I have (ill or well) expounded & faithfully recorded a heap of the most curious facts regarding a simple plant that have been brought to light for many years.

The whole thing is however a dry record of singular structures, & sinks down to the level of the dullest descriptive account of dead matter, beside your jolly dancing facts anent orchid life & bee-life. I have looked at an Orchid or two since reading the Orchid book & feel that I never should have made out one of your points, even had I limitless leisure zeal & material— I am a dull dog, a very dull dog.— I may content myself with the per contra reflection that you could not (be dull enough to) write a "Genera plantarum", which is just about what I am best fitted for.—[9] I feel I have a call that way, & you the other.

The dismal fact you quote of hybrid transitions between *Verb. Thapsus* & *nigra* (or whichever two it was) & its bearing on my practice of lumping species through intermediate specimens, is a very horrible one; & would open my eyes to my own blindness if nothing else could.[10] I have long been prepared for such a case, though I once wrote much against its probability—[11] I feel tolerably sure I must have encountered many such, but have not the tact to discern them, when under my nose: & I hence feel as if all my vast experience in the field has been thrown away. Your Orchid book has pretty well convinced me that such cases must be abundant,[12] & they only tend further to disturb our ideas of physiological versus structural species.[13] Perhaps my intermediates between *Habenaria chlorantha* & *bifolia* (of which I retain a lively recollection) were of this hybrid nature.[14] Certain it is that I had only to look for Hybrid orchids at the in Switzerland to find *two* different sorts. & numerous specimens of one of them.[15]

Huxley seems to have made short work of Owen at Cambridge.[16] the latter H. says "trailed his coat"—![17] Otherwise the meeting seems to me to have been dull enough, but cheery & friendly as far as sociability goes

Ever Yours affec | J D Hooker

I don't think it can be worth while returning the Melastomas but will enquire.[18]

DAR 101: 59–60, 86

[1] The date is established by the relationship to the letter to J. D. Hooker, 6 October [1862], and by reference to a note in DAR 205.8: 4, dated 14 October 1862, which records some of the observations made in this letter (see n. 5, below); the intervening Sunday fell on 12 October.

[2] Augé de Lassus 1861. In his letter of 26 September [1862], CD had asked Hooker for his opinion concerning irritability in the insectivorous plant *Drosera.*

[3] Letter to J. D. Hooker, 6 October [1862].

[4] In his letter to Hooker of [18 September 1862], CD asked Hooker to send him '2 or 3 single Balsams', so that he could trace the vascular bundles as part of his study of the homologies of floral parts; 'balsam' is a common name for plants of the genus *Impatiens*.

[5] In his letter to Hooker of 22 [August 1862], CD requested information about species in which there were differently coloured sets of anthers or pollen. Hooker apparently refers to the Commelinaceae. There is a note in DAR 205.8: 4, dated 14 October 1862, which records Hooker's identification of two sets of stamens in *Cassia.*

[6] Daniel Oliver was librarian and assistant in the herbarium at the Royal Botanic Gardens, Kew.

[7] Hooker was preparing a monograph on the Angolan plant *Welwitschia mirabilis* (J. D. Hooker 1863a). In his letter of 16 September 1862, Hooker informed CD that he intended to display the plant at the 1862 meeting of the British Association for the Advancement of Science, held in Cambridge from 1 to 8 October, and that his wife, Frances Harriet Hooker, planned to accompany him.

[8] See letter to J. D. Hooker, 6 October [1862]. Hooker alludes to William Shakespeare's *Hamlet*, 1.2.133–4: 'How weary, stale, flat, and unprofitable Seem to me all the uses of this world!'

[9] For CD's concern at the amount of time Hooker spent in preparing *Genera plantarum* (Bentham and Hooker 1862–83), see the letter to J. D. Hooker, 9 February [1862], and the letter to Asa Gray, 21 April [1862].

[10] See letter to J. D. Hooker, 6 October [1862]. In a section of J. D. Hooker 1853, which is headed 'Species vary in a state of nature more than is usually supposed' (pp. xii–xvii), Hooker stated:

> The result of my observations is, that differences of habit, colour, hairiness, and outline of leaves, and minute characters drawn from other organs than those of reproduction, are generally fallacious as specific marks, as being attributable to external causes, and easily obliterated under cultivation. It has hence been my plan to group the individuals of a genus which I assume after

careful examination to contain many species whose limits I cannot define, that the species shall have the same relative value as those have of allied genera whose specific characters are evident . . . with all, the tendency would be to regard dubious species as varieties, to take enlarged views of the range and variation of species, and to weigh characters not only *per se*, but with reference to those which prevail in the Order to which the species under consideration belong.

Hooker employed this principle in his systematic work, and was inclined to combine in one species individuals previously considered to represent distinct species (see, for example, *Correspondence* vol. 5, letter from Asa Gray, 22 May 1855 and n.6).

[11] In Hooker 1853, p. xv, Hooker stated:

Hybridization has been supposed by many to be an important element in confusing and masking species. Nature, however, seems effectually to have guarded against its extensive operation and its effects in a natural state, and as a general rule the genera most easily hybridized in gardens, are not those in which the species present the greatest difficulties. With regard to the facility with which hybrids are produced, the prevalent ideas on the subject are extremely erroneous.

[12] See, for example, *Orchids*, pp. 72 and 88–9, where CD questioned the classification by some botanical authorities of *Ophrys apifera* and *O. arachnites*, and of *Habenaria bifolia* and *H. chlorantha*, as two varieties of one species, rather than as separate species.

[13] In his review of *Origin*, Thomas Henry Huxley distinguished between what he termed 'morphological' species, 'distinctly definable from all others, by certain constant . . . morphological peculiarities', and 'physiological' species ([T. H. Huxley] 1860, p. 543). He stated that the best way to distinguish between two true physiological species and two varieties was to attempt hybridisation: true species would either be infertile *inter se* or would produce infertile offspring, whereas two varieties would give rise to fertile progeny ([T. H. Huxley] 1860, pp. 552–3). However, CD had come to the conclusion, in view of the hybridisation experiments carried out on *Verbascum* by Karl Friedrich von Gärtner, and of his own crossing experiments with *Primula sinensis*, that 'the lessened fertility of the first union of the offspring of two forms is no sure criterion of specific distinctness' ('Illegitimate offspring of dimorphic and trimorphic plants', p. 436). See also Appendix VI, letter to T. H. Huxley, 14 [January 1862], and letter to Asa Gray, 9 August [1862] and n. 13.

[14] See n. 12, above. See also letter to J. D. Hooker, 19 [July 1860] and n. 4 (*Correspondence* vol. 8).

[15] See letter from J. D. Hooker, [24 July 1862].

[16] Hooker refers to Thomas Henry Huxley and Richard Owen, who had engaged in intense debate concerning the differences in brain anatomy between humans and anthropoid apes, at the 1862 meeting of the British Association, held in Cambridge from 1 to 8 October. For Huxley's account of the exchanges, see his letter to CD of 9 October 1862.

[17] The phrase is used to describe one who seeks an argument (*OED*).

[18] See letter to J. D. Hooker, 6 October [1862].

To George Bentham 13 October [1862][1]

Down. | *Bromley.* | *Kent. S.E.*
Oct 13th

Dear Bentham

You sent me some time ago Hist. notes on Cultivated Plants by Targioni-Tozzetti, & which I wish occasionally to quote. I hate quoting an anonymous work. Would you have the kindness to allow me to add your name,—that is of course if you have no objection. Otherwise I must say a "Writer".—[2]

If not much trouble could you tell me the *volume* & *year* & *first page* of Hort. Journal, in which it appears, as none of these very necessary elements appear.—

Pray forgive me troubling you & believe me | Yours very sincerely | Ch. Darwin

Royal Botanic Gardens, Kew (Bentham letters: 715)

[1] The year is established by the relationship to the letter from George Bentham, 15 October 1862.
[2] There is an annotated presentation copy of Bentham's anonymous synoptical review of Targioni Tozzetti 1853 ([Bentham] 1855) in the Darwin Pamphlet Collection–CUL. [Bentham] 1855 is frequently cited in chapters 9 and 10 of *Variation*; CD's 'Journal' (Appendix II) records that he began to work on the section of *Variation* dealing with 'Facts of variation of Plants', on 7 October 1862.

To Daniel Oliver 13 October [1862][1]

Down. | *Bromley.* | *Kent. S.E.*
Oct. 13th

Dear Oliver

You must be a clair-voyant or something of that kind to have sent me such useful plants. Twenty-five years ago I described in my Father's garden 2 forms of L. flavum (thinking it case of mere variation);[2] from that day to this I have several times looked, but never saw the 2^{d} form till it arrived from Kew.— Virtue is never its own reward: I took paper this summer to write to you to ask you to send me flowers, that I might beg plants of this Linum, if you had the other form, & refrained from not wishing to trouble you.[3] But I am now sorry I did; for I have hardly any doubt that L. flavum never seeds in any garden that I have seen, because one form alone is cultivated by slips.—

Secondly I raised a lot of plants from Kew seed marked "L. Austriacum", but certainly different from the flowers you sent, which no doubt, as Lecoq says,[4] are dimorphic. The Kew seedlings 112 in number were not dimorphic & were all self-fertile & I strongly suspect were L. usitatissimum: now it would be of great use to me to know whether you keep seed at Kew of any other *blue* Linums,, besides L. perenne, Austriacum & usitatissimum; for my plants were *not* the two former, & I could thus perhaps know what they were.—[5]

Many thanks for your caution about stigmas of long-styled L. perenne: I knew & think I said that position was due to twisting of style; but I will look more carefully at what period twisting takes place.—[6]

You sent me some most curious plants: what in name of Heaven is name of the genus of which you sent me 3 species, & especially of pinkish-purple flower with long thin nectary: I suppose you sent them from seeing their relation like in orchids to visits of insects.[7] If a greenhouse plant I will get this plant.—

I certainly thought that you were author of the excellent review of the orchid book, especially from passage about unisexuality in high plants.[8] The Reviewer (is it a secret, who?) I daresay is quite right about arrangement of Book; but I hardly know with my materials that I could have made it better.[9] I wish he would criticize the last chapter; I sh^{d} like to hear what a good hand would say to it.—[10]

I have not yet had time to examine Lythrum flowers; most cordial thanks for them—[11] | Yours very sincerely | C. Darwin[12]

DAR 261 (DH/MS 10: 66)

[1] The year is established by the reference to [J. D. Hooker] 1862d (see nn. 8–10, below).

[2] There is a note in CD's 'Questions & Experiments' notebook, under the heading '*Shrewsbury*', which states 'Linum flavum *put in spirits [*interl*] which plant seeds?' (DAR 206: 14; see also *Notebooks*, p. 505); this notebook was probably opened in mid-1839 (*Notebooks*, p. 487). CD's father, Robert Waring Darwin, had resided at The Mount, Shrewsbury. CD had referred to this case in 'Dimorphic condition in *Primula*', p. 96 (*Collected papers* 2: 62–3).

[3] Having discovered in 1861 that certain species of *Linum* were dimorphic, CD carried out crossing experiments in the summer of 1862 on *L. perenne* and *L. grandiflorum*; the results from these experiments are given in 'Two forms in species of *Linum*', which was read before the Linnean Society of London on 5 February 1863. Oliver had probably either read or heard about CD's letter to J. D. Hooker, 6 October [1862] (see nn. 6, 8, and 11, below), in which CD referred to his research on *Linum*. CD described *L. flavum* in 'Two forms in species of *Linum*', p. 81 (*Collected papers* 2: 104), noting: 'I have not been able to try any experiments on this species'.

[4] Lecoq 1854–8, 5: 325. There is a heavily annotated copy of this work in the Darwin Library–CUL (see *Marginalia* 1: 488–95).

[5] In 'Two forms in species of *Linum*', p. 82 (*Collected papers* 2: 104), CD noted that not all species of *Linum* were dimorphic, and described his experiments on plants raised from seed sent from Kew, 'incorrectly named *L. Austriacum*'; he did not, however, identify the species.

[6] See letter to J. D. Hooker, 6 October [1862] and n. 12. No letter from Oliver discussing this point has been found.

[7] CD apparently refers to the flowers of two species of *Impatiens* that had in fact been sent at his own request by Joseph Dalton Hooker, together with some flowers of a species of Loasaceae (see letters from J. D. Hooker, [12 October 1862] and [18 October 1862]). See also letter to J. D. Hooker, 14 [October 1862].

[8] In his letter to Oliver of 24 July [1862], CD had suggested that Oliver should write a review of *Orchids* for the *Natural History Review* (see also letter from Oliver, 28 July 1862); CD had subsequently asked Hooker to thank Oliver 'for his clear & favourable notice' (see letter to J. D. Hooker, 6 October [1862]). However, it was Hooker who wrote the review ([J. D. Hooker] 1862d; see letter from J. D. Hooker, 7 November 1862). The review began with a statement that one of the 'prominent differences between the animal and plant kingdoms' was that 'whereas unisexuality is the rule amongst the highest orders of animals, and hermaphroditism becomes more frequent as we descend in the scale, the contrary is the case with plants' (p. 371). CD's work, Hooker continued, had shown that 'amongst many plants apparent and real hermaphroditism are totally different things', and that 'before reasoning further on the subject, we must begin again not only to observe, but also to experiment.'

[9] In his review of *Orchids* ([J. D. Hooker] 1862d, p. 373), Hooker argued that the book suffered from a lack of concise headings and from a degree of arbitrariness in the division of the chapters; it was, he stated, as if 'the conception of putting forth the treatise as a separate work was an after-thought'. He also noted that it was an arrangement that would 'not recommend itself to the general reader, who thus loses sight of the grand divisions of the Order as well as of the subject.'

[10] In chapter 7 of *Orchids*, CD discussed the homologies of orchid flowers, and concluded with remarks on natural selection and on the importance of cross-fertilisation (*Orchids*, pp. 286–360). Hooker concluded his review ([J. D. Hooker 1862d], p. 376) by noting that he had preferred giving an extended synopsis of CD's observations on British orchids to 'reviewing the very extensive and intricate chapters devoted to foreign Orchids, the homologies of Orchid flowers, and general considerations'; this, Hooker stated, was because CD's observations on British orchids could be 'repeated by any observer and extended by many', and because such a procedure gave 'a better idea of the completeness of the work', in spite of the fact that the 'other chapters' were 'by far the more interesting and important'.

[11] In the letter to J. D. Hooker, 6 October [1862], CD had asked Hooker for dried flowers of *Lythrum Graefferi* and *L. thymifolia*, if they were readily available in the herbarium at the Royal Botanic Gardens, Kew; Oliver was an assistant in the herbarium. In 'Three forms of *Lythrum salicaria*', p. 190 (*Collected papers* 2: 124), CD referred to his having seen two dried flowers of *L. thymifolia* from Kew.

[12] No reply to this letter has been found. However, on the reverse of the letter Oliver noted: 'Seed of *Linums* dib. | L. hyssopifolia *dib.* **if riper**'. In his letter to CD of 13 September 1862, Oliver reported that he had 'given directions about saving seed of *Lythrum hyssopifolium*'.

From B. J. Sulivan 13 October [1862][1]

Board of Trade. S.W.
Oct 13

My dear Darwin

I was very sorry to hear of your being so unwell.[2] You must *not think* of our coming to see you unless quite well enough.[3] It must also depend on Wickham[4] remaining *next* week in Town. He comes up today with his family having gone down from us on Thursday—& may have to go to France on Saturday. One day this week he has to be at Portsmouth so that unless he remains longer in London he cannot manage one day having to prepare for their trip to Continent where he expects to remain two or three years.

If he should remain another week and *you are well enough* we might be able to go to you about Wednesday 22nd for one night—[5]

with best regards | Believe me very sincly Yours | B. J. Sulivan

DAR 177: 277

[1] The year is established by reference to the visit to Down House on 21 October 1862, of Sulivan, John Clements Wickham, and Arthur Mellersh (see n. 5, below).
[2] CD's letter has not been found.
[3] Sulivan had suggested a reunion in London between CD and two other former officers of HMS *Beagle*; instead, CD had invited them to visit Down House (see letters from B. J. Sulivan, 27 September [1862] and 2 October [1862]).
[4] John Clements Wickham.
[5] Emma Darwin's diary (DAR 242) records that Sulivan, Wickham, and Arthur Mellersh visited Down House on 21 October 1862.

To J. D. Hooker 14 [October 1862]

Down Bromley Kent
14th

My dear Hooker

Your letter is a mine of wealth.[1] But first I must scold you: I cannot abide to hear you abuse yourself, even in joke, & call yourself a stupid dog. You in fact thus abuse me; because for long years I have looked up to you as the man whose opinion I have valued more on any scientific subject than any one else in the world. I continually marvel at what you know & at what you do. I have been looking at the Genera, & of course cannot judge at all of its real value; but I can judge of amount of condensed facts under each family & genus.—[2]

I am glad you know my feeling of not being able to judge about one's own work; but I suspect that you have been overworking. I shd. think you could not give too much time to Wellwitchia (I spell it different every time I write it); at least I am sure in animal kingdom monographs cannot be too long on the osculant groups.—[3]

Hereafter I shall be excessively glad to read paper about Aldrovanda; & am very much obliged for reference.[4] It is pretty to see how the caught flies support Drosera, where nothing else can live. I answer your Query on separate slip.[5]

Thanks about plant with 2 kinds of anthers. I presume (if an included flower was a Cassia) that Cassia is like Lupines but with some stamens still more rudimentary.—[6] If I hear I will return the 3 Melastomateds; I do not want them & indeed have cuttings; I am very low about them, & have wasted *enormous* labour over them & cannot yet get a glimpse of the meaning of the parts.[7]

I wish I knew any Botanical collector, to whom I could apply for seeds *in the native land* for any Heterocentron or Monchætum: I have raised plenty of seedlings from your plants; but, I find in other cases that from a *homomorphic* union, one generally gets solely the parent form.—[8] Do you chance to know of any Botanical collector in Mexico or Peru?

Here is a pretty job: I thought Oliver[9] had sent me the flowers of Impatiens, as they are so beautifully adapted for insect fertilisation:[10] I did not guess that they were Impatiens & after looking at them threw them away! But anyhow I must not now indulge myself with looking after vessels & homologies. Some future time I will indulge myself. By the way sometime I want to talk over the alternation of organs in flowers with you; for I think I must have quite misunderstood you that it was not explicable.[11]

I found out the Verbascum case by pure accident, having transplanted one for experiment, & finding it to my astonishment utterly sterile.[12] I formerly thought with you about rarity of natural hybrids;[13] but I am beginning to change, viz Oxlips (not quite proven),[14] Verbascum,—Cistus (not quite proven)[15] Ægilops triticoides (beautifully shown by Godron)[16] Weddell[17] & your orchids,[18] & I daresay many others recorded.—

Your letters are one of my greatest pleasures in life, but I earnestly beg you *never* to write, unless you feel somewhat inclined; for I know how hard you work. As I work only in morning, it is different with me & is only a pleasant relaxation. You will never know how much I owe to you for your constant kindness & encouragement.

Yours affectionately | C. Darwin

Incomplete[19]
Endorsement: 'Oct/62'
DAR 115.2: 166

[1] Letter from J. D. Hooker, [12 October 1862].

[2] The first part of the first volume of *Genera plantarum* (Bentham and Hooker 1862–83) was published on 7 August 1862 (Stearn 1956, p. 130); CD's lightly annotated copy of this work is in the Darwin Library–CUL (see *Marginalia* 1: 51–2).

[3] Hooker was preparing a monograph on the Angolan plant *Welwitschia mirabilis* (J. D. Hooker 1863a). In a letter to Hooker that is now missing, CD had compared Hooker's work on *Welwitschia* to his own

extensive work on barnacles (L. Huxley ed. 1918, 2: 24); his comment here may have been intended to reiterate that comparison.

4 Augé de Lassus 1861 is cited in *Insectivorous plants*, p. 321. See letter from J. D. Hooker, [12 October 1862] and n. 2.

5 CD's note has not been found; it apparently addressed a query raised by Hooker in response to CD's letter of 26 September [1862], in which CD gave an account of his experiments on the insectivorous plant *Drosera rotundifolia*. Hooker's letter has not been found, but see the letter to J. D. Hooker, 6 October [1862].

6 See letter from J. D. Hooker, [12 October 1862] and n. 5. In a note dated 14 October 1862 (DAR 205.8: 4), CD recorded: 'Hooker says Cassia (Leguminosae) has different anthers. ['*I think*' *del*] I saw one like Lupine *but more rudimentary [*interl*], & believe some of Comelyneae here.—' CD had learned from Vaucher 1841, 2: 213, that lupines exhibit two differently coloured sets of anthers; there are observational notes on the different sets of anthers in *Lupinus nanus*, dated 5 August 1862, in DAR 76: 92, and further notes in DAR 48: 49 v.

7 See letter from J. D. Hooker, [12 October 1862] and n. 18. In October 1861, CD had begun to investigate the occurrence of two different sets of stamens in the flowers of the Melastomataceae, the structure and colour of the stamens facing the petals in many species differing from that of those facing the sepals (see *Correspondence* vol. 9, letter to J. D. Hooker, 17 November [1861] and n. 14). He suspected that the Melastomataceae might exhibit a novel form of dimorphism, and continued to work on the family throughout 1862 and 1863 without ultimately being able to account for the two sets of stamens (see *Cross and self fertilisation*, p. 298 n., and *ML* 2: 292–302). CD's notes from these experiments are in DAR 205.8.

8 CD had observed this phenomenon in his crossing experiments performed in 1861 and 1862 with the dimorphic *Primula sinensis* (see 'Illegitimate offspring of dimorphic and trimorphic plants', pp. 413–4).

9 Daniel Oliver.

10 Hooker had sent CD the flowers at his own request, so that he might trace their vascular bundles as part of his study of the homologies of plant parts (see letter from J. D. Hooker, [12 October 1862] and n. 4, and letter to Daniel Oliver, 13 October [1862] and n. 7).

11 See letter to J. D. Hooker, 11 September [1862] and n. 1; see also letter from J. D. Hooker, 16 September 1862, and letter to J. D. Hooker, [18 September 1862].

12 See letter from J. D. Hooker [12 October 1862], and letter to J. D. Hooker, 6 October [1862] and n. 14.

13 See letter from J. D. Hooker, [12 October 1862] and n. 11.

14 CD was keen to test the commonly held view that oxlips were the hybrid offspring of primroses and cowslips. He initially maintained that the various forms were varieties descended from a common parent (*Natural selection*, pp. 128–33, and *Origin*, pp. 49–50), but as a result of his work on dimorphism in *Primula*, he came to distrust the experimental evidence against the occurrence of hybridisation, stating that further experiments were 'absolutely necessary' ('Dimorphic condition in *Primula*', pp. 93–4; see also *Collected papers* 2: 60–1); CD carried out such experiments between 1862 and 1867 (see the experimental notes in DAR 108 and DAR 157a), as a result of which he concluded that the common cowslip, the primrose, and the Bardfield oxlip were distinct species, but that the common oxlip was a hybrid between the cowslip and the primrose ('Specific difference in *Primula*').

15 Between 7 October and 11 December 1862, CD wrote a draft section of *Variation* in which he discussed 'Facts of variation of Plants', appearing in the published form as chapters 9 and 10 (*Variation* 1: 305–72; see 'Journal' (Appendix II)). In *Variation* 1: 336, CD noted that 'undoubted species of Verbascum, Cistus, Primula, Salix, &c., frequently cross in a state of nature.'

16 Godron 1859, 2: 168. CD's annotated copy of this work is in the Darwin Library–CUL (see *Marginalia*, 1: 331–5). In *Variation* 1: 313 n., CD referred to Dominique Alexandre Godron's 'careful experiments' showing that *Ægilops triticoides* was 'a hybrid between wheat and *Æ. ovata*'.

17 Weddell 1852. CD cited Hugh Algernon Weddell's account of 'naturally produced' hybrids between *Aceras anthropomorpha* and *Orchis galeata* in *Orchids*, p. 19 n.

[18] Hooker had described his observation of naturally occurring orchid hybrids in his letter to CD of [24 July 1862].
[19] See n. 5, above.

To Charles Lyell 14 October [1862][1]

Down. | Bromley. | Kent. S.E.
Oct 14th

My dear Lyell

I return Jamieson's capital letter.[2] I have no comments, except to say that he has removed all my difficulties & that now & for ever more I give up & abominate Glen Roy & all its belongings.— It certainly is a splendid case & wonderful monument of the old Ice Period.—[3] You ought to give a woodcut.—[4] How many have blundered over those horrid shelves![5]

That was a capital paper by Jamieson in last G. Journal:[6] I was never before fully convinced of the land glacialisation of Scotland before; though Chambers tried hard to convince me.—[7]

I must say I differ rather about Ramsay's paper;[8] perhaps he pushes it too far. It struck me the more from remembering some years ago marvelling what cd be the meaning of such a multitude of Lakes in Finland & other northern districts. Ramsay wrote to me & I suggested that he ought to compare mountainous Tropical regions with Northern regions.— I could not remember many lakes in any mountainous Tropical country.[9] When Tyndall talks of every valley in Switzerland being formed by glacier, he seems to forget there are valleys in the Tropics;[10] & it is monstrous in my opinion the accounting for glacial period in Alps by greater height of mountains, & their lessened heights, if I understand, by glacial erosion.—[11] Ne sutor ultra crepidam, I think applies in this case to him.—[12] I am hard at work on "Variation under Domestication".[13]

Ever yours | C. Darwin

P.S. I am rather overwhelmed with letters at present, & it has just occurred to me that perhaps you will forward my note to Mr Jamieson; as it will show that I entirely yield. I do believe every word in my Glen Roy paper is false—

[Enclosure][14]

Ellon, Aberdeensh.
15, Aug. 1862

My Dear Sir Charles,

I have been again at Lochaber this summer, and have come away more convinced than ever of my last years views of the Glen Roy Lakes. I made a careful measurement with a good spirit level of the height of the lowest line near the mouth of Glen Roy and make it 854 feet above the mean sea level—starting from an advance mark at Bridge of Roy. Another measurement I made about six miles east—up Glen Spean—indicates a *rise* of about a foot in the mile fully in the line as you go east. Circumstances prevented me getting to the E. end of L. Laggan, but I hope to get there at some future time— meanwhile my recollection of that locality leads me to think that this slight elevation to Eastward will be borne out there also—

The Glen Arkaig glacier has been the main agent in blocking up Glen Gluoy & the mouth of Glen Spean but has been aided in the latter case by the neighbouring ice streams in particular by one from the longest corry on the N. flank of the Ben Nevis ridge called Corry N. Youn.[15]

I also discovered the most beautiful set of moraines that I believe will be found in the United Kingdom.!

Good instances of Moraines have been always a desiratum in this country but I think these will satisfy everybody that sees them. They are the work of the old Glen Treig glacier but at such a distance from the mouth of the glen as to have eluded the observation of former visitors. I noticed part of them from the shoulder of a hill last year & it was one of my objects this season to revisit them— They have a sweep of several miles, forming a sort of huge semi circle—tier within tier—in many places almost as regular as a railway embankment— the inner one is composed in many parts *wholly* of large blocks of syenite, often from 10 to 15 feet in length, without the intermixture of earthy matter, which makes it very striking— these have been pushed by the glacier before it in certain places off the syenitic mass *up hill* & left *upon the gneiss* on the N. side of Glen Spean— The retreat of the ice has left a smooth space like a bay within these moraines almost free of blocks—

These moraines are as fine a sight as the Glen Roy lines themselves & to a geologist— as interesting. their fine preservation is certainly a treat.

—Another object of my visit as you are aware was to ascertain the relation of the Glen Roy lakes to the great submergence— with this view I examined the lower part of Glen Nevis which as it opens out upon an arm of the sea I thought most suitable to throw light upon the question— The result was that nothing appeared to indicate that Glen Nevis had been occupied by the sea since the glacier left it— The moranic hillocks even at its mouth seem to have remained undisturbed by any marine action, with this exception, that the great 40 foot terrace of the West Coast can be *distinctly traced up to the mouth of the glen* and apparently extending *across* it up to Bannavie[16] & fringing the bottom of the moranic heaps with a flat surface of shingle— This would seem to place the Glen Roy glacier lake period after the great submergence and before the era of the 40 foot old coast line of Argyleshire— This was what I expected— But I don't know how to reconcile it with Chamber's account of the Glen Iorsa moraine in Arran being slung over upon the old beach for I should think the Glen Nevis glacier would have protruded longer than most others in Scotland[17]

—Half a mile below Fort William I discovered a famous section of this old 40 foot terrace or ancient coast line— its height at the mouth of Glen Nevis is 40— or 43 feet above the mean sea-level— this section contained a bed of shells—some of which are *arctic*— an hour or twos search yielded me the following;

Modiola modiolus—	abundant
Astarte elliptica—	numerous
" compressa	4 specimens
Pecten opercularis	several
Nucula.	1. app. N. nucleus–
Pilidium fulvum.	a few
Anomia ephippium	several
Puncturella noachina	2.
Trophon clathratus	numerous
" scalariforme.	1. imperfect
Lacuna vincta.	numerous
Littorina expansa	several
(Lochus expansus, Brown & Smith)	
" littorea	1. large specimen
" rudis	5. some of these perhaps young of L. littorea
Natica clausa	8. or 9.
Trochus cinerarius	several
" tumidus	2. or 3.
Buccinum undatum.	a few broken
Bela trevelliana.	1. specimen
Bela. undetermined sp.	7.
Rissoa striata.	2.
Chiton.	fragment of a large species
Echinus.	some plates & spines

The prevalence of Littorinæ & Lacuna are good evidence of a coast or shore line— the shells seemed to have been mostly all dead ones, several of them are pierced by a boring mollusk—. all the species range into shallow water except the Puncturella, Pilidium, & Nucula, which are in small numbers and had been drifted— we often get such thrown up on our coast here—

The prevailing littorina was as you see the Arctic variety of L. littoralis— and all the naticus were N. *clausa*.

This would seem to shew that the arctic shells lingered in our seas till a later period than has been hitherto supposed.

In 1860 I found at Otter ferry at Loch Fyne just a little above high water mark a sandy clay with shells which I am inclined to think may also belong to this 40 feet old coast line as the group was so similar to those at Fort William— & I half suspect the shelly bed in the Kyles of Bute may also belong to it & also those got at Loch Lomond wh. would embrace many of Jordanhill's arctic clyde species—[18] The species I got at Otter were.

Tellina proxima—	not of large size—several
Astarte compressa—	1. complete

Cardium.	small species
Saxicava arctica..	1. valve of large variety
Mya truncata.	several
Pildium fulvum.	3.
Littorina littoralis.	2. or 3.
" rudis.	1. or 2
" littorea.	1. or 2. small
Trochus cinerarius.	1. small
Margantia undulata.	1. damaged
Trophon clathratus.	1.
Lacuna vincta.	2.
Echinus.	a few plates & spines.

I think it is evident that the Glen Nevis glacier must have extended down further than this section at Fort William when that of Glen Treig or Glen Arkaig were so much developed. so this shell bed must be later than that period— the glaciers from Ben Lomond I think w^d. have also prevented the accumulation of any shelly beds along Loch Lomond & w^d. have destroyed previous ones— ditto *probably* at Kyles of Bute—

But if the elevation has increased a foot in the mile from W to E. as I have hinted this 40 feet terrace of the West coast would correspond to beds of greater elevation on the east— so that it is probably older than the 25. to 30 foot raised beaches at Falkirk. Forth, &c.

—The undetermined species of Bela or Mangelia which I got in this Fort William section seems different from any British species in Forbes & Hanley—[19] neither can I find any agreeing with it in Searles Woods Crag Mollusca.[20]

—I enclose a drawing of it— It a good deal resembles the *Fusus proprinquus* in miniature. but with the canal less produced. I w^d. take it very kind if you could give me any hint of what it is.

I am very ill off for books on Arctic shells with good figures— Wood's Crag Mollusca is almost my sole refuge. I have searched the Catalogues of the Geol. Soc. library[21] but cannot find *Gould's Report on the Invertebrata of Massachusetts.*[22] nor. Moller[23] nor Middendorf[24] nor Loven.[25] nor any other. I think the Council sh^d. get some of these for the use of members

—There will be a parcel of books sent to me from the library in a few days— if you could lend me a look of Gould's Report if you have it, it would be a great favour & could be sent in my parcel by M^r. Jones.[26]

I enclose a little map shewing the disposition of the Glen Treig Moraines & the direction of the glacial striæ in Glen Arkaig &c. & position of shelly section of old coast line near Fort William.

The sloping terraces which Darwin had a difficulty about in the lower part of Glen Spean I believe to have been accumulated in a lower lake retained by the protrusion of the Glen Nevis glacier after that of Glen Arkaig had shrunk out of the mouth of Glen Spean— His buttresses or terraces above the highest line near the outfalls I believe to be of the nature of side moraines or moraine matter accumulated in lateral pools between the ice & the hill at a period antecedent to the lakes—[27] It would oblige me if you would shew him this communication if you think he would be interested in it.

Please excuse the haste in which I have had to write the above & believe me to be | with the greatest respect & esteem | Your very obed. serv^{t}. | Thos. F. Jamieson.

Sir. Charles Lyell. FRS

P.S. Mr R. Chambers had some word of perhaps going to Lochaber in Sept. if you should see him you might tell him of these moraines as I have no doubt he w^{d}. like to see them if he goes.[28]

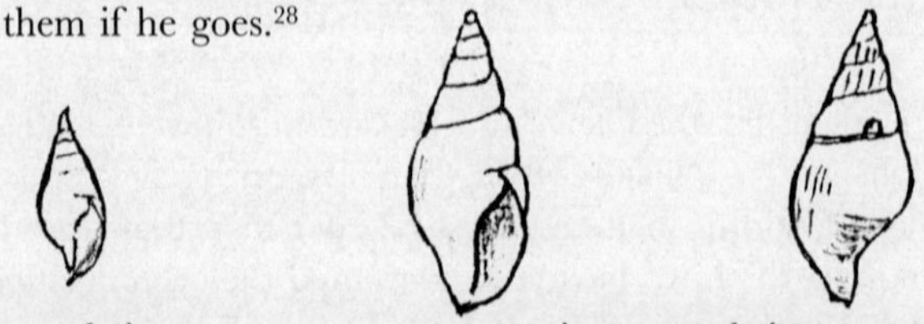

natural size twice natural size

Shell smooth, white, rather thin, of about six turns, outer lip thin; spin tapering, apex rather obtuse; volutions slightly convex— traces of spiral striæ on the base of the shell & some irregular slight plications on the 3rd. & 4.th turns.

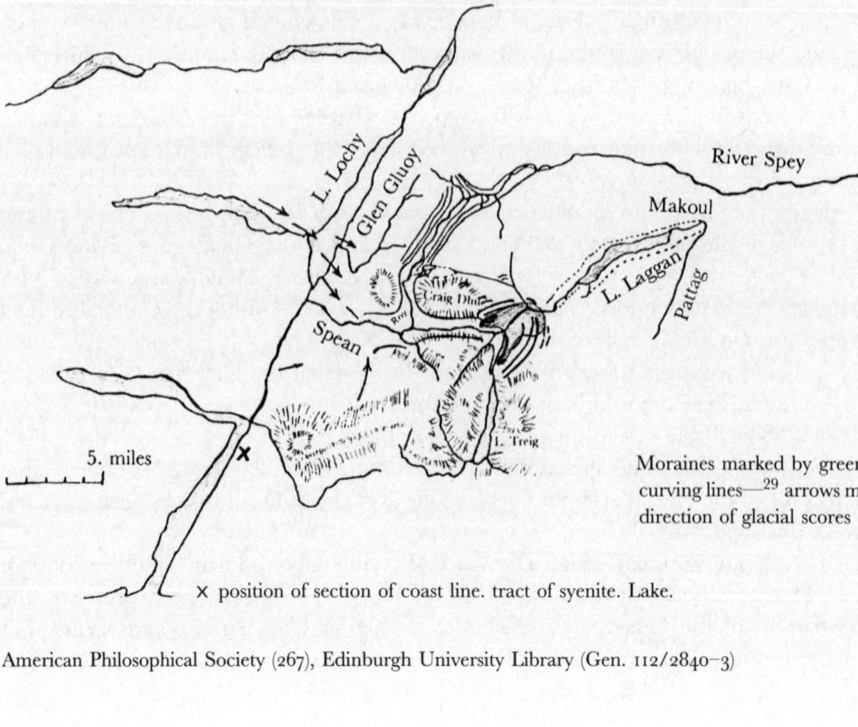

Moraines marked by green curving lines—[29] arrows mark direction of glacial scores

× position of section of coast line. tract of syenite. Lake.

American Philosophical Society (267), Edinburgh University Library (Gen. 112/2840–3)

[1] The year is established by the reference to Jamieson 1862 (see n. 6, below).

[2] See enclosure. No letter from Lyell containing the enclosure has been found; Lyell may have lent CD the letter from Jamieson when CD visited him in London on 30 September 1862 (see 'Journal' (Appendix II)).

[3] In 'Parallel roads of Glen Roy', published in 1839, CD argued that the so-called 'roads', a series of terraces running parallel to each other along the sides of Glen Roy, in Lochaber, Scotland, were formed by the action of the sea during successive periods of elevation of the landmass. However, Thomas Francis Jamieson, having visited the area in August 1861, concluded that the 'roads' represented the successive shorelines of a former lake that had been trapped by ice-flows (see *Correspondence* vol. 9, letter from T. F. Jamieson, 3 September 1861). While CD at first appeared to have conceded defeat on the question, stating that his paper had been 'one long gigantic blunder' (see *Correspondence* vol. 9, letter to Charles Lyell, 6 September [1861]), subsequent letters indicate that he was reluctant to abandon his own explanation (see *Correspondence* vol. 9, Appendix IX).

[4] Lyell planned to discuss the Glen Roy formation in the forthcoming *Antiquity of man* (see C. Lyell 1863a, pp. 252–64). He illustrated his account with three woodcuts (*ibid.*, plate II (facing p. 252), p. 254, and p. 255).

[5] For an account of the principal attempts to explain the 'parallel roads' of Glen Roy, see Rudwick 1974.

[6] Jamieson 1862.

[7] In Chambers 1853 and 1861, Robert Chambers upheld evidence for the effects of large-scale glaciation in Europe, dismissing CD's favoured explanation of many of the phenomena in terms of the action of icebergs; however, no letters from Chambers seeking to convince CD of this belief have been found. CD and Chambers may have discussed this point in April 1861, when CD visited Chambers at his London home (see *Correspondence* vol. 9, letter to Robert Chambers, 30 April [1861]). There is a presentation copy of Chambers 1861 in the Darwin Pamphlet Collection–CUL.

[8] Lyell's letter discussing Ramsay 1862 has not been found. Lyell was highly critical of Ramsay's controversial theory that many European and American rock-basins, now containing lakes, owed their origin to glacial erosion (Davies 1969, pp. 305–6). CD, by contrast, told Ramsay in his letter of 5 September [1862] that, so far as he could judge, the theory '*must* be right to a large extent, possibly wholly'.

[9] See letter from A. C. Ramsay, 26 August 1862, and letter to A. C. Ramsay, 5 September [1862].

[10] John Tyndall was a supporter of Ramsay's theory; in Tyndall 1862, he argued that the extent of glacial action in Switzerland was so great that almost all the features of the present landscape were glacial in origin (see Davies 1969, p. 308).

[11] CD was opposed to the view, advocated by Charles Lyell, that the elevation and subsidence of landmasses was responsible for the climatic changes of the glacial period (see letter to A. C. Ramsay, 5 September [1862] and nn. 8 and 9).

[12] Erasmus, *Adagia*, 1.4.16. The passage translates 'The shoemaker should not go beyond his last'. Desiderius Erasmus gave the aphorism with the comment: 'Men should not attempt what they are not qualified by education or natural aptitude to perform, nor should they discourse on matters they do not understand' (Stevenson ed. 1949, p. 2098). CD apparently refers to the fact that Tyndall, who was professor of natural philosophy at the Royal Institution, was primarily a physicist, rather than a geologist.

[13] According to CD's 'Journal' (Appendix II), he began work on 7 October 1862 on the section of *Variation* dealing with 'Facts of variation of Plants', appearing in the published form as chapters 9 and 10 (*Variation* 1: 305–72).

[14] The enclosure has been identified from the nature of CD's response, and by reference to Lyell's annotation to CD's letter, which states: 'Darwin on Jamieson revisit to Glen Roy'. Although subsequent letters from Jamieson to Lyell also discuss subjects related to Jamieson's Glen Roy theory, none is concerned directly with his second visit to Glen Roy in August 1862 (see the letters from T. F. Jamieson to Charles Lyell, 27 August 1862, 29 August 1862, 24 September 1862, 29 September 1862, 7 October 1862, and 8 October 1862, all of which are in Edinburgh University Library, Gen.

112: 2844–58). In addition, this is the only letter of the period that Jamieson asked Lyell to forward to CD. See also letter from Charles Lyell, 20 August 1862.

15 The modern spelling of this feature is 'Coire nan Eòin' (grid reference NN 221738).

16 The modern spelling is 'Banavie' (*Ordnance Survey gazetteer*).

17 Chambers 1861, p. 20.

18 The geologist James Smith, commonly known as 'Smith of Jordanhill', carried out extensive researches in the 1830s in regard to the raised beaches of the 'newer pliocene' (in modern terminology, Pleistocene) period that were common on the west coast of Scotland; most of the beaches studied by Smith were in the Clyde basin. Noting that some of the molluscs in those deposits were currently very rare or extinct off the coast of Scotland, Smith identified many of them as species currently living in more Arctic regions. On the basis of his findings, he maintained that the climate of the 'newer Pliocene' period had been much colder than at present (see *DNB* and J. Smith 1839a). He provided a 'Catalogue of shells from the newer Pliocene deposits in the British Islands' in J. Smith 1839b, pp. 93–7, including a separate list of those that were either extinct or that were only found in Arctic seas.

19 Forbes and Hanley 1853.

20 S. V. Wood 1848–61.

21 Jamieson refers to the *Catalogue of the books and maps in the library of the Geological Society of London* (London, 1846), the *Supplemental catalogue of the books, maps, sections and drawings in the library of the Geological Society of London* (London, 1856), and the *Alphabetical supplement to the classified catalogues of the library of the Geological Society of London: additional books and maps, 1854–59* (London, 1860).

22 A. A. Gould 1841.

23 Möller 1846.

24 Middendorf 1848–75.

25 Lovén 1846.

26 A. A. Gould 1841. Thomas Rupert Jones was assistant secretary at the Geological Society of London (*DNB*).

27 CD expressed his concerns on these points in the letter to Charles Lyell, [15 September 1861] (*Correspondence* vol. 9), which Lyell forwarded to Jamieson. Jamieson had replied somewhat differently to CD's queries in his letter to Charles Lyell of 19 September 1861 (*ibid.*, Appendix IX).

28 See n. 7, above.

29 The moraines, drawn in green on the original map, are represented here by long dashed lines. The parallel roads of Glen Roy, originally drawn in red, are represented by shorter dashed lines.

From B. J. Sulivan 14 October [1862][1]

Board of Trade. S.W.
Oct 14th.

My dear Darwin

I think Tuesday 21st. if you are well enough would be the best day for Wickham[2] he does not go till Friday 24th. we could go down by 3.30. or 4.40. train—& I could get back in time by 11.29 train on Wednesday but I shall see Wickham & Mellersh on Friday & will then settle with them.

Yours very sincly | B. J. Sulivan

DAR 177: 278

1 The year is established by reference to the visit to Down House on 21 October 1862, of Sulivan, John Clements Wickham, and Arthur Mellersh (Emma Darwin's diary (DAR 242)).

2 See letters from B. J. Sulivan, 27 September [1862], 2 October [1862], and 13 October [1862].

To H. W. Bates 15 October [1862][1]

Down. | Bromley. | Kent. S.E.
Oct 15th

My dear Mr Bates

I want to hear a little news of you & your Book, & how you & it go on.—[2] We have had a wretched summer & have returned home about a fortnight.—[3]

One of my poor Boys, Leonard, was fearfully ill for two months from effect of Scarlet fever & on our journey to sea-side, Mrs Darwin sickened with the fever & we were detained 3 weeks at Southampton.[4] My health has suffered considerably, but I am now slowly at work again.—

When at leisure pray let me have a line, telling me what you have been doing.—

By the way the other day a Mr Edwin Brown of Burton sent me Procs. of N. Ent. Soc. with a letter,[5] in which he tells me that he is working at a genealogical classification of genus Carabus.— In answer I told him that you had thought of something of the kind.[6]

Pray believe me | Yours very sincerely | C. Darwin

Cleveland Health Sciences Library (Robert M. Stecher collection)

[1] The year is established by the relationship to the letter from H. W. Bates, 17 October 1862.

[2] Bates 1863.

[3] The Darwin family spent September 1862 in Bournemouth, returning to Down House on 30 September (see 'Journal' (Appendix II)).

[4] Leonard Darwin became ill with scarlet fever on 12 June 1862 (see Emma Darwin's diary (DAR 242)). Emma Darwin became ill with scarlet fever in Southampton on 13 August, while on the way to Bournemouth with CD and Leonard; they remained in Southampton until 1 September 1862 (see 'Journal' (Appendix II)).

[5] Edwin Brown's letter has not been found. A lightly annotated copy of the *Proceedings of the Northern Entomological Society* for 28 July 1862, printed lithographically in handwritten script, is preserved in the Darwin Pamphlet Collection–CUL; a paper by Brown 'on the mutability of specific or race forms' (E. Brown 1862) is printed on pp. 7–18. Brown argued that while CD had 'dwelt ingeniously and satisfactorily upon one cause for the alteration of forms of life', namely, natural selection, he had almost lost sight of other causes, including 'the direct influences of climate and food, and the accumulative effects of those apparently causeless individual variations that take place at every generation'.

[6] CD's letter to Brown has not been found. CD had been impressed by Bates's discussion of the distribution of beetles of the genus *Carabus* in South America, and of its significance for CD's views respecting the migration of species during the Pleistocene glacial period (see *Correspondence* vol. 9, letter from H. W. Bates, 28 March 1861, and letter to H. W. Bates, 4 April [1861], and this volume, letter from H. W. Bates, 30 April 1862). In his letter to Bates of 4 May [1862], CD had suggested that this case, supported by another, 'would be worth a paper'.

From George Bentham 15 October 1862

Royal Gardens Kew
Oct 15/62

My dear Darwin

The abstract of Prof Targioni-Tozzetti's work is in the Journal of the Horticultural Society v. IX (1855) p. 133[1]

You are at perfect liberty to mention me as having made the abstract—but if I recollect right the facts are all given on the authority of Targioni [2]

Can we hope to have any communication from you this winter for the Linnean Society?

Yours very sincerely | George Bentham

DAR 160.1: 153

[1] In his letter to Bentham of 13 October [1862], CD had asked for permission to refer to Bentham's authorship of an anonymous synoptical review of Targioni Tozzetti 1853 ([Bentham] 1855). There is an annotated presentation copy of the review in the Darwin Pamphlet Collection–CUL.

[2] CD referred to Bentham's authorship of the review in *Variation* 1: 306 n.

From C. C. Babington 16 October 1862

Cambridge
16 Oct. 1862

Dear Darwin

I am sorry that the very full occupation of the late meeting totally prevented me from attending to your wishes at the proper time.[1]

I believe however that now our Curator, Stratton, has sent you some of the seeds that you want and I hope that you will succeed in raising plants from them.[2]

We have none of the others that you want.

Yours very truly | Charles C. Babington—

DAR 160.1: 5

[1] See letter to C. C. Babington, 2 September [1862]. Babington was one of the three local secretaries for the annual meeting of the British Association for the Advancement of Science, held in Cambridge from 1 to 8 October 1862 (*Report of the 32d meeting of the British Association for the Advancement of Science held at Cambridge*, p. xxiii).

[2] See letter from C. C. Babington, 16 September 1862. James Stratton was the curator of the Cambridge University Botanic Garden (Walters 1981).

To Asa Gray 16 October [1862][1]

Down Bromley Kent
Oct. 16th

My dear Gray

I have not indulged myself for some little time in writing to you, though I have to thank you for two very pleasant notes, & for some pages of Silliman with several notices which I was glad to see.[2] Lythrum salicaria is coming out so clear (though all seed not counted) that I do not care much for other species; but I shall be *very glad* of seed of Nesæa.[3] My geese are always at first Swans; but I cannot help going on marvelling at Lythrum. It is a consolation to me for I am utterly routed, beaten, "whipped" by those odious Melastomatads; yet I feel sure there is something very curious to be made out about them.[4]

On getting your last note, I looked at Rothrock's observations on Houstonia;[5] they are capital on the structure; but when I began to think how I could put the case for Gard. Chron, I failed; for the observations are not sufficient about reciprocal fertilisation.[6] It would be a pity to spoil a good case; by aid a few experiments, if your seed will grow, I could make a paper & give all his facts.[7]

Now that we are at home again, I have begun dull steady work on "Variation under Domestication"; but alas & alas pottering over plants is much better sport.—[8]

By the way at Bournemouth, for the want of something else to do, I worked a bit at my old friend Drosera: I took to testing all sorts of fluids, which are not corrosive & do not, I believe, act on ordinary organic compounds, but do act on the nervous system of animals; & I declare I am coming to the conclusion that plants or at least Drosera, must have something closely analogous to nervous matter.[9] It was pretty to see effect of acetate of strychnine, how it stopped all movement; & how acetate of morphia greatly dulled & retarted movement. I think I shall some day pursue this subject.—[10]

Another little point has interested me, viz finding such a number of natural hybrids between two species of Verbascum; & linking V. thapsus & lychnitis closely together. They are all utterly sterile. This fact has given Hooker, to whom I told it, a fit of the horrors.—[11]

So we are all come to you next summer! Alas my days for moving anywhere are come to an end.[12]

Many thanks for sending the article in Daily News, which we read aloud in Family conclave.[13] Our verdict was, that the N. was fully justified in going to war with the S.; but that as soon as it was plain that there was no majority in the S. for ReUnion, you ought, after your victories in Kentucky & Tennessee, to have made peace & agreed to a divorce. How curious it is that you all seem to believe that you can annex the South; whilst on this side of the Atlantic, it is the almost universal opinion that this is utterly impossible. If I could believe that your Presidents proclamation would have any effect, it would make a great alteration in my wishes;[14] I would then run the risk of your seizing on Canada (I wish with all my heart it was an independent country) & declaring war against us. But slavery seems to me to grow a more hopeless curse. How detestably the special correspondent of the Times writes on the subject; the man has not a shade of feeling against slavery.[15] This war of yours, however it may end, is a fearful evil to the whole world; & its evil effect will, I must think, be felt for years.— I can see already it has produced wide spread feeling in favour of aristocracy & Monarchism: no one in England will speak for years in favour of the people governing themselves.

Well good night.— Do not be indignant with me & do not let M[rs] Gray be more indignant than she can help.—[16] Good Night & farewell | Yours cordially | Ch. Darwin

N.B | Do you chance know anything of Mr Floy of N. York[17] who sent in 1846 Hort. Soc. Journal vol I to Lindley ears of wild maize & says he cultivated it & saw Bracts decreasing.[18] De Candolle doubts story.[19] Is he trustworthy ie Mr Floy?)

Gray Herbarium of Harvard University (81)

1 The year is established by the relationship to the letter from Asa Gray, 22 September 1862.

2 Letters from Asa Gray, 5 September 1862 and 22 September 1862. 'Silliman's journal' was the name commonly given to the *American Journal of Science and Arts*, founded by Benjamin Silliman. The pages referred to have not been found. Gray had already sent CD part of the July 1862 number of the journal (see letter to Asa Gray, 28 July [1862]); the subsequent number (September 1862) contained several notes by Gray in the section headed 'Scientific intelligence. III. Botany and zoology'. CD may refer particularly to a synoptical review of the latest numbers of the *Journal of the Proceedings of the Linnean Society (Botany)* (A. Gray 1862f), in which Gray briefly noticed CD's paper, 'Dimorphic condition in *Primula*'.

3 CD refers to his crossing experiments carried out on *Lythrum salicaria* in the summer of 1862, the notes from which are in DAR 27.2. In his letter of 9 August [1862], CD had asked Gray to make observations on, and, if possible, to send seed of American species of *Lythrum* and the related plant *Nesea verticillata.* See also letters from Asa Gray, 5 September 1862 and 22 September 1862.

4 See letter to J. D. Hooker, 14 [October 1862] and n. 7.

5 Gray's pupil, Joseph Trimble Rothrock, had carried out crossing experiments on CD's behalf on the dimorphic plant, *Houstonia caerulea*, the results from which Gray sent to CD in his letter of 4 August 1862.

6 In his letter to CD of 22 September 1862, Gray suggested that CD draw up an account of *Houstonia* for the *Gardeners' Chronicle and Agricultural Gazette.*

7 Gray had sent CD seed of *Houstonia* (see letter to Asa Gray, [3–]4 September [1862] and n. 10). CD included Rothrock's observations and experimental results in *Forms of flowers*, pp. 132, 254.

8 The Darwins returned from their holiday in Bournemouth on 30 September 1862; CD began work on the section of *Variation* dealing with 'Facts of variation of Plants' on 7 October 1862 (see 'Journal' (Appendix II)).

9 CD had begun his investigations with the insectivorous plant *Drosera rotundifolia* while 'idling and resting' at the Sussex home of his sister-in-law, Sarah Elizabeth Wedgwood, in July 1860 (*Autobiography*, p. 132, and *Correspondence* vol. 8). For an account of the trials made on *Drosera* in Bournemouth in 1862, see the letter to J. D. Hooker, 26 September [1862]. There are notes in DAR 54: 29–49 detailing CD's experiments with this species between 14 and 26 September 1862.

10 CD did not again work extensively on the subject of insectivorous plants until 1872 (*LL* 3: 322); his findings were published in 1875 as *Insectivorous plants.*

11 See letter to J. D. Hooker, 6 October [1862], and letter from J. D. Hooker, [12 October 1862].

12 See letter from Asa Gray, 22 September 1862.

13 Gray had sent a copy of D[icey] 1862, which discussed the American Civil War, with his letter to CD of 22 September 1862.

14 On 22 September 1862, Abraham Lincoln issued a preliminary emancipation proclamation, announcing that from 1 January 1863, all slaves in the states still in rebellion would be freed (McPherson 1988, p. 557).

15 Charles Mackay became the North American correspondent for *The Times* in 1862 (Brogan ed. 1975, p. xvi).

16 Like Asa Gray, Jane Loring Gray was a fervent supporter of the Republican Party and the war effort (Dupree 1959, pp. 307–8 and 329). See also letter to Asa Gray, 22 January [1862] and n. 8.

17 James Floy.

18 [Lindley] 1846.

19 A. de Candolle 1855, pp. 951–2.

From E. Schweizerbart'sche Verlagsbuchhandlung[1] 16 October 1862

Stuttgart
den 16 Oct. 1862

⟨Hoch⟩verehrter Herr!

Sie werden ohne Zweifel längst erwartet haben über das ⟨For⟩tschreiten der beiden Uebersetzungen Kenntniss zu erhalten;[2] ⟨d⟩er Druck verzögerte sich gegen Erwarten, aber jedenfalls sollen auch die beiden anderen Lieferungen des "Origin" binnen etwa 5 Wochen ausgegeben werden.[3]

Indem ich nun anliegend die beiden Schriften
über die Entstehung der Arten,
über die Einrichtung zur Befruchtung der Orchideen
in je 3 Exemplaren Ihnen sende, hoffe ich durch die Ausführung Ihre Zufriedenheit zu erhalten. Sollten Sie noch weitere Ex. davon wünschen, so bitte mir es zu sagen.[4]

Ich füge der heutigen Sendung ein neues Werk bei: *Die Pflanzenblätter in Natur Druck* (Morphologie d. Blätter) von Reuss, 1^e^ Lieferung.[5] das Ganze soll 42 Blätter umfassen; ich glaube dass Sie Interesse daran finden und würde mich freuen, wenn Sie das engl. Publikum bei Gelegenheit darauf aufmerksam machen könnten.

Neulich erzählte mir Jemand einen Fall, der Sie interessieren wird; ein Mädchen wurde Schwanger und beschuldigte einen jungen Mann der Vaterschaft, dieser bestritt solches, fügte aber bei, wenn das Kind an Hand und Fuss je 10 finger, Zehen habe, wie er, so wolle er der Vater seyn; als das Kind geboren war, zeigte es an Händen u. Füssen je 10 Glieder mehr oder weniger entwickelt!

Mit grösster Hochachtung verharre Ihr
E Schweizerbart[6]

DAR 177: 70

[1] For a translation of this letter, see Appendix I. The letter was written by Christian Friedrich Schweizerbart, who was head of the Stuttgart publishing firm E. Schweizerbart'sche Verlagsbuchhandlung (see n. 6, below).

[2] Bronn trans. 1862 and 1863. See letter from E. Schweizerbart'sche Verlagsbuchhandlung, 11 July 1862.

[3] The second German edition of *Origin* (Bronn trans. 1863) was initially issued in three parts, the publication of which was announced on 6 October, 17 November, and 19 December 1862. The publication of the first German edition of *Orchids* (Bronn trans. 1862) was announced on 20 October 1862. See *Börsenblatt für den Deutschen Buchhandel* 29 (1862): 2083, 2195, 2447, 2735.

[4] There is a copy of the second German edition of *Origin* (Bronn trans. 1863) in the Darwin Library–CUL, and there are two copies of Bronn's translation of *Orchids* (Bronn trans. 1862) in the Darwin Library–Down.

[5] The publication of the first part of Reuss 1862–70 was announced on 26 September 1862 (*Börsenblatt für den Deutschen Buchhandel* 29 (1862): 2010). There is a copy of this work in the Darwin Library–Down.

[6] Although C. F. Schweizerbart purchased the publishing firm of E. Schweizerbart'sche Verlagsbuchhandlung from his uncle Wilhelm Emanuel Schweizerbart in 1841, he continued to use the signature 'E. Schweizerbart' in business communications (*Jubiläums-Katalog*, pp. x–xi).

From H. W. Bates 17 October 1862

King St Leicester
17 Oct 1862

My Dear Mr Darwin

I was as pleased as a child this morning to have a letter from you again.[1] Mr Wallace two months ago told me of the illness in your house & knowing the distress you would be in I did not like to trouble you with letters. You do not say whether Mrs Darwin & Leonard are quite recovered.[2]

Regarding my book. I am thoroughly ashamed of myself after so much bragging at the beginning not to have finished the work after 12 months employed on it.[3] I think I told you that it would be only external stimulus that would impel me on with it I felt so disinclined to write. I hoped, however, having once commenced, a liking for the task would set in but it has not been so. I have been working & bodging against inclination ever since April last. You will be glad to hear that now, 620 pages are finished out of the 700 of which the work is to consist. Two thirds of the M.S. have been delivered to Murray:[4] after the last receipt M. writes "it keeps up to the mark". With the autumnal weather a better activity has arisen & I am writing rapidly.

I must plead as a little excuse for slowness, an interruption caused by re-writing & passing through the press of my memoir in Linnæan Transactions.[5] This took me 3 hours a day for six weeks in June to August. It fills 72 pages quarto. I took great pleasure in this. I shall get sharply scolded, I expect, for expense caused to the society in altering proofs; but I wanted to make the treatise worthy of the high honour of its place in the Transactions.[6] It was written only as a review of species with ordinary introductory remarks for the "Proceedings". I feel now that the memoir does not express my thoughts with the force & clearness that I think I could impart were I allowed to rewrite again the whole. I shall be most anxious to have your deliberate criticism on it I should think it is now about ready for distribution. I have matter for another better treatise on the origin of species out of local varieties.

Mr Edwin Brown is manager in a large bank at Burton.[7] I have known him 21 years: he was my earliest Naturalist friend. I have always looked on him as a man of extraordinary intellectual ability. I have given him my notions on Carabi. He is amassing material (specimens) at a very great expense.[8] He has never travelled: this is a great deficiency for the relations of species to closely allied species & varieties cannot, I think, be thoroughly understood without personal observation in different countries. He has very little leisure & perhaps will not be able to devote the enormous time & anxious thought to the subject which are required to work it out.[9]

It was a damper not to see my book advertised in Murray's lists for November, December.[10] Should you advise me to recommend him to begin printing? The correction of proof would sooner help me to finish than hinder me. There are 3 or 4 more engravings required, which he will put in hand doubtless when he has the additional M.S. which I shall send him in 5 or 6 days.

Please give my kind regards to Mrs Darwin & family
Yours sincerely | H W Bates

DAR 160.1: 71

[1] Letter to H. W. Bates, 15 October [1862].
[2] Leonard and Emma Darwin were both ill with scarlet fever during the summer of 1862 (see letter to A. R. Wallace, 20 August [1862]).
[3] Bates 1863. In his letter of 4 April [1861] (*Correspondence* vol. 9), CD had encouraged Bates to write an account of his travels as a naturalist and collector in the Amazon basin. Bates's letter announcing his intention to do so has not been found, but see the letter from H. W. Bates, [before 25 September 1861], and the letter to H. W. Bates, 25 September [1861] (*Correspondence* vol. 9).
[4] John Murray.
[5] Bates 1862a.
[6] The part of the *Transactions of the Linnean Society of London* in which Bates 1862a appeared was published on 13 November 1862 (Raphael 1970, p. 70)
[7] See letter to H. W. Bates, 15 October [1862]. Edwin Brown was manager of the Burton, Uttoxeter, and Ashbourne Union Bank in Burton-on-Trent, Staffordshire (*Banking almanac* 1865).
[8] The contents of Brown's extensive natural history collections at the time of his death are described in the *Entomologist's Monthly Magazine* 13 (1876–7): 116–7, 257–8.
[9] Although Brown amassed 'a large and valuable collection' of Carabidae, he only published one short paper on the subject (Brown 1869; see *Entomologist's Monthly Magazine* 13 (1876–7): 116–7).
[10] See, for example, *Athenæum*, 4 October 1862, p. 421.

From T. F. Jamieson to Charles Lyell 17 October 1862

Ellon Aberdeensh.
17 Oct 62

My Dear Sir Charles,

I return you Mr. Darwin's too flattering letter and am glad to find that he thinks his difficulties about Glen Roy sufficiently answered.[1]

With regard to the lakes I think Ramsay's idea is good altho he may push it too far.[2] It seems to me intelligible only on the supposition that whenever you have a glacially formed lake in a rock basin there must have been a spot of soft or more easily eroded rock otherwise how would the scooping commence.— Dr. Hooker says that glaciers sometimes *raise* their beds by accumulating debris underneath.[3]

Your explanation of the larger lakes not being filled up owing to the occupation of the hollow by the ice seems to me imperfect in as much as it does not account for the original formation of the hollow—.[4]

Your case of Zurich is very interesting.[5]

I think Ramsay should have shown that glaciers have the immense scooping power which he requires— as that is begging a good deal of the question.

I should like to know also how he explains the absence of lakes in the Himmalaya seeing how large the glaciers were there.

Your suggestion of a subsidence occurring in part of the course of a long valley ought I think to account for some lakes. and I think lines of fracture would account for others— Ramsay's arguments against these latter seem to me not conclusive. for altho the original yawning chasm may have been removed by denudation yet the deepseated fractures would have a tendency to gape again during fresh movements of disturbance—[6]

However I am very much inclined to think that a great many lakes have been formed in the way Ramsay suggests— It is difficult otherwise to account for their immense number in those northern countries which have been so much ice worn.

I am glad to see that Mr Darwin thinks Dr Tyndall's explanation of the glacial period insufficient—[7] for I cannot see how it will answer for all that he supposes. Really will no astronomer come to the rescue and give us a lift with this great ice affair.[8]

I am | My Dear Sir Charles | Your very obed. serv | Thos. F. Jamieson.

P.S. I hope M^r^. King will publish his lists of Clyde &c shells—giving those of each locality separate with the comparative prevalence of each species—[9] I have often urged M^r^. Smith of Jordanhill to make out accurate *local* lists but in vain.[10]

T. F. J.

Edinburgh University Library (Gen. 112/2859)

[1] CD had asked Lyell to forward to Jamieson his letter of 14 October [1862].

[2] The reference is to Andrew Crombie Ramsay's controversial theory, detailed in Ramsay 1862, that many European and American rock-basins, now containing lakes, owed their origin to glacial erosion (see letter to Charles Lyell, 14 October [1862] and nn. 8 and 9).

[3] J. D. Hooker 1854b, 2: 121 n.

[4] Lyell's letter has not been found; however, Lyell was the leading critic of Ramsay's theory (see Davies 1969, pp. 305–6).

[5] In Ramsay 1862, pp. 192–3, Ramsay argued that the absence of freshwater strata in the Alps intermediate in age between the close of the Miocene and the start of the Pleistocene glacial epoch, constituted further evidence that the Alpine lakes did not exist before the glacial epoch. Lyell had apparently told Jamieson of evidence he had collected showing that there were strata of the sort referred to by Ramsay on the shores of the lake at Zürich; Lyell cited this case as evidence against Ramsay's theory in C. Lyell 1863a, pp. 314–6.

[6] In his paper, Ramsay had argued against the possibility of the great Alpine lakes being the result of structural features associated with synclines, or local subsidence, or fissures along fault lines (Ramsay 1862, pp. 188–90).

[7] See letter to Charles Lyell, 14 October [1862] and n. 10.

[8] This is apparently a reference to CD's comment that John Tyndall, who was primarily a physicist, was going beyond his expertise in writing about glacial phenomena (see letter to Charles Lyell, 14 October [1862] and n. 12).

[9] The reference is probably to the geologist, Samuel William King, a friend of Lyell's, who had helped him with a number of geological investigations (*DNB*). Lyell had apparently sent Jamieson one of his letters from King at the end of September, in which King described a shelly deposit that had a bearing on Jamieson's views concerning the raised beaches of Scotland and the lower level of the Scottish landmass in the recent geological past (see the letter from Jamieson to Lyell of 29 September 1862, which is in Edinburgh University Library, Gen. 112: 2853–4). No publication by King on this subject has been found.

[10] The Scottish geologist and antiquary, James Smith (commonly referred to as 'Smith of Jordanhill'), had published extensively on the raised beaches of the west coast of Scotland, and especially those in the Clyde basin (*DNB*, *RSCSP*). However, his catalogues of shells, published in J. Smith 1839b, pp. 89–97, were not arranged by location.

To Friedrich Rolle 17 October [1862]

Down. Bromley. Kent.
Oct. 17th

Dear &. Respected Sir

On my return home after a long absence I found the first number of your work; & yesterday I received the second number.—[1] I suppose I owe this present to your kindness, & I sincerey thank you. I am, I am, I am sory, to say, a very poor german scholar & have as yet only skinned through the first number & turned ower the pages of the second.

I can see that there will be much to interess me in this second number; For I am now writing on "Variation and Domestication".—[2] Will you permit me to say that I regret to see that You do not give many foot-notes with References.

You have conferred much honnour on me, by undertaking the illustration and enlargement of my views.[3] You will do excellent service in my opinion by getting the subject more discussed & understood; for thus alone will the truth be discovered. Many who criticise what I have written do not at all understand the subject.—

A new german edition of the "Origin" with a good many additions, translated by the lamented Prof. Bronn just before his death, is either now published or will by published immediately:[4] Also a translation of my small work on "Orchids";[5] the last chapter of wich contain some general remarks on species.—

With sincer respect & my best thanks I beg leave to remain | Dear Sir | your obliged servant | Ch. Darwin.

Copy[6]
Endorsement: '(r. 20 October 1862.)'
Forschungsinstitut und Natur-Museum Senckenberg, Frankfurt

[1] Rolle 1863 was initially issued in four parts; the publication of the first two was announced on 8 September and 27 October 1862, and that of the remaining two parts on 19 January 1863 (*Börsenblatt für den Deutschen Buchhandel* 29 (1862): 1862, 2258; 30 (1863): 113). CD's annotated copies of all four parts of this work are in the Darwin Pamphlet Collection–CUL. It appears that the first part of the work was forwarded to CD before his return from Bournemouth on 30 September, and that he subsequently lent it to Charles Lyell (see letter to Charles Lyell, 1 October [1862] and n. 2).

[2] In the second part of Rolle 1863 the author discussed several questions pertaining to domesticated plants and animals, including the importance of external influences on producing variation, acclimatisation, crossing, and the origins of domesticated species. CD began work on the section of *Variation* dealing with 'Facts of variation of Plants' on 7 October 1862 (see 'Journal' (Appendix II)); he cited Rolle's account of the history of peach cultivation in *Variation* 2: 308.

[3] Rolle 1863 was written at the suggestion of the Frankfurt publisher, Friedrich Emil Suchsland, as a popular exposition of CD's views on the species question. In translation its title reads: 'Charles Darwin's theory of the origin of species in the plant and animal kingdoms and its application to the history of creation'. In the foreword to the book (Rolle 1863, p. [ii]), Rolle referred to CD's approval of his efforts as expressed in this letter: 'Nicht das geringste des bisherigen Lohnes der Arbeit aber war der briefliche Ausdruk offener Anerkennung des Strebens und Leistens von Herrn Chs. Darwin selbst.' [The greatest compensation for this work, moreover, came from Mr Charles Darwin himself, who in a letter expressed his candid recognition of the endeavour and the result.]

[4] Bronn trans. 1863. Heinrich Georg Bronn, who translated both *Origin* and *Orchids* into German, died on 5 July 1862 (see letter from E. Schweizerbart'sche Verlagsbuchhandlung, 11 July 1862).
[5] Bronn trans. 1862.
[6] The copy is in Rolle's hand and includes his errors of transcription; it may have been prepared on behalf of Francis Darwin, who, after his father's death, borrowed letters or copies of letters from many of CD's correspondents, in order to prepare *LL* and *ML* (see Martin and Uschmann 1969, p. 70).

To Walter White 17 [October – November 1862?][1]

Down
17th

Dear Sir

I have received Vol I of Hort. Transact. & am much obliged but it is not the right volume; I believe that I wrote Vol I **second series.**[2] Have you not 2d series? Will you kindly pass this note on to Linn. Soc to see if they have the Second series; & if so ask them to send it by Delivery Coy on Wednesday morning to enclosed address. Will you kindly inform me whether either Soc. has this second series? for I must somehow get it— I am sorry to cause so much trouble—

In Haste | Yours faithfully | C. Darwin

Librarian Royal Soc | or | Linnean Socy.[3]

Wellcome Institute for the History of Medicine Library

[1] The date range is conjectured from the reference to volume 1 of the second series of the *Transactions of the Horticultural Society of London* (see n. 2, below).
[2] CD's earlier letter to White, who was the librarian of the Royal Society, has not been found. Between 7 October and 11 December 1862, CD wrote a draft section of *Variation* in which he discussed 'Facts of variation of Plants', appearing in the published form as chapters 9 and 10 (*Variation* 1: 305–72; see 'Journal' (Appendix II)). In his discussion he cited on several occasions Robert Thompson's papers on the varieties of apricot, gooseberry, and cherry, which were published in volume 1 of the second series of the *Transactions of the Horticultural Society of London* (Thompson 1831, 1832a, and 1832b).
[3] The librarian of the Linnean Society of London was Richard Kippist.

To H. W. Bates 18 October [1862][1]

Down. | Bromley. | Kent. S.E.
Oct 18th

My dear Mr Bates

Many thanks for your very friendly note.—[2] I am not *at all* surprised at the time which your Book has taken.[3] Everyone finds the same thing. It is dull work, but must be borne. People differ about being hurried by Proofs: I hate it, so shd recommend you not to be in too great a hurry. Murray[4] would be an excellent adviser when you had better begin. After M.S. sent in, proofs seldom arrive for first fortnight, & then the flow is steady. I am most anxious that yours shd be a real good Book; so do not hurry over proofs—; yet I have known some very few men correct too much, & flatten their style.— I find it good to correct in pencil & read aloud, & if it sounds well, not to plague more over it. But you write a capital style.

I am very glad you have finished paper for Linn. Soc.[5] When I go to London, I must enquire for the Part; for it is not sent till enquired for.[6] You need not ask me to read it carefully; I shall be sure to do that.— not having been in London I have not seen a paper which you told me would appear in Annals.[7]

I am glad to hear about Mr Brown: I thought he was something remarkable![8]

Mrs Darwin & my Boy are well;[9] but my children are cause of never ending anxiety to me.—

This is a stupid note, but I am tired with a long day's work.

With hearty good wishes for your work through the press— Dear Bates | Yours very sincerely | C. Darwin

Cleveland Health Sciences Library (Robert M. Stecher collection)

[1] The year is established by the relationship to the letter from H. W. Bates, 17 October 1862.

[2] Letter from H. W. Bates, 17 October 1862.

[3] Bates 1863.

[4] John Murray.

[5] Bates 1862a.

[6] CD refers to the third part of volume 23 of the *Transactions of the Linnean Society of London*, in which Bates 1862a appeared; the part was published on 13 November 1862 (Raphael 1970, p. 70).

[7] Bates 1862b. See letter from H. W. Bates, 19 May 1862 and n. 7.

[8] Edwin Brown. See letter to H. W. Bates, 15 October [1862], and letter from H. W. Bates, 17 October 1862.

[9] Emma and Leonard Darwin had been ill with scarlet fever during the summer (see 'Journal' (Appendix II)).

From J. D. Hooker [18 October 1862][1]

Royal Gardens Kew
Saturday.

Dr Darwin

Do not send back the Melastomads[2]

Masdevallia is beautifully in flower, do you want it? flower? or whole plant?[3]

I put flowers of curious Loasaceous plant in the box with *Impatiens*— I saw bees hard at work on it; but not on *Impatiens*. I also put a Cassia flower in.[4]

R. Spruce Esq^e. at Quito (or elsewhere) care of HRM Consul Guayaquil—would send you by post Melastomaceæ seeds, or Messrs Herbst & Co nurserymen Rio de Janeiro.[5]

Ever Yours affec | J D Hooker

I have ordered Bonafuss on Maize.[6]

DAR 101: 63

[1] The date is established by the relationship to the letter to J. D. Hooker, 14 [October 1862], and the letter from J. D. Hooker, 25 October 1862; the intervening Saturday was 18 October 1862.

2 See letter to J. D. Hooker, 14 [October 1862] and n. 7.
3 In *Orchids*, p. 168, CD described the peculiar closed structure of the flowers in *Masdevallia fenestrata*, with their 'two minute, lateral, oval windows'; he stated that he had 'failed to understand' how insects could remove or insert pollinia in order to effect pollination in this species. In his letter to J. D. Hooker, 25 February [1862], CD had asked to borrow a plant of the species when next in flower (see also the letter to J. D. Hooker, 30 [June 1862], and letters from J. D. Hooker, 3 March 1862 and 2 July 1862).
4 See letter from J. D. Hooker, [12 October 1862], and letter to J. D. Hooker, 14 [October 1862].
5 In the letter to J. D. Hooker, 14 [October 1862], CD asked Hooker for the names of people to whom he could apply for seeds of *Heterocentron* or *Monochaetum* in their native South America. The reference is to Richard Spruce. The nursery of Messrs Herbst & Co. was probably established by Hermann Carl Gottlieb Herbst. Although CD wrote to Spruce to ask him for information concerning the Melastomataceae (see letter from A. R. Wallace, 2 January 1864 (*Calendar* no. 4378), and letter from Richard Spruce, 15 April 1869 (*Calendar* no. 6697)), his letter has not been found.
6 Bonafous 1836. CD was preparing the section of *Variation* dealing with 'Facts of variation of Plants' (see 'Journal' (Appendix II)); he cited Bonafous 1836 in his account of varieties of maize (*Variation* 1: 320).

From B. J. Sulivan 18 October [1862][1]

Board of Trade S.W.
Oct 18.

My dear Darwin

We will come on Tuesday D.V. by the 4.10 train from Victoria St.[2] I hope you are really well enough. I trust it will not be too much for you & do not care about sending for us as we can take a cab if your carriage is not there.

Yours very sincerely | B. J. Sulivan

Wickham was to leave on Wednesday for Paris but has put it off till Thursday

DAR 177: 276

1 The year is established by reference to the visit to Down House on 21 October 1862, of Sulivan, John Clements Wickham, and Arthur Mellersh (Emma Darwin's diary (DAR 242)).
2 See letters from B. J. Sulivan, 27 September [1862], 2 October [1862], 13 October [1862], and 14 October [1862].

From W. E. Darwin 21 October [1862][1]

Bank.
Oct 21.

My Dear Father,

I send off by the early post today, some pods of the L. styled Lythrum, from a plant growing in a clump as I put on the envelope.[2]

The Long styled that I first sent you (wrongly named short-styled) was growing quite by itself some yards from any other plant, and a hundred to two hundred from any clump. The plant itself as far as I can remember—was not an unhealthy plant, but the difference in size and number of pods is very marked. So that comparing a branch of each they look almost like different species, one is so fat and tightly packed in its pods compared to the other; the situations did not differ much, one along side a stream, the other along an all-the-year-round wet ditch.[3]

I will count the seeds soon. 6 of each as you say—[4] Maud Atherley seems much the same, all Mr Atherley can say is that if anything his hopes are rather more than his fears.[5] it seems an attack on the brain, all Sunday and Sunday Night, she was screaming most terribly, taking brandy every two hours which she has to do still.

I expect I have seen the last of the exhibition, as I shall not be able to get away this Saturday Certainly, or next most probably.[6]

Mama was wonderful in London—[7] | your affect son | W E Darwin

DAR 162.1: 93

CD ANNOTATIONS
1.1 I send ... envelope. 1.2] *crossed pencil*
3.1 I *will* ... London— 5.1] *crossed ink*
Top of letter: 'Distance of single Plant' *pencil*

[1] The year is established by the reference to the International Exhibition held in London in 1862 (see n. 6, below).

[2] CD had apparently asked William, in a letter which is now missing, to send him seed-pods from a wild specimen of each of the three sexual forms of the trimorphic plant *Lythrum salicaria*, so that he could establish their relative fecundity. CD's notes on the seed-pods sent in this letter (from a long-styled plant), with a tally of the numbers of seeds, are in DAR 27.2 (ser. 2): 55. Having verified the results by comparison with immature plants in his own garden, CD published them in 'Three forms of *Lythrum salicaria*', p. 173 (*Collected papers* 2: 109–10); see also the notes in DAR 27.2 (ser. 2): 51, 53.

[3] No letter enclosing such specimens has been found. However, there are notes in DAR 27.2 (ser. 2): 45 r. that are headed: 'Long styled wild (marked short, so was this not *separate* plant?) I think so & hence infertility'. At the bottom of the page CD noted the lower tally of seeds from the pods of this plant compared to that recorded in DAR 27.2 (ser. 2): 55, writing in explanation: 'Effects of distance from other plants'. These seed-pods were probably sent at the same time as the seed-pods from mid-styled specimens of *Lythrum salicaria* (see CD's notes in DAR 27.2 (ser. 2): 45 v.), and as those from short-styled specimens (see CD's notes, dated 'Oct 8th & 13th', in DAR 27.2 (ser. 2): 56).

[4] There are notes in William's botanical notebook (DAR 117: 50), dated 2 November 1862, recording his tally of seeds in one pod of long-styled *Lythrum.*

[5] Maud Atherley was the daughter of George Atherley, William's partner in the Southampton and Hampshire Bank, Southampton.

[6] The International Exhibition opened in South Kensington, London on 1 May 1862 (*The Times*, 2 May 1862, pp. 11–12). It was due to close in September; however, by popular demand it was kept open until the end of October (*Athenæum*, 20 September 1862, p. 374).

[7] According to her diary (DAR 242), Emma Darwin travelled to London on 16 October 1862 and visited the International Exhibition on 17 and 18 October, before returning to Down House on 20 October. She had been ill with scarlet fever during the summer (see 'Journal' (Appendix II)).

To John Lubbock 23 October [1862][1]

Down
Oct. 23^d^.

My dear Lubbock

I had intended writing to you on my return home, but two days afterwards I had a bad attack, which kept me in house for week or two & stopped me going to Cambridge, as I had much wished & where I sh^{d}. have seen you.—[2]

I very much wish to see you; but I do not know what to do. Two days ago three officers of the Beagle came here to dinner;[3] I took every possible precaution, but it made me very ill with violent shaking & vomiting till the early morning; & Could not even wish them goodbye next morning. Forgive these wretched details on my health, for I cannot bear that you sh^{d}. think me indifferent of your friendship or ungrateful for your kindness about my son William.—[4]

It is not worth your while, but if you could any day come & dine & sleep here & let me go away for an hour after dinner & retire to my room at 9 o clock I do not think it would hurt me; or if weather sh^{d}. ever come fine again would you spare time, if so inclined, to come on a Sunday to our early dinner, which would do, for your lunch.—

Whether I see you or not, I hope you will believe in my sincere attachment.

Yours very truly | Ch. Darwin

DAR 263

[1] The year is established by the relationship to the letter from John Lubbock, 25 October 1862.

[2] The Darwins spent September 1862 on holiday in Bournemouth (see 'Journal' (Appendix II)); following their return to Down House, CD developed eczema (see letter to J. D. Hooker, 6 October [1862]). In 1862, the British Association for the Advancement of Science held its annual meeting in Cambridge from 1 to 8 October.

[3] Emma Darwin's diary (DAR 242) records that Arthur Mellersh, Bartholomew James Sulivan, and John Clements Wickham, three of CD's former shipmates from the *Beagle*, dined at Down House on 21 October 1862.

[4] In 1861, Lubbock helped to arrange William Erasmus Darwin's partnership with George Atherley in the Southampton and Hampshire Bank, Southampton (see *Correspondence* vol. 9).

To W. E. Darwin [25 October 1862][1]

My dear William.—

one word to say that if a Sunday be fine I sh^{d}. be very glad if you would test your observation on mid-styled & see what difference is in pods.—[2] You might get $\frac{1}{2}$ dozen pods from other plant. Also, if then you could step distance from the old long-styled to nearest other plant & look what form that is.— I shall not publish this year & shall work out whole case very carefully.—[3]

We are most sincerely sorry about Maude A.—[4]

In Haste. Your | C. Darwin

Etty came home yesterday very brisk.—[5]

Poor Mamma is unwell with very feverish cold.—[6]

Saturday. Down—

DAR 210.6: 106

[1] The date is established by the relationship to the letters from W. E. Darwin, 21 October [1862] and 28 October 1862; the intervening Saturday was 25 October 1862.

[2] The letter containing William's comments on the seed-pods of the mid-styled form of *Lythrum salicaria* has not been found; however, see the letter to W. E. Darwin, 30 [October 1862].

[3] William was assisting his father by collecting seed-pods from wild plants of *Lythrum salicaria* (see letter from W. E. Darwin, 21 October [1862]); however, CD had concluded that he would have to perform a further 126 crosses before he could publish his results (see letter to J. D. Hooker, 27 [October 1862] and n. 11). CD's paper, 'Three forms of *Lythrum salicaria*', was read before the Linnean Society of London on 16 June 1864; he reported observations based on William's specimens on page 173 of the published paper (*Collected papers* 2: 109–10).

[4] Maud Atherley. See letter from W. E. Darwin, 21 October [1862] and n. 5.

[5] Emma Darwin's diary (DAR 242) records that Henrietta Emma Darwin returned to Down House on 22 October 1862, and that on 24 October she was 'languid in [the] m[ornin]g as before'.

[6] On 25 October 1862, Emma Darwin recorded in her diary (DAR 242) that she took to her bed feeling 'feverish' with a 'bad cold'.

From J. D. Hooker 25 October 1862

Royal Gardens Kew
Oct 25/62

Dr. Darwin

One short note only to say, that I sent as, directed, the other day, the *Masdevallia*[1] & plants of *Desmod. gyrans* & the *sensitive plant*.[2] These last are tender sort of things, though **very easy** to manage—with plenty of humidity & heat—both thrive best with us standing *over* the tank in the hot Victoria House.[3] I should suggest for your Experiments—standing the pot in a pan of warmed water—& a bellglass. I feel sure you must let both recover well & grow a bit before beginning for scarce any packing will prevent their suffering from the Journey. Tell me if they are packed well. I gave stringent orders about it.

Shall I grow some more Desmodiums & sensitives for you?

We are all well, but my wife complaining of headaches. Miss Henslow got safe to St Albans, & had no attack at all for a fortnight, but is bad again now (& still there)[4] I was very averse to her going, but she would.

What a nice book Parrotts Ararat is; it is refreshing to read his simple faith in the ark being still under the snow![5] Wife saw Lyells yesterday all well.[6]

Ever yours affec | J D Hooker

DAR 101: 64–5

CD ANNOTATION

Top of letter: 'Potato'[7] *brown crayon*

[1] CD's letter has not been found. In his letter of 18 [October 1862], Hooker offered to send CD a flower or plant of *Masdevallia* from the Royal Botanic Gardens, Kew.

[2] In September 1862, CD had carried out a series of experiments on the effects of various substances (chiefly poisons and narcotics) in modifying the responsiveness to various stimuli of the insectivorous plant *Drosera rotundifolia* (see letter to J. D. Hooker, 26 September [1862]), and had extended his interest to a number of other plant species, including *Dionaea*, *Linum perenne*, *Oxalis acetosella* and *Trifolium repens* (see the notes in DAR 54: 64–5 and DAR 60.1: 1–3). In the missing letter to Hooker,

CD had apparently asked for specimens of the telegraph plant, *Desmodium gyrans*, and the sensitive plant, *Mimosa pudica*, in order to carry out similar experiments on them (see letter to J. D. Hooker, 27 [October 1862] and n. 5). The account in Henslow 1837, pp. 165–6, of irritability in these species, is annotated in CD's copy with suggestions for experiments and observations; CD's copy of the work is in the Darwin Library–CUL (see *Marginalia* 1: 369–71).

[3] Victoria House was a hot-house with a circular tank, 36 ft in diameter, built in 1852 at the Royal Botanic Gardens, Kew, to house the giant waterlily, *Victoria regia* (Bean 1908, p. 44).

[4] One of Frances Harriet Hooker's aunts, probably Anne Frances Henslow, had been unwell when she stayed with the Hookers in August (see letter from J. D. Hooker, 20 August 1862 and n. 4).

[5] Friedrich von Parrot's *Reise zum Ararat* (Parrot 1834), was translated into English in 1845 (Parrot 1845).

[6] Charles and Mary Elizabeth Lyell.

[7] See letter to J. D. Hooker, 27 [October 1862].

From John Lubbock 25 October 1862

15, Lombard Street. E.C.
25 Oct. 1862

My dear Mr. Darwin

I am very sorry to hear so bad an account of your health, as I had hoped that you were better.[1] F. Galton said that you were coming to Cambridge, but I felt that that was too good to be true.[2]

I should dearly like to see you & have a good talk. Tomorrow, Next Sunday & the one after are engaged, so if you are well enough to see me I will come & dine with you, but had better go back home perhaps afterwards.

Friday next would suit me & you might let me know at the last moment if you felt well enough; I would hold myself disengaged & could come, or not, according to your feelings when the time arrived.[3] Pray however do not overdo yourself.

You do not mention Mrs Darwin or the children, so I hope that they are well.[4]

Believe me always | Yours affectly | John Lubbock

P.S. Herbert Norman has found a celt of the "drift" type at Cudham.[5] It is a surface specimen.

DAR 170: 32

CD ANNOTATIONS

Top of letter: '2° 31′ in 10′ [*del* 'not'] sensitive' *pencil*[6]

Verso of last page: '10
10
16
9
11
6
10
10
6
—
88' *ink*[7]

[1] See letter to John Lubbock, 23 October [1862].

[2] Francis Galton was CD's cousin. In 1862, the British Association for the Advancement of Science held its annual meeting in Cambridge from 1 to 8 October.

[3] Emma Darwin recorded in her diary (DAR 242) that Lubbock dined at Down House on Friday 31 October 1862; she also noted: 'Ch. attack of sickness in night but not so bad'.

[4] Emma and Leonard Darwin had been ill with scarlet fever during the summer (see letters to John Lubbock, 21 August [1862] and 2 September [1862]).

[5] Herbert George Henry Norman lived at Oakley, a village near to Down; Cudham is a village about two miles south-east of Down. Lubbock was actively engaged in prehistoric archaeology, and had recently written an article for the *Natural History Review* on the celts, or prehistoric flint implements, found by Jacques Boucher de Perthes, together with fossil remains of extinct species, in the Drift deposits of the Somme Valley (Lubbock 1862c). Lubbock cited Norman's discovery of a celt in Lubbock 1865, pp. 274–5, noting that it had been found 'near Greenstreet Green, a locality which is interesting as having produced remains, not only of the mammoth, but also of the musk ox.'

[6] CD was studying the effects of various substances (chiefly poisons and narcotics) in modifying the responsiveness of plants. These annotations probably relate to CD's experiment on *Mimosa pudica*, carried out on 27 October 1862; CD's notes from the experiment record (DAR 209.2: 86): 'I gave it at 2°. 30′, thirty drops of Ether for 10′ no effect; added 20 drops more & left it for 30′, was then quite sensitive & petioles became depressed—'. See also letter from J. D. Hooker, 25 October 1862 and n. 2, and letter to J. D. Hooker, 27 [October 1862] and n. 5.

[7] The tabulation probably represents a tally of seeds harvested from seed-pods of *Lythrum salicaria*; in October 1862, CD made several such tallies of the seed-pods resulting from his crossing experiments with this species (see the notes in DAR 27.2 (ser. 2)).

From Asa Gray 27 October 1862

Cambridge. [Massachusetts]
27 Oct. 1862.

My Dear Darwin

This heavy mail for you is merely for the purpose of carrying a *30 cents* stamp for Leonard, so you must distribute the contents to oblige him.[1] Do not prepay the continental letters, unless required, as I think is not the case.

Enclosed is a *cent* stamp, the like of which is new to me, & perhaps to the young gentleman. Tell him, also, that I have to-day bought stamps on envelopes, of 12, 20 & 24 cts. which I shall make do duty—like the present 30 ct in carrying my letters, and then they will go into his collection.— These make up all his desiderata, except the 90^{cts}—which I never saw, but I will invest in this, whenever I have something heavy to send.

I enclose seeds of *Nesæa*—near *Lythrum*, likely to be in the dimorphous or trimorphous way.[2]

I wait for Capt. Anderson, of Cunard Steamer, that I may send you 3 Cypripedia & Mitchella.[3]

A second notice of your Orchid book, in Sill. Jour. is mostly made up of obs. on our Ophrydeæ, and *Cypripedium*.[4] I hope to send you the sheets soon.

No scientific news— no letter of yours to answer. I am *bound* to write to you, to utilize the three envelopes before me.—

Presidents Emancipation proclamation is working well on the whole.[5] Our Courage does not fail, and I think will not.

Ever Yours cordially | Asa Gray

DAR 165: 121

CD ANNOTATION

End of letter: 'Duke of Argyl | is Fragaria virginiana much larger than F. vesca— Do you know the wild Pine Strawberry or Carolina or Surinam Strawberry | ['Maize to plant | Bonafous' *del*] | Bates Paper | Macmillan— Horace formerly spec. change | Bates paper—My [*illeg*] & Literary I'[6] *ink*

[1] The enclosures have not been identified. Leonard Darwin had written at Gray's suggestion to tell him which American postage stamps he wanted for his collection (see letter from Asa Gray, 22 September 1862 and n. 3).

[2] CD had requested seeds of *Nesaea verticillata* in his letter to Gray of 9 August [1862].

[3] James Anderson was captain of the Cunard line's trans-Atlantic steamer *Africa* (*Men of the time*). See also letter from Asa Gray, 5 September 1862 and nn. 4 and 5.

[4] Gray refers to the follow-up article to his review of *Orchids* (A. Gray 1862b), which was published in the *American Journal of Science and Arts* (commonly known as 'Silliman's journal' after its founder, Benjamin Silliman). See also letter from Asa Gray, 4 and 13 October 1862 and n. 4.

[5] On 22 September 1862, Abraham Lincoln issued a preliminary emancipation proclamation, serving notice that from 1 January 1863 all slaves in those states still in rebellion would be freed (McPherson 1988, p. 557).

[6] See letter to Asa Gray, 23 November [1862].

To J. D. Hooker 27 [October 1862]

Down Bromley Kent

27th

My dear Hooker

I am sorry to hear about Miss Henslow, & not a very good account of Mrs Hooker.[1] Our Henrietta has been failing a little in the old way again (do not allude to it) & some of the others are yet very delicate.[2] It is a heavy burthen to bear; but I will scribble a bit to you about plants, & forget things. The plants all arrived quite safe.[3] Masdevallia turns out nothing wonderful; I was merely stupid about it; I am not the less obliged for its loan, for if I had lived till 100 years old, I shd. have been uneasy about it.[4] It shall be returned the first day I send to Bromley.

I have steamed the other plants, & made the Sensitive plant very sensitive, & shall soon try some experiment on it.[5] But after all it will only be amusement. Nevertheless, if not causing too much trouble, I shd. be very glad of a few young plants of this & Hedysarum in summer; for this kind of work takes no time & amuses me much.— Have you seeds of Oxalis sensitiva, which I see mentioned in Books.[6] By the way what a fault it is in Henslow's Botany that he gives hardly any references: he alludes to great series of experiments on absorption of poisons by roots, but where to find them I cannot guess.[7] Possibly the all knowing Oliver may know.—[8] I can plainly see that glands of Drosera from rapid power, almost instantaneous, of absorption & power of movement, give enormous advantage for such experiments. And someday I will enjoy myself with a good set to work; but it will be great advantage if I can get some preliminary notion on other sensitive plants, & on roots.—[9]

Oliver said he would speak about some seed of Lythrum hyssopifolium being preserved for me.—[10] By the way I am rather disgusted to find I cannot publish this year on Lythrum salicaria: I must make 126 additional crosses!![11] All that I expected is true, but I have plain indications of much higher complexity. There are 3 pistils of different structure & functional power & 3 kinds of pollen of different structure & functional power, & I strongly suspect altogether five kinds of pollen, all different in this one species![12]

By any chance have you at Kew, any odd *varieties* of the common Potato: I want to grow a few plants of every var. to compare flowers, leaves fruit &c, as I have done with Peas &c.—[13]

I am crawling on with my book on variation; but I have had some bad attacks lately. I fear I cannot come to Kew; yet I wish it much to look at specimen (if such you have) of wheat & maize &c; & I sh$^{d.}$ enjoy it so much.—[14]

Farewell. C. Darwin

Endorsement: 'Oct/62'
DAR 115.2: 167

[1] CD refers to Frances Harriet Hooker and one of her aunts, probably Anne Frances Henslow (see letter from J. D. Hooker, 25 October 1862).

[2] Throughout 1860 and 1861, Henrietta Emma Darwin had been ill with a fever diagnosed as a form of typhus (see *Correspondence* vols. 8 and 9). On 13 October 1862, Emma Darwin recorded in her diary (DAR 242): 'Etty poorly & weak', and on 24 October: 'Henrietta languid in m[ornin]g as before'. Horace Darwin had been seriously ill earlier in the year, and Emma and Leonard had been ill with scarlet fever during the summer (see 'Journal' (Appendix II)). On 13 October, Emma recorded in her diary: 'all children poorly', and on 25 October she recorded that she herself was 'feverish' with a 'bad cold'.

[3] See letter from J. D. Hooker, 25 October 1862.

[4] CD had been anxious to see again the orchid species *Masdevallia fenestrata*, since he had previously been unable to explain how fertilisation was effected in this species (see letter from J. D. Hooker, [18 October 1862] and n. 3). His notes, dated 24 October 1862, on the specimen sent from the Royal Botanic Gardens, Kew, are in DAR 70: 105–6. CD published his observations on this specimen in *Orchids* 2d ed., pp. 135–7, but reiterated his earlier statement that he had failed to understand 'how insects perform the act of fertilisation'.

[5] See letter from J. D. Hooker, 25 October 1862 and n. 2. CD's notes, dated 27 October – 5 November 1862, on his experiments with the sensitive plant, *Mimosa pudica*, are in DAR 209.2: 86; CD exposed the plant to ether and chloroform and observed the effects on the depression of the petiole. These notes also contain a reference to the other plant sent by Hooker, the telegraph plant (*Hedysarum gyrans* or *Desmodium gyrans*), which states: 'I see Hedysarum gyrans drops its leaves downwards at night.' In his copy of Henslow 1837, pp. 165–6, CD noted with regard to *M. pudica*: 'try this with Ether', and with regard to *D. gyrans*: 'Does Desmodium gyrans *sleep*'; CD's copy of the work is in the Darwin Library–CUL (see *Marginalia* 1: 369–71).

[6] From 17 to 20 October 1862, CD carried out experiments on the effects of ether, chloroform, and morphine in modifying the responsiveness of *Oxalis acetosella* (see the experimental notes in DAR 54: 65 and DAR 60.1: 1); he may have intended to extend the experiments to *O. sensitiva.*

7 In Henslow 1837, pp. 168–9, John Stevens Henslow discussed what he called the 'Sensibility' of plants, noting:

> When corrosive poisons are imbibed into their system, they destroy the tissue much in the same way as in the animal frame; but when narcotic poisons are imbibed, although they kill the plants, they do not appear to have produced any derangement or disorganisation in their tissue. But it has been argued that, as these latter poisons act upon the nervous system of animals, we may suspect something analogous to this system to exist in vegetables also.

Henslow went on to note that 'A long list has been given of substances which act as poisons on plants'. This passage is marked with a marginal line in the annotated copy of Henslow 1837 in the Darwin Library–CUL (see *Marginalia* 1: 369–71). CD had recently come to the conclusion that some plants 'must have diffused matter . . . closely analogous to the nervous matter of animals' (see letter to J. D. Hooker, 26 September [1862]).

8 Daniel Oliver was the librarian at the Royal Botanic Gardens, Kew, and one of the compilers of the annual botanical bibliography in the *Natural History Review*; CD had repeatedly commented on Oliver's extensive bibliographic knowledge (see letters to J. D. Hooker, 11 June [1862] and 6 October [1862]).

9 CD did not again work extensively on *Drosera rotundifolia* until 1872 (*LL* 3: 322); his experiments on the species consituted the main body of *Insectivorous plants*, which was published in 1875. CD used species of *Mimosa*, *Desmodium*, and *Oxalis* (see nn. 5 and 6, above) in a further study, published in 1880 as *Movement in plants*.

10 See letter from Daniel Oliver, 13 September 1862, and letter to Daniel Oliver, 13 October [1862] and n. 12.

11 CD had carried out ninety-four crosses with *Lythrum salicaria* in the summer of 1862; however, he decided to carry out further crosses in 1863 in order to be sure of his results (see 'Three forms of *Lythrum salicaria*', p. 179 (*Collected papers* 2: 114), letter to Asa Gray, 9 August [1862], and letter to W. E. Darwin, [25 October 1862]). CD's paper, 'Three forms of *Lythrum salicaria*', was read before the Linnean Society of London on 16 June 1864.

12 In terms of structure, there are only three kinds of stamen in the different forms of flower of *Lythrum salicaria*; however, in 'Three forms of *Lythrum salicaria*', p. 188 (*Collected papers* 2: 122), CD observed that the pollen produced by the long and short stamens of the mid-styled form was less potent than the pollen produced by the structurally identical stamens of the other two forms. Thus, functionally speaking, there were, he claimed, five kinds of pollen in this species ('Three forms of *Lythrum salicaria*', p. 169; *Collected papers*, p. 106).

13 CD was preparing the section of *Variation* dealing with 'Facts of variation of Plants' (see 'Journal' (Appendix II)); he discussed the variation and cultivation of the potato in *Variation* 1: 330–1. In the summer of 1855, CD grew in his garden forty-six varieties of pea, and varieties of many other vegetables (see *Correspondence* vol. 5; see also the experimental notes in DAR 46.2: 3–31, 37–41). CD discussed pea varieties and his own observations on them in *Variation* 1: 326–30.

14 Cultivated and wild varieties of wheat and other cereals are discussed in *Variation* 1: 312–20; varieties of maize are discussed in *Variation* 1: 320–2. CD did not visit the Royal Botanic Gardens, Kew, until 11 February 1863 (Emma Darwin's diary (DAR 242)).

To John Lubbock 27 [October 1862][1]

Down *Bromley* your letter went to Dover. Kent
27th

My dear Lubbock.

Many thanks for your note.[2] I hope & expect to be well enough on Friday to see you. I presume 7° oclock will be the best hour for you.—[3] If I shd. be bad, I would send over to Lamas[4] on Friday afternoon. I cannot say much for several of us; for we have had a touch of Influenza[5]

Ever yours very sincerely | C. Darwin

Smithsonian Institution Libraries (Special collections)

[1] The date is established by the relationship to the letter from John Lubbock, 25 October 1862.
[2] Letter from John Lubbock, 25 October 1862.
[3] Emma Darwin's diary (DAR 242) records that Lubbock dined at Down House on Friday 31 October 1862.
[4] 'Lamas' was Lubbock's house in Chislehurst, Kent.
[5] Emma Darwin recorded in her diary (DAR 242) on 25 October 1862: 'I feverish & bad cold | in bed'; on 26 October she wrote: 'in bed—'.

From W. E. Darwin 28 October 1862

Southampton
Oct 28. 1862.

My Dear Father,

I managed to get over on Sunday to the Lythrum, but the floods prevented me examining the clumps as to whether I could distinguish the MS. from the LS. & S.S. Lythrum— But it must be all rubbish. I managed to measure the distance of the *first* L.S. from other plants. though I think some plants must have been broken down or cut near it.[1]

One side there was a LP. Lythrum, and a very shabby one, distant 14 y^{ds}. And on the same side of it there was a Lythrum, but I could not distinguish what it was, (as the pods were all gone) distant 10 yd^{s}. And on the other side there was a middle S. Lythrum distant 9. y^{ds}.

I have not seen Mr Atherley yet; he has written to ask me to go over and see him tomorrow, I suppose I shall see poor Mrs Atherley too.[2]

I should not be surprised if he means to go away for some long time

Your affect son | W E Darwin

DAR 162.1: 94

CD ANNOTATIONS
1.1 I . . . rubbish. 1.3] *crossed ink*
1.4 *first* LS.] *altered to* '*first* Long Styled' *ink*; 'first sent & imperfectly fertile'[3] *added ink*
2.1 LP.] *altered to* 'Long styled [*above del* 'Pistilled']' *ink*
2.3 there was . . . 9. y^{ds}. 2.4] '& 100 yards from any clump.'[4] *added ink*
3.1 I have . . . time 4.1] *crossed ink*

[1] William was assisting his father by making observations on the trimorphic plant, *Lythrum salicaria* (see letter to W. E. Darwin, [25 October 1862]); he refers to the short-styled, mid-styled, and long-styled forms.
[2] George Atherley was William's partner in the Southampton and Hampshire Bank, Southampton. The Atherleys' daughter, Maud, died in October 1862 (see letter to W. E. Darwin, 30 [October 1862]).
[3] See letter from W. E. Darwin, 21 October [1862] and n. 3.
[4] See letter from W. E. Darwin, 21 October [1862].

From Henrietta Emma Darwin[1] [29 October 1862]

My cat Greysy deserted her 2nd kitten Bullsig when it was a month old, I suppose because she had no milk, as it was always thin & she neglected it so that we heard its remarkably loud mews from the cellar over the house. Bullsig always wished to play with his mother whilst she hated & repulsed him; till the night before her confinement when she let him get into the basket prepared for her & touzle her as much as he liked, though he didn't try to suck her. We saved the kitten she had next day & ever after he used to suck her & stay in the basket at the same time with his younger brother. She was now a very good mother, & kept them very fat.—

My kitten Miss Legge I took from its mother when she was a month old & cd not eat meat, so that it was a great deprivation. At Hartfield where we were,[2] there was an old cat & kittens & about 2 days after we got there she adopted this cat & sucked it with the others. She had this mother for a fortnight & then nearly 3 weeks interval till we came to Shanklin where there were two old cats giving milk & 3 kittens.[3] She took to both these cats & played with the kittens. One day I observed her snowzling in the stomach of the kittens, & constantly beginning to purr as soon as she thought she had found the tit. She did not leave off till its whole stomach was quite wet.—

our Ilkley[4] cat used to suck its own stomach when it was happy just like a child does its thumb. It did it a great part of the day, & used to purr over it as if it was so good.—

A memorandum
DAR 162: 68

CD ANNOTATIONS
2.8 She ... quite wet.— 2.9] 'Strong experience affecting sucking instinct' *pencil*
Top of letter: 'Henrietta on Cats.— | Oct. 29/62/' *pencil*

[1] CD's annotations suggest that Henrietta's account of these incidents was intended as possible evidence for a planned treatment of instinct; however, her testimony on these points was not referred to by CD in any publication. In her manuscript autobiography (DAR 246: 11–12), Henrietta recalled that, as a child, she would sit 'for long hours' watching her cats, and 'sympathising with the cat's admiration of her kittens'.

[2] Henrietta refers to the period from 2 to 16 July 1858 when the Darwin children stayed at The Ridge in Hartfield, Sussex, the home of their aunt, Sarah Elizabeth Wedgwood (Emma Darwin's diary (DAR 242); see also n. 3, below).

[3] The Darwins were on holiday in Shanklin on the Isle of Wight from 27 July to 13 August 1858 (see *Correspondence* vol. 7, Appendix II).

[4] The Darwins stayed in Ilkley, Yorkshire, in 1859 (see *Correspondence* vol. 7, Appendix II).

To W. E. Darwin 30 [October 1862][1]

Down
30th

My dear William

I write only just to thank you about the Lythrum. I do not feel at all so sure, that what you thought about the mid-styled looking different is rubbish:[2] I am sure that it yields more seed. I think the seed of long-styled are rather larger than of the two other forms.—[3]

What a melancholy thing poor little Mauds death is. I am truly sorry for the Atherleys.[4] It is not at all likely that you will have any opportunity, but if you have say how sincerely we sympathise with them.— I fear if Mr A. goes away for some time, you will be much confined to your work.— How does your health keep.? Poor Etty has not been very well lately, & it makes her low.[5]

John Lubbock comes here to dinner tomorrow; Heaven knows how I shall stand it.—[6] Herbert Norman found the other day a very perfect Celt near Cudham:[7] & a grand Mammoth's tusk has just been found at Greenstreet Green.—[8]

Our greatest piece of news is that Lizzie at her own wish is to go to school at Kensington after Xmas;[9] & if Horace is able to go, we shall part with Miss Ludwig.—[10] Etty is going on Saturday,, if well enough as I very much hope she will be, to stay for a few days at the Bonham Carters.[11]

Thanks for Saturday Review; it was a nice little article.—[12]

Farewell | My dear old fellow | Your affect | C. Darwin

Months hence will do about counting seed; you will find it tedious work.—[13]

DAR 210.6: 107

[1] Dated by the relationship to the letter from W. E. Darwin, 28 October 1862.

[2] See letter to W. E. Darwin, [25 October 1862] and n. 2, and letter from W. E. Darwin, 28 October 1862.

[3] CD reported this observation in 'Three forms of *Lythrum salicaria*', p. 172 (*Collected papers* 2: 108).

[4] Maud Atherley was the daughter of George Atherley, William's banking partner in the Southampton and Hampshire Bank, Southampton.

[5] Henrietta Emma Darwin. See letter to J. D. Hooker, 27 [October 1862] and n. 2.

[6] The excitement of conversation with guests usually resulted in CD's suffering attacks of vomiting and chills during the night. He had suffered a severe attack on 21 October 1862, following the visit to Down House of three former shipmates from HMS *Beagle* (see letter to John Lubbock, 23 October [1862]). Following Lubbock's visit to Down House on 31 October, Emma Darwin recorded in her diary (DAR 242): 'Ch. attack of sickness in night but not so bad'.

[7] Herbert George Henry Norman. See letter from John Lubbock, 25 October 1862 and n. 5.

[8] Greenstreet Green is a village about two miles north-east of Down. See letter from John Lubbock, 25 October 1862, n. 5.

[9] Elizabeth Darwin started at a school in Kensington run by Miss Buob, on 27 January 1863 (see Emma Darwin's diary (DAR 242), and the letter from Emma Darwin to W. E. Darwin, [29 October 1862], in DAR 219.1: 63).

[10] The Darwins' governess, Camilla Ludwig, was on an extended visit to her family in Hamburg, having apparently been sent away in early June, on full pay, in order to separate her from Horace Darwin. The Down surgeon, Stephen Paul Engleheart, was concerned that Horace's attachment to her might have been exacerbating the illness from which he had been suffering earlier in the year. See the letters from Emma Darwin to William Erasmus Darwin, [2 March 1862], [27 May 1862], and [6 November 1862], in DAR 219.1: 49, 57, 64; see also Emma Darwin's diary (DAR 242), and CD's Classed account book (Down House MS).

[11] CD refers to the family of Joanna Maria Bonham-Carter, whose house, Ravensbourne, was in the village of Keston, two miles north-west of Down. On Saturday 1 November 1862, Emma Darwin

recorded in her diary (DAR 242) that Henrietta had been 'better but languid in m[ornin]g all week', but there is no record of a visit to the Bonham-Carters on that day. Henrietta visited and stayed with the Bonham-Carters on more than one occasion in November and December 1862, having become friendly with Elinor Mary Bonham-Carter (see Emma Darwin's diary (DAR 242), and the letters from Emma Darwin to W. E. Darwin, [19 November 1862] and [2 December 1862?], in DAR 219.1: 67–8).

12 William had probably sent CD a copy of the anonymous review of *Orchids* that appeared in the *Saturday Review* of 18 October 1862 (Anon 1862); there is a copy of the review in CD's Scrapbook of reviews (DAR 226.1).

13 See letter from W. E. Darwin, 21 October [1862] and n. 4, and letter to W. E. Darwin, [25 October 1862].

From John Lubbock 30 October 1862

London
30 Oct 1862

My dear Mr. Darwin

I am looking forward with real pleasure to the prospect of a talk with you.[1]

If however you do not feel up to it, please send me word to the *Bickley Station* & *not to Chiselhurst* as I shall not go home.[2]

Hoping to see you so soon I will only add that I am always

Your affectn— | John Lubbock

DAR 170: 33

1 Emma Darwin's diary (DAR 242) records that John Lubbock dined at Down House on Friday 31 October 1862.

2 Lubbock worked in London and lived in Chislehurst, Kent. Bickley railway station, situated between Chislehurst and Bromley, was probably the one used by Lubbock in travelling to and from work (*Post Office directory of the six home counties* 1862).

From C. W. Crocker 31 October 1862

South St. Chichester
Oct 31st. 62

Dear Sir

I am ashamed to have allowed your letter to have remained so long un-answered but the fact is I have as long as possible deferred acknowledging even to myself the amount of *real* information I have gained from experiment during the last year.[1] It has indeed been a year of trial to me and mine, God grant we may not have another like it.[2] Leaving family matters out of the question I have still had no end of difficulties to contend against. In the first place the only bit of ground I could get was ten minutes walk away from my living house, and even for that I had to pay a rent which to one in my situation is something serious. The garden when I got it contained nothing but a few Snap-dragons, (of which anon) The next difficulty was how to obtain plants I could not afford to buy but was not ashamed to beg. I went to Kew early in the season, but could not see Dr Hooker who was good enough to say before I left that he would give any plants I might require.[3] I may perhaps be over-modest, but some day I shall summon courage to ask him for some. I obtained at the time some ferns for a Wardean case[4] and two or three conifers, but no herbaceous plants. I know from experience how strict their rules are and will not press upon them.— I have been very kindly aided by the curator of the Liverpool and some other botanic gardens from whom I hardly expected it.[5] Their presents have however for the most part been in seeds and therefore this years work has been to raise them.

My Linum grandiflorum has been very fine and confirmed your statements with regard to it. L. flavum in my case has come *true* but not many plants of it.[6] As I had no means of pushing it on in heat in the spring I did not expect to see it flower this season. It will I have no doubt stand the winter and come into bloom early next year.

The Antirrhinums, being the only thing I had to my hand early in the season, were put through a series of little experiments.[7] The lip closes firmly enclosing stamens and stigma and one might naturally expect it to be ordinarily self-fertilized and perhaps it may be. The common honey-bee never goes to it so far as I have been able to observe the humble bee however (or some large sp. of bee I am afraid to speak upon entomological subjects to you) is very fond of it. He invariably settles on the lower lip and if his weight is not sufficient to force it down he puts one leg against the upper and pushes it open. His proboscis goes in and his head comes out covered with pollen, and off he goes to the next. I castrated some flowers marked them and as might have been expected they produced seed as though nothing had been the matter. The next thing was to find if they would do so when the bees were excluded. They have done so but not so perfectly, the details are still unrecorded but I shall weigh seed and I shall try the experiment again.

For my next experiment I was obliged to fall back upon seeds. and somehow the Plantagos took my fancy.[8] I had seen a var. of P. major which produces a rosette of leaves instead of a spike of flowers. What induced the plant to do this I wanted to know. I took one and planted him in good soil, took care that he had no struggle for existance, fed him on the fat of the land, and hoped for the result. It produced no less than 180 very fine spikes of flowers, instead of ten or a dozen, and three times the usual size, the foliage fine and not inclined to keep flat upon the ground (which is a selfish habit in the family no doubt intended to keep as much space as possible to themselves should they require it) But the only modification was that the lower bracts were *very* much developed, being quite leaves 2 inches long at the base of spike.— I went out for a walk one day and to my infinite disgust I found a plant of P. lanceolata which without any artificial help had produced a rosette of leaves at the top of the flower shoot instead of flowers. True the plant was luxuriant and growing by the side of the canal. I took it up planted it in my garden and found that it produced a few flowers among the leaves. I have carefully raised plants from these seeds to see if they will exhibit the same character next year to any extent. The rosette I pegged down to see if it would root and make a plant of itself it has not done so yet but it is as fresh as ever although it was three months ago that I brought it home. I have since received one or two curious varieties of Plantago from Liverpool and Mr. Borrer's gardener[9] and shall follow them up, though I dont see what they are to prove.— It is astonishing how selfish P. media is, its leaves are pressed quite hard upon the soil and anything covered by them actually rots.

A good many little experiments have been dead failures but they have not disheartened me, they have only whet my appetite and roused my temper to the extent of saying that I would not be beaten next year if my health be spared so long. Among other things I tried to hybridize Tropæolum canariense with the dark var of T. Major without results. All the Orchids hereabouts I have compared with your observations and I need not tell you with what result. I did not expect to find anything different I only wanted to understand the subject properly.—[10]

Did I tell you that I think there is a *possibility* of finding an intermediate form in the common Primrose. Out of a handful I gathered (one from each plant) last April—13 were long styled—29 short styled, and there was one decidedly intermediate— It may have been only an accidental form as I had only the one flower. Next season I shall examine them as they are gathered.[11]

I have not heard how the series of experiments I commenced at Kew upon the deterioration of vegetables and serials is going on.[12] Prof. Oliver I have seen several times since but not to spend much time with him.[13] I suppose you know of those which have been conducted by Mr. Lawes at Harpenden for some years past.[14] I have lately had a friend of mine stopping with me for a week who was engaged there for some months previously and I have been much interested in his descriptions of their wonderful experiments.

Trusting that I may have a better account to give of work done next year I must now bring this long and I fear uninteresting letter to a close with the hope that it will find you in better health.

I remain, Sir, | yours most respectfully | C. W. Crocker

C. Darwin Esq.[r]

DAR 76 (ser. 2): 84a–d

CD ANNOTATIONS

1.4 It has . . . like it. 1.5] *scored brown crayon*
1.9 The next . . . beg. 1.10] *scored brown crayon*
2.4 It will . . . may be. 3.4] *crossed ink*; 'Dichogamy' *added brown crayon*; 'C. W. Crocker Nov 1. 1862' *added ink*
3.4 The common] *opening square bracket, ink*
3.9 I castrated . . . again. 3.13] *scored brown crayon*
4.1 For my . . . fancy. 4.2] *scored brown crayon*
4.1 For my . . . require it) 4.9] *crossed ink*
4.19 I have . . . prove.— 4.21] *scored brown crayon*
5.1 A good . . . me, 5.2] *scored brown crayon*
5.4 I tried . . . results. 5.5] *cross in margin brown crayon*
5.4 Tropæolum cananariense] *underl brown crayon*

6.3 13 were ... intermediate— 6.4] *scored brown crayon*
7.1 I have ... him. 7.3] *scored brown crayon*

[1] CD's letter has not been found. Crocker had made numerous observations and experiments on CD's behalf during his time as foreman of the propagating department at the Royal Botanic Gardens, Kew (see *Correspondence* vols. 8 and 9). Following his retirement in the winter of 1861–2, he had offered to continue to assist CD with botanical experiments. See letters from C. W. Crocker, 17 February 1862, [before 13 March 1862], 13 March 1862, 22 April 1862, and 17 May 1862.

[2] Crocker's father, Charles Crocker, died on 6 October 1861 (*DNB*); his mother died on 27 April 1862 (*Sussex Advertiser*, 6 May 1862).

[3] Joseph Dalton Hooker was assistant director at the Royal Botanic Gardens, Kew.

[4] Wardian case: 'a close-fitting case with glass sides and top for growing small ferns and other moisture-loving plants' (*OED*).

[5] The curator of the Liverpool Botanic Gardens was John Simpson Tyerman (R. Desmond 1994).

[6] CD had mentioned that *Linum flavum* and *L. grandiflorum* were dimorphic in 'Dimorphic condition in *Primula*', p. 96 (*Collected papers* 2: 62–3), a copy of which he had sent to Crocker (see letter from C. W. Crocker, 17 February 1862, and Appendix III). CD may have asked Crocker for assistance with the crossing experiments that he planned to carry out in 1862 (see letter from C. W. Crocker, [before 13 March 1862]). In his paper, 'Two forms in species of *Linum*', which was read before the Linnean Society of London on 5 February 1863, CD reported (p. 81) with regard to *L. flavum*: 'I have not been able to try any experiments on this species; but a careful observer, Mr. W. C. Crocker, intends proving their reciprocal fertility next summer' (see also *Collected papers* 2: 104).

[7] These are the snapdragons referred to earlier in the letter. CD cited Crocker's observations on this species in *Cross and self fertilisation*, p. 363.

[8] In 'Dimorphic condition of *Primula*', p. 95 (see *Collected papers*, p. 62), CD mentioned that several species of the North American *Plantago* were dimorphic.

[9] Charles Green was gardener to the botanist William Borrer (R. Desmond 1994).

[10] CD sent Crocker a presentation copy of *Orchids* (see letter from C. W. Crocker, 17 May 1862, and Appendix IV).

[11] CD and Crocker had discussed the existence of a mid-styled form in the Chinese primrose, *Primula sinensis*, earlier in the year (see letters from C. W. Crocker, 22 April 1862 and 17 May 1862).

[12] In his letter to Asa Gray of 25 April [1860] (*Correspondence* vol. 8), CD mentioned Hooker's attempts to 'degenerate our culinary vegetables', in experiments conducted at the Royal Botanic Gardens, Kew.

[13] In addition to his duties as librarian and assistant in the herbarium at the Royal Botanic Gardens, Kew, Daniel Oliver was professor of botany at University College London.

[14] Crocker refers to John Bennet Lawes, founder of the Rothamsted Agricultural Experimental Station near Harpenden, Hertfordshire (*DNB*).

From Andrew Smith [November? 1862][1]

last seven months more than usually out of order in fact I was regarded by my medical attendants at one time as in sate of great danger—. violent rheumatic fever extending to the Heart was the illness and I dare say you will ere this have noticed from my writing that my fingers are not yet as they were before the attack.

I have just finished reading your most interesting volumes on the Orchids[2] how you must have laboured— you will leave a name for yourself which were be named with admiration long after you and I shall have ceased to exist I think you will give me credit for wishing you every good thing to which a person with claims such as you have may fairly expect.

My own health does not enable me to do much still I go on amusing myself principally in enquiries about the natives of South Africa, their languages &c. also in endeavouring to trace if any connection between them and the population of Northern Africa is to found.[3] I think I have made out some points of interest but the whole subject is one of such difficulty that I almost despair of any ones being able to ascertain how far the various tribes are related to each other.

With reference to a question of yours I do not recollect having seen that the natives of Australia in times of scarcety eat of the vegetable production natives of the soil, but I have no doubt they do in fact all people who do not cultive the ground for their food trust not a little to what nature offers them.— the Bushmen when the young bulbs of a certain Gladiolus are in force and the larva of ants in season get quite independent & even fat. the whole year round these wild Hottentots have roots &. on which they depend for a portion of their diet[4]

I am | My Dear Darwin | Yours most faithfully | Andrew Smith

Incomplete
DAR 177: 184

[1] Dated by the reference to CD's work on cultivated plants (see n. 4, below).

[2] *Orchids* was offered for sale on 15 May 1862 (Freeman 1977, p. 112).

[3] Smith, formerly superintendent and director-general of the Army Medical Department, retired in 1858 because of ill health. He served in South Africa between 1821 and 1837 and wrote many papers on the origin and history of bushmen (*DNB*).

[4] Between 7 October and 11 December 1862, CD prepared a draft of the part of *Variation* dealing with 'Facts of variation of Plants' (see 'Journal' (Appendix II)); in the introduction to the section, he included 'a few general remarks on the origin of cultivated plants' (*Variation* 1: 306–12. CD countered the arguments of some botanists that for cultivated plants like cereals to have been 'noticed and valued as objects of food', their original state must have closely resembled their present one; he pointed to 'the many accounts given by travellers of the wretched food collected by savages', and cited Smith on the consumption by the natives of South Africa of roots, leaves, and other sources of meagre nutrition during times of scarcity or famine (*Variation* 1: 307–8). CD made a related inquiry in his letter to J. D. Hooker, 4 November [1862].

From J. D. Hooker 2 November 1862

Kew
Nov 2/62

Dear Darwin

Oliver has I believe written to you about *Lythrum* seeds,[1] if not to be had elsewhere I will get it from continent by letter, if you will tell me when you want them.

The *Masdevallia* is all safe.[2] We have no Potatoes at all, I wish we had!—sometimes we grow a few on waste ground.— not a cob of Indian corn ripened this shady year.[3]

I will have the *Desmodium* & *Mimosa* plants raised for you. & *Oxalis sensitiva* also—.[4] I am stupified at the idea of your 5 forms of Lythrum & ∞ *intercrossings* to make, your work looks to me like a "Binomial theorem" if you remember that delightful mathematical affair that always got bigger by terms & never ended.[5]

Asa Gray seems busy with *Cypripedium*— ours are coming into flower, so if you want a fight I can I hope supply you with weapons!.[6] Did I tell you how **deeply** pleased I was with Grays notice of my Arctic Essay, it was awfully good of him, for I am sure he must have seen several blunders.[7] He tells me that Dr Dawson is down on me[8] & I have a very nice Lecture on Arctic & Alp plants from D^r^. D with sup critique on the Arctic Essay which he did not see till afterwards.[9] He has found some mare's nests,[10] in my Essay & *one* very venial blunder in the tables—[11] he seems to **hate** Darwinism—[12] he accuses me of overlooking the Geolog. facts & dwells much on my overlooking subsidence of temp. America during glacial period,—& my asserting a subsidence of Arctic America, which never entered into my head—[13] I wish however, if it would not make your head ache too much, you would just look over my first 3 pages & tell me if I have outraged any Geological facts or made any oversights.[14] I expounded the whole thing twice to Lyell[15] before I printed it with map & tables, intending to get (& thought I had) his imprimatur for all I did & said,: but when here three nights ago I found he was as ignorant of my having written an Arctic Essay as could be! & so I suppose he either did not take it in, or thought it of little consequence. Hector approved of it in toto.[16] I need hardly say that I set out on Biological grounds & hold myself as independent of theories of subsidence as you do of the opinions of Physicists on heat of Globe!! I have written a long screed to Dawson

By the way do you see the Athenæum notice of L. Bonapartes Basque & Finnish language—[17]is it not possible that the Basques are Finns left behind after Glacial period, like the Arctic plants!.[18] I have often thought this theory would explain the Mexican & Chinese national affinities

I am plodding away at Welwitschia by night & Genera Plantarum by day—[19]We had a very jolly dinner at Club on Thursday.[20]

We are all well. | Ever yours | J D Hooker

DAR 101: 66–7, 70

CD ANNOTATIONS
1.1 Oliver . . . them. 1.3] *cross in margin, pencil*
3.1 I will . . . also—. 3.1] *cross in margin, pencil*
4.8 he seems . . . period,— 4.10] *cross in margin, pencil*
5.2 is it . . . affinities 5.4] *scored pencil*

[1] Daniel Oliver's letter has not been found, but see the letter to Daniel Oliver, 13 October [1862] and n. 12, and the letter to J. D. Hooker, 4 November [1862] and n. 9. In his letter to Hooker of 27 [October 1862], CD mentioned that Oliver had promised to arrange for seed of *Lythrum hyssopifolia* to be reserved for him at the Royal Botanic Gardens, Kew.

[2] See letter to J. D. Hooker, 27 [October 1862].

[3] See letter to J. D. Hooker, 27 [October 1862] and nn. 13 and 14.

[4] See letter to J. D. Hooker, 27 [October 1862] and nn. 5 and 6.

[5] See letter to J. D. Hooker 27 [October 1862].

[6] During the summer, Asa Gray had carried out at CD's request, observations and experiments on several species of the orchid genus *Cypripedium*, sending CD his notes on the subject (see letters to Asa Gray, 10–20 June [1862] and 1 July [1862]). In a letter to Hooker of 13 October 1862, which is at the Royal Botanic Gardens, Kew (Asa Gray letters: 307), Gray reported that he had decided to publish the substance of his observations in A. Gray 1862b, rather than waiting until he had repeated the experiments the following year. Referring to CD's observations in *Orchids* on the pollination mechanisms of British species of *Cypripedium*, Gray continued:

> Darwin's hypothesis is wrong for our species. If I could draw, I would publish a set of figures to illustrate my view of *Cypripedium* fertilization,— and a pretty paper it would make.

See also letter to J. D. Hooker, 4 November [1862] and n. 10.

[7] Gray reviewed Hooker's paper on the distribution of Arctic plants (J. D. Hooker 1861a) in the July number of the *American Journal of Science and Arts* (A. Gray 1862d).

[8] John William Dawson discussed Hooker's views on palaeogeography in a letter to Gray of 19 September 1862, which is preserved at the Gray Herbarium of Harvard University (see Sheets-Pyenson 1992, p. 15). Gray apparently forwarded this letter to Hooker with his own letter of 15 October 1862, which is at the Royal Botanic Gardens, Kew (Asa Gray letters: 308); in his letter, Gray reported: 'your confident expectation that somebody will yet cut your Arctic Essay up by the roots is likely to be verified!' He had, he stated, lent Dawson a copy of J. D. Hooker 1861a, 'which he *mastered* and returned to me in a *week or so*.' He continued: 'You see you blunder about the distribution of common Canadian plants—as well as in geology! Expect a severe blast from Canada this winter.'

[9] Hooker refers to Dawson's lecture on 'Alpine and Arctic plants', delivered before the Young Men's Christian Association of Montreal in February 1862 (Dawson 1862a). No copy of Dawson 1862a has been located that includes a supplementary critique of J. D. Hooker 1861a. Hooker probably refers to Dawson's review of the work in the October 1861 issue of *Canadian Naturalist and Geologist* (Dawson 1862b); a version of this review may also have been issued with some copies of Dawson 1862a.

[10] Mare's nest: 'an illusory discovery, esp. one that is much vaunted and displays foolish credulity' (*OED*).

[11] Dawson 1862b, p. 343; the reference is to J. D. Hooker 1861a, p. 275.

[12] While praising Hooker's paper as 'a most valuable summary of the Arctic Flora', Dawson stated (Dawson 1862b, p. 335):

> [geologists will] find themselves obliged to dissent . . . from the assumption, for it is nothing more, of the unlimited variation of species in a Darwinian sense, which pervades the paper, notwithstanding the positive geological testimony to the permanence of several of these throughout a great lapse of geological time.

In conclusion, Dawson criticised Hooker's treatment of the 'intricate' problem of the geographical distribution of Arctic plants (pp. 343–4), stating:

> We believe that he has not sufficiently weighed some of the elements for its solution, and has been disposed instead to lean on the hypothesis, which however specious and apparently useful in explaining difficulties, has not yet been proved by a single tangible fact, that under certain circumstances two real species may spring from one.

[13] Dawson 1862b, pp. 335, 341–3.

[14] CD's annotated copy of the number of the *Transactions of the Linnean Society of London* in which J. D. Hooker 1861a appeared is in the Darwin Library–CUL.

[15] Charles Lyell.

[16] James Hector was surgeon and geologist to the government exploring expedition of British North America between 1857 and 1860 (*DNB*).

[17] The anonymous review of Louis Lucien Bonaparte's privately printed pamphlet (Bonaparte 1862) appeared in the *Athenæum*, 1 November 1862, pp. 559–61.

[18] In J. D. Hooker 1861a, Hooker sought to explain the current distribution of Arctic plants on the basis of migrations that had occurred during and after the Pleistocene glacial period.
[19] J. D. Hooker 1863a and Bentham and Hooker 1862–83.
[20] The Philosophical Club of the Royal Society of London met once a month on a Thursday (Bonney 1919, p. 2).

To J. D. Hooker 3 November [1862]

Down Bromley Kent
Nov. 3^d.

My dear Hooker

I am going to give you two bothers.— (1st) Can you give me reference to Vol. (which I suppose I can get from Linn. Soc^y.) in Hooker's Bot. Journal, in which Planchon gives monograph of Linum, & states that several species have long & short pistils.—[1]

(2nd) Can you supply me with *seeds* of any of enclosed list of plants, for experiment;[2] I know it is mere chance if you have any: what I want most is any of the species of Oxalis & of the Boragineæ, especially Alkanna: it would hardly cost you a minute to look to reference in Prodromus, in the list sent me in a letter by Alp. De Candolle.[3]

So much for business. In a note the other day Asa Gray speaks of the Reviews of the Orchis Book in Gardeners' Chronicle, as written by you.[4] Is this possible? I assumed that they were by Lindley.[5] They are gorgeous, but too strong. Nevertheless, on chance of their being yours I could not resist rereading them. If by you, it was too bad your not telling me; for I declare I value a word of praise from you more than from rest of world. But somehow I do not think they are by you. Well might A. Gray say "how they praise you."— You must sometime tell me. I do not think you could possibly have spared time.—

I have been trying a little the Mimosa; but I hurt its constitution with too much chloroform. Desmodium won't move,[6] & I have sent it friend's Hot house.—[7] These experiments will be nice little amusement for me; after my dull daily work. Now I am compiling on vegetables & fruit-trees; & awful work it is drawing any conclusions from my mass of references & notes.—[8]

Do you remember the scarlet Leschenaultia formosa, with the sticky margin *outside* the indusium; well this is the stigma, at least I find the pollen-tubes here penetrate & no where else.[9] What a joke it would be if stigma is always exterior;[10] & this by far greatest difficulty in my crossing notions sh^d. turn out a case eminently requiring insect aid, & consequently almost inevitably insuring crossing. By the way have you any other Goodenia, which you could lend me besides Leschenaultia & Scævola, of which I have seen enough.—[11]

I had long letter the other day from Crocker of Chichester;[12] he has real spirit of experimentalist, but has not done much this summer.

Do you know whether there are *two* Revd Prof. Haughtons at Dublin; one of this name has made a splendid *medical* discovery of nicotine counteracting strychnine

& tetanus;[13] Can it be my dear friend? if so, he is at full liberty for the future to sneer & abuse me to his heart's content.[14]

I had a nice letter two or three weeks ago from Asa Gray, who seems as politically rabid as ever; he says $\frac{9}{10}$ of property & $\frac{4}{5}$ of northern men (or some such proportion) may be destroyed before, as he hopes, the war will be given up. He owns it is a far tougher job than he anticipated.[15]

Farewell. I hope M[rs] Hooker[16] is going on pretty well. We are rather brighter. Farewell | my dear old friend | C. Darwin

I hope to Heaven Masdevallia got safe home.—[17]

Endorsement: '/62'
DAR 115.2: 171

[1] Planchon 1847–8 appeared in volumes 6 and 7 of the *London Journal of Botany*. After 1848, the *London Journal of Botany* was continued as *Hooker's Journal of Botany and Kew Garden Miscellany* (*BUCOP*); both journals were edited by Hooker's father, William Jackson Hooker. CD cited Planchon 1847–8 in 'Two forms in species of *Linum*', p. 81 (*Collected papers* 2: 103–4), which was written between 11 and 21 December 1862 (see 'Journal' (Appendix II)), and read before the Linnean Society of London on 5 February 1863.

[2] The enclosure has not been found. CD required some of the plants for his experiments on heterostyly (see n. 3, below), and others for his experiments on plant sensitivity and movement (see letter to J. D. Hooker, [10–]12 November [1862] and n. 19).

[3] The words 'in the list' were added as an afterthought: CD means that he had included on his list (see n. 2, above) a reference to Candolle and Candolle 1824–73, 10: 94, which Alphonse de Candolle had sent him in his letter of 13 June 1862. For CD's interest in heterostyly in *Oxalis* and *Alkanna*, see the letter to Alphonse de Candolle, 17 June [1862] and n. 3.

[4] See letter from Asa Gray, 4 and 13 October 1862. Hooker acknowledged his authorship of the review ([J. D. Hooker] 1862c) in his letter to CD of 7 November 1862.

[5] In a letter of 14 September [1862], CD had thanked John Lindley, who was the principal editor of the *Gardeners' Chronicle and Agricultural Gazette* (*DNB*), for having reviewed *Orchids* so favourably.

[6] See letter to J. D. Hooker, 27 [October 1862] and n. 5.

[7] The reference is probably to George Henry Turnbull of The Rookery, Down, who allowed CD to use his hot-house while preparing *Orchids* (see *Orchids*, p. 158 n.).

[8] Between 7 October and 11 December 1862, CD wrote a draft section of *Variation* in which he discussed 'Facts of variation of Plants', appearing in the published book as chapters 9 and 10 (*Variation* 1: 305–72; see 'Journal' (Appendix II)).

[9] CD began to experiment on the pollination mechanism of *Leschenaultia* in April 1860 (see *Correspondence* vol. 8, letter to J. D. Hooker, 12 March [1860] and n. 1), noting that *L. formosa* seemed to have an effective mechanism for preventing cross-pollination, the pollen being 'shed in the early bud' and 'there shut up round the stigma within a cup or indusium' (see *Correspondence* vol. 9, letter to *Journal of Horticulture*, [17 May 1861]). He had initially suspected that the two viscid surfaces on the outside of the indusium might in fact be stigmatic surfaces, thus requiring insect agency to effect pollination (see *Correspondence* vol. 8, letter from J. D. Hooker, 26 April [1860]), but Hooker convinced him that the stigma was *inside* the indusium (*Correspondence* vol. 8, letter to J. D. Hooker, [28 April 1860]). However, CD carried out further investigations that encouraged him to maintain that pollination involved removal of the pollen from the indusium by insects, thus allowing *cross*-pollination (see *Correspondence* vol. 8, and *Correspondence* vol. 9, letter to *Journal of Horticulture*, [17 May 1861]). See also n. 10, below. CD's experimental notes recording the observations given here, which are dated 29–31 October 1862, are in DAR 265. CD subsequently published an account of his observations on *L. formosa*, made during 1860 and 1862, in a letter to the *Gardeners' Chronicle and Agricultural Gazette*, 9 September 1871 (*Collected papers* 2: 162–5).

[10] CD told Hooker in his letter of 1 May [1862] that he had discovered the stigma to be exterior in *Leschenaultia biloba*, and Hooker subsequently confirmed the observation in that and another species (see letters from J. D. Hooker, [16 May 1862] and [17 May 1862]).
[11] CD refers to the Goodeniaceae; for his work on *Scaevola*, see *Correspondence* vol. 8.
[12] Letter from C. W. Crocker, 31 October 1862.
[13] Haughton 1862.
[14] Samuel Haughton, professor of geology at Dublin University, was one of the earliest critics of the theory of natural selection. In his presidential address to the Geological Society of Dublin on 9 February 1859 (Haughton 1860a), Haughton criticised CD and Alfred Russel Wallace's paper announcing the theory (Darwin and Wallace 1858); he subsequently criticised *Origin* in a review that appeared in the *Natural History Review* ([Haughton] 1860b).
[15] See letter from Asa Gray, 22 September 1862.
[16] Frances Harriet Hooker.
[17] CD had borrowed a specimen of the orchid *Masdevallia fenestrata* from the Royal Botanic Gardens, Kew (see letter to J. D. Hooker, 27 [October 1862]).

From J. B. Jukes 3 November 1862

Geological Survey of Ireland, | Office, 51, Stephen's Green, Dublin,
Nov 3rd *1862*

My dear Darwin

You know of course how the inaccuracy of some little trifling incidental statement in a book reflects its authority in the minds of some people who never take the trouble to compare the crooked grain of sand with the symmetrical mountain.

I heard your Origin of Species laughed at the other day because you assume that the young bird pecks its way out of the shell with its own beak.—[1] Certainly this seems to me an impossibility looking at the way in w^{h}. the young bird is coiled up in the shell with its beak almost under its wing (see for instance the plate in Rymer Jones Outlines An: Kingd)[2]

—Others present supposed that the old bird broke the egg for the young one to come out, & my wife tells me that that was the way all the old henwives told her it was done.—

Your critic however said that on applying to an old woman on the point she laughed at him and said "why Lord bless you, Sir, the young un' grows too big for the shell & busts it" & that does seem to me the most reasonable supposition—

Your statement at p 87 w^{d}. therefore either need revision or the facts on which it is based verification.—[3]

Believe me | Very truly yours | J. Beete Jukes

DAR 168: 92

[1] Jukes refers to *Origin*, p. 87 (see n. 3, below); the same passage appears in the third edition of *Origin* (p. 92).
[2] Jones 1841, p. 630 (fig. 292).
[3] *Origin*, p. 87. In his copy of the third edition of *Origin*, which is in the Rare Books Room–CUL, CD added opposite this passage (p. 92): 'all right | Jukes says I am laughed at about Chickens breaking eggs. See Dixon Poultry p. 213.' The reference is to Dixon 1848, a copy of which is in the Darwin Library–CUL. The passage remained unchanged in the fourth edition of *Origin* (p. 99).

To W. E. Darwin 4 [November 1862][1]

Down
4th

My dear William

There were only 5 seeds in packet sent; these were all perfectly good—[2]

But do not bother yourself & count anymore; from long practice I can do it very quickly: I have counted about 30 pods of wild—20 of home-grown; besides my artificially fertilised pods.[3] So I have had lots of counting! Next year I shall make 136 more crosses, so I shall have counting with a vengeance next year.—[4]

Good Bye | My dear old man | C. D

DAR 210.6: 105

[1] The date is established by the relationship to the letter from W. E. Darwin, 21 October [1862], and to the letter to W. E. Darwin, 30 [October 1862], and by reference to dated notes by CD and William, recording the numbers of seeds counted in the pods of *Lythrum salicaria* (see nn. 2 and 3, below).

[2] CD had asked William to count the seeds from six pods of each of the three forms of *Lythrum salicaria*, collected from plants growing wild near Southampton (see letter from W. E. Darwin, 21 October [1862], and letter to W. E. Darwin, 30 [October 1862]). On 2 November 1862, William recorded in his Botanical notebook (DAR 117: 50) that he had counted the seeds in one pod of the long-styled form, excluding 'those evidently void or those shrivelled on the convex side'.

[3] In October 1862, William had sent his father pods of each of the three forms of *Lythrum salicaria*, collected from plants growing wild near Southampton (see letter from W. E. Darwin, 21 October [1862] and nn. 2 and 3). CD's tally of the seeds from thirty-one of these pods is in DAR 27.2 (ser. 2): 45, 55–6; some of these notes are dated 8 and 13 October. In addition, his tally of the seeds from twenty-one pods collected from plants grown in his garden and pollinated by insects, is in DAR 27.2 (ser. 2): 53, dated 17 October 1862. CD's results from the crosses made with this species in 1862 are also recorded in DAR 27.2 (ser. 2).

[4] See letter to J. D. Hooker, 27 [October 1862] and n. 11.

To J. D. Hooker 4 November [1862]

Down Bromley Kent
Nov 4th

My dear Hooker

I have read the pages attentively (with even very much more admiration than the first time) & cannot imagine what makes D^{r}. D. accuse you of asserting a subsidence, of Arctic America.[1] No doubt there was a subsidence in N. America during glacial period & over a large part, but to maintain that the subsidence extended over nearly whole breadth of continent or lasted during whole glacial period, I do not believe he can support.— I suspect much of evidence of subsidence during glacial period there will prove false, as it largely rests on ice action, which is becoming, as you know, to be viewed as more & more subaerial.[2] If Dawson has published criticisms, I sh^{d} like to see them.[3] I have heard he is rabid against me, & no doubt partly in consequence against anything you write in my favour. (& never was anything published more favourable than the Arctic paper). Lyell had difficulty in preventing

Dawson reviewing the Origin on hearsay, without having looked at it.[4] No spirit of fairness can be expected from so biassed a judge.—

All I can say is that your few first pages have impressed me far more this reading than the first time. Can Scandinavian portion of Flora be so potent, from having been preserved in the corner, warmed by Gulf stream, & from now alone representing the entire circumpolar flora, during the warmer preglacial period? From the first I have not been able to resist impression (shared by Asa Gray, whose Review on you pleased me much) that during Glacial period there must have been almost entire extinction in Greenland;[5] for depth of sea does not favour former southerly extension of land there: I must suspect that plants have largely been introduced by sea-currents, which bring so much wood from n. Europe.— But here we shall split as wide as the poles asunder.—[6] All the world could not persuade me, if it tried, that yours is not a grand essay; I do not quite understand whether it is this essay that Dawson has been "down on".— What a curious notion about glacial climate & Basques & Finns! Are the Basques mountaineers, I hope so. I am sorry I have not seen Athenæum, but I now take in the Parthenon.[7] By the way I have just read with much interest Max Muller; the last part about first origin of language seems the least satisfactory part.—[8]

Pray thank Oliver *heartily* for his heap of references on Poisons.—[9] How the Devil does he find them out? I must not indulge with Cypripedium: Asa Gray has made out pretty clearly that at least in some cases the act of fertilisation is effected by small insects being forced to crawl in & out of flower in a particular direction; & perhaps I am quite wrong that it is ever effected by proboscis.[10]

I retract so far that if you have the rare C. *hirsutissimum*, I sh^{d} very much like to examine a cut *single* flower; for I saw one at a Flower Show, & as far as I could see, it seemed widely different from other forms.—[11]

Farewell | Ever yours | C. Darwin.

P.S. Answer this if by chance you can.— I remember distinctly having read in some book of Travels, I am nearly sure in Australia, an account of the Natives during Famines, trying & cooking in all sorts of ways, various vegetable productions & sometimes being injured by them.. Can you remember any such account? I want to find it.[12] I thought it was in Sir. G. Grey, but it is **not**.—[13] Could it have been in Eyre's Book?[14]

Endorsement: '/62'
DAR 115.2: 168

[1] In his letter of 2 November 1862, Hooker asked CD to examine his paper on the distribution of Arctic plants (J. D. Hooker 1861a) to see whether the criticisms put forward by John William Dawson were justified. CD's annotated copy of the number of the *Transactions of the Linnean Society of London* in which J. D. Hooker 1861a appeared is in the Darwin Library–CUL.

[2] CD refers to the increasing popularity of the idea that many of the surface features of European geology were due to the action of land-ice. During the mid-nineteenth century, most British geologists (including CD), although considering valley glaciers to have been geologically important during the Pleistocene glacial period, had rejected the idea of a Europe-wide ice-sheet; it was believed instead

that most of the geological phenomena of Europe were due to the submergence of much of the landmass under the sea, and to the action of icebergs. This consensus had recently been shaken by the controversial work of Andrew Crombie Ramsay (Ramsay 1862), John Tyndall (Tyndall 1862), and Thomas Francis Jamieson (Jamieson 1862; see Davies 1969, pp. 301–13). CD's reading of Ramsay 1862 and Jamieson 1862 had largely convinced him that land-ice had been more extensive and geologically significant during the glacial period than was commonly supposed (see letter to A. C. Ramsay, 5 September [1862], and letter to Charles Lyell, 14 October [1862]).

3 See letter from J. D. Hooker, 2 November 1862 and n. 9.

4 The reference has not been identified; however, CD and Charles Lyell discussed Dawson's reaction to *Origin* in their correspondence in 1860 (see *Correspondence* vol. 8).

5 In his paper, Hooker sought to explain the general paucity of the Greenland flora on the hypothesis that 'the polar region was once occupied by the Scandinavian flora', and that the cold of the glacial epoch drove the vegetation southwards, causing the temperate flora of southern Greenland to be 'driven into the sea'. Following 'the return of the heat', Hooker argued, Greenland would have been repopulated by the migration of Arctic species surviving in the south of the peninsula (J. D. Hooker 1862a, p. 254). On reading Hooker's paper, CD concluded that 'during the coldest part of Glacial period, Greenland must have been quite depopulated', suggesting that it must then have been repopulated by 'accidental means of transport by ice & currents' (see letter to J. D. Hooker, 25 February [1862]; see also letter from J. D. Hooker, 3 March 1862, and letter to J. D. Hooker, 7 March [1862]). In his review of Hooker's essay (A. Gray 1862d, p. 145), Asa Gray espoused a similar view. See also letter to Asa Gray, 28 July [1862] and n. 11.

6 Since the 1840s, CD and Hooker had debated the merits of invoking former land-bridges or other means of dispersal to account for the present geographical distribution of animals and plants (see *Correspondence*, especially vols. 3 and 6). Whereas Hooker was favourably disposed to the idea of land-bridges, CD believed there was insufficient geological evidence in most cases, arguing instead for trans-oceanic migration. Between 1855 and 1857, he conducted an extensive series of experiments on the possible means by which organisms could be transported across oceans, but failed to convince Hooker of his position (see *Correspondence* vols. 5 and 6).

7 See letter from J. D. Hooker, 2 November 1862 and n. 17.

8 Max Müller 1861. Concerning Friedrich Max Müller's views on the origins of language, see the letter from Asa Gray, 4 and 13 October 1862, n. 3.

9 In his letter to Hooker of 27 [October 1862], CD suggested that Daniel Oliver might be able to provide a reference for experiments on the absorption of poisons by plants. No letter from Oliver on this subject has been found, but see the letter from J. D. Hooker, 2 November 1862.

10 See letter from J. D. Hooker, 2 November 1862 and n. 6. In *Orchids*, pp. 274–5, CD concluded that *Cypripedium* must be pollinated by an insect inserting its proboscis into one of the two lateral entrances at the base of the labellum, directly over one of the two lateral anthers, and thus either placing the pollen onto the flower's own stigma, or carrying it away to another flower. In 'Fertilization of orchids', pp. 155–6 (*Collected papers* 2: 152), CD stated:

> Prof. Asa Gray, after examining several American species of *Cypripedium*, wrote to me . . . that he was convinced that I was in error, and that the flowers are fertilized by small insects entering the labellum through the large opening on the upper surface, and crawling out by one of the two small orifices close to either anther and the stigma.

Gray detailed his hypothesis in A. Gray 1862b, p. 428.

11 See letter to Asa Gray, 10–20 June [1862] and n. 17. There is an undated note in DAR 70: 120 that states:

> Cypripedium [*underl brown crayon*] hirsutissimum w$^{d.}$ be worth experimenting, on account of testing my remarks, supposing that I am correct that upper anther is soldered to labellum.

12 CD was preparing a draft of the part of *Variation* dealing with 'Facts of variation of Plants' (see 'Journal' (Appendix II)); in the introduction to the section, he included 'a few general remarks on the

origin of cultivated plants' (*Variation* 1: 306–12). CD countered the arguments of some botanists that for cultivated plants like cereals to have been 'noticed and valued as objects of food', their original state must have closely resembled their present one; he pointed to 'the many accounts given by travellers of the wretched food collected by savages' (*Variation* 1: 307), and continued, without giving a reference: 'I have read an account of the savages of Australia cooking, during a dearth, many vegetables in various ways, in the hopes of rendering them innocuous and more nutritious.'

[13] Grey 1841.

[14] Eyre 1845.

To Asa Gray 6 November [1862][1]

Down Bromley Kent
Nov. 6th

My dear Gray

When your note of Oct 4th & 13th, (chiefly about Max Muller) arrived I was nearly at the end of the same book & had intended recommending you to read it![2] I quite agree that it is extremely interesting, but the latter part about *first* origin of language much the least satisfactory.[3] It is a marvellous problem. I have heard, whether truly or not I do not know, but the book has rather given me the same impression, that he is dreadfully afraid of not being thought strictly orthodox. He even hints at truth of Tower of Babel![4] I thus accounted for covert sneers at me, which he seems to get the better of towards the close of the book.—[5] I cannot quite see how it will forward "my cause" as you call it; but I can see how anyone with literary talent (I do not feel up to it) could make great use of the subject, in illustration. What pretty metaphors you would make from it![6] I wish some one would keep a lot of the most noisy monkeys, half free, & study their means of communication!

A book has just appeared here, which will, I suppose make a noise, by Bishop Colenso, who, judging from extracts smashes most of the old Testament.—[7] Talking of Books, I am in middle of one which pleases me, though it is very innocent food, viz "Miss Cooper's Journal of Naturalist". *Who is she?*[8] She seems a very clever woman & gives a capital account of the battle between *our* & *your* weeds.[9] Does it not hurt your Yankee pride that we thrash you so confoundedly. I am sure Mrs. Gray will stick up for your own weeds. Ask her whether they are not more honest downright good sort of weeds.—[10] The Book gives an extremely pretty picture of one of your villages; but I see your autumn, though so much more gorgeous than ours, comes on sooner, & that is one comfort.

I am glad to hear that you have sent off your account of orchids to Newhaven; let me have a copy if you can, for I see no periodicals.—[11] I wish you had an active pupil in the country: it would be curious to block up with cotton or something the holes on each side of the sterile anther in Cypripedium; & then if pollen were at all disturbed it would show that little insects had entered by the toe.[12] I shall be very glad indeed of Mitchella, & of seed, if possible, of Nessæa. I am more than ever interested in Lythrum; the seed of my 88 crossed flowers prove truth of diagram, if you remember it;[13] but there is something more, mid-styled is in addition moderately fertile with half its own stamens, & I must make many more crosses, & shall not publish this year—[14] The case, I think, is worth any labour.—

I did not know, & yet doubt, that it was Hooker who reviewed me so gorgeously in Gardeners Ch^le. but I have asked him.—[15]

Now for two questions: please give me *reference* to your notice of Gourds affecting each others fruit;[16] & secondly on the movements of the tendrils.—[17]

Also do you know of any treatise descriptive of all your vars. of Maize, if so give me title.—; if I fail to get here, could you help me to half-a-dozen grains of the most marked varieties of Maize.—[18] What a good book "Downing on Fruit" is.—[19]

I am crawling steadily on & today have been compiling all about peaches & nectarines; & a curious case it is.—[20]

I sh^d. like to try a few experiments on your Tendrils;[21] I wonder what would be good & easy plant to raise in pot.

Farewell, my dear Gray. God help your poor country, though perhaps you scorn our pity. Farewell, my good friend | C. Darwin

Gray Herbarium of Harvard University (78)

[1] The year is established by the relationship to the letter from Asa Gray, 4 and 13 October 1862.

[2] Max Müller 1861; see letter from Asa Gray, 4 and 13 October 1862.

[3] Concerning Friedrich Max Müller's views on the origin of language, see the letter from Asa Gray, 4 and 13 October 1862, n. 3.

[4] Max Müller 1861, p. 125. The city and 'tower' of Babel are described in Gen. 11: 1–9 as being the scene of the 'confusion of languages' (*OED*).

[5] CD may refer to Max Müller's argument that humans and animals are separated by the barrier of language (Max Müller 1861, p. 340):

> Man speaks, and no brute has ever uttered a word. Language is our Rubicon, and no brute will dare to cross it. This is our matter of fact answer to those who speak of development, who think they discover the rudiments at least of all human faculties in apes, and who would fain keep open the possibility that man is only a more favoured beast, the triumphant conqueror in the primeval struggle for life. Language is something more palpable than a fold of the brain, or an angle of the skull. It admits of no cavilling, and no process of natural selection will ever distill significant words out of the notes of birds or the cries of beasts.

Towards the end of the book, however, Max Müller used Darwin's concept of natural selection to account for the development of human language (p. 371).

[6] CD had been impressed by Gray's use of metaphor in his three-part review of *Origin* published in the *Atlantic Monthly* in 1860 ([A. Gray] 1860), telling him that he was 'a complex cross of Lawyer, Poet, Naturalist, & Theologian' (see *Correspondence* vol. 8, letter to Asa Gray, 10 September [1860]).

[7] The reference is to volume 1 of John William Colenso's critical examination of the Old Testament (Colenso 1862–79); Colenso was bishop of Natal in southern Africa. Extracts from the volume appeared in a review in the *Parthenon*, 1 November 1862, pp. 833–4; CD subscribed to the *Parthenon* at this time (see preceding letter).

[8] Cooper 1855. Susan Augusta Fenimore Cooper was the daughter of the American novelist James Fenimore Cooper, and had been his amanuensis in the years before his death; Cooper 1855 was based on a journal she kept during her father's last years (*DAB*). On 19 August 1862, the *Journal of Horticulture and Cottage Gardener* published the first in a series of extracts from this book (*Journal of

Horticulture and Cottage Gardener n.s. 3 (1862): 400); CD's annotated copies of this journal are in the Darwin Library–CUL. CD had previously noted an earlier edition of this work (Cooper 1850) in his list of 'Books to be Read', 1837–51 (see *Correspondence* vol. 4, Appendix IV, *119: 24).

9 Cooper 1855, 1: 117–23.

10 CD refers indirectly to Jane Loring Gray's strong opinions about the attitudes of the British in regard to the American Civil War (see, for example, letter to Asa Gray, 22 January [1862] and n. 8).

11 CD refers to Gray's follow-up article to his review of *Orchids*, which appeared in the November 1862 number of the *American Journal of Science and Arts* (A. Gray 1862b); the journal was published in New Haven, Connecticut. See letter from Asa Gray, 4 and 13 October 1862.

12 See preceding letter, n. 10. Gray had reiterated a promise to send CD specimens of American species of *Cypripedium* in his letter of 4 and 13 October 1862.

13 See letter to Asa Gray, [3–]4 September [1862].

14 See letter to J. D. Hooker, 27 [October 1862] and nn. 11 and 12.

15 [J. D. Hooker] 1862c. See letter to J. D. Hooker, 3 November [1862] and n. 4. CD's annotated copy of the *Gardeners' Chronicle* is in the Cory Library, Cambridge Botanic Garden; CD kept in a separate parcel his copies of the numbers in which [J. D. Hooker] 1862c appeared (see DAR 222 and DAR 75: 1–12).

16 CD refers to Gray's observations, given before the American Academy of Arts and Sciences on 9 February 1858, which were reported in the *Proceedings of the American Academy of Arts and Sciences* 4 (1860): 21–2 (see also n. 17, below). CD was preparing a draft of the part of *Variation* dealing with 'Facts of variation of Plants' (see 'Journal' (Appendix II)).

17 A. Gray 1858b. CD's annotated copy of the volume of the *Proceedings of the American Academy of Arts and Sciences* in which this paper was published is in the Darwin Library–CUL. CD had previously read this article (see *Correspondence* vol. 9, letter to Asa Gray, 17 February [1861] and n. 15); in his *Autobiography*, p. 129, he recalled that it was this that led him to study climbing plants. See also n. 21, below.

18 CD was preparing a draft of the part of *Variation* dealing with 'Facts of variation of Plants' (see 'Journal' (Appendix II)); he discussed maize in *Variation* 1: 320–3.

19 CD frequently cited Downing 1845 in *Variation*; there is an annotated copy of the work in the Darwin Library–CUL (see *Marginalia* 1: 205–7).

20 CD refers to his draft of *Variation* 1: 336–44.

21 In his *Autobiography*, p. 129, CD recalled of Gray:

> He sent me seeds, and on raising some plants I was so much fascinated and perplexed by the revolving movements of the tendrils and stems, which movements are really very simple, though appearing at first very complex, that I procured various other kinds of Climbing Plants, and studied the whole subject.

CD carried out numerous experiments in 1863 and 1864, and his paper, 'Climbing plants', was read before the Linnean Society of London on 2 February 1865.

From J. D. Hooker 7 November 1862

Royal Gardens Kew
Nov 7/62

Dear Old Darwin

I assure you it was not my fault! I worried Lindly over & over again to notice your Orchid: book in Chronicle, by the very broadest hints man could give. At last he said, "really I cannot, you must do it. for me," & so I did—volontiers—[1] Lindley felt that he ought to have done it himself, & my main effort was to write it "a la Lindley" & in this alone I have succeeded—that people all think it is exactly Lindley's style!!! which diverts me vastly. The fact is between ourselves I fear that poor L. is breaking up— he said that he could not fix his mind on your book.[2]

Miss L.[3] told my wife[4] the other day, that he had twice lately lost all consciousness of outer world, once in a shop. He works himself beyond his mental or physical powers.—[5] His left arm is nearly disabled by an affection of the thumb (ligaments?) & the right is grumbling—all this entre nous. And now my dear Darwin, I may as well make a clean breast of it, & tell you that I wrote the Nat. Hist. Review notice too—[6]to me a very difficult task, & one I fancied I failed in, comparatively. of this you are no judge & can be none, you told me to tell Oliver it pleased you, & so I am content & happy.[7]

I am greatly relieved by your letter this morning about my Arctic Essay,[8] for I had been conjuring up some egregious blunder (like the granitic plains of Patagonia)—[9] Certes after what you have told me of Dawson, he will not like the letter I wrote to him days ago in which I told him that it was impossible to entertain a strong opinion against Darwinian hypothesis without its giving rise to a mental twist when viewing matters in which that hypothesis was or might be involved—[10] I told him I felt that this *was* so with me when I opposed *you*, & that all minds are subject to such obliquities!— the Lord help me, & this to an L.L.D & Principal of a College!—[11] I proceeded to discuss *his* Geology with the effrontery of a novice; &, thank God, I urged the very argument of your letter about evidence of subsidence, viz, not all submerged at once, & glacial action being subærial & not oceanic— Your letter hence was a relief, for I felt I was hardly strong enough to have launched out as I did to a professed Geologist.

The main part of Dawsons criticism is I suppose in the sup to pamphlet herewith sent,[12] but A Gray tells me to expect a *blast* from Canada this winter.[13]

I quite see, & feel the force of your impression that Greenland may have been repeopled from Scandinavia; but if so I should think not by oceanic currents,—[14] I have speculated on the probability of there having been a post glacial arctic Norwego-Greenlandian connection, which would account for the strong fact, that temperate Greenland is as Arctic as Arctic Greenland is, a fact, to me, of astounding force.— I do confess, that such a Northern migration would thus fill Greenland as it is filled, in so far as the whole Flora (temperate & arctic) would be Arctic—but then the same plants should have gone to the other polar Islands, & above all, so many common Arctic Scandinavian plants should not be absent in Greenland. still less should whole Nat Ords be absent, & above all the Arctic Leguminosæ.— It is difficult (as I have told Dawson) to conceive of the force with which arguments drawn from the absence of certain familiar ubiquitous plants strike the Botanists.— I would not throw overboard altogether Sea-transport & water transport, but I cannot realize their giving rise to such anomalies, in the distribution, as Greenland presents.

So too I have always felt the force of your objection, that Greenland should have been depopulated in the Glacial period,[15] but then reflected that vegetation now ascends I forget how high (above 1000 ft) in Disco[16] in 70°. & that even in a glacial ocean there may always have been lurking places for the few hundred plants Greenland now possesses

Supposing Greenland were re-peopled from Scandinavia over ocean way, why should Carices be the chief things brought? why should there have been no *Leguminosæ* brought, no plants but high arctic— why no *Caltha palustris*—which *gilds* the marshes of Norway & *paints* the house tops of Iceland. In short to my eyes the trans oceanic migration would no more make such an assemblage than special creations will account for representative species.— & no "*ingenious wriggling*[17] ever satisfied me that it would!— there there.—

Say the word when you wish for *Oxalis sensitiva*.[18]

We have a *Cyprid. hirsutiss.* wh. we hope will flower in spring.[19]

I dined with Henry Christy last night who has just returned from celt hunting with Lartet, amongst Basques— they are Pyreneans.[20] Lubbock was there & told me that my previous speculation was one of Von Bærs,[21] & that the Finns are supposed to have made the Kœkken middens[22]

I read Max Muller a year ago—& quite agree, first part is excellent, last on origin of language fatuous & feeble, as a Scientific argument[23]

Incomplete[24]
DAR 101: 68–9, 73–4

[1] In his letter of 3 November [1862], CD asked Hooker whether he was the author of the three-part review of *Orchids* that appeared in the *Gardeners' Chronicle and Agricultural Gazette* in August and September 1862 ([J. D. Hooker] 1862c). John Lindley, who was the principal editor of the magazine (*DNB*), was a leading orchid specialist.

[2] During the last few years before his death in 1865, Lindley reportedly suffered from 'gradual softening of the brain' (*DNB*).

[3] The reference is probably to Lindley's unmarried daughter, Barbara.

[4] Frances Harriet Hooker.

[5] As a young man, Lindley had assumed responsibility for his father's heavy debts, and, partly as a result of financial necessity, he undertook 'ever more tasks and duties without relinquishing those he already had' (*DSB*).

[6] [J. D. Hooker] 1862d.

[7] CD had initially believed the review ([J. D. Hooker] 1862d) to have been written by Daniel Oliver (see letter to J. D. Hooker, 6 October [1862] and n. 10, and letter to Daniel Oliver, 13 October [1862]).

[8] See letter to J. D. Hooker, 4 November [1862].

[9] Hooker refers to an error in J. D. Hooker 1844–7, p. 212, where he wrote of 'the granitic formation of Patagonia'. Although no correspondence between CD and Hooker on this point has been found, CD's copy of the work, which is in the Darwin Library–CUL, has the word 'granitic' underlined, and '?' marked in the margin (see *Marginalia* 1: 388–92).

[10] John William Dawson. See letter from J. D. Hooker, 2 November 1862, and letter to J. D. Hooker, 4 November [1862].

[11] Dawson was principal of McGill University, Montreal.

[12] Hooker refers to Dawson's pamphlet *Alpine and Arctic plants* (Dawson 1862a). No copy of the pamphlet has been located that includes a supplementary critique of J. D. Hooker 1861a. Hooker probably refers to Dawson's review of the work in the *Canadian Naturalist and Geologist* for October 1862 (Dawson 1862b); a version of this review may also have been issued with some copies of Dawson 1862a.

[13] Dawson discussed Hooker's views on palaeogeography in a letter to Asa Gray of 19 September 1862 that is at the Gray Herbarium of Harvard University (see Sheets-Pyenson 1992, p. 15). Gray had

apparently forwarded this letter to Hooker with his own letter of 15 October 1862 (Royal Botanic Gardens, Kew, Asa Gray letters: 308), warning him: 'Expect a severe blast from Canada this winter.'

14 See letter to J. D. Hooker, 4 November [1862] and nn. 5 and 6.

15 See letter to J. D. Hooker, 4 November [1862] and n. 5.

16 Disko Island and Disko Bay are on the west coast of Greenland.

17 Hooker had light-heartedly accused CD of having a very 'elastic' theory as early as 1854 (see *Correspondence* vol. 5, letter to J. D. Hooker, 7 July [1854]), and his ability to 'wriggle out' of theoretical difficulties had become a joke between the two friends (see, for example, *Correspondence* vol. 6, letter from J. D. Hooker, [16 November 1856], and letter to J. D. Hooker, 18 November [1856], *Correspondence* vol. 8, letter to J. D. Hooker, 7 May [1860], and *Correspondence* vol. 9, letters to J. D. Hooker, 27 [March 1861] and 7 November [1861]).

18 See letter to J. D. Hooker, 27 [October 1862] and n. 6.

19 The reference is to the orchid *Cypripedium hirsutissimum* (see letter to J. D. Hooker, 4 November [1862]).

20 See letter to J. D. Hooker, 4 November [1862] and n. 7. Henry Christy and Edouard Amant Isidore Hippolyte Lartet excavated together at a number of prehistoric sites in Europe in the early 1860s, most notably at the Vézère valley caves in the Dordogne (*DNB*). Their 1862 itinerary has not been identified.

21 See letter from J. D. Hooker, 2 November 1862. The references are to John Lubbock and Karl Ernst von Baer; the work by Baer has not been identified.

22 Lubbock had given an account of the prehistoric 'Kjökkenmöddings', or kitchen middens of Denmark, in Lubbock 1861.

23 Max Müller 1861. See letter to J. D. Hooker, 4 November [1862] and n. 8.

24 Some indication of the content of the missing portion of this letter is given by CD's reply (letter to J. D. Hooker, [10–]12 November [1862]).

To *Gardeners' Chronicle* [before 8 November 1862][1]

Will any one learned in Peas have the kindness to tell me whether Knight's Tall Blue and White Marrows were raised by Knight himself?[2] If so, I presume that they are the offspring of the crosses described by him in the Philosophical Transactions for 1799.[3] I find that the name "Knight" tacked to a Pea is not a guarantee that the sort was of his production. I will beg permission to ask one other question. Has any one who has saved seed Peas grown close to other kinds observed that the succeeding crop came up untrue or crossed?[4] This certainly occurs rarely if ever; yet from what I have observed on the manner of fructification of the Pea, I should have expected that such crossing would occasionally happen, as in the case of Dwarf Kidney Beans, of which fact I gave a striking example in your Paper of October 25, 1857.[5]

Charles Darwin, Down, Kent.

Gardeners' Chronicle and Agricultural Gazette, 8 November 1862, p. 1052

1 The letter was published in the issue of 8 November 1862 (see also *Collected papers* 2: 70).

2 CD was preparing a draft of the part of *Variation* dealing with 'Facts of variation of Plants' (see 'Journal' (Appendix II)); he discussed Thomas Andrew Knight's pea varieties in *Variation* 1: 326, 329–30.

3 Knight 1799; CD discussed Knight's experiments in *Variation* 2: 129.

4 See letter from Adam Fitch, 18 November 1862.

[5] CD was mistaken about the date; his letter was published in the *Gardeners' Chronicle and Agricultural Gazette* on 24 October 1857 (see *Correspondence* vol. 6, letter to *Gardeners' Chronicle*, 18 October [1857]).

From Asa Gray 10 November 1862

Cambridge, Mass
Nov. 10, 1862

Dear Darwin

Here is a new stamp for L. D.—tho' not *postage*.[1]

And I shall put this in an envelope embossed with a *20 cent* postage stamp. I have really nothing to write this week. I trust I shall receive to-morrow in time for foreign post, some copies of sheets of notices in Nor. Amer. Journal,—in which there are two articles upon which I wish your opinion.— One of them is continuation of remarks on your far-famed Orchid-book.—with the substance of my notes on our species of Ophryd. & Cypripedium.[2]

I am waiting for Capt. Anderson to come to this port that I may send you Cypripediums, &c—for your study next spring.[3] If he does not come over in next Cunarder, I fear I shall have to consign my package to Kew.— —where they may get out of the way before Hooker can turn them over to you.[4] We are now wintry with precocious snow; but we still expect a short Indian summer.

I have to thank you for yours of the 16 Oct.—which has been lying a fortnight here.[5] As you do not speak of your family, I conclude they are doing very well.

It is just as I thought about Rothrock's notes on *Houstonia*.[6] All his projected experiments came to nothing, as I thought.—not well carried out.

It is refreshing to me that you find the Special Correspondent of the Times *detestable*.[7]

Your comments upon our affairs always show such a good spirit, that you need not fear even my wife's "indignation".[8]

We are sorry that you suffer in England; but you must blame the rebels for it, not us,—and your Manchester people should have looked earlier to India for cotton.[9]

You dont see, as you would if here—the total impossibility of coming to any terms of peace with the South, based on their independence. Before that can be they or *we* must be thoroughly beaten. You can't be expected to see too,—what seems plain to me, that you English would give us no end of trouble, if we attempted a piece-meal existence. We must be strong enough to keep any Old-world power at bay. Then we shall behave pretty well, on the whole,—surely so when the North is dominant and is fairly treated. "Siezing on Canada".[10] What do we want of Canada? When the South was aggressive and making Slave States, we often looked to the *peaceful* acquisition of Canada as desirable as a counterpoise— But when we had "changed all that"—and *it is* changed, a⟨nd⟩ slavery limited, past all do⟨ubt,⟩—however the combat ends— we no longer have use or need of Canada. If we get set up again, we have work enough at home, & our hands full for years— we shall be strong for defence but weak for aggression. The ill-feeling to England will die out when we are well able to defend ourselves and our home interests.

It does seem that all England *wishes* us to be weak and divided,—perhaps that is good national policy. But the more that is so, the more necessary it is for us to vindicate our integrity, at whatever cost. Let us have it out now, even at the cost of 10 times what it has cost so far.

I never thought anything of American institutions for England.[11] Aristocracy is a natural & needful appendage to Monarchy. You work out your own type—and you will liberalize fast enough,—and leave us to do ours. We'll make it do,—with some jangling.—

I wish we could be *shut up*, like the Japanese of old,—for 10 or 20 years,—[12] —only with a weekly ⟨ma⟩il from you and Dr. Hooker, ⟨Fa⟩re-well.— Ever Yours cordially | Asa Gray.

Incomplete[13]
DAR 165: 122

CD ANNOTATION
Top of letter: '(Please return as I shall have to answer)' *ink*[14]

1 Leonard Darwin had written to Gray at his suggestion to tell him which North American postage stamps he most wanted for his collection (see letter from Asa Gray, 22 September 1862 and n. 3).

2 Gray refers to the November 1862 number of the *American Journal of Science and Arts*, which contained the follow-up article to his review of *Orchids* (A. Gray 1862b), in which he gave an account of his observations on several North American species of Ophrydeae and *Cypripedium*. The second item to which Gray refers was probably his brief notice on dimorphism in the stamens and pistils of flowers (A. Gray 1862e), published in the same number of the *American Journal of Science and Arts*.

3 James Anderson was captain of the Cunard line's trans-Atlantic steamer *Africa* (*Men of the time*). In his letter to CD of 5 September 1862, Gray promised to send live roots of three American species of the orchid genus *Cypripedium* and seeds of *Mitchella* (see also letter from Asa Gray, 27 October 1862 and n. 3).

4 Joseph Dalton Hooker was assistant director of the Royal Botanic Gardens, Kew.

5 Letter to Asa Gray, 16 October [1862].

6 See letter to Asa Gray, 16 October [1862] and n. 5. Joseph Trimble Rothrock had been Gray's assistant and student at the Lawrence Scientific School, Harvard University, until he enlisted in the Union army in July 1862 (*DAB*). In his letter of 4 August 1862, Gray sent CD an abstract of Rothrock's notes from his observations and experiments on the dimorphic plant *Houstonia caerulea*, carried out in the summer of 1862 at CD's request. For Gray's opinion of Rothrock's work, see also the letter from Asa Gray, 22 September 1862.

7 See letter to Asa Gray, 16 October [1862]. The reference is to Charles Mackay.

8 Jane Loring Gray. See letter to Asa Gray, 16 October [1862] and n. 16.

9 Manchester was the centre of the cotton textile industry in Britain. After the start of the American Civil War in 1861, the Union navy established a blockade of the Confederate states, cutting off the supply of cotton to Europe; supply reached its lowest level in 1862. In July, the *Economist* reported that the time when 'mills must stop and Lancashire must starve from an actual exhaustion of the whole supply of raw materials' might be 'very near at hand' (*Economist*, 5 July 1862, p. 729). For accounts of the Lancashire cotton famine, see Arnold 1864 and Longmate 1978.

10 See letter to Asa Gray, 16 October [1862].

[11] In his letter to Gray of 16 October [1862], CD commented that the American Civil War had retarded the prospect of democratic reform in England, and produced 'wide spread feeling in favour of aristocracy & monarchism'.

[12] The Japanese Tokugawa regime instituted the Sakoku, or closed country policy, in 1639; it was maintained until 1854 (Jansen ed. 1989).

[13] There was a postscript to this letter, which has not been found; it was apparently a response to CD's queries about maize, included in his letter of 16 October [1862] (see letter to Asa Gray, 26[–7] November [1862], and letter from Asa Gray, 24 November 1862).

[14] This is a note to Hooker, to whom CD sent this letter (see letter to J. D. Hooker, 24 [November 1862]).

To J. D. Hooker [10–]12 November [1862][1]

Down Bromley Kent
Nov 12th.

My dear Hooker

What a long & interesting letter you have sent me, & you so hard-worked.—[2]

So you did write the Review in Gardeners Chrone: Once or twice I doubted whether it was Lindley;[3] but when I came to a little slap at R. Brown,[4] I doubted no longer. You arch-rogue!— I do not wonder you have deceived others also. Perhaps I am a conceited dog; but if so, you have much to answer for: I never received so much praise, & coming from you I value it, much more than from any other.—

I am extremely sorry to hear about Lindley, & I am glad you told me, for I was going to have written to him to ask him about any odd Potatoes in Hort. Gardens.[5] I have thought of late that Gardeners' Chronicle has been rather flat. It is wonderful for my purposes, what a lot of matter the Gardeners' Chron. has contained during last 20 years.—[6]

I return by this Post Dawson's lecture, which seems to me interesting, but with nothing new.[7] I think he must be rather conceited with his "If Dr. Hooker had known this & that, he would have said so & so".— It seems to me absurd in Dawson assuming that N. America was under sea during *whole* Glacial period.—[8] Certainly Greenland is a most curious & difficult problem. But as for the Leguminosæ, the case, my dear fellow, is as plain as a pikestaff, as the seeds are so very quickly killed by sea water.[9] Seriously it would be a curious experiment to try vitality in salt-water of the plants which ought to be in Green-land:[10] I forget, however, that it would be impossible, I suppose, to get hardly any except the Caltha,[11] & if ever I stumble on that plant in seed I will try it.

I wish to Heaven some one would examine the rocks near sea-level at south point of Greenland & see if they are well scored; that would tell something;[12] but then subsidence might have brought down higher rocks to present sea-level. I am much more willing to admit your Norwego-Greenland connecting land than most other cases, from the nature of rocks in Spitzbergen & Bear Isld.—[13] You have broached & thrown a lot of light on a splendid problem, which some day will be solved. It rejoices me to think that when a boy I was shown an erratic boulder in Shrewsbury & was told by a clever old gentleman, that till the world's end no one would ever guess how it came there.—[14] It makes me laugh to think of D^r. Dawson's indignation at your sentence about "obliquity of vision":[15] by Jove he will try & pitch into you some day.—

Good Night for the present.—[16]

I am particularly obliged for sentence about Sikhim people in Famines, & I find nearly same in your Journal: it is a little point which interests me, & I shall quote it.[17]

(To return for a moment to Glacial period, you might have asked Dawson whether Ibex, Marmot &c &c were carried from mountain to mountain in Europe on floating ice; & whether musk ox got to England on iceberg; yet England has subsided, if we trust to good evidence alone of shells *more* during glacial period, than America is known to have done.)[18]

Hearty thanks about seeds; I thought it useless to send my list to seedsman first, for besides 2 or 3 I sh^d. think there was no chance of getting them. Add, if you can, seeds of any Stylidium (for excitable column) of easy greenhouse culture.—[19] You write "say the word when you wish for Oxalis sensitiva".[20] Do you mean you could let me have some plants now? if so I sh^d. be delighted for I can try little experiments at odds & ends of time; *but could I keep plant sensitive now in greenhouse & study*? If so send them to G. Snow Nag's Head, Boro', by midday on any Thursday.[21]

Some day I will amuse myself by sending diagram with explanation on Lythrum, which you can study or burn as you like—[22] Good Heaven how your friends work you with M.S. (we will say nothing about another friend on seeds, plants, facts &c &c) For Heaven sake instill a word of caution into Tyndalls ears; I saw extract that valleys of Switzerland were wholly due to Glaciers.[23] He cannot have reflected on valleys in Tropical countries. The grandest valleys I ever saw were in Tahiti. Again if I understand, he supposes that Glaciers wear down whole mountain ranges, thus lower their height, decrease the temperature & decrease the glaciers themselves. Does he suppose whole of Scotland has been thus worn down? Surely he must forget oscillation of level would be more potent one way or another during such enormous lapses of time. It would be hard to believe any mountain range has been so long stationary.—

I suppose Lyell's book will soon be out:[24] I was very glad to see in News Paper that Murray sold 4000; what a sale![25]

To recur to Orchids, I hear that D. of Argyle has been praising my book in Edinburgh R. which is a good joke, as you told me he could not understand it.—[26]

I am now working on cultivated plants & rather like my work;[27] but I am horribly afraid I make the rashest remarks on value of differences; I trust to a sort of instinct, & God knows can seldom give any reason for my remarks. Lord in what a medley the origin of cultivated plants is in.— I have been reading on Strawberries & I can find hardly two Botanists agree what are the wild forms; but I pick out of horticultural books here & there queer cases of variation inheritance &c. &c.—

What a long letter I have scribbled; but you must forgive me, for it a great pleasure thus talking to you.—

Farewell | Ever yours | C. Darwin

Did you ever hear of "Condy's Ozonised Water"? I have been trying it with, I think *extraordinary* advantage to comfort at least,—a tea-spoon in water 3 or 4 times. a day.— If you meet any poor dyspeptic devil like me suggest it.—[28]

Endorsement: '/62'
DAR 115.2: 169

[1] CD initially dated the letter '10th', but later deleted the date and wrote '12th'; the text of the letter confirms that it was written over more than one day (see n. 16, below).

[2] Letter from J. D. Hooker, 7 November 1862.

[3] In his letter to CD of 7 November 1862, Hooker admitted having written the anonymous review of *Orchids* that had appeared in the *Gardeners' Chronicle and Agricultural Gazette* ([J. D. Hooker] 1862c), and which CD had assumed to be by the magazine's principal editor, John Lindley.

[4] Lindley and Robert Brown had quarrelled in the late 1820s and early 1830s, since which time Lindley had been extremely critical of Brown's work (Mabberley 1985).

[5] In his letter of 7 November 1862, Hooker told CD that he feared Lindley was 'breaking up'. CD was preparing a draft of the part of *Variation* dealing with 'Facts of variation of Plants' (see 'Journal' (Appendix II)), and had unsuccessfully sought 'odd *varieties*' of potato from Hooker, with the intention of growing a few plants of each for comparison (see letter to J. D. Hooker, 27 [October 1862], and letter from J. D. Hooker, 2 November 1862). Lindley was secretary of the Horticultural Society of London (Fletcher 1969).

[6] Lindley was principal editor of the *Gardeners' Chronicle and Agricultural Gazette* (*DNB*); CD's annotated copy of the magazine from 1841 to 1871 is in the Cory Library, Cambridge Botanic Garden. In DAR 222 there is a 'List of the numbers of special interest to Darwin and kept by him in separate parcels', and CD's abstracts of the journal are in DAR 75: 1–12.

[7] Dawson 1862a. See letter from J. D. Hooker, 7 November 1862 and n. 12.

[8] See letter from J. D. Hooker, 2 November 1862 and n. 13.

[9] See letter from J. D. Hooker, 7 November 1862.

[10] Between 1855 and 1857, CD had carried out a number of experiments in order to establish the means by which the transoceanic dispersal of plants might occur; he immersed seeds in salt water for periods of time, and then tested their ability to germinate (see *Correspondence* vols. 5 and 6).

[11] See letter from J. D. Hooker, 3 March 1862.

[12] CD refers to glacial striations, the occurrence of which at the southernmost part of Greenland would provide strong support for his view that there must have been 'almost entire extinction in Greenland' during the Pleistocene glacial period as a result of the extreme cold (see letter to J. D. Hooker, 4 November [1862]).

[13] In his letter to CD of 7 November 1862, Hooker mentioned having considered 'the probability of there having been a post glacial arctic Norwego-Greenlandian connection'. Unlike Hooker, CD was generally opposed to the invocation of former land-bridges to account for the present geographical distribution of plants and animals, on the grounds that there was insufficient geological evidence in most cases. See letter to J. D. Hooker, 4 November [1862] and n. 6.

[14] The reference is to the Shrewsbury naturalist, Richard Cotton (*Autobiography*, p. 52).

[15] See letter from J. D. Hooker, 7 November 1862. The reference is to John William Dawson.

[16] See n. 1, above.

[17] In his letter of 4 November [1862], CD asked Hooker for a reference to Australian aborigines preparing and eating poisonous plants during times of famine. Hooker's reply of 7 November 1862 is incomplete; however, the missing portion of Hooker's letter evidently contained information regarding starving inhabitants of a Sikkim village eating arum roots, after pounding and fermenting them to reduce their poisonous effects (see *Variation* 1: 307). A similar account appears in Hooker's *Himalayan journals* (J. D. Hooker 1854b, p. 49).

[18] See n. 8, above.
[19] CD enclosed with his letter to Hooker of 3 November [1862] a list of species of which he required seed; the list has not been found. CD had recently begun to experiment on plant sensitivity and movement (see letters to J. D. Hooker, 26 September [1862], 27 [October 1862], and 3 November [1862], and letter from J. D. Hooker, 25 October 1862).
[20] See letter from J. D. Hooker, 7 November 1862 and n. 18.
[21] George Snow operated a carrier service between Down and London every Thursday (Freeman 1978).
[22] See letter to J. D. Hooker, 24 [November 1862] and n. 15.
[23] The extract from Tyndall 1862 has not been identified.
[24] Charles Lyell's book, *Antiquity of man* (C. Lyell 1863a), was not published until 6 February 1863 (C. Lyell 1863b, p. [vii]).
[25] The trade sale referred to was held on 4 November 1862 (*Athenæum*, 8 November 1862, p. 595).
[26] [G. D. Campbell] 1862. See letter from J. D. Hooker, 16 September 1862 and n. 4.
[27] See n. 5, above.
[28] This disinfectant liquid was recommended to CD for bathing and drinking by George Chichester Oxenden (see letters from G. C. Oxenden, 21 June 1862 and 17 September [1862]).

From John Scott[1] 11 November 1862

Edinburgh | Botanic Gardens
⟨Nov. 11th, 1862.⟩

To Charles Darwin Esq.
Sir,

I take the liberty of addressing you for the purpose of directing your attention to an error in one of your ingenious explanations of the structural adaptations of the Orchidaceæ in your late work. This occurs in the genus *Acropera*, two species of which ⟨you⟩ assume to be unisexual, and so far as known represented by male individuals only.[2] Theoretically you have no doubt assigned good grounds for this view, nevertheless, experimental observations that I am now making, have already convinced me of its fallacy. And I thus hurriedly—and as you may think prematurely—direct your attention to it—before I have seen the final result of my own experiments, that you might have the longer time for re-considering the structure of this genus for another edition of your interesting book—if, indeed, it be not already called for.[3] I am, furthermore, induced to communicate the results of my yet imperfect experiments in the believe that the actuating principle of your late work is the elicitation of truth, and that you will gladly avail yourself of *this* even at the sacrifi⟨ce⟩ ⟨of much⟩ ingenious theoretical argumentation.

Since I have had an opportunity of perusing your ⟨work on⟩ Orchid Fertilisation, my attention h⟨as⟩ been ⟨particul⟩arly ⟨direc⟩ted ⟨to⟩ the curiously constructed floral organs ⟨of⟩ Acropera. I unf⟨or⟩tu⟨nate⟩ly have as yet had only a few flowers for experiment⟨al⟩ en⟨quiry,⟩ ⟨otherwise⟩ my remarks might have been clearer and more satisfact⟨ory⟩. ⟨Such as they are,⟩ however, I respectfully, lay before you, with a full assur⟨ance of their ver⟩acity, and I sincerely trust that as such you will receive them.

Your observations, appear to have been chiefly directed to the ⟨*A. lu*⟩*teola*, mine to the A. Loddiegesii, which, however, as you remark is ⟨in⟩ a ⟨very simi⟩lar structural condition with the former;[4] having the same ⟨narrow⟩ stigmatic chamber, abnormally developed placenta, &c. In regard ⟨to the⟩ former point—contraction of stigmatic chamber—I may remark that it does not appear to be absolutely necessary that the pollen-masses penetrate this chamber for effecting fecundation. Thus a raceme was produced upon a plant of A. Loddigesii in the Botanic Gardens here lately; upon this I left only six flowers, these I attempted to fertilise, but with two only of the six, have I been successful. I succeeded in forcing a single pollen-mass into the stigmatic chamber of one of the latter, but I failed to do this on the other; however, by inserting a portion of a pedicel with a pollinium attached, I caused the latter to adhere, with a gentle press—to the mouth of the stigmatic chamber. Both of these, as I have already remarked, are, nevertheless, fertilised, one of them I have cut off for examination—and its condition I will presently describe, the other is still upon the plant and promises fair to attain maturity. In regard to the other four flowers, I may remark that though similarly fertilised—part having pollinia inserted, others merely attached—they all withered and dropped off without the least swelling of the ovary. Can it be then that this is really a *Mono-synoicous* species?—part of the flowers male others truly hermaphrodite.[5]

In making longitudinal sections of the fertilised ovary ⟨before⟩ mentioned, I found the basal portion entirely destitute of ⟨ovules⟩ their place being substituted by transparent cellular ram⟨ification⟩ of the placentæ. As I traced the placentæ upwards the ovules bec⟨ame⟩ ⟨gr⟩ad⟨ually⟩ ⟨more abundant⟩ towards its apex. A transverse section ⟨n⟩e⟨ar⟩ the apex of ⟨th⟩e ovary, however, still exhibited a more than ordinary ⟨plac⟩ental development—i.e. congenerically considered—each end giving ⟨off two⟩ branches, which meet each other in the centre of the ovary, the ovules being irregularly and sparingly disposed upon their surfaces.

In regard to the mere question of fertilisation, then I am perfectly satisifed, but there are other points which require further elucidation. Among these I may particularly refer to the contracted stigmatic chamber, and the slight viscidity ⟨of it⟩s disk. ⟨The latter⟩, however, may be a consequence of uncongenial conditions ⟨as you do not⟩ mention particularly its examination by any author ⟨in its⟩ natural habitat.— If such be the case, the contracted stigmatic chamber will offer no real difficulty, should the viscous exudations be only sufficient to render the mouth adhesive. For as I have already shown the pollen-tubes may be emitted in this condition, and effect fecundation without being in actual contact with the stigmatic surface, as occurs pretty regularly in the fertilisation of the *Stapelias*, for example. But indeed, your own discovery of the independent germinative capabilities of the pollen-grains of certain Orchidaceæ is sufficiently illustrative of this.[6]

I may also refer to the peculiar abnormal condition, that many at least of the ovaries present on a comparative examination of their placentæ; and of which I beg to suggest the following explanation, though it is yet founded on limited observations. In examining certain young ovaries of *A. Loddiegesii*, I found some of them filled with the transparent membranous fringes of more or less distinctly cellular matter, which from your description of the ovaries of *luteola*, appears to differ simply in the greater development in the former species. Again in others I found small mammillary bodies, which appeared to be true ovules, though I could not perfectly satisfy myself as to the existence of the micropyle, or nucleus. I unfortunately neglected to apply any chemical test. The fact, however, that in certain of the examined ovaries few or none of the latter bodies occurred—the placenta al⟨on⟩e ⟨being⟩ developed in an irregular membranous form, taken in conjunct⟨ion⟩ ⟨wi⟩th the results of my experiments—before alluded to—on ⟨their⟩ fe⟨rtilisation⟩, leads ⟨m⟩e ⟨to⟩ infer *that two sexual conditions are presented by ⟨the⟩ flowers of this plant* In short that many of the ova⟨ries⟩ are now ⟨nor⟩mal⟨ly⟩ abortive, though Nature occasionally makes futile efforts for ⟨their⟩ per⟩fect development, in the production of ovuloid bodies, these ⟨then I regard as the⟩ male flowers. The others that are still capable of ⟨fertilisation, and⟩ likewise possessing male organs, are hermaphro⟨dite,⟩ and must as I think from the results of your comparative examinations present a somewhat different condition. As it can scarcely be supposed that ovules in the condition you describe, could ever be fertilised.

Th⟨is is⟩ at least the most plausible explanation I can offer for the different results ⟨in⟩ my experiments ⟨on⟩ the fertilisation of apparently similar morphologically constructed flowers, others may however occur to you. For there is not, as ⟨in⟩ the *Catasetum* any external change visible in the respective unisexual and bisexual flowers.[7] And yet it would appear from your researches that the ovules of *Acropera* are in a more highly atrophied condition than occurs in *Catasetum*,[8] though as you likewise remark M. Neuman has never succeeded in fertilising *C. tridentatum*.[9] If there be not then, an arrangement of the reproductive structures, such as I have indicated, how can the different results in M. Neuman's experiments and mine be accounted for? However as you have examined many flowers of both *A. luteola* and Loddigesii, such a difference in the ovulary or placental structures could scarcely have escaped your observation. But be this as it may the—to me at least—demonstrated fact still remains, that certain flowers of A. Loddigesii are capable of fertilisation and that though *[*there*]* are good grounds for supposing that important physiological changes are going on in the sexual phenomena of this species, there is no evidence whatever, for supposing that ex⟨terna⟩l morphological changes have so masqued certain individuals, as to prevent their recognition.

I would now, Sir, in conclusion beg you to excuse me for this infringement upon your valuable time, as I have been induced to write ⟨you in the⟩ believe that you have had negative results from other exper⟨iments⟩ before you ventured to propose your theoretical explanation, and consequently that you have been unknowingly led into error. I will ⟨contin⟩ue as opportunities present themselves to examine the many peculiarities you have pointed out in this as well as others of the Orchid ⟨family.⟩ And at present I am looking forward ⟨with⟩ anxiety for the ⟨maturation⟩

of the ovary of A. Loddigesii which will bear testi⟨mony⟩ ⟨to the⟩ veracity of the remarks I have ventured to lay before you

I remain | Sir | Your obedient servant | John Scott

To *Charles Darwin* Esq.

DAR 177: 77

CD ANNOTATIONS

1.1 I take ... them. 2.7] *crossed pencil*
3.5 I may ... Gardens 3.7] *scored pencil*
3.8 these I ... successful. 3.9] *scored pencil*
3.12 I caused ... fertilised, 3.14] *scored red crayon*
3.18 Can ... hermaphrodite. 3.20] *scored pencil*
4.2 basal] *underlined pencil*
4.4 A ... surfaces. 4.8] *scored pencil*
4.5 apex] *underlined pencil*
5.2 Among ... this. 5.13] *scored pencil*
6.11 the placenta ... flowers. 6.18] *scored pencil*
7.11 such a ... observation. 7.12] *scored pencil*
7.13 that certain ... recognition. 7.18] *scored pencil*
8.1 I would ... you 8.9] *crossed pencil*; 'Apex means Flower end.—'[10] *ink*
Top of letter: 'Acropera | Gongora | Stanhopea | &c' *blue crayon*

[1] The manuscript has deteriorated considerably. Missing words have been transcribed from a copy of the letter, now in DAR 147, which was made for the use of Francis Darwin in editing CD's correspondence (see *ML* 2: 302–6).

[2] *Acropera luteola* and *A. loddigesii* are discussed in *Orchids*, pp. 203–10. At the start of his account (p. 203), CD reported: 'this genus for a long time remained the opprobrium of my work. All the parts seemed determinately contrived that the plant should never be fertilized.' He considered that he had partly solved the 'mystery' by the inference that *Acropera luteola* was dioecious, and that the specimen from which he had seen flowers was male. He concluded (p. 209):

> What the female or hermaphrodite form may prove to be— whether resembling in most respects the male, or whether it be at present named and masked as some distinct genus—it is impossible to say.

[3] The second edition of *Orchids* was not published until 1877; however, CD referred to Scott's observations and experiments on this genus in 1869, in 'Fertilization of orchids'. This paper comprised revised and additional notes on the subject, keyed by page numbers to the first edition of *Orchids.* Under the heading '*Sexes of* Acropera *Not Separated*' ('Fertilization of orchids', p. 153; *Collected papers* 2: 150) he reported:

> I have committed a great error about this genus, in supposing that the sexes were separate. Mr. J. Scott, of the Royal Botanic Garden of Edinburgh, soon convinced me that it was a hermaphrodite, by sending me capsules containing good seed, which he had obtained by fertilizing some flowers with pollen from the same plant.

[4] *Orchids*, p. 203 n.

[5] In *ML* 2: 303, Francis Darwin replaced the word 'Mono-synoicous' with the editorial insertion '[andromonœcious]'; Scott apparently intended to combine the senses of 'monoecious' and 'synoicous'. In 1863, the latter (now obsolete) term was reported to be applied to plants in which male and female flowers are found 'on the same *receptacle*' (see *OED*); however, Scott apparently used it simply as a synonym for 'hermaphrodite'.

[6] *Orchids*, p. 324 n.

[7] In *Orchids*, pp. 231–48, CD argued that *Catasetum tridentatum* was the male form of an orchid, the female and hermaphrodite forms of which were so different as to have previously been assigned to distinct genera (as *Monachanthus viridis* and *Myanthus barbatus*, respectively). Discussing the genus *Catasetum* more generally (p. 247), CD noted: 'The separation of the sexes is unknown in other Orchids, excepting as we shall see probably in the allied genus Cycnoches and in the before-given case of Acropera.'
[8] *Orchids*, p. 238.
[9] *Orchids*, p. 236 n.; the reference is to Louis Neumann, a gardener at the Muséum d'Histoire Naturelle, Paris.
[10] See letter from John Scott, 15 November [1862] and n. 8.

From Hugh Falconer 12 November [1862][1]

21 Park Crescᵗ.
12 Noᵛʳ.

My Dear Darwin

I have sent you a copy of a paper on the contested affinities of *Plagiaulax*[2]

It is not in your line of reading—much—but oblige me by glancing an eye over it.

Am I never likely to see you again—

Yours Ever | H. Falconer

DAR 164: 9

[1] The year is established by the reference to Falconer 1862 (see n. 2, below).
[2] Falconer 1862.

From J. D. Hooker 12 November 1862

Kew
Nov 12 *1862*

Dʳ Darwin

I have found out that Haughton is the identical one & indivisible with our Nat. & prejudiced reviewer of the Origin.[1] He is certainly a man of large capacity, & that is his special quality— he has taken to the Medical classes lately & turned his attention to poisons,—[2] My own impression as regards his strychnine case[3] is that it is only a deduction from Dʳ. Andersons discovery, some 8 years ago, that one Narcotic &c counteracts another, I do not know where Andersons paper was published, but can enquire & let you know—[4]

Haughton is as I am given to understand a man without any faculties of imagination or discovery.—but a plodder & applies with many horse power brains. He will grapple any subject on a moments notice, & the only thing to be deplored is that some of his friends would not put him up to a wrinkle or two on Theology, & let him get his steam up— wouldn't he just break things!—

I send the Maize book by Carrier today, & it will go on from Nag's head tomorrow by your Carrier.[5]

I have a little yesterday from our West Africa collector, he has been across Du Chaillus country & says his accounts are *all false*.[6] That impudent liar Burton (another Geogr. Soc. protegee) has in a public despatch) filched away all poor Mann's credit for the ascent of the Cameroons, calls it his Expedition, planned & carried out by him, & calls Mann his volunteer associate.[7] I never read any thing so gross in my life— Poor Mann had set his heart on this thing for 2 years, had failed the first time, & was actually leaving F. Po. for the ascent when Burton arrived at F. Po as Consul, did leave & had ascended the Mt. several weeks before Burton following him was at its foot; having prepared the way & provided guides & every thing— I am quite disgusted, but hardly know how to act, I dislike & despise the Geogr. Soc. way of going so much that I do not like to bring the matter forward there, & as to having a quarrel with Burton, we all know what it is to touch pitch.[8]

I have some more matters (in your letter) to write about, but they are not at hand here (B. House) where I am examining.[9]

Ever Yours | J D Hooker

DAR 101: 75–6

CD ANNOTATIONS[10]
Top of first page: '(Acropera)' *ink*
End of letter: 'Hooker | L Büchner'[11] | on Mann | Welwitschia' *ink*

[1] See letter to J. D. Hooker, 3 November [1862] and n. 14. Hooker refers to Samuel Haughton, author of a critical review of *Origin* that was published in the *Natural History Review* ([Haughton] 1860b).

[2] Haughton had been professor of geology at the University of Dublin since 1851; he entered the university's medical school in 1859, at the age of 38, in order to obtain a better knowledge of anatomy for his palaeontological researches. He graduated in 1862 and was appointed medical registrar of the school (*DNB*).

[3] Haughton 1862.

[4] The reference is probably to Anderson 1848, a paper documenting a 'Case of Recovery from a poisonous dose of Strychnia'; it was published in the *Monthly Journal of Medical Science* 8 (1848): 566–74. In the paper, Thomas Anderson, professor of chemistry at the university of Glasgow, and an expert on the chemistry of narcotics, stated (p. 569):

> The question may be raised as to how far the action of the strychnia may have been affected by the habitual use of opium . . . In this case, it is just possible that intoxication [with opium] may have prevented the immediate access of symptoms [of strychnine poisoning], exactly as it is known to do in the case of poisoning by opium.

[5] Bonafous 1836 (see letter from J. D. Hooker, [18 October 1862] and n. 6). Hooker refers to George Snow, who operated a weekly carrier service between Down and London.

[6] Gustav Mann was botanical collector for the Royal Botanic Gardens, Kew, on the Niger expedition led by William Balfour Baikie (R. Desmond 1995, p. 433). The veracity of Paul Belloni Du Chaillu's account of his explorations and adventures in equatorial Africa (Du Chaillu 1861) had been repeatedly questioned and was the cause of much public controversy (see Vaucaire 1930, pp. 125–41).

[7] Burton 1862. Richard Francis Burton was appointed British consul in Fernando Po, West Africa, in March 1862. He had previously received Royal Geographical Society support for a projected expedition to cross the Arabian peninsula, and for an exploring expedition into Central Africa (*DNB*).

[8] Hooker sought to secure Mann's reputation in a paper read before the Linnean Society on 5 November 1863 (J. D. Hooker 1863c). He apparently began the paper at the end of 1862; in a letter to

Thomas Henry Huxley of [before 19 December 1862] (The Huxley Papers, College Archives, Imperial College, London, V.3.78), Hooker wrote: 'I am making a précis of our poor German collector, G. Mann's, West African letters, to contradict Burton's assertions'. In his account of Mann's ascent of the Cameroon mountains (J. D. Hooker 1863c, p. 174), Hooker stated:

> Some of the physical characters of the group have been described in a memorandum transmitted to the Secretary of State for Foreign Affairs by Consul Burton, who accompanied Mr. Mann on his second visit to this group. The account there given of this adventurous expedition seeming to imply that it was one planned and conducted by Consul Burton, to which Mr. Mann had attached himself, I have been desired by Mr. Mann to publish the accompanying statement of the facts of the case as communicated by himself

The subsequent report demonstrated, as Hooker noted, that Mann began his exploration 'nine months before Consul Burton arrived on the coast of Africa'.

9 Hooker probably refers to examinations at University College London, where he was examiner in botany (*Medical directory* (1862): 200). University College was one of several organisations to have its offices at Burlington House, Piccadilly (*Post Office London directory* 1861).

10 See letter to J. D. Hooker, 18 [November 1862].

11 Ludwig Büchner.

To John Scott 12 November [1862][1]

Down. | *Bromley.* | *Kent. S.E.*
Nov. 12th

Dear Sir

I thank you most sincerely for your kindness in writing to me, & for very interesting letter.[2] Your fact has surprised me greatly, & has alarmed me not a little, for if I am in error about Acropera I may be in error about Catasetum.[3] Yet when I call to mind the state of the placentæ in A. luteola, I am astonished that they should produce ovules. You will see in my book that I state that I did not look at the ovarium of A. Loddegesii.[4] Would you have the kindness to send me word, which end of the ovarium is meant by *apex* (that nearest the flower?) for I must try & get this species from Kew & look at its ovarium.

I shall be extremely curious to hear whether the fruit, which is now maturing, produces a large number of *good* & plump seed; perhaps you may have seen the ripe capsules of other Vandeæ & may be able to form some conjecture what it ought to produce. In the young unfertilised ovaria of many Vandeæ, there seemed an infinitude of ovules. In desperation it occurs to me as just possible, as almost every thing in nature goes by gradation, that a properly male flower might occasionally produce a few seeds, in same manner as female plants sometimes produce a little pollen. All your remarks seem to me excellent & very interesting & I again thank you for your kindness in writing to me.

I am pleased to observe that my description of the structure of Acropera seems to agree pretty well with what you have observed. Does it not strike you as very difficult to understand how insects remove the pollinia & carry them to the stigmas?— Your suggestion that the mouth of the stigmatic cavity may become charged with viscid matter & thus secure the pollinia, & that the pollen-tubes may then protrude, seems very ingenious & new to me; but it would be very anomalous in Orchids, i.e. as far as I have seen. No doubt, however, though I tried my best, that I shall be proved wrong in many points. Botany is a new subject to me.—

With respect to the protrusion of pollen-tubes, you might like to hear (if you do not already know the fact) that, as I saw this summer, in the little imperfect flowers of Viola & Oxalis, which never open, that the pollen-tubes always come out of the pollen-grain, *whilst still in the anthers*, & direct themselves in a beautiful manner to the stigma seated at some little distance.—[5]

I hope that you will continue your very interesting observations, & I beg leave to remain | Dear Sir | Yours faithfully & obliged | Charles Darwin

DAR 93 (ser. 2): 7–10

[1] The year is established by the relationship to the letter from John Scott, 11 November 1862.
[2] Letter from John Scott, 11 November 1862.
[3] See letter from John Scott, 11 November 1862 and nn. 2, 3, and 7.
[4] *Orchids*, p. 210.
[5] CD's notes on his observations of *Viola canina* and *Oxalis acetosella*, made in May and June 1862, are preserved in DAR 111: 3–5, 44. See also letter to J. D. Hooker, 30 May [1862], letter to W. E. Darwin, [31 May 1862], and letter to Asa Gray, 10–20 June [1862].

From Robert Swinhoe 12 November 1862

18 Royal Avenue Terrace | Chelsea— SW.
12 Novr. 1862.

My dear Sir,

I have just returned from the country, and been engaged in unpacking some of the bones from Formosa—[1]In the large collections I have made I have many things that confirm your theory of species which I will bring to your notice so soon as my series are arranged.[2] At the present my chief object in writing to you is to forward you a specimen of the ordinary domestic Pigeon of China. It is unfortunately somewhat albino, but in form and proportions it is quite typical of the bird reared in the dovecots of China—[3]

I have further a remark to make regarding *Anatidae*— you questioned me some time ago as to whether the hybrid between the Muscovy and Chinese domestic Duck ever interbred.[4] Notwithstanding what Pallas states to the contrary in his Zoograph—Ross—Asiat.[5] I am confident that a thorough race has been produced between these two species, which though constantly bred in and in still preserves its characters, viz those of being much larger than the common chinese Duck, and of having the legs placed more central under the belly giving it a more goose-like appearance. This race has a smooth head, *never* wattled, and is usually black, has in Amoy (China) the name of *Aw-ah* or Black Duck, as distinguishing it from the *Ah* or Duck proper and the Liengtasu ah (Dragon-headed Duck) or Muscovy.

I remain, | Your's very truly, | Robert Swinhoe

P.S. The Pigeon is from Foochow. (Foochow)[6]

Charles Darwin, Esq^{e}.

DAR 177: 326

[1] Swinhoe was an amateur naturalist, who had served with the British Foreign Office in the Far East since 1854, latterly as Vice-Consul of Formosa (now called Taiwan). He returned to London in September 1862, bringing with him a large collection of specimens that included examples of sixteen new bird species from Formosa (P. B. Hall 1987).

[2] Swinhoe's name appears on CD's presentation list for *Origin* (see *Correspondence* vol. 8, Appendix III). Although this is the earliest known letter between Swinhoe and CD, they evidently corresponded from as early as 1855, when Swinhoe sent CD specimens of bird skins (see *Correspondence* vol. 5, CD memorandum, [December 1855], and *Variation* 1: 132 n.1); they had subsequently continued to correspond about pigeons (see *Correspondence* vol. 9, letter to P. L. Sclater, 4 May [1861]).

[3] CD referred to Swinhoe having sent him a 'dovecot-pigeon' from Foochow, China, in *Variation* 1: 186; see also n. 6, below.

[4] CD's letter has not been found; however, he wrote the section of *Variation* dealing with ducks in May 1861 (see *Correspondence* vol. 9, Appendix II).

[5] Pallas 1811–31, 2: 257.

[6] Swinhoe apparently repeated the word 'Foochow' in order to make the word clear; on the first occasion he had altered the spelling.

To Hugh Falconer 14 November [1862][1]

Down
Nov. 14th

My dear Falconer

I have read your paper with extreme interest, and I thank you for sending it, though I should certainly have carefully read it, or anything with your name, in the Journal.[2] It seems to me a masterpiece of close reasoning: although of course not a judge of such subjects, I cannot feel any doubt that it is conclusive. Will Owen answer you: I expect that from his arrogant view of his own position he will not answer.[3] Your paper is dreadfully severe on him; but perfectly courteous and polished as the finest dagger.[4] How kind you are towards me: your first sentence has pleased me more than perhaps it ought to do, if I had any modesty in my composition.[5] By the way after reading the first whole paragraph, I reread it not for matter, but for style; and then it suddenly occurred to me that a certain man once said to me, when I urged him to publish some of his miscellaneous wealth of knowledge, "Oh, he could not write,—he hated it", &c.[6] You false man, never say that to me again. Your incidental remark on the remarkable specialisation of Plagiaulax (which has stuck in my gizzard ever since I read your first paper[7]) as bearing on the number of preceding forms is quite new to me, and of course is in accordance to my notions a most impressive argument.[8] I was, also, glad to be reminded of teeth of camel and tarsal bones.[9] Descent from an intermediate form, Ahem![10]

Well, all I can say is that I have not been for a long time more interested with a paper than with yours. It gives me a demoniacal chuckle to think of Owen's pleasant countenance when he reads it.

I have not been in London since end of September; when I do come I will beat up your quarters if I possibly can;[11] but I do not know what has come over me: I am worse than ever in bearing any excitement. Even talking of an evening for less than two hours has twice recently brought on such violent vomiting and trembling;[12] that I dread coming up to London. I hear that you came out strong at Cambridge, and am heartily glad you attacked the Australian Mastodon.[13] I never did or could believe in him. I wish you would read my little Primula paper in Linn. Journ., Vol. VI. Botany (No. 22), p. 77 (I have no copy which I can spare)[14] as I think there is a good **chance** that you may have observed similar cases. This is my real hobby-horse at present. I have retested this summer the functional difference of the two forms in Primula and find all strictly accurate.[15] If you should know of any cases analogous, pray inform me. Farewell my good and kind friend.

Yours very sincerely | Ch. Darwin.

P.S. I am going to send a paper soon to Linn. Soc. on the genus Linum, like Primula.[16]

Copy
DAR 144: 27

[1] The year is established by the relationship to the letter from Hugh Falconer, 12 November [1862].

[2] Falconer sent CD a copy of Falconer 1862 with his letter of 12 November [1862]; it was published in the *Quarterly Journal of the Geological Society of London*.

[3] In R. Owen 1859 (republished as R. Owen 1860a), Richard Owen had disputed Falconer's identification of two fossil species of the marsupial genus *Plagiaulax* as herbivorous rodents (Falconer 1857), arguing instead that they were predatory carnivores. Falconer's paper was a forthright critique of Owen's position, in which he claimed to have answered '*seriatim*' all Owen's objections to his own view of the affinities of *Plagiaulax* (Falconer 1862, pp. 363, 365).

[4] In his paper, Falconer personalised the dispute between Owen and himself, emphasising that it arose 'from different methods having been followed by the observers in dealing with the evidence' (Falconer 1862, p. 350). He stated that he considered himself 'bound, in the interest of science,' either to support his previous interpretation, 'or frankly to admit the correction, if . . . found to be in error'. He continued:

> I am further impelled by my sense of self-respect, as an observer, to consider whether—apart from the conclusions—I have fallen into such errors of observation and description as would necessarily be implied, should Professor Owen's manner of viewing the objects prove correct

In his conclusion (p. 365), having countered Owen's arguments, Falconer again contrasted his own and Owen's methodology, stating: 'The case is of sufficient interest and importance to test the sufficiency of the respective modes of analysis.'

[5] Referring to *Origin* in a footnote (Falconer 1862, p. 348 n.), Falconer's opening sentence reads: 'One of the most accurate observers and original thinkers of our time has discoursed with emphatic eloquence on the imperfection of the geological record.'

[6] The reference has not been traced.

[7] In his former paper (Falconer 1857, p. 276), Falconer had suggested that *Plagiaulax* furnished a 'crucial test' of the generalisations made by Owen, William Benjamin Carpenter, and others, that the degree of specialisation within a group like the Mammalia increased over geological time. In particular, he noted, these naturalists argued that the earliest Mammalia usually possessed the full complement of teeth, while forms characteristic of later times were 'remarkable for the special suppression of these organs'. In opposition to this view, Falconer argued that, while *Plagiaulax* was 'the oldest well-ascertained

herbivourous mammal yet discovered', it represented 'the most specialized exception ... from the rule to be met with in the whole range of the Marsupialia, fossil or recent.' This argument had implications for CD's view that natural selection led to a constant tendency to specialisation of function (see Bowler 1976 and Ospovat 1981). In his abstract of the number of the *Quarterly Journal of the Geological Society of London* in which Falconer 1857 appeared (DAR 75: 21), CD noted: 'Falconer on Plagiaulax ... p 276 a most specialized form.' See also letter to Hugh Falconer, [7 March 1857?] (*Calendar* no. 3791).

8 In his conclusion (Falconer 1862, p. 365), Falconer stated:

> If ... *Plagiaulax* be regarded through the medium of the view advocated with such power by Darwin, through what a number of intermediate forms must not the genus have passed before it attained the specialized condition in which the fossils come before us!

See also n. 9, below.

9 See Falconer 1862, p. 353. CD refers to the functional correlation, established by Georges Cuvier, between the upper canine-shaped incisors of the camel and the bones of the ankle or tarsus (Cuvier 1812, 1: 64). In his abstract of the number of the *Quarterly Journal of the Geological Society of London* in which Falconer 1862 appeared (DAR 75: 24), CD noted:

> Falconer p. 353. Excellent on connection (by Descent) of canines & bones of leg in camels. p. 365. on antiquity of plagiaullax, judged by *its specialization [*above del* 'antiquity of']

10 CD apparently refers to the fact that Falconer, who had previously been opposed to the theory of the transmutation of species, appeared to give countenance to such a view by writing of *Plagiaulax* passing through 'intermediate forms' (see n. 8, above). See also letter to J. D. Hooker, 24 [November 1862] and n. 12.

11 See letter from Hugh Falconer, 12 November [1862]. Falconer had regretted not seeing CD when he visited London on 29 and 30 September (see letter from Hugh Falconer, 4 October 1862).

12 See letter to John Lubbock, 23 October [1862], and letter to W. E. Darwin, 30 [October 1862].

13 At the annual meeting of the British Association for the Advancement of Science, held in Cambridge in October 1862, Owen read a paper entitled 'On a tooth of *Mastodon* from Tertiary marls, near Shanghai, China', in which he made reference to *Mastodon Australis*, a species he had proposed in 1844 on the basis of a single fossil molar tooth. The report of the meeting in the *Parthenon*, 11 October 1862, p. 754, stated that, following Owen's presentation, Falconer, who was an authority on fossil elephants,

> disputed the possible or even probable former existence of the elephant in Australia, contending, in an elaborate argument, that the data of its asserted discovery were worthless, and strongly protesting against the third repetition of a statement which rested on so slight a foundation.

The report concluded:

> Professor Owen gracefully acknowledged the loose and possibly inaccurate data upon which the discovery was claimed, and as, during the twenty years which had elapsed ... no other specimens or corroborative facts had been offered, he must consent to abandon the *Mastodon Australis* as indefensible, and its former existence in Australia as an unsupported proposition.

No abstract of Owen's paper appeared in the *Report of the 32d meeting of the British Association for the Advancement of Science held at Cambridge* (Transactions of the sections), nor was it included on the 'List of papers of which abstracts have not been received' (pp. 195–6).

14 'Dimorphic condition in *Primula*'.

15 See letter to J. D. Hooker, 23 June [1862] and n. 4.

16 CD wrote his paper, 'Two forms in species of *Linum*', between 11 and 21 December 1862 (see 'Journal' (Appendix II)); it was read before the Linnean Society of London on 5 February 1863.

From J. D. Hooker [15 and] 20 November [1862][1]

Royal Gardens Kew
Saturday

Dr Darwin

I thought you might care to see these W. Ireland Soundings, showing the bank tailing off to the South, burn it when done with, as I have a copy in Nautical Journal[2]

I am horrified at your telling me that I have made an allusion to R Brown that might seem Lindleyitish, for certainly that was the last point in which I should have tried to write like him.[3] I had indeed forgotten that I made any allusion to Brown In apeing Lindly I went no further, to my knowledge, than trying to treat the subject as I thought he might could would or should have, & setting out with a rather clumsy force of wording, which he delights in. I have not reread any of it but the last column of the last article, where I was at first exceedingly puzzled to recover my meaning, (which is not very like Lindley!) I mean where I allude to the bearing of the whole thing on your theory.[4]

What a capital article Bates has made in his paper in Linn. Trans.[5] I have written to him about it, calling attention to the weakness of the paragraph at p. 508 about the effect of physical causes on variation.—[6]

What a poor paper Murrays is! he does not in the least see how he is playing into your hands, from sheer ignorance of your hand, which he thinks he quite understands—[7] nevertheless it is a sort of Entomological paper I am glad to see in the Transactions.

I was too late to send the London journal of Botany with Bonafuss on Thursday, I am trying to find a copy of vol VII. for Linn. Soc.—[8] They have put me on council of R.S. "heu me miserum!"[9] & with Owen too, for my sins.[10] I wish you would come up on Monday & substitute Falconer's[11] name for his!— There is a talk about organising an opposition to his election, but I suppose it will come to nothing.[12]

By the way I see you are alluded to in no less than 3 of the papers in Linn. Trans!—[13] I do not think you are conceited, but really I do think you have a good right to be so. I have said nothing of your writing but what I verily think & believe, & I find people are fast coming round to my way of thinking.[14]

Nov 20th | I send letter written 4 days ago[15]

I have no recollection of applying N.S. to Polynesians, though I dare say I have alluded to so obvious a deduction— none but a German would dig out such a passage & attach importance to it (*if it exists in print*)[16] The idea is of course not new to the future author of "The Aristocracy," or Darwin in all in all."[17]

I have caused Tyndall to modify extensively his pseudo geology, but do not know to what extent— He is awfully wrong about Valleys.[18]

I have not seen D of Argylls review— he did not understand it the *least little bit* about Orchids when I saw him nor understand fully the Origin.[19] I shall hunt up a clue I have to his proceedings I do know something of the origin of his book on Church matters at time of Scotch disruption.[20]

Would you like our dried strawberries we have all the wild forms from insular localities &c.[21] Oliver reminds me of curious remark on sexualism of strawberries by an American, he says it is alluded to at length in Technologist.[22] He also tells me to call your attention to 2 states of *Epilobium angustifolium*[23]

I am ashamed of so long neglecting Oxalis sensitiva I have a plant for you but this is worst season.[24]

Our *admirable* (perfect) governess is so ill with Lungs, that we must part with her, & she has no home poor thing—[25] My wife is in pack of troubles, I take her today to get tooth pulled, a worse affair for me than for her! for she behaves *shockingly* ill; & I am afraid to have her chloroformed indeed I doubt if dentist will do it—as I must tell him her heart action is not what it should be.[26]

Yours perplexed | J D Hooker

DAR 101: 71–2, 79

CD ANNOTATIONS
1.1 I thought . . . Journal 1.3] *crossed pencil*
3.2 weakness . . . variation.— 3.3] *scored pencil*

1 The year is established by the relationship to the letter to J. D. Hooker, [10–]12 November [1862]; the Saturday following that date, but preceding 20 November, was 15 November.

2 The enclosure has not been found. Hooker apparently sent CD a map of the deep sea soundings made to the west of Ireland in June, July, and August 1862 by HMS *Porcupine*, under the superintendence of Richard Hoskyn. Hooker also refers to the account given in the *Nautical Magazine and Naval Chronicle* for November 1862 (Hoskyn 1862); his error in citing the title of the periodical may be explained by the fact that its subtitle was 'a journal of papers on subjects connected with maritime affairs'. Hooker evidently considered these soundings relevant to the ongoing debate between himself and CD, over the possible former existence of land-bridges connecting existing islands and continents (see, for example, letter from J. D. Hooker, 7 November 1862, and letters to J. D. Hooker, 4 November [1862] and [10–]12 November [1862]).

3 Hooker refers to the botanists Robert Brown and John Lindley, and to his anonymous three-part review of *Orchids* ([J. D. Hooker] 1862c). See letter to J. D. Hooker, [10–]12 November [1862].

4 [J. D. Hooker] 1862c, p. 910.

5 Bates 1862a. The third issue of volume 23 of the *Transactions of the Linnean Society of London*, in which Bates 1862a appeared, was published on 13 November 1862 (Raphael 1970). There is an annotated copy of this number of the journal in the Darwin Library–CUL.

6 Hooker refers to the following passage (Bates 1862a, p. 508):

> I think the facts of similar variation in two already nearly allied forms do sometimes show that they have been affected in a similar way by physical conditions. A great number of insects are modified in one direction by a seaside habitat. I found, also, the general colours of many widely different species affected in a uniform way in the interior of the South American continent.

In his letter to Henry Walter Bates of 13 November 1862, which is reproduced in Bates 1892, pp. xlvi–xlvii, Hooker stated that he could not see how Bates applied '*physical conditions* here, *logically*, as a cause different in kind and operation from *natural conditions* elsewhere.' If the 'physical conditions' of the sea caused species change, Hooker argued, they could only do so by natural selection rather than by direct action, and were thus identical to 'natural conditions'. CD, Bates, and Hooker had corresponded on this subject earlier in the year (see letters from J. D. Hooker, [10 March 1862], 17 March 1862, and [23 March 1862], and letters to J. D. Hooker, 14 March [1862], 18 March [1862], and 26 [March 1862]).

7 Murray 1862. In his paper, Andrew Murray discussed the Coleoptera of Old Calabar in Africa, in the light of Edward Forbes's theory that the South American continent was 'at one time united or in close proximity to Western Africa', a view that Murray tentatively supported. Hooker refers particularly to the following statement of Murray's (p. 453):

> I am willing to assume for the present that Europe and North America have been at some period united, and that the affinity between *Anophthalmus Bilimekii* and *A. Tellkampfii* may be explained, on Mr. Darwin's theory, by assuming them to be the product of the same or of allied *Trechi* which have wandered into the caves, and that the like conditions have impressed a like form on their offspring. This latter assumption is of course not Mr. Darwin's; for he repudiates the idea of physical conditions making much, if any, impress upon life. I confess I am still a believer in that exploded heresy. I can see no other way of explaining the existence of these allied blind insects in caves so widely separated; and if it applies to the caves, it may equally apply to any district with well marked physical conditions.

CD lightly annotated his copy of the paper (see n. 5, above), including in his manuscript index for this number of the journal the comment: '[p.] 455 Distribution of Beetles'.

8 Hooker refers to his sending CD a copy of Bonafous 1836 by the Down carrier service, operated every Thursday by George Snow. In his letter of 3 November [1862], CD requested that Hooker inform him of the number of the volume of the *London Journal of Botany* in which Planchon 1847–8 appeared, so that he might borrow it from the library of the Linnean Society of London.

9 This commonly used phrase translates: 'Alas! wretched me'.

10 Richard Owen and Hooker were formally elected to the council of the Royal Society of London at the anniversary meeting of the society on 1 December 1862 (*Proceedings of the Royal Society of London* 12 (1863): 299).

11 Hugh Falconer.

12 Thomas Henry Huxley conducted a correspondence with the secretary of the Royal Society, William Sharpey, protesting against the proposal to elect Owen to the council of the society. Writing to Sharpey on 13 November 1862 (University College London, MS Archives, Add MS 227/4 no. 122), Huxley stated that he was 'greatly and not pleasantly astonished to find Owen's name' on the list of the new council of the society. He continued, referring to the so-called 'Hippocampus controversy':

> Considering the position of the controversy between us is, as you know very well, such, that one of the two of us is guilty of wilful & deliberate falsehood, I did not expect to find the Council of the Royal Society throwing even a feather's weight into the scale against me— But of the fact that Owen's election into the Council at this time will be viewed & used in that light there cannot be a doubt—

13 Bates 1862, Murray 1862, and Blanford 1862. In his paper, Bates invoked the theory of natural selection to account for the phenomenon of mimicry in Amazonian butterflies, arguing that the case offered 'a most beautiful proof of the truth of the theory' (Bates 1862a, p. 513). CD's copy of this paper (see n. 5, above) is heavily annotated. On Murray 1862, see n. 7, above. In his paper on the genus *Tanalia*, Henry Francis Blanford argued not only that many of the supposed species were merely varieties of one species, but also that this genus and a number of related genera should merely be considered 'sections of the genus *Melania*'. Referring to the structure of the operculum, on which alone the generic distinction was based (Blanford 1862, pp. 609–10), he stated:

> Indeed, accepting the views of Mr. Darwin, we might regard the group as affording an instance of variable structure in an organ usually constant, the tendency to vary having survived generic and specific differentiation and obtaining to some extent in coexistent forms of the same species.

CD marked this passage with a marginal line in his copy of the paper, also adding a reference to it and to Bates 1862a in his manuscript index to this number of the journal, with the comment: 'Laws of Variation & Variability'.

14 See letter to J. D. Hooker, [10–]12 November [1862].

[15] The second part of this letter was written on a separate sheet of paper; however, since the two parts were sent under the same cover and include only one salutation and valediction, they have been treated as a single letter.

[16] See letter to J. D. Hooker, 18 [November 1862] and n. 4. Hooker refers to Ludwig Büchner.

[17] Hooker refers to his somewhat whimsical suggestion that the development of an aristocracy was the necessary consequence of natural selection (see letters from J. D. Hooker, [19 January 1862], [31 January – 8 February 1862], and [23 March 1862]).

[18] John Tyndall. See letter to J. D. Hooker, [10–]12 November [1862] and n. 23.

[19] [G. D. Campbell] 1862. See letter to J. D. Hooker, [10–]12 November [1862] and n. 26, and letter to J. D. Hooker, 18 [November 1862].

[20] Hooker refers to the disruption of the Church of Scotland in 1843, when more than 450 ministers seceded over the issue of the church's relationship with the civil government, forming the Free Church of Scotland. The seventh duke of Argyll, John Douglas Edward Henry Campbell, was involved in trying to resolve the issue in the House of Lords, and his son, George Douglas Campbell (later eighth duke of Argyll) published two pamphlet letters seeking to avert the disruption (see Buchanan 1849 and Cameron *et al.*, eds. 1993). Hooker apparently refers to a polemical essay on the ecclesiastical history of Scotland written by the eighth duke (G. D. Campbell 1848), and may be implying that it was not entirely of Argyll's own authorship. The attribution of [G. D. Campbell] 1862 is confirmed by the *Wellesley index* 1: 511–12.

[21] See letter to J. D. Hooker, [10–]12 November [1862]. Hooker refers to specimens in the herbarium of the Royal Botanic Gardens, Kew.

[22] Hooker refers to Wray 1861a, in which the author contrasted the 'scientific culture of the strawberry' in the United States, where the tendency to dioecism of the cultivated strawberry was made the basis for horticultural practice, with strawberry growing in Britain.

[23] See letter to Daniel Oliver, 23 [November 1862].

[24] See letter to J. D. Hooker, [10–]12 November [1862] and n. 20.

[25] The governess has not been identified; she was appointed in June 1862 (see letter from J. D. Hooker, 19 [June 1862] and n. 4).

[26] Frances Harriet Hooker had been ill earlier in the year, causing Hooker to fear that she may have inherited a heart condition from her father, John Stevens Henslow (see letter from J. D. Hooker, 28 June 1862).

From John Scott 15 November [1862][1]

Botanic Gardens Edinburgh
Novr 15th—

Sir.

I had the honour to receive yours of the 12th.[2] I am pleased to see that the four suggestions &c. I ventured in elucidating the structure of Acropera have been considered worthy of your notice.[3] I was afraid that, they—unaccompanied as they were by either specimens or drawings, and communicated by one of whom you had no knowledge—would ever remain unacknowledged. And I can assure you, Sir, I felt deeply ashamed at my presumption, on reflecting after the letter had gone, now however I can only thank you—and that cordially for your unexpected attention. I may state that since I wrote you, I have mentioned the experiments I have been making to M^{r}. McNab—the Curator of the Gardens, and under whom, I am—[4]and requested his permission to communicate any specimens I might think of interest to you. He has kindly granted me this, and I will be most happy to communicate any specimens I observe in any way elucidating your views. I will soon then instead of attempting the *dissection of the now fast swelling ovary of Acropera—transmit it unbroken to you* that you may form the required deduction. I will, however, continue my experiments upon Acropera, and others; the results you shall have for your consideration.

My attention is at present principally engaged with that much controverted subject *Vegetable Parthenogenesis*—. This I am experimenting largely upon, and I trust ere long to show, that "Parthenogenesis in plants", is *not* as Karsten expresses it "thrust aside".[5] Professor Balfour[6] occasionally gives my experiments a look, and is not a little surprised with some of my results. As yet I have only recorded one instance of Parthenogenesis in a paper I read before the Botanical Society here,[7] on the Nature and Peculiarities of the Fern-spore—as illustration of certain views I there proposed and which I am now carefully—and in more detail—working out.

But I must discontinue this digression and return to the chief topic of my communication. And first regarding the application of the term *apex* to the ovary I applied it as you inferred to the portion nearest the flower.[8] I will be glad to hear the results of your examinations on this point as the suggestion I offered ⟨*two pages missing*⟩ have been substituted by **females** Can cultivation effect this? as botanists appear to regard many of the Palms dioecious which present this phenomena in cultivation.

In your application of this phenomena to Acropera, however—there seems to be no difficulty. For may it not be supposed in accordance with your interesting views that the change from the unisexual to the bisexual condition is not yet perfected? and the appearance of fertilised ovules thus be accounted for. As hinted before, however, you will no doubt still observe a difficulty in the way, viz; the contracted stigmatic chamber &c. unless you are inclined to accept the suggestion of the protrusion of pollen-tubes, independent of direct contact with stigma. This difficulty seems to be a necessary consequence of the fact that when female flowers substitute male they are usually quite similar to normal female flowers. This, however, may perhaps be neutralised, should the fertilised ovaries produce only a few seeds.

Your discovery of the independent development of the pollen-tubes in imperfect flowers of Viola & Oxalis is quite new to me.[9] As soon as I have an opportunity I will make experiments on *Acropera* for the purpose of testing its qualities for self-development of pollen-tubes.

The mode by which its natural fertilisation is effected is most assuredly a complete mystry to me.

I remain | Sir | Yours very respectfully | John Scott

To Charles Darwin Esq.

Incomplete[10]
DAR 177: 78

CD ANNOTATION
3.2 regarding . . . flower. 3.3] *scored pencil*

[1] The year is established by the reference to Scott 1862a (see n. 7, below).
[2] Letter to John Scott, 12 November [1862].
[3] See letter from John Scott, 11 November 1862, and letter to John Scott, 12 November [1862].
[4] James McNab was the curator of the Royal Botanic Garden, Edinburgh, where Scott was foreman of the propagating department (R. Desmond 1994).
[5] Karsten 1861, p. 93.
[6] John Hutton Balfour was Regius keeper of the Royal Botanic Garden, Edinburgh, and professor of botany at the University of Edinburgh (*DNB*).
[7] Scott read his paper (Scott 1862a) before the Botanical Society of Edinburgh on 12 June 1862. There is an annotated copy of this paper in the Darwin Pamphlet Collection–CUL.
[8] See letter from John Scott, 11 November 1862, and letter to John Scott, 12 November [1862].
[9] See letter to John Scott, 12 November [1862].
[10] Some indication of the contents of the missing portion of this letter is given by CD's reply (see letter to John Scott, 19 November [1862]).

To P. G. King 16 November [1862][1]

Down. | Bromley. | Kent. S.E.
Nov. 16th

My dear King

I was much pleased to get your letter;[2] only about a fortnight since I was asking all I could about you, for Wickham, Sulivan & Mellersh came down here; & much I enjoyed seeing these good & true old Beagle friends.[3] I wish you had told me a bit more about yourself & family; for the remembrance of old days when we used to sit & talk on the booms of the Beagle, will always to the day of my death make me glad to hear of your happiness & prosperity.— I have no proper carte, but I send a photograph of myself made 3 or 4 years ago by my eldest son (now a partner in a Bank,—this shows how old I am) & which is a good likeness of me.—[4] Thanks for your information about sheep; I remember being puzzled about their degeneracy.—[5] But to the main purpose of your letter, I grieve to say that my health is so indifferent, I cannot stand seeing at present anyone here. Twice lately I could not resist seeing old friends (once was when Wickham & Co came here) & the excitement made me so ill afterwards, that I have been advised not to do so again.[6] I am well enough in the mornings & when I keep quiet. I *must* write to your Brother to this effect.[7]

I seldom go to London, which is 16 or 17 miles distant, & so rarely see FitzRoy, but if I do I will not fail to give your message.[8]

Farewell my old friend, may all prosperity attend you.

Yours very sincerely | Ch. Darwin

Do you remember Syms Covington? I had a letter about a year ago from Twofold Bay to say he was dead.—[9]

Mitchell Library, State Library of New South Wales (ZML MSS 3447/2)

[1] The year is established by the relationship to the letter from P. G. King, 16 September 1862. See also n. 3, below.

[2] Letter from P. G. King, 16 September 1862.

[3] Like King, Arthur Mellersh, Bartholomew James Sulivan, and John Clements Wickham had been officers on board HMS *Beagle* during its surveying voyage of 1831–6. Emma Darwin's diary (DAR 242) records that they visited Down House on 21 October 1862.

[4] King had requested a photograph of CD in his letter of 16 September 1862. William Erasmus Darwin, who became a partner in the Southampton and Hampshire Bank, Southampton in 1861 (see *Correspondence* vol. 9), had been interested in photography since 1857 (see *Correspondence* vol. 6, letter to W. E. Darwin, [before 11 September 1857] and n. 3). A photograph of CD taken by William in April 1861, is reproduced as the frontispiece to *Correspondence* vol. 9 (see *Correspondence* vol. 9, letter to Asa Gray, 11 April [1861] and n. 19).

[5] See letter from P. G. King, 16 September 1862.

[6] For an account of CD's illness following the visit of Mellersh, Sullivan, and Wickham, see the letter to John Lubbock, 23 October [1862]. CD also refers to the visit of John Lubbock to Down House on 31 October, following which Emma Darwin recorded in her diary (DAR 242): 'Ch. attack of sickness in night'.

[7] In his letter to CD of 16 September 1862, King had expressed a wish that his younger brother, John, might visit CD while in England.

[8] Robert FitzRoy was commander of the *Beagle* during its surveying voyage of 1831–6. See letter from P. G. King, 16 September 1862.

[9] The letter has not been found. Syms Covington became CD's servant and assistant on the *Beagle* in 1833. After the voyage, he continued in CD's service until 1839, when he emigrated to Australia; he died in 1861 (Freeman 1978).

To Ludwig Büchner 17 November [1862][1]

Down. | *Bromley.* | *Kent. S.E.*
Nov. 17th

Respected Sir!

I beg leave to return you my sincere thanks for your great kindness in sending me your last work "Aus Natur und Wissenschaft."[2] I have read the chapter, in which you treat of my work, & you have indeed conferred the most distinguished honour on me.[3] I venture to thank you, also, on account of the good which you will effect in getting the subject widely discussed, & the truth finally discovered, which I have much at heart.—

I am, like so many Englishmen a very poor proficient at Languages, & German is to me, excepting the simplest descriptions, extremely difficult; but I hope to make intelligible to myself some of your valuable volume.

With my best thanks, I have the honour to remain, Respected Sir | Your obliged servant | Charles Darwin

Cleveland Health Sciences Library (Robert M. Stecher collection)

[1] The year is established by the reference to Büchner 1862 (see n. 2, below).

[2] Büchner 1862. There is a copy of this work in the Darwin Library–CUL.

[3] Büchner 1862 is a collection of several of Büchner's papers and reviews, which had originally been published between 1856 and 1860. CD refers to Büchner's review of *Origin*, first published as Büchner 1860, but republished in Büchner 1862, pp. 245–53. In his initial review, Büchner, who was a leading member of the intellectual circle of scientific materialists in Germany, described *Origin* sympathetically;

however, in Büchner 1862, pp. 252–3, he added a paragraph in which he suggested that CD's emphasis on the theory of natural selection might be too one-sided, and that it should be supplemented by a greater consideration of the effect of direct external influences on organisms. On the reaction of the German scientific materialists to Darwinism, see F. Gregory 1977.

From Asa Gray 17 November 1862

Cambridge, Mass.
Nov. 17, 62.

Dear Darwin

Nothing to say.— Seeds from a different region of *Nesæa verticillata* have come in. I send a few—to make surer,—[1] and I put upon you the posting of the enclosures,—and the turning over of the *12^cts^* stamp to Leonard, with my compliments—[2]

Yes—a note about *Cypripedium insigne*, just flowering in conservatory. Orifice in front very large—different from our species.— The lateral basal orifice under anther pretty large & free. Beard at base of the labellum inside very short and sparse.—

Altogether more room to move into and out of flower, and in it. Fertilization I guess is done by larger insects passing bodily into the cavity.[3]

But the stigma is *smooth*! not a vestige of the *rasping* arrangement which I found in our indigenous species.

What is the use of *genera* if the structure is not same throughout? Verily, I shall begin to credit your theory of *accidents*![4]

Ever Yours | A. Gray

I rejoice at the removal of *Mc Lellan*,—lacks *decision*.[5]

DAR 165: 123

CD ANNOTATIONS
0.1 Cambridge ... compliments— 1.4] *crossed ink*
Top of letter: '(Cypripedium)' *red crayon*

[1] At CD's request, Gray had sent seeds of *Nesaea verticillata* with his letter to CD of 27 October 1862; CD considered the plant likely to be trimorphic.

[2] Leonard Darwin had written to Gray at his suggestion to tell him which North American postage stamps he most wanted for his collection (see letter from Asa Gray, 22 September 1862 and n. 3).

[3] At CD's request, Gray had examined the pollination mechanisms in a number of North American species of the orchid genus *Cypripedium*, concluding that, for the North American species, the account given by CD in *Orchids* was incorrect (see, for example, letter to Asa Gray, 10–20 June [1862] and n. 16). *C. insigne*, which is not a native North American species, was one of the species on which CD had based his account (see *Orchids*, p. 273). In *Orchids*, pp. 274–5, CD suggested that *Cypripedium* must be pollinated by an insect inserting its proboscis into one of the two lateral entrances at the base of the labellum, directly over one of the two lateral anthers, and thus either placing the pollen onto the flower's own stigma, or carrying it away to another flower. However, Gray suggests here that *C. insigne* supported his view that the flowers were pollinated by insects entering the labellum through the large opening on the upper surface, and crawling out by one of the two lateral orifices (see A. Gray 1862b).

[4] Unlike CD, Gray preferred to interpret natural selection as evidence of design in nature. They had corresponded extensively on the subject in 1860 and 1861 (see *Correspondence* vols. 8 and 9).

[5] George Brinton McClellan was removed from command of the Union's army of the Potomac on 7 November 1862 (McPherson 1988, pp. 562 and 570).

From Adam Fitch 18 November 1862

Vicarage | Thornton Steward | Bedale
Novr 18th. 1862

Sir,

I have just met with your queries in the Gardeners' Chronicle, respecting peas—[1]I venture therefore to trouble you with a letter in answer.

My late Rector D^{r}. Davy Master of Caius was a most intimate friend of M^{r}. Knights, who was in the habit of sending his new productions, if very good, to the Doctors country residence in Norfolk—[2] On one occasion I recollect seeing some peas in bloom which came from M^{r}. Knight, the tall Marrow; and I noticed a large proportion of coloured blossom—in fact the blossom of the common field pea— I noticed this to my Rector, who told me the Peas were a cross by Knight between the common Prussian Blue and the common field pea—[3] When Curate of Cottenham & Willingham having resided in the two Parishes for 20 years,[4] I had a very favourite pale pea—Greens tall Marrow. It was my great crop & of course was surrounded by different kinds of peas—most being more or less in blossom at the same time— I never saw any change—or anything approximating to a cross—but at end of 20 years the seed appeared to be as true as at commencement.[5] I tried the same kind here but our West Winds which are most violent, compelled me to give it up.[6]

As a practical gardener I would venture to suggest that the pea producer's on a large scale, are not sufficiently careful in gathering or in threshing & thus too frequently you find a mixture. This was the case this last season with Hairs Dwrf Marrow a most desirable variety

Incomplete
DAR 77: 166–7

CD ANNOTATIONS
2.9 Greens tall Marrow.] *underl brown crayon*
2.11 but . . . commencement. 2.12] *scored brown crayon*
Top of letter: 'Revd Adam Fitch.' *ink*; 'Ch. 3.'[7] *brown crayon*

[1] See letter to *Gardeners' Chronicle*, [before 8 November 1862].

[2] Fitch refers to the horticulturalist Thomas Andrew Knight, and to the physician Martin Davy, master of Gonville and Caius College, Cambridge, from 1803 to 1839 (*DNB*). Davy was rector of Cottenham, Cambridgeshire, between 1827 and 1839, during part of which time Fitch served as his curate (*Alum. Cantab.*); Davy's country residence was Heacham Lodge, Norfolk (*DNB*).

[3] CD reported this information in *Variation* 1: 326.

[4] Fitch was curate of Cottenham, Cambridgeshire from 1832, and curate of the neighbouring parish of Willingham from 1841 to 1849.

[5] CD reported this account in *Variation* 1: 329–30.

[6] From 1849, Fitch was vicar of Thornton Steward, Yorkshire (*Alum. Cantab.*).

[7] This is a reference to CD's 'big book' on species, originally drafted between 1856 and 1858; the third chapter was entitled, 'On the possibility of all organisms crossing, & on susceptibility of the reproductive system to external agencies' (see *Natural selection*, pp. 33–91). Having published *Origin* as an abstract of his 'big book', CD planned to revise the longer manuscript for publication (see *Correspondence* vol. 7, letter to John Murray, 22 December [1859]). In the event, only the first part of the projected work was published, broadly based on the first two chapters of the original manuscript (*Variation*). However, in *Cross and self fertilisation* CD dealt with many of the topics originally covered in the third chapter of his 'big book', and several of the other notes and letters in CD's portfolios of materials used in preparing the work (DAR 76–9) are annotated 'Ch. 3'. In *Cross and self fertilisation*, p. 161, CD referred the reader to the evidence given in *Variation* (including Fitch's) concerning the true breeding exhibited by varieties of common pea grown in close proximity over many generations.

To J. D. Hooker 18 [November 1862]

Down Bromley Kent
18th.

My dear Hooker

Very many thanks for Bonafous, which has been very useful to me; but I greatly fear will not be worth your Library having so magnificently purchased.[1] Thanks, also, for information about Haughton, on which I was simply curious; he must have, as you say, "a many-horse power of brains".—[2] what a shame about Mann & Burton; it will be a horrid bore for you to get into a dispute with such a man as Burton.—[3]

I received the other day a German Book, by Dr L. Bückner in which it is said that the distinguished Botanist Hooker first applied natural selection to the replacement of races of men,—the ruder races as Polynesians &c yielding to the civilised Europeans.—[4] I cannot remember reading this.—

I hope your awesome labours on Wellwitschia are drawing to a close.— Take care it does not run on into a case of Barnacles, & consume years instead of months.[5]

Strange to say I have only one little bother for you today, & that is to let me know about what month flowers appear in Acropera Loddigesii & luteola; for I want extremely to beg a few more flowers; & if I knew time I would keep memorandum to remind you.— Why I want these flowers is (& I am much alarmed), that Mr J. Scott of Bot. Garden of Edinburgh, (do you know anything of him?) has written me a very long & clever letter, in which he confirms most of my observations;[6] but tells me that with much difficulty he managed to get pollen into orifice or as far as mouth of orifice of six flowers of A. *Loddigesii* (the ovarium of which I did not examine) & two pods set; one he gathered & saw a very few ovules, as he thinks, on the large & mostly rudimentary placentæ. I shall be most curious to hear whether the other pod produces a good lot of seed. He says he regrets that he did not test the ovules with chemical agents: does he mean Tincture of Iodine?[7] He suggests that in state of nature the viscid matter may come to very surface of stigmatic chamber, & so pollen-masses need not be inserted. This is possible, but I shd. think improbable. Altogether the case is very odd, & I am very uneasy; for

I cannot hope that A. Loddigesii is hermaphrodite & A. luteola the male of same species.— Whenever I can get Acropera, would be very good time for me to look at Vanda in spirits, which you so kindly preserved for me.—[8]

Ever yours | C. Darwin

The D. of Argylls Review is very clever, but not convincing to me.—[9]

P.S. I am in middle of Bates' paper;[10] it is a very admirable & is worth labour (& that not slight) of careful reading— The remarks in the systematic part excellent on formation of species from vars.—[11] It is a pity the title did not more plainly tell contents.[12] Most wonderful the mimetic resemblances!

Endorsement: 'Nov/62'
DAR 115.2: 170

[1] Bonafous 1836. See letter from J. D. Hooker, 12 November 1862 and n. 5.

[2] Samuel Haughton. See letter to J. D. Hooker, 3 November [1862], and letter from J. D. Hooker, 12 November 1862.

[3] CD refers to Richard Francis Burton and Gustav Mann. See letter from J. D. Hooker, 12 November 1862 and nn. 7 and 8.

[4] Büchner 1862, pp. 249–50; there is a copy of this work in the Darwin Library–CUL. Although Ludwig Büchner cited J. D. Hooker 1855–60, the reference has not been traced. However, Büchner may have misunderstood a passage in which Hooker discussed the impact of the 'progress of civilization' on the struggle for existence between native and introduced plants (J. D. Hooker 1855–60, pp. civ–cv). See also letter to Ludwig Büchner, 17 November [1862].

[5] Hooker was preparing a monograph on the Angolan plant *Welwitschia mirabilis* (J. D. Hooker 1863a; see letter from J. D. Hooker, 2 November 1862). In a letter that is now missing (L. Huxley ed. 1918, 2: 24), CD had compared Hooker's work on *Welwitschia* to his own extensive work on barnacles, carried out between 1846 and 1854 (see *Correspondence* vol. 4, Appendix II); the allusion had been repeated on several occasions in their correspondence (see letter from J. D. Hooker, [19 January 1862], and letter to J. D. Hooker, 14 [October 1862]).

[6] Letter from John Scott, 11 November 1862.

[7] 'Tincture of Iodine' is used to check for the presence of starch grains in plant cells.

[8] See letter from J. D. Hooker, 20 August 1862 and n. 12. CD wished to compare a seed-capsule of *Acropera Loddigesii*, which John Scott had promised to send him, with a capsule from another species of the Vandeae, the orchid tribe to which *Vanda* belongs (see following letter).

[9] [G. D. Campbell] 1862; there is an annotated copy of this publication in the Darwin Pamphlet Collection–CUL. In his review, the duke of Argyll, George Douglas Campbell, discussed *Orchids* in conjunction with five works on miracles and the supernatural. He argued that miracles were not supernatural, since they did not involve the suspension or violation of natural laws, but that they were superhuman, since natural laws were in such cases applied by the divine will in a way that was analogous to human contrivance, but more powerful. Argyll used this interpretation of divine action to account for the design apparent in the contrivances by which orchids are pollinated, contrasting it with CD's explanations in terms of natural selection, which he considered 'the vaguest and most unsatisfactory conjectures' (p. 394).

[10] Bates 1862a.

[11] Henry Walter Bates's paper comprised a discursive introduction (Bates 1862a, pp. 495–515), followed by a taxonomic study of the Amazonian Heliconidae (pp. 515–64). Having described the formation of species from varieties in general terms in his introduction (pp. 500–2), Bates referred the reader to his systematic descriptions of the individual species for 'the details of variation'.

[12] Bates's paper was entitled, 'Contributions to an insect fauna of the Amazon Valley. *Lepidoptera: Heliconidæ*.' In the paper, Bates invoked the theory of natural selection to account for the phenomenon of

mimicry in Amazonian butterflies, arguing that the case offered 'a most beautiful proof of the truth of the theory' (Bates 1862a, p. 513).

To John Scott 19 November [1862][1]

Down. | Bromley. | Kent. S.E.
Nov. 19th

Dear Sir

I am much obliged for your letter which is full of interesting matter.[2] I shall be very glad to look at the capsule of the Acropera when ripe & pray present my thanks to Mr McNab.[3] I shd. like to keep it, till I could get a capsule of some other member of the Vandeæ for comparison; but ultimately all the seeds shall be returned in case you would like to write any notice on subject:[4] It was as I said only "*in desperation*" that I suggested that the flower might be a male & yet occasionally capable of producing a few seeds.[5] I had forgotten Gärtner's remark;[6] in fact I know only odds & ends of Botany & you know far more. One point makes the above view more probable in Acropera than in other cases,—viz the presence of rudimentary placentæ or testæ; for I cannot hear that these have been observed in other male plants. They do not occur in male Lychnis dioica;[7] but next spring I will look to male Holly Flowers.—[8]

I fully admit difficulty of similarity of stigmatic chamber in the two Acroperas.[9] As far as I remember the blunt end of pollen-mass would not easily even stick in the orifice of the chamber. Your view may be correct about abundance of viscid matter, but seems rather improbable.[10]

Your facts about female flowers occurring when males alone ought to occur is new to me: if I do not hear that you object, I will quote the Zea case on your authority in what I am now writing on the varieties of the Maize.—[11]

I am glad to hear that you are working on the most curious subject of Parthenogenesis.[12] I formerly *fancied* that I observed female Lychnis dioica seeded without pollen.[13] I send by this post a paper on Primula, which may interest you.[14] I am working on this subject, & if you shd. ever observe any analogous case I shd be glad to hear. I have added another very clever pamphet by Prof. Asa Gray.—[15] Have you a copy of my Orchis Book? if you have not, & would like one, I shd. be pleased to send one.—

I plainly see that you have the true spirit of an Experimentalist & good observer. Therefore I ask, whether you have ever made any trials on relative fertility of *varieties* of plants (like those I quote from Gärtner on the varieties of Verbascum)[16] I much want information on this head, & on those marvellous cases (as some Lobelias & Crinum & Passiflora) in which a plant can be more easily fertilised by the pollen of another *species* than by its own *good* pollen.—[17]

I am compelled to write in haste. With many thanks for your kindness. | Dear Sir | Yours very faithfully | Ch. Darwin.

AL & Copy[18]
DAR 93 (ser. 2): 11–14, DAR 147: 431

[1] The year is established by the relationship to the letter from John Scott, 15 November [1862].
[2] Letter from John Scott, 15 November [1862].
[3] James McNab. See letter from John Scott, 15 November [1862].
[4] No publication by Scott on this subject has been found.
[5] See letter to John Scott, 12 November [1862].
[6] The reference to the remark by Karl Friedrich von Gärtner has not been identified; John Scott apparently referred to this remark in the missing portion of his letter of 15 November [1862].
[7] See letter to J. D. Hooker, 30 May [1862] and n. 10.
[8] CD had been interested in the 'gradation in sexes' in species of holly for some time, and had unsuccessfully sought assistance from Asa Gray earlier in the year (see letters to Asa Gray, 21 April [1862] and 10–20 June [1862], and letter from Asa Gray, 18 May 1862).
[9] See letter from John Scott, 15 November [1862].
[10] See letter from John Scott, 11 November 1862. See also preceding letter.
[11] CD was preparing a draft of the part of *Variation* dealing with 'Facts of variation of Plants' (see 'Journal' (Appendix II)). The case to which CD refers was apparently described in the missing portion of the letter from John Scott, 15 November [1862]; see also letter from John Scott, [20 November – 2 December 1862]. Scott described this case, in which female flowers appeared on a male panicle of *Zea mays*, in Scott 1864a; CD referred to the paper in *Variation* 1: 321.
[12] See letter from John Scott, 15 November [1862].
[13] In April 1839, John Stevens Henslow had told CD that, while *Lychnis dioica* was generally dioecious, the male parts in female flowers (and vice versa) were only 'very slightly abortive', and that a 'bed of female flowers will sometimes produce a few seeds' (*Notebooks*, p. 434). However, although CD planned to examine the pollination mechanism in this species (see *Notebooks*, p. 499), and later carried out experiments on the effects of external conditions (see *Correspondence* vol. 6, letter to J. S. Henslow, 16 June [1856]), no notes have been found describing the observation mentioned here. See also *Correspondence* vol. 9, letter to Daniel Oliver, 7 December [1861] and n. 5.
[14] 'Dimorphic condition in *Primula*'.
[15] A. Gray 1861.
[16] 'Dimorphic condition in *Primula*', p. 91 (*Collected papers* 2: 58–9). CD discussed Karl Friedrich von Gärtner's experiments on *Verbascum*, in which crosses between differently coloured varieties of the same or of different species were found to produce less seed than the parallel crosses between similarly coloured varieties, in *Origin*, pp. 270–1. In September 1861, he told Joseph Dalton Hooker that he had decided to test Gärtner's experiments, stating: 'I do not think any experiment can be more important on Origin of species; for if he is correct, we certainly have what Huxley calls new physiological species arising' (*Correspondence* vol. 9, letter to J. D. Hooker, 28 September [1861]; see also Appendix VI). CD unsuccessfully sought the requisite specimens from a number of botanical acquaintances (see *Correspondence* vol. 9, letters to J. D. Hooker, 18 October [1861], 23 October [1861], and this volume, letter to C. C. Babington, 20 January [1862]), and also attempted to use a wild specimen for the purpose (see letter to W. E. Darwin, 4 [July 1862] and n. 4). Scott carried out the suggested experiments in 1863 (see *Correspondence* vol. 11), publishing his results in Scott 1868; CD cited Scott's work in *Variation* 2: 106–7.
[17] See *Origin*, pp. 249–51.
[18] The last paragraph is missing from the original, and has been transcribed here from the copy made for the use of Francis Darwin in editing his father's letters for publication.

To H. W. Bates 20 November [1862][1]

Down Bromley Kent
Nov. 20th

Dear Bates

I have just finished after several reads your Paper.[2] In my opinion it is one of the most remarkable & admirable papers I ever read in my life. The mimetic cases are truly marvellous & you connect excellently a host of analogous facts. The illustrations are beautiful & seem very well chosen; but it would have saved the reader not a little trouble, if the name of each had been engraved below each separate figure;[3] no doubt this would have put the engraver[4] into fits, as it would have destroyed beauty of Plate. I am not at all surprised at such a paper having consumed much time. I rejoice that I passed over whole subject in the Origin, for I sh^d. have made a precious mess of it.[5] You have most clearly stated & solved a wonderful problem.—

No doubt with most people this will be the cream of the paper; but I am not sure that all your facts & reasoning on variation & on the segregation of complete & semi-complete species is not really more, or at least as valuable a part.—[6] I never conceived the process nearly so clearly before; one feels present at the creation of new forms.— I wish, however, you had enlarged a little more on the pairing of similar varieties; a rather more numerous body of facts seems here wanted.[7]

Then again what a host of curious miscellaneous observations there are,—as on related sexual & individual variability you give; these will some day, if I live, be a treasure to me.—[8]

With respect to mimetic resemblance being so common with insects; do you not think it may be connected with their small size; they cannot defend themselves;—they cannot escape by flight at least from Birds; therefore they escape by trickery & deception?[9]

I have one serious criticism to make & that is about title of paper; I cannot but think that you ought to have called prominent attention in it to the mimetic resemblances.—[10] Your paper is too good to be largely appreciated by the mob of naturalists without souls; but rely on it, that it will have *lasting* value, & I cordially congratulate you on your first great work. You will find, I sh^d. think, that Wallace will fully appreciate it.—[11]

How gets on your Book?—[12] Keep your spirits up. A Book is no light labour. I have been better lately & working hard; but my health is very indifferent. How is your health?

Believe me Dear Bates | Yours very sincerely | Ch. Darwin

Excuse my poor M.S. paper.—

Cleveland Health Sciences Library (Robert M. Stecher collection)

[1] The year is established by the relationship to the letter from H. W. Bates, 24 November 1862.

[2] Bates 1862a appeared in the third part of volume 23 of the *Transactions of the Linnean Society of London*, which was published on 13 November 1862 (Raphael 1970); there is a copy of this part of the journal in the Darwin Library–CUL, in which Bates's paper is heavily annotated.

[3] The two colour plates of Amazonian mimetic butterflies published in Bates 1862a (plates LV and LVI), were keyed by numbers to identifications and descriptions at the end of the paper (pp. 564–6). CD added the species names to the figures in his copy of the paper.

[4] The plates were drawn and engraved by Edward W. Robinson, a noted entomological artist.

[5] In his paper, Bates invoked the theory of natural selection to account for the phenomenon of mimicry in Amazonian butterflies, arguing that the case offered 'a most beautiful proof of the truth of the

theory' (Bates 1862a, p. 513). CD included a summary of Bates's findings in the fourth edition of *Origin*, published in 1866 (*Origin* 4th ed., pp. 502–6).

6 In the introduction to his paper, Bates argued that 'many of the now distinct species of *Heliconidæ* have arisen from local varieties, segregated from the variations of preexisting widely disseminated species' (Bates 1862a, p. 501). In particular, he noted that in some species the varieties presented 'all the different grades between simple individual differences and well-marked local varieties or races, which latter cannot be distinguished from true species, when two or more of them are found coexisting in the same locality without intercrossing'. For the details of these cases, Bates referred the reader to his systematic descriptions of the various species (pp. 515–64). See also letter to J. D. Hooker, 18 [November 1862] and n. 11.

7 In Bates 1862a, p. 501, Bates stated:

> The process of the creation of a new species I believe to be accelerated in the *Ithomiæ* and allied genera by the strong tendency of the insects, when pairing, to select none but their exact counterparts: this also enables a number of very closely allied ones to exist together, or the representative forms to live side by side on the confines of their areas, without amalgamating.

In his copy of the paper, CD wrote at this point 'What proof?', but added 'See further on'. Bates's reply has not been found, but see the letter to H. W. Bates, 15 December [1862].

8 See, for example, Bates 1862a, p. 502. CD frequently cited Bates's work in his account of sexual selection in insects in *Descent* 1: 309–15 and 341–423; however, he did not cite this paper.

9 In Bates 1862a, p. 507, Bates noted:

> It may be asked, why are mimetic analogies so numerous and amazingly exact in insects, whilst so rare and vague in the higher animals? The only answer that I can suggest is, that insects have perhaps attained a higher degree of specialization, after their type, than most other classes: this seems to be shown by the perfection of their adaptive structures and instincts.

In *Origin* 4th ed., p. 506, following his discussion of Bates's work on mimicry, CD suggested that the 'much greater frequency of mockery with insects than with other animals' was 'probably the consequence of their small size', and continued:

> insects cannot defend themselves, excepting indeed the kinds that sting, and I have never heard of an instance of these mocking other insects, though they are mocked: insects cannot escape by flight from the larger animals; hence they are reduced, like most weak creatures, to trickery and dissimulation.

10 Bates's paper was entitled: 'Contributions to an insect fauna of the Amazon Valley. *Lepidoptera*: *Heliconidæ*.' See also letter to J. D. Hooker, 18 [November 1862] and n. 12.

11 Bates and Alfred Russel Wallace were close friends, and had travelled together to South America in 1848 (see, for example, Wallace 1905). In his paper, Bates cited an instance of bird mimicry communicated to him by Wallace (Bates 1862a, p. 507 n.).

12 Bates 1863. See also letter from H. W. Bates, 17 October 1862.

From John Scott [20 November – 2 December 1862][1]

seed.— P. *vulgaris elatior* and *veris*—carefully protected from cross-impregnation.[2] But I never succeeded—as not a few aver they have—in raising *oxlips* from *cowslips* or *primroses*. The explanation of these cases where trustworthy authorities, have raised these very distinct forms from one plant, is I believe found in mediate or immediate crosses with the seed-producing plants. As you remark further experiments are absolutely necessary to show that these forms may originate in an independent and ungraduated manner from *vulgaris*. The permanence and stability—so to speak—of these three forms when raised from seed—as the majority of the offspring, at least, always represent their immediate parent, seems to demonstrate well-established characteristics, and inconsistent with the laws governing the reproduction of variations in the *Phanerogamic Class* of plants. These laws, however appear to me—not a little capricious in their mode of action in the Vegetable Kingdom—different I believe from what has place in the Animal—and present great difficulties to me in fully accepting your captivating theory of the origin of species by Natural Selection. Time however, does not permit me here to specify these difficulties, there are other points for present consideration.

In regard to the substitution of female flowers for male in *Zea*, it may be necessary to specify particularly that in the case I have given these were truly products of the *male* panicle.[3] As in the axils of the upper leaves small monoecious spikes are occasionally produced, bearing a few male flowers at their base, and female on the upper part. I mention this now as you regard it as worthy of your recording; to prevent any equivocation to those who may have observed the latter spikes; and, therefore, in the case of a non-specific statement, suppose that those which are in reality female spikes, were taken advantage of.

I am extremely sorry that I can afford you no information whatever on the relative fertility of varieties of plants.[4] I may mention, here, however, though you may already be aware of it an experiment which I made in illustration of the fertility of hybrids. This was made upon *Clivia cyrtanthiflorum*, which was raised by crossing the *C. nobilis* with C. *miniata*. I find it to be perfectly fertile with its own pollen as well as that of either parent.

I am in the same predicament in regard to your other question on the greater facilities certain species present for fertilisation by other pollen than their own. I have only one experiment I can give you on this point and it is quite inconclusive. I have more than one season fertilised flowers of *Tacsonia pinnasistipula* in the Gardens here,[5] yet I have rarely succeeded in getting any fruit to set. During the latter part of this autumn another unnamed species flowered. I impregnated a flower of the former species—the only one upon the plant at the time—the ovary began to swell and promised fair for some time, eventually however it dropped off. But this I think was due to its being so late in the season, and the cold damp state of the atmosphere—the plant being grown in a cool greenhouse. I believe however if I had opportunities for experiment that it will present a very analogous case with *Passiflora*. In the course of the ensuing season I will commence a series of experiments on those interesting questions.[6]

And now before closing allow me to express my most sincere thanks for the honour you would have done in presenting me with a copy of your interesting work on Orchids, which I would indeed have highly valued.[7] I have now, however, a copy of it in my possession, otherwise I would have gladly availed myself of your unmerited kindness to me in this likewise.

I remain | Sir | Your obedient Servt—

John Scott.

P.S. | Should you not be able to get a capsule of Vande⟨ae⟩ for comparison, I will have an Oncidium in flower, whic⟨h⟩ I will impregnate, and send along with Acropera.[8] J. S.

Incomplete[9]
DAR 177: 79

CD ANNOTATIONS
1.1 seed.— . . . plants. 1.11] *crossed ink*
2.2 products . . . panicle. 2.3] *cross in margin, brown crayon*
3.5 I find . . . parent. 3.6] *scored brown crayon; cross in margin, brown crayon*

1 The date range is established by the relationship to the letter to John Scott, 19 November [1862] (see nn. 2, 3, and 7, below), and to the letter to John Scott, 3 December [1862].

2 CD mentioned in his letter to Scott of 19 November [1862] that he would send a copy of 'Dimorphic condition in *Primula*'. In this paper, CD discussed the much-disputed question of whether the primrose (*Primula vulgaris*) and the cowslip (*P. veris*) were distinct species or varieties of the same species. On the former hypothesis, some authors had claimed that the common oxlip was a hybrid of the primrose and cowslip; CD, by contrast, had formerly held the view that the various forms were varieties descended from a common parent (*Natural selection*, pp. 128–33, and *Origin*, pp. 49–50). However, as a result of his work on dimorphism in *Primula* he came to distrust the evidence relating to this point, arguing that further experiments were 'absolutely necessary' ('Dimorphic condition in *Primula*', pp. 93–4; *Collected papers* 2: 60–1). Scott gave an account of his experiments on various species of *Primula* in Scott 1864c. See also letter from John Scott, 7 January [1864] (*Calendar* no. 4382).

3 See letter to John Scott, 19 November [1862] and n. 11.

4 See letter to John Scott, 19 November [1862].

5 Scott was foreman of the propagating department at the Royal Botanic Garden, Edinburgh (R. Desmond 1994).

6 Scott reported the results from his experiments on *Tacsonia*, together with the results from a series of experiments on *Passiflora*, in Scott 1864b.

7 See letter to John Scott, 19 November [1862].

8 CD had asked Scott to send him, when ripe, a seed-capsule of *Acropera Loddigesii*, noting that he would like to keep it until he could obtain a capsule of another member of the Vandeae for comparison (see letter to John Scott, 19 November [1862]).

9 Some indication of the contents of the missing portion of this letter is given by CD's reply (see letter to John Scott, 3 December [1862]). Scott apparently also enclosed with this letter a list giving details of various species of *Primula* (see letter to John Scott, 3 December [1862] and n. 2, and letter from John Scott, 6 December [1862] and n. 3); however, the list has not been found.

To T. F. Jamieson 21 November 1862

Down. | *Bromley.* | *Kent. S.E.*
November 21— 1862

My dear Sir

I have great pleasure in expressing my strong opinion on your qualifications, as far as knowledge of the natural sciences is concerned, to give Lectures on Agriculture.[1] I have read with admiration all your papers on the more recent geological changes in Scotland; & I can with entire truth say that according to my power of judging I have never read more able productions.[2] I may add that the more recent geological phenomena are obviously those most connected with Agriculture. From our frequent correspondence, I am well aware that your knowledge of the several branches of Natural History is very considerable.

With my cordial wishes for that success, which I am sure you are well entitled to, I remain | My dear Sir | Yours sincerely | Charles Darwin

To | Th. F. Jamieson Esqre—

National Library of Scotland (MS 5406: 171–2)

[1] In a letter that is now missing, Jamieson had evidently asked CD for a reference; he was appointed Fordyce lecturer on agriculture at the University of Aberdeen in 1862 (*Roll of the graduates of the University of Aberdeen*).

[2] Jamieson had written a number of important papers on the Pleistocene geology of Scotland (Jamieson 1858, 1860a, 1860b, and 1862). In particular, CD had been greatly impressed by Jamieson's explanation of the so-called 'parallel roads' of Glen Roy, in Lochaber, Scotland, later detailed in Jamieson 1863 (see, for example, letter to Charles Lyell, 14 October [1862] and n. 3).

From Edward Blyth 23 November 1862

Calcutta
Novr. 23/62—

My dear Sir,

I suppose you have heard that Sir C. Wood has disallowed my pension, although so strongly and flatteringly recommended by the Viceroy in Council—[1] But it matters not. I will win it yet.[2] Nobody has told me of it, nor do I require to be told that the Earl of E. and K. (the representative of the old Scottish kingly family of Bruce) is piqued and nettled at being thus coolly snubbed by the Yorkshire Baronet.[3] Ld Elgin held a grand levee on the 10th ult^{o}., at which I attended; and as I passed him to make the usual bow on such occasions, he said nothing, of course, nor to anybody else in such a crowded assemblage, but I think that I understood the expression of his countenance. At all events, I revisit England immediately,—as I trust, by the screw steamer which leaves this on the 15th. prox^{o}., proceeding viâ the Cape.[4] I have many reasons to prefer that route; among which are, that I can take a lot of living animals with me, for the Z. Gns,[5] and because I wish to visit Capetown and examine its museum. It is understood also that I go home to combat Sir Charles on the spot, as likewise to recruit my health and strength, after so long a residence within the tropics. I am to be well supplied with funds, supposed to come (and perhaps they do so) from the Asiatic Socie⟨ty,⟩ but I cannot help suspecting that there are 'wheels within wheels' in this matter— I am exceedingly well supported here, and have no lack of friends in England, both in the House & out of it;[6] among whom I surely reckon upon you,[7] and as surely upon my personal friend, our renowned financier, M^{r}. Laing,[8] at whose princely table I have had the honour to dine, & who in fact knows me well. He is likely to prove a nice thorn, in the House of Commons, in the side of the present Indian minister; but I doubt if the latter will be able to maintain his official position much longer, despite the support of Ld P.,[9] in face of the tremendous opposition which is fast rising against him from every quarter— Sir C. undoubtedly does possess one transcendant talent, which is that of making himself enemies, in which he is most preeminently successful. But enough of all this. I hope and expect to be able to shake you by the hand soon, if you will allow me the honour and the pleasure. Meanwhile, I may remark that I forgot to mention to you in my last letter, that there are no soundings (in ordinary seaman's parlance, i.e. with the usual instruments,) between the Andamáns & the main, to the eastward; but, that a string of islands, rocks, and shoals, extend from Cape Negrais[10] to Acheen Head in Sumatra; which may have once been continuous land.[11]

Yours ever Sincerely, Ed. Blyth—

DAR 160.2: 204, DAR 205.2 (Letters)

CD ANNOTATIONS

1.19 among whom ... minister; 1.22] *crossed ink*

1.31 but, ... land. 1.33] '18.'[12] *added in margin, brown crayon*

[1] Charles Wood was the secretary of state for India; the viceroy and governor-general of India was James Bruce, earl of Elgin and Kincardine (*DNB*). Blyth had been curator of the museum of the Asiatic Society of Bengal since 1841, and in 1855, had applied to the society's Court of Directors for a pension to be awarded to him after a certain number of years' service; his memorial was forwarded to the British government in 1856, but met with no success. In July 1861, following the deterioration of Blyth's health, the society sent a second memorial to the government (Blyth 1875, pp. ix–xi). The earl of Elgin argued that Blyth's application had claims to consideration as a 'special case' (the pension had been refused on the grounds that grants from public revenues were limited to those in the 'direct service of the government'), and he stated (*Journal of the Asiatic Society of Bengal* 31 (1862): 430):

> It is the case ... of a man of science, who has devoted himself for a very small salary in duties in connexion with the Asiatic Society, a body aided by and closely identified with the Government of India from which the public have derived great advantage. Mr. Blyth may truly be said to have been, in a great measure, the creator of the Natural History Museum, which has hitherto supplied the place of a Public Museum in the metropolis of India ... [I]f, under such circumstances, Mr. Blyth should after twenty years' service, be compelled to retire from ill-health, it would not be creditable to the Government that he should be allowed to leave without any retiring pension

[2] A pension of £150 was granted to Blyth in 1863 (see *Correspondence* vol. 11, letter from Edward Blyth, 21 September 1863, and *Proceedings of the Asiatic Society of Bengal* 33 (1864): 73).

[3] Bruce was the eighth earl of Elgin and the twelfth earl of Kincardine. Wood, who was the member of parliament for Halifax, Yorkshire, was a baronet (*DNB*).

[4] Blyth's departure was precipitated by his ill health. In the circumstances, the council of the society agreed to give him a year's leave with full pay (Blyth 1875, p. xii, and *Journal of the Asiatic Society of Bengal* 32 (1863): 32).

[5] The reference is to the zoological gardens in Regent's Park, London, run by the Zoological Society of London, of which Blyth was a fellow.

[6] Blyth reportedly owed his pension mainly to 'the untiring efforts made in London' by Hugh Falconer and Proby Thomas Cautley (Blyth 1875, p. xii).

[7] During the 1850s, Blyth became one of CD's most important correspondents, providing him with a vast amount of information on the plants and animals of India (see *Correspondence* vols. 5–7).

[8] Blyth refers to Samuel Laing, financial minister in India, on the council of the governor-general (*DNB*).
[9] Henry John Temple, Viscount Palmerston, was the prime minister.
[10] Cape Negrais is on the south coast of Burma.
[11] Blyth's letter has not been found; however, CD and Blyth had apparently discussed the occurrence of amphibians in the Andaman Islands. In *Origin*, pp. 392–3, CD had written:

> With respect to the absence of whole orders on oceanic islands, Bory St. Vincent long ago remarked that Batrachians (frogs, toads, newts) have never been found on any of the many islands with which the great oceans are studded. I have taken pains to verify this assertion, and I have found it strictly true.

In the fourth edition, published in 1866, CD repeated the above passage (*Origin* 4th ed., pp. 467–8), but continued:

> with the exception of New Zealand, New Caledonia, the Andaman Islands, and perhaps the Salomon Islands and the Seychelles. But I have already remarked that it is doubtful whether New Zealand ought to be classed as an oceanic island; and this is still more doubtful with respect to the Andaman and Salomon groups.

[12] This is the number of CD's portfolio of notes on the means of dispersal of plants and animals.

To Asa Gray 23 November [1862][1]

Down Bromley Kent
Nov. 23^d

My dear Gray

I have been rather slack in obeying my boy Leonard's instructions to thank you most sincerely for a number of stamps with which your envelope of Oct. 27th was plaistered (the enclosures were duly posted) & for a precious Canada stamp previously sent.—[2] He believes he is by far the richest boy in the whole school in N. American stamps, & expects to get wonderful treasures by trading with some sent by you in duplicate.—[3] The Nesæa seeds came all safe, & are as great a treasure to me as the stamps to Leonard, & stronger I cannot put it.[4] I see you say you have sent 2^d notice on Orchid Book to Silliman;[5] I have not received the first, which I sh^d. think must be out, & which I sh^d. very much like to possess.—[6] By the way in the last Edinburgh R^w. there is an article by Duke of Argyll on "Supernaturalism" in which Orchids are brought in:[7] it is clever; but I do not see that it really removes any of the difficulties of Theology.—

In the last Macmillan there is a little Review on Max Muller,—on the origin of language; (by my Brother-in-law, H. Wedgwood & his daughter) which I think is worth looking at.[8] This letter is a sort of Literary Intelligencer; for I am going to tell you that Bates' paper on Butterflies of Amazonia, in last part of Linn. Transactions. is well worth some labour in studying, though out of your special line.—[9] The mimetic cases are really wonderful, & no one has brought so clearly before my mind the process of segregation of varieties into species. But I doubt whether you will have time to go into paper carefully enough to appreciate it. Lyell's book, which I am very curious to see is not yet out; but Murray sold 4000! copies the other day at his auction.—[10]

I have nearly finished a long chapter on the simple facts of the variation of a few of our cultivated plants,[11] & I sh^{d}. be very much obliged for an answer to one question. Is the fruit of the wild Fragaria Virginiana much larger, (twice or thrice?) than that of F. vesca?[12] and secondly do you know anything of the F. grandiflorus, said to have come from Surinam, & often called the Carolina strawberry. How far south does F. virginiana range?

Do you remember about my Boy, Horace, on the natural selection of coward adders?[13] I must tell you that the other day he overheard me talking about species; & afterwards he came to me, with his eyes open with astonishment & asked "Did people *formerly* really believe that animals & plants never changed? I answered oh yes. "Well then what did they say about the kinds of cabbages & peas in the Garden?" I answered that these were all due to man's agency. "But do not wild plants vary". I answered that they varied within certain fixed but unknown limits. To this he shrugged his shoulders with pity for the poor people who *formerly* believed in such conclusions.— I believe Horace is a prophetic type, as Agassiz would say, of future naturalists.—[14]

I see in Times today the great news of Macclellan'ds dismissal from your army—.— Good God what will be the end of all?—[15]

Ever my dear Gray | Yours most truly
C. Darwin

Gray Herbarium of Harvard University (49)

[1] The year is established by the relationship to the letter from Asa Gray, 27 October 1862. From Gray's reply, it appears that this letter and the letter to Gray of 26[–7] November [1862] were posted in the same envelope (see letter from Asa Gray, 29 December 1862).

[2] See letter from Asa Gray, 27 October 1862 and n. 1. Since June 1862, Gray had regularly sent Leonard Darwin North American postage stamps for his collection.

[3] Leonard Darwin had begun to attend Clapham Grammar School, in south London, in January 1862; however, he became ill with scarlet fever in June 1862, and apparently did not return to the school until January 1863 (see letter to W. E. Darwin, 26 April [1862], n. 2).

[4] See letter from Asa Gray, 27 October 1862 and n. 2.

[5] CD refers to Gray's follow-up article to his review of *Orchids* (A. Gray 1862b), which was published in the *American Journal of Science and Arts* in November 1862; the journal was commonly known as 'Silliman's journal' after its founder, Benjamin Silliman. Gray had expressed a hope in his letter of 27 October 1862 that he would be able to send proof-sheets of the review 'soon'; he sent them with his letter to CD of 10 November 1862.

[6] CD made this statement under a misapprehension: he later told Gray that he had 'confounded one Review with another—' (*Correspondence* vol. 11, letter to Asa Gray, 19 January [1863]). Gray had sent CD a copy of A. Gray 1862a with his letter of 15 July [1862]; although the review is listed in the List of reviews (DAR 261 (DH/MS* 8: 6–18)) that served as CD's index to his collection of reviews of his own books, it is absent from the Darwin Pamphlet Collection–CUL.

[7] [G. D. Campbell] 1862; there is an annotated copy of this work in the Darwin Pamphlet Collection–CUL. See letter to J. D. Hooker, 18 [November 1862], n. 9.

[8] Hensleigh and Frances Julia Wedgwood's review of Max Müller 1861 ([H. Wedgwood and F. J. Wedgwood] 1862) appeared in *Macmillan's Magazine* for November 1862. The attribution to F. J. Wedgwood is based on the reference in the letter from Emma Darwin to William Erasmus Darwin

of [13 November 1862] (DAR 219.1: 69). In addition, the *Wellesley index* reports that a cheque was paid to F. J. Wedgwood for a contribution to this number of the magazine.

9 Bates 1862a.

10 CD refers to Charles Lyell's *Antiquity of man* (C. Lyell 1863a), which was not published until 6 February 1863 (C. Lyell 1863b, p. [vii]). John Murray's annual trade sale was held on 4 November 1862 (*Athenæum*, 8 November 1862, p. 595).

11 According to his 'Journal' (Appendix II), CD prepared a draft of the part of *Variation* dealing with 'Facts of variation of Plants' between 7 October and 11 December 1862. The material formed chapters 9 and 10 of the published work (*Variation* 1: 305–72).

12 Gray's reply to these questions was apparently included in the missing portion of the letter from Gray of 29 December 1862; however, in the section on strawberries in *Variation* 1: 351–4, CD stated that he was informed by Gray that the fruit of *Fragaria virginiana* 'is only a little larger' than that of the common wood strawberry, *F. vesca* (*Variation* 1: 351 n.).

13 Horace Darwin. See letter to Asa Gray, [3–]4 September [1862].

14 CD refers to Louis Agassiz's concept of a 'prophetic type', the first-appearing representative of a group that reached its full development at a later period (Agassiz 1857–62 1: 116–18; CD's annotated copy of this work is in the Darwin Library–CUL).

15 The leader in *The Times* of 22 November 1862 (pp. 8–9) reported that on 7 November 1862, George Brindle McClellan had been removed from command of the Union's army of the Potomac (see also McPherson 1988, pp. 562 and 570).

To Daniel Oliver 23 [November 1862][1]

Down Bromley Kent
23^{d}.

Dear Oliver

I am *very curious* to hear about Epilobium angustifolium,[2] both on account of fact itself, & for following odd psychological case.— I knew plant well between 20 & 30 years ago in my Father's grounds;[3] well, this summer it flashed across my mind that there was something dimorphic in it.[4] I tried my best, but could remember no vestige of particulars; yet I was very near sending to my sister for a lot of flowers;[5] but as I could hardly remember anything of flower, except its colour, I thought it too foolish.— If it really is dimorphic I shall always look at my flash of memory like one of those cases of persons in a fever who have temporarily remembered a language learnt in infancy & ever after forgotten.—

If you have pretty good reason to think it dimorphic, ask Hooker to put it down in list of seeds required, if I can possibly get it.[6] Seeds are much better than roots, as same form may spread by suckers.— For several reasons (Clarkia elegans) I sh^{d}. very much like to experiment on this plant.[7]

About strawberries, do you refer to American plan of planting 6 rows of "*pistillates*" & one row of *hermaphrodites*?— If something else kindly inform me, as I have written a very little on strawberries.[8] Thank Hooker for very kind offer of dried plants;[9] but I hate dried plants; I can make nothing of them & I profoundly pity all you Botanists.—

Will you ask Hooker (to whom I shall be writing before long, for I never give him a long holiday) if he gets the missing vol. of Bot. Journal for Linn. Soc. kindly to send it me (per Railway to care of Down Postman) *first* & then I will send it in his name to the Soc; for I want pretty soon to consult it, before I write my paper on Linum for Linn. Soc^{y}.—[10]

Ever | Yours sincerely | Ch. Darwin

DAR 261 (DH/MS 10: 57)

[1] Dated by the relationship to the letter from J. D. Hooker, [15 and] 20 November [1862].
[2] See letter from J. D. Hooker, [15 and] 20 November [1862].
[3] Robert Waring Darwin had lived at The Mount, Shrewsbury, Shropshire (Freeman 1978).
[4] CD had had a similar recollection in the summer with regard to *Linum flavum* (see letter to Daniel Oliver, 13 October [1862] and n. 2). See also n. 7, below.
[5] Susan Elizabeth Darwin still lived at The Mount (Freeman 1978).
[6] Joseph Dalton Hooker; with his letter to Hooker of 3 November [1862], CD had sent a list of seeds of plants he wanted for his experiments.
[7] While staying in Bournemouth in September 1862, CD had corresponded with Oliver concerning a plant with two differently coloured sets of anthers, a feature that CD considered might be 'a good guide to dimorphism'. He initially thought that the plant was either an *Epilobium* or a *Clarkia*, subsequently identifying it, with Oliver's help, as *Clarkia elegans* (see letter to Daniel Oliver, 2 September [1862] and n. 7, and letter from Daniel Oliver, 4 September 1862).
[8] CD was preparing a draft of chapter 10 of *Variation*, in which he included a section on variation in cultivated strawberries (*Variation* 1: 351–4). Hooker had told CD that Oliver had reminded him of a 'curious remark on sexualism of strawberries by an American … alluded to at length in [the] Technologist' (Wray 1861a); see letter from J. D. Hooker, [15 and] 20 November [1862] and n. 22. Wray 1861a was reprinted in the *Gardeners' Chronicle and Agricultural Gazette* for 3 August 1861 (Wray 1861b), a copy of which CD kept in a separate parcel (see DAR 222 and DAR 75: 1–12; CD's annotated copy of the *Gardeners' Chronicle* is at the Cory Library, Cambridge Botanic Garden); he cited Wray 1861b in support of the observation that several English hermaphrodite varieties of strawberry 'when cultivated in rich soils under the climate of North America' commonly produced plants with 'separate sexes' (*Variation* 1: 353 n.). CD also cited Wray 1861b on this point in *Forms of flowers*, p. 293 n.
[9] See letter from J. D. Hooker, [15 and] 20 November [1862].
[10] Hooker was trying to acquire a copy of volume 7 of the *London Journal of Botany* for the library of the Linnean Society of London (see letter from J. D. Hooker, [15 and] 20 November [1862]); CD wished to borrow the volume since it contained part of a paper on the genus *Linum* (Planchon 1847–8) that he needed to consult. CD cited Planchon 1847–8 in 'Two forms in species of *Linum*', p. 81 (*Collected papers* 2: 103–4), which was written between 11 and 21 December 1862 (see 'Journal' (Appendix II)), and read before the Linnean Society on 5 February 1863.

From H. W. Bates 24 November 1862

King St Leicester
Nov 24 1862

My Dear Mr Darwin

I am now well repaid for the labour of writing the paper in Linn. Trans. having received your warm & discriminating approval.[1] Dr Hooker has also written a most friendly letter on the subject valuing the treatise highly.[2] Mr Wallace & I discussed the matter last time I was in London & he thinks me entirely right.[3] As for the "mob" of Naturalists, I don't care much about them.[4] There is only one man in Europe who will be able to follow my reasonings species by species & variety by variety; this is Dr C. Felder of Vienna & I shall be glad to hear from him upon the subject.[5] As to ordinary Entomologists they cannot be considered scientific men but must be ranked with collectors of postage stamps & crockery.

I wish to have more criticisms from you on the subject, when you are quite at leisure (what an idea!).— I am quite sure I could now put the case much more strongly. But you say the argument is quite clear so I must rest contented. I now believe one of the forms of Leptalis (L. argochloe) is not an immediate descendant from L. Theonoe, as the other varieties are, but has originated (in a similar way) somewhere westward in Andean Valleys & has wandered to the Amazons. It would be tedious here to give reasons. Does it occur to you that a great deal may be unexpectedly learnt by thus thoroughly going into one small group of Natural objects? The more I study them the more I am surprised at the wonderful revelations which spring from them: much more than is explained in my treatise. ⟨*two pages missing*⟩

I am vain enough to wish to see the paper noticed in some widely circulated publication, think'g it may introduce my book.[6] Do you know anything of Sir J. Herschel or his address?[7] I hear he writes those capital monthly summaries of science in the "Cornhill" & would like to send him a copy.[8]

Incomplete[9]
DAR 160.1: 72

CD ANNOTATION
2.7 Does ... objects? 2.8] *cross in margin, ink*

[1] Bates 1862a. See letter to H. W. Bates, 20 November [1862].
[2] Joseph Dalton Hooker's letter to Bates of 13 November 1862 is reproduced in Bates 1892, pp. xlvi–xlvii. See also letter from J. D. Hooker, [15 and] 20 November [1862].
[3] Alfred Russel Wallace; see letter to H. W. Bates, 20 November [1862].
[4] See letter to H. W. Bates, 20 November [1862].
[5] Cajetan von Felder was an authority on Brazilian butterflies (*OBL*).
[6] Bates was preparing an account of his travels in the Amazon region of South America (Bates 1863).
[7] CD had known John Frederick William Herschel for many years (see *Correspondence* vol. 2); however, his name is not in CD's Address book (Down House MS). Herschel lived at Collingwood House, Hawkhurst, Kent (*Post Office directory of the six home counties* 1862).
[8] Herschel contributed monthly summaries of recent scientific discoveries to the *Cornhill Magazine* between September 1862 and March 1863 (*Wellesley index*); however, Bates 1862a was not discussed in any of his notices.
[9] Some indication of the contents of the missing portion of the letter is given by CD's reply (see letter to H. W. Bates, 25 November [1862]).

From C. W. Crocker 24 November 1862

Chichester.
Novr. 24th. 62

Dear Sir

I should think that you would be able to manage Begonia frigida and its allies, at this season of the year, much better in the window of a room where a fire is kept every day than in a greenhouse where the frost is kept out and nothing more.[1] I believe you would find no difficulty in growing them in your study window, only taking the precaution of moving them away from the windows in severe frosts. B. frigida flowers nearly or quite all the year round; I expect it could only be procured from Kew; the nurserymen would hardly think it showy enough to keep. I left a stock of young ones when I came away. I believe its most nearly allied species would be the little group to which B. Dregei, B. parvifolia, and B. Natalensis belong. My poor old friend Dr. Klot of Berlin, who parted Begonia up into some 30 or 40 genera, would I think put all of them into his group "Augustia"—[2] There can be but little doubt that B. frigida would hybridize with those mentioned. As far as I have been enabled to judge I should say that the genus Begonia is too comprehensive to get all the members to cross with each other.— Thus I could never get the B. Xanthina group (to which most of the coloured leaved species now so popular belong) to cross with the upright caulescent sp. Not more than 4 or 6 species have been used in procuring these painted leaved vars.—[3] B. frigida *does* produce normal male & female flowers as well as the hermaphrodite ones.

I did not know of the Peloria Tropæolum, I have never seen it.[4]

Since I wrote to you last I have reason to think that my Antirrhinum experiment was more successful than I anticipated.[5] Still one must not build too much upon a single experiment. I refer to those flowers which were simply protected with netting without being otherwise touched. I believe that not more than half the flowers were fertilised; and even these might have been done by insect agency—for the flowers of the Snap-dragon are inhabited by a number of very minute insects which no net could keep out, but still large enough to carry a few grains of pollen about.

It is very rarely indeed that a flower misses producing seed if left to nature and therefore as several were left unfertilised in my netted plant I shall try again with great hope of proving what could hardly be anticipated. Should it prove to be sterile unless artificially aided why where can we look for a plant which would be self-fertilised they would surely be exceptions. But I am perhaps looking ahead too hopefully— Do you know of any one who has got the Peloria form of Antirrhinum? I have never seen it.—[6] Many thanks for your hint relative to dimorphism among my Plantains I shall look diligently for it—though I have hitherto seen nothing which led me to expect it.[7] I was much pleased to hear of your success with the Pelargonium[8]

I have to thank you very much for offer of aid in the way of plants, but I hope and trust, at the same time, that nothing I said looked like complaining[9] I hate a grumbler and should despise myself if I thought I had been weak enough to do it. The man who suits my taste is the one who makes the most of the opportunities he has (and everybody has opportunities of some kind). I'll tell you how I have acted this year. I bought and begged some annual seeds last spring these I have grown on, have harvested my crop, and as soon as the seed is clean, mean to exchange it with some nurseryman for plants. My income is larger than it was at Kew and I am not afraid to spend a little upon a hobby which is as dear as life itself to me.[10] I would as soon be dead as have no plants, even though they be weeds.

I think I shall be all right for Hollyhocks next year.—[11] I have got some Scarlet Geraniums upon which I am trying influence of pollen from long and short stamens, but my experiments are quite young.— I last year collected seed of all the kinds of Aquilegia I could get hold of with a vague sort of idea that something would come of it for there are several distinct colours among them. They'll flower next year. But what a job I had to raise them! I was obliged to sow in the open ground (I have since bought two frames) and the seeds laid for four blessed months without one germinating. I gave them up for lost but I have now got them strong enough to stand the winter I trust.

In conclusion allow me to say that there is nothing in the world which could give me greater pleasure than to answer to the best of my ability any question you could can put— the more of these questions there are the better I shall be pleased— at the same time you may trust my honesty to say that I dont know if I have any doubt upon the subject.

Trusting you will often give me the happiness of answering anything upon which I can be of the very least assistance to you, I remain, dear Sir, | yours most respectfully | C. W. Crocker

C. Darwin Esq^{e}.

As there is not much in flower just now I have been amusing myself with the Ivy—perhaps the most useful plant in all our flora so far as birds and the insect world is concerned. It flowers when everything else has done and ripens its fruit when other supplies of food are nearly exhausted. I find lots of little variations which nobody has thought it worth while to notice, as for instance in general form of inflorescence (not of individual flowers), in time of blooming, &c. Some persist in making only one terminal umbel—while others produce as usual many, it is the case all over the plant.

DAR 161.2: 259

CD ANNOTATION

4.5 they . . . exceptions.] *cross in margin, ink*

[1] CD was preparing a draft of the part of *Variation* dealing with 'Facts of variation of Plants' (*Variation* 1: 305–72; see 'Journal' (Appendix II)), in which he included a brief discussion of variation in cultivated flowers (*Variation* 1: 364–72). CD cited a case (pp. 365–6) of a 'monstrous' *Begonia frigida* grown at the Royal Botanic Gardens, Kew, with, untypically, a number of hermaphrodite flowers that also showed other structural changes. Crocker, who until recently had been foreman of the propagating department of the Royal Botanic Gardens, Kew, had observed 'that seedlings from the *normal* flowers produced plants which bore, in about the same proportion as the parent-plant, hermaphrodite flowers having inferior perianths' (*Variation* 1: 366; see Crocker 1861). CD wished to obtain one of the 'monstrous' offspring for crossing purposes (see letter to J. D. Hooker, 12 [December 1862], and letter to John Scott, 19 December [1862]), and had apparently written to Crocker to ask about the practicalities; the letter to Crocker has not been found.

[2] Crocker refers to Klotzsch 1854, in which begonias were divided into forty-two genera (*Taxonomic literature*). Crocker probably met Johann Friedrich Klotzsch, who was curator of the herbarium at the University of Berlin from 1833 to 1860, while working as a gardener near Berlin.

[3] In Crocker 1861, Crocker had discussed the restricted parentage of the popular variegated hybrids of Begonia, expressing a fear that the 'present passion for plants with coloured foliage will drive some of our old favourites out of cultivation', and suggesting that nurserymen should attempt to hybridise the variegated varieties with those from other groups. CD kept separately his copy of the number of the *Gardeners' Chronicle and Agricultural Gazette* that contained Crocker 1861, and which is now at the Cory Library, Cambridge Botanic Garden (see the 'List of the numbers of special interest to Darwin and kept by him in separate parcels' in DAR 222).

[4] In his letter to CD of 31 October 1862, Crocker had mentioned conducting experiments on species of the genus *Tropaeolum*. CD's reply has not been found, but in the letter to Isaac Anderson-Henry, 20 January [1863] (*Correspondence* vol. 11), he mentioned his belief that there was 'a *peloric* and common variety of *Tropæolum*'.

[5] See letter from C. W. Crocker, 31 October 1862 and n. 7.

[6] Following Crocker's discussion of *Antirrhinum* in his letter to CD of 31 October 1862, CD may have suggested to him that he should attempt to obtain seed from the peloric flowers of *Antirrhinum* as Karl Ludwig Willdenow had done (see *Correspondence* vol. 9, letter to *Journal of Horticulture*, [before 18 June 1861]). Such experiments would parallel CD's own experiments on pelargoniums (see n. 8, below).

[7] See letter from C. W. Crocker, 31 October 1862 and n. 8. On the basis of CD's information, Crocker carried out a number of observations on and crossing experiments with *Plantago*, publishing his results in Crocker 1864; CD cited Crocker's work in *Forms of flowers*, p. 306 n. See also *Correspondence* vol. 11, letter from C. W. Crocker, 1 May 1863.

[8] CD had apparently told Crocker of his success in producing fertile seed from the normally sterile peloric flowers of pelargoniums (see letter to M. T. Masters, 8 July [1862], and letter to John Scott, 19 December [1862]).

[9] See letter from C. W. Crocker, 31 October 1862.

[10] After leaving the Royal Botanic Gardens, Kew (see n. 1, above), Crocker worked as a journalist and cathedral official in Chichester, Sussex (*Gardeners' Chronicle and Agricultural Gazette*, 7 March 1868, p. 242).

[11] See letter from C. W. Crocker, 17 May 1862 and n. 3.

From Asa Gray 24 November 1862

Cambridge. [Massachusetts]
Nov. 24, 1862.

My Dear Darwin.

As another peculiar stamp is waiting for Leonard's collection, and I have notes &c—to fill it, here goes for an early response to yours of the 6th inst.[1]

About Max Müller. Surely you can't wonder that the attempt to account for the "*first origin of* language" or of anything else, should be the "least satisfactory".[2]

The *use* that I fancied could be made of Max Muller's book,—or rather of history of language—is something more than *illustration*, but only a *little* more,—i.e. you may point to analogies of development & diversification of language—of no value at all in evidence in support of your theory, but good & pertinent as rebutting objections, urged against it.[3]

Bishop Colenso's book will make a noise in England, indeed. I have only read the notice in the Athenæum[4]

You detest the spirit of the *Times* quoad U.S.[5] The Athenæum is just as bad in its little penny-trumpet way, every chance it can get from the very first. Can you be much surprised that we return dislike with interest. But we are pleased to find there sensible & fair writers,—such as Cairnes & Mill.[6]

No, dear Darwin, we don't scorn your joining in the prayer that we daily offer that "God would help our poor country", and I know and appreciate your honest & right feeling.

I see also, from the English papers I read how you must picture us as in the extreme of turmoil and confusion and chaos.— But if you were here, you would open your eyes to see everything going on quietly, hopefully, and comfortably as possible. I suppose we do not appreciate our miseries. We accept our misfortunes and adversities, but mean to retrieve them,—and would sink all that we have before giving up. We work hard, & persevere, and expect to come out all right,—to lay the foundations of a better future, no matter if they be laid in suffering. That will not hurt us *now*, and may bring great good hereafter.

I never saw, and have scarcely heard of Miss Cooper's book you ask after.[7] She is the daughter of late J. Fenimore Cooper, the novelist.[8] The village she describes must be Cooperstown—in the county adjacent to that in which I was brought up,—a region, which every time I visit it, I say it is the the fairest of lands, and the people the happiest.[9]

Oh—as to the weeds,—Mrs. Gray says she allows that our weeds give up to yours. Ours are modest, woodland, retiring things, and no match for the intrusive, pretentious, self-asserting foreigners.[10] But I send you seeds of one native weed which corrupted by bad company—is as nasty and troublesome as any I know, viz. *Sycios angulatus*,—also of a more genteel Cucurbitacea, *Echinocystis lobata* (the larger seeds). Upon these, especially upon the first, I made my observations of tendrils coiling to the touch.[11] *Put the seeds directly into the ground*; they will come up in spring, in moist garden soil. My observations were made on a warm sunny day. I doubt if you have warmth and sunshine enough in England to get up a *sensible* movement.

My note about them is in Proceed. Amer. Acad. 4, p. 98, reprinted in Sill. Jour. March, 1859, p. 277.[12] I must own, that upon casually taking them up since, I never have obtained such *very good results*, as upon 2 days of Aug. 1858.

Upon gourds affecting each others fruits, I have made *no observations* at all. I have only *referred* to that,—as a well-known thing—at least, of common repute here,—and then referred to *maize*, where the soft *sweet-corn*, when fertilized by hard yellow-corn, the grain so fertilized takes the character of the fertilizer. My note about it is in Acad. Proceed. vol. IV, I think.—[13] You have the vols, (which I have not in reach now), & can find it by the Index. It does not amount to much. Nothing on Maize I know of except Bonafous' folio volume.[14] I am going to get & send you grains of 4 or 5 sorts of maize. About the involucrate form, I wrote in my last.[15]

Whenever I post to you a Boston *Semi Weekly* Advertizer, please to send it on to D^r^. Boott.[16]

Ever, dear Darwin | Yours cordially | Asa Gray

DAR 165: 124

CD ANNOTATIONS
1.1 As . . . day. 9.8] *crossed pencil*

10.1 My note . . . 1858. 10.3] *scored brown crayon*
11.4 the grain . . . I think.— 11.5] *scored brown crayon*
Top of letter: 'Tendrils' *brown crayon*

[1] Letter to Asa Gray, 6 November [1862]. Since June 1862, Gray had regularly sent Leonard Darwin North American postage stamps for his collection.
[2] Max Müller 1861. See letter to Asa Gray, 6 November [1862] and n. 3.
[3] See letter from Asa Gray, 4 and 13 October 1862, and letter to Asa Gray, 6 November [1862].
[4] See letter to Asa Gray, 6 November [1862] and n. 7. Gray refers to volume 1 of Colenso 1862–79, and to the review published in the *Athenæum*, 1 November 1862, pp. 553–4.
[5] See letter to Asa Gray, 16 October [1862] and n. 15. Gray refers to the American Civil War.
[6] The Irish political economist John Elliot Cairnes was the author of *The slave power* (Cairnes 1862), which made a great impression in both Britain and America, and has been described as 'the most powerful defence of the cause of the Northern states ever written' (*DNB*). John Stuart Mill wrote a number of articles in support of the Union cause in the American Civil War, a copy of one of which Gray had sent to CD earlier in 1862 (Mill 1862; see letter from Asa Gray, 6 March [1862] and n. 8).
[7] Cooper 1855. See letter to Asa Gray, 6 November [1862] and n. 8.
[8] James Fenimore Cooper.
[9] Cooperstown is a town in New York State (Seltzer 1952); Asa Gray was brought up in the nearby villages of Sauquoit and Paris Furnace, New York State (Dupree 1959, pp. 4–5).
[10] Jane Loring Gray. See letter to Asa Gray, 6 November [1862] and nn. 9 and 10.
[11] See letter to Asa Gray, 6 November [1862] and nn. 17 and 21.
[12] CD's annotated copy of the volume of the *Proceedings of the American Academy of Arts and Sciences*, in which A. Gray 1858b was published, is in the Darwin Library–CUL. A. Gray 1858b was reprinted in the *American Journal of Science and Arts* (A. Gray 1859), commonly known as 'Silliman's journal' after its founder, Benjamin Silliman.
[13] Gray refers to his observations, given before the American Academy of Arts and Sciences on 9 February 1858, which were reported in the *Proceedings of the American Academy of Arts and Sciences* 4 (1860): 21–2.
[14] Bonafous 1836.
[15] Gray refers to the postscript, now missing, to his letter of 10 November 1862 (see letter to Asa Gray, 26[–7] November [1862]).
[16] Gray refers to the American botanist Francis Boott, who had been resident in London for many years (Dupree 1959).

To J. D. Hooker 24 [November 1862]

Down Bromley Kent
24th

My dear Hooker.

I have just received enclosed for you,[1] & I have thought that you would like to read the latter half of A. Grays letter to me, as it is political & nearly as mad as ever in our English eyes.—[2] You will see how the loss of the power of bullying is in fact the sore loss to the men of the north from disunion.—

I return with thanks Bates' letter, which I was glad to see.[3] It was very good of you writing to him; for he is a man, who evidently wants encouragement.— I have now finished his paper (but have read nothing else in the volume);[4] it seems to me admirable. To my mind the act of segregation of varieties into species was never so plainly brought forward; & there are heaps of capital miscellaneous observations.

I hardly know why I am a little sorry, but my present work is leading me to believe rather more in the direct action of physical conditions.—[5] I presume I regret it, because it lessens the glory of Natural selection, & is so confoundedly doubtful.— Perhaps I shall change again when I get all my facts under one point of view, & a pretty hard job this will be.—

Thanks for soundings off Ireland, which I have been glad to see.—[6]

What a pleasant letter your last was.— When shall you begin your great work "The Aristocracy"! By the Lord how it would sell![7] I noted your opening sentence in Review of Orchids on my *second* reading, as laughably like Lindley.—[8] In that Review, it seemed to me that you laid rather too much stress on importance of crossing with respect to origin of species; yet certainly it is very important in keeping forms stable.—

What a thing it would be if Owen were excluded by the Soc^{y}. from the Council:[9] I do not think he is fit man; but I hate him so I sh^{d}. not like to vote against him, as I should never know whether I did it honestly. And what a sting it would be to put in good honest Falconer![10] I doubt whether I could stand excitement of public meeting; but I must come soon to London & try what I can do.—

Have you seen Falconer's paper in last Geolog. Journal versus Owen;[11] he pays me an extraordinary compliment, but what is far better, I think I see he is slowly coming round about permanence of species—[12] In some M.S. of his, which I have seen, this is pretty clear; but he does not like N. Selection.[13] Parts of Falconer's paper strike me as quite admirably written.—

I am sorry to hear about your Governess—good & bad they are the trouble of one's life.[14] You must **not** write so often, for it *must* be as great a bore to you to write as it is a very great pleasure to me to receive them.

Ever yours affec^{ly}— | C. Darwin

Asa Gray's letter to you is fine excuse for writing.— I enclose Lythrum salicaria Diagram.—[15] Study it or burn it. In my opinion it is a very curious case of generation.—

You must read, if you want to understand, the side description of parts & M.S. at bottom of diagram.—

[Enclosure]

King St Leicester
17 Nov 1862

My Dear Dr Hooker

I need not say how glad I am to have your good opinion, (given so promptly) on my little essay.[16] To tell you the truth I bestowed extra pains on it because I considered that what I had previously done was not sufficient to merit the high estimation which yourself and Mr Darwin so very kindly placed upon it. As to being satisfied with the treatment of the Society I shall consider myself a lucky man to escape a severe scolding (I know Kippist will inflict it) for putting it to so much expense.[17] Mr Busk[18] rather pressed me towards the last & I finished in a hurry, consequently

I had to rewrite about 2 pages besides other alterations after the treatise was put in type. The printer was also very careless in altering some of the type in some sheets, which had to be changed so I am afraid the expense altogether will be great.

You hit me on what I know is a weak point.[19] You will recollect our discussion (by letter) last winter & will have perceived how much I have been influenced by your teaching; for I have abandoned the notion that physical conditions on the individuals have had anything to do with the production of those close imitations figured.[20] But I find it difficult to abandon the idea of some effect being produced directly on individuals by the action of physical conditions. If half a dozen beetles belonging to different genera show brassy varieties when living under the sea air; if a number of butterflies equally independent of each other have their orange colour changed into brown in the interior of the S. American continent & if many different plants become changed in a similar way on the sea side or on a mountain it seems to me that all have been operated upon by local physical conditions. But your remarks I believe will tend to change my opinion for I can see, by their light, that Selection may have been after all the cause of the establishment of the varietal forms.

I have thought that when a species first migrated, say to the seaside, the effects of sea-air would be visible in a generation or two & therefore that the maritime variety would be due to direct action. If, however, it required a very great number of generations to effect the change we now see, why then of course Nat. Select. must have played a part. You would say, as to a succulent maritime var. of a plant, that its original parent (its condition before migrating to sea-side) varied in a succulent direction quite as much in the interior of the land as on the sea side, but that the sea-side habitat favoured the first germs of succulence & led them on generation after generation; whereas the interior habitat neglected them or favoured the opposite tendency.

I have no doubt you have facts to show that a plant like those we are discussing originally showed (in the first one or two generations) no more tendency to succulence on the sea side than in the interior & therefore I give it up.— I shall be most anxious to have Mr Darwins opinion on the essay.[21] My book is getting on I have had 6 sheets of proof down already.[22]

Yours sincerely | H W Bates

Does Linn. Soc. allow coloured plates to author's 25 copies? Some societies do not.

Endorsement: 'Nov/'62'
DAR 115.2: 173; Royal Botanic Gardens, Kew (Hooker letters 2: 46)

[1] CD probably refers to proof-sheets of part of the November 1862 number of the *American Journal of Science and Arts*, in which A. Gray 1862b and 1862e were published. Asa Gray enclosed a set of proof-sheets for CD with his letter of 10 November 1862, and apparently also enclosed a set for Hooker (see letter to Asa Gray, 26[–7] November [1862]). Hooker discussed A. Gray 1862e in his reply to this letter (see letter from J. D. Hooker, 26 November 1862 and n. 2).

[2] See letter from Asa Gray, 10 November 1862 and n. 14.

[3] See enclosure.

[4] Bates 1862. The third part of volume 23 of the *Transactions of the Linnean Society of London*, in which this paper appeared, was published on 13 November 1862 (Raphael 1970); in his letter to CD of [15 and] 20 November [1862], Hooker discussed several other papers in this part of the journal.

[5] In his letter to CD of [15 and] 20 November [1862], Hooker mentioned his having criticised Bates's endorsement of the direct influence of physical causes on variation. CD was preparing a draft of the part of *Variation* dealing with 'Facts of variation of Plants' (see 'Journal' (Appendix II)). He concluded his discussion of varieties of maize by examining the facts concerning the changes effected by Europe's climate on newly imported American varieties; he noted that after a few generations, some of the seeds became rounded, approaching the common European forms, and stated that this afforded 'the most remarkable instance' known to him of 'the direct and prompt action of climate on a plant' (*Variation* 1: 322). In his discussion of the 'laws of variation' in *Origin*, pp. 131–70, CD noted that the 'external conditions of life', such as climate and food, seemed to have 'induced some slight modifications' of structure (p. 167); however, he emphasised his view that it was 'the steady accumulation, through natural selection, of such differences, when beneficial to the individual', that gave rise to 'all the more important modifications of structure' (p. 170).

[6] See letter from J. D. Hooker, [15 and] 20 November [1862] and n. 2.

[7] Letter from J. D. Hooker, [15 and] 20 November [1862] and n. 17.

[8] The reference is to John Lindley, in whose style Hooker had written a review of *Orchids* for the *Gardeners' Chronicle and Agricultural Gazette* ([J. D. Hooker] 1862c; see letters from J. D. Hooker, 7 November 1862 and [15 and] 20 November [1862], and letter to J. D. Hooker, [10–]12 November [1862]). The opening sentence (p. 789) reads: 'It is not unadvisedly that we have so long delayed reviewing the work that heads this article.' CD's annotated copy of the *Gardeners' Chronicle* is in the Cory Library, Cambridge Botanic Garden; CD kept in a separate parcel his copy of the numbers in which [J. D. Hooker] 1862c appeared (see DAR 222 and DAR 75: 1–12).

[9] See letter from J. D. Hooker, [15 and] 20 November [1862] and nn. 10 and 12. Richard Owen was elected to the council of the Royal Society at the anniversary meeting of the society on 1 December 1862 (*Proceedings of the Royal Society of London* 12: 299).

[10] Hugh Falconer. See letter from J. D. Hooker, [15 and] 20 November [1862].

[11] Falconer 1862.

[12] Falconer began his paper, 'On the disputed affinity of the mammalian genus *Plagiaulax* from the Purbeck beds', by referring to CD as 'One of the most accurate observers and original thinkers of our time' (Falconer 1862, p. 348). See also letter to Hugh Falconer, 14 November [1862] and nn. 7 and 9.

[13] CD refers to a portion of the manuscript of Falconer 1863 that he had seen (see letter to Hugh Falconer, 1 October [1862]).

[14] See letter from J. D. Hooker, [15 and] 20 November [1862] and n. 25; the governess has not been identified.

[15] The diagram has not been found; however, see the diagrams in the letter to W. E. Darwin, [2–3 August 1862], and those in the letter to Asa Gray, [3–]4 September [1862].

[16] Hooker's letter to Bates of 13 November 1862 is reproduced in Bates 1892, pp. xlvi–xlvii. See also letter from J. D. Hooker, [15 and] 20 November [1862].

[17] In his letter to Bates of 13 November 1862, Hooker, who was a vice-president of the Linnean Society of London, expressed a hope that Bates had found the society to have 'done justice' to Bates 1862 (Bates 1892, p. xlvi; see also n. 4, above). Bates refers to the fact that he had incurred additional expense for the society by making extensive alterations to the proofs of Bates 1862 (see letter from H. W. Bates, 17 October 1862); Richard Kippist was librarian of the Linnean Society.

[18] George Busk was one of the secretaries of the Linnean Society.

[19] In his letter to Bates of 13 November 1862, Hooker had criticised Bates's attempt to account for some of the variations in species of Lepidoptera by the direct action of 'physical conditions' (Bates 1892, pp. xlvi–xlvii). See also letter from J. D. Hooker, [15 and] 20 November [1862] and n. 6.

[20] CD, Bates, and Hooker had corresponded on this subject earlier in the year (see letters from J. D. Hooker, [10 March 1862], 17 March 1862, and [23 March 1862], and letters to J. D. Hooker, 14 March [1862], 18 March [1862], and 26 [March 1862]).
[21] See letter to H. W. Bates, 20 November [1862].
[22] Bates 1863.

To *Journal of Horticulture* [before 25 November 1862][1]

Will any of your correspondents who have attended to the history of the Strawberry, kindly inform me whether any of the kinds now, or formerly, cultivated have been raised from a cross between any of the Woods or Alpines with the Scarlets, Pines, and Chilis?[2] Also, whether any one has succeeded in getting any good from a cross between the Hautbois and any other kind? I am aware that Mr. Williams, of Pitmaston, succeeded in getting some sterile hybrids from the Hautbois and Woods;[3] but whether these were ever at all largely propagated, I cannot find out. I am, also, aware that Mr. Knight and Mr. Williams raised many seedlings by crossing Scarlets, Pines, and Chilis;[4] but what I want to know is, whether any one has crossed these three latter kinds with the Wood or Alpine. I should feel greatly indebted to any one who would take the trouble to inform me on this head.—[5]

C. Darwin, Down, Bromley, Kent.

Journal of Horticulture, Cottage Gardener, and Country Gentleman, n.s. 3 (1862): 672

[1] The letter was published in the issue of 25 November 1862. There is a draft of this letter in DAR 96: 9.
[2] CD was preparing a draft of the part of *Variation* dealing with 'Facts of variation of Plants' (*Variation* 1: 305–72; see 'Journal' (Appendix II)); varieties of cultivated strawberries are discussed on pp. 351–4.
[3] John Williams's observations are recorded in Knight 1824, p. 294; CD cited them in *Variation* 1: 352 n.
[4] CD cited Knight 1820 on this point in *Variation* 1: 351 n.
[5] See letter from R. T. Clarke, [after 25 November 1862]. CD's letter also elicited a number of replies in the *Journal of Horticulture*: a letter signed 'D., *Deal*' was published in the issue of 9 December 1862, p. 721; a letter from William Smith of York was published in the issue of 30 December 1862, p. 779; and one from Isaac Anderson-Henry was published in the issue of 20 January 1863, pp. 45–6. CD cited the first two of these letters in *Variation* 1: 352 n. Smith sent with his letter 'three small runners' of a cross between a wild Hautbois and 'a fine seedling . . . obtained between British Queen and Black Prince', stating: 'If you thought it might in any way interest Mr. Darwin, it is at your option to send him one or all.' The editors sent CD the specimens (see letter to G. W. Johnson or Robert Hogg, 20 December [1862]).

To H. W. Bates 25 November [1862][1]

Down Bromley Kent
Nov. 25

Dear Bates

I sh^d. think it was not necessary to get a written agreement.— I have never had one from Murray.[2] I suppose you have a letter with terms; if not, I sh^d. think you had better ask for one to prevent misunderstandings.— I *think* Sir C. Lyell told me he had not any formal agreements.—[3] I am heartily glad to hear that your Book is progressing.— Could you find some place, even foot-note (though these are in nine cases out of ten objectionable) where you could state, as fully as your materials permit, all the facts about similar varieties pairing,—at a guess how many you caught, & how many now in your collection.—[4] I look at this fact as very important: if not in your book, put it somewhere else, or let me have cases.—

I entirely agree with you on enormous advantage of thoroughily studying one group.—[5] I sh^{d}. doubt Sir J. Herschel reading or reviewing nat. History; his address is "Collingwood Hawkhurst, Kent".—[6]

I have already drawn Asa Gray's attention to your paper; but I fear it is out of his line, as he contributes only Bot. Reviews to N. American Journal.—[7] I will see whether a suggestion to one of Editors of Nat. Hist. R. will do any good;[8] but as you are aware it is very unusual to review papers.—

I wish I had spare strength or time to review your paper; but in truth I have lost such months of time this whole summer that I must work on my own work when well enough.—[9]

I really have no criticism to make; style seems to me very good & clear; but I much regret, that in title or opening passage that you did not blow loud trumpet about what you were going to show.—[10] Perhaps paper would have been better more divided into sections with Headings— Perhaps you might have given somewhere rather more of a summary on the process of segregation of varieties & not referred your readers to the descriptive part, excepting such readers as wanted minute detail.—[11]

But these are trifles; I consider your paper as a most admirable production in every way. Whenever I come to variation under natural conditions (my head for months has been exclusively occupied with domestic varieties) I shall have to study & restudy your paper & no doubt shall then have to plague you with questions.—[12]

I am heartily glad to hear that you are well.—

I have been compelled to write in a hurry, so excuse me, & believe me, Yours very sincerely | C. Darwin

Cleveland Health Sciences Library (Robert M. Stecher collection)

[1] The year is established by the relationship to the letter from H. W. Bates, 24 November 1862.

[2] Bates apparently asked CD for advice on this point in the missing portion of his letter of 24 November 1862. Having encouraged Bates to write an account of his travels as a naturalist in South America (Bates 1863; see *Correspondence* vol. 9, letter to H. W. Bates, 4 April [1861]), CD had acted as an intermediary between Bates and his own publisher, John Murray (see letter to John Murray, 28 January [1862]).

[3] John Murray had been Charles Lyell's publisher for many years, and it was as a result of Lyell's advice and intervention that *Origin* was published by John Murray (see *Correspondence* vol. 7, letter to Charles Lyell, 28 March [1859], and letter to J. D. Hooker, 30 March [1859]). CD may refer to Lyell's response, now missing, to CD's letter of 28 March [1859] (*Correspondence* vol. 7) asking for advice about agreeing terms with Murray.

[4] None of this information is provided in Bates 1863, but see the letter to H. W. Bates, 15 December [1862] and nn. 2 and 3.

[5] See letter from H. W. Bates, 24 November 1862.

[6] John Frederick William Herschel. See letter from H. W. Bates, 24 November 1862 and n. 8.

[7] In his letter of 24 November 1862, Bates had expressed a hope that someone might review his paper on South American mimetic butterflies (Bates 1862a) in order to stimulate advance interest in Bates 1863. CD refers to Asa Gray, who regularly wrote botanical reviews and notices for the *American Journal of Science and Arts* (see letters to Asa Gray, 23 November [1862] and 26[–7] November [1862]).

[8] The *Natural History Review* was edited by a committee of eleven co-editors, under the effectual control of Thomas Henry Huxley. CD suggested to John Lubbock, who was one of the editors, that he might review Bates 1862a; however, Lubbock declined, suggesting CD should write the review himself (see letter from John Lubbock, 15 December 1862).

[9] At Lubbock's suggestion, CD reviewed Bates 1862a for the *Natural History Review* ('Review of Bates on mimetic butterflies'; see letter to John Lubbock, 16 [December 1862]). CD recorded in his 'Journal' (Appendix II) that much time was 'wasted' in the summer of 1862 due to Leonard Darwin's illness with scarlet fever.

[10] Bates's paper was entitled, 'Contributions to an insect fauna of the Amazon Valley. *Lepidoptera: Heliconidæ*.' In the paper, Bates invoked the theory of natural selection to account for the phenomenon of mimicry in Amazonian butterflies, arguing that the case offered 'a most beautiful proof of the truth of the theory' (Bates 1862a, p. 513).

[11] Bates 1862a, pp. 500–1. See letter to H. W. Bates, 20 November [1862] and n. 6.

[12] CD was working on the manuscript of *Variation*, which dealt with variation under domestication (see 'Journal' (Appendix II)). *Variation* was originally intended to comprise the first volume of a three-volume work on natural selection, each volume having a separate title (see *Correspondence* vol. 7, letter to John Murray, 22 December [1859]); however, the remaining two parts were never published.

From Daniel Oliver 25 November 1862

Royal Gardens Kew
25. XI. 62

My dear Sir

Two forms of *Epilobium angustifolium* are described in the English Floras—a short & long capsuled form.—[1] But my attention was attracted to their possible dimorphism by a M.S. note of M^r. Babington's—in the herb. of the late M^r. Borrer.—[2] Mr Bab^n. writes (in 1841) stating that he had received specimens of the *E. macrocarpum* described by Stephens (Ann. Nat. Hist. viii. 170)[3] & that he thinks it is *E. angustifolium* producing seed for—he says "I suspect that the 'short turgid pods' of *E. angustifolium* do not ripen their seed".—[4]

More about these forms is written in the same vol. of the Annals.[5]

I must look again at the paper on Strawberries in the Museum copy of the Technologist[6] & if it seem worthwhile have it sent down to you along with the Gardens copy of Hooker's Journal containing the *Linum* paper.[7]

I shewed your note to D^r. Hooker.[8]

Very sincerely yours | Danl. Oliver

Chas. Darwin Esq

DAR 111 (ser. 2): 61–2

CD ANNOTATIONS

Top of letter: 'The case w^d. concern me if **everywhere** the two forms appear— | See about Babington | case of Stellaria (?) graminea.—'[9] *pencil*

End of letter, ink: Vol. 8

p. 246—further information on do.[10]

p. 401 }
p. 403 }

[1] See letter from J. D. Hooker, [15 and] 20 November [1862], and letter to Daniel Oliver, 23 [November 1862].

[2] William Borrer died in January 1862, leaving his herbarium to the Royal Botanic Gardens, Kew (*DNB*); the reference is to Charles Cardale Babington.

[3] Stephens 1841. There is an annotated copy of the volume of the *Annals and Magazine of Natural History* containing this paper in the Darwin Library–CUL.

[4] In his paper, Henry Oxley Stephens claimed that the name *Epilobium angustifolium* had been applied to two distinct species, the second of which he proposed to call *E. macrocarpum.* William Allport Leighton, who supported this identification, sent Borrer his descriptions for comment, before publishing an account in December 1841 (Leighton 1841). In a letter from Babington to Borrer of 12 November 1841, which is reproduced in A. M. Babington ed. 1897, pp. 284–5, Babington stated:

> From what you said in your letter I have again examined the *Epilobium angustifolium*, but am unable to see any real distinction between the two forms . . . I presume that Leighton is not prepared to separate the two plants, or we should have heard more about it before this time.

[5] Leighton 1841 and 1842. There is an annotated copy of the volume of the *Annals and Magazine of Natural History* containing this paper in the Darwin Library–CUL. See also letter to W. A. Leighton, 26 November [1862].

[6] Wray 1861a. See letter to Daniel Oliver, 23 [November 1862] and n. 8.

[7] Oliver refers to Planchon 1847–8, which appeared in volumes 6 and 7 of the *London Journal of Botany*; after 1848 the journal was called *Hooker's Journal of Botany and Kew Garden Miscellany* after its editor, William Jackson Hooker (*BUCOP*). See letter to Daniel Oliver, 23 [November 1862] and n. 10.

[8] Joseph Dalton Hooker. See letter to Daniel Oliver, 23 [November 1862].

[9] See letter from C. C. Babington, 17 January 1862, and letter to C. C. Babington, 20 January [1862].

[10] These are references to additional articles on *Epilobium angustifolium* in volume 8 of the *Annals and Magazine of Natural History* (see n. 3, above).

From Richard Trevor Clarke [after 25 November 1862][1]

Welton Place | near Daventry

My dear Sir

I have just read your letter of enquiry about Strawberries in the Cottage Gardener.[2]

I have been at them for years and here follows all I have been able to ascertain. Myatts British Queen with hautbois pollen produces luxuriant robust very florid plants with numerous anthers, apparently fertile They were however barren though fertized from other strawberries, with one exception in a plant which bears a few dry ill formed fruit from this I have raised *one* solitary plant but have not yet fruited it— Alpine and wild crossed with Myatt and Keens Seedling also barren with, also, one exception The excepted plant can scarcely be called fertile the fruit swells bearing a few seeds one or two on each fruit protruding in a curious way from the fleshy receptacle. In this case also I have raised one seedling not yet adult. Hautbois x^{d} by Myatt produced seedlings so exactly resembling the mother that I doubt the cross. They are however all but one real hermaphrodites full of anthers and most prolific.—[3]

Amongst a batch of crossed seedlings between various sorts in which experiment I remember using Hautbois pollen and having HB plants in the vicinity of the parents, I found a very interesting plant The flavour is strongly musky and the general appearance of the fruit very peculiar. Now this plant in the second year of fruiting that is to say this spring sent up a spike of bloom from a distinct crown (the centre one) nearly, nay almost exactly resembling the hautbois in appearance and flavour.

I must observe this individual closely next season and should be happy to send you a fruiting plant in a pot

In great haste very truly yours | R Trevor Clarke

DAR 161.2: 166

CD ANNOTATIONS
2.6 Myatt] 'Pine class' *added ink*
2.6 Keens] 'Pine class' *interl ink*
3.1 Amongst . . . fruiting 3.5] '1st Letter'[4] *added pencil*
3.1 Amongst . . . flavour. 3.7] *crossed brown crayon; marked with cross, pencil*
3.5 that is . . . Clarke 5.1] 'First Letter' *added pencil*
Top of letter: '1st Note' *ink*

1 Dated by the relationship to the letter to the *Journal of Horticulture*, [before 25 November 1862].
2 Letter to the *Journal of Horticulture*, [before 25 November 1862]; the *Journal of Horticulture, Cottage Gardener and Country Gentleman* had, until 1861, been called *The Cottage Gardener, Country Gentleman's Companion, and Poultry Chronicle.*
3 CD cited Clarke's observations in *Variation* 1: 352.
4 See letter from R. T. Clarke, [after 27 November 1862].

To Asa Gray 26[–7] November [1862][1]

Down Bromley Kent
Nov. 26th

My dear Gray

The very day after my last letter yours of Novr. 10th & the Review in Silliman, which I feared might have been lost, reached me.[2] We were all very much interested by the political part of your letter: in some odd way one never feels that information & opinions printed in a newspaper come from a living source; they seem dead, whereas all that you write is full of life.— Many thanks for P.S. about maize; if the husked form had been the aboriginal, it would surely have not varied so readily; there must be some mistake in statement of Indian, quoted by Aug. St. Hilaire.—[3] The Reviews interested me profoundly;[4] you rashly ask for my opinion & you must consequently endure a long letter.

First for Dimorphism:[5] I do not *at present* like the term "Diœcio-dimorphism"; for I think it gives quite false notion, that the phenomena are connected with a separation of the sexes.[6] Certainly in Primula there is unequal fertility in the two forms, & I suspect this is case with Linum; & therefore I felt bound in Primula paper to state that it might be a step towards dioicous condition;[7] though I believe there are no dioicous forms in Primulaceæ or Linaceæ. But the three forms in Lythrum convince me that the phenomenon is in no way necessarily connected with any tendency to separation of sexes.[8] The case seems to me in result or function to be almost identical with what old C. K. Sprengel called "dichogamy",[9] & which is so frequent in truly hermaphrodite groups; namely the pollen & stigma of each flower being mature at different periods. If I am right it is very advisable not to use term "diœcious" as this at once brings notion of separation of sexes.— I hope you will be able to attend a little to Plantago; I can hardly understand the sentence in your article.[10] In which form does stigma project in bud (this occurs in long-styled Lythrum, but is not then fertilised)?[11] is the short-styled (i.e. your long-stamened) really sterile? You will think that I am in the most unpleasant, contradictory, fractious humour, when I tell you that I do not like your term of "precocious fertilisation" for your second class of dimorphism.[12] If I can trust my memory, the state of corolla, of stigma & pollen-grains is different from state of parts in bud; that they are in a condition of special modification. But upon my life I am ashamed of myself to differ so much from my betters on this head.— The *temporary* theory which I have formed on this class of Dimorphism, just to guide experiment, is that the *perfect* flowers can only be perfectly fertilised by insects & are in this case abundantly crossed; but that the flowers are not always, especially in early spring, visited enough by insects, & therefore the little imperfect self-fertilising flowers are developed to ensure a sufficiency of seed for present generations.[13] Viola canina is sterile, when not visited by insects, but when so visited forms plenty of seed.[14] I infer from *structure* of 3 or 4 forms of Balsamineæ that these require insects; at least there is almost as plain adaptation to insects as in Orchids.—[15] I have Oxalis acetosella ready in pots for experiment next spring;[16] & I fear this will upset my little theory; unless I can as Hooker says "Oh you will wriggle out of anything".—[17] Campanula carpathica, as I proved this summer, is absolutely sterile if insects are excluded. Specularia speculum is fairly fertile when enclosed; & this seemed to me to be effected by the frequent closing of the flower; the inward angular folds of corolla corresponding with the clefts of the open stigma, & in this action pushing pollen from outside of stigma on to its surface.[18] Now can you tell me, does Spec. perfoliata close its flower like S. speculum with angular inward folds; if so, I am smashed without some fearful "wriggling".— Are the *imperfect* flowers of your Specularia the early or the later ones? very early or very late? It is rather pretty to see importance of closing of flowers of Sp. speculum.—[19]

I entirely agree with you in your remarks on the part which crossing plays.[20] I was much perplexed by Oliver's remarks in N. Hist Review of the Primula case, on the lower plants having sexes more often separated than in the higher plants,—so exactly the reverse of what takes place in animals.—[21] Hooker in Review of Orchids repeats this remark.[22] There seems to me much truth in what you say, & it did not occur to me, about no improbability of specilisation in *certain* lines in lowly organised beings. I could hardly doubt that the Hermaphroditic state is the aboriginal one. But how is it in the conjugation of Confervæ—is not one of the two individuals here in fact male & the other female??[23] I have been much puzzled by this contrast in sexual arrangements between plants & animals. Can there be anything in following consideration. By *roughest* calculation about $\frac{1}{3}$ of British *genera* of aquatic plants belong to Linnean classes of Mono- & Diœcia; whilst of terrestrial plants (the aquatic genera being substracted) only $\frac{1}{13}$ of genera belong to these two classes. Is there any truth in this fact generally? Can aquatic plants, being confined to a small area or small community of individuals, require more free crossing & therefore have separate sexes?

But to return to one point; does not Alph. Decandolle say that aquatic plants taken as a whole are lowly organised compared with terrestrial;[24] & may not Oliver's remark on separation of sexes in lowly organised plants stand in some relation to their being frequently aquatic? Or is this all rubbish?

I have left myself little room for orchids & indeed I have little to say except to express my admiration at clearness & ingenuity with which you explain & describe all the forms.[25] It seems to me all excellently done, & has interested me beyond measure.— Do your Platantheras smell sweetest at night; this I suspect is clear guide that moths are the fertilisers.— I have been especially interested by case of P. psycodes, more especially since the D. of Argyll's contemptuous remarks on my case of Angræcum, which in action seems analogous to your case.[26]

But by far the most wonderful is the case of G. tridentata; I hope you will confirm so remarkable a physiological fact.[27] If I understand rightly the rostellum alone is penetrated,—the part primordially of a stigmatic nature. In this observation you have anticipated an experiment, which I mean to try, whether pollen-tubes will penetrate the great rostellum of Cattleya.—[28] I daresay you are quite right about self-fertilisation being much commoner than I thought with orchids.[29] Did I tell you that I have found in Neottia nidus avis that this ensues, if in course of few days the flowers are not visited by insects.?[30]

Your observations on Cypripedium seem excellent; & I daresay I am wholly wrong;[31] it seems to me now more likely that small insects should lick juice off hairs with jaws or short proboscis, than with long proboscis. How curious about the little bristles on the stigma! What a magnificent compliment you end your Review with![32] You & Hooker seem determined to turn my head with conceit & vanity (if not already turned) & make me an unbearable wretch.—

With most cordial thanks, my good & kind friend | Farewell | C. Darwin

P.S. | In my last letter, I mentioned Bates' paper:[33] he is a man of lowly origin, of great force of character, & wonderfully self-educated, but constitutionally of low spirits & poor & under unpleasant circumstances of life. Could you induce any of your Zoological co-editors, just to notice his paper (& if so inform me);[34] it would be a good & charitable deed, for it would encourage & please a man, that wants & deserves encouragement.

What a fearfully long letter I have written!

P.S. 2d. Would you be so kind as to tell me whether Fragaria vesca & Virginiana differ much *Botanically* for I cannot make out that any one has succeeded in crossing them.—[35]

I have just had long letter from Hooker on part which crossing plays in Nature; I must consider it well, & see if it alters my notions.—[36]

Gray Herbarium of Harvard University (50)

[1] The year is established by the relationship to the letter from Asa Gray, 10 November 1862. CD refers in his second postscript to having received the letter from J. D. Hooker, 26 November 1862; since Joseph Dalton Hooker's letter could not have arrived at Down before 27 November 1862, the second postscript must have been added on the latter date. From Gray's reply (see letter from Asa Gray, 9 December 1862), it appears that this letter was sent in the same envelope as the letter to Asa Gray, 23 November [1862]; however, since the two letters have separate salutations and valedictions, they have been treated separately.

[2] In his letter of 10 November 1862, Gray had enclosed proof-sheets of a portion of the November 1862 number of the *American Journal of Science and Arts* (commonly known as 'Silliman's journal', after its founder Benjamin Silliman), which included Gray's follow-up article to his review of *Orchids* (A. Gray 1862b). See also letter to Asa Gray, 23 November [1862].

[3] CD refers to a postscript to the letter from Asa Gray, 10 November 1862, that is now missing. In his letter of 16 October [1862], CD had asked Gray about a report that, under cultivation, the bracts of wild maize had been found to decrease in size. CD here refers to the claim of a young Guarany Indian, reported by Auguste Saint-Hilaire, that *Zea mays* grew wild in the humid forests of his native Paraguay (A. de Candolle 1855, 2: 951). In *Variation* 1: 320–1, CD stated:

> A peculiar kind [of maize], in which the grains, instead of being naked, are concealed by husks as much as eleven times in length, has been stated on insufficient evidence to grow wild in Brazil. It is almost certain that the aboriginal form would have had its grains thus protected; but the seeds of the Brazilian variety produce, as I hear from Professor Asa Gray, and as is stated in two published accounts, either common or husked maize; and it is not credible that a wild species, when first cultivated, should vary so quickly and in so great a degree.

[4] See n. 2, above. In addition to A. Gray 1862b, the November 1862 number of the *American Journal of Science and Arts* included A. Gray 1862e (see n. 5, below).

[5] A. Gray 1862e.

[6] The phenomenon now known as heterostyly, the occurrence of which CD had described in his paper 'Dimorphic condition in *Primula*', had earlier been termed 'dioecio-dimorphism' by Gray and John Torrey (Torrey and Gray 1838–43, 2: 38–9). See also letters from Asa Gray, [10 July 1860] (*Correspondence* vol. 8) and 11 October 1861 (*Correspondence* vol. 9). Gray used the term in his discussion of dimorphism in the *American Journal of Science and Arts* (A. Gray 1862e, p. 419).

[7] 'Dimorphic condition in *Primula*', p. 95 (*Collected papers* 2: 61–2). In the conclusion to his later paper, 'Two forms in species of *Linum*', p. 83 (*Collected papers* 2: 105), CD stated:

> That in some cases this dimorphism may be a step towards a complete separation of the sexes, I will not dispute; but good reasons could be assigned to show that there is no necessary connexion between reciprocal dimorphism and a tendency to dioecious structure. Although good is gained by the inevitable crossing of the dimorphic flowers, yet numerous other analogous facts lead me to conclude that some other quite unknown law of nature is here dimly indicated to us.

[8] In 'Three forms of *Lythrum salicaria*', which was read before the Linnean Society on 16 June 1864, CD stated:

As some authors consider reciprocal dimorphism to be the first stage toward dioeciousness, the difficulty of understanding how a trimorphic plant like *Lythrum salicaria* could become dioecious should be noticed; and as dimorphism and trimorphism are so closely allied, it is not probable that either state is necessarily in any way related to a separation of the sexes—though it may occasionally lead to this end.

[9] C. K. Sprengel 1793. CD's heavily annotated copy of this work is in the Darwin Library–CUL (*Marginalia* 1: 774–85).

[10] In A. Gray 1862e, p. 419, Gray stated that two principal kinds of dimorphism had been noticed by botanists. The first of these, he noted, he had called '*diœcio-dimorphism*' (now called heterostyly). Before discussing the second kind, which he called 'precocious fertilization' (now called cleistogamy), he commented:

The diœcio-dimorphous species of *Plantago* had seemed to confuse this case with the next. That is, the short-stamened flowers appeared to be fertilized in the closed flower, and the long-stamened and generally sterile plants therefore to be generally useless . . . a recent observation on a single specimen . . . shows the top of the style projecting from the tip of the closed corolla. This refers the case to the same category with *Houstonia*, *Primula*, &c.

See also letter from Asa Gray, 29 December 1862, and *Correspondence* vol. 11, letter to Asa Gray, 19 January [1863].

[11] See 'Three forms of *Lythrum salicaria*', p. 170 (*Collected papers* 2: 106).

[12] See n. 10, above. See also letter to Asa Gray, 23[–4] July [1862].

[13] See also letter to Daniel Oliver, 12 [April 1862].

[14] CD's notes of his observations on *Viola canina*, made in May 1862, are in DAR 111: 3–5. See also letter to J. D. Hooker, 30 May [1862], letter to W. E. Darwin, [31 May 1862], and letter to Asa Gray, 10–20 June [1862]. Gray wrote on CD's letter 'But *Viola* goes on all summer'; see letter from Asa Gray, 29 December 1862.

[15] CD refers to several species of Balsaminaceae sent to him by Hooker in October 1862 (see letter to Daniel Oliver, 13 October [1862], letter to J. D. Hooker, 14 [October 1862], and letter from J. D. Hooker, [18 October 1862]).

[16] See letter to Asa Gray, 10–20 June [1862] and n. 24. CD had cultivated plants of *Oxalis acetosella* with a view to testing a possible case of heterostyly, but later decided that the species was not heterostyled (see letters to Daniel Oliver, 20 [April 1862] and 24 April [1862], and letter to J. D. Hooker, 30 May [1862]). There are notes on his experiments on the cleistogamic flowers of *Oxalis acetosella*, made in the spring of 1863, in DAR 109 (ser. 2): 6.

[17] See letter from J. D. Hooker, 7 November 1862, n. 17.

[18] CD's notes of his observations on *Campanula carpathica* and *Specularia speculum*, dated 4 October 1862, are in DAR 79: 113. These observations are recorded in *Cross and self fertilisation*, p. 174.

[19] CD had repeatedly asked Gray for information regarding *Specularia speculum* since learning from him that it had cleistogamic flowers (see letters from Asa Gray, 11 October 1861 (*Correspondence* vol. 9), 2–3 July 1862, 18–19 August 1862, and 5 September 1862, and letters to Asa Gray, 10–20 June [1862], 23[–4] July [1862], and [3–]4 September [1862]). CD marked the last three sentences of this paragraph with a marginal line, apparently in order to draw Gray's attention to his request.

[20] In his review, Gray criticised a point in Daniel Oliver's anonymous review of CD's paper, published in the *Natural History Review* ([Oliver] 1862c). Oliver had referred in his review (p. 236; see also p. 238) to two kinds of dimorphism: one that was 'apparently favourable to variation, marked primarily by a partial or complete separation of the sexes' (i.e., heterostyly), and one that was 'conservative, and unfavourable to variation', with the sexual organs of certain flowers to some extent enclosed and sealed up (i.e., cleistogamy). In A. Gray 1862e, p. 420, Gray commented:

we were somewhat surprised at finding that the reviewer of Darwin's *Primula*-paper in the Natural History Review . . . regards the separation of sexes, and therefore cross-fertilization, as favouring variation, and self-fertilization as necessarily inimical to it. This probably comes from not considering that while close-breeding tends to keep a given form true—in virtue of the ordinary likeness of offspring to parents—it equally and in the same way tends to perpetuate a variation once originated from that form, and also, along with selection (natural or artificial), to educe and further develope or confirm said variety. On the other hand, free cross-breeding of incipient varieties *inter se* and with their original types is just the way to blend all together, to repress all salient characteristics as fast as the mysterious process of variation originates them, and fuse the whole into a homogeneous form.

21 In considering the apparent tendency in *Primula* towards a separation of the sexes, Oliver raised the question 'why did they ever become hermaphrodite?' ([Oliver] 1862c, p. 238). He went on to say:

> While we may . . . suggest that certain species are tending to a separation of the sexes, we must not forget that arguments may be advanced to shew that it is not impossible but that they may be striving towards more perfect hermaphroditism, especially if we bring to mind the evidence . . . furnished by the 'Geological Record.' This evidence does certainly appear in favour of a greater predominance of unisexual forms at an early period than obtains at the present day.

Gray responded to this in A. Gray 1862e, p. 420, stating:

> on morphological grounds, we should look upon hermaphroditism, rather than the contrary, as the normal or primary condition of flowers, and enquire how and why so many became diclinous . . . Forms which are low in the scale as respects morphological completeness may be high in the scale of rank founded on specialization of structure and functions.

See also letter to Daniel Oliver, 15 April [1862] and n. 3.

22 [J. D. Hooker] 1862d, p. 371.

23 CD refers to filamentous green algae.

24 The reference is probably to A. de Candolle 1855, a heavily annotated copy of which is in the Darwin Library–CUL (see *Marginalia* 1: 106–53).

25 CD refers to Gray's follow-up article to his review of *Orchids* (A. Gray 1862b), in which he detailed a number of his observations on North American orchid species.

26 In his article, Gray noted that *Platanthera psycodes* had a long and curved nectary (A. Gray 1862b, p. 425). In his review of *Orchids*, George Douglas Campbell, eighth duke of Argyll, had questioned CD's conclusion that a moth with an unusually long proboscis must be responsible for the pollination of *Angraecum sesquipedale*, in view of its similarly long nectary ([G. D. Campbell] 1862, pp. 394–5; see *Orchids*, pp. 197–203).

27 Gray had introduced his description of *Gymnadenia tridentata* with the comment: 'we hesitate to bring forward our too scanty observations until another summer affords an opportunity to test them' (A. Gray 1862b, p. 426); he confirmed his observations in A. Gray 1863a, pp. 293–4. See also letters from Asa Gray, 18–19 August 1862 and 22 September 1862. CD made undated notes referring to Gray's account of this species (DAR 70: 8, 17), and cited Gray's observations in 'Fertilization of orchids', p. 147 (*Collected papers* 2: 144).

28 CD suggested to John Scott that he might carry out this experiment on *Cattleya*, in the letter to John Scott, 3 December [1862].

29 In *Orchids*, CD sought to demonstrate that the 'main object' of the various 'contrivances by which Orchids are fertilised' was to ensure cross-fertilisation. He noted only one exception (p. 359). In A. Gray 1862b, p. 426, Gray repeated his earlier description (A. Gray 1862c, p. 259–60) of the occurrence of self-pollination in *Platanthera hyperborea* and *Gymnadenia tridentata*, noting: '*Natura non agit saltatim*, and is more flexible and diversified in her ways than we are apt to think: many other cases of occasional or habitual self-fertilization may be expected among Orchids.' CD included the two species mentioned by Gray on an undated list of 'self-fertilisers' that is now in DAR 70: 167, and included a modified discussion of the occurrence of self-pollination in orchids in *Orchids* 2d ed., pp. 288–93.

30 See letter to Asa Gray, [3–]4 September [1862] and n. 9.

31 A. Gray 1862b, pp. 427–8. See also letter from J. D. Hooker, 2 November 1862, and letter to J. D. Hooker, 4 November [1862].

32 Gray concluded his article by expressing his gratitude to CD for 'having brought back teleological considerations into botany' (A. Gray 1862b, p. 428). He continued (p. 429):

> In this fascinating book on the fertilization of Orchids and in his paper explaining the meaning of dimorphism in hermaphrodite flowers, Mr. Darwin,—who does not pretend to be a botanist—has given new eyes to botanists, and inaugurated a new era in the science.

33 Bates 1862a. See letter to Asa Gray, 23 November [1862].

34 Gray was one of the botanical editors of the *American Journal of Science and Arts*; Gray reviewed Bates 1862a in the journal in September 1863 (A. Gray 1863b). See also letter from H. W. Bates, 24 November 1862, and letter to H. W. Bates, 25 November [1862].

35 CD was preparing a draft of the part of *Variation* dealing with 'Facts of variation of Plants' (*Variation* 1: 305–72; see 'Journal' (Appendix II)). In his account of strawberries (*Variation* 1: 351–4), CD noted that the European and American species could be crossed 'with some difficulty', but considered it 'improbable that hybrids sufficiently fertile to be worth cultivation will ever be thus produced.' He continued (p. 352): 'This fact is surprising, as these forms structurally are not widely distinct, and are sometimes connected in the districts where they grow wild, as I hear from Professor Asa Gray, by puzzling intermediate forms.'

36 See following letter.

From J. D. Hooker 26 November 1862

Royal Gardens Kew
Nov 26/62

Dr. Darwin

I return A Grays letter with a thousand thanks,[1] I am very glad to see it—sorry too— how odd it is that men in his position have not learnt by experience that they are no judges of contemporaneous events.

He has made a great blunder in his criticism on Oliver, who was aghast (poor man) till I pointed it out.—[2] breeding in does *not* favor variation, as he supposes,— he mistakes the "perpetuation of *a* variety", for the "propagation of variation"— which is a totally different thing— close breeding will of course tend to preserve & multiply *a* variety in as much as every individual is a variety— (he forgets that the type is a myth.) whereas crossing tends to variation by *adding* differences to preexisting ones. I have several times caught A. Gray in lacking precision of thought—logic in short.

I shall look out for the Cypripeds for you if they come through us[3] Au reste, his whole letter breathes an accursed spirit of jealousy of our strength, & Americas weakness.

The more I reflect, the more sure I am that America will never settle untill she has the equivalent of an Aristocracy (used in best sense) wherefrom to chuse able Governors & statesmen.[4] There is no more certain fruits of your doctrines than this—that the laws of nature lead infallibly to an aristocracy, as the only security for a *settled condition of improvement*—[5] What has prevented America having one of same sort hitherto?, but the incessant pouring in of democratic* elements from the West—which has prevented the sorting of the masses, & frustrated all good effects of Natural Selection. *By a democracy in bad sense I mean a tendency to reduce the better to the worse level

By the way when you have any difficulties such as believing too much in action of physical conditions,[6] you must do as the parsons tell their flocks—come to **me** or some other wise & discreet &c &c &c— you may as well talk to me of expressing the Glory of the Almighty as of ditto to Natural Selection. I *am* a jolly good neophyte. I do however calmly think that there is still amongst us some confusion of ideas between "action of Physical causes" & "Effects of Physical causes".[7]

I am still very strong in holding to impotence of crossing with respect to *origin* of species— I regard variation as so illimitable in *[*animals*]*— You must remember that it is neither crossing nor N. Selection that has **made** so many divergent human individuals, but simply **variation**: Nat. Sel. no doubt has hastened the process, intensified it so to speak, has regulated the lines places &c &c &c. in which & to which the races have run & led, & the number of each & so forth;—but, given a pair of individuals with power to propagate, & infinite span to procreate in, so that not one be lost,—or that in short Nat. Sel. is not called on to play a part at all & I maintain that after n generations you will have extreme individuals as totally unlike one another as if Nat Sel. had extinguished half— If once you hold that Nat. Sel. can *make* a difference, ie *create* a character, your whole doctrine tumbles to the ground— N.S. is as powerless as physical causes to make a variation;—the law that "like shall *not* produce like" is at the bottom of all, & is as inscrutable as life itself. This it is that Lyell & I feel you have failed to convey with force enough to us & the public:[8] & this is at the bottom of half the infidelity of the scientific world to your doctrine. You have not, as you ought, begun by attacking old false doctrines, that "like does produce like" the first chapter of your book should have been devoted to this & to nothing else. But there is some truth I now see in the objection to you, that you make N.S. the "Deus ex machina."[9] for you do somehow seem to do it.—by neglecting to dwell on the *facts* of infinite incessant variation,— Your 8 children are really all **totally** unlike one another they agree **Exactly** *in no one property* how is this? you answer that they display the inherited differences of different progenitors—well—but go back, & back & back in time & you are driven at last to your original pair for origin of differences, & logically you must grant, that the differences between the original ♂ & ♀ of your species were = the su⟨m⟩ of the extreme differences between the most dissimilar existing individuals of your species!—or that the latter varied from some inherent law that had them. Now am not I a cool fish to lecture you so glibly—

No steps will be taken regarding Owen & I am glad of it— let his wickedness find him out of itself.[10]

I shall read Falconers paper with great interest.[11]

I shall not send Oxalis this weather—*without you wish it*.— it is of no value, but would disappoint you, I fear, these sensitive things want fine warm sunny weather[12]

It is no bore to write to you God knows, it is jolly good fun & what a relief from *Welwitschia*![13] I shall look at Lythrum tomorrow your drawing I mean, & description, a thousand thanks for it, I wanted much to know it.[14]

Huxley has just sent me his No 1. working mens lectures.[15] The one only wise good & conservative thing I ever did was to hold out against lecturing for love or money or fame— it is equally admirable—whether you call it a horribly selfish act—or a 'cute sense of my own inability—, or piece of confounded lazyness—to all which motives I plead proudly guilty. & am your | dear friend

J D Hooker

PS. | Another Box of Welwitschia has arrived at Lisbon en route for me— I am fainting away—

DAR 101: 61–2, 77–8

CD ANNOTATIONS
5.6 "action] *underl brown crayon*
5.6 "Effects] *underl brown crayon*
6.10 If once ... ground— 6.12] *scored brown crayon*
6.12 N.S. is ... itself. 6.14] *scored brown crayon*

[1] Letter from Asa Gray, 10 November 1862; see letter to J. D. Hooker, 24 [November 1862] and n. 1.

[2] Gray had apparently sent Hooker, through CD, proof-sheets of the November 1862 number of the *American Journal of Science and Arts*, in which Gray's notice on dimorphism in the stamens and pistils of flowers was published (A. Gray 1862e; see letter to J. D. Hooker, 24 [November 1862] and n. 1). For Gray's comments on Daniel Oliver's anonymous review of 'Dimorphic condition in *Primula*' ([Oliver] 1862c), and for CD's reaction, see the preceding letter and nn. 20 and 21. See also letter to J. D. Hooker, [after 26] November [1862].

[3] Hooker refers to Gray's comment, in his letter to CD of 10 November 1862, that he might have to send specimens of the orchid *Cypripedium* to the Royal Botanic Gardens, Kew, 'where they may get out of the way before Hooker can turn them over to you.'

[4] See letter from Asa Gray, 10 November 1862 and n. 11.

[5] Hooker refers to his somewhat whimsical suggestion that the development of an aristocracy was the necessary consequence of natural selection (see letters from J. D. Hooker, [19 January 1862], [31 January – 8 February 1862], and [23 March 1862]). See also letter to J. D. Hooker, 24 [November 1862].

[6] See letter to J. D. Hooker, 24 [November 1862] and n. 5.

[7] See letter from J. D. Hooker, [15 and] 20 November [1862], and letter to J. D. Hooker, 24 [November 1862].

[8] CD, Hooker, and Charles Lyell had corresponded extensively on this point (see *Correspondence* vol. 7, letter from J. D. Hooker, [20 December 1859], and *Correspondence* vol. 8, letter from J. D. Hooker, 8 June 1860, letters to J. D. Hooker, 29 [May 1860], 5 June [1860], and 12 [June 1860], letters from Charles Lyell, 15 June 1860 and 30 September 1860, and letters to Charles Lyell, 6 June [1860], 14 [June 1860], 17 June [1860], 28 [September 1860], and 3 October [1860]). See also *Correspondence* vol. 9, letter to Charles Lyell, 21 August [1861]. For Lyell's view of natural selection and the question of 'the variety-making power', see also L. G. Wilson ed. 1970, pp. 55–6, 327–46, 458–9, and 515–16. CD responded to these criticisms in *Variation* 1: 6.

[9] *Deus ex machina*: 'A power, event, person, or thing that comes in the nick of time to solve a difficulty; providential interposition' (*OED*).

[10] Richard Owen. See letter from J. D. Hooker, [15 and] 20 November [1862], and letter to J. D. Hooker, 24 [November 1862].

[11] Falconer 1863. See letter to J. D. Hooker, 24 [November 1862].

[12] In his letter of 27 [October 1862], CD asked Hooker to provide him with specimens of *Oxalis sensitiva* for use in experiments on the sensitive reactions of plants. See also letters to J. D. Hooker, 3 November

[1862] and [10–]12 November [1862], and letters from J. D. Hooker, 2 November 1862, 7 November 1862, and [15 and] 20 November [1862].

13 See letter to J. D. Hooker, 24 [November 1862]. Hooker refers to the work for his monograph on *Welwitschia mirabilis* (J. D. Hooker 1863a), which involved prolonged microscopical examination (see letters from J. D. Hooker, 20 August 1862 and [12 October 1862]). See also L. Huxley ed. 1918, 2: 23–6.

14 See letter to J. D. Hooker, 24 [November 1862] and n. 15.

15 Hooker refers to the first part of T. H. Huxley 1862c, which was a report of the first in Thomas Henry Huxley's course of six lectures to working men on 'Our knowledge of the causes of the phenomena of organic nature', delivered in the Museum of Practical Geology, London, in November and December 1862 (see letter from T. H. Huxley, 10 October [1862] and n. 3). The first lecture was entitled 'The present condition of organic nature'.

To William Allport Leighton 26 November [1862][1]

Down. | Bromley. | Kent. S.E.
Nov. 26th

My dear Sir

Will you forgive me troubling you.— Prof. Oliver has called my attention to your papers published so long ago as 1842 on Epilobium angustifolium.—[2] He suggests as possible that these forms may be reciprocally connected like the two forms of Primula. I much doubt this fact; but as I am working on Dimorphism, I shd. very much like to grow the two forms to experiment on.— Is there any chance of your being able to anyhow procure & send me roots of the two forms; it would be a great kindness if you could.— Anyhow pray excuse me venturing to trouble you.—

With my hopes that your health is good,[3] I beg leave to remain, my dear Sir | Yours very faithfully | Ch. Darwin

address for Parcel
C. Darwin
Care of Down Postman
Bromley—
Kent
per Rail.

Milton D. Forsyth, Jr (private collection)

1 The year is established by the relationship to the letter from Daniel Oliver, 25 November 1862.

2 Leighton 1841 and 1842. See letter from Daniel Oliver, 25 November 1862.

3 Leighton had been a friend of CD's at Shrewsbury School and at the University of Cambridge (Browne 1995).

To J. D. Hooker [after 26] November [1862][1]

Down
Nov. 20th

My dear Hooker

Your last letter has interested me to an extraordinary degree, & your truly parsonic advice "some other wise & discreet person" &c, amused us not a little.—[2] I will put a "concrete" case to show what I think A. Gray believes about crossing & what I believe.—[3] If 1000 pigeons were bred together in cage for ten 1000 years, their number not being allowed to increase by *chance* killing, then from mutual intercrossing no varieties would arise; but if each pigeon were a self-fertilising hermaphrodite a multitude of varieties would arise. This I believe is common effect of crossing, viz the obliteration of incipient varieties.[4] I do not deny that when two marked varieties have been produced; their crossing will produce a third or more *intermediate* varieties. Possibly or probably with domestic varieties, with strong tendency to vary, the act of crossing tends to give rise to *new* characters; & thus a third or more races, not strictly intermediate, may be produced. But there is heavy evidence against *new* characters arising from crossing wild forms; only intermediate races are then produced.— Now do you agree thus far? if not, it is no use arguing, we must come to swearing, & I am convinced I can swear harder than you. ∴ I am right. Q.E.D.—

If the number of 1000 pigeons were prevented increasing, not by chance killing, but by, say, all the shorter-beaked birds being killed, then the **whole** body would come to have longer beaks. Do you agree?

Thirdly, if 1000 pigeons were kept in hot country, & another 1000 in cold country, & fed on different food & confined in different size aviary & kept constant in number by *chance* killing, then I sh^d. expect as rather probable that after ten 1000 years, the two bodies would differ slighty in size, colour & perhaps other trifling characters. This I sh^d. call the *direct* action of physical conditions.[5] By this action I wish to imply that the innate vital forces are somehow led to act rather differently in the two cases. Just as heat will allow or cause two elements to combine, which otherwise would not have combined.— I sh^d. be especially obliged if you would tell me what you think on this head.—[6]

But the part of your letter which fairly pitched me head over heels with astonishment; is that where you state that every single difference which we see might have occurred without any selection. I do & have always fully agreed; but you have got right round the subject & viewed it from an entirely opposite & new side, & when you took me there, I was astounded. When I say I agree, I must make proviso, that under your view, as now, each form long remains adapted to certain fixed conditions & that the conditions of life are in long run changeable; & 2^d, which is more important that each individual form is a self-fertilising hermaphrodite, so that each hair-breadth variation is not lost by intercrossing. Your manner of putting case would be even more striking than it is, if the mind could grapple with such numbers— it is grappling with eternity— think of each of a thousand seeds bringing forth its plant, & then each a thousand. A globe stretching to furthest fixed star would very soon be covered. I cannot even grapple with idea even with races of dogs, cattle, pigeons or fowls; & here all must admit & see the accurate strictness of your illustration.—

Such men, as you & Lyell thinking that I make too much of a Deus of N. Selection is conclusive against me.—[7] Yet I hardly know how I could have put in, in all parts of my Book, stronger sentences. The title, as you once pointed out, might have been better.[8] No one ever objects to agriculturalists using the strongest language about their selection; yet every breeder knows that he does not produce the modification which he selects. My enormous difficulty for years was to understand adaptation, & this made me, I cannot but think rightly, insist so much on N. Selection. God forgive me for writing at such length; but you cannot tell how much your letter has interested me, & how important it is for me with my present Book in hand to try & get clear ideas.[9] Do think a bit about what is meant by direct action of physical conditions. I do not mean *whether* they act; my facts will throw some light on this. I am collecting all cases of "*bud*-variations in contradistinction to 'seed-variation'" (do you like this term for what *some* gardeners call "sports"):[10] these eliminate all effect of crossing.— Pray remember how much I value your opinion, as the clearest & most original I ever get.—

Very sincere thanks to you & Oliver for the Books—: Planchon has been very useful:[11] they shall be *all* returned by Rail the first day, probably Tuesday, on which I send to Bromley.—

I see plainly Wellwitschia will be a case of Barnacles.—[12]

Please do **not** send Oxalis sensitiva.[13]

I have another plant to beg, but I write on separate paper, as more convenient for you to keep.[14] I meant to have said before, as excuse for asking for so much from Kew; that I have now lost *two* seasons, by accursed nurserymen not having right plants, & sending me the wrong instead of saying that they did not possess.—

Ever yours | My dear Hooker | C. Darwin

Endorsement: '/1862'
DAR 115.2: 172

[1] Although CD wrote 'Nov. 20th', this letter is evidently a reply to the letter from J. D. Hooker, 26 November 1862 (see n. 2, below).

[2] See letter from J. D. Hooker, 26 November 1862.

[3] See letter from J. D. Hooker, 26 November 1862 and n. 2. See also letter to Asa Gray, 26[–7] November [1862] and nn. 20 and 21.

[4] On CD's views respecting blending inheritance, see Vorzimmer 1963.

[5] See letter to J. D. Hooker, 24 [November 1862].

[6] Hooker's reply to this letter has not been found; however, see the letter to J. D. Hooker, 12 [December 1862].

[7] Charles Lyell. See letter from J. D. Hooker, 26 November 1862 and nn. 8 and 9.

[8] The reference has not been traced, but see *Correspondence* vol. 9, letter to Charles Lyell, 21 August [1861].

[9] CD had resumed work on *Variation* in the spring, after several months spent writing *Orchids* (see 'Journal' (Appendix II)).

[10] CD prepared a draft of chapter 11 of *Variation*, 'On bud-variation, and on certain anomalous modes of reproduction and variation' (*Variation* 1: 373–411), between 21 December 1862 and 23 January 1863 (see 'Journal' (Appendix II)).

[11] CD refers to Bonafous 1836, volumes 6 and 7 of the *London Journal of Botany* (containing Planchon 1847–8), and possibly to volume 1 of the *Technologist* (containing Wray 1861a), sent to him from the Royal Botanic Gardens, Kew (see letter to J. D. Hooker, 18 [November 1862], and letter from Daniel Oliver, 25 November 1862). Daniel Oliver was librarian at the Royal Botanic Gardens, Kew (*List of the Linnean Society of London* 1862).

[12] See letter to J. D. Hooker, 18 [November 1862] and n. 5, and letter from J. D. Hooker, 26 November 1862 and n. 13.
[13] See letter from J. D. Hooker, 26 November 1862 and n. 12.
[14] The enclosure has not been found.

From R. T. Clarke [after 27 November 1862][1]

Dear Mr Darwin

With this I send the strawberries. Hybrid wild, (Vesca + Myatt) Hybrid Hautbois, a chance seedling, and the plant which sent up a crown of true hautbois.[2] The Six plants dug up with balls of earth were raised from (as I thought) a *very female* hautbois by pollen of Myatt^s British Queen. These seedl^gs. were not to be distinguished from ordinary Hautbois, save that they are quite hermaphrodite, bearing freely. I think *one* was inclined to female developement. Hyb Hautbois 2^d generation, is the offspring of a nearly barren seedling from Myatt × Hautbois. Has not yet flowered.—

At the last moment I am trying to cram in a plant of Fragaria Lucida, a remarkable looking wild sort from, I think, California introduced by Mlle. Vilmorin.[3] Enclosed also two or three plants of Cheiranthus, what shall I say the wild yellow Gillyflower of old walls, crossed by Cheiranthus alpinus, forms a very pretty, *very* fragrant garden flower. The brassicaceous smell of alpinus has imparted a peculiar piquance to the scent of the ordinary wall flower. | R T C

DAR 161.2: 167

CD ANNOTATIONS

1.2 and the ... hautbois 1.3] *underl brown crayon*; '(see ['from' *del*] other note)'[4] *added ink*, *square brackets in MS*
Top of first page: '1.' *brown crayon*
Top of verso of first page: '(2' *brown crayon*

[1] The date is conjectured from the relationship to the letter from R. T. Clarke, [after 25 November 1862] (see n. 2, below).
[2] Clarke had written, in response to CD's letter to the *Journal of Horticulture*, [before 25 November 1862], informing CD of various crossing experiments that he had carried out on strawberries, and offering to send him a plant of a hybrid individual that had produced a crown of the Hautbois variety (see letter from R. T. Clarke, [after 25 November 1862]). CD apparently wrote to Clarke accepting his offer, but no such letter has been found. See also letter from R. T. Clarke, [April? 1863] (*Correspondence* vol. 11).
[3] The reference is to Elisa de Vilmorin, who named the Californian strawberry *Fragaria lucida*, and who began to cultivate it in 1855 (Gay 1857, p. 201).
[4] See letter from R. T. Clarke, [after 25 November 1862].

From Robert Monsey Rolfe, Lord Cranworth 28 November 1862

Holwood | Bromley
28 Novr 1862.

My dear Mr Darwin

At this time of the year I become your debtor as my Almoner for the parish of

Downne—[1] I therefore enclose a cheque of which you will perhaps at your leisure acknowledge the receipt—

I was pleased to hear from Lady Cranworth a short time ago that she had seen & conversed with you & that you appeared very well—[2]

I hope that Mrs Darwin is now well, though I believe she was not so when Lady C. called—[3]

Believe me | yours very truly | Cranworth

DAR 161: 229

[1] Almoner: 'An official distributor of the alms of another' (*OED*). CD was treasurer of the Down Friendly Club and the Down Coal and Clothing Club (Freeman 1978); in addition, he apparently distributed Cranworth's annual subscriptions to a day school and a Sunday school in Down (see *Correspondence* vol. 7, letter to John Lubbock, 14 December [1859]). CD's Account book–cash account (Down House MS) records a payment of £10 10*s.* on 30 November 1862 under the heading 'Ld Cranworth Subscription for several Down Charities'; the receipt of Cranworth's cheque is recorded in CD's Account book–banking account (Down House MS) on 3 December 1862. The Down Coal and Clothing Club account book (Down House MS) records a payment of £2 2*s.* from Lord Cranworth on 29 November 1862.

[2] Laura Rolfe, Lady Cranworth. In a letter to William Erasmus Darwin of [13 November 1862] (DAR 219.1: 66), Emma Darwin wrote: 'They had a visit from Lady Cran[worth]. when I was away & a little flirtation with her always pleases your father'; according to her diary (DAR 242), Emma was in London from 8 to 12 November.

[3] Emma Darwin's diary (DAR 242) records that she had been 'feverish' with a 'bad cold' at the end of October.

From Arthur Mellersh 30 November [1862][1]

16 Queens Terrace | S^t^. Johns Wood | London. N.W.
Nov. 30.

My dear Darwin,

I am going out to be Captain of the Forte Flagship to Admiral Warren on the E. Coast of S. America, and You think there is anything I can do for You in the fossil way I shall be happy to do it if you will tell me.[2] Sulivan was telling me that there is a bed of fossil bones at the Gallegos which he wished to examine but had not an opportunity, if any small vessel of the squadron should be going down there I may be able to procure some if You think them worth having—[3] I should be much obliged if You would tell me the right name of those fine birds which are found in Patagonia something like a guirea fowl, of which I shot the only specimen obtained, and only preserved the head and legs.[4] A large landed proprietor in Sussex who has a large enclosed tract of heath has promised to endeavour to naturalize them if I will send some over which I shall endeavour to do, and as a lady who formerly lived at Montevideo told me that she kept some tame and that they did well I hope to get them home alive. I hope your indisposition was only temporary the night we passed at your house, but I was sorry to leave you poorly after passing such a pleasant evening.[5] I saw Admiral FitzRoy on Friday, who was finishing his book as he told me, and looked worn as if he had been closely confined to his work.[6] I leave England in a few days and if you think it worth while to write I shall be here until Thursday—

Hoping some day to meet you again I am my dear Darwin with compliments to Mrs. Darwin, Yours very truly | A. Mellersh

DAR 171.1: 145

[1] The year is established by reference to Mellersh's appointment to HMS *Forte* (see n. 2, below).

[2] Mellersh, who had been a junior officer on HMS *Beagle* during its surveying voyage of 1831–6, was appointed on 12 November 1862 to the command of HMS *Forte*, flagship of Richard Laird Warren, the British navy's commander-in-chief on the south-east coast of America (*Navy list*, January 1863).

[3] Bartholomew James Sulivan provided CD with an account of his fossil find at Rio Gallegos, Patagonia, in 1845 (see *Correspondence* vol. 3, letter from B. J. Sulivan, 13 January – 12 February 1845).

[4] The reference has not been identified; however, Mellersh probably refers to a member of the Tinamidae, three species of which CD mentioned as having been abandoned on the plains of Patagonia in *Journal of researches*, p. 131. See also *Zoology* 3: 119–20.

[5] Mellersh dined with CD at Down House on 21 October 1862, in company with Sulivan and John Clements Wickham, also former shipmates from HMS *Beagle* (see 'Journal' (Appendix II)). Following the meal, CD was so ill that he was unable to 'wish them goodbye next morning' (see letter to John Lubbock, 23 October [1862]).

[6] FitzRoy 1863. Robert FitzRoy commanded HMS *Beagle* during the surveying voyage of 1831–6, when Mellersh and CD were on board.

To Pickard & Stoneman[1] 1 December [1862][2]

Down. | Bromley. | Kent. S.E.
December 1st

Sir

I shd. be very much obliged if you would answer me the following questions. You can return this note with answers which will save you trouble & I enclose envelope.—[3]

Would a case 36 inches long & **24** inches high act for **stove** plants, & price in varnished deal?[4]

Answer[5]

How many gallons of water does the pan of the 48 inch case hold?

Answer.[6]

The case will stand in window which comes to within 18 inches of floor.—

Do you think 20 inches would be good height for legs; or what height do you generally recommend?

Answer.[7]

Sir | Your obedt. sevt | Ch. Darwin

American Philosophical Society (283)

[1] The recipient is identified by an entry in CD's Account book–cash accounts (Down House MS) recording a payment to 'Pickard' of £4 14*s*. for a 'Glass Plant case' (see n. 4, below). Pickard &

Stoneman was a firm of cabinet makers located at 10 Spencer Street, Shoreditch, London (*Post Office London directory* 1861).

[2] The year is established by the inquiry concerning a case for '**stove**' plants; CD ordered a case from Pickard & Stoneman on 11 December 1862 (see n. 1, above, and letter to J. D. Hooker, 12 [December 1862]).

[3] The answers were written in the spaces provided (see nn. 5, 6, and 7, below).

[4] CD required the plant case so that he could work with temperature-sensitive plants like *Oxalis sensitiva* (see letter from J. D. Hooker, 26 November 1862, and letter to J. D. Hooker, [after 26] November [1862]).

[5] The recipient wrote: 'Yes— Refer to Price List'. The enclosed price list has not been found.

[6] The recipient wrote: 'About 4 Gallons'.

[7] The recipient wrote: 'We consider for *that* height of Window, the legs Should be 18 inches—'.

To *Journal of Horticulture* [before 2 December 1862][1]

As you have been so obliging as to insert my query on the crossing of Strawberries,[2] perhaps you will grant me the favour to insert two or three other questions, for the chance of some one having the kindness to answer them. I am writing a book on "Variation under Domestication," in which I treat chiefly on animals; but I wish to give some few facts on the changes of cultivated plants.[3]

1st. The fruit of the wild Gooseberry is said to weigh about 5 dwts. (I am surprised that it is so heavy), and from various records I find that towards the close of the last century the fruit had doubled in weight; in 1817, a weight of 26 dwts. 17 grs. was obtained; in 1825, 31 dwts. 13 grs.; in 1841, "Wonderful" weighed 32 dwts. 16 grs.; in 1845, "London" reached the astonishing weight of 36 dwts. 16 grs., or 880 grains. I find in the "Gooseberry Register" for 1862,[4] that this famous kind attained only the weight of 29 dwts. 8 grs., and was beaten by "Antagonist." Will any one have the kindness to inform me whether it is authentically known that the weight of 36 dwts. 16 grs., has, since the year 1845, been ever excelled?

2nd. Is any record kept of the diameter attained by the largest Pansies? I have read of one above 2 inches in diameter, which is a surprising size compared with the flowers of the wild Viola tricolor, and the allied species or varieties.[5]

3rd. How early does any variety of the Dahlia flower? Mr. Salisbury, writing in 1808, shortly after the first introduction of this plant into England, speaks of their flowering from September, or the end of September, to November.[6] Whereas, Mr. J. Wells, in Loudon's "Gardener's Magazine" for 1828, states that some of his dwarf kinds began flowering in June.[7] I presume the end of June. Do any of the varieties now regularly flower as early as June? Have any varieties been observed to withstand frost better than other varieties?[8]

If any one will give me information on these small points, I shall feel greatly obliged.— Chas. Darwin, Down, Bromley. Kent.

Journal of Horticulture, Cottage Gardener, and Country Gentleman n.s. 3 (1862): 696

[1] The letter was published in the issue of 2 December 1862.

[2] Letter to *Journal of Horticulture*, [before 25 November 1862].
[3] CD prepared a draft of the part of *Variation* dealing with 'Facts of variation of Plants' between 7 October and 11 December 1862 (*Variation* 1: 305–72; see 'Journal' (Appendix II)).
[4] *Gooseberry Grower's Register* (1862): 192, 210. CD's copy of this volume of the journal is in the Darwin Library–CUL (see *Marginalia* 1: 335).
[5] In his discussion of the pansy in *Variation* 1: 368–9, CD stated that the great number and recent origin of the varieties of this plant had made it seem worth studying, 'more especially from the great contrast between the small, dull, elongated, irregular flowers of the wild pansy, and the beautiful, flat, symmetrical, circular, velvet-like flowers, more than two inches in diameter, magnificently and variously coloured, which are exhibited at our shows.' However, he noted that upon closer inquiry he had found that, despite their modern origin, there was much doubt about the parentage of the different varieties.
[6] Salisbury 1808.
[7] *Gardener's Magazine* 3 (1828): 179. The reference is to Joseph Wells. The passage was marked by CD in his copy of this volume of the *Gardener's Magazine*, which is in the Darwin Library–CUL. John Claudius Loudon founded the *Gardener's Magazine, and register of rural & domestic improvement* in 1826 and edited it until his death in 1843 (*DNB*).
[8] In *Variation* 1: 370, CD cited information on this point provided by 'Mr. Grieve'. No such letter has been found, but the reference is probably either to Peter Grieve or James Grieve (R. Desmond 1994).

From T. H. Huxley 2 December 1862

Jermyn St.
Dec 2nd | 1862

My dear Darwin

I send you by this post three of my working men's Lectures—now in course of delivery[1] As you will see by their prefatory notice I was asked to allow them to be taken down in shorthand for the use of the audience but I have no interest in them & do not desire or intend that they should be widely circulated[2]

Some time hence may be, I may revise & illustrate them—and make them into a book as a sort of popular exposition of your views—or at any rate of my version of your views[3]

There really is nothing new in them nor anything worth your attention—but if on glancing over them at any time you should see anything to object to—I should like to know—

I am very hard worked just now—six Lectures a week & no end of other things—but as vigorous as a three year old—[4] Somebody told me you had been ill—but I hope it was fiction—and that you & Mrs Darwin[5] and all your belongings are flourishing—

Ever | yours faithfully | T H Huxley

DAR 166.2: 296

[1] CD's annotated copies of the six parts of T. H. Huxley 1862c (see n. 2, below) are in the Darwin Library–CUL. Beginning on Monday 10 November 1862, Huxley delivered a series of six weekly lectures to working men, with the title 'On our knowledge of the causes of the phenomena of organic nature', at the Museum of Practical Geology, Jermyn Street, London (L. Huxley 1900, 1: 206).

2 The publisher Robert Hardwicke had arranged with Huxley to publish the substance of his lectures from shorthand notes taken by J. Aldous Mays. The lectures were published in six parts (T. H. Huxley 1862c), the paper covers of each of which bore the announcement:

> The Publisher of these interesting Lectures having made an arrangement for their publication with Mr. J. A. Mays, the Reporter, begs to append the following Note from Professor Huxley:— "Mr. J. Aldous Mays, who is taking Shorthand Notes of my Lectures to Working Men, has asked me to allow him, on his own account, to print these Notes for the use of my audience. I willingly accede to this request, but as I have no leisure to revise the Lectures, or to make alterations in them, beyond the correction of any important error of fact, it must be clearly understood, that the Notes can be properly comprehended only by those who are acquainted with the context of the oral discourse. T. H. HUXLEY."

The lectures were reissued in bound form the following year (T. H. Huxley 1863b), with the title on the publisher's binding: 'Huxley's Lectures on Origin of Species'.

3 In the introduction to his lectures, Huxley stated that they were intended to present *Origin* 'in a true light' (T. H. Huxley 1862c, p. 5; see also letter from T. H. Huxley, 10 October [1862] and n. 3). Despite the fact that Huxley 'more than once' set about the task of revising these lectures (T. H. Huxley 1893–4, 2: vii), he did not do so, and they were reproduced, unaltered, in his *Collected essays* (T. H. Huxley 1893–4, 2: 303–474).

4 Huxley probably refers to his lectures as professor of natural history at the Government School of Mines, Jermyn Street, London (*DSB*); he gave fifty lectures there in the 1861–2 session (*Medical directory* 1862, p. 243).

5 Emma Darwin.

From Robert Swinhoe 2 December 1862

18 Royal Avenue Terrace | Chelsea. SW.
2 Decr. 1862.

My dear Sir,

I have just turned out another domestic animal which I had laid aside for you—a Guinea Pig.[1] It is of the Chinese breed, and may perhaps be interesting to you for comparison with the European bred animals. I heard it asserted the other day at Cambridge that the domestic Guinea-pig would not interbreed with the Rock Cavy, it's supposed parent-stock, and that consequently a new species had been formed by *artificial* selection. Have you heard of this asserted fact?[2] The person that gave it forth at Cambridge had it I believe on good authority. I should like much to hear your opinion on this subject.

I remain, | Dear Sir, | Your's very truly,
Robert Swinhoe

Charles Darwin, Esqre.

DAR 177: 327

1 Swinhoe, who had previously supplied CD with animal specimens, had recently returned from the Far East, bringing with him a large natural history collection (see letter from Robert Swinhoe, 12 November 1862).

2 No letter to Swinhoe on this point has been found.

From Patrick Matthew 3 December 1862

Gourdie hill, Errol, N.B.
Dec. 3 /62

Dear Sir,

When in London last summ⟨er⟩ it was only for a few days, engrossed with mercantile concerns so that I could not bestow that attention to scientific thought that I should have liked; I also learned from Prof. Huxley,[1] that by coming up to London you were sometimes rendered unwell. I would have been sorry to bring you from home least I might do you injury, & therefore did not reply to your letter.[2] I also could not but feel that I was an intrud⟨er⟩ & that there existed in scientific me⟨n⟩ a strong vis inertiæ & retiring inclin⟨ation⟩ which I had no right to disturb, ⟨more⟩ especially as I believed I could ⟨be of⟩ no service in advancing your p⟨resent⟩ pursuit. While you have been making advances in vegetable science, I have been attempting to promote a ⟨be⟩tter system of land occupancy by the ⟨f⟩armer—that there might be protection ⟨of⟩ property created by the farmer in ⟨e⟩nriching the vegetable mould. This is a question of the highest importance to the British Empire & Race. My line lies more in the political & social, Your's in tracing out the admirably balanced scheme of Nature all linked together in dependant connection—the vital endowed with a variation-power in accommodation to material change. Altho' this is a grand field for contemplation, yet am I tired of ⟨it⟩— of a world where my sympathies ⟨are⟩ intended to be bounded almost ⟨exclu⟩sively to my own race & family. ⟨I am⟩ not satisfied with my existence ⟨ ⟩ to devour & trample upon my ⟨fellow⟩ creature. I cannot pluck a flower without regarding myself a destroyer. At present we feel some enjoyment in tracing out the scheme of Nature. Since I have paid attention to the progress of discovery, so much has been done that comparatively little remains to do. What will become of man when all the great facts of material & vital science are pointed out? We may be satisfied that we have lived in the great age of discovery & in the country & of the Race in which & by whom these discoveries have been made. Man cannot advance much higher. A reaction such as attended Babylonian, Egyptian, Grecian & Roman civilizations must soon ensue. The same powers that have reached high civilization cannot support it. Fall we must.

We have had a very bleak & unpleasant summer in Scotland, yet another season may be more propitious. Change of air & scene if the change is not too great acts a salutory part in the human constitution & a journey to Scotland might next summer be of service to you or any of your family. You mentioned you had a Son unwell. I hope he soon recovered.[3] Should you think of a jaunt to Scotland I would be most happy in pointing out the little I know of the character of the country. There is something in the change of place which stimulates mental conception.

I enclose one or two pieces which I have been amusing myself with,[4] & remain | Dear Sir | Yours truly | P. Matthew.

Charles Darwin Esq.

DAR 171: 91

[1] Thomas Henry Huxley.
[2] Letter to Patrick Matthew, 13 June [1862].
[3] Leonard Darwin was ill with scarlet fever during the summer of 1862 (see letter to Patrick Matthew, 13 June [1862] and n. 3).
[4] The enclosures have not been found.

To John Scott 3 December [1862][1]

Down Bromley Kent
Dec. 3[d]

Dear Sir

What a capital observer you are! & how well you have worked the Primulas. All your facts are new to me.[2] It is likely that I overrate the interest of the subject; but it seems to me that you ought to publish a paper on the subject.—[3] It would, however, greatly add to value, if you were to cover up any of the forms having pistil & anthers of same height, & prove that they were *fully* self-fertile. The occurrence of dimorphic & *non*-dimorphic species in same genus is quite the same as I find in Linum.—[4] Have any of the forms of Primula, which are *non*-dimorphic been propagated for some little time by *seed* in garden?— I suppose not: I ask because I find in P. Sinensis, a *third*, rather fluctuating form, apparently due to culture, with stigma & anthers of same height.—[5] I have been working successive generations homorphically of this Primula & think I am getting curious results: I shall probably publish next autumn;[6] & if you do not (but I hope you will) publish yourself previously, I sh[d]. be glad to quote in abstract some of your facts.— But I repeat that I hope you will yourself publish——

Hottonia is dimorphic, with pollen of very different sizes in the two forms.—[7]

I think you are mistaken about Siphocampylus; but I feel rather doubtful in saying this to so good an observer:[8] In Lobelia the closed pistil grows rapidly & pushes out the pollen, & then the stigma expands; & the flower in *function* is monoicous;[9] from appearance I believe this is case with your plant.— I hope it is so; for this plant can hardly require a cross, being in function monoicous; so that dimorphism in such a case would be a heavy blow to understanding its nature or good, in all other cases.—

I see few periodicals; where have you published on Clivia?[10] I suppose that you did not actually count the seeds in the hybrids in comparison with those of the parent-forms; but this is almost necessary after Gärtner's observations—[11]

I **very much** hope you will make a good series of comparative trials on the same plant of Tacsonia.—[12]

I have raised 700–800 seedlings from cowslips, artificially fertilised with care; & they presented not a hairs-bredth approach to Oxlips.—[13] I have now seed in pots from Cowslip fertilised by pollen of Primrose, & I hope they will grow;[14] I have also got fine seedlings from seed of wild Oxlips; so I hope to make out this case.—[15]

You speak of difficulties on Natural Selection;[16] there are indeed plenty; if ever you have spare time (which is not likely, as I am sure you must be a hard worker)

I sh^d^ be very glad to hear difficulties from one who has observed so much as you have.— The majority of criticisms on the "Origin" are in my opinion not worth the paper they are printed on. Sir C. Lyell is coming out with what, I expect, will prove really good remarks.—[17] Pray do not think me intrusive; but if you would like to have any Book I have published, such as my Journal of Researches or the Origin, I sh^d^ esteem it a compliment to be allowed to send it.—

Will you permit me to suggest one experiment, which I sh^d^ much like to see tried, & which I *now* wish the more from an extraordinary observation by Asa Gray (in number just out of Sillimans N. American Journal) on Gymnenadia tridentata.—[18] Namely to split the labellum of a Cattleya, or of some allied orchis,—remove caudicle from pollen-mass (so that no lose grains are about) & put it carefully into the large tongue-like Rostellum, & see if pollen-tubes will penetrate, or better, see if capsule will swell— *Similar* pollen-mass ought to be put on true stigmas of 2 or 3 other flowers of same plant for comparison.— It is to discover whether Rostellum yet retains some of *its primordial function* of being penetrated by pollen-tubes.—

You will be sorry you ever entered into correspondence with me.— But do not answer till at leisure, & as briefly as you like.— My hand-writing, I know, is dreadfully bad—

Excuse this scribbling paper, as I can write faster on it, & I have a rather large correspondence to keep up.—

with sincere thanks for all your very interesting information— I remain | Dear Sir | Yours very faithfully | Ch. Darwin

Many thanks about capsule of Orchis for comparison with that of Acropera—[19]

DAR 93 (ser. 2): 60–3

[1] The year is established by the relationship to the letter from John Scott, 6 December [1862].

[2] See letter from John Scott, [20 November – 2 December 1862]; some of Scott's comments on the genus *Primula* were included in the missing portion of that letter. In a note dated 3 December 1862 (see Appendix VI), CD referred to a list, provided by Scott, containing information about species of *Primula*; the list has not been found, but was apparently sent with Scott's letter of [20 November – 2 December 1862] (see letter from John Scott, 6 December [1862] and n. 3).

[3] After further research, Scott wrote a paper entitled 'Observations on the functions and structure of the reproductive organs in the *Primulaceæ*' (Scott 1864c), which was communicated by CD to the Linnean Society of London in 1864.

[4] See 'Two forms in species of *Linum*', pp. 82–3 (*Collected papers* 2: 104–5). According to his Journal (Appendix II), CD wrote this paper between 11 and 21 December 1862.

[5] In 'Illegitimate offspring of dimorphic and trimorphic plants', p. 414, CD recalled that his interest in the equal-styled variety of *Primula sinensis* first arose in 1862, following his observation of some anomalous flowers in a long-styled plant, descended from a self-pollinated long-styled parent. He reported that upon subsequent examination of the plants 'in several small collections', he had discovered that the equal-styled variety was 'not rare' (*ibid*., p. 415). CD's notes on these observations, dated 1–30 March and 24 April 1862, are in DAR 108: 56–66.

[6] CD carried out crossing experiments with *Primula sinensis* in 1861 (see 'Dimorphic condition in *Primula*', pp. 87–8; see also *Collected papers* 2: 54–6); he subsequently raised plants from the seeds resulting from those experiments in which plants were crossed with their own-form pollen (what

CD called 'homomorphic unions' and later renamed as 'illegitimate unions'). With this second generation of plants CD carried out further homomorphic crosses in late January and February 1862, the results of which were published in 'Illegitimate offspring of dimorphic and trimorphic plants', pp. 410–18 (see also the experimental notes in DAR 108: 15–18, 26–8, 34–9). The experiments were designed to investigate the phenomena of cross and hybrid sterility (see Appendix VI and 'Illegitimate offspring of dimorphic and trimorphic plants', pp. 431–7). 'Illegitimate offspring of dimorphic and trimorphic plants' was read before the Linnean Society on 20 February 1868.

7 See letter from M. S. Wedgwood, [6 August 1862] and n. 2.

8 The letter from John Scott, [20 November – 2 December 1862], is incomplete; Scott evidently commented on *Siphocampylus* in the portion that is now missing.

9 CD discussed this case in *Origin*, p. 98.

10 Scott mentioned some crossing experiments that he had carried out with species of *Clivia* in his letter to CD of [20 November – 2 December 1862]; he had not, however, published an account of his experiments (see letter from John Scott, 6 December [1862]).

11 CD refers to Karl Friedrich von Gärtner's practice of assessing the degree of inter-sterility of two plants by the number of seeds produced from the hybrid cross, as compared with the number produced by pure-type parents (Gärtner 1849, pp. 207–15). CD's heavily annotated copy of Gärtner 1849 is in the Darwin Library–CUL (see *Marginalia* 1: 256–98). CD described Gärtner's practice in *Origin*, p. 247.

12 See letter from John Scott, [20 November – 2 December 1862].

13 See letter from John Scott, [20 November – 2 December 1862] and n. 2. CD was interested in the commonly held view that oxlips were the hybrid offspring of primroses and cowslips. He initially maintained that the various forms were varieties descended from a common parent (*Natural selection*, pp. 128–33, and *Origin*, pp. 49–50), but as a result of his work on dimorphism in *Primula*, he came to distrust the experimental evidence against the occurrence of hybridisation, stating that further experiments were 'absolutely necessary' ('Dimorphic condition in *Primula*', pp. 93–4; see also *Collected papers* 2: 60–1). In a note in his Experiment book (DAR 157a, p. 75), dated 21 April 1862, CD reported that since 1859 he had grown several generations of artificially pollinated cowslips (amounting to 765 plants), in order to ascertain whether they would vary under changed conditions of culture, without a 'vestige of a sign of modification'. These findings were reported in 'Specific difference in *Primula*', p. 442.

14 CD began his crossing experiments with cowslips and primroses in April 1862 (see the dated notes in DAR 157a, pp. 76–7 and DAR 108: 70). He continued to carry out similar experiments between 1863 and 1867, the results of which are recorded in DAR 108 and in 'Specific difference in *Primula*', pp. 438–40.

15 There are notes on these seedlings, dated 27 April 1862, in DAR 157a, pp. 75–7. CD's crossing experiments with oxlips, carried out over succeeding years, are detailed in DAR 108 and in 'Specific difference in *Primula*', pp. 443–5. CD eventually concluded that the common cowslip, the primrose, and the Bardfield oxlip were distinct species, but that the common oxlip was a hybrid between the cowslip and the primrose.

16 See letter from John Scott, [20 November – 2 December 1862].

17 CD refers to Charles Lyell's *Antiquity of man* (C. Lyell 1863a), which was published on 6 February 1863 (C. Lyell 1863b, p. vii).

18 In A. Gray 1862b, p. 426, Asa Gray reported that pollen falling onto certain viscid parts of the rostellum in the orchid *Gymnadenia tridentata*, readily sent pollen tubes into it, and that those parts of the rostellum appeared 'to act as stigmas'. In the letter to Asa Gray, 26[–7] November [1862], CD mentioned that this observation anticipated an experiment he was intending to try with *Cattleya*. Gray's paper appeared in the November issue of the *American Journal of Science and Arts*, commonly known as 'Silliman's journal' after its founding editor, Benjamin Silliman.

19 See letter from John Scott, [20 November – 2 December 1862] and n. 8.

From James Dwight Dana 4 December 1862

New Haven,
Dec. 4, 1862

My dear Mr Darwin:

A year and a half ago I partially completed a letter to you in reply to your kind words which greeted me soon after my arrival in the country.—[1] I have been delaying ever since then against my inclination, with the hope of being able soon to report that I was in a condition to read your work.[2] Many long months, and now, even years, have passed by, and still your book remains unopened. You see that I have been gaining and doing some work in the Geological Manual which I trust will have reached you before you have the reading of this note.[3] But I have worked to great disadvantage, 1 to 3 hours a day and often none at all and thus, have gradually pushed through the labor to the end. I am now resuming my duties in the University.[4] But one hours intercourse with the Students in the lecture room, is a day's work for me. Thus you will yet pardon my seeming neglect of your work. In my geology, I had a chapter partly prepared on the question whether the origination of species was a subject within the range of Dynamical Geology,—taking sides, I confess, against you—: but I omitted it entirely because I could not study up the subject to the extent that was necessary to do it justice. I have however expressed an opinion on this point in the geology; and this you will excuse: for my pesuasions are so strong that I could not say less.—[5] You will perhaps be the more interested in the work because of its American Character.—[6]

I have thus far had nothing to do, since the summer of 1859, with the editing of the Jour. Sci. although wholly charged with it before then. I hope soon to take hold again.[7]

I shall take great pleasure in hearing from you; and if a photograph of yourself could be added to your letter it would enhance greatly the pleasure.

Although so long silent, there is no failing of esteem and admiration on the part of your friend | James D. Dana

Charles Darwin Esq—

Dana Family Papers, Manuscripts and Archives, Yale University Library

[1] See *Correspondence* vol. 8, letter to J. D. Dana, 30 July [1860]. In 1859 Dana suffered a 'serious nervous breakdown' and travelled to Europe to convalesce; he never fully recovered, but was ultimately able to continue his work (*DAB*).

[2] CD sent Dana a copy of *Origin* in 1859, and had been anxious to know his views on the subject, but expressed a wish that Dana should not read it until he was well enough to do so (see *Correspondence* vol. 7, letters to J. D. Dana, 11 November [1859] and 30 December [1859], and *Correspondence* vol. 8, letter to J. D. Dana, 30 July [1860]).

[3] The preface of Dana's *Manual of geology* was dated 1 November 1862 (Dana 1863, p. ix), and the work was included on the 'List of New Books' in the 1 December 1862 issue of the *American Publishers' Circular* (*American Publishers' Circular* 8 (1862): 131). However, the title-page bears the year '1863', and the *Manual of geology* was apparently not published in Britain until January 1863 (*Publishers' Circular* 26 (1863): 9). CD did not receive his copy of the book, which is now in the Darwin Library–Down,

until February 1863 (see *Correspondence* vol. 11, letters to J. D. Dana, 7 January [1863] and 20 February [1863], and letter from J. D. Dana, 5 February 1863).

4 Dana was Silliman Professor of geology at Yale University (*DSB*).

5 At the conclusion of his section on 'historical geology', Dana argued that, while it was right for science to seek for natural causes, geology suggested 'no theory of natural forces' that could account for the origination of species (Dana 1863, p. 602). Referring in particular to palaeontology, he continued:

> It has brought to light no facts sustaining a theory that derives species from others, either by a system of evolution, or by a system of variations of living individuals, and bears strongly against both hypotheses ... Geology appears to bring us directly before the Creator ... and leads to no other solution of the great problem of creation, whether of matter or of species of life, than this:— DEUS FECIT.

On Dana's changing views on evolution and natural selection, see Sanford 1965.

6 Dana 1863 was entitled *Manual of geology: treating of the principles of the science with special reference to American geological history.*

7 Dana had been one of the editors of the *American Journal of Science and Arts* since 1846 (*DSB*).

From John Scott 6 December [1862][1]

Botanic Gardens [Edinburgh]
Decr. 6th.

Sir.

I should have been glad to publish—as you desire me, a paper on the Primulas, had my *materials* for such been sufficient.[2] This however is not the case. My observations have been entirely confined to the examination of the relative length of the sexual organs, and I have never tested their comparative fertility. In regard to this point in the few non-dimorphic forms, I have given in my list, one only—P. Scotica—is cultivated here.[3] The plants of this—as you suppose—have not been raised from seeds with us, but introduced from their native habitats. It—non-dimorphism—cannot therefore, in these cases as in P. Sinensis which you mention,—be an effect of Cultivation, as I find it the same in native specimens of P. *Scotica*, and the few others similarly conditioned were native specimens likewise.[4]

I have just heard of a young *M.D.* who intends bringing forward some objections to your views on Primula.[5] I have not been able to ascertain what these are, but I shall make it a point to do so on Thursday first when I expect to meet him at the Botanical Society's Meeting. I am fully anticipating that they are founded on the conditions of P. Scotica—as he has frequently been North and gathered it. if so, however, your case of Linum may meet all the difficulties he will urge.—[6]

By the way, I may here state that I intend to read a short paper, on the *Irritability* &c. of *Drosera* to the Society on Thursday.[7] I am very much puzzled with the capricious appearance of this phenomena in such widely different parts of the Vegetable Kingdom. So far as I remember in my overhasty perusal of your 'Origin', you do not refer to this point. May I be permitted to ask you, if you explain its *isolated appearances*, in the same manner as the electric phenomena of fishes? of which I have taken a note.—[8]

When I first observed the difference in the length of styles in Siphocampylus, I thought it might be analogous to Lobelia, and that the shorter styled flowers were not yet fully developed. This I find is not the case. I now enclose a few withered flowers showing stigma expanded with styles scarcely longer than stamens. In referring to this case, however, it was merely in analogical illustration of those forms of Primula, where the length of stigma varies, which as I hinted might in them as in Siphocampylus, be the first indication of dimorphism.[9] The existence—on the authority of Lindley, and of which you may perhaps be aware, as I have not yet ascertained the species—of a structurally dioecious Lobelia,[10] furthermore induced me to mention Siphocampylus, as I supposed such a state consequent on analogous changes, to those you have demonstrated in Primula. I may, however, after all be in an error about it, and I now merely state these points; which I thought justified my inference. I have made no notice of Clivia, nor have I counted the seeds.[11] This I will now, have an opportunity of doing, as the hybrid & one of its parents are at present in flower

You kindly ask me if there is any of your books I would like.[12] I can assure you Sir, that any of them would be a highly valued boon, but since you ask me to mention one, I will indeed feel deeply obliged and honoured of your presenting me with a copy of your Journal, which I have less frequently seen than the Origin.

The experiments you suggest on Cattleya shall be attended to as soon as an opportunity presents itself.[13] At present, unfortunately there are none of them in flower. **Lælia** *anceps*, however, will be in flower shortly, and I will experiment upon it. It shall indeed afford me a pleasure to make any experiments you may be pleased to suggest. I will be glad to here of many more.

I will try and state my difficulties with Natural Selection in some future letter.[14] I may here state, however, that it is not the mere question of variability, but the reproductive inheritance which presents difficulties to me. I send, by post a copy of my paper treating upon the reproduction of varieties of Ferns, on which I will be glad to hear any remarks, as I believe the relations of the sexual organs in the higher plants greatly infacilitates the reproduction of individual variations.[15] That—in fact—the facilities for such are inversely proportionate, to the more or less intimate association of these organs. The most favourable condition for such being presented by Ferns, where the **genitalia** are the products of a single organ, and the least favourable by the diœcious Phænogams where these are produced by distinct individuals.

This subject I may state has engaged me for sometime back, and I now am inclined to think, that the views you advocate in your Orchid Book, are directly opposed to those I have adopted in my paper on the Ferns, where I apply it to like phenomena in the Higher Plants. That in short—all unfavourable to your views of variation, might—according to mine—regard that unknown great good derived from the union of distinct individuals which you speak of;[16] as a natural institution for the prevention of any continuous flux in the forms of Vegetative Life. Inasmuch, as in this kingdom—it will rarely happen that two individuals having like differences will fertilise each other. And thus though each may individually present a departure from the normal specific type, the resultant of this combined action, will not blend their idyosyncrasies—if I may so speak. This immediate change in the genitive conditions, having a tendency to call into energetic action the somewhat latent specific formative power, and so efface the individual tendencies of both.

Time, however, does not at present permit me to enter fully and clearly, into these matters, which in fact are the basis of my difficulties with Natural Selection. With many thanks for the attention you have been pleased to honour me with, I remain

Sir | Your obedient Servt | John Scott

DAR 108: 182a–d

CD ANNOTATIONS
0.1 Botanic ... fertility. 1.4] *crossed brown crayon*
1.4 In ... are, 2.2] *crossed red crayon*
1.7 from seeds ... mention, 1.9] *scored brown crayon*
1.9 as I ... likewise. 1.11] *scored brown crayon*
2.2 I shall ... note.— 3.7] *crossed ink*
3.5 May I ... fishes? 3.6] *cross in margin, brown crayon*
4.3 now enclose] *underl brown crayon*; *cross brown crayon*
6.3 I will ... more. 6.5] *cross brown crayon*
7.10 and the ... individuals. 7.11] *scored brown crayon*; *cross in margin, brown crayon*
Top of letter: '*John Scott*' *ink*

[1] The year is provided by the reference to Scott 1862b (see n. 7, below).

[2] See letter to John Scott, 3 December [1862] and nn. 2 and 3.

[3] Scott apparently sent CD a list containing information on several species of *Primula* with his letter of [20 November – 2 December 1862]. CD refers to such a list in a note, dated 3 December 1862, discussing the origins of dimorphism (DAR 205.7: 162). The list has not been found; however, Scott included a list of dimorphic and non-dimorphic species of *Primula* in Scott 1864c, p. 80.

[4] See letter to John Scott, 3 December [1862] and n. 5.

[5] This individual has not been identified. See also letter from John Scott, 17 December [1862].

[6] See letter to John Scott, 3 December [1862] and n. 4.

[7] Scott read his paper, 'On the propagation and irritability of *Drosera* and *Dionaea*', before the Botanical Society of Edinburgh on 11 December 1862 (Scott 1862b).

[8] In *Origin* 3d ed., pp. 212–13, CD discussed the 'serious difficulty' presented to his theory by the presence of electric organs in only a small number of species of fish, several of which were 'widely remote in their affinities'. He observed that in such cases of 'two very distinct species furnished with apparently the same anomalous organ', some 'fundamental difference' could generally be found, and continued:

> I am inclined to believe that ... natural selection, working for the good of each being and taking advantage of analogous variations, has sometimes modified in very nearly the same manner two parts in two organic beings, which beings owe but little of their structure in common to inheritance from the same ancestor.

[9] Scott had apparently mentioned this point in the missing portion of his letter of [20 November – 2 December 1862] (see letter to John Scott, 3 December [1862] and n. 8). Scott omitted to enclose the flowers with this letter (see letter to John Scott, 11 December [1862], and letter from John Scott, 17 December [1862]).

[10] See, for instance, Lindley 1853, p. 692.

[11] See letter to John Scott, 3 December [1862] and n. 10.
[12] See letter to John Scott, 3 December [1862].
[13] See letter to John Scott, 3 December [1862] and n. 18.
[14] See letter to John Scott, 3 December [1862] and n. 16.
[15] See Scott 1862a, p. 214. There is an annotated copy of this paper in the Darwin Pamphlet Collection–CUL. See also letter from John Scott, 15 November [1862].
[16] One of CD's objects in *Orchids* had been to show that the flowers of orchids were adapted for cross-pollination (p. 1). CD concluded the volume (p. 360) by referring to the probability that 'some unknown great good is derived from the union of individuals which have been kept distinct for many generations'.

To T. H. Huxley 7 December [1862][1]

Down. | Bromley. | Kent. S.E.
Dec. 7th

My dear Huxley

I was on point of adding to an order to Williams & Norgate for your Lectures, when they arrived, & much obliged I am.[2] I have read them with interest & they seem to me very good for their purpose & capitally written as is everything which you write. I suppose every-book now-a-days requires some pushing, so that if you do not wish these Lectures to be extensively circulated,[3] I suppose they will not; otherwise I shd. think they would do good & spread a taste for the Natural Sciences. Anyhow I have liked them; but I get more & more, I am sorry to say, to care for nothing but natural history; & chiefly, as you once said, for the mere species question.[4] I think I liked no III the best of all.[5] I have often said & thought that the process of scientific discovery was identical with every day thought, only with more care; but I never succeeded in putting the case to myself with one-tenth of the clearness with which you have done.—[6] I think your second Geological section will puzzle your non-scientific readers; anyhow it has puzzled me, & with the strong middle line, which must represent either a line of stratification or some great mineralogical change, I cannot conceive how your statement can hold good.—[7]

I am very glad to hear of your "three-year old" vigour, but I fear with all your multifarious work that your Book on Man will necessarily be delayed.—[8] You bad man you say not a word about Mrs. Huxley, of whom my wife & self are always truly anxious to hear.—[9]

My dear Huxley | Ever yours very truly | C. Darwin

I see in Cornhill mag. a notice of work by Cohn which apparently is important on the *contractile* tissue of Plants.[10] You ought to have it reviewed.[11] I have ordered it, & must try & make out, if I can, some of the accursed German, for I am much interested in subject and experimented a little on it this summer, and came to conclusion that plants must contain some substance most closely analogous to the supposed diffused nervous matter in the lower animals: or as, I presume, it would be more accurate to say with Cohn, that they have contractile tissue.[12]

ALS & Copy[13]
DAR 145: 227, Imperial College of Science, Technology and Medicine Archives (Huxley 5: 179)

[1] The year is established by the relationship to the letter from T. H. Huxley, 2 December 1862.

[2] Huxley sent CD the first three numbers of T. H. Huxley 1862c on 2 December 1862 (see letter from T. H. Huxley, 2 December 1862). CD commonly bought his books from the London booksellers and publishers Williams and Norgate, who specialised in scientific literature (CD's Classed account book (Down House MS), *Modern English biography*). See also letter from J. D. Hooker, 26 November 1862.

[3] CD refers to the fact that Huxley had not personally arranged for his lectures to be published, but that they had been published from shorthand notes, without any revisions, by an enterprising publisher. See letter from T. H. Huxley, 2 December 1862 and n. 2.

[4] The reference has not been identified.

[5] The third of Huxley's lectures, delivered on 24 November 1862, was headed 'The method by which the causes of the present and past conditions of organic nature are to be discovered.— The origination of living beings' (T. H. Huxley 1862c, pp. 53–82).

[6] T. H. Huxley 1862c, pp. 53–68. On CD's views concerning scientific method, see Ghiselin 1969 and Ruse 1975.

[7] T. H. Huxley 1862c, p. 40 and fig. 5. CD refers to Huxley's argument that in continuously stratified rocks the upper stratum in one locality might be deposited before the lower stratum in a second locality. Huxley used this case as an example of his doctrine that the stratigraphic identity of deposits in different areas did not necessarily indicate identity in the dates of deposition (T. H. Huxley 1862c, pp. 39–42). See also letter from T. H. Huxley, 6 May 1862, and letter to T. H. Huxley, 10 May [1862].

[8] See letter from T. H. Huxley, 2 December 1862. Huxley had been working intermittently on *Evidence as to man's place in nature* (T. H. Huxley 1863a) since early in the year (A. Desmond 1994, pp. 301, 304).

[9] CD refers to Henrietta Anne Huxley and Emma Darwin.

[10] CD refers to a review in the November 1862 number of the *Cornhill Magazine* ([Herschel] 1862, pp. 853–5), evidently of Cohn 1860, although the reviewer gave the date as '1862'. There is an annotated copy of the number of the *Abhandlungen der Schlesischen Gesellschaft für vaterländsche Cultur: Abtheilung für Naturwissenschaften und Medicin* containing this paper in the Darwin Pamphlet Collection–CUL.

[11] Huxley was editor-in-chief of the *Natural History Review* (L. Huxley ed. 1900, 1: 209–10). No review of Cohn 1860 appeared in the *Natural History Review*; however an abstract was published in the *Annals and Magazine of Natural History* 3d ser. 11 (1863): 188–202.

[12] See, for example, letter to Daniel Oliver, [17 September 1862], and letter to J. D. Hooker, 26 September [1862].

[13] Part of the original letter is missing. The text in the final paragraph from line 4, 'subject', to line 7, 'tissue.', has been taken from the copy made for Francis Darwin's use in editing his father's letters for publication.

To T. W. Woodbury 7 December [1862][1]

Down,
Dec. 7th.

Dear Sir

I cannot think of any means of aiding you with respect to Bees from the E. Indies.[2] You must be sanguine to hope to keep a Tropical insect through our winters; but I fully admit that no one can foresee what will happen in such cases. Mr Wallace was only touring in Devonshire,[3] his address is

A. R. Wallace, Esq[re]
5, Westbourne Grove Terrace
London W

The only man, whom I can suggest to communicate with is

E. Blyth, Esqre[4]
Asiatic Soc
Calcutta

He is the Curator and an excellent Naturalist, but sadly overworked; yet very obliging. He is not an Entomologist; but attends much to domesticated animals.

I shall be truly grateful for the piece of comb: be so kind as to address it as follows to my Brother's House[5]

Ch. Darwin, Esqre
6 Queen Anne Street
Cavendish Square
London

To be left till called for

I am also truly obliged to your great kindness in getting my query on the variation of the Bee circulated in Germany; I have been very glad to see the answers.[6]

My dear Sir | yours sincerely obliged | Ch. Darwin.

Copy
DAR 148: 374

[1] The year is established by the relationship to the letter from T. W. Woodbury, 17 March 1863 (*Correspondence* vol. 11; see nn. 2 and 5, below).

[2] Woodbury's letter has not been found; however, he was interested in the possibility of naturalising non-native species of hive bee, and in August had sought CD's assistance in procuring bees from West Africa (see letter from T. W. Woodbury, 9 August 1862, letter to A. R. Wallace, 20 August [1862], and letter to J. D. Hooker, 22 [August 1862]). Woodbury believed that the large Indian bee species, *Apis dorsata*, would be 'very valuable' if it could be naturalised (see *Correspondence* vol. 11, letter from T. W. Woodbury, 17 March 1863).

[3] Alfred Russel Wallace returned to Britain from the Malay archipelago at the end of March 1862; in August, following a period of illness, he left London for a holiday in Devon (see letter from A. R. Wallace, 8 August 1862, and Brackman 1980, pp. 236–42). Woodbury resided in Exeter, Devon (*Post Office directory of Devonshire* 1856).

[4] Edward Blyth was the curator of the museum of the Asiatic Society of Bengal from 1841 to 1862, and in the 1850s had provided CD with much information on the plants and animals of India (*Correspondence* vols. 5–7, *DNB*). CD had evidently not yet received Blyth's letter of 23 November 1862, reporting his imminent return to Britain.

[5] Erasmus Alvey Darwin. See also letter from E. A. Darwin, 14 December [1862?]. CD refers to a piece of 'partially completed artificial comb' (see *Correspondence* vol. 11, letter from T. W. Woodbury, 17 March 1863).

[6] See letter to the *Journal of Horticulture*, [before 10 June 1862], and letter to *Bienen Zeitung*, 18 June 1862.

From Asa Gray[1] 9 December 1862

Cambridge. [Massachusetts]
Dec. 9, 1862.

⟨My de⟩ar Darwin

This is to let you know ⟨that⟩ I have put in charge ⟨of o⟩ur kind Capt. *James ⟨A⟩nderson*, of this weeks ⟨M⟩.M. Steamer (Africa) a small box, wrapped as a paper parcel $1\frac{1}{2}$ foot long. 5×5 inches or so—addressed to you—which he will forward from Liverpool—as you wish—containing *Mitchella* (I hope both sorts) & *Cypripedium acaule*, *spectabile*, & *arietinum*. Pot & keep them cold till spring, & then let grow; they will surely flower.[2]

I hope you are all well now, and will have a merry *Christmas*— *We* do not expect that, but are hopeful. The Lord will remember us in due time, and we shall see some of the oppressed free.—

Adieu | Yours ever | A. Gray

DAR 165: 125

[1] This note was apparently enclosed with a letter from Gray to Joseph Dalton Hooker, dated 8 December 1862 (Royal Botanic Gardens, Kew (Asa Gray Letters: 312)); in the letter Gray asked Hooker to forward to CD an enclosure concerning the species mentioned here.

[2] James Anderson had been in the service of the Cunard line since 1851, and commanded fourteen different steamers belonging to that fleet in the Mediterranean and Atlantic trades (*Modern English biography*). Gray had been waiting since October to send 'patches' of the dimorphic plant *Mitchella repens* and live roots of North American species of the orchid genus *Cypripedium* (see letters from Asa Gray, 5 September 1862, 4 and 13 October 1862, and 27 October 1862). See also letter to James Anderson, 23 December [1862].

From Julius von Haast 9 December 1862

(Copy)[1]
Lake Ohau. N.Z.
December 9th. 1862.

Dear Sir

Our common friend Dr J. D. Hooker tells me he did me the honour to send you my last letter & extracts and that you were interested in them, yet at the same time communicates to me that you wish for a specimen of our native rat & frog.[2] I had the pleasure to send you a month ago two productions of my pen (addressed to the British Museum) for the perusal of which I hope you will find a spare moment.[3] You will observe that I took the liberty to call one of our Alpine giants after you as a feeble tribute from the Southern hemisphere to the author of 'Origin of species'.[4]

I am very sorry that I have no specimen of the native rat, it being almost extinct, but I shall do my best to procure one for you; The frog exists only in one or two small creeks at Coromandel in the northern island, and I shall write instantly to one of my Auckland friends, Mr. Chs. Petschler to procure one, and to send it directly to you.[5] There are some more highly interesting animals in N.Z. quite unknown to science, as for instance a small quadruped in the rivers forming this lake, & as no doubt I shall be able to procure some, I shall do myself the honour to send them to you for examination & description.[6] Dr. Hooker tells me that you ask him if I had your 'Origin of Species', if ever in my life I could be induced to tell an untruth, it would be here the case, because I should consider it the highest compliment to receive a work like yours from the hand of the Author, and on returning to Ch.

Church, I shall present my copy to our embryo of a library, so that I can then fairly say, I do not possess it.[7]

You will see in my address as Presi^{t}. of the Phil. Inst^{t}. of Canterbury, that I tried to explain in few words to the members & public at large, the great object of your work, so as to preserve them against the prejudices of bigoted people.[8] It will perhaps interest you when I tell you, that the 'Origin of Species' was my travelling companion during my last journey, in the N.Z. Alps, taking always a book with me, the careful study of which, the long evenings & rainy days in a tent afford the best opportunity, and I need scarcely say that very often I forgot hunger & fatigue, cold & storm in its perusal, & sometimes I was only roused from its study by the falling of an avalanche, or the howling of the storm.[9] If you wish me to make any observations on the subject of natural history, I shall be most happy to fulfil your desires. You will allow me to give you one instance which came under my knowlege, how animals in order to preserve their offspring, adapt themselves to circumstances: the 'Casarca Variegata', the 'Paradise duck' of the setlers, builds its nest along the bank of rivers, on the ground, but several instances have been observed at the Arowenui bush between Lyttelton and Timaru on the east coast of this island, that these webfooted birds when disturbed from their nests, have built new ones on the top of high trees, bringing afterwards their young ones on their backs down to the water; This occurrance has been observed by many respectable people so that there is not the least doubt about its truthfulness, & have not the deductions from such a change in the habit of an animal a very high bearing on the confirmation of your theories? Any parcel for me if entrusted to the care of M^{r}. J. Marshman our Provin^{l}. Agent, 16 Charing Cross London, will reach me safely.[10] Having hurt my right hand slightly, by a fall with a horse, you will excuse my employing one of my travelling companions to write this. Hoping that you will find time to send me a few lines,

Believe me my dear Sir | very sincerely yours | Julius Haast

Ch^{s}. Darwin Esqr | ect FRS | London.

LS
DAR 166.1: 1

CD ANNOTATIONS
1.1 Our . . . species'. 1.7] *crossed ink*
2.1 I am] *after opening square bracket, ink*
2.2 The frog . . . desires. 3.11] *crossed ink*
Top of letter: 'Change in Habits in nest in Duck. | New Zealand Vertebrata.' *ink*

1 Haast initially wrote this letter on 9 December 1862, enclosing or intending to enclose it with a letter to Joseph Dalton Hooker of 10 December 1862. Hooker received Haast's letter in mid-April 1863, but either he lost Haast's letter to CD, or Haast failed to enclose it as promised (see *Correspondence* vol. 11, letters from J. D. Hooker, 20 April 1863 and [30 April 1863], and letter to J. D. Hooker, 23 April [1863]). Fearing, however, that the entire body of his correspondence sent from Lake Ohau at this time had gone astray, Haast sent Hooker copies of his letters to CD and Hooker, together

with two covering letters, in March 1863 (see *ibid.*, letter from Julius Haast, 5 March 1863). These arrived in mid-June 1863, and Hooker transmitted to CD the letter from Julius Haast, 5 March 1863 (*ibid.*), together with this copy of the letter from Julius Haast, 9 December 1862 (see *ibid.*, letter from J. D. Hooker, 19 June 1863). For reasons of clarity, the copy of Haast's letter has been reproduced here, in addition to being reproduced as an enclosure to the letter from Julius Haast, 5 March 1863 (*ibid.*).

2 See the enclosure to the letter from J. D. Hooker, 20 September 1862. See also letter to J. D. Hooker, 21 [September 1862] and nn. 6 and 7.

3 J. F. J. von Haast 1862a and 1862b (see *Correspondence* vol. 11, letter to Julius von Haast, 22 January 1863). There are annotated copies of these publications in the Darwin Pamphlet Collection–CUL.

4 J. F. J. von Haast 1862a, p. 127. Mount Darwin is at the northern end of the Malte Brun Range, in the Southern Alps of New Zealand. Haast explained his practice of naming mountains in a letter to William Jackson Hooker of 9 June 1862: 'When beginning with the survey of the Southern Alps of New Zealand, hitherto entirely unknown, I proposed myself to create a kind of Pantheon or Walhalla for my illustrious contemporaries amongst those never-trodden peaks and glaciers' (H. F. von Haast 1948, p. 213).

5 See letter to J. D. Hooker, 21 [September 1862] and nn. 6 and 7. Charles Petschler has not been identified.

6 In J. F. J. von Haast 1861, p. 135, Haast had stated: 'The native rat (Mus rattus) is the only known indigenous land quadruped'. However, in J. F. J. von Haast 1862b, p. 6, Haast had discussed the possible existence of two further indigenous quadrupeds. One was a badger- or otter-like quadruped, 'called by the natives Kaureke', which he believed probably still existed in the lakes and rivers of the Southern Alps. The second was a smaller, nocturnal quadruped, traces of which Haast had found 'in the river bed of the Hopkins, the stream which forms Lake Ohou [Ohau]'.

7 Haast refers to the Philosophical Institute of Canterbury, which he founded in the summer of 1862 (H. F. von Haast 1948, pp. 220–30).

8 In J. F. J. von Haast 1862b, p. 7, Haast presented his discussion of *Origin* as 'a tribute to its illustrious author', stating that this was 'the great work of the age' in natural history. While asserting that CD's theories were not altogether new, Haast claimed that CD's 'great merit' was that he had not only dealt with the subject in a 'true philosophical spirit' but had also collected 'a great mass of facts, which throw new light upon this inexhaustible source of inquiry'.

9 As geologist for the province of Canterbury, Haast had spent the period from January to May 1862 in the Southern Alps of New Zealand, carrying out the regular work of the geological survey of the province, together with a search for gold-bearing deposits (H. F. von Haast 1948, pp. 199–219).

10 John Marshman was agent to the Canterbury Emigration Office, 16 Charing Cross, London (*Post Office London directory* 1863).

To John Scott 11 December [1862][1]

Down Bromley Kent
Decr. 11th.—

Dear Sir

I have read your paper with much interest.[2] You ask for remarks on matter, which is alone really important, shall you think me impertinent (I am sure I do not mean to be so) if I hazard a remark on style, which is of more importance than some think? In my opinion (whether or no worth much) your paper would have been much better if written more simply & less elaborated,—more like your letters. It is a golden rule always to use, *if possible*, a short old Saxon word. Such a sentence as "so purely dependent is the incipient plant on the specific morphological tendency"[3]—does not sound to my ears like good mother English—it wants translating.—

Here & there you might, I think, have condensed some sentences: I go on plan of thinking every *single* word which can be omitted without actual loss of sense as a *decided* gain.— Now perhaps you will think me a muddling intruder; Anyhow it is the advice of an old hackneyed writer who sincerely wishes you well.—

Your remarks on the two sexes counteracting variability in product of the one is new to me.[4] But I cannot avoid thinking that there is something unknown & deeper in seminal generation. Reflect on the long succession of embryological changes in every animal. Does a bud ever produce cotyledons or embryonic leaves? I have been much interested by your remarks on inheritance at corresponding ages; I hope you will, as you say, continue to attend to this.[5] Is it true that female plants always produce female by parthenogenesis?[6] If you can answer this I sh^d. be glad; it bears on my Primula work: I thought on subject but gave up investigating what had been observed, because female Bee by parthenogenesis produces *males* alone. Your paper has told me much that in my ignorance was quite new to me.—

Thanks about P. Scotica. If any *important* criticisms are made on Primula to Bot. Soc. I sh^d. be glad to hear them.[7] If you think fit, you may state that I repeated the crossing experiments on P. Sinensis & Cowslip with the same result this Spring as last year—indeed with rather more marked difference in fertility of the two crosses.[8] In fact had I then proved Linum case I would not have wasted time in repetition.— I am determined I will at once publish on Linum.—[9]

We seem predestined to work on same subjects: for 3 or 4 summers I have worked *hard* at Drosera & Dionæa & have almost a volume of materials, which I suppose some day I shall publish; ie if ever I have time to work my materials into shape.[10] Irritability of plants has been a hobby-horse to me. I **suspect** it will turn out that irritability of same nature, only intensified in Dionæa &c. is very common with plants.—[11] If your paper on Drosera is published, I sh^d. be grateful for a copy.—[12] Two German papers have been written on subject, but I now forget where; & there is good French paper on Structure, which I daresay you know[13] I was right to be cautious in supposing you in error about Siphocampylus (no flowers were enclosed):[14] I hope that you will make out whether the pistil presents two *definite* lengths; I shall be astounded if it does.—

I do not fully understand your objections to N. Selection;[15] if I do, I presume they would apply with *full* force to, for instance, Birds. Reflect on modification of Arab-Turk Horse with our English Race-Horse.—[16]

I have had satisfaction to tell my publisher to send my Journal & Origin to your address.—[17]

I suspect with your fertile mind, you would find it far better to experiment on your own choice; but if on reflection you would like to try some which interest me, I sh^d. be truly delighted, & in this case would write in some detail.[18] If you have means, to repeat Gärtners experiments on vars. of Verbascum or on Maize (see Origin) such experiments would be preeminently important.—[19] I could never get vars. of Verbascum.[20] I could suggest experiments on Potatoes analogous with case of Passiflora; even this case of Passiflora,, often as it has been repeated, might be with advantage repeated.[21]

I have worked like a slave (having counted about 9000 seeds) on Melastomas on meaning of the two sets of very different stamens, & as yet have been shamefully beaten, & I now cry for aid.—[22]

I could suggest what I believe very good scheme (at least D^r. Hooker thought so) for systematic degeneration of culinary plants & so find out their origin; but this would be laborious & work of years.—[23]

I am tired, so pray believe me, yours very faithfully | Ch. Darwin

P.S. I have been thinking that if you do not complete your beginning on the non-dimorphic Primula, I should like *extremely* to do so & would of course fully acknowledge your work.[24] What I sh^d. do would be to fertilise a dozen or score of flowers with own pollen & another score of flowers with pollen of others, the most different, & count the seed of each capsule.—

Now could you aid me (unless you resolve to do the work yourself) & procure me from any Edinburgh nursery-garden ½ a dozen plants (or fewer, if so many could not be got) & ½ dozen of any of the other non-dimorphic species. They could be sent in pots in box addressed. "C. Darwin care of Down Postman, Bromley Kent" & I could pay by P.O.— They give or lend me all plants at Kew; but they are very weak in Primulas. I am sick of ordering plants at London nurseries; I so often get wrong thing. Could you aid me in this?—

DAR 93 (ser. 2): 37, 49–52

1 The year is provided by the relationship to the letter from John Scott, 6 December [1862].

2 Scott 1862a. See letter from John Scott, 6 December [1862]. CD's annotated copy of this paper is in the Darwin Pamphlet Collection–CUL.

3 Scott 1862a, p. 216.

4 In his paper, Scott aimed, by analysing the nature of fern spores, to shed new light on their 'apparently anomalous properties', and particularly on 'that peculiar facility afforded by spores for the reproduction and perpetuation of any accidental variation of the parts upon which they originate' (Scott 1862a, p. 210). Scott contrasted the reproductive organs in ferns with those in higher plants. Since in the latter the embryo was 'the *modified resultant* of two originally distinct organs' (p. 214), there would, he claimed, 'necessarily be a greater tendency to efface any individual peculiarities of these than would have been the case, had the embryo been the product of a single organ.' See also letter from John Scott, 6 December [1862] and n. 15.

5 Scott 1862a, pp. 217–8. Scott prefaced his remarks by quoting CD's statement (*Origin*, p. 13): 'That at whatever period a peculiarity first appears, it tends to appear in the offspring at a corresponding age, *though sometimes earlier*.'

6 Scott 1862a, p. 219.

7 CD refers to the Botanical Society of Edinburgh (see letters from John Scott, 6 December [1862] and 17 December [1862]).

8 See letter to J. D. Hooker, 23 June [1862], n. 4.

9 Having discovered in 1861 that certain species of *Linum* were dimorphic, CD carried out crossing experiments in the summer of 1862 on *L. perenne* and *L. grandiflorum*. According to his 'Journal' (Appendix II), CD wrote his paper, 'Two forms in species of *Linum*', between 11 and 21 December 1862.

[10] See letter from John Scott, 6 December [1862]. In 1860, CD began to experiment on the sensitivity to various substances of the insectivorous plants, *Drosera rotundifolia* and *Dionaea muscipula* (see *Correspondence* vol. 8). He had hoped to continue and complete the experiments in the summer of 1861, but subsequently decided to postpone them (see *Correspondence* vol. 9, letter to J. D. Hooker, 4 February [1861], and letter to Daniel Oliver, 11 September [1861]). He carried out further experiments in May and September 1862 (see DAR 54: 29–49, 74–5). However, he did not again work extensively on this subject until 1872 (*LL* 3: 322); his findings were published in 1875 as *Insectivorous plants*.

[11] See also letter to J. D. Hooker, 26 September [1862].

[12] Scott read his paper on *Drosera* and *Dionaea* (Scott 1862b) before the Botanical Society of Edinburgh on 11 December 1862. CD cited Scott's paper in *Insectivorous plants*, pp. 1 n.–2 n.

[13] In a brief overview of the literature on insectivorous plants in *Insectivorous plants*, p. 1 n., CD referred to five papers on *Drosera rotundifolia* published by German and French writers before 1862: Milde 1852 and Nitschke 1861a and 1861b (in *Botanische Zeitung*), and Grönland 1855 and Trécul 1855 (in *Annales des Sciences Naturelles* (*Botanique*)).

[14] See letter to John Scott, 3 December [1862], and letter from John Scott, 6 December [1862]. See also letter from John Scott, 17 December [1862].

[15] See letter from John Scott, 6 December [1862].

[16] CD apparently refers to the fact that an objection to natural selection based on the effect of blending inheritance on incipient varieties, would be particularly telling in cases where individuals paired for life (as with birds). CD had cited the gradual improvement of 'the whole body of English racehorses' by selection and training in *Origin* p. 35, noting that they had come to surpass the parent Arab stock. See also Appendix VI.

[17] John Murray. See letter to John Scott, 3 December [1862], and letter from John Scott, 6 December [1862].

[18] See letter from John Scott, 6 December [1862].

[19] The references are to *Origin*, pp. 269–71, and Gärtner 1844 and 1849; CD's extensively annotated copies of the latter works are in the Darwin Library–CUL (see *Marginalia* 1: 248–98). CD considered the experiments particularly important because of their bearing on an objection raised against natural selection by Thomas Henry Huxley (see letter to T. H. Huxley, 18 December [1862] and nn. 8–12, and Appendix VI).

[20] See, for example, *Correspondence* vol. 9, letters to J. D. Hooker, 28 September [1861], 18 October [1861], 23 October [1861], and 1 November [1861], and this volume, letter to C. C. Babington, 20 January [1862], and letter to W. E. Darwin, 4 [July 1862] and n. 4.

[21] The reference is to the fact that in *Passiflora* it had been found that plants could be 'much more easily fertilised by the pollen of a distinct species, than with its own' (*Natural selection*, p. 400; see also *Origin*, p. 251). Scott reported the results from a series of experiments on sterility and hybridisation in *Passiflora*, begun in 1863, in Scott 1864b.

[22] See letter to J. D. Hooker, 14 [October 1862], n. 7.

[23] Hooker discussed this project with CD during a visit to Down House in April 1860 (see *Correspondence* vol. 8, letter to Asa Gray, 25 April [1860]). See also letter from C. W. Crocker, 31 October 1862 and n. 12.

[24] Scott had apparently sent CD an account of his experiments on *Primula* in the missing portion of his letter of [20 November – 2 December 1862]). After further research, Scott wrote a paper on the subject (Scott 1864c), which CD later cited in *Forms of flowers*.

To J. D. Hooker 12 [December 1862]

Down Bromley Kent
12th

My good old friend,

How kind you have been to give me so much of your time! Your letter is of real use & has been & shall be well considered.[1] I am much pleased to find that we do not differ as much as I feared. I begin my book with saying that my chief object is to show inordinate scale of variation;[2] & I have especially studied all sorts of little variations of the individual. On crossing I cannot change; the more I think, the more reason I have to believe that my conclusion would be agreed to by all practised breeders.[3] I, also, greatly doubt about variability & domestication being at all necessarily correlative; but I have touched on this in Origin.—[4] Plants being identical under very different conditions has always seemed to me a very heavy argument against what I call direct action.

I think perhaps I will take case of 1000 pigeons as means to sum up my volume.[5] I will not discuss other points; but as I have said I shall recur to your letter. But I must just say that if sterility be allowed to come into play—if long-beaked be in least degree sterile with short beak, my whole case is altered.[6] By the way my notions on hybridity are becoming considerably altered by my dimorphic work: I am now strongly inclined to believe that sterility is at first a selected quality to keep incipient species distinct.[7] If you have looked at Lythrum, you will see how pollen can be modified merely to favour crossing; with equal readiness it could be modified to prevent crossing.—[8] It is this which makes me so much interested with dimorphism &c.— One word more: when you pitched me head over heels by your new way of looking at the back side of Variation,[9] I received assurance & strength by considering monsters,—due to law, but so horribly strange as they are. I looked at some Plates;[10] the monsters were alive till at least when born. They differ at least as much from parent as any one mammal from another.—

I have just finished a long weary chapter on simple facts of variation of cultivated Plants; & am now refreshing myself with paper on Linum for Linn. Soc.[y].—[11]

I paid Bonafous & other Books to London, but could not pay to Kew.—[12]

I have just ordered one of the Glass cases, which are warmed by dish of hot-water twice a day, & I hope I shall then be able to keep Oxalis sensitiva.[13] I see a book mentioned which I have ordered Cohn on *contractile* tissue in plants;[14] I suspect he has been at work like, but far fuller, mine on Drosera.[15] I am reading Dutrochet's work, which seems extremely clever, but, I know not why, does not convince me about the swelling of cells by endosmose of fluid & *of oxygen*.![16]

If you know, do please tell me who is John Scott of Bot. Garden of Edinburgh; I have been corresponding largely with him: he is no common man.—[17] I enclose one other request to be added to awful number, which you already have.—[18] I shall be anxious to know whether I can have a Begonia frigida with the strange flowers to cross.[19]

You once told me that I sh[d]. be executed for Origin in Edinburgh;[20] but I received the other day Diploma of Medical Soc. with signatures of 37 Edinburgh big-wig medicals; so I must be rather up with at least heretical Doctors.[21]

I hardly know whether the enclosed letter of Gray's is worth sending, for as it is not answered, *I must have it back*. & for References.— —[22]

My dear Hooker | Yours ever most truly | C. Darwin.

Endorsement: 'Dec/62'
DAR 115.2: 176

[1] Hooker's letter, presumably a reply to the letter to J. D. Hooker, [after 26] November [1862], has not been found.

[2] According to his 'Journal' (*Correspondence* vol. 8, Appendix II), CD wrote a draft of the introduction to *Variation* in the spring of 1860. In the published version he began by stating his intention to provide, 'under the head of each species', those facts that he had been able to collect or observe 'showing the amount and nature of the changes which animals and plants have undergone whilst under man's dominion, or which bear on the general principles of variation' (*Variation* 1: 1).

[3] See letter from J. D. Hooker, 26 November 1862, and letter to J. D. Hooker, [after 26] November [1862].

[4] CD discussed variation under domestication in the first chapter of *Origin* (pp. 7–43). He began by observing (p. 7) that individuals of the same variety or sub-variety of older cultivated plants and animals 'generally differ much more from each other than do the individuals of any one species or variety in a state of nature', suggesting that this greater variability was 'simply due to our domestic productions having been raised under conditions of life not so uniform as, and somewhat different from, those to which the parent-species have been exposed under nature.' In particular, he gave some countenance to the view, suggested by Thomas Andrew Knight, that 'this variability may be partly connected with excess of food.'

[5] See letter to J. D. Hooker, [after 26] November [1862], and *Variation* 2: 420–4.

[6] CD made a note to this effect on 18 December 1862; the note is preserved in DAR 205.7: 163 (see Appendix VI).

[7] See Appendix VI.

[8] In his letter to J. D. Hooker, 24 [November 1862], CD sent a diagram illustrating what he believed to be the sexual relations between the three forms of flower in *Lythrum salicaria*.

[9] See letter from J. D. Hooker, 26 November 1862, and letter to J. D. Hooker, [after 26] November [1862].

[10] Possibly a reference to the plates in Geoffroy Saint-Hilaire 1832–7, an extensively annotated copy of which is in the Darwin Library–CUL (see *Marginalia* 1: 306–16).

[11] According to his 'Journal' (Appendix II), CD wrote a draft chapter on 'Facts of variation of Plants', which became chapters 9 and 10 of *Variation*, between 7 October and 11 December 1862. The journal also records that he wrote his paper, 'Two forms in species of *Linum*', between 11 and 21 December.

[12] The reference is to copies of Bonafous 1836 and volumes 6 and 7 of the *London Journal of Botany* that had been sent to CD from the Royal Botanic Gardens, Kew, possibly also with volume 1 of the *Technologist* (see letter from J. D. Hooker, [15 and] 20 November [1862], and letter from Daniel Oliver, 25 November 1862).

[13] See letter to Pickard & Stoneman, 1 December [1862]. CD required the plant case so that he could work with temperature-sensitive plants like *Oxalis sensitiva* (see letter from J. D. Hooker, 26 November 1862, and letter to J. D. Hooker, [after 26] November [1862]).

[14] Cohn 1860; see letter to T. H. Huxley, 7 December [1862] and n. 10.

[15] CD had begun his experiments with the insectivorous plant *Drosera rotundifolia* in 1860 (see *Correspondence* vol. 8). He had hoped to continue and complete the experiments in the summer of 1861, but subsequently decided to postpone them (see *Correspondence* vol. 9, letter to J. D. Hooker, 4 February [1861], and letter to Daniel Oliver, 11 September [1861]). After experimenting further with the plant in September 1862, CD had come to the conclusion that it must have 'diffused matter, in organic connection, closely analogous to the nervous matter of animals', in which view Cohn had preceded him (see letter to J. D. Hooker, 26 September [1862], and letter to T. H. Huxley, 7 December [1862]).

[16] CD refers to Dutrochet 1837. According to his library catalogue (DAR 240), CD owned a copy of this work; however, only the atlas of plates is in the Darwin Library–CUL. See also *Correspondence* vol. 4, Appendix IV, *119: 2v. In Dutrochet 1837, 1: 534–73, René Joachim Henri Dutrochet discussed

the mechanism by which he believed sensitive plants to move. He explained the movements in mechanical terms, distinguishing two antagonistic causes for the depression and elevation of petioles. He claimed that the former was caused by the flexing of fibrous tissues owing to oxygenation, while the latter was caused by the distention of cellular tissue owing to 'endosmis'. Dutrochet had pioneered the study of what he termed 'endosmis' and 'exosmis' (in modern terms, osmosis and diffusion). See *DSB.*

17 John Scott was foreman of the propagating department at the Royal Botanic Garden, Edinburgh; he began to correspond with CD on 11 November 1862 (see letter from John Scott, 11 November 1862).

18 The enclosure has not been found; it apparently contained a request for specimens of the wild gooseberry (see letter from J. D. Hooker, [14 December 1862] and n. 10).

19 See letter from C. W. Crocker, 24 November 1862 and n. 1.

20 The reference has not been identified.

21 See Appendix IX. The Royal Medical Society of Edinburgh is a student society, founded in 1737, and closely associated with the Edinburgh Medical School (J. Gray 1952).

22 The reference is apparently to the letter from Asa Gray, 24 November 1862 (see letter from J. D. Hooker, [14 December 1862]).

To William Forsell Kirby 12 December [1862]

Down. | Bromley. | Kent. S.E.
December 12th

Dear Sir

I am very much obliged for your extremely kind present of your Manual of European Butterflies.—[1] From all that I have heard I can well believe that it must have been a difficult task. It pleases me much to hear that you are not a believer in the immutability of species,—a doctrine perfectly adapted to stop philosophical research.—[2] I trust that you some day will write on variation in Butterflies, & express your beliefs on the subject of species.— I do not know whether you are any relation of that great man, Kirby the author of that admirable work on Entomology;[3] but if you are you have in truth good right to success; & if you are not, your name ought to be an omen of success.—

With my best wishes & thanks | Pray believe me | Dear Sir | Yours faithfully | Ch. Darwin

I have just observed that your note is dated *Nov.* 7th but was received only this day.—

Postmark: DE 13 62
American Philosophical Society (Getz 3854)

1 Kirby 1862 was Kirby's first work of importance, and brought him into public notice (*Entomologist's Record and Journal of Variation* 24 (1912): 315).

2 The letter from Kirby has not been found; however, he was reportedly 'an early disciple of Darwin' (*Entomologist's Record and Journal of Variation* 24 (1912): 315).

3 CD refers to the Suffolk clergyman, William Kirby, who was co-author with William Spence of *Introduction to entomology* (Kirby and Spence 1815–26); there is a copy of Kirby and Spence 1815–26 in the Darwin Library–CUL. W. F. Kirby was the son of a Leicester banker, and was apparently not a relation of the older entomologist (*Entomologist's Record and Journal of Variation* 24 (1912): 315).

From A. C. Ramsay 13 December 1862

London
13 Decr 1862

My dear Sir

By this post I send you the 3^{d} edition of our Catalogue of rocks. It is a kind of manual.[1]

I have read Jamiesons paper on the Parallel Roads.[2] It is to be read at the Society January 21st. It is excellent & not long. Might I venture to refer it to you.[3] It would give you very little trouble, I am sure. If you would rather not have it I must refer it to someone else less able to judge of its merits.

In the Catalogue I now send, you may possibly be interested in the passages p 1 to 3, bottom of p 13 & p 14, p 19 also top $\frac{1}{2}$ of p 21,[4] and the bottom of page 185 where you will find a remarkable quotation from the writings of old Peter Merian of Basle on the subject of fracture & regelation of glacier ice, though not in these express terms.[5]

Ever truly | And^{w} C Ramsay

DAR 176: 10

[1] Ramsay refers to a catalogue of the rock specimens in the Museum of Practical Geology, Jermyn Street, London (Ramsay *et al.* 1862). Ramsay was on the staff of the Geological Survey of Great Britain and was lecturer in geology at the Government School of Mines, both of which were part of the Jermyn Street establishment (Geikie ed. 1895). The catalogue was intended as the first in a series of publications 'to illustrate the several branches of science which are taught in the Government School of Mines' (Ramsay *et al.* 1862, p. ii); in his introduction Ramsay explained how the collections had been arranged for didactic purposes.

[2] Jamieson 1863. See n. 3, below.

[3] Ramsay was president of the Geological Society; in his letter to Ramsay of 14 December [1862], CD agreed to referee Jamieson 1863 for publication in the *Quarterly Journal of the Geological Society of London.* In his paper, Thomas Francis Jamieson offered an explanation for the series of parallel terraces on the sides of Glen Roy in Lochaber, Scotland, contradicting that given by CD in his 1839 paper, 'Parallel roads of Glen Roy'. CD had fully accepted Jamieson's explanation (see letter to A. C. Ramsay, 5 September [1862]).

[4] Ramsay refers to a number of passages relating to the origin of metamorphic rocks, and, in particular, to the phenomenon of cleavage. He cited a suggestion of CD's from *South America*, based on his observations in the Andes, that

> "*foliation may be the extreme result of the process of which cleavage is the first effect*;" or, in other words, that the process of re-arrangement of particles in the rocks began with cleavage, and ended in their entire crystalline re-arrangement in the same lines, thus producing *foliation.*

Ramsay, on the contrary, held that there was no 'necessary connection betwen cleavage and foliation.'

[5] Ramsay *et al.* 1862, pp. 185–6. The reference is to Merian 1843, p. 156. Ramsay noted that Peter Merian's theory that glaciers moved by repeated fracturing, followed by rapid regelation under pressure, was similar to that recently proposed by the physicist John Tyndall.

From E. A. Darwin 14 December [1862?][1]

Dec 14

There is a mysterious box come for you, marked Glass but with a kind of gridiron lid as if it had something alive inside—size a thin quarto.[2] I have no doubt you have received Prices circulars but I enclose one. I quite forget his address which I shall be obliged for[3]

E D

DAR 105 (ser. 2): 12

[1] The year is conjectured from the reference to a package delivered to Erasmus's house and from the reference to the circulars for Price 1863–4 (see nn. 2 and 3, below).

[2] The reference is probably to the piece of 'partially completed artificial comb' that CD had directed Thomas White Woodbury to send to his brother Erasmus's address (see letter to T. W. Woodbury, 7 December [1862] and n. 5).

[3] The reference is probably to John Price, who had been a contemporary of Erasmus's and CD's at Shrewsbury School. Price's twelve-part miscellaneous work, *Old Price's remains* (Price 1863–4), was published between April 1863 and March 1864; CD's copy is in the Darwin Library–CUL. Price lived at 38 Watergate Street, Chester (Price 1863–4; see also CD's Address book (Down House MS)).

From J. D. Hooker [14 December 1862][1]

Kew
Sunday

Dr Darwin

I always like to see what A Gray says, though I must say it is sadly unsatisfactory to me.[2] When writing to him the other day, I broke the Ice, in so far as to say, that I avoided alluding to the War, because I never knew any one to be right in his ideas or prognostications of a war *in progress*, & no belligerent ever knew what he was bringing about. Events **alone** can justify a war, motives may excuse, but never justify however good— the whole thing is as much an anomaly in civilization as duelling. Of course (I added) we both deplore it, but we are not children, & have something better to do than to hold pocket handkerchiefs to one anothers eyes across the Atlantic.

I read Max Muller, liked some parts, thought the *last* eminently unphilosophical, & concluded that it contained as little grist for your mill as could well be, considering how fertile the subject might be made[3]

Mann of W. Africa is sending you bees & honey comb as I requested.[4]

I will see to the Begonia &c *toute de suite*[5]

Think again over your assumption of long-beaked pigeons being or not being in any degree sterile with short.[6] *They must* be one or the other— there is no such thing as Equality.. it is inconceivable,—hence there is no such thing as *Chance*; & Nat. Seln. is the Sword of Damocles hanging over your own head, if you make a slip in your premisses

I have read, with delight, the note on *Lythrum*, you sent me some weeks ago—[7] its consequences are of the *most* prolific order to your doctrine

I know nothing of Dutrochet & Cohns works, & am becoming a miserable ignoramus.—[8] I once asked about John Scott for you & was told he was a very smart fellow—nothing more was known—or said.[9]

We have not wild Gooseberry nor do I know where to get it in this country, but will write to St Petersburgh— it is a Scandinavian plant.[10]

Ever yours affec— | J D Hooker

Dear Willy comes home on Thursday, he keeps at the *very* bottom of his school poor boy.[11]

DAR 101: 83–4

CD ANNOTATIONS
5.2 there is . . . sent me 6.1] *scored brown crayon*
8.1 We . . . plant. 8.2] *scored brown crayon*
End of letter: 'Not Worth'[12] *brown crayon*; '(Teucrium | campanulatum)'[13] *ink, del ink*; '(Huxley Lectures)'[14] *ink*

[1] Dated by the relationship to the letter to J. D. Hooker, 12 [December 1862], and the letter from J. D. Hooker, [21 December 1862]; the intervening Sunday was 14 December.
[2] CD apparently enclosed Asa Gray's letter of 24 November 1862 with his letter to J. D. Hooker, 12 [December 1862]. Hooker and Gray held radically different views on the American Civil War, and had tacitly agreed not to discuss the matter in their letters (see letter from J. D. Hooker, [19 January 1862], and letter from Asa Gray, 18 February 1862).
[3] Max Müller 1861. See letter from Asa Gray, 24 November 1862 and nn. 2 and 3.
[4] Gustav Mann. See letter from J. D. Hooker, 20 August 1862 and nn. 8 and 9.
[5] See letter to J. D. Hooker, 12 [December 1862].
[6] See letters to J. D. Hooker, [after 26] November [1862] and 12 [December 1862].
[7] See letter to J. D. Hooker, 24 [November 1862].
[8] See letter to J. D. Hooker, 12 [December 1862]. The references are to Cohn 1860 and Dutrochet 1837.
[9] In his letter to Hooker of 12 [December 1862], CD inquired about John Scott, who was foreman of the propagating department at the Royal Botanic Garden, Edinburgh. He had previously made inquiries in his letter to Hooker of 18 [November 1862].
[10] See letter to J. D. Hooker, 12 [December 1862] and n. 18. According to his 'Journal' (Appendix II), CD finished writing a draft of the section of *Variation* dealing with 'Facts of variation of Plants' (*Variation* 1: 305–72) on 11 December 1862; the gooseberry is discussed on pp. 354–6. See also letter to *Journal of Horticulture*, [before 2 December 1862].
[11] Hooker's oldest son, William Henslow Hooker, was 9 years old.
[12] See letter to J. D. Hooker, [21 December 1862].
[13] CD discussed the incidence of peloric flowers in *Teucrium campanulatum* in *Variation* 2: 345, citing Moquin-Tandon 1841, p. 192; there is an annotated copy of this work in the Darwin Library–CUL.
[14] T. H. Huxley 1862c. See letter to J. D. Hooker, [21 December 1862].

To A. C. Ramsay 14 December [1862][1]

Down. | Bromley. | Kent. S.E.
Dec. 14th

Dear Ramsay

Many thanks for Catalogue: I have read with interest parts referred to & other parts—[2] I have been pleased to find some of Volcanic specimens have been honoured by being enumerated.—[3]

With respect to Glen Roy paper, I shall be happy to undertake it.[4] It will in fact be no trouble, as I have corresponded with Mr J. on subject & know most of his facts & arguments, & know, alas, too well that I am everlastingly smashed

Ever Dear Ramsay | Yours sincerely | C. Darwin

DAR 261 (DH/MS 9: 5)

[1] The year is established by the relationship to the letter from A. C. Ramsay, 13 December 1862.

[2] Ramsay *et al.* 1862. See letter from A. C. Ramsay, 13 December 1862. CD's copy of this work has not been found.

[3] CD refers to some of the volcanic rock specimens collected during the *Beagle* voyage, which he had presented to the Museum of Practical Geology (see Ramsay *et al.* 1862, pp. 220–9; see also Porter 1985, p. 996).

[4] In his letter of 13 December 1862, Ramsay asked CD to referee Thomas Francis Jamieson's paper on the geology of Glen Roy in Lochaber, Scotland (Jamieson 1863), for possible publication in the *Quarterly Journal of the Geological Society of London*. Jamieson's paper was sent to CD for review immediately after it was read to the society on 21 January 1863; although CD's report was received by the society on 26 January 1863, it has not been preserved in the archives (see 'Papers read', Geological Society of London, archives, COM P 3/1).

To H. W. Bates 15 December [1862][1]

Down. | *Bromley.* | *Kent. S.E.*
Dec. 15th

Dear Bates

Many thanks for the paper & references on pairing of vars,—a subject which interests me much.— If I do not hear to contrary I will keep the paper sent, though I do not see in it much on pairing of varieties.[2]

I would not on any account give you trouble to copy the passages out of Zoologist;[3] when I come to subject I can consult the volume in London.—

I am sorry for all your trouble & delay about artists &c; but it is a law of nature that they will give trouble & it is of no use fighting against a law of nature.[4]

Good luck to you. | Yours very sincerely | C. Darwin

By any chance did you ever come across Indians, hardly pressed by famine,— who were thus compelled to cook in *new* ways & try *new* vegetable productions?—[5] If by any odd chance, you have, I sh^d^ be glad to hear; otherwise do *not* write to say that you do not know.—

Cleveland Health Sciences Library (Robert M. Stecher collection)

[1] The year is established by the relationship to the letter to H. W. Bates, 25 November [1862] (see n. 2, below).

[2] In his letter to Bates of 25 November [1862], CD asked for information on the Lepidoptera, including 'facts about similar varieties pairing', and an estimate of how many he had caught and how many were in his collection. The paper and references sent by Bates are now missing. In the catalogue of his collection of offprints, CD listed a paper of Bates's with the title 'On Non-Crossing of Varieties', recording that it was in the *Transactions of the Entomological Society of London* for 1858 (Vorzimmer 1964); the offprint corresponding to this description is absent from the Darwin Pamphlet Collection–CUL, but it is likely that it was Bates 1857, which comprised notes on a wide range of South American Lepidoptera.

[3] CD refers to the *Zoologist: a popular miscellany of natural history*; the reference may be to some of Bates's own communications. In addition to several more formal papers, the *Zoologist* carried numerous letters from Bates between 1850 (volume 8) and 1857 (volume 15), describing his travels and collecting activities in the Amazonian region.

[4] Bates was in the final stages of preparing for publication an account of his travels as a naturalist in South America (Bates 1863). In his preface (p. vi), Bates acknowledged the assistance of three artists in preparing the illustrations: Edward W. Robinson, Joseph Wolf, and Johann Baptist Zwecker.

[5] See letter to J. D. Hooker, 4 November [1862] and n. 12.

From John Lubbock 15 December 1862

London
15 Dec. 1862

My dear M^r. Darwin

I did not reply to your note at once because I had not then finished reading Bates' paper.[1]

It is certainly very good, but as you yourself suggested I can't see any reason why you should not, & several why you should, write the article for us yourself.[2] I have two short articles in our forthcoming number,[3] I have to do one on the Swiss lakes for our next,[4] besides two papers for the Linnæan & a lecture at the Royal Institution.[5] So you see my hands are already too full.

But you have the subject so completely at your fingers end, & have no doubt studied the question of the mimetic resemblances so much & thought out its bearings on Natural Selection, that it would give you very little trouble (comparatively) to write two or three pages about it.

Do see whether you can't do this for us.

I hope you have all kept well.

Believe me, dear Mr Darwin, always | Yours affectionately | John Lubbock

C Darwin Esq

DAR 170: 34

[1] CD's letter has not been found; however, he had evidently written to suggest that Lubbock, who was one of the editors of the *Natural History Review*, might review Bates 1862a in the journal (see the letter to H. W. Bates, 25 November [1862] and n. 8).

[2] CD reviewed Bates 1862a in the April 1863 number of the *Natural History Review* ('Review of Bates on mimetic butterflies').

[3] Lubbock 1863a and [Lubbock] 1863b.

[4] Lubbock visited Switzerland in July and early August 1862, when he made a study of the recently discovered remains of prehistoric lake-dwellings or 'Pfahlbauten' (see letter from John Lubbock, 23 August 1862 and nn. 3 and 5). He had previously described these remains from written reports in a paper for the January 1862 number of the *Natural History Review* (Lubbock 1862b), but did not publish an account of his own observations in the journal. He did, however, present some of his observations in a lecture at the Royal Institution of Great Britain on 27 February 1863 (Lubbock 1863c); see also Lubbock 1865, pp. 119–70.

[5] Lubbock 1863c, 1863d, and 1863e.

From J. B. Innes 16 December [1862][1]

Milton Brodie | Forres | NB.
Decr. 16th.

Dear Darwin,

When I left Downe I could not find Johnny's Savings Bank book.[2] It had been put away with other papers and has only now turned up. Will you be so kind as to have it made up and returned to me.

I hope its absence has not increased the trouble you and the other managers are so kind as to take on behalf of depositors.[3]

I am glad to have a reason for writing to you and to be able to wish you and yours a happy Christmas and New year. Stephens has not mentioned you in any very recent letter wherefore I conclude you are all well.[4] I hope William continues to like his occupation and finds it as golden as he could anticipate.[5] Henry Lubbock seems to hang on to the paternal nest very firmly at present.[6] John must be a great loss to Lady Lubbock but his own party increased too fast to remain with comfort.[7] I am sorry to hear Ring's wife is so ill.[8] She has been a good wife and brought up her children far better than most— I had a very cheerful letter from Knight Bruce a few days ago. He is still without a house and has just returned to his Father's at Roehampton from Versailles where he has been all the summer.[9]

We lead a very quiet life here. My wife[10] is quite as well as she was in England and has several times gone out to dinner some 4 or 5 miles, which she could not, or would not do at Downe. Our change has been of the greatest advantage to Johnny's health. He has grown quite stout and robust as well as tall, and has not had an hour's illness, indeed I am thankful to say that we have had no need for a doctor in the house since we have been here. We want a cook very badly but it is not much use to ask you as I don't think they grow abundantly in your soil. There is no special natural history that has come under my unscientific observation, except that I saw a white rabbit with black tips to his ears on a muir where only brown and occasionally a black one commonly dwell. What do you say to wheat being grown from oats in the second year? Do you trust it enough to try it for yourself?[11]

With all our best regards | Believe me | Dear Darwin | Yours faithfully | J Brodie Innes—

DAR 167: 10

[1] The year is established by the relationship to the letter to J. B. Innes, 22 December [1862].

[2] The reference is to Innes's fifteen-year-old son, John William Brodie, and to the Bromley Savings Bank (see n. 3, below). Innes, who was the perpetual curate of Down (that is, the Anglican incumbent of the parish), had removed with his family to Scotland at the beginning of the year (see letters from J. B. Innes, 2 January [1862] and 19 February [1862]).

[3] CD was one of the trustees of the Bromley Savings Bank, High Street, Bromley (see *Correspondence* vol. 11, letter from J. T. Austen, 30 May and 3 June 1863). An account of the bank, which was founded in 1816 'to protect the Savings of the poor' (see *Correspondence* vol. 11, letter from J. T. Austen, 30 May and 3 June 1863), is given in Horsburgh 1980, pp. 303–4.

[4] Thomas Sellwood Stephens was Innes's curate at Down (*Post Office directory of the six home counties* 1862); he was evidently responsible for discharging Innes's parochial duties following the latter's removal to Scotland (see n. 2, above).

[5] CD's eldest son, William Erasmus Darwin, became a partner in the Southampton and Hampshire Bank, Southampton, in 1861 (see *Correspondence* vol. 9).

[6] Henry James Lubbock, the second son of John William and Harriet Lubbock, continued to live with them at High Elms, near Down, until 1869, when he was 31 years old (see letter to J. B. Innes, 18 October 1869, *Calendar* no. 6942).

[7] The Lubbocks' eldest son, John, moved from High Elms to Chislehurst, a village about five miles north of Down, on 19 August 1861 (John Lubbock's diary (British Library, Add. Ms. 62679: 64 r.); John Lubbock's fourth child, Norman, was born in 1861 (Hutchinson 1914).

[8] The reference is to Abraham Ring, a gardener resident in Down, and to his wife Charlotte (Census returns 1861 (Public Record Office, RG9/462: 72)).

[9] Lewis Bruce Knight Bruce had formerly resided in Keston, a village two miles north of Down (*Post Office directory of the six home counties* 1851, 1855, 1859). His father, James Lewis Knight Bruce, lived at Roehampton Priory, Surrey (*Post Office London directory* 1862).

[10] Eliza Mary Brodie Innes.

[11] Innes probably refers to an extract that appeared in *The Times*, 10 December 1862, p. 7, quoting a letter from a Northamptonshire farmer, William Cowper, to the *Berkshire Chronicle*. Cowper stated as a 'positive fact' that he had grown both wheat and barley from oats, and cited several other such reports. See also letter to J. B. Innes, 22 December [1862] and n. 9.

To John Lubbock 16 [December 1862][1]

Down. | Bromley. | Kent. S.E.
16th

My dear Lubbock

You are right, I am a selfish dog.— I am aghast at all you have to do.[2] In a day or two I will do my best (but I confess to sighs & groans) to give your readers a brief notion of Bates' paper; how far I shall succeed, I know not.—[3]

I wonder what you are going to Lecture on at Royal. Instn—[4]

Most truly yours | C. Darwin

DAR 263

[1] The date is established by the relationship to the letter from John Lubbock, 15 December 1862.

[2] See letter from John Lubbock, 15 December 1862.

[3] Lubbock was one of the editors of the *Natural History Review*; CD's review of Bates 1862a appeared in the April 1863 number of the journal ('Review of Bates on mimetic butterflies').

[4] Lubbock's lecture, 'On the ancient lake-habitations of Switzerland' (Lubbock 1863c), was delivered at the Royal Institution of Great Britain on 27 February 1863.

From John Scott 17 December [1862][1]

Botanic Gardens | Edinburgh
Dec. 17th.

Sir,

I have the honour to acknowledge, the due receipt of your Journal and Origin,[2] for which I now return my most sincere thanks, in humble and grateful acknowledgement of the entirely unmerited kindness you have done me. And this not only in presenting me with these inestimable works, but also in condescending, even to recognise the observations of one entirely unknown, a young and ardent admirer of Science, who is yet standing beyond her pale. I would also most particularly thank you for the valuable remarks you have made in your last, inculcating upon me a due regard for simplicity of style.[3] Would that some of those—able and qualified, friends ⟨*half a line excised*⟩ forward with me, and point out my ⟨*half a line excised*⟩ being perhaps ⟨*nine lines excised*⟩ disadvantages in regard to this point. I feel insuperable difficulties in expressing my ideas, and I am thus frequently forced to do this in a style very different from what I like. The paper on Ferns which I ventured to send you, was with another trifling exception or so, my first attempt, and I can assure you their style now annoys me very much.[4] Indeed, if it had not been, that I was anxious to draw your attention to the explanation I had proposed for that singular facility the Ferns present for reproducing variations— I would never thought of sending it to you. I am however, deeply impressed with the validity of this view on its special application, and I believe it will be found perfectly accordant when applied generally to like phenomena in the Vegetable Kingdom. I may be in an error, and such a proposition as the following, may cause you to smile—viz: that the facilities for the reproduction of variations in plants, is inversely dependent on the less or more mediate relation of the sexual organs— nevertheless, I think, I could show good grounds for my believe—though you may perhaps explain it very differently—by a cursory review of the Vegetable Kingdom, from this stand-point. I will not, however, even if desirable, attempt this at present, but confine myself to some of the other points ⟨*nine lines excised*⟩

I have not been at all successful in further illustrating the subject of inheritance at corresponding ages.[5] I have been looking principally after cases for experimenting upon amongst the Ferns, and when I wrote my paper, I was looking forward to spores which were then being produced upon an accidentally variegated portion of a frond; since that I have observed another such case. From both of these I got a few spores, but I have not succeeded in raising a single plant, not one of them even germinated

If you are not already aware of the fact, however, a case in point is presented by the variegated varieties of Begonias—in at least some of them for I have only raised a single batch of seedlings. Being curious to see the results of two very marked varieties of these, I crossed them, and got a number of good seeds. After these germinated, I found that the early leaves presented no resemblance whatever, to the parents, nor did they do so before the fourth or more generally the sixth leaf, when their immediate parentage was betrayed. I know nothing of the history of these Begonias at present, otherwise, I would have given you fuller particulars regarding them, for they evidently betray in their young state an earlier progenitor, from which they have inherited their incipient characters. If you have not already information on this point, much I believe might be derived from the Messrs Veitch.[6] It is a point however, that I will not neglect, to test, as opportunities present themselves.

I was led into an error in regard to the statement, that female plants, always produced females by parthenogenesis,[7] by taking it upon the authority of a friend, having then seen Karsten's paper only;[8] and a short abstract of Braun's by Prof. Balfour.[9] I was therefore very sorry, on consulting authors on the subject to find them stating that both sexes were produced. The subject, however, yet requires a fair testing, as we find such flatly contradictory statements made by authors on the subject, as applied to plants. There is one fact, however, that still makes me cling to the idea, viz that *Cœlobgyne ilicifolia*, has always produced female plants. Now in my opinion, if all the seeds that it has produced had been fertilised, as Karsten seems to suppose,[10] the male plants ought likewise to have appeared amongst the seedlings. Their absence therefore still induces me to regard it as an excellent illustration of the doctrine. It does not necessarily happen that though he has observed hermaphrodite flowers upon the plant, they should always be so. I have made a great number of experiments on the subject this season, and got apparently perfect seeds from a number of different plants, yet I have only succeeded in germinating those belonging the *Lychnis dioica*, and *Melandryum prætense*. I am anxious to see the produce of them, as I was exceedingly careful in guarding against all pollinic influence. I have also raised the *Selaginella denticulata*, and I am trying a number of others.[11]

I completely overlooked the flowers of Siphocampylus which I ought to have enclosed in my last.[12] I now send you a few withered flowers, to show you the state of the short-styled flowers. The lengths, however, so far as I have yet observed are very fluctuating, but I will attend to this point, when the plants flower again.

I will be highly gratified if you favour me with an account of the experiments, you would like to have performed, and I can assure you if it is in my power to do them, it will afford me a great pleasure.[13] I would much rather that you would do this; than be allowed to work on in the blind, unsystematic manner, it must necessarily be if left to myself. As I have had so little experience yet in these as in others, I will therefore be glad if you communicate all which would be of interest to you, and I hope the mere circumstance of years of experimental enquiry, being necessary for the attainment of definite results, will not deter you from communicating such. If it is within my power to perform them, it would afford me a pleasure to carry out your schemes.

With my best thanks for your kindness to me, | I remain | Sir | Yours respectfully, John Scott

P.S. I have just received a letter from a friend, whom I wrote regarding P. Scotica—[14] I was not able to get plants from the Messrs Lawson's—they are quite out of it; and, I have no faith, in any of the others, here—[15] they are something like your London nurseryman— My friend has, however, promised to send me a few plants; which I will forward to you. That these may be correct, I shall take the following precautions,—viz: replace them by plants that I have had under my eye last summer, so that there may be no fear of mistakes in those you get. We have not many plants of P. Scotica in the gardens here at present,—but perhaps we may get a few in the spring—otherwise I should have asked Mr. McNab, to favour you with a few.[16] I will not therefore, let him know anything about them, and I hope you will take no notice of having received them from me, as it might be a cause of offence to him, if he saw or heard it. I have no other non-dimorphic species that I could send you, nor so far as I am aware, are there any of them in cultivation at present. The P. *pusilla*, has been introduced, but now I believe lost. If you have not the P. *farinosa*, however, I will try and send you a few plants of it along with the P. *Scotica*, as being near the latter, and a truly dimorphic species. The comparative fertility of these two is very remarkable, when left to Natural Agencies. I observe nearly all the flowers of P. Scotica produce a greater or less number of seeds, in some capsules, I have found 120 seeds, while in P. farinosa few capsules are produced, and these contain only a very few seeds. I have rarely found in one of them 20 seeds. If the former were artificially fertilised; I believe it will prove, as you suspect perfectly self-fertile. I will make a careful series of experiments upon the plants I have here in the course of the ensuing season, in accordance with your directions, and you shall have the results, to compare with those you perform.[17]

By the way I spoke to the young gentleman at the Society's Meeting regarding his objections to your views on Primula, and it turns out that instead of being opposed, he is quite favourable, so far as he has examined, but he does not appear to have done much amongst them.[18] He has not observed P. Scotica.

I am not sure how many of the varieties of Verbascum, we have in the Gardens here, but there are a few of them, which I will try this season. Passiflora we have, and I will try and get a few varieties of Maize. I know not, where I could get those that Gartner experimented upon. Can you give me any information regarding them?[19]

In my next I expect to have a little information on the fertilisation of *Loelia*, which is just coming into flower.[20] I have fertilised two flowers properly, which are evidently fecundated, as the perianth has now become quite flaccid. I also applied a single pollen-mass to the viscid surface of the tongue-like rostellum of two other flowers, but these give no signs yet, of being affected, though the act was performed at the same time. If I have a few more flowers, to experiment upon, I will cut some of them for examination; but as I have so few at present I think it will be better, to see whether the capsules swell or not.

I am afraid lest the pollen-tubes instead of penetrating the rostellum—pass along its surface, down sides of stigmatic cavity, and penetrate stigmas proper. Could you not suggest some preventive for such an occurrence? For perhaps such an objection might occur to some, though there may be small probability of its taking place. I thought of applying a thickish solution of gum, or what would you suggest? i.e. if you think it necessary to apply anything.

Incomplete
DAR 177: 80

CD ANNOTATIONS
1.1 I have … case. 2.5] *crossed ink*
Top of letter: '(*John Scott*)' *ink*

[1] The year is established by the reference to Scott 1862a (see n. 4, below).
[2] See letter to John Scott, 11 December [1862] and n. 17.
[3] See letter to John Scott, 11 December [1862].
[4] Scott had enclosed a copy of Scott 1862a with his letter of 6 December [1862]; the earlier papers to which he refers are Scott 1862c and 1862d.
[5] See letter to John Scott, 11 December [1862] and n. 5.
[6] Scott refers to the Chelsea father and son nurserymen James and James Veitch.
[7] Scott 1862a, p. 219. See letter to John Scott, 11 December [1862].
[8] Karsten 1861.
[9] Braun 1859. John Hutton Balfour was keeper of the Royal Botanic Garden, Edinburgh.
[10] Karsten 1861.
[11] Scott later discussed the sexuality of the higher Cryptogams, and particularly of *Selaginella*, in Scott 1864d.
[12] See letter from John Scott, 6 December [1862], and letters to John Scott, 3 December [1862] and 11 December [1862].
[13] See letter to John Scott, 11 December [1862].
[14] See letter to John Scott, 11 December [1862]; this individual has not been identified.
[15] Charles Lawson conducted the Edinburgh nursery, Peter Lawson & Son.
[16] James McNab was curator of the Royal Botanic Garden, Edinburgh.
[17] See letter to John Scott, 3 December [1862] and nn. 3 and 6.
[18] See letter from John Scott, 6 December [1862] and n. 5.
[19] The reference is to Karl Friedrich von Gärtner (see letter to John Scott, 11 December [1862] and nn. 19 and 20). Scott repeated a number of Gärtner's crossing experiments with *Verbascum* in 1863 (see *Correspondence* vol. 11, letter to John Scott, 7 November [1863]). CD cited some of Scott's findings in *Variation* 2: 106–7; Scott published his full results in Scott 1868.
[20] See letter from John Scott, 6 December [1862].

To T. H. Huxley 18 December [1862][1]

Down Bromley Kent
Dec. 18th

My dear Huxley

I have read nos IV & V. They are simply perfect.[2] They ought to be largely advertised; but it is very good in me to say so; for I threw down nor IV with this reflexion. "What is the good of my writing a thundering big book, when everything is in this green little book so despicable for its size"? In the name of all that is good & bad I may as well shut up shop altogether.[3] You put capitally & most simply & clearly the relation of animals & plants to each other at p. 122.—[4]

Be careful about Fantails; their tail-feathers are **fixed** in radiating position but they can depress & elevate them:[5] I remember in a pigeon-book seeing withering contempt expressed at some naturalist for not knowing this important point![6]

p. 111. seems a little too strong, viz 99 out of hundred, unless you except plants[7]

p. 118. you say the answer to varieties when crossed being at all sterile is "absolutely a negative".[8] Do you mean to say that Gärtner lied, after experiments by the hundred (& he a hostile witness) when he showed that this was the case with Verbascum & with Maize (& here you have *selected* races):[9] does Kölreuter lie when he speaks about the vars. of Tobacco.[10] My God is not the case difficult enough, without its being, as I must think, falsely made more difficult? I believe it is my own fault—my d—d— candour; I ought to have made ten-times more fuss, about these most careful experiments.[11] I did put it stronger in 3^d. Edition of Origin[12]

If you have a new Edit. do consider your *second* Geological section:[13] I do not dispute the truth of your statement; but I maintain that in almost every case the gravel would graduate into the mud; that there would not be a hard **straight** line between the mass of gravel & mud. That the gravel in crawling inland, would be separated from the underlying beds by oblique lines of stratification. A nice idea of the difficulty of geology your section would give to a working man!— Do show your section to Ramsay & tell him what I say,[14] & if he thinks it a fair section for a beginner, I am shut up & "will for ever hold my tongue".[15]

Good Night— | Yours very sincerely | C. Darwin

My 2^d son George (who is no fool & his Master says would be a wrangler at Cambridge, if *now* examined)[16] is reading your Lectures & likes them very much— I asked him whether he understood the Geological bit; he answered he thinks he could have done so, *without* that second diagram.—[17]

Imperial College of Science, Technology and Medicine Archives (Huxley 5: 186)

[1] The year is established by the reference to T. H. Huxley 1862c (see n. 2, below).

[2] CD refers to the published version of Huxley's lectures to working men at the Museum of Practical Geology (T. H. Huxley 1862c, pp. 83–132; see letter from T. H. Huxley, 2 December 1862 and nn. 1 and 2). The fourth and fifth lectures were delivered on 1 and 8 December respectively, and were entitled: 'The perpetuation of living beings, hereditary transmission and variation' and 'The conditions of existence as affecting the perpetuation of living beings'. CD's lightly annotated copy of the lectures is in the Darwin Library–CUL (see *Marginalia* 1: 425).

[3] In the introduction to his lectures, Huxley stated that they were intended to present *Origin* 'in a true light' (T. H. Huxley 1862c, p. 5). See also letter from T. H. Huxley, 10 October [1862].

[4] The passage to which CD refers (T. H. Huxley 1862c, p. 122) is part of a section in which Huxley sought to demonstrate that the 'CONDITIONS OF EXISTENCE . . . exercise an influence which is exactly comparable to that of artificial selection.' In considering the influence on an organism of 'the state of the rest of organic creation', Huxley divided organic beings into '*indirect opponents*' (or '*rivals*'), '*direct opponents*' (or '*enemies*'), and '*helpers*'. CD marked this section in his copy of the lectures with a marginal line, noting 'very good on Relation of all Beings in Struggle for life'.

[5] T. H. Huxley 1862c, p. 103; CD noted this correction in his copy of the work.

[6] Dixon 1851, pp. 90–1; there is an annotated copy of this work in the Darwin Library–CUL (see *Marginalia* 1: 199–201). The relationship between naturalists and breeders, and their differing views of pigeons, is discussed in Secord 1981, pp. 166–70.

[7] CD refers to Huxley's statement that the crossing of two hybrids would, 'in ninety-nine cases out of a hundred', result in 'no offspring at all' (T. H. Huxley 1862c, p. 111). In his copy of the lectures CD noted that this statement was 'too strong'.

[8] CD wrote 'p. 118' in error; the statement appears on p. 113, and is underlined in CD's copy of the lectures. CD and Huxley had a long-standing disagreement over Huxley's statement, in his review of *Origin* ([T. H. Huxley] 1860, pp. 552–5), that a significant, if inconclusive test of whether two individuals represented distinct 'physiological' species was to attempt to hybridise them: distinct species would often either be infertile *inter se* or produce infertile offspring, whereas varieties of the same species would give rise to fertile progeny. Huxley argued that the lack of positive evidence

that any group of animals had, 'by variation and selective breeding', given rise to another group which was 'even in the least degree infertile with the first', was the weakest point in CD's hypothesis of natural selection ([T. H. Huxley] 1860, pp. 567–8). Huxley reiterated the first of these points in T. H. Huxley 1862c, pp. 108–113 (and the second on pp. 146–50). See also Appendix VI.

9 For the results of Karl Friedrich von Gärtner's experiments on *Verbascum* and *Zea*, see Gärtner 1844 and 1849. CD's heavily annotated copies of these works are in the Darwin Library–CUL (see *Marginalia* 1: 248–98). There is also an abstract of Gärtner 1849 in DAR 116. In *Origin*, p. 270, CD explained that Gärtner and Joseph Gottlieb Kölreuter were 'hostile witnesses' since they in all other cases considered 'fertility and sterility as safe criterions of specific distinction.'

10 Kölreuter 1761–6; CD's heavily annotated copy of this work is in the Darwin Library–CUL (see *Marginalia* 1: 458–71).

11 In *Origin*, pp. 269–71, CD described the crossing experiments carried out by Gärtner and Kölreuter on varieties of *Verbascum* and *Nicotiana* respectively, citing them as important evidence that in some cases varieties of the same species are found to be somewhat sterile when crossed.

12 Partly in response to Huxley's criticisms, but also in the light of his work on dimorphism, CD made a number of minor alterations throughout the chapter on hybridism in the third edition of *Origin*, serving to increase the force of his argument against the view espoused by Huxley (see Peckham 1959, pp. 424–74). In introducing Gärtner's and Kölreuter's experiments, he changed the sentence: 'But it seems to me impossible to resist the evidence of the existence of a certain amount of sterility in the few following cases' to read: 'But it is impossible . . .' (*Origin* 3d ed., p. 292). CD's annotated copies of the six editions of *Origin* are in the Rare Books Room–CUL.

13 See letter to T. H. Huxley, 7 December [1862] and n. 7, and T. H. Huxley 1862c, p. 40 and fig. 5. Although Huxley 'more than once' set about the task of revising these lectures (T. H. Huxley 1893–4, 2: vii), he did not do so, and they were reproduced, unaltered, in his *Collected essays* (T. H. Huxley 1893–4, 2: 303–474).

14 Andrew Crombie Ramsay was a colleague of Huxley's at the Government School of Mines, Jermyn Street, London.

15 'Let him speak now, or else hereafter for ever hold his peace' (*Book of common prayer*, Solemnization of matrimony, 'Banns').

16 George Howard Darwin was a student at Clapham Grammar School, the then headmaster of which was Alfred Wrigley; he graduated as second wrangler at Trinity College, Cambridge, in 1868 (*DNB*).

17 See n. 13, above.

From John Lubbock 18 December 1862

15, Lombard Street. E.C.
18 Dec 62

Dear M^r. Darwin

Hurrah, you are a brick & I shall read your article with the greatest interest.[1]

My lecture is to be on the "Pfahlbauten".[2]

Y^rs. affec | John Lubbock

C Darwin Esqr

DAR 170: 35

1 See letter from John Lubbock, 15 December 1862, and letter to John Lubbock, 16 [December 1862].

2 Lubbock refers to his lecture at the Royal Institution of Great Britain on the ancient lake-habitations of Switzerland (Lubbock 1863c): the dwellings ('Pfahlbauten') were built on stilts or piles ('Pfahl').

To John Scott 19 December [1862][1]

Down Bromley Kent
Dec. 19th

Dear Sir

There is no occasion to thank me; it is simply a pleasure to me to aid (if aid it can be called) anyone who really loves science; & you conferred a great favour on me about Acropera.—[2] You have taken my little criticisms in the spirit in which they were intended.[3] You have reason, in my opinion, to be proud of your paper.—[4] By odd chance reading last night some short Lectures just published by Prof. Huxley, I find your observation, independently arrived at by him, on the confluence of two sexes causing variability.—[5] I must write briefly.— Difference or similarity of buds & ovules is beyond my tether: I have spoken of them in deference to others as identical, but I have always *suspected* some fundamental difference.[6]

Talking of buds, if you can give me any cases of what Gardeners sometimes call sports, & which I shall call "bud-variation" I shall be grateful, as I want to collect all cases like moss—on common rose,—nectarine on Peach.—[7]

I did not at all know about foliage of Begonia.—[8]

I shall be very glad indeed of P. farinosa & Scotica.: *if in pots in open ground I suppose they will do*???[9]

I can tell you nothing about vars of Verbascum, except that I have in vain tried to get white & yellow vars. of same species.[10]

If Lælia should set (as does not from what you say appear probable) seed-capsules with pollen put on Rostellum: then if you were to examine one specimen & see whether pollen-tubes penetrated Rostellum it wd., I shd. think, suffice.—[11]

In my experiments, I have found net, (such as bit enclosed) distended with hoops or fastened on square frame of sticks, answer perfectly in excluding insects, during flowering season, & not injure fertility of plants. Now there ought to be no false delicacy between us followers of science; if any pecuniary assistance is required to purchase pieces of net & pay for making, nets, grant me the favour to apply to me.— Excuse Brevity, as I am tired. I will append about experiments separately.—[12]

Yours very faithfully | C. Darwin

[Enclosure 1]

Experiments

To the best of my judgment, no subject is so important in relation to Theoretical Natural Science, in several respects, & likewise in itself deserving investigation; as the effects of changed or unnatural conditions, or of changed structure on the reproductive system.

Under this point of view the relation of well-marked, but undoubted varieties in fertilising each other, requires far more experiments than have been tried.

See in Origin the brief abstract of Gärtner on Verbascum & Zea.—[13]

Mr. W. C. Crocker, lately foreman at Kew & a very good observer is going at my suggestion to work vars of Hollyhocks.[14] Climate wd be too cold, I suppose, for vars.

of Tobacco.[15] I began on Cabbage, but immediately stopped from Early shedding of their pollen causes too much trouble.[16] Your knowledge would suggest some.

On same principle it would be well to test peloric flowers, with their own pollen; & with pollen of regular flowers; & try pollen of peloric on regular flowers,—seeds being counted in each case.

I have now got one seedling from many crosses of a peloric Pelargonium by peloric pollen; I have 2 or 3 seedlings from a peloric flower by pollen of regular flower.— I have ordered a peloric Antirrhinum & the peloric Gloxinia, but I much fear I shall never have time to try them.[17]

The Passiflora-cases are truly wonderful, like the Crinum cases (see Origin).[18] I have read in German paper that some varieties of Potatoes (name not given) cannot be fertilised by own pollen, but can by pollen of other *vars*; well worth trying[19]

Again fertility of any monster flower, which is pretty regularly produced; I have got the wonderful Begonia frigida from Kew, but doubt whether I have heat to set its seeds.[20] If an unmodified Celosia could be got; it would be well to test with the modified Cockscomb. There is a var. of Columbine with simple petals without nectaries &c &c.—[21]

I never could think what to try; but if one could get hold of a long cultivated plant which crossed with a distinct species & yielded a very small number of seed; then it would be highly good to test comparatively the wild parent form & its varying offspring with this third species: for instance if a polyanthus would cross with some species of Primula, then to try a wild cowslip with it.— I believe hardly any Primulas have ever been crossed. If we knew & could get parent of Carnation it would be very good for this end.—

(Any member of the Lythraceæ, raised from seed ought to be well looked after for dimorphism: I have wonderful facts, the result of experiment, on Lythrum salicaria.—[22]

[Enclosure 2]

Melastomatads.[23]

Have you any species with two sets of stamens *widely dissimilar.* It would be better not to try Heterocentron roseum(?) or mexicanum(?) with purple flowers—or Monochætum ensiferum; for I have tried many experiments on these with no sure result, but with plain indications that there is something odd about them; for seedlings produced by the two kinds of pollen differed wonderfully in their rate of growth & in setting seed &c—[24]

If you have any such & are inclined to operate; first observe whether the pistil retains the same position from time flower opens till it withers: if it does *not* move, merely try both pollens & mark capsules. The Counting the seed is awful labour, but I would undertake it. (N.B. in all experiments it is quite necessary to wrap up well each ripe capsules in *separate* paper.)

If pistil moves, experiment is more troublesome, for *both* pollens sh$^{d.}$ be applied to the stigma at the *two* ages & positions.[25] I am completely baffled, but I strongly suspect there is something very strange in these Melastomas. There is nothing odd, as I know by trial, in such genera as Centradenia & Rhexia.[26]

DAR 93 (ser. 2): 35–6, 64–5, 80

[1] The year is established by the relationship to the letter from John Scott, 17 December [1862].

[2] See letter from John Scott, 17 December [1862]. Scott had initiated the correspondence with CD in November when he wrote concerning the sexual forms of the orchid *Acropera* (see letter from John Scott, 11 November 1862). Earlier in December, CD had arranged for copies of *Journal of researches* and the third edition of *Origin* to be sent to Scott (see letters to John Scott, 3 December [1862] and 11 December [1862]).

[3] See letter to John Scott, 11 December [1862], and letter from John Scott, 17 December [1862].

[4] Scott had sent CD a copy of Scott 1862a (see letter from John Scott, 6 December [1862]); it is preserved in the Darwin Pamphlet Collection–CUL.

[5] See Scott 1862a, p. 214, and Huxley 1862, p. 90–2. See also letter from John Scott, 17 December [1862].

[6] CD discussed the resemblances of buds and ovules in chapter 11 of *Variation*, a chapter he began writing in December 1862 (see n. 7, below). He concluded by stating that the facts he had collected showed in how 'close and remarkable' a manner the germ of a fertilised seed and the small cellular mass forming a bud resemble each other in function and in their capacity for variation (*Variation* 1: 411).

[7] CD began writing a draft of chapter 11 of *Variation*, 'On bud-variation, and on certain anomalous modes of reproduction and variation', on 21 December 1862 (see 'Journal' (Appendix II)). He cited observations by Scott of this phenomenon in *Variation* 1: 385.

[8] See letter from John Scott, 17 December [1862] and n. 6.

[9] Scott offerred to send CD specimens of *Primula scotica* and *P. farinosa* in his letter of 17 December [1862].

[10] See letter from John Scott, 17 December [1862] and n. 19. Regarding CD's attempts to obtain specimens of *Verbascum*, see the letter to W. E. Darwin, 4 [July 1862] and n. 4.

[11] See letter from John Scott, 17 December [1862] and n. 20.

[12] See letter to John Scott, 11 December [1862], and letter from John Scott, 17 December [1862].

[13] *Origin*, pp. 270–1.

[14] See letters from C. W. Crocker, [before 13 March 1862] and 24 November 1862.

[15] CD refers to the possibility of Scott repeating a series of crossing experiments on varieties of the tobacco plant (*Nicotiana*), which had originally been carried out by Joseph Gottlieb Kölreuter. CD reported Kölreuter's findings in *Origin*, pp. 271–2, discussing their significance with regard to hybrid sterility.

[16] In DAR 76: 40, there are dated notes from CD's crossing experiments with cabbages in May and June 1862. CD continued to carry out similar experiments with cabbages between 1863 and 1871 (see DAR 76: 41–4 and DAR 78: 139–56); he discussed his findings in *Cross and self fertilisation*, pp. 98–103.

[17] CD refers to the crossing experiments, begun in May 1862, with the normally sterile peloric flowers of pelargoniums (see letter to Daniel Oliver, 8 June [1862], letter to Asa Gray, 1 July [1862], and letters to M. T. Masters, 8 July [1862] and 24 July [1862]). CD's notes from these experiments are in DAR 51 (ser. 2): 4–9, 12–13. In *Variation* 2: 70, CD described crossing experiments on peloric flowers of *Antirrhinum major*; his notes from these experiments, dated 1863–5, are in DAR 51 (ser. 2): 18–23.

[18] *Origin*, pp. 250–1.

[19] CD refers to an article entitled 'German pamphlets on the potato disease, and its remedies', which appeared in the *Gardeners' Chronicle and Agricultural Gazette*, 21 March 1846, pp. 181–3. The article provided an abstract of Albert 1845, and included the observations noted here in a translation of the appendix, 'Upon the regeneration, or profitable propagation of potatoes from seeds', by 'Inspector' Tinzmann. CD cited Tinzmann's observations in *Variation* 2: 137. CD's annotated copy of the *Gardeners'*

Chronicle is in the Cory Library, Cambridge Botanic Garden; he noted the article cited here in his abstract of the journal (DAR 75: 1–12).

20 See letter to J. D. Hooker, 12 [December 1862], and letter from J. D. Hooker, [14 December 1862].

21 CD discussed the modified forms of the columbine and the cock's-comb (*Celosia cristata*) in *Variation* 1: 365.

22 In the summer of 1862, CD made extensive crossing experiments with the trimorphic plant, *Lythrum salicaria* (see letter to Daniel Oliver, 29 [July 1862]); the observational notes from this work are preserved in DAR 27.2. CD's paper, 'Three forms of *Lythrum salicaria*', was read before the Linnean Society of London on 16 June 1864.

23 CD began to investigate the occurrence of two sets of stamens in the flowers of the Melastomataceae in October 1861, suspecting that the family might exhibit a novel form of dimorphism (see *Correspondence* vol. 9, letter to J. D. Hooker, 17 November [1861] and n. 14, and this volume, letter to George Bentham, 3 February [1862]). He continued to work on the family throughout 1862 and 1863 without ultimately being able to account for the two sets of stamens (see *Cross and self fertilisation*, p. 298 n., and *ML* 2: 292–302). CD's notes from these experiments are in DAR 205.8. CD mentioned his work on the Melastomataceae in his letter to Scott of 11 December [1862], and sought Scott's assistance with his experiments.

24 CD had begun experimenting with *Heterocentron roseum* in October 1861; his observational and experimental notes on this species, dated October 1861 – January 1863, are in DAR 205.8: 45–7, 49–53, 56. He began crossing experiments on *Monochaetum ensiferum* in February 1862 and continued to work on the species until May 1863 (see DAR 205.8: 22–41). His experimental notes on *H. mexicanum*, dated June – October 1862, are in DAR 205.8: 19 v., 54. CD was also seeking to obtain specimens of these genera from their native South America (see letter to J. D. Hooker, 14 [October 1862], and letter from J. D. Hooker, [18 October 1862]).

25 Having noticed that in *Monochaetum ensiferum* there were changes over time in the positions of the pistils and stamens, CD had speculated that the pistil might react differently to the two kinds of pollen produced by this species, depending on its age and position (see letter to Asa Gray, 16 February [1862]).

26 CD's observational and experimental notes on *Rhexia glandulosa* (dated June and July 1862) and on various species of *Centradenia* (dated January – July 1862) are in DAR 205.8: 14–21. In addition, Asa Gray had, at CD's request, supplied him with observations on *Rhexia Virginica* (see letter from Asa Gray, 4 August 1862).

To George William Johnson or Robert Hogg[1] 20 December [1862][2]

Down. | *Bromley.* | *Kent. S.E.*
Dec. 20th

Dear Sir

I am extremely much obliged to you for your kindness in sending me the Strawberry Hybrid, which I have planted.[3] When the communication appears in the Journal will be in ample time for my purpose.—[4]

Dear Sir | Yours faithfully | Charles Darwin

C. H. Hughes-Johnson, private collection

1 The correspondent is identified by the reference to his having sent CD a strawberry hybrid (see n. 3, below); Johnson and Hogg were the editors of the *Journal of Horticulture.*

2 The year is established by the reference to William Smith's letter to the *Journal of Horticulture* (see n. 3, below).

[3] The reference is to three runners of a hybrid strawberry, which were sent to Johnson and Hogg by William Smith of York, in response to CD's letter to the *Journal of Horticulture* of [before 25 November 1862]. Smith requested that the editors send CD the runners if they thought he might be interested in them (*Journal of Horticulture* n.s. 3 (1862): 779). No letter on this subject from Hogg or Johnson has been found. Smith's letter to the *Journal of Horticulture* was published in the number for 30 December 1862, with the editorial comment: 'Mr. Darwin, to whom we forwarded all the runners, is very much obliged, and has planted them.'

[4] This appears to be a reference to Smith's letter to the *Journal of Horticulture* (see n. 3, above).

From J. D. Hooker [21 December 1862][1]

Kew
Sunday

My dear Darwin

I have nothing worth bothering you with a letter about, but feel impelled to write—apropos of nothing! I saw Erasmus the other evening looking as young & blithe as anything—[2] I never saw him looking half so well. He told me you were but poorly. I am longing to run down & see you, but not yet for some weeks, I have promised to go to Paris with Bentham for a week or so after 15th. January about Gen. Pl.[3] I throttled off Welwitschia at the Linnæan the other night. there was a good meeting & they all seemed pleased— I have sent it to press.[4]

If you want any information about Australia & its savages, a capital man is now about Kew who has collected plants well & lived amongst & observed natives shrewdly from K.G.S. to Sharks bay, & W. up to Wide bay.[5]

What shall I do with the African bees & comb?— If I hear nothing I shall send it by next Thursday's Carrier; so do not write on purpose.[6]

I am actually reading de Tocquevilles Democracy in America,[7] it appears to me a most able book, though I do not at all agree with it. (bigger fool you, you may so, & double big fool I am to say so) but I cannot help it. He assumes that D. in America was a success— Now I never regarded America as having cohesion enough to be pronounced either a success or a failure: there has been hitherto far too much freedom of motion there, too little "struggle for existence"—to develop any settled Govt. at all, & it is impossible to predicate what shape the existing *introduced* form of Govt. would take in 100 years, even if this war had not stepped in to confound all calculations. Democracy has *persisted* in America, because there has been no cause for its overthrow—just as Monarchies might persist indefinitely (though they persist under much greater disadvantages. Specialization I conceive to be a dominant law governing everything, & I cannot see how either a Democracy or a Republican form of Govt can resist the effects of Natural Selection— In short I regard a pure Democracy as visionary as a country peopled by one invariable species.— This with me is no question of what is good or bad, but of what must ever be.—& I do hold that a Govt must always eventually get into the hands of an individual, or a family, or a class—or there is no truth in Nat: Selection. Q.E.D. as you say.

My boy Willy[8] is back from school, a standing protest against the "Origin of Species" I do not know one quality of my wifes[9] or my own that he inherits. He cares for no one thing in life, I mean naturally, as Music. or Nat Hist. or even seeing things, has neither Memory Sensation or judgement;—"Like I was" he is very young of his age, & I suspect there is a great deal in that— Per contra, he is a very good amiable generous & most conscientious child, extremely fond of us all & of his brothers & sisters especially—& is only not docile because it is impossible to fix his attention. I cannot find out any one thing he is learning at School, he can give no account of himself or his time whatever, of course he is at bottom of his classes, but that I do not mind. he has no pride, nor mischief, no vice,—no emulation no pursuit & no taste—& is quite indifferent to lollypops!— In the others I can trace everywhere family vices & virtues— The most wonderful thing to me is that not one has an atom of ear, & whereas I could sing in tune at 3 years old, I do not find that these have the smallest ear for music, or care—to hear it.

Ever Yours affec | Jos D Hooker

I send Dawsons most unsatisfactory letter[10]

DAR 101: 80–2

[1] The date is established by the reference to J. D. Hooker 1863a (see n. 4, below); the Sunday following 18 December 1862 was 21 December.

[2] CD's elder brother, Erasmus Alvey Darwin, lived at 6 Queen Anne Street, London.

[3] Hooker refers to George Bentham and to *Genera plantarum* (Bentham and Hooker 1862–83). Hooker and Bentham left for a ten-day trip to Paris on 17 January 1863 (Jackson 1906, p. 193).

[4] J. D. Hooker 1863a. Hooker read the first part of his monograph on the Angolan plant *Welwitschia* before the Linnean Society of London on 18 December 1862, concluding at a subsequent meeting of the society on 16 January 1863. The number of the *Transactions of the Linnean Society of London* in which the paper appeared was published on 30 January 1863 (Raphael 1970, p. 76).

[5] The reference is to the botanical collector Augustus Frederick Oldfield, who had returned to Britain from Australia earlier in the year (see letter from J. D. Hooker, [27 or 28 December 1862] and R. Desmond 1994). In his letter of 4 November [1862], CD had asked Hooker for assistance in tracing a reference to Australian aborigines preparing and eating poisonous plants during times of famine (see also letter to J. D. Hooker, [10–]12 November [1862]).

[6] The bees and honeycomb had been acquired for CD in West Africa by the botanical collector Gustav Mann (see letters from J. D. Hooker, 20 August 1862 and nn. 8 and 9, and [14 December 1862]). A weekly carrier service between Down and London was operated by George Snow.

[7] Tocqueville 1836. In 1862 Henry Reeve published a new edition of his translation of Tocqueville 1836 (Reeve trans. 1862).

[8] William Henslow Hooker.

[9] Frances Harriet Hooker.

[10] The letter from John William Dawson has not been found. However, see letters from J. D. Hooker, 2 November 1862 and 7 November 1862, and letters to J. D. Hooker, 4 November [1862] and [10–]12 November [1862]. See also letter to J. D. Hooker, 24 December [1862].

To J. D. Hooker [21 December 1862][1]

Down

Sunday evening

My dear Hooker

I send just a line to thank for Begonias & the very curious Oxalis, which arrived quite safe.— I am low about the Plant=case, & cannot keep it hot enough.—[2]

It is not at all worth while to write about wild Gooseberry, unless you wish to have it in garden.—[3]

I keep obstinate about crossing & could argue till doomsday, but will not bother you.—[4]

I infer from G. Chronicle that you have read your Wellwitschia paper & I heartily wish you joy;[5] for it is great satisfaction finishing a job. It is certainly the greatest pleasure about a book. I inferred from one of your notes that you did not think much of Huxley's Lectures;[6] they seem to me *capital*; perhaps not deserving of *such* a man's time, but otherwise, as it seems to me, excellent.—

I have finished my Linum paper[7] & an abstract of Bates' paper[8] for N. Hist Review—thank God—& today have begun to think of arrangement of my concluding chapters on Inheritance, Reversion—Selection & such things & am fairly paralysed how to begin & how to end & what to do with my huge piles of materials[9]

Ever yours affect^{ly} | C. Darwin

Endorsement: 'Dec 22/62'
DAR 115.2: 174

1 The date is established by the reference to CD having finished his paper, 'Two forms in species of *Linum*', and to his resumption of work on *Variation* (see nn. 7 and 9, below).

2 See letter to J. D. Hooker, 12 [December 1862] and nn. 13 and 19, and letter from J. D. Hooker, [14 December 1862] and n. 5.

3 See letter from J. D. Hooker, [14 December 1862] and nn. 10 and 12.

4 See letter from J. D. Hooker, [14 December 1862] and n. 6.

5 The *Gardeners' Chronicle* of 20 December 1862, p. 1194, reported that Hooker had read his paper on *Welwitschia* (J. D. Hooker 1863a) before the Linnean Society of London on 18 December. See also preceding letter and n. 4.

6 T. H. Huxley 1862c. See letter from J. D. Hooker, 26 November 1862. For CD's opinion of Huxley's lectures, see the letters to T. H. Huxley, 7 December [1862] and 18 December [1862].

7 According to his 'Journal' (Appendix II), CD began his paper 'Two forms in species of *Linum*' on 11 December 1862.

8 'Review of Bates on mimetic butterflies' was published in the April 1863 number of the *Natural History Review*. See also letters from John Lubbock, 15 December 1862 and 18 December 1862, and letter to John Lubbock, 16 [December 1862].

9 According to his 'Journal' (Appendix II), CD began to prepare a draft of chapter 11 of *Variation*, 'On bud-variation and on certain anomalous modes of reproduction and variation' (*Variation* 1: 373–411) on 21 December 1862.

From Francis Boott 22 December [1862][1]

24 Gower St.
Dec 22.

My Dear Darwin

I have just rec^{d} the enclosed from Gray, who in alluding to the War tells me they have no idea of giving up. tho' the state of their finance I think must suggest the necessity of it sooner or later.[2] I have just rec^{d} a bill at the exchange of $146\frac{1}{2}$—[3]

I have never thanked you for your work on the Orchidiæ—which I read with wonder & admiration.[4] It confirms my faith that White of Selborne bequeathes his mantle to you—[5]

I do not know whether Dear Henslow saw your sagacious interpretation, but I am quite sure that it adds to the joys which he & White & Linneus have in Heaven, for I cannot believe such pure spirits have ceased to exist,[6]

I hope you are well | Yrs sincerely | F. Boott

C. Darwin Esq[re]

DAR 160: 251

[1] The year is established by the relationship to the letter from Francis Boott, 26 December 1862.

[2] Asa Gray's letter has not been found; it was probably returned to Boott.

[3] Boott, was the son of a wealthy New England merchant and had inherited 'a competency' (*DAB* s.v. Boott, Kirk; *DNB*). In order to avert the financial crisis precipitated by the American Civil War, the Union government abandoned the gold standard at the end of 1861, and issued large numbers of Treasury notes ('greenbacks') in the course of 1862, prompting fears of ruinous inflation (MacPherson 1988, pp. 442–50). Following the abandonment of the gold standard, to which the other major trading nations continued to subscribe, the price of foreign exchange became a function of the price of gold as measure in 'greenbacks' (Kessel and Alchian 1967, p. 15). Boott quotes a figure of $\$146\frac{1}{2}$ in 'greenbacks' for \$100 in gold, representing 46% inflation since the introduction of 'greenbacks'.

[4] Boott's name appears on CD's presentation list for *Orchids* (see Appendix IV).

[5] Boott refers to Gilbert White, the author of *The natural history and antiquities of Selborne* (White 1789), who was renowned for his careful observation of detail. For CD's reading of this work, see *Correspondence* vol. 4, Appendix IV, *119: 6v and 119: 13a. CD's annotated copies of White 1825 and 1843 are in the Darwin Library–CUL (see *Marginalia* 1: 868–70).

[6] Boott refers to John Stevens Henslow, CD's former mentor and professor of botany at Cambridge University, who died on 16 May 1861 (see *Correspondence* vol. 9), and to the originator of the Linnean system of classification, Carolus Linnaeus.

To J. B. Innes 22 December [1862][1]

Down. | *Bromley.* | *Kent. S.E.*
Dec. 22[d]

Dear Innes

We were very glad to hear so good an account of your son & of M[rs]. Innes; for dining out I know is a triumph of strength for her.[2] As the Highlands do so much good, I wish some honest old gentlemean would leave us an estate, for I am sure we all want doing good to.[3] Both our younger Boys are delicate & whether they will be fit for school, I am sure I hardly know.[4] As for myself I doubt whether I shall ever dine out again, so M[rs]. Innes has clearly beaten me.—

The next time I send to the Bank at Bromley, I will send the Book & return it to you when I get it back.—[5] Poor M[rs] Ring is a dying woman:[6] I don't remember any other piece of news. The Clubs go on well & everything else in the parish, as far as I know. At last Coal Club there was a brilliant attendance of four members.—[7] The man (name unknown) from Clapham who bought M[r] Ainslies house, must be as odd a man as M[r] A;[8] for he never came to see it; & enquired anxiously whether there was a place for a single cow & was astounded when he saw all the ranges of stalls. He bought everything in the House & amongst other things a large Box full of personal letters. He opened one & put it back with a *laugh*, "saying this could never have been meant for a stranger to see."—

I do not believe a word about the wheat story,[9] which has been repeated at intervals for a century; but when carefully tried (as it has been) has always failed. How it arises I know not.

With my wife's very kind remembrances to Mrs Innes & yourself, Believe me | Dear Innes | Always yours very truly | C. Darwin

American Philosophical Society (3872)

[1] The year is established by the references to Leonard and Horace Darwin (see n. 4, below).

[2] The references are to Innes's wife, Eliza Mary Brodie, and his son, John William Brodie (see letter from J. B. Innes, 16 December [1862]).

[3] Innes, the perpetual curate of Down, removed to Scotland early in 1862, having in 1861 inherited an entailed estate at Milton Brodie, near Forres, Morayshire, from his cousin Eliza Brodie Dunn.

[4] Leonard and Horace Darwin had both been seriously ill in 1862 (see 'Journal' (Appendix II)). In June, Leonard was sent home from Clapham Grammar School, South London, suffering from scarlet fever (see letter from Charles Pritchard, 17 June [1862]). During his convalescence he was tutored by George Varenne Reed, and returned to school in January 1863 (see CD's Classed account book (Down House MS)). Horace, having apparently been tutored by Reed in the autumn of 1861, did not return until the autumn of 1862; he did not attend Clapham Grammar School until 1865 (see CD's Classed account book (Down House MS), *Correspondence* vol. 11, letter from G. V. Reed, 12 January 1863, and Notes on Horace Darwin, p. 3 (Cambridge University Library, Cambridge Scientific Instrument Company Archives, Box 3)).

[5] See letter from J. B. Innes, 16 December [1862] and n. 3.

[6] Charlotte Ring.

[7] CD was treasurer of the Down Coal and Clothing Club, a charity designed 'to encourage saving Habits in the Poor' by adding subscriptions from the gentry to the money contributed by the members (see *Correspondence* vol. 5, letter to G. H. Turnbull, 28 October [1854], and *Correspondence* vol. 7, letter from John Innes, 9 January [1858–9], n. 2). Innes had previously been treasurer of the club and retained a close interest in its affairs (see *Correspondence* vol. 4, letter to John Innes, [8 May 1848], and J. R. Moore 1985, pp. 468–9). CD was also treasurer for thirty years of the Down Friendly Club, which he helped to found in 1852 (Freeman 1978).

[8] Robert Ainslie, who resided at Tromer Lodge, Down, from 1845 until 1858, had angered CD by his illegal altering of the road outside his house, by his refusal to pay his Church rates, and by his mistreatment of his horses (see *Correspondence* vols. 3 and 7, and letter to [Local landowner], [1866], *Calendar* no. 4963). Innes had considered purchasing Tromer Lodge from Ainslie in 1860 (see *Correspondence* vol. 8, letter to John Innes, 18 July [1860]). It was, however, purchased in the summer of 1862 by Robert Haswell, who in 1863 attracted local notoriety when convicted of smoking in a first-class railway carriage (see letter from J. B Innes to T. S. Stephens, [before 5 May 1862], the letter from Henrietta Emma to William Erasmus Darwin, dated [22 February 1863], in DAR 210.6: 109, and *The Times*, 20 February 1863, p. 11).

[9] See letter from J. B. Innes, 16 December [1862] and n. 11, and Weissenborn 1837. CD wrote in the margin of his copy of the volume of the *Magazine of Natural History* containing this paper (Darwin Library–CUL): 'Try single grains in pots placed in triangle'. See also his Questions & Experiments notebook (DAR 206: 15 v.; *Notebooks*, p. 506). Weissenborn's report was cited approvingly by Robert Chambers in his controversial evolutionary work, *Vestiges of the natural history of creation* (see [Chambers] 1844, pp. 220–2 and [Chambers] 1845, pp. 111–12).

To James Anderson 23 December [1862][1]

Down. | *Bromley.* | *Kent. S.E.*
Dec. 23[d]

Dear Sir

I have just heard from Prof. Asa Gray that you have had the extreme kindness to bring over for me a box with live plants for experiment.[2] I thank you sincerely.— Will you have the goodness to send the box per Rail addressed

"Ch. Darwin Esq[e]"
"care of Down Postman"
"Bromley"
"Kent"

Dear Sir, Yours truly obliged | Ch. Darwin

Autographia, San Rafael, California

[1] The year is established by the relationship to the letter from Asa Gray, 9 December 1862 (see n. 2, below).

[2] See letter from Asa Gray, 9 December 1862. Anderson was captain of the Cunard trans-Atlantic steamer *Africa*.

To Thomas Rivers 23 December [1862][1]

Down. | *Bromley.* | *Kent. S.E.*
Dec. 23[d]

Sir

I do not know whether you will forgive a stranger addressing you. My name may possibly be known to you.— I am now writing a book on the Variation of Animals & Plants under domestication; & there is one little piece of information, which it is more likely that you could give me, than any man in the world, if you can spare half an hour from your professional labours & are inclined to be so kind.[2] I am collecting all accounts of what some call "Sports", that is, of what I shall call "bud-variations",[3] ie a moss-rose suddenly appearing on a provence rose—a nectarine on a peach &c &c.— Now what I want to know, & which is not likely to be recorded in print, is whether very slight difference, too slight to be worth propagating, thus appear suddenly by *buds*. As every one knows in raising seedlings you may have every gradation from individuals identical with the parent, to slight varieties, to strongly marked varieties.— Now does this occur with buds? Or do only rather strongly marked varieties thus appear at rare intervals of time by buds? I sh[d]. be most grateful for information.—[4]

I may add that if you have observed in your enormous experience any remarkable "bud-variations" & could spare time to inform me, & allow me to quote them on your authority, it would be the greatest favour. I feel sure that these "bud-variations" are most interesting to anyone endeavouring to make out, what little can be made out on the obscure subject of variation.

I hope that you will forgive the liberty which I have taken, & I remain, with much respect, Dear Sir | Yours faithfully | Charles Darwin

P.S. I have read with *greatest* interest your communication in the last G. Chron; & venture to express an earnest hope that they will be followed by others.—[5]

Sotheby Sale (23–4 July 1987)

[1] The year is established by the reference to Rivers 1862 (see n. 5, below).

[2] Rivers, who was a nurseryman, was the author of several books on the cultivation of roses and fruit-trees, and a regular contributor to gardening journals (*DNB*).

[3] According to his 'Journal' (Appendix II), CD began working on a draft of chapter 11 of *Variation*, 'On bud-variation, and on certain anomalous modes of reproduction and variation' (*Variation* 1: 373–411), on 21 December 1862.

[4] Rivers's reply has not been found; however, see the letter to Thomas Rivers, 28 December [1862].

[5] Rivers 1862. Rivers concluded his article (p. 1196):

> I have thus far given a portion of my experience in raising seedling fruits; if I have not carried the subject to a satiating length, I will continue a relation of my experiments with seedling Apples, Plums, Cherries, Strawberries, Rasberries, and Currants.

Further articles by Rivers on seedling fruits appeared in the *Gardeners' Chronicle* for 1863 (pp. 27, 76, 244–5, and 364); CD kept the relevant numbers of the journal in separate parcels (see DAR 222 and DAR 75: 1–12; CD's annotated copy of the *Gardeners' Chronicle* is at the Cory Library, Cambridge Botanic Garden).

To J. D. Hooker 24 December [1862]

Down Bromley Kent
Dec. 24th.—

My dear Hooker

Many thanks for sight of Dawson's letter:[1] he does not show cockishness towards you. I shd like to hear what evidence he could advance that when the country was first upheaved after his later submergence that climate was not glacial.[2] He is contradictory about me; for if I make others study the limits of species it is enough.—

We shd. indeed rejoice to see you here, whenever you can come; but you would find it dull, for I must talk but little,—yet that little would be a real enjoyment to me.—[3] I am glad to hear of your Paris scheme; for it will be enjoyable.—[4] As I suppose you will see Naudin I will enclose a memorandum of enquiry:[5] I have been for some time wishing to write to him; but scrupled. Can you put this memorandum with any other papers, so as not to forget it; you would do me *great* service, if you would master the question & interest Naudin.—

Please send comb & Bees as proposed. *What shall I do about paying Mann??*[6]

If you think of it ask the Australian Traveller whether he ever came across savages starving, & ever knew of their trying various plans to make vegetables eatible.[7] Ask him whether the Australians blush (Oh oh I forgot that they were black); it is an odd theory I have never met in notice by anyone of Expression in Savages; expression is one of my hobby-horses; I have got some funny notions on subject.—[8] 3^d. ask him whether they take any pain in breeding dogs or get them crossed with European Dogs.—[9]

And now I am going to tell you a *most* important piece of news!! I have almost resolved to build a small hot-house: my neighbours really first-rate gardener has suggested it & offered to make me plans & see that it is well done, & he is a really a clever fellow, who wins lots of prizes & is very observant.[10] He believes that we sh^d succeed with a little patience; it will be grand amusement for me to experiment with plants.—

I like to hear your notions about America; I think Asa Gray would consider them two or three degrees more atrocious than mine.[11] Slavery draws me one day one way & another day another way. But certainly the Yankees are utterly detestable towards us.— What a new idea of Struggle for existence being necessary to try & purge a government! I daresay it is very true.

By Jove you must write your book on Aristocracy—[12] I read De Tocqueville some years ago with *great* interest.[13] Your boy must give you much anxiety for the future:[14] but how good it is that he has conscientiousness; this is a whole volume to itself. What an extraordinary combination of character he has. I sh^d. think he would certainly alter. As for musical ear Emma declares it sometimes comes late; our Willy[15] had *none*, now he has a good one!

Yours affection^ly | C. Darwin

Endorsement: '/62'
DAR 115.2: 177

1 With his letter to CD of [21 December 1862], Hooker had enclosed a letter, now missing, that he had received from John William Dawson.

2 CD refers to the geology of North America. See letter from J. D. Hooker, 2 November 1862, and letter to J. D. Hooker, 4 November [1862].

3 See letter from J. D. Hooker, [21 December 1862].

4 See letter from J. D. Hooker, [21 December 1862].

5 Neither CD's 'memorandum' nor Charles Victor Naudin's reply have been found. However, a number of letters from CD to Naudin are referred to in Blaringhem 1913, p. 94 n.; one of the letters is described as being accompanied by an explanatory note in the hand of George Bentham (see *Correspondence* vol. 11, letter from J. D. Hooker, 24 January 1863), and may be the missing memorandum. CD had probably written to Naudin concerning the latter's work on plant hybridisation (see letter from C. V. Naudin, 26 June 1862). See also *Correspondence* vol. 11, letter to C. V. Naudin, 7 February 1863.

6 Gustav Mann. See letter from J. D. Hooker, [21 December 1862] and n. 6.

7 Augustus Frederick Oldfield. See letter from J. D. Hooker, [21 December 1862] and n. 5.

8 CD's work on the expression of emotions in man and animals (*Expression*) was not published until 1872; it included a discussion of 'Blushing in the various races of man' (pp. 316–22).

9 Oldfield's written answers to CD's queries were enclosed with Hooker's letter to CD of [27 or 28 December 1862], but have not been found. However, in *Variation* 2: 215, CD wrote:

> Mr. Oldfield, who has seen so much of the aborigines of Australia, informs me that "they are all very glad to get a European kangaroo dog, and several instances have been known of the father killing his own infant that the mother might suckle the much-prized puppy."

10 The reference is to John Horwood, gardener to CD's neighbour George Henry Turnbull, who superintended the building of the hot-house in late January and early February 1863 (see *Correspondence*

vol. 11, letters to J. D. Hooker, 13 January [1862] and 15 February [1863], and letter to George Henry Turnbull, [16? February 1863]). CD had recently attempted, unsuccessfully, to grow plants for experimental purposes in a heated glass case (see letter to J. D. Hooker, [21 December 1862] and n. 2).

11 See letter from J. D. Hooker, [21 December 1862]. Hooker and Asa Gray held radically different views on the American Civil War, and had tacitly agreed not to discuss the matter in their letters (see letter from J. D. Hooker, [19 January 1862], and letter from Asa Gray, 18 February 1862).

12 CD refers to Hooker's whimsical suggestion that he would write a book arguing that aristocracy was the result of natural selection (see letters from J. D. Hooker, [23 March 1862], [15 and] 20 November [1862], and 26 November 1862).

13 See letter from J. D. Hooker, [21 December 1862]. CD read Tocqueville 1836 in February 1849 (see *Correspondence* vol. 4, Appendix IV, 119: 226).

14 William Henslow Hooker. See letter from J. D. Hooker, [21 December 1862].

15 Emma and William Erasmus Darwin.

From Mary Butler [before 25 December 1862][1]

Sudbrook Park | Petersham
Wednesday

My dear Mr. Darwin

We were all relieved & made happy by Mrs. Darwins account of you and I wish that I could go to you now and have the very great pleasure of being once more amongst you all, but Lady Drysdale some time since, made me promise not to leave Sudbrook till after Christmas—[2] We are a quiet sociable party here, & the absence of even one would make some difference in the arrangements of the house. I have an interest much at heart just now, which I fear you will not be able to assist—willing to do so I am certain you will be— You remember Mr Thom—who excited yr. admiration by the several victories which he achieved over Brandy, Opium Tobacco—& himself!! he has been the steadiest of men ever since Clever, well educated, highly principled—modest!—[3] For some years he has been nominally Sub Editor of the Home News (from which Mr. Robert Bell derives the revenue) but really the sole Manager of the Paper, which has an extensive circulation, & is said to be extremely well conducted—[4]

The Sedentary life in a damp office in the City has so completely undermined his health that he is obliged to give up his employment, & has no prospect of meeting with a suitable one in this Country—so that as a last resource he is going to Queensland—at the age of 33—to spend the remainder of his days amongst Cows & Sheep (he scarcely knows one from the other) in a strange Country—where he has not even a friend; I feel for him deeply

There are many persons to whom the services of such a man would be valuable—for his abilities are excellent—& he has the highest testimonials as to character, whilst both his appearance & manner are prepossessing— It has struck me as *just possible* that you may know of some place to fit him. He has been trying to get into the Constabulary but has not the proper interest to give any hope of success— he tried for the Secretaryship to an Hospital & found that there were Six hundred Candidates!!— An Inspector of Schools he once thought of, but of that there is no chance, from the Government Interest requisite— Could you speak a good word for him in some influential quarter? he would do you no discredit I believe in any way, for he is really a superior & meritorious man?

We have felt a good deal of anxiety about M^{r}. Smyth—who was thrown *penniless* upon the kind family here—but M^{r}. Tennant of Glasgow has given him an appointment in Trinidad where he has a prospect of becoming a Planter & doing well—[5] he sailed for the West Indies a fortnight ago.— My dear M^{r}. Darwin I will not excuse myself for writing all this to you, I scarcely believe in the possibility of your having it in your power to befriend poor Mr Thom, but at all events I am assured of your most kind sympathy—[6]

My best love to M^{rs}. Darwin. I dont like the idea of your long beard. M^{r}. Davenport who is here—wears one from the same cause, but he has benefited wonderfully from the frequent use of the Turkish Bath—& is beginning to look perfectly handsome—[7]

Always Sincerely & affectionately Yours | Mary Butler

DAR 160: 392

[1] Dated by the references to Christmas, to Sudbrook Park (see n. 2, below), and to J. P. Thom's difficulties (see n. 6, below).

[2] CD first met Mary Butler at Edward Wickstead Lane's hydropathic establishment at Moor Park, near Farnham, Surrey, in February 1859; they met again in October 1859 while attending Edmund Smith's hydropathic establishment at Ilkley Wells, Otley, Yorkshire (see *Correspondence* vol. 7), and again in April 1860, when Butler visited Down House for a week (Emma Darwin's diary (DAR 242)). In 1862, Lane took over a hydropathic establishment at Sudbrook Park, Petersham, Surrey (Metcalfe 1906, pp. 56–7; *Post Office directory of the six home counties* 1862). Lady Elizabeth Drysdale was Lane's mother-in-law, and lived with him (*Emma Darwin* (1904) 2: 184). No letters from Emma Darwin to Butler have been found.

[3] CD made the acquaintance of J. P. Thom in the course of hydropathic treatment at Lane's establishment, and considered him a 'very sensible nice young man' (*Correspondence* vol. 7, letter to W. D. Fox, 24 June [1858]).

[4] Robert Bell is reported to have edited the colonial newspaper *Home news* 'with assiduity' (*DNB*).

[5] Mr Tennant and Mr Smyth have not been identified. Mr Tennant is also mentioned in the letter to Mary Butler, 11 September [1859] (*Correspondence* vol. 7).

[6] CD's Account book–banking accounts (Down House MS) records payment on 1 January 1863 of £20 under the entry 'Mr Thom Present'. See also *Correspondence* vol. 11, letter from J. P. Thom, 14 January 1863.

[7] Mr Davenport has not been identified. Emma had been encouraging CD to grow a beard (see letter to W. E. Darwin, 4 [July 1862]). The condition referred to by Butler may be eczema, from which CD had begun to suffer earlier in the year (see letter to J. D. Hooker, 30 [June 1862]).

From Francis Boott 26 December 1862

Gower St.
Dec. 26. 1862

My Dear Darwin

My faith is an orthodoxy of its own, embracing a much wider revelation than the church admits of, & the reverence I feel for those "who still sway our spirits from their irons",[1] rests with a fond complacency on you as one of Natures true interpreters. I also hold to the privilege of openly professing worship to all who have enlightened & quickened my spirit. I have not been in a church but once for the last 25 years, & I never yet comprehended the possibility of prayer beyond that emotional habit of the mind, when one contemplates the beauty of nature or the glories of genius. I place you in a nitch in my self-consecrated Temple, associated with a few of my Divinities. Your altar close to that of Linneus & Gilbert White,—& the Dear Hookers[2]—& not far from those imputed sinners Byron, Burns & Charles Lambe[3]—& those earlier Saints Milton, Shakespeare[4] & that still earlier Bard whose spirit was more simply divine than the church tries to make it.—[5] I make this confession to justify my last note to you, & you are too liberal to object to the worship of any man.[6]

But my especial object in writing is to beg you not to take the trouble of sending me any American Newspapers. If Dear Gray asks you to do so, silently appropriate them to the only use I make of them.[7] I never look at them, nor do I read a syllable about this horrid war. I cannot see any desirable issue to it or to the fate of the poor blacks. I have had a nephew shot thro' the abdomen,[8] & my last remittance was at the rate of exchange $146\frac{1}{2}$—[9] These evils I consent to, for I have no power to redress them, but I cannot read details of hopeless bloodshed—

Many happy new years to you & Mrs Darwin & your family— | Yrs sincerely | F. Boott

Charles Darwin Esqre | &c &c

DAR 160: 253

1 The source of the quotation has not been identified.
2 Carolus Linnaeus, Gilbert White, William Jackson Hooker, and Joseph Dalton Hooker. See also letter from Francis Boott, 22 December [1862].
3 George Gordon Byron, Robert Burns, and Charles Lamb.
4 John Milton and William Shakespeare.
5 The reference has not been traced.
6 See letter from Francis Boott, 22 December [1862]. CD's reply has not been found
7 Asa Gray had apparently enclosed copies of the *Boston Advertiser* with his letter to CD of 24 November 1862, requesting that he send them on to Boott. CD may have forwarded them with his reply to Boott's letter of 22 December [1862], which has not been found. See also letter to J. D. Hooker, 29 [December 1862].
8 Boott's nephew has not been identified.
9 See letter from Francis Boott, 22 December [1862] and n. 3.

To *Journal of Horticulture* [before 27 December 1862][1]

If any of your readers have kept Penguin Ducks, and will have the kindness to observe one little point, and communicate the result, I should be greatly obliged.[2] On examining the skeleton, I find that certain bones of the leg are longer than in the other breeds. I formerly kept these birds alive, and as far as I dare trust my memory, they could run considerably faster than other Ducks.[3] Is this the case? It would, perhaps, be a good way to test their running powers to call the two kinds, when hungry, from a distance to their food, and see which arrived first.—

Charles Darwin, Down, Bromley, Kent.

Journal of Horticulture, Cottage Gardener, and Country Gentleman n.s. 3 (1862): 797

[1] The letter was published in the issue of 27 December 1862.

[2] According to his 'Journal', CD prepared a draft of the section of *Variation* relating to ducks (*Variation* 1: 276–87) between 16 and 31 May 1861 (see *Correspondence* vol. 9, Appendix II).

[3] CD discussed the osteological peculiarities distinctive of penguin ducks in *Variation* 1: 281–6, noting that this was 'the most remarkable of all the breeds' of domestic duck (p. 281), and stating that the elongation of the femur and metatarsus, relative to the same bones in the wild duck, was 'no doubt connected with its peculiar upright manner of walking' (p. 284).

From J. D. Hooker [27 or 28 December 1862][1]

K⟨ew⟩
⟨ ⟩

Dr Darwin

I send Mr Oldfields answer to your questions— he has misunderstood me about *Lythrum*.[2]

Do you know that I very much agree with J. E. Gray about collecting—& its interest[3] Our greatest interest in Nat. Hist specimens is the knowledge we obtain thereby of function, affinity, distribution &c—but there is a much more catholic view of specimens than this, which none but a very few of the very ablest ⟨naturalists have e⟩ntertained ⟨*two words missing*⟩ higher interest ⟨ ⟩ & which ⟨inclu⟩des a love of all ⟨the⟩ artistic pleasure they afford, the pleasure of classyfying by eye, & not by knowledge, the historic interest attached to them, & the pleasure of being successful in obtaining, as well as the love of *possessing*, without which much of all the rest is nothing. R. Brown[4] was one of those few rare great men, who loved specimens for all that Gray loved them for, & did not despise the springs of that love; & who besides loved them for all that we should. His fault was, not that he loved too wisely, but too much, & was cursedly selfish.

Henslow was another.[5] This interest is neither less useful nor good nor worthy of cultivation because it is often perverted. it made Brown selfish; it frittered away Henslows energies, it obfuscates all Grays good qualities, & it leads to many vices; but then look at the grist it brings to the Mill-Scientific—& look on the other-hand to the effects of the love of specimens for the higher objects alone— in $\frac{99}{100}$ cases it leads to nothing, in some others to indolence, jealousy, very narrow-mindedness & as many evils of a less docile nature What a mess poor J. E. G. has made of Rowland Hill, & what a Jesuitical letter is his last;[6] how du Chaillu will chuckle.[7]

By the way—now don't despise me—I am collecting Wedgewoods simply & solely because they are pretty & I love them— I have not even a Grayan Excuse, they afford me pleasure—voila tout—

I saw Boott yesterday, first time for ages—looks more old & thin, but was in great force.[8] F. Palgrave is to be married on Tuesday, to a Miss Gaskell, who rejoices in the name Cecil;—he was author of "Passionate Pilgrim" & so I was all wrong together, in every conceivable way[9]

I should like to turn the water-spout of Herbert Spencers abstract philosophy on the subject of Nat: Selection as applied to Politics, Govt, & Society:—[10] By the way what splendid books Jeremy Bentham might have written had he clearer notions & read the Origin.[11]

I will take care of your note for Naudin.[12]

Ever yours affec | J D Hooker

Nothing to pay for Bees.[13]

DAR 101: 93–5

1 Dated by the reference to John Edward Gray's letter to the *Athenæum* (see n. 6, below), and by the mention of the forthcoming marriage between Francis Turner Palgrave and Cecil Grenville Milnes Gaskell (see n. 9, below).

2 Augustus Frederick Oldfield's communication has not been found; however, see the letter to J. D. Hooker, 24 December [1862] and n. 9.

3 John Edward Gray discussed the value and purpose of collecting in the introduction to his *Hand catalogue of postage stamps* (J. E. Gray 1862c). See Gunther 1975, p. 192.

4 Robert Brown was keeper of the botanical collections at the British Museum from 1827 to 1858.

5 John Stevens Henslow was professor of botany at Cambridge from 1825 to 1861, with responsibility for the University Botanic Garden.

6 In the introduction to his *Hand catalogue of postage stamps*, Gray had claimed to be the first to propose 'the system of a small uniform rate of postage to be pre-paid by stamps' (J. E. Gray 1862c, p. viii). Repeated in the *Athenæum*, 13 December 1862, p. 768, the claim elicited a letter of complaint from Rowland Hill, the instigator of the penny postage, who was angered by the implication that his scheme had originated with Gray (*Athenæum*, 20 December 1863, p. 806). In a letter to the *Athenæum*, 27 December 1863, pp. 845–6, Gray sought to distance himself from any such claim, and to confirm the independence, though not the priority, of Hill's scheme. The controversy continued into January 1863 with an exchange of letters in the *Athenæum* between Gray and the publisher Charles Knight. See Gunther 1975, pp. 192–4.

7 The reference is to Gray's attack during the summer of 1861 on *Explorations & adventures in equatorial Africa* by the explorer Paul Belloni Du Chaillu (Du Chaillu 1861). In a series of letters to the *Athenæum*, and in papers in the *Annals and Magazine of Natural History* and the *Proceedings of the Zoological Society of London*, Gray challenged the accuracy of Du Chaillu's account of his travels and the identification and description of the animals he collected, sparking a public controversy that spilled over into the pages of *The Times* and other periodicals. See *Correspondence* vol. 9, letter to Edward Cresy, 28 May [1861], Gunther 1975, pp. 132–4, and Rupke 1994, pp. 314–22.

8 The physician and botanist, Francis Boott, was seventy years old; he had resigned as treasurer of the Linnean Society of London in May 1861 (*DNB*).

9 Hooker's cousin, the poet and critic Francis Turner Palgrave, married Cecil Grenville Milnes Gaskell on 30 December 1862 (*Gentleman's Magazine* n.s. 14 (1863): 231). *The passionate pilgrim* was a pseudonymous romantic novel (Thurstan 1858) about which CD and Hooker had exchanged comments in 1858 (see *Correspondence* vol. 7, letter to J. D. Hooker, 12 [October 1858]).

10 Hooker refers to his whimsical suggestion that he would write a book arguing that aristocracy was the result of natural selection (see letter from J. D. Hooker, [21 December 1862], and letter to J. D. Hooker, 24 December [1862]). The sixth number of Herbert Spencer's *First principles* (Spencer 1860–2) was

published in June 1862, completing the first volume of his projected five-part series entitled 'A System of philosophy'. The volume presented Spencer's philosophy in abstract terms, and the remaining four divisions promised to deal with its application to biology, psychology, sociology, and morality. CD had subscribed to the *First principles* (see *Correspondence* vol. 8, letter to Herbert Spencer, 2 February [1860]), but the last two numbers of his unannotated copy, which is in the Darwin Library–CUL, are uncut.

11 The reference is to the utilitarian writer on jurisprudence and economics, Jeremy Bentham.

12 In his letter of 24 December [1862], CD had sent Hooker a 'memorandum of enquiry' for the French botanist and plant hybridiser, Charles Victor Naudin, whom he expected Hooker to see on his forthcoming visit to Paris.

13 See letter to J. D. Hooker, 24 December [1862] and n. 6.

To W. B. Tegetmeier 27 [December 1862]

Down Bromley Kent
27th

My dear Sir

I thank you sincerely for your letter,[1] & am heartily glad to hear of R.S. making so good a move.[2] I am, however, not sanguine of success.— The present plan is to try whether any existing breeds happen to have acquired accidentally any degree of sterility;—but to this point hereafter.[3] The enclosed M.S. will show what I have done & know on the subject. Please at some future time carefully return the MS. to me.[4] If I were going to try again, I would prefer Turbit with Carrier or Dragon.—

I will suggest an analogous experiment, which I have had for two years in my Experimental book with "be sure & try".[5] but which as my health gets yearly weaker & weaker & my other work increases, I suppose I shall never try. Permit me to add that if 5£ would cover expences of experiment, I shd. be delighted to give it & you could publish result *if there be any result*.[6] I crossed Spanish Cock (your bird) & white Silk hen & got plenty of eggs & chickens; but two of these *seemed* to be quite sterile.[7]

I was then sadly overdone with work but have ever since much reproached myself, that I did not preserve & carefully test the procretive power of these hens.— Now if you are inclined to get a Spanish Cock & a couple of *white* Silk hens, I shall be most grateful to hear whether the offspring breed well; they will prove, I think, not hardy; if they shd. prove sterile,, which I can hardly believe, they will anyhow do for the pot.—

If you do try this; how would it be to put a silk cock to your curious silky Cochin Hen; so as to get a *big* Silk breed; it would be curious if you could get silky fowl with bright colours— I believe a silk **hen** crossed by any other breed never give silky feather. A cross from Silk Cock & Cochin Silk Hen ought to give silky feathers & probably bright colours.—

I have been led lately from experiments (not published) on Dimorphism to reflect much on sterility from Hybridism & *partially* to change the opinion given in Origin.[8] I have now letters out enquiring on following point, implied in the experiment, which seems to me well-worth trying, but too laborious ever to be attempted.[9] I would ask every Pigeon & Fowl Fancier, whether they have ever observed *in the same breed*, a cock A paired to a hen B, which did not produce young. Then I would get cock A & match it to a hen of its nearest blood; & hen B to its nearest blood. I would then match the offspring of A (viz a, b, c, d, e) to the offspring of B, (viz f, g, h, i, j)—& all these children which were fertile together should be destroyed until I found, one, (say *a*) which was not quite fertile with (say *i*). Then *a* & *i* sh^d. be preserved & paired with their parents A & B, so as to try & get two families, which would not unite together; but the members **within** each family being fertile together. This would probably be quite hopeless; but he who could effect this, would, I believe, solve the problem of Sterility from Hybridism.—

If you sh^d ever hear of *individual* fowls or pigeons which are sterile together, I sh^d. be very grateful to hear of case. It is parallel case to those recorded of a man *not* impotent long living with a woman who remained childless; the husband died & the woman married again & had plenty of children. *Apparently* (by no means certainly) this first man & woman were dissimilar in their sexual organisation.[10] I conceive it possible that their offspring (if both had married again & both had children would be sexually dissimilar like their parents or sterile together.—

Pray forgive my dreadful writing; I have been very unwell all day, & have no strength to rewrite this scrawl.— I am working slowly on, & I suppose in 3 or 4 months shall be ready for M.S. of Fowls.[11]

My dear Sir | Yours very sincerely | C. Darwin

I am sure I do not know whether any human being could understand or read this shameful scrawl.—

Endorsement: 'Decr. 1862.'
New York Botanical Garden Library (Charles Finney Cox collection)

1 Tegetmeier's letter has not been found, but see the following letter.

2 On 1 December 1862, the council of the Royal Society of London resolved to grant £10 to Tegetmeier for 'experiments on the cross-breeding of pigeons' (Royal Society, Council minutes, 1 December 1862). Between 1863 and 1865, Tegetmeier carried out a series of crosses between varieties of pigeon designed to test the fertility of their hybrids; he tested the products of three generations without finding 'any trace of sterility' (see letters from W. B. Tegetmeier, 7 July 1863 (*Correspondence* vol. 11) and 13 March 1865, *Calendar* no. 4785). CD reported Tegetmeier's results in *Variation* 1: 192.

3 See n. 8, below.

4 The enclosure has not been found; it was evidently a list of the crosses between different pigeon varieties carried out by CD and others (see *Correspondence* vol. 11, letter to W. B. Tegetmeier, 19 February [1863]). The list is probably that referred to in *Variation* 1: 192 n., which CD did not think 'worth publishing'. CD began keeping pigeons for experimental purposes in April 1855, and in August 1856 he began crossing all kinds, 'to see whether crosses are fertile' (see *Correspondence* vol. 5, letter to W. E. Darwin, [25 April 1855], and *Correspondence* vol. 6, letter to W. B. Tegetmeier, 30 August [1856]); the records of his crosses in 1856 and 1857 are in DAR 205.7: 166–89. See also CD's Catalogue of Down specimens (Down House MS). Tegetmeier apparently did not return the list until 1865 (see letter to W. B. Tegetmeier, 6 March [1865], *Calendar* no. 4779, and letter from W. B Tegetmeier, 13 March 1865, *Calendar* no. 4785).

5 This reference has not been found in CD's Experiment book (DAR 157a), but see n. 7, below.

6 CD sent £5 5*s.* to cover the cost of the experiments in his letter to Tegetmeier of 9 July [1863] (*Correspondence* vol. 11). Tegetmeier carried out the crosses between 1863 and 1865 (see letters from

John Lubbock. Photograph, *c.* 1865.
(Courtesy of the College Archives, Imperial College of Science, Technology and Medicine, London.)

Hooker Glacier, New Zealand. Engraving, 1862.
(From *Illustrated Australian Mail*, 21 July 1862, p. 136.
Courtesy of the Special Collections, Baillieu Library, University of Melbourne.)

W. B. Tegetmeier, 7 July 1863 (*Correspondence* vol. 11) and 13 March 1865, *Calendar* no. 4785). He published an account of the experiments in Tegetmeier 1867, p. 224; CD reported Tegetmeier's observations in *Variation* 1: 242.

7 The reference is to crossing experiments that CD carried out in 1859 and 1860 with a male Spanish fowl provided by Tegetmeier (see *Correspondence* vol. 7, letters to W. B. Tegetmeier, 16 November [1858], 27 [November 1858], and 24 December [1858]). By crossing different varieties, CD hoped to produce a reversion to the coloration of the presumed ancestral breed. The results from these experiments are given in his Experiment book (DAR 157a), pp. 41–2, 49–50, and in *Variation* 1: 240–2, 2: 67.

8 See Appendix VI.

9 The letters referred to have not been found, but see Appendix VI.

10 CD referred to this point in his notes on hybridity (see Appendix VI).

11 In June 1861, CD had sent the manuscript of a draft chapter on domestic fowls, prepared for inclusion in *Variation*, for Tegetmeier's comments (see *Correspondence* vol. 9, letter to W. B. Tegetmeier, 14 June [1861]). In the event, Tegetmeier did not read and return the manuscript with his comments until 1865 (see letter to W. B. Tegetmeier, [7 April 1865], *Calendar* no. 4806).

To T. H. Huxley 28 December [1862][1]

Down Bromley Kent
Dec. 28th

My dear Huxley

I return enclosed: if you write thank Mr. Kingsley for thinking of letting me see the sound sense of an Eastern potentate.[2] All that I said about the little book is strictly my opinion; it is in every way excellent & cannot fail to do good the wider it is circulated. Whether it is worth your while to give up time to it, is another question for you alone to decide; that it will do good for the subject is beyond all question.[3] I do not think a dunce exists, who could not understand it; & that is a bold saying after the extent to which I have been misunderstood.

I did not understand what you required about sterility: assuredly facts given do not go nearly so far. We differ so much that it is no use arguing. To get the degree of sterility you expect in recently formed varieties seems to me simply hopeless.[4] It seems to me almost like those naturalists who declare they will never believe that one species turns into another till they see every stage in process.—

I have heard from Tegetmeier & have given him the results of my crosses of the birds which he proposes to try, & have told him how alone I think the experiment could be tried with faintest hope of success.[5] Namely to get if possible case of two birds which when paired were unproductive, yet neither impotent; for instance I had this morning I had a letter with case of Hereford Heifer which seemed to be after repeated trials sterile with one particular far from impotent Bull, but not with another Bull..—[6] But it is too long a story— it is to attempt to make two strains, both fertile, & yet sterile when one of one strain is crossed with one of the other strain. But the difficulty & distinction of the fertile individuals would be beyond calculation.—

As far as I see Tegetmeiers plan would simply test whether two existing breeds are now in any slight degree sterile; which has already been largely tested: not that I dispute good of retesting.

You must have had a very pleasant little tour in Isle of Wight & I hope it has done you a world of good.[7] We are both heartily glad of the good account of M^{rs}. Huxley[8]

I am tired—so good night | Ever yours very truly | C. Darwin

[Enclosure]

Eversley
De^{c}. 20/62.

My Dear Huxley

I send you this, as a hint of the effect on my natural theology Darwin has had on me.[9] If you think it worth, send it on to the good man, as it may please him.

"Once on a time, in Tartary, there was a jolly old heathen miscreant of a Khan, who was given to worshipping a horse's scull, & other devotions of a rudimentary nature.

And there came to him two bronzes, Moollahs, or other sort of missionaries, animated with a pious desire of converting him to their faith; but as they worshipped two different Deities, they hated each other accordingly, as in duty bound, & each believed the other was going to Gehanna.

Well. The old Kahn was frank enough with them. He confest that he had no great respect for his horse's scull; that he had totally failed in obtaining from it any rational answer, several times, when he was at a great pinch; & that on the whole, he was ready to take up with any other deity, provided the said deity was wise enough. He demanded therefore of the two Moollahs, w^{h}. of their deities was the cleverest.

Then the first Moollah said, 'Oh Khan, worship my God. He is so wise, that he made all things."

'Wah!' said the Khan 'him a great sultan. He is a wise builder.' But what can thy God do, oh Moollah number two?'

Then said the second Moollah, 'Oh Khan, it is a light thing for a God to make all things. A God who could not do that would not be good enough for a Samoiede who eats blubber, or a Tom-goose who digs mammoth bones. May their mothers graves be defiled! But, Oh Kahn, my God is a God indeed; *For he is so wise, that he makes all things make themselves*."

"'Wah Wah! said the Kahn. "He is the Sultan of all sultans; He is the wisest of all Master-builders. He is the God for me henceforth, if he be wise enough to make things make themselves."

Verbum sat sapienti.[10]

Imperial College of Science, Technology and Medicine Archives (Huxley 5: 189, 19: 209–12)

1 The year is established by the reference to T. H. Huxley 1862c (see n. 3, below).

2 See enclosure. Huxley had apparently sent Charles Kingsley's letter with a letter to CD that has not been found.

3 T. H. Huxley 1862c. See letters to T. H. Huxley, 7 December [1862] and 18 December [1862]. The lectures were reproduced, unaltered, in Huxley's *Collected essays* (T. H. Huxley 1893–4, 2: 303–474).

[4] See letter to T. H. Huxley, 18 December [1862]; Huxley had evidently answered CD's objections in a letter that has not been found.
[5] The letter from William Bernhard Tegetmeier has not been found, but see the preceding letter.
[6] The letter has not been found, but see the letter to W. B. Tegetmeier, 28 [December 1862].
[7] Huxley had evidently told CD of this trip in a letter that has not been found.
[8] Henrietta Anne Huxley.
[9] Kingsley outlined his view of the significance for natural theology of the theory of evolution by natural selection in the preface to his *Westminster sermons* (Kingsley 1874, pp. v–xxxii). He entreated his readers to study *Orchids* if they wished to see how little evolution by natural selection affected 'the old theory of design, contrivance, and adaptation' (p. xxv), and asserted that, whether or not Darwin's theory was true, *Orchids* was 'a most valuable addition to natural Theology.' He repeated the view expressed in this letter in the following words (p. xxvii): 'We knew of old that God was so wise that He could make all things: but, behold, He is so much wiser than even that, that He can make all things make themselves.'
[10] 'A word is enough for a wise man' (H. P. Jones ed. 1900, p. 123).

To Thomas Rivers 28 December [1862][1]

Down. | *Bromley.* | *Kent. S.E.*
Dec. 28th

My dear Sir

Permit me to thank you cordially for your most kind letter.[2] For years I have read with interest every scrap which you have written in Periodical & abstracted in M.S. your book on Roses,[3] & several times I thought I would write to you, but did not know whether you would think me too intrusive. I shall indeed be truly obliged for any information you can supply me on bud-variation or sports.[4] When any extra difficult point occurs to me in my present subject (which is a mass of difficulties), I will apply to you; but I will not be unreasonable. It is most true what you say that any one to study well the physiology of the life of plants, ought to have under his eye a multitude of plants. I have endeavoured to do what I can by comparing statements by many writers & observing what I could myself. Unfortunately few have observed like you have done. As you are so kind I will mention one other point on which I am collecting facts, namely the effect produced on the stock by the graft; thus it is *said* that the purple-leaved Filbert affects the leaves of the common hazel on which it is grafted (I have just procured a plant to try) so variegated Jessamine is *said* to affect its stock.[5] I want these facts partly to throw light on the marvellous Laburnum Adami—Trifacial oranges &c. That Laburnum case seems one of the strangest in physiology: I have now growing splendid, *fertile* yellow Laburnums (with long racemes like the so-called Waterer's Laburnum) from seed of yellow flowers on the L. adami.—[6]

I do so wish I could accept your invitation; I will see in Spring what I can do; but I suffer severely from ill-health of a very peculiar kind, which prevents me from all mental excitement, which is always followed by spasmodic sickness, & I do not think I could stand conversation with you, which to me would be so full of enjoyment. I send a little pamphet on a subject on which I am at work, & on which I shall soon publish some much more striking cases.— These cases show how little we yet know on fertilisation.—[7]

To a man like myself who is compelled to live a solitary life & sees few persons, it is no slight satisfaction to hear that I have been able at all interest by my books, observers like yourself.—

As I shall publish on my present subject, I presume within a year, it will be of no use your sending me the shoots of Peaches & nectarines which you so kindly offer; I have recorded your facts.—[8]

Permit me again to thank you cordially; I have not often in my life received a kinder letter.— | My dear Sir | Yours sincerely | Ch. Darwin

Sotheby Sale (23–4 July 1987)

[1] The year is established by the relationship to the letter to Thomas Rivers, 23 December [1862].

[2] Rivers's letter has not been found.

[3] CD's reading notebooks record that he read Rivers 1837 in 1840 (see *Correspondence* vol. 4, Appendix IV, *119: 22v and 119: 8a); however, his abstract of the work has not been found. Rivers regularly contributed to horticultural journals like the *Gardeners' Chronicle and Agricultural Gazette* and *Journal of Horticulture* (*DNB*). There is also a copy of Rivers 1848 in the Darwin Library–CUL.

[4] See letter to Thomas Rivers, 23 December [1862]. Rivers's answer to this letter has not been found; however, see *Correspondence* vol. 11, letter to Thomas Rivers, 7 January [1863].

[5] Rivers's reply has not been found. However, his observations on the cases described by CD, and on one other case of this type, are cited in *Variation* 1: 394–5. See also *Correspondence* vol. 11, letter to Thomas Rivers, 7 January [1863].

[6] CD had for many years been interested in the origin of the hybrid laburnum, *Cytisus Adami*, branches of which had reportedly been found to bear the leaves and flowers of both the parent species (*C. purpureus* and *C. laburnum*, the common laburnum) in addition to those of the hybrid form (see *Correspondence* vols. 4–6). In *Variation* 1: 387–97, CD discussed at length the evidence for and against the different theories that had been advanced to account for the hybrid form, dwelling particularly on the possibility that it was either an ordinary hybrid, formed by seed, or that it was what he called a 'graft-hybrid'. He cited the observations referred to here in *Variation* 1: 388. The analogous case of the trifacial orange is discussed in *Variation* 1: 391.

[7] CD refers to 'Dimorphic condition in *Primula*', although Rivers's name does not appear on CD's presentation list for this paper (see Appendix III). CD also refers to his experiments in 1862 on dimorphism in species of *Linum* and trimorphism in *Lythrum salicaria*, and to his further experiments with dimorphic species of *Primula*. His results from these experiments were given in 'Two forms in species of *Linum*', and 'Three forms of *Lythrum salicaria*', which were published in 1863 and 1864, respectively, and 'Illegitimate unions of dimorphic and trimorphic plants', which was not published until 1868 (*General index to the Journal of the Linnean Society*, p. vi).

[8] In his letter to Rivers of 23 December [1862], CD had asked for accounts of what he called 'bud-variations', citing in particular the example of a nectarine appearing on a peach tree. CD discussed such examples in *Variation* 1: 374–5, but he did not cite information from Rivers; however, he did include a number of observations by Rivers on the relationship between almonds, peaches, and nectarines in *Variation* 1: 338–40 (see also *Correspondence* vol. 11, letter to Thomas Rivers, 11 January [1863], 15 January [1863], and 17 [January 1863]). *Variation* was not published until 1868.

To W. B. Tegetmeier 28 [December 1862]

28th

My dear Sir

You will see in *next* Nor. of Cottage Gardener, or in one subsequent, a question about the running powers of Penguin Ducks.—[1] Could you aid me in getting this

Query answered by some private individual; or by its insertion in the Field, which I do not take in.—[2]

This morning I had letter from a great squire & breeder who tells me of a Hereford Heifer put *several* times to one of his Bulls (a *sure* calf-getter) & never produced, but now she has been put to another Bull has conceived.[3]

I shall beg him to save calf & put her to the 1st. Bull; & see if special sterility has descended. but with these big animals, years must elapse—

Yours very sincerely | C. Darwin

Endorsement: 'Decr 1862'
New York Botanical Garden Library (Charles Finney Cox collection)

1 See letter to *Journal of Horticulture*, [before 27 December 1862].
2 Tegetmeier was a regular contributor to the natural history columns of the *Field* (Richardson 1916, pp. 140–3); he communicated a query to the journal on the point raised by CD, which was published on 3 January 1862 in the section entitled 'Notes and questions on natural history' (*Field* 21 (1863): 29).
3 This letter has not been found. In *Variation* 2: 162, CD acknowledged communications on this phenomenon from three individuals who were landed gentry: namely Thomas Campbell Eyton, William Waring, and 'Mr. Wicksted' (probably Charles Wicksted). The letter was probably from Eyton, who was the head of a renowned Shropshire land-owning family, and a childhood friend of CD's. Eyton was an authority on Hereford cattle (*DNB*), and author of two volumes of the *Herd book of Hereford cattle* (Eyton 1846–53).

From J. D. Hooker [before 29 December 1862][1]

P.S.[2] I am very glad to see the Sat. Review of D. of Argylls review,[3] it is *perfect*, & if S.R. always wrote like that instead of *out-corresponding* the Times Correspondent, or indulging in flatulent & pompous language I would take it in.

I read the Edinb Review, & wondered; of course I saw that the Duke had strained at a gnat & swallowed a Camel in accepting the superhuman & rejecting the supernatural.—or rather had found the desiderated difference between *tweedledum* & *tweedledee*, but I could not have convicted him so neatly & concisely as the Saturday has.[4]

But what on earth has your orchid book to do there— he evidently wrote 2 reviews, clipped them both, & made a most uncommon bad *one* out of them[5]

Incomplete
DAR 101: 85

1 Dated by the reference to CD's having sent Hooker a copy of the issue of the *Saturday Review* containing [Parker] 1862 (see n. 3, below).
2 It is not known whether this letter was the postscript to one of the extant letters from Hooker, or to a letter which has not been found (see letter to J. D. Hooker, 29 [December 1862]).
3 Hooker refers to [Parker] 1862, published in the *Saturday Review* on 15 November 1862, which criticised the views advanced by George Douglas Campbell, eighth duke of Argyll, in [G. D. Campbell] 1862. It is not known when CD sent Hooker the magazine, but see the letter to J. D. Hooker, 29 [December 1862], and the letter from J. D. Hooker, [31 December 1862].

[4] In his article, Argyll argued that miracles were not supernatural, since they did not involve the suspension or violation of the laws of nature, but that they were superhuman, since natural laws were in such cases applied by the divine will in a way that was analogous to, but more powerful than, human contrivance. In his review, Henry Parker argued that Argyll's distinction did not hold true, since it was 'simply to avoid ambiguity' that the word supernatural was not also applied to 'the agency of man as standing in opposition to nature in its narrower sense' ([Parker] 1862, p. 590). He asserted that God's 'coming in and applying natural laws to effect a particular purpose, just as men apply them', still constituted a violation of the laws of nature, and that without 'the notion of a special interposition', however remote, there could be no such thing as a 'true miracle'.

[5] [G. D. Campbell] 1862 reviewed *Orchids* in conjunction with five works on miracles and the supernatural.

To Hugh Falconer 29 December [1862][1]

Down
Dec. 29th.

My dear Falconer

Heaven only knows when I shall be able to come to London.[2] If I could I would not have troubled you with this query, which I send for the *chance* of your having any facts out of the stores of your knowledge. Have you met with any cases of what gardeners call "sports" and what I shall call "*bud-variations*";[3] i.e., a bud suddenly assuming a new character, such as a moss-rose on a provence, or a nectarine on a peach tree. I much wish to collect all authentic cases. I especially ask you because Sir R. Schonburgk (no good authority) states that such bud-variations occurred rather often with flowers from warmer temperate regions grown in hot St. Domingo.[4] Can you aid me? Little or great changes would be all gratefully received. There are two or three other points on which I want to talk with you; so that whenever I can get to London, I must beat up your quarters.

My dear Falconer | Yours most truly | Ch. Darwin

Copy
DAR 144: 28

[1] The year is established by the reference to '*bud-variations*' (see n. 3, below).

[2] Falconer had for some time been anxious to see CD, having failed to do so when CD visited London in September (see letter to Hugh Falconer, 14 November [1862] and n. 11).

[3] According to his 'Journal' (Appendix II), CD began preparing a draft of chapter 11 of *Variation*, 'On bud-variation, and on certain anomalous modes of reproduction and variation', on 21 December 1862.

[4] Schomburgk 1857, p. 132.

From Asa Gray 29 December 1862

Cambridge. [Massachusetts]
Dec. 29, 1862

My Dear Darwin

A happy New Year to you and yours,—and another stamp for Leonard![1] Also some corn (maize) of a few distinct sorts.[2]

To-day a student of Agassiz (entomological)[3] brought up to show me a small butterfly received from Canada, "with a singular disease" The student, who last summer studied your book and made some observations,—of course detected the nature of the disease. viz—a pollinium of *Platanthera* I think *Hookeri*, neatly affixed by its disk to *each eye*;—and both had lost some pollen-packets from the upper part; showing it had done work before being captured.

It is a *day-flyer*. I judge from structure it is either P. Hookeri or P. orbiculata, and from size rather the former.— And the head is just about the size to catch both discs of the former—as it has done—but to small to have done that with P. orbiculata.

The young man will give me the name of the butterfly in a few days,—when you shall have it.[4]

I thank you heartily for your long and interesting letter of Nov. 23.–26.[5] I have not had the time nor the spirits to write to you,—and have no time now, tho' the spirits are better.

Edinb. Review & Argyle's article has not come to me yet.[6] When it does I shall read it with interest; I heard it well spoken of by a very fair and good judge the other day.—who had no idea who wrote it.

McMillan I do not see, without going to libraries to look it up.[7] As Agassiz does nothing at ⟨*section missing*⟩

I do not at all object to your criticisms on my *Dimorphous* notes in Sill. Jour.[8]

My object was not so much to commend the terms I had used, as to note that we had here long ago observed (tho' we did not comprehend the meaning of) these facts.—and to say that I thought the terms not bad even now. I know nothing of Sprengel's *'dichogamy'*.[9] Where?

"Dichogamous" then every way better, as well as older term.— extend the meaning and use it.

As to Plantago, my words "closed corolla" show it was the short-stamened & long-styled I was speaking of, as that alone *closes the corolla* But the whole of that 2^{d} paragraph was introduced in proof, and as short as possible—hence partly the obscurity.— I saw no sense, and the printers as you see made bad work with it.[10]

"Precocious fertilization" I incline to stand by.—[11] In Violets & Impatiens, Lespedeza, &c &c— I believe that the condition is not so much "special modification" as it is *arrest of development*.

While it is the earlier flowers in Abronia, Nyctaginia, Pavonia hastata, Ruellia &c[12]—that are thus fertilized in the bud,—and partly so in Impatiens, in Viola sagittata, cucullata, rotundifolia, &c— the vernal flowers are the ordinary,—but the peculiar closed ones are produced all summer long.

(I cant tell you about *Specularia*, as I could get no plants of it last summer.[13]

The main object of my note was to correct Oliver, who, I suspect, took his cue from Hooker—who, I believe, now came to understand this matter as you and I do.—[14]

We have, at least, a clear & consistent notion about it, and I believe it is a correct one.

The various suggestive matters in your long letter I must let drop— —only answering your question in P.S.—about Fragaria[15]

Incomplete
DAR 109: 85, DAR 165: 126

CD ANNOTATIONS
0.3 My Dear ... sorts 1.2] *crossed ink*
2.1 To-day] *after opening square bracket, ink*; 'Orchids' *ink*; 'Platanthera' *red crayon*
5.1 I thank ... nothing at 7.2] *crossed ink*
8.1 I do ... use it 10.2] *crossed ink*
13.1 While ... &c 13.2] *cross in margin, red crayon*
15.1 The main ... Fragaria 17.2] *crossed ink*
Top of letter: 'Return this'[16] *ink*

[1] Since June 1862, Gray had regularly sent North American postage stamps for Leonard Darwin's collection.
[2] Gray had promised to send seeds of several North American varieties of maize in response to CD's request for specimens (see letter to Asa Gray, 6 November [1862], and letter from Asa Gray, 24 November 1862).
[3] Gray refers to Louis Agassiz, professor of natural history at Harvard University, and Samuel Hubbard Scudder (see *Correspondence* vol. 11, letter from Asa Gray, 27 January 1863).
[4] See the enclosure to the letter from Asa Gray, 27 January 1863 (*Correspondence* vol. 11).
[5] Letters to Asa Gray, 23 November [1862] and 26[–7] November [1862].
[6] [G. D. Campbell] 1862; see letter to Asa Gray, 23 November [1862] and n. 7.
[7] Gray refers to a review of Max Müller 1861 in *Macmillan's Magazine* ([J. Wedgwood and F. J. Wedgwood] 1862), which CD had recommended to him (see letter to Asa Gray, 23 November [1862] and n. 8).
[8] In his letter to Gray of 26[–7] November [1862], CD had expressed a fear that his extensive criticisms of the botanical terminology used in A. Gray 1862e would cause Gray to think him 'in the most unpleasant, contradictory, fractious humour'.
[9] C. K. Sprengel 1793. See letter to Asa Gray, 26[–7] November [1862] and n. 9.
[10] See letter to Asa Gray, 26[–7] November [1862] and n. 10, and A. Gray 1862e, p. 419.
[11] See letter to Asa Gray, 26[–7] November [1862] and n. 12.
[12] CD enclosed this letter with his letter to J. D. Hooker of 13 January [1863] (*Correspondence* vol. 11), asking him to rewrite the names of the four genera listed here (see *Correspondence* vol. 11, letter from J. D. Hooker, [15 January 1863]).
[13] See letter to Asa Gray, 26[–7] November [1862] and nn. 18 and 19.
[14] Joseph Dalton Hooker and Daniel Oliver (see letter to Asa Gray, 26[–7] November [1862] and nn. 20–2).
[15] See letters to Asa Gray, 23 November [1862] and n. 12, and 26[–7] November [1862] and n. 35.
[16] See n. 12, above.

To J. D. Hooker 29 [December 1862]

Down
29th

My dear Hooker

The spirit moves me to write a line or two, though I have nothing particular to say & what is very strange nothing to ask. I heartily wish you joy that Wellwitschia is finished & no doubt it will be a great work.[1] Your Scotch simile made me laugh;[2] but by the Lord how true it is: I have often said that when this or that job is finished, I will give myself a holiday; but alas & alas the days of even half-holidays are over with me;—a holiday is simply an unendurable bore.—

The Genera plantarum are reviewed in the last Parthenon, by some one who thinks you have neglected the treasures of the B. Museum.[3] The Reviewer thinks you a disgraced man; for he says you are "notoriously tinged with Darwinism".

Here is an odd chance; my nephew Henry Parker, an Oxford classical swell & Fellow of Oriel came here this evening; & I asked him whether he knew who had written the little article in the Saturday, smashing the Duke of A. which we liked;[4] & after a little hesitation he owned he had. I never knew that he wrote in Saturday; & was it not an odd chance? I sh^{d}. like now to read the Saturday again, & wish I had thought of asking you to return it.—[5]

I have just received a cargo of orchids & Mitchella in excellent state from Asa Gray.[6] He sends me American newspapers which I never read, & tells me to forward them to D^{r}. Boott, who writes to me not to send them, as he won't read them, but that I must not say so to Gray;[7] & what the deuce I am to do to stop the poor fellow having trouble of posting them, I know not.—

Ever yours affect | C. Darwin

Endorsement: '/Dec/62'
DAR 115.2: 175

1 J. D. Hooker 1863a. See letter from J. D. Hooker, [21 December 1862], and n. 2, below.
2 The reference has not been identified. However, it seems likely that Hooker made this comment in a letter that is now missing, written in answer to the letter to J. D. Hooker, [21 December 1862]; the incomplete letter from J. D. Hooker, [before 29 December 1862], may be the postscript to such a letter.
3 The anonymous review of Bentham and Hooker 1862–83 appeared in the *Parthenon*, 27 December 1862, pp. 1102–3.
4 [Parker] 1862 and [G. D. Campbell] 1862. See letter from J. D. Hooker, [before 29 December 1862] and nn. 3 and 4.
5 CD had sent Hooker his copy of the issue of the *Saturday Review* containing [Parker] 1862 (see letter from J. D. Hooker, [before 29 December 1862]).
6 See letter from Asa Gray, 9 December 1862, and letter to James Anderson, 23 December [1862].
7 See letter from Asa Gray, 24 November 1862, and letter from Francis Boott, 26 December 1862.

To G. H. K. Thwaites 29 December [1862]

Down. | Bromley. | Kent. S.E.
Dec. 29th

My dear Mr Thwaites

Can you give me any cases of what some gardeners call "Sports", but what I shall call "bud-variations", ie when a leaf-bud assumes suddenly a new character—like moss-rose on Provence &c.— I sh^{d} be very grateful for any authentic cases.—[1] Why I more particularly ask you, is that Sir R. Schomburghk says that at St Domingo, introduced flowers from the warmer temperate regions were there particularly apt to sport in this manner.[2]

I wrote some time ago to you about Cinchona being dimorphic like Primula;[3] I have now better reason to believe that this is the case, & I I have proof that some of these dimorphic plants are *absolutely* sterile with their own-form pollen.—[4]

You mentioned two Ceylon genera (I have your note but names forgotten; one was allied to Menyanthes)[5] Would you haye kindness to examine them & see that *both* forms produce pollen & if you will, compare size of pollen. Or if you prefer could you send me the two forms of these genera in a letter, well dried, but not much pressed & named & then I could compare by soaking the pollen. The flowers sh[d] not be old.[6]

I believe that you will forgive me for troubling you, & I remain | Yours very sincerely | Ch. Darwin

I am hard at work at a book on "Variation under Domestication".[7]

Endorsement: '1862'
Cleveland Health Sciences Library (Robert M. Stecher collection)

1 According to his 'Journal' (Appendix II), CD began working on chapter 11 of *Variation*, 'On bud-variation, and on certain anomalous modes of reproduction and variation', on 21 December 1862.
2 Schomburgk 1858.
3 See letter to G. H. K. Thwaites, 15 June [1862].
4 See letter to Asa Gray, 14 July [1862] and nn. 11 and 12.
5 See letter from G. H. K. Thwaites, 15 May 1862, and letter to G. H. K. Thwaites, 20 June [1862].
6 See *Correspondence* vol. 11, letter from G. H. K. Thwaites, 17 February 1863, and letter to G. H. K. Thwaites, 30 March [1863].
7 CD had begun work on *Variation*, the first part of a projected three-part work on natural selection, in January 1860 (*Correspondence* vol. 8); it was not published until 1868.

From J. D. Hooker [31 December 1862][1]

Kew
Wednesday

D[r] Darwin

It is rather jolly this writing about matters non-scientific— let's give up Science when you have done the 3 vols & take to gossip.[2] I quite agree with you that a holiday is an unendurable bore,[3] but depend on it that is because we have no vices to indulge in, & if you will only join me in some good vice, such as talking about & writing about what will do no good to our neighbours & some harm to ourselves—we shall get on capitally, & scratch away. As luck would have it—I put aside the Ducal critique for a more careful reading with the Article itself.— I read the article itself, & in turn forgot I had put aside the smasher, which by a curious coincidence I stumbled upon **yesterday**, the day you wrote to me! so here it is. I congratulate you on so clever logical & acute a relative—[4] Have I not met him at Down? many years ago.

I saw Mr Froud[5] yesterday for first time at my Cousins wedding— what a singularly magnetic man he is to look at & talk (2 words) to; I think Frank P. has married a nice girl of a nice family.[6]

I have a great mind to send the Parthenon a s⟨creed⟩[7] —The only 7 times I ap⟨plied⟩ for information at b⟨ritish museum⟩ I got *none whatever* ⟨and on one⟩ occasion the attendant could not find the Natural Order *Cruciferæ*!— The last time I went I found the invaluable plants of *Loureiro*,[8] the only authentic scraps for identifying genera which it is impossible to make out by descriptions, in such a state of dirt, disorder & confusion that I came away determining never to try there again.

Ever Yrs affec | J D Hooker

DAR 101: 96–7

[1] Dated by the relationship to the letter to J. D. Hooker, 29 [December 1862] (see nn. 3 and 4, below), and by the reference to the wedding of Hooker's cousin, Francis Turner Palgrave (see n. 6).

[2] *Variation* was intended to be the first part of a three-volume work on natural selection (see *Correspondence* vol. 7, letter to T. H. Huxley, 16 December [1859], and *Correspondence* vol. 8, letter to Asa Gray, 28 January [1860]).

[3] See letter to J. D. Hooker, 29 [December 1862].

[4] See letter to J. D. Hooker, 29 [December 1862] and n. 5. Hooker refers to a critique of George Douglas Campbell's review of *Orchids* ([G. D. Campbell] 1862), published in the *Saturday Review* by CD's nephew, Henry Parker ([Parker] 1862). CD had sent Hooker his copy of the critique before realising its authorship.

[5] James Anthony Froude.

[6] Hooker refers to the marriage of his cousin, F. T. Palgrave, to Cecil Grenville Milnes Gaskell on 30 December 1862 (see *Gentleman's Magazine* n.s. 14 (1863): 231).

[7] Hooker refers to a review, published in the *Parthenon*, 27 December 1862, pp. 1102–3, of part 1 of Bentham and Hooker 1862–83 (see letter to J. D. Hooker, 29 [December 1862]). The reviewer criticised the authors for having neglected the materials in the British Museum.

[8] The reference is to João de Louriero, whose herbarium had been acquired by the British Museum (*Taxonomic literature*).

APPENDIX I
Translations of letters

From Imperial Zoological Society of Acclimatisation[1] 10 January 1862

Paris,
10 January 1862

Sir and dear colleague,

I have the honour of informing you that the Imperial Zoological Society of Acclimatisation will be meeting on Friday the 14th of February at 3 o'clock precisely, in 19, rue de Lille, and will proceed with the annual elections of the board and of a third of the Council.[2]

According to regulations, absent members are invited to send their ballot to the General Secretary at the Society headquarters before the ⟨ ⟩ February (see opposite).

Please accept, sir and dear colleague, the expression of my highest esteem.

The General Secretary, | Count d'Éprémesnil.[3]

DAR 96: 11 v.

[1] For the transcription of this letter in its original French, see pp. 10–11.

[2] The Société d'Acclimatation had as its aim the acclimatisation of exotic species of plants and animals to Europe (Geoffroy Saint-Hilaire 1857, Osborne 1987). The society was founded by Isidore Geoffroy Saint-Hilaire in 1854. CD was admitted to membership at the meeting of the society held on 17 February 1860 (*Bulletin de la Société Impériale Zoologique d'Acclimatation* 7 (1860): 100).

[3] Jacques Louis Raoul Duval d'Eprémesnil.

From H. G. Bronn[1] [before 11 March 1862][2]

⟨ ⟩
⟨ ⟩

Most honoured Sir,

After having been strenuously occupied last year with zoological work, a letter I received from Herr Schweizerbart brought me back to natural selection, since he informs me that the German edition of the "Origin of Species" (which, by the way, was not very large) will be out of print within the year,[3] so that a new edition would have to be arranged. Permit me therefore to address the humble inquiry to you whether a *new* English edition is also imminent, and, if so, whether it will contain changes and additions—or whether (even if such is not the case) you would like to make alterations and additions to the German edition.[4] Perhaps the larger work in several volumes, which was to follow the smaller one, is also about to appear?[5]

I have read about a dozen reviews of your book in German, Dutch, English, and American journals, some favourable and many unfavourable. However, they have not altered *my own* opinion.[6]

1) I see in your theory the only natural route to the final solution of the enigma of creation.

2) Yet, your theory contradicts the *current* state of our knowledge of the formation of *organic matter* from inorganic, of the *vitality* of organic matter and its organisation into organic *form* **without** *previous influence*

3). Our knowledge in this regard, however, is negative, and it *cannot be maintained* that future discoveries will not give us the kind of positive knowledge needed to be able to accept your theory. Even if your theory cannot be accepted at present, one cannot reject it *for all time*!

Among the many opinions of your **theory**, I have not found one that agrees with my own, but I consider mine correct. All judgments about your **book**, however, were favourable

Not long ago you sent me an article on Primula, for which I thank you cordially.[7] I have read it with much interest, just as I generally study carefully all of the observations and phenomena connected to your theory. Many stimulate considerations that I might otherwise have overlooked.

Please accept the assurance of my highest esteem. I have the honour to remain | your humble servant

H. G. Bronn

DAR 160.3: 319

[1] For the transcription of this letter in its original German and CD's annotations, see pp. 109–11.

[2] Dated by the relationship to the letter to H. G. Bronn, 11 March [1862].

[3] Bronn had translated *Origin* into German soon after its publication in November 1859. The German edition (Bronn trans. 1860) was published in Stuttgart by the firm of E. Schweizerbart'sche Verlagsbuchhandlung, under the direction of Christian Friedrich Schweizerbart. See *Correspondence* vol. 8, letter to T. H. Huxley, 2 [February 1860], and letters to H. G. Bronn, 4 February [1860] and 5 October [1860], and letter from H. G. Bronn, 13 [or 15] October [1860].

[4] The first German edition of *Origin* (Bronn trans. 1860) had been translated from the second English edition (see *Correspondence* vol. 8, letter to H. G. Bronn, 4 February [1860], and this volume, letter to H. G. Bronn, 25 April [1862]). A third English edition was issued in April 1861, and it was from this that the second German edition (Bronn trans. 1863) was prepared (see letter to H. G. Bronn, 11 March [1862]).

[5] CD had originally planned to publish his account of the origin of species by means of natural selection as a multi-volume treatise (see *Correspondence* vols. 6 and 7), but was persuaded to postpone this project in favour of producing an 'abstract' of his material, *Origin* (see, for example, *Correspondence* vol. 7, letter to Charles Lyell, 18 July [1858]). The only part of the projected larger work that was published during CD's lifetime was *Variation* (1868).

[6] For a list of the reviews of *Origin* that appeared in 1860, see *Correspondence* vol. 8, Appendix VII. Bronn included a final chapter in his translation of *Origin* that presented his own views of CD's theory (Bronn trans. 1860, pp. 495–520). See also Bronn 1860a, and *Correspondence* vol. 8, letter to

H. G. Bronn, 14 July [1860], and letter from H. G. Bronn, 13 [or 15] October 1860. For a discussion of Bronn's attitude towards CD's theory, see Junker 1991.

[7] The reference is to CD's paper, 'Dimorphic condition in *Primula*'. Bronn's name is included on CD's list of people to whom presentation copies of the paper were to be sent (see Appendix III).

From H. G. Bronn[1] 27 March 1862

Heidelberg
27 March 1862

Dear Sir,

I am very much obliged for your favourable reaction and kind answer to my last letter,[2] in which you write that you are willing to support the new translation of your "Origin of Species" with contributions, if you could be granted some time to do so. This can easily be done, for I myself cannot begin with the work before the first of May— by the way, it would not be very difficult and should not take up much time, as long as I know which passages in the original version have been amended since the first translation.

My main reason for writing to you now is to inform you that Mr Schweizerbart would also be ready to translate your work on the fertilisation of orchids,[3] and therefore would be grateful if you would be good enough to obtain for him copies of the pertinent illustrations, or the blocks with the woodcuts, or the *plates*. I was also pleased to see from your letter that you are still pursuing your larger work.[4]

In asking you kindly to accept the assurance of my sincere esteem, I have the honour to remain your most obedient servant

H. G. Bronn

DAR 160.3: 320

[1] For the transcription of this letter in its original German, see pp. 138–9.

[2] See letter from H. G. Bronn, [before 11 March 1862], and letter to H. G. Bronn, 11 March [1862].

[3] In the letter to H. G. Bronn, 11 March [1862], CD had asked whether Christian Friedrich Schweizerbart, head of the publishing firm E. Schweizerbart'sche Verlagsbuchhandlung, might be interested in publishing a translation of *Orchids*.

[4] See letter to H. G. Bronn, 11 March [1862] and n. 5.

From H. G. Bronn[1] 19 May 1862

Heidelberg
19 May 1862

Sir,

I have had the honour of receiving the packets that you were kind enough to send me either directly or through Mr Schweizerbart;[2]—and I thank you sincerely not only for the additions to the latest English edition of your Origin of Species, but even more especially for the manner in which you made it easy to find the pages and the lines where a change or an addition is to be made in the new German edition, which will be going to press any day now.[3] I must however comment on a passage in the latest English edition that is not in the previous one.

Pag. 142: Geology shows us, at least within the whole imm⟨ense⟩ tertiary period, that the number of species of shells and probably, *of mammals*, has not greatly or at all increased.

Is there not some printing error there? The ⟨ ⟩ do not exist at the beginning of the tertiary period ⟨ ⟩ number and only a few orders; they become more numerous such that in the miocene stratum they appear to be even more varied and more abundant than today. At least the number of species found together in the same bed at the same locality (at Mayance, Sansan etc)[4] is sometimes greater than that to be found nowadays in the whole of Europe. There is no doubt that the number of mammals, of which only about a dozen species are known in the pre-tertiary beds, has increased considerably perhaps more than any other class precisely in the tertiary period. I beg you therefore, my dear Sir, to let me know how this passage should be understood?[5]

The pages of your book on the Orchids made me realize that it will be of great importance for science and that it will provide new facts relating to a family of plants of the greatest interest, the interpretation of which it will revise.[6] Though perhaps not all readers of the "Origin" would be sufficiently botanically minded to be interested in it, there will certainly be a great number of botanists who do not possess a copy of the Origin but who will buy the Orchids. This is what I wrote to Mr. Schweizerbart when I sent him the proofs; however I have asked him also to consult a genuine botanist, such as, for example, Mr. *v. Mohl*.[7] Afterwards he will have to make an account of his expenses; and I think that he will himself inform you of his decision within no more than 18–24 days—since I have just been informed that he has left for the Leipzig fair, a journey that every publisher must make every year and which lasts 2 to 3 weeks.[8]

I have told you that I do not support your theory because, in spite of all the advantages it would have for science, it is still in opposition to some fundamental facts of science (the first development of an animal from inorganic matter), but nevertheless I am convinced that in the long run *it* will lead us on the path of truth.[9] It is possible that, if I have enough time, I will add to the translation some interesting observations on the variability of species with respect to some characters that are believed to be the most constant.[10]

Please accept, Sir, the expression of my highest esteem, with which I have the honour, Sir, of being your obedient servant

H. G. Bronn

DAR 160.3: 321

[1] For the transcription of this letter in its original French and CD's annotations, see pp. 212–14. Although Bronn had previously corresponded with CD in German, he began writing in French after learning that CD found it more difficult to read German than French (see letter to H. G. Bronn, 11 March [1862]).

[2] See letter to H. G. Bronn, 25 April [1862]. Christian Friedrich Schweizerbart was the head of E. Schweizerbart'sche Verlagsbuchhandlung, the publishing firm that published the German translation of *Origin*, and was preparing to publish a translation of *Orchids*.

[3] Bronn was preparing a second German edition of *Origin* (Bronn trans. 1863) from the third English edition. The manuscript alterations and corrections sent by CD with his letter of 25 April [1862] have not been found. However, those changes incorporated in the second German edition of *Origin* that do not occur in the third English edition are given in Appendix VIII. Bronn trans. 1863 was initially published in three parts, whose publication was announced on 6 October, 17 November, and 19 December 1862, respectively (*Börsenblatt für den Deutschen Buchhandel* 29 (1862): 2083, 2447, and 2735).

[4] Mayence is the French name for the German city of Mainz. Bronn refers to the rich Miocene fauna found at both Eppelsheim, a village south of Mainz, and at Sansan, a village near Auch in the south of France (see Bronn 1858, p. 473).

[5] No reply from CD to this query has been found, and in Bronn trans. 1863, p. 152, the sentence is translated without modification. However, Bronn added a footnote that reads: 'Bekanntlich hat sich die Säugthier-Welt fast ganz erst im Laufe der Tertiär-Zeit entwickelt.' [As is commonly known, mammals developed almost exclusively during the Tertiary period.] CD annotated the sentence in his copy of the third edition of *Origin*, and in the fourth edition published in 1866, he amended it to read: 'But geology shows us, that from an early part of the long tertiary period the number of species of shells, and that from the middle part of this same period the number of mammals, has not greatly or at all increased' (*Origin* 4th ed., p. 151). CD's copies of his own books are in the Rare Books Room–CUL.

[6] CD sent Bronn the first half of the proof-sheets of *Orchids* at the end of April (see letter to H. G. Bronn, 25 April [1862]); Bronn's name also appears on CD's list of presentation copies for this work (see Appendix IV).

[7] See letter to H. G. Bronn, 25 April [1862]. Hugo von Mohl, professor of botany at Tübingen, was a specialist on plant anatomy.

[8] See letter from E. Schweizerbart'sche Verlagsbuchhandlung, 7 June 1862. Leipzig was a major centre for the book trade in the nineteenth century; the book fair formed part of the annual Easter fair (*Börsenblatt für den Deutschen Buchhandel* 29 (1862): 1172–3, and *EB*).

[9] See also letter from H. G. Bronn, [before 11 March 1862]. For a discussion of Bronn's attitude towards CD's theory, see Junker 1991.

[10] Bronn had appended a chapter commenting on *Origin* to his translation of the first edition (see *Correspondence* vol. 8). He added no further remarks to the second German edition, stating that, as the latest English edition (*Origin* 3d ed.) contained several responses to his original comments, he felt obliged to leave his chapter unchanged in the second German edition (Bronn trans. 1863, p. 525 n.).

From E. Schweizerbart'sche Verlagsbuchhandlung[1] 7 June 1862

Stuttgart
7 June 1862

Most honoured Sir,

You will have heard from Herr Hofrat Bronn that the copy of your latest edition of the "Origin", which you had the kindness to have sent to me has arrived safely and has been sent to Mr Bronn.[2] On this occasion I must again praise the great kindness with which you have assisted us in organising a new printing of 1000 copies of the German edition of your famous book, for which assistance I am most grateful.[3]

Mr Bronn also informed me that you were kind enough to send him a copy of the book on orchids, and also that you would be so good as to arrange with your publisher, Mr Murray that he should send me for the price of £10— the plates of the illustrations that are inserted in the text;[4] I accept your offer gladly and herewith request that you ask Mr Murray to send me good castings of the plates as soon as possible and to inform me to whom I should pay the amount, or whether he would like a London draft.

I shall add your portrait to the 2nd. German edition, from a photograph that I had sent to me from London for this purpose, since I believe it will make a welcome gift to all those who admire the book.[5]

Please accept, most honoured Sir, the assurance of the fullest respect of your most faithful | E. Schweizerbart.[6]

DAR 177: 68

[1] For the transcription of this letter in its original German, see pp. 235–6. The letter was written by Christian Friedrich Schweizerbart, who was head of the Stuttgart publishing company E. Schweizerbart'sche Verlagsbuchhandlung (see n. 6, below).

[2] See letter to H. G. Bronn, 25 April [1862], and letter from H. G. Bronn, 19 May 1862.

[3] Schweizerbart refers to CD's offer to assist in incorporating new material from the third English edition of *Origin* into the new (second) German edition of the book (Bronn trans. 1863). See letters to H. G. Bronn, 11 March [1862] and 25 April [1862].

[4] Schweizerbart had agreed, through Bronn, to CD's suggestion that he might publish a German translation of *Orchids* (see letter to H. G. Bronn, 11 March [1862], and letter from H. G. Bronn, 27 March 1862); he refers to John Murray, CD's British publisher.

[5] The photograph included as the frontispiece to the second German edition of *Origin* (Bronn trans. 1863) is that taken by Maull & Polyblank *circa* 1857 (see the frontispiece to *Correspondence* vol. 8; the photograph reproduced there is of a print prepared by Maull & Fox from the original negative).

[6] The publishing firm E. Schweizerbart'sche Verlagsbuchhandlung was founded by Wilhelm Emanuel Schweizerbart; in 1841 he sold the firm to his nephew, Christian Friedrich Schweizerbart, who continued using the signature 'E. Schweizerbart' in business communications (*Jubiläums-Katalog* pp. x–xi).

From Alphonse de Candolle[1] 13 June 1862

Geneva
13 June 1862.

My dear Sir

I wish to thank you very much for sending me your article *On the dimorphic condition of Primula*.[2] It is very intriguing and rightly gives one cause for reflection, like everything you publish. I have instructed my son to prepare an extract for the Bibliothèque universelle (archives des sciences) and in fact he has edited it for one of the next issues.[3]

The main fact, that of superior fertility by crossing of the two least similar forms, seems to me to be connected in physiology with only one known fact, that of the higher fertility and more vigorous offspring of individuals not related to each other, while in-breeding is unfavourable. This is quite mysterious—from a theoretical point of view, one would sooner have expected the opposite—but it is a fact.

I have not noted whether you sowed the seeds obtained from the different crossings of your Primula. One would like to know whether the two forms reappear in equal proportions from seeds resulting from different crosses, or whether a given kind of seed produces one form. You are right when you say that the same plant keeps its form of flowers from year to year, but what happens generation after generation?

The double form of the Boraginaceae struck me while I was writing about them for the *Prodromus*.[4] I have noted the fact in this work for certain species, for example *Alkanna hispidissima* DC. Prodr. X p. 94, but I never carried out such interesting experiments as you did. One should also note in ⟨ ⟩ from the family Campanulaceae late flowers ⟨ ⟩ hardly any, which are quite fertile, and in ⟨ ⟩ as far as I remember two types of flowers, where the least ⟨ ⟩ are the most fertile, but these are not ⟨ ⟩ analogous and no experiments have been performed on the ⟨ ⟩

Will we soon have the great work that you announce as providing detailed evidence of facts mentioned in your book on the origin of species?[5] I await it with great impatience. In conclusion, after reading your book three or four times, sometimes entirely, sometimes partially, my view is close to Asa Gray's.[6] I like your theory. It delights my mind. It is the only one that makes sense of very obscure questions, unapproachable by other paths—but we need proofs for it, especially regarding *natural selection*. The general hypothesis of indefinite transmission across centuries of forms with more or less marked modifications seems preferable to any other, but I am uncertain that natural selection is the means for it. There are so many factors that for a long time keep forms the same from generation to generation or that cause them to revert! It is so rare for a new form to be preserved without the protection of man! I know of no proven instance of the latter case. There are some, probably, but none has been proved, as far as I know. These grave questions on which you have thrown so much light, have been occupying me while preparing a review of the genus *Quercus* for the *Prodromus*[7] and an analysis of Heer's works on the tertiary flora for the Bibliothèque universelle (May 1862).[8] My intention was to treat *Quercus* ⟨ ⟩ on species, by means of materials ⟨ ⟩ of its forms. Variations of the ⟨ ⟩ are astonishing. The result has been a very large ⟨ ⟩ of supposed species by various authors. This has led me again to the question of the origin of forms: I laughed heartily when I reread Linnaeus's definition (Philos. bot. n. 157): *Species tot numeramus, quot diversæ formæ in principio sunt creatæ*.[9] How much the good man thought he knew about things we do not know a century later! I have nowhere discovered the date and exact place of origin of perhaps a thousand forms of *Quercus* which I group in about two hundred species and which I would reduce still further, if I could have as many specimens of foreign forms as I have of forms from Europe and the United States. The main forms of Q. Robur L. (which I consider as *one* species) are prior to the separation of Ireland from Great Britain and consequently of the latter from the Continent. So the pedunculate and sessile forms have been struggling for quite some time in all the forests of Europe without one overthrowing the other. The American ones are less aggressive. If I decide to publish the work I have begun to prepare on *Quercus* I will have the honour of sending it to you.[10]

The completion of the Dicotyledons for the Prodromus has occupied me for some time. I hope, or rather, I wish to have sufficient strength to take up botanical geography once again, after publishing volumes XV and XVI with the help of various collaborators. Unfortunately for me the years are advancing and my strength is diminishing.

Please accept, dear Sir, the expression of my high regard and all my devotion. | Alph. de Candolle

DAR 161.1: 10

[1] For the transcription of this letter in its original French, see pp. 248–50.
[2] Candolle's name is on CD's presentation list for his paper 'Dimorphic condition in *Primula*' (see Appendix III).
[3] C. de Candolle 1862. Candolle refers to his son, Casimir de Candolle.
[4] Candolle and Candolle 1824–73. There is a note, dated 13 June 1862, recording this observation and reference, in DAR 110 (ser. 2): 25a.
[5] Candolle refers to CD's explanation in *Origin*, p. 2, that the work was an abstract, without references or authorities. CD stated: 'No one can feel more sensible than I do of the necessity of hereafter publishing in detail all the facts, with references, on which my conclusions have been grounded; and I hope in a future work to do this.' The only part of the planned three-part work that was published during CD's lifetime was *Variation* (1868).
[6] CD had sent Candolle a copy of Asa Gray's pamphlet on *Origin* in which Gray argued that natural selection and natural theology were not inconsistent with one another if one took the view that natural selection operated in accordance with divine purpose (A. Gray 1861a; for CD's presentation list for this pamphlet, see *Correspondence* vol. 9, Appendix III). In the discussion of CD's theory given in A. de Candolle 1862b, p. 60 (see n. 10, below), Candolle noted that many apparent anomalies that were repugnant and embarrassing ('répugnent et embarrassent') for the theory of special creation (like the rudimentary nipples of male mammals), were, by contrast, brought under a general law by the theory of evolution from a common ancestor. He continued (p. 61):

> je trouve naturel que des hommes fort éloignés des idées matérialistes, ayant même une tendance prononcée vers d'autres opinions, comme le docteur Hooker, M. Asa Gray et le professeur Heer, préfèrent la théorie de l'évolution et s'attachent plus ou moins aux doctrines ou aux études par lesquelles on s'efforce de la démontrer. [I find it natural that men who distance themselves from materialist ideas, and who may even have a definite tendency towards other opinions, like Dr Hooker, Mr Asa Gray and Professor Heer, prefer the theory of evolution and attach themselves more or less to those doctrines and studies by which attempts are made to demonstrate it.]

The references are to Joseph Dalton Hooker and Oswald Heer. On Gray's response to CD's theory, see also Dupree 1959.
[7] Candolle and Candolle 1824–73, 16, pt 2: 1–109.
[8] Candolle's paper (A. de Candolle 1862a) discussed Heer 1855–9 and Heer 1861a.
[9] Linnaeus 1751, aphorism 157: 'We count as many species as there were forms created in the beginning' (Stafleu 1971, p. 63).
[10] Candolle published two articles on the genus *Quercus* (oaks) in 1862 (A. de Candolle 1862b and 1862c). There are annotated copies of these publications in the Darwin Pamphlet Collection–CUL. The first paper (A. de Candolle 1862c) described a newly identified defining characteristic in acorns, and proposed a new division of the genus. Candolle probably refers to the second paper (A. de

Candolle 1862b), entitled 'Étude sur l'espèce à l'occasion d'une révision de la famille des Cupulifères' [Study on species occasioned by a revision of the family of Cupuliferae], in which he discussed the grouping and historical origins of the members of the oak family, concluding with an extensive commentary on CD's theory (A. de Candolle 1862b, pp. 57–68). Although Candolle considered that evolution from a common ancestor was 'l'hypothèse la plus naturelle' [the most natural hypothesis] (p. 66), which explained otherwise inexplicable facts, he also observed that there was no direct proof of the theory, and recommended a cautious approach, especially in view of the vast periods of time that would be required to accomplish species change by such a process.

To *Bienen Zeitung*[1] 18 June 1862

It would please me greatly, if the Reverend Dzierzon,[2] or another experienced correspondent of the Bienenzeitung, would be kind enough to clarify whether or not a noticeable difference occurs in the ordinary bee (apis mellifera) kept in various regions of Germany. An observant naturalist and clergyman, as well as gardener, claimed a few years ago that certain broods of bees were smaller than others, and that the bees differed in behaviour.[3] This clergyman stated further that the wild bees in certain English forests were smaller than the usual tame ones. Mons. Godson, a learned French naturalist said, also, that in southern France the bees were larger than elsewhere, and that in comparing certain hives a slight difference in the colour of the hair can be detected.[4]

I hope that some experienced observers who have seen bees in various parts of Germany will establish to what extent the foregoing reports are well founded.[5]

Bromley, Kent, England 18/6 62.
Charles Darwin.

Bienen Zeitung 18 (1862): 145

[1] For the transcription of this letter in its original German, see pp. 257–8. CD sent a slightly different version of this inquiry to the *Journal of Horticulture* (letter to *Journal of Horticulture*, [before 10 June 1862]). It was Thomas White Woodbury, one of the editors of the bee section of the *Journal of Horticulture*, who forwarded CD's letter to the editors of the *Bienen Zeitung* (*Journal of Horticulture and Cottage Gardener* n.s. 3 (1862): 463).

[2] The reference is to the renowned German beekeeper, Johannes Dzierzon.

[3] The reference has not been traced.

[4] 'Godson' is a misspelling; the reference is to Godron 1859, 1: 459. There is an annotated copy of this work in the Darwin Library–CUL (see *Marginalia* 1: 331).

[5] A reply from Dzierzon was published in the *Bienen Zeitung* 18 (1862): 145–6. CD's query also elicited a response from Georg Kleine (*Bienen Zeitung* 18 (1862): 206–7). These replies were translated into English and published in the *Journal of Horticulture and Cottage Gardener* n.s. 3 (1862): 463–4 and 642–3. In *Variation* 1: 298, referring to Dzierzon's reply to this letter, CD stated: 'The great apiarian Dzierzon . . . says that in Germany bees of some stocks are decidedly dark, whilst others are remarkable for their yellow colour. Bees also seem to differ in habits in different districts'. CD also summarised Kleine's reply (p. 298, n. 59): 'though there is some variability in colour, no constant or perceptible differences can be detected in the bees of Germany'.

From H. G. Bronn[1] 21 June 1862

Heidelberg
21 June 1862

Sir,

I have just received a letter from Mr. Schweizerbart telling me that Mr. Treviranus is willing to translate your excellent volume on the Orchids.[2] You could not have found a better translator to introduce your book into Germany,[3] and I would have congratulated you with all my heart. But I am distressed to tell you that the translation *is done* up to *page* **306**.[4]

Mr Schweizerbart had asked me to do it. I replied that I would be happy to do so if he could not find a translator whose good name as a botanist would be advantageous for the success of the book, and on condition that he would entrust the work to me quickly enough so that I could finish it before the end of June since I was already committed for the months of July and August. He replied that for his part he wished to receive the translation as soon as possible, and so I have worked on it day and night, in order to respond to *your* wish to see a German translation of your Orchids and to *his* intention to publish this translation soon![5]

I had consented to do the translation as I considered this work as a complement to the "Origin";[6] I hoped in this way to reach a more complete understanding of your views than through a passive reading of the book, and I thought thus to teach myself in a more fundamental manner. I had brought to me, from the hothouse of our botanical garden, some species of Orchids that were in bloom, in order to study and compare them with your descriptions. After having begun the translation, I soon felt my mind stretched to the highest degree by the interest that reading your observations and the conclusions that you have drawn from them continually inspires in me. And now, I wanted to write to you about some minor difficulties that transpired during the course of my translation, the list of which you find attached here.[7] But with all the interest that your researches on the sexual relations of hermaphroditic plants have inspired in me, researches that are equally important for my zoological studies, I admit that I am perfectly aware that I cannot replace a translator-botanist of Mr. Treviranus's merit, in whose hands any German edition would be bound to gain considerably vis-à-vis my own.

Mr Schweizerbart will have already started the printing, if he has not been awaiting copies of the woodcuts.[8] The 'Origin' is in press.[9]

Please excuse me, Sir, if this time my intentions have not been so advantageous either for your publications, or it must be said for science itself, as I had wished and hoped by charging myself with a task that is, at least for myself, one of the most attractive and instructive.

Please accept, Sir, the expression of my distinguished consideration, with which I have the honour to be | Your | very devoted | H. G. Br⟨onn⟩

[Enclosure]

Darwin Orchids

pag. 22, fg. III: (**1′**) is missing in the figure[10]

pag. 171: **india-rubber-band**: what is this, I cannot find this term anywhere

pag. 254: **cocket-hat**: nor this![11]

pag. 272, fg. XXXI: (**a′**) is missing in the figure.[12]

pag. 285 **Homopterous Insects**. we understand (*in Germany*) in a general sense under this name a division of Hemiptera or Rhynchota, and although there are some very distinctive forms (Fulgora), I do not know whether the often fantastic forms of tropical *Orthopteras* would not serve your comparison a great deal better?[13]

Anther-case: is there a botanical term for this expression? I still do not always know whether this is the *actual* covering of the anthers or the cavity that surrounds them?

pag. 104, line 4: **renewed** should be **removed**??[14]

(pag 292, fg. XXXII. and following) **group of spiral-vessels**: can this expression be rendered by *Spiralgefäss-Bündel*, i.e. *bundle of spiral vessels*,—or can *a group* contain several *bundles*? (we are very accustomed to the use of the word Gefäss-*Bündel*)

—In the description on page 292 and following, is there always a distinction between

group of vessels and	}	this might be very important
vessels ?		for the "general reader" to know.

—Finally, it would appear to me, that it would be very useful, again for the general reader, to have a very small table (only a single page) of the composition of the whole family of Orchids, its tribes, sub-tribes and most important genera (following Lindley?[15]) in order to give a fair idea of the affinities of genera, whose isolated names he will encounter. The botanist has no need for this.

DAR 70: 2, DAR 160.3: 318

[1] For the transcription of this letter in its original French and CD's annotations, see pp. 265–8.

[2] Bronn refers to Christian Friedrich Schweizerbart, head of E. Schweizerbart'sche Verlagsbuchhandlung, the Stuttgart publishing firm that was preparing to publish a German edition of *Orchids* (see letter from H. G. Bronn, 27 March 1862, and letter from E. Schweizerbart'sche Verlagsbuchhandlung, 7 June 1862). CD had apparently written on 18 June 1862 to ask Ludolph Christian Treviranus, professor of botany at the University of Bonn, whether he would be prepared to translate *Orchids* for this edition. However, having learned that Bronn had nearly completed a translation of the book, CD wrote to inform Treviranus of this on 24 June. See letter from F. H. G. Hildebrand, 14 July 1862, and *Correspondence* vol. 11, letter from L. C. Treviranus, 12 February 1863. CD's letters to Treviranus of June 1862 have not been found.

[3] Treviranus was well known in Germany for his botanical publications. He was familiar with CD's views as they pertained to botanical problems. In his review of Joseph Dalton Hooker's *Flora Tasmaniæ* (J. D. Hooker 1855–60), where Hooker discussed CD's views on species as they related to the question of the geographical distribution of plants, Treviranus was critical of CD's ideas about the advantages to plants of cross-fertilisation (see Treviranus 1861; see also Junker 1989, pp. 143–4). CD was apparently unaware of Treviranus's review: there is no copy in the Darwin Pamphlet Collection–CUL, and CD made no mention of it in his extant correspondence. For Treviranus's views on *Orchids*, see Treviranus 1863a, and *Correspondence* vol. 11, letter from L. C. Treviranus, 12 February 1863.

[4] The English edition of *Orchids* has 365 pages. Since CD did not send Bronn the sheets of the first half of the book until the beginning of May, Bronn apparently translated more than 300 pages in only six weeks (see letter to H. G. Bronn, 25 April [1862]).

[5] Bronn's German translation of *Orchids* (Bronn trans. 1862) was published on 20 October 1862 (*Börsenblatt für den Deutschen Buchhandel* 29: 2195). See also letter from E. Schweizerbart'sche Verlagsbuchhandlung, 16 October 1862.

[6] Bronn translated both the first and second German editions of *Origin* (Bronn trans. 1860 and 1863).

[7] See enclosure.

[8] CD had arranged for Schweizerbart to purchase electrotype plates of the illustrations for *Orchids*, made from John Murray's original woodcuts (see letters to John Murray, 13 June [1862] and 20 [June 1862]).

[9] Bronn refers to the second German edition of *Origin* (Bronn trans. 1863; see letters from H. G. Bronn, [before 11 March 1862], 27 March 1862, and 19 May 1862). It was published in three parts, appearing on 6 October, 17 November, and 19 December 1862 (*Börsenblatt für den Deutschen Buchhandel* 29: 2083, 2447, and 2735); however, the title-page bore the date '1863'. There is a copy of the work in the Darwin Library–CUL.

[10] In the key to figure 3, given in *Orchids*, p. 23, the label 'l' is said to stand for 'labellum', and 'l′' for 'guiding plate on the labellum'; there is, however, no l′ shown on the figure. The error was corrected for the second edition (*Orchids* 2d ed., p. 18).

[11] Bronn mistranscribed the term: it read 'cocked-hat'.

[12] In the key to figure 31 (*Orchids*, p. 272), the label 'a' is said to stand for 'anther', and 'a′' for 'rudimentary shield-like anther'; there is, however, no a′ shown on the figure. The error was corrected for the second edition (*Orchids* 2d ed., p. 227).

[13] In the last paragraph of his chapter on the Catasetidae (*Origin*, p. 285), CD stated:

> The flowers of Orchids, in their strange and endless diversity of shape, may be compared with the great vertebrate class of Fish, or still more appropriately with tropical Homopterous insects, which seem to us in our ignorance as if modelled by the wildest caprice.

In his own copy of *Orchids* (Darwin Library–CUL), CD marked this final paragraph for deletion. However, the paragraph including the comparison between orchids and homopterous insects is maintained in the second edition (*Orchids* 2d ed., pp. 224–5).

[14] The sentence Bronn refers to reads:

> In England the flowers are much visited by insects: during the wet and cold season of 1860 a friend in Sussex examined five spikes bearing eighty-five expanded flowers: of these, fifty-three had the pollinia removed, and thirty-two had them in place; but as many of the latter were immediately beneath the buds, ultimately a larger proportion would almost certainly have been renewed.

In his own copy of *Orchids* (Darwin Library–CUL), CD deleted the word 'renewed' in pencil and wrote 'removed' in the margin. In the second edition, the word was again misprinted, this time as 'remowed' (*Orchids* 2d ed., p. 102).

[15] John Lindley was the leading authority on orchid taxonomy. Bronn refers to the table listing the various tribes of orchids and their genera in Lindley 1853, pp. 181–3, which he included in his translation (Bronn trans. 1862, p. vi).

From C. V. Naudin[1] 26 June 1862

Muséum | d'Histoire naturelle, | Culture. | Paris,
26 June 1862.

To Mr C. Darwin, Member of the Royal Society &c.
Sir,

I have received, through the agency of the publisher Mr J. Murray, a copy of your learned work on the fertilisation of Orchids by insects that you had been kind enough to send me;[2] I have read it with the utmost interest, and I wish to convey to you my most heart-felt and sincere thanks.

You know that I have been occupied for many years with a similar subject, the observation of hybrids in the vegetable kingdom.[3]

I think that I have reached, if not the definitive solution to this problem, at least something that approaches it very closely. By the end of the year I shall be in a position to publish a number of observations that I hope will convince my readers about the *usual* termination of hybrid generations. It will be for me both a duty and a pleasure to send you, in return, a copy of my work.[4]

In the meantime, please accept, Sir, with my thanks, the expression of my esteem and my most devoted sentiments, | Your most obedient servant, | Ch. Naudin | aide-naturaliste at the Muséum.

DAR 172.1: 6

[1] For a transcription of this letter in its original French and CD's annotations, see pp. 272–3.

[2] *Orchids* was published by John Murray on 15 May 1862 (Freeman 1977, p. 112). Naudin's name appears on CD's presentation list for the volume (see Appendix IV).

[3] In the 1840s, Naudin had become interested in hybridisation as a means of establishing taxonomic relationships between different forms. In the 1850s he became increasingly interested in its significance in regard to species formation, and from 1854 he carried out extensive crossing experiments in the botanic garden of the Muséum d'Histoire Naturelle in Paris (Olby 1985, pp. 47–8, and Rheinberger 1983, p. 201). CD owned a copy of one of Naudin's papers on hybridisation (Naudin 1858), which is now in the Darwin Pamphlet Collection–CUL, and had read two further papers by Naudin that dealt with the subject (Naudin 1852 and 1856; see *Correspondence* vol. 4, Appendix IV, *128: 155, 157, 167). In Naudin 1852, which CD had apparently read by November 1855 (see *Correspondence* vol. 7, letter to J. D. Hooker, 6 November [1855] and n. 2), the author maintained that the present range of species had been derived from a smaller number of ancestral forms by a process analogous to that of artificial selection, by which domestic varieties had been derived. After the publication of *Origin*, Naudin's colleague at the Muséum, Joseph Decaisne, claimed that Naudin had anticipated CD's views in this paper (see *Correspondence* vol. 7, letter to J. D. Hooker, 23 [December 1859]); CD subsequently corresponded on the subject with another of Naudin's colleagues, Armand de Quatrefages (see *Correspondence* vol. 9, letter to J. L. A. de Quatrefages de Bréau, 25 April [1861] and n. 7).

[4] In December 1861, Naudin submitted an account of his researches into plant hybridisation to the Académie des Sciences for consideration for the *prix des sciences physiques*. The essay was awarded the prize in December 1862, and the second part of it, containing his conclusions, was published the following year (Naudin 1863). The material on Cucurbitaceae used in this paper was covered in greater detail in a separate paper (Naudin 1862). There are offprints of both papers in the Darwin Pamphlet Collection–CUL. See also *Correspondence* vol. 11, letter to C. V. Naudin, 3 February 1863. Naudin's prize-winning essay was not published in full until 1865 (Naudin 1865); there is an annotated reprint of this paper in the Darwin Library–CUL (see *Marginalia* 1: 638–9). In the essay, Naudin described the results of hybridisation experiments with sixty different species of plant, the offspring from which he had propagated for as many as five generations. In his conclusions, Naudin discussed the significance of his discovery that whereas the first generation from each of his crosses consisted of individuals that all resembled each other, the subsequent generations often comprised individuals that were markedly different from each other, and that, over a number of generations, the hybrid forms reverted to the parental types. Naudin explained this in terms of his theory of specific essences and

non-blending heredity, arguing: 'Tous ces faits vont s'expliquer naturellement par la *disjonction des deux essences spécifiques* dans le *pollen* et dans les *ovules* de l'hybride' [All these facts are naturally explained by the disjunction of the two specific essences in the pollen and ovules of the hybrid] (Naudin 1865, p. 150). He also stated that he had not been studying hybrids for long enough to have formed a settled opinion on the question of whether there were any exceptions to his law of reversion, by which certain hybrids might become new species, but that he doubted this was so (Naudin 1865, pp. 156–7). See also Guédès 1975.

From E. Schweizerbart'sche Verlagsbuchhandlung[1] 11 July 1862

Stuttgart
11 July 1862

Most honoured Sir,

I do not know whether you have already been informed that German science has suffered a great loss that is particularly telling for both you and me,— our dear friend Hofrat Bronn died suddenly from a heart attack a few days ago while he was drinking his afternoon coffee as usual in the Schrieder Hotel!—[2]

I heard this sad news while I was away from home; I hurried immediately to Heidelberg to present my condolences in person to his family.[3] Mr Bronn leaves a widow, one daughter and two sons who arrived all the way from Marseille where they are employed in business.

Bronn was so immensely industrious that not only was the new edition of the translation of the "Origin" prepared for the press,[4] but also the translation of the book on Orchids was finished so quickly, that we are only waiting for the plates before going to press. I would be most obliged if you would be so kind as to ask Mr Murray to send the plates as soon as possible.[5]

As far as the observations of Mr Treviranus are concerned, perhaps their publication should be effected at the same time, if he intends to publish them at all and his manuscript is ready.[6] Do you know anything about him?

Mr Bronn recently sent the manuscript of the additions with which you supplied him and that you wished to be returned for the American edition;[7] I enclose them with this letter.

I will not fail to send you a copy of the translation as soon as it is ready;[8] unfortunately, the proofs can no longer be corrected by Mr Br., so I will employ a competent proofreader for this purpose.

I would be most obliged if you should wish to send me later your new work in progress.[9]

With the most profound respect and reverence | Your most humble servant | E. Schweizerbart[10]

DAR 177: 69

[1] For the transcription of this letter in its original German, see pp. 315–16. The letter was written by Christian Friedrich Schweizerbart, who was head of the Stuttgart publishing firm E. Schweizerbart'sche Verlagsbuchhandlung (see n. 10, below).

2 Heinrich Georg Bronn died on 5 July 1862, aged 62 (*Quarterly Journal of the Geological Society of London* 19 (1863): xxxiii).
3 Bronn was professor of zoology at the University of Heidelberg.
4 The second German edition of *Origin* (Bronn trans. 1863) was initially issued in three parts, whose publication was announced on 6 October, 17 November, and 19 December 1862, respectively (*Börsenblatt für den Deutschen Buchhandel* 29 (1862): 2083, 2447, and 2735).
5 Bronn trans. 1862. CD arranged for Schweizerbart to buy electrotype plates of the illustrations for *Orchids* from John Murray, CD's British publisher (see letter from E. Schweizerbart'sche Verlagsbuchhandlung, 7 June 1862, and letter to John Murray, 13 June [1862]). The German translation of *Orchids* was published on 20 October 1862 (*Börsenblatt für den Deutschen Buchhandel* 29 (1862): 2195).
6 CD had apparently written to ask the German botanist Ludolf Christian Treviranus whether he would be prepared to translate *Orchids* into German, only to discover that Bronn had already almost completed a translation (see letter from H. G. Bronn, 21 June 1862 and n. 2). The observations referred to by Schweizerbart have not been identified; Treviranus included a detailed critique of *Orchids* in his review of CD's views on dichogamy, published in the *Botanische Zeitung* in January 1863 (Treviranus 1863a), and in his 'Supplementary observations on the fertilisation of some orchids' published in August 1863 (Treviranus 1863b). Treviranus sent CD a copy of Treviranus 1863a with his letter of 12 February 1863 (*Correspondence* vol. 11); there are lightly annotated copies of Treviranus 1863a and 1863b in the Darwin Pamphlet Collection–CUL.
7 See letter to H. G. Bronn, 30 June [1862].
8 See letter to H. G. Bronn, 30 June [1862].
9 The reference is to *Variation*, on which CD had been working intermittently since January 1860 (see *Correspondence* vol. 8, 'Journal' (Appendix II)); the work was not published until 1868.
10 Although C. F. Schweizerbart purchased the publishing firm of E. Schweizerbart from his uncle Wilhelm Emanuel Schweizerbart in 1841, he continued to use the signature 'E. Schweizerbart' in business communications (*Jubiläumskatalog*, pp. x–xi).

From Armand de Quatrefages[1] [after 11 July 1862][2]

My dear Colleague

It is too late this ⟨year⟩ to reexamine the new information concerning the various races of ⟨silk⟩ worms but next year towards the month of April the hatchings will start again and then I will attempt to respond ⟨to⟩ your wishes—[3] Only ⟨ ⟩ *be kind enough* **to remind me of my promise**. ⟨I⟩ mistrust my ⟨memory⟩ and would be very unhappy if you were to take for negligence ⟨or⟩ ill-will that ⟨which is⟩ the result of a mental failure. This is why I am writing to you without delay, my distractions causing me recently to forget to send you a reference that unfortunately I know is necessary.*[4]

You are right to say that we greatly differ. But perhaps less than you think. I believe that during the present geological period species have remained unchanged. I feel that we lack sufficient data to make judgments about what existed before that. Therefore I abstain rather than do battle. But above all—I can say—I seek truth and science. Everything that serves to extend this double domain has my deep interest. Were there to be a hint of poetry in the endeavours attempted, this would not alarm me. In short, while preferring Cuvier I regard Geoffroy highly and

Buffon is in my view a very great man.[5] I apply the same criteria of judgment to my contemporaries— That is all.

I shall send you soon a small book on metamorphoses.[6] Its conclusion expresses more or less closely the ideas to which I was just alluding. It will therefore disagree with your views. But you can be absolutely sure that a doctrinal dispute, as far as I am concerned, will have no influence on the esteem that you deserve for all efforts made to enlarge the domain of the human spirit. As to my personal independence, I believe it is proved by the introduction to my *Unité de l'espèce humaine*. I am told it is that which prevented its translation into English.[7]

Farewell my dear colleague. If I can be useful to you, please rely on me— | De Quatrefages

Incomplete[8]
DAR 175: 7

[1] For the transcription of this letter in its original French, see pp. 316–17.

[2] Dated by the relationship to the letter to Armand de Quatrefages, 11 July [1862], and by the references to Quatrefages 1861 and 1862 (see nn. 6 and 7, below). See also *Correspondence* vol. 11, letter from Armand de Quatrefages, 29 March 1863.

[3] See letter to Armand de Quatrefages, 11 July [1862].

[4] The original letter included a symbol like an asterisk at this point. It was apparently intended to refer CD to a marginal reference; the broad left margin of this page of the letter has been excised.

[5] Quatrefages refers to the prolonged controversy earlier in the century between Etienne Geoffroy Saint-Hilaire and Georges Cuvier, both professors at the Muséum d'Histoire Naturelle in Paris, that culminated in 1830 with a famous debate at the Académie des Sciences. Part of the substance of the controversy related to the question of species change, Geofroy Saint-Hilaire arguing for environmentally driven transmutation, while Cuvier argued for the permanence of species. Geoffroy Saint-Hilaire drew heavily on the philosophical approach of Georges Louis Leclerc, comte de Buffon, who, in his later writings, classified mammals and birds into natural families, each of which he believed had descended by 'degeneration' from a single parent stock, modifications being the result of adaptations to changes in the environment (Appel 1987).

[6] Quatrefages 1862. There is an annotated copy of this publication in the Darwin Library–CUL (see *Marginalia* 1: 691).

[7] Quatrefages 1861, pp. 1–10. The introduction to the volume details the religious and political basis of arguments for and against the doctrine that all human races belong to one species. Quatrefages proposed that the question was settled by the scientific evidence of the complete interfertility between the different races. There is an annotated copy of this work in the Darwin Library–CUL (see *Marginalia* 1: 693–4).

[8] See n. 4, above.

From Edouard Claparède[1] 6 September 1862

Cologny near Geneva
6 Sept. 62.

Sir,

I am truly ashamed to take up my pen so tardily in order to thank you for the pleasant letter with which you were kind enough to honour me.[2] I have been ill for seven or eight months following an attack of typhoid fever and consequently

I have fallen behind in a great number of things. One of the first things I read, however, was your work on the fertilisation of orchids, the wonderful *contrivances* of which greatly excited my interest.—[3] I am pleased that the analysis of your theory of natural selection that I inserted in the *Revue Germanique* has met with your approval.[4] I published it with all the *love* generated by a conception that is in my eyes the greatest step forward of which the natural sciences can boast in our era and yet I believe I have examined your ideas dispassionately, casting aside all partiality.

You also thank me, Sir, for my assistance to Mlle. Royer.[5] I would have preferred you not to know about this, for I have to say that I am sorry to see your work translated by this person for whom I otherwise profess considerable admiration. Her translation is heavy, indigestible, sometimes incorrect and the notes that accompany it will certainly not be to your taste. I have used all my influence on Mlle. Royer to persuade her to confine herself purely to the role of translator, but my efforts have not been rewarded with success. I must however say in Mlle. Royer's favour that she has removed without exception all the notes that I judged absurd and scientifically meaningless. On the other hand she has printed a very large number (most of those that elucidate her translation) that had not been submitted to me at all.

Mlle. Royer is a singular individual, whose attractions are not those of her sex. However, the semi-masculine education that she has given herself by means of hard work has primarily derived from an exclusively deductive school of philosophy and her mode of thinking reflects this.[6] In translating your work, she had the idea of introducing corrections devised by herself, corrections that would have strangely and disagreeably surprised you. However I have succeeded in dissuading her from doing this by showing her that ⟨ ⟩ was disrespectful to you.— ⟨The⟩ nature of these corrections was truly interesting in showing how the methods of a mind like that of Mlle. Royer are opposed to the progress of the natural sciences. Let me cite two examples.

In the chapter on the instincts of bees, Mlle. Royer throughout her translation had replaced the term three-sided pyramid (for the base of the honeycomb cells) with that of six-sided pyramid, because she believed bees could not complete a hexagonal prism other than from a hexagonal base.[7] It had simply not occurred to her, before she introduced such a significant amendment, to glance at a honeycomb.

The second example is of the same nature. Mlle. Royer had conceived nothing better in her translation than to trace the descent of all electric fish from a common ancestor with an electric organ. Given that she has no more knowledge of zoology than she has of comparative anatomy, I have had much difficulty in getting her to understand that you had your reasons for not expressing an idea as simple as that. I did however succeed in convincing her more or less with a description of the electric organs of the torpedo, the gymnotus, the malapterurus, the mormyrus, and the nerves that make their way there that these organs, although histologically identical, are, however, certainly not morphologically homologous.[8]

So, however imperfect Mlle. Royer's translation may be, and however out of place certain portions of her preface and notes, I congratulate myself nevertheless for having prevented her from disfiguring your work more completely. But if the great work on species whose publication you envisage in the not-distant future comes, as I hope, to be published, I would wish it to have a translator better versed in the natural sciences and less desirous of bringing their own personality to the fore.[9]

Some time ago I took the liberty of sending you a memoir on the evolution of spiders and a little later another on oligochetes.[10] Both, especially the former consider problems of morphological homologies that I hope will be of interest to you.

Your obedient servant | Ed. Claparède

DAR 161.1: 149

[1] For the transcription of this letter in its original French, see pp. 398–400.

[2] See letter to Edouard Claparède, [*c.* 16 April 1862]. However, either the extant draft of that letter is incomplete, or CD must have sent a subsequent letter to Claparède (see n. 4, below).

[3] Claparède's name appears on CD's presentation list for *Orchids* (see Appendix IV).

[4] CD praised Claparède 1861 in his letter to Edouard Claparède, [*c.* 16 April 1862]. There is an annotated copy of this review in the Darwin Pamphlet Collection–CUL.

[5] Clémence Auguste Royer translated and edited the first French edition of *Origin*, which was published on 31 May 1862 (Royer trans. 1862 and *Bibliothèque Générale de l'Imprimerie et de la Libraire* 2d ser. 6 (pt 3): 341).

[6] As a young woman, Royer had obtained certificates to teach music and languages, and, after reading works by radical Enlightenment authors, devoted herself to acquiring a knowledge of science and philosophy; having educated herself through a rigorous programme of reading, she instituted in 1858 a course in philosophy for women, extending it in 1859 to include science (Harvey 1987, pp. 150–2).

[7] CD discussed the cell-making instinct of hive-bees in chapter seven of the third edition of *Origin*, which is entitled 'Instinct' (pp. 245–56); he referred to 'the three-sided pyramidal bases of the cell of the hive-bee' on page 246. In Royer trans. 1862, the passage is translated accurately (p. 323).

[8] CD discussed the origin of electrical organs in fish in a chapter of *Origin* entitled 'Difficulties on theory', pointing out that they occurred 'in only about a dozen fishes', several of which were 'widely remote in their affinities' (*Origin* 3d ed., p. 212). Consequently, CD argued, the occurrence of electrical organs in these fish could not be explained as a product of inheritance from a common ancestor, but might be explained as the result of independent adaptations (*Origin* 3d ed., p. 213). Royer added a lengthy footnote to CD's discussion of this question in Royer trans. 1862, pp. 277 n.–279 n. She began by suggesting that all the ancient ancestral forms from which the existing species of electric fish were descended, themselves possessed electrical organs, but that the organs had only been preserved in certain lines. However, Royer also noted that a number of serious objections could be raised against this hypothesis, including the considerable differences between the various electrical organs and the very distant affinities of the existing species of electric fish. In consequence, she proposed an explanation for the independent origin of electrical organs, based on the localised concentration of the capability of all muscle tissue to produce weak electric currents.

[9] In the introduction to *Origin*, CD described the book as an abstract of a larger work on natural selection, which he anticipated would take him 'two or three more years to complete' (p. 1). He intended to publish the work in three parts (see *Correspondence* vol. 7, letter to John Murray, 22 December [1859]); however, the only part ever completed was *Variation*, published in 1868. The remainder of the first draft of CD's 'big book' on species was published posthumously as *Natural selection.*

[10] Claparède 1862a and 1862b. There are lightly annotated copies of these works in the Darwin Pamphlet Collection–CUL.

From Alphonse de Candolle[1] 18 September 1862

My dear Sir

I received and read with great interest the volume on Orchids that you were kind enough to send me.[2] It is so full of facts that one would like to be able to read it next to each plant mentioned in order to understand the details better, but there are many general ideas that relate to your theories as a whole and it is this aspect that interested me the most. I have just sent to the editor of the scientific part of the Bibliothèque universelle a few pages to announce the work.[3] I did not dare go into details, but every reader should be alerted to see the book if interested.

The pollen masses attached to insects have in the past given rise to curious mistakes. I cannot remember which botanists had taken them for cryptogams! As far as I can remember, Wydler devoted an article to pointing out the mistake in some German journal—twenty-five or thirty years ago.[4]

The last issues of the Bulletin de la Société botanique de France contain articles on hybridisation by Lecoq[5] and on dimorphism in Orchids by Duchartre[6] that touch upon the subjects with which you are occupied.

I send you my best wishes for your health since with such a wealth of good observations it would be unfortunate if you could not publish any more.[7]

My work on Quercus is finished.[8] The descriptive section in the Prodromus will be the main thing,[9] but I intend to publish beforehand two little pamphlets which I shall take the opportunity to send you.[10]

In the meantime, please accept the expression of my highest regard | Alph. de Candolle

Geneva 18 Sept. 1862.

DAR 161.1: 11

1 For the transcription of this letter in its original French, see p. 418.
2 Candolle's name appears on the presentation list for *Orchids* (see Appendix IV).
3 A. de Candolle 1862d.
4 The article by Heinrich Wydler has not been identified. See also Baehni 1955, p. 119 n. 13.
5 Lecoq 1862.
6 Duchartre 1862.
7 See letter to Alphonse de Candolle, 17 June [1862] and n. 4.
8 See letter from Alphonse de Candolle, 13 June 1862.
9 Candolle and Candolle 1824–73, 16, pt 2: 1–109.
10 A. de Candolle 1862b and 1862c; there are annotated copies of these works in the Darwin Pamphlet Collection–CUL. See also letter from Alphonse de Candolle, 13 June 1862 and n. 10.

From Emile M. J. M. P. Goubert[1] 20 September 1862

London
20 Sept 1862

Sir,

I have taken the liberty of introducing myself to you without having the honour and privilege of meeting you. I would indeed have been very pleased to visit your palaeontological collection, if you had consented to permit me to do so.

Be that as it may, I should like to seize the opportunity to propose to exchange material with you. In return for English fossils from whatever area you please, provided that they were identified and their localities indicated, I could send you some beautiful series of French and German species from any stratum that you wanted. I would like especially to draw your attention to my numerous duplicate tertiary specimens from Paris and Mainz from the lowest strata, nearly all identified by M. Deshayes.[2] Furthermore, I can offer you particularly fossils from the Cretaceous of Picardy, from the Jurassic of Normandy, from the Permian and Carboniferous schists of Germany.

I have published the details of some very pretty coralline species and others that I would be pleased to send you, with the corresponding pamphlets.[3]

In any case, will you, my ⟨dear⟩ and much honoured colleague, accept the tribute of my highest esteem | Emile Goubert

Paris, rue du Cherche-midi 23 | professor at the collège Louis le grand | member of the Société géologique and the Soc. Botanique of France, of the Academy of Moscow and that of Hanau, etc

DAR 165: 80

[1] For the transcription of this letter in its original French, see p. 423.
[2] Gérard Paul Deshayes.
[3] The paper on corals to which Goubert refers is probably Goubert and Zittel 1861.

From E. Schweizerbart'sche Verlagsbuchhandlung[1] 16 October 1862

Stuttgart
16 Oct. 1862

Highly honoured Sir,

No doubt you have long been expecting to hear word of the progress of the two translations;[2] contrary to expectations the printing has been held up, but in any case the two other installments of the *Origin* will be out within about 5 weeks.[3]

In sending you the three enclosed copies of both works

über die Entstehung der Arten,
über die Einrichtung zur Befruchtung der Orchideen

I hope that their production will meet with your approval. Should you want additional copies, please let me know.[4]

I am adding to today's shipment a new work: *Die Pflanzenblätter in Natur Druck* (Morphologie d. Blätter) by Reuss, 1st installment.[5] The whole work will comprise 42 folios; I think that you will find it interesting and I would be pleased if you could at your convenience bring it to the attention of the English public.

Recently I was told of a case that you might find interesting; a girl became pregnant and accused a young man of being the father; he denied the accusation but added that he was prepared to be the father provided that the child had ten fingers and toes on each hand and foot as he had. When the child was born, it had on each hand and foot ten digits, more or less developed.

With the most profound respect I remain yours | E. Schweizerbart[6]

DAR 177: 70

[1] For the transcription of this letter in its original German, see p. 473. The letter was written by Christian Friedrich Schweizerbart, who was head of the Stuttgart publishing firm E Schweizerbart'sche Verlagsbuchhandling (see n. 6, below).

[2] Bronn trans. 1862 and 1863. See letter from E. Schweizerbart'sche Verlagsbuchhandlung, 11 July 1862.

[3] The second German edition of *Origin* (Bronn trans. 1863) was initially issued in three parts, the publication of which was announced on 6 October, 17 November, and 19 December 1862. The publication of the first German edition of *Orchids* (Bronn trans. 1862) was announced on 20 October 1862. See *Börsenblatt für den Deutschen Buchhandel* 29 (1862): 2083, 2195, 2447, 2735.

[4] There is a copy of the second German edition of *Origin* (Bronn trans. 1863) in the Darwin Library–CUL, and there are two copies of Bronn's translation of *Orchids* (Bronn trans. 1862) in the Darwin Library–Down.

[5] The publication of the first part of Reuss 1862–70 was announced on 26 September 1862 (*Börsenblatt für den Deutschen Buchhandel* 29 (1862): 2010). There is a copy of this work in the Darwin Library–Down.

[6] Although C. F. Schweizerbart purchased the publishing firm of E. Schweizerbart'sche Verlagsbuchhandlung from his uncle Wilhelm Emanuel Schweizerbart in 1841, he continued to use the signature 'E. Schweizerbart' in business communications (*Jubilums-Katalog*, pp. x–xi).

APPENDIX II
Chronology 1862

This appendix contains a transcription of Darwin's 'Journal' for the year 1862. Darwin commenced his 'Journal' in August 1838 and continued to maintain it until December 1881. In this small notebook, measuring 3 inches by $4\frac{1}{2}$ inches, Darwin recorded the periods he was away from home, the progress and publication of his work, and important events in his family life.

The version published by Sir Gavin de Beer as 'Darwin's Journal' (1959) was edited before the original 'Journal' had been found and relied upon a transcription made by an unknown copyist. The original, now in the Darwin Archive in the Cambridge University Library (DAR 158), reveals that the copyist did not clearly distinguish between the various types of entries it contains.

From 1845 onward, Darwin recorded all that pertained to his work (including his illnesses, since these accounted for time lost from work) on the left-hand pages of the 'Journal', while the periods he was away from home, and family events, were noted on the right-hand pages. In order to show clearly Darwin's deliberate separation of the types of entries he made in his 'Journal', the transcription has the left- and right-hand pages labelled, with the left-hand page of an opening preceding the right-hand page.

All alterations, interlineations, additions, and the use of a different ink or pencil have been noted. In addition, the editors have inserted relevant additional information throughout this transcription of the 'Journal' for 1862. These interpolations are enclosed in square brackets to distinguish them from Darwin's own entries.

[Left]

1862

[2–4 April. Went to London.][1]

[3 April. Read 'Three sexual forms of *Catasetum tridentatum*' to the Linnean Society of London.][2] April 28th[3] finished Orchis Book[4] Besides odd times for several years,[5] this Book has cost me 9 months, if I do not count Torquay;[6] but in this time I did Primula paper:[7] 2d Edit German of Origin[8] & experiments &c.[9] Say 10 months counting $\frac{1}{2}$ time at Torquay.

May 15th. Orchid Book published.[10]

Much time wasted June & July from Leonards illness.[11]

Finished Silk-worms Geese &c[12]

Oct. 7th Facts of variation of Plants;[13]

Dec. 11. finished long Chapter—[14] Paper on Linum.[15]
Dec. 21. Bud-Variation[16]

[Right]
1862
[1 February. John Lubbock came.][17]
[14 February. Sarah Elizabeth Wedgwood and Mary Susan Parker visited.][18]
[15 February. Lunch at John Lubbock's house with Joseph Dalton Hooker and George and Ellen Busk.][19]
[19 February. Went to London. Visited dentist.][20]
Feb. March & April. Horace ill[21]
[30 March. John Lubbock and Joseph Dalton Hooker came.][22]
[6 April. John Lubbock, George Busk, and Thomas Henry Huxley came.][23]
[17–21 April. Joseph Dalton Hooker visited.][24]
[18–21 April. Henry Walter Bates visited.][25]
[6–9 May. Went to London. Visited London International Exhibition. Called on Hugh Falconer and Charles Lyell.][26]
May 15 to 22d. Leith Hill Place[27]
June, July Augt Leonard very ill Scarlet Fever[28]
Aug 12 to Southampton,[29] Emma Scarlet Fever
Sept 1 to Bournemouth.
[29 September. Went to London. Stayed at Erasmus Alvey Darwin's house.][30]
Sept. 30 [Called on Charles Lyell.][31] Home
[21 October. John Clements Wickham, Bartholomew James Sulivan, and Arthur Mellersh came.][32]
[31 October. John Lubbock dined.][33]

1 Emma Darwin's diary (DAR 242).
2 Letter to George Bentham, 30 March [1862], and letter to Charles Lyell, 1 April [1862].
3 *After del* 'May'.
4 *Orchids* was published by John Murray on 15 May 1862 (Freeman 1977, p. 112). For CD's presentation list for this work, see Appendix IV.
5 'several' *interl*.
6 The Darwins spent July and August 1861 at Torquay in Devon (see *Correspondence* vol. 9, Appendix II).
7 CD read his paper, 'Dimorphic condition in *Primula*', before the Linnean Society of London on 21 November 1861. For CD's presentation list for this work, see Appendix III.
8 The second German edition of *Origin* (Bronn trans. 1863) was initially published in three parts, whose publication was announced on 6 October, 17 November, and 19 December 1862, respectively (*Börsenblatt für den Deutschen Buchhandel* 29 (1862): 2083, 2447, and 2735). In April 1862, CD marked on a copy of the third English edition all the additions and corrections made since the second English edition, from which the first German edition had been prepared; in addition he compiled a few new additions and corrections for incorporation in the second German edition (see letter to H. G. Bronn, 25 April [1862]). See also Appendix VIII.
9 During the early part of 1862, CD carried out numerous experiments in relation to his interest in dimorphism in flowers (see letter to Alphonse de Candolle, 17 June [1862], n. 2).

[10] *Marked with cross, pencil.* See n. 4, above.

[11] Leonard Darwin contracted scarlet fever in June 1862 (Emma Darwin's diary (DAR 242); letter to W. E. Darwin, 13 [June 1862]).

[12] 'Geese &c' *interl*. CD was preparing the manuscript of *Variation*, published in 1868. He refers to material that was eventually incorporated in chapter 8.

[13] CD refers to material on cultivated plants to be included in chapters 9 and 10 of *Variation*.

[14] See n. 12, above, and letter to J. D. Hooker, 12 [December 1862].

[15] CD read his paper, 'Two forms in species of *Linum*', to the Linnean Society of London on 5 February 1863 (see *Correspondence* vol. 11).

[16] This material eventually formed chapter 11 of *Variation*.

[17] Letter from Emma Darwin to W. E. Darwin, [3 February 1862] (DAR 219.1: 48).

[18] Letter to W. E. Darwin, 14 February [1862].

[19] Letter from John Lubbock, 13 February 1862. See also letter to W. E. Darwin, 14 February [1862], and letter to J. D. Hooker, [before 15 February 1862].

[20] Letter to W. E. Darwin, 14 February [1862].

[21] See also letter to W. E. Darwin, 14 February [1862].

[22] Emma Darwin's diary (DAR 242).

[23] Emma Darwin's diary (DAR 242); letter from T. H. Huxley, 6 May 1862.

[24] Emma Darwin's diary (DAR 242); letter from J. D. Hooker, [15 April 1862].

[25] Emma Darwin's diary (DAR 242); letter to H. W. Bates, 16 April [1862].

[26] Emma Darwin's diary (DAR 242); letters to H. W. Bates, 9 May [1862], and to J. D. Hooker, 9 May [1862].

[27] Leith Hill Place, six miles from Dorking, Surrey, was the home of CD's sister, Caroline Sarah Wedgwood, and her husband, Josiah Wedgwood III.

[28] 'Scarlet fever' *interl*. See n. 11, above.

[29] Emma Darwin's diary (DAR 242) gives the date as 13 August 1862; see also letter to W. B. Tegetmeier, 12 August [1862].

[30] Emma Darwin's diary (DAR 242).

[31] See letter to Hugh Falconer, 1 October [1862].

[32] Emma Darwin's diary (DAR 242); letter to John Lubbock, 23 October [1862].

[33] Emma Darwin's diary (DAR 242).

APPENDIX III
Presentation list for 'Dimorphic condition in *Primula*'

Darwin read his paper on the two forms of *Primula* before the Linnean Society of London on 21 November 1861; it was published in the number of the *Journal of the Proceedings of the Linnean Society (Botany)* issued on 1 March 1862 (*General index to the Journal of the Linnean Society*, p. vi). In addition, however, Darwin requested permission from the council of the Linnean Society, to have fifty copies of 'Dimorphic condition in *Primula*' printed at his own expense, 'to give to gardeners & others' who had been sending him specimens (see *Correspondence* vol. 9, letter to George Bentham, 24 November [1861]). No reply is extant, but an entry in CD's Classed account book (Down House MS) for 3 March 1862, records payment of £1 10*s.* 6*d.* to Taylor & Francis, the printers of the *Journal of the Proceedings of the Linnean Society*, under the heading 'Primula Paper'. CD began distributing the offprints by the middle of January 1862; the presentation list reproduced here presumably identifies most of the recipients of these copies.

During his work on *Primula*, Darwin relied heavily on the help of private individuals from whom he received specimens or notes and observations, and who subsequently became regular correspondents. Consequently, the names on the presentation list include a number of gardeners, amateur botanists, and plant collectors. Darwin also included in the list translators of his works, family members and friends, and British and foreign scientists whom he wished to keep apprised of his work. Many of Darwin's closest friends and colleagues, such as Joseph Dalton Hooker, Daniel Oliver, and Thomas Henry Huxley, are omitted from the list; as fellows of the Linnean Society, they would have received a copy of the paper in the society's journal.

The manuscript list is in DAR 210.11: 20 v., and is reproduced below as close to its original format as possible, with all alterations, interlineations, additions, and the use of a different ink or pencil noted. The symbol ✔ denotes a brown crayon tick; ✓, a tick marked in ink; ✔ ✔, a tick marked in blue crayon; and ✓ ✓, a pencil tick. The symbol ✘ denotes a cross marked in brown crayon.

The list is followed by a biographical register, arranged alphabetically, identifying each recipient and, where possible, giving an indication as to why Darwin would have sent them a presentation copy of 'Dimorphic condition in *Primula*'. Individuals not otherwise mentioned in the volume are only identified here and are not included in the main Biographical register and index to correspondents.

Primula Paper

✓ Brown Sequard ✗
Look at Orchis list[1]
✔ Rev. H. Dombraine, Deal.
✔ A. Murray
Mr Horwood
✓ Dr. Lindley
✔ Rev A. Rawson
✔✔ King ? Hollyhock man.[2]
✓ Veitch.
✓ 3 copies Asa Gray[3]
✓ Sir H Holland ✗
✔✔ Eliz. Wedgwood
✓ William ✗[4]
✓ Mr. Cattell
✓ H. C. Watson[5] ✓✓
✓ Mr D. Beaton. care of Editor✗ of Hort. Journ
& Editor
✓ Mr W. Wooler Geneva H. Darlington
H. Fawett Bodenham Salisbury[6]
✓ Naudin
✓ E. Cresy ✗
✓ Decaisne
Winckler Dutch Translator
✔✔ Bronn.✗
✔✔ Alph. De Candolle—
✓ Prof. Babington ✗
The Botanist who sent me Astrantia ✗[7]
✓ Prof Dickie Aberdeen ✗.—
✓ Prof. Balfour.✗
✓ James Campbell Hendon H Grove House[8]
✔✔ Mr W. Borrer— Henfield Sussex[9]

✓ Lecoq[10]
✓ W. H. Dunnett Esq
East. H. Farm
Dedham — Essex.[11]
✓ I Anderson Henry E
Hay Lodge. Trinity | Edinburgh.[12]
Erasmus[13]
✓ W. Hugh. Gower. Kew G.
✓ Sir P. Egerton ✗
✓ Dr Lindley[14]
J Bateman
C. W Crocker
Dr. Weddell
✔✔— Leighton Shrewsbury[15] ✗
Naudin Decaisne[16]
Prof. Harvey T. C. Dublin

[1] Written in blue crayon. CD refers to his presentation list for *Orchids* (see Appendix IV).

[2] CD published queries relating to variation and crossing of hollyhocks in the *Journal of Horticulture* in 1861, and also corresponded with Charles William Crocker on the topic (see *Correspondence* vol. 9). CD may have confused the name 'King' with Donald Beaton's description of a plant breeder as 'the king of all the British cross-breeders' in the *Cottage Gardener*, 26 June 1860, p. 194 (see *Correspondence* vol. 9, letter to the *Journal of Horticulture,* [17 May 1861]).

[3] 'copies' was interlined in ink.

[4] William Erasmus Darwin.

[5] Written in pencil.

[6] Deleted in ink. CD misspelled the name 'Fawcett'.
[7] This individual has not been identified.
[8] 'H' deleted in ink.
[9] 'M^{r}' deleted in ink.
[10] Written in pencil.
[11] 'W. H. . . . Essex.' circled in ink.
[12] 'I Anderson . . . Edinburgh.' circled in ink.
[13] Erasmus Alvey Darwin.
[14] Deleted in ink.
[15] Written in pencil.
[16] 'Naudin Decaisne' deleted in ink.

Biographical register of recipients of 'Dimorphic condition in *Primula*'

Anderson-Henry, Isaac (1800–84). Scottish solicitor and horticulturist. Bred new and rare plant varieties in his garden at Hay Lodge, Edinburgh. (R. Desmond 1994; *Gardeners' Chronicle and Agricultural Gazette*, 22 March 1873, p. 399.) There is no extant correspondence between Anderson-Henry and CD until 1863, but Andrew Murray described Anderson-Henry as a 'great hybridiser' in his letter to CD of 3 May 1860 (*Correspondence* vol. 8). Anderson-Henry later conducted experiments for CD, particularly on hybrids (see *Correspondence* vol. 11).

Babington, Charles Cardale (1808–95). Botanist and archaeologist. Professor of botany at the University of Cambridge, 1861–95; an expert on plant taxonomy. (*DNB*, *DSB*.) Babington was a helpful source of seeds and specimens throughout 1862. CD cited Babington 1851 in *Orchids*, p. 320 n.

Balfour, John Hutton (1808–84). Physician and botanist. Professor of botany and Regius keeper of the Royal Botanic Garden, Edinburgh, 1845–79. (*DNB*, *DSB*.) Balfour cultivated exotic orchids at the Botanic Garden in Edinburgh; CD wrote to him in 1861 requesting specimens (see *Correspondence* vol. 9, letter to George Gordon, 4 July [1861]).

Bateman, James (1811–97). Horticulturist. Cultivated tropical plants, and became a leading orchid-collector. (R. Desmond 1994.) Bateman supplied CD with plant specimens (see letter from Robert Bateman, [28 January 1862], and *Orchids*, pp. 114 n., 158 n., and 197).

Beaton, Donald (1802–63). Gardener and regular columnist for the *Journal of Horticulture*. (R. Desmond 1994.) Beaton provided CD with information about plant varieties and hybridisation (see *Correspondence* vol. 9).

Borrer, William (1781–1862). Botanist and friend of William Jackson Hooker. Borrer endeavoured to cultivate every critical British species and many hardy exotic plants. (*DNB*.) CD had hoped to obtain his assistance in procuring specimens, but Borrer died on 10 January 1862 (see *Correspondence* vol. 9, letter to J. D. Hooker, 18 [December 1861], and this volume, letter from J. D. Hooker, [19 January 1862]).

Bronn, Heinrich Georg (1800–62). German palaeontologist. Professor of natural science at Heidelberg, 1833. (*DSB*, *NDB*.) Bronn translated the first German edition of *Origin* (see *Correspondence* vol. 8), and *Orchids* (see letter to H. G. Bronn, 11 March [1862]).

Brown-Séquard, Charles Edouard (1817–94). French physiologist. Practised medicine in France, the United States, and England. Physician, National Hospital for the Paralysed and Epileptics, 1860–3. Pioneer in endocrinological and neurological research. (*DBF*, *DSB*.) CD heard that Brown-Séquard was to review the new French translation of *Origin* (see letter to C. E. Brown-Séquard, 2 January [1862], and letter from C. E. Brown-Séquard, 13 January 1862).

Campbell, James. Of The Grove, Church End, Hendon, Middlesex. (*Post Office directory of the six home counties* 1862.) CD corresponded with a 'Mr Campbell', who agreed to do some experimental work for him, in 1861 (see *Correspondence* vol. 9, letters to T. H. Huxley, 22 October [1861] and 31 October [1861]).

Candolle, Alphonse de (1806–93). Swiss botanist. Professor of botany and director of the botanic gardens, Geneva, 1835–50. (*DSB*.) CD had been greatly interested by Candolle's work on the geographical distribution of plants (A. de Candolle 1855), and had sent him a presentation copy of *Origin* (see *Correspondence* vol. 8) and of Asa Gray's review pamphlet (A. Gray 1861) (see *Correspondence* vol. 9).

Cattell, John (*fl.* 1840s). Florist, nurseryman, and seedsman in Westerham, Kent. (R. Desmond 1994, *Post Office directory of the six home counties* 1832.) A note in DAR 108: 66 indicates that Cattell provided CD with *Primula* seeds or seedlings.

Cresy, Edward (1823–70). Surveyor and civil engineer. Principal assistant clerk at the Metropolitan Board of Works. (*DNB*.) Cresy, a family friend of the Darwins, had assisted CD with his experiments on insectivorous plants in the autumn of 1860 (see *Correspondence* vol. 8). He was given presentation copies of the second and third editions of *Origin* (*Correspondence* vol. 9).

Crocker, Charles William (1832–68). Foreman of the propagating department, Royal Botanic Gardens, Kew, 1857–62. (R. Desmond 1994.) Crocker carried out experiments for CD on pollination and insect agency (see *Correspondence* vol. 9); he was a regular correspondent in 1862.

Darwin, Erasmus Alvey (1804–81). CD's brother. Lived in London. (*Alum. Cantab.*, Freeman 1978.) CD was in regular contact with Erasmus and usually stayed with him when visiting London.

Darwin, William Erasmus (1839–1914). CD's eldest child. Banker in Southampton. (*Alum. Cantab.*) Since joining the bank in 1861, William had spent some of his leisure time carrying out botanical experiments; CD encouraged this interest in botany (see *Correspondence* vol. 9).

Decaisne, Joseph (1807–82). French botanist. Professor of plant cultivation, Muséum d'Histoire Naturelle, Paris, 1850. (*DBF*, *NBU*.) Decaisne was a friend of Joseph Dalton Hooker's to whom CD had sent a presentation copy of *Origin* (see *Correspondence* vol. 8).

Dickie, George (1812–82). Scottish botanist. Professor of botany, Aberdeen, 1860–77. (R. Desmond 1994, *DNB*.) Dickie had supplied CD with orchid specimens (see *Correspondence* vol. 9, letter to George Dickie, [5 July 1861], and *Orchids*, p. 152).

Dombrain, Henry Honywood (1818–1905). Clergyman and botanist. Perpetual curate of St George's, Deal, Kent. (*Clergy list*, R. Desmond 1994.) Dombrain was a regular contributor to the *Journal of Horticulture*, a journal to which CD subscribed; CD responded in the journal to a query from Dombrain about variation in *Auricula* (see *Correspondence* vol. 9, letter to the *Journal of Horticulture*, [before 18 June 1861]).

Dunnett, William H. Seedsman and farmer. (*Post Office directory of the six home counties* 1862.)

Egerton, Philip de Malpas Grey- (1806–81). Of Oulton Park, Cheshire. Tory MP and palaeontologist who specialised in fossil fish. (*DNB*, Sarjeant 1980.) Egerton was a prominent member of several London scientific societies and was a long-standing acquaintance of CD's. He and CD had served together on the council of the Royal Society in 1856 (see *Correspondence* vol. 6, letter to J. D. Hooker, 8 April [1856] and n. 7). Egerton was believed to be the author of the 'squib' on *Origin* published in *Punch* in May 1861 (see *Correspondence* vol. 9, Appendix VIII).

Fawcett, Henry (1833–84). Statesman. Blinded in a shooting accident in 1858. (*DNB*.) Fawcett had reviewed *Origin*, staunchly defending CD's methods of reasoning (Fawcett 1860; see *Correspondence* vol. 8, letter to T. H. Huxley, 5 [December 1860]).

Gower, William Hugh (1835–94). Foreman, Royal Botanic Gardens, Kew, until 1865. Principally known for his knowledge of orchids. (R. Desmond 1994.) Gower had provided CD with information about pollination experiments carried out on *Victoria regia*, the giant waterlily at Kew, and with specimens for CD's work on dimorphism (see *Correspondence* vol. 9).

Gray, Asa (1810–88). American botanist. Fisher Professor of natural history, Harvard University, 1842–73. (*DAB*, *DSB*.) Throughout their extensive correspondence, CD often wrote of his botanical work, seeking Gray's advice, and asking him to carry out experiments on American species. Gray superintended the production of the American edition of *Origin* (see *Correspondence* vol. 8).

Harvey, William Henry (1811–66). Irish botanist. Professor of botany, Trinity College, Dublin, 1856–66. (*DNB*, *DSB*.) Harvey was a good friend of Joseph Dalton Hooker's (see L. Huxley ed. 1918). CD regarded Harvey as 'a first-rate Botanist' (see *Correspondence* vol. 8, letter to Charles Lyell, 18 [and 19 February 1860]). Harvey had evinced a specific interest in hybridism among Primulaceae (see *Correspondence* vol. 8, letter to J. D. Hooker, 20 May [1860]).

Henry, Isaac Anderson. *See* Anderson-Henry, Isaac.

Holland, Henry (1788–1873). Physican. Distant relative of the Darwins and Wedgwoods. Physician in ordinary to Queen Victoria, 1852. President of the Royal

Institution for many years. (*DNB*, *Physicians*.) CD consulted Holland on matters of family health.

Horwood, John Gardener to George Henry Turnbull of Down, Kent. (Census returns 1861 (Public Record Office, RG9/462: 70).) Horwood helped CD with some of his observations and supplied him with orchid specimens and varieties of dimorphic plants (see *Correspondence* vol. 9 and *Orchids*, p. 158 n.).

Lecoq, Henri (1802–71). French naturalist and vulcanologist. Professor in science faculty, Preparatory School of Medicine and Pharmacy, Clermont-Ferrand, from 1854. (*Grande encyclopédie*, Sarjeant 1980.) CD purchased Lecoq's *Étude sur la géographie botanique de l'Europe* (Lecoq 1854–8) in 1861, and found therein some useful references to polymorphic forms and to the relationship between plant colour and climate (see *Correspondence* vol. 9, letters to J. D. Hooker, 18 [December 1861] and 28 [December 1861]). CD's annotated copy is in the Darwin Library–CUL.

Leighton, William Allport (1805–89). Botanist, clergyman, and antiquary. (R. Desmond 1994, *DNB*.) Leighton had been a boyhood friend of CD's; the two kept in touch, corresponding on botanical subjects. Towards the end of 1862, CD requested specimens from Leighton to help with his work on dimorphism (see letter to W. A. Leighton, 26 November [1862]).

Lindley, John (1799–1865). Botanist and horticulturist. Professor of botany at London University (later University College London), 1828–60. Editor of the *Gardeners' Chronicle* from 1841. (*DNB*, *DSB*.) Lindley was an authority on orchid taxonomy, whom CD cited frequently in *Orchids*; he also supplied CD with orchid specimens (see *Correspondence* vol. 9 and *Orchids*, p. 158). The *Gardeners' Chronicle* was an important forum for the discussion of queries that arose in the course of CD's work.

Murray, Andrew (1812–78). Entomologist and botanist. Assistant secretary to the Royal Horticultural Society of London from 1860. (*Entomologist's Monthly Magazine* 14 (1878): 215–16, Gilbert 1977.) Murray published a critical review of *Origin* (Murray 1860), founding several of his objections on entomological examples, about which he and CD corresponded prior to the review's publication (see *Correspondence* vol. 8).

Naudin, Charles Victor (1815–99). French botanist. Aide-naturaliste at the Muséum d'Histoire Naturelle from 1854. Experimented widely on plants, particularly on acclimatisation and hybridity. (*DSB*.) CD mentioned Naudin in the historical preface to the third edition of *Origin* as one of those who had previously published on the modification of species; CD was made aware of this work of Naudin's in December 1859 (see *Correspondence* vol. 7).

Rawson, Arthur (1818–91). Perpetual curate of Trinity Church, Bromley, Kent, 1843–82. A keen gardener who bred florists' flowers, particularly pelargoniums. (*Alum. Cantab.*, R. Desmond 1994.)

Veitch, James (1815–69). Nurseryman. In partnership with his father James Veitch (1792–1863) as the owners of a nursery in Chelsea. (R. Desmond 1994.) Veitch

Veitch, James, cont.
frequently provided CD with orchid specimens (see *Correspondence* vol. 9 and *Orchids*, pp. 158 n., 214 n.)

Watson, Hewett Cottrell (1804–81). Botanist, phytogeographer, and phrenologist. Published guides to the distribution of British plants. (*DNB*, *DSB*.) CD made extensive use of Watson's botanical works; the two had corresponded about various aspects of CD's work on species over a number of years (see *Correspondence* vols. 6–9).

Weddell, Hugh Algernon (1819–77). British-born botanist and physician who worked in France. Specialist on South American flora. (R. Desmond 1994.) In 1861, CD had been referred to some cases of dimorphism mentioned in Weddell's published works (see *Correspondence* vol. 9).

Wedgwood, Sarah Elizabeth (Elizabeth) (1793–1880). Emma Darwin's sister. Resided at the Ridge, Hartfield, Sussex, 1847–62. She moved to London before settling in Down in 1868. (*Burke's peerage* 1980, *Emma Darwin*.)

Winkler, Tiberius Cornelis (1822–97). Dutch geologist and palaeontologist. Translated *Origin* into Dutch (1860). (*NNBW*, Sarjeant 1980.)

Wooler, William A. Landowner, colliery-owner, and agriculturalist of Halliwell House, Heighington, County Durham. (*Northern Echo*, 5 May 1891.) Wooler corresponded with CD on domesticated fowls and rabbits and on cultivated forms of fruit and vegetables while CD was working on the manuscript of *Variation* (see *Correspondence* vol. 9).

APPENDIX IV
Presentation list for *Orchids*

Darwin began preparing an account of the pollination mechanisms of orchids during a family holiday in Torquay in July and August 1861 (see *Correspondence* vol. 9, Appendix II). His original intention had been to present the work before the Linnean Society of London, but by September the paper was already longer than envisioned; in a letter to his publisher John Murray on 21 September [1861] (*Correspondence* vol. 9), Darwin announced:

> I have just finished a very long paper intended for Linn. Soc.[y] . . . & yesterday for the first time it occurred to me that **possibly** it might be worth publishing separately, which would save me trouble & delay.— The facts are new & have been collected during 20 years & strike me as curious. Like a Bridgewater Treatise the chief object is to show the perfection of the many contrivances in Orchids.

Having provided a brief account of the probable audience for such a book ('persons who care for Nat. History, **but no others**'), its size ('*at most* 135 pages'), and the minimum terms upon which he would be willing to offer the manuscript for publication ('any terms . . . but I would not publish on my sole risk'), Darwin expressed uncertainty about the idea: 'it might injure sale of my future larger Books.— In fact I am utterly in doubt.—'

Murray did not share Darwin's trepidation, replying, 'I have no hesitation in offering to print & publish the work including Illustrations at my own sole cost & risque giving you one half of the profits of every Edition' (*Correspondence* vol. 9, letter from John Murray, 23 September 1861). *Orchids* was published by John Murray in May 1862.

Despite Darwin's reiterated assertion that the work would be too dull for the general public, its appeal at best confined to persons with a strong interest in natural history, *Orchids* was carefully planned so as to attract and maintain the interest of non-botanists; CD even asked Henrietta Emma Darwin to read it in proofs to see whether she understood it, though Emma Darwin reported: 'she finds it not easy' (see the letter from Emma to William Erasmus Darwin, dated [2 March 1862], in DAR 219.1: 49). The introduction to the work includes a section defining the botanical terms with which the reader would need to be familiar in order to understand the text, 'In case any one should look at this treatise who has never attended to Botany' (*Orchids*, p. 5). Darwin's account of the first orchid treated, *Orchis mascula*, is set out pedagogically, and he assured his readers that, while the details might be difficult to understand, if they persisted and had the 'patience to make out this first case', the succeeding cases would be 'easily intelligible' (*Orchids*, p. 9). A detailed, fold-out illustration of *O. mascula* (facing p. 18) can be referred to as the reader proceeds.

Darwin was also interested in the appearance of the product, asking Murray: 'Do you think of little Book with Cloth Back or Pamphlet in paper?' (*Correspondence* vol. 9, letter to John Murray, 24 September [1861]). In October he was worried about the size of the book, asking Murray to use 'large type & lines wider apart' so that it should not look 'ridiculously small' (*Correspondence* vol. 9, letter to John Murray, 5 October [1861]). In addition, Darwin was concerned about the lettering on the volume's spine, the retail price (feeling that ten shillings was too much), and the colour of the cover-cloth—red would be 'too gaudy for a grave volume' (letter to John Murray, 2 May [1862]).

On 10 February 1862, Darwin delivered the bulk of his manuscript to Murray via a servant (see letter to John Murray, 9 [February 1862]). He asked Murray to urge the printers, William Clowes & Sons, to print it '*quickly*', declaring: 'I am incapable of changing my work & want to get on with my other Books.—' Darwin was soon submerged by a tide of proofs, and by the end of March he was 'consumedly sick' of correcting them (see letter to W. E. Darwin, 14 February [1862], and letter to J. D. Hooker, 22 [March 1862]). *Orchids* was offered for sale on 15 May 1862 (Freeman 1977, p. 112), the print-run being 1500 copies. The volume comprises 365 pages with thirty-six illustrations, and bears an embossed gilt impression of a *Cycnoches* orchid on a plum-coloured cloth covering; it sold for nine shillings. As Darwin recorded in his 'Journal' (Appendix II), 'Besides odd times for several years, this Book has cost me 9 months'.

Darwin's presentation list for *Orchids* is in DAR 210.11: 20, and is reproduced below, as close to its original format as possible. Excluding Darwin's 'self', sixty-five individuals and three institutions are listed. The list reflects Darwin's acknowledgment of the assistance he received from colleagues and enthusiasts, his hope of future help, and his gratitude to those who had provided him with assistance, information, and specimens. The list was written in ink, with question marks added in pencil. All further alterations, interlineations, and additions are noted. Following the list is a biographical register, arranged alphabetically, identifying the recipients and giving an indication as to why Darwin might have sent them a presentation copy of *Orchids*. Those individuals to whom Darwin wished to send presentation copies of both 'Dimorphic condition in *Primula*' and *Orchids* are given a full entry in the biographical register that concludes Appendix III, with a cross-reference below.

Orchis Book.

Asa Gray	G. Maw
G. B. Sowerby ?[1]	
W. E. Darwin ?	
Rucker	
Lady Dorothy[2]	

Georgina Tollet ?
E. Cresy
Sir H. Holland
Duke of Argyll. ?[3]
D^r^. Weddell— Poitiers ?
Prof.. Andersson, Sec. Aca Stockholm [4]
Prof. Braun. Berlin
Sturr.[5]
Prof. Harvey.[6]
D^r^. Falconer
George Darwin
Alph. De Candolle
F. Smith.
F. Walker Esq

Royer	Royal Soc.[9]
J. R. Greene ?	Brown-Sequard ?
Bentham	Crocker
Wallace	Naudin
Oliver	Masters
Hooker	Lindley
Bates	Bond
Claparede ?	self
Lyell	Rodgers[10]
Huxley	Rolleston[11] ?
Carpenter ?	Ludwig[12]
Babington	Jukes
Erasmus ?[7]	Westwood
Rev W. B. Clarke	Stainton
Balfour	Parfitt
Dickie	Veitch
Bateman	Parker
Gordon	Fremond[13] ?
A. G. More	Horwood
H. C. Watson	Turnbull
Wallis	Bronn
Malden	Lubbock
Oxenden	Bot. Soc. de France[14]
Jamieson	D^r^. Boot[15]
Linn Soc.[8]	

[1] Name deleted in pencil.
[2] Dorothy Frances Nevill.
[3] George Douglas Campbell, eighth duke of Argyll.
[4] 'Prof..' and 'Sec. Aca.' interlined.

[5] CD misspelled Dionys Stur's name.
[6] 'Prof.' written over '12'
[7] Erasmus Alvey Darwin; deleted in pencil.
[8] Linnean Society of London.
[9] Royal Society of London.
[10] CD misspelled John Rogers's name.
[11] Deleted in pencil.
[12] Deleted in pencil.
[13] Deleted in pencil. CD misspelled Charles Fermond's name.
[14] Société Botanique de France.
[15] CD misspelled Francis Boott's name.

Biographical register of recipients of *Orchids*

Andersson, Nils Johan (1821–80). Swedish botanist. Explored the Americas on the *Eugenie* expedition, 1851–3. (*SBL.*) While Andersson was working at the Royal Botanic Gardens, Kew, in 1857, CD had sent him queries on the colouring of horses in Norway (see *Correspondence* vol. 6, letter to J. D. Hooker, 11 September [1857]). CD subsequently sent him a presentation copy of *Origin* (see *Correspondence* vol. 8, Appendix III).

Babington, Charles Cardale. *See* Appendix III.

Balfour, John Hutton. *See* Appendix III.

Bateman, James. *See* Appendix III.

Bates, Henry Walter (1825–92). Naturalist. Explored the Amazon region, 1848–59. Provided the first comprehensive explanation of the phenomenon subsequently known as 'Batesian mimicry'. (*DNB*, *DSB.*) After his return from the Amazon, Bates began a correspondence with CD in which Bates related his observations in the field to arguments put forward in *Origin*. Bates sent CD his papers on South American butterflies (Bates 1861 and 1862a). CD was interested in Bates's work on mimicry as providing evidence for the operation of natural selection, and was instrumental in the publication of Bates's 'travels' (Bates 1863). See *Correspondence* vols. 8–10.

Bentham, George (1800–84). Botanist. President of the Linnean Society, 1861–74. (*DNB*, *DSB.*) Bentham, a long-standing correspondent of CD's, had helped him with his botanical work over many years. After Bentham heard CD's paper, 'Dimorphic condition in *Primula*', at the Linnean Society, the two corresponded about dimorphism (see *Correspondence* vol. 9). They also discussed CD's work on orchids, and Bentham supplied CD with a specimen of *Orchis pyramidalis* (see *Correspondence* vol. 9, letter to George Bentham, 22 June [1861]). CD cited Bentham 1858 in *Orchids*, p. 47.

Bond, Frederick (1811–89). Entomologist and ornithologist. Collector of British Lepidoptera. (*Entomologist's Monthly Magazine* 25 (1888–9): 384, Gilbert 1977.) Bond sent CD specimens of moths with pollinia attached to their bodies (see *Correspondence* vol. 9, letter to George Bentham, 22 June [1861], and *Orchids*, pp. 35, 88, 91).

Boott, Francis (1792–1863). American physician and botanist. Practised medicine in London. Treasurer of the Linnean Society, 1856–61. (R. Desmond 1994, *DNB*.) Boott admired *Origin*, though unable to accept its tenets wholesale (see *Correspondence* vol. 8, letter from Francis Boott, 29 February 1860, and letter to Asa Gray, 8 March [1860]). He was an expert on the taxonomy of *Carex*, the sedges, and sent CD part of his work on the genus (Boott 1858–67) in January 1862 (see letter from Francis Boott, 27 January 1862).

Braun, Alexander Carl Heinrich (1805–77). Professor of botany and director of the botanic garden at the University of Berlin, 1851–77. His main studies were on plant morphology. Deeply influenced by *Naturphilosophie*. (*DSB*, *NDB*.) CD's only former contact with Braun or his work seems to have been his reading of two of Braun's papers (Braun 1853a and 1853b) in 1854 (see *Correspondence* vol. 5).

Bronn, Heinrich Georg. *See* Appendix III.

Brown-Séquard, Charles Edouard. *See* Appendix III.

Campbell, George Douglas, 8th duke of Argyll (1823–1900). Whig politician and author of works on science, religion, and politics. President of the Royal Society of Edinburgh in 1861. A defender of the concept of design in nature. (*DNB*.) Campbell had discussed *Origin* in his presidential address to the Royal Society of Edinburgh in December 1860 (G. D. Campbell 1860; see *Correspondence* vol. 9, letter to T. H. Huxley, 1 April [1861] and n. 2). CD also sent him a copy of Asa Gray's review pamphlet (A. Gray 1861; see *Correspondence* vol. 9, Appendix III).

Candolle, Alphonse de. *See* Appendix III.

Carpenter, William Benjamin (1813–85). Physician and physiologist. Registrar of the University of London, 1856–79. (*DNB*, *DSB*.) CD and Carpenter had corresponded on a wide variety of topics over many years, ever since Carpenter had carried out microscopic examinations on CD's South American geological specimens (see *Correspondence* vol. 3, *et seq.*). CD sent Carpenter a presentation copy of *Origin*; Carpenter wrote two reviews of the work ([Carpenter] 1860a and 1860b; see *Correspondence* vol. 8). CD cited Carpenter 1854 in *Orchids*, p. 236 n.

Claparède, Jean Louis René Antoine Edouard (Edouard) (1832–71). Swiss naturalist and invertebrate zoologist. Professor of zoology and comparative anatomy at the Academy of Geneva. (*IBN*, Gilbert 1977.) CD's only former contact with Claparède or his work seems to have been his reading of a paper by Claparède on the morphology of arthropod eyes (Claparède 1860; see *Correspondence* vol. 8, letter from John Lubbock, [after 28 April 1860?]).

Clarke, William Branwhite (1798–1878). Divine and geologist. Emigrated to Australia in 1839, where he carried out extensive geological surveys. (*DNB*, *DSB*.) Clarke wrote to CD in 1861 about some points raised in *Origin* touching Australian geology (see *Correspondence* vol. 9, letter from W. B. Clarke, [August 1861]). CD subsequently asked him to carry out an experiment regarding insect pollination (see *Correspondence* vol. 9, letter to W. B. Clarke, 25 October [1861], and this volume, letter from W. B. Clarke, 16 January 1862).

Cresy, Edward. *See* Appendix III.

Crocker, Charles Willliam. *See* Appendix III.

Darwin, Erasmus Alvey. *See* Appendix III.

Darwin, George Howard (1845–1912). CD's second son. (*DNB*, *DSB*.) George may have helped CD with observations for *Orchids*; in June 1862, subsequent to the publication of *Orchids*, he made numerous observations on the habits of insects visiting various species of orchids (see DAR 70: 13–14, 30, 32–6; see also letter to J. D. Hooker, 30 [June 1862], and letter to Asa Gray, 1 July [1862]).

Darwin, William Erasmus. *See* Appendix III.

Dickie, George. *See* Appendix III.

Falconer, Hugh (1808–65). Palaeontologist and botanist. Superintendent of the Calcutta Botanic Garden and professor of botany, Calcutta Medical College, 1848–55. (*DNB*, *DSB*.) CD sent Falconer, a long-time friend and correspondent, presentation copies of *Origin* and of Asa Gray's review pamphlet (A. Gray 1861; see *Correspondence* vols. 8 and 9), hoping to induce him to become 'less fixed' in his belief in the immutability of species (*Correspondence* vol. 7, letter to Hugh Falconer, 11 November [1859]).

Fermond, Charles (1810 – *c.* 1875). French pharmacist and botanist. (*DBF*.) CD was provided with several references to Fermond's work by Daniel Oliver in 1861 (see *Correspondence* vol. 9, letters from Daniel Oliver, [before 8 November 1861] and 8 November 1861), and read Fermond's review of the literature on plant fertilisation (Fermond 1859, now in the Darwin Pamphlet Collection–CUL).

Gordon, George (1801–93). Scottish botanist, geologist, and clergyman. Minister of Birnie, near Elgin, 1832–89. (R. Desmond 1994, Sarjeant 1980.) Gordon supplied CD with specimens of *Goodyera repens* (see *Correspondence* vol. 8, letter to George Gordon, 17 September [1860], and *Orchids*, p. 112 n.), and assisted him in his attempts to acquire specimens of *Corallorhiza* (see *Correspondence* vol. 9, letters to George Gordon, 22 June [1861] and 4 July [1861]).

Gray, Asa. *See* Appendix III. CD thanked Gray in *Orchids*, p. 123 n. for his examination of *Spiranthes gracilis* and *S. cernua*.

Greene, Joseph Reay. Irish naturalist. One of the editors of the *Natural History Review*. Studied Protozoa and coelenterates. (*Catalogue of graduates, University of Dublin*; L. Huxley ed. 1900.) Greene probably sent CD a copy of his *Manual of the sub-kingdom Cœlenterata* (the second part of Greene 1859–61) in 1861 (see *Correspondence* vol. 9, letter to Joseph Reay Greene?, 24 September [1861]).

Harvey, William Henry. *See* Appendix III.

Holland, Henry. *See* Appendix III.

Hooker, Joseph Dalton (1817–1911). Botanist. Friend and confidant of CD. Assistant director of the Royal Botanic Gardens, Kew, 1855–65. (*DNB*, *DSB*.) Hooker assisted CD in procuring specimens, carrying out experiments, and confirming observations. His assistance is acknowledged in *Orchids*, p. 158; CD cited J. D. Hooker 1854a in *Orchids*, p. 139, and J. D. Hooker 1855–60 in *Orchids*, p. 172.

Horwood, John. *See* Appendix III.

Huxley, Thomas Henry (1825–95). Naturalist and physiologist. Friend and supporter of CD. Professor of natural history at the Royal School of Mines since 1857. (*DNB*, *DSB*.) Huxley's support for his species theory was important to CD (see *Correspondence* vols. 6 and 7); CD had given Huxley presentation copies of both first and second editions of *Origin* and of the German translation (Bronn trans. 1860). CD was concerned about Huxley's continuing insistence that the theory would only be completely validated by the demonstration that new 'physiological' species had arisen by means of selection (see *Correspondence* vol. 9, letter to J. D. Hooker, 28 September [1861] and n. 12, and Appendix VI).

Jamieson, Thomas Francis (1829–1913). Scottish geologist and agriculturalist. Appointed Fordyce lecturer in agricultural research at the University of Aberdeen in 1862. (*Geological Magazine* 50 (1913): 332–3, Sarjeant 1980.) CD asked Jamieson for assistance in procuring specimens of *Listera cordata* and *Habenaria viridis* (see *Correspondence* vol. 9, letter from T. F. Jamieson, 13 June 1861). The two subsequently conducted an extensive correspondence about the formation of the 'parallel roads' of Glen Roy during which CD formed a high opinion of Jamieson's abilities (see *Correspondence* vol. 9 and Appendix IX).

Jukes, Joseph Beete (1811–69). Geologist. Director of the Irish branch of the Geological Survey, 1850–69. (*DNB*, *DSB*.) Jukes was one of the '*practical* geologists' from whom CD was keen to have support for his theory (see *Correspondence* vol. 8, letter to Andrew Murray, 28 April [1860], and letter to Charles Lyell, 25 February [1860]).

Lindley, John. *See* Appendix III.

Lubbock, John (1834–1913). Banker, politician, and naturalist. Friend, neighbour, and active supporter of CD. (*DNB*, *DSB*.) CD and Lubbock frequently corresponded about their scientific work (see *Correspondence* vols. 8 and 9).

Ludwig, Camilla. Governess to the Darwin children. Joined the household in 1860. Translated German works for CD. (CD's Account books (Down House MS).)

Lyell, Charles (1797–1875). Scottish geologist. Friend of CD and one of his scientific mentors. (*DNB*, *DSB*.) Lyell had encouraged CD to publish an abstract of his views on species (*Origin*; see *Correspondence* vol. 7); Lyell's opinion of his scientific work was important to CD (see *Correspondence* vols. 7–9). Lyell also sent CD specimens of 'spurless' *Orchis pyramidalis* (see *Correspondence* vol. 9, letters to Charles Lyell, 6 August [1861] and 13 [August 1861], and *Orchids*, p. 41).

Malden, Bingham Sibthorpe (1830–1906). Curate of St George-the-Martyr, Canterbury, 1858–63. (*Alum. Cantab.*, *Crockford's*.) Malden provided CD with several orchid specimens including *Orchis fusca* and *Peristylus viridis* (see *Correspondence* vol. 9, letter to B. S. Malden, 15–16 June [1861], and *Orchids*, pp. 43, 78).

Masters, Maxwell Tylden (1833–1907). Physician. Lecturer on botany at St George's Hospital, 1855–68. (R. Desmond 1994.) CD had corresponded with Masters about variability in plants (see *Correspondence* vol. 8, and this volume, letter from M. T. Masters, 17 March 1862). CD had also been in contact with

Masters, Maxwell Tylden, cont.

Masters's father, the nurseryman, William Masters (see *Correspondence* vols. 4 and 8).

Maw, George (1832–1912). Tile manufacturer, geologist, botanist, and antiquarian. (R. Desmond 1994, Sarjeant 1980.) CD was impressed by Maw's review of *Origin* for the *Zoologist* (Maw 1861; see *Correspondence* vol. 9, letters to George Maw, 13 July [1861] and 19 July [1861]).

More, Alexander Goodman (1830–95). Botanist and ornithologist. (R. Desmond 1994, Gilbert 1977, *Natural Science* 6 (1895): 351.) More carried out observations of orchids for CD and provided him with fresh specimens (see *Correspondence* vols. 8 and 9, and *Orchids*, pp. 67, 95 n., and 99).

Naudin, Charles Victor. *See* Appendix III.

Nevill, Dorothy Frances (1826–1913). Owned a notable garden at Dangstein in Sussex where she cultivated orchids, nepenthes, and other tropical plants. (R. Desmond 1994.) Nevill provided CD with orchid specimens from her collection (see *Correspondence* vol. 9, letter to D. F. Nevill, 19 November [1861], and *Orchids*, p. 158 n.).

Oliver, Daniel (1830–1916). Librarian and assistant in the herbarium of the Royal Botanic Gardens, Kew. Professor of botany at University College London, 1861–88. (R. Desmond 1994, *List of the Linnean Society of London* 1859–91.) Oliver provided CD with specimens and information regarding the literature on orchids, and carried out observations for CD (see *Correspondence* vol. 9, and *Orchids*, p. 158 n.).

Oxenden, George Chichester (1797–1865). Author of satiric verse and parodies. Resided at Broome Park, near Canterbury, Kent. (*Alum. Cantab.*, *Burke's peerage* 1895.) Oxenden provided CD with a number of orchid specimens and carried out observations for CD (see *Correspondence* vol. 9, letter to A. G. More, 4 June 1861, and *Orchids*, pp. 31 n., 61).

Parfitt, Edward (1820–93). Naturalist and gardener. Librarian to the Devon and Exeter Institute, 1861–93. (R. Desmond 1994, *DNB*.) CD wrote to Parfitt in 1860, having read a notice by him in the *Entomologist's Weekly Intelligencer*, 3 October 1857, pp. 3–4, describing the adherence of pollinia from the bee orchis, *Ophrys apifera*, to the proboscis of a moth (see *Correspondence* vol. 8, letter to H. T. Stainton, 11 June [1860]). Parfitt subsequently sent CD specimens of the moth *Anthrocera trifolii* with pollinia attached, and CD concluded that the pollinia were not from the bee orchis (see *Orchids*, p. 36 n.)

Parker, Robert (*fl.* 1850s–80s). In partnership with Benjamin Samuel Williams, ran a nursery in Holloway, London, 1854–61. He opened his own nursery in Tooting, Surrey, in 1861. (R. Desmond 1994.) Parker's nurseries were noted for developing successful techniques for cultivating orchids (see Stewart ed. 1992). CD thanked Parker for sending him an 'extremely valuable series of forms' in *Orchids*, p. 158 n. See also *Correspondence* vol. 9, letter to J. D. Hooker, 17 [July 1861].

Rogers, John (1806–67). Barrister. Resided in Sevenoaks Weald, Kent. (*Modern English biography*, *Post Office directory of the six home counties* 1862.) Rogers provided

CD with information on *Myanthus* and *Monachanthus* (see letter from John Rogers, 22 January 1862, and *Orchids*, pp. 158 n., 236 n.).

Rolleston, George (1829–81). Physician. Linacre professor of anatomy and physiology, Oxford, 1860–81. (*DNB*.) CD had sent Rolleston presentation copies of the third edition of *Origin* and of Gray's review pamphlet (A. Gray 1861; see *Correspondence* vol. 9, letter to George Rolleston, 2 March [1861], and letter from George Rolleston, 1 September 1861, and Appendixes III and VII).

Royer, Clémence Auguste (1830–1902). French author and economist. Studied natural science and philosophy in Switzerland. (*Dictionnaire universel des contemporains*.) Royer translated *Origin* into French in 1862 (see *Correspondence* vol. 9, letter to John Murray, 10 September [1861], and this volume, letter to Asa Gray, 10–20 June [1862]).

Rucker, Sigismund. East and West India broker with premises in Great Tower Street, City of London. Resided at West Hill, Wandsworth. (*Post Office London directory* 1861, *Post Office London suburban directory* 1860.) Rucker is acknowledged several times in *Orchids* for having 'repeatedly' sent CD orchid specimens (see *Orchids*, pp. 158 n., 214 n., 231 n., and 249 n.; see also *Correspondence* vol. 9, letter to John Lindley, 16 November [1861]).

Smith, Frederick (1805–79). Entomologist in the zoological department of the British Museum from 1849. Specialised in the Hymenoptera. (*Entomologist* 12 (1879): 89–92, Gilbert 1977.) CD had corresponded with Smith about various entomological subjects for several years, and sometimes sent insects to him for identification (see *Correspondence* vols. 6–9). CD sent him a presentation copy of *Origin* (see *Correspondence* vol. 8). CD thanked Smith for sending him a humble-bee specimen carrying orchid pollinia in *Orchids*, p. 164.

Sowerby, George Brettingham (1812–84). Conchologist and artist. Illustrated numerous works on fossils. (*DNB*, Sarjeant 1980.) Sowerby prepared the illustrations for *Orchids* (see *Correspondence* vol. 9, and *Orchids*, p. vi).

Stainton, Henry Tibbats (1822–92). Entomologist. Founder and editor (1855–74) of the *Entomologists' Annual* and of the *Entomologist's Weekly Intelligencer* (1856–61). (*DNB*.) CD had long regarded Stainton as a 'distinguished Entomologist' (see *Correspondence* vol. 5, letter to W. E. Darwin, [25 April 1855]). CD occasionally read the *Entomologist's Weekly Intelligencer* and had a letter published in it in 1860 (see *Correspondence* vol. 8).

Stur, Dénes Rudolf Jossef (Dionys) (1827–93). Hungarian stratigrapher, palaeobotanist, and vertebrate palaeontologist. Joined the staff of the Imperial-Royal Geological Survey of Vienna in 1850. (*BHGW*, Sarjeant 1980, *Quarterly Journal of the Geological Society of London* 50 (1894): Proceedings, pp. 55–6.) CD's only former contact with Stur or his work seems to have been reading Stur's monograph on *Astrantia* (Stur 1860; see *Correspondence* vol. 9, letter to J. D. Hooker, 15 January [1861]), a copy of which is in the Darwin Library–CUL.

Tollet, Georgina. Daughter of George Tollett (1767–1855), of Betley Hall, Staffordshire. (*Burke's landed gentry* 1846.) Family friend of the Darwins and Wedgwoods.

Tollet, Georgina, cont.

(Freeman 1978.) Edited the manuscript of *Origin* (see *Correspondence* vol. 7).

Turnbull, George Henry. Resided at the Rookery, Down. (*Post Office directory of the six home counties* 1862.) CD thanked Turnbull in *Orchids*, p. 158 n. for allowing him to use his hot-house and for providing 'some interesting Orchids'.

Veitch, James. *See* Appendix III.

Walker, Francis (1809–74). Entomologist. Expert on Hymenoptera and Diptera. Described the Chalcididae (parasitic Hymenoptera) collected by CD on the *Beagle* voyage. (*Entomologist's Monthly Magazine* 11 (1874): 140–1, Gilbert 1977.) Walker, the author of numerous catalogues of the insect collections at the British Museum, was an acquaintance of Frederick Smith (see above), and also identified insects for CD (see *Correspondence* vol. 9, letter from Frederick Smith, 1 February 1861, and letter to Francis Walker, 19 June [1861]).

Wallace, Alfred Russel (1823–1913). Naturalist. Collector in the Amazon regions, 1848–52, and in the Malay Archipelago, 1854–62. Independently formulated a theory of natural selection in 1858. (*DNB*, *DSB*.) Since the joint publication of their papers on natural selection (Darwin and Wallace 1858; see *Correspondence* vol. 7), CD had remained in correspondence with Wallace, sending him a presentation copy of *Origin* (see *Correspondence* vol. 8, Appendix III), and discussing some of the reviews of the work (see *Correspondence* vols. 8 and 9).

Wallis, William (1813/14–93). Surgeon of Hartfield, Sussex, *c.* 1845–91. District medical officer, East Grinstead Union, *c.* 1851–88; surgeon, Hartfield Self-supporting Dispensary, *c.* 1871–6; registrar of births and deaths for the parishes of Hartfield and Withyam, *c.* 1862 – *c.* 1882. Collector of orchids. (*Medical directory* 1847–94, *Post Office directory of the six home counties* 1845–86.) Wallis provided CD with numerous specimens including the rare *Malaxis paludosa* (see *Correspondence* vol. 8, and *Orchids*, p. 130 n.).

Watson, Hewett Cottrell. *See* Appendix III.

Weddell, Hugh Algernon. *See* Appendix III. Weddell's work on hybrids of *Aceras* (Weddell 1852) is cited in *Orchids*, p. 19 n.

Westwood, John Obadiah (1805–93). Entomologist and palaeographer. Hope professor of zoology at the University of Oxford, 1861–93. Entomological referee for the *Gardeners' Chronicle*. (*DNB*, Gilbert 1977.) CD admired Westwood's work, nominating him for the Royal Medal in 1855 (see *Correspondence* vol. 5). Westwood provided CD with specimens of bees with orchid pollen-masses attached (see *Correspondence* vol. 8, letters to J. O. Westwood, 25 June [1860] and 9 July [1860], and *Orchids*, p. 35).

APPENDIX V

Reports from the Scottish press on Thomas Henry Huxley's Edinburgh lectures on the 'relation of man to the lower animals'

On 4 and 7 February 1862, Thomas Henry Huxley lectured at the Philosophical Institution of Edinburgh on the subject of the 'relation of man to the lower animals'. He wrote to Darwin on 13 January telling him of the enthusiastic reaction of his audience, and sending a report of the second lecture that had appeared in the *Edinburgh Evening Courant*. The following day, a fiercely critical leading article was published in the evangelical newspaper, the *Witness*, which had previously published relatively neutral reports of Huxley's lectures (see *Witness*, 9 January 1862, p. 3, and 11 January 1862, p. 3). The article attacked not only Huxley's lectures, but also the directors of the Philosophical Institution itself for allowing them even to take place. Huxley sent Darwin a copy of the piece with his letter of 20 January 1862, noting that it had been followed by 'another more violent more scurrilously personal & more foolish' (see *Witness*, 18 January 1862, pp. 2–3). Darwin was 'much amused' by the article, stating that such abuse was 'as good as praise' (letter to T. H. Huxley, 22 January [1862]).

By way of response to the criticism of his lectures in the Scottish press, Huxley wrote a letter to the *Scotsman*, which was published on 24 January 1862. In the letter, he contrasted the coverage of his lectures in the *Witness*, which he cited as an example of the 'noisy injustice and unreason' he had expected to receive from the less 'respectable' parts of the Scottish press, with the candid and courteous, though nonetheless hostile, review that had appeared in the *Week*. (On 25 January 1862, the *Witness* published a final attack on Huxley, in response to his letter to the *Scotsman*, under the headline 'Professor Huxley's compliment to the "Witness"'.)

Darwin received a copy of Huxley's letter to the *Scotsman* through Charles Lyell, and subsequently obtained from Huxley the article from the *Week*, pronouncing its author 'a good man' (see letters to T. H. Huxley, 2 February [1862] and 6 February [1862]).

As a supplement to the correspondence about Huxley's lectures, the report from the *Edinburgh Evening Courant*, the articles from the *Witness* and the *Week*, and Huxley's letter to the *Scotsman* are all printed below, in chronological order of their publication.

Edinburgh Evening Courant, 11 January 1862
The relation of man to the lower animals.

Yesterday evening Professor Huxley, of the Government School of Mines, London, delivered the second of two lectures, "On the structural relations of man to the other forms of animal life." He stated that in his first lecture he had endeavoured to illustrate the proposition that, whatever system of organs were considered, man, if compared with the highest of the apes, approached that animal far more nearly than the lowest apes approached the highest apes. On the former occasion he considered the external forms and proportions, and the chief characteristics of the bony skeleton. He had now to continue and conclude the subject by taking into consideration the structure and relative forms and proportions of two sets of organs, on which more stress had been laid than any others in the discussion of these questions—namely, first, the hand and foot; and second, the brain. The learned Professor accordingly proceeded, with the aid of numerous diagrams and anatomical preparations (including the skeleton and preserved brain of a chimpanzee, lent by Professor Goodsir from the Anatomical Museum), to show the relation of the human to the gorilla and the chimpanzee hand—pointing out the close resemblance of the gorilla's hand to man's; the less perfect resemblance in the hand of the chimpanzee, where the thumb was very short; and the apparent absence of a thumb in some of the lower species, in which, however, its rudiments were found on dissection. He went on to show the relations of the foot, remarking that apes were improperly designated by Cuvier quadrumana; for, taking bone for bone, and proportion for proportion, in the structure of the lower limb of the ape, it was a true foot, on precisely the same plan as the human foot. It had, however, a prehensile faculty which man's foot had altogether lost by enclosure in boots. The great toe, he pointed out, was only rudimentary in the orang's foot, and altogether there was a greater difference between the orang's and the gorilla's foot than between the gorilla's foot and man's. Speaking of the brain, he said it had been to him a matter of great surprise to find that there was so little to be seen, so far as anatomy went, between the brain of the higher apes and that of man. They might be sure his business here was not in any way to lower the dignity of human nature, or to lead his hearers to suppose that they were simply apes, but to state anatomical facts; and considering that, without doubt, the brain was the organ of the mind, and that there was great gulf fixed between the highest apes and the lowest of mankind in intellectual faculty, it was a very remarkable fact that, excepting in point of mass, there should be so very little difference discoverable. The minimum weight of the human brain was considered to be 31oz., while the largest brain of any of the apes was 17 or 18—perhaps not more than 15oz. The maximum weight of the human brain was about 64oz., or double the minimum; so that, on the scale of proportion, there was no greater interval between the gorilla and lowest man and lowest man and highest man. In speaking of these matters, we yet spoke in total ignorance of the human race as it existed in the earlier epochs of the world's history. The time over which our knowledge of the structure of man extended was brief, viewed as geologic time, and not by the insignificant standard of civil history. It was only within the last few years or months that we had obtained clear and unmistakeable evidences of man in even the most recent of the geological formations. They had, however, discovered the traces, not of man, but of his works, in hatchets and other implements which, it was proved to demonstration, belonged to man as the contemporary of the great mammoth, and other animals that existed about the time of the great Drift. Yet, while no positive evidence of that period had been found, while even no fragment of a skull had been found, in connection with these implements, there was a most singular indication of what they might hope to find in geological records, if pushed further. Some time ago there was discovered in a cave, in the valley of Neanderthal, in Germany, a most remarkable human skull, associated with bones of the ordinary human size and proportion, but in which the mass of the brain must have been marvellously smaller than even the lowest Australian brain, and there were other marks about the skull, all of which had an approximation to the ape type. No one had been able to ascertain what its geologic age was, but he believed it was the first discovered relic of a race and variety of mankind older than any of which we had at present cognisance. It would be hazardous in the last degree for any cautious thinker to say how close the structure of the human and the simian brain might one day be found to approximate. Whatever system of organs they might study, he asserted that the same result would hold good, that the higher apes approached man more closely than the lower did to the higher apes. One of the daily papers, in noticing his first lecture, stated that it consisted of an advocacy of the views of Mr Darwin. He might be excused for saying that that was not quite the case. It had been his province in these two lectures simply to state facts. But he should be wanting in his duty did he state these facts baldly without drawing attention to what seemed to be their clear indication. He must not be misunderstood as saying that the differences between man and the apes were slight or insignificant. The very reverse of that was true. So far as our knowledge of existing creation went, no intermediate forms existed which bridged over the interval between the ape and the man. The highest ape and lowest man were as distinct from one another as were any two genera of allied families. But so was the orang from the gorilla, the lemur from the orang, and so on down to the rhinoceros and the horse. So that they might see to what conclusion all he had said tended; and therefore he gave a tacit amount of praise to the sagacity of his reporter. His conclusion was this, that if any physical causation could be discovered by which the existence of such animals as the gorilla, the orang, the lemur, the rhinoceros or the horse, was to be accounted for, this causation was also amply sufficient to account for the origin of man so far as he was an anatomical structure, for the organisation of man was not more different from that of the animals which approached to him most nearly, than that of the lemurs and other lower forms were from one another. Or finally, if Mr Darwin's hypothesis or any other would account for the origin of the gorilla, the orang, and the lemur from a common stock, then to the eye of an anatomist the origin of man presented no difficulty on account of any gap between him and the animals immediately below him. But, on the other hand, if they asked him whether he conceived Mr Darwin's hypothesis to be correct, and whether it would account for the origin of these animals from a common stock, then he replied frankly that Mr Darwin's hypothesis lacked one great link in the chain of proof.

He had, he thought, satisfactorily proved that what he termed selection must and did occur in nature. He has also proved that the superfluity of this selection was competent to produce from one common stock forms which differed as much from one another as did our genera or species from the animals most nearly approaching us. But he had not yet proved that any form of selection could produce from one common stock forms which differed from one another so far that they would not breed with one another, which was one of the chief distinctions of species in nature. Mr Huxley went on to say that he simply viewed Mr Darwin's hypothesis as a key which helped to solve all sorts of riddles, and which he would not throw away so long as he had no other. Being the only key before him he adopted it, because it enabled the naturalist to group together a vast number of phenomena which would otherwise remain isolated and without meaning. The learned Professor concluded by an eloquently expressed analogy between the Alps upheaved from the lower strata of the earth's crust and towering above the plains, and man standing high above the lower animals, but sprung from the same level as his humbler compeers. The lecturer was listened to with deep attention by the audience.

Witness, 14 January 1862

The philosophical institution and Professor Huxley.

It is far from our design to express either opposition or indifference to this admirable Institution, and its many costly schemes and arrangements for the intellectual advancement of various classes in our city. Its title might have been objected to as rather pretentious, had not the assumption of it by almost every small "Mutual Improvement Society" in our cities, towns, and villages, been a familiar fact, and had not Philosophy been in our days not only brought from the lofty region where she "commerced with the skies," and with such transcendent spirits of earth as Socrates and Plato, down to the academic groves and guarded cloisters in which the disciples of these great men give a smattering of their own knowledge, and act as a sort of pupil-teachers; but also led forth from these comparatively select haunts, into the public streets, and, to the tune of "the march of intellect in the nineteenth century," marshalled to the market-place, where she was required to explain her profound secrets in a way that should be intelligible to the indiscriminate throng, and whence she was afterwards taken to preside over an aspiring "Infant School." It may be doubted whether, along with a change in her seat, range, and votaries, there has not also been a change upon Philosophy herself; and whether, in so far as she has been popularized and cheapened, she has not also been made to forego her high character and mission. Moreover, every art and every branch of knowledge are now dignified with the sounding name of Philosophy; and all societies that pursue either of them rejoice in the name of "Philosophical." The city Institution with which we are about to deal is unquestionably far better entitled than any of these to such a designation, though we cannot altogether regard its membership as constituted by so many Platos, or its studies and exercises as purely, or even prominently, philosophical. The ticket of membership is not a passport into the temple of wisdom, much less a certificate of philosophical attainment. We should, indeed, rejoice to see the monopoly of philosophy really broken up, and the whole of our city population entering upon the hitherto fenced field, digging for the long hidden and interdicted treasures, and proving themselves a great congregation of profound sages; but the requisite leisure, genius (including inclination as well as capacity), and training, are lacking, we fear, with many in such a multitude, and the Institution cannot supply what is wanting. It is with various classes in Edinburgh a favourite Institution, and in certain respects has been eminently successful; whilst there is also the promise of a signal increase for its future activity and prosperity. Still, scarcely enough has been done to justify the select title conferred at its birth, and its progress can hardly be defined a march of intellect in and through Philosophy. The Institution is very valuable for its library, which is already large and choice,—for its reading-rooms, fully provided with the freshest stock of current periodical literature, and placed in almost hourly communication with the telegraphs of Europe,—and for the ancient and foreign language classes regularly taught in connection with it. Its capital recommendation, however, in the estimate of our fellow-citizens, is, that it furnishes courses of lectures, delivered annually during winter. Nor can it be denied that the most of these are of very high merit and value, the services of men—most of them celebrated, and all of them perfectly well-known in the country—having been engaged for this important department. Still, there is, with the view of meeting the tastes of all the members, and of avoiding exhaustiveness and monotony, so great a variety of subjects, chosen, too, out of all the fields of knowledge, that the members can neither be taught nor trained in any one branch or section of "Philosophy." The syllabus of a single session's prelections resembles the index to an encyclopædia, and gives us a lively idea of the abrupt, irregular, and incoherent way in which the lecturers sweep from pole to pole of the intellect, and alight at all points of the compass for an hour's superficial survey.

Our intention, however, is not to notice the defects obviously attaching to the rapid, complete, and grotesque change of subjects in the different courses of one winter's lectures, though these defects call for prompt consideration and amendment. We have a simple question, yet one of no small importance and interest to the Christian public, to put regarding the Directors' meaning and motives in engaging Professor Huxley to appear on the Institution's platform, and exhibit his anti-scriptural and most debasing theory of the origin and kindred of Man. That theory is viewed by all theologians, and by the great body of Christians of all denominations, as standing in blasphemous contradiction to Bible narrative and doctrine, whilst almost all the masters in philosophy and science utterly repudiate it as untenable and absurd. Was it then indispensable to the creed or the culture of the members—especially the youthful ones—of the Institution, that the advocate of the vilest and beastliest paradox ever vented in ancient or modern times amongst Pagans or Christians, should be invited, for a handsome consideration, to expound and defend it? Have those members made such a prodigious progress in Philosophy, and attained to such a sublime intellectual elevation, that, in order to keep them duly humble, or simply to furnish the most striking of contrasts, they must be told that they have each and all directly sprung from *apes*, and that the gorilla of our day is their elder brother, though some sinister chance has doomed him as yet to a shorter thumb and lighter brains, and though hitherto the speechless and reason-lacking animal has never stepped forward to assert his relationship and his priority, and to claim the first and largest share in the inheritance of humanity? What conceivable good end could be gained by bringing this Professor and his ignoble hobby upon the platform of the Edinburgh Institution? We wonder greatly that he and the Directors, along with the shadow of the creature hailed by them as a brother, were not entirely left to themselves in the lecture-room. On behalf of the dignity of humanity, the members ought unanimously to have deserted the Hall—if, indeed, they should not have adopted a more emphatic mode of protesting against the foul outrage committed upon them individually, and upon the whole species as "made in the likeness of God!" It seems, however, that the strange proceedings of the Directors in hiring a person to parade the monkey as the father and the brother of all men, were fully sanctioned by the members of the "Philosophical," who also applauded the lecturer, and did so with extra vigour and enthusiasm in those passages which most explicitly and bluntly asserted the original and absolute identity of man and ape. It is surprising that, at the close of the lectures, the hearers refrained from forming themselves into a "Gorilla Emancipation Society," and from concerting some prompt measures for humanizing and civilizing their unfortunate brother, as a significant earnest of what they intend to do for the whole fraternity of apes. It may be easier, however, for the kindred-loving members of the "Philosophical," when aiming at one common level for them and the gorilla, to step down to his position, than to bring him up to theirs—to degrade and sink themselves to his state, than to elevate him to theirs; though, fortunately for the grand distinctions of humanity, no man, by any amount of moral and intellectual self-abnegation, can cast himself down into identity of nature or destiny with even the highest type of an irrational animal: he may, in a serious sense, fall *beneath* the brutes, but he cannot place himself *beside* them. Sceptical followers of science have frequently assured us that the poor negro could not, at least for several ages, and unless his race intermixed with the European, be civilized or Christianized, and that all attempts to bring him, or any less remote descendant than his great-great-great-great-grandchild, within the pale of a Church were hopeless; but now the science ("falsely so-called") of sceptics leads them with an outrageous inconsistency which yet is fully characteristic, to rush into the opposite extreme, and to hold that their dearly beloved monkey brother is capable of a speedy assumption of all the grand distinctions of man.

It will be strange if our fellow-citizens outside the "Philosophical" do not press the pertinent question why Edinburgh and one of her leading Institutions should have been compromised by the two exhibitions of last week in Queen Street Hall. Are we next to have the author of the "Vestiges" engaged by the Directors to state and maintain, on the Institution's platform, his theory of all terrestrial existence? This would be far fairer, and to the youth of our city less injurious, since this theory has now been many years before the public, and has fallen dead within the circle of the sciences, every one of which inflicted upon it blows both disfiguring and fatal, until its whole body, from top to toe, was one ugly gash; but Professor Huxley's theory has as yet the recommendation of novelty; nor has there been time for its complete demolition,—at least the conclusive facts and arguments against it are comparatively unknown among our intelligent young men. Will the "Philosophical" by and by produce in the Hall some apostle of Mormonism? Even this would be a less offensive, mischievous, and inexcusable exhibition than was made in the recent two lectures by Professor Huxley. Let us not be misunderstood as denying or disliking, or seeking either to curtail or hamper, the rights of free inquiry, on the part of the lecturer, or of any person who may also publish to the world, by speech or writing, the course and issue of his unfettered speculations. But we do emphatically deny that the platform of such an Institution as the Edinburgh "Philosophical" is the proper place for broaching and defending such suspicious novelties as are the pets of Professor Huxley; and we cannot but reprobate severely the conduct of the Directors in inviting, or even permitting, him to mount his hobby there. When our young men have for months,—with the assurance of receiving no harm to their Biblical faith, but great instruction and delight for their minds,—been listening to admirable lectures on history, poetry, and music, unexpectedly and without warning this anti-Scriptural theory regarding the origin of man is insinuated, and to many an inexperienced and unarmed hearer its novelty may render it so seductive as to lead him astray from Revealed Truth. Had the Directors such and other, equally sad contingencies before their eyes when they engaged the Professor to propound and advocate the brotherhood of man with monkey? We have in this paper dealt exclusively with the share which the Directors of the Institution took in last week's unseemly exhibitions. We shall next examine these exhibitions themselves, and discuss Professor Huxley's odious, absurd, and mischievous theory, that we have all sprung from apes.

Week, 17 January 1862

Men and apes.

In the second of the two lectures which he recently delivered before the Philosophical Institution here, Professor Huxley declined to acknowledge that those lectures were to be taken as advocating Mr Darwin's theory of the Origin of Species. They were, he said, a mere statement of facts which any man could verify for himself. Was this expected to be taken seriously? Does any one assemble facts without assuming, consciously or unconsciously, principles which regulate the manner in which they are collected? Do scientific lecturers, known to take a particular interest in a given theory, give statements of facts which have no relation to any theory whatever? Is there no such thing as arranging facts so as to make them suggest the theory which they seem to the lecturer to support? Surely this was precisely what Mr Huxley did, and what he had no need to disown. He thinks, he tells us, that Mr Darwin's theory, though not quite made out, is more strongly supported by a number of facts than most persons are disposed to admit. He thinks it has hardly had fair play. He comes down accordingly, and delivers two lectures, in which he arranges his facts, to the best of his ability, in the way that will most strongly exhibit the doctrine which may, he thinks, be the true explanation of them. He accompanies his facts, as he produces them, with a commentary which secures that no one shall escape the question—'What now do these facts suggest?' and that no one shall remain ignorant of the kind of reply which the lecturer is inclined to give. This is just one way of putting a case; the strongest of all ways if well managed, and a perfectly fair one, if the object is sufficiently avowed, and the advocate does not affect an impartiality which he does not feel. Although Mr Huxley objected to his lectures being regarded as advocating Mr Darwin's theory, he did not conceal his own tendencies in that direction. We accuse him therefore not of anything disingenuous, but of squeamishness, in drawing a distinction where there was no difference. And we have made these remarks simply in order to vindicate our own fairness in animadverting farther on the lectures. The truth is, that the question, how far the lectures made a plausible case for Darwin's theory, in one of its applications, is the main question, we had well-nigh said the only one, which has any interest for those who heard the lectures, or for those who discuss them.

Mr Darwin maintains the hypothesis that all species of living things may have been developed by slow degrees and divergencies, under the operation of a principle of "Natural Selection" which he explains, from one or at the most from very few original simple forms. In so far as the application of the principle to man is concerned, it evidently points towards a common origin for him with the higher apes, as his nearest cousins. Mr Huxley's lectures were chiefly occupied with facts which illustrate the physical analogies between man and the animals, and in particular between man and the apes. It is this hypothesis, and this least welcome application of it, which, designedly or undesignedly, the lectures suggest. Now we are to make no ungrounded insinuations as to the views, or the motives either of Mr Darwin or of Professor Huxley. Neither of them, as far as we are aware, has expressed himself adversely to the idea of creative power being concerned, at some point, in the order of the world. Neither of them has denied that there may be an element in man which cannot be accounted for by the hypothesis of transmutation, however fitly that may account for his physical framework. Nor do we maintain that either the creative energy of God, or the immateriality of man is necessarily excluded by their hypothesis, so far as they have developed it. At the same time, it goes in that direction, that is to say it tends to remove difficulties out of the way of those who desire to introduce an atheistic view of the universe, and a materialistic view of man. It is this circumstance which has undoubtedly made the hypothesis so welcome to a number of persons. It is this circumstance, also, which disposes those who are otherwise minded to examine it with particular care. An interest is felt, certainly, on other than scientific grounds, in the question whether nothing beyond natural law is concerned in the production of all species of living things past and present, and whether the origin of man is physically identical with that of every other species, and in particular of the apes. Men ask how does this comport either with Natural or Revealed Religion; and they are quite right in asking the question. As for Mr Darwin and Professor Huxley, however (of whom the one advances hypothesis, and the other without precisely affirming, thinks there is probably a great deal in it), they profess to take their stand simply on the fair construction of ascertained facts. They have been looking, as they tell us, for truth, as Nature presents it to their eyes, and have no wish to go one step beyond her findings. It is quite right that they should ask us to accompany them into this field, and we do not see why anybody should feel the least hesitation in complying.

Although Professor Huxley's "facts" pointed principally to the question about man, we shall make a few remarks in the first place on the more general question as to the Origin of Species. What is the state of that question? And what has Mr Darwin done to throw light upon it? The state of the question is this. Since the beginning of the world it has been apparent to men's eyes that there is a very considerable resemblance among various species of animals. We refer to the animal kingdom alone, for convenience sake. As observation became more minute and scientific, it became clear that analogies the most remarkable and extensive connected together the different species, as types of organised being; so that each holds its place on a system which not merely comprehends but unites all. By-and-bye the facts of geology made men aware that the past had teemed with life, in forms subject to the same analogies, and capable of being ranked in the same system with those which now exist; that these had succeeded one another, some becoming extinct and others appearing; and that, generally speaking, the more highly organised and endowed forms of life appear later in the series. It is this fact of the succession of species on the earth, and the appearance of a certain order (though not rigorously systematic) in the succession, which is the new element supplied by geology to this question of species. Long ago these facts became the starting points for hypothesis. Could no natural causes be imagined or discovered to explain the origin and succession of species? It appears, on the one hand, that species have become extinct, and have originated, in times of which the rocks preserve the record. It is known, on the other hand, that there is a certain plasticity in organised life. The forms of animals and their instincts do yield somewhat to the presence of altered circumstances, and certain changes may thus be superinduced by various means on the original type. May it not be, then, that species has produced species? May it not be that at some time, under some new influence, the manner and form of life of a given species has altered a little, or even more than a little; and so, by little and little, by one divergence and another, the types have multiplied, and the vast catalogue of species has been filled up? Guesses of this kind were so obvious that they could not fail to occur to speculative men from the dawn of zoological science. But the answer which science, truly so called, gives to them is equally simple and obvious. It is granted that it is not inconceivable that species should produce species, as individual produces individual—like, and yet another. It is conceivable that it might be so; not in the sense of our being able to represent to ourselves how it should be so, but in this sense, that no man can demonstrate, *a priori*, that such an arrangement is impossible. But the question is one of fact. Is there any evidence of such a thing ever or anywhere taking place, as a new species being physically derived from an old one? After the most heedful examination of all facts adduced, science has answered and answers, No. There are curious facts about the variations observable or producible, especially in domestic animals. But all these, even in the extreme cases, sport within the limits of the specific character proper to each kind, and tend to return from all eccentricities that would obscure that character. There is no such change as estranges the creatures produced from the old kind or constitutes them into a new specific type, and you have only to turn them loose, free from the artificial influences, in order that this may appear undeniably in their progeny. As far as can be judged from all evidence bearing on the question, the species is as true a natural unity, with as strict and incommunicable an identity, as any separate *self* in the whole range of beings. All attempts to break down this position have failed, simply and entirely failed. Therefore all attempts to account for the origin of species on the assumption that one species can produce, as its progeny, another which will prove diverse from the first, proceed in defiance of all that nature teaches us about species. Proceeding in this way any dream may be propounded. The legitimate object of a hypothesis is to propound a possible cause or mode of origin, *conformable to the certain knowledge we already have.* This is the rigorous condition of an admissible hypothesis in the case; and this is the reason why all hypotheses of the kind to which we have referred have been generally rejected by scientific men.

Now, the fundamental mistake of Mr Darwin's book was this, that professing to account for the Origin of Species by transmutation, it spent its force entirely away from the main point. In so far as it dwelt upon the similitudes which obtain between the various species of organised beings, it simply added some new illustration of what was perfectly well known before, but nothing that was in the least fitted to alter the balance of argument or evidence on the question. In so far again as it dwelt on the changes which living species may be made to undergo by careful breeding, it still only gave some new illustrations of that which, as to its whole argumentative force, was known before. There was much entertaining detail about pigeons and other impressible creatures. But, after all, it amounted to as much and no more than was patent two thousand years ago to the Roman when he recognised and proceeded on the distinction between the Thessalian and the African horse; or to the Greek, when he bred the Spartan and the Molossian hound. There are variations within the limits of the specific character, but there is no origination of new species. The attempt to produce evidence to establish that there is such a thing, was not only a failure, but we may fairly say a confessed failure. On what, then, was the strength of the work spent? On an endeavour to show that, *provided you admit that the species in its natural process of generations is capable of altering into a new species*, there are natural causes at work ("natural selection") which would tend to educe the capability into fact, which might sometimes—perhaps often—favour the evolution of the possible change. This is *ignoratio elenchi* with a vengeance. Why, if the hypothesis be admitted, it will be easy enough to point out or imagine causes that might set changes agoing. It needs no wizard to succeed in making this appear likely; and Mr Darwin succeeds, so pleasantly and so ingeniously, that many an unwary reader imagines that he is proving his theory. All he is proving is, that if you admit the fundamental principle of his theory, it might possibly work in practice. He gets into some serious difficulties even as to this, but it is hardly worth one's while to advert to them. He has done nothing to alter the state of the argument as to the identity and permanence of species, as determined by all the proper evidence upon the point to which we have access. Yet this was the point, the one point, with which he was bound to grapple in advocating an Origin of Species by transmutation. His agreeable but overrated book has therefore left the real question where it found it.

We have dwelt on this, a little too long perhaps, because a precisely similar vice pervades Professor Huxley's manner of assembling and and presenting the "facts," which, he says, we are to verify if we choose. We are, in the judgment of those who heard the lectures, whether the impression which they seeme fitted and intended to leave, was not of this kind;—that there is something in the course of recent discovery which tends towards obliterating the old distinctions between man and the lower animals—something which requires us to look on the relations between the two in a new light—something which calls on us to be prepared to see, ere long, "the differences between man and the apes swept away, and their relationship established." Certainly the reports of the lectures, though not *verbatim,* sufficiently bear out this representation of their tendency. And the *something* which calls upon us thus to modify our position, is the information which men of science can now give us of the close structural analogy between man and the higher anthropomorphoid apes, especially our ugly friend—the gorilla. Now, we do not dispute that the structural similarities in question were in the main fairly stated by Professor Huxley; though stating them from his own point of view, and under the influence of his own theoretic tendency, he failed we think to estimate soundly the relative value of the resemblances and the diversities. The resemblances are, however, extremely striking, interesting, and worthy to be observed and studied. But yet the plain truth is, (and Professor Huxley, if he would study logic, could not fail to see it), that there is nothing in any or all of them that goes to establish between man and the apes any relationship different from that which everybody has seen, confessed, and accepted, ever since man and ape were first compared as to their physical structure. The argument, if it is meant for an argument, and if it proposes to persuade us to believe anything we did not believe before, is a mere blunder.

From the days of Galen, from the days of Aristotle, from time immemorial, it has been accepted truth that man's physical organisation is analogous to that of the lower animals. Whatever marks he bears of another higher nature combined with the material, whatever dignity is stamped on him as an heir of immortality, the manner of his bodily organisation is animal. Anatomically, chemically, physiologically, the similarity has been constantly assumed, and constantly verified. There is no doubt at all, therefore, in anybody's mind that if you class man according to his physical characters he belongs to the animal kingdom. Neither is there any doubt that of all animals the apes approach him nearest in form. The relationship is therefore already acknowledged. All the vertebrata are so far like him, for he is one of them. All the mammalia are still more like him, for he is one of them. Still more nearly do the quadrumana approach him. This we all knew. What has Mr Huxley found out? That some of these apes approach him a little nearer still. Why, suppose that to-morrow we discover some irrational creature whose physical form, judging from that alone, would rank him even in the same genus with ourselves, does Mr Huxley suppose that this would alter materially the old conditions of relationship? The only point in the matter which any one is concerned either to affirm or deny (except of course in so far as every one is concerned to have the real truth ascertained) is identity of species, inferring a probable or actual common origin. The Professor does not assert anything like this of the gorilla; he hardly expects it, we suppose, in any irrational creature that is to turn up. Then if it be so, all the affinities on which he dwells are simply interesting and instructive facts, illustrating that close connection of ours with the animal kingdom which is admitted on all hands.

We are very far from thinking that it is an unimportant matter whether our zoological affinities with the quadrumana are or are not overrated, in such statements of "facts" as Professor Huxley's. We believe that the overstatement of the affinity is unwholesome, and frequently proceeds from unwholesome mental tendencies. If therefore we were considering that question, we should have something to say upon the point. Setting aside all higher criterion, and keeping to what is physical, we should maintain that a very high value ought to be assigned not merely to the diversities of anatomical detail to which Professor Huxley had to advert, but to the readjustment of every portion of the framework, with a view to a new posture and mode of progression, which we find in passing from the apes to man. So thorough and pervading a re-adjustment, the subtle influence of which pervades the whole structure, the true foot, the muscles of the calf and thigh, the relation of the thigh-bone to the pelvis, the obliquity of the latter, the curvatures of spine, the disposition of the internal organs as to bulk and weight, the position and setting on of the head, with the absence either of ligature or of powerful muscles to hold it up, the arrangement of the carotids, and the like—all these we should assert to constitute a diversity, which fairly demands to be recognised by a very decided chasm in the classification. But when our affinities to the animal kingdom are brought forward as pointing to some new and portentous "relationship," such discussions of detail are out of place, and only obscure the true ground of argument. Since men began to think they have owned that, by the body, they belong to the animal kingdom. The true distinction of man is grounded on his double nature, which really constitutes him into a new kingdom of spiritual and moral being, as distinct from the animal kingdom as that is from any other. As to his bodily connection with the animal kingdom, that being palpable and acknowledged, it really matters nothing at all to the question of man's distinction and superiority, whether certain animals are classified a little nearer him, or a little further off. One naturalist, as Cuvier, may assign man to the separate order of the *Bimana*, and place him there by himself. Another, as Bury de Saint Vincent, fired we may suppose with sentiments of liberty, equality, and fraternity, and indignant that man should assume to "bear like the Turk, no brother near the throne," may give him for an *adlatus* the Ourang-outang; straining a point to place among the *Bimana* that four-handed creature. What does it matter? Suppose that of a number of animals, all more or less like man in structure, some are shown to be a little liker than was supposed, does that alter anything in the state of the argument, or call for any change in the general convictions of men? Man remains a distinct species even in physical character. On a wider view of what his nature includes, he stands divided from all the others by a sovereignty of superior nature, which not the widest range of zoological classification can adequately measure. And it is worth noticing that the apes, which approach in form so nearly, are not so remarkable for animal intelligence as many creatures whose form is less human.

While, therefore, we thank Professor Huxley for his lectures, as turning attention to many interesting facts, we must look upon it as a mere private bewilderment of his own that he regards these as either so surprising or so subversive. If he is to make progress in the direction towards which he has set his face, he must apply himself to break down that principle of the reality and constancy of specific difference, which has hitherto been nibbled at in vain. He will not break it down by any amount of evidence as to the analogies which extend from species to species, with a wonderful order in variety. Nay, in this way he will not even advance his cause one step. All this is but new illustration of an accepted truth, in which the accepted doctrine of species reposes. There would still be some scientific difficulties in the way of establishing any generic relationship between man and the animals, even if the doctrine of the constancy of species were overthrown. There would still remain also all the difficulties arising from revelation. We do not now consider how either class of difficulties bears on the argument. It is enough to say in the meantime that Mr Darwin and Professor Huxley, both alike, mistake the question, and have failed to lay the very foundation of their case.

Scotsman, 24 January 1862

Professor Huxley and his critics

The Government School of Mines,
Jermyn Street, London,
January 21, 1862.

Sir,— In accepting the Invitation to lecture in Edinburgh with which the directors of the Philosophical Institution honoured me last year, I was fully aware of the unusual difficulty of the task I undertook.

For my acquaintance with the conditions of the enterprise led me to see the necessity of striving towards the attainment of three very different objects—the first and the most important of these being the discharge of my obligation to the Directors and of my duty to myself, to state fully and fairly those conclusions at which I, as a man of science, had arrived, after no brief or hasty study of a great scientific question.

My second aim was, not only to avoid all unnecessary offence to persons whose views differed from my own, but to present the doctrines I espouse in such a manner as to win at least a fair and favourable consideration even from those who had been taught to regard them with disfavour.

And, in the last place, while feeling assured of the just and reasonable dealing of the respectable part of the Scottish press, I naturally hoped for noisy injustice and unreason from the rest, seeing, as I did, the best security for the dissemination of my views through regions which they might not otherwise reach, in the certainty of a violent attack by an Edinburgh paper for which the striking abilities of the lamented Hugh Miller once earned an honest fame, but which since his death has swiftly sunk to its now universally recognised position of a regenerate "Satirist" or Pharisaic "Age."

The fulfilment of these my aspirations has afforded me unfeigned satisfaction. For, much as I was obliged to condense my lectures, the accusation that their meaning or purport could be doubted has yet to be brought against me.

Nor has any report which I have seen affirmed that my audience listened with other than the kindest and most uninterrupted attention. And if I now recal with genuine pleasure the unmixed applause which fell upon my ears, it is not, I trust, because that recollection solaces a petty vanity, but because it bids me continue in the faith on which I acted—that a man who speaks out honestly and fearlessly that which he knows, and that which he believes, will always enlist the good-will and the respect, however much he may fail in winning the assent, of his fellow-men.

Thus, my first ends have assuredly been attained; and who that has read in the *Witness* the articles (exceptional, even in its columns, I trust) directed against my doctrines, my audience, the directors of the Philosophical Institution, and myself, can accuse me of having failed in reaching the third?

I do not trouble you with this letter, however, for the mere purpose of expressing my justifiable satisfaction at a success, whose sole importance consists in the conclusive evidence which it affords of the feebleness of the hold which the bigots really have over the minds of your countrymen; nor will you suspect me of any intention of replying directly or indirectly to the *Witness*. I should not even feel justified in asking permission to occupy your well-filled columns with a detailed reply to criticisms of such a totally different stamp, as those which are to be found in the pages of the *Week* of Friday last.

But I desire to be allowed to state that all such comments as these last are very welcome, and will receive my most earnest and respectful attention. I am not without hope that when my intention of publishing these well-abused lectures in full is carried into effect, the scope and weight of my argument will wear a different aspect to my candid, though hostile critic; but, in the meanwhile, let me thank my unknown adversary for his practical demonstration, that even the strongest dissent from, and opposition to, the views held by my revered friend Mr Darwin, and by myself (so largely, in these matters, a mere disciple of that great naturalist) are perfectly compatible with the courtesy of a gentleman, the language of a scholar, the knowledge of a man of science, and even (incredible as this may seem to the *Witness*) with the genuine piety of an orthodox Christian.— I am &c.

Thomas Henry Huxley, F.R.V., | Professor of Natural History in the Government Schools of Mines.

APPENDIX VI
Notes on the causes of cross and hybrid sterility

In June 1861, having completed a series of reciprocal crossing experiments on dimorphic species of *Primula*, Darwin wrote the first in an important series of dated theoretical notes in which he began to revise his former views on the question of the causes of cross and hybrid sterility. Darwin wrote notes on this subject intermittently throughout 1862, the last in the series being dated 4 January 1863. The notes are preserved in his portfolio of notes on hybridism (DAR 205.7: 155–65). Since Darwin subsequently reverted to his earlier views on cross and hybrid sterility, only briefly referring to the ideas mentioned in these notes in his published work (see *Origin* 4th ed., pp. 310–14, and *Variation* 2: 185–8), and since his notes shed light on the letters in this volume, they have been reproduced here.

In *Origin*, Darwin dedicated an entire chapter to 'Hybridism', noting at the outset that the theory of natural selection accorded a particular importance to the subject of hybrid sterility. He began (p. 245):

> The view generally entertained by naturalists is that species, when intercrossed, have been specially endowed with the quality of sterility, in order to prevent the confusion of all organic forms. ... On the theory of natural selection the case is especially important, inasmuch as the sterility of hybrids could not possibly be of an advantage to them, and therefore could not have been acquired by the continued preservation of successive profitable degrees of sterility. I hope, however, to be able to show that sterility is not a specially acquired or endowed quality, but is incidental on other acquired differences.

Darwin's sense of the importance of the subject was increased by the response of Thomas Henry Huxley to the theory of natural selection. From January 1860, Huxley repeatedly stated his view, first privately, then publicly, that natural selection could never be considered a *vera causa* for the origin of species until artificial selection had been shown to be capable of producing varieties of a species that were cross sterile (T. H. Huxley 1860a and 1860b). Darwin took this criticism seriously (see *Correspondence* vol. 8, letter to T. H. Huxley, 11 January [1860]), but attacked Huxley's tendency to imply that sterility was a 'universal & infallible criterion of species' (see *Correspondence* vol. 8, letter to Charles Lyell, 12 [February 1860]). To demonstrate that this was not the case, he repeatedly referred to the crossing experiments of Karl Friedrich von Gärtner, already discussed in *Origin*, in which similarly coloured varieties of the same or different species of *Verbascum* were found to be less cross sterile than differently coloured varieties. Darwin considered the experiments so significant in countering Huxley's objection that he sought to replicate them himself (see *Correspondence* vol. 9, letter to J. D. Hooker, 28 September [1861]), telling Joseph Dalton Hooker: 'I do not think any experiment can be more important on Origin of species; for if he [Gärtner] is correct, we certainly have what Huxley calls new physiological species arising.' However, Darwin could not obtain the requisite plants, and ultimately had to entrust the task to John Scott (see letter to W. E. Darwin, 4 [July 1862], and letter to John Scott, 11 December [1862] and n. 19).

The subject of hybrid sterility continued to be the source of serious disagreement between Huxley and Darwin in 1862. When, in January, Huxley reiterated his views in his lectures at the Philosophical Institution of Edinburgh, telling Darwin that he had taken his 'old line about the infertility difficulty' (see letter from T. H. Huxley, 13 January 1862), Darwin once again referred him to Gärtner's experiments, but added: 'Do oblige me by reading latter half of my Primula paper in Lin. Journal for it leads me to suspect that sterility will hereafter have to be largely viewed as an acquired or *selected* character.—a view which I wish I had had facts to maintain in the Origin.—' (see letter to T. H. Huxley, 14 [January 1862]).

Darwin began to suspect that cross and hybrid sterility might be the result of natural selection following his experiments on dimorphic species of *Primula* in the spring of 1861. In these experiments, Darwin carried out all four possible crosses between the two forms of flower in each species, namely two crosses between like forms (which he called 'homomorphic', or, later, 'illegitimate' crosses) and two between different forms (which he called 'heteromorphic', or 'legitimate' crosses). He discovered that the former crosses were much less fertile than the latter, and concluded that dimorphism occurred in order to favour cross-pollination, a finding in accord with his current work on the pollination of orchids, and with his long-standing conviction that an occasional cross was a 'law of nature'.

As the first of the dated notes reproduced below indicates, Darwin immediately began to see in these results implications for the question of hybrid sterility. In September 1861 he wrote to report his findings in *Primula* to Asa Gray, and to ask for details of any analogous cases of dimorphism, telling him: 'This subject interests me much, so do help me if you can; for I have some *very faint* hopes that it may throw some light on Hybridisation' (*Correspondence* vol. 9, letter to Asa Gray, 16 September [1861]). When, shortly afterwards, he presented his results in a paper for the Linnean Society of London ('Dimorphic condition in *Primula*', p. 94; *Collected papers* 2: 61), he was more explicit:

> The simple fact of two individuals of the same undoubted species, when homomorphically united, being as sterile as are many distinct species when crossed, will surprise those who look at sterility as a special endowment to keep created species distinct. Hybridizers have shown that individual plants of the same species vary in their sexual powers, so far that one individual will unite more readily than another individual of the same species with a distinct species. Seeing that we thus have a groundwork of variability in sexual power,

and seeing that sterility of a peculiar kind has been acquired by the species of *Primula* to favour intercrossing, those who believe in the slow modification of specific forms will naturally ask themselves whether sterility may not have been slowly acquired for a distinct object, namely, to prevent two forms, whilst being fitted for distinct lines of life, becoming blended by marriage, and thus less well adapted for their new habits of life.

The implications of this view of hybrid sterility were far-ranging, not only in addressing Huxley's objection to natural selection, but also (as Darwin's notes make clear) in addressing the issues of blending inheritance, and the role of geographic isolation in speciation. However, it was chiefly in regard to Huxley's objection that Darwin began to introduce his ideas to a number of correspondents in 1862. Having broached the subject with Huxley in January, he did not raise it directly for some months, merely hinting at the nature of his findings (see, for example, the letter to J. D. Hooker, 7 March [1862]). Meanwhile, he pursued an extensive research programme on dimorphism, repeating his earlier experiments on *Primula veris*, extending his experiments on *P. sinensis* to a second generation, experimenting on several species of *Linum*, and carrying out less detailed experiments on other dimorphic species. As a result, he became still more impressed by the degree of intra-specific sterility that could be obtained in dimorphic plants, telling Asa Gray that each of the two forms of *Linum grandiflorum* was so sterile with itself that, '[t]aking sexual power as the criterion of difference the two forms of this one species may be said to be generically distinct' (see letter to Asa Gray, 14 July [1862]). Soon afterwards, while occupied with a new enthusiasm for the trimorphic plant *Lythrum salicaria*, CD wrote to Gray with regard to his work on heterostyly: 'to those who *already believe in change of species*, these facts will modify to certain extent whole view of Hybridity—' (see letter to Asa Gray, 9 August [1862]).

In December 1862, as Darwin reviewed his experiments on heterostyly, and discussed blending inheritance and the causes of variation with Hooker (see letters to J. D. Hooker, 24 [November 1862], [after 26] November [1862], and 12 [December 1862], and letters from J. D. Hooker, [15 and] 20 November [1862] and 26 November 1862), he once more ventured to disclose his private views on the origin of cross and hybrid sterility. In a letter dated 12 [December 1862], he told Hooker:

> By the way my notions on hybridity are becoming considerably altered by my dimorphic work: I am now strongly inclined to believe that sterility is at first a selected quality to keep incipient species distinct. If you have looked at Lythrum, you will see how pollen can be modified merely to favour crossing; with equal readiness it could be modified to prevent crossing.— It is this which makes me so much interested with dimorphism &c.—

Shortly afterwards, Darwin received another unpleasant surprise in the form of the published version of one of Huxley's recently delivered lectures to working men (T. H. Huxley 1862c; see letter to T. H. Huxley, 18 December [1862]). In the lecture, Huxley reiterated his views concerning natural selection, suggesting that the weakest point in Darwin's hypothesis was the lack of positive evidence that any groups of animals known to have been produced 'by selective breeding from a common stock', had become mutually sterile (T. H. Huxley 1862c, pp. 108–13; see also pp. 146–50). Darwin was exasperated by Huxley's criticisms, and especially by his claim that 'the answer to varieties when crossed being at all sterile' was 'absolutely a negative', which Darwin considered to be plainly contradicted by Gärtner's hybridising experiments. 'My God', he continued, 'is not the case difficult enough without its being, as I must think, falsely made more difficult?' After a further exchange of letters (now missing), Darwin told his friend in a letter of 28 December [1862]: 'We differ so much that it is no use arguing. To get the degree of sterility you expect in recently formed varieties seems to me hopeless.'

Darwin's changing views about the selective origin of cross and hybrid sterility, together with Huxley's repeated criticisms, prompted him to seek further experimental evidence on the question, especially in regard to animals. Some time in December, Darwin sent out letters to a number of animal breeders inquiring about the occurrence of intra-specific sterility, although none of this correspondence has been found (see letters to W. B. Tegetmeier, 27 [December 1862] and 28 [December 1862]). In particular, he apparently suggested a breeding experiment to pigeon and fowl fanciers, designed to test whether hybrid sterility could be artificially selected among individuals of a single species. If successful, the experiment would, Darwin believed, 'solve the problem of Sterility from Hybridism' (see letter to W. B. Tegetmeier, 27 [December 1862]). It would also have fully met Huxley's objection to the acceptance of natural selection as a *vera causa*. Indeed, Huxley had suggested something similar in his letter to Darwin of 20 January 1862, stating:

> I have told my students that I entertain no doubt that twenty years experiments on pigeons conducted by a skilled physiologist instead of a mere breeder—would give us physiological species sterile *inter se* from a common stock—(& in this if I mistake not I go further than you do yourself) and I have told them that when these experiments have been performed I shall consider your views to have a complete physical basis—and to stand on as firm ground as any physiological theory whatever—

Darwin discussed the proposed experiment with Huxley in his letter of 28 December [1862], telling him of his hope that the pigeon fancier and naturalist William Bernhard Tegetmeier would carry it out. At the start of December, Tegetmeier had obtained a grant of £10 from the Royal Society for 'experiments on the cross-breeding of pigeons' (Royal Society, Council minutes, 1 December 1862). He originally planned the experiments to test 'whether any existing breeds happen to have acquired accidentally any degree of sterility', a point that Darwin had previously tested, but Darwin encouraged him to carry out the experiments according to his own plan of selecting for sterility (see letter to W. B. Tegetmeier, 27 [December

1862]). In the event, however, Tegetmeier's experiments produced no sterile offspring (*Variation* 1: 192).

In these letters of December 1862, Darwin began for the first time to commit himself more publicly to the idea that hybrid sterility was a selected quality. However, the series of notes reproduced here ends abruptly with a note dated 4 January 1863, in which Darwin twice rejects the selective origin of sterility: 'It will not do'. Moreover, between 1 April and 16 June 1863, Darwin wrote a draft of the section of *Variation* dealing with 'Crossing and Sterility': in the published version (*Variation*, chapters 25–9), he reverted to his earlier views on cross and hybrid sterility. Although Darwin revised his manuscript before it was finally published in January 1868, there is no evidence to suggest that the first draft differed from the published version in this regard.[1]

In *Variation* 2: 185–9, Darwin referred to his former interest in the idea of the selective origin of cross and hybrid sterility, emphasising not only the evidence from heterostyled plants, but also the evidence that sterility had been acquired through natural selection for 'other and widely different purposes, as with neuter insects in reference to their social economy' (p. 185). However, he cited three objections to this view. Firstly he noted that in cases of speciation where the incipient species were geographically isolated, cross and hybrid sterility had evidently been obtained without selection, since under such circumstances there could be no selective advantage attached to sterility. Secondly he pointed out that selection could not account for those instances where, in reciprocal crosses, 'the male element of one form should have been rendered utterly impotent on a second form, whilst at the same time the male element of this second form is enabled freely to fertilise the first form' (p. 186), since selection ought to act equally on each of the reciprocal crosses. Lastly, and most importantly, he argued that a 'great difficulty' lay in 'the existence of many graduated steps from slightly lessened fertility to absolute sterility.' He continued:

> It may be admitted, on the principle above explained, that it would profit an incipient species if it were rendered in some slight degree sterile when crossed with its parent-form or with some other variety; for thus fewer bastardised and deteriorated offspring would be produced to commingle their blood with the new species in process of formation. But he who will take the trouble to reflect on the steps by which this first degree of sterility could be increased through natural selection to that higher degree which is common to so many species, and which is universal with species which have been differentiated to a generic or family rank, will find the subject extraordinarily complex. After mature reflection it seems to me that this could have been of no direct advantage to an individual animal to breed badly with another individual of a different variety, and thus leave few offspring; consequently such individuals could not have been preserved or selected.

While Darwin considered that the case might be 'somewhat different' with plants (p. 187), he pointed out that since cross and hybrid sterility followed 'the same general laws in the vegetable as in the animal kingdom, it is improbable, though apparently possible, that with plants crossed species should have been rendered sterile by a different process' (p. 188). His conclusion was that cross and hybrid sterility had arisen 'incidentally . . . in connection with other and unknown changes in their organisation.'

Darwin's notes from 1861–3 include discussions of these potential objections, although it is clear that the third was uppermost in his mind during this period. The 'mature reflection' which Darwin mentions in *Variation* clearly refers to the internal dialogue recorded in these notes.

[DAR 205.7: 155]

June 6th. 61/ The case of Primula just opposite to Species; similar forms are here fully fertile dissimilar sterile; with Primula the similar are sterile & the dissimilar fertile.

Keep in Hybrid Chapt

As we see sterility acquired merely to prevent intermarriage, & this can hardly be disputed by those who consider all above facts & who believe in change of species; so may we not ask whether sterility may not have been acquired to prevent blending of two incipiently divergent forms. As one became adapted for aquatic life would it not have been an advantage that it shd. not have become blended with the truly terrestrial form— This first change may have been infinitely slight, the degree which my facts on Primula perhaps explain Gartner's ill-success in crossing Primrose & Cowslip.—[2] Did not Lecoq succeed.[3] I have succeeded with greatest ease.—[4]

If we take broad view of good of Divergence— good of sterility follows—

In neuter insects sterility from loss of sexual instinct.

[DAR 205.7: 156]

Hybridism

Take case of Pigeons paired for life—with all equal fixed fertility—hardest case.— Assume that country will support a given number— —But that with some sensible modification of structure & habit, a new species might be found & be supported in additional numbers— Each variety leading in this direction wd be preserved; but crossing with old form would constantly drag back the incipient form— now suppose a bird, varying slightly in desired direction was born, which when paired with its nearest ally, was fully fertile but was sterile with the old form; then its offspring which chanced to pair with its like would constantly be propagated, but those which propagated with the old would die out.— we shd have fewer of intermediate form, & more of the pure variety.

Take case of a pigeon paired with similar var, producing offspring, fully fertile with own var. & not fully fertile (in either way) with distinct var. would this new little family A have advantage in long run over other similar vars, which had no such tendency to sterility B? Many of A^{s} would die out from sterility; a few would pair & leave pure own vars & these would slowly increase. Of B all the gradation would live, yet occasionally some purely matched vars would be produced. **Occasionally perhaps** *all B^{s} would match with other vars & then new variety would be lost*— so it would be with *A*!!! Perhaps matched pairs if sterile would separate— My gretest difficulty with animals why mere sexual predeliction does not by itself suffice; perhaps passion too strong.— If no matching for life case w^{d} be clear: prepotency clear with plants.—

[DAR 205.7: 157–8]

Sept. 24— 1861. (After considering Primula.)

The creationist who sees organisms from arctic & tropical countries or from 2 continents, which never could meet, sterile together, must admit that incidental from endowed sterility to prevent forms which do come into contact from blending— So with sterility of Hybrid progeny.—

Hence on view of sterility acquired by nat. Selection no difficulty from such facts.—

In animals hard to see why sterility sh^{d} have been acquired, why not mere instinctive dislike, which does prevail, why is sterility superadded.— If animal born with slightly different habits of life & at same time by chance some sterility concurred, it would be great advantage.

In plants, where no instinctive choice such sterility necessary.—

Perhaps cases of Birds which pair for life gretest difficulty. how c^{d} nat. selection come in? Yes it could if new var. paired with old it would be lost: if paired with similar var it would be preserved.— Sterility must supervene at first formation of Var.— When we see how almost universal some sterility is, can we believe *always* acquired? Must be incidental.— When whole body of a species changes like our Race-Horse, no use in sterility—[5] if one body changed in England & another in America, such 2 new forms would have no tendency to sterility if acquired through nat. selection.—

Gartner Bastard s. 165| on individuals refusing to cross.—[6] *[seeing]* Primula & Linum & neuter Hymenoptera & Neuroptera no difficulty in sterility being acquired, if this view would explain facts.— It would, I think, apply to all offsets of one species in same country. The Verbascum & Tobacco case shows incidental in some cases.—[7] (All the facts about Lobelia—Passiflora &c & Crinum crossing more readily with *another species* (this cannot be dimorphism) shows how easily reproductive system is affected by condition of life & rendered self sterile.).[8] (Such cases as Rawson's Gladiolus must be different case, ie. good of crossing with distant individual.[9]

I suspect it will come to this that generally when a new form is selected within the same country with its parent, from early start sterility must be selected.— The vast diversity in degree of sterility is compatible with its ensuing from various causes—correlation, as well as selection.— Cases are on record of women sterile with one man & not with another.

Nov. 18th 1861. We must look at selection for own (say) equation from stedily going on; but always retarded till sterility is at same time gained.— I suspect this sterility must be habitually acquired before instinctive dislike acquired; otherwise this instinctive dislike would have been potent by itself & sterility would have been superfluous. Instinctive dislike perhaps does come much into play—see multiple origin of domestic animals— Those that pair for life seems gretest difficulty.—

With respect to /slow/ modifications like our Race-Horses. Tendency to sterility may be inherited from former process of selection.

[DAR 205.7: 159–60]

Feb. 26— 62.— It is true as I said in Origin no good to creature to be sterile;[10] but I forgot that this was only means to end— To keep vars. apart highest good, as we see by man by fences &c.— In plants it is prepotency of pollen. which is probably aimed at.— Though the sterility of parents & hybrids would be of good.—

In animals my difficulty is that instincts would have sufficed; but one never knows how nature will work.— Certainly instincts are not invariably strong enough to keep species apart— Hooded *[illeg]*— Black /fans/ Woodpecker &c &c—

One can clearly see that under general point of view good to keep vars apart— If the hybrids could breed they would drag back the new var. to primordial state.—

Hence variation so slow as structural & physiological changes have to go together.—

In all cases when animals pair & live in herd under one Lord, prepotency of semen. (if it ever comes in) cannot apply.—

Nature acts as a man would if two of his best breeds united he would castrate the mongrel— As nature does in sterile—or neuter Hymenoptera & Neuroptera— Border flowers of Compositæ Umbelliferae— Hydrangeæ are rendered sterile: they perhaps as these cases due to laws of growth.—

An pairing animal could not be selected, even if it varied in right way, until it chanced to /exercise/ either instinctive (& that would not occur early enough) or physiological aversion to unite with parent form| Perhaps gretest first a little varieties & sub-vars formed| this would be lost, unless it were separated by sterility from parent or other races.—

[DAR 205.7: 161]

Nov. 18/62/ The fact proved by such a mass of evidence that crossing two individuals distinct, & probably longer kept distinct the better, adds to vigour & fertility of offspring, shows that there must be some excessively slight difference in pollen of every individual of same species.— Under unnatural conditions we find that the pollen gets so wonderfully altered (though effectual in itself) that as in Lobelia, Potato, Passiflora—Crinum &c the flower is quite sterile with its own pollen & requires that of another. This throws flood of light on dimorphism & makes it not improbable (seeing the individual difference of pollen) that sterility may be selected to keep vars. distinct.

[DAR 205.7: 162]

Dec. 3^{d}. 1862. The structure of animals & plants, many decisive experiments & common experience shows that good is derived from crossing one individual with another of a distinct family or race.— Now if for instance, pollen from a distinct plant produces more numerous and more vigorous seedlings, than flowers own pollen on the same stigma, then two pollens must certainly be in some slight degree differentiated. Hence some differentiation of pollen is commonest Phenomenon.— So must the stigma or ovules, for if pollen on 1st. plant will be effective on the 2^{d} plant.— The differentiation of female organ in same plant is carried to extremes in certain plants under culture, as Lobelia, Crinum, &c.— We see thus see that dimorphism may readily arise & we can understand why so variable within same genus; how easily lost under culture as in P. Sinensis, & in structure in P. auricula & Siberica (see J. Scott list).[11] But if we once admit, the *[easy]* of differentiation of pollen, *we can understand hybridity*— P. Sinensis shows how this sterility affects offspring.— N. Selection only collects & forms this sterility—no doubt it is always in some degree correlated with changed structure.—

[DAR 205.7: 163]

Perhaps important— Dec. 18— 62.

[With] formation of marsh var. of a Plant, it w^{d} not be necessary that it sh^{d} be sterile with land var, only that it produced rather more seed with marsh var. & the result would be that the marsh var. having slight advantage of habit & greter fertility would gain the end. This tendency which is in fact all that occurs in more fertile hybrids or *[illeg]* then might be selected or go on aggregating by simple effects of conditions.—

Suppose a short-beaked var & short beaked var, when paired produced on a verge a few more percentage of young ones than a heterogeneous pair—this would clearly be advantage—(like seed in Wheat field.) & so onwards.— Those which produced most young when matched would necessarily prevail— subject to other contingencies.—

A closely interbred dog, I have heard several cases, w^{d} not breed with its own kind, but did freely with mongrel— This serves to show how easily generative system modified. Possibly or *not* some foal- getter may be perfectly fertile with certain individual females, but not with all.—[12]

[DAR 205.7: 164]

Jan 4th 1863. An animal, say, becoming adapted for aquatic life & surrounded by so many of its parent terrestrial forms, that in course of few generations all such vars. chanced to get crossed by them; would be dragged back from its favourable variation. But if a variation arose, which was sterile with ordinary form, then although even so large a proportion united with ordinary form & were ultimately lost by sterility; yet if even so few united with similar form, these would be kept pure; but then *chances* are supposed that they would unite with ordinary form & so would be lost by sterility. It will not do. The sterility must supervene from being kept close to uniform condition of life for ages.— If sexual disinclination supervened all would go well; but then why does sterility supervene, & why in Hybrid & not always in first cross— — **Will not do.**—

N.B Sexual disinclination *must* **be an** acquired instinct; for how else could it arise?

[DAR 205.7: 165]

Make pollen of one var. of Cabbage prepotent over another or one var. of Verbascum slightly prepotent over another. It may be in some cases incidental. But Whenever a new species has arisen, there has not always been another (as in such case as change of our Race-horse from Arab); this is a difficulty.— So in case of 2 species developed in 2 distant quarters of world.— Again how are Hybrids themselves rendered sterile: this must be incidental. for I will never believe that Nat. Selection was so blundering as to make the offspring sterile, & not the parents; moreover relation of sterility of two parents & Hybrid child not strictly correlated— N.B we have sterility of another kind induced in neuter insects of several kinds— We could on this view understand no sterility in vars. selected by man.— Perhaps understand the extreme variability in degree of sterility of Species.— Gärtners Verbascum case, & Maize & Kölreuters tobacco, must be correlation.— The case of Lobelia, Crinum & Passiflora, shows that under culture, the potentiality of pollen & stigma varies. It may be correlation in some cases & Selection in others: or increased by selection; but then the extreme & increasing degree of sterility as shown by different genera never crossing must be wholly due to correlation.—

Mem. Reciprocal Union how could these be explained? What in case of Linum; may there not be case with one form being dimorphic & not another

MANUSCRIPT ALTERATIONS AND COMMENTS

[DAR 205.7: 155]
1.1 similar . . . fertile. 1.3] *cross in margin, blue crayon*
2.1 Keep . . . Chapt] *added red crayon*
4.1 If we . . . instinct. 5.1] *added pencil*

[DAR 205.7: 156]
2.3 might] *above del* 'w^{d}'
2.4 Each] *after del* 'As long as these new forms crossed'
2.5 with] *after del* 'would'
3.1 case of] 'of' *interl*
3.3 family A] 'A' *interl*
3.4 sterility B] 'B' *interl*
3.4 ; a few] *after del* 'would die out'
3.5 & these would slowly increase] *interl*
3.7 so it would be with *A*!!! 3.8] *interl*

[DAR 205.7: 157–8]
1.1 (After considering Primula.)] *square brackets in MS*
7.5 (All the . . . sterile.). 7.7] *square brackets in MS*

7.6 easily] *after del* 'much'
7.7 (Such] *square bracket in MS*

[DAR 205.7: 159–60]
7.1 pairing] *interl*
7.3 aversion] *after del* 'abho'
7.5 or] *over* '&'

[DAR 205.7: 161]
1.2 distinct] *after del* 'long'
1.5 (though . . . itself)] *parentheses over commas*
1.8 (seeing] *parenthesis over comma*

[DAR 205.7: 162]
1.7 So] *after del* 'This diff. under culture is'
1.9 as] *after del* 'when'
1.9 We] *altered from* 'we', *after del* 'Again'
1.9 thus] *after del* 'in whole'
1.12 *[*easy*/]* *after del* 'd'

[DAR 205.7: 163]
2.4 which is . . . or *[illeg]* 2.5] *interl*
3.1 short beaked] 'short' *above del* 'long'
3.4 subject] *after del* '& of these'
4.3 be] *after del* 'not'

[DAR 205.7: 164]
1.5 united] *after del* 'was'
1.7 supposed] *interl*

[DAR 205.7: 165]
1.3 But] *interl*
1.3 (as] parenthesis *over* ';'
1.4 such case as] *interl*
1.8 two] *after del* 'p'
1.12 & Maize] *interl*
1.15 & increasing] *after del* 'degree of'
2.1 Mem. . . . another] *added blue crayon*
2.2 being] *after del* 'having'

[1] M. J. Kottler has suggested that CD's rejection of the selective origin of cross and hybrid sterility was occasioned by his work on the trimorphic plant *Lythrum salicaria* (Kottler 1985, pp. 402–5), but this does not appear to be the case. Having carried out ninety crosses between the three forms of this plant in 1862, the results of which had evidently not shaken his growing belief in the selective origin of cross and hybrid sterility, Darwin resolved that, in order to make his results fully reliable, he should carry out a further 136 crosses the following year (see letter to W. E. Darwin, 4 [November 1862]). However, Darwin did not count the seeds from the crosses made in 1863 until April 1864, long after he had drafted the chapters of *Variation* dealing with crossing and sterility. His comments on this subject in 'Three forms of *Lythrum salicaria*', p. 186 (*Collected papers* 2: 120), were thus apparently part of a wider rejection of the selective origin of cross and hybrid sterility.

[2] Gärtner 1849, p. 721. There is a heavily annotated copy of this work in the Darwin Library–CUL (see *Marginalia* 1: 256–98).

[3] Lecoq 1845, p. 195. There is an annotated copy of this work in the Darwin Library–CUL (see *Marginalia* 1: 495–6).

[4] CD refers to the crossing experiments he had carried out on *Primula veris* and *P. vulgaris* in the spring of 1862.

[5] CD refers to the fact that, by a 'process of selection, and by careful training, the whole body of English racehorses have come to surpass in fleetness and size the parent Arab stock' (*Origin*, p. 35).

[6] Gärtner 1849.

[7] In *Origin*, pp. 270–1, CD described crossing experiments carried out by Karl Friedrich von Gärtner on *Verbascum* and by Joseph Gottlieb Kölreuter on *Nicotiana*. Gärtner found that 'yellow and white varieties of the same species of Verbascum when intercrossed produce less seed, than do either coloured varieties when fertilised with pollen from their own coloured flowers.' Kölreuter observed that 'one variety of the common tobacco is more fertile, when crossed with a widely distinct species, than are the other varieties'.

[8] In *Origin*, pp. 249–51, CD described instances where cultivated species of *Crinum*, *Lobelia*, and *Hippeastrum* had been observed to be 'far more easily fertilised by the pollen of another and distinct species, than by their own pollen.'

[9] CD refers to the observations of Arthur Rawson, that a number of varieties of gladiolus descended from a 'well-known old hybrid', were found not to set seed with pollen of their own variety, but would set seed freely with pollen from any other variety (see *Variation* 2: 139–40).

[10] In the introduction to his chapter on hybridism in *Origin*, CD stated (p. 245):

> The importance of the fact that hybrids are very generally sterile, has, I think, been much underrated by some late writers. On the theory of natural selection the case is especially important, inasmuch as the sterility of hybrids could not possibly be of any advantage to them, and therefore could not have been acquired by the continued preservation of successive profitable degrees of sterility.

[11] John Scott's list of the crossing experiments that he had carried out on species of *Primula*, which was probably included with Scott's letter to CD of [20 November – 2 December 1862], has not been found. However, see the letter from John Scott, 6 December [1862].

[12] CD apparently asked a number of his correspondents for examples of this sort (see letters to W. B. Tegetmeier, 27 [December 1862] and 28 [December 1862]).

APPENDIX VII
Reviews of *Orchids*

This brief list identifies the reviews of *Orchids* mentioned in Darwin's correspondence between May and December 1862. In addition, the list includes reviews not mentioned in the correspondence but referred to in Darwin's own manuscript list of reviews or in the 'Scrapbook of reviews' (see below). Some of the reviews were sent as presentation copies to Darwin; others he obtained himself. Darwin frequently wrote on his copies the name of the author, when known, and the date of publication. In many, he also made marginal annotations.

The manuscript headed 'List Reviews of Origin of Sp & of C. Darwins Books' (DAR 261 (DH/MS* 8: 6–18)) served as Darwin's index to his collection of reviews, each being assigned a number. Although most of the items in Darwin's collection are now in the Darwin Pamphlet Collection–CUL (where Darwin's own numbering has been preserved), a number of them have been lost (see, for example, nn. 3 and 7, below). Darwin's manuscript list includes reviews of *Orchids* that are not mentioned in the correspondence; their bibliographic details have been transcribed here since these reviews were evidently known to Darwin.

In addition to his main review collection, Darwin maintained a collection of shorter reviews, mainly from newspapers; these were ultimately pasted into two bound scrapbooks. The first volume of this 'Scrapbook of reviews' (DAR 226.1) contains several reviews of *Orchids* that are also included here.

The following list is arranged chronologically according to the month of publication. For those reviews that were unsigned, the author's name, if known, is given in square brackets with the source of the identification supplied in a note. If the review is in the Darwin Pamphlet Collection–CUL or in the 'Scrapbook of reviews' (DAR 226.1), this information follows the entry.

May 1862

[Leifchild, John R.][1] Review of *Orchids. Athenæum*, 24 May 1862, pp. 683–5. [Scrapbook of reviews (DAR 226.1).]

June 1862

[Anon.] Review of *Orchids. Parthenon*, 7 June 1862, pp. 177–9. [Scrapbook of reviews (DAR 226.1).]

[Berkeley, Miles Joseph.][2] Fertilization of orchids. *London Review and Weekly Journal of Politics and Science*, 14 June 1862, pp. 553–4. [Scrapbook of reviews (DAR 226.1).]

July 1862

[Anon.] Review of *Orchids. Literary Churchman*, 16 July 1862. [Scrapbook of reviews (DAR 226.1).]

G[ray], A[sa.] Review of *Orchids. American Journal of Science and Arts* 2d ser. 34: 138–44.[3]

August 1862

Tegetmeier, William Bernhard. Review of *Orchids. Weldon's Register of Facts and Occurrences relating to Literature, the Sciences, and the Arts*, 1 August 1862, pp. 38–9. [Scrapbook of reviews (DAR 226.1).]

August–September 1862

[Hooker, Joseph Dalton.][4] Review of *Orchids. Gardeners' Chronicle and Agricultural Gazette*, 23 August 1862, pp. 789–90; 13 September 1862, p. 863; 27 September 1862, p. 910.[5]

October 1862

[Anon.] *The British and Foreign Medico-Chirurgical Review or Quarterly Journal of Practical Medicine and Surgery* 30 (1862): 312–18. [Darwin Pamphlet Collection–CUL.]

[Anon.] Mr. Darwin's Orchids. *Saturday Review*, 18 October 1862, p. 486. [Scrapbook of reviews (DAR 226.1).]

[Hooker, Joseph Dalton.][6] Review of *Orchids. Natural History Review* n.s. 2: 371–6.[7]

November 1862

G[ray], A[sa.] Fertilization of orchids through the agency of insects. *American Journal of Science and Arts* 2d ser. 34: 420–9.[8]

[Campbell, George Douglas.][9] The supernatural. *Edinburgh Review* 116: 378–97. [Darwin Pamphlet Collection–CUL.]

[1] See letter to John Murray, 13 June [1862] and n. 10.

[2] See letter from J. D. Hooker, 28 June 1862.

[3] Gray sent CD a copy of this review (see letter to Asa Gray, 28 July [1862]), and it is listed in CD's list of reviews; however, the review is absent from the Darwin Pamphlet Collection–CUL.

[4] See letter from J. D. Hooker, 7 November 1862.

[5] CD included this review in his list of reviews. His collection of the *Gardeners' Chronicle* is now in the Cory Library of the Cambridge Botanic Garden. The three numbers in which the review was published are listed on the 'List of numbers of special interest to Darwin and kept by him in separate parcels' (DAR 222); see also his manuscript index to the journal in DAR 75: 1–12.

[6] See letter from J. D. Hooker, 7 November 1862.

[7] CD's unbound copies of the *Natural History Review* are in the Darwin Library–CUL.

[8] Gray sent CD a copy of this review (see letter to Asa Gray, 26[–7] November [1862]), and it is listed in CD's list of reviews; however, the review is absent from the Darwin Pamphlet Collection–CUL.

[9] See *Wellesley index* 1: 511–12.

APPENDIX VIII

Additions and corrections to the second German edition of *Origin*

In March 1862, Heinrich Georg Bronn wrote to Darwin stating his intention to prepare a second German edition of *Origin* (Bronn trans. 1863); he asked whether a new English edition containing corrections and additions was imminent, or whether Darwin would like to make any such changes to the existing German edition (Bronn trans. 1860; see letter from H. G. Bronn, [before 11 March 1862]). Since the publication of the first German edition, a third English edition had appeared, published in April 1861, containing, as Darwin told Bronn, 'a considerable number of small corrections & a few of importance' (see letter to H. G. Bronn, 11 March [1862]). Darwin had sent Bronn some of these alterations in manuscript for inclusion in the first German edition, but he considered it important that the remainder be included in the new edition; in his letter to Bronn of 25 April [1862], he mentioned that he was sending a set of sheets of the third English edition, marked with the alterations from the second edition. In addition, Darwin referred to 'a few new M.S. additions & corrections', which he was he sending with the purpose of 'enlarging the parts which have been most criticised'.

The original manuscript of these additions and corrections has not been found, although they were returned to Darwin for possible use in a new American edition of *Origin* (see letter from E. Schweizerbart'sche Verlagsbuchhandlung, 11 July 1862). (No American edition incorporating further alterations was published until 1870 (see Freeman 1977, pp. 85–6).) However, the editors have compiled a list of the changes that are incorporated in the second German edition of *Origin*, but that do not occur in the third English edition: these changes should correspond to the additional alterations sent by Darwin to Bronn. Many of these additions and corrections were noted in pencil by Darwin in his copy of the third English edition, which is now at CUL; most of them were also ultimately incorporated in the fourth English edition, which appeared in 1866.

The changes and additions have been translated into English, and are keyed to the third English edition by page, paragraph, and line number.

Page xiv, par. 1, line 1, insert before 'Geoffroy':[1]

Etienne

Page xiv, n., lines 2–5, delete 'It is curious . . . in 1794'.[2]

Page xiv, n., line 7, insert after 'long afterwards.':[3]
He has pointedly remarked (Goethe als Naturforscher, von Dr. Karl Meding, s. 34) that the future question for naturalists will be how, for instance, cattle got their horns, and not for what they are used.

Page xiv, n., line 10, delete '(as we shall immediately see)'.

Page xiv, n., line 11, delete 'in the years 1794–5'.

Page xviii, par. 3, line 9, insert after 'continued reproduction."':[4]

A well-known French botanist, M. Lecoq, writes in 1854 ('Etudes sur Géograph. Bot.,' tom. I. p. 250), 'On voit que nos recherches sur la fixité ou la variation de l'espèce, nous conduisent directement aux idées émises par deux hommes justement célèbres, Geoffroy Saint-Hilaire et Goethe.' Some other passages scattered through M. Lecoq's large work, make it a little doubtful how far he extends his views on the modification of species.

Page xix, par. 1, line 2, insert after 'and clearness.':[5]

Von Baer, towards whom all zoologists feel so profound a respect, expressed about the year 1859 (see Prof. Rudolph Wagner, 'Zoologisch-Anthropologische Untersuchungen,' 1861, s.51) his conviction, chiefly grounded on the laws of geographical distribution, that forms now perfectly distinct have descended from a single parent-form.

Page xix, par. 4, line 2, insert after '1860':[6]
and the third edition in April 1861.

Page 12, par. 1, line 18, substitute for 'animals':[7]
ruminants

Page 18, par. 2, line 15, insert after 'England':[8]
and Germany

Page 19, par. 1, line 8, insert after 'parent.':[9]
This conclusion, as well as the specific distinction between the humped and common cattle, may indeed be looked at as almost established by the recent admirable researches of Professor Rütimeyer.

Page 19, par. 1, line 16, insert after '(Gallus bankiva)':[10]
Having kept nearly all the English kinds alive, having bred and crossed them, and examined their skeletons, I have come to a similar conclusion,—the grounds of which will be given in a future work.

Page 46, par. 2, lines 22–4, substitute for 'but then . . . kinds of flowers.':[11]
In just some of these cases it seems that the generation of two different flowers takes place gradually. This is indicated by the intermediate forms between the two main kinds of flowers in certain campanula and violet species.

Page 52, par. 1, line 12, insert after 'varieties.':[12]
But I have to state that according to some more recent observations of mine on the sexual relations of the long- and short styled kinds of this genus, Primula vulgaris and

Primula veris seem to be two distinct species. This is corroborated by experiments which are still in progress.

Page 80, par. 1, line 4, insert after 'other cases.':[13]

In Australia, the imported hive-bee is rapidly exterminating the small, stingless native bee.

Page 93, par. 1, line 20, insert after 'other males.':[14]

The males of certain hymenopterous insects have been frequently seen by that inimitable observer M. Fabre, fighting for a particular female, who sits by an apparently unconcerned beholder of the struggle, and then retires with the conqueror. By the way,

Page 99, par. 2, line 26, to page 100, par. 1, line 5, substitute for 'The difference . . . hive-bees.':[15]

It is certain that the hive-bee sucks the nectar out, because I have long since observed many of these bees (although only possible in the autumn) suck the nectar through holes in the base of the corolla which I believe the small species of humble-bee have bitten.

Page 151, par. 1, line 12, insert after 'appreciable by us.':[16]

In a far-fetched sense, however, the conditions of life may be said, not only to cause variability, but likewise to include natural selection; for it depends on the nature of the conditions whether this or that variety shall be preserved. But we see in selection by man, that these two elements of change are essentially distinct; the conditions under domestication causing the variability, and the will of man, acting either consciously or unconsciously, accumulating the variations in certain definite directions.

Page 162, par. 2, lines 6–7, insert after 'deafness in':[17]

some white

Page 171, par. 1, lines 6–7, insert after 'wattle of':[18]

English

Page 179, par. 1, line 4, insert after 'barb-pigeon':[19]

(p. 32)

Page 182, par. 2, line 15, insert after 'full-grown animal.':[20]

I have myself recently bred a foal from a bay mare (offspring of a Turcoman horse and a Flemish mare) by a bay English race-horse; this foal when a week old was marked on its hinder quarters and on its forehead with numerous, very narrow, dark zebra-like bars, and its legs were feebly striped; all the stripes soon disappeared completely.

Page 198, par. 1, line 7, delete 'and I could have given no answer'.[21]

Page 222, par. 1, line 3, substitute for 'on high authority':[22]

on the high authority of Joh. Müller

Page 222, par. 1, line 4, insert after 'organ, the':[23]

human

Page 235, par. 1, lines 27–8, substitute for 'do not suppose that domestic rabbits have ever':[24]
can hardly suppose that domestic rabbits have

Page 235, par. 1, line 32, insert after 'confinement.':[25]
Yet, as the French translator of this book has noted, it is common to keep the tamest rabbits which cause least trouble, so that breeding plays a role here too.

Page 236, par. 1, line 21, substitute for 'in the same way . . . under a hen.':[26]
for I am informed by Captain Hutton that the young chickens of the parent-stock, the Gallus bankiva, when reared in India under a hen, are at first excessively wild. So it is with young pheasants reared in England under a hen.

Page 264, par. 1, lines 8–9, substitute for 'and not':[27]
whilst man works

Page 264, par. 1, lines 10-16, delete ', a perfect . . . natural selection'.[28]

Page 269, par. 1, line 26, substitute for 'such good reason':[29]
some reason

Page 275, par. 2, line 5, delete 'and with P. versicolor'.[30]

Page 275, par. 2, line 6, substitute for 'three':[31]
two closely related species of

Page 275, par. 2, line 7, insert after 'common,':[32]
and

Page 275, par. 2, lines 7–8, substitute for 'and the Japan':[33]
along with the Japanese P. versicolor

Page 275, par. 2, line 9, insert after 'England.':[34]
From the experiments lately made on a large scale in France, it seems that two such distinct species as the hare and rabbit, when they can be got to breed together, produce offspring almost perfectly fertile.

Page 276, par. 2, lines 18–21, substitute for 'again there. . . distinct species.':[35]
again I have lately acquired decisive evidence that the crossed offspring from the Indian humped and common cattle are *inter se* perfectly fertile; and from the observations by Rütimeyer on their important osteological differences, these two forms must be regarded as good and distinct species - as good as any in the world.

Page 310, par. 1, line 2, insert after 'preserved.':[36]
Some of the many kinds of animals which live on the beach between high and low water mark seem to be rarely preserved.

Page 334, par. 3, line 3, substitute for 'not one oceanic island':[37]
not one truly oceanic island (with the exception of New Zealand, Svalbard, and the adjacent Bear Island, if these can be called truly oceanic islands)

Page 363, par. 3, line 15, insert after 'life.': [38]
It is no valid objection to my views, although we know too little of the condition of life to offer any explanation, that certain Brachiopods have remained unaltered

from the earliest geological period; the fresh-water molluscs, subjected to less competition than the shells of the great oceans, have remained in nearly the same condition.

Page 364, par. 1, line 22, insert after 'geological research.':[39]
Bronn has dealt with this topic better and in more depth than any other author.

Page 400, par. 1, line 28, insert after 'epoch.':[40]
This view has been supported by three referees: Prof. Asa Gray, Dr. Hooker, and Prof. Oliver.

Page 403, par. 2, lines 15–16, substitute for 'If one . . . evidence':[41]
From facts lately communicated to me by the Rev. W. B. Clarke, it appears also that there are clear traces

Page 405, par. 1, line 18, insert after 'Africa.':[42]
Dr. Hooker has also lately shown that several of the plants, living on the upper parts of the lofty island of Fernando Po and on the neighbouring Cameroon mountains in the Gulf of Guinea, are closely related to those on the mountains of Abyssinia, and likewise to those of temperate Europe. This is one of the most astonishing facts ever recorded in the distribution of plants.

Page 407, par. 2, lines 14–15, insert after 'now concerned.':[43]
The whole problem of what will have occurred is excessively complex. The probable existence before the Glacial period of a pleistocene equatorial flora and fauna, fitted for a hotter climate than any now existing, must not be overlooked. This old equatorial flora will have been almost wholly destroyed, and the two pleistocene sub-tropical floras, commingled and reduced in number, will then have formed the equatorial flora. There will also probably have been during the Glacial period great changes in the precise nature of the climate, in the degree of humidity, &c.; and various animals and plants will have migrated in different proportions and at different rates. So that altogether during the Glacial period the inhabitants of the tropics must have been greatly disturbed in all their relations of life.

Page 407, par. 2, line 26, insert after 'extent.':[44]
The chief difficulty is to understand how they can have escaped entire annihilation. It must not be overlooked that, as the cold will have come on very slowly, it is almost certain that many of the inhabitants of the tropics will have become in some degree acclimatised.

Page 408, par. 1, line 22, substitute for 'But I do not doubt':[45]
We might of course speculate on the land having been formerly higher than at present in various parts of the tropics, where temperate forms apparently have crossed; but as the lines of migration have been so numerous, such speculation would be rash. So I am forced to believe

Page 408, par. 1, line 23, insert after 'tropics':[46]
especially of India

Page 409, par. 1, line 1, insert before 'Thus,':[47]

So again, on the island of Fernando Po, Mr. Mann found temperate European forms first beginning to appear at the height of about five thousand feet. On the mountains of Panama, at the height of only two thousand feet, Dr. Seemann found the vegetation like that of Mexico, with forms of the torrid zone harmoniously blended with those of the temperate. So that under certain conditions of climate it is certainly possible that strictly tropical forms might have co-existed for an indefinitely long period mingled with temperate forms.

At one time I had hoped to find evidence that the tropics in some part of the world had escaped the chilling effects of the Glacial period, and had afforded a safe refuge for the suffering tropical productions. We cannot look to the peninsula of India for such a refuge, as temperate forms have reached nearly all its isolated mountain-ranges, as well as Ceylon; we cannot look to the Malay archipelago, for on the volcanic cones of Java we see European forms, and on the heights of Borneo temperate Australian productions. If we look to Africa, we find that not only some temperate European forms have passed through Abyssinia along the eastern side of the continent to its southern extremity; but we now know that temperate forms have likewise travelled in a transverse direction from the mountains of Abyssinia to Fernando Po, aided perhaps in their march by east and west ranges, which there is some reason to believe traverse the continent. But even granting that some one large tropical region had retained during the Glacial period its full warmth, the supposition would be of no avail, for the tropical forms therein preserved could not have travelled to the other great tropical regions within so short a period as has elapsed since the Glacial epoch. Nor are the tropical productions of the whole world by any means of so uniform a character as to appear to have proceeded from any harbour of refuge.

The eastern plains of tropical South America apparently have suffered least from the Glacial period; yet even here there are on the mountains of Brazil a few southern and northern temperate and some Andean forms, which it appears must have crossed the continent from the Cordillera; and some forms on the Silla of Caraccas, which must have migrated from the same great mountain-chain. But Mr. Bates, who has studied with such care the insect-fauna of the Guiano-Amazonian region, has argued with much force against any recent refrigeration in this great region; for he shows that it abounds with highly peculiar Lepidopterous forms, thus apparently contradicting the belief in much recent extinction near the equator. How far his facts can be explained on the supposition of the almost entire annihilation during the Glacial period of a pleistocene equatorial fauna adapted for greater heat than any now prevailing, and the formation of the present equatorial fauna by the commingling of two former subtropical faunas, I will not pretend to say.

Page 410, par. 1, lines 19–25, substitute for 'Something ... north.':[48]

The Neilgherrie mountains in India, however, offer a partial exception; for here, as I hear from Dr. Hooker, Australian forms are rapidly sowing themselves and becoming naturalised.

Page 411, par. 2, line 5, insert after 'regions.':[49]

It is extremely difficult to understand how a vast number of peculiar forms confined to the tropics could have been therein preserved during the coldest part of the Glacial period. The number of forms in Australia, which are related to European temperate forms, but which differ so greatly that it is impossible to believe that they could have been modified since the Glacial period, perhaps indicates some much more ancient cold period, even as far back as the miocene age, in accordance with the recent speculations of certain geologists. I do not pretend to indicate the exact lines and means of migration, or the reason why certain species and not others have migrated; why certain species have been modified and have given rise to new groups of forms, and others have remained unaltered.

Page 422, par. 1, lines 16-19, substitute for 'Madeira . . . Harcourt.':[50]

Almost every year, as I am informed by Mr. E. V. Harcourt, many European and African birds are blown to Madeira; this island is inhabited by 99 kinds, of which one alone is peculiar, though very closely related to a European form; and three or four other species are confined to this island and to the Canaries.

Page 423, par. 2, line 5, insert after 'mammals.':[51]

Although New Zealand is here spoken of as an oceanic island, it is in some degree doubtful whether it should be so ranked; it is of large size, and is not separated from Australia by a profoundly deep sea: from its geological character and the direction of its mountain-ranges, the Rev. W. B. Clarke has lately maintained that this island, as well as New Caledonia, should be considered as appurtenances of Australia.

Page 424, par. 2, lines 6–10, substitute for 'I have . . . agency.':[52]

In the meantime Dr. Hochstetter has discovered a frog in the mountains of New Zealand which above all, and this is highly surprising, is related to a South American form. But frozen and revitalizable frogs have been found embedded in glaciers. It even seems possible that a frog or its spawn has been carried there from the islands of the South Pole on one of the large icebergs in the Antarctic oceans, from where the highly peculiar plant forms have originated which Australia, New Zealand and the southern tip of America have in common.

Page 425, par. 1, line 25, substitute for 'thought':[53]

known

Page 425, par. 1, line 30, insert after 'world':[54]

but this island, as before remarked, can only be classed as oceanic with some doubt

Page 443, par. 1, line 14, substitute for 'country':[55]

area

Page 445, par. 1, line 8, insert after 'explained.':[56]

(This is not to say that there be no other explanation of the subordination of character. We know that Mr. Maw has raised the objection against our theory that also minerals and even elementary matter can be classified into groups and subgroups where there is of course no genealogical succession. But the view developed above, explains the classification of organic bodies, and no other explanation has ever been put forward.)

Page 456, par. 2, lines 8–9, insert after 'separating them.':[57]

As soon as the three Orchidean forms, Monachanthus, Myanthus, and Catasetum, which had previously been ranked as three distinct genera, were known to be sometimes produced on the same plant, they were immediately considered as varieties; but now I have been able to show that they really constitute the male, female, and hermaphrodite forms of the same species.

Page 456, par. 2, line 20, to page 457, par. 1, line 3, delete 'As soon ... single species.'[58]

Page 460, par. 1, line 1, insert before 'As':[59]

A number of cases of analogical or adaptive resemblance are very remarkable. Here, I will only discuss one such case which is less eyecatching than the outward resemblance between sea-mammals and fish, between flying possums and flying squirrels, etc. Bates has recently reported how some species of one genus and even varieties within one species of butterflies in the large Amazona valley mimick to such perfection the appearances of species of completely different genera or subfamilies that they can only be distinguished by the most careful examination. It is a further noteworthy fact that the mocking species have rarely been very successful in their struggle for survival whereas the mocked often have. Mr. Bates concludes that the imitators gradually come to their present appearance through natural selection thereby escaping a threat by hiding behind the mask of the more common and successful species.)

[1] Bronn trans. 1863, p. 2.
[2] CD retained this sentence in *Origin* 4th ed., p. xiv, n., but modified it slightly.
[3] Bronn trans. 1863, p. 2 n. This sentence also appears in *Origin* 4th ed., p. xiv, n.
[4] Bronn trans. 1863, pp. 8–9. This passage also appears in *Origin* 4th ed., p. xx.
[5] Bronn trans. 1863, p. 9. This passage also appears in *Origin* 4th ed., pp. xx.
[6] Bronn trans. 1863, p. 10.
[7] Bronn trans. 1863, p. 22.
[8] Bronn trans. 1863, p. 28.
[9] Bronn trans. 1863, p. 29. This sentence also appears in *Origin* 4th ed., p. 19.
[10] Bronn trans. 1863, p. 29. This sentence also appears in *Origin* 4th ed., p. 20.
[11] Bronn trans. 1863, p. 56. This whole paragraph was extensively rewritten in *Origin* 4th ed., p. 47.
[12] Bronn trans. 1863, pp. 61–2.
[13] Bronn trans. 1863, p. 89. This sentence also appears in *Origin* 4th ed., p. 87.
[14] Bronn trans. 1863, p. 102. A similar sentence to this also appears in *Origin* 4th ed., p. 100.
[15] Bronn trans. 1863, p. 109. A modified version of the original sentence appears in *Origin* 4th ed., p. 107.
[16] Bronn trans. 1863, pp. 160–1. This passage also appears in *Origin* 4th ed., p. 160.
[17] Bronn trans. 1863, p. 171. The original sentence was rewritten in *Origin* 4th ed., p. 171.
[18] Bronn trans. 1863, p. 180.
[19] Bronn trans. 1863, p. 187.
[20] Bronn trans. 1863, p. 191. This sentence also appears in *Origin* 4th ed., p. 192.

21 A similar deletion occurs in *Origin* 4th ed., p. 208.
22 Bronn trans. 1863, p. 229. A similar substitution occurs in *Origin* 4th ed., p. 242.
23 Bronn trans. 1863, p. 229. This addition also appears in *Origin* 4th ed., p. 242.
24 Bronn trans. 1863, p. 242. This substitution also occurs in *Origin* 4th ed., p. 256.
25 Bronn trans. 1863, p. 242. The original sentence was rewritten in *Origin* 4th ed., p. 256.
26 Bronn trans. 1863, p. 243. This passage also appears in *Origin* 4th ed., p. 257.
27 Bronn trans. 1863, p. 271. This substitution also occurs in *Origin* 4th ed., p. 289.
28 This deletion also occurs in *Origin* 4th ed., p. 289.
29 Bronn trans. 1863, p. 276. The original sentence was rewritten in *Origin* 4th ed., p. 289.
30 This deletion also occurs in *Origin* 4th ed., p. 300.
31 Bronn trans. 1863, p. 281. The original sentence was removed in *Origin* 4th ed., p. 300.
32 Bronn trans. 1863, p. 281. The original sentence was removed in *Origin* 4th ed., p. 300.
33 Bronn trans. 1863, p. 281. The original sentence was removed in *Origin* 4th ed., p. 300.
34 Bronn trans. 1863, p. 281. This sentence also appears in *Origin* 4th ed., p. 300.
35 Bronn trans. 1863, p. 282. This sentence, with further additions, also appears in *Origin* 4th ed., p. 301.
36 Bronn trans. 1863, p. 315. This sentence also appears in *Origin* 4th ed., p. 347.
37 Bronn trans. 1863, p. 339. A similar substitution occurs in *Origin* 4th ed., pp. 372–3.
38 Bronn trans. 1863, p. 366. A similar sentence also appears, in a different part of the text, in *Origin* 4th ed., p. 402.
39 Bronn trans. 1863, p. 367. The original sentence was rewritten in *Origin* 4th ed., p. 402.
40 Bronn trans. 1863, p. 401.
41 Bronn trans. 1863, p. 403. This substitution also occurs in *Origin* 4th ed., p. 443.
42 Bronn trans. 1863, p. 405. This passage, with further additions, also appears in *Origin* 4th ed., p. 445.
43 Bronn trans. 1863, pp. 407–8. This passage also appears in *Origin* 4th ed., pp. 447–8.
44 Bronn trans. 1863, p. 408. A similar passage also appears, with further alterations, in *Origin* 4th ed., p. 448.
45 Bronn trans. 1863, p. 409. A similar passage also appears in *Origin* 4th ed., p. 449.
46 Bronn trans. 1863, p. 409. A similar addition also appears in *Origin* 4th ed., p. 449.
47 Bronn trans. 1863, p. 409–11. This passage also appears, with slight modifications, in *Origin* 4th ed., pp. 450–1.
48 Bronn trans. 1863, p. 412. This sentence also appears in *Origin* 4th ed., p. 453.
49 Bronn trans. 1863, p. 413. This passage, with additions and modifications, also appears in *Origin* 4th ed., p. 454.
50 Bronn trans. 1863, p. 423. This substitution also occurs in *Origin* 4th ed., p. 465.
51 Bronn trans. 1863, p. 424. This sentence also appears in *Origin* 4th ed., p. 466.
52 Bronn trans. 1863, p. 426.
53 Bronn trans. 1863, p. 427. This substitution also occurs in *Origin* 4th ed., p. 469.
54 Bronn trans. 1863, p. 427.
55 Bronn trans. 1863, p. 444.
56 Bronn trans. 1863, p. 445. A new passage was added at this point in the text in *Origin* 4th ed., p. 488.
57 Bronn trans. 1863, pp. 456–7. This sentence also appears in *Origin* 4th ed., p. 500.
58 A similar deletion occurs in *Origin* 4th ed., p. 500.
59 Bronn trans. 1863, p. 460. CD included several paragraphs on this subject in *Origin* 4th ed., pp. 503–6.

APPENDIX IX
Diplomas presented to Charles Darwin

In 1862, Darwin received two diplomas from institutions that had honoured him. Although neither of these is a letter in the conventional sense, they clearly represent significant communications between Darwin and the insitutions involved. Moreover, the citations in the diplomas provide valuable indications of those aspects of Darwin's work that had been considered worthy of honour. In view of this, and of the fact that the diplomas were referred to in Darwin's correspondence during 1862, they have been included here, together with translations. Footnotes are keyed to both the original text and the translation. The diplomas are followed by a biographical register, arranged alphabetically, identifying the persons mentioned in the diplomas. Individuals not otherwise mentioned in the volume are only identified here and are not included in the main biographical register.

From University of Breslau [before 10 February 1862][1]

Q. D. B. V.[2]
Summis auspiciis
serenissimi ac potentissimi principis
Guilelmi
Regis Borussiae caet.[3]
regis ac domini nostri iustissimi et clementissimi
eiusque auctoritate regia
Universitatis Litterarum Vratislaviensis
Rectore Magnifico
Christophilo Iulio Braniss[4]
Philosophiae Doctore et Professore Publico Ordinario
ex decreto Ordinis Medicorum
promotor legitime constitutus
Ioannes Carolus Leopoldus Barkow[5]
Med. et Chir. Doctor et Professor Publicus Ordinarius Regi
a Consiliis Medicis Ordinis Regii de Aquila Rubra Eques
Ordinis Medicorum H. T. Decanus
Carolo Darwin
Artium Liberalium Magistro Societatis Geologicae
et Societatis Regiae Londinensis membro clarissimo
originis specierum in regno animalium et vegetabilium

investigatori solertissimo viro doctissimo
de zoologia et de physiologia comparata
generali egregie merito
Doctoris Medicinae et Chirurgiae
honores et privilegia
in solemnibus Universitatis Litterarum Vratislaviensis
ante quinquaginta annos conditae
D. IV. M. AUGUSTI A. MDCCCLXI
honoris causa rite contulit
collataque
publico hoc diplomate
medicorum ordinis obsignatione comprobato
declaravit.

Dr. F. C. L. Barkow | L.T.D.[6]

[Translation]

May God turn it to good[2]
Under the august patronage of the most serene and most powerful
Prince William, King of Prussia[3] etc.,
our most just and most merciful king and lord,
and by his royal authority, at the University of Breslau,
in the High Rectorship of Christlieb Julius Braniss,[4]
Doctor of Philosophy and Public Ordinary Professor,
by a decree of the Faculty of Medicine,
Johann Carl Leopold Barkow,[5]
Doctor of Medicine and Surgery and Public Ordinary Professor,
Medical Counsellor to the King, Knight of the Royal Order of the Red Eagle,
present Dean of the Faculty of Medicine,
being officially appointed to confer degrees,
at the solemn festival of the University of Breslau fifty years after its foundation,
on the 4th of August 1861, duly conferred, in honour of his great deserts,
the honours and privileges of a Doctor of Medicine and Surgery
on Charles Darwin,
Master of Arts, a most distinguished member of the Geological Society
and of the Royal Society of London,
investigator of the origin of species in the animal and vegetable kingdoms,
a man of the greatest skill and learning
in zoology and general comparative physiology;
and the conferment was announced by this public diploma,
confirmed by the seal of the Faculty of Medicine.

Dr. F. C. L. Barkow | L.T.D.[6]

DAR 229

[1] CD was awarded his honorary doctorate in medicine and surgery by the Faculty of Medicine of the University of Breslau on 4 August 1861. However, the diploma and the covering letter from the University of Breslau, dated 4 August 1861 (*Correspondence* vol. 9), were not sent until early in 1862. They were both transmitted to CD via the Prussian embassy in London (see letters from Maurice Alberts, 10 February 1862 and 13 February 1862, and letter to Maurice Alberts, [after 13 February 1862]).
[2] 'Q.D.B.V.': 'Quod Deus bene vertat'.
[3] William I.
[4] Christian Julius Braniss.
[5] Hans Karl Leopold Barkow.
[6] 'L.T.D.': 'Litterarum Doctor' (doctor of letters).

From Edinburgh Royal Medical Society[1] [before 12 December 1862][2]

Societas Medica Edinburgeria
A.D. MDCCXXXVII CONSTITUTA ET REGIA AUCTORITATE
A.D. MDCCLXXIX CONFIRMATA
Omnibus ad quos hæc pervenerint
SALUTEM.

Spectatissimum ornatissimumque Virum Carolum Darwin inter Primates, nempe Honorarios, nostræ Societatis adscripsimus; quippe cujus felicis ingenii, laudum, animique ad optimum, quod que parati certiores facti fuerimus: cujus rei, quo major sit fides manibus nostris societatisque Sigillo munitas hasce Literas lubentissime donamus

Edinburgi, Anno, MDCCCLXI

Jacobus Crichton Browne Praeses.
Alexander Crum Brown M.A., B.Sc., Praeses
Gulielmus M. Watson. Praeses.
Jacobus Pettigrew— Praeses.[3]

Thomas R Fraser Sec. Hon.
Joannes H. Balfour olim. Præses
Gulielmus Rutherford Soc. Ext.
Jacobus Gerhard Reid Soc Ext.
Jacobus D. Gillespie M.D. F.R.C.S.E. olim Praes.
R. Christison Soc. Honor.
Joannes H. Bennett Soc. Hon. olim Præses.
Joannes Goodsir. Olim Præs.
J. Matthews Duncan M.D. Soc: Extr.
P. D. Handyside MD. Soc Reg Ed Soc. Soc Reg Med. Ed. Soc.
Georgius Hearder Soc. Ord.

Joannes Bayldon B.Sc. olim Praeses
Alex. R. Simpson, M.D. olim Præses.
Joannes Young MD. Soc. Extr.
Joannes Sibbald M.D. Olim Praeses.
T. Grainger Stewart M.D. F.R.C.P.E. olim Praeses—
Archibaldus Dickson, Soc. Extr.
Joannes Duncan M.A. Sec Hon.
Dyce Duckworth. Musæi Curator.
Andreas Pow. M.D olim Præs.
H. Cleghorn M.D. F.L.S.
F. de Chaumont MD. olim Praeses.
Gulielmus Anderson. M.D. Soc. Extr.

Alexander J. Macfarlan Quæstor
et olim Præs.
R. J. B. Cunninghame. Curator.
Henricus Season Wilson Olim Præses:
Alexander Dickson M.D. Olim Præses
Henricus Season Wilson M.D.
Olim Præses:

Andreas Smart olim Praeses
Gulielmus R. Sanders MD.
olim Praeses.
Josephus Bell MD. LRCS.
olim Praeses.
Doyle M. Shaw Olim Præs:
James Clark Wilson Soc: ord

[Translation]

Edinburgh Medical Society
FOUNDED A.D. MDCCXXXVII
AND CHARTERED A.D. MDCCLXXIX
To all to whom these present shall come
GREETINGS.

We enroll the most distinguished and most honoured man Charles Darwin among the very foremost of the honorary members of our society, since we are most certain of his happy genius, renown and his mind prepared for whatsoever is of the best. Wherefore, for a greater confirmation, we give these presents most willingly guaranteed under our hands and the seal of the society.

Edinburgh, in the year MDCCCLXI

James Crichton Browne President.
Alexander Crum Brown M.A., B.Sc., President.
William M. Watson. President.
James Pettigrew— President.[3]

Thomas R Fraser
Honorary Secretary
John H. Balfour
Former President
William Rutherford
Extraordinary Fellow
James Gerhard Reid
Extraordinary Fellow
James D. Gillespie M.D.
F.R.C.S.E. Former President
R. Christison
Honorary Fellow
John H. Bennett
Honorary Fellow Former President.
John Goodsir.
Former President
J. Matthews Duncan M.D.
Extraordinary Fellow

John Bayldon B.Sc.
Former President
Alex. R. Simpson, M.D.
Former President.
John Young MD.
Extraordinary Fellow
John Sibbald M.D.
Former President.
T. Grainger Stewart M.D.
F.R.C.P.E. Former President—
Archibald Dickson,
Extraordinary Fellow
John Duncan M.A.
Honorary Secretary
Dyce Duckworth.
Curator of the Museum.
Andrew Pow. M.D
Former President

P. D. Handyside MD.
Fellow of the Royal Society of Edinburgh
Fellow of the Edinburgh Royal Medical Society.
George Hearder
Ordinary Fellow
Alexander J. Macfarlan
Treasurer and Former President
R. J. B. Cunninghame.
Curator.
Henry Season Wilson
Former President:
Alexander Dickson M.D.
Former President
Henry Season Wilson M.D.
Former President

H. Cleghorn M.D. F.L.S.
F. de Chaumont MD.
Former President.
William Anderson. M.D.
Extraordinary Fellow
Andrew Smart
Former President
William R. Sanders MD.
Former President
Joseph Bell MD. LRCS.
Former President.
Doyle M. Shaw
Former President
James Clark Wilson
ordinary Fellow

DAR 229

[1] The Royal Medical Society of Edinburgh is a student medical society, founded in 1737, which was closely associated with the Edinburgh Medical School (J. Gray 1952). As a student at Edinburgh University, CD had been an ordinary member of the society, and had attended 'pretty regularly' (*LL* 1: 40, 3: 373). In his autobiography (*LL* 1: 40), CD stated that he thought it was owing to his attendance at this society and to his having visited the Royal Society of Edinburgh as a youth, that he felt 'the honour of being elected ... an honorary member of both these Societies, more than any other similar honour'.

[2] Although the diploma is dated '1861' (see also n. 3, below), it was not sent to CD until December 1862 (see letter to J. D. Hooker, 12 [December 1862]).

[3] Browne, Brown, Watson, and Pettigrew were the presidents of the society for the session 1860–1 (J. Gray 1952, p. 322).

Biographical register of persons mentioned in the diplomas

Anderson, William (1848–1902). Physician. Resident physician at the Royal Infirmary, Edinburgh, until 1865. Physician at General Hospital, Birmingham; settled in Richmond, Surrey, in 1868. (*Medical directory* 1862–1902.)

Balfour, John Hutton (1808–84). Physician and botanist. Professor of botany, Glasgow University, 1841–5. Professor of botany and Regius keeper of the Royal Botanic Garden, Edinburgh, 1845–79. FRS 1856. (*DNB*, *DSB*.)

Barkow, Hans Karl Leopold (1798–1873). Anatomist and physiologist. Appointed as professor of anatomy at Rostock, 1826; at Breslau, 1835. Director of the anatomical institute at Breslau from 1845. (*ADB*.)

Bell, Joseph (1837–1911). Scottish surgeon. Surgeon at Royal Infirmary, Edinburgh, and consulting surgeon at Royal Hospital for Sick Children. Editor of the *Edinburgh Medical Journal*, 1873–96. (*WWW*.)

Bennett, John Hughes (1812–75). Physician and physiologist. Physician, Edinburgh Royal Dispensary. Pathologist, Edinburgh Royal Infirmary. Professor of Institutes of Medicine, Edinburgh University, 1848. Editor, *London and Edinburgh Monthly Journal of Medical Science*. (*DNB*.)

Blaydon, John (*fl.* 1856–63). Physician and botanist. President of the Edinburgh Royal Medical Society, 1856–7. BSc, London University, 1860; MB, 1862. Professor of botany and materia medica, Royal Agricultural College, Cirencester, 1863. (J. Gray 1952, *Historical record*, *Post office directory of Gloucestershire* 1863.)

Braniss, Christian Julius (1792–1873). Philosopher. Professor at Breslau from *c.* 1829. Author of works on the history of philosophy. (*ADB*.)

Brown, Alexander Crum (1838–1922). Scottish chemist. Extramural lecturer in chemistry, Edinburgh University, 1863–9; professor, 1869–1908. Manager, Edinburgh Royal Infirmary. President, Chemical Society of London, 1891–3. FRS 1879. (*DNB*, *DSB*.)

Browne, James Crichton (1840–1938). Scottish physician and psychologist. Medical director, West Riding Asylum, Wakefield, 1866. Lord chancellor's visitor in lunacy, 1875–1922. Co-editor of *Brain*, 1878–85. Knighted, 1886. FRS 1883. (*DNB*.)

Chaumont, Francis Stephen Bennet François de (1833–88). Military surgeon. Assistant surgeon with Army Medical Department, 1854–65; surgeon, 1865–88. Served with the Rifle Brigade in the Crimea in 1855. Professor of military hygiene, Army Medical School, Netley, 1869–88. (*Edinburgh Academy register* 1914, *Medical directory* 1862–88.)

Christison, Robert (1797–1882). Scottish physician. Professor of medical jurisprudence, Edinburgh University, 1822–32; of clinical medicine, 1832–55; of materia medica and therapeutics, 1832–77. President of the Royal College of Physicians of Edinburgh, 1839 and 1848. President of the Royal Society of Edinburgh, 1868–73. President of the British Medical Association, 1875. Physician in ordinary to the queen in Scotland, 1848. Created baronet, 1871. (*DNB*.)

Cleghorn, Hugh Francis Clark (1820–95). Physician and forester. Appointed to Madras Medical Service, Mysore, 1842. Became professor of botany, Madras, in 1852, and Conservator of Forests, Madras, in 1856. Returned to Scotland *c.* 1869. (R. Desmond 1994, Goring 1992.)

Cunynghame, Robert James Blair (Blair) (1840/1–1903). Scottish surgeon. Resident physician, Edinburgh Royal Infirmary. Medical officer, Edinburgh New Town Dispensary, 1868–80. Surgeon, Edinburgh Ear Dispensary, 1876–1903. Examiner and conservator of museum, Royal College of Surgeons of Edinburgh, 1877–1903. Physician, Edinburgh Royal Hospital for Sick Children, 1880–3. Superintendent of Statistics, Registrar General's Department in Scotland, 1884–1903. President, Royal College of Surgeons of Edinburgh, 1892. (J. Gray 1952, *Medical directory* 1865–.)

Dickson, Alexander (1836–87). Scottish botanist. Professor of botany at Trinity College, Dublin, 1866–8; at Royal College of Science, Dublin, 1867–8. Professor of botany, Glasgow University, 1868–79. Professor of botany, Edinburgh

University, and Regius keeper of the Royal Botanic Garden, Edinburgh, 1879–87. (R. Desmond 1994, *DNB.*)

Dickson, Archibald (1837–1902). Scottish physician. MD, Edinburgh University, 1863. Medical officer at Edinburgh New Town Dispensary, 1866–72. Brother of Alexander Dickson; succeeded to his Lanarkshire estate in 1887. (*Burke's landed gentry* 1921, *List of graduates of the University of Edinburgh* 1889, *Medical directory* 1872.)

Duckworth, Dyce, 1st baronet (1840–1928). Physician. MD, Edinburgh University, 1863. Physician at St Bartholomew's Hospital, London, 1883–1905; joint lecturer on medicine, 1890–1901. Knighted, 1886. Created baronet, 1909. (*DNB.*)

Duncan, James Matthews (1826–90). Scottish physician. Physician, Edinburgh Royal Infirmary, 1861. President of the Obstetrical Society of Edinburgh, 1873–5. Lecturer on midwifery and obstetric physician, St Bartholomew's Hospital, London, 1877–90. FRS 1883. (*DNB.*)

Duncan, John (1839–99). Scottish surgeon. MD, University of Edinburgh, 1862. Appointed surgeon to the Royal Infirmary, Edinburgh, 1877. Instrumental in the development of expertise in the surgical applications of electricity. (J. Gray 1952, *Medical directory* 1880, *NUC.*)

Fraser, Thomas Richard (1841–1920). Scottish pharmacologist. Assistant physician, Edinburgh Royal Infirmary, 1869–74. Professor of materia medica and clinical medicine at Edinburgh University, 1877–1917. President of the Royal College of Physicians of Edinburgh, 1900–2. Knighted, 1902. Physician in ordinary to the King in Scotland, 1907. FRS 1877. (*DNB.*)

Gillespie, James Donaldson (1822/3–91). Scottish surgeon. Extramural lecturer, Edinburgh University, 1851. Surgeon and lecturer in clinical surgery, Edinburgh Royal Infirmary, 1851–91. Medical Officer, Edinburgh New Town Dispensary. Surgeon, Gillespie's and Donaldson's Hospitals, Edinburgh. Examiner, Royal College of Surgeons of Edinburgh, *c.* 1865–91; president, 1869. President, Royal Medical Society of Edinburgh, 1846–7. (J. Gray 1952, *Medical directory* 1863–, *Medical directory for Scotland* 1857.)

Goodsir, John (1814–67). Curator of the Edinburgh University museum and demonstrator in anatomy, 1843–6. Professor of anatomy at Edinburgh, 1846–67. FRS 1846. (*DNB, DSB.*)

Handyside, Peter David (d. 1881). Scottish surgeon and anatomist. Extramural lecturer in anatomy, Edinburgh University, 1846. Examiner and lecturer in anatomy, Royal College of Surgeons of Edinburgh, 1863–81. President, Royal Medical Society of Edinburgh, 1828–9. (J. Gray 1952, *Medical directory* 1863–.)

Hearder, George Jonathan (1839/40–94). Physician. Assistant medical officer, Worcester County and City Lunatic Asylum, *c.* 1864–8. Medical superintendent, Joint Counties Lunatic Asylum, Carmarthen, Wales, 1869–94. President, South Wales and Monmouth Branch of the British Medical Association, 1873. (*Medical directory* 1863–.)

Macfarlan, Alexander Johnstone (1837/8–69). Scottish physician and botanist. Curator, Botanical Society of Edinburgh. President, Royal Medical Society of

Macfarlan, Alexander Johnstone, cont.
Edinburgh, 1859–60; treasurer, 1861. Partner in family firm, J. F. Macfarlan & Co., chemists, druggists and chemical manufacturers. (R. Desmond 1994, J. Gray 1952, *Post Office Edinburgh and Leith directory* 1868.)

Pettigrew, James Bell (1834–1908). Scottish anatomist. Croonian lecturer, Royal Society of London, 1860. House surgeon, Edinburgh Royal Infirmary, 1861. Assistant, Hunterian museum, Royal College of Surgeons of England, 1862–7. Curator, museum of the Royal College of Surgeons of Edinburgh, 1869; lecturer in physiology, 1873. Chandos professor of medicine and anatomy and dean of the medical faculty, St Andrews University, 1875. Represented Glasgow and St Andrews Universities on the General Medical Council, 1877–86. FRS 1869. (*DNB*.)

Pow, Andrew (d. 1863). Scottish physician. MD, University of Edinburgh, 1857. Resident physician, Royal Infirmary, Edinburgh. (*Medical directory* 1861–5.)

Reid, James Gerhard (*fl.* 1861). Extraordinary fellow of the Edinburgh Royal Medical Society.

Rutherford, William (1839–99). Scottish physiologist. Professor of physiology, King's College, London, 1869. Fullerian professor of physiology, Royal Institution of London, 1871. Professor of physiology, Edinburgh University, 1874–99. Co-editor of *Journal of Anatomy and Physiology*, 1875–6 and *Journal of Physiology*, 1878. (*DNB*.)

Sanders, William Rutherford (1828–81). Scottish physician. Conservator of the Royal College of Surgeons, Edinburgh, 1853. Lecturer in medicine from 1853. Appointed physician to the Edinburgh Royal Infirmary, 1861. Appointed chair of pathology at University of Edinburgh, 1869. (*DNB*.)

Shaw, Doyle Money (1830–1918). Naval surgeon. Served as surgeon on various naval vessels and flagships, 1854–83. Inspector general of Royal Navy hospitals at Plymouth, Malta, and Haslar, 1884–92. Knighted, 1911. (*WWW*.)

Sibbald, John (1833–1905). Scottish physician concerned with the treatment of mental illness. Medical superintendent, Argyll District Asylum, 1862–70. Deputy commissioner of lunacy for Scotland, 1870–8; commissioner, 1878–99. Knighted, 1899. (J. Gray 1952, *Medical directory* 1862, *WWW*.)

Simpson, Alexander Russell (1835–1916). Scottish physician. Assisted at the practice of his uncle, James Gray Simpson, 1st baronet (*DNB*), 1858–65. Practised in Glasgow, 1865–70. Dean of the faculty of medicine and professor of midwifery and the diseases of women and children, Edinburgh University, 1870–1905. Knighted, 1906. (J. Gray 1952, *Medical directory* 1862–70, *WWW*.)

Smart, Andrew (1825–1911). Scottish physician. MD, Edinburgh University, 1862. Physician at the Royal Infirmary, Royal Public Dispensary, and Royal Medical Asylum in Edinburgh. Assistant professor of the practice of medicine, Edinburgh University; and of clinical medicine, Royal Infirmary, Edinburgh, 1866. (*List of graduates of the University of Edinburgh* 1889, *Medical directory* 1870–1912.)

Stewart, Thomas Grainger (1837–1900). Scottish physician. House physician at the Edinburgh Royal Infirmary, and lecturer on pathology at Surgeons'

Hall, 1862–9. Resigned these posts to become junior ordinary physician in the infirmary and lecturer in clinical medicine, and, later, in the practice of physic. Appointed professor of the practice of physic, Edinburgh University, 1876. Knighted, 1894. (*DNB*, J. Gray 1952.)

Watson, William M'Culloch (1839/40–89). Scottish medical practitioner. Resident physician, Edinburgh Royal Infirmary. President, Royal Medical Society of Edinburgh, 1860–1. (*British Medical Journal* (1889), pt 1: 1260; J. Gray 1952; *Medical directory* 1863–.)

William I (1797–1888). King of Prussia, 1861; German Emperor, 1871. (*ADB*.)

Wilson, Henry Season (*fl.* 1861–1900). Anatomist and physician. Demonstrator in anatomy, Cambridge University, 1868–76. Demonstrator, London Hospital Medical College, 1877. Physician, Isleworth, Middlesex, 1881. (*Alum. Cantab.*, *Medical directory* 1865–.)

Wilson, James Clark (1842–77). Scottish physician. MD, Edinburgh University, 1863. Physician in Ayr. (*Edinburgh Academy register* 1914, *List of graduates of the University of Edinburgh* 1889.)

Young, John (1835–1902). Scottish physician and geologist. Physician at the Royal Infirmary and Royal Lunatic Asylum, Edinburgh, until 1866. Worked on the Royal Geological Survey, 1860–6. Keeper of the Hunterian Museum, University of Glasgow, 1866. Professor of natural history and Honeyman Gillespie lecturer on geology, University of Glasgow, 1866–1901. (*Medical directory* 1861–5, *WWW*.)

MANUSCRIPT ALTERATIONS AND COMMENTS

The alteration notes and comments are keyed to the letter texts by paragraph and line numbers. The precise section of the letter text to which the note applies precedes the square bracket. The changes recorded are those made to the manuscript by CD or his amanuensis; readers should consult the Note on editorial policy in the front matter for details of editorial practice and intent. The following terms are used in the notes as here defined:

del	deleted
illeg	illegible
interl	interlined, i.e., inserted between existing text lines
omitted	omitted by the editors to clarify the transcription
over	written over, i.e., superimposed

To John Brodie Innes [3] January [1862]
1.1 arrived] *after del* 'has'

To Société Impériale Zoologique d'Acclimation [after 10 January 1862]
1.1 what] *after del* 'how much'
1.2 Imperiale de Acclimatation] *interl*
1.2 the annual] *added*

To Henry Walter Bates 13 January [1862]
2.3 or] *over* '&'
3.4 have] *above del* 'are'

To Thomas Henry Huxley 14 [January 1862]
1.3 You] *over illeg*
4.3 partially] *interl*
4.4 (as I have to day)] *parentheses over commas*
4.7 in Lin. Journal] *interl*

To Ludwig Rütimeyer 15 [and 16] January [1862]
2.3 to Ld. Tankerville] *interl*

To Joseph Dalton Hooker 16 January [1862]
1.2 at a time] *interl*
4.4 was ... what] *added*
10.1 with curious address] *interl*

To Asa Gray 22 January [1862]
1.1 curious] *after del* 've'
1.6 there] *above del* 'there'
4.3 insult!] '!' *over comma*
6.1 big] *before omitted point*
6.5 the same] 'the' *interl*
6.6 the Melastomas;] 'the' *added*

To T. H. Huxley 22 January [1862]
2.3 hear that] 'that' *over* 'it'
2.3 you will touch on it.] *interl*
2.7 such] *after del illeg*
3.1 me,] *added*

To John Lubbock 23 January [1862]
1.7 to me] *interl*

To J. D. Hooker 25 [and 26] January [1862]
1.7 making] *interl*
2.4 subject:] *interl*
2.6 get] *after del illeg*
2.8 the lower] 'the' *interl*
3.4 2 or 3] *interl*
3.5 book] *interl*
4.1 very] *above del* 'very'
4.3 that Owen] 'O' *over* 'o'
6.4 at] *before omitted point*

To William Walmisley Baxter 26 January [1862]
0.3 Dear] 'D' *over illeg*

To John Murray 28 January [1862]
1.3 generalizer] 'z' *over* 's'
1.4 animals;] ';' *over full stop*
1.5 but Entomology . . . forte.—] *added*
1.7 several months ago] *interl*
1.12 Chapters] *interl*
2.2 whom] *after del* 'ask'
2.7 common] *interl*

To J. D. Hooker 30 January [1862]
2.1 Saturday,] *after del illeg*

To George Bentham 3 February [1862]
2.4 of pollen] *interl*
3.4 is exactly opposed to] *above del* 'would agree with'
3.8 own] *interl*
3.18 the name] 'the' *interl*
3.20 (which would be best)] *interl*
3.22 differently in different plants. 3.23] *interl*
3.25 viz] *interl*

To Charles Kingsley 6 February [1862]
1.2 Argyle] 'A' *over* 'a'
3.4 a naked] *after del comma*
3.7 was a hairy beast.] *interl*
3.8 as] *over illeg*
4.2 &] *over illeg*

To J. D. Hooker 9 February [1862]
1.4 (as . . . mortals)] *parentheses over commas*
1.5 in case] *after del illeg*
1.11 & forms] *interl*
3.1 You allude] *after del illeg*
4.1 glad] *interl*

To John Murray 9 [February 1862]
1.1 (Monday)] *interl*
1.2 fully] *interl*
4.1 *quickly*,] *after del* 'pretty'

To Ludwig Rütimeyer 11 February [1862]
2.5 new &] *interl*

To Maurice Alberts [after 13 February 1862]
1.1 beg leave] *above del* 'have the honour'
1.1 your letter & the] *interl*
1.3 Univ] *after del* 'Body'

To William Erasmus Darwin 14 February [1862]
1.1 that] *interl*
1.4 that you have got one;] *above del* 'it'
2.12 the] *interl*
2.13 Compositæ] *interl*
3.1 constant] *interl*
3.2 Skimp] 'S' *over* 's'
3.2 a day] *after del* 'of'
3.3 sobbing,] *interl*
3.4 during] *interl*
3.5 accompanied] *after del* 'not'
3.6 probably] *after del* 'was'

To Asa Gray 16 February [1862]
2.7 basal] *above del* 'ba lower'
2.7 ring] *after del* 'u'
3.3 the pistil] 'the' *over illeg*
3.4 prove] *above del* 'prove'

To J. B. Innes 24 February [1862]
1.3 (except] *parenthesis over comma*
3.2 account of] 'of' *over* 'I'
4.3 our] *interl*
4.5 (to whom] *parenthesis over comma*

To J. D. Hooker 25 February [1862]
1.3 there] *interl*
1.6 with] *after del* 'of'
1.9 & of your Lebanon case] *interl*
1.12 whole] *interl*
2.5 going back &] *above del* 'going'
2.6 also is] *interl*
2.7 What] *after del* 'is.'
2.8 Flora of] *interl*
2.9 point] *interl*
2.13 any] *above del* 'ay'
3.2 transplanting] *interl*
4.3 this next summer] *interl*
5.2 to Linn. Soc:] *interl*

To J. D. Hooker 7 March [1862]
2.1 Orchid] *interl*
3.2 at the time] *interl*
4.1 Arctic] *interl*
4.3 of dispersal] *interl*
4.5 their] *over illeg*
4.6 from want . . . modification] *added*
4.7 as you] *interl*
4.7 (considering . . . north) 4.8] *parentheses over commas*
4.7 direction of] *interl*
4.9 to there] 'to' *interl*
6.3 in Primula] *interl*

To J. D. Hooker 14 March [1862]
4.3 as in woodpecker] *interl*
4.7 for] *over* 'on'

To Asa Gray 15 March [1862]
2.6 Natural Selection] *before omitted point*
3.3 sent)] *parenthesis over comma*
3.5 compare the] 'the' *added*
3.5 the part] 'the' *interl*
3.8 & stigma] *interl*
3.9 & give me a sketch.] *interl*
5.1 Again . . . me.] *added*

To J. D. Hooker 18 March [1862]
1.2 Castle] 'C' *over* 'c'
2.4 or plant] *interl*
2.6 seems] *after del illeg*
2.8 owing to Orchids] *interl*
2.9 the] *over* 'an'
2.10 of an animal] *interl*
2.11 shade] *before omitted point*
2.11 produce] *altered from* 'produces'
2.11 on a] 'a' *interl*
3.2 also] *interl*
3.9 possibly] *added*
3.10 *without selection*] *interl*
4.7 serious] *after del* 'very'

To Richard Kippist 18 March [1862]
1.2 have,] *added*
1.2 will soon] 'will' *interl*
1.3 it,] *interl*

To Richard Kippist 18 March [1862]
1.4 the whole] *above del* 'all'

To William Bernhard Tegetmeier 18 March [1862]
1.2 ashamed] [1]'a' *over* 's'
2.4 one] 'o' *over illeg*

To J. D. Hooker 22 [March 1862]
2.1 3^{d}.] *before omitted point*

To J. D. Hooker 26 [March 1862]
1.1 must be the] *above del* 'is'
1.4 direct] *interl*
2.2 with] *added*
2.6 to see] *interl*
2.6 (& yourself)] *interl*
2.9 2nd] *over* '3rd'

To Charles Lyell 1 April [1862]
1.1 I must] 'I' *interl*
1.2 -theory] *interl*
1.5 outlet at] *interl*
1.5 **first**] *interl*
1.6 began] *after del comma*
1.8 was discovered by Milne] *above del* 'is seated'
1.10 along all] *above del* 'in'
2.2 ?] *over* '.—'

To J. D. Hooker 9 [April 1862]
1.3 hours,] *above del* 'drops'
1.6 some] *after del* 'when'
1.6 long] *interl*

To John Murray 9 April [1862]
0.2 April 9th] *above del* 'March'

To Daniel Oliver 12 [April 1862]
2.1 (pollen of)] *interl*
2.2 much] *interl*
2.5 of Primrose] *interl*
2.6 the reverse] 'the' *interl*
2.6 "long"] *above del* 'short'
2.7 is the] 'the' *altered from* 'this'
2.7 size] *interl*
2.8 number of ovules] 'number of' *interl*
2.8 look at] *before omitted point*
3.6 latter cases] *before del caret*
3.6 from what . . . Violets] *interl*
3.7 the imperfect] *interl*
3.10 provisional] *interl*
4.2 (& so does Vaucher)] *interl and square brackets in MS*
4.6 this letter] *interl*
5.1 insect] *interl*
5.2 Dielytra] 'ie' *over* 'ei'
5.2 Adlumia] 'A' *over* 'ad'
5.2 Adlumia &c &c 5.3] *interl*
5.4 the stamens] 'the' *added*
5.4 to] *above del* 'on'
5.4 the other] 'the' *interl*
5.5 covered] *after del* 'ru'
5.5 This] *after del* 'Next'
5.6 & the pistil is straight] *interl*
5.7 & allies] *interl*
5.8 secreting] *interl*
5.8 only] *interl*
5.9 the pistil] 'the' *interl*
6.4 at once] *interl*
6.5 (but flowers . . . fertile)] *interl*

7.4 as you say,] *interl*
8.3 rejoice] *after omitted point*
9.1 also] *interl*
10.8 groups] *interl before del* 'families', *above del* 'genera'

To Daniel Oliver 15 April [1862]
1.4 more than] *interl*
3.3 some] *interl*
4.2 make] *above del* 'have'
4.2 of] *after del* 'made'
4.3 suspect] *above del* 'believe'
6.1 N. America] *interl*

To H. W. Bates 16 April [1862]
4.2 "Shortlands] *before del* ' " '

To Charles Edouard Brown-Séquard 16 April [1862]
1.6 to give it, 1.7] *interl*

To Daniel Oliver 20 [April 1862]
1.5 my] *interl*

To Asa Gray 21 April [1862]
1.4 many] *after del* 'several'
1.6 has not] *after del* 'Secondly'
2.1 of my] *interl*
2.3 too severely] *interl*
3.1 North] 'N' *altered from* 'n'
4.2 yet] *interl*

To Daniel Oliver 24 April [1862]
1.3 useful] *interl*

To Heinrich Georg Bronn 25 April [1862]
1.8 new] *interl*
1.10 almost] *interl*
2.7 by Post to you] *interl*
2.10 Schweizerbart] 'w' *over* 'ei'
2.15 Schweizerbart] 'w' *over* 'ew'

To W. E. Darwin 26 April [1862]
2.1 Evening] 'E' *over* 'e'

To T. H. Huxley 30 April [1862]
1.3 & I] '&' *over* '—'

To J. D. Hooker 1 May [1862]
1.2 it] *interl*
1.5 highly] *interl*

To J. B. Innes 1 May [1862]
1.7 , also,] *interl*

To John Murray
1.1 to Printers] *interl*

To H. W. Bates 4 May [1862]
1.4 by] *below del* 'with'
2.2 closely allied] *above del* 'the'
2.4 is] *after del* 'of'
3.2 Owen] 'O' *over* 'o'
3.2 before very long. 3.3] *interl*
3.4 personal] *interl*

To H. W. Bates 9 May [1862]
1.2 almost certainly] *interl*
1.2 In] *before omitted point*
2.1 mimetic] *interl*

To J. D. Hooker 9 May [1862]
1.8 much] *interl*
2.1 &] *over* 'ha'
5.4 correct] *after del illeg*
6.1 I] *after del illeg*
6.2 !!!] *over* '.—'

To T. H. Huxley 10 May [1862]
2.2 "your] *interl*
2.3 look] *altered from* 'looked'
2.4 of your Address] *interl*
3.2 so-called] *interl*
3.4 on non] *above del* 'non-'
3.5 enough] *interl*
4.1 Address] 'A' *over* 'a'
4.2 dead] *after del illeg*

To William Darwin Fox 12 May [1862]
1.3 so-called] *interl*
1.4 your] *over illeg*
1.6 quite] *interl*
1.7 Turkeys?] '?' *over* '.—'
2.1 &] *interl*

To Philip Lutley Sclater 12 May [1862]
1.4 can you] *interl*
1.6 recent] *interl*

To P. L. Sclater 14 May [1862]
1.9 are] *altered from* 'is'

To J. D. Hooker 15 [May 1862]
1.1 I known] 'I' *interl*

1.1 that] *after del* 'fo'
1.1 was] *altered from* 'way'
1.3 two] *altered from* 'to'
1.5 must] *after del* 'I'
1.6 does it] *interl*
2.1 at Linn.] 'at' *over illeg*
2.6 on the same plant] *interl*
2.8 me] *after del* 'see'
2.8 though] *interl*
2.9 all] *interl*
2.9 rising] *after del illeg*
2.10 anthers.] *full stop over semicolon*
2.13 (*mind*, ... *leisure*)] *square brackets in MS*
2.17 the] *over* 'my'
5.2 Boothii] 'oo' *altered from* 'ho'
5.2 smallish] *interl*

To W. D. Fox [17 May 1862]
1.1 here,] *comma over semicolon*
2.3 &c] *interl*

To J. D. Hooker [18 May 1862]
1.4 collect] *above del* 'Catch'
1.6 diversity] *above del* 'diversity'
1.11 been] *interl*
2.3 my] *interl*
2.3 am] *interl*
2.6 degree to] 'to' *over* 'at'
2.8 has a] *interl*

To J. D. Hooker 30 May [1862]
1.2 of Heterocentron 1.3] *interl*
1.5 which I have seen,] *interl*
1.12 included] *interl*
1.12 empty)] *parenthesis over comma*
1.15 the anther] 'the' *interl*
1.16 are] *after del* 'with'
5.1 opened] *before del comma*
5.3 , like that of Primula,] *interl*
6.1 which] *before del* 'p'
6.2 the stigma] 'the' *interl*
6.2 & no pollen 6.3] *interl*
6.3 fertile] *interl*
6.3 sterile] *interl*
7.1 (single flower)] *above del illeg*

To W. E. Darwin [31 May 1862]
4.2 & odorata] *interl*

To Daniel Oliver 8 June [1862]
1.6 tissue or] *interl*
2.1 is] *after del* 'wa'
3.2 published anything which I] *interl*

To Asa Gray 10–20 June [1862]
1.2 May] *interl*
1.3 value;] *after del illeg*
2.2 of] *over illeg*
4.17 "Blood's] *after del* 'for a'
4.17 Envelope, 1, 3,] *commas over full stops*
8.5 (See p. 8 at back of p. 5.)] *square brackets in MS*
8.5 (This ... place) 8.6] *square brackets in MS*
9.4 other notes.] 'notes.' *interl*
9.9 it] *interl*
10.2 ardent] *interl after del* 'an'
10.3 Deist] *after del interl* 'ardent'
10.5 & good] *interl*
10.5 these] *after del illeg*
12.2 unlike that of] *above del* 'unlike'
13.2 grains collected] 'grains' *interl*
14.1 & he gives me facts] *interl*
14.4 . . . . . .] *interl*

To Daniel Oliver [before 11 June 1862]
1.2 seen,] *comma over full stop*
1.2 minute] *interl*
3.1 Müller] *after del* 'he'
4.3 allied] *interl*

To J. D. Hooker 11 June [1862]
1.2 (who is here)] *interl*
1.5 the change)] 'the' *interl*
1.9 marry] *above del* 'marry'
4.1 a paper] *after del illeg*
5.1 (with] *parenthesis over* '—'
5.2 not] *added*

To W. E. Darwin 13 [June 1862]
2.1 some little] *interl*

To Leonard Horner 13 June [1862]
1.5 us] *over* '&'

To John Murray 13 June [1862]
1.6 ; you will probably know] *interl*
3.2 but] *interl*
3.5 (an admirable Journal)] *interl*
3.5 of] *interl*
3.9 light] *interl*

To George Henry Kendrick Thwaites 15 June [1862]
1.2 D^r.] *added*
2.2 *Ceylon*] *above del* '*the*'

2.2 , making] *over* '.—'
2.3 There is no need of castration] *interl*

To Alphonse de Candolle 17 June [1862]
1.4 on several plants &] *interl*
1.10 that of] *added*
2.6 regards] *above del* 'is with respect to'
2.7 at] *interl*
3.5 & admiration 3.6] *interl*

To G. H. K. Thwaites 20 June [1862]
2.3 (I have been ... diœcious.)] *square brackets in MS*
2.7 the pollen-grains] *after del* 'wit'
5.2 forms] *interl*
5.3 Primula] *before omitted point*

To J. D. Hooker 23 June [1862]
1.4 world:] *altered from* 'word:'

To Nicholas Trübner 23 June [1862]
1.2 at once] *interl*

To James Brown Gibson [after 29 June 1862]
1.2 to my] *after del* '& ha'
1.2 in the endeavour] *interl after del interl* 'with respect'
1.3 the endeavour to find] *interl after del interl* 'find' *above del* 'to write'
2.1 not *[illeg]*] *interl*

To H. G. Bronn 30 June [1862]
7.1 = "Gummi"] *interl*
7.1 or hinge] *after del* 'of hin'
9.2 small] *added*
12.5 as one petal or one anther 12.6] *interl*

To J. D. Hooker 30 [June 1862]
1.6 (but ... sells)] *interl*

To Asa Gray 1 July [1862]
1.1 tonight] *after del* 'toda'
1.2 (& have no other object)] *interl*
1.4 to write] 'to' *interl*
1.7 bridal] *interl*
1.10 without permission first granted] *interl*
2.2 it is] 'is' *interl*
2.5 probosces.] *altered from* 'proboscis.'
2.8 viz to] 'to' *interl*
2.8 exterior] 'ex' *over* 'ba'
2.11 they] *interl*
2.13 a wrong] *above del* 'wrng'
2.16 strongly] *after del* 'as'

To John St Barbe [before 3 July 1862]
1.2 if I ... suit me.] *interl*
1.3 it] *above del* 'my request'
1.4 for Guaranteed] *after del* 'to have what'; 'for' *over* 'in'
1.4 or Preference 1.5] *interl*
1.6 some of] *interl*
1.6 & of] *interl*
1.7 Will] *after del* 'Can'
1.7 shares] *after del* 'other'
1.7 there are] *after del* 'can'
1.9 sh^{d}.] 'd' *above* 'd'
1.9 perhaps] *above del* 'probably'
1.11 Preference or Guaranteed] *interl*
1.12 *perpetual*] *before interl del* 'or *[two words illeg]*'
1.12 (though ... stock) 1.13] *added*
1.13 said to be] *interl*
1.15 the best 1.6] *interl*

To George Maw 3 July [1862]
1.22 (though] *after del* '&'
1.23 in lesser degree] *interl*
1.24 or] *interl*
1.25 even] *interl*
1.29 Some] *interl*
1.30 a quasi marsupial] *interl*
1.34 (I must add)] *parentheses over commas*
2.4 As] *over* 'I'

To W. E. Darwin 4 [July 1862]
1.5 stigma] *before del* ')'
2.1 of hardly ... pollen-tube: 2.2] *interl*
2.4 mere] *interl*
2.5 that] *above del* 'may not'
2.9 expose stigma;] ';' *over* '.—'
2.9 the stigma 2.10] 'the' *over illeg and* 'stigma' *added*
4.1 well,] *after del* 'so'
4.2 is not] 'is' *interl*
5.2 horses)] *parenthesis over comma*

To Maxwell Tylden Masters 8 July [1862]
2.6 wild] *interl*
2.6 peloric] 'el' *over illeg*
2.7 Does] 'D' *over illeg*
4.1 sterility] *after del* 'hybridit'

To H. G. Bronn 11 July 1862
1.2 "Mormodes"*] *interl*
2.6 after the movement] *interl*
2.7 anther.] *full stop altered from comma*

2.7 There] *after del* 'which is protected'
2.7 either] *after del* 'touch'
2.8 anther or] *interl*
2.9 flirted] *after del* 'flied'
2.9 probably] *after del* 'to'
5.1 most of] *above del* 'all'
5.2 until lately] *interl*
5.3 are] *interl*

To Armand de Quatrefages 11 July [1862]
1.6 any facts] *above del* 'them'

From Down Friendly Society to Bromley Savings Bank [before 14 July 1862]
1.4 C. Darwin] *after del* 'Trustee'

To Asa Gray 14 July [1862]
1.1 have] *interl*
1.2 (the Postage Stamp collector)] *interl*; *square brackets in MS*
2.5 end to] 'to' *over* 'as'
2.6 when] *above del* 'how'
2.7 that] *before del* 'in'
2.7 the structure] 'the' *over* 'al'
4.1 has] *interl*
4.5 pollen of] *after del* 'two'
4.5 division of the] *interl*
4.7 ; nor . . . exserted.] *interl*
4.11 of this one species] *interl*

To W. E. Darwin [after 14 July 1862]
1.1 (but] *parenthesis over comma*

To John St Barbe [16 July 1862]
1.1 I enclose cheque] *interl*
1.3 pay into B. of E 1.4] *above del* 'invest'
1.4 . . . . . .] *below del interl* '(in figures and writing)'
1.4 enclosed] *interl*
1.4 for 122] *interl*
1.5 (figures] *square bracket in MS*
1.6 which must . . . Commission 1.6] *interl*

To Asa Gray 23[–4] July [1862]
1.1 your] *over* 'on'
1.24 too] *after del* 'more'
3.1 of thin paper] *interl*
4.3 advances] *after del* 'that'
5.3 news-] *interl*
7.2 diversity] *after del* 'of'
9.2 observations!] *after del* 'letter'
9.5 to me] *interl*
9.7 actually] *interl*

To W. E. Darwin [24 July 1862]
1.1 powdered meat] *interl*

To M. T. Masters 24 July [1862]
3.2 some few] 'few' *interl*

To J. D. Hooker 26 July [1862]
7.1 orchids] *after del* 'are'
7.2 & flower] *interl*

To Asa Gray 28 July [1862]
3.8 that] *over illeg*

To W. E. Darwin [2–3 August 1862]
2.2 labour, by] *after del* 'the'
2.3 a grand] 'a' *interl*
2.6 your] *after del* 'or'
3.3 only] *interl*
3.5 the short-styled] 'the' *interl*
4.1 my] *interl*
5.1 case?] '?' *over* ':'
5.3 from flower to flower.] *interl*
6.1 that the] *interl*
9.1 two] *after del* 'one'
12.8 on] *over* 'of'

To Asa Gray 9 August [1862]
2.2 $6^{s}\cdot 8^{d}$, instead of 9^{s}.)] *interl*
3.2 merely] *above del* 'mery'
3.4 believe,] *before omitted point*
3.5 distinct] *interl*
4.1 hyssopifolia] *altered from* 'hyssipofolia'
4.2 & I have looked to your manual 4.3] *interl*
7.3 two] *above del* 'the union'
7.3 (from two homorphic unions)] *interl*

To Alfred Russel Wallace 20 August [1862]
3.4 increase] *above del* 'change'

To Asa Gray 21 August [1862]
1.3 well] *after del* 'pretty'
2.7 first] *interl*
4.4 will] *after del* '&'

To J. D. Hooker 22 [August 1862]
2.7 case] *below del* 'fact'
2.11 but] *above del* 'that'
2.16 weak] *interl*
3.2 of dissected object 3.3] *interl*
4.2 & in] *added*
5.3 emit] *after del illeg*

To Charles Lyell 22 August [1862]
2.7 &c] *interl*
2.9 plainly] *interl*
2.11 or] *over* '.—'
3.3 the statement] *interl*
3.4 probably true.] *interl above del full stop*
3.5 at all] *added*
3.5 contrary] *interl*
4.2 almost] *interl*
4.4 (or during all successive periods)] *interl*
4.6 e.g.] *interl*
5.1 at] *above del* 'on'
6.1 seem to] *interl*
6.2 sea] *above del* 'water'
6.7 rate of] *interl*
7.2 from each other 7.3] *interl*

To Camilla Ludwig 26 August [1862]
2.3 been forced to] *interl*
2.4 engage] *altered from* 'engaged'

To John Lubbock 2 September [1862]
2.5 bitten] *after del illeg*
2.9 probosces.] *altered from* 'proboscis.'

To John Lubbock [3 September 1862]
1.3 stupid] *after del illeg*

To Asa Gray [3–]4 September [1862]
1.8 must] *interl*
1.12 horn-like] *interl*
1.12 its structure] 'its' *interl*
1.13 apparently] *interl*
2.2 (why . . . things?) 2.4] *square brackets in MS*
2.7 it] *interl*
2.16 or] *over illeg*
3.2 (but . . . blunder) 3.3] *square brackets in MS*
7.1 little] *interl*
7.1 Horace] *interl*

To Andrew Crombie Ramsay 5 September [1862]
1.20 possibly] *above del* 'easily'
2.6 be due] 'be' *interl*
2.13 several] *interl*
3.4 recent] *interl*

To W. E. Darwin [10? September 1862]
1.1 good] *after del* 'muc'
2.1 much] *interl*
2.3 London] *interl*

To J. D. Hooker 11 September [1862]
1.5 some time] *interl*
1.7 of outer whorl] *interl*
1.7 perhaps] *interl*
1.8 so few, only two, sub-bundles] *interl after del* 'three sub- [*interl*] bundles of vessels in all the 8 [*interl*] main bundles, excepting'
1.12 confluent] *interl*
1.13 stamens] 't' *over* 'p'
2.4 he had] 'he' *interl*
2.9 flower,] *comma added*
2.9 explain] *after del* '&'
3.6 now] *interl*
3.7 I saw] 'I' *over* 'M'
3.9 Thanks . . . Mann &c &c.— 3.10] *interl*
5.1 (except] *parenthesis over* ';'
7.2 in] *over* '—'

To W. D. Fox 12 September [1862]
2.2 from] 'fr' *over* 'w'

To Daniel Oliver 14 September [1862]
1.10 species] *interl*
1.11 *unopened*] *interl*

To Daniel Oliver [17 September 1862]
1.2 shall] *above del* 'will'
1.2 great] *after del* 'care.'
1.6 the case] 't' *over illeg*
4.3 this] *interl*
4.3 plant was 4.4] *altered from* 'plants were'
4.4 to various] 'to' *interl*

To J. D. Hooker [18 September 1862]
1.7 also] *interl*
1.8 Sweet Pea,] *above del* 'Lupine'
1.9 vessel] *interl*
1.10 the vessels] *after del* 'in Crucifer'
1.10 of the carpel-leaf often] *interl*
1.11 sepals.] *above del* 'calyx.'
2.1 for] *over* '&'
2.2 had] *interl*
2.3 every] *above del* 'the'

To W. D. Fox 20 [September 1862]
1.1 & pleasant] '&' *over comma*
2.4 had] *interl*

To J. D. Hooker 21 [September 1862]
1.8 to] *over ampersand*
2.6 &c] *over colon*
3.2 few] *interl*

To J. D. Hooker 26 September [1862]
1.3 in] *after del* 'but'
1.3 organic] *interl*
1.17 most of] *interl*
1.18 on the] 'the' *interl*
1.21 partially] *interl*
1.23 possesses] *after del* 'at least'
1.23 at least in some degree 1.24] *above del* 'closely'
7.1 but if] *after del* '&'
7.2 be given] 'be' *interl*
13.11 & killed the glands 13.12] *interl*
13.17 tentacles] *interl*
13.21 was excessively] 'was' *interl*
13.23 Poison from adders'] *after del* '[*illeg*] caused no movement & stopped for some [*illeg*] subsequent action of meat'
13.26 Poison from] *after del* 'Poison from'

To Charles Lyell 1 October [1862]
2.2 long] *interl*
3.1 soon] *interl*
3.2 Proboscidean] *altered from* 'Probiscidean'
4.2 form] *altered from* 'force'

To J. D. Hooker 6 October [1862]
4.2 open] *interl*
4.2 (i.e.] *parenthesis over comma*
4.5 6 or 7] *interl*
4.6 Some] *over* 'L.'
4.7 a **tri**] 'a' *above del* 'the'
4.8 form] *interl*
4.8 or graduates] *interl*
5.6 Kew,] *comma over* ';'
11.1 (N.B . . . return.)] *square brackets in MS*
12.3 have] *after del* 'of'

To George Bentham 13 October [1862]
3.1 Pray] *altered from* 'Pary'

To Daniel Oliver 13 October [1862]
1.6 plants] *after del* 'fl'
1.8 any] *after del* 'most'
2.3 & were all self-fertile 2.4] *interl*
4.3 thin] *interl*
4.3 I suppose] *after del illeg*
4.3 like in orchids] *interl*
5.4 would] 'w' *over illeg*

To J. D. Hooker 14 [October 1862]
1.1 first] *after del* 'I'
3.3 I answer . . . slip.] *added*
4.2 some] *interl*
4.4 about] *interl*
5.3 in other cases] *interl*
6.5 alternation] *below del* 'alteration'
7.4 triticoides] *interl*

To Charles Lyell 14 October [1862]
2.2 the land] *interl*
2.2 before;] ';' *over* ':'
3.5 many] *interl*
3.5 mountainous] *interl*

To H. W. Bates 15 October [1862]
2.1 for two months] *interl*

To Asa Gray 16 October [1862]
2.4 fertilisation.] *interl*
7.5 to have] 'to' *over* 'm'
9.2 Hort. Soc. Journal vol I] *interl*

To Walter White 17 [October – November 1862?]
1.1 & am much obliged] *interl*
1.2 Vol I] *interl*
1.5 series?] '?' *over* ';'

To H. W. Bates 18 October [1862]
1.5 adviser] *interl*
1.7 men] *interl*
2.3 I shall be sure to do that.—] *interl*

To John Lubbock 23 October [1862]
2.3 & Could . . . morning. 2.4] *interl*
3.1 you] *after del* 'at'
3.2 o clock] *interl*
4.1 I hope] 'I' *over* 'h'

To W. E. Darwin [25 October 1862]
1.4 nearest] *interl*

To J. D. Hooker 27 [October 1862]
2.1 the other] 'the' *interl*
3.5 structure &] *interl*
3.6 structure &] *interl*
3.6 altogether] *interl*
3.7 different] *interl*
4.1 you] *above del illeg*
4.2 every] *above del* 'Every'
5.2 (if such you have) 5.3] *parentheses over commas*

To John Lubbock 27 [October 1862]
0.1 your letter went to Dover.] *interl*
1.3 to Lamas] *interl*
1.3 afternoon] *above del* 'morning'

To W. E. Darwin 30 [October 1862]
1.2 what] *after del* 'y'

To J. D. Hooker 3 November [1862]
1.1 (1st)] *interl*
2.4 reference] *after del* 'given'
2.4 in the list] *interl*
3.1 business. In] '. In' *altered from* ': in'
7.1 Revd] *interl*
7.3 Can] *altered from* 'can'
8.1 two or three] *interl*
8.2 northern] *interl*

To W. E. Darwin 4 [November 1862]
2.2 20] *added before del* '18'

To J. D. Hooker 4 November [1862]
1.1 (with] *parenthesis over comma*
1.4 to] *interl*
1.9 partly] *interl*
2.2 first time.] *above del* 'last.'
2.3 now alone] *interl*
2.6 on you] *interl*
2.9 n.] *interl*
2.10 could] *after del* 'w'
2.13 glacial climate &] *interl*
2.13 climate] *after del* 'acti'
2.13 Are . . . so. 2.14] *interl*
3.1 How . . . them out? 3.2] *interl*
3.4 & out of] *interl*
3.4 in a] 'a' *interl*
4.1 the] *added*
4.2 a cut] 'a' *interl*
6.3 trying &] *added*
6.4 & sometimes being injured by them.] *interl*
6.5 Could . . . Book? 6.6] *added*

To Asa Gray 6 November [1862]
1.2 of the] 'the' *over illeg*
1.2 same book] *interl*
1.4 I have] *after del* 'As'
1.6 strictly] *after del comma*
3.3 the country:] 'the' *interl*
3.3 or something] *interl*
6.1 , if . . . title.— 6.2] *interl*
6.3 of Maize.] *interl*

To J. D. Hooker [10–]12 November [1862]
0.2 12th.] *after del* '10th'
3.4 a lot] 'a' *over* 'al'
4.6 pikestaff,] *comma over* ';'
5.2 & see if they are well scored] *interl*
5.4 land] *interl*
8.1 (To return . . . done.) 8.5] *square brackets in MS*
8.4 to good] *after del* 'only'
8.4 alone] *interl*
9.4 Oxalis] 'O' *over illeg*
9.7 G. Snow] *interl*
10.7 thus] *interl*
11.1 News] *interl*
12.2 R.] *over full stop*
12.2 a good] 'a' *interl*
13.3 in what] 'in' *interl*

To John Scott 12 November [1862]
1.8 its] *after del* 'the'
2.2 & plump] *interl*
2.2 perhaps] *after del* 'but was'
2.2 ripe] *interl*
3.5 may] *interl*
4.1 the protrusion] 'the' *interl*
4.3 always] *interl*

To J. D. Hooker 18 [November 1862]
2.1 by Dr L. Bückner] *interl*
2.3 ruder races] 'races' *interl*
2.3 yielding to] *above del* 'by'
2.4 cannot] *after del* 'ha'
4.2 flowers appear in] *interl*
4.2 Loddigesii] *after del* 'lute'
4.5 (do . . . him?)] *square brackets in MS*
4.7 or as far as mouth of orifice 4.8] *interl*
4.11 the other] 'the' *interl*

To John Scott 19 November [1862]
1.2 the capsule] *after del* 'seeds'
1.2 when ripe] *interl*
1.11 male plants] 'male' *above del* 'female'
1.11 male Lychnis] *after del* 'fem'
2.2 of pollen-mass] *interl*
4.2 I observed] *interl*
4.6 Book?] '?' *over* ';'
5.4 some] *interl*
5.4 & Crinum 5.5] *interl*

To H. W. Bates 20 November [1862]
1.4 the] *interl*
1.5 the] *added*
1.8 passed] *after del* 'slurred'
2.2 on the] 'on' *added*
2.5 enlarged] *after del* 'g'
2.6 a rather] 'a' *interl*

3.1 miscellaneous] *interl*
3.1 there are] *interl*
3.2 be] *interl*

To Asa Gray 23 November [1862]
1.3 (the enclosures . . . posted)] *square brackets in MS*
1.12 of the] *interl*
2.1 Review] *after del illeg*
2.3 Intelligencer] 'I' *altered from* 'i'
2.4 in last . . . Transactions. 2.5] *interl*
2.4 part] *below del* 'vol.'
2.7 process of] *interl*
3.3 fruit] *above del* 'fruit'
3.3 the wild] *interl*
4.8 for] *over* 'on'
4.9 conclusions.—] *after del* 'things'
5.1 from your army— 5.2] *interl*

To Daniel Oliver 23 [November 1862]
1.2 psychological] 'lo' *interl*
1.7 flash of] *interl*
1.8 one of] *interl*
2.1 it down] *below del* 'the plant'
2.3 For several . . . this plant. 2.4] *added*
4.3 (per Railway . . . Postman)] *square brackets in MS*

To J. D. Hooker 24 [November 1862]
1.3 loss of the] *interl*
2.3 have read] 'have' *above del* 'as'
2.5 forward] *above del* 'up'
2.5 capital] *interl*
5.3 in Review of Orchids] *interl*
6.1 Owen] 'O' *altered from* 'o'
7.3 permanence of] *interl*
8.2 to write 8.3] *interl*
8.3 me to] *interl*

To H. W. Bates 25 November [1862]
1.1 not] *interl*
1.1 a written] 'a' *interl*
1.3 misunderstandings] *altered from* 'mistakes'
1.4 he] *interl*
1.7 similar] *interl*
1.8 &] *over* 'or'
1.9 put it] *interl*
3.1 already] *interl*
3.3 one of Editors of] *interl*

To Asa Gray 26[–7] November [1862]
1.3 some] 's' *over* 'o'
1.4 a living] 'a' *interl*
2.4 in Primula paper 2.5] *interl*
2.7 phenomenon] *altered from* 'phenomena'
2.7 is in] 'is' *above del* 'are'
2.15 long-styled] *interl*
2.18 your second] 'y' *over* 't'
2.19 the state] 'the' *interl*
2.20 parts in] *interl*
2.24 in this case] *altered from* 'thus'
2.24 that the] 'the' *altered from* 'these'
2.24 flowers] *interl*
2.26 self-fertilising] *interl*
2.28 Balsamineæ 2.29] *after del* 'Impat'
2.34 effected] *after del* 'partially'
2.35 of corolla] *interl*
2.36 open] *interl*
2.38 inward] *interl*
3.8 of the two individuals] *interl*
3.11 *genera* of] *interl*
3.15 or] *over* '&'
3.15 & therefore . . . sexes 3.16] *interl*
5.2 explain &] *interl*
9.1 Bates'] *after del* 'Wallace'

To William Allport Leighton 26 November [1862]
1.6 roots of] *interl*

To J. D. Hooker [after 26] November [1862]
1.3 about crossing] *interl*
1.4 together] *interl*
1.7 viz] *interl*
2.1 , not] *before del* 'by not'
3.2 & confined in different size aviary] *interl*
3.2 in number] *interl*
4.2 might] *above del* 'would'
4.4 & new] *interl*
4.8 individual] *interl*
4.8 a self-fertilising] 'a' *interl*
4.9 of putting case 4.10] *interl*
4.12 very] *interl*
4.14 & see] '&' *over* 'the'
5.4 ever] *above del* 'ever'
5.9 has] *interl*
5.9 with my] *above del* 'at'
5.12 in contradistinction to 'seed-variation'" 5.13] *interl*
5.13 (do you . . . "sports")] *square brackets in MS*
5.13 term] *interl*
5.14 these] *over illeg*
5.15 as] *over* '—'
6.1 Oliver] 'O' *over* 'o'
9.4 the] *interl*

To Pickard & Stoneman 1 December [1862]
2.2 ?] *over* '.—'
4.1 many gallons of] *above del* 'much'

To John Scott 3 December [1862]
1.2 the interest] 'the' *interl*
1.2 the subject] 'the' *interl*
1.4 any] *above del* 'many'
1.4 having] *above del* 'with'
1.6 the same] *after del* 'analo'
1.7 propagated] *after del illeg*
1.8 garden?] '?' *over full stop*
1.12 (but] *over* ',—'
1.13 yourself publish— 1.14] *interl*
2.1 Hottonia] *after del* 'In those primulas with ster.'
2.1 the two] 'the' *interl*
3.3 the pollen] 'the' *interl*
3.5 so;] *interl*
4.2 the seeds] 'the' *interl*
4.2 those of] *interl*
5.1 make] *above del* 'try'
5.1 the] *interl*
5.2 of] *over* 'on'
6.3 fertilised] *interl*
6.4 Oxlips;] ';' *over* ':'
7.3 difficulties] *after del caret*
7.7 the] *interl*
8.4 Cattleya,] *comma over caret*
8.4 of some] *interl*
8.5 (so] *parenthesis over comma*
8.6 the large] 'the' *interl*
8.7 true] *interl*
8.8 of same . . . comparison.] *interl*

To T. H. Huxley 7 December [1862]
1.7 get more] 'm' *over* 't'

To John Scott 11 December [1862]
2.3 *decided*] *interl*
2.3 intruder] *after del* 'an'
3.2 that] *interl*
4.1 on Primula . . . Soc. 4.2] *interl*
4.3 & Cowslip] *interl*
4.3 the same] 'the' *interl*
4.3 as last year 4.4] *interl*
4.4 of the two crosses.] *interl*
6.3 Arab-Turk Horse] 'Horse' *interl*
8.5 such experiments] *interl*
8.5 I could . . . Verbascum. 8.6] *interl*
8.7 this] *altered from* 'the'
8.7 case of Passiflora,] *above del* 'latter'
8.7 might] *after del* 'experiment'
8.8 with advantage] *interl*
9.1 on meaning of the 9.2] *above del* 'with very'
9.2 stamens] *above del* 'anthers'
9.3 I now] 'I' *over* 'n'
10.1 could] *after omitted point*
10.2 out] *interl*

To J. D. Hooker 12 [December 1862]
2.6 at first] *interl*
2.7 If] *above del* 'Have'
2.7 have] *interl*
2.7 how] *interl*
2.13 the monsters] *above del* 'they'
2.13 at least] *interl*
4.1 & other Books] *interl*
6.3 request] *above del* 'query'
6.4 a Begonia] 'a' *interl*
8.1 the] *interl*
8.1 letter] *interl*
8.2 & for References.—] *interl*

To H. W. Bates 15 December [1862]
1.1 pairing of vars] *above del* 'crossing'

To John Lubbock 16 [December 1862]
1.2 (but . . . groans)] *parentheses over commas*
2.1 at] *over* '.—'

To T. H. Huxley 18 December [1862]
1.5 & most simply & clearly 1.6] *interl*
2.1 -feathers] *interl*
4.3 he showed] *after del* 'sh'
4.4 with Maize] 'with' *interl*
4.8 of Origin 4.9] *interl*
5.1 a new] 'a' *interl*
5.2 case] *interl*
5.4 inland] *above del* 'inward'

To John Scott 19 December [1862]
5.2 species.] *before omitted point*
6.1 (as does . . . probable)] *parentheses over commas*
6.1 you] *added*
6.1 say] *interl*
7.5 nets,] *interl*
12.2 vars of] *interl*
12.3 vars. of] *interl*
12.4 causes . . . trouble.] *interl*
13.2 flowers,—] *comma over* ';'
14.1 one] *interl*
16.4 var. of] *interl*
16.4 with] *after del* 'var.'
17.1 long cultivated] *interl*

17.2 crossed] *after del* 'could'
17.2 very] *interl*
17.3 comparatively] *interl*
17.6 If we ... end.— 17.7] *added*
18.1 raised from seed] *interl*
20.3 The Counting] 'The' *interl*
20.3 the seed] 'the' *interl*
20.4 (N.B. ... paper.) 20.5] *square brackets in MS*
20.5 well] *interl after del interl* 'firmly'
20.5 ripe] *interl*
21.2 & positions.] *interl*

To J. D. Hooker [21 December 1862]
5.2 to think] *after del* 'wri'

To Thomas Rivers 23 December [1862]
1.5 &] *over* '.—'
1.7 suddenly appearing] *interl*
1.8 nectarine] *after del* 'pea'

To J. D. Hooker 24 December [1862]
2.6 ; you ... Naudin.— 2.7] *added*
4.3 eatible] 'a' *over illeg*
4.3 (Oh ... black) 4.4] *added*; *square brackets in MS*
4.6 3^{d}. ... Dogs.— 4.7] *added*
5.3 suggested it &] *interl*
5.4 a clever] 'a' *added*

To W. B. Tegetmeier 27 [December 1862]
1.4 M.S.] *after del* 'dreadfully rough'
1.5 on the] *after del* '(and long noted)'
1.5 the subject] 'the' *interl*
1.5 carefully] *interl*
1.6 with] *above del* '&'
2.2 as] *over illeg*
2.2 my health] *after del* 'I'
3.1 with work] *interl*
3.2 power] *above del* 'power'
3.4 breed] *above del* 'breed'
3.5 not hardy] *above del* 'tender'
4.1 it be] 'it' *interl*
4.1 silky] *interl*
4.4 Silk Hen] 'Silk' *interl*
5.2 the opinion] 'the' *interl*
5.7 cock] *interl*
5.7 to] *interl*
5.8 I would then match] *above del* 'Of'
5.9 & all ... which] *above del* 'if these little letters'
5.9 children] *after del* 'grand'
5.9 together should 5.10] *above del* 'they sh^{d}. all'
5.11 Then *a* & *i*] *above del* 'These'
5.14 this,] *interl*
6.1 fowls] *above del* 'fowles'
6.2 *not* impotent 6.3] *interl*
6.4 plenty of] *interl*
6.5 first] *interl*
6.6 (if both ... children] *interl*
6.7 like their parents] *interl*
6.7 together] *over* '.—'
7.1 day,] *comma altered from full stop*
7.1 & have ... scrawl.— 7.2] *added*

To T. H. Huxley 28 December [1862]
2.1 assuredly ... so far. 2.2] *interl*
3.6 far from impotent] *interl*
3.7 Bull.] *interl*

To Thomas Rivers 28 December [1862]
1.3 in M.S.] *interl*
1.6 extra] *interl*
1.13 -leaved] *interl*
1.14 the common] 'the' *interl*
1.18 the so-called] 'the' *interl*
2.2 severely] *interl*
2.6 These cases ... fertilisation.— 2.7] *added*
4.2 of Peaches & nectarines] *interl*
5.1 in my life] *interl*

To W. B. Tegetmeier 28 [December 1862]
3.1 & see ... descended. 3.2] *added*

To J. D. Hooker 29 [December 1862]
2.1 Genera] 'G' *over* 'g'
2.3 Darwinism] *before omitted point*
4.1 excellent] *above del* 'fine'

To G. H. K. Thwaites 29 December [1862]
3.4 of these genera] *interl*
3.5 much] *interl*

BIBLIOGRAPHY

The following bibliography contains all the books and papers referred to in this volume by author–date reference or by short title. Short titles are used for some standard reference works (e.g., *DNB, OED*), for CD's books, and for editions of his letters and manuscripts (e.g., *Descent, LL, Notebooks*). Works referred to by short titles are listed in alphabetical order according to the title; those given author–date references occur in alphabetical order according to the author's surname. Notes on manuscript sources are given at the end of the bibliography.

Adams, Henry. 1918. *The education of Henry Adams. An autobiography.* Massachusetts: Massachusetts Historical Society.

ADB: *Allgemeine deutsche Biographie.* Under the auspices of the Historical Commission of the Royal Academy of Sciences. 56 vols. Leipzig: Duncker and Humblot. 1875–1912.

Agassiz, Jean Louis Rodolphe. 1857–62. *Contributions to the natural history of the United States of America.* 4 vols. Boston and London.

Ainslie, Robert. 1865. *Discourses delivered in Christ Church, New Road, Brighton.* London.

Albert, W. 1845. *Authentische Thatsachen über die Verjüngung der Kartoffeln aus Samenkörnern, mit Berücksichtigung der unter denselben jetzt herrschenden Krankheiten.* Magdeburg.

Allan, Mea. 1967. *The Hookers of Kew, 1785–1911.* London: Michael Joseph.

Altholz, Josef L. 1989. *The religious press in Britain, 1760–1900.* New York, Westport, Conn., and London: Greenwood Press.

Alum. Cantab.: *Alumni Cantabrigienses. A biographical list of all known students, graduates and holders of office at the University of Cambridge, from the earliest times to 1900. Part II. From 1752 to 1900.* Compiled by J. A. Venn. 6 vols. Cambridge: Cambridge University Press. 1940–54.

Alum. Oxon.: *Alumni Oxonienses: the members of the University of Oxford, 1715–1886.* By Joseph Foster. 4 vols. Oxford. 1888.

Anderson, Thomas. 1848. Case of recovery from a poisonous dose of strychnia; with observations on the tests for the organic alkalies. *Monthly Journal of Medical Science* 8: 566–74.

Anon. 1862. Mr. Darwin's orchids. *Saturday Review* 14 (1862): 486.

Appel, Toby A. 1987. *The Cuvier–Geoffroy debate: French biology in the decades before Darwin.* New York and Oxford: Oxford University Press.

Army list: *The army list.* London. 1815–81.

Arnold, R. Arthur. 1864. *The history of the cotton famine, from the fall of Sumter to the passing of the Public Works Act.* London.

Ashworth, J. H. 1935. Charles Darwin as a student in Edinburgh, 1825–1827. *Proceedings of the Royal Society of Edinburgh* 55 (1934–5): 97–113.

'"Auditory-sac" of Cirripedes': On the so-called "auditory-sac" of Cirripedes. *Natural History Review* n.s. 3: 115–16 (*Collected papers* 2: 85–7). By Charles Darwin. 1863.

Augé de Lassus, M. 1861. Analyse du mémoire de Gaetan monti sur l'*Aldrovandia*, suivie de quelques observations sur l'irritabilité des *follicules* de cette plante. *Bulletin de la Société Botanique de France* 8: 519–23.

Auld, R. C. 1888. The wild cattle of Great Britain. *American Naturalist* 22: 498–509.

Aust. dict. biog.: *Australian dictionary of biography: 1788–1850; 1851–1890.* Edited by Douglas Pike and Bede Nairn. 6 vols. Melbourne: Melbourne University Press. 1966–76.

Autobiography: *The autobiography of Charles Darwin 1809–1882. With original omissions restored.* Edited with appendix and notes by Nora Barlow. London: Collins. 1958.

Babington, Anna Maria, ed. 1897. *Memorials journal and botanical correspondence of Charles Cardale Babington.* Cambridge.

Babington, Charles Cardale. 1843. *Manual of British botany, containing the flowering plants and ferns arranged according to the natural orders.* London.

——. 1851. *Manual of British botany, containing the flowering plants and ferns arranged according to the natural orders.* 3d ed. London.

——. 1856. *Manual of British botany, containing the flowering plants and ferns arranged according to the natural orders.* 4th ed. London.

Baehni, Charles. 1955. Correspondance de Charles Darwin et d'Alphonse de Candolle. *Gesnerus* 12 (1955): 109–56.

Baker, Herbert G. 1965. Charles Darwin and the perennial flax—a controversy and its implications. *Huntia* 2: 141–61.

Banking almanac: *The banking almanac, directory, yearbook and diary.* Edited by D. Morier Evans. London. 1844–1919.

Barlow, Nora, ed. 1967. *Darwin and Henslow. The growth of an idea.* Berkeley and Los Angeles: University of California Press.

Barnhart, John Hendley. 1965. *Biographical notes upon botanists. Compiled by John Hendley Barnhart ... and maintained in the New York Botanical Garden Library.* 3 vols. Boston, Mass.

Bartholomew, John. n.d. *The survey gazetteer of the British Isles including summary of 1951 census.* 9th ed. (reprint). Edinburgh: John Bartholomew & Son.

——. [1864?] *Philips' atlas of Australia, including a general map of the world; a series of maps, constructed from the latest and best authorities.* Edited by William Hughes. London and Liverpool.

Bartholomew, Michael. 1975. Huxley's defence of Darwin. *Annals of Science* 32: 525–35.

——. 1976. The non-progress of non-progression: two responses to Lyell's doctrine. *British Journal for the History of Science* 9: 166–74.

Bateman, James. [1843.] *The Orchidaceæ of Mexico & Guatemala.* London.

Bates, Henry Walter. 1857. Notes on South American butterflies. [Read 7 September

1857.] *Transactions of the Entomological Society of London* n.s. 5 (1858–61): 1–11.

Bates, Henry Walter. 1861. Contributions to an insect fauna of the Amazon valley. [Read 5 March and 24 November 1860.] *Transactions of the Entomological Society of London* n.s. 5 (1858–61): 223–8, 335–61.

——. 1862a. Contributions to an insect fauna of the Amazon valley. *Lepidoptera: Heliconidæ*. [Read 21 November 1861.] *Transactions of the Linnean Society of London* 23 (1862): 495–566.

——. 1862b. Contributions to an insect fauna of the Amazon valley. *Coleoptera: Longicornes*. *Annals and Magazine of Natural History* 3d ser. 9: 117–24, 396–405, 446–58.

——. 1863. *The naturalist on the River Amazons, a record of adventures, habits of animals, sketches of Brazilian and Indian life, and aspects of nature under the equator, during eleven years of travel.* London.

——. 1892. *The naturalist on the River Amazons. A record of adventures, habits of animals, sketches of Brazilian and Indian life, and aspects of nature under the equator, during eleven years of travel. With a memoir of the author by Edward Clodd.* Reprint edition. London.

'Beagle' diary: *Charles Darwin's diary of the voyage of H.M.S. "Beagle"*. Edited by Nora Barlow. Cambridge: Cambridge University Press. 1933.

Bean, W. J. 1908. *The Royal Botanic Gardens, Kew: historical and descriptive.* London, Paris, New York, Toronto and Melbourne: Cassell and Company.

Beck, Richard. 1865. *A treatise on the construction, proper use, and capabilities of Smith, Beck, and Beck's achromatic microscopes*. London.

Beddoe, John. 1862. On the relation of temperament and complexion to disease. *British Medical Journal* 1 (1862): 431–4.

Bénézit, Emmanuel. 1976. *Dictionnaire critique et documentaire des peintres, sculpteurs, dessinateurs et graveurs de tous les temps et de tous les pays par un groupe d'écrivains spécialistes français et étrangers.* Revised edition. 10 vols. Paris: Librairie Gründ.

[Bentham, George.] 1855. Historical notes on the introduction of various plants into the agriculture and horticulture of Tuscany. *Journal of the Horticultural Society of London* 9: 133–81.

——. 1858. *Handbook of the British flora; a description of the flowering plants and ferns indigenous to, or naturalized in, the British Isles.* London.

——. 1862a. Notes on *Caryophylleæ*, *Portulaceæ*, and some allied orders. *Journal of the Proceedings of the Linnean Society of London (Botany)* 6: 55–77.

——. 1862b. Anniversary address. *Journal of the Proceedings of the Linnean Society of London (Botany)* 6: lxvi–lxxxiii.

Bentham, George and Hooker, Joseph Dalton. 1862–83. *Genera plantarum ad exemplaria imprimis in herbariis kewensibus servata definita.* 3 vols. London.

Berg, Carlos. 1895. Carlos Germán Conrado Burmeister. Reseña Biográfica. *Anales del Museo Nacional de Buenos Aires* 2d ser. 1: 316–57.

[Berkeley, Miles Joseph.] 1862. Fertilization of orchids. *London Review and Weekly Journal of Politics, Arts and Sciences* 4: 553–4.

BHGW: *Biographisch-literarisches Handwörterbuch zur Geschichte der exacten Wissenschaften enthaltend Nachweisungen über Lebensverhältnisse und Leistungen von Mathematikern, Astronomen, Physikern, Chemikern, Mineralogen, Geologen usw.* By Johann Christian Poggendorff. Vols. 1–4, Leipzig: Johann Ambrosius Barth. 1863–1904. Vol. 5, Leipzig and Berlin: Verlag Chemie. 1926.

Bibby, Cyril. 1959. *T. H. Huxley. Scientist, humanist and educator.* London: Watts.

Binney, Edward William. 1862. On some fossil plants, showing structure, from the lower coal-measures of Lancashire. *Quarterly Journal of the Geological Society of London* 18: 106–12.

Blagg, C. J. 1902. *A history of the North Staffordshire hounds and country 1825 to 1902.* London: Sampson Low, Marston & Company.

Blake, Charles Carter. 1861a. On the discovery of *Macrauchenia* in Bolivia. *Annals and Magazine of Natural History* 3d ser. 7: 441–3. [Reprinted in *Geologist* 4: 354–5.]

——. 1861b. On the distribution of Mastodon in South America. *Geologist* 4: 469–72.

——. 1862a. Past life in South America. *Geologist* 5: 323–30.

——. 1862b. Fossil monkeys. *Geologist* 5: 81–6.

——, ed. 1863. *A manual of zoology.* By Henri Milne-Edwards. Translated by Robert Knox. 2d ed. London.

Blanford, Henry Francis. 1862. On the specific identity of the described forms of *Tanalia. Transactions of the Linnean Society of London* 23: 603–10.

Blaringhem, Louis. 1913. La notion d'espèce et la disjonction des hybrides, d'après Charles Naudin (1852–1875). *Progressus rei botanicæ* 4: 27–108.

Blaschko, Mary D. 1958. *Sidcot school: register of old scholars, 1808–1958.* Winscombe, Somerset: Sidcot School.

BLC: *The British Library general catalogues of printed books to 1975.* 360 vols. and supplement (6 vols.). London. 1979–88.

Bliss, Trudy, ed. 1953. *Thomas Carlyle: letters to his wife.* London: Victor Gollancz.

Blyth, Edward. 1875. *Catalogue of mammals and birds of Burma . . . With a memoir, and portrait of the author.* Hertford. [*Journal of the Asiatic Society of Bengal* pt 2, extra number, August 1875.]

Bollaert, William. 1860. *Antiquarian, ethnological and other researches in New Granada, Equador, Peru and Chile.* London.

Bonafous, Matthieu. 1836. *Histoire naturelle, agricole et économique du maïs.* Paris and Turin.

Bonaparte, Louis Lucien. 1862. *Langue Basque et langues Finnoises.* London.

Bonham-Carter, Victor. 1960. *In a liberal tradition: a social biography 1700–1950.* London: Constable.

Bonney, Thomas George. 1919. *Annals of the Philosophical Club of the Royal Society, written from its minute books.* London: Macmillan and Co.

Boott, Francis. 1858–67. *Illustrations of the genus* Carex. 4 pts. London.

Bowler, Peter John. 1976. *Fossils and progress. Palaeontology and the idea of progressive evolution in the nineteenth century.* New York: Science History Publications.

Brackman, Arnold C. 1980. *A delicate arrangement. The strange case of Charles Darwin and*

Alfred Russel Wallace. New York: Times Books.

Braun, Alexander Carl Heinrich. 1853a. Das Individuum der Pflanze in seinem Verhältniss zur Species. Generationsfolge, Generationswechsel und Generationstheilung der Pflanze. *Abhandlungen der Königlichen Akademie der Wissenschaften zu Berlin*, Physikalische Abhandlungen, pp. 19–122.

——. 1853b. Considerations on the phenomenon of rejuvenescence in nature, especially in the life and development of plants. Translated by A. Henfrey. In *Botanical and physiological memoirs*, edited by Arthur Henfrey. London.

——. 1859. Uber Polyembryonie und Keimung von Caelebogyne. Ein Nachtrag zu der Abhandlung über Parthenogenesis bei Pflanzen. *Abhandlungen der Königlichen Akademie der Wissenschaften zu Berlin* (1860), pp. 109–263.

Brent, Peter. 1981. *Charles Darwin. "A man of enlarged curiosity"*. London: Heinemann.

Brewster, David. 1862. The facts and fancies of Mr. Darwin. *Good Words* (1862), pp. 3–9.

British postal guide: *British postal guide; containing the chief public regulations of the Post Office, with other information. Published quarterly by command of the postmaster general*. London. 1856–79.

Brogan, Hugh ed. 1975. The Times *reports. The American Civil War. Extracts from* The Times, *1860–1865*. London: Times Books.

Brongniart, Adolphe Théodore. 1839. Note sur les poils collecteurs des Campanules et sur le mode de fécondation de ces plantes. *Annales des Sciences Naturelles (Botanique)* 2d ser. 12: 244–7.

Bronn, Heinrich Georg. 1860. Review of *Origin of species*. *Neues Jahrbuch für Mineralogie, Geognosie, Geologie und Petrefakten-Kunde* (1860), pp. 112–16. [Translated in Hull 1973, pp. 120–4.]

——, trans. 1860. *Charles Darwin, über die Entstehung der Arten im Thier- und Pflanzen-Reich durch natürliche Züchtung, oder Erhaltung der vervollkommneten Rassen im Kampfe um's Daseyn*. Stuttgart.

——. 1861. Essai d'une réponse à la question de prix proposée en 1850 par l'Académie des Sciences pour le concours de 1853, et puis remise pour celui de 1856. *Supplément aux Comptes Rendus Hebdomadaires des Séances de l'Académie des Sciences* 2: 377–918.

——, trans. 1862. *Charles Darwin, über die Einrichtungen zur Befruchtung Britischer und ausländischer Orchideen durch Insekten und über die günstigen Erfolge der Wechselbefruchtung*. Stuttgart.

——, trans. 1863. *Charles Darwin, über die Entstehung der Arten im Thier- und Pflanzen-Reich durch natürliche Züchtung, oder Erhaltung der vervollkommneten Rassen im Kampfe um's Daseyn*. 2d ed. Stuttgart.

Brown, Edwin. 1862. On the mutability of specific or race forms. *Proceedings of the Northern Entomological Society*, 28 July 1862, pp. 7–18. [Reprinted in the *Zoologist* 20 (1862): 8249–52.]

Brown, Philip Arthur Henry. 1982. *London publishers and printers, c. 1800–1870*. London: British Library.

Browne, Janet. 1980. Darwin's botanical arithmetic and the "principle of divergence", 1854–1858. *Journal of the History of Biology* 13: 53–89.

——. 1983. *The secular ark: studies in the history of biogeography.* New Haven and London: Yale University Press.

——. 1995. *Charles Darwin. Voyaging. Volume 1 of a biography.* New York: Alfred A. Knopf.

Brown-Séquard, Charles Edouard. 1860. Hereditary transmission of an epileptiform affection accidentally produced. [Read 2 February 1860.] *Proceedings of the Royal Society of London* 10 (1859–60): 297–8.

Buchanan, Robert. 1849. *The ten years' conflict: being the history of the disruption of the Church of Scotland.* 2 vols. Glasgow, Edinburgh, and London.

Büchner, Friedrich Karl Christian Ludwig. 1860. Eine neue Schöpfungstheorie. *Stimmen der Zeit* 2: 356–60.

——. 1862. *Aus Natur und Wissenschaft. Studien, Kritiken und Abhandlungen.* Leipzig.

Buckle, Henry Thomas. 1857–61. *History of civilization in England.* 2 vols. London.

BUCOP: *British Union-catalogue of periodicals: a record of the periodicals of the world, from the seventeenth century to the present day, in British libraries.* Edited by James D. Stewart *et al.* 4 vols. and supplement. London: Butterworths scientific publications. 1955–62.

Buller, Walter. 1870. On the New Zealand rat. *Transactions and Proceedings of the New Zealand Institute* 3: 1–4.

Burke's landed gentry: *Burke's genealogical and heraldic history of the landed gentry.* By John Burke. 1–18 editions. London: Burke's Peerage Ltd. 1836–1972.

Burke's peerage: *Burke's peerage and baronetage.* 1–105 editions. London: Burke's Peerage (Genealogical Books). 1826–1980.

Burmeister, Karl Hermann Konrad. 1853. *Reise nach Brasilien, durch die Provinzen von Rio de Janeiro und Minas geraës.* Berlin.

Burton, Richard Francis. 1862. Account of the ascent of the Camaroons Mountain, in Western Africa. *Proceedings of the Royal Geographical Society of London* 6: 238–48.

Butterworths medical dictionary: *Butterworths medical dictionary.* 2d ed. Editor-in-chief Macdonald Critchley. London and Boston: Butterworth & Co. 1978.

Cairnes, John Elliot. 1862. *The slave power: its character, career, & probable designs: being an attempt to explain the real issues involved in the American contest.* London.

Calendar: *A calendar of the correspondence of Charles Darwin, 1821–1882. With supplement.* Edited by Frederick Burkhardt and Sydney Smith. 2d ed. Cambridge: Cambridge University Press. 1994.

Cameron, Nigel M. de S. *et al.*, eds. 1993. *Dictionary of Scottish church history & theology.* Edinburgh: T. & T. Clark.

Campbell, George Douglas. 1848. *Presbytery examined. An essay, critical and historical, on the ecclesiastical history of Scotland since the Reformation.* London.

——. 1850. On a fossiliferous deposit underlying basalt in the island of Mull. *Report of the 20th meeting of the British Association for the Advancement of Science held at Edinburgh* (1851), Transactions of the sections, pp. 70–1.

Campbell, George Douglas. 1851. On tertiary leaf-beds in the Isle of Mull. *Quarterly Journal of the Geological Society of London* 7: 89–103.

——. 1860. Opening address, session 1860–1. [Read 3 December 1860.] *Proceedings of the Royal Society of Edinburgh* 4 (1857–62): 350–77.

[——.] 1862. The supernatural. *Edinburgh Review* 116: 378–97.

Campbell, Ina Erskine, ed. 1906. *George Douglas, eighth duke of Argyll . . . (1823–1900): autobiography and memoirs.* 2 vols. London: John Murray.

Candolle, Alphonse de. 1855. *Géographie botanique raisonnée; ou exposition des faits principaux et des lois concernant la distribution géographique des plantes de l'époque actuelle.* 2 vols. Paris and Geneva.

——. 1862a. De la flore Européenne et de la configuration des continents à l'époque tertiaire, d'après l'ensemble des travaux de M. le professeur Heer. *Bibliothèque Universelle et Revue Suisse. Archives des sciences physiques et naturelles* n.s. 14: 48–77.

——. 1862b. Étude sur l'espèce à l'occasion d'une révision de la famille des Cupulifères. *Bibliothèque Universelle et Revue Suisse. Archives des sciences physiques et naturelles* n.s. 15: 211–37, 326–65.

——. 1862c. Note sur un nouveau caractère observé dans le fruit des chênes et sur la meilleure division à adopter pour le genre Quercus. *Bibliothèque Universelle et Revue Suisse. Archives des sciences physiques et naturelles* n.s. 15: 89–101.

——. 1862d. Charles Darwin. On the various contrivances, etc . . . Sur les divers moyens par lesquels les orchidées indigènes de la Grande-Bretagne et exotiques sont fécondées par les insectes et sur les avantages du croisement dans les fécondations. *Bibliothèque Universelle et Revue Suisse. Archives des sciences physiques et naturelles* n.s. 15: 173–6.

Candolle, Augustin Pyramus de and Candolle, Alphonse de. 1824–73. *Prodromus systematis naturalis regni vegetabilis, sive enumeratio contracta ordinum generum specierumque plantarum huc usque cognitarum, juxta methodi naturalis normas digesta.* 17 vols. Paris.

Candolle, Casimir de. 1862. Charles Darwin. On the two forms, etc.— Sur les deux formes ou l'état dimorphe dans les espèces du genre Primula et sur leurs rapports sexuels remarquables. *Bibliothèque Universelle et Revue Suisse. Archives des sciences physiques et naturelles* n.s. 14: 192–6.

Carlyle, Thomas. 1858–65. *History of Friedrich II of Prussia, called Frederick the Great.* 4 vols. London.

Carpenter, William Benjamin. 1854. *Principles of comparative physiology.* 4th ed. London.

[——.] 1860a. Darwin on the origin of species. *National Review* 10: 188–214.

[——.] 1860b. Review of *Origin of species. British and Foreign Medico-Chirurgical Review* 25: 367–404.

Carroll, P. Thomas, ed. 1976. *An annotated calendar of the letters of Charles Darwin in the library of the American Philosophical Society.* Wilmington, Del.: Scholarly Resources Inc.

Catalogue de la Bibliothèque nationale: *Catalogue général des livres imprimés de la Bibliothèque Nationale.* 231 vols. Paris: Imprimerie nationale. 1897–1981.

Catalogue of graduates, University of Dublin: *A catalogue of graduates who have proceeded to degrees in the University of Dublin, from the earliest recorded commencements to July, 1866: with supplement to December 16, 1868.* Dublin. 1869.

Catalogue of the library of the British Museum (Natural History): *Catalogue of the books, manuscripts, maps and drawings in the British Museum (Natural History).* 5 vols. London: Trustees of the British Museum. 1903–15.

Cawood, John. 1979. The magnetic crusade: science and politics and early Victorian Britain. *Isis* 70: 493–518.

CDEL: *A critical dictionary of English literature and British and American authors, living and deceased, from the earliest accounts to the middle of the nineteenth century.* By S. Austin Allibone. 3 vols. Philadelphia and London. 1859–71. Supplement by John Foster Kirk. 2 vols. Philadelphia and London. 1891.

[Chambers, Robert.] 1844. *Vestiges of the natural history of creation.* London.

[——.] 1845. *Explanations: a sequel to 'Vestiges of the natural history of creation'.* London.

——. 1853. On glacial phenomena in Scotland and parts of England. *Edinburgh New Philosophical Journal* 54 (1852–3): 229–81.

——. 1861. *Edinburgh papers [pt 3]. Ice and water: a review of the superficial formation.* London and Edinburgh.

Chemical manufactories' directory: *The chemical manufactories' directory of England.* London: 1867–1955.

Claparède, Edouard. 1859. Zur Morphologie der zusammengesetzten Augen bei den Arthropoden. *Zeitschrift für wissenschaftliche Zoologie* 10: 191–214. [Reprinted in *Annals and Magazine of Natural History* 3d ser. 6 (1860): 455–7.]

——. 1860. Zur Morphologie der zusammengesetzten Augen bei den Arthropoden. *Zeitschrift für wissenschaftliche Zoologie* 10: 191–214. [Reprinted in *Annals and Magazine of Natural History* 3d ser. 6 (1860): 455–7.]

——. 1861. M. Darwin et sa théorie de la formation des espèces. *Revue Germanique Française et Étrangère* 16: 523–59; 17: 232–63.

——. 1862a. Recherches sur l'évolution des araignées. *Memoire auquel la Société des Arts et Sciences d'Utrecht.*

——. 1862b. Recherches anatomiques sur les oligochètes. *Mémoires de la Société de Physique et d'Histoire Naturelle de Genève* 16: 217–91.

Clark, Samuel. 1979. *Social origins of the Irish Land War.* Princeton, N.J. and Guildford, Surrey: Princeton University Press.

Clarke, William Branwhite. 1861a. On the relative positions of certain plants in the coal-bearing beds of Australia. *Quarterly Journal of the Geological Society of London* 17: 354–62.

——. 1861b. *Recent geological discoveries in Australasia.* [Read before the Philosophical Society of New South Wales on 20 November 1861.] 2d ed. with notes and addenda. Sydney.

——. 1861c. On the Carboniferous and other geological relations of the Maranoa district in Queensland in reference to a discovery of zoological fossils in Wollombilla Creek and Stoney Creek, West Maitland. [Read 30 December 1861.] *Transactions and Proceedings of the Royal Society of Victoria* 6 (1865): 32–42.

Clergy list: *The clergy list . . . containing an alphabetical list of the clergy.* London. 1841–.

'Climbing plants': On the movements and habits of climbing plants. By Charles Darwin. [Read 2 February 1865.] *Journal of the Proceedings of the Linnean Society of London (Botany)* 9 (1866): 1–118.

Climbing plants: *The movements and habits of climbing plants.* By Charles Darwin. London. 1875.

Cohn, Ferdinand Julius. 1860. Contractile Gewebe im Pflanzenreiche. [Read 1 November 1860.] *Abhandlungen der Schlesischen Gesellschaft für vaterländische Cultur. Abtheilung für Naturwissenschaften und Medicin* 1 (1861): 1–48.

Coke, Edward. 1628. *The first part of the institutes of the lawes of England. Or, a commentarie upon Littleton, not the name of a lawyer onely, but of the law it selfe.* London.

Colenso, John William. 1862–79. *The Pentateuch and book of Joshua critically examined.* 5 vols. London.

Collected letters of the Carlyles: *The collected letters of Thomas and Jane Welsh Carlyle.* Edited by Charles Richard Sanders *et al.* 21 vols. Durham, N.C.: Duke University Press. 1970–93.

Collected papers: *The collected papers of Charles Darwin.* Edited by Paul H. Barrett. 2 vols. Chicago and London: University of Chicago Press. 1977.

Colonial Office list: *The Colonial Office list.* London. 1862–1925.

Complete peerage: *The complete peerage of England, Scotland, Ireland, Great Britain, and the United Kingdom, extant, extinct, or dormant.* By George Edward Cokayne. New edition, revised and much enlarged. Edited by the Hon. Vicary Gibbs and others. 12 vols. London: St Catherine Press. 1910–59.

Concordance: *A concordance to Darwin's Origin of species, first edition.* Edited by Paul H. Barrett *et al.* Ithaca and London: Cornell University Press. 1981.

Condy, Henry Bollmann. 1862. *Air and water: their impurities and purification.* London.

Cooper, Susan Augusta Fenimore. 1850. *Rural hours.* 2 vols. London.

Cooper, [——.] 1855. *Journal of a naturalist in the United States.* 2 vols. London.

Coral reefs: *The structure and distribution of coral reefs. Being the first part of the geology of the voyage of the Beagle, under the command of Capt. FitzRoy, R.N. during the years 1832 to 1836.* By Charles Darwin. London. 1842.

Correns, C. 1916. Friedrich Hildebrand. *Berichte der deutschen botanischen Gesellschaft* 34 (Schluss-heft): 28–49.

Correspondence: *The correspondence of Charles Darwin.* Edited by Frederick Burkhardt, Sydney Smith, and others. Cambridge: Cambridge University Press. 1985–.

Corsi, Pietro. 1985. Recent studies on Italian reactions to Darwin. In *The Darwinian heritage*, edited by David Kohn. Princeton: Princeton University Press in association with Nova Pacifica.

County families: *The county families of the United Kingdom; or, a royal manual of the titled & untitled aristocracy of Great Britain & Ireland.* By Edward Walford. London. 1860–1920.

Crocker, Charles William. 1861. Begonias. *Gardeners' Chronicle and Agricultural Gazette*, 14 December 1861, p. 1092.

Crocker, Charles William. 1862. The lesser celandine, Ranunculus ficaria. *Gardener's Weekly Magazine, and Floricultural Cabinet* 4: 68–70.

——. 1864. Plantago lanceolata. *Gardeners' Chronicle and Agricultural Gazette*, 26 March 1864, pp. 293–4.

Crockford's: *The clerical directory, a biographical and statistical book of reference for facts relating to the clergy and the Church.* London. 1858. Continued as *Crockford's clerical directory etc.* London. 1860–.

Cross and self fertilisation: *The effects of cross and self fertilisation in the vegetable kingdom.* By Charles Darwin. London. 1876.

Curry, Leonard P. 1968. *Blueprint for modern America: nonmilitary legislation of the first Civil War congress.* Nashville: Vanderbilt University Press.

Cuvier, Jean Léopold Nicolas Frédéric (Georges). 1812. *Recherches sur les ossemens fossiles de quadrupèdes, où l'on rétablit les caractères de plusieurs espèces d'animaux que les révolutions du globe paroissent avoir détruites.* 4 vols. Paris.

DAB: *Dictionary of American biography.* Under the auspices of the American Council of Learned Societies. 20 vols., index, and 10 supplements. New York: Charles Scribner's Sons. London: Oxford University Press. 1928–95.

Dallas, James. 1881. *Catalogue of the Library of the Geological Society of London.* London.

——. 1921. *The history of the family of Dallas and their connections and descendants from the twelfth century.* Edinburgh: T. and A. Constable.

Dana, James Dwight. 1850. On the degradation of the rocks of New South Wales and formation of valleys. *American Journal of Science and Arts* 2d ser. 9: 289–94.

——. 1863. *Manual of geology: treating of the principles of the science with special reference to American geological history, for the use of colleges, academics, and schools of science.* Philadelphia and London.

Darwin, Charles and Wallace, Alfred Russel. 1858. On the tendency of species to form varieties; and on the perpetuation of varieties and species by natural means of selection . . . communicated by Sir Charles Lyell . . . and J. D. Hooker. [Read 1 July 1858.] *Journal of the Proceedings of the Linnean Society of London (Zoology)* 3 (1859): 45–62

Darwin, Francis. 1914. William Erasmus Darwin, 1839–1914. *Christ's College Magazine* 29: 16–23.

——. 1916. Memoir of Sir George Darwin. In *Scientific papers*, by George Howard Darwin. 1916. Cambridge: Cambridge University Press.

——. 1920. *Springtime and other essays.* London: John Murray.

Darwin pedigree: *Pedigree of the family of Darwin.* Compiled by H. Farnham Burke. Privately printed. 1888. [Reprinted in facsimile in *Darwin pedigrees*, by Richard Broke Freeman. London: printed for the author. 1984.]

Daubeny, Charles Giles Bridle. 1860. Remarks on the final causes of the sexuality of plants, with particular reference to Mr. Darwin's work 'On the origin of species by natural selection'. *Report of the 30th meeting of the British Association for the Advancement of Science held at Oxford* (1861), Transactions of the sections, pp. 109–10.

Daubeny, Charles Giles Bridle. 1867. *Miscellanies: being a collection of memoirs and essays on scientific and literary subjects, published at various times.* 2 vols. Oxford and London.

Davies, Gordon Leslie. 1969. *The earth in decay. A history of British geomorphology 1578–1878.* London: Macdonald Technical and Scientific.

Dawson, John William 1862a. *Alpine and Arctic plants: a lecture delivered before the Young Men's Christian Association of Montreal, February, 1862.* Montreal.

——. 1862b. Review of Hooker's outlines of the distribution of Arctic plants. *Canadian Naturalist and Geologist, and proceedings of the Natural History Society of Montreal* 7: 334–44.

DBF: *Dictionnaire de biographie Française.* Under the direction of J. Balteau, M. Barroux, M. Prevost, and others. 17 vols. and 7 fascicles of vol. 18 (A–Laglenne). Paris: Libraire Letouzey et Ané. 1933–95.

DBI: *Dizionario biografico degli Italiani.* Edited by Alberto M. Ghisalberti. 44 vols. (A–Farina). Rome: Istituto della Enciclopedia Italiana. 1960–94.

DBS: *Dictionnaire historique & biographique de la Suisse.* Under the direction of Marcel Godet, Henri Türler, and Victor Attinger. 7 vols. and supplement. Neuchatel: Administration du Dictionnaire Historique et Biographique de la Suisse. 1921–34.

De Filippi, Filippo. 1861. Osservazioni zoologiche. *Archivio per la Zoologia, l'Anatomia e la Fisiologia* 1: 200–27.

Dempster, W. J. 1983. *Patrick Matthew and natural selection: nineteenth century gentleman-farmer, naturalist and writer.* Edinburgh: Paul Harris publishing.

Descent: *The descent of man, and selection in relation to sex.* By Charles Darwin. 2 vols. London. 1870–1.

Desmond, Adrian. 1982. *Archetypes and ancestors: palaeontology in Victorian London, 1850–1875.* London: Blond & Briggs.

——. 1994. *Huxley: the devil's disciple.* London: Michael Joseph.

Desmond, Adrian and Moore, James. 1991. *Darwin.* London: Michael Joseph.

Desmond, Ray. 1994. *Dictionary of British and Irish botanists and horticulturists including plant collectors, flower painters and garden designers.* New edition, revised and completely updated. London: Taylor & Francis and the Natural History Museum. Bristol, Pa.: Taylor & Francis.

——. 1995. *Kew: the history of the Royal Botanic Gardens.* London: Harvill Press with the Royal Botanic Gardens, Kew.

D[icey], E[dward]. 1862. The outlook of the war. *Daily News*, 4 September 1862, p. 3. Abridged and reprinted from *Macmillan's Magazine* 6 (1862): 408–20.

Dickens, Charles. 1853. *Bleak House.* London.

Dictionary of artists in America: *The New-York Historical Society's dictionary of artists in America 1564–1860.* By George C. Grove and David H. Wallace. New Haven: Yale University Press. London: Oxford University Press. 1957.

Dictionary of geological terms: *Dictionary of geological terms.* Revised edition. New York: Anchor Books/Doubleday. 1976.

Dictionary of watercolour artists: *The dictionary of watercolour artists up to 1920.* By H. L. Mallalieu. 2d ed. 3 vols. Woodbridge, Suffolk: Antique Collectors Club. 1986.

Dictionnaire des parlementaires français: *Dictionnaire des parlementaires français, comprenant tous les membres des assemblées françaises et tous les ministres français.* Edited by Adolphe Robert, Edgar Bourloton, and Gaston Cougny. 5 vols. Paris. 1889–91.

Dictionnaire universel des contemporains: *Dictionnaire universel des contemporains contentant toutes les personnes notables de la France et des pays étrangers.* Edited by Louis Gustave Vapereau. Paris. 1858. 3d ed., 1865. 4th ed., 1870. 5th ed., 1880. 6th ed., 1893.

'Dimorphic condition in *Primula*': On the two forms, or dimorphic condition, in the species of *Primula*, and on their remarkable sexual relations. By Charles Darwin. [Read 21 November 1861.] *Journal of the Proceedings of the Linnean Society of London (Botany)* 6: 77–96. [*Collected papers* 2: 45–63.]

Dixon, Edmund Saul. 1848. *Ornamental and domestic poultry: their history and management.* London.

——. 1851. *The dovecote and the aviary: being sketches of the natural history of pigeons and other domestic birds in a captive state, with hints for their management.* London.

DNB: *Dictionary of national biography.* Edited by Leslie Stephen and Sidney Lee. 63 vols. and 2 supplements (6 vols.). London: Smith, Elder, and Co. 1885–1912. *Dictionary of national biography 1912–85.* Edited by H. W. C. Davis, J. R. H. Weaver, and others. 8 vols. London: Oxford University Press. 1927–90.

DNZB: *A dictionary of New Zealand biography.* Edited by G. H. Scholefield. 2 vols. Wellington: Department of Internal Affairs, New Zealand. 1940.

Dodd, Victor. 1983. *Beemasters of the past.* Hebden Bridge: Northern Bee Books.

Downing, Andrew Jackson. 1845. *The fruits and fruit trees of America.* London.

Drejer, Salomon Thomas Nicolai. 1840–1. Revisio critica Caricum borealium in terris sub imperio Danico jacentibus inventarum. *Naturhistorisck Tidsskrift* 3: 423–80.

——. 1844. *Symbolae Caricologicæ ad synonymiam Caricum extricandam stabiliendamque et affinitates naturales eruendas.* Hafniæ.

Drouet, Henri. 1859. Coléoptères açoréens. *Revue et Magasin de Zoologie pure et appliquée* 2d ser. 11: 243–59.

DSB: *Dictionary of scientific biography.* Edited by Charles Coulston Gillispie. 14 vols., supplement, and index. New York: Charles Scribner's Sons. 1970–80.

Du Chaillu, Paul Belloni. 1861. *Explorations & adventures in equatorial Africa; with accounts of the manners and customs of the people, and of the chace of the gorilla, crocodile, leopard, elephant, hippopotamus, and other animals.* London.

Duchartre, Pierre Etienne Simon. 1862. Note sur le polymorphisme de la fleur chez quelques orchidées. *Bulletin de la Société Botanique de France* 9: 113–25.

Dun, Finlay. 1881. *Landlords and tenants in Ireland.* London.

Dupree, Anderson Hunter. 1959. *Asa Gray, 1810–1888.* Cambridge, Mass: Belknap Press of Harvard University.

Dutrochet, René Joachim Henri. 1837. *Mémoires pour servir à l'histoire anatomique et physiologique des végétaux et des animaux.* Paris.

Dyer, Isaac Watson. 1928. *A bibliography of Thomas Carlyle's writings and ana.* Portland, Maine: The Southworth Press.

EB: *The Encyclopædia Britannica. A dictionary of arts, sciences, literature and general information.*

11th ed. 29 vols. Cambridge: Cambridge University Press. 1910–11.

Edinburgh Academy register: *The Edinburgh Academy register. A record of all those who have entered the school since its foundation in 1824*. Edinburgh: Edinburgh Academical Club. 1914.

Egerton, F. Clement C. 1957. *Angola in perspective: endeavour and achievement in Portuguese West Africa*. London: Routledge & Kegan Paul.

Ehrenberg, Christian Gottfried. 1848. Beobachtungen über das gewöhnlich in der Atmosphäre unsichtbar getragene formenreiche Leben. *Bericht über die zur Bekanntmachung geeigneten Verhandlungen der Königlichen Preussischen Akademie der Wissenschaften zu Berlin* (1848): 325–45, 349–62, 370–81.

——. 1849. Über weitere atmosphärische mikroskopische Verhältnisse während der berliner Cholerazeit von 1848 und über schalenlose Infusorien der Atmosphäre. *Bericht über die zur Bekanntmachung geeigneten Verhandlungen der Königlichen Preussischen Akademie der Wissenschaften zu Berlin* (1848): 91–7.

Elliot, Daniel Giraud. 1865. *A monograph of the Tetraoninae, or family of the grouse*. New York.

Emma Darwin (1904): *Emma Darwin, wife of Charles Darwin. A century of family letters*. Edited by Henrietta Litchfield. 2 vols. Cambridge: privately printed by Cambridge University Press. 1904.

Emma Darwin: *Emma Darwin: a century of family letters 1792–1896*. Edited by Henrietta Litchfield. Revised edition. 2 vols. London: John Murray. 1915.

ENZ: *An encyclopedia of New Zealand*. Edited by A. H. McLintock. 3 vols. Wellington, New Zealand: R. E. Owen. 1966.

Erslew, Thomas Hansen. 1843–68. *Almindeligt Forfatter-Lexicon for Kongeriget Danmark med tilhørende Bilande*. With supplement. 6 vols. Copenhagen.

Evans, George Eyre. 1897. *Vestiges of Protestant dissent: being lists of ministers, sacramental plate, registers, antiquities, and other matters pertaining to most of the churches (and a few others) included in the National Conference of Unitarian, Liberal Christian, Free Christian, congregations*. Liverpool.

Evans, Glyn. 1975. *The life of beetles*. London: George Allen & Unwin.

Eve, A. S. and Creasey, C. H. 1945. *Life and work of John Tyndall*. London: Macmillan & Co.

Expression: *The expression of the emotions in man and animals*. By Charles Darwin. London. 1872.

Eyre, Edward John. 1845. *Journals of expeditions of discovery into central Australia, and overland from Adelaide to King George's Sound, in the years 1840–1*. 2 vols. London.

Eyton, Thomas Campbell. 1846–53. *The herd book of Hereford cattle*. 2 vols. London.

Falconer, Hugh. 1857a. Description of two species of the fossil mammalian genus Plagiaulax from Purbeck. *Quarterly Journal of the Geological Society of London* 13: 261–82.

——. 1857b. On the species of *Mastodon* and *Elephant* occurring in the fossil state in Great Britain. Part 1. Mastodon. *Quarterly Journal of the Geological Society of London* 13: 307–60.

Falconer, Hugh. 1862. On the disputed affinity of the mammalian genus *Plagiaulax*, from the Purbeck beds. *Quarterly Journal of the Geological Society of London* 18: 348–69.

——. 1863. On the American fossil elephant of the regions bordering the Gulf of Mexico, (*E. columbi*, Falc.); with general observations on the living and extinct species. *Natural History Review* n.s. 3: 43–114.

Farley, John. 1977. *The spontaneous generation controversy from Descartes to Oparin.* Baltimore and London: Johns Hopkins University Press.

Fawcett, Henry. 1860. A popular exposition of Mr. Darwin on the origin of species. *Macmillan's Magazine* 3 (1861): 81–92.

Felder, Cajetan Friedrich von and Felder, Rudolf. 1862. Specimen faunae lepidopterologicae riparum fluminis Negro superioris in Brasilia septentrionali. *Wiener Entomologische Monatschrift* 6: 65–80, 109–26.

Fermond, Charles. 1859. Faits pour servir à l'histoire générale de la fécondation chez les végétaux. *Bulletin de la Société Botanique de France* 6: 749–51.

Ferris, Norman B. 1977. *The* Trent *affair: a diplomatic crisis.* Knoxville: The University of Tennessee Press.

'Fertilization of orchids': Notes on the fertilization of orchids. By Charles Darwin. *Annals and Magazine of Natural History* 4th ser. 4 (1869): 141–59. [*Collected papers* 2: 138–56.]

Fielding, K. J. 1978. Froude's revenge, or the Carlyles and Erasmus A. Darwin. *Essays and Studies* n.s. 31: 75–97.

Filmer, J. L. 1977. The Norman family of Bromley Common. *Bromley Local History* 2: 16–24.

FitzRoy, Robert. 1863. *The weather book: a manual of practical meteorology.* London.

Fletcher, Harold Roy. 1969. *The story of the Royal Horticultural Society, 1804–1968.* London: Oxford University Press for the Royal Horticultural Society.

Flett, John Smith. 1937. *The first hundred years of the Geological Survey of Great Britain.* London: His Majesty's Stationery Office.

Forbes, Edward. 1846. On the connexion between the distribution of the existing fauna and flora of the British Isles, and the geological changes which have affected their area, especially during the epoch of the Northern Drift. *Memoirs of the Geological Survey of Great Britain, and of the Museum of Economic Geology in London* 1: 336–432.

——. 1850. Report on the investigation of British marine zoology by means of the dredge. Part 1. The infra-littoral distribution of marine invertebrata on the southern, western, and northern coasts of Great Britain. *Report of the 20th meeting of the British Association for the Advancement of Science held at Edinburgh* (1851), pp. 192–263.

Forbes, Edward and Hanley, Sylvanus. 1853. *A history of British Mollusca and their shells.* 4 vols. London.

Foreign Office list: *The Foreign Office list.* London. 1852–1965.

Forms of flowers: *The different forms of flowers on plants of the same species.* By Charles Darwin. London. 1877.

Fossil Cirripedia (1851): *A monograph on the fossil Lepadidæ, or, pedunculated cirripedes of Great Britain.* By Charles Darwin. London. 1851.

Foster, Joseph. 1885. *Men-at-the bar: a biographical hand-list of the members of the various inns of court, including Her Majesty's judges, etc.* London.

Foster, Michael and Lankester, Edwin Ray, eds. 1898–1903. *The scientific memoirs of Thomas Henry Huxley.* 4 vols. and supplement. London: Macmillan and Co. New York: D. Appleton.

Freeman, Richard Broke. 1977. *The works of Charles Darwin: an annotated bibliographical handlist.* 2d ed. Folkestone, Kent: William Dawson and Sons. Hamden, Conn.: Archon Books, Shoe String Press.

——. 1978. *Charles Darwin: a companion.* Folkestone, Kent: William Dawson and Sons. Hamden, Conn.: Archon Books, Shoe String Press.

Gage, Andrew Thomas. 1938. *A history of the Linnean Society of London.* London: Linnean Society of London.

Garner, Robert. 1861. On the structure of the *Lepadidæ. Report of the 30th Meeting of the British Association for the Advancement of Science held at Oxford*, Transactions of the sections, p. 130.

Gärtner, Karl Friedrich von. 1844. *Versuche und Beobachtungen über die Befruchtungsorgane der vollkommeneren Gewächse und über die natürliche und künstliche Befruchtung durch den eigenen Pollen.* Pt 1 of *Beiträge zur Kenntniss der Befruchtung der vollkommeneren Gewächse.* Stuttgart.

——. 1849. *Versuche und Beobachtungen über die Bastarderzeugung im Pflanzenreich.* Stuttgart.

Gay, Jacques Etienne. 1857. Recherches sur les caractères de la végétation du fraisier et sur la distribution géographique de ses espèces, avec la description de deux nouvelles. *Annales des Sciences Naturelles (Botanique)* 4th ser. 8: 185–208.

Geikie, Archibald, ed. 1895. *Memoir of Sir Andrew Crombie Ramsay.* London.

General index to the Journal of the Linnean Society: *General index to the first twenty volumes of the Journal (Botany), and the botanical portion of the Proceedings, November 1838 to June 1886, of the Linnean Society.* London. 1888.

Geoffroy Saint-Hilaire, Isidore. 1832–7. *Histoire générale et particulière des anomalies de l'organisation chez l'homme et les animaux.* 3 vols. and atlas. Paris.

——. 1854–62. *Histoire naturelle générale des règnes organiques, principalement étudiée chez l'homme et les animaux.* 3 vols. Paris.

——. 1857. Discours d'ouverture du président de la société. *Bulletin de la Société Impériale Zoologique d'Acclimatation* 4 (1857): xxv–xxxiv.

Geology of the 'Beagle': *Geological observations on coral reefs, volcanic islands, and on South America.* By Charles Darwin. London. 1851.

Gerstaecker, Carl Eduard Adolph. 1858. Die Chilenischen Arten der Gattung Carabus. *Linnaea Entomologica* 12: 417–58.

Ghiselin, Michael T. 1969. *The triumph of the Darwinian method.* Berkeley and Los Angeles: University of California Press.

Gilbert, Pamela. 1977. *A compendium of the biographical literature on deceased entomologists.* London: British Museum (Natural History).

Gillespie, Neal C. 1977. The duke of Argyll, evolutionary anthropology, and the art of scientific controversy. *Isis* 68: 40–54.

Gingins-La-Sarra, Frédéric Charles Jean de. 1823. *Mémoire sur la famille des violacées.* Geneva.

Godron, Dominique Alexandre. 1859. *De l'espèce et des races dans les êtres organisés et spécialement de l'unité de l'espèce humaine.* 2 vols. Paris.

Goethes Werke: *Goethes Werke. Herausgegeben im Auftrage der Großherzogin Sophie von Sachsen.* 4 pts in 142 vols. Weimar. 1887–1912.

Goring, Rosemary, ed. 1992. *Chambers Scottish biographical dictionary.* Edinburgh: W. & R. Chambers.

Gorman, W. Gordon. 1910. *Converts to Rome. A biographical list of the more notable converts to the Catholic Church in the United Kingdom during the last sixty years.* New and enlarged edition. London: Sands & Co.

Götzinger, Wilhelm Lebrecht. 1812. *Schandau und seine Umgebungen, oder Beschreibung der sächischen Schweiz.* 2d ed. Dresden.

Goubert, Emile and Zittel, Carl Alfred von. 1861. Description des fossiles du coral-rag de Glos. *Journal de Conchyliologie* 9: 192–208.

Gould, Augustus Addison. 1841. *Report on the invertebrata of Massachusetts, comprising the mollusca, crustacea, annelida and radiata.* Boston.

Gould, John. 1856. On a new turkey, Meleagris mexicana. *Proceedings of the Zoological Society of London* 24: 61–3.

Grande encyclopédie: *La grande encyclopédie inventaire raisonné des sciences, des lettres et des arts.* Edited by F. Camille Dreyfus and others. 31 vols. Paris. 1886–1902.

Gratiolet, Louis Pierre. [1854.] *Mémoire sur les plis cérébraux de l'homme et des primates.* Paris.

Gray, Asa. 1848. *A manual of the botany of the northern United States.* Cambridge, Mass.

——. 1848–9. *Genera Floræ Americæ Boreali-Orientalis illustrata. The genera of the plants of the United States.* 2 vols. Vol. 1: Boston, New York, and London. Vol. 2: New York.

——. 1856. *Manual of the botany of the northern United States: second edition; including Virginia, Kentucky, and all east of the Mississippi: arranged according to the natural system.* New York.

——. 1858a. *Introduction to structural and systematic botany, and vegetable physiology, being a fifth and revised edition of the botanical text-book.* New York, Chicago, and London.

——. 1858b. Note on the coiling of tendrils. [Read 12 August 1858.] *Proceedings of the American Academy of Arts and Sciences* 4 (1860): 98–9.

——. 1858–9. Diagnostic characters of new species of phænogamous plants, collected in Japan by Charles Wright, botanist of the U.S. North Pacific Exploring Expedition . . . With observations upon the relations of the Japanese flora to that of North America, and of other parts of the northern temperate zone. [Read 14 December 1858 and 11 January 1859.] *Memoirs of the American Academy of Arts and Sciences* n.s. 6 (1857–9): 377–452.

——. 1859. On the coiling of tendrils. *American Journal of Science and Arts* 2d ser. 27: 277–8.

[Gray, Asa.] 1860. Darwin on the origin of species. *Atlantic Monthly* 6: 109–16, 229–39; Darwin and his reviewers. 6: 406–25.

——. 1861. *Natural selection not inconsistent with natural theology. A free examination of Darwin's treatise on the origin of species, and of its American reviewers.* Boston and London.

——. 1862a. Review of *Orchids. American Journal of Science and Arts* 2d ser. 34: 138–44.

——. 1862b. Fertilization of orchids through the agency of insects. *American Journal of Science and Arts* 2d ser. 34: 420–9.

——. 1862c. Enumeration of the plants of Dr. Parry's collection in the Rocky Mountains in 1861. *American Journal of Science and Arts* 2d ser. 33: 404–11, 34: 249–61, 330–41; 35 (1863): 137.

——. 1862d. Review of *Outlines of the distribution of Arctic plants*, with a map. By Jos. D. Hooker. *American Journal of Science and Arts* 2d ser. 34: 144–8.

——. 1862e. Dimorphism in the genitalia of flowers. *American Journal of Science and Arts* 2d ser. 34: 419–20.

——. 1862f. Review of *Journal of the Proceedings of the Linnean Society, Botany*. Nos. 22, 23. *American Journal of Science and Arts* 34: 285–6.

——. 1863a. Structure and fertilization of certain orchids. *American Journal of Science and Arts* 36: 292–4.

——. 1863b. Variation and mimetic analogy in Lepidoptera. *American Journal of Science and Arts* 36: 285–90.

Gray, James. 1952. *History of the Royal Medical Society, 1737–1937*. Edited by Douglas Guthrie. Edinburgh: Edinburgh University Press.

Gray, John Edward. 1862a. On the skull of the Japanese pig (Sus pliciceps). [Read 28 January 1862.] *Proceedings of the Scientific Meetings of the Zoological Society of London* (1862): 13–17.

——, ed. 1862b. *Catalogue of the bones of Mammalia in the collection of the British Museum.* London.

——. 1862c. *A hand catalogue of postage stamps for the use of collectors.* London.

Grayson, Donald K. 1985. The first three editions of Charles Lyell's *The geological evidences of the antiquity of man. Archives of Natural History* 13: 105–21.

Greene, Joseph Reay. 1859–61. *Manual of the animal kingdom.* Pt 1: Protozoa. Pt 2: Cœlenterata. London.

Gregory, Frances W. 1975. *Nathan Appleton: merchant and entrepreneur 1779–1861.* Charlottesville, Va.: University Press of Virginia.

Gregory, Francis Thomas. 1862. Expedition to the north-west coast of Australia. *Journal of the Royal Geographical Society* 32: 372–429.

Gregory, Frederick. 1977. *Scientific materialism in nineteenth century Germany.* Dordrecht and Boston: D. Reidel Publishing Company.

Grey, George. 1841. *Journals of two expeditions of discovery in north-west and western Australia, during the years 1837, 38, and 39.* 2 vols. London.

Gris, Arthur. 1859. Note sur quelques cas remarquables de pélorie dans le genre *Zingiber*. *Bulletin de la Société Botanique de France* 6: 346–8.

Grönland, Johannes. 1855. Note sur les organes glanduleux du genre *Drosera*. *Annales des Sciences Naturelles (Botanique)* 4th ser. 3: 297–303.

——. 1858. Notice nécrologique sur M. Neumann. *Revue Horticole* (1858): 622–4.

Gross, Charles G. 1993. Hippocampus minor and man's place in nature: a case study in the social construction of neuroanatomy. *Hippocampus* 3: 403–16.

Gruber, Jacob W. and Thackray, John C. 1992. *Richard Owen commemoration: three studies*. London: Natural History Museum.

Guédès, Michel. 1975. Charles Naudin: l'évolutionnisme sans transformisme. *Histoire et Nature* 7: 77–88.

Gunther, Albert E. 1975. *A century of zoology at the British Museum through the lives of two keepers, 1815–1914*. Folkestone, Kent: William Dawson and Sons.

Haast, H. F. von. 1948. *The life and times of Sir Julius von Haast, explorer, geologist, museum builder*. Wellington, New Zealand: privately published.

Haast, John Francis Julius von. 1861. *Report of a topographical and geological exploration of the western districts of the Nelson Province, New Zealand*. Nelson, New Zealand.

——. 1862a. Notes on the geology of the province of Canterbury, N.Z. *New Zealand Government Gazette. Province of Canterbury*, 24 October 1862, pp. 121–31.

——. 1862b. *Address delivered by Mr. Haast at the dinner of the Philosophical Institute of Canterbury*. Canterbury, New Zealand: Philosophical Institute of Canterbury.

——. 1864a. Notes to a sketch-map of the province of Canterbury, New Zealand, showing the glaciation during the Pleistocene and recent periods as far as explored. [Read 7 December 1864.] *Quarterly Journal of the Geological Society of London* 21 (1865): 133–7.

——. 1864b. Notes on the mountains and glaciers of the Canterbury province, New Zealand. *Journal of the Royal Geographical Society* 34: 87–96.

——. 1867. Notes on the geology of the province of Canterbury, N.Z., principally in reference to the deposits of the glacial epoch at the western base of the Southern Alps. *Quarterly Journal of the Geological Society of London* 23: 342–52.

Hall, John N. 1991. *Trollope. A biography*. Oxford: Clarendon Press.

Hall, Philip B. 1987. Robert Swinhoe (1836–1877), FRS, FZS, FRGS: a Victorian naturalist in Treaty Port China. *Geographical Journal* 153: 37–47.

Hamberger, Georg Christoph and Meusel, Johann Georg. 1796–1834. *Das gelehrte Teutschland, oder Lexikon der jetzt lebenden teutschen Schriftsteller*. 23 vols. in 24. Lemgo.

Harbottle, Thomas Benfield. 1902. *Dictionary of quotations (classical)*. 2d ed. London: Swan Sonnenschein & Co. New York: Macmillan Co.

Harvey, Joy. 1987. "Strangers to each other": male and female relationships in the life and work of Clémence Royer. In *Uneasy careers and intimate lives: women in science 1789–1979*, edited by Pnina G. Abir-Am and Dorinda Outram. New Brunswick and London: Rutgers University Press.

Haughton, Samuel. 1860a. Presidential address. *Journal of the Geological Society of Dublin* 8 (1857–60): 137–56.

[——.] 1860b. Biogenesis. *Natural History Review* 7: 23–32.

——. 1862. On the use of nicotine in tetanus, and cases of poisoning by strychnia.

Dublin Quarterly Journal of Medical Science 34: 172–86.

Heer, Oswald. 1855–9. *Flora tertiaria Helvetiae. Die Tertiäre Flora der Schweiz.* 3 vols. Winterthur.

——. 1857. Ueber die fossilen Pflanzen von St. Jorge in Madeira. [Read 5 November 1855.] *Neue Denkschriften der allgemeinen schweizerischen Gesellschaft für die gesammten Naturwissenschaften* 15 (1857): 1–40.

——. 1860. *Untersuchungen über das Klima und die Vegetationsverhältnisse des Tertiärlandes.* Winterthur.

——. 1861a. *Recherches sur le climat et la végétation du pays tertiaire.* Translated from the German by Charles Théophile Gaudin. Winterthur, Geneva, and Paris.

——. 1861b. Über die fossilen Calosomen. *Neues Jahrbuch für Mineralogie, Geognosie, Geologie und Petrefakten-Kunde* (1861): 52–8.

——. 1862. On the fossil flora of Bovey Tracey. *Philosophical Transactions of the Royal Society of London* 152: 1039–86.

Heinecke, P. M. 1985. The parish of Chelsfield. *Bromley Local History* 7: 62–8.

Henslow, John Stevens. 1837. *Descriptive and physiological botany.* (Dionysius Lardner's *Cabinet cyclopædia.*) London.

Heron, Robert. 1835. Notes on the habits of the pea-fowl. *Proceedings of the Zoological Society of London* (1833–5), pt 3: 54.

Herringshaw, Thomas W. 1909–14. *Herringshaw's national library of American biography.* 5 vols. Chicago: American Publishers' Association.

[Herschel, John Frederick William.] 1862. On the existence of muscles in plants. In 'Our survey of literature and science'. *Cornhill Magazine* 6: 853–5.

Heuzé, Gustave. 1899. Les Vilmorin. *Revue Horticole* 71: 453–9.

Hildebrand, F. 1861. Die Verbreitung der Coniferen in der Jetztzeit und in den früheren geologischen Perioden. *Verhandlungen des naturhistorischen Vereines der preussischen Rheinlande und Westphalens* n.s. 8: 199–384.

Historical record: *University of London: the historical record (1836–1912). Being a supplement to the calendar completed to September 1912.* London: University of London Press. 1912.

History of Hampshire: *A history of Hampshire and the Isle of Wight.* Edited by H. Arthur Doubleday *et al.* 5 vols. London: Archibald Constable and Company. 1900–12.

Holland, Henry. 1855. *Medical notes and reflections.* 3d ed. London.

[——.] 1859. Life and organisation. *Edinburgh Review* 109: 227–63.

——. 1862. *Essays on scientific and other subjects from the Edinburgh and Quarterly Reviews.* London.

Hooker, Joseph Dalton. 1844–7. *Flora Antarctica.* Pt 1 of *The botany of the Antarctic voyage of H.M. Discovery Ships* Erebus *and* Terror *in the years 1839–1843, under the command of Captain Sir James Clark Ross.* 2 vols. in 1. London.

——. 1849. *The rhododendrons of Sikkim-Himalaya; being an account, botanical and geographical of the rhododendrons recently discovered in the mountains of eastern Himalaya, from drawings and descriptions made on the spot, during a government botanical mission to that country.* Edited by William Jackson Hooker. London.

Hooker, Joseph Dalton. 1853. *Introductory essay to the flora of New Zealand.* London.

——. 1853–5. *Flora Novæ-Zelandiæ.* Pt 2 of *The botany of the Antarctic voyage of H.M. Discovery Ships* Erebus *and* Terror, *in the years 1839–1843, under the command of Captain Sir James Clark Ross.* 2 vols. London.

——. 1854a. On the functions and structure of the rostellum of *Listera ovata. Philosophical Transactions of the Royal Society of London* 144: 259–63.

——. 1854b. *Himalayan journals; or, notes of a naturalist in Bengal, the Sikkim and Nepal Himalayas, the Khasia mountains, &c.* 2 vols. London.

——. 1855–60. *Flora Tasmaniæ.* Pt 3 of *The botany of the Antarctic voyage of H.M. Discovery Ships* Erebus *and* Terror, *in the years 1839–1843, under the command of Captain Sir James Clark Ross.* 2 vols. London.

——. 1859. *On the flora of Australia, its origin, affinities, and distribution; being an introductory essay to the Flora of Tasmania.* London. [Reprinted from J. D. Hooker 1855–60.]

——. 1861a. Outlines of the distribution of Arctic plants. [Read 21 June 1860.] *Transactions of the Linnean Society of London* 23 (1862): 251–348.

——. 1861b. An account of the plants collected by Dr. Walker in Greenland and Arctic America during the expedition of Sir Francis M'Clintock, R.N., in the Yacht 'Fox'. *Journal of the Proceedings of the Linnean Society of London (Botany)* 5: 79–88.

——. 1862a. On the cedars of Lebanon, Taurus, Algeria, and India. *Natural History Review* n.s. 2: 11–18.

——. 1862b. On the vegetation of Clarence Peak, Fernando Po; with descriptions of the plants collected by Mr. Gustav Mann on the higher parts of that mountain. [Read 7 March 1861.] *Journal of the Proceedings of the Linnean Society of London (Botany)* 6: 1–23.

[——.] 1862c. Review of *Orchids*, by Charles Darwin. *Gardener's Chronicle and Agricultural Gazette*, 23 August 1862, pp. 789–90, 13 September 1862, p. 863, and 27 September 1862, p. 910.

[——.] 1862d. Review of *Orchids*, by Charles Darwin. *Natural History Review* n.s. 2: 371–6.

——. 1863a. On *Welwitschia*, a new genus of *Gnetaceæ.* [Read 11 December 1863.] *Transactions of the Linnean Society of London* 24 (1864): 1–48.

——. 1863b. Enumeration of the mountain flowering plants and ferns collected by M. Gustav Mann, government botanist, during his various ascents of the Camaroons Mountains, of Clarence Peak, Fernando Po, and of the peak of San Thomé. Appendix 3 in *Abeokuta and the Camaroons Mountains. An exploration*, by Richard F. Burton. 2 vols. London.

——. 1863c. On the plants of the temperate regions of the Cameroons Mountains and islands in the Bight of Benin; collected by Mr. Gustav Mann, government botanist. [Read 5 November 1863.] *Journal of the Proceedings of the Linnean Society (Botany)* 7 (1864): 171–240.

——. 1864–7. *Handbook of the New Zealand flora: a systematic description of the native plants of New Zealand and the Chatham, Kermadec's, Lord Auckland's, Campbell's, and MacQuarrie's Islands.* 2 pts. London. 1864–7.

Hooker, Joseph Dalton. 1903. *A sketch of the life and labours of Sir William Jackson Hooker, . . . late director of the Royal Gardens of Kew.* Oxford: Clarendon Press.

Hooker, Joseph Dalton and Thomson, Thomas. 1858. *Præcursores ad Floram Indicam*: being sketches of the natural families of Indian plants, with remarks on their distribution, structure, and affinities. [Read 3 March 1857.] *Journal of the Proceedings of the Linnean Society (Botany)* 2 (1858): 1–29, 54–103, 163–80.

Hooker, William Jackson. 1821. *Flora Scotica; or a description of Scottish plants, arranged both according to the artificial and natural methods.* 2 pts. London.

Hopkins, William. 1852a. Presidential address. *Quarterly Journal of the Geological Society of London* 8: xxi–lxxx.

——. 1852b. On the influence of various configurations of land and sea, and oceanic currents, on the earth's superficial temperature. *Quarterly Journal of the Geological Society of London* 8: 63–92.

——. 1860. Physical theories of the phenomena of life. *Fraser's Magazine* 61: 739–52; 62: 74–90. [Reprinted in Hull 1973.]

Horsburgh, E. L. S. 1980. *Bromley, Kent: from the earliest times to the present century, compiled from materials collected from original sources by various hands.* Reprint edition. Chislehurst, Kent: Lodgemark Press.

Hoskyn, Richard. 1862. Report on the deep sea soundings to the westward of Ireland,—made in H.M.S. "Porcupine," in June, July, and August, 1862. *Nautical Magazine and Naval Chronicle* 31: 561–9.

Hull, David L. 1973. *Darwin and his critics: the reception of Darwin's theory of evolution by the scientific community.* Cambridge, Mass.: Harvard University Press.

Humboldt, Alexander von and Bonpland, Aimé Jacques Alexandre. 1819–29. *Personal narrative of travels to the equinoctial regions of the new continent, during the years 1799–1804.* 3d ed. London.

Hutchinson, Horace Gordon. 1914. *Life of Sir John Lubbock, Lord Avebury.* 2 vols. London: Macmillan and Co.

Huxley, Leonard, ed. 1900. *Life and letters of Thomas Henry Huxley.* 2 vols. London: Macmillan and Co.

——, ed. 1918. *Life and letters of Sir Joseph Dalton Hooker.* 2 vols. London: John Murray.

[Huxley, Thomas Henry.] 1860. Darwin on the origin of species. *Westminster Review* n.s. 17: 541–70.

——. 1861a. On a new species of *Macrauchenia* (*M. Boliviensis*). [Read 21 November 1860.] *Quarterly Journal of the Geological Society of London* 17: 73–84.

——. 1861b. On the zoological relations of man with the lower animals. *Natural History Review* n.s. 1: 67–84.

——. 1862a. On fossil remains of man. *Notices of the Proceedings at the Meetings of Members of the Royal Institution* 3 (1858–62): 420–2. [Reprinted in Foster and Lankester eds. 1898–1903, 2: 509–11.]

——. 1862b. The anniversary address. *Quarterly Journal of the Geological Society of London* 18: xl–liv.

Huxley, Thomas Henry. 1862c. *On our knowledge of the causes of the phenomena of organic nature.* 6 pts. London.

——. 1862d. *Address delivered at the anniversary meeting of the Geological Society of London on the 21st of February, 1862.* London.

——. 1863a. *Evidence as to man's place in nature.* London and Edinburgh.

——. 1863b. *On our knowledge of the causes of the phenomena of organic nature. Being six lectures to working men, delivered at the Museum of Practical Geology.* London.

——. 1893–4. *Collected essays.* 7 vols. London.

IBN: *Index bio-bibliographicus notorum hominum.* Pt C: *Corpus alphabeticum sectio generalis.* Edited by Jean-Pierre Lobies and Otto and Wolfram Zeller. 60 vols. (A–Durau, Pierre). Osnabrück: Biblio Verlag. 1974–92.

'Illegitimate offspring of dimorphic and trimorphic plants': On the character and hybrid-like nature of the offspring from the illegitimate unions of dimorphic and trimorphic plants. By Charles Darwin. [Read 20 February 1868.] *Journal of the Linnean Society of London (Botany)* 10 (1869): 393–437.

Index animalium: *Index animalium sive index nominum quae ab* A.D. MDCCLVIII *generibus et speciebus animalium imposita sunt.* By Charles Davies Sherborn. 10 vols. Cambridge: Cambridge University Press (vol. 1). London: British Museum. 1902–32.

Index biographique de l'Académie des sciences: *Index biographique de l'Académie des sciences du 22 décembre 1666 au 1er octobre 1978.* Paris: Gauthier-Villars. 1979.

Insectivorous plants: *Insectivorous plants.* By Charles Darwin. London. 1875.

Jackson, Benjamin Daydon. 1906. *George Bentham.* London: J. M. Dent & Co. New York: E. P. Dutton & Co.

Jamieson, Thomas Francis. 1858. On the Pleistocene deposits of Aberdeenshire. *Quarterly Journal of the Geological Society of London* 14: 509–32.

——. 1860a. On the drift and rolled gravel of the north of Scotland. *Quarterly Journal of the Geological Society of London* 16: 347–71.

——. 1860b. On the occurrence of Crag strata beneath the boulder-clay in Aberdeenshire. *Quarterly Journal of the Geologial Society of London* 16: 371–3.

——. 1862. On the ice-worn rocks of Scotland. *Quarterly Journal of the Geological Society of London* 18: 164–84.

——. 1863. On the parallel roads of Glen Roy, and their place in the history of the Glacial period. *Quarterly Journal of the Geological Society of London* 19: 235–59.

Jansen, Marius B. ed. 1989. *The Cambridge history of Japan. Volume 5. The nineteenth century.* Cambridge: Cambridge University Press.

Jenyns, Leonard. 1862. *Memoirs of the Rev. John Stevens Henslow, late rector of Hitcham, and professor of botany in the University of Cambridge.* London.

Jervis, James. [1945.] *W. B. Clarke: "The father of Australian geology".* Sydney: Royal Australian Historical Society.

Johnson, Henry. 1862. On the distorted skulls found at Wroxeter (Salop), with a mechanico-chemical explanation of the distortion. *Proceedings of the Royal Society of London* 12: 149–50.

Jones, Hugh Percy, ed. 1900. *A new dictionary of foreign phrases and classical quotations.* London: Charles William Deacan and Co.

Jones, Thomas Rymer. 1841. *A general outline of the animal kingdom, and manual of comparative anatomy.* London.

Journal of researches: *Journal of researches into the geology and natural history of the various countries visited by H.M.S. Beagle, under the command of Captain FitzRoy, R.N. from 1832 to 1836.* By Charles Darwin. London. 1839.

Journal of researches 2d ed.: *Journal of researches into the natural history and geology of the countries visited during the voyage of H.M.S. Beagle round the world, under the command of Capt. FitzRoy, R.N.* Second edition, corrected, with additions. By Charles Darwin. London. 1845.

Jubiläums-Katalog: *Jubiläums-Katalog der E. Schweizerbart'schen Verlagsbuchhandlung (Erwin Nägele) G.m.b.H., Stuttgart, 1826–1926.* Stuttgart: E. Schweizerbart. 1926.

Jukes, Joseph Beete. 1862. On the mode of formation of some of the river-valleys in the south of Ireland. *Quarterly Journal of the Geological Society of London* 18: 378–403.

Junker, Thomas. 1989. *Darwinismus und Botanik. Rezeption, Kritik und theoretische Alternativen im Deutschland des 19. Jahrhunderts.* Stuttgart: Deutscher Apotheker Verlag.

——. 1991. Henrich Georg Bronn und die *Entstehung der Arten. Sudhoffs Archiv Zeitschrift für Wissenschaftsgeschichte* 75: 180–208.

Kaplan, Fred. 1983. *Thomas Carlyle: a biography.* Cambridge: Cambridge University Press.

Karsten, Gustave Karl Wilhelm Hermann. 1861. On the sexual life of plants, and parthenogenesis. *Annals and Magazine of Natural History* 3d ser. 8: 81–9, 200–9.

Kessel, Reuben A. and Alchian, Armen A. 1967. Real wages in the north during the Civil War: Mitchell's data reinterpreted. In *The economic impact of the American Civil War*, edited by Ralph Andreano. 2d ed. Cambridge, Mass.: Schenkman Publishing Company.

Keynes, Margaret. 1943. *Leonard Darwin, 1850–1943.* Cambridge: privately printed at Cambridge University Press.

Keynes, Richard Darwin, ed. 1979. *The Beagle record. Selections from the original pictorial records and written accounts of the voyage of H.M.S. Beagle.* Cambridge: Cambridge University Press.

——, ed. 1988. *Charles Darwin's* Beagle *diary.* Cambridge: Cambridge University Press.

Kingsley, Charles. 1874. *Westminster sermons.* London.

Kirby, William and Spence, William. 1815–26. *An introduction to entomology: or elements of the natural history of insects.* London.

Kirby, William Forsell. 1862. *A manual of European butterflies, on the plan of Stainton's Manual of British butterflies and moths.* London and Edinburgh.

Klotzsch, Johann Friedrich. 1854. Begoniaceen-Gattungen und Arten. [Read 2 March 1854.] *Abhandlungen der königlichen Akademie der Wissenschaften zu Berlin* 39 (1855): 121–255.

Knight, Thomas Andrew. 1799. An account of some experiments on the fecundation of vegetables. *Philosophical Transactions of the Royal Society of London* (1799), pp. 195–204.

——. 1820. Upon the variations of the scarlet strawberry (*Fragaria Virginiana*) when propagated by seeds. [Read 4 August 1818.] *Transactions of the Horticultural Society of London* 3 (1820): 207–10.

——. 1824. An account of some mule plants. [Read 6 May 1823.] *Transactions of the Horticultural Society of London* 5 (1824): 292–6.

Knoll, Elizabeth. 1986. The sciences of language and the evolution of mind: Max Müller's quarrel with Darwinism. *Journal of the History of the Behavioral Sciences* 22: 3–22.

Koch, Wilhelm Daniel Joseph. 1843–4. *Synopsis florae Germanicae et Helveticae, exhibens stirpes phanerogamas rite cognitas.* 2d ed. 2 vols. Frankfurt and Leipzig.

Kohn, David. 1985. Darwin's principle of divergence as internal dialogue. In *The Darwinian heritage*, edited by David Kohn. Princeton: Princeton University Press in association with Nova Pacifica.

Kolenati, Friedrich A. 1845–6. *Meletemata entomologica.* St Petersburg.

Kölreuter, Joseph Gottlieb. 1761–6. *Vorläufige Nachricht von einigen das Geschlecht der Pflanzen betreffenden Versuchen und Beobachtungen.* Leipzig.

Kraatz, Ernst Gustave. 1860. Ueber die europäischen Hirschkäfer. *Berliner Entomologische Zeitschrift* 4 (1859–60): 68–75, 265–75.

Kritsky, Gene. 1991. Darwin's Madagascan Hawk Moth prediction. *American Entomologist* 37: 206–9.

Krohn, August David. 1859. Beobachtungen über den Cementapparat und die weiblichen Zeugungsorgane einiger Cirripedien. *Archiv für Naturgeschichte* 25, pt 1: 355–64.

Kunth, Karl Sigismund. 1815. Considérations générales sur la famille des Cypéracées. *Mémoires du Muséum d'Histoire Naturelle* 2: 147–53.

——. 1833–50. *Enumeratio plantarum omnium hucusque cognitarum, secundum familias naturales disposita, adjectis characteribus, differentiis et synonymis.* 5 vols. Stuttgart and Tübingen.

Lang, David. 1989. *A guide to the wild orchids of Great Britain and Ireland.* 2d ed. Oxford: Oxford University Press.

Larousse: *Larousse du XXe siècle.* Under the direction of Paul Augé. 6 vols. Paris: Librairie Larousse. 1928–33.

Latreille, Pierre André. 1802. *Histoire naturelle des Fourmis, et recueil de mémoires et d'observations sur les Abeilles, les Araignées, les Faucheurs, et autres insectes.* Paris.

Lecoq, Henri. 1854–8. *Études sur la géographie botanique de l'Europe et en particulier sur la végétation du plateau central de la France.* 9 vols. Paris.

——. 1862. De l'espèce et de ses croisements dans le genre *Mirabilis. Bulletin de la Société Botanique de France* 9: 217–35.

Leidy, Joseph. 1858. Notice of remains of extinct Vertebrata, from the valley of the Niobrara River, collected during the exploring expedition of 1857, in Nebraska, under the command of Lieut. G. K. Warren, U.S. Top. Eng., by Dr. F. V. Hayden, geologist to the expedition. *Proceedings of the Academy of Natural Sciences of Philadelphia*, pp. 20–9.

[Leifchild, John R.] 1862. Review of *Orchids*, by Charles Darwin. *Athenæum*, 24 May 1862, pp. 683–5.

Leighton, William Allport. 1841. On *Epilobium angustifolium*, Linn., and species which have been confounded with it. *Annals and Magazine of Natural History* 8 (1842): 246–8. [Published in issue of December 1841.]

——. 1842. On *Epilobium angustifolium*, Linn., and species which have been confounded with it. *Annals and Magazine of Natural History* 8: 401–4.

Lewis, George Cornewall. 1862. *Historical survey of the astronomy of the ancients*. London.

Lindley, John. 1830. *An introduction to the natural system of botany: or, a systematic view of the organisation, natural affinities, and geographical distribution, of the whole vegetable kingdom; together with the uses of the most important species in medicine, the arts, and rural or domestic economy*. London.

——. 1836. *A natural system of botany; or, a systematic view of the organisation, natural affinities, and geographical distribution, of the whole vegetable kingdom; together with the uses of the most important species in medicine, the arts, and rural or domestic economy*. 2d ed. London.

[——.] 1846. A note upon the wild state of maize, or Indian corn. *Journal of the Horticultural Society of London* 1: 114–17.

——. 1853. *The vegetable kingdom; or, the structure, classification, and uses of plants, illustrated upon the natural system*. 3d ed. London.

——. 1859. *A synopsis of the British flora arranged according to the natural orders. Containing vasculares or flowering plants*. 3d ed. London.

——. 1862a. West African tropical orchids. [Read 7 November 1861.] *Journal of the Proceedings of the Linnean Society of London (Botany)* 6: 123–40.

[——.] 1862b. Review of *Genera plantarum* vol. 1, pt 1. *Gardeners' Chronicle and Agricultural Gazette*, 6 September 1862, pp. 838–9.

Lindsay, Jean. 1979. *The Trent & Mersey Canal*. Newton Abbot, Devon: David & Charles.

Lindsay, Lionel. 1968. *Conrad Martens, the man and his art*. Revised edition. Sydney: Angus and Robertson.

Link, Heinrich Friedrich. 1821. *Die Urwelt und das Alterthum, erläutert durch die Naturkunde*. Berlin.

Linnaeus, Carl. 1751. *Philosophia botanica in qua explicantur fundamenta botanica cum definitionibus partium, exemplis terminorum, observationibus rariorum, adjectis figuris æneis*. Stockholm.

——. 1792. *Prælectiones in ordines naturales plantarum*. Hamburg.

List of fellows of the Zoological Society: *A list of the fellows and honorary, foreign, and corresponding members and medallists of the Zoological Society of London*. London. 1893.

List of the Geological Society of London: *List of the Geological Society of London*. London. 1864–1965.

List of graduates of the University of Edinburgh: *Alphabetical list of graduates of the University of Edinburgh from 1859 to 1888 (both years included) with historical appendix (including present and past office-bearers) and separate lists of honorary graduates and graduates with honours*. Edinburgh. [1889.]

List of the Linnean Society of London: *List of the Linnean Society of London*. London. 1805–1939.

List of the Microscopical Society: *List of the officers and members of the Microscopical Society of London*. London. 1862.

Lister, Raymond. 1981. *George Richmond: a critical biography*. London: Robin Garton.

Littleton, Thomas. *c*. 1481. *Tenores novelli*.

Living Cirripedia (1851): *A monograph on the sub-class Cirripedia, with figures of all the species. The Lepadidæ; or, pedunculated cirripedes*. By Charles Darwin. London. 1851.

LL: *The life and letters of Charles Darwin, including an autobiographical chapter*. Edited by Francis Darwin. 3 vols. London. 1887–8.

Longmate, Norman. 1978. *The hungry mills. The story of the Lancashire cotton famine 1861–5*. London: Temple Smith.

Lorenz, Otto, ed. 1867–71. *Catalogue général de la librairie Française pendant 25 ans (1840–1865)*. 4 vols. Paris.

Loudon, John Claudius. 1850. *Loudon's hortus Britannicus. A catalogue of all the plants indigenous, cultivated in, or introduced to Britain*. New ed. London.

Lovén, Sven Ludvig. 1846. *Index Molluscorum litora Scandinaviæ occidentalia habitantium*. Holmiæ.

Lowe, Richard Thomas. 1857–68. *A manual flora of Madeira and the adjacent islands of Porto Santo and the Desertas*. Vol. 1. London.

Lubbock, John. 1861. The Kjökkenmöddings: recent geologico-archæological researches in Denmark. *Natural History Review* n.s. 1: 489–504.

——. 1862a. On the geologico-archæological discoveries in Denmark, Switzerland, and France. [Read 19 February 1862.] *Report of the Proceedings of the Geological and Polytechnic Society of the West Riding of Yorkshire* 4 (1860–2): 238–73.

——. 1862b. On the ancient lake habitations of Switzerland. *Natural History Review* n.s. 2: 26–51.

——. 1862c. On the evidence of the antiquity of man, afforded by the physical structure of the Somme Valley. *Natural History Review* n.s. 2: 244–69.

——. 1862d. On two aquatic *Hymenoptera*. *Report of the 32d meeting of the British Association for the Advancement of Science held at Cambridge*, Transactions of the sections, pp. 110–11.

——. 1863a. North American archæology. *Natural History Review* n.s. 3: 1–26.

[——.] 1863b. Review of *Prehistoric man: researches into the origin of civilisation in the Old and New World*, by Daniel Wilson. *Natural History Review* n.s. 3: 26–30.

——. 1863c. On the ancient lake-habitations of Switzerland. [Read 27 February 1863.] *Proceedings of the Royal Institution of Great Britain* 4 (1866): 29–40.

——. 1863d. On the development of *Chloëon* (Ephemera) *dimidiatum.—Part 1*. [Read 15 January 1863.] *Transactions of the Linnean Society of London* 24 (1864): 61–78.

——. 1863e. On two aquatic *Hymenoptera*, one of which uses its wings in swimming.

[Read 7 May 1863.] *Transactions of the Linnean Society of London* 24 (1864): 135–42.

Lubbock, John. 1865. *Pre-historic times, as illustrated by ancient remains, and the manners and customs of modern savages.* London and Edinburgh.

Lucas, Prosper. 1847–50. *Traité philosophique et physiologique de l'hérédité naturelle dans les états de santé et de maladie du système nerveux avec l'application méthodique des lois de la procréation au traitement général des affections dont elle est le principe.* 2 vols. Paris.

Lund, Peter Wilhelm. 1839–40. Coup-d'oeil sur les espèces éteintes de Mammifères du Brésil; extrait de quelques mémoires présentés à l'Académie royale des Sciences de Copenhague. *Annales des Sciences Naturelles (Zoologie)* 2d ser. 11 (1839): 214–34; 12 (1839): 205–8; 13 (1840): 310–19.

Lupton, I. J. and Walsh, John Henry. 1861. *The horse, in the stable and the field: his varieties, management in health and disease, anatomy, physiology, etc. etc.* London.

Lyell, Charles. 1830–3. *Principles of geology, being an attempt to explain the former changes of the earth's surface, by reference to causes now in operation.* 3 vols. London.

——. 1853. *Principles of geology; or, the modern changes of the earth and its inhabitants considered as illustrative of geology.* 9th ed. London.

——. 1855. *A manual of elementary geology: or, the ancient changes of the earth and its inhabitants as illustrated by geological monuments.* 5th ed. London.

——. 1863a. *The geological evidences of the antiquity of man; with remarks on theories of the origin of species by variation.* London.

——. 1863b. *The geological evidences of the antiquity of man; with remarks on theories of the origin of species by variation.* 2d ed., revised. London.

Lyell, Katherine Murray, ed. 1881. *Life, letters and journals of Sir Charles Lyell, Bart.* 2 vols. London.

——, ed. 1890. *Memoir of Leonard Horner, consisting of letters to his family and from some of his friends.* 2 vols. London: privately printed.

Mabberley, D. J. 1985. *Jupiter botanicus: Robert Brown of the British Museum.* London: British Museum (Natural History). Braunschweig: J. Cramer.

McCoy, Frederick. 1847. On the fossil botany and zoology of the rocks associated with the coal of Australia. *Annals and Magazine of Natural History* 20: 145–57, 226–36, 298–312.

——. 1861. Remarks on a series of fossils collected at Wollumbilla, and transmitted by Rev. W. D. Clarke, of Sydney. [Read 30 December 1861.] *Transactions and Proceedings of the Royal Society of Victoria* 6 (1865): 42–6.

MacMinn, Ney *et al.* 1990. *Bibliography of the published writings of John Stuart Mill. Edited from his manuscript with corrections and notes.* Bristol: Thoemmes.

McPherson, James M. 1988. *Battle cry of freedom. The Civil War era.* New York and Oxford: Oxford University Press.

Magnus, P. 1891. Johannes Grönland. *Verhandlungen des botanischen Vereins der Provinz Brandenburg* 33: il–li.

Marginalia: *Charles Darwin's marginalia.* Edited by Mario A. di Gregorio with the assistance of N. W. Gill. Vol. 1. New York and London: Garland Publishing. 1990.

Marshall, John. 1861. On the brain of a young chimpanzee. *Natural History Review* n.s. 1: 296–315.

Martin, Gerald P. R. and Uschmann, Georg. 1969. *Friedrich Rolle 1827–1887, ein Vorkämpfer neuen biologischen Denkens in Deutschland.* Leipzig: Johann Ambrosius Barth.

Martin, William Charles Linnaeus. 1837. Observations on three specimens of the genus *Felis* presented to the Society by Charles Darwin, Esq. *Proceedings of the Zoological Society of London* 5: 3–4.

Masters, Maxwell Tylden. 1862. Vegetable morphology: its history and present condition. *British and Foreign Medico-Chirurgical Review* 29: 202–18.

——. 1863. On the existence of two forms of Peloria. *Natural History Review* 3: 258–62.

——. 1869. *Vegetable teratology, an account of the principal deviations from the usual construction of plants.* London.

Matthew, Patrick. 1831. *On naval timber and arboriculture; with critical notes on authors who have recently treated the subject of planting.* London and Edinburgh.

Maw, George. 1861. Review of *Origin of species* and other works. *Zoologist* 19: 7577–611.

Max Müller, Friedrich. 1861. *Lectures on the science of language delivered at the Royal Institution of Great Britain in April, May, and June, 1861.* London.

Maximovicz, Karl Ivanovich. 1859. Primitiae florae amurensis. Versuch einer Flora des Amurlandes. *Mémoires Présentés à l'Académie Impériale des Sciences de St.-Petersbourg par divers savants et lus dans ses Assemblées* 9: 1–504.

Medical directory: *The London & provincial medical directory, inclusive of the medical directory for Scotland, and the medical directory for Ireland, and general medical register.* London. 1861–9. *The medical directory . . . , including the London and provincial medical directory, the medical directory for Scotland, the medical directory for Ireland.* London. 1870–1905.

Medical directory for Scotland: *The medical directory for Scotland.* London. 1852–60.

Medical who's who: *The medical who's who.* London: London & Counties Press Association. 1913–27.

Meehan, Thomas. 1862. On the uniformity of relative characters between allied species of European and American trees. *Proceedings of the Academy of Natural Sciences of Philadelphia* 14: 10–13.

Men of the time: *Men of the time: a dictionary of contemporaries, containing biographical notices of eminent characters of both sexes.* 7th ed. London. 1868.

Merian, Peter. 1843. Ueber die Theorie der Gletscher. *Bericht über die Verhandlungen der Naturforschenden Gesellschaft in Basel* 5: 110–60.

Metcalfe, Richard. 1906. *The rise and progress of hydropathy in England and Scotland.* London: Simpkin, Marshall, Hamilton, Kent & Co.

Michalet, Louis Eugène. 1860. Sur la floraison des *Viola* de la section *Nomimium*, de l'*Oxalis acetosella* et du *Linaria spuria. Bulletin de la Société Botanique de France* 7: 465–9.

Middendorf, Aleksandr Fedorovich. 1848–75. *Reise in den äussersten Norden und Osten*

Sibiriens während der Jahre 1843 und 1844 mit allerhöchster Genehmigung auf Veranstaltung der kaiserlichen Akademie der Wissenschaften zu St. Petersburg ausgeführt und in Verbindung mit vielen Gelehrten herausgegeben. 4 vols. St Petersburg.

Mieder, Wolfgang *et al.*, eds. 1992. *A dictionary of American proverbs.* New York and Oxford: Oxford University Press.

Milde, Karl August Julius. 1852. Ueber die Reizbarkeit der Blätter von *Drosera rotundifolia*. *Botanische Zeitung* 10: 540.

Mill, John Stuart. 1862. The contest in America. *Fraser's Magazine for Town and Country* 65: 258–68.

Milne, David. 1849. On the parallel roads of Lochaber, with remarks on the change of relative levels of sea and land in Scotland, and on the detrital deposits in that country. [Read 1 March and 5 April 1847.] *Transactions of the Royal Society of Edinburgh* 16 (1849): 395–418.

ML: *More letters of Charles Darwin: a record of his work in a series of hitherto unpublished letters.* Edited by Francis Darwin and Albert Charles Seward. 2 vols. London: John Murray. 1903.

Modern English biography: *Modern English biography containing many thousand concise memoirs of persons who have died since the year 1850.* By Frederick Boase. 3 vols. and supplement (3 vols.). Truro: printed for the author. 1892–1921.

Moffat, C. B., ed. 1898. *Life and letters of Alexander Goodman More . . . with selections from his zoological writings.* With a preface by Frances M. More. Dublin.

Möller, Hans Peter Christian. 1842. *Index Molluscorum Grœnlandiæ.* Hafniæ.

Monnier, Auguste. 1833. Note sur quelques espèces du genre *Viola*. *Archives de Botanique* 1: 412–15.

Montagu, Ashley, ed. 1968. *On the origin of species or, the causes of the phenomena of organic nature. By Thomas H. Huxley.* Ann Arbor: The University of Michigan Press.

Monteiro, Joachim John. 1875. *Angola and the River Congo.* 2 vols. London.

Moore, Charles. 1861. On the zones of the *Lower Lias* and the *Avicula contorta* zone. *Quarterly Journal of the Geological Society of London* 17: 483–516.

——. 1870. Australian Mesozoic geology *and* palæontology. [Read 10 November 1869.] *Quarterly Journal of the Geological Society of London* 26 (1870): 226–61.

Moore, James R. 1985. Darwin of Down: the evolutionist as squarson-naturalist. In *The Darwinian heritage*, edited by David Kohn. Princeton: Princeton University Press in association with Nova Pacifica.

Moquin-Tandon, Horace Bénédict Alfred. 1841. *Eléments de tératologie végétale, ou, histoire abrégée des anomalies de l'organisation dans les végétaux.* Paris.

Morell, John Daniel. 1862. *An introduction to mental philosophy, on the inductive method.* London.

Movement in plants: *The power of movement in plants.* By Charles Darwin. London. 1880.

Müller, Daniel Ernst. 1857. Ueber die Befruchtung der incompleten Blumen einiger *Viola*-Arten. *Botanische Zeitung* 15: 729–33.

Murray, Andrew. 1860. On Mr Darwin's theory of the origin of species. [Read 20 February 1860.] *Proceedings of the Royal Society of Edinburgh* 4 (1857–62): 274–91.

Murray, Andrew. 1862. On the geographical relations of the *Coleoptera* of Old Calabar. *Transactions of the Linnean Society* 23: 449–55.

Natural selection: *Charles Darwin's Natural Selection; being the second part of his big species book written from 1856 to 1858*. Cambridge: Cambridge University Press. 1975.

Naudin, Charles Victor. 1849–52. Melastomacearum quæ in Musæo Parisiensi continentur. Monographicæ descriptionis et secundum affinitateo distributionis tentamen. *Annales des Sciences Naturelles (Botanique)* 3d ser. 12 (1849): 196–284; 13 (1849): 25–39, 126–59, 273–303, 347–62; 14 (1850): 53–76, 118–65; 15 (1851): 43–79, 276–345; 16 (1851): 83–246; 17 (1852): 305–82; 18 (1852): 85–154, 257–94.

——. 1852. Considérations philosophiques sur l'espèce et la variété. *Revue Horticole* 4th ser. 1: 102–9.

——. 1856. Nouvelles recherches sur les caractères spécifiques et les variétés des plantes du genre *Cucurbita*. *Annales des Sciences Naturelles (Botanique)* 4th ser. 6: 5–73.

——. 1858. Observations concernant quelques plantes hybrides qui ont été cultivées au muséum. *Annales des Sciences Naturelles (Botanique)* 4th ser. 9: 257–78.

——. 1862. Cucurbitacées cultivées au Muséum d'Histoire Naturelle en 1862. Description d'espèces nouvelles et de quelques formes hybrides obtenues de plantes de cette famille. *Annales des Sciences Naturelles (Botanique)* 4th ser. 18: 159–208.

——. 1863. Nouvelles recherches sur l'hybridité dans les végétaux. Mémoire couronné par l'Académie des sciences en 1862. *Annales des Sciences Naturelles (Botanique)* 4th ser. 19: 180–203.

——. 1865. Nouvelles recherches sur l'hybridité dans les végétaux. Mémoire présenté a l'Académie des Sciences par M. Ch. Naudin en décembre 1861 et couronné dans la séance du 29 décembre 1862. *Nouvelles Archives du Muséum d'Histoire Naturelle* 1: 25–176.

Navy list: *The navy list*. London. 1815–70.

NBU: *Nouvelle biographie universelle depuis les temps les plus regulés jusqu'à nos jours, avec les renseignements bibliographiques et l'indication des sources a consulter*. Edited by Jean Chrétien Ferdinand Hoefer. 46 vols. Paris. 1852–66.

NDB: *Neue deutsche Biographie*. Under the auspices of the Historical Commission of the Bavarian Academy of Sciences. 17 vols. (A–Moller). Berlin: Duncker and Humblot. 1953–94.

Nevill, Dorothy. [1910.] *Under five reigns*. Edited by Ralph Nevill. London: Methuen & Co.

Nicholas, F. W. and Nicholas, J. M. 1989. *Charles Darwin in Australia*. Cambridge: Cambridge University Press.

Nitschke, Theodor Rudolf Joseph. 1861a. Morphologie des Blattes von *Drosera rotundifolia* L. *Botanische Zeitung* 19: 145–51.

——. 1861b. Anatomie des Sonnenthaublattes (*Drosera rotundifolia* L.). *Botanische Zeitung* 19: 233–5, 241–6, 252–5.

NNBW: *Nieuw Nederlandsch biografisch Woordenboek*. Edited by P. C. Molhuysen, P. J. Blok, and K. H. Kossman. 10 vols. Leiden: A. W. Sijthoff's Uitgevers-Maatschappij. 1911–37.

Notebooks: *Charles Darwin's notebooks, 1836–1844. Geology, transmutation of species, metaphysical enquiries.* Transcribed and edited by Paul H. Barrett, Peter J. Gautrey, Sandra Herbert, David Kohn, and Sydney Smith. Cambridge: Cambridge University Press for the British Museum (Natural History). 1987.

NSTC: *Nineteenth century short title catalogue.* 54 vols. Newcastle-upon-Tyne: Avero Publications. 1984–95.

NUC: *The national union catalog. Pre-1956 imprints.* 685 vols. and supplement (vols. 686–754). London and Chicago: Mansell. 1968–81.

OBL: *Österreichisches biographisches Lexikon 1815–1950.* Edited by Leo Santifaller, Eva Obermayer-Marnach, and others. 9 vols. and 3 fascicles of vol. 10 (A–Schmid). Vienna: Österreichischer Akademie der Wissenschaften. 1957–92.

O'Donoghue, Freeman. 1908–25. *Catalogue of engraved British portraits preserved in the Department of Prints and Drawings in the British Museum.* 6 vols. London: British Museum.

OED: *The Oxford English dictionary.* 2d ed. Prepared by J. A. Simpson and E. S. C. Weiner. First edited by James A. H. Murray, Henry Bradlet, W. A. Craigie, and C. T. Onions. Combined with *A supplement to the Oxford English dictionary*, edited by R. W. Burchfield. 20 vols. Oxford: Clarendon Press. 1989.

Olby, Robert. 1985. *Origins of Mendelism.* Chicago and London: University of Chicago Press.

Oliver, Daniel. 1862a. On the distribution of northern plants. *Proceedings of the Royal Institution of Great Britain* 3 (1858–62): 431–3.

——. 1862b. The Atlantis hypothesis in its botanical aspect. *Natural History Review* n.s. 2: 149–70.

[——.] 1862c. On the two forms, or dimorphic conditions, in the species of *Primula*, and on their remarkable sexual relations. By Charles Darwin. *Natural History Review* n.s. 2: 235–43.

[——.] 1862d. Botanical bibliography. *Natural History Review* n.s. 2: 416–85.

Orchids: *On the various contrivances by which British and foreign orchids are fertilised by insects, and the good effects of intercrossing.* By Charles Darwin. London. 1862.

Orchids 2d ed.: *On the various contrivances by which British and foreign orchids are fertilised by insects, and the good effects of intercrossing.* Revised, etc. By Charles Darwin. London. 1877.

Ordnance Survey gazetteer: *The Ordnance Survey gazetteer of Great Britain.* 3d ed. Southampton: Ordnance Survey. London and Basingstoke: Macmillan Press. 1992.

Origin: *On the origin of species by means of natural selection, or the preservation of favoured races in the struggle for life.* By Charles Darwin. London. 1859.

Origin 2d ed.: *On the origin of species by means of natural selection, or the preservation of favoured races in the struggle for life.* By Charles Darwin. London. 1860.

Origin 3d ed.: *On the origin of species by means of natural selection, or the preservation of favoured races in the struggle for life.* With additions and corrections. By Charles Darwin. London. 1861.

Origin 4th ed.: *On the origin of species by means of natural selection, or the preservation of favoured races in the struggle for life.* With additions and corrections. By Charles Darwin. London. 1866.

Osborne, Michael Andrew. 1987. The Société Zoologique d'Acclimatation and the new French Empire: the science and political economy of economic zoology during the Second Empire. (PhD dissertation: University of Wisconsin).

Ospovat, Dov. 1981. *The development of Darwin's theory. Natural history, natural theology, and natural selection, 1838–1859.* Cambridge: Cambridge University Press.

Owen, Richard. 1844. Description of a fossil molar tooth of a mastodon discovered by Count Strzlecki in Australia. *Annals and Magazine of Natural History* 14: 268–71.

——. 1859. Palæontology. *The Encyclopædia Britannica, or dictionary of arts, sciences, and general literature* 17: 91–176. *Encyclopaedia Britannica*. 8th ed. Edited by T. S. Traill. 21 vols. and index. London.

——. 1860a. *Palæontology or a systematic summary of extinct animals and their geological relations.* Edinburgh.

[——.] 1860b. Review of *Origin of species* and other works. *Edinburgh Review* 111: 487–532.

——. 1861a. *Palæontology or a systematic summary of extinct animals and their geological relations.* 2d ed. Edinburgh.

——. 1861b. On the cerebral characters of man and the ape. *Annals and Magazine of Natural History* 3d ser. 7: 456–8.

——. 1862a. On the aye-aye (*Chiromys*, Cuvier; *Chiromys madagascariensis,* Desm.; *Sciurus madagascariensis,* Geml., Sonnerat; *Lemur psilodactylus,* Schreber, Shaw). *Transactions of the Zoological Society of London* 5 (1862–6): 33–101.

——. 1862b. On the Hyracotherian character of the lower molars of the supposed *Macacus* from the Eocene sand of Kyson, Suffolk. *Annals and Magazine of Natural History* 3d ser. 10: 240.

——. 1862c. On the zoological significance of the cerebral and pedial characters of man. *Report of the 32d meeting of the British Association for the Advancement of Science held at Cambridge*, Transactions of the sections, pp. 116–18.

——. 1862d. On the homologies of the bones of the head of the *Polypterus niloticus. Report of the 32d meeting of the British Association for the Advancement of Science held at Cambridge*, Transactions of the sections, p. 118.

——. 1862e. On the characters of the aye-aye, as a test of the Lamarckian and Darwinian hypothesis of the transmutation and origin of species. *Report of the 32d meeting of the British Association for the Advancement of Science held at Cambridge*, Transactions of the sections, pp. 114–16.

Owen, Richard Startin, ed. 1894. *The life of Richard Owen . . . With the scientific portions revised by C. Davies Sherborn; also an essay on Owen's position in anatomical science by the Right Hon. T. H. Huxley.* 2 vols. London.

Oxenden, George Chichester. 1862. *Railway Horace.* London.

Oxford Latin dictionary: *The Oxford Latin dictionary.* 2 vols. Oxford: Clarendon Press. 1968–82.

Page, William, ed. 1908–32. *The Victoria county history of Kent.* 3 vols. London: Archibald Constable and Company.

Palgrave, Gwenllian F. 1899. *Francis Turner Palgrave: his journals and memories of his life.* London.

Pallas, Pyotr Simon. 1811–31. *Zoographia Rosso-Asiatica. Sistens omnium animalium in extenso imperio Rossico et adjacenibus maribus observatorum recensionem . . . anatomen atque icones plurimorum.* 3 vols. Petropoli.

Pancaldi, Giuliano. 1991. *Darwin in Italy: science across cultural frontiers.* Translated by Ruey Brodine Morelli. Bloomington and Indianapolis: Indiana University Press.

'Parallel roads of Glen Roy': Observations on the parallel roads of Glen Roy, and of other parts of Lochaber in Scotland, with an attempt to prove that they are of marine origin. By Charles Darwin. *Philosophical Transactions of the Royal Society of London* (1839), pt 1: 39–81. [*Collected papers* 1: 89–132.]

[Parker, Henry.] 1862. The *Edinburgh review* on the supernatural. *Saturday review of politics, literature, science, and art*, 15 November 1862, pp. 589–90.

Parris, Henry. 1965. *Government and the railways in nineteenth-century Britain.* London: Routledge & Kegan Paul. Toronto: University of Toronto Press.

Parrot, J. J. F. W. (Friedrich) von. 1834. *Reise zum Ararat.* 2 vols. in 1. Berlin.

——. 1845. *Journey to Ararat.* London.

Pascoe, Francis Polkinghorne. 1862. Note on Xenocerus semiluctuosus. *Transactions of the Entomological Society of London* 3d ser. 1 (1862–4), Proceedings, pp. 71–2.

Pasteur, Louis. 1861. Mémoire sur les corpuscules organisés qui existent dans l'atmosphère, examen de la doctrine des générations spontanées. *Annales des Sciences Naturelles (Zoologie)* 4th ser. 16: 5–98.

Peckham, Morse. 1959. *The origin of species by Charles Darwin. A variorum text.* Philadelphia: University of Pennsylvania Press.

Pengelly, William. 1862. The lignites and clays of Bovey Tracey, Devonshire. *Philosophical Transactions of the Royal Society of London* 152: 1019–38.

Physicians: *The roll of the Royal College of Physicians of London.* Vol. 3: 1801–25. By William Munk. 2d ed., revised and enlarged. London. 1878. Vol. 4: *Munk's roll, vol. 4: Lives of the fellows of the Royal College of Physicians of London 1826–1925.* Compiled by G. H. Brown. London: Royal College of Physicians. 1955.

Pictet de la Rive, François Jules. 1860. Sur l'Origine de l'espèce. *Bibliothèque Universelle et Revue Suisse. Archives des sciences physiques et naturelles* n.s. 7: 233–55.

Pine, L. G. 1972. *The new extinct peerage 1884–1971: containing extinct, abeyant, dormant & suspended peerages with genealogies and arms.* London: Heraldry Today.

Planchon, Jules Emile. 1847–8. Sur la famille des Linées. *London Journal of Botany* 6 (1847): 588–603; 7 (1848): 165–86, 473–501, 507–28.

Plarr, Victor Gustave. 1930. *Plarr's lives of the fellows of the Royal College of Surgeons of England.* Revised by Sir D'Arcy Power. 2 vols. London: Simpkin Marshall.

Porter, Duncan M. 1985. The *Beagle* collector and his collections. In *The Darwinian heritage*, edited by David Kohn. Princeton: Princeton University Press in association with Nova Pacifica.

Post Office directory of Birmingham: *Post Office directory of Birmingham, Warwickshire, and part of Staffordshire*. 1845–1928.

Post Office directory of Cambridge, Norfolk, and Suffolk: *Post Office directory of Cambridge, Norfolk, and Suffolk*. London. 1853–.

Post Office directory of the county of Durham: *The Post Office directory of the county of Durham, and the principal towns and adjacent places in Northumberland, Cumberland & Westmorland*. London. 1873–.

Post Office directory of Devonshire: *Post Office directory of Devonshire and Cornwall*. London. 1856–1926.

Post Office directory of Gloucestershire: *Post Office directory of Gloucestershire, with Bath, Bristol, Herefordshire, and Shropshire*. London. 1856–85.

Post Office directory of Hampshire, Dorsetshire, and Wiltshire: *Post Office directory of Hampshire, Dorsetshire, and Wiltshire*. London. 1848–1939.

Post Office directory of Lincolnshire: *Post Office directory of Lincolnshire*. London. 1849–76.

Post Office directory of Northamptonshire: *Post Office directory of Berkshire, Northamptonshire, Oxfordshire, with Bedfordshire, Buckinghamshire, and Huntingdonshire*. London. 1847–77.

Post Office directory of Northumberland: *Post Office directory of Westmoreland, Cumberland, Northumberland, and Durham*. London. 1858.

Post Office directory of the six home counties: *Post Office directory of the six home counties, viz., Essex, Herts, Kent, Middlesex, Surrey and Sussex*. London. 1845–.

Post Office directory of Somerset and Devon: *The Post Office directory of Somerset and Devon, with Bristol*. London. 1866.

Post Office Edinburgh and Leith directory: *Post Office Edinburgh and Leith directory*. Edinburgh. 1833–.

Post Office London directory: *Post Office London directory*. London. 1802–.

Post Office London suburban directory: *The Post Office London suburban directory*. London. 1860–1918.

Pouchet, Félix Archimède. 1858. Note sur des proto-organismes végétaux et animaux, nés spontanément dans l'air artificiel et dans le gaz oxygène. *Comptes rendus Hebdomadaires des Séances de l'Académie des Sciences* 47: 979–84.

——. 1859. *Hétérogénie ou traité de la génération spontanée*. Paris.

Prestwich, Grace Anne. 1899. *Life and letters of Sir Joseph Prestwich*. Edinburgh and London.

Prete, Frederick R. 1990. The conundrum of the honey bees: one impediment to the publication of Darwin's theory. *Journal of the History of Biology* 23: 271–90.

Price, John. 1863–4. *Old Price's remain's; præhumous, or during life*. 12 pts. London.

Pritchard, Ada, ed. 1897. *Charles Pritchard. Memoirs of his life*. London.

Pritchard, Michael. 1986. *A directory of London photographers 1841–1908*. Watford: ALLM Books.

Provincial medical directory: *The provincial medical directory*. London. 1847.

Quatrefages de Bréau, Jean Louis Armand de. 1861. *Unité de l'espèce humaine*. Paris.

——. 1862. *Métamorphoses de l'homme et des animaux*. Paris.

Rachootin, Stan P. 1985. Owen and Darwin reading a fossil: *Macrauchenia* in a boney light. In *The Darwinian heritage*, edited by David Kohn. Princeton: Princeton University Press in association with Nova Pacifica.

Ramsay, Andrew Crombie. 1862. On the glacial origin of certain lakes in Switzerland, the Black Forest, Great Britain, Sweden, North America, and elsewhere. [Read 5 March 1862.] *Quarterly Journal of the Geological Society of London* 18: 185–204.

Ramsay, Andrew Crombie *et al.* 1862. *A descriptive catalogue of the rock specimens in the Museum of Practical Geology, with explanatory notices of their nature and mode of occurrence in place.* 3d ed., revised, and partly re-written. London.

Raphael, Sandra. 1970. The publication dates of the *Transactions of the Linnean Society of London*, series 1, 1791–1875. *Biological Journal of the Linnean Society* 2: 61–76.

Record of the Royal Society of London: *The record of the Royal Society of London for the promotion of natural knowledge.* 4th ed. London: Royal Society. 1940.

Reeve, Henry, trans. 1862. *Democracy in America.* New ed. 2 vols. London.

Reeve, Lovell, ed. 1863–6. *Portraits of men of eminence in literature, science, and art, with biographical memoirs.* 4 vols. London.

Rehbock, Philip F. 1983. *The philosophical naturalists. Themes in early nineteenth-century British biology.* Madison, Wis. and London: University of Wisconsin Press.

Report of the Royal Commission on Herring: *Report of the Royal Commission on the Operation of the Acts Relating to Trawling for Herring on the Coasts of Scotland.* [House of Commons Parliamentary papers, Session 1863, 28: 139–74.]

Reports by the juries: *International Exhibition, 1862. Reports by the juries on the subjects in the thirty-six classes into which the exhibition was divided.* London. 1863.

Reuss, Georg Christian. 1862–70. *Die Pflanzenblätter in Naturdruck mit der botanischen Kunstsprache für die Blattform.* Stuttgart.

'Review of Bates on mimetic butterflies': Review of 'Contributions to an insect fauna of the Amazon Valley', by Henry Walter Bates. By Charles Darwin. *Natural History Review* n.s. 3 (1863): 219–24. [*Collected papers* 2: 87–92.]

Rheinberger, Hans-Jörg. 1983. Naudin, Darwin, Nägeli: Bemerkungen zu den Vererbungsvorstellungen des 19. Jahrhunderts. *Medizin Historisches Journal* 18: 198–212.

Richardson, Edmund William. 1916. *A veteran naturalist; being the life and work of W. B. Tegetmeier.* London: Witherby & Co.

Ritvo, Harriet. 1992. Race, breed, and myths of origin: Chillingham cattle as ancient Britons. *Representations* 39: 1–22.

Rivers, Thomas. 1837. *The rose amateur's guide; containing ample descriptions of all the fine leading varieties of roses, regularly classed in their respective families, their history and modes of culture.* London.

——. 1848. *The miniature fruit garden, or the culture of pyramidal fruit trees; with instructions for root pruning, &c.* 3d ed. London.

——. 1862. Seedling fruits. *Gardeners' Chronicle and Agricultural Gazette*, 20 December 1862, pp. 1195–6.

Rivière, Auguste. 1866. Observations sur la fécondation et la germination des

Orchidées, à l'occasion d'un Lœlia hybride présenté en fleurs, le 25 août 1865. *Journal de la Société Impériale et Centrale d'Horticulture* 12: 268–91.

Rogers, Mary Eliza. [1861.] *Domestic life in Palestine*. London.

Roll of the graduates of the University of Aberdeen: *Roll of the graduates of the University of Aberdeen 1860–1900*. By William Johnston. Aberdeen. 1906.

Roll of the graduates of the University of Glasgow: *A roll of the graduates of the University of Glasgow. From 31st December, 1727. To 31st December, 1897*. Compiled by W. Innes Addison. Glasgow. 1898.

Rolle, Friedrich. 1863. *Ch. Darwin's Lehre von der Entstehung der Arten im Pflanzen- und Thierreich in ihrer Anwendung auf die Schöpfungsgeschichte*. Frankfurt.

[Rorison, Gilbert.] 1861. *The three barriers: notes on Mr. Darwin's "Origin of species"*. Edinburgh and London.

Royal Society catalogue of scientific papers: *Royal Society catalogue of scientific papers (1800–1900). Compiled and published by the Royal Society of London*. 19 vols. and index (3 vols.). London and Cambridge. 1867–1925.

Royer, Clémence Auguste, trans. 1862. *De l'origine des espèces ou des lois du progrès chez les êtres organisées par Ch. Darwin*. Paris.

Rudwick, Martin. 1974. Darwin and Glen Roy: A "great failure" in scientific method? *Studies in History and Philosophy of Science* 5: 97–185.

Rupke, Nicolaas A. 1994. *Richard Owen: Victorian naturalist*. New Haven and London: Yale University Press.

Ruse, Michael. 1975. Darwin's debt to philosophy: an examination of the influence of the philosophical ideas of John F. W. Herschel and William Whewell on the development of Darwin's theory of evolution. *Studies in History and Philosophy of Science* 6: 159–81.

Salisbury, Richard Anthony. 1808. Observations on the different species of *Dahlia*, and the best method of cultivating them in Great Britain. [Read 5 April 1808.] *Transactions of the Horticultural Society of London* 1 (1812): 84–98.

Sanford, William F., Jr. 1965. Dana and Darwinism. *Journal of the History of Ideas* 26: 531–46.

Sarjeant, William A. S. 1980. *Geologists and the history of geology: an international bibliography from the origins to 1978*. 5 vols. and supplement (2 vols., 1987). London: Macmillan.

SBI: *Scandinavian biographical index*. Edited by Laureen Baillie. 4 vols. London: K. G. Saur. 1994.

SBL: *Svenskt biografiskt lexikon*. Edited by Bertil Boëthius, Erik Grill, Birgitta Lager-Kromnow, and Göran Nilzén. 27 vols. and 5 fascicles of vol. 28 (A–Pederby). Stockholm: Albert Bonnier and P. A. Norstedt. 1918–94.

Scherren, Henry. 1905. *The Zoological Society of London. A sketch of its foundation and development and the story of its farm, museum, gardens, menagerie and library*. London: Cassell and Co.

Schlegel, Hermann. 1862. The Sumatran elephant. *Natural History Review* n.s. 2: 72–8.

Schmidt, August and Voigt, Bernhard Friedrich. 1824–56. *Neue Nekrolog der Deutschen.* 30 vols.

Schomburgk, Robert Hermann. 1837. On the identity of three supposed genera of orchideous epiphytes. [Read 15 November 1836.] *Transactions of the Linnean Society of London* 17 (1837): 551–2.

——. 1857. Description of a remarkable spike or bunch of fruits of the fig banana (*Musa sapientum*), *var.* [Read 2 June 1857.] *Proceedings of the Linnean Society (Botany)* 2 (1858): 130–2.

Schrempp, Gregory. 1983. The re-education of Friedrich Max Müller: intellectual appropriation and epistemological antinomy in mid-Victorian evolutionary thought. *Man* n.s. 18: 90–110.

Schroeder van der Kolk, Jacob Ludwig Conrad and Vrolik, Willem. 1849. Ontleedkundige nasporingen over de hersenen van den chimpansé. *Verhandelingen der eerste klasse van het Koninklijk-Nederlandsche Instituut van wetenschappen, letterkunde en schoone kunsten te Amsterdam* 3d ser. 1 (1848–50): 263–78.

——. 1862. Note sur l'encéphale de l'orang-outang. *Natural History Review* n.s. 2: 111–17.

Sclater, Philip Lutley. 1860. On the black-shouldered peacock of Latham (*Pavo nigripennis*). *Proceedings of the Scientific Meetings of the Zoological Society of London* 28: 221–2.

Scott, John. 1862a. Remarks on the nature and peculiarities of the fern-spore. [Read 12 June 1862.] *Transactions of the Botanical Society [of Edinburgh]* 7 (1863): 352–70.

——. 1862b. On the propagation and irritability of *Drosera* and *Dionæa.* [Read 11 December 1862.] *Transactions of the Botanical Society [of Edinburgh]* 7 (1863): 429–30. [Abstract of paper.]

——. 1862c. On the homologies of the sporiferous appendices of Anemia, &c. [Read 13 February 1862.] *Transactions of the Botanical Society [of Edinburgh]* 7 (1863): 300–5.

——. 1862d. On *Marsilea Macropus* (the Nardoo Plant,) which had germinated in the Edinburgh Botanic Garden. [Read 13 March 1862.] *Transactions of the Botanical Society [of Edinburgh]* 7 (1863): 335–6. [Abstract of paper.]

——. 1864a. Remarks on the sexual changes in the inflorescence of *Zea Mays.* [Read 10 December 1864.] *Transactions of the Botanical Society [of Edinburgh]* 8 (1866): 55–62.

——. 1864b. Notes on the sterility and hybridization of certain species of *Passiflora*, *Disemma*, and *Tacsonia.* [Read 16 June 1864.] *Journal of the Linnean Society (Botany)* 8 (1865): 197–206.

——. 1864c. Observations on the functions and structure of the reproductive organs in the *Primulaceæ.* [Read 4 February 1864.] *Journal of the Linnean Society (Botany)* 8 (1865): 78–126.

——. 1864d. Remarks on the sexuality of the higher Cryptogams, with a notice of a hybrid Selaginella. [Read 10 March 1864.] *Transactions of the Botanical Society [of Edinburgh]* 8 (1866): 144–51.

——. 1868. On the reproductive functional relations of several species and varieties of Verbasca. *Journal of the Asiatic Society of Bengal* 36 (pt 2): 145–74.

Scriba, Heinrich Eduard. 1831–43. *Biographisch-literarisches Lexikon der Schriftsteller des Großherzogthums Hessen im 1. Viertel des 19. Jahrhunderts.* 2 vols. [Darmstadt.]

Scudder, Samuel H. 1861. On the genus Raphidophora, Serville; with descriptions of four species from the caves of Kentucky, and from the Pacific coast. *Proceedings of the Boston Society of Natural History* 8 (1861–2): 6–14.

Secord, James A. 1981. Nature's fancy: Charles Darwin and the breeding of pigeons. *Isis* 72: 163–86.

——. 1989. Behind the veil: Robert Chambers and *Vestiges*. In *History, humanity and evolution: essays for John C. Greene*, edited by James R. Moore. Cambridge: Cambridge University Press.

——, ed. 1995. *Vestiges of the natural history of creation and other evolutionary writings.* By Robert Chambers. Chicago and London: University of Chicago Press.

Seemann, Berthold Carl. 1852–7. *The botany of the voyage of H.M.S. Herald, under the command of Captain Henry Kellett, R.N., C.B., during the years 1845–51.* London.

——. 1853. *Narrative of the voyage of H.M.S. Herald during the years 1845–51, under the command of Captain Henry Kellett, R.N., C.B.; being a circumnavigation of the globe, and three cruizes to the Arctic regions in search of Sir John Franklin.* 2 vols. London.

——. 1862. *Viti: an account of a government mission to the Vitian or Fijian Islands in the year 1860–1.* Cambridge and London.

——. 1865–73. *Flora Vitiensis: a description of the plants of the Viti or Fiji islands with an account of their history, uses, and properties.* London.

Seltzer, Leon E., ed. 1952. *The Columbia Lippincott gazetteer of the world.* New York: Columbia University Press.

Sheets-Pyenson, Susan. 1992. *Index to the scientific correspondence of John William Dawson.* Stanford in the Vale, Oxfordshire: British Society for the History of Science.

Shotwell, Walter Gaston. 1923. *The civil war in America.* 2 vols. London: Longmans, Green and Co.

Smith, Benjamin E. 1894. *The cyclopædia of names: a pronouncing and etymological dictionary of names in geography, biography, mythology, history, ethnology, art, archæology, fiction, etc., etc., etc.* London.

Smith, Charles Card. 1903. Memoir of William Sumner Appleton, A.M. *Proceedings of the Massachusetts Historical Society* 2d ser. 17: 516–31.

Smith, James. 1839a. On the climate of the newer pliocene tertiary period. [Read 24 April 1839.] *Proceedings of the Geological Society of London* 3 (1842): 118–19.

——. 1839b. On the last changes in the relative levels of the land and sea in the British Islands. *Memoirs of the Wernerian Natural History Society* 8: 49–113.

Smith, James Edward and Sowerby, James. 1790–1814. *English botany; or, coloured figures of British plants, with their essential characters, synonyms, and places of growth.* 36 vols. London.

Smith, William. 1826. On *Dahlias*. [Read 5 December 1826.] *Transactions of the Horticultural Society of London* 7 (1830): 141–67.

Solier, Antoine Joseph Jean. 1848. Observations sur les genres *Procrustes*, *Procerus*,

Carabus et *Calosoma* formant la famille des Carabiens de M. Brullé. *Studi entomologici* 1: 49–62.

South America: *Geological observations on South America. Being the third part of the geology of the voyage of the Beagle, under the command of Capt. Fitzroy, R.N. during the years 1832 to 1836*. By Charles Darwin. London. 1846.

'Specific difference in *Primula*': On the specific difference between *Primula veris*, Brit Fl. (var. *officinalis* of Linn.), *P. vulgaris*, Brit. Fl. (var. *acaulis*, Linn.), and *P. elatior*, Jacq.; and on the hybrid nature of the common Oxlip. With supplementary remarks on naturally-produced hybrids in the genus *Verbascum*. By Charles Darwin. [Read 19 March 1868.] *Journal of the Linnean Society (Botany)* 10 (1869): 437–54.

Spence, James. 1861. *The American Union; its effect on national character and policy, with an inquiry into secession as a constitutional right, and the causes of the disruption*. London.

Spencer, Herbert. 1860–2. *First principles*. London.

Sprengel, Christian Konrad. 1793. *Das entdeckte Geheimniss der Natur im Bau und in der Befruchtung der Blumen*. Berlin.

Sprengel, Kurt Polycarp Joachim, ed. 1825–8. *Systema vegetabilium*. By Carl Linnaeus. 16th ed. 5 vols. Göttingen.

Stafleu, Frans A. 1971. *Linnaeus and the Linnaeans. The spreading of their ideas in systematic botany, 1735–1789*. Utrecht: A. Oosthoek's Uitgeversmaatschappij N.V. for the International Association for Plant Taxonomy.

Stearn, William T. 1956. Bentham and Hooker's *Genera Plantarum*: its history and dates of publication. *Journal of the Society for the Bibliography of Natural History* 3 (1953–60): 127–32.

Stenton, Michael. 1976. *Who's who of British members of Parliament, 1832–1885*. Brighton: Hanover Press.

Stephens, Henry Oxley. 1841. On *Epilobium angustifolium*, and species which have been confounded with it. *Annals and Magazine of Natural History* 8 (1842): 170–1. [Published in issue of November 1841.]

Steudel, Ernst Gottlieb. 1841. *Nomenclator botanicus seu: synonymia plantarum universalis, enumerans ordine alphabetico nomina atque synonyma tum generica tum specifica et a Linnaeo et a recentioribus de re botanica scriptoribus plantis phanerogamis imposita*. 2d ed. Stuttgart and Tübingen.

Stevenson, Burton, ed. 1949. *Stevenson's book of proverbs, maxims and familiar phrases*. London: Routledge and Kegan Paul.

Stewart, Joyce, ed. 1992. *Orchids at Kew*. London: Her Majesty's Stationery Office.

Stocking, George Ward, Jr. 1987. *Victorian anthropology*. New York: Free Press. London: Collier Macmillan.

Sutton, R. J. 1966. *The stamp collector's encyclopaedia*. 6th ed. Revised by K. W. Anthony. London: Stanley Paul.

Targioni Tozzetti, Antonio. 1853. *Cenni storici sulla introduzione di varie piante nell' agricoltura ed orticoltura toscana*. Florence.

Taxonomic literature: *Taxonomic literature. A selective guide to botanical publications and collections with dates, commentaries and types*. By Frans A. Stafleu and Richard S. Cowan.

2d ed. 7 vols. Utrecht: Bohn, Scheltema & Holkema. 1976–88.

Tegetmeier, William Bernhard. 1858. On the formation of the cells of bees. *Report of the 28th meeting of the British Association for the Advancement of Science held at Leeds*, Transactions of the sections, pp. 132–3.

——. 1862. Darwin on orchids. *Weldon's register of facts and occurrences relating to literature, the sciences, and the arts*, 1 August 1862, pp. 38–9.

Thompson, Robert. 1831. A report upon the varieties of apricot cultivated in the garden of the Horticultural Society. [Read 15 February 1831.] *Transactions of the Horticultural Society of London* 2d ser. 1 (1831–5): 56–75.

——. 1832a. A report upon the best varieties of the gooseberry, cultivated in the garden of the Horticultural Society, at Chiswick. *Transactions of the Horticultural Society of London* 2d ser. 1 (1831–5): 218–45.

——. 1832b. A report upon the principal varieties of the cherry cultivated in the garden of the Society. [Read 16 October 1832.] *Transactions of the Horticultural Society of London* 2d ser. 1 (1831–5): 248–94.

'Three forms of *Lythrum salicaria*': On the sexual relations of the three forms of *Lythrum salicaria.* By Charles Darwin. [Read 16 June 1864.] *Journal of the Linnean Society (Botany)* 8 (1865): 169–96. [*Collected papers* 2: 106–31.]

'Three sexual forms of *Catasetum tridentatum*': On the three remarkable sexual forms of *Catasetum tridentatum*, an orchid in the possession of the Linnean Society. By Charles Darwin. [Read 3 April 1862.] *Journal of the Proceedings of the Linnean Society of London (Botany)* 6: 151–7. [*Collected papers* 2: 63–70.]

Thurstan, Henry J., pseud. [Francis Turner Palgrave]. 1858. *The passionate pilgrim; or, Eros and Anteros.* London.

Thwaites, George Henry Kendrick. 1858–64. *Enumeratio plantarum Zeylaniæ: an enumeration of Ceylon plants, with descriptions of the new and little-known genera and species, observations on their habitats, uses, native names, etc.* London.

Tocqueville, Alexis Charles Henri Clerel de. 1836. *De la démocratie en Amérique.* 4th ed. 2 vols. Paris.

Torrey, John, ed. 1824. A monograph of the North American species of *Carex.* By Lewis D. de Schweinitz. *Annals of the Lyceum of Natural History of New-York* 1: 283–374.

——. 1836. Monograph of North American Cyperaceæ. *Annals of the Lyceum of Natural History of New-York* 3 (1828–36): 239–448.

——. 1843. *A flora of the state of New-York, comprising full descriptions of all the indigenous and naturalized plants hitherto discovered in the state; with remarks on their economical and medicinal properties.* Pt 2 of *Natural History of New York.* 2 vols. New York and Boston.

Torrey, John and Gray, Asa. 1838–43. *A flora of North America: containing abridged descriptions of all the known indigenous and naturalized plants growing north of Mexico; arranged according to the natural system.* 2 vols. New York and London.

Trécul, Auguste Adolphe Lucien. 1855. Organisation des glandes pédicellées des feuilles du *Drosera rotundifolia. Annales des Sciences Naturelles (Botanique)* 3: 303–11.

Treviranus, Ludolf Christian. 1861. Review of *Flora Tasmaniæ* . . . by Joseph Dalton Hooker. *Botanische Zeitung* 19: 133–5, 142–4.

——. 1863a. Ueber Dichogamie nach C. C. Sprengel und Ch. Darwin. *Botanische Zeitung* 21: 1–7, 9–16.

——. 1863b. Nachträgliche Bemerkungen über die Befruchtung einiger Orchideen. *Botanische Zeitung* 21: 241–3.

Trollope, Anthony. 1862. *Orley Farm*. 2 vols. London.

Tuckerman, Edward. 1843. *Enumeratio methodica Caricum quarundam. Species recensuit et secundum habitum pro viribus disponere tentavit*. Schenectadiæ.

Turner, Gerard L'E. 1989. *The great age of the microscope: the collection of the Royal Microscopical Society through 150 years*. Bristol and New York: Adam Hilger.

Turrill, W. B. 1959. *The Royal Botanic Gardens, Kew, past and present*. London: Herbert Jenkins.

'Two forms in species of *Linum*': On the existence of two forms, and on their reciprocal sexual relation, in several species of the genus *Linum*. By Charles Darwin. [Read 5 February 1863.] *Journal of the Proceedings of the Linnean Society (Botany)* 7 (1864): 69–83. [*Collected papers* 2: 93–105).]

Tyndall, John. 1862. On the conformation of the Alps. *The London, Edinburgh, and Dublin Philosophical Magazine and Journal of Science*. 4th ser. 24: 169–73.

Unger, Franz. 1860. *Die versunkene Insel Atlantis*. Vienna.

Variation: *The variation of animals and plants under domestication*. By Charles Darwin. 2 vols. London. 1868.

Vaucaire, Michel. 1930. *Paul du Chaillu: gorilla hunter. Being the extraordinary life and adventures of Paul du Chaillu*. New York and London: Harper & Brothers.

Vaucher, Jean Pierre Etienne. 1803. *Histoire des conferves d'eau douce, . . . suivie de l'histoire des trémelles et des ulves d'eau douce*. Geneva.

——. 1841. *Histoire physiologique des plantes d'Europe ou exposition des phénomènes qu'elles présentent dans les diverses périodes de leur développement*. 4 vols. Paris.

Veterinary directory: *The veterinary directory; or, annual register of the veterinary practitioners throughout the United Kingdom*. Edinburgh. 1861.

Veterinary register: *The register of veterinary surgeons*. London. 1884–.

Vilmorin, Jacques de. 1918. Notice biographique sur J. Gay. *Bulletin de la Société Botanique de France* 65: 78–82.

Volcanic islands: *Geological observations on the volcanic islands, visited during the voyage of H.M.S. Beagle, together with some brief notices on the geology of Australia and the Cape of Good Hope. Being the second part of the geology of the voyage of the Beagle, under the command of Capt. Fitzroy, R.N. during the years 1832 to 1836*. By Charles Darwin. London. 1844.

Vorzimmer, Peter James. 1963. Charles Darwin and blending inheritance. *Isis* 54: 371–90.

——. 1964. The development of Darwin's evolutionary thought after 1859. 2 vols. (PhD thesis: University of Cambridge).

Wahlenberg, Göran. 1802–3. Inledning til Caricographien. *Kongl. Vetenskaps Academiens nya Handlingar* 23 (1802): 225–47, 301–29; 24 (1803): 66–98, 138–70. [Partly translated as: A monograph of the genus *Carex*. *Annals of Botany* 2 (1806): 182–44.]

Walker, David M. 1980. *The Oxford companion to law.* Oxford: Clarendon Press.

Walker, Francis. 1839. *Monographia Chalciditum.* London.

Wallace, Alfred Russel. 1853. *A narrative of travels on the Amazon and Rio Negro.* London.

——. 1905. *My life: a record of events and opinions.* 2 vols. London: Chapman & Hall.

Wallich, G. C. 1870. *Eminent men of the day. Scientific series.* London.

Wallich, Nathaniel. 1830–1. *Plantæ Asiaticæ rariores; or, descriptions and figures of a select number of unpublished East Indian plants.* 2 vols. London.

Walsh, John Henry ('Stonehenge'). 1859. *The dog in health and disease.* London.

Walters, S. M. 1981. *The shaping of Cambridge botany. A short history of whole-plant botany in Cambridge from the time of Ray into the present century.* Cambridge: Cambridge University Press.

Warwick-Haller, Sally. 1990. *William O'Brien and the Irish Land War.* Dublin: Irish Academic Press.

Watkins, A. H. 1977. William Baxter: Bromley antiquary. *Bromley Local History* 2: 30–5.

Watt, Robert. 1824. *Bibliotheca Britannica; or a general index to British and foreign literature.* 4 vols. Edinburgh and London.

Weddell, Hugh Algernon. 1849. *Histoire naturelle des quinquinas ou monographie du genre Cinchona suivie d'une déscription du genre Cascarilla et de quelques autres plantes de la même tribu.* Paris.

——. 1852. Description d'un cas remarquable d'hybridité entre des orchidées de genres différents. *Annales des Sciences Naturelles (Botanique)* 3d ser. 18: 5–10.

Wedgwood, Barbara and Wedgwood, Hensleigh. 1980. *The Wedgwood circle, 1730–1897: four generations of a family and their friends.* London: Studio Vista.

Wedgwood, Hensleigh. 1866. *On the origin of language.* London.

[Wedgwood, Hensleigh and Wedgwood, Frances Julia.] 1862. The origin of language. The imitative theory and Mr. Max Müller's theory of phonetic types. *Macmillan's Magazine* 7 (1863): 54–60. [Published in issue of November 1862.]

Weissenborn, W. 1837. On the alleged transformation of Avèna sativa into Secàle cereàle. *Magazine of Natural History* n.s. 1: 574–7.

Wellesley index: *The Wellesley index to Victorian periodicals 1824–1900.* Edited by Walter E. Houghton and others. 5 vols. Toronto: University of Toronto Press. London: Routledge & Kegan Paul. 1966–89.

Westwood, John Obadiah. 1860. Mr. Darwin's theory of development. *Annals and Magazine of Natural History* 3d ser. 5: 347–8. [Reprinted from *Gardeners' Chronicle*, 11 February 1860, p. 120.]

Whately, Richard. 1831. *Introductory lectures on political economy, being part of a course delivered in Easter term, 1831.* London.

——. 1854. *On the origin of civilisation. A lecture by his grace the archbishop of Dublin to the Young Men's Christian Association.* London.

Whewell, William. 1854. *Of the plurality of worlds: an essay. Also a dialogue on the same subject.* 2d ed. London.

White, Gilbert. 1789. *The natural history and antiquities of Selborne, in the county of Southampton.* 2 vols. London.

——. 1825. *The natural history of Selborne.* New ed. 2 vols. London.

——. 1843. *The natural history of Selborne.* New ed., with notes by Leonard Jenyns. London.

White, Henry and Newton, Thomas W. 1878. *A catalogue of the library of the Museum of Practical Geology and Geological Survey.* London.

Whitworth, Robert Percy, ed. 1866. *Bailliere's New South Wales gazetteer and road guide, containing the most recent and accurate information as to every place in the colony.* Sydney, Melbourne, and London.

[Wilberforce, Samuel.] 1860. Review of *Origin of species*, by Charles Darwin. *Quarterly Review* 108: 225–64.

Willdenow, Karl Ludwig, ed. 1797–1824. *Species plantarum exhibentes plantas rite cognitas ad genera relatas cum differentiis specificis, nominibus trivialibus, synonymis selectis, locis natalibus, secundum systema sexuale digestas.* By Carl Linnaeus. 4th ed. 6 vols. Berlin.

——. 1805. *Caricologia, sive descriptiones omnium specierum Carici, in usum excursionum botanicarum pro amicis seorsim impressa* . . . Berlin.

Williamson, George C., ed. 1903–5. *Bryan's dictionary of painters and engravers.* New ed. 5 vols. London: George Bell & sons.

Willis, J. C. 1973. *A dictionary of the flowering plants and ferns.* 8th ed. Revised by H. K. Airy Shaw. Cambridge: Cambridge University Press.

Wilson, David Alec. 1929. *Carlyle to threescore-and-ten (1853–1865).* London: Kegan Paul, Trench, Trubner & Co. New York: E. P. Dutton & Co.

Wilson, Leonard G., ed. 1970. *Sir Charles Lyell's scientific journals on the species question.* New Haven and London: Yale University Press.

Winkler, Tiberius Cornelis, trans. 1860. *Het ontstaan der soorten door middel van de natuurkeus, of het bewaard blijven van bevoorregte rassen in den strijd des levens.* [*Origin*], by Charles Darwin. Translated from the English. Haarlem.

Wollaston, Alexander Frederick Richmond. 1921. *Life of Alfred Newton, professor of comparative anatomy, Cambridge University, 1866–1907. With a preface by Sir Archibald Geikie.* London: John Murray.

Wollaston, George Buchanan. 1855. Various notes on British Orchideæ. *Phytologist* n.s. 1 (1855–6): 225–7.

Wood, Christopher. 1978. *The dictionary of Victorian painters.* 2d ed. Woodbridge, Suffolk: Antique Collectors' Club.

Wood, J. C. 1973. *A dictionary of British animal painters.* Leigh-on-Sea: F. Lewis.

Wood, Searles Valentine. 1848–61. *A monograph of the Crag Mollusca, with descriptions of shells from the upper Tertiaries of the British Isles.* 2 vols. London.

——. 1861–77. *A monograph of the Eocene Mollusca, or, descriptions of shells from the older Tertiaries of England. Bivalves.* 3 pts. and supplement. London.

Woodward, Horace B. 1907. *The history of the Geological Society of London.* London: Geological Society.

Wooler, William A. 1860. The varieties of Polyanthus best for early bedding. *Cottage Gardener, Country Gentleman's Companion, and Poultry Chronicle*, 29 May 1860, p. 133.

Wray, Leonard. 1861a. On the scientific culture of the strawberry. *Technologist. A monthly record of science applied to art and manufacture* 1: 44–50.

——. 1861b. Scientific culture of the strawberry. *Gardeners' Chronicle and Agricultural Gazette*, 3 August 1861, pp. 716–17. [Reprint of Wray 1861a.]

Wright, Thomas. 1872. *Uriconium. A historical account of the ancient Roman city, and of the excavations made upon its site at Wroxeter, in Shropshire, forming a sketch of the condition and history of the Welsh border during the Roman period.* London and Shrewsbury.

WWW: *Who was who. A companion to* Who's who. 8 vols. and cumulated index, 1897–1990. London: Adam & Charles Black. 1920–91.

Zoology: *The zoology of the voyage of H.M.S. Beagle, under the command of Captain FitzRoy, R.N., during the years 1832 to 1836. Published with the approval of the Lords Commissioners of Her Majesty's Treasury.* Edited and superintended by Charles Darwin. 5 pts. London. 1838–43.

NOTES ON MANUSCRIPT SOURCES

The majority of the manuscript sources cited in the footnotes to the letters are located either in the Darwin Archive, Cambridge University Library, or at Down House Museum, Kent. Further details about the Darwin Archive are available from the *Handlist of Darwin papers at the University Library Cambridge* (Cambridge: Cambridge University Press, 1960) and the unpublished supplementary handlist available at the library; a new catalogue of the papers is currently being prepared. Further details about the manuscripts at Down House are available from Philip Titheradge, ed. *The Charles Darwin Memorial at Down House, Downe, Kent*, revised ed. ([Downe: Down House Museum], 1981) and from the curator (The Curator, Down House Museum, Downe, Kent, BR6 7JT). In addition, there are a number of named sources that are commonly used in the footnotes: for each of these, the editors have provided brief descriptive notes.

CD's Account books (Down House MS). This series of seventeen account books begins on 12 February 1839, a fortnight after CD and Emma's marriage, and ends with CD's death. The books contain two sets of accounts. From the start, CD recorded his *cash account* according to a system of double-entry book-keeping. On each left-hand page he recorded credits (i.e., withdrawals from the bank, either in the form of cash paid to himself or cheques drawn for others), and on each right-hand page, he recorded debits (i.e., cash or cheques paid to others). CD also recorded details of his *banking account* from the start, but only noted them down in a single column at the bottom of the left-hand page of his cash account. In August 1848, however, he began a system of detailing his banking account according to double-entry book-keeping, in a separate chronological section at the back of each account book. On the left, he recorded credits to the account in the form of income (i.e., investments, rent, book sales, etc.). On the right, he recorded debits to the account (i.e., cash or cheque withdrawals).

CD's Classed account books (Down House MS). This series of four account books, covering the years 1839–81, runs parallel to CD's Account books. For each year, September – August (after 1867, January – December), CD divided his expenditure into different classified headings; in addition, he made a tally for the year of his income, expenditure, cash in hand, and money in the bank. From 1843, CD also compiled at the back of each book a separate account of the total expenditure under the various headings in each year, and from 1844 he added a full account of his income in each year, and of capital invested and 'paid' up. The classification of expenditure varied over time, but in 1862 the classes were: 'Miscellaneous', 'Personal', 'Governess', 'House', 'William & Boys',

'Coals & Firewood', 'Beer', 'Wine', 'Man-Servants', 'Repairs. Furniture', 'Medical attendance', 'Stationery, Postage newspapers', 'Books', 'Science', 'Garden', 'Cows, Pigs, Field', 'Stable', 'Travelling Expenses', 'Taxes. Rates', 'Charities', and 'Gifts. Annual Subscriptions'.

CD's Experimental notebook (DAR 157a). This notebook contains notes on some, though by no means all of the experiments carried out between 13 November 1855 (with some back references) and 20 May 1868, although the majority of the notes date from before 1863. In many cases only the details of the experiment attempted are given, often with cross-references to results recorded in CD's portfolios of notes. The notebook also contains a number of letters to CD.

CD's Investment book (Down House MS). This book records under each of CD's investments the income received during the period 1846–81.

CD's Library catalogue (DAR 240). This manuscript catalogue of CD's scientific library was compiled by Thomas W. Newton, assistant librarian of the Museum of Practical Geology, London, in August 1875. Additions to the catalogue were subsequently made by Francis Darwin (who inherited most of his father's scientific library) and by H. W. Rutherford, who apparently used this catalogue as a basis for compiling his *Catalogue of the library of Charles Darwin now in the Botany School, Cambridge* (Cambridge: Cambridge University Press, 1908). However, there are items listed in this manuscript catalogue that do not appear in Rutherford's published catalogue, and which must have been dispersed subsequently to being listed.

Darwin's copy of the *Gardeners' Chronicle*: account of numbers and index to marks & memoranda (DAR 222). This is a bound manuscript index to CD's collection of the *Gardeners' Chronicle and Agricultural Gazette*, 1841–71 (which is at the Cory Library, Cambridge Botanic Garden). It comprises a list showing which numbers of the journal are missing from CD's copy, together with a 'List of the numbers of special interest to Darwin and kept by him in separate parcels'. The latter details the page-numbers and subjects of the articles that were of particular interest to (and annotated by) CD in the numbers that he kept separately. The handwriting of the list is unknown, but probably dates from the late nineteenth century.

Down Coal and Clothing Club account book (Down House MS). CD was for some years treasurer of this charitable organisation. The account book records subscriptions made by honorary subscribers between 1841–76; between 1848 and 1869 the entries are in CD's handwriting. For the years 1841–8 and 1868–76, there is also a statement of expenditures, though not in CD's handwriting.

Emma Darwin's diary (DAR 242). This comprises a series of small pocket diaries, in which Emma recorded details of the health of family members, trips made by herself, CD, and the children, school holidays, and visits to Down by others. The collection at CUL comprises diaries for the years 1824, 1833–4, 1839–45, and 1848–96.

H. E. Litchfield's autobiography (DAR 246). This unfinished autobiography, written in 1926 on forty-two loose leaves, and chiefly concerning Henrietta Emma Darwin's childhood, has never been published.

List of reviews (DAR 226 (DH/MS* 8: 6–18)). This manuscript, headed 'List Reviews of Origin of Sp & of C. Darwins Books', was CD's working index to his collection of the reviews of his own books. It corresponds approximately to the review collection in the Darwin Pamphlet Collection–CUL, but includes some items that were dispersed subsequently to being listed.

Scrapbook of reviews (DAR 226.1 and 226.2). Many of the reviews contained in these volumes bear CD's annotations and thus were evidently collected by CD. However, the scrapbook seems to have been assembled by Francis Darwin: the tables of contents are in the handwriting of H. W. Rutherford, an assistant at Cambridge University Library who acted as a copyist for Francis on several occasions (see *ML*, 1: x, and Francis Darwin, ed. *The foundations of the Origin of Species. Two essays written in 1842 and 1844 by Charles Darwin* (Cambridge: Cambridge University Press, 1909)). In addition, the scrapbook is identified as Francis's in a note (DAR 226.1:132a) made in 1935 by Arthur Keith, whose appeal led to the purchase of Down House as a Darwin memorial (see Arthur Keith, *An autobiography* (London: Watts & Co., 1950)). DAR 226.1 bears the inscription 'Reviews of C. Darwin's works' on the spine, and contains, among others, reviews of *Origin* and *Orchids*; DAR 226.2 is inscribed: 'Reviews. Descent. Expression. Insect. Pl. Eras. D.'

W. E. Darwin's botanical notebook (DAR 117). This notebook contains observational and experimental notes on plants made by William, often in consultation with CD. The first observation bears the date 13 July 1862, and, although the date of the last observation is 26 June 1870, most of the notes were made between 1862 and 1864. The notebook originally contained letters from CD, but these were subsequently removed. William entered notes made from botanical textbooks in a separate notebook (DAR 234).

W. E. Darwin's botanical sketchbook (DAR 186: 43). This sketchbook, which contains entries dated 1862–72, was evidently begun in parallel to William's botanical notebook. It contains ink drawings of various parts of plants, and of sections, together with descriptions, which in some cases are very extensive.

BIOGRAPHICAL REGISTER

This list includes all correspondents and all persons mentioned in the letters and notes whom the editors have been able to identify. Dates of letters to and from correspondents are given in chronological order. Letters to the correspondents are listed in roman type; letters from correspondents in italic type; third-party letters are listed in italic type with the name of the recipient given in square brackets. Following the register, a list of biographical sources referred to in the entries is given. These works are also listed in the main bibliography.

Adams, Charles Francis (1807–86). American diplomat. Elected to the Massachusetts legislature as a Whig, 1840; led the state in opposition to slavery. Elected to Congress in 1858 and again in 1860 as a Republican. American minister to London, 1861–8. Published the papers and a biography of his father, the former president, John Quincey Adams (*DAB*). (*DAB*.)

Adams, John Couch (1819–92). Astronomer and mathematician. Co-discoverer of the planet Neptune. Fellow and tutor, St John's College, Cambridge, 1843–52; Pembroke College from 1853. Lowndean professor of astronomy and geometry, Cambridge University, 1858; director of the Cambridge Observatory, 1861. FRS 1849. (*DNB*, *DSB*.)

Addington, Henry, 1st Viscount Sidmouth (1757–1844). Statesman. MP, Devizes, 1783. Speaker of the House of Commons, 1783–1801; chancellor of the Exchequer, 1801; president of the council, 1805; home secretary, 1812. (*DNB*.)

Agassiz, Jean Louis Rodolphe (Louis) (1807–73). Swiss-born zoologist. Professor of natural history, Neuchâtel, 1832–46. Emigrated to the United States in 1846. Professor of natural history, Harvard University, 1847–73. Established the Museum of Comparative Zoology at Harvard in 1859. Foreign member, Royal Society, 1838. (*DAB*, *DSB*.)

Ainslie, Robert (1802/3–76). Independent unitarian minister and religious writer. Resident at Pond House (later renamed Tromer Lodge), Down, 1845–58. Minister of New Court Chapel, Carey Street, London. Secretary of the London City Mission and of the Congregational Board of Education. Minister of Christ Church, Brighton, 1860–74. (Ainslie 1865; *Calendar* no. 12842, letter from O. A. Ainslie, 23 November 1880; *Correspondence* vol. 3, letter to Susan Darwin, 3[–4] September 1845; *Correspondence* vol. 7, letter to W. E. Darwin 14 [May 1858]; G. E. Evans 1897; *The Times*, 23 August 1876, p. 16.)

Albert Francis Charles Augustus Emmanuel, prince-consort of England (1819–61). Married Queen Victoria in 1840. A patron of the arts and sciences.

Chancellor of Cambridge University, 1847. Instrumental in the inception and organisation of the Great Exhibition, 1849–51. President of the British Association for the Advancement of Science, 1859. FRS 1840. (*DNB.*)

Albert, W. (*fl.* 1846). German civil servant and author. Chief administrative officer of a rural district. Author of a work on potato cultivation. (*Gardeners' Chronicle and Agricultural Gazette*, 21 March 1846, pp. 181–3.)

Alberts, Maurice. Prussian diplomat. Secretary and chef de chancellerie in the Prussian embassy in London, 1859 (or before) – 1863; director of the Chancery, 1863–7. (*Foreign office list*, 1858–68.)
10 February 1862, *13 February 1862*, [after 13 February 1862]

Allen, Catherine (1765–1830). Married James Mackintosh in 1798. (*Emma Darwin.*)

Anderson, James (1824–93). Naval officer. Entered merchant navy, 1840. Entered service with the Cunard line in 1851; commanded fourteen different steamers belonging to that fleet. Commanded the *Great Eastern* for the Atlantic Telegraph Company expeditions, 1865–6. Managing director of the Electric Telegraph Company, 1866–93. Fellow of the Geological Society, 1870. Knighted, 1866. (*Modern English biography.*)
23 December [1862]

Anderson, Thomas (1819–74). Chemist to the Highland and Agricultural Society of Scotland, 1848. Professor of chemistry, University of Glasgow, 1852. (*DNB.*)

Anderson-Henry, Isaac (1800–84). Scottish lawyer and horticulturist. Practised as a solicitor in Edinburgh from 1834 or earlier; member of the Incorporated Society of Solicitors in Supreme Courts. Retired from practice and took the additional name of Henry on his marriage to the heiress of the entailed estates of Woodend, Perthshire, in 1861. Owned a garden in Edinburgh known for the rarity of some of its contents. Interested in plant hybridisation and acclimatisation; became one of the foremost hybridisers of his day. Joined the Caledonian Horticultural Society, and the Linnean Society of London; president of the Botanical Society of Edinburgh, 1867–8. (*County families* 1878; *Gardeners' Chronicle and Agricultural Gazette*, 22 March 1873, p. 399; *Post Office Edinburgh and Leith directory* 1834–64; *Proceedings of the Linnean Society of London* (1886–7), pp. 42–4.)

Andrew, John Albion (1818–67). American statesman. Admitted to the bar, 1840. Elected to the Massachusetts legislature as a Republican, 1857. Governor of Massachusetts, 1860–6. An ardent anti-slavery campaigner. (*DAB.*)

Appleton, Daniel (1785–1849). New York publisher. In partnership with his son, William Henry Appleton, founded the New York publishing firm D. Appleton & Co. in 1838. (*DAB.*)

Appleton, Thomas Gold (1812–84). Boston essayist, poet, and artist. A staunch supporter of his friend John Albion Andrew. Actively interested in the growth and improvement of Boston; a trustee of the Athenaeum, the Public Library, and the Museum of Fine Arts. (*DAB.*)
24 April [1862], *28 June [1862]*

Appleton, William Henry (1814–99). New York publisher. Partner in the firm D. Appleton & Co., 1838–48. Became head of the firm in 1848 on the retirement of his father, Daniel Appleton. D. Appleton & Co. was the medium through which the works of British scientists such as CD, Thomas Henry Huxley, and Herbert Spencer reached the American public. (*DAB*.)

Appleton, William Sumner (1840–1903). American antiquary and numismatist. Half-brother of Thomas Gold Appleton. Founding member of the Boston Numismatic Society. Appointed to the commission to attend the assay of the United States Mint, 1867. Commissioner of Boston records, 1875–92. Established the arms and seal of Harvard University, 1885. (C. C. Smith 1903.)

Argyll, duchess of. *See* Campbell, Elizabeth Georgiana.

Argyll, duke of. *See* Campbell, George Douglas.

Arnold, Arthur (1833–1902). Historian and politician. Appointed assistant commissioner and subsequently government inspector of public works, 1863. Published *History of the cotton famine* in 1864. Edited the *Echo*, an evening newspaper, 1868–75. MP for Salford, 1880–5. Alderman, London County Council, 1889–1902; chairman, 1895–6. Knighted, 1895. (*DNB*.)

Arrowsmith, John (1790–1873). Geographer and map-maker. For many years aided his uncle, Aaron Arrowsmith (*DNB*) in the construction of his large collection of maps and charts. Established his own business in 1823. Founding fellow of Royal Geographical Society, 1830; for many years a member of the society's council. Many of his smaller maps appeared in the journals of the Royal Geographical Society. (*DNB*.)

Ashburton, Lord. *See* Baring, Alexander Hugh; Baring, Francis; and Baring, William Bingham.

Atherley, Francis Henry (1831–97). Army officer. Captain, Rifle Brigade, 1855–8. Elected fellow of Zoological Society, 1861. (*List of fellows of the Zoological Society*, *Modern English biography*.)

Atherley, George (1819–83). Banker. William Erasmus Darwin's partner in the Southampton and Hampshire Bank, Southampton. (*Alum. Cantab.*, *Banking almanac* 1861.)

Atherley, Maud (d. 1862). Daughter of George Atherley. (Letter from Emma Darwin to William Erasmus Darwin, [September 1862] (DAR 219); letter from W. E. Darwin, 21 October [1862]; letter to W. E. Darwin, 30 [October 1862].)

Augé de Lassus, M. (d. 1863). French botanist. Receiver of finances at Poligny in the department of Jura. Member of the Société botanique de France, 1858. Made several botanical discoveries in the department of Var. (*Bulletin de la Société Botanique de France* 5 (1858): i; 10 (1863): 216.)

Austen, John Thomas (1794–1876). Clergyman. Rector of West Wickham, Kent, from 1848. (*Modern English biography*.)

Babington, Anna Maria. Daughter of John Walker of the Madras civil service. Married Charles Cardale Babington in 1866. (A. M. Babington ed. 1897.)

Babington, Charles Cardale (1808–95). Botanist and archaeologist. Greatly involved in the natural history activities at Cambridge for more than forty years. A founding member of the Entomological Society in Cambridge and the Cambridge Antiquarian Society. Professor of botany, Cambridge University, 1861–95. An expert on plant taxonomy. FRS 1851. (*DNB*, *DSB*.)
17 January 1862, 20 January [1862], *30 January 1862*, 1 February [1862], *22 May 1862*, 2 September [1862], *16 September 1862*, *16 October 1862*

Baer, Karl Ernst von (1792–1876). Estonian naturalist. Professor of anatomy at Königsberg University, 1817; of zoology, 1826–34. Professor of zoology at the Academy of Sciences, St Petersburg, 1834–67. Carried out embryological researches and, in 1826, discovered the mammalian egg. Propounded the influential view that development proceeds from the general to the specific. Foreign member, Royal Society, 1854. (*DSB*, *NDB*.)

Baikie, William Balfour (1825–64). Naval surgeon, naturalist, and philologist. Assistant surgeon, Royal Navy, 1848. Assistant surgeon, Haslar Hospital, 1851–4. Surgeon and naturalist on the Niger expedition, 1854; succeeded to the command of the vessel during the voyage. Undertook a second Niger expedition in 1857. Explored West Africa until his death in Sierra Leone. (R. Desmond 1994, *DNB*.)

Baker, Samuel C. and Charles N. Dealers in birds and live animals, with premises at 3 Halfmoon Passage, Gracechurch Street, and 15a Beaufort Street, Chelsea, London, and at the Rue de la Faisanderie, avenue de l'Impératrice, Paris. (*Post Office London directory* 1861.)

Balfour, John Hutton (1808–84). Physician and botanist. Professor of botany, Glasgow University, 1841–5. Professor of botany and Regius keeper of the Royal Botanic Garden, Edinburgh, 1845–79. FRS 1856. (*DNB*, *DSB*.)
14 January 1862

Baly, Joseph Sugar (1817–90). Physician and entomologist. Surgeon at Kentish Town and Camden Town dispensary, London; then at Warwick, 1868–90. Medical health officer, Leamington Priors, 1873–90. A leading authority on phytophagous Coleoptera. (*Modern English biography*.)

Banks, Joseph (1743–1820). Botanist. Travelled around the world on HMS *Endeavour*, 1768–71. President of the Royal Society, 1778–1820. Created baronet, 1781; privy councillor, 1791. FRS 1766. (R. Desmond 1994, *DNB*.)

Baring, Alexander Hugh, 4th Baron Ashburton (1835–89). Statesman. MP for Thetford, 1857–67. Succeeded to the peerage in 1868. (*Modern English biography*.)

Baring, Francis, 3d Baron Ashburton (1800–68). Statesman. MP for Thetford, 1832–41 and 1848–57. Succeeded to the peerage in 1864. (*Modern English biography*.)

Baring, William Bingham, 2d Baron Ashburton (1799–1864). Statesman. Conservative MP for North Staffordshire, 1837–41; for Thetford, 1841–8. Succeeded to the peerage in 1848. FRS 1854. (*DNB*.)

Bartholomew, John (1831–93). Scottish cartographer. Assistant to August Heinrich Petermann (*Modern English biography*), 1853–6. Succeeded his father as cartographer at Chambers Street, Edinburgh, 1856. (*Modern English biography*.)

Bartlett, Abraham Dee (1812–97). Zoologist. Taxidermist in London, 1834–52. Superintendent of the natural history department of the Crystal Palace, 1852–9; of the Zoological Society's gardens, Regent's Park, 1859–97. (*Modern English biography.*)

Bateman, James (1812–97). Landed gentleman and horticulturist. Made his fortune from engineering, iron, and banking. Succeeded to his father's estates in Staffordshire and Westmoreland in 1857. Cultivated tropical plants, especially orchids. Employed collectors to send him orchids from Central America, on the basis of which he published *The Orchidaceæ of Mexico and Guatemale* (1843); pioneered techniques of cool orchid cultivation. Created a remarkable garden at Biddulph Grange, Staffordshire. President, North Staffordshire Field Club, 1865–70. FRS 1838. (*DNB*, R. Desmond 1994.)

[1 February 1862]

Bateman, Robert (1842–1922). Landed gentleman. Third son of James Bateman. (*Burke's landed gentry* 1937.)

[28 January 1862]

Bates, Frederick (1829–1903). Brewer and naturalist. Younger brother of Henry Walter Bates. Managing director of a brewing company in Leicester. A keen entomologist, particularly interested in the Heteromera, publishing many papers and forming an extensive collection. Also formed collections of Cicindelidae and British Coleoptera. Cultivated orchids for some years. (*Entomologist's Monthly Magazine* 39 (1903): 286–7; Gilbert 1977.)

Bates, Henry (d. 1870). Father of Frederick and Henry Walter Bates. Hosiery manufacturer in Leicester. (*DNB* s.v. Bates, Henry Walter.)

Bates, Henry Walter (1825–92). Naturalist. Undertook a joint expedition to the Amazon with Alfred Russel Wallace, 1848–50; continued to explore the region alone until 1859. Provided the first comprehensive scientific explanation for the phenomenon subsequently known as Batesian mimicry. Published *The naturalist on the River Amazons* in 1863. Assistant secretary, Royal Geographical Society, 1864–92. FRS 1881. (*DNB*, *DSB*.)

6 January 1862, *11 January 1862*, 13 January [1862], *25 January 1862*,
31 January [1862], 27 [February 1862], 27 February [1862], 16 April [1862],
[17 April 1862], *30 April 1862*, 4 May [1862], 9 May [1862], *19 May 1862*,
11 June [1862], *14 June 1862*, 15 October [1862], *17 October 1862*,
18 October [1862], 20 November [1862], *24 November 1862*,
25 November [1862], 15 December [1862]

Baxter, William Walmisley (b. 1829). Chemist and druggist. Succeeded his father in business at 40 High Street, Bromley, Kent, 1857 – *c.* 1899. (*Post Office directory of the six home counties* 1857–1903, Watkins 1977.)

26 January [1862]

Bayly, Nicholas Paget (1814–79). Australian farmer and sheep breeder. Owned 14,000 acres of property in Mudgee; involved in a controversy in the 1860s over the comparative merits of Mudgee and Victorian wools. (*Aust. dict. biog.*)

Beaton, Donald (1802–63). Scottish gardener interested in hybridisation. A regular contributor to the *Cottage Gardener*, *Gardener's Magazine*, and *Journal of Horticulture*. An expert in bedding schemes, Beaton was gardener to William Fowle Middleton of Shrubland Hall, Suffolk, *c.* 1840–52. (*Cottage Gardener*, 28 November 1854, pp. 153–8; R. Desmond 1994.)

Beck, Barbara Van (b. 1629). Born Barbara Ursler. Woman with 'a completely hairy face'; she was exhibited in London in 1663. (O'Donoghue 1908–25, *Variation* 2: 4.)

Beck, Richard (1827–66). Manufacturing optician. Nephew of Joseph Jackson Lister (*DNB*), whose design for achromatic objective lenses transformed microscope technology. Apprenticed to James Smith (Turner 1989), a mathematical instrument-maker in London. Entered into partnership as Smith & Beck, 1847 – *c.* 1857; as Smith, Beck & Beck, *c.* 1857–65; as R. & J. Beck, 1865–6. Elected member of the Microscopical Society, 1855. (*Modern English biography*; Turner 1989, p. 171.)

Beddoe, John (1826–1911). Physician and anthropologist. Began medical practice in Bristol in 1857. Physician, Bristol Royal Infirmary, 1862–73; consulting physician, Bristol Children's Hospital, 1866–1911. Began active researches on ethnology during an extended European tour prior to setting up practice. An authority on the physical characteristics of living European races. Founder member of the Ethnological Society. President of the Anthropological Society, 1869–70. FRS 1873. (*DNB*.)

Bell, Charles (1774–1842). Anatomist who investigated the nervous system and the expression of emotions in humans. Co-owner of and principal lecturer at the Great Windmill Street School of Anatomy, 1812–25. Surgeon at the Middlesex Hospital, 1812. Professor of surgery, Edinburgh University, 1836. Knighted, 1831. FRS 1826. (*DNB*, *DSB*.)

Bell, Marion (1787–1876). Wife of Charles Bell. Published *Letters of Sir Charles Bell* in 1870. (*Modern English biography*.)

Bell, Robert (1800–67). Popular Irish writer and journalist. Edited the colonial newspaper *Home News*. (*DNB*.)

Bennet, Charles Augustus, 5th earl of Tankerville (1776–1859). Father of Charles Augustus Bennet, sixth earl of Tankerville. (*Burke's peerage* 1970.)

Bennet, Charles Augustus, 6th earl of Tankerville (1810–99). Styled Lord Ossulston, 1822–59. Tory MP for North Northumberland, 1832–59. Succeeded as sixth earl in 1859. Lord Steward of the Household to Queen Victoria, 1867–8. (*Burke's peerage* 1980, *Modern English biography*, Stenton 1976.)
[9 February 1862]

Bentham, George (1800–84). Botanist. Pursued his botanical studies at the Royal Botanic Gardens, Kew. President of the Linnean Society, 1861–74. Published *Genera plantarum* (1862–83) with Joseph Dalton Hooker. FRS 1862. (*DNB*, *DSB*.)
3 February [1862], 30 March [1862], *15 May 1862*, 13 October [1862], *15 October 1862*

Bentham, Jeremy (1748–1832). Utilitarian philosopher and writer on jurisprudence. (*DNB*.)

Bentham, Mary Louisa. *See* Chesnel, Mary Louisa de.

Berkeley, Miles Joseph (1803–89). Clergyman and botanist. Perpetual curate of Apethorpe and Wood Newton, Northamptonshire, 1833–68. An expert on British fungi. FRS 1879. (*DNB*, *DSB*.)

Bienen Zeitung

18 June 1862

Bigelow, George Tyler (1810–78). Jurist and state senator in Boston, Massachusetts. Chief justice, Boston. (*IBN*, *NUC*.)

Binney, Edward William (1812–81). Solicitor. Palaeobotanist. A founder of the Manchester Geological Society, 1838; president, 1857–9, 1865–7. FRS 1856. (R. Desmond 1994, *DNB*.)

Black, Allan A. (1832–65). Botanist. First curator of the herbarium at the Royal Botanic Gardens, Kew, 1853–64. Superintendent, Botanic Garden, Bangalore, 1864. (R. Desmond 1994.)

Blake, Charles Carter (1840? – after 1887). Anthropologist, palaeontologist, and comparative anatomist. Protégé of Richard Owen, *c.* 1855–63. Lecturer on zoology, London Institution, 1862–3. Founding member of the Anthropological Society of London, 1863; assistant secretary, 1863–6; curator and librarian, 1863–7; honorary secretary and treasurer, 1868; honorary fellow, 1868–73. Fellow of the Geological Society, 1863. (*Anthropologia* 1 (1873–5); Blake 1862a; Blake ed. 1863; *CDEL*; Correspondence with Richard Owen (Natural History Museum); Gorman 1910; *Journal of the Anthropological Society of London* 6 (1868): cviii, 7 (1869): lxxx; *List of the fellows of the Anthropological Society of London* 1865, 1869; *List of the Geological Society of London* 1865–8; *Medical directory* 1866–81; Subscription records, Royal Anthropological Institute; *Wellesley index*.)

13 January 1862

Blanford, Henry Francis (1834–93). Meteorologist and geologist. Served on the Geological Survey of India, 1855–62. Professor at the Presidency College, Calcutta, 1862–72. Meteorological reporter to the government of India. FRS 1880. (*DNB*, Sarjeant 1980.)

Blomefield, Leonard. *See* Jenyns, Leonard.

Blood, Daniel O. & Co. United States mail company based in Philadelphia. Issued stamps for local use between 1841 and *c.* 1860. (*NUC*; Sutton 1966, p. 41.)

Blyth, Edward (1810–73). Owned an unsuccessful druggist's business in Tooting. Wrote zoological articles and edited zoological works before being appointed curator of the museum of the Asiatic Society of Bengal in 1841. He retired from this post in 1862 and returned to England, where he continued to write on zoology and on the origin of species. (*DNB*, *DSB*.)

23 November 1862

Boll, Ernst Friedrich August (1817–68). German theologian, botanist, and geologist. Edited *Archiv des Vereins der Freunde der Naturgeschichte*, 1847–68. Author of

Geognosie der deutschen Ostseeländer zwischen Eider und Oder (1846). (*ADB*, Sarjeant 1980.)

Bollaert, William (1807–76). Chemist and ethnologist. Chemical assistant at the Royal Institution, then chemist and assayist for Peruvian silver mines. Travelled extensively in South America, Central America, and Texas, publishing geographical and ethnological observations made during his explorations. An associate member of the Royal Geographical Society and a fellow of the Ethnological and Anthropological Societies of London. (*Journal of the Royal Geographical Society* 47 (1877): 148–150, *Modern English biography*.)

Bolle, Carl August (1821–1909). German dendrologist and ornithologist resident in Berlin; collector of plants. (*Taxonomic literature*.)

Bonafous, Matthieu (1793–1852). French agriculturist. Settled in Turin and became director of the Agricultural Institute. Published works on the cultivation of cereal crops. Studied and translated works on silkworms, compiling a catalogue of nearly 2000 published works on the subject. (*DBF*, *Index biographique de l'Académie de sciences*.)

Bonaparte, Louis Lucien (1813–91). French statesman and philologist. Nephew of Napoleon I, emperor of France (*EB*). Born and educated in England; went to France after the 1848 revolution. Deputy for the Seine in the Legislative Assembly, 1849. Returned to England after 1870 and concentrated on his interest in philology. (*EB*.)

Bonham-Carter, Elinor Mary (1837–1923). Daughter of Joanna Maria Bonham-Carter. Married Albert Venn Dicey (*DNB*) in 1872. (*Burke's landed gentry* 1965.)

Bonham-Carter, Joanna Maria (1791/2–1884). Widow of John Bonham-Carter, for many years MP for Portsmouth. Resided at Ravensbourne, Keston, two miles north-west of Down, 1853–84. Mother of Elinor Mary Bonham-Carter. (Bonham-Carter 1960, *Burke's landed gentry* 1965.)

Bonpland, Aimé Jacques Alexandre (Aimé) (1773–1858). French traveller and botanist. Accompanied Alexander von Humboldt on his South American travels, 1799–1804. (*DBF*, *EB*.)

Booth, John (1779–1857). Breeder of cattle. Eldest son of Thomas Booth. Came into his father's estate at Killerby, Yorkshire, and part of his short-horn herd, in 1819. (*DNB* s.v. Booth, Thomas.)

Booth, Richard (1788–1864). Breeder of livestock. Second son of Thomas Booth; succeeded to his father's estate at Warlaby, Yorkshire, in 1835. (*DNB* s.v. Booth, Thomas.)

Booth, Thomas (d. 1835). Farmer and breeder of livestock. Owner of estates at Killerby and Warlaby in Yorkshire. Originated the 'Booth' strain of shorthorn cattle in 1790. According to the *DNB*, 'About the period of 1814 he was considered to be the most enterprising and skilful improver of cattle in his district, if not of his day.' (*DNB*, *EB*.)

Boott, Francis (1792–1863). American physician and botanist resident in London. Had a successful medical practice in London, 1825–32. Lecturer on botany at

Boott, Francis, cont.
the Webb Street School of Medicine, London, 1825. Secretary, Linnean Society, 1832–9; treasurer, 1856–61. (R. Desmond 1994, *DNB*.)
27 January 1862, 22 December [1862], 26 December 1862

Borrer, William (1781–1862). Wealthy botanist who sought to cultivate every critical British species, together with many hardy exotic plants. Had considerable knowledge of the British genera *Salix*, *Rubus*, and *Rosa*. Contributed to William Jackson Hooker's *British flora* (1830). FRS 1835. (R. Desmond 1994, *DNB*.)

Boucher de Crèvecoeur de Perthes, Jacques (Jacques Boucher de Perthes) (1788–1868). Customs official and archaeologist. Director of customs at Abbeville, France. Investigated the evidence of what became known as Stone Age cultures discovered at Abbeville. (*DSB*.)

Bradford, Lord. *See* Bridgeman, George Augustus Frederick Henry.

Brent, Bernard Philip (d. 1867). Bird fancier and author. Studied pigeon breeding in France and Germany. (*CDEL*.)
15 July 1862

Brewster, David (1781–1868). Scottish physicist who specialised in optics. Invented the kaleidoscope in 1816. Edited the *Edinburgh encyclopædia*, 1807–29. Assisted in the establishment of the British Association for the Advancement of Science in 1831. Principal of the United College of St Salvatore and St Leonard, St Andrews University, 1838–59. Vice-chancellor, Edinburgh University, 1859–68. Knighted, 1832. FRS 1815. (*DNB*, *DSB*.)

Bridgeman, George Augustus Frederick Henry, 2d earl of Bradford (1789–1865). Succeeded to earldom in 1825. Resident at Weston Park, Staffordshire, and Castle Bromwich, Warwickshire. (*Burke's peerage* 1862, 1963.)

Bridgeman, John Robert Orlando (1831–97). Third son of George Augustus Frederick Henry Bridgeman. Married Marianne Caroline Clive in 1862. Domestic chaplain to his father, 1859–65. Rector of Weston-under-Lizard, Staffordshire, 1859–97. Rural dean of Brewood, 1869–85. (*Alum. Cantab.*)

Brodie (d. 1873). The Darwin children's nurse at 12 Upper Gower Street and Down House, 1842–51. (Freeman 1978.)

Brodie Innes, Eliza Mary. *See* Innes, Eliza Mary Brodie.

Brodie Innes, John. *See* Innes, John Brodie.

Brodie Innes, John William. *See* Innes, John William Brodie.

Bromley Savings Bank
[before 14 July 1862]

Brongniart, Adolphe Théodore (1801–76). French palaeobotanist and taxonomist. A founder of the *Annales des Sciences Naturelles*, 1824. Professor of botany, Muséum d'Histoire Naturelle, 1833. Foreign member, Royal Society, 1852. (*DBF*, *DSB*.)

Bronn, Heinrich Georg (1800–62). German palaeontologist. Professor of natural science at Heidelberg, 1833. Translated and superintended the first German edition of *Origin* (1860) and of *Orchids* (1862). (*DSB*, *NDB*.)

[before 11 March 1862], 11 March [1862], *27 March 1862*, 25 April [1862], *19 May 1862*, *21 June 1862*, 30 June [1862], 11 July 1862

Brown, Edwin (1819–76). Bank manager and naturalist. Employed by the Burton, Uttoxeter, and Ashbourne Union Bank in Burton-upon-Trent, Staffordshire, 1834; manager, 1861. Member of the Entomological Society of London, 1849; fellow of the Geological Society and the Royal Geographical Society; member of the Botanical Society of London. A life-long friend and patron of Henry Walter Bates; Bates's letters to him from the Amazon were published in the *Zoologist*, 1849–59. (*Banking almanac* 1865; Bates 1892, p. 20; R. Desmond 1994; *Entomologist* 9 (1876): 240; *Entomologist's Monthly Magazine* 13 (1876–7): 116–7, 257–8.)

Brown, Robert (1773–1858). Scottish botanist. Naturalist to the expedition surveying the coast of Australia, 1801–5; published descriptions of the plants he collected. Librarian, Linnean Society, 1805. Librarian to Joseph Banks, 1810–20. Keeper of the botanical collections, British Museum, 1827–58. FRS 1811. (*DNB*, *DSB*.)

Browne, William Alexander Francis (1805–85). Physician. President of the Plinian Society, Edinburgh, when CD was a member. Physician at Stirling, 1830. Superintendent of Montrose lunatic asylum; superintendent of Crichton Institute at Dumfries, 1839. Commissioner in lunacy for Scotland, 1857–70. (*Modern English biography*.)

Brown-Séquard, Charles Edouard (1817–94). French physiologist. Practised medicine in France and the United States before accepting a professorship at Virginia Medical College, Richmond, in 1854. Returned to Paris in 1855, but moved to England in 1858. Physician, National Hospital for the Paralysed and Epileptics, 1860–3. Professor of physiology and pathology, Harvard Medical School, 1864–7. Professor of medicine, Collège de France, 1878–94. Conducted pioneering research in neurology and endocrinology. FRS 1860. (*DBF*, *DSB*.)

2 January [1862], *13 January 1862*, 16 April [1862]

Bruce, James, 8th earl of Elgin and 12th earl of Kincardine (1811–63). Statesman. Liberal-conservative MP for Southampton, 1841. Succeeded to earldoms in 1841. Governor of Jamaica, 1842; governor-general of Canada, 1846–54; postmaster-general, 1859. Governor-general and viceroy of India, 1861. (*DNB*.)

Bruce, James Lewis Knight (1791–1866). Judge. Called to the bar, 1817. Appointed king's counsel, 1829. MP for Bishop's Castle, 1831–4. Appointed a vice-chancellor of the court of chancery in 1841. Member of privy council, 1842. Chief judge in bankruptcy. Lord justice of appeal, 1851–66. Knighted, 1841. (*DNB*.)

Bruce, Lewis Bruce Knight (1820–1906). Barrister and expert on beekeeping. Son of James Lewis Knight Bruce. Admitted at Lincoln's Inn, 1842. Justice of the peace for Surrey and Middlesex. Resident at Heathfield, Keston, Kent, *c.* 1851 – *c.* 1859; friend of CD's. Succeeded to family estate of Roehampton Priory, Surrey, in 1866. (*Alum. Oxon.*, *Burke's landed gentry* 1906, *Correspondence* vol. 7, *Post Office directory of the six home counties*.)

Buchanan, Christina Laura (d. 1853). Daughter of James Smith of Jordanhill. Married Walter Buchanan in 1851. (*Burke's landed gentry* 1863.)

Buchanan, Walter (1797–1877). Merchant. Owned an estate at Shandon, Dumbartonshire. Liberal MP for Glasgow, 1857–65. Married Christina Louisa Smith in 1851. (Stenton 1976.)

Büchner, Friedrich Karl Christian Ludwig (Ludwig) (1824–99). German physician and philosopher of science. Lecturer in medicine at Tübingen, 1854; forced to resign following the publication of *Kraft und Stoff* in 1855. Saw CD's work as a confirmation of his own materialist philosophy. (*ADB*, *DSB*, *NDB*.)
17 November [1862]

Buckle, Henry Thomas (1821–62). Historian of civilisation. (*DNB*.)

Buffon, Georges Louis Leclerc, comte de (1707–88). French naturalist and scientific administrator. FRS 1739. (*DBF*, *DSB*.)

Buob, Miss (*fl.* 1863–7). Schoolmistress. Ran a private school at 2 Courtland Terrace, Kensington, London, *c.* 1862 – *c.* 1867. (Letter from Emma Darwin to William Erasmus Darwin, [29 October 1862] (DAR 219.1: 63); letter to W. E. Darwin, 30 [October 1862]; *Post Office London directory* 1862–8.)

Burke, Robert O'Hara (1821–61). Irish-born explorer and police officer. Served as an officer in the Austrian army and the Irish Mounted Constabulary before migrating to Australia in 1853, where he joined the police force. Superintendent of police in the Castlemaine district, 1858. In 1860, commissioned by the Royal Society of Victoria to lead the exploring expedition to traverse the continent from south to north. (*Aust. dict. biog.*)

Burmeister, Karl Hermann Konrad (1807–92). German zoologist, ethnographer, and geologist. Published extensively on entomology. Professor of zoology, University of Halle, 1837–61. Travelled in Brazil, 1850–2; and Argentina, 1857–60. Director of the Museo Nacional in Buenos Aires, 1861–80. (Berg 1895, *NBU*, Sarjeant 1980.)

Burns, Robert (1759–96). Scottish poet. (*DNB*.)

Burton, Richard Francis (1821–90). Soldier, diplomat, explorer, and writer. Joined the Indian army in 1842. Adopting an assumed identity, pilgrimaged to Mecca in 1853. Participated in the search for the source of the Nile, 1856–9. British consul, Fernando Po, West Africa, 1861–5; Santos, Brazil, 1865–9; Damascus, 1869–71; Trieste, 1872. In later years, in England, devoted himself to literary scholarship. (*DNB*.)

Busk, Ellen. Married George Busk in 1843. (*DNB* s.v. Busk, George.)

Busk, George (1807–86). Naval surgeon and naturalist who served on the hospital ship at Greenwich, 1832–55. President of the Microscopical Society, 1848–9. Hunterian professor of comparative anatomy, Royal College of Surgeons, 1856–9. Specialised in palaeontology and in the study of Bryozoa. FRS 1850. (*DNB*, *DSB*.)
1 April 1862

Butler, Mary (d. 1865?). Patient at Moor Park and Ilkley Wells hydropathic

establishments. Became a friend of CD's and visited the Darwins at Down. Probably sister of Richard Butler, Vicar of Trim, Ireland, 1819–62 (*Modern English biography*). (*Burke's landed gentry* 1879; *Correspondence* vol. 7, letters to Mary Butler, 20 February [1859] and 11 September [1859], and letter to John Murray, 14 November [1859].)
[before 25 December 1862]

Byron, Lord. *See* Byron, George Gordon.

Byron, George Gordon, 6th baron (1788–1824). Poet. (*DNB*.)

Cairnes, John Elliot (1823–75). Irish political economist. Appointed Whately professor of political economy, Trinity College, Dublin, in 1856. Professor of political economy and jurisprudence, Queen's College, Galway, 1859. Professor of political economy, University College London, 1866. (*DNB*.)

Campbell, Elizabeth Georgiana (d. 1878). Daughter of George Granville Leveson-Gower, second duke of Sutherland (*Burke's peerage* 1980). Married George Douglas Campbell in 1844. (*Burke's peerage* 1980; *DNB* s.v. Campbell, George Douglas.)

Campbell, George Douglas, 8th duke of Argyll (1823–1900). Whig politician and author of works on science, religion, and politics. A defender of the concept of design in nature. Succeeded to dukedom in 1847. Chancellor of St Andrews University, 1851. President of the Royal Society of Edinburgh, 1861. In government, privy seal, 1853–5, 1859–60, and 1860–6; postmaster general, 1855–8 and 1860; secretary of state for India, 1868–74. FRS 1851. (*DNB*.)

Campbell, James. Merchant in Glasgow. Married Willielma Hooker in 1862. (*The Times*, 9 September 1862, p. 1)

Campbell, John Douglas Edward Henry, 7th duke of Argyll (1777–1847). Father of George Douglas Campbell. Whig MP for Argyllshire, 1799–1822. Succeeded to dukedom in 1839. FRS 1819. (*Complete peerage*.)

Candolle, Alphonse de (1806–93). Swiss botanist whose home was a centre of botanical activity. Son of Augustin Pyramus de Candolle. Professor of botany and director of the botanic gardens, Geneva, 1835–50. After 1850, concentrated on his own research. Foreign member, Royal Society, 1869. (*DSB*.)
13 June 1862, 17 June [1862], *18 September 1862*

Candolle, Augustin Pyramus de (1778–1841). Swiss botanist. Professor of natural history, Academy of Geneva, 1816–35. Foreign member, Royal Society, 1822. (*DSB*.)

Candolle, Anne Casimir Pyramus (Casimir) de (1836–1918). Swiss botanist. Son of Alphonse de Candolle; assistant and colleague of his father. Published monographs of several families of plants. Foreign member, Linnean Society of London. (*DBS*, *Proceedings of the Linnean Society of London* (1918–19): 51–2.)

Carey, John (1797–1880). Botanist. Travelled to the United States in 1830. Worked with Asa Gray in North Carolina in 1841. Returned to England in 1852. Contributed sections on *Carex* and *Salix* to Asa Gray's *Manual of the botany of the northern United States* (1848). (R. Desmond 1994.)

Carlyle, Thomas (1795–1881). Essayist and historian. (*DNB*.)

Carpenter, Louisa. Married William Benjamin Carpenter in 1840. (*DNB* s.v. Carpenter, William Benjamin.)

Carpenter, William Benjamin (1813–85). Physician and physiologist. Fullerian professor of physiology at the Royal Institution, 1844–56; professor of forensic medicine at University College London; physiology lecturer, London Hospital, 1845–56. Registrar of the University of London, 1856–79. FRS 1844. (*DNB*, *DSB*.)

Carter, James, & Co. Seedsmen of High Holborn, London. Founded by James Carter (*c.* 1797–1885). (R. Desmond 1994.)

Cautley, Proby Thomas (1802–71). Army officer, irrigation engineer, and palaeontologist. Planned and constructed the Ganges Canal, 1840–54. With Hugh Falconer, carried out extensive geological and palaeontological research in the Siválik range of northern India, for which they were jointly awarded the Wollaston Medal of the Geological Society in 1837. Knighted, 1854. (*DNB*.)

Chambers, Robert (1802–71). Publisher, writer, and geologist. Partner, with his brother William Chambers (*DNB*), in the Edinburgh publishing company W. & R. Chambers. Joint editor of *Chambers's Journal* from 1832. Anonymous author of *Vestiges of the natural history of creation* (1844). (*DNB*, *DSB*.)

Chesnel, Mary Louisa de (1797–1865). Sister of George Bentham. (Jackson 1906.)

Christy, Henry (1810–65). Ethnologist, archaeologist, geologist, and banker. Partner in the London banking house, Christy & Co.; a director of the London Joint-Stock Bank. Travelled for ethnological purposes in the East, 1850; Scandinavia, 1852–3; North America, Cuba, and Mexico, 1856–7. Joined the Geological Society of London in 1858; subsequently excavated caves in the Vézère valley in southern France with Edward Lartet. (*DNB*.)

Claparède, Jean Louis René Antoine Edouard (Edouard) (1832–71). Swiss physician and naturalist. Professor of comparative anatomy, Academy of Geneva, 1862. Specialised in invertebrate anatomy, histology, embryology, and evolution. One of the first Swiss naturalists to endorse CD's theory. (*DSB*.)

[*c.* 16 April 1862], *6 September 1862*

Clarke, Richard Trevor (1813–97). Army officer and horticulturist. Honorary lieutenant-colonel in the Northampton and Rutland Infantry Militia, 1862. Conducted hybridisation experiments on cotton and begonias. Awarded a gold medal by the Cotton Supply Association of Manchester for his contribution to cotton culture. Active member of the Royal Horticultural Society, serving for many years on the council and scientific committee; awarded the society's Veitchian Medal, 1894. (*Army list*; R. Desmond 1994; *Gardeners' Chronicle*, 17 April 1897, p. 263.)

[after 25 November 1862], *[after 27 November 1862]*

Clarke, William Branwhite (1798–1878). Anglican clergyman and geologist. Emigrated to Australia in 1839, where he carried out extensive geological surveys. Discovered gold in New South Wales in 1841. In clerical charge of the country from Paramatta to the Hawkesbury river, 1839–44; of Campbelltown, 1844–7;

and of Willoughby parish (St Leonard's, Sydney), 1847–70. FRS 1876. (*DNB*, *DSB*, Jervis [1945].)

16 January 1862, 21 January 1862, 20 June 1862, 20 September 1862

Clerk, George, 6th baronet (1787–1867). Statesman. Succeeded to the baronetcy in 1798. MP for Midlothian, 1811–32 and 1835–7; for Stamford, 1838–47 and 1847–52. Lord of the Admiralty, 1819–27. Secretary to the Treasury, 1834–5 and 1841–5. Vice-president of the Board of Trade and master of the Royal Mint, 1845–6. President of the Zoological Society, 1862–7. FRS 1819. (*DNB*, *Modern English biography*.)

Clive, Edward, 1st earl of Powis (1754–1839). Eldest son of Robert Clive (*DNB*). Tory MP for Ludlow, 1774–94. Governor of Madras, 1798–1803. Lord lieutenant of Shropshire, 1804–39. (*DNB*.)

Clive, Marianne Caroline (d. 1930). Niece of Georgina Tollet. Married John Robert Orlando Bridgeman in 1862. (*Burke's landed gentry* 1862, *Burke's peerage* 1980.)

Clodd, Edward (1840–1930). Banker and author. Clerk at the London Joint Stock Bank, 1862–72; appointed secretary in 1872; retired, 1915. Joined the unconventional Century Club in 1877, and the Folk-Lore Society in 1878. Helped to found the Johnson Club in 1884, and the Omar Khayyám Club in 1892. He developed a remarkable circle of friends, and his books, the first of which was published in 1873, were very popular. (*DNB*.)

Clowes, William & Sons. Printers. William Clowes (1807–83), eldest son of William Clowes (1779–1847), joined his father's printing business in 1823 and the name of the firm was changed to William Clowes & Sons in 1846. Printed the offical catalogue of the Great Exhibition of 1851. Introduced improvements in type-music printing. Printers to John Murray. (*DNB*.)

Cochrane, Louisa Harriet (d. 1902). Married Thomas Barnes Cochrane, later eleventh earl of Dundonald, in 1847. (*Burke's peerage* 1980.)

Cochrane, Thomas Barnes, 11th earl of Dundonald (1814–85). Army captain. Representative peer for Scotland, 1879–85. (*Burke's peerage* 1980.)

Cohn, Ferdinand Julius (1828–98). German botanist and bacteriologist. Established as a *privatdozent* at the University of Breslau; professor, 1872. Founded the first institute for plant physiology, at Breslau, in 1866. Worked on unicellular algae, and argued for the identity of the contractile contents of plant and animal cells. Foreign member, Royal Society, 1897. (*DSB*, *NDB*.)

Coke, Edward (1552–1634). Jurist. Speaker of the House of Commons, 1592–3; attorney-general, 1593–4; chief justice of the Common Pleas, 1606; chief justice of the King's Bench, 1613. His *Institutes* (1628), consisting of a translation and reprint of Thomas Littleton's *Tenures* (*c.* 1481) with a commentary, the text of various statutes with commentaries, and a treatise on the jurisdiction of different courts, had a profound influence on early modern English law. (*DNB*.)

Coke, Thomas William, earl of Leicester of Holkham (1752–1842). Whig politician and landowner. Came into estates in 1776. Elected MP for Norfolk

Coke, Thomas William, earl of Leicester of Holkham, cont.
in same year. Retired as father of the House of Commons in 1833. Created earl of Leicester of Holkham and Viscount Coke, 1837. Bred Southdown variety of sheep and Devon cattle. Improved Suffolk breed of pigs by crossing with Neapolitan. (*DNB.*)

Colburn, Henry (d. 1855). London publisher who published the first edition of CD's *Journal of researches* (1839). (*DNB.*)

Colenso, John William (1814–83). Divine, mathematician, and missionary. Fellow, St John's College, Cambridge, 1837; college tutor, 1842–6. Mathematical tutor, Harrow, 1839–42. Vicar of Forncett St Mary, Norfolk, 1846–53. Bishop of Natal, 1853–83. Published *Critical examination of the Pentateuch* (1862–79), in which he argued that these books were post-exile forgeries. Excommunicated in 1863, his possession of the see was confirmed by the law courts in 1866. (*DNB.*)

Condy, Henry Bollmann (*fl.* 1859–1900). Manufacturing chemist and author of works on sanitation. By 1857, had developed and patented 'Condy's fluid', a disinfectant solution of alkaline permanganates that could be taken internally or used externally. Manufactured the fluid at his works in Battersea, *c.* 1867–97. (*Chemical manufactories' directory* 1867–1901, Condy 1862, *NUC*, *Post Office London directory* 1854–66.)

Cooper, James Fenimore (1789–1851). American novelist. (*DAB.*)

Cooper, Susan Augusta Fenimore (1813–94). Author and philanthropist. Daughter of James Fenimore Cooper, for whom she acted as an amenuensis until his death in 1851. Wrote and edited books on the countryside, of which the most successful was *Rural hours* (New York, 1850). (*BLC*, *DAB.*)

Cotton, Richard (d. 1839). Shrewsbury naturalist. (*Correspondence* vol. 1, Freeman 1978.)

Covington, Syms (1816?–61). Became CD's servant on HMS *Beagle* in 1833 and remained with him as assistant, secretary, and servant until 1839. Emigrated to Australia in 1839. (Freeman 1978.)

Cowper, Charles (1807–75). Statesman and farmer. Secretary of the Church and Schools Lands Corporation, New South Wales, Australia. Between 1856 and 1865, Cowper held a number of posts in various administrations, serving as colonial secretary between 1860 and 1863. Appointed agent-general for New South Wales, 1870. (*DNB.*)

Cowper, William (*fl.* 1854–64). Farmer and grazier of Wappenham, Northamptonshire. (*Post Office directory of Northamptonshire* 1854, 1864.)

Cranworth, Lady. *See* Rolfe, Laura.

Cranworth, Lord. *See* Rolfe, Robert Monsey.

Cresy, Edward (1823/4–70). Surveyor and civil engineer, resident at Ham Moor, Black Boy Bridge, Chertsey, Surrey. Son of Edward Cresy (1792–1858), the architect and civil engineer who advised CD about the purchase and improvement of Down House. As a young man, worked as an architectural draftsman in his father's office. Assisted his father in preparing his *Cyclopædia of* *civil engineering*, in 1845. Assistant surveyor under the Commissioners of Sewers, 1849; afterwards engineer. Principal assistant clerk at the Metropolitan Board of Works, 1859. Architect to the fire brigade, 1866. Founder member of the Geologists' Association, 1858; president, 1864–5; vice-president, 1865–70. (*Annual Report of the Geologists' Assocation* 1859–70; Census returns 1861 (Public Record Office, RG9/422: 118a); *DNB* s.v. Cresy, Edward, (1792–1858); *Engineer* 30 (1870): 409).

8 January [1862 or 8], *19 May 1862*, *13 September 1862*, 15 September [1862]

Cresy, Eliza. Translator and mother of Edward Cresy. Resident at Riverhead, near Sevenoaks, Kent. In 1814, married the architect Edward Cresy (1792–1858). (*DNB* s.v. Cresy, Edward, (1792–1858); *Post Office directory of the six home counties* 1862.)

Cresy, Mary (b. 1820/1). Wife of Edward Cresy. (Census returns 1861 (Public Record Office, RG9/422: 118a).)

Crocker, Charles (1797–1861). Poet. Worked as a shoemaker for twenty years; bookseller, 1839–45. Sexton of Chichester Cathedral, 1845; later, also bishop's verger. Father of Charles William Crocker. (*DNB*, *Journal of Horticulture* n.s. 14 (1868): 206–7.)

Crocker, Charles William (1832–68). Gardener. Spent time in Germany as a gardener at the summer residence of Victoria, Crown Princess of Prussia, near Berlin. Foreman of the propagating department, Royal Botanic Gardens, Kew, 1857–62. Moved to his native Chichester in 1862 where he worked as a journalist and a cathedral official. (R. Desmond 1994, *Journal of Horticulture* n.s. 14 (1868): 206–7.)

17 February 1862, *[before 13 March 1862]*, *13 March 1862*, *22 April 1862*, *17 May 1862*, *31 October 1862*, *24 November 1862*

Crocker, Mrs Charles (1791/2–1862). Wife of Charles Crocker and mother of Charles William Crocker. (*Sussex Advertiser*, 6 May 1862.).

Currey, Frederick (1819–81). Mycologist. Called to the bar, 1844. Secretary, Linnean Society, 1860–80; vice-president and treasurer, 1880–1. One of the editors of the *Natural History Review.* FRS 1858. (*Alum. Cantab.*, *DNB.*)

12 June 1862, *3 July 1862*

Cuvier, Jean Léopold Nicolas Frédéric (Georges) (1769–1832). French systematist, comparative anatomist, and palaeontologist. Professor of natural history, Collège de France, 1800–32; professor of comparative anatomy, Muséum d'Histoire Naturelle, 1802–32. Permanent secretary of the Académie des Sciences from 1803. Foreign member, Royal Society, 1806. (*DBF*, *DSB.*)

Dallas, James (1853–1916). Museum curator and librarian. Son of William Sweetland Dallas (*Modern English biography*), assistant secretary, librarian, and curator of the Geological Society of London. Assistant in the library of the Geological Society, 1872; in the library and museum, 1878–9. Fellow of the Linnean Society, 1884. Local secretary, Society of Antiquaries. (Dallas 1921, *List of the Linnean Society*, Woodward 1907.)

Dana, James Dwight (1813–95). American geologist and zoologist. Naturalist with the United States Exploring Expedition to the Pacific, 1838–42. An editor of the *American Journal of Science and Arts* from 1846. Professor of geology, Yale University, 1856–90. Foreign member, Royal Society, 1884. (*DAB*, *DSB*.)
4 December 1862

Daniell, William Freeman (1818–65). Surgeon and botanist. Assistant surgeon to the British army in West Africa, 1841–53. Sent plants to the Royal Botanic Gardens, Kew. (R. Desmond 1994, *DNB*.)

Darwin, Caroline Sarah. *See* Wedgwood, Caroline Sarah.

Darwin, Elizabeth (1847–1926). CD's daughter. (*Darwin pedigree*.)

Darwin, Emily Catherine (1810–66). CD's sister. Resided at The Mount, Shrewsbury, until she married Charles Langton in 1863. (*Darwin pedigree*.)

Darwin, Emma (1808–96). Youngest daughter of Josiah Wedgwood II. Married CD, her cousin, in 1839. (*Emma Darwin*.)
[January 1862], 28 June [1862]

Darwin, Erasmus (1731–1802). CD's grandfather. Physician, botanist, and poet. Advanced an evolutionary theory similar to that subsequently propounded by Lamarck. FRS 1761. (*DNB*, *DSB*.)

Darwin, Erasmus Alvey (1804–81). CD's brother. Qualified as a physician but never practised. Admitted to Christ's College, Cambridge, 1822; at Edinburgh University, 1825–6. Lived in London from 1829. (*Alum. Cantab.*, Freeman 1978.)
[April – May? 1862], 20 June [1862], 21 June [1862], *1 July [1862], 2 July [1862], 14 December [1862?]*

Darwin, Francis (1848–1925). CD's son. Botanist. Qualified as a physician but did not practise. BA, Trinity College, Cambridge, 1870. Collaborated with CD on several projects, 1875–82. Lecturer in botany, Cambridge University, 1884; reader, 1888–1904. Edited CD's letters (*LL* (1887) and *ML* (1903)). President of the British Association, 1908. Knighted, 1913. FRS 1882. (*Alum. Cantab.*, *DNB*, *DSB*.)

Darwin, George Howard (1845–1912). CD's son. Mathematician. BA, Trinity College, Cambridge, 1868; fellow, 1868–78. Called to the bar in 1872 but did not practise. Plumian professor of astronomy and experimental philosophy, Cambridge University, 1883–1912. President of the British Association, 1905. Knighted, 1905. FRS 1879. (*Alum. Cantab.*, *DNB*, *DSB*.)
[12 June 1862], [after 5 August 1862]

Darwin, Henrietta Emma (1843–1927). CD's daughter. Married Richard Buckley Litchfield (Freeman 1978) in 1871. Assisted CD with some of his work. Edited *Emma Darwin* (1904). (*Burke's landed gentry* 1952, Freeman 1978.)
[29 October 1862]

Darwin, Horace (1851–1928). CD's son. Civil engineer. BA, Trinity College, Cambridge, 1874. Apprenticed to an engineering firm in Kent before returning to Cambridge in 1875 to design and make scientific instruments. Founder and director of the Cambridge Scientific Instrument Company. Mayor of Cambridge, 1896–7. Knighted, 1918. FRS 1903. (*Alum. Cantab.*, *DNB*.)

Darwin, Leonard (1850–1943). CD's son. Attended the Royal Military Academy, Woolwich. Commissioned in the Royal Engineers, 1870; major, 1889. Served on several scientific expeditions, including those to observe the transit of Venus in 1874 and 1882. Instructor in chemistry and photography, School of Military Engineering, Chatham, 1877–82. Staff, Intelligence Department, War Office, 1885–90. Liberal Unionist MP, Lichfield division of Staffordshire, 1892–5. President of the Royal Geographical Society, 1908–11, and of the Eugenics Education Society, 1911–28. Chairman, Bedford College, London University, 1913–20. (M. Keynes 1943, *WWW*.)

Darwin, Robert Waring (1766–1848). CD's father. Physician. Had a large practice in Shrewsbury and resided at The Mount, which he built *circa* 1796–8. Son of Erasmus Darwin (*DNB*) and his first wife, Mary Howard. Married Susannah, daughter of Josiah Wedgwood I (*DNB*), in 1796. FRS 1788. (Freeman 1978.)

Darwin, Sacheverel Charles (1844–1900). CD's first cousin, once removed. Naval officer. Entered British navy in 1858. Appointed midshipman to HMS *Phœbe*, 1862. Obtained his lieutenant's commission, 1866. Promoted to Captain, 1885. Promoted to retired rear-admiral, 1899. (*Colburn's United Service Magazine* (1862), pt 3: 447; *Darwin pedigree*, p. 18; *The Times*, 8 January 1900, p. 4.)

Darwin, Susan Elizabeth (1803–66). CD's sister. Lived at The Mount, Shrewsbury, the family home, until her death. (*Darwin pedigree*.)

Darwin, Susannah (1765–1817). CD's mother. Daughter of Josiah Wedgwood I. Married Robert Waring Darwin in 1796. (*Darwin pedigree*.)

Darwin, William Erasmus (1839–1914). CD's eldest child. Banker. Attended Rugby School. BA, Christ's College, Cambridge, 1862. Partner in the Southampton and Hampshire Bank, Southampton, 1861. Chairman of the Southampton Water Company. Keen amateur photographer. (*Alum. Cantab.*, F. Darwin 1914.)

14 February [1862], 26 April [1862], [8 May 1862], [31 May 1862], *June 1862*, 13 [June 1862], 4 [July 1862], 9 July [1862], *14 July 1862*, [after 14 July 1862], [24 July 1862], *1 August 1862*, *2 August [1862]*, [2–3 August 1862], *5 August 1862*, [2–3 September 1862], [10? September 1862], *21 October [1862]*, [25 October 1862], *28 October 1862*, 30 [October 1862], 4 [November 1862]

Daubeny, Charles Giles Bridle (1795–1867). Chemist and botanist. Professor of chemistry, Oxford University, 1822–55; professor of botany, 1834; and rural economy, 1840. An active supporter of the British Association for the Advancement of Science. FRS 1822. (*DNB*, *DSB*.)

5 July 1862

Davy, Humphry (1778–1829). Chemist. Professor of chemistry at the Royal Institution, 1802–13. President of the Royal Society, 1820–7. FRS 1803. (*DNB*, *DSB*.)

Davy, Martin (1763–1839). Physician. Master of Gonville and Caius College, Cambridge, 1803–39. Rector of Cottenham, Cambridgeshire, 1827–39. FRS 1801. (*DNB*.)

Dawson, John William (1820–99). Canadian geologist and educationist. Investigated the geology of maritime provinces with Charles Lyell in 1846 and 1852. Superintendent of education for common schools in Nova Scotia, 1850. Appointed principal and professor of geology at McGill University in 1855. Knighted, 1884. FRS 1862. (*DNB*, *DSB*.)

Decaisne, Joseph (1807–82). French botanist. Botanist at the Jardin des Plantes, Paris, 1824. Professor of statistical agriculture, Collège de France, 1848. Professor of plant cultivation, Muséum d'Histoire Naturelle, 1850. Foreign member, Royal Society, 1877. (*DBF*, *NBU*.)

De Filippi, Filippo (1814–67). Italian zoologist, embryologist, and geologist. Professor of zoology and director of the Museum of Zoology, Turin, 1847. Travelled as naturalist to a diplomatic and scientific mission to Persia, 1862; and with a scientific voyage of global circumnavigation, 1865–7. In 1855, advocated belief in the limited transmutation of species. His lecture on 'Man and the Monkeys', delivered in Turin in January 1864, initiated the public debate on Darwin's work in Italy. (Corsi 1983, *DBI*, Pancaldi 1991.)

Delamer, Eugene Sebastian. *See* Dixon, Edmund Saul.

Deshayes, Gérard Paul (1797–1875). French palaeontologist and conchologist. Collected and studied living and fossil molluscs. Prepared a catalogue of the bivalve molluscs in the British Museum, 1853–4. (*DBF*, *DSB*.)

Dickens, Charles John Huffam (1812–70). Novelist. (*DNB*.)

Dickie, George (1812–82). Scottish botanist. Lecturer in botany, King's College, Aberdeen, 1839–49. Professor of natural history, Queen's College, Belfast, 1849–60. Professor of botany, Aberdeen University, 1860–77. Specialised in the study of marine algae and described many of the collections sent to the Royal Botanic Gardens, Kew. FRS 1881. (R. Desmond 1994, *DNB*.)
30 May 1862

Dixon, Edmund Saul (1809–93). Clergyman and poultry-fancier. Rector of Intwood with Keswick, Norfolk, 1842–93. Author of books on the history and management of poultry. From 1854, published under the pseudonym Eugene Sebastian Delamer. (*Alum. Cantab.*, *Modern English biography*.)

Dominy, John (1816–91). Gardener at the nurseries of James and James Veitch in Exeter (1834–41) and Chelsea (*c.* 1846–80). Grew the first known artificially produced orchid hybrid, *Calanthe dominii*. Hybridised nepenthes and fuchsias. (R. Desmond 1994.)

Doubleday, Henry (1808–75). Entomologist and ornithologist. Grocer and insurance agent in Epping, Essex. Noted for his work on the systematics of Lepidoptera. (R. Desmond 1994, *DNB*, *Post Office directory of the six home counties* 1862.)

Douglas, William Alexander Anthony Archibald, 11th duke of Hamilton (1811–63). Succeeded to dukedom in 1852. Held the offices of knight marischal of Scotland, lord-lieutenant of Lanarkshire, and colonel of the Lanarkshire Militia. Major-commandant of the Glasgow Yeomanry, 1849–57. Grandmaster of the society of freemasons. (*DNB*.)

Down Friendly Society. Founded with CD's help, probably in 1852. CD acted as the society's treasurer for thirty years. The annual general meeting was held at Down House. (Freeman 1978.)
[before 14 July 1862]

Downing, Andrew Jackson (1815–52). American landscape gardener, architect, and horticulturalist. Commissioned to design the grounds for the White House and the Smithsonian Institution in Washington in 1851. (*DAB.*)

Drejer, Salomon Thomas Nicolai (1813–42). Danish botanist. Worked at the veterinary school, Copenhagen. Edited *Flora Danica* (1841–2). (*Taxonomic literature.*)

Drouet, Henri (1829–1900). French administrator and zoologist. Went on a scientific voyage to the Azores, French Guyana, and Angola, in 1854. Became principal private secretary to the prefect of Vienna in 1861. Adviser to the prefectures of Ardennes, Vaucluse, and the Côte-d'Or, 1863–4. Sometime inspector of prison services in Algeria. (*DBF.*)

Drummond, James (1784?–1863). Scottish botanist. Curator, Botanic Garden, Cork, 1809–29. Emigrated to Australia in 1829. Superintendent of the government gardens, Western Australia. (*Aust. dict. biog.*, *DNB.*)

Drysdale, Elizabeth (1781/2–1882). Daughter of John Pew of Hilltown, Kinkudbrightshire. Married William Copland of Colliston, Dumfries; widowed, 1808. Married Sir William Drysdale (1781–1847), for many years treasurer of the city of Edinburgh. Mother of John James Drysdale (1816–92), a leading homeopathic doctor. Mother-in-law of the hydropathic specialist Edward Wickstead Lane. (*Emma Darwin* (1904), 2: 184; *Modern English Biography* s.v. Drysdale, John James.)

Du Chaillu, Paul Belloni (1835–1903). French explorer. Undertook an exploring expedition to Central Africa, 1856–60. Collected rare birds and mammals, bringing to America the first gorillas seen there. His controversial account of the journey, *Explorations and adventures in equatorial Africa* (1861), engendered scepticism about his claims; however, subsequent travellers confirmed many of his findings. Revisited Africa, 1863–5. Resided in New York, 1867–71. (*DAB.*)

Duchartre, Pierre Etienne Simon (1811–94). French botanist. Specialised in plant structure. Professor of botany, Agricultural Institute, Versailles, 1849–52; professor of botany at the Faculty of Sciences, Paris, 1861. (Barnhart 1965, *DBF.*)

Duck, George (b. 1794/5). Carpenter, grocer, and landlord of the 'George' public house in Down. (Census returns 1861 (Public Record Office, RG9/462: 73), Freeman 1978, *Post Office directory of the six home counties* 1862.)

Dunalley, Lord. *See* Prittie, Henry Sadleir.

Dundonald, earl of. *See* Cochrane, Thomas Barnes.

Dundonald, Lady. *See* Cochrane, Lousia Harriet.

Dunn, Eliza Brodie (d. 1861). Wife of Henry Joseph Brodie Dunn. Cousin of John Brodie Innes. (*County families* 1864, *Gentleman's Magazine* (1861), pt 2: 215.)

Dunn, Henry Joseph Brodie (1797/8–1861). Landowner of Milton Brodie, near Forres, Scotland. (*Gentleman's Magazine* (1861), pt 2: 98.)

Durand, Elie Magloire (Elias) (1794–1873). French-born pharmacist and botanist. Pharmacist in the French army, 1813–14. Opened a drug store in Philadelphia in 1825, introducing many foreign medicines to the United States. Elected to the Philadelphia Academy of Natural Sciences, 1825; to the American Philosophical Society, 1854. Vice-president of the College of Pharmacy, 1844. (*DAB*, *Taxonomic literature*.)

Dutrochet, René Joachim Henri (1776–1847). French military medical officer, 1808–9. Abandoned medical practice and devoted himself to natural science. Elected corresponding member of the Académie des sciences, 1819; full member, 1831. Primarily studied plant physiology and is noted for his work on the phenomena of osmosis and diffusion. (*DBF*, *DSB*.)

Dyne, Musgrave James Bradley (b. 1827). Army officer of Gore Court, Sittingbourne, Kent. Matriculated, Merton College, Oxford, 1845. Major, 2d (The Queen's) Dragoon Guards, 1861. (*Alum. Oxon.*; *Army list*, September 1862; *Burke's landed gentry* 1862.)

Dzierzon, Johannes (1811–1906). German Catholic priest in Karlsmarkt and authority on bees. (*NDB*, *NUC*.)

Edmands, John Wiley (1809–77). American merchant and congressman. Representative in Congress for Massachusetts, 1853–5. Founded a public library at Newton, Massachusetts. Presided at the banquet held in Boston in honour of Charles Wilkes, captain of the *San Jacinto*, for his part in the '*Trent* affair'. (*Boston Post*, 27 November 1861, p. 2; Herringshaw 1909–14.)

Edwards, George. Farmer and horse-breeder in Bromley, Kent. (CD's Address book (Down House MS), *Post Office directory of the six home counties* 1862.)

Ehrenberg, Christian Gottfried (1795–1876). German zoologist, comparative anatomist, and microscopist. Studied the development of coral reefs and worked extensively on infusoria. Appointed professor of medicine at Berlin University, 1839. Foreign member, Royal Society, 1837. (*DSB*, *NDB*.)

Elgin, earl of. *See* Bruce, James.

Engleheart, Stephen Paul (1831/2–85). Surgeon. Member of the Royal College of Surgeons of England, 1859; licentiate of the Royal College of Surgeons of Edinburgh, 1860. Surgeon in Down, Kent, 1861–70. Medical officer, 2d District, Bromley Union, 1863–70; Divisional Surgeon of Police, 1863–70. Resident in Shelton, Norfolk, 1870–81; in Old Calabar, Nigeria, 1882–5. (*Medical directory* 1861–86, *Post Office directory of the six home counties* 1862.)

Eprémesnil, Jacques Louis Raoul Duval, comte d' (d. 1891). French property-owner. Member of the Conseil générale de l'Eure, 1862. Founder member of the Société Impériale Zoologique d'Acclimatation, 1854; secretary general, 1854–71; vice-president, 1872–87; honorary vice-president, 1887–91. (*Bulletin de la Société Impériale Zoologique d'Acclimatation* 2 (1855): xxiv, (1891), pt 1: 109–11; *Catalogue de la Bibliothèque Nationale*; *Liste générale des membres de la société et des sociétés affiliées et agrégées au 16 Mai 1862*, p. 34.)

Erasmus, Desiderius (1466–1536). Dutch scholar and theologian. (*EB*.)

Evans, John (1812–61). American physician and geologist. Undertook exploratory work in Washington and Oregon in 1851, and in the White River Badlands, now South Dakota and Nebraska, in 1853. Travelled with the Chiriqui exploring expedition to Central America in 1860. (Sarjeant 1980.)

Eyre, Edward John (1815–1901). Explorer and colonial governor. Emigrated to Australia in 1833, where he took up sheep farming. Explored the Australian interior. Lieutenant-governor of New Zealand, 1846–53; of St Vincent, West Indies, 1854–60. Acting governor of Jamaica, 1861; governor in chief, 1864. Brutally repressed the Morant Bay uprising in 1865; recalled, 1866. (*Aust. dict. biog.*, *DNB*.)

Eyton, Thomas Campbell (1809–80). Shropshire naturalist. Friend and Cambridge contemporary of CD. Author of several works on natural history. On coming into possession of the family estate at Eyton, Shropshire, in 1855, he built a museum in which he formed a collection of skins and skeletons of European birds. (*DNB*.)

[after 19 May 1862?]

Falconer, Hugh (1808–65). Palaeontologist and botanist. Superintendent of the botanic garden, Saharanpur, India, 1832–42. Superintended the arrangement of Indian fossils for the British Museum in 1844. Superintendent of the Calcutta Botanic Garden and professor of botany, Calcutta Medical College, 1848–55. FRS 1845. (*DNB*, *DSB*.)

7 May [1862], [8 May 1862], *24–7 September [1862]*, 1 October [1862], *3 October 1862*, *4 October 1862*, *12 November [1862]*, 14 November [1862], 29 December [1862]

Fargo, William George (1818–81). Expressman. In 1844, became messenger for Wells & Company, of which he was one of three owners. Secretary of the American Express Company in 1850. Wells, Fargo & Company established the first express to and from California in 1852. Mayor of Buffalo, New York, 1862–6. (*DAB*.)

Felder, Cajetan von (1814–94). Austrian jurist and entomologist. Elected to the first Viennese district council in 1848. Mayor of Vienna, 1868. Made a special study of Lepidoptera and had an extensive private collection of specimens. (*OBL*.)

Felder, Rudolf (1842–71). Austrian entomologist. Son of Cajetan von Felder. Used his father's extensive entomological collections as the basis for his studies and publications. (*OBL*.)

Felton, Cornelius Conway (1807–62). American philologist. Eliot Professor of Greek literature, Harvard University, 1834–60. President of Harvard University, 1860–2. (*DAB*.)

Fermond, Charles (1810 – *c.* 1875). French pharmacist and botanist. Head pharmacist at the Salpetrière hospital, Paris, 1837. His botanical studies encompassed the germination of plants, their movements, flowering, fertilisation, hybridisation, and the production of monstrosities. (*DBF*.)

Ffinden, George Skertchley. Vicar of Down, 1871–1911. (Freeman 1978.)

Fields, James Thomas. *See* Ticknor & Fields.

Filippi, Filippo De. *See* De Filippi, Filippo.

Fitch, Adam (1807–85). Clergyman. Curate of Cottenham, Cambridgeshire, 1832; of Willingham, Cambridgeshire, 1841–9. Vicar of Thornton Steward, Bedale, Yorkshire, 1849–85. (*Alum. Cantab.*)
18 November 1862

Fitch, Walter Hood (1817–92). Botanical artist at the Royal Botanic Gardens, Kew, from 1841. Produced over 10,000 published drawings. Illustrated Joseph Dalton Hooker's *Rhododendrons of Sikkim-Himalaya* (1849–51) and *Illustrations of Himalayan Plants* (1855). (R. Desmond 1994, *Proceedings of the Linnean Society of London* (1891–2): 68.)

FitzRoy, Robert (1805–65). Naval officer, hydrographer, and meteorologist. Commander of HMS *Beagle*, 1828–36. Tory MP for Durham, 1841–3. Governor of New Zealand, 1843–5. Superintendent of the dockyard at Woolwich, 1848–50. Chief of the meteorological department at the Board of Trade, 1854. Rear-admiral, 1857; vice-admiral, 1863. FRS 1851. (*DNB*, *DSB*.)

Flower, William Henry (1831–99). Anatomist and zoologist. Curator of the Hunterian Museum, Royal College of Surgeons, 1861–84; Hunterian professor of comparative anatomy and physiology, Royal College of Surgeons, 1870–84. Director of the Natural History Museum, London, 1884–98. President of the Zoological Society of London, 1879–99. President of the British Association for the Advancement of Science, 1889. Knighted, 1892. FRS 1864. (*DNB*.)

Floy, James (1806–63). American clergyman, writer, and horticulturalist. (*DAB*.)

Forbes, Edward (1815–54). Zoologist, botanist, and palaeontologist. Naturalist on board HMS *Beacon*, 1841–2. Appointed professor of botany, King's College, London, and curator of the museum of the Geological Society in 1842. Palaeontologist with the Geological Survey, 1844–54. Professor of natural history, Edinburgh University, 1854. FRS 1845. (*DNB*, *DSB*.)

Fox, Samuel William Darwin (b. 1841). Clergyman. Eldest son of William Darwin Fox. Curate of St Paul's, Manningham, 1865–7; of St Augustine's, Halifax, 1867–74. Rector of St Peter's, Lymm, 1874–8. Secretary of the Parish Mission Society, 1878–87. Vicar of St Paul's, Maidstone, 1887–91; of Christ Church, Bridlington Quay, 1891–1900; of St Mark's, Manningham, 1900–8. (*Alum. Oxon.*, *Crockford's*.)

Fox, Sylvanus Bevan (1825–1912). Quaker dentist, beekeeper, and writer. Licentiate of dental surgery, Royal College of Surgeons of England, 1863. Dentist in Exeter, 1865 (or earlier) – 1908. Dental surgeon, Exeter Dental Hospital, 1880 (or earlier); consultant dental surgeon, 1889–1908. A contributor to the beekeeping section of the *Cottage Gardener*, 1860–1; of the *Journal of Horticulture*, 1861–73. Member and sometime councillor of the Devonish Association for the Advancement of Science and Art. (Blaschko 1958, *Medical directory* 1865–1911.)

Fox, William Darwin (1805–80). Clergyman. CD's second cousin. A friend of CD's at Cambridge who shared his enthusiasm for entomology. Maintained an active interest in natural history throughout his life and provided CD with much information. Rector of Delamere, Cheshire, 1838–73. Spent the last years of his life at Sandown, Isle of Wight. (*Alum. Cantab.*)

12 May [1862], [17 May 1862], 12 September [1862], 20 [September 1862]

Frederick II (1712–86). Known as 'Frederick the Great'. King of Prussia, 1740–86. (*EB.*)

Froude, James Anthony (1818–94). Historian. Editor of *Fraser's Magazine*, 1860–74. Regius professor of modern history, Oxford University, 1892. (*DNB.*)

Gallio, Junius Annaeus (d. A.D. 65). Elder brother of the philosopher Seneca. Proconsul of Achaea, *c.* A.D. 51–3. (*EB.*)

Galton, Francis (1822–1911). Traveller, statistician, and scientific writer. CD's cousin. Explored in south-west Africa, 1850–2. Carried out various hereditarian inquiries. Founder of the eugenics movement. FRS 1860. (*DNB*, *DSB.*)

Gardeners' Chronicle

[before 8 November 1862]

Garibaldi, Giuseppe (1807–82). Italian general and revolutionary. (*EB.*)

Garner, Robert (1808–90). Naturalist and surgeon in Stoke-upon-Trent. Surgeon to the North Staffordshire Infirmary, 1834. President of the Staffordshire Branch of the British Medical Association. Fellow of the Linnean Society, 1856. President of the North Staffordshire Naturalists' Field Club, 1878. Specialised in anatomical work on invertebrates and vertebrates. (*Modern English biography*, Plarr 1930.)

Gärtner, Karl Friedrich von (1772–1850). German physician and botanist. Practised medicine in Calw, Germany, from 1802. Contracted an eye ailment in the course of his microscopical investigations, forcing his retirement from medical practise. From about 1824, he devoted his time to the study of plant hybridisation. (*DSB*, *NDB.*)

Gaskell, Cecil Grenville Milnes (d. 1890). Eldest daughter of James Milnes Gaskell, MP. Married Francis Turner Palgrave in 1862. (Palgrave 1899.)

Gay, Jacques Etienne (1786–1864). Swiss-born botanist and civil servant. Moved to Paris in 1811; appointed to the office of the senate. Appointed secretary to the Comité des pétitions. Carried out extensive research in descriptive botany. Instrumental in the foundation of the Société botanique de France, 1854. (*DBF*, Vilmorin 1918.)

Geoffroy Saint-Hilaire, Etienne (1772–1844). French zoologist. Professor of zoology, Muséum d'Histoire Naturelle, 1793. Devoted much attention to embryology and teratology. (*DBF*, *DSB.*)

Geoffroy Saint-Hilaire, Isidore (1805–61). French zoologist. Succeeded his father, Etienne Geoffroy Saint-Hilaire, as professor of zoology at the Muséum d'Histoire Naturelle in 1841. Continued his father's work in teratology. Became professor of zoology at the Sorbonne, Paris, in 1850. (*DBF*, *DSB.*)

Gerrard, Edward (1811–1910). Attendant at the British Museum, 1841–96. Preserved and registered bottled animals. Compiled a catalogue of osteological specimens at the British Museum. (Gunther 1975; letter from J. E. Gray, 21 February 1862.)

Gerstaecker, Carl Edvard Adolf (1828–95). German entomologist and zoologist. Curator of the entomological collection at the Natural History Museum, Berlin, 1856. Lecturer at the Agricultural College, Berlin, 1864–74; professor of zoology, 1874–6. (*NDB*.)

Gibson, James Brown (1805–68). Physician. Entered military service in 1826 as a hospital assistant; later promoted to surgeon. Served in the Crimean War. Director-general of the Army Medical Department, 1860–7. Knighted, 1865. (*DNB*.)

[after 29 June 1862]

Gingins-La-Sarra, Frédéric Charles Jean de (1790–1863). Swiss historian and botanist. Pupil of Augustin Pyramus de Candolle. Founded the Société d'Histoire de la Suisse Romande in 1837. (*Grande encyclopédie*, *Taxonomic literature*.)

Glaisher, James (1809–1903). Astronomer and meteorologist. Assistant, Cambridge University observatory, 1833–5. In charge of the magnetic and meteorological department at the Royal Observatory, Greenwich, 1838–74. Secretary, Royal Meteorological Society, 1850–72. FRS 1849. (*DNB*, *DSB*.)

Godron, Dominique Alexandre (1807–80). French botanist, zoologist, and ethnologist. Became head of the Nancy science faculty in 1854, where he established a natural history museum and a botanic garden. (*DBF*.)

Goethe, Johann Wolfgang von (1749–1832). German poet and naturalist. (*DSB*, *NDB*.)

Götzinger, Wilhelm Lebrecht (1758–1818). German clergyman and geographer. Published on the history of a number of areas in Saxony and on Lutheran theology. His publications were used in the redefinition of the Saxony borders during the Napoleonic wars. (*ADB* s.v. Götzinger, Max Wilhelm; Hamberger and Meusel 1796–1834; *NDB* s.v. Götzinger, Max Wilhelm.)

Goubert, Emile M. J. M. P. (d. 1867). French physician, botanist, and geologist. Became a member of the Société botanique de France while still a student, 1857/8. Professor of physical sciences, Lycée Impérial Louis-le-Grand, Paris. Author of botanical and geological papers, medical works, and of a work on science and religion. His herbarium comprised five or six thousand French plants, together with about a thousand Russian plants, and a collection of mosses and seaweeds. (*BLC*; *Bulletin de la Société Botanique de France* 5 (1858): vi, 14 (1867): 144, 240; letter from E. M. J. M. P. Goubert, 20 September 1862; Lorenz ed. 1868.)

20 September 1862

Gould, Augustus Addison (1805–66). American physician and conchologist. Practised medicine in Boston, Massachusetts. Co-author, with Louis Agassiz, of the *Principles of zoology* (1848). Described the shells collected during the United States Exploring Expedition, 1838–42. An active member of the Boston Society of Natural History. (*DAB*, *DSB*.)

Gould, John (1804–81). Ornithologist and artist. Taxidermist to the Zoological Society of London, 1826–81. Described the birds collected on the *Beagle* and *Sulphur* expeditions. Published forty-one illustrated folio volumes. FRS 1843. (*DNB*, *DSB*.)

Gower, William Hugh (1835–94). Foreman, Royal Botanic Gardens, Kew, until 1865. Principally recognised for his knowledge of orchids. (R. Desmond 1994.)

Gratiolet, Louis Pierre (1815–65). French anatomist and anthropologist. Laboratory assistant, Muséum d'Histoire Naturelle, Paris, 1842–53. Lectured on anatomy at the museum, 1844–50. Deputy professor of zoology, Muséum d'Histoire Naturelle, 1862–3; professor, 1863–5. (*DSB.*)

Gray, Asa (1810–88). American botanist. Fisher Professor of natural history, Harvard University, 1842–73. Devoted much time to the Harvard botanic garden and herbarium. Foreign member, Royal Society, 1873. (*DAB*, *DSB.*)

22 January [1862], 16 February [1862], *18 February 1862*, *6 March [1862]*, 15 March [1862], *31 March [1862]*, 21 April [1862], *18 May 1862*, *[2 June 1862]*, 10–20 June [1862], *[late June 1862]*, 1 July [1862], *2–3 July 1862*, 14 July [1862], *15 July [1862]*, *21 July 1862*, 23[–4] July [1862], 28 July [1862], *29 July 1862*, *4 August 1862*, 9 August [1862], *18–19 August 1862*, 21 August [1862], [3–]4 September [1862], *5 September 1862*, *22 September 1862*, *4 and 13 October 1862*, 16 October [1862], *27 October 1862*, 6 November [1862], *10 November 1862*, *17 November 1862*, 23 November [1862], *24 November 1862*, 26[–7] November [1862], *9 December 1862*, *29 December 1862*

Gray, Jane Loring (b. 1821/2). Wife of Asa Gray. (Dupree 1959.)

Gray, John Edward (1800–75). Naturalist. Assistant keeper of the zoological collections at the British Museum, 1824; keeper, 1840–74. President, Botanical Society of London, 1836–56. FRS 1832. (R. Desmond 1994, *DNB.*)

28 January 1862, *29 January 1862*, *1 February 1862*, *21 February 1862*

Green, Charles (d. 1886). Gardener to William Borrer. (R. Desmond 1994.)

Greene, Joseph Reay. Irish naturalist. One of the editors of the *Natural History Review.* Studied Protozoa and coelenterates. (*Catalogue of graduates, University of Dublin*; L. Huxley ed. 1900.)

Gregory, Augustus Charles (1819–1905). Explorer and surveyor in Australia. First commissioner of crown lands, Queensland, 1859–63; surveyor-general, 1859–79. Appointed to the Legislative Council in 1882. Strong critic of the government. President of the Australasian Association for the Advancement of Science, 1895. Knighted, 1903. (*Aust. dict. biog.*)

Gregory, Francis Thomas (1821–88). Surveyor and explorer in Australia. Assistant surveyor, Western Australia, 1847–9; staff surveyor, 1849. Commanded an expedition to the Gascoigne River in 1858. Campaigned for the establishment of a colony on the north-west coast of Australia. Awarded the founder's medal of the Royal Geographical Society, 1863. Brother of Augustus Charles Gregory. (*Aust. dict. biog.*)

Grey, George (1812–98). Army officer, explorer, and colonial governor. Governor of South Australia, 1841–5; of New Zealand, 1845–53 and 1861–8; of Cape Colony, 1854–61. Settled in New Zealand, 1870–94. Prime minister of New Zealand, 1877–9. Knighted, 1848. (*Aust. dict. biog.*, *DNB.*)

Grieve, James (*c.* 1840–1924). Scottish nurseryman. Hybridised pansies for Messrs Dickson and Co., Edinburgh, 1859–95. Nurseryman, Edinburgh, 1896. (R. Desmond 1994.)

Grieve, Peter (1812–95). Gardener to Brinsley Butler, 4th earl of Lanesborough (*Complete peerage*) and to Edward Richard Bengon (*Alum. Cantab.*) at Culford Hall, 1847. Author of works on pelargoniums. (R. Desmond 1994.)

Gris, Jean Baptiste Arthur (Arthur) (1829–72). French botanist. Botanical assistant, Muséum d'Histoire Naturelle, 1858. Succeeded Adolphe Théodore Brongniart as professor of botany at the muséum. Studied diverse aspects of plant anatomy and physiology; in collaboration with Brongniart, studied the flora of New Caledonia. (*DBF*, *Grande encyclopédie.*)

Grönland, Johannes (1824–91). German botanist. Worked as a chemist before going to Paris in 1853, where he became Pierre Vilmorin's botanical assistant. Member of the Société botanique de France, 1854. Published several papers on lichens and argued for the hybrid origin of plants such as wheat. Left Paris in 1871, after the Franco-Prussian war. Worked at the Botanical and Agricultural Chemistry Research Station at Dahme, 1872–91. (Magnus 1891.)

Guillaumin et cie. French publishers. Founded in 1833 by Gilbert Urbain Guillaumin (1801–64) to publish works on political economy and business. Published the first French translation of *Origin* (1862). (*DBF*, *NUC.*)

Günther, Albrecht Carl Ludwig Gotthilf (Albert Charles Lewis) (1830–1914). German-born zoologist. Began his association with the British Museum in 1857 when he was charged with completing the museum's catalogues of amphibia, reptiles, and fish; officially joined the staff in 1862. Assistant keeper of the zoological department, 1872–5; keeper, 1875–95. Edited the *Record of Zoological Literature*, 1865–9. FRS 1867. (*DNB*, *NDB.*)

Gurney, Hudson (1775–1864). Antiquary and verse-writer. Inherited a fortune from his father. MP for Shaftesbury, 1812; for Newton, Isle of Wight, 1816–31. High sheriff of Norfolk, 1835. Fellow of the Society of Antiquaries, 1818; vice-president, 1822–46. FRS 1818. (*DNB*, *Modern English biography.*)

Haast, John Francis Julius (Julius) von (1822–87). German-born explorer and geologist. Travelled to New Zealand in 1858 to report on the prospects of German emigration for the English ship-owners Willis Gann and Company. Explored the western districts of Nelson province in 1859 at the request of the provincial government. Appointed provincial geologist in Canterbury province, 1861. Conducted the first geological survey of Canterbury province, 1861–8. Became a British national in 1861. Founded the Canterbury Museum in 1861 and the Philosophical Institution of Canterbury in 1862. One of the founders of Canterbury Collegiate Union, 1871. Progessor of geology, Canterbury College, 1876–87. Member of the New Zealand senate, 1879–87. Knighted, 1886. FRS 1867. (*DNZB.*)

9 December 1862

Halifax, Lord. *See* Wood, Charles.

Hamilton, duke of. *See* Douglas, William Alexander Anthony Archibald.

Hardwicke, Robert (1823–75). London printer and publisher. One of the founders of the Quekett microscopical club, 1865. Published *Hardwicke's Science Gossip*, 1865–75. (*Modern English biography.*)

Hardy, Charles (1803–85). Clergyman. Curate of the Subdeanery Church in Chichester, 1827–32. Vicar of North and South Hayling, Hampshire, 1832–80. (*Alum. Cantab.*, *Crockford's.*)

Hardy, William. Bailiff and steward at Chillingham Castle, Northumberland, seat of Charles Augustus Bennet, earl of Tankerville. (*Post Office directory of Northumberland* 1858.)

Harris, George E. Tailor at 20 Gloucester Street, Bloomsbury, London. (Letter from G. E. Harris, 3 March 1862.)
3 March 1862

Harvey, William Henry (1811–66). Irish botanist. Colonial treasurer in Cape Town, 1836–42. Keeper of the herbarium, Trinity College, Dublin, from 1844; professor of botany, 1856–66. FRS 1858. (*DNB*, *DSB.*)

Haswell, Robert. Moved to Tromer Lodge, Down, in 1862. Formerly resident in Clapham. (Letter from Henrietta Emma Darwin to William Erasmus Darwin, [22 February 1863] (DAR 210.6: 109); letter from J. B. Innes to T. S. Stephens, [before 5 May 1862]; letter to J. B. Innes, 22 December [1862]; *The Times*, 20 February 1863, p. 11.)

Haughton, Samuel (1821–97). Clergyman, mathematician, and palaeontologist. Professor of geology at Dublin University, 1851–81. Became registrar of the medical school after graduating in medicine in 1862. President of the Royal Irish Academy, 1887. FRS 1858. (R. Desmond 1994, *DNB*, Sarjeant 1980.)

Hawthorn, Sarah. Eldest daughter of the clergyman, Robert Hawthorn of Stapleford, Cambridgeshire. Married Leonard Jenyns in 1862. (*DNB* s.v. Jenyns, Leonard.)

Headland, Edward (1803–69). Surgeon practising in London. Member of the Royal College of Surgeons, 1848. President of the Medical Society of London. The leading general practitioner of his time. One of the first practitioners to charge for his services rather than for medicine. (*Modern English biography.*)

Hector, James (1834–1907). Scottish geologist. Surgeon and geologist on the government exploring expedition to the western parts of British North America, 1857–60. Geologist to the provincial government of Otago, New Zealand, 1861–5. Director of the Geological Survey of New Zealand, 1865. Director of the meteorological and weather department of the New Zealand Institute, of the Colonial Museum, and of the botanical garden, Wellington, 1866–1903. Knighted, 1887. FRS 1866. (*DNZB.*)

Heer, Oswald (1809–83). Swiss botanist, palaeontologist, and entomologist. An expert on Tertiary flora. Lecturer on botany, University of Zürich, 1834–5; director of the botanic garden, 1834; associate professor, 1835–52; professor of botany and entomology, 1852–83. (*DSB*, *NDB.*)

Henry, Isaac Anderson. *See* Anderson-Henry, Isaac.

Henslow, Anne Frances (d. 1863). Eldest daughter of John Prentis Henslow and sister of John Stevens Henslow. (*Gentleman's Magazine* n.s. 15 (1863): 520; Jenyns 1862, pp. 4–5.)

Henslow, John Stevens (1796–1861). Clergyman, botanist, and mineralogist. CD's teacher and friend. Professor of mineralogy, Cambridge University, 1822–7; of botany, 1825–61. Extended and remodelled the Cambridge botanic garden. Curate of Little St Mary's Church, Cambridge, 1824–32; vicar of Cholsey-cum-Moulsford, Berkshire, 1832–7; rector of Hitcham, Suffolk, 1837–61. (*DNB*, *DSB*.)

Henslow, Leonard Ramsay (1831–1915). Eldest son of John Stevens Henslow. Curate of Hitcham, Suffolk, 1854–6; of Bangor-Monachorum, Flintshire, 1856–60; of Great Chart, Kent, 1860–3. Rector of St Mary Magdalene, Pulham, Norfolk, 1863–70; of Zeals, Wiltshire, 1870–1915. (*Alum. Cantab.*)

Herbert, Edward, 2d earl of Powis (1785–1848). Tory MP for Ludlow, 1806–39. Assumed the surname Herbert, in lieu of Clive, in 1807. Succeeded to the earldom in 1839. (*Complete peerage*, *DNB*.)

Herbert, Edward James, 3d earl of Powis (1818–91). Styled Viscount Clive, 1839–48. Conservative MP for North Shropshire, 1843–8. Lieutenant-colonel, South Shropshire Yeomanry, 1848. High steward of Cambridge University, 1863–91. Succeeded to the earldom in 1848. (*Alum. Cantab.*, *Complete peerage*.)

Herbert, Percy Egerton (1822–76). Army officer. Second son of Edward Herbert, second earl of Powis. Served in the Kaffir and Crimean Wars. Deputy quartermaster-general at Horse Guards, 1860–5; at Aldershot, 1865–7. Major-general, 1868. MP for Ludlow, 1854–60; for South Shropshire, 1865–76. Knighted, 1869. (*Burke's peerage* 1980, *DNB*.)

Herbst, Hermann Carl Gottlieb (*c.* 1830–1904). Botanist and nurseryman. Director of the Botanic Gardens, Rio de Janeiro. Probably established the nursery company, Herbst & Co., Rio de Janeiro. Nurseryman in Richmond, Surrey, *c.* 1864–84. (R. Desmond 1994, *Gardener's Magazine* 47 (1904): 222.)

Herbst & Co. Nurserymen of Rio de Janeiro. Probably established by Hermann Carl Gottlieb Herbst. (Letter from J. D. Hooker, [18 October 1862].)

Heron, Robert (1765–1854). Whig politician. MP for Grimsby, 1812–18; Peterborough, 1819–47. Succeeded to a baronetcy upon his father's death in 1805. (*DNB*.)

Herschel, John Frederick William (1792–1871). Astronomer, mathematician, chemist, and philosopher. Member of many learned societies. Carried out astronomical observations at the Cape of Good Hope, 1834–8. Master of the Royal Mint, 1850–5, retiring after a bout of illness. Created baronet, 1838. FRS 1813. (*DNB*, *DSB*.)

Higgins, John (1796–1872). Land agent. Born in Shrewsbury but settled in Alford, Lincolnshire, in 1819. Agent for CD's farm at Beesby, Lincolnshire. Crown agent and receiver for the northern district of England. Assistant enclosure commissioner for England and Wales. (*Post Office directory of Lincolnshire* 1849–68.)

Hildebrand, Friedrich Hermann Gustav (1835–1915). German botanist. Lecturer in Bonn, 1860. Professor of botany, University of Freiburg, 1868–1907. Primarily interested in ecological aspects of botany. (Barnhart 1965, Junker 1989.)
14 July 1862

Hill, Richard (1795–1872). American entomologist. (Gilbert 1977, *NUC*.)

Hill, Rowland (1795–1879). Inventor of the penny postage. Invented the rotatory printing-press and established the penny postage in 1840. Dismissed from the Post Office, 1842. Chairman of the Brighton Railway, 1843–6. Secretary to the post-master general, 1846; to the Post Office, 1854–64. Knighted, 1860. FRS 1857. (*DNB*.)

Hofmann, August Wilhelm von (1818–92). German organic chemist. Student of Justus von Liebig (*DSB*) at the University of Giessen, 1836; personal assistant to Liebig, 1843. Director, Royal College of Chemistry, London, 1845–65. Professor of chemistry, Berlin University, 1865–92. FRS 1851. (*DSB*, *NDB*.)
27 June 1862

Hogg, Robert (1818–97). Horticulturist and pomologist. Gained practical horticultural training with Peter Lawson in Edinburgh, and Hugh Ronalds (R. Desmond 1994) in Brentford, after which he studied in Paris, Belgium, and Germany. Partner in a nursery in Brompton Park, London, 1842–51. Co-editor of the *Cottage Gardener* (later *Journal of Horticulture*), 1855–95. (R. Desmond 1994.)
20 December [1862]. *See also* Johnson, George William.

Holland, Henry, 1st baronet (1788–1873). Physician. Distant relative of the Darwins and Wedgwoods. Physician in ordinary to Prince Albert, 1840; to Queen Victoria, 1852. President of the Royal Institution for many years. Created baronet, 1853. FRS 1815. (*DNB*, *Physicians*.)
[3–14 January 1862], 15 January [1862], [21 January 1862?], 30 January [1862], [1 or 8 February 1862], 26 March [1862], [c. April 1862], 19 May [1862]

Hooker, Brian Harvey Hodgson (b. 1860). Fifth child of Frances Harriet and Joseph Dalton Hooker. (Allan 1967.)

Hooker, Charles Paget (1855–1933). Physician and surgeon. Son of Frances Harriet and Joseph Dalton Hooker. Trained at St Bartholomew's Hospital, London; became a licentiate of the Royal College of Physicians and the Royal College of Surgeons, Edinburgh, in 1879, before being appointed to the staff of the Hertfordshire General Infirmary. Worked at Cottishall Cottage Hospital, Norfolk, 1880–5; at Cirencester Cottage Hospital, Gloustershire, 1885–1912. (Allan 1967, *Medical directory* 1880–1900, *Medical who's who* 1914.)

Hooker, Frances Harriet (1825–74). Daughter of John Stevens Henslow. Married Joseph Dalton Hooker in 1851. (*DNB* s.v. Hooker, Joseph Dalton.)

Hooker, Isabella Whitehead (1819–80). Daughter of Mr Smith, a land surveyor from Glasgow. Married William Dawson Hooker in 1839. (Allan 1967.)

Hooker, Joseph Dalton (1817–1911). Botanist. Worked chiefly on taxonomy and plant geography. Son of William Jackson Hooker. Friend and confidant of CD's. Accompanied James Clark Ross on his Antarctic expedition, 1839–43,

Hooker, Joseph Dalton, cont.

and published the botanical results of the voyage. Appointed palaeobotanist to the Geological Survey of Great Britain in 1846. Travelled in the Himalayas, 1847–50. Assistant director, Royal Botanic Gardens, Kew, 1855–65; director, 1865–85. Knighted, 1869. FRS 1847. (*DNB*, *DSB*.)

[1 January 1862], *[30 December 1861 or 6 January 1862]*, 16 January [1862], *[19 January 1862]*, 25 [and 26] January [1862], *[25 January 1862]*, 30 January [1862], *[31 January – 8 February 1862]*, *[8 February 1862]*, 9 February [1862], [before 15 February 1862], *[before 15 February 1862]*, 25 February [1862], *[26 February 1862?]*, *27 February 1862*, [1 March – 15 May 1862], *3 March 1862*, 7 March [1862], *[10 March 1862]*, 14 March [1862], *17 March 1862*, 18 March [1862], 22 [March 1862], *[23 March 1862]*, *[23–5 March 1862]*, 26 [March 1862], *[after 26 March 1862?]*, *[7 April 1862]*, 9 [April 1862], *[15 April 1862]*, 1 May [1862], *[5 May 1862]*, 9 May [1862], 15 [May 1862], *[16 May 1862]*, *[17 May 1862]*, [18 May 1862], *23 May 1862*, *[29 May 1862]*, 30 May [1862], *9 June 1862*, 11 June [1862], *19 [June 1862]*, 23 June [1862], *28 June 1862*, 30 [June 1862], *2 July 1862*, *10 July 1862*, *[24 July 1862]*, 26 July [1862], *20 August 1862*, 22 [August 1862], *[26–31 August 1862]*, 11 September [1862], *16 September 1862*, [18 September 1862], *20 September 1862*, 21 [September 1862], 26 September [1862], 6 October [1862], *[12 October 1862]*, 14 [October 1862], *[18 October 1862]*, *25 October 1862*, 27 [October 1862], *2 November 1862*, 3 November [1862], 4 November [1862], *7 November 1862*, [10–]12 November [1862], *12 November 1862*, *[15 and] 20 November [1862]*, 18 [November 1862], 24 [November 1862], *26 November 1862*, [after 26] November [1862], 12 [December 1862], *[14 December 1862]*, *[21 December 1862]*, [21 December 1862], 24 December [1862], *[27 or 28 December 1862]*, *[before 29 December 1862]*, 29 [December 1862], *[31 December 1862]*

Hooker, William Dawson (1816–40). Ornithologist and entomologist. Eldest son of William Jackson Hooker. MD, Glasgow University, 1838. Professor of materia medica, Anderson's College, Jamaica. (R. Desmond 1994, *Roll of the graduates of the University of Glasgow*.)

Hooker, William Henslow (1853–1942). Eldest child of Frances Harriet and Joseph Dalton Hooker. (Allan 1967.)

Hooker, William Jackson (1785–1865). Botanist. Father of Joseph Dalton Hooker. Regius professor of botany, Glasgow University, 1820. Director of the Royal Botanic Gardens, Kew, 1841–65. Knighted, 1836. FRS 1812. (*DNB*, *DSB*.)

Hooker, Willielma (1840–79). Daughter of William Dawson and Isabella Whitehead Hooker. Married James Campbell in 1862. (Allan 1967.)

Hope, Frederick William (1797–1862). Entomologist and clergyman. Curate of Frodesley, Shropshire. President of the Entomological Society, 1835 and 1846. Founded a professorship of zoology at Oxford University in 1849, and in the same year donated his entomological and other collections. FRS 1834. (*DNB*, Gilbert 1977.)

Hopkins, William (1793–1866). Mathematician and geologist. Tutor in mathematics at Cambridge University. President of the Geological Society, 1851–3. Specialised in quantitative studies of geological and geophysical questions. FRS 1837. (*DNB*, *DSB*.)

Horace [Quintus Horatius Flaccus] (65–8 B.C.). Roman poet. (*EB*.)

Horner, Anne Susan (1789–1862). Married Leonard Horner in 1806. (Freeman 1978.)

Horner, Leonard (1785–1864). Scottish geologist and educationist. Founded the Edinburgh School of Arts in 1821. Warden of University College London, 1827–31. Inspector of factories, 1833–56. A promoter of science-based education at all social levels. President of the Geological Society, 1846 and 1860–2. Father-in-law of Charles Lyell. FRS 1813. (*DNB*, *DSB*.)
13 June [1862]

Horwood, John (b. 1824). Gardener to George Henry Turnbull of Down, Kent. (Census returns 1861 (Public Record Office, RG9/462: 70); *Orchids*, p. 158 n.)

Hoskyn, Richard (*fl.* 1840–62). Naval officer and surveyor. Master in Royal Navy, 1840. During 1862, carried out deep sea soundings to the west of Ireland in HMS *Porcupine*. (Hoskyn 1862, *Navy list* 1862.)

Humboldt, Friedrich Wilhelm Heinrich Alexander (Alexander) von (1769–1859). Prussian naturalist and traveller. Official in the Prussian mining service, 1792–6. Explored equatorial South America, Mexico, and the United States, 1799–1804. Travelled in Siberia, 1829. Foreign member, Royal Society, 1815. (*DSB*, *NDB*.)

Hunter-Blair, Jane Anne Eliza. Married Philip Lutley Sclater in 1862. (*Gentleman's Magazine* n.s. 13 (1862): 630.)

Huxley, Henrietta Anne (1825–1915). Married Thomas Henry Huxley in 1855. (Freeman 1978.)

Huxley, Noel (1856–60). Eldest child of Henrietta Anne and Thomas Henry Huxley. Died of scarlet fever. (L. Huxley ed. 1900.)

Huxley, Thomas Henry (1825–95). Zoologist. Assistant surgeon on HMS *Rattlesnake*, 1846–50. Lecturer on natural history, Royal School of Mines, 1854; professor, 1857. Naturalist to the Geological Survey of Great Britain, 1855. Fullerian professor of physiology, Royal Institution, 1863–7. Hunterian professor of comparative anatomy, Royal College of Surgeons, 1863–9. Editor-in-chief of the *Natural History Review* from 1861. President of the Royal Society, 1883–5. Served on several commissions of inquiry and administration. Privy councillor, 1892. FRS 1851. (*DNB*, *DSB*.)
13 January 1862, 14 [January 1862], *20 January 1862*, 22 January [1862], 2 February [1862], 6 February [1862], 30 April [1862], *6 May 1862*, 10 May [1862], *9 October 1862*, *10 October [1862]*, *2 December 1862*, 7 December [1862], 18 December [1862], 28 December [1862]

Innes, Eliza Mary Brodie. Married John Innes (later John Brodie Innes) in 1847. (*Burke's landed gentry* 1879.)

Innes, John Brodie (1817–94). Perpetual curate of Down, 1846–68; vicar, 1868–9. Left Down at the start of 1862 after inheriting an entailed estate at Milton Brodie, near Forres, Scotland; in 1861, changed his name to Brodie Innes as a condition of the legacy. Priest in charge of Milton Brodie Mission and general licentiate of the diocese of Moray, 1861. Chaplain to the Bishop of Moray, 1861–80 and 1886–94. (*Clergy list*, *County families* 1864, *Crockford's*, Freeman 1978, J. R. Moore 1985.)

2 January [1862], [3] January [1862], *19 February [1862]*, 24 February [1862], 1 May [1862], *[before 5 May 1862]*, *5 May [1862]*, *16 December [1862]*, 22 December [1862]

Innes, John William Brodie (1848–1923). Barrister and novelist. Son of John Innes. Called to the bar, 1876. Advocate of the Scottish bar, 1888. Interested in antiquarian research, romance, demonology, witchcraft, and criminology. (*Alum. Cantab.*, Freeman 1978.)

Jamieson, Thomas Francis (1829–1913). Scottish geologist and agriculturalist of Ellon, Aberdeenshire. Factor on estates for many years; later took the farm of Mains, Waterton. Fordyce lecturer on agricultural research, University of Aberdeen, 1862. Notable for his researches on Scottish Quaternary geology and geomorphology. Fellow of the Geological Society, 1862. (*Geological Magazine* 50 (1913): 332–3, *Roll of the graduates of the University of Aberdeen*, Sarjeant 1980.)

24 March 1862, *17 October 1862*, 21 November 1862

Janson, Edward Westley (1822–91). Natural history agent, publisher, and entomologist. Authority on British Coleoptera. Curator of the Entomological Society, 1850–63; librarian, 1863–74. (*Entomologist* 24 (1891): 252, Gilbert 1977.)

Jenyns, Leonard (1800–93). Naturalist and clergyman. Brother-in-law of John Stevens Henslow. Vicar of Swaffham Bulbeck, Cambridgeshire, 1828–49. Settled near Bath in 1850. Founder and first president of the Bath Natural History and Antiquarian Field Club in 1855. Member of many scientific societies. Described the *Beagle* fish specimens. Adopted the name Blomefield in 1871. (R. Desmond 1994, *DNB*.)

22 January [1862], 24 January [1862], 24 May [1862], *28 May 1862*

Johnson, George Henry Sacheverell (1808–81). Scholar and clergyman. Brother of Henry Johnson. Savilian professor of astronomy, Oxford University, 1839–42; Whyte professor of moral philosophy, 1842–5. Dean of Wells, 1854–81. FRS 1838. (*DNB*, *Salopian Shreds and Patches* 5 (1882): 2.)

Johnson, George William (1802–86). Barrister and writer on gardening. Barrister, Gray's Inn, 1836. Professor of moral and political economy at the Hindoo college, Calcutta, 1839–42. Founded the *Cottage Gardener* (later *Journal of Horticulture*) in 1848. (R. Desmond 1994, *DNB*.)

20 December [1862]. *See also* Hogg, Robert.

Johnson, Henry (1802/3–81). Physician. A contemporary of CD's at Shrewsbury School and Edinburgh University. Senior physician, Shropshire Infirmary. Member of Royal College of Physicians of London, 1859. Founder member and honorary secretary of the Shropshire and North Wales Natural History and Antiquarian Society, 1835–77. (*Medical directory*, *Salopian Shreds and Patches* 5 (1882): 2.)

30 September 1862, *8 October 1862*

Jones, Thomas Rupert (1819–1911). Geologist and palaeontologist. Assistant-secretary, Geological Society, 1851–62; responsible for editing the *Journal of the Geological Society*. Lecturer on geology, Royal Military College, 1858; professor, 1862. Later professor of geology at the Staff College, Sandhurst; retired in 1880. President of the Geologists' Association, 1879–81. An authority on Phyllopoda, Ostracoda, and Foraminifera. FRS 1872. (*DNB*.)

Jones, Thomas Rymer (1810–80). Zoologist. Professor of comparative anatomy, King's College, London, 1836–74. Fullerian professor of physiology at the Royal Institution, 1840–2. FRS 1844. (*DNB*.)

Journal of Horticulture

[before 10 June 1862], [before 15 July 1862], [before 22 July 1862], [before 25 November 1862], [before 2 December 1862], [before 27 December 1862]

Jukes, Joseph Beete (1811–69). Geologist. Geological surveyor of Newfoundland, 1839–40. Naturalist aboard HMS *Fly* in the survey of the north-east of Australia, 1842–6. Geologist with the Geological Survey of Great Britain working in North Wales, 1846–50. Director of the Irish branch of the Geological Survey, 1850–69. FRS 1853. (*DNB*, *DSB*.)

25 May 1862, *30 May 1862*, *3 November 1862*

Karsten, Gustave Karl Wilhelm Hermann (Hermann) (1817–1908). German botanist. Lecturer on botany at the University of Berlin. Travelled in South America, 1848–56. Professor of botany, University of Vienna, 1866. (*NDB*.)

Kincardine, earl of. *See* Bruce, James.

King, John (1820–95). Australian landowner and farmer. Resident in Gippsland, Victoria, Australia. Brother of Philip Gidley King. (*Aust. dict. biog.*)

King, Philip Gidley (1817–1904). Australian farmer and mining company manager. Midshipman, HMS *Beagle*, 1831–6. Returned to Australia in 1836. Entered the service of the Australian Agricultural Company in 1842; superintendent of stock, 1851. New South Wales manager of the Peel River Land and Mineral Company, 1852–81. (*Aust. dict. biog.*)

16 September 1862, 16 November [1862]

King, Samuel William (1821–68). Traveller, entomologist, and geologist. Rector of Saxlingham Nethergate, Norfolk, 1851. Friend of Charles Lyell, whom he assisted in geological investigations. (*DNB*.)

Kingsley, Charles (1819–75). Author and clergyman. Lecturer on English literature, Queen's College, London, 1848–9. Professor of modern history, Cambridge

Kingsley, Charles, cont.
University, 1860–9. Rector of Eversley, Hampshire, 1844–75. Chaplain to the Queen, 1859–75. (*Alum. Cantab.*, *DNB.*)
31 January 1862, 6 February [1862]

Kippist, Richard (1812–82). Botanist. Librarian of the Linnean Society of London, 1842–80. Specialist on Australian plants. (R. Desmond 1994, *DNB.*)
18 March [1862], 18 March [1862]

Kirby, William (1759–1850). Clergyman and entomologist. One of the authors of the Bridgewater Treatises (1833–6) and, in collaboration with William Spence, co-author of *Introduction to entomology* (1815–26). Vicar of Barham, Suffolk, 1797–1850. One of the original fellows of the Linnean Society, 1788. Honorary president of the Entomological Society, 1837. FRS 1818. (*Alum. Cantab.*, *DNB.*)

Kirby, William Forsell (1844–1912). Entomologist. Assistant at the Royal Dublin Society, 1867–79. Assistant in the zoological department of the British Museum, 1879. (R. Desmond 1994, Gilbert 1977.)
12 December [1862]

Kleine, Georg (1806–94). Beekeeper in Lüethorst, Hanover. (Gilbert 1977.)

Klotzsch, Johann Friedrich (1805–60). German botanist. Curator of the herbarium, University of Berlin, 1833–60. (*ADB*, *Taxonomic literature.*)

Knight, Charles (1791–1873). Publisher and journalist. Wrote works of popular instruction; edited the *Plain Englishman* and other popular periodicals. Publisher to the Society for the Diffusion of Useful Knowledge, 1829–46. Campaigned for the removal of stamp duty on newspapers. (*DNB.*)

Knight, Thomas Andrew (1759–1838). Botanist and horticulturist. Correspondent to the Board of Agriculture from 1795. President of the Horticultural Society of London, 1811–38. Carried out important work on the phenomenon now known as geotropism. FRS 1805. (R. Desmond 1994, *DSB.*)

Koch, Wilhelm Daniel Joseph (1771–1849). German physician and botanist. Professor of botany, Erlangen University, 1824. (*ADB.*)

Kolenati, Friedrich August Rudolph (1812–64). Bohemian zoologist and botanist at the Polytechnic Institute of Brünn. Travelled in Russia, 1842–5. (*Taxonomic literature.*)

Kölreuter, Joseph Gottlieb (1733–1806). German botanist. Keeper of the natural history collections, Imperial Academy of Sciences, St Petersburg, 1755–61. Professor of natural history and director of the court gardens at Karlsruhe, Baden, 1761–86. Carried out extensive hybridisation experiments on plants. (*DSB*, *NDB.*)

Kraatz, Ernst Gustav (1831–1909). German entomologist. An authority on the Coleoptera. Co-founder of the Berliner Entomologische Gesellschaft in 1856. Founded the first purely entomological museum, the Deutsches Entomologisches National Museum. Fellow of the Entomological Society of London, 1876–1907. (*Entomologist's Monthly Magazine* 2d ser. 46 (1910): 21–2, Gilbert 1977.)

Krohn, August David (1803/4–91). Russian invertebrate zoologist and embryologist. (Freeman 1978.)

Kunth, Karl Sigismund (1788–1850). German botanist. Between 1813 and 1829, worked in Paris on the plants collected by Alexander von Humboldt on his travels in South America. Appointed professor of botany, Berlin University, 1829. (*ADB.*)

Laing, Samuel (1812–97). Railway magnate, politician, and author. Fellow of St John's College, Cambridge, 1834. Called to the bar, 1837. Secretary of the railway department of the Board of Trade, 1842. Chairman and managing director of the London, Brighton, and South Coast Railway Company, 1848. Chairman of the Crystal Palace Company, 1852. Indian finance minister, 1860. Published popular accounts of science, anthropology, and human origins. (*DNB.*)

Lamarck, Jean Baptiste Pierre Antoine de Monet de (1744–1829). French naturalist. Held various botanical positions at the Jardin du Roi, 1788–93. Professor of zoology, Muséum d'Histoire Naturelle, 1793. Believed in spontaneous generation and in the progressive development of animal types; widely known for his theory of transmutation. (*DSB.*)

Lamb, Charles (1775–1834). Essayist. (*DNB.*)

Landsborough, William (1825–86). Explorer and prospector. Emigrated from Scotland to New South Wales, Australia, in 1841. Made several exploratory excursions to the Australian interior. In 1861, led a search for the ill-fated expedition party led by Robert O'Hara Burke and William John Wills. First recorded explorer to cross the continent from north to south. Police magistrate and commissioner of crown lands in Carpentaria, 1865–70. (*Aust. dict. biog.*)

Lane, Edward Wickstead (1823–89). Physician. Proprietor of a hydropathic establishment at Moor Park, near Farnham, Surrey, 1854 (or before) – 1860; at Sudbrook Park, Surrey, 1860–79. Practised in Harley Street, London, 1879–89. Member of the Faculty of Advocates, the Botanical Society, and the Speculative Society, Edinburgh. Author of works on hydropathy. (Freeman 1978, *Medical directory* 1858–90, *Post Office directory of the six home counties* 1851–62.)

Langton, Charles (1801–86). Rector of Onibury, Shropshire, 1832–40. Left the Church of England in 1840. Resided at Maer, Staffordshire, 1840–6; at Hartfield Grove, Sussex, 1847–62. Married Emma Darwin's sister, Charlotte Wedgwood, in 1832. After her death, married CD's sister, Emily Catherine Darwin, in 1863. (*Alum. Oxon.*, *Emma Darwin.*)

Langton, Charlotte (1797–1862). Emma Darwin's sister. Married Charles Langton in 1832. Resided at Maer, Staffordshire, 1840–6; at Hartfield Grove, Sussex, 1847–62. (*Emma Darwin.*)

Lartet, Edouard Amant Isidore Hippolyte (1801–71). French palaeontologist and prehistorian. Trained as a lawyer but never practised. Began palaeontological research in the 1830s, and became interested in the question of the antiquity of the human species in the late 1850s; carried out investigations in 1860 that provided proof of the contemporaneity of man with extinct animal species. Named professor of palaeontology, Muséum d'Histoire Naturelle in 1869, but ill health prevented him from fulfilling his duties. (*DSB*, Sarjeant 1980.)

Latreille, Pierre André (1762–1833). French naturalist. Professor of entomology, Muséum d'Histoire Naturelle, 1829. (*DSB*, *NBU*.)

Lawes, John Bennet (1814–1900). Agricultural chemist. Developed the British industrial production of organic and chemical fertilisers; founded the Rothamsted Agricultural Experiment Station near Harpenden, Hertfordshire, in 1843. FRS 1854. (*DNB*, *DSB*.)

Lawson, Charles (1794–1873). From 1821 managed the Edinburgh nursery Peter Lawson & Son. (R. Desmond 1994.)

Lawson, Peter (*fl.* 1770s–1821). Founded the Edinburgh nursery Peter Lawson & Son in 1770. (R. Desmond 1994.)

Le Blanc, Sarah Jane (1804–64). George Bentham's sister. (Jackson 1906.)

Lecoq, Henri (1802–71). French naturalist and vulcanologist. Professor of natural history, University of Clermont-Ferrand, and director of the town's botanic garden, 1826–54. Taught at the Preparatory School of Medicine and Pharmacy, Clermont-Ferrand, from 1840; professor, science faculty, from 1854. Published widely on botany, agriculture, and meteorology. (*Grande encyclopédie*, Sarjeant 1980.)

Leicester, earl of. *See* Coke, Thomas William.

Leidy, Joseph (1823–91). American physician and naturalist. Demonstrator of anatomy, Franklin Medical College, 1846. Visited Europe in 1848 and 1850. Professor of anatomy, University of Pennsylvania, 1853–91. President of the Academy of Natural Sciences of Philadelphia, 1881–91. Published extensively on the fossil remains of extinct vertebrates of North America. (*DAB*, *DSB*.)

Leifchild, John R. (b. 1815). Author and reviewer. Son of John Leifchild, a leading evangelical minister (*DNB*). Visiting government commissioner to coal-mines in Newcastle and Durham. Author of works on mining and coal. Reviewed *Orchids* in the *Athenæum* in 1862. (*BLC*, *CDEL*, *NUC*, *Wellesley index*.)

Leighton, William Allport (1805–89). Botanist, clergyman, and antiquary. School-fellow of CD's in Shrewsbury, 1817. Curate of St Giles's, Shrewsbury, 1846–8. For many years, edited the *Transactions of the Shropshire Archaeological Society*. Published *The flora of Shropshire* (1841). (R. Desmond 1994, *DNB*, Freeman 1978.) 26 November [1862]

Leverrier, Urbain Jean Joseph (1811–77). Astronomer and politician. Elected to the Académie des Sciences, and appointed to the chair of astronomy at the Sorbonne, 1846. The first to postulate the existence of the planet Neptune. Elected as representative to the French legislative assembly, 1849; senator, 1852. Director of the Observatory, 1854–70 and 1873–7. (*Dictionnaire des parlementaires français*.)

Lewis, George Cornewall, 2d baronet (1806–63). Statesman and author. Poor Law commissioner for England and Wales, 1839–47. MP for Herefordshire, 1847–52; for Radnor boroughs, 1855–63. Editor of the *Edinburgh Review*, 1852–5. Chancellor of the Exchequer, 1855–8. Succeeded to the baronetcy in 1855. Home secretary, 1859–61. (*DNB*, *Modern English biography*.)

Lincoln, Abraham (1809–65). Republican president of the United States, 1861–5, at the time of the American Civil War. (*DAB*.)

Lindley, Barbara (*fl.* 1867). Younger daughter of John Lindley. Assisted her father in the illustration of some of his later works. Married Edmund Thompson, fellow of Christ's College, Cambridge, and rector of Clipstare, in 1867. (*DNB* s.v. Lindley, John; Pine 1972.)

Lindley, John (1799–1865). Botanist and horticulturist. Assistant in Joseph Banks's library, 1819. Garden clerk, Horticultural Society of London, 1822–7; assistant secretary, 1827–41; vice-secretary, 1841–58; honorary secretary, 1858–63. Lecturer on botany, Apothecaries' Company, 1836–53. Professor of botany, London University (later University College London), 1828–60. Editor of the *Gardeners' Chronicle* from 1841. FRS 1828. (R. Desmond 1994, *DNB*, *DSB*.)
14 September [1862]

Link, Heinrich Friedrich (1767–1851). German naturalist and philosopher. Professor of zoology, botany, and chemistry, University of Rostock, 1792–1811; of chemistry and botany, University of Breslau, 1811–15. Professor of botany and director of the botanic garden at the University of Berlin, 1815. Interested in the philosophical foundations of the natural sciences. Foreign member, Royal Society, 1842. (*DSB*, *NDB*.)

Linnaeus, Carolus (Carl von Linné) (1707–78). Swedish botanist and zoologist. Professor of medicine, University of Uppsala, 1741; court physician, 1747. Proposed a system for the classification of the natural world and reformed scientific nomenclature. FRS 1753. (*DSB*.)

Litchfield, Henrietta Emma. *See* Darwin, Henrietta Emma.

Littleton, Thomas (1402–81). Jurist. King's serjeant, 1455. Justice of the Common Pleas, 1466. Author of a treatise on legal tenure, *Tenures* (*c.* 1481), which had a profound influence on English property law. (*DNB*.)

Liveing, Catherine (d. 1888). Daughter of Rowland Ingram, rector of Little Ellingham, Norfolk (*Alum. Cantab.*) Married George Downing Liveing, 1860. (*DNB*, s.v. Liveing, George Downing.)

Liveing, George Downing (1827–1924). Chemist. Fellow, St John's College, Cambridge, 1853–60 and 1880–1924; president, 1911–24. Professor of chemistry, Staff and Royal Military Colleges, Sandhurst, 1860. Professor of chemistry, Cambridge University, 1861–1908. FRS 1879. (*Alum. Cantab.*, *DNB*.)

Loring, Charles Greely (d. 1867). Lawyer in Boston. Father of Jane Loring Gray. (Dupree 1959.)

Loudon, John Claudius (1783–1843). Landscape gardener and horticultural writer. Travelled in northern Europe, 1813–15; in France and Italy, 1819–20. Designed public gardens and cemeteries in England and Scotland. Founded and edited the *Gardener's Magazine*, 1826–43, and the *Magazine of Natural History*, 1828–36. (R. Desmond 1994, *DNB*.)

Louriero, João de (1717–91). Portuguese missionary and naturalist in Mozambique, Goa (India), and Chochin-China (Vietnam). (*Taxonomic literature*.)

Lovén, Sven (1809–95). Swedish marine biologist. Travelled to Spitsbergen and northern Norway, 1836–7. Curator of the invertebrate section of the Museum of Natural History, Stockholm, 1841. Studied shell banks on the west coast of Sweden, providing evidence that an arctic sea had once covered much of the present Scandinavian land-mass. (*Catalogue of the library of the British Museum (Natural History)*, *DSB*.)

Lowe, J. Beekeeper in Edinburgh. (Letter to *Journal of Horticulture*, [before 10 June 1862].)

Lowe, Richard Thomas (1802–74). Clergyman and botanist. English chaplain in Madeira, 1832–54. Rector of Lea, Lincolnshire, 1854–74. Published a flora of Madeira (1857–72). (R. Desmond 1994, *DNB*.)

Lowell, John Amory (1798–1881). Boston businessman and philanthropist. Trustee of the Lowell Institute, Boston. Member of the Harvard Corporation. Fellow of the American Academy of Arts and Sciences. A generous supporter of botanical studies. (Dupree 1959, *NUC*, Rossiter 1971.)

Lubbock, Ellen Frances (d. 1879). Married John Lubbock in 1856. (*Burke's peerage* 1970.)
[January 1862]

Lubbock, Harriet (1810–73). Married John William Lubbock in 1833.

Lubbock, Henry James (1838–1910). Banker. Second son of John William Lubbock. High sheriff for the county of London, 1897. (*WWW*.)

Lubbock, John, 4th baronet and 1st Baron Avebury (1834–1913). Banker, politician, and naturalist. Son of Harriet and John William Lubbock. A partner in the family bank. MP for Maidstone, Kent, 1870 and 1874; for London University, 1880–1900. Studied entomology and anthropology. An active supporter of CD's theory of natural selection. Succeeded to the baronetcy in 1865. Privy councillor, 1890. Created Baron Avebury in 1900. FRS 1858. (*DNB*, *DSB*.)
6 January 1862, 23 January [1862], *27 January 1862*, *29 January 1862*, *7 [February] 1862*, *13 February 1862*, *17 April 1862*, *15 May 1862*, 21 August [1862], *23 August 1862*, 2 September [1862], [3 September 1862], 5 September [1862], 23 October [1862], *25 October 1862*, 27 [October 1862], *30 October 1862*, *15 December 1862*, 16 [December 1862], *18 December 1862*

Lubbock, John William, 3d baronet (1803–65). Astronomer, mathematician, and banker. CD's neighbour in Down. First vice-chancellor of London University, 1837–42. Partner in the family bank, 1825. Treasurer and vice-president of the Royal Society, 1830–5 and 1837–48. Succeeded to the baronetcy in 1840. FRS 1829. (*DNB*, *DSB*.)

Lubbock, Montagu (b. 1842). Younger brother of John Lubbock. (*Burke's peerage* 1970.)

Lubbock, Norman (1861–1926). Fourth child of John and Ellen Frances Lubbock. (*Burke's peerage* 1980.)

Lucas, Prosper (1805–85). French physician and medical writer interested in heredity. (*NUC*.)

Ludwig, Camilla. German governess. Governess to the Darwin family, 1860–3. Translated German works for CD. Married Reginald Saint Pattrick, vicar of Sellinge, Kent (*Alum. Oxon.*). (CD's Classed account book (Down House MS); letter from R. S. Pattrick, 19 October 1881 (*Calendar* no. 13416).)
26 August [1862]

Ludwig, Louisa. German schoolmistress. Acted as governess to the Darwin family for periods in 1862 and 1864. Sister of Camilla Ludwig. (CD's Classed account book (Down House MS); letter to Camilla Ludwig, 26 August [1862].)

Lund, Peter Wilhelm (1801–80). Danish naturalist and speleologist. Carried out studies of South American caves. (Sarjeant 1980.)

Lupton, James Irvine (1830/1–1900). Veterinarian and author on veterinary subjects. Member of the Royal College of Veterinary Surgeons, 1859; fellow, 1877. In practice at Kingburgh, Warwickshire, 1860; at Richmond, Surrey, 1865 (or before) – 1900. (*Alum. Oxon.*, *CDEL*, *Veterinary directory* 1861, *Veterinary register* 1901.)

Lyell, Charles, 1st baronet (1797–1875). Scottish geologist. Scientific mentor and friend of CD's. Uniformitarian geologist whose *Principles of geology* (1830–3) and *Elements of geology* (1838) appeared in many editions. Professor of geology, King's College, London, 1831–3. President of the Geological Society, 1834–6 and 1849–50. Travelled widely and published accounts of his trips. Knighted, 1848; created baronet, 1864. FRS 1826. (*DNB*, *DSB*.)
[26–31 March 1862], 1 April [1862], *20 August 1862*, 22 August [1862], 1 October [1862], 14 October [1862], *17 October 1862*

Lyell, Leonard, 1st baron (1850–1926). Eldest son of Henry Lyell, Charles Lyell's brother, and Katharine Murray Lyell. MP for Orkney and Shetland, 1885–1900. Created baronet, 1894; peer, 1914. (*Burke's peerage* 1980.)

Lyell, Mary Elizabeth (1808–73). Eldest child of Leonard Horner. Married Charles Lyell in 1832. (Freeman 1978.)

Macarthur, Hannibal Hawkins (1788–1861). Australian farmer, politician, and businessman. Married Anna Maria King, Philip Parker King's sister, in 1812. (*Aust. dict. biog.*, *DNB*.)

MacCarthy, Charles Justin (1811–64). Statesman. Auditor-general of Ceylon, 1847; colonial secretary, 1851; governor, 1860–4. Knighted, 1857. (*Modern English biography*.)

McClellan, George Brinton (1826–85). American soldier. General-in-chief of the Union army, 1862–3. Unsuccessful Democratic candidate for the presidency, 1864. Governor of New Jersey, 1878–81. (*DAB*.)

McCoy, Frederick (1823–99). Naturalist and geologist. Employed in arranging the collection in the Woodwardian Museum, Cambridge University, 1846–54. Professor of mineralogy and geology, and curator of the museum, Queen's College, Belfast, 1849–54. Professor of natural science, University of Melbourne, 1854–99. Palaeontologist to the Australian Geological Survey, 1856. Director of the National Museum of Natural History and Geology, Melbourne, 1857–99. An ardent opponent of natural selection. FRS 1880. (*Aust. dict. biog.*, *DNB*.)

McGilvray, Anne (1851–62). Daughter of Joseph Dalton Hooker's sister, Maria McGilvray. (Allan 1967.)

McGilvray, Maria (1819–89). Sister of Joseph Dalton Hooker. (Allan 1967.)

Mackay, Charles (1814–89). Poet and journalist. Editor, *Illustrated London News*, 1852–9. Established the *London Review and Weekly Journal of Politics, Literature, Art and Science* in 1860. Special correspondent for *The Times* in New York, 1862–5. (*DNB.*)

McKinlay, John (1819–72). Explorer and land prospector. Emigrated from Scotland to New South Wales, Australia, in 1836. In 1861, led a search for the ill-fated expedition party led by Robert O'Hara Burke and William John Wills. McKinlay's was the second recorded expedition to cross the continent from south to north. (*Aust. dict. biog.*)

Mackintosh, James (1765–1832). Philosopher and historian. Professor of law and general politics at the East India Company College, Haileybury, 1818–24. Married Catherine Allen in 1798. (*DNB.*)

Mackintosh, Mary. American-born wife of Emma Darwin's cousin, Robert Mackintosh. Sister of Thomas Gold Appleton. (Wedgwood and Wedgwood 1980.)

Mackintosh, Robert (1806–64). Brother of Frances Wedgwood, Emma Darwin's sister-in-law. (Wedgwood and Wedgwood 1980.)

Mackintosh, Ronald. Son of Robert Mackintosh. (Wedgwood and Wedgwood 1980.)

Macleay, William Sharp (1792–1865). Naturalist and diplomat. Originated the circular or quinary system of classification. Commissary judge, Havana, 1830–7. Emigrated to Sydney, Australia, in 1839, where he devoted himself to the study of Australian natural history. Made large collections of Australian insects and marine fauna, and supervised the cultivation of exotic plants in his private gardens. Trustee of the Australian Museum, 1841–62. (*Aust. dict. biog.*, *DNB.*)

McNab, James (1810–78). Botanist. Collected plants in North America, 1834. Superintendent, Caledonian Horticultural Society, 1835. Curator of the Royal Botanic Garden, Edinburgh, 1849. President of the Edinburgh Botanical Society, 1872. (R. Desmond 1994.)

Malden, Bingham Sibthorpe (1830–1906). Clergyman. Curate of Clayton, Sussex, 1854–8; of St George-the-Martyr, Canterbury, 1858–63; of Pattingham, Staffordshire, 1863–70. Vicar of Sheldwich, Kent, 1870–1906. (*Alum. Cantab.*)

Mann, Gustav (1836–1916). German-born explorer and botanist. Gardener at the Royal Botanic Gardens, Kew, 1859. Botanical collector to the Niger Expedition, 1859–62. Worked for the Indian Forest Service, 1863–91. (R. Desmond 1994; R. Desmond 1995.)

Markham, Richard Anthony. *See* Salisbury, Richard Anthony.

Marshall, John (1818–91). Anatomist and surgeon. Demonstrator of anatomy, University College London, *c.* 1845; professor of surgery, 1866; later, Holme professor of clinical surgery. Consulting surgeon to University College Hospital, 1884. Professor of anatomy, Royal Academy, 1873–91. Pioneered operations to excise varicose veins. Instrumental in the discovery that cholera is a water-borne disease. FRS 1857. (*DNB*.)

Marshman, John (*fl.* 1862–7). Agent at the Canterbury Emigration Office, 16 Charing Cross, London, *c.* 1862–7. (*Post Office London directory*.)

Martens, Conrad (1801–78). Landscape painter. Joined HMS *Beagle* in Montevideo in 1833 and served as draughtsman until 1834. Settled in Australia in 1835. (*Aust. dict. biog.*, Bénézit 1976.)

20 January 1862

Martin, William Charles Linnaeus (1798–1864). Writer on natural history. Resided in Lee, Kent. Superintendent of the museum of the Zoological Society of London, 1830–8. Wrote works on domestic animals. (*DNB*, *Gentleman's Magazine* 1 (1864): 536, *Modern English biography*.)

Mason, James Murray (1798–1871). American politician and diplomat. United States senator from 1847 to 1861, when his state, Virginia, seceded from the Union. Appointed Confederate diplomatic commissioner to England in 1861; his detention with John Slidell by the United States Navy, created a tense diplomatic situation between Britain and the United States. (*DAB*.)

Masson, Victor (1807–79). Paris publisher. Joined the publishing firm of Crochard in 1838, becoming the sole proprietor in 1846. Published elegant editions of scientific works. (*Dictionnaire universel des contemporains*.)

Masters, Maxwell Tylden (1833–1907). Physician and botanist. In general practice at Peckham from 1856. Lecturer on botany at St George's Hospital, 1855–68. Editor of the *Gardeners' Chronicle*, 1865–1907. Specialised in the study of plant teratology. FRS 1870. (*DNB*.)

26 February [1862], *17 March 1862*, *[c. 15 May 1862]*, 8 July [1862], *12 July 1862*, 24 July [1862]

Masters, William (1796–1874). Nurseryman in Canterbury. Founded the Canterbury Museum, 1823; honorary curator, 1823–46. Conducted hybridisation experiments on passion-flowers. Designed the formal gardens at Walmer Castle. Father of Maxwell Tylden Masters. (R. Desmond 1994.)

Matthew, Patrick (1790–1874). Scottish gentleman farmer. Author of works on political and agricultural subjects. Advanced a theory of natural selection in the 1830s. (Dempster 1983, R. Desmond 1994.)

13 June [1862], *3 December 1862*

Maull & Fox. Photographic artists of Piccadilly and Cheapside, London, from 1879 until after 1908. (Pritchard 1986.)

Maull & Polyblank. Photographers of Piccadilly and Gracechurch Street, London, 1856–65. A partnership between Henry Maull and George Henry Polyblank. (*Post Office London directory* 1860, Pritchard 1986.)

Maull, Henry. *See* Maull & Polyblank.

Maw, George (1832–1912). Tile manufacturer, geologist, botanist, and antiquarian. Had a notable garden at Benthall Hall, Broseley, Shropshire. Wrote on the geology of western England and north Wales. Travelled to Morocco and Algeria

Maw, George, cont.
with Joseph Dalton Hooker in 1871 and independently in 1873, writing on the geology of these countries. (R. Desmond 1994, Sarjeant 1980.)
30 June 1862, 3 July [1862], *7 July 1862*

Max Müller, Friedrich (1823–1900). German-born orientalist and philologist. Published an edition of the 'Rigveda', the most important of the sacred books of the Brahmans (1849–73). Moved to Paris in 1845; settled in Oxford in 1848 after fleeing the revolution in France. Deputy Taylorian professor of modern European languages, Oxford University, 1850–4; professor, 1854–68; professor of comparative philology, 1868–75. Curator of the Bodleian library, 1856–63 and 1881–94. (*DNB.*)

Maximowicz, Karl Ivanovich (Carl Johann) (1827–91). Russian botanist and explorer. Travelled in eastern Asia, 1853–7; China and Japan, 1859–64. Curator of the St Petersburg herbaria. (*Taxonomic literature.*)

Mays, J. Aldous. Reporter and shorthand writer. (Freeman 1978.)

Meehan, Thomas (1826–1901). English-born botanist, horticulturist, and author. Gardener at the Royal Botanic Gardens, Kew, 1845–8. Emigrated to the United States in 1848, where he worked as a gardener. Established a nursery in Germantown, Pennsylvania, *c.* 1853. Editor, *Gardener's Monthly*, 1859–87; *Meehan's Monthly*, 1891–1901. Botanist on the Philadelphia state board of agriculture, 1877–1901. Elected to the Academy of Natural Sciences of Philadelphia, 1860; to the American Philosophical Society, 1871. (Baker 1965, *DAB.*)

Mellersh, Arthur (1812–94). Naval officer. Midshipman and mate on HMS *Beagle*, 1825–36. Served off the coast of Syria, then in command of HMS *Rattler* in the Burma campaign in 1852. Served off the coast of China in the 1850s, suppressing piracy. Served in the Caribbean and South America before retiring in 1864. (*Modern English biography*; *The Times*, 28 September 1894, p. 4.)
30 November [1862]

Merian, Peter (1795–1883). Swiss geologist and chemist. Established geological and mineralogical collections at the University of Basel. (Sarjeant 1980.)

Meyer, Bernhard (1767–1836). German ornithologist and botanist. Formed an ornithological collection for the museum at the Senckenberg Society for Natural Sciences. Made ornithological expeditions to Holland, Switzerland, and Denmark, 1805–27. (Schmidt and Voigt 1824–56, Scriba 1831–43.)

Michalet, Louis Eugène (1826–62). French botanist and magistrate at Besançon. (*Taxonomic literature.*)

Middendorf, Aleksandr Fedorovich (1815–94). Russian biogeographer. Taught zoology at Kiev University, 1839–40. Travelled to Kola Peninsula with Karl Ernst von Baer, 1839. Commissioned by the St Petersburg Academy of Sciences to travel in northern and eastern Siberia, 1843–4; published a description and analysis of his materials, 1848–75. Member of the St Petersburg Academy of Sciences, 1845; permanent secretary, 1852–65. His Siberian journey led to the establishment of the Russian Geographical Society. (*DSB.*)

Milde, Karl August Julius (Julius) (1824–71). German botanist. Taught at secondary schools in Breslau from 1851, conducting botanical research in his spare time. (*ADB*.)

Mill, John Stuart (1806–73). Philosopher and political economist. (*DNB*, *DSB*.)

Milne, David. *See* Milne-Home, David.

Milne-Home, David (1805–90). Scottish advocate and geologist. Studied earthquakes and the parallel roads of Glen Roy. Founder of the Scottish Meteorological Society. (*DNB* s.v. Milne, Sir David (1763–1845); *Modern English biography*; Sarjeant 1980.)

Milton, John (1608–74). Poet. (*DNB*.)

Milton, John. Tea-dealer and grocer at 10 Great Marylebone Street West, London. (*Post Office London directory* 1861.)

Mitchell, David William (1813–59). Zoologist. Secretary of the Zoological Society of London, 1847–59; director of the Jardin d'Acclimatation, Paris, 1859. (*Modern English biography*, Scherren 1905.)

Moggridge, John Traherne (1842–74). Entomologist and botanist. Wintered in Menton, France, and studied the flora of the area. (R. Desmond 1994, Gilbert 1977.)

Mohl, Hugo von (1805–72). German biologist. Professor of physiology, Bern, 1832–5; of botany, Tübingen, 1835–72. Known for his work on the microscopic anatomy of plants and for his contributions to knowledge of the plant cell. Foreign member, Royal Society, 1868. (*DSB*.)

Möller, Hans Peter Christian (1810–45). Danish army officer, whaling inspector, naturalist, and author. Author of *Index Molluscorum Grœnlandiæ* (Hafniæ, 1842). (Erslew 1843–58, *SBI*.)

Monnier, Auguste (b. 1800). French botanist. Professor of botany, Nancy. (Barnhart 1965, *Taxonomic literature*.)

Monteiro, Joachim John (*fl.* 1858–75). Mining engineer and zoologist, resident in Angola from 1858. Involved in trading and mining operations; engaged by the Western African Malachite Copper Mines Company to develop a concession at Bembe, 1858. Author of papers on African ornithology. Associate of the Royal School of Mines; corresponding member of the Zoological Society. (*CDEL*, Egerton 1957, Monteiro 1875, *Royal Society catalogue of scientific papers*.)

Moore, Charles (1815–81). Bookseller and geologist. Resident in Bath from 1853, from where he studied the stratigraphy of Somerset. His collection of fossils formed the basis of the geological museum of the Bath Royal Literary and Scientific Institute. Fellow of the Geological Society, 1854. (*DNB*, Sarjeant 1980.)

Moore, Charles (1820–1905). Botanist. Gardener at the Trinity College Botanic Garden, Dublin, 1835–8. Gardener at the Royal Botanic Gardens, Kew, 1847. Emigrated to Sydney, Australia, in 1848. Toured the South Pacific on HMS *Havannah*, 1850. Director of the Botanic Gardens, Sydney, 1848–96. (R. Desmond 1994.)

Moquin-Tandon, Horace Bénédict Alfred (Alfred) (1804–63). French botanist

Moquin-Tandon, Horace Bénédict Alfred (Alfred), cont.
and naturalist. Professor of zoology, Marseille, 1829–33. Professor of botany and director of the Jardin des plantes, Toulouse, 1833–53. Professor of natural history and director of the Jardin des plantes, faculty of medicine, Paris, 1853–63. (*Dictionnaire universel des contemporains*, *NBU*.)

More, Alexander Goodman (1830–95). Naturalist. Lived on the Isle of Wight before moving to Ireland in the 1860s. Studied the flora of Ireland, publishing *Cybele Hibernica* in 1866. In his zoological work, concentrated mainly on birds. Assistant, Dublin Natural History Museum, 1867–81; curator, 1881–7. (R. Desmond 1994, Gilbert 1977, Moffat ed. 1898, *Natural Science* 6 (1895): 351.)
18 May 1862, 7 June 1862, 26 September 1862, 1 October 1862

Morell, John Daniel (1816–91). Clergyman, philosopher, and inspector of schools. Congregational minister, Gosport, 1842–5. His *Historical and critical view of the speculative philosophy of Europe in the nineteenth century* (1846) led to his appointment as inspector of schools, a post which he held from 1848 until 1876. Published numerous works on grammar and philosophy. (*DNB*.)

Morgan, Charles (1814/15–94). Surgeon. Member of the Royal College of Surgeons, 1837. Licentiate of the Society of Apothecaries, 1839. In practice in Farnborough, Kent, 1845–9. In partnership with Edward Augustus Williams in Bromley, Kent, 1849–65. (*Medical directory*, *Post Office directory of the six home counties*.)

Morlot, Charles Adolphe (1820–67). Swiss stratigrapher and archaeologist. Appointed professor of geology and mineralogy, Lausanne, 1851. Conservator of the Bern Archaeological Museum. Wrote on the Tertiary and Quaternary geology of Austria, Switzerland, and Denmark. Originator of the term 'Quaternary'. (*ADB*, Sarjeant 1980.)

Morton, Thomas (1764?–1838). Dramatist. Author of *Speed the plough* (1798). (*DNB*.)

Mueller, Ferdinand Jakob Heinrich (Ferdinand) von (1825–96). German-born explorer and botanist. Emigrated to Australia in 1847. Government botanist, Victoria, 1852. Botanist to the North West Australia Expedition, 1855–7. Director of the Botanical Gardens, Melbourne, 1857–73. President of the Australian Association for the Advancement of Science, 1890. FRS 1861. (*Aust. dict. biog.*, R. Desmond 1994.)

Müller, Daniel Ernst (*fl.* 1819–29). German forester. Made a study of insect damage to conifers in the 1820s. (*ADB*.)

Müller, Friedrich Max. *See* Max Müller, Friedrich.

Murchison, Roderick Impey, 1st baronet (1792–1871). Geologist. Noted for his work on the Silurian system. President of the Geological Society, 1831–3 and 1841–3; of the British Association for the Advancement of Science, 1846; of the Royal Geographical Society, 1843–58. Director-general of the Geological Survey of Great Britain, 1855. Knighted, 1846; created baronet, 1866. FRS 1826. (*DNB*, *DSB*.)

Murray, Andrew (1812–78). Entomologist and botanist. Practised law in Edinburgh until 1860, when he was appointed assistant secretary to the Royal

Horticultural Society of London. In entomology, specialised on the Coleoptera; in botany, on the Coniferae. An expert on crop-destroying insects. (*Entomologist's Monthly Magazine* 14 (1878): 215–16, Gilbert 1977.)

10 April [1862], *12 April 1862*

Murray, John (1808–92). Publisher and author of guide-books. CD's publisher from 1845. (*DNB.*)

28 January [1862], *30 January [1862]*, 9 [February 1862], 9 April [1862], 2 May [1862], 13 June [1862], 18 [June 1862], 20 [June 1862], 24 August [1862], 7 October [1862?]

Natural History Review

[before 10 October 1862]

Naudin, Charles Victor (Charles) (1815–99). French botanist. Entered the herbarium staff at the Muséum d'Histoire Naturelle and became professor of zoology at the Collège Chaptal, Paris, 1846. Obliged to resign almost immediately due to a severe nervous disorder. Appointed aide-naturaliste at the Muséum d'Histoire Naturelle, 1854. Experimented widely on plants, particularly on acclimatisation and hybridity. Published a theory of transmutation based on hybridisation. (*DSB*, *Taxonomic literature*.)

26 June 1862

Neuberg, Joseph (1806–67). German-born manufacturer, and friend of Thomas Carlyle. In business in Hamburg and Nottingham; retired early. Became a British citizen in 1845. Became friends with Carlyle in 1848 and then worked as his 'volunteer secretary' from 1849. Travelled as Carlyle's guide across some of the battlefields of Frederick II of Prussia, 1852 and 1858. Translated some of Carlyle's works into German. (Kaplan 1983, *Modern English biography*.)

Neumann, Louis (*fl.* 1849–97). French gardener. Gardener at the Muséum d'Histoire Naturelle, 1849 (or before) – 1878 (or after). Member of the Société Centrale d'Horticulture de France, 1863; secretary, 1865–6. Editor of *Le Nouveau Jardinier Illustré*, 1865–93. (*Annuaire de la Société [Nationale d'Horticulture de France]* (1878): lxvi; (1884): 60; (1893): 109, *Catalogue de la Bibliothèque Nationale*, Grönland 1858, *Journal de la Société Impériale et Centrale d'Horticulture de France* 11 (1865): 15, 44; 12 (1866): 25, Rivière 1866, p. 284.)

Nevill, Dorothy Frances (1826–1913). Daughter of Horatio Walpole, third earl of Orford; married Reginald Henry Nevill in 1847 (*Burke's peerage* 1914). Developed a notable garden at Dangstein in Sussex, where she cultivated orchids, nepenthes, and other tropical plants; employed thirty-four gardeners. (R. Desmond 1994, *WWW*.)

[before 22 January 1862], *[before 22 January 1862]*, 22 January [1862], *[c. 14 March 1862]*, *[16? May 1862]*

Newman, Edward (1801–76). Naturalist. Founder of the Entomological Club, the precursor of the Entomological Society, 1826. Natural history editor of the *Field*, 1858–76. (R. Desmond 1994, *DNB*, Gilbert 1977.)

6 April 1862

Newman, Henry Wenman (1788–1865). Army officer and landowner. Joined the South Gloucestershire Militia in 1814; captain, 1820; lieutenant-colonel commandant, 1854–60. Succeeded to his father's estates at Thornbury Park, Gloucestershire, in 1829. (*Burke's landed gentry* 1871, *Modern English biography*.)

Newton, Alfred (1829–1907). Zoologist and ornithologist. Travelled extensively throughout northern Europe and North America on ornithological expeditions, 1854–64. Professor of zoology and comparative anatomy at Cambridge University, 1866–1907. FRS 1870. (*DNB*.)

Newton, Edward (1832–97). Colonial administrator. Brother of Alfred Newton. Colonial secretary, Mauritius, 1859–63; auditor-general, 1863–8. Colonial secretary, 1868–77. Lieutenant-governor and colonial secretary, Jamaica, 1877–83. Knighted, 1887. (*Alum. Cantab.*)

Newton, Thomas W. (1822–1902). Assistant librarian at the Museum of Practical Geology, London, 1860–95. (*BLC*, Flett 1937.)

Nitschke, Theodor Rudolf Joseph (1834–83). German botanist. Professor of botany, University of Münster, 1867. (*Taxonomic literature*.)

Norman, George Warde (1793–1882). Writer on finance. Merchant in the Norway timber trade, 1810–30. A director of the Bank of England, 1821–72. Exchequer bill commissioner, 1831; public works loan commissioner, 1842–76. A director of the Sun Insurance Office, 1830–64. A founder member of the Political Economy Club, 1821. Succeeded to his father's estate at the Rookery, Bromley Common, Kent, in 1830. A family friend of the Darwins'. (*Burke's landed gentry* 1965, *DNB*, Freeman 1978, *Post Office directory of the six home counties* 1859.)

Norman, Herbert George Henry (b. 1838). Barrister. Son of Henry Norman of Oakley, Kent; nephew of George Warde Norman. BA, Christ Church, Oxford, 1859. Entered Lincoln's Inn, London, 1860. Called to the bar, 1863. (*Alum. Oxon.*, Filmer 1977, Foster 1885, *Post Office directory of the six home counties* 1862.)

Northumberland, duke of. *See* Percy, Algernon.

Oldfield, Augustus Frederick (1820–87). Botanist and ethnologist. Collected plants in Australia and Tasmania for Ferdinand von Mueller and others during the 1850s. Placed botanical information and specimens at the disposal of Joseph Dalton Hooker and George Bentham. Acquired extensive knowledge of the Australian aborigines. Returned to England in 1862. Became blind *c.* 1867. (R. Desmond 1994, *Papers and Proceedings of the Royal Society of Tasmania* (1909): 24–6.)

Oliver, Daniel (1830–1916). Botanist. Assistant in the herbarium of the Royal Botanic Gardens, Kew, 1858; librarian, 1860–90; keeper, 1864–90. Professor of botany, University College London, 1861–88. FRS 1863. (R. Desmond 1994, *List of the Linnean Society of London* 1859–91.)

[4–8 February 1862], *10 April 1862*, 12 [April 1862], *14 April 1862*,
15 April [1862], 20 [April 1862], *23 April 1862* 24 April [1862], *14 May 1862*,
8 June [1862], [before 11 June 1862], 24 July [1862], *28 July 1862*,
29 [July 1862], 2 September [1862], *4 September 1862*, *13 September 1862*,

14 September [1862], [17 September 1862], 13 October [1862], 23 [November 1862], *25 November 1862*

Osborne & Whitehead. Butcher's shop in Down, Kent. (*Post Office directory of the six home counties* 1862.)

Osborne, Alfred James (b. 1832/3). Butcher in Down, Kent. (Census returns 1861 (Public Record Office, RG9/462: 71).)

Overton, Frederick Arnold (b. 1862). Grandson of William Darwin Fox. Educated at Exeter College, Oxford. (*Alum. Oxon.*; *Darwin pedigree*; letter to W. D. Fox, 16 September [1862].)

Overton, Harriet Emma (b. 1837). Daughter of William Darwin Fox. Married Samuel Charlesworth Overton (*Crockford's*), curate of Hackness, Yorkshire, in 1861. Mother of Frederick Arnold Overton. (*Darwin pedigree*, pp. 16–17; *Gentleman's Magazine* n.s. 11 (1861): 558.)

Owen, Richard (1804–92). Naturalist. Assistant conservator of the Hunterian Museum, Royal College of Surgeons, 1827; Hunterian professor of comparative anatomy and physiology, 1836–56. Superintendent of the natural history departments, British Museum, 1856–84; championed the establishment of the Natural History Museum, South Kensington, in 1881. President of the British Association for the Advancement of Science, 1858. Described the *Beagle* fossil mammal specimens. Knighted, 1884. FRS 1834. (*DNB*, *DSB*, Rupke 1994.)

Oxenden, George Chichester (1797–1865). Author of satiric verse and parodies, and orchid-fancier. Son of Henry Oxenden, seventh baronet (*Burke's peerage* 1895). Lived at the family seat at Broome Park, near Canterbury, Kent. (*Alum. Cantab.*, *Burke's peerage* 1895.)

15 May 1862, 26 May 1862, [before 30 May 1862], 30 May [1862], 31 May [1862], 4 June [1862], [before 21 June 1862], 21 June 1862, 8 July [1862], 8 July 1862, 4 August [1862?], 11 September 1862, 17 September [1862], 27 September [1862]

Paget, James, 1st baronet (1814–99). Surgeon. Assistant surgeon at St Bartholomew's Hospital, London, 1847; surgeon, 1861–71. Arris and Gale Professor of anatomy and surgery at the Royal College of Surgeons, 1847–52. Lectured on physiology in the medical school, St Bartholomew's, 1859–61; on surgery, 1865–9. Appointed surgeon-extraordinary to Queen Victoria, 1858; serjeant-surgeon, 1877. Created baronet, 1871. FRS 1851. (*DNB*.)

Palgrave, Francis (1788–1861). Historian. Author of several works on the medieval history of England and on antiquarian topics. Deputy-keeper of the Queen's records, 1838–61. Married Joseph Dalton Hooker's aunt, Elizabeth Turner, in 1823. Knighted, 1832. FRS 1821. (*DNB*.)

Palgrave, Francis Turner (1824–97). Poet, critic, and educationist. Assistant private secretary to William Ewart Gladstone (*DNB*), 1846. Joined the department of education in 1848; examiner and assistant secretary, 1855–84. Art critic for the *Saturday Review.* Professor of poetry, Oxford University, 1885–95. (*DNB*.)

Pallas, Pyotr Simon (1741–1811). German naturalist and geographer. Travelled widely in the Russian empire. Foreign member, Royal Society, 1764. (*ADB*, *DSB*.)

Palmerston, Lord. *See* Temple, Henry John.

Pamplin, William (1806–99). Botanist, bookseller, and publisher. Son of the nurseryman, William Pamplin (1768–1844; R. Desmond 1994); assisted his father until 1839. Botanical bookseller in Soho, London, 1839–62. Retired to Wales and tried to establish the North Wales Central Botanic Gardens. Published the *Phytologist*, 1855–63. Associate of the Linnean Society, 1830. Member of the Botanical Society of London. (R. Desmond 1994, *Modern English biography.*)

Panizzi, Anthony (1797–1879). Italian exile in Britain. Assistant librarian of the British Museum, 1831; principal librarian, 1856–66. (*DNB.*)

Papillon, Frances Margaret (d. 1885). Sister of George Chichester Oxenden. Married Thomas Papillon in 1825. (*Burke's landed gentry* 1894, *Burke's peerage* 1895.)

Papillon, Thomas (1803–83). Brother-in-law of George Chichester Oxenden. Resided at Crowhurst Park, Sussex. Married Frances Margaret Oxenden in 1825. (*Alum. Oxon.*, *Burke's landed gentry* 1894.)

Parker, Henry (1788–1856). Physician at the Shropshire Infirmary. Married Marianne Darwin in 1824. (*Darwin pedigree*, *Provincial medical directory.*)

Parker, Henry (1828–92). Scholar and author. CD's nephew. Son of Henry and Marianne Parker. Fellow of Oriel College, Oxford, 1851–85. Author of works on the fine arts. (*Alum. Oxon.*, *CDEL.*)

Parker, Marianne (1798–1858). CD's eldest sister. Married Henry Parker in 1824. (*Darwin pedigree.*)

Parker, Mary Susan (1836–93). CD's niece. Daughter of Henry and Marianne Parker. (*Darwin pedigree.*)

Parkes, Edmund Alexander (1819–76). Physician and author of works on tropical diseases. Served as assistant surgeon in India, 1842–5. Special professor of clinical medicine at University College London, and physician to the University College Hospital, 1849. Professor of hygiene at the Army Medical School at Chatham, 1860. FRS 1861. (*DNB.*)

8 April 1862, 28 April 1862, 29 June 1862

Parrot, Johann Jakob Friedrich Wilhelm (Friedrich) von (1791–1841). German physician and traveller. Professor of physiology and pathology, University of Dorpat, 1821. Made the first modern ascent of Mount Ararat in 1829. (*ADB*, *BHGW.*)

Pascoe, Francis Polkinghorne (1813–93). Entomologist. Surgeon in the Royal Navy until 1843. Devoted himself to science, forming a large entomological collection, later given to the Natural History Museum, South Kensington. Although a believer in evolution, he persistently opposed the theory of natural selection. (R. Desmond 1994, *DNB*, Gilbert 1977.)

Pasteur, Louis (1822–95). French chemist and microbiologist. Professor of chemistry, Strasbourg University, 1849–54. Professor of chemistry and dean of the science faculty, Lille University, 1854–7. Administrator and director of scientific studies, Ecole Normale, Paris, 1857–67; director of the laboratory of physiological chemistry, 1867–88. Renowned for his work on fermentation and for experiments providing evidence against the theory of spontaneous generation. Foreign member, Royal Society, 1869. (*Dictionnaire universel des contemporains*, *DSB*.)

Pellew, George (1793–1866). Clergyman and theologian. Dean of Norwich, 1828–66; Rector of Great Chart, Kent, 1852–66. (*DNB*.)

Pengelly, William (1812–94). Mathematical tutor and geologist. An expert on the geology of Devon; explored the plant-bearing deposits of Bovey Tracey, Brixham Cave and Kent's Hole at Torquay. Honorary secretary of the Torquay Natural History Society, 1851–90. President of the Devonshire Association for the Advancement of Literature, Science, and Art, 1867–8. FRS 1863. (*DNB*, Sarjeant 1980.)

Percy, Algernon, 4th duke of Northumberland and 1st Baron Prudhoe (1792–1865). Succeeded to the title in 1847. Fellow of the Astronomical and Geological Societies; president of the Royal Institution; trustee of the British Museum. Developed the gardens at Syon House, Isleworth, Middlesex. FRS 1818. (*DNB*, R. Desmond 1994.)

Petschler, Charles (*fl.* 1862–5). Resident of Auckland, New Zealand. Friend of Julius von Haast. (Letter from Julius Haast, 9 December 1862; H. F. von Haast 1948, p. 446.)

Pickard & Stoneman. Firm of cabinet makers, located at $3\frac{1}{2}$ New Norfolk Street, Shoreditch, London, 1853–5, and at 10 Spencer Street, Shoreditch, London, 1855–85. (*Post Office London directory*.)

1 December [1862]

Pictet de la Rive, François Jules1809–72). Swiss zoologist and invertebrate palaeontologist. Professor of zoology, University of Geneva, 1835. (Gilbert 1977, Sarjeant 1980.)

Planchon, Jules Emile (1823–88). French botanist. Assistant in the herbarium, Royal Botanic Gardens, Kew, 1844–8. Professor of botany at the Institute of Horticulture, Ghent, Belgium, 1849–51; at the School of Medicine and Pharmacy, Nancy, 1851–3. Professor of botany at the faculty of sciences at Montpellier, 1853–81; director of the school of pharmacy, 1856. (R. Desmond 1994, *Grande encyclopédie*.)

Plato (427–347 B.C.). Athenian philosopher. (*EB*.)

Polyblank, George Henry. *See* Maull & Polyblank.

Pouchet, Félix Archimède (1800–72). French biologist and naturalist. Director of the Muséum d'Histoire Naturelle, Rouen. Prolific author and populariser of science. Adversary of Louis Pasteur in the debate over spontaneous generation. (*Dictionnaire universel des contemporains*, *DSB*.)

Powis, earl of. *See* Clive, Edward; Herbert, Edward; and Herbert, Edward James.

Prestwich, Joseph (1812–96). Geologist and businessman. Entered the family wine business in London in 1830; became proprietor in 1842. Professor of geology, Oxford University, 1874–88. Fellow of the Geological Society, 1833; president, 1870. An expert on the Tertiary geology of Europe. Prominent in studies of human prehistory. Knighted, 1896. FRS 1853. (*DNB*, *DSB*.)

Price, John (1803–87). Welsh scholar, school teacher, and naturalist. Educated at Shrewsbury School (1818–22) and St John's College, Cambridge. Assistant master, Shrewsbury, 1826–7. Headmaster of the junior department at Bristol College, then classics principal at the Liverpool High School, before settling in Chester. A founding member of the Chester Natural Science Society. (*Alum. Cantab.*, *Eagle* 15 (1888): 169–72, *Modern English biography*.)

Pritchard, Charles (1808–93). Clergyman and astronomer. Headmaster of Clapham Grammar School, 1834–62, where he erected an observatory. Hulsean lecturer, Cambridge University, 1867. Savilian professor of astronomy, Oxford University, 1870–93. President of the Royal Astronomical Society, 1886. FRS 1840. (*Alum. Cantab.*, *DNB*, *DSB*.)
17 June [1862]

Prittie, Henry Sadleir, 3d Baron Dunalley of Kilboy (1807–85). Sheriff of County Tipperary, Ireland, 1840. Succeeded to the peerage in 1854. (*Burke's peerage* 1980.)

Pugh, Miss. Governess of the Darwin children, January 1857 – January 1859. Resident in Kew, Surrey, in 1862. Later became insane and was confined to an asylum. (CD's classed account book (Down House MS), Emma Darwin's diary (DAR 242), Freeman 1978, H. E. Litchfield's autobiography (DAR 246: 36–7).)

Quatrefages de Bréau, Jean Louis Armand de (Armand de Quatrefages) (1810–92). French zoologist and anthropologist. Undertook a study of invertebrates and, in 1844, accompanied Henri Milne-Edwards (*DSB*) on a zoological excursion to the coast of Sicily. Awarded chair of natural history at the Lycée Henri IV, Paris, 1850; professor of the natural history of man, Muséum d'Histoire Naturelle, 1855. Foreign member, Royal Society, 1879. (*DSB*.)
11 July [1862], *[after 11 July 1862]*

Ramsay, Andrew Crombie (1814–91). Geologist with the Geological Survey of Great Britain, 1841; director for England and Wales, 1862; director-general, 1871–81. Professor of geology, University College London, 1847–52; lecturer on geology at the Government School of Mines, 1852–71. Knighted, 1881. FRS 1849. (*DNB*, *DSB*.)
17 February 1862, 18 February [1862], *26 August 1862*, 5 September [1862], *13 December 1862*, 14 December [1862]

Reed, George Varenne (1816–86). Anglican clergyman. Curate of Hayes, Kent, 1837–9; of Tingewick, Buckinghamshire, 1839–54. Rector of Hayes, 1854–86. Tutor to George Howard, Francis, Leonard, and Horace Darwin. (*Alum. Cantab.*, Freeman 1978.)

Reeks, Trenham (1823/4–79). Mineralogist. Worked in the laboratory of the Museum of Economic Geology, London, in 1840. Registrant of the School of Mines, London, 1851–79. Curator and librarian of the Jermyn Street institution, London, 1851–99. (*Modern English biography*.)

Reeve, Henry (1813–95). Man of letters. Began a literary career in London in 1834. Joined the staff of *The Times* in 1840; resigned in 1855. Editor of the *Edinburgh*

Review, 1855–95. Member of the Society of Antiquaries, 1852; vice-president, 1879–82. (*DNB*.)

Reuss, Georg Christian (*fl.* 1869). German botanist at Ulm. (*Taxonomic literature*.)

Richmond, George (1809–96). Portrait painter. Until about 1846, worked almost entirely in crayon and water-colour, then began to paint in oils. Painted portraits of Darwin and Wedgwood family members. (*DNB*.)

Ring, Abraham (b. 1824/5). Gardener. Resident in Down, Kent. (Census returns 1861 (Public Record Office, RG9/462: 72).)

Ring, Charlotte (b. 1826/7). Wife of Abraham Ring. Resident in Down, Kent. (Census returns 1861 (Public Record Office, RG9/462: 72).)

Rivers, Thomas (1798–1877). Nurseryman in Sawbridgeworth, Hertfordshire. Succeeded to family business (founded 1720) in 1827. Obtained best collection of roses in Britain, and published books on the subject. Author of works on fruit trees. One of founders of British Pomological Society, 1854. Contributed widely to gardening journals. (*DNB*.)

23 December [1862], 28 December [1862]

Robarts, Lubbock & Co. London bank in which John William Lubbock, his son John Lubbock, and Abraham John Robarts (*WWW*) were partners. (*Banking almanac*, *DNB*, *WWW*.)

Robinson, Edward W. (1835–77). London-based entomological artist, steel-engraver, and landscape painter. Exhibited, 1859–76. Worked in Britain, France, Switzerland, and northern Italy. Illustrated the *Entomologists Annual*, 1857–74, and many of the papers in the *Transactions of the Entomological Society of London*, the *Journal of the Linnean Society*, the *Journal of Entomology*, and the *Annals and Magazine of Natural History*. Illustrated Bates 1863. (*Dictionary of watercolour artists*, *Entomologist's Monthly Magazine* 14 (1877): 118–19.)

Rogers, John (1807–67). Barrister. Resided at River Hill House, Sevenoaks Weald, Kent. Keen orchid grower, who supplied CD with information on orchids. FRS 1839. (*Modern English biography*; *Orchids*, p. 236 n.; *Post Office directory of the six home counties* 1862.)

22 January 1862

Rogers, Mary Eliza (*fl.* 1861–5). Poet and writer on Palestine. Sister of Edward Thomas Rogers (*Modern English biography*), British consul at Damascus. (*CDEL*.)

Rolfe, Laura (1787–1868). Married Robert Monsey Rolfe in 1845. (*DNB* s.v. Rolfe, Robert Monsey.)

Rolfe, Robert Monsey, 1st Baron Cranworth of Cranworth (1790–1868). Statesman and jurist. MP for Penryn and Falmouth, 1832–9. Created Baron Cranworth of Cranworth, 1850. Lord chancellor, 1852–8 and 1865–6. (*DNB*.)

28 November 1862

Rolle, Friedrich (1827–87). German geologist, palaeontologist, and natural history dealer. Associate of the Hofmineralien-Cabinet, Vienna, 1859–62. Returned to Bad Homburg from Vienna in 1862. Author of *Ch. Darwin's Lehre von der Entstehung der Arten und ihre Anwendung auf die Schöpfungsgeschichte* (1863) and *Der Mensch, seine*

Rolle, Friedrich, cont.
Abstammung und Gesittung im Lichte der Darwin'schen Lehre von der Art-Entstehung und auf Grundlage der neuen geologischen Entdeckungen dargestellt (1866). (*ADB*, Martin and Uschmann 1969, Sarjeant 1980.)
17 October [1862]

Rolleston, George (1829–81). Physician and anatomist. A physician to the British Civil Hospital at Smyrna, 1855–6. Physician to Radcliffe Infirmary, Oxford, and Leeds reader in anatomy at Christ Church College, Oxford, 1857. Linacre professor of anatomy and physiology, Oxford University, 1860–81. Fellow of Merton College, Oxford, 1872. FRS 1862. (*DNB*.)

Rorison, Gilbert (1821–69). Clergyman. Curate of St John's Episcopal Church, Edinburgh, 1844–5; of St James's, Leith, 1845–6. Incumbent of St Peter's, Peterhead, 1846–69. (*Roll of the graduates of the University of Glasgow*.)

Ross, Andrew (1798–1859). Optician and instrument maker. Collaborated with Joseph Jackson Lister (Turner 1989) in manufacturing objective lenses to Lister's new design, *c.* 1837–41. Founder member of the Microscopical Society of London, 1841. Prize-winner for his microscopes at the Great Exhibition in 1851. Together with Lister, he 'transformed the microscope from a toy to a scientific instrument of immense importance'. (Turner 1989.)

Ross, James Clark (1800–62). Naval officer and polar explorer. Joined the navy in 1812. Discovered the northern magnetic pole in 1831. Employed on the magnetic survey of Britain, 1838. Commander of an expedition to the Antarctic, 1839–43; and of a search expedition for John Franklin (*DNB*), 1848–9. Rear-admiral, 1856. Knighted, 1843. FRS 1828. (*DNB*.)

Ross, John (1777–1856). Naval officer and Arctic explorer. Entered the service of the East India Company in 1794. Joined the Navy as a midshipman in 1799; commander, 1812. Leader of the expeditions in search of the north-west passage in 1818 and 1829–33. Consul in Stockholm, 1839–46. Led a private expedition to search for John Franklin (*DNB*), 1850–1. Knighted 1834. (*DNB*.)

Ross, Thomas (d. 1870). Optician and scientific instrument maker at 2 and 3 Featherstone Buildings, High Holborn, London. Son of Andrew Ross. Took over his father's business in 1859. Elected member of the Microscopical Society of London, 1859. (*Post Office London directory* 1861, Turner 1989.)

Rothrock, Joseph Trimble (1839–1922). Physician, botanist, and forester. Entered the Lawrence Scientific School, Harvard University, in 1860 where he was a student-assistant to Asa Gray. Enlisted in the 131st Pennsylvania Infantry in 1862; commissioned captain in the 20th Pennsylvania Cavalry in 1863; honourably discharged, 1864. Professor of botany, Pennsylvania State Agricultural College, 1867. Botanist and surgeon to the government survey in Colorado, New Mexico, and California, 1873. Professor of botany, University of Pennsylvania, 1877–1904. (*DAB*.)

Royer, Clémence Auguste (1830–1902). French author and economist. Studied natural science and philosophy in Switzerland. In Lausanne in 1859, founded a course on logic aimed at women. Translated *Origin* into French in 1862. (*Dictionnaire universel des contemporains.*)

Rucker, Sigismund. East and West India broker with premises in Great Tower Street, City of London. Resided at West Hill, Wandsworth. (*Post Office London directory* 1861, *Post Office London suburban directory* 1860.)

Russell, John, 1st Earl Russell (1792–1878). Statesman. Liberal prime minister, 1846–52 and 1865–6. Home secretary, 1835; secretary of state for foreign affairs, 1852–3; for the colonies, 1839 and 1853–5. Foreign secretary under Lord Palmerston, 1859–65. Created Earl Russell of Kingston Russell in 1861. President of the Royal Historical Society, 1872–8. FRS 1847. (*DNB.*)

Rütimeyer, Karl Ludwig (Ludwig) (1825–95). Palaeozoologist and geographer. Professor of zoology and comparative anatomy, University of Basel, 1855; rector, 1865; professor in the medical and philosophical faculties, 1874–93. Made significant contributions to the natural history and evolutionary palaeontology of ungulate mammals. (*DSB.*)
15 [and 16] January [1862], 11 February [1862]

Sabine, Edward (1788–1883). Astronomer, geophysicist, and army officer. Entered army in 1803; major-general, 1856; general, 1870. Astronomer to expeditions in search of a north-west passage, 1818 and 1819–20. Appointed one of three scientific advisers to the Admiralty, 1828. General secretary of the British Association for the Advancement of Science, 1838–52 and 1853–9. Foreign secretary of the Royal Society, 1845–50; treasurer, 1850–61; president, 1861–71. Knighted, 1869. FRS 1818. (*DNB*, *DSB.*)

St Barbe, John. Banker. Manager of the Charing Cross branch, Union Bank of London, 1860–70. (*Banking almanac* 1860–70.)
[before 3 July 1862], [16 July 1862]

Saint-Hilaire, Augustin François César Prouvençal (Auguste) de (1779–1853). French naturalist. Surveyed the flora and fauna of Brazil, 1816–22. Appointed professor at the faculty of sciences, Paris, 1830. (*DSB.*)

Salisbury, Richard Anthony (1761–1829). Botanist. In 1785, changed his name from Markham to Salisbury as the condition of a gift of £10,000. Secretary of the Horticultural Society of London, 1805–16. FRS 1787. (R. Desmond 1994, *DNB.*)

Sanders, Charles Henry Martyn (b. 1862). Grandson of William Darwin Fox and son of Eliza Ann Sanders. (*Darwin pedigree*, p. 16; *Gentleman's Magazine* n.s. 12 (1862): 638.)

Sanders, Eliza Ann (1836–74). Daughter of William Darwin Fox. Married Henry Martyn Sanders (*Crockford's*), Vicar of Skidby, Yorkshire. Mother of Charles Henry Martyn Sanders. (*Darwin pedigree*, pp. 15–16.)

Schlegel, Hermann (1804–84). German naturalist. Director of the National Museum of the Netherlands in Leiden. (*ADB*, Sarjeant 1980.)

Schomburgk, Robert Hermann (1804–65). German-born explorer, diplomat, and naturalist. Explored British Guiana under the direction of the Royal Geographical Society, 1831–5. Government commissioner in charge of surveying

Schomburgk, Robert Hermann, cont.
and establishing the boundaries of British Guiana, 1841–3. British consul, San Domingo, 1848; Bangkok, 1857–64. Knighted, 1844. FRS 1859. (*DNB*.)

Schroeder van der Kolk, Jacob Ludwig Conrad (1797–1862). Dutch physician, anatomist, physiologist, and zoologist. Professor of anatomy and physiology, University of Utrecht, 1827–62. Inspector of lunatic asylums, 1842–62; involved in reforms in the care of the mentally ill. (*DSB*.)

Schweizerbart, Christian Friedrich (1805–79). German publisher. Nephew of Wilhelm Emanuel Schweizerbart. Director of E. Schweizerbart'sche Verlagsbuchhandlung of Stuttgart, 1841–67, publishers of the German translation of *Origin* (1860) and of *Orchids* (1862). (*Jubiläums-Katalog*.) *Per pro* E. Schweizerbart'sche Verlagsbuchhandlung, Stuttgart:
7 June 1862, *11 July 1862*, *16 October 1862*

Schweizerbart, Wilhelm Emanuel (1785–1870). German publisher. Uncle of Christian Friedrich Schweizerbart. Founded E. Schweizerbart'sche Verlagsbuchhandlung in Stuttgart in 1826. Publisher of many German scientific works. Retired in 1841. (*Jubiläums-Katalog*.)

E. Schweizerbart'sche Verlagsbuchhandlung, Stuttgart. German publishing company.
7 June 1862, *11 July 1862*, *16 October 1862*. *See also under* Schweizerbart, Christian Friedrich.

Sclater, Philip Lutley (1829–1913). Lawyer and ornithologist. Founder of *Ibis*, 1858; editor, 1858–65 and 1878–1912. Secretary of the Zoological Society of London, 1859–1903. FRS 1861. (*DSB*, Scherren 1905.)
12 May [1862], 14 May [1862]

Scott, John (1836–80). Scottish botanist. Gardener at Chatsworth House, Derbyshire, before becoming foreman of the propagating department at the Royal Botanic Garden, Edinburgh, in 1859. Through CD's patronage, emigrated to India in 1864 and began a professional career as curator of the Calcutta botanic garden. Fellow of the Linnean Society, 1873. Carried out numerous experiments on CD's behalf. (R. Desmond 1994, Freeman 1978, *Transactions of the Botanical Society of Edinburgh* 14 (1883): 160–1.)
11 November 1862, 12 November [1862], *15 November [1862]*, 19 November [1862], *[20 November – 2 December 1862]*, 3 December [1862], *6 December [1862]*, 11 December [1862], *17 December [1862]*, 19 December [1862]

Scott, Walter, 1st baronet (1771–1832). Scottish novelist and poet. (*DNB*.)

Scudder, Samuel Hubbard (1837–1911). American entomologist and invertebrate palaeontologist. Assistant to Louis Agassiz, 1862–4. Assistant librarian of Harvard University, 1879–82. Carried out notable studies of fossil arthropods, especially insects. (*DAB*, Sarjeant 1980.)

Sedgwick, Adam (1785–1873). Geologist and clergyman. Woodwardian professor of geology, Cambridge University, 1818–73. Canon of Norwich, 1834–73.

President of the Geological Society, 1831; and of the British Association for the Advancement of Science, 1833. FRS 1821. (*DNB*, *DSB*.)

Seemann, Berthold Carl (1825–71). German-born traveller and botanist. Studied botany at the Royal Botanic Gardens, Kew, 1844–6. Naturalist to HMS *Herald*, 1847–51. Commissioned in 1860 to report on the Fiji Islands before their cession and published a botanical catalogue of the islands. Travelled in Venezuela, 1864; in Nicaragua, 1866–7. (R. Desmond 1994, *DNB*.)
24 April 1862

Shakespeare, William (1564–1616). Poet and dramatist. (*DNB*.)

Silliman, Benjamin (1779–1864). American chemist, geologist, and mineralogist. Professor of chemistry and natural history, Yale University, 1802–53. Founder and first editor of the *American Journal of Science and Arts*, 1818. (*DAB*, *DSB*.)

Silliman, Benjamin, Jr (1816–85). American chemist. Teaching assistant to his father, Benjamin Silliman, 1837. An editor of the *American Journal of Science and Arts*, 1838–85. Appointed professor of practical chemistry, Yale University, 1846; succeeded his father as professor of chemistry and natural history in 1853. Professor of chemistry in the medical department, University of Louisville, Kentucky, 1849–54. (*DAB*, *DSB*.)

Sinclair, Andrew (1794–1861). Scottish surgeon, naturalist, and colonial administrator. Assistant surgeon, Royal Navy, 1822–32; surgeon, 1835–43. Collected botanical and zoological specimens for the British Museum and Royal Botanic Gardens, Kew. Settled in New Zealand, 1843; colonial secretary, 1844–56; became a member of the legislative council in 1844. Elected a fellow of the Linnean Society, 1857. In 1858, began collecting material for Joseph Dalton Hooker's *Handbook of the New Zealand flora* (J. D. Hooker 1864–7). Drowned on Julius von Haast's geological surveying expedition to the headwaters of the Rangitata river, Canterbury province. (*DNZB*.)

Slaney, Robert Aglionby (1792–1862). Reformer. MP for Shrewsbury, 1826–35, 1837–41, and 1847–62. Commissioner on health of towns, 1843–6. High sheriff of Shropshire, 1854. Fell through the floor at the opening of the International Exhibition and later died. Father-in-law of Thomas Campbell Eyton. (*DNB*.)

Slidell, John (1793–1871). American politician and diplomat. United States senator for Louisiana, 1853–61. Appointed Confederate diplomatic commissioner to France in 1861; his detention, along with James Murray Mason, by the United States Navy in November 1861 caused a tense diplomatic situation to develop between the United States and Britain. (*DAB*.)

Smith, Andrew (1797–1872). Army surgeon. Stationed in South Africa, 1821–37. Principal medical officer at Fort Pitt, Chatham, 1837; deputy inspector-general, 1845. Director-general, Army Medical Department, 1853–8. An authority on South African zoology. Knighted, 1859. FRS 1857. (*DNB*.)
[November? 1862]

Smith, Beck & Beck. Instrument makers at 6 Coleman Street, and Pear Tree Cottage, Holloway Road, London; among the leading manufacturers of microscopes

Smith, Beck & Beck, cont.
in London. The partners, James Smith (d. 1870), Richard Beck (1827–66), and Joseph Beck (1829–91), were all members of the Microscopical Society of London. (*Post Office London directory* 1861; Turner 1989, p. 171.)

Smith, Edmund (1804–64). Surgeon and hydropathic doctor. Proprietor of Ilkley Wells hydropathic establishment, Ilkley, Yorkshire, 1858–64. (*Medical directory* 1848–65; Metcalfe 1906, p. 107.)

Smith, Frederick (1805–79). Entomologist. Employed in the zoological department of the British Museum from 1849. Specialised in the Hymenoptera. Elected president of the Entomological Society of London in 1862. (*Entomologist* 12 (1879): 89–92, Gilbert 1977.)
27 June [1862], 28 June 1862

Smith, James (1782–1867). Scottish antiquarian, numismatist, and geologist. Known as Smith of Jordanhill. Partner in the firm of West India merchants, Leitch and Smith. Made major early contributions to Quaternary geology; also wrote on the geology of Madeira, Gibraltar, and Portugal. FRS 1830. (*DNB*, Sarjeant 1980.)

Smith, James Edward (1759–1828). Botanist. Purchased the library and collections of Carolus Linnaeus in 1784. Founded the Linnean Society of London in 1788; served as president until his death. Knighted, 1814. FRS 1785. (*DNB*, *DSB*.)

Smith, Jane Charlotte (d. 1864). Daughter of James Smith of Jordanhill. (*Burke's landed gentry* 1886.)

Smith, John (1798–1888). Scottish gardener and pteridologist. Gardener at the Royal Botanic Garden, Edinburgh, 1818; at the Royal Botanic Gardens, Kew, 1822. Curator, Royal Botanic Gardens, Kew, 1842–64. (R. Desmond 1994, *Taxonomic literature*.)

Smith, Sabina Douglas Clavering. Daughter of James Smith of Jordanhill. Married Robert Paisley (d. 1881) of St. Ninian's, County Stirling, in 1878. (*Burke's landed gentry* 1914.)

Smith, William (*c.* 1804–28). Scottish gardener. Under-gardener in the arboretum department at the Chiswick gardens of the Horticultural Society of London. (R. Desmond 1994, W. Smith 1826.)

Smith, William (*fl.* 1862). Resident of York. (*Journal of Horticulture*, 30 December 1862, p. 779.)

Snow, George (1820/1–85). Coal dealer of Down, Kent. Operated a weekly carrier service between Down and London. (Census returns 1862 (Public Record Office RG9/462: 72), Down churchyard, *Post Office directory of the six home counties* 1862.)

Société Impériale Zoologique d'Acclimatation.
10 January 1862, [after 10 January 1862]

Solier, Antoine Joseph Jean (1792–1851). French entomologist. (Gilbert 1977, *NUC*.)

Sowerby, James (1757–1822). Naturalist and artist. Published several illustrated works on natural history, most famously *English botany* (1790–1814). (*DNB*, *DSB*.)

Spence, James (b. 1816). Writer on political subjects. Resident of Liverpool. Author of works advocating the cause of the Confederate states in the American Civil War, 1861–4. (*CDEL*, *NUC*.)

Spence, William (1783–1860). Entomologist. A founder of the Entomological Society, 1833; president, 1847. Author, with William Kirby, of the celebrated *Introduction to entomology* (1815–26). FRS 1834. (*DNB*, Gilbert 1977.)

Spencer, Herbert (1820–1903). Philosopher. Civil engineer on the railways, 1837–41 and 1844–6. In 1844, became sub-editor of the *Pilot*, a newspaper devoted to the suffrage movement. Sub-editor of the *Economist*, 1848–53. From 1852, author of papers on evolution and numerous works on philosophy and the social sciences. (*DNB*, *DSB*.)

Sprague, Isaac (1811–95). American landscape painter. Served as artist-assistant to John James Audubon (*DAB*) on an ornithological expedition along the Missouri River, 1843. Settled in Cambridge, Massachusetts; carried out work for Asa Gray from 1845. (*Dictionary of artists in America*, Dupree 1959.)

Sprengel, Christian Konrad (1750–1816). German botanist who studied the pollination of flowers. Rector of the Great Lutheran Town School, Spandau, where he taught languages and natural science, 1780–94. Private tutor in Berlin. Published his major work on the pollination of flowers in 1793. (*ADB*, *DSB*.)

Sprengel, Kurt Polycarp Joachim (1766–1833). German botanist and physician. Professor of medicine at Halle, 1787–97; of botany, 1797–1833. (*Taxonomic literature*.)

Spruce, Richard (1817–93). Botanist and traveller. Master at St Peter's School, York, until 1846. Collected plants in South America, 1849–64. Interested in mosses and liverworts. (R. Desmond 1994, *DNB*.)

Stanhope, Philip Henry, 5th Earl Stanhope (1805–75). Historian and politician. Styled Viscount Mahon from 1816 until 1855 when he succeeded to the earldom. Tory MP for Wootton Bassett, 1830–2; for Hertford, 1832–3 and 1835–52. Under-secretary for foreign affairs, 1834–5; secretary to the board of control for India, 1845–6. Instrumental in the founding of the National Portrait Gallery and the Historical Manuscripts Commission. President of the Society of Arts, 1846–75. Resided at the family seat at Chevening in Kent. FRS 1827. (*DNB*.)
28 July 1862

Stephens, Henry Oxley (1816–81). Surgeon and botanist. Surgeon, St Peter's Hospital, Bristol, 1849–57. Superintendent of the Bristol Lunatic Asylum, 1857–71. (R. Desmond 1994, *Medical directory* 1847–82.)

Stephens, Thomas Sellwood (b. 1825). Clergyman. Curate of Wanstead, Essex, 1853–9; of Down, Kent, 1859–67. Rector of St Erme, Cornwall, 1867–1904. (*Alum. Oxon.*, *Crockford's*.)
[before 5 May 1862]

Steudel, Ernst Gottlieb (1783–1856). Swiss physician and botanist. Began medical practise in Esslingen, Switzerland, in 1806. Senior consultant, 1828–56. (*ADB*.)

Stevens, Samuel (1817–99). Natural history agent and collector of British Coleoptera and Lepidoptera. Active member of the Zoological Society of London. Alfred Russel Wallace's agent, 1848–63. (Wallace 1905, 1: 266.)

Stilwell, John G. & Thomas. Naval agents of 22 Arundel Street, London.

Stratton, James (d. 1864). Gardener. Curator of the Cambridge University Botanic Garden, 1850–64. (Walters 1981.)

Stuart, John McDouall (1815–66). Explorer, surveyor, and estate agent. Emigrated from Scotland to South Australia in 1838. In 1858, made the first of three attempts to cross the continent from south to north; succeeded, the first person to do so, in 1862. Awarded the Royal Geographical Society's gold medal in 1861. (*Aust. dict. biog.*)

Suchsland, Friedrich Emil (1808–1903). German publisher and book-dealer. Purchased the publishing company of Johann Christian Hermann'sche Verlagsbuchhandlung, Frankfurt-am-Main, in 1836. Published Friedrich Rolle's *Ch. Darwin's Lehre von der Entstehung der Arten*, 1863. (*Börsenblatt für den Deutschen Buchhandel* 3 (1836): 830; 27 (1860): 2208–9; 36 (1869): 434.)

Sulivan, Bartholomew James (1810–90). Naval officer and hydrographer. Lieutenant, HMS *Beagle*, 1831–6. Surveyed the Falkland Islands, 1838–9. Commander of HMS *Philomel*, 1842–6. Resided in the Falkland Islands, 1848–51. Commanded HMS *Lightning* in the Baltic, 1854–5. Naval officer in the marine department of the Board of Trade, 1856–65. Admiral, 1877. Knighted, 1869. (*DNB*.)

27 September [1862], 2 October [1862], 13 October [1862], 14 October [1862], 18 October [1862]

Swinhoe, Robert (1836–77). Official in the British foreign service. Attached to the British consulate in Hong Kong, 1854; Amoy, China, 1855. British vice-consul, Taiwan, 1860–5; consul, 1865–73. Acting consul, Amoy, 1865–71; Ning-po, 1871–3. Consul, Ning-po, 1873–5. Collected plants and animals in the Far East. FRS 1876. (Hall 1987.)

12 November 1862, 2 December 1862

Sydney, Lord. *See* Townsend, John Robert.

Tait, Robert (*fl.* 1845–75). London portrait painter, resident at 14 Queen Anne Street, Cavendish Square. Exhibited at the Royal Academy, 1845–75. Best known for his work 'Thomas and Jane Carlyle in the Drawing Room of their House in Cheyne Row'. (*Post Office London directory* 1861, C. Wood 1978.)

Tankerville, Lord. *See* Bennet, Charles Augustus.

Targioni Tozzetti, Antonio (1785–1856). Italian botanist. Professor of botany and materia medica at the S. Maria hospital, Florence. Director of the medical garden and professor of chemistry at the Academy of Fine Arts. (*Taxonomic literature*.)

Taylor & Francis. Printers and publishers at Red Lion Court, Fleet Street and 3 Robinhood Court, Shoe Lane, London. The firm was founded by Richard Taylor (*DNB*), becoming Taylor & Francis after William Francis (*DNB* s.v. Taylor, Richard) joined the firm in 1852. (*DNB*, *Post Office London directory* 1862.)

Taylor, Tom (1817–80). Dramatist, journalist, and civil servant. Author of *Our American cousin*, first performed at Laura Keene's theatre, New York, 1858. (*DNB*.)

Tegetmeier, William Bernhard (1816–1912). Editor, journalist, and naturalist. Pigeon fancier and an expert on fowls. Secretary of the Apiarian Society of London. (Richardson 1916.)
18 March [1862], 20 June 1862, 12 August [1862], 27 [December 1862], 28 [December 1862]

Temple, Henry John, 3d Viscount Palmerston (1784–1865). Statesman. Succeeded to peerage in 1802. Foreign secretary, 1830–41 and 1846–51. Home secretary, 1852–5. Prime minister, 1855–8 and 1859–65. FRS 1853. (*DNB*.)

Theophrastus (*c.* 372–287 B.C.). Greek philosopher, historian, and botanist. Native of Lesbos. Studied in Athens under Aristotle; succeeded him as head of the Peripatetic school. His most important works are two botanical treatises. (*EB*.)

Thom, J. P. (b. 1838/9). Journalist. Patient at Edward Wickstead Lane's hydropathic establishment in the 1850s. Sub-editor of the colonial newspaper *Home News*. Emigrated to Queensland, Australia, in 1863. (*Correspondence* vol. 7, this volume, letter from Mary Butler, [before 25 December 1862], and *Correspondence* vol. 11, letter from J. P. Thom, 14 January 1863.)

Thompson, Robert (1798–1869). Scottish gardener. Worked at the Horticultural Society of London's Chiswick gardens, 1824–69, where he became superintendent of the fruit department. (R. Desmond 1994.)

Thomson, Thomas (1817–78). East India company surgeon and botanist. Curator of the Asiatic Society's museum, Calcutta, 1840. Superintendent of the Calcutta botanic garden and professor of botany at the Calcutta Medical College, 1854–61. Collaborated with Joseph Dalton Hooker on various publications and joined him on his Himalayan expedition in 1849. FRS 1855. (R. Desmond 1994, *DNB*.)

Thwaites, George Henry Kendrick (1812–82). Botanist and entomologist. Superintendent of the Peradeniya botanic gardens, Ceylon, 1849; director, 1857–80. FRS 1865. (R. Desmond 1994, *DNB*.)
15 May 1862, 15 June [1862], 20 June [1862], 29 December [1862]

Ticknor & Fields. Boston publishing house, founded in 1832 by William Davis Ticknor (*DAB*); James Thomas Fields (*DAB*) joined the company in 1854. Published the journal *Atlantic Monthly* as well as the works of many leading British and American writers. (*DAB*.)

Ticknor, William Davis. *See* Ticknor & Fields.

Tinzmann, 'Inspector' (*fl.* 1845). German civil servant and author. Contributor to Albert 1845. (*Gardeners' Chronicle and Agricultural Gazette*, 21 March 1846, pp. 181–3.)

Tocqueville, Alexis Charles Henri Clerel (Alexis) de (1805–59). French statesman and political writer. Studied law in Paris, 1823–6. Visited the United States, 1831–2, and England in 1833. Wrote his major work, *De la démocratie en Amérique* between 1832 and 1834. Wrote numerous works on democracy, prison reform, French colonial possessions, and the abolition of slavery. (*NBU*.)

Tollet, George (1767–1855). Landowner and agricultural reformer. Changed his name from Embury in 1796, on succeeding to the estate of Betley Hall, Staffordshire. Justice of the peace and deputy lieutenant of Staffordshire. A close friend of Josiah Wedgwood II. (*Burke's landed gentry* 1879 s.v. Wicksted of Betley Hall, Freeman 1978.)

Tollet, Georgina. Author. Daughter of George Tollet. A close friend of the Wedgwoods and Darwins. Edited the manuscript of *Origin*. (Freeman 1978.)
17 May [1862]

Torrey, John (1796–1873). American botanist and chemist. Collected and described plants from various government explorations. Professor of chemistry at the College of Physicians and Surgeons, New York City, 1827–54. Assayer of the United States Mint in New York, 1853–73. (*DAB*, *DSB*.)

Townshend, John Robert, 3d Viscount Sydney and 1st earl Sydney (1805–90). MP for Whitchurch, 1826–31. Captain, West Kent Militia, 1827; West Kent Yeomanry, 1830. Succeeded to earldom in 1831. Lord lieutenant of Kent, 1856–90. His principal residence, Frognal, Foots Cray, was about five miles from Down. (*Complete peerage*.)

Trécul, Auguste Adolphe Lucien (1818–96). French botanist. Appointed by the Muséum d'Histoire Naturelle and the French Ministry of Agriculture to study nutrition and collect plants in the United States, 1848–50. Continued his researches into the anatomy, physiology, and organogeny of plants. Awarded the legion of honour, 1867. (*Dictionnaire universel des contemporains*, *Index biographique de l'Académie des sciences*.)

Treviranus, Ludolph Christian (1779–1864). German botanist. Professor of botany, University of Breslau, 1816–30; University of Bonn, 1830–64. (*ADB*, *DSB*.)

Trollope, Anthony (1815–82). Novelist and post-office official. (*DNB*.)

Trübner, Johann Nicolaus (Nicholas) (1817–84). German-born publisher. Worked for the Longman publishing company, 1843–51. Entered into partnership with Thomas Delf and David Nutt, and established a successful London publishing house. Interested in philology, philosophy, religion, and oriental studies. (*ADB*, *DNB*.)
23 June [1862]

Tuckerman, Edward (1817–86). American botanist. Lecturer in history at Amherst College, Massachusetts, 1854; professor of botany, 1858–86. An expert on lichens. (*DAB*.)

Turnbull, George Henry. Resided at The Rookery, Down. (*Post Office directory of the six home counties* 1862.)

Turner, Elizabeth. Aunt of Joseph Dalton Hooker. Married the historian Francis Palgrave in 1823. (*DNB* s.v. Palgrave, Francis.)

Tyerman, John Simpson (*c.* 1830–89). Botanist and conchologist. Foreman, Francis Dickson & Co. nursery, Chester, 1858–61. Curator, Liverpool Botanic Gardens, 1861–71. (R. Desmond 1994.)

Tyndall, John (1820–93). Irish physicist, lecturer, and populariser of science.

Professor of natural philosophy at the Royal Institution, London, 1853; superintendent of the Royal Institution, 1867–87. Scientific adviser to Trinity House and the Board of Trade, 1866–83. FRS 1852. (*DNB*, *DSB*.)

Unger, Franz (1800–70). Austrian botanist and palaeobotanist. Professor of plant anatomy and physiology, University of Vienna, 1849–66. (*ADB*, *DSB*.)

Vaucher, Jean Pierre Etienne (1763–1841). Swiss botanist and divine. Parish priest, Geneva, 1797–1822. Honorary professor of botany, Geneva University, 1798–1807; of ecclesiastical history, 1807–39; rector, 1818–21. Founder member of the Société de Physique et d'Histoire Naturelle de Genève. Worked on conjugation and spore formation in algae. (*DSB*.)

Veitch, James (1792–1863). Nurseryman. In partnership with his son James Veitch (1815–69), as proprietors of nurseries in Chelsea and Exeter. (R. Desmond 1994.)

Veitch, James (1815–69). Nurseryman. In partnership with his father James Veitch (1792–1863). (R. Desmond 1994.)

Victor Emmanuel II (1820–78). King of Sardinia and first king of Italy. (*EB*.)

Vilmorin, Elisa de (d. 1866). French horticulturalist. Born Elisa Bailly. Married Pierre Louis François Léveque de Vilmorin in 1841. President of the Paris seed firm Vilmorin-Andrieux et Cie, 1860–6. Published papers on the breeding of strawberries. (Barnhart 1965, Heuzé 1899, *Journal of Horticulture* 40 (1868): 134.)

Vilmorin, Pierre Louis François Léveque de (1816–60). French botanist. President of the Paris seed firm Vilmorin-Andrieux et Cie from 1843. Published papers on the breeding of cereal grains, potatoes, sugar-beet, and flowers. (*DSB*, *NUC*.)

Virgil [Publius Vergilius Maro] (70–19 B.C.). Roman poet. (*EB*.)

Vrolik, Willem (1801–63). Dutch anatomist and physiologist. Appointed to the chair of the medical faculty in Groningen, 1828. Member of the Atheneum Illustre, 1831–61. Member of numerous Dutch and international societies, including the second dividion of the Royal Netherlands Institute and the Academy of Sciences. (*NNBW*.)

Wahlenberg, Göran (1780–1851). Swedish botanist and homoeopathist. Travelled extensively in Europe, 1799–1814, studying the geographical distribution of plants. Professor of medicine and botany, University of Uppsala, 1828. (*DSB*, *SBL*.)

Walker, Charles Vincent (1812–82). Electrical engineer. Secretary, London Electrical Society, 1843. Editor of *Electrical Magazine*, 1845–6. Electrician to the South-Eastern Railway Company, 1845–82. Instrumental in the development of submarine telegraphy. Member of the British Meteorological Society, 1850; secretary, 1857–64; president, 1869–70. FRS 1855. (*DNB*, *Quarterly Journal of the Meteorological Society* 9 (1883): 99–100.)

Walker, Francis (1809–74). Entomologist. Specialised in the small parasitic Hymenoptera, the chalcids; also an authority on the Diptera. Described the Chalcididae CD collected on the *Beagle* voyage. Catalogued a number of the insect collections in the British Museum. (*Entomologist's Monthly Magazine* 11 (1874): 140–1, Gilbert 1977.)

Walker, Frederick (1820?–66). Police officer. Emigrated from England to New South Wales, Australia, in 1844 and joined the police force. Clerk of petty sessions at Tumut, 1847. Held positive views on racial harmony, and recruited and trained aboriginals for service in the police force. In 1861, he was commissioned to search for the ill-fated expedition party led by Robert O'Hara Burke and William John Wills. (*Aust. dict. biog.*)

Wall, Susan (b. 1842/3). Married Leonard Ramsay Henslow in 1862. Granddaughter of Henry Addington, 1st Viscount Sidmouth. (*Burke's peerage* 1980, *Gentleman's Magazine* 13 (1862): 489.)

Wallace, Alfred Russel (1823–1913). Naturalist. Collected natural history specimens in the Amazon, 1848–52; in the Malay Archipelago, 1854–62. Independently formulated a theory of natural selection in 1858. Lecturer and author of works on protective coloration, mimicry, and zoogeography. Considered a controversial figure because of his views on spiritualism, vaccination, land nationalisation, women's rights, and socialism. FRS 1893. (*DNB*, *DSB*.)
7 April 1862, *23 May 1862*, 24 [May 1862], *8 August 1862*, 20 August [1862], *[after 20 August 1862]*, *30 September 1862*

Wallich, Nathaniel (1786–1854). Danish-born botanist and surgeon. Surgeon at Serampore, India, 1807–13. Superintendent of the Calcutta botanic garden, 1817–46. Made great collections of plants in India and Burma. FRS 1829. (R. Desmond 1994, *DNB*.)

Walsh, John Henry (1810–88). Surgeon, journalist, and sports writer. Surgeon at the Worcester Eye Infirmary and lecturer on surgery and anatomy at the Aldersgate Street School of Medicine. Writing under the pseudonym 'Stonehenge', he published works on dogs, horses, guns, sports, domestic economy, medicine, and cookery. Editor of the *Field* from 1857. (*DNB*, *Medical directory* 1862.)

Walshe, Walter Hayle (1812–92). Physician. Professor of morbid anatomy, University College London, 1841–6; Holme professor of clinical medicine and physician to the hospital, 1846–8; professor of the principles and practice of medicine, 1848–62. (*DNB*.)

Ward, Nathaniel Bagshaw (1791–1868). Botanist and physician. Inherited his father's medical practice in Whitechapel, London. Examiner in botany for the Society of Apothecaries, 1836–54. Invented the 'Wardian' case for the transport of plant specimens. An original member of the Botanical Society of Edinburgh and a founding member of the Microscopical Society. (R. Desmond 1994, *DNB*.)

Waring, William (1818–1904). Landowner. Principal landowner of Chelsfield, a village four miles east of Down. Owned Chelsfield and Hewitts, two of the four estates in the parish. Justice of the peace. (*Burke's landed gentry* 1939, Heinecke 1985, *Post Office directory of the six home counties* 1862.)

Warren, Richard Laird (1806–75). Naval officer. Joined the British navy in 1822; captain, 1839; rear-admiral, 1858. Commander-in-chief, east coast of South America, 1861–4; Sheerness, 1869–70. Admiral, 1870. (*Modern English biography*, *Navy list*.)

Waterhouse, George Robert (1810–88). Naturalist. Curator of the Zoological Society of London, 1836–43. Assistant in the mineralogical and geological branch of the natural history section of the British Museum, 1843; keeper, 1851–7; keeper of the department of geology, 1857–80. Founding member of the Entomological Society, 1833; president, 1849–50. Described CD's mammalian and entomological specimens from the *Beagle* voyage. (*DNB*, Gilbert 1977.)

Watson, Hewett Cottrell (1804–81). Botanist, phytogeographer, and phrenologist. Published various guides to the distribution of British plants. Edited the *Phrenological Journal*, 1837–40. Thereafter, returned to the study of botany, to which he devoted the rest of his life. Collected plants in the Azores, 1842. An early supporter of the idea of the progressive development of plant species. (*DNB*, *DSB*.)

8 [August 1862]

Watson-Taylor, George Graeme (1816–65). Owner of the Mediterranean island of Monte Cristo. Purchased Monte Cristo in 1852, having been ordered to the Mediterranean for his health. In 1860, with his wife, Victorine, convicted by a Sardinian court of sedition, but granted remission of penalties. Fled to Britain and sought compensation from the Sardinian government following the sacking of Monte Cristo by Giuseppe Garibaldi's troops. (*Alum. Cantab.*, *Burke's landed gentry* 1868, *Hansard's Parliamentary Debates* 3d ser. 167 (1862): 545–608.)

Watson-Taylor, Victorine. Married George Graeme Watson-Taylor in 1847. (*Burke's landed gentry* 1868.)

Webb, Jonas (1796–1862). Agriculturist and livestock breeder. Began business as a farmer in Babraham, Cambridgeshire, in 1822. Devoted himself to breeding southdown sheep; a leading exhibitor, 1840–62. Developed an important herd of shorthorn cattle, 1838–62. (*DNB*.)

Weddell, Hugh Algernon (1819–77). British-born botanist and physician. Assistant naturalist to the Muséum d'Histoire Naturelle, Paris, 1850–7. Specialist on South American flora. (R. Desmond 1994.)

Wedgwood, Alfred Allen (1842–92). Son of Hensleigh and Frances Mackintosh Wedgwood. (Freeman 1978.)

Wedgwood, Caroline Sarah (1800–88). CD's sister. Married Josiah Wedgwood III, her cousin, in 1837. (*Darwin pedigree*.)

Wedgwood, Emma. *See* Darwin, Emma.

Wedgwood, Ernest Hensleigh (1838–98). Civil servant. Clerk in the office of the secretary of state for the colonies, 1860; assistant private secretary, 1868–72. Second class clerk, 1872; first class clerk, 1876–96. Son of Hensleigh and Frances Wedgwood. (*Alum. Cantab.*, *Colonial Office list* 1896.)

Wedgwood, Frances Julia (Snow) (1833–1913). Novelist, biographer, historian, and literary critic. Daughter of Hensleigh and Frances Wedgwood. (*Burke's peerage* 1980, Wedgwood and Wedgwood 1980.)

Wedgwood, Frances (1800–89). Second child of James Mackintosh and Catherine Allen. Married Hensleigh Wedgwood in 1832. (Freeman 1978.)

Wedgwood, Hensleigh (1803–91). Philologist and barrister. Emma Darwin's brother. Fellow, Christ's College, Cambridge, 1829–30. Metropolitan police magistrate at Lambeth, 1832–7. Registrar of metropolitan carriages, 1838–49. An original member of the Philological Society, 1842. Published the first volume of his *Dictionary of English etymology* in 1859. Married Frances Mackintosh in 1832. (*DNB.*)

Wedgwood, James Mackintosh (1834–64). Eldest son of Hensleigh and Frances Wedgwood. Held a position in the Colonial Office. (Freeman 1978, Wedgwood and Wedgwood 1980.)

Wedgwood, Josiah, I (1730–95). Master-potter. Founded the Wedgwood pottery works at Etruria, Staffordshire. Grandfather of CD and Emma Darwin. Greatly interested in experimental chemistry. Contributed several papers on the measurement of high temperature to the Royal Society's *Philosophical Transactions.* Actively associated with scientists and scientific societies. FRS 1783. (*DNB*, *DSB.*)

Wedgwood, Josiah, II (1769–1843). Master-potter of Etruria. Resident at Maer Hall, Staffordshire. Whig MP for Stoke-on-Trent, 1832–4. Emma Darwin's father. (*Burke's peerage* 1980, *Emma Darwin.*)

Wedgwood, Josiah, III (1795–1880). Master-potter. Partner in the Wedgwood pottery in Staffordshire until 1844, when he moved to Leith Hill Place, Surrey. Emma Darwin's brother. Married Caroline Sarah Darwin, his cousin, in 1837. (Freeman 1978.)

Wedgwood, Katherine Elizabeth Sophy (1842–1911). Daughter of Caroline and Josiah Wedgwood III. CD's niece. (*Emma Darwin.*)
4 [August 1862]

Wedgwood, Lucy Caroline (1846–1919). Daughter of Caroline and Josiah Wedgwood III. CD's niece. (*Darwin pedigree.*)
4 [August 1862]

Wedgwood, Margaret Susan (1843–1937). Daughter of Caroline and Josiah Wedgwood III. CD's niece. Married Arthur Charles Vaughan Williams (Freeman 1978) in 1869. Mother of Ralph Vaughan Williams (*DNB*). (*Emma Darwin.*)
[before 4 August 1862], 4 [August 1862], *[6 August 1862]*

Wedgwood, Sarah Elizabeth (Elizabeth) (1793–1880). Emma Darwin's sister. Resided at Maer Hall, Staffordshire, until 1847, then at The Ridge, Hartfield, Sussex, until 1862. She moved to London before settling in Down in 1868. (*Emma Darwin*, Freeman 1978.)

Wedgwood, Susannah. *See* Darwin, Susannah.

Weissenborn, W. (*fl.* 1837–8). Author of Weissenborn 1837.

Wells, Henry (1805–78). American expressman. Established Livingstone, Wells and Pomeroy's, a mail service operating between Albany and Buffalo, in 1843. In 1844, established Wells & Company, operating between Buffalo and Detroit with William George Fargo as messenger. In 1850, the merger of Wells & Company with two other companies created the American Express Company of which Wells and Fargo were presidents, 1850–68. The American Express Company absorbed the Pacific Union Express Company in 1868. Wells retired as president in 1873. (*DAB*.)

Wells, Joseph. Gardener to William Wells. (R. Desmond 1994.)

Wells, William (1760–1847). Art collector. With the aid of his gardener Joseph Wells, created a garden at Redleaf, Kent, 'exploiting the natural features with sensitive skill'. (R. Desmond 1994.)

Westwood, Eliza (d. 1882). Married John Obadiah Westwood in 1839. Accompanied him on archaeological tours. (*DNB* s.v. Westwood, John Obadiah.)

Westwood, John Obadiah (1805–93). Entomologist and palaeographer. An active member of the Entomological Society of London. Hope professor of zoology, Oxford University, 1861–93. Entomological referee for the *Gardeners' Chronicle*. Awarded the Royal Medal of the Royal Society in 1855. (*DNB*, Gilbert 1977.)

14 May 1862, 1 August 1862

Whately, Richard (1787–1863). Clergyman and scholar. Principal of St Alban Hall, Oxford, 1825–31. Drummond professor of political economy, Oxford University, 1829–31. Archbishop of Dublin, 1831–63. Wrote extensively on theology, philosophy, and political economy. (*DNB*.)

Whewell, William (1794–1866). Mathematician, historian, and philosopher of science. Master of Trinity College, Cambridge, 1841–66. Professor of moral philosophy, Cambridge University, 1838–55. FRS 1820. (*DNB*, *DSB*.)

White, Adam (1817–79). Naturalist. Official in the zoological department of the British Museum, 1835–63. Described some of the insect specimens collected on the *Beagle* voyage. (*DNB*, Gilbert 1977.)

White, Gilbert (1720–93). Naturalist and clergyman. Author of *The natural history and antiquities of Selborne* (1789). (*DNB*, *DSB*.)

White, Henry (1812–80). Historical and educational writer. Published several important historical textbooks. Superintended the compilation of the *Royal Society catalogue of scientific papers*, 1858–80. In conjunction with Thomas W. Newton, prepared the *Catalogue of the library of the museum of practical geology* (1878). (*DNB*.)

White, Walter (1811–93). Miscellaneous writer. Cabinetmaker in Reading until 1834. Travelled in America, 1834–9. Attendant in the library of the Royal Society, 1844; assistant secretary and librarian, 1861–84. Author of many travel books. (*DNB*, *Modern English biography*.)

17 [October – November 1862?]

Whitworth, Robert Percy (1831–1901). Journalist and author. Emigrated from Lancashire to Sydney, Australia, in 1855. Began work on a series of gazetteers of various Australian colonies in 1864. (*Aust. dict. biog.*)

Wickham, John Clements (1798–1864). Naval officer and magistrate. First-lieutenant on HMS *Beagle*, 1831–6; commander, 1837–41, surveying the Australian coast. Settled in Australia in 1842. Police magistrate in New South Wales, 1843–57; government resident, 1853. Left Australia and took his family to southern France in 1862. (*Aust. dict. biog.*, R. Desmond 1994.)

Wicksted, Charles (1796–1870). Landowner and breeder. Son of George Tollet and childhood friend of CD's. Took the name Wicksted in 1814. High-sheriff of Cheshire, 1822. Inherited Betley Hall, Staffordshire, in 1855; continued living at Shakenhurst, Worcestershire, the home inherited by his wife to which he had moved *c.* 1844. A noted foxhunter and breeder of hounds. (Blagg 1902, *Burke's landed gentry* 1875, Freeman 1978, *Post Office directory of Birmingham* 1864.)

Wight, Alexander (1815/16–71). Banker. Manager of the Charing Cross Branch of the Union Bank of London, 1844 (or earlier) – 1860. General Manager of the Manchester and County Bank, 1862–71. (*Banking almanac*, 1845–72.)

Wilberforce, Samuel (1805–73). Clergyman. Rector of Brightstone, Isle of Wight, 1830–40; of Alverstoke, Hampshire, 1840–3. Chaplain to Prince Albert, 1841. Appointed dean of Westminster in 1845. Bishop of Oxford, 1845–69. Opposed CD's theory of evolution. FRS 1845. (*DNB*, Sarjeant 1980.)

Wilkes, Charles (1798–1877). American naval officer and explorer. Commanded the United States Exploring Expedition, 1838–42, exploring the American north-west coast, the Antarctic coastline, and the islands of the Pacific Ocean. From 1843 until 1861, on special service, preparing the information collected by the expedition for publication. Took command of the *San Jacinto* in 1861 and sparked off the '*Trent* affair' by forcibly removing two Confederate diplomatic commissioners from a British ship. (*DAB*.)

Willdenow, Karl Ludwig (1765–1812). German botanist and apothecary. Gave up his apothecary's shop when he became professor of natural history at the Royal College of Medicine and Surgery, Berlin, in 1798. Appointed chief botanist of the Berlin Academy of Science and curator of the botanical gardens in 1801. Professor of botany, University of Berlin, 1809–12. (*ADB*, *DSB*.)

Williams, Edward Augustus (1800/1–75). Surgeon. Member of the Royal College of Surgeons, 1824. Licentiate of the Society of Apothecaries, 1823. In practice in Bromley, Kent, 1829; in partnership there with Charles Morgan, 1849–65. Attended various members of the Darwin household, 1845–63. (CD's Classed account book (Down House MS), *London Medical Gazette* 4 (1829): 297, *Medical directory*, *Post office directory of the six home counties*.)

Williams, John (*fl.* 1806–54). Writer on climate, viticulture, and pomology; resident at Pitmarston House, Worcester. Corresponding member of the Horticultural Society of London, *c.* 1818. (*Post Office directory of Birmingham* 1854; *Transactions of the Horticultural Society of London* 2 (1817): 324, 3 (1820): 249; Watt 1824.)

Williams and Norgate. London booksellers and publishers who specialised in foreign scientific literature. The business was a partnership between Edmund Sydney Williams (1817–91) and Frederick Norgate. Publishers of the *Natural History Review*. (*Modern English biography*.)

Wills, William John (1834–61). Explorer, surveyor, and astronomer. Studied medicine in London before emigrating to Australia in 1852. Assistant at the astronomical and magnetical observatories in Melbourne. In 1860, accompanied

Robert O'Hara Burke's ill-fated expedition attempting to cross the continent from south to north (*Aust. dict. biog.*)

Wilson, William James Erasmus (Erasmus) (1809–84). Surgeon. Taught anatomy and physiology at University College London, 1831–6. Established Sydenham College 1836. Entered private practice in 1840, specialising in the treatment of skin diseases. Founded and subsequently held the chair of dermatology at the Royal College of Surgeons, 1869–77. President, Royal College of Surgeons, 1881. Interested in the study of Egyptian antiquities; president of the Biblical Archaeology Society. Knighted, 1881. FRS 1845. (*DNB*.)

Winkler, Tiberius Cornelius (1822–97). Dutch geologist and palaeontologist. Curator of the Teylor Museum, Haarlem, from 1864. Translated *Origin* into Dutch (1860). (*NNBW*, Sarjeant 1980.)

Wolf, Joseph (1820–99). German-born animal painter and illustrator. Apprenticed and worked as a lithographer. Studied art at Darmstadt and Antwerp. Emigrated to London, 1848. Employed at the British Museum. Illustrated many important books and periodicals on natural history. (*DNB*, J. C. Wood 1973.)

Wollaston, George Buchanan (1814–99). Collector of ferns. Resided at Bishop's Well, Chislehurst in Kent. (R. Desmond 1994, *Post Office directory of the six home counties* 1862.)

Wood, Charles, 1st Viscount Halifax (1800–85). Statesman. MP for Halifax, 1832–65. Chancellor of the exchequer, 1846. President of board of control, 1852; First Lord of the Admiralty, 1855. Secretary of state for India, 1859–66. Created Viscount Halifax, 1866. (*DNB*.)

Wood, Searles Valentine (1798–1880). Palaeontologist. Officer in the East India Company's navy, 1811–25. Became a partner in a bank at Hasketon, near Woodbridge. Curator of the Geological Society of London, 1838–9; fellow, 1839. Studied the fossils of the East Anglian Crag, specialising in Eocene fossil Mollusca. (*DNB*, Woodward 1907.)

18 February 1862

Woodbury, Thomas White (1818–71). Beekeeper and journalist. Resided at 17 Lower Mount Radford Terrace, Exeter, Devon. Son of W. H. Woodbury (*Modern English biography*), a linguist and part-proprietor of the *Exeter and Plymouth Gazette*. Took an active part in the management of his father's newspaper, before retiring to live on private means. After 1850, devoted himself to the study of bees. Introduced Ligurian bees to Britain in 1859, and developed the 'Woodbury hive', incorporating a movable frame around the bee space; a contributor to the bee-keeping section of the *Journal of Horticulture*, 1859–71. (Dodd 1983, *Journal of Horticulture* n.s. 21 (1871): 97, 113; *Post Office directory of Devonshire* 1856.)

9 August 1862, 7 December [1862]

Woodfall and Kinder. Printing firm with premises in Angel Court, Skinner Street, London, 1853–64, and elsewhere, 1864–1900 (or later). Henry Dick Woodfall, the eldest son of George Woodfall (*DNB*), became a partner in the firm in 1840. (P. A. H. Brown 1982, *DNB*.)

Wooler, William Alexander (1813–91). Landowner, colliery-owner, and agriculturist of Halliwell House, Heighington, County Durham. Speculated in land; owned several collieries; director of the Middleton and Dinsdale Gas Works. Founder and honorary treasurer of the County Durham and North Yorkshire Horse and Dog Show; instrumental in founding the South Durham and North Yorkshire Chamber of Agriculture. Interested in all aspects of agriculture and sanitation; reported to be the originator of the system, later adopted by the government, of slaughtering diseased animals to prevent the spread of rinderpest. (*North Eastern Daily Gazette*, 5 May 1891; *Northern Echo*, 5 May 1891; *North Star*, 5 May 1891; *Post Office directory of the county of Durham* 1873.)
5 July 1862

Wray, Leonard Hume (1816–1901). Sugar-planter and horticultural writer. For sixteen years a sugar-planter in Jamaica, Bengal, and the Straits Settlements. Author of *The practical sugar planter* (London, 1848) and *The Zulu-Zafir lanphee, or "Sweet Reed"* (London, 1854–7). (*Catalogue of the library of the British Museum (Natural History), CDEL.*)

Wright, Edward Perceval (1834–1910). Irish naturalist. Edited the *Natural History Review*, 1854–66. Appointed director of the university museum in Dublin, 1857. Lecturer in zoology, Trinity College, Dublin, 1858–68; professor of botany and keeper of the herbarium, 1869. Resigned the chair in 1904 due to ill health, but continued to superintend the herbarium. (*DNB.*)

Wrigley, Alfred (1817–98). Mathematician. Professor of mathematics at the Royal Military College, Addiscombe, 1841–61. Headmaster of Clapham Grammar School, 1862–82. (*Alum. Cantab., Modern English biography.*)

Wydler, Heinrich (1800–83). Swiss botanist. Keeper of the de Candolle botanical collections in University of Geneva, 1830–4. Taught in schoools in Zürich and Bern. After his marriage in 1840, settled in Strasbourg and devoted himself to his botanical studies. (*ADB.*)

Wyman, Jeffries (1814–74). American anatomist and ethnologist. Hersey Professor of anatomy, Harvard University, 1847–74. Curator of the Peabody Museum of American Archaeology and Ethnology, Harvard, 1866–74. (*DAB*, *DSB.*)

Zittel, Karl Alfred von (1839–1904). German geologist and palaeontologist. After completing his studies, worked as a volunteer in the Geologische Reichsanstalt, Vienna, 1862–3. Professor of mineralogy, geognosy, and palaeontology, Karlsruhe Technical College, 1863–6; at the University of Munich from 1866. Spent twenty years producing his *Handbuch der Palaeontologie* (1876–93), a systematic study of the organic fossil record. (*DSB.*)

Zwecker, Johann Baptist (1814–76). German-born painter and illustrator. Emigrated to England in 1850. Illustrated books by travellers and explorers. Contributed to illustrated newspapers. (*Modern English biography*, Williamson ed. 1903–5.)

Bibliography of biographical sources

ADB: *Allgemeine deutsche Biographie.* Under the auspices of the Historical Commission of the Royal Academy of Sciences. 56 vols. Leipzig: Duncker and Humblot. 1875–1912.

Ainslie, Robert. 1865. *Discourses delivered in Christ Church, New Road, Brighton.* London.

Albert, W. 1845. *Authentische Thatsachen über die Verjüngung der Kartoffeln aus Samenkörnern, mit Berücksichtigung der unter denselben jetzt herrschenden Krankheiten.* Magdeburg.

Allan, Mea. 1967. *The Hookers of Kew, 1785–1911.* London: Michael Joseph.

Alum. Cantab.: *Alumni Cantabrigienses. A biographical list of all known students, graduates and holders of office at the University of Cambridge, from the earliest times to 1900. Part II. From 1752 to 1900.* Compiled by J. A. Venn. 6 vols. Cambridge: Cambridge University Press. 1940–54.

Alum. Oxon.: *Alumni Oxonienses: the members of the University of Oxford, 1715–1886.* By Joseph Foster. 4 vols. Oxford. 1888.

Army list: *The army list*. London. 1815–81.

Aust. dict. biog.: *Australian dictionary of biography: 1788–1850; 1851–1890.* Edited by Douglas Pike and Bede Nairn. 6 vols. Melbourne: Melbourne University Press. 1966–76.

Babington, Anna Maria, ed. 1897. *Memorials journal and botanical correspondence of Charles Cardale Babington.* Cambridge.

Baker, Herbert G. 1965. Charles Darwin and the perennial flax—a controversy and its implications. *Huntia* 2: 141–61.

Banking almanac: *The banking almanac, directory, yearbook and diary.* Edited by D. Morier Evans. London. 1844–1919.

Barnhart, John Hendley. 1965. *Biographical notes upon botanists. Compiled by John Hendley Barnhart . . . and maintained in the New York Botanical Garden Library.* 3 vols. Boston, Mass.

Bates, Henry Walter. 1863. *The naturalist on the River Amazons, a record of adventures, habits of animals, sketches of Brazilian and Indian life, and aspects of nature under the equator, during eleven years of travel.* London.

——. 1892. *The naturalist on the River Amazons. A record of adventures, habits of animals, sketches of Brazilian and Indian life, and aspects of nature under the equator, during eleven years of travel. With a memoir of the author by Edward Clodd.* Reprint edition. London.

Bénézit, Emmanuel. 1976. *Dictionnaire critique et documentaire des peintres, sculpteurs, dessinateurs et graveurs de tous les temps et de tous les pays par un groupe d'écrivains spécialistes français et étrangers.* Revised edition. 10 vols. Paris: Librairie Gründ.

Berg, Carlos. 1895. Carlos Germán Conrado Burmeister. Reseña Biográfica. *Anales des Museo Nacional de Buenos Aires* 2d ser. 1: 316–57.

BHGW: *Biographisch-literarisches Handwörterbuch zur Geschichte der exacten Wissenschaften enthaltend Nachweisungen über Lebensverhältnisse und Leistungen von Mathematikern, Astronomen, Physikern, Chemikern, Mineralogen, Geologen usw.* By Johann Christian

Poggendorff. Vols. 1–4, Leipzig: Johann Ambrosius Barth. 1863–1904. Vol. 5, Leipzig and Berlin: Verlag Chemie. 1926.

Blake, Charles Carter. 1862a. Past life in South America. *Geologist* 5: 323–30.

——, ed. 1863. *A manual of zoology*. By Henri Milne-Edwards. Translated by Robert Knox. 2d ed. London.

Blagg, C. J. 1902 *A history of the North Staffordshire hounds and country 1825 to 1902*. London: Sampson Low, Marston & Company.

Blaschko, Mary D. 1958. *Sidcot school: register of old scholars, 1808–1958*. Winscombe, Somerset: Sidcot School.

BLC: *The British Library general catalogue of printed books to 1975*. 360 vols. and supplement (6 vols.). London. 1979–88.

Bonham-Carter, Victor. 1960. *In a liberal tradition: a social biography 1700–1950*. London: Constable.

Brown, Philip Arthur Henry. 1982. *London publishers and printers, c. 1800–1870*. London: British Library.

Burke's landed gentry: *Burke's genealogical and heraldic history of the landed gentry*. By John Burke. 1–18 editions. London: Burke's Peerage Ltd. 1836–1972.

Burke's peerage: *Burke's peerage and baronetage*. 1–105 editions. London: Burke's Peerage (Genealogical Books). 1826–1980.

Butterworths medical dictionary: *Butterworths medical dictionary*. 2d ed. Editor-in-chief Macdonald Critchley. London and Boston: Butterworth & Co. 1978.

Calendar: *A calendar of the correspondence of Charles Darwin, 1821–1882. With supplement.* Edited by Frederick Burkhardt and Sydney Smith. 2d ed. Cambridge: Cambridge University Press. 1994.

Catalogue de la Bibliothèque nationale: *Catalogue général des livres imprimés de la Bibliothèque Nationale.* 231 vols. Paris: Imprimerie nationale. 1897–1981.

Catalogue of the library of the British Museum (Natural History): *Catalogue of the books, manuscripts, maps and drawings in the British Museum (Natural History)*. 5 vols. London: Trustees of the British Museum. 1903–15.

CDEL: *A critical dictionary of English literature and British and American authors, living and deceased, from the earliest accounts to the middle of the nineteenth century.* By S. Austin Allibone. 3 vols. Philadelphia and London. 1859–71. Supplement by John Foster Kirk. 2 vols. Philadelphia and London. 1891.

Chemical manufactories' directory: *The chemical manufactories' directory of England*. London: 1867–1955.

Clergy list: *The clergy list . . . containing an alphabetical list of the clergy*. London. 1841–.

Colonial Office list: *The Colonial Office list*. London. 1862–1925.

Complete peerage: *The complete peerage of England, Scotland, Ireland, Great Britain, and the United Kingdom, extant, extinct, or dormant.* By George Edward Cokayne. New edition, revised and much enlarged. Edited by the Hon. Vicary Gibbs and others. 12 vols. London: St Catherine Press. 1910–59.

Correspondence: *The correspondence of Charles Darwin.* Edited by Frederick Burkhardt, Sydney Smith, and others. Cambridge: Cambridge University Press. 1985–.

Corsi, Pietro. 1985. Recent studies on Italian reactions to Darwin. In *The Darwinian heritage*, edited by David Kohn. Princeton: Princeton University Press in association with Nova Pacifica.

County families: *The county families of the United Kingdom; or, a royal manual of the titled & untitled aristocracy of Great Britain & Ireland.* By Edward Walford. London. 1860–1920.

Crockford's: *The clerical directory, a biographical and statistical book of reference for facts relating to the clergy and the Church.* London. 1858. Continued as *Crockford's clerical directory etc.* London. 1860–.

DAB: *Dictionary of American biography.* Under the auspices of the American Council of Learned Societies. 20 vols., index, and 10 supplements. New York: Charles Scribner's Sons; London: Oxford University Press. 1928–95.

Dallas, James. 1921. *The history of the family of Dallas and their connections and descendants from the twelfth century.* Edinburgh: T. and A. Constable.

Darwin pedigree: *Pedigree of the family of Darwin.* Compiled by H. Farnham Burke. Privately printed. 1888. [Reprinted in facsimile in *Darwin pedigrees*, by Richard Broke Freeman. London: printed for the author. 1984.]

Darwin, Francis. 1914. William Erasmus Darwin, 1839–1914. *Christ's College Magazine* 29: 16–23.

DBF: *Dictionnaire de biographie Française.* Under the direction of J. Balteau, M. Barroux, M. Prevost, and others. 17 vols. and 7 fascicles of vol. 18. (A–Laglenne). Paris: Libraire Letouzey et Ané. 1933–95.

DBI: *Dizionario biografico degli Italiani.* Edited by Alberto M. Ghisalberti. 44 vols. (A–Farina). Rome: Istituto della Enciclopedia Italiana. 1960–94.

DBS: *Dictionnaire historique & biographique de la Suisse.* Under the direction of Marcel Godet, Henri Türler, and Victor Attinger. 7 vols. and supplement. Neuchatel: Administration du Dictionnaire Historique et Biographique de la Suisse. 1921–34.

Dempster, W. J. 1983. *Patrick Matthew and natural selection: nineteenth century gentleman-farmer, naturalist and writer.* Edinburgh: Paul Harris publishing.

Desmond, Ray. 1994. *Dictionary of British and Irish botanists and horticulturists including plant collectors, flower painters and garden designers.* New edition, revised and completely updated. London: Taylor & Francis and the Natural History Museum. Bristol, Pa.: Taylor & Francis.

——. 1995. *Kew: the history of the Royal Botanic Gardens.* London: Harvill Press with the Royal Botanic Gardens, Kew.

Dictionary of artists in America: *The New-York Historical Society's dictionary of artists in America 1564–1860.* By George C. Grove and David H. Wallace. New Haven: Yale University Press. London: Oxford University Press. 1957.

Dictionary of watercolour artists: *The dictionary of watercolour artists up to 1920.* By H. L. Mallalieu. 2d ed. 3 vols. Woodbridge, Suffolk: Antique Collectors Club. 1986.

Dictionnaire des parlementaires français: *Dictionnaire des parlementaires français, comprenant tous les membres des assemblées françaises et tous les ministres français.* Edited by Adolphe Robert, Edgar Bourloton, and Gaston Cougny. 5 vols. Paris. 1889–91.

Dictionnaire universel des contemporains: *Dictionnaire universel des contemporains contenant toutes les personnes notables de la France et des pays étrangers.* Edited by Louis Gustave Vapereau. Paris. 1858. 3d ed., 1865. 4th ed., 1870. 5th ed., 1880. 6th ed., 1893.

DNB: *Dictionary of national biography.* Edited by Leslie Stephen and Sidney Lee. 63 vols. and 2 supplements (6 vols.). London: Smith, Elder, and Co. 1885–1912. *Dictionary of national biography 1912–85.* Edited by H. W. C. Davis, J. R. H. Weaver, and others. 8 vols. London: Oxford University Press. 1927–90.

DNZB: *A dictionary of New Zealand biography.* Edited by G. H. Scholefield. 2 vols. Wellington: Department of Internal Affairs, New Zealand. 1940.

Dodd, Victor. 1983. *Beemasters of the past.* Hebden Bridge: Northern Bee Books.

DSB: *Dictionary of scientific biography.* Edited by Charles Coulston Gillispie. 14 vols., supplement, and index. New York: Charles Scribner's Sons. 1970–80.

Dupree, Anderson Hunter. 1959. *Asa Gray, 1810–1888.* Cambridge, Mass.: Belknap Press of Harvard University.

EB: *The Encyclopædia Britannica. A dictionary of arts, sciences, literature and general information.* 11th ed. 29 vols. Cambridge: Cambridge University Press. 1910–11.

Egerton, F. Clement C. 1957. *Angola in perspective: endeavour and achievement in Portuguese West Africa.* London: Routledge & Kegan Paul.

Emma Darwin (1904): *Emma Darwin, wife of Charles Darwin. A century of family letters.* Edited by Henrietta Litchfield. 2 vols. Cambridge: privately printed by Cambridge University Press. 1904.

Emma Darwin: *Emma Darwin: a century of family letters 1792–1896.* Edited by Henrietta Litchfield. Revised edition. 2 vols. London: John Murray. 1915.

ENZ: *An encyclopaedia of New Zealand.* Edited by A. H. McLintock. 3 vols. Wellington, New Zealand: R. E. Owen. 1966.

Erslew, Thomas Hansen. 1843–58. *Almindeligt Forfatter-Lexicon for Kongeriget Danmark med tilhørende Bilande.* With supplement. 6 vols. Copenhagen.

Evans, George Eyre. 1897. *Vestiges of Protestant dissent: being lists of ministers, sacramental plate, registers, antiquities, and other matters pertaining to most of the churches (and a few others) included in the National Conference of Unitarian, Liberal Christian, Free Christian, congregations.* Liverpool.

Filmer, J. L. 1977. The Norman family of Bromley Common. *Bromley Local History* 2: 16–24.

Flett, John Smith. 1937. *The first hundred years of the Geological Survey of Great Britain.* London: His Majesty's Stationery Office.

Foreign Office list: *The Foreign Office list.* London. 1852–1965.

Foster, Joseph. 1885. *Men-at-the bar: a biographical hand-list of the members of the various inns of court, including Her Majesty's judges, etc.* London.

Freeman, Richard Broke. 1978. *Charles Darwin: a companion.* Folkestone, Kent: William Dawson and Sons. Hamden, Conn.: Archon Books, Shoe String Press.

Gilbert, Pamela. 1977. *A compendium of the biographical literature on deceased entomologists.* London: British Museum (Natural History).

Gorman, W. Gordon. 1910. *Converts to Rome. A biographical list of the more notable converts to the Catholic Church in the United Kingdom during the last sixty years.* New and enlarged edition. London: Sands & Co.

Grande encyclopédie: *La grande encyclopédie inventaire raisonné des sciences, des lettres et des arts.* Edited by F. Camille Dreyfus and others. 31 vols. Paris. 1886–1902.

Grönland, Johannes. 1858. Notice nécrologique sur M. Neumann. *Revue Horticole* (1858): 622–4.

Gruber, Jacob W. and Thackray, John C. 1992. *Richard Owen commemoration: three studies.* London: Natural History Museum.

Gunther, Albert E. 1975. *A century of zoology at the British Museum through the lives of two keepers, 1815–1914.* Folkestone, Kent: William Dawson and Sons.

Haast, H. F. von. 1948. *The life and times of Sir Julius von Haast, explorer, geologist, museum builder.* Wellington, New Zealand: privately published.

Hall, Philip B. 1987. Robert Swinhoe (1836–1877), FRS, FZS, FRGS: a Victorian naturalist in Treaty Port China. *Geographical Journal* 153: 37–47.

Hamberger, Georg Christoph and Meusel, Johann Georg. 1796–1834. *Das gelehrte Teutschland, oder Lexikon der jetzt lebenden teutschen Schriftsteller.* 23 vols. in 24. Lemgo.

Heinecke, P. M. 1985. The parish of Chelsfield. *Bromley Local History* 7: 62–8.

Herringshaw, Thomas W. 1909–14. *Herringshaw's national library of American biography.* 5 vols. Chicago: American Publishers' Association.

Heuzé, Gustave. 1899. Les Vilmorin. *Revue Horticole* 71: 453–9.

Hoskyn, Richard. 1862. Report on the deep sea soundings to the westward of Ireland,— made in H.M.S. "Porcupine", in June, July, and August, 1862. *Nautical Magazine and Naval Chronicle* 31: 561–9.

Huxley, Leonard, ed. 1900. *Life and letters of Thomas Henry Huxley.* 2 vols. London: Macmillan and Co.

IBN: *Index bio-bibliographicus notorum hominum.* Pt C: *Corpus alphabeticum sectio generalis.* Edited by Jean-Pierre Lobies and Otto and Wolfram Zeller. 60 vols. (A–Durau, Pierre). Osnabrück: Biblio Verlag. 1974–92.

Index biographique de l'Académie des sciences: *Index biographique de l'Académie des sciences du 22 décembre 1666 au 1er octobre 1978.* Paris: Gauthier-Villars. 1979.

Jackson, Benjamin Daydon. 1906. *George Bentham.* London: J. M. Dent & Co. New York: E. P. Dutton & Co.

Jervis, James. [1945.] *W. B. Clarke: "The father of Australian geology".* Sydney: Royal Australian Historical Society.

Jubiläums-Katalog: *Jubiläums-Katalog der E. Schweizerbart'schen Verlagsbuchhandlung (Erwin Nägele) G.m.b.H., Stuttgart, 1826–1926.* Stuttgart: E. Schweizerbart. 1926.

Junker, Thomas. 1989. *Darwinismus und Botanik. Rezeption, Kritik und theoretische Alternativen im Deutschland des 19. Jahrhunderts.* Stuttgart: Deutscher Apotheker Verlag.

Kaplan, Fred. 1983. *Thomas Carlyle: a biography.* Cambridge: Cambridge University Press.

Keynes, Margaret. 1943. *Leonard Darwin, 1850–1943.* Cambridge: privately printed at Cambridge University Press.

Larousse: *Larousse du XXe siècle*. Under the direction of Paul Augé. 6 vols. Paris: Librairie Larousse. 1928–33.

List of fellows of the Zoological Society: *A list of the fellows and honorary, foreign, and corresponding members and medallists of the Zoological Society of London*. London. 1893.

List of the Geological Society of London: *List of the Geological Society of London*. London. 1864–1965.

List of the Linnean Society of London: *List of the Linnean Society of London*. London. 1805–1939.

LL: *The life and letters of Charles Darwin, including an autobiographical chapter*. Edited by Francis Darwin. 3 vols. London. 1887–8.

Lorenz, Otto, ed. 1867–71. *Catalogue général de la librairie Française pendant 25 ans (1840–1865)*. 4 vols. Paris.

Magnus, P. 1891. Johannes Grönland. *Verhandlungen des botanischen Vereins der Provinz Brandenburg* 33: il–li.

Medical directory: *The London & provincial medical directory, inclusive of the medical directory for Scotland, and the medical directory for Ireland, and general medical register*. London. 1861–9. *The medical directory . . . , including the London and provincial medical directory, the medical directory for Scotland, the medical directory for Ireland*. London. 1870–1905.

Medical who's who: *The medical who's who*. London: London & Counties Press Association. 1913–27.

Metcalfe, Richard. 1906. *The rise and progress of hydropathy in England and Scotland*. London: Simpkin, Marshall, Hamilton, Kent & Co.

ML: *More letters of Charles Darwin: a record of his work in a series of hitherto unpublished letters*. Edited by Francis Darwin and Albert Charles Seward. 2 vols. London: John Murray. 1903.

Modern English biography: *Modern English biography containing many thousand concise memoirs of persons who have died since the year 1850*. By Frederic Boase. 3 vols. and supplement (3 vols.). Truro: printed for the author. 1892–1921.

Moffat, C. B., ed. 1898. *Life and letters of Alexander Goodman More . . . with selections from his zoological writings*. With a preface by Frances M. More. Dublin.

Monteiro, Joachim John. 1875. *Angola and the River Congo*. 2 vols. London.

Moore, James R. 1985. Darwin of Down: the evolutionist as squarson-naturalist. In *The Darwinian heritage*, edited by David Kohn. Princeton: Princeton University Press in association with Nova Pacifica.

Navy list: *The navy list*. London. 1815–70.

NBU: *Nouvelle biographie universelle depuis les temps les plus regulés jusqu'à nos jours, avec les renseignements bibliographiques et l'indication des sources a consulter*. Edited by Jean Chrétien Ferdinand Hoefer. 46 vols. Paris. 1852–66.

NDB: *Neue deutsche Biographie*. Under the auspices of the Historical Commission of the Bavarian Academy of Sciences. 17 vols. (A–Moller). Berlin: Duncker and Humblot. 1953–94.

NNBW: *Nieuw Nederlandsch biografisch Woordenboek*. Edited by P. C. Molhuysen, P. J.

Blok, and K. H. Kossmann. 10 vols. Leiden: A. W. Sijthoff's Uitgevers-Maatschappij. 1911–37.

NUC: *The national union catalog. Pre-1956 imprints.* 685 vols. and supplement (vols. 686–754). London and Chicago: Mansell. 1968–81.

OBL: *Österreichisches biographisches Lexikon 1815–1950.* Edited by Leo Santifaller, Eva Obermayer-Marnach, and others. 9 vols. and 3 fascicles of vol. 10 (A–Schmid). Vienna: Österreichischer Akademie der Wissenschaften. 1957–92.

O'Donoghue, Freeman. 1908–25. *Catalogue of engraved British portraits preserved in the Department of Prints and Drawings in the British Museum.* 6 vols. London: British Museum.

Orchids: *On the various contrivances by which British and foreign orchids are fertilised by insects, and the good effects of intercrossing.* By Charles Darwin. London. 1862.

Palgrave, Gwenllian F. 1899. *Francis Turner Palgrave: his journals and memories of his life.* London.

Pancaldi, Giuliano. 1991. *Darwin in Italy: science across cultural frontiers.* Translated by Ruey Brodine Morelli. Bloomington and Indianapolis: Indiana University Press.

Physicians: *The roll of the Royal College of Physicians of London.* Vol. 3: 1801–25. By William Munk. 2d ed., revised and enlarged. London. 1878. Vol. 4: *Munk's roll, vol. 4: Lives of the fellows of the Royal College of Physicians of London 1826–1925.* Compiled by G. H. Brown. London: Royal College of Physicians. 1955.

Pine, L. G. 1972. *The new extinct peerage 1884–1971: containing extinct, abeyant, dormant & suspended peerages with genealogies and arms.* London: Heraldry Today.

Plarr, Victor Gustave. 1930. *Plarr's lives of the fellows of the Royal College of Surgeons of England.* Revised by Sir D'Arcy Power. 2 vols. London: Simpkin Marshall.

Post office directory of Birmingham: *Post office directory of Birmingham, Warwickshire, and part of Staffordshire.* 1854–1928.

Post Office directory of the county of Durham: *The Post Office directory of the county of Durham, and the principal towns and adjacent places in Northumberland, Cumberland & Westmorland.* London. 1873–.

Post Office directory of Devonshire: *Post Office directory of Devonshire and Cornwall.* London. 1856–1926.

Post Office directory of Lincolnshire: *Post Office directory of Lincolnshire.* London. 1849–76.

Post Office directory of Northamptonshire: *Post Office directory of Berkshire, Northamptonshire, Oxfordshire, with Bedfordshire, Buckinghamshire, and Huntingdonshire.* London. 1847–77.

Post Office directory of Northumberland: *Post Office directory of Westmoreland, Cumberland, Northumberland, and Durham.* London. 1858.

Post Office directory of the six home counties: *Post Office directory of the six home counties, viz., Essex, Herts, Kent, Middlesex, Surrey and Sussex.* London. 1845–.

Post Office directory of Somerset and Devon: *The Post Office directory of Somerset and Devon, with Bristol.* London. 1866.

Post Office Edinburgh and Leith directory: *Post Office Edinburgh and Leith directory.* Edinburgh. 1833–.

Post Office London directory: *Post Office London directory.* London. 1802–.

Post Office London suburban directory: *The Post Office London suburban directory*. London: 1860–1918.

Pritchard, Michael. 1986. *A directory of London photographers 1841–1908*. Watford: ALLM Books.

Provincial medical directory: *The provincial medical directory*. London. 1847.

Richardson, Edmund William. 1916. *A veteran naturalist; being the life and work of W. B. Tegetmeier*. London: Witherby & Co.

Rivière, Auguste. 1866. Observations sur la fécondation et la germination des Orchidées, à l'occasion d'un Lœlia hybride présenté en fleurs, le 25 août 1865. *Journal de la Société Impériale et Centrale d'Horticulture* 12: 268–91.

Roll of the graduates of the University of Aberdeen: *Roll of the graduates of the University of Aberdeen 1860–1900*. By William Johnston. Aberdeen. 1906.

Roll of the graduates of the University of Glasgow: *A roll of the graduates of the University of Glasgow. From 31st December, 1727. To 31st December, 1897*. Compiled by W. Innes Addison. Glasgow. 1898.

Royal Society catalogue of scientific papers: *Royal Society catalogue of scientific papers (1800–1900). Compiled and published by the Royal Society of London*. 19 vols. and index (3 vols.). London and Cambridge. 1867–1925.

Rupke, Nicolaas A. 1994. *Richard Owen: Victorian naturalist*. New Haven and London: Yale University Press.

Sarjeant, William A. S. 1980. *Geologists and the history of geology: an international bibliography from the origins to 1978*. 5 vols. and supplement (2 vols., 1987). London: Macmillan.

SBI: *Scandinavian biographical index*. Edited by Laureen Baillie. 4 vols. London: K. G. Saur. 1994.

SBL: *Svenskt biografiskt lexikon*. Edited by Bertil Boëthius, Erik Grill, Birgitta Lager-Kromnow, and Göran Nilzén. 27 vols. and 5 fascicles of vol. 28 (A–Pederby). Stockholm: Albert Bonnier and P. A. Norstedt. 1918–94.

Scherren, Henry. 1905. *The Zoological Society of London. A sketch of its foundation and development and the story of its farm, museum, gardens, menagerie and library*. London: Cassell and Co.

Schmidt, August and Voigt, Bernhard Friedrich. 1824–56. *Neue Nekrolog der Deutschen*. 30 vols.

Scriba, Heinrich Eduard. 1831–43. *Biographisch-literarisches Lexikon der Schriftsteller des Großherzogthums Hessen im 1. Viertel des 19. Jahrhunderts*. 2 vols. [Darmstadt.]

Smith, Charles Card. 1903. Memoir of William Sumner Appleton, A.M. *Proceedings of the Massachusetts Historical Society* 2d ser. 17: 516–31.

Smith, William. 1826. On *Dahlias*. [Read 5 December 1826.] *Transactions of the Horticultural Society of London* 7 (1830): 141–67.

Stenton, Michael. 1976. *Who's who of British members of Parliament, 1832–1885*. Brighton: Hanover Press.

Sutton, R. J. 1966. *The stamp collector's encyclopaedia*. 6th ed. Revised by K. W. Anthony. London: Stanley Paul.

Taxonomic literature: *Taxonomic literature. A selective guide to botanical publications and collections with dates, commentaries and types*. By Frans A. Stafleu and Richard S. Cowan. 2d ed. 7 vols. Utrecht: Bohn, Scheltema & Holkema. 1976–88.

Turner, Gerard L'E. 1989. *The great age of the microscope: the collection of the Royal Microscopical Society through 150 years*. Bristol and New York: Adam Hilger.

Variation: *The variation of animals and plants under domestication*. By Charles Darwin. 2 vols. London. 1868.

Veterinary directory: *The veterinary directory; or, annual register of the veterinary practitioners throughout the United Kingdom*. Edinburgh. 1861.

Veterinary register: *The register of veterinary surgeons*. London. 1884–.

Vilmorin, Jacques de. 1918. Notice biographique sur J. Gay. *Bulletin de la Société Botanique de France* 65: 78–82.

Wallace, Alfred Russel. 1905. *My life: a record of events and opinions*. 2 vols. London: Chapman & Hall.

Walters, S. M. 1981. *The shaping of Cambridge botany. A short history of whole-plant botany in Cambridge from the time of Ray into the present century*. Cambridge: Cambridge University Press.

Watkins, A. H. 1977. William Baxter: Bromley antiquary. *Bromley Local History* 2: 30–5.

Watt, Robert. 1824. *Bibliotheca Britannica; or a general index to British and foreign literature*. 4 vols. Edinburgh and London.

Wedgwood, Barbara and Wedgwood, Hensleigh. 1980. *The Wedgwood circle, 1730–1897: four generations of a family and their friends*. London: Studio Vista.

Wellesley index: *The Wellesley index to Victorian periodicals 1824–1900*. Edited by Walter E. Houghton and others. 5 vols. Toronto: University of Toronto Press. London: Routledge and Kegan Paul. 1966–89.

Williamson, George C., ed. 1903–5. *Bryan's dictionary of painters and engravers*. New ed. 5 vols. London: George Bell & sons.

Wood, Christopher. 1978. *The dictionary of Victorian painters*. 2d ed. Woodbridge, Suffolk: Antique Collectors' Club.

Wood, J. C. 1973. *A dictionary of British animal painters*. Leigh-on-Sea: F. Lewis.

Woodward, Horace B. 1907. *The history of the Geological Society of London*. London: Geological Society.

WWW: *Who was who. A companion to* Who's who. 8 vols. and cumulated index, 1897–1990. London: Adam & Charles Black. 1920–91.

INDEX

The dates of letters to and from Darwin's correspondents are listed in the Biographical register and index to correspondents and are not repeated here. Darwin's works are indexed under the short titles used throughout this volume and listed in the bibliography.

Table

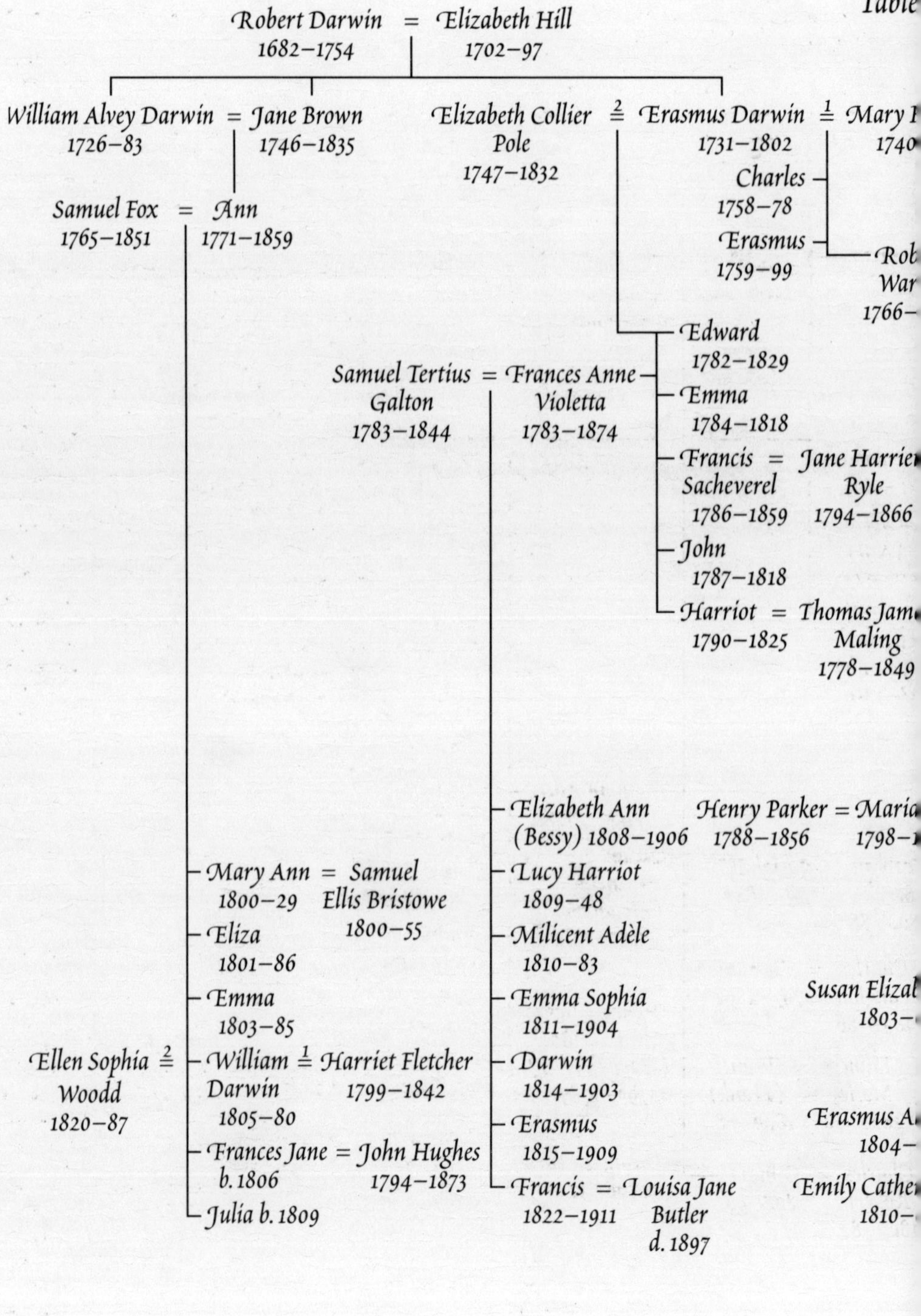
Robert Darwin = Elizabeth Hill
1682–1754
1702–97
William Alvey Darwin = Jane Brown
1726–83
1746–1835
Elizabeth Collier Pole
1747–1832
Erasmus Darwin
1731–1802
Mary
1740
Charles
1758–78
Erasmus
1759–99
Rob
War
1766–
Samuel Fox = Ann
1765–1851
1771–1859
Edward
1782–1829
Samuel Tertius Galton = Frances Anne Violetta
1783–1844
1783–1874
Emma
1784–1818
Francis Sacheverel = Jane Harriet Ryle
1786–1859
1794–1866
John
1787–1818
Harriot = Thomas James Maling
1790–1825
1778–1849
Elizabeth Ann (Bessy) 1808–1906
Henry Parker = Maria
1788–1856
1798–
Mary Ann = Samuel Ellis Bristowe
1800–29
1800–55
Lucy Harriot
1809–48
Eliza
1801–86
Milicent Adèle
1810–83
Emma
1803–85
Emma Sophia
1811–1904
Susan Elizab
1803–
Ellen Sophia Woodd
1820–87
William Darwin
1805–80
Harriet Fletcher
1799–1842
Darwin
1814–1903
Erasmus
1815–1909
Erasmus A
1804–
Frances Jane = John Hughes
b. 1806
1794–1873
Francis = Louisa Jane Butler
1822–1911
d. 1897
Emily Cathe
1810–
Julia b. 1809